Natural History of West Indian Reptiles and Amphibians

UNIVERSITY PRESS OF FLORIDA
Florida A&M University, Tallahassee
Florida Atlantic University, Boca Raton
Florida Gulf Coast University, Ft. Myers
Florida International University, Miami
Florida State University, Tallahassee
New College of Florida, Sarasota
University of Central Florida, Orlando
University of Florida, Gainesville
University of North Florida, Jacksonville
University of South Florida, Tampa
University of West Florida, Pensacola

Boa nebulosa, one of two endemic West Indian species formerly considered subspecies of widely distributed Neotropical *B. constrictor*. Photograph by Robert Powell.

Natural History of West Indian Reptiles and Amphibians

Robert W. Henderson and Robert Powell

University Press of Florida

GAINESVILLE · TALLAHASSEE · TAMPA · BOCA RATON · PENSACOLA · ORLANDO · MIAMI · JACKSONVILLE · FT. MYERS · SARASOTA

14 13 12 11 10 09 6 5 4 3 2 1

Library of Congress Cataloging-in-Publication Data
Henderson, Robert W., 1945–
Natural history of West Indian reptiles and amphibians /
Robert W. Henderson and Robert Powell.
p. cm.
Includes bibliographical references and index.
ISBN 978-0-8130-3394-5 (alk. paper)
1. Reptiles—West Indies. 2. Amphibians—West Indies.
I. Powell,
Robert, 1948– II. Title.
QL656.5.A1H46 2009
597.909729—dc22 2009017139

The University Press of Florida is the scholarly publishing agency for the State University System of Florida, comprising Florida A&M University, Florida Atlantic University, Florida Gulf Coast University, Florida International University, Florida State University, New College of Florida, University of Central Florida, University of Florida, University of North Florida, University of South Florida, and University of West Florida.

University Press of Florida
15 Northwest 15th Street
Gainesville, FL 32611-2079
http://www.upf.com

Funding to assist in publication of this book was generously provided by the following organizations:

The Falconwood Foundation through The Conservation Agency, Jamestown, Rhode Island

The Herpetologists' League

To our wives
(good sports that they are)
Rose MacLaren Henderson
and
Beverly Jean Powell

Contents

List of Species ix

List of Illustrations xix

Acknowledgments xxiii

Introduction 1

Methods 2
Where We Are Today 5
A Current Overview of Natural History 17
Future Emphases 20

Part I. Amphibia

Section 1. Anura

1. Family Aromobatidae 25
2. Family Bufonidae 26
3. Family Eleutherodactylidae 33
4. Family Hylidae 84
5. Family Leptodactylidae 91
6. Family Microhylidae 95
7. Family Ranidae 96
8. Family Strabomantidae 98

Part II. Reptilia

Section 1. Squamata (lizards)

9. Family Anguidae 101
10. Family Gekkonidae 108
11. Family Gymnophthalmidae 112
12. Family Iguanidae 114
13. Family Leiocephalidae 137
14. Family Phyllodactylidae 153
15. Family Polychrotidae 157
16. Family Scincidae 292
17. Family Sphaerodactylidae 294
18. Family Teiidae 319
19. Family Xantusiidae 335

Section 2. Squamata (amphisbaenians)

20. Family Amphisbaenidae 337
21. Family Cadeidae 341

Section 3. Squamata (snakes)

22. Family Boidae 343
23. Family Colubridae 354
24. Family Dipsadidae 356
25. Family Leptotyphlopidae 373
26. Family Natricidae 375
27. Family Tropidophiidae 376
28. Family Typhlopidae 382
29. Family Viperidae 390

Section 4. Testudines

30. Family Emydidae 394
31. Family Pelomedusidae 399
32. Family Testudinidae 400

Section 5. Crocodilia

33. Family Alligatoridae 403
34. Family Crocodylidae 404

Literature Cited 407

Index to Scientific Names of Amphibians and Reptiles 487

Contents

Species

Amphibia (197)

Anura (197)

Family Aromobatidae (1)

Allobates chalcopis

Family Bufonidae (13)

Bufo cataulaciceps
Bufo empusus
Bufo florentinoi
Bufo fluviaticus
Bufo fractus
Bufo fustiger
Bufo guentheri
Bufo gundlachi
Bufo lemur
Bufo longinasus
Bufo marinus
Bufo peltocephalus
Bufo taladai

Family Eleutherodactylidae (161)

Eleutherodactylus abbotti
Eleutherodactylus acmonis
Eleutherodactylus adelus
Eleutherodactylus albipes
Eleutherodactylus alcoae
Eleutherodactylus alticola
Eleutherodactylus amadeus
Eleutherodactylus amplinympha
Eleutherodactylus andrewsi
Eleutherodactylus antillensis
Eleutherodactylus aporostegus
Eleutherodactylus apostates
Eleutherodactylus armstrongi
Eleutherodactylus atkinsi
Eleutherodactylus audanti
Eleutherodactylus auriculatoides
Eleutherodactylus auriculatus
Eleutherodactylus bakeri
Eleutherodactylus barlagnei
Eleutherodactylus bartonsmithi
Eleutherodactylus blairhedgesi
Eleutherodactylus bothroboans
Eleutherodactylus bresslerae
Eleutherodactylus brevirostris
Eleutherodactylus brittoni
Eleutherodactylus caribe
Eleutherodactylus casparii
Eleutherodactylus cavernicola
Eleutherodactylus chlorophenax
Eleutherodactylus cochranae
Eleutherodactylus cooki
Eleutherodactylus coqui
Eleutherodactylus corona
Eleutherodactylus counouspeus
Eleutherodactylus cubanus
Eleutherodactylus cundalli
Eleutherodactylus cuneatus
Eleutherodactylus darlingtoni
Eleutherodactylus dimidiatus
Eleutherodactylus diplasius
Eleutherodactylus dolomedes
Eleutherodactylus eileenae
Eleutherodactylus emiliae
Eleutherodactylus eneidae
Eleutherodactylus erythroproctus
Eleutherodactylus etheridgei
Eleutherodactylus eunaster
Eleutherodactylus flavescens
Eleutherodactylus fowleri
Eleutherodactylus furcyensis
Eleutherodactylus fuscus
Eleutherodactylus glamyrus
Eleutherodactylus glandulifer
Eleutherodactylus glanduliferoides
Eleutherodactylus glaphycompus
Eleutherodactylus glaucoreius
Eleutherodactylus goini
Eleutherodactylus gossei
Eleutherodactylus grabhami

Eleutherodactylus grahami
Eleutherodactylus greyi
Eleutherodactylus griphus
Eleutherodactylus gryllus
Eleutherodactylus guanahacabibes
Eleutherodactylus guantanamera
Eleutherodactylus gundlachi
Eleutherodactylus haitianus
Eleutherodactylus hedricki
Eleutherodactylus heminota
Eleutherodactylus hypostenor
Eleutherodactylus iberia
Eleutherodactylus inoptatus
Eleutherodactylus intermedius
Eleutherodactylus ionthus
Eleutherodactylus jamaicensis
Eleutherodactylus jasperi
Eleutherodactylus jaumei
Eleutherodactylus johnstonei
Eleutherodactylus juanariveroi
Eleutherodactylus jugans
Eleutherodactylus junori
Eleutherodactylus karlschmidti
Eleutherodactylus klinikowskii
Eleutherodactylus lamprotes
Eleutherodactylus leberi
Eleutherodactylus lentus
Eleutherodactylus leoncei
Eleutherodactylus limbatus
Eleutherodactylus limbensis
Eleutherodactylus locustus
Eleutherodactylus lucioi
Eleutherodactylus luteolus
Eleutherodactylus maestrensis
Eleutherodactylus mariposa
Eleutherodactylus martinicensis
Eleutherodactylus melacara
Eleutherodactylus melatrigonum
Eleutherodactylus michaelschmidi
Eleutherodactylus minutus
Eleutherodactylus monensis
Eleutherodactylus montanus
Eleutherodactylus nortoni
Eleutherodactylus notidodes
Eleutherodactylus nubicola
Eleutherodactylus olibrus
Eleutherodactylus orcutti
Eleutherodactylus orientalis
Eleutherodactylus oxyrhyncus
Eleutherodactylus pantoni
Eleutherodactylus parabates
Eleutherodactylus paralius
Eleutherodactylus parapelates
Eleutherodactylus patriciae
Eleutherodactylus paulsoni
Eleutherodactylus pentasyringos
Eleutherodactylus pezopetrus
Eleutherodactylus pictissimus
Eleutherodactylus pinarensis
Eleutherodactylus pinchoni
Eleutherodactylus pituinus
Eleutherodactylus planirostris
Eleutherodactylus poolei
Eleutherodactylus portoricensis
Eleutherodactylus principalis
Eleutherodactylus probolaeus
Eleutherodactylus rhodesi
Eleutherodactylus richmondi
Eleutherodactylus ricordii
Eleutherodactylus riparius
Eleutherodactylus rivularis
Eleutherodactylus rogersi
Eleutherodactylus ronaldi
Eleutherodactylus rucillensis
Eleutherodactylus rufifemoralis
Eleutherodactylus ruthae
Eleutherodactylus schmidti
Eleutherodactylus schwartzi
Eleutherodactylus sciagraphus
Eleutherodactylus semipalmatus
Eleutherodactylus simulans
Eleutherodactylus sisyphodemus
Eleutherodactylus sommeri
Eleutherodactylus staurometopon
Eleutherodactylus symingtoni
Eleutherodactylus tetajulia
Eleutherodactylus thomasi
Eleutherodactylus thorectes
Eleutherodactylus toa
Eleutherodactylus tonyi
Eleutherodactylus turquinensis
Eleutherodactylus tychathrous
Eleutherodactylus unicolor
Eleutherodactylus varians
Eleutherodactylus varleyi
Eleutherodactylus ventrilineatus
Eleutherodactylus warreni
Eleutherodactylus weinlandi
Eleutherodactylus wetmorei
Eleutherodactylus wightmanae
Eleutherodactylus zeus
Eleutherodactylus zugi

Family Hylidae (13)

Hyla cinerea
Hyla squirella
Hypsiboas heilprini
Osteopilus brunneus
Osteopilus crucialis
Osteopilus dominicensis
Osteopilus marianae
Osteopilus pulchrilineatus
Osteopilus septentrionalis
Osteopilus vastus
Osteopilus wilderi
Scinax ruber
Scinax cf. *x-signatus*

Family Leptodactylidae (3)

Leptodactylus albilabris
Leptodactylus fallax
Leptodactylus validus

Family Microhylidae (1)

Gastrophryne carolinensis

Family Ranidae (3)

Rana catesbeiana
Rana grylio
Rana sphenocephala

Family Strabomantidae (2)

Pristimantis euphronides
Pristimantis shrevei

Reptilia (539)

Squamata (lizards) (364)

Family Anguidae (26)

Celestus agasepsoides
Celestus anelpistus
Celestus badius
Celestus barbouri
Celestus costatus
Celestus crusculus
Celestus curtissi
Celestus darlingtoni
Celestus duquesneyi
Celestus fowleri
Celestus haetianus
Celestus hewardii
Celestus macrotus
Celestus maculatus
Celestus marcanoi
Celestus microblepharis
Celestus molesworthi
Celestus occiduus
Celestus sepsoides
Celestus stenurus
Celestus warreni
Diploglossus delasagra
Diploglossus garridoi
Diploglossus montisserrati
Diploglossus nigropunctatus
Diploglossus pleii

Family Gekkonidae (6)

Gekko gecko
Hemidactylus angulatus
Hemidactylus garnotii
Hemidactylus mabouia
Hemidactylus palaichthus
Hemidactylus turcicus

Family Gymnophthalmidae (3)

Bachia heteropus
Gymnophthalmus pleii
Gymnophthalmus underwoodi

Family Iguanidae (13)

Cyclura carinata
Cyclura collei
Cyclura cornuta
Cyclura cychlura
Cyclura lewisi
Cyclura nubila
Cyclura onchiopsis
Cyclura pinguis
Cyclura ricordii

Cyclura rileyi
Cyclura stejnegeri
Iguana delicatissima
Iguana iguana

Family Leiocephalidae (25)

Leiocephalus altavelensis
Leiocephalus barahonensis
Leiocephalus carinatus
Leiocephalus cubensis
Leiocephalus cuneus
Leiocephalus endomychus
Leiocephalus eremitus
Leiocephalus greenwayi
Leiocephalus herminieri
Leiocephalus inaguae
Leiocephalus loxogrammus
Leiocephalus lunatus
Leiocephalus macropus
Leiocephalus melanochlorus
Leiocephalus onaneyi
Leiocephalus personatus
Leiocephalus pratensis
Leiocephalus psammodromus
Leiocephalus punctatus
Leiocephalus raviceps
Leiocephalus rhutidira
Leiocephalus schreibersii
Leiocephalus semilineatus
Leiocephalus stictigaster
Leiocephalus vinculum

Family Phyllodactylidae (8)

Phyllodactylus hispaniolae
Phyllodactylus pulcher
Phyllodactylus sommeri
Phyllodactylus wirshingi
Tarentola albertschwartzi
Tarentola americana
Tarentola crombiei
Thecadactylus rapicauda

Family Polychrotidae (157)

Anolis acutus
Anolis aeneus
Anolis agueroi
Anolis ahli
Anolis alayoni
Anolis alfaroi
Anolis aliniger
Anolis allisoni
Anolis allogus
Anolis altavelensis
Anolis altitudinalis
Anolis alumina
Anolis alutaceus
Anolis anfiloquioi
Anolis angusticeps
Anolis argenteolus
Anolis argillaceus
Anolis armouri
Anolis bahorucoensis
Anolis baleatus
Anolis baracoae
Anolis barahonae
Anolis barbatus
Anolis barbouri
Anolis bartschi
Anolis bimaculatus
Anolis birama
Anolis bremeri
Anolis breslini
Anolis brevirostris
Anolis brunneus
Anolis carolinensis
Anolis caudalis
Anolis centralis
Anolis chamaeleonides
Anolis chlorocyanus
Anolis christophei
Anolis chrysops
Anolis clivicola
Anolis coelestinus
Anolis confusus
Anolis conspersus
Anolis cooki
Anolis cristatellus
Anolis cupeyalensis
Anolis cuvieri
Anolis cyanopleurus
Anolis cybotes
Anolis darlingtoni
Anolis delafuentei
Anolis desechensis
Anolis desiradei
Anolis distichus
Anolis dolichocephalus
Anolis equestris

Anolis ernestwilliamsi
Anolis etheridgei
Anolis eugenegrahami
Anolis evermanni
Anolis extremus
Anolis fairchildi
Anolis ferreus
Anolis fowleri
Anolis fugitivus
Anolis garmani
Anolis garridoi
Anolis gingivinus
Anolis grahami
Anolis griseus
Anolis guafe
Anolis guamuhaya
Anolis guazuma
Anolis gundlachi
Anolis haetianus
Anolis hendersoni
Anolis homolechis
Anolis imias
Anolis incredulus
Anolis inexpectatus
Anolis insolitus
Anolis isolepis
Anolis juangundlachi
Anolis jubar
Anolis kahouannensis
Anolis koopmani
Anolis krugi
Anolis leachii
Anolis lineatopus
Anolis litoralis
Anolis lividus
Anolis longiceps
Anolis longitibialis
Anolis loysianus
Anolis luciae
Anolis lucius
Anolis luteogularis
Anolis luteosignifer
Anolis macilentus
Anolis marcanoi
Anolis marmoratus
Anolis marron
Anolis maynardii
Anolis mestrei
Anolis monensis
Anolis monticola
Anolis noblei
Anolis nubilis
Anolis occultus
Anolis oculatus
Anolis olssoni
Anolis opalinus
Anolis ophiolepis
Anolis oporinus
Anolis paternus
Anolis pigmaequestris
Anolis placidus
Anolis pogus
Anolis poncensis
Anolis porcatus
Anolis porcus
Anolis pulchellus
Anolis pumilus
Anolis quadriocellifer
Anolis reconditus
Anolis rejectus
Anolis richardii
Anolis ricordii
Anolis rimarum
Anolis roosevelti
Anolis roquet
Anolis rubribarbus
Anolis ruibali
Anolis rupinae
Anolis sabanus
Anolis sagrei
Anolis schwartzi
Anolis scriptus
Anolis semilineatus
Anolis sheplani
Anolis shrevei
Anolis singularis
Anolis smallwoodi
Anolis smaragdinus
Anolis spectrum
Anolis strahmi
Anolis stratulus
Anolis terraealtae
Anolis terueli
Anolis toldo
Anolis trinitatis
Anolis valencienni
Anolis vanidicus
Anolis vermiculatus
Anolis vescus
Anolis wattsi
Anolis websteri
Anolis whitemani

Family Scincidae (4)

Mabuya lineolata
Mabuya mabouya
Mabuya macleani
Mabuya sloanii

Family Sphaerodactylidae (97)

Aristelliger barbouri
Aristelliger cochranae
Aristelliger expectatus
Aristelliger hechti
Aristelliger lar
Aristelliger praesignis
Aristelliger reyesi
Gonatodes albogularis
Gonatodes daudini
Sphaerodactylus altavelensis
Sphaerodactylus argivus
Sphaerodactylus argus
Sphaerodactylus ariasae
Sphaerodactylus armasi
Sphaerodactylus armstrongi
Sphaerodactylus asterulus
Sphaerodactylus beattyi
Sphaerodactylus becki
Sphaerodactylus bromeliarum
Sphaerodactylus caicosensis
Sphaerodactylus callocricus
Sphaerodactylus celicara
Sphaerodactylus cinereus
Sphaerodactylus clenchi
Sphaerodactylus cochranae
Sphaerodactylus copei
Sphaerodactylus corticola
Sphaerodactylus cricoderus
Sphaerodactylus cryphius
Sphaerodactylus dacnicolor
Sphaerodactylus darlingtoni
Sphaerodactylus difficilis
Sphaerodactylus dimorphicus
Sphaerodactylus docimus
Sphaerodactylus elasmorhynchus
Sphaerodactylus elegans
Sphaerodactylus elegantulus
Sphaerodactylus epiurus
Sphaerodactylus fantasticus
Sphaerodactylus gaigeae
Sphaerodactylus gilvitorques
Sphaerodactylus goniorhynchus
Sphaerodactylus inaguae
Sphaerodactylus intermedius
Sphaerodactylus kirbyi
Sphaerodactylus klauberi
Sphaerodactylus ladae
Sphaerodactylus lazelli
Sphaerodactylus leucaster
Sphaerodactylus levinsi
Sphaerodactylus macrolepis
Sphaerodactylus mariguanae
Sphaerodactylus microlepis
Sphaerodactylus micropithecus
Sphaerodactylus monensis
Sphaerodactylus nicholsi
Sphaerodactylus nigropunctatus
Sphaerodactylus notatus
Sphaerodactylus nycteropus
Sphaerodactylus ocoae
Sphaerodactylus oliveri
Sphaerodactylus omoglaux
Sphaerodactylus oxyrhinus
Sphaerodactylus parkeri
Sphaerodactylus parthenopion
Sphaerodactylus parvus
Sphaerodactylus perissodactylius
Sphaerodactylus phyzacinus
Sphaerodactylus pimienta
Sphaerodactylus plummeri
Sphaerodactylus ramsdeni
Sphaerodactylus randi
Sphaerodactylus rhabdotus
Sphaerodactylus richardi
Sphaerodactylus richardsoni
Sphaerodactylus roosevelti
Sphaerodactylus ruibali
Sphaerodactylus sabanus
Sphaerodactylus samanensis
Sphaerodactylus savagei
Sphaerodactylus scaber
Sphaerodactylus schuberti
Sphaerodactylus schwartzi
Sphaerodactylus semasiops
Sphaerodactylus shrevei
Sphaerodactylus siboney
Sphaerodactylus sommeri
Sphaerodactylus sputator
Sphaerodactylus storeyae
Sphaerodactylus streptophorus
Sphaerodactylus thompsoni
Sphaerodactylus torrei
Sphaerodactylus townsendi

Sphaerodactylus underwoodi
Sphaerodactylus vincenti
Sphaerodactylus williamsi
Sphaerodactylus zygaena

Family Teiidae (24)

Ameiva alboguttata
Ameiva ameiva
Ameiva atrata
Ameiva auberi
Ameiva chrysolaema
Ameiva cineracea
Ameiva corax
Ameiva corvina
Ameiva desechensis
Ameiva dorsalis
Ameiva erythrocephala
Ameiva exsul
Ameiva fuscata
Ameiva griswoldi
Ameiva lineolata
Ameiva major
Ameiva maynardii
Ameiva pleii
Ameiva pluvianotata
Ameiva polops
Ameiva taeniura
Ameiva wetmorei
Cnemidophorus vanzoi
Kentropyx borckiana

Family Xantusiidae (1)

Cricosaura typica

Squamata (amphisbaenians) (17)

Family Amphisbaenidae (15)

Amphisbaena bakeri
Amphisbaena barbouri
Amphisbaena caeca
Amphisbaena carlgansi
Amphisbaena caudalis
Amphisbaena cayemite
Amphisbaena cubana
Amphisbaena fenestrata
Amphisbaena gonavensis
Amphisbaena hyporissor
Amphisbaena innocens
Amphisbaena leali
Amphisbaena manni
Amphisbaena schmidti
Amphisbaena xera

Family Cadeidae (2)

Cadea blanoides
Cadea palirostrata

Squamata (snakes) (148)

Family Boidae (13)

Boa nebulosa
Boa orophias
Corallus cookii
Corallus grenadensis
Epicrates angulifer
Epicrates chrysogaster
Epicrates exsul
Epicrates fordii
Epicrates gracilis
Epicrates inornatus
Epicrates monensis
Epicrates striatus
Epicrates subflavus

Family Colubridae (4)

Chironius vincenti
Mastigodryas bruesi
Pantherophis guttatus
Tantilla melanocephala

Family Dipsadidae (50)

Alsophis antiguae
Alsophis antillensis
Alsophis manselli
Alsophis rijgersmaei
Alsophis rufiventris
Alsophis sajdaki
Alsophis sanctonum
Alsophis sibonius

Arrhyton ainictum
Arrhyton dolichura
Arrhyton procerum
Arrhyton redimitum
Arrhyton supernum
Arrhyton taeniatum
Arrhyton tanyplectum
Arrhyton vittatum
Borikenophis portoricensis
Borikenophis sanctaecrucis
Borikenophis variegatus
Caraiba andreae
Clelia clelia
Clelia errabunda
Cubophis cantherigerus
Cubophis caymanus
Cubophis fuscicauda
Cubophis ruttyi
Cubophis vudii
Haitiophis anomalus
Hypsirhynchus ater
Hypsirhynchus callilaemus
Hypsirhynchus ferox
Hypsirhynchus funereus
Hypsirhynchus melanichnus
Hypsirhynchus parvifrons
Hypsirhynchus polylepis
Hypsirhynchus scalaris
Ialtris agyrtes
Ialtris dorsalis
Ialtris haetianus
Ialtris parishi
Liophis cursor
Liophis juliae
Liophis ornatus
Liophis perfuscus
Magliophis exiguus
Magliophis stahli
Tretanorhinus variabilis
Uromacer catesbyi
Uromacer frenatus
Uromacer oxyrhynchus

Family Leptotyphlopidae (9)

Leptotyphlops asbolepis
Leptotyphlops bilineatus
Leptotyphlops breuili
Leptotyphlops calypso
Leptotyphlops carlae
Leptotyphlops columbi
Leptotyphlops goudoti
Leptotyphlops leptepileptus
Leptotyphlops pyrites

Family Natricidae (1)

Nerodia clarkii

Family Tropidophiidae (27)

Tropidophis bucculentus
Tropidophis canus
Tropidophis caymanensis
Tropidophis celiae
Tropidophis curtus
Tropidophis feicki
Tropidophis fuscus
Tropidophis galacelidus
Tropidophis greenwayi
Tropidophis haetianus
Tropidophis hardyi
Tropidophis hendersoni
Tropidophis jamaicensis
Tropidophis maculatus
Tropidophis melanurus
Tropidophis morenoi
Tropidophis nigriventris
Tropidophis pardalis
Tropidophis parkeri
Tropidophis pilsbryi
Tropidophis schwartzi
Tropidophis semicinctus
Tropidophis spiritus
Tropidophis stejnegeri
Tropidophis stullae
Tropidophis wrighti
Tropidophis xanthogaster

Family Typhlopidae (42)

Ramphotyphlops braminus
Typhlops agoralionis
Typhlops anchaurus
Typhlops annae
Typhlops anousius
Typhlops arator

Typhlops biminiensis
Typhlops capitulatus
Typhlops catapontus
Typhlops caymanensis
Typhlops contorhinus
Typhlops dominicanus
Typhlops epactius
Typhlops eperopeus
Typhlops geotomus
Typhlops golyathi
Typhlops gonavensis
Typhlops granti
Typhlops guadeloupensis
Typhlops hectus
Typhlops hypomethes
Typhlops jamaicensis
Typhlops lumbricalis
Typhlops monastus
Typhlops monensis
Typhlops naugus
Typhlops notorachius
Typhlops paradoxus
Typhlops perimychus
Typhlops platycephalus
Typhlops proancylops
Typhlops pusillus
Typhlops richardii
Typhlops rostellatus
Typhlops satelles
Typhlops schwartzi
Typhlops sulcatus
Typhlops sylleptor
Typhlops syntherus
Typhlops tasymicris
Typhlops tetrathyreus
Typhlops titanops

Family Viperidae (2)

Bothrops caribbaeus
Bothrops lanceolatus

Testudines (8)

Family Emydidae (5)

Trachemys decorata
Trachemys decussata
Trachemys scripta
Trachemys stejnegeri
Trachemys terrapen

Family Pelomedusidae (1)

Pelusios castaneus

Family Testudinidae (2)

Chelonoidis carbonaria
Chelonoidis denticulata

Crocodilia (3)

Family Alligatoridae (1)

Caiman crocodilus

Family Crocodylidae (2)

Crocodylus acutus
Crocodylus rhombifer

Illustrations

Figures

Frontispiece. *Boa nebulosa*

1. Map of the West Indies xx
2. Map of the Lesser Antilles xxi
3. Number of natural history publications by decade 4
4. Family Aromobatidae, *Allobates chalcopis* 25
5. Family Bufonidae, *Bufo lemur* 26
6. Family Eleutherodactylidae, *Eleutherodactylus bakeri* and *E. oxyrhyncus* 33
7. Family Hylidae, *Osteopilus septentrionalis* 84
8. Family Leptodactylidae, *Leptodactylus validus* 91
9. Family Microhylidae, *Gastrophryne carolinensis* 95
10. Family Ranidae, *Rana catesbeiana* 96
11. Family Strabomantidae, *Pristimantis shrevei* 98
12. Family Anguidae, *Celestus badius* 101
13. Family Gekkonidae, *Hemidactylus mabouia* 108
14. Family Gymnophthalmidae, *Gymnophthalmus underwoodi* 112
15. Family Iguanidae, *Cyclura lewisi* 114
16. Family Leiocephalidae, *Leiocephalus lunatus* 137
17. Family Phyllodactylidae, *Thecadactylus rapicauda* 153
18. Family Polychrotidae, *Anolis allogus* 157
19. Family Scincidae, *Mabuya mabouya* 292
20. Family Sphaerodactylidae, *Sphaerodactylus macrolepis* 294
21. Family Teiidae, *Ameiva ameiva* 319
22. Family Xantusiidae, *Cricosaura typica* 335
23. Family Amphisbaenidae, *Amphisbaena bakeri* 337
24. Family Cadeidae, *Cadea blanoides* 341
25. Family Boidae, *Corallus cookii* 343
26. Family Colubridae, *Mastigodryas bruesi* 354
27. Family Dipsadidae, *Borikenophis portoricensis* 356
28. Family Leptotyphlopidae, *Leptotyphlops breuili* 373
29. Family Natricidae, *Nerodia clarkii* 375
30. Family Tropidophiidae, *Tropidophis wrighti* 376
31. Family Typhlopidae, *Typhlops granti* 382
32. Family Viperidae, *Bothrops caribbaeus* 390
33. Family Emydidae, *Trachemys stejnegeri stejnegeri* 394
34. Family Pelomedusidae, *Pelusios castaneus* 399
35. Family Testudinidae, *Chelonoidis carbonaria* 400
36. Family Alligatoridae, *Caiman crocodilus* 403
37. Family Crocodylidae, *Crocodylus acutus* 404

Tables

1. Families, genera, and number of species of frogs and reptiles that occur in the West Indies 6
2. Presence or absence of species of *Ameiva*, *Alsophis*, *Liophis*, and the mongoose on Lesser Antillean islands 9
3. Associations of West Indian frogs, lizards, and snakes with 13 habitat variables resulting from human impact 13

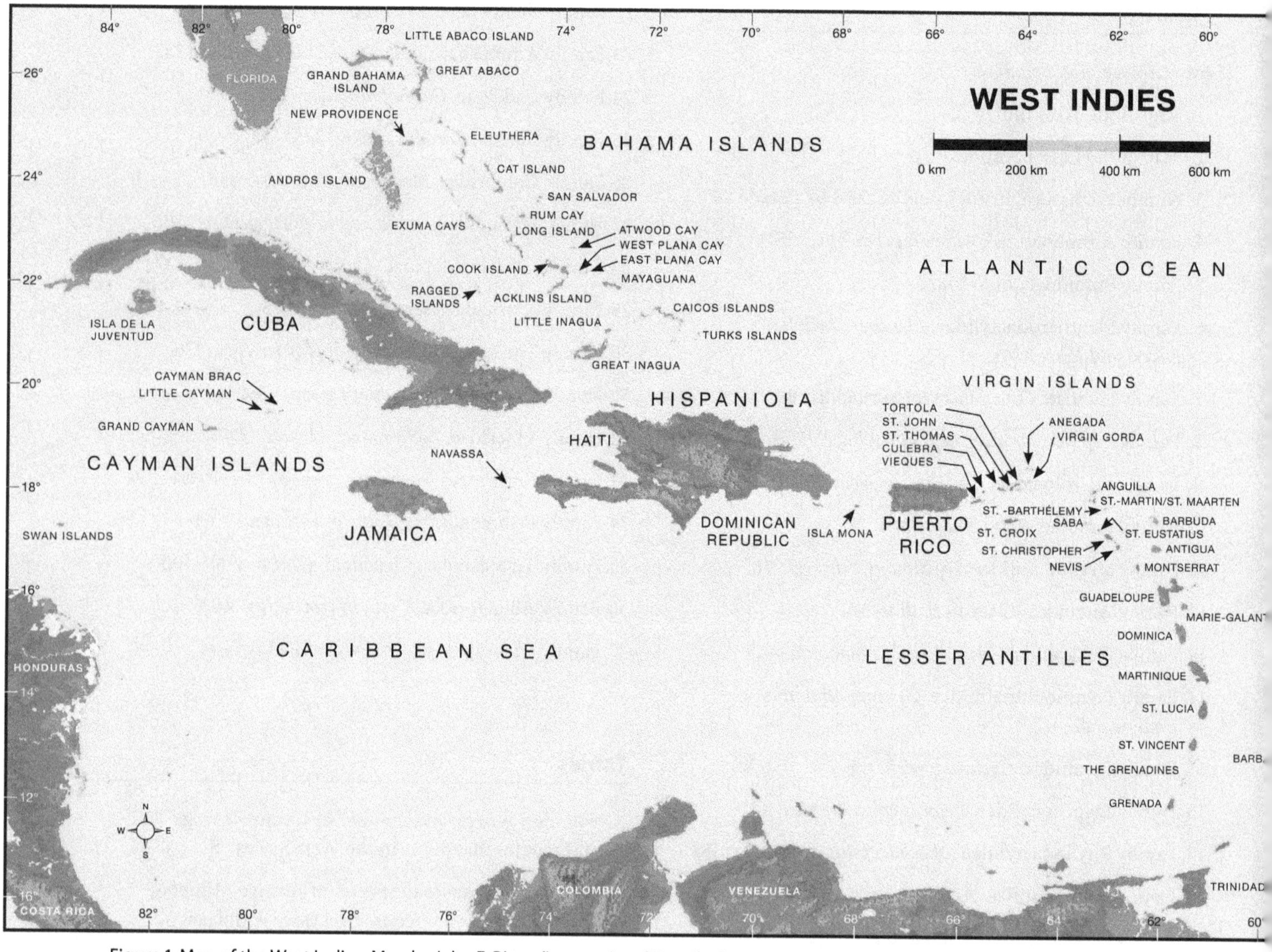

Figure 1. Map of the West Indies. Map by John F. Binns (International Reptile Conservation Foundation).

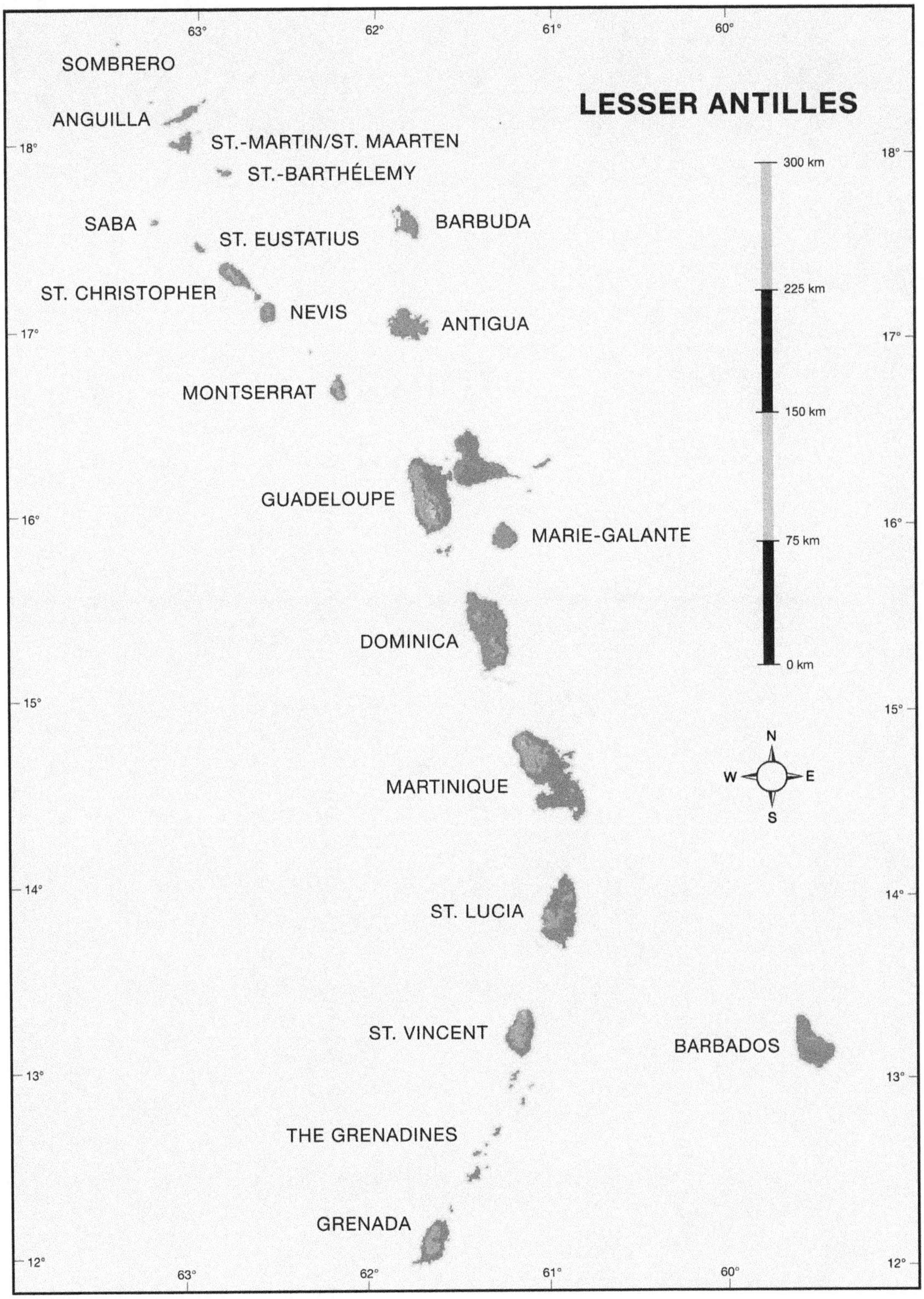

Figure 2. Map of the Lesser Antilles. Map by John F. Binns (International Reptile Conservation Foundation).

Acknowledgments

We have amassed an incredible debt of gratitude to all those who provided information for the species accounts that follow: Kraig Adler, Allison Alberts, Brittany S. Barker, Aaron M. Bauer, Breck Bartholomew, Craig S. Berg, John F. Binns, Jon Boone, Michel Breuil (French West Indies), Rafe M. Brown, Frederick J. Burton (Cayman Islands), Ryan Calsbeek, William E. Cooper, Brian I. Crother, Jennifer Daltry, Kevin de Queiroz, Mark de Silva (Grenadines), Luis M. Díaz (Cuba), William E. Duellman, Taura Ebanks, Arthur D. (Sandy) Echternacht (Cayman Islands), Alberto R. Estrada (Cuba), Eladio Fernández (Hispaniola), Angela Fields (documented the presence of *Anolis sagrei* on Barbados, in litt. to J. B. Losos, 14.IV.2008), Axel Fläschendräger, Ansel Fong Grillo (Cuba), Matthew E. Gifford, Richard E. Glor, Stephen R. Goldberg (parasites), Tandora Grant (iguanas), Traci D. Hartsell, S. Blair Hedges, Courtney Higgins, Karim V. D. Hodge (Anguilla), John B. Iverson, Arlington James (Dominica), Thomas A. Jenssen, Rafael L. Joglar (Puerto Rico), Charles R. Knapp, Fred Kraus (identified the snail taken by an *Anolis cristatellus* on Guana Island, British Virgin Islands), James D. (Skip) Lazell, Manuel Leal, Allen R. Lewis, Jonathan B. Losos, Gregory C. Mayer, Colin McCarthy, Adam Mitchell, Numi Mitchell, John C. Murphy, Luis O. Nieves (Puerto Rico), Kristiina Ovaska, John S. Parmerlee, Jr., Gad Perry, Renata Platenberg (U.S. Virgin Islands), Louis Porras (Bahamas), Alan Resetar, Javier Rodríguez-Robles, Lourdes Rodríguez Schettino (Cuba), Alejandro J. Sánchez Muñoz, Michael E. Seidel (turtles), Brian E. Smith, Peter J. Stafford, John B. Thorbjarnarson (crocodilians), Daniel S. Townsend, Susan C. Valentine-Cooper, Nicolas Vidal, David Wareham, Heiko Werning, and Byron Wilson (Jamaica).

Luis Diaz, Ansel Fong Grillo, and Lourdes Rodríguez Schettino did more than merely provide reprints. In many instances, they manually scanned and sent copies of publications we otherwise would not have seen or been able to include, resulting in an underestimation of the contributions to natural history made by Cuban biologists. Along those same lines, Aaron Bauer, Axel Fläschendräger, Matthew E. Gifford, Richard E. Glor, S. Blair Hedges, and Jonathan B. Losos went far beyond the call of duty to dig out and provide access to materials in their private collections and, in some instances, to unpublished information that contributed significantly to the completion of our task. Charles R. Bursey and Stephen R. Goldberg reviewed the taxonomy of parasites in amphibians and reptiles to ensure the highest degree of accuracy possible. Because they worked from incomplete manuscripts and did not see the entire volume at any time, however, any remaining errors are ours alone. Guy Poitras and Marla E. Allen compensated for our linguistic limitations by translating French publications. In addition, Susan Otto, Milwaukee Public Museum, and Farrukh Hasan, Avila University, facilitated an inordinate number of interlibrary loans, and the staff of the Linda Hall Library of Science, Engineering, and Technology (Kansas City, Missouri) was immensely helpful to RP.

Allison C. Alberts, Brian I. Crother, Matthew E. Gifford, Richard E. Glor, Tandora Grant, S. Blair Hedges, James (Skip) Lazell, Jonathan B. Losos, Richard Thomas, and one anonymous reviewer read at least sections of the manuscript in various stages of completion. Their comments were invariably helpful and contributed greatly to the quality of the final product.

John F. Binns (International Reptile Conservation Foundation) prepared the maps. John F. Binns, Michel Breuil, Joseph Burgess, Mark de Silva, Arthur C. (Sandy) Echternacht, Eladio Fernández, Axel Fläschendräger, S. Blair Hedges, Hinrich Kaiser, John Magor, Luis O. Nieves, Kristiina Ovaska, John S. Parmerlee, Jr., Mel José Rivera, Alejandro J. Sánchez Muñoz, and Rod D. Wittenberg granted permission to use their photographs.

Over the past 30+ years, Henderson's fieldwork has been funded through a variety of sources: Milwaukee Public Museum, Windway Foundation, Puelicher Foundation, Milwaukee County Zoological Gardens, Zoological Society of Milwaukee, American Philosophical Society, Robert Bourgeois Memorial Fund, Robert Bourgeois Family, The Conservation Agency (through grants from the Falconwood Foundation), Chicago Zoological Society, Museo de Historia Natural (Dominican Republic), Gulf + Western Dominicana, and the National Science Foundation (through Robert Powell). Powell's fieldwork over 20+ years has been funded by Avila University, The Conservation Agency, and a series of REU (Research Experiences for Undergraduates) grants from the National Science Foundation. We are grateful to all for their support.

We also thank the various governmental agencies on many islands that issued permits and often facilitated our work. We also thank everyone who made us welcome, inviting us to stay in their homes or merely making us feel as though we were at home (Lazy Lagoon, Maffiken Apartments, King's Well Hotel, El Momo Cottages).

We have had the good fortune to spend many pleasurable hours conducting fieldwork in the West Indies with many good companions: Arijana Barun, Craig Berg, John Binns, Bill Buck, Fred Burton, Andy Caballero, Ellen Censky, Sandy Echternacht, Doug Eifler, Julio Féliz, Henry Fitch, Brian Fontenot, Neil Ford, Joel Friesch, Glenn Gerber, Matt Gifford, Robert Halley, Mary Haskins, Ky Henderson, Rose Henderson, Karim Hodge, Sixto Incháustegui, James Johnson, Jane Jolley, Martin Jolley, Keet Kopecky, Nathan Kraucunas, Samantha Larimer, Amy Lathrop, Skip Lazell, Carlos Martínez, Tim McCarthy, John Murphy, David Nieves, Luis Nieves, Sascha Oerter, James Oland, José Ottenwalder, John Parmerlee, Gad Perry, Renata Platenberg, Bev Powell, Mike Powell, Andy Price, Rich Sajdak, Alejandro Sánchez, Aaron Savit, Andreas Schubert, Albert Schwartz, Don Smith, Michael Smith, Roland Sosa, Amy Sproston, Warren Steiner, Jil Swearingen, Jaime Villa, Al Winstel, Tom Zanoni (who insists that biology is the study of plants and their "parasites" instead of the study of animals and their food), many Avila College (now Avila University) students, and participants in the Avila REU programs, a number of whom have since become colleagues and friends. Each contributed to the success of our projects and we cherish the adventures shared and the times spent searching, recording, complaining, and laughing. We hope to do it for another 30 years.

Introduction

The islands, especially the shores, are teeming with lizards of every color, of every variety of marking, and of all sizes.
—Frederick Ober (1879)

One usually has to take only a few steps from the airplane to the ornamental vegetation on the airport grounds to see the first anole.
—Fläschendräger and Wijffels (2009)

Our final tally of 737 currently recognized species of frogs and non-avian reptiles (excluding the marine turtles) in the West Indies might be outdated by the time this book is published. Based on an approximate total of 700 species, Hedges (2006) determined that the West Indian islands harbor 3.0% of the world's amphibians and 6.3% of the world's known reptiles, while comprising only 0.15% of Earth's land area. He further predicted that the total number of species would most likely exceed 800, with more than 1,000 a possibility. Coupled with the great number of species are the amazingly high percentages of endemism, 99% for frogs and 93% for reptiles (Hedges, 1996). Also, because the numbers of indigenous West Indian mammals (about 60 species) and birds (fewer than 300 species) are quite modest, amphibians and reptiles are the most abundant and arguably the most important terrestrial animals. Among others, the smallest known frog, lizard, and snake (*Eleutherodactylus iberia*, *Sphaerodactylus ariasae*, *Leptotyphlops carlae*, respectively) live in the West Indies. Iguanas (both *Iguana* spp. and *Cyclura* spp.) are the largest native herbivores in many island ecosystems and, as such, are intimately implicated in the well-being of vegetative communities and the dispersal of many plants (e.g., Hartley et al., 2000). However, the islands support not only a remarkably species-rich herpetofauna, but one in which many species occur at phenomenally high densities (e.g., *Sphaerodactylus macrolepis*, to 67,600/ha on Guana Island, British Virgin Islands; Rodda et al., 2001). Because we still know so little about so many of the frogs and reptiles of the islands, the West Indies collectively represent a natural laboratory that will continue to provide an unending stream of questions for many decades to come.

Since 1991, when the precursor to this volume emanated from a "conviction that as much information as possible concerning West Indian amphibians and reptiles should be brought between two covers" (Schwartz and Henderson, 1991; = "S&H"), our knowledge of those remarkable animals has increased dramatically. In addition, the number of recognized West Indian species has increased from 585 to 737 since 1990, when S&H ceased collecting information. The incidence of introduced species also has risen considerably, often with unpredictable effects on the native animals with which they interact. In addition, concerns for natural habitats that are rapidly retreating to meet human needs have led to an awareness that many species and populations are declining, at least some have disappeared, and others are in danger of doing so in the near future. On the other hand, increased interactions between humans and the native wildlife of West Indian islands have stimulated research in environments altered to varying degrees by humans (conditions that will apply to an increasingly larger percentage of West Indian species over time). This, in turn, has led to an appreciation of the ecological versatility evident in so many species. A large percentage of the region's amphibians and some of its reptiles are intimately associated with pristine (or formerly pristine) habitats. These species are unlikely to adjust to large-scale alterations in those habitats, and are suffering from the never-ending onslaught of growing human populations that desire a standard of living that necessitates the exploitation of natural resources. However, many others not only survive but may thrive in conditions that arise as a consequence of humans extracting resources or building infrastructures (e.g., Henderson and Powell, 2001; Powell and Henderson, 2008c).

The book by S&H had two principal goals: systematic and ecological. In regard to the former, new species have been discovered at a surprisingly constant rate, and new relationships, phylogenetic and biogeographic, have been elucidated by many studies, at least some of them triggered by the need to apply new paradigms to a taxonomy that had been quite "traditional" in its reliance on alpha methodologies, slight geographic variants often captured by Albert Schwartz's discriminating eyes, and a subspecies concept that was never adequately defined and has fallen largely into disfavor. In this volume, we do not stress systematics or taxonomy, other than to embrace newly discovered and acknowledged species, which, along with those that have been recognized for many years, we use as the basis of our individual accounts. In addition, unlike the predecessor to this work, we chose not to give type localities or to list holotypes; instead, we refer readers to a recent checklist (Powell et al., 1996) and subsequent updates (Powell and Henderson, 1999, 2003) that provide full citations for all original descriptions of species-level taxa in the region. In addition, valuable taxonomic information may be gleaned from earlier checklists and updates (Schwartz and Thomas, 1975; Schwartz et al., 1978; Schwartz and Henderson, 1988; Hedges and Thomas, 1989a).

Driven by that twofold emphasis, S&H provided detailed, standardized descriptions of species (with efforts to include all variations, taxonomically recognized or not), lists of published illustrations, and distribution maps. Although the descriptions were drawn largely from previously published works, the lists of illustrations and the maps for the first time provided a newcomer to the West Indies with effective means of familiarization with the herpetofauna. Paralleling our overall increase in awareness, all these elements obviously have changed, in some instances substantially. In addition to new species, some populations (some recognized as subspecies) that were subsumed in descriptions of species as then understood are now known to represent species-level taxa on their own, and, in a few instances, populations previously thought to be distinct on the basis of color or morphology are now known to be nothing more than variable individuals or populations. Illustrations of many species have been published since 1991. Some of those had not been illustrated previously, or the only available illustrations were line drawings of specific characters rather than of entire individuals. Distribution maps generated today might look different from those in the 1991 volume. In some instances, we have discovered new populations; in others, new populations have become established largely as a consequence of human-mediated introductions. Nevertheless, and despite the importance of the descriptions (including those of anuran calls), lists of illustrations, and distribution maps in 1991, we omit them from this volume, referring the reader instead to the forthcoming field guide to amphibians and reptiles of the region (S. Blair Hedges, in prep.), which will provide photographs of nearly every taxon (and many variants) and maps that illustrate distributions, albeit without dots indicating specific localities.

Methods

We herein defined the West Indies as the Greater Antilles (Cuba, Jamaica, Hispaniola, Puerto Rico, and associated satellite islands) and Lesser Antilles (the island arc extending from Sombrero south to, and including, Grenada), plus the Bahamas, Turks and Caicos, Cayman Islands, and Swan Islands (Figures 1 and 2). A total of 737 species of frogs and reptiles has been documented on more than 600 West Indian islands, and still more islands await basic survey work. We have opted to provide accounts only for those species represented by breeding populations; therefore, we omitted introduced populations recorded for islands but no longer extant (e.g., *Hyla crucifer* on Cuba; R. I. Crombie in G. Peters, 1974) and what we considered "waif" species (often a single individual of a species that arrived on an island, usually through human-mediated circumstances). Examples of waif species in the West Indies from outside the region are *Ophisaurus ventralis* on Grand Cayman (Schwartz and Henderson, 1991), *Phrynops geoffroanus* on Anguilla (K. V. D. Hodge, in litt., 11.XII.2008), and *Crocodylus intermedius* on Grenada (Groome, 1970); examples from within the region (i.e., from other West Indian islands) include *Anolis sagrei* on St. Lucia (C. Cox, in litt., 24.VIII.2006 to D. Simberloff) and *A. bimaculatus* on Dominica (A. James, pers. comm.).

What we provide in this volume is a broad view of distributions and detailed natural history data, all organized by species. For each account, we note the distribution of the entire species, often with little regard for subspecies beyond the number currently accepted. Under "natural history," we include information pertaining to ecology, physiology, parasitology, behavior, and anything else we concluded was pertinent. We have excluded, except where they include mention of topics pertinent to natural history, phylogenetic and purely systematic studies. In addition, with the occasional exception of selectively extracted natural history data, we have omitted mention of karyological, electrophoretic, or genetic studies (including considerations of DNA or RNA sequences), since they more logically apply to systematic studies, and we also chose to exclude information on morphology (e.g., Herrel et al., 2008; Sanger et al., 2008), venom composition (but did include data on snakebites), color variation as it relates to, for example, habitat type (e.g., Thorpe, 2002; Thorpe and Stenson, 2003), and detailed analyses of frog vocalizations (e.g., Estrada, 1992; Alonso Bosch and Rodríguez Gómez, 2003c; Díaz and Cádiz, 2008) or behavioral displays, such as those of anoles and iguanas (although we did mention components of those displays).

However, in recognition of the changing realities that affect many species and their habitats and the fact that the West Indies are particularly rich in species of amphibians (Stuart et al., 2008) and reptiles (Collen et al., 2008) threatened with extinction, we have explicitly included references to the conservation status of species, a topic that was barely acknowledged in 1991. For Hispaniola, Puerto Rico, and Cuba, we include older references regarding conservation status (Aquino, 1983; Johnson, 1988; Moreno, 1991d; Vales et al., 1998) to illustrate the rapidly deteriorating circumstances of some species. For example, 19 species of Cuban amphibians were listed in Vales et al. (1998) as "vulnerable"; the acknowledged status of 17 has subsequently deteriorated to "endangered" or "critically endangered" (one remains "vulnerable" and the other is listed as being of "least concern" after new studies led to taxonomic changes).

When data for animals in nature were available, we selectively supplemented them with observations made on captive animals. For the most part, we omitted entirely observations made in geographical areas outside of the West Indies for species with ranges extending beyond the region (e.g., *Osteopilus septentrionalis*, *Rana catesbeiana*, *Iguana iguana*, *Ramphotyplops braminus*, *Crocodylus acutus*). Where we have included either captive or extralimital data, they are annotated accordingly. We have sometimes included apparently contradictory information for historical purposes or to emphasize the reality that conditions that apply in some situations are not necessarily applicable across a species' range or at all times of year.

We also have avoided generalizations. Information pertaining to a genus is not included unless explicitly applied to a particular species. For example, lizards in the genus *Anolis* are generally diurnal (although some are known to exploit the night-light niche; see also Z. Vogel, 1965a), and females in both *Anolis* (H. M. Smith et al., 1972, 1973) and *Sphaerodactylus* (Kluge, 1995) nearly always produce and lay one egg at a time (although some female anoles may contain two fully developed shelled eggs and deposit them in short order under circumstances such as those following extreme drought or stress; see, for example, Stamps, 1976b). Such general statements are noted only when specifically stated in a cited reference or if we have personally made the observation.

In some instances, if an observation could not be attributed to a particular island for those species with multi-island distributions, or, in the case of polytypic species, to a particular subspecies, we were inclined to omit the observation and reference. Also, we generally made no references to color or pattern (or variation by habitat) except in the specific context of behavior or some other category (e.g., "move slowly and rely on crypsis").

Consequently, each account includes sections on distribution, habitat, various additional aspects of natural history, and conservation status, often based on the IUCN Red Book designation (IUCN, 2001, 2008), a CITES (Convention on International Trade in Endangered Species of Wild Fauna and Flora) Appendix listing (UNEP-WCMC, 2008), or inclusion in a state or national list of endangered species (e.g., USFWS, 2008). Aspects of natural history might include abundance, activity, behavior, biomass, competition, diet and foraging, dispersal, growth, home range, movement, parasites, population size and density, predation, reproduction, sex ratio, size, tail autotomy, thermal biology, and miscellany. For anoles, we also include the topic of ecomorphology (see definition in Williams, 1972) in which we list formal ecomorphological designations based on actual data as well as similar categories that have been used in various ways to describe morphological attributes associated with particular microhabitats. When data for a particular category were unavailable, that category was omitted. The only exception was for conservation status, which we left in place with the annotations "unknown" or "not assessed" in order to emphasize the reality that we simply do not know the status of many species, and omitting mention of status might imply that those for which data are omitted are not in need of conservation.

When data were available only for populations outside the West Indies, we either state that explicitly or indicate the lack of data pertaining to West Indian populations, a decision made on a species-by-species basis. Sources for all data are cited, and, at the end of some accounts, under the topic of "additional pertinent literature," we cite accounts in the *Catalogue of American Amphibians and Reptiles* (published by the Society for the Study of Amphibians and Reptiles), the most recent editions of which (since about 1990) provide essentially complete reviews of the literature pertaining to each species.

For the most part, information was gleaned from our best efforts to thoroughly survey the pertinent literature and was supplemented with unpublished data cited in the 1991 volume and additional information accumulated in the course of our own fieldwork (indicated by our initials) or solicited from and contributed by colleagues. In some instances, secondary citations are given in order to acknowledge the actual source of an observation, particularly when these emanated from the "gray literature" (e.g., reports, abstracts, etc.) or when cited as "pers. comm." (personal communication) or "in litt." (from written correspondence). Despite our best efforts, we have undoubtedly missed some original contributors, especially when cited in compendia such as S&H, the edited volumes of Lourdes Rodríguez Schettino (*The Iguanid Lizards of Cuba* in 1999 and *Anfibios y Reptiles de Cuba* in 2003b), or the volume on Cuban frogs by Díaz and Cádiz (2008). Regardless of the source, we cite the authority for each piece of information instead of providing a cumulative list at the end of each account. However, although citations are provided for the data in each account, we have sometimes varied in the detail included and often omitted sample sizes. Consequently, we urge readers to consult the original sources before citing them in their own studies.

Many taxonomic changes have occurred since the publication of S&H. We either accepted those changes or, based on advice from trusted colleagues, decided to take a "wait and see" approach and opted for a more established taxonomy. Recent changes are noted under "Remarks" at the ends of relevant accounts. Recent papers that proposed significant taxonomic changes include Faivovich et al. (2005), Utiger et al. (2005), Frost et al. (2006), Grant et al. (2006), Le et al. (2006), Bickham et al. (2007), Burbrink and Lawson (2007), Gamble et al. (2007, 2008), Heinicke et al. (2007), Vidal et al. (2007, 2008), Hedges et al. (2008a, 2009), and Zaher et al. (2009). One up-to-date list providing the current status of the taxonomy pertaining to West Indian species is available via the online database "Caribherp" (Hedges, 2008a), although we have chosen to recognize the iguanian families not acknowledged in that list.

After completing S&H, Schwartz intimated that, in retrospect, RWH should have written all the natural history sections, and he should have written all the descriptions, as those were their respective areas of interest. Nevertheless, 18 years later, Al Schwartz is herein cited more than any other single senior author (when we exclude the IUCN Red List accounts). Why? Because, although natural history/ecology/behavior was not the primary interest or focus of his research, he recognized the value of observations on, for example, habitat, diet, and reproduction, and often included them in papers usually written to describe new taxa or elucidate the relationships within particular

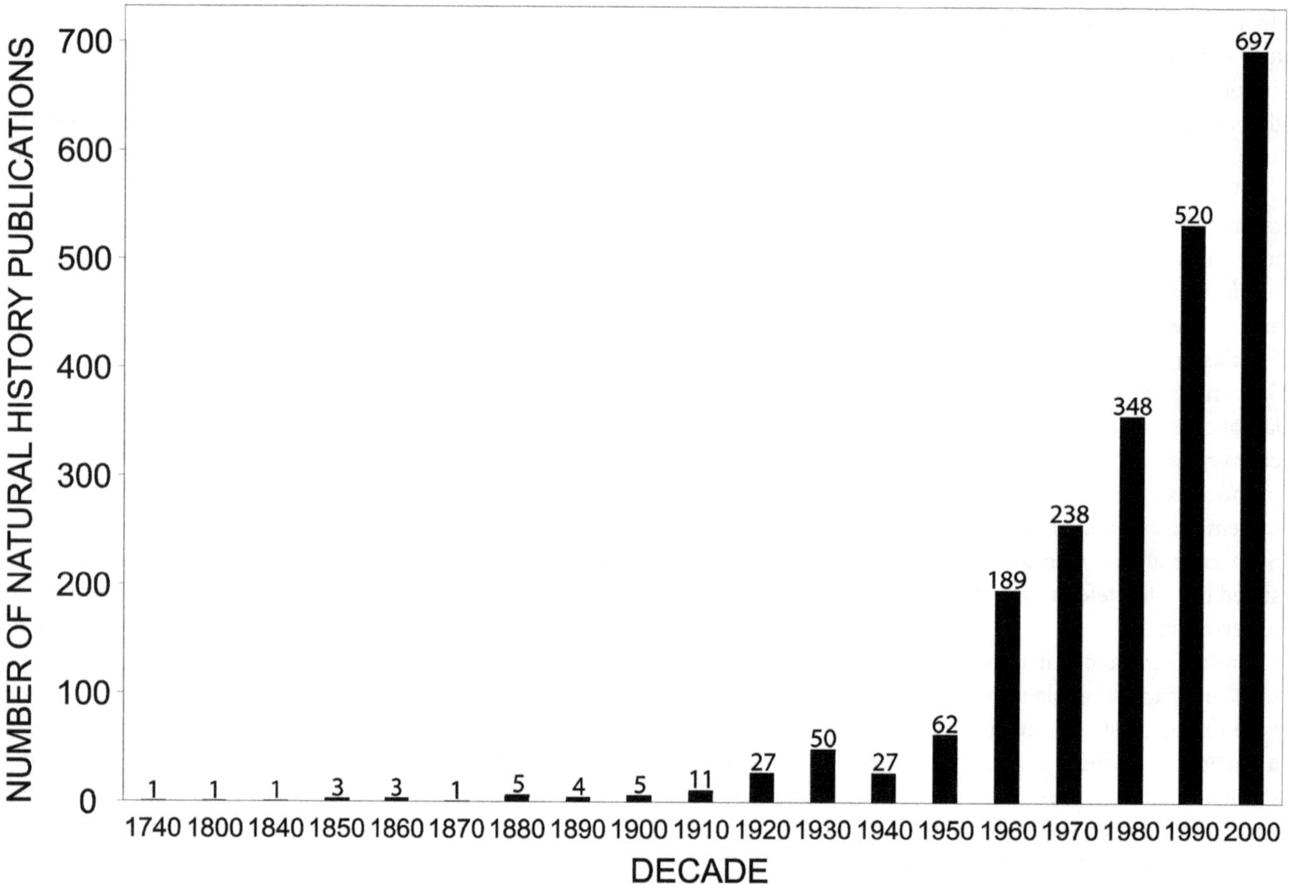

Figure 3. The number of publications produced by decade (1740–49 . . . 2000–09) that include some natural history information relevant to West Indian amphibians and reptiles. The 1910s and 1920s largely represent the efforts of Thomas Barbour (with E. R. Dunn and K. P. Schmidt contributing in the 1920s); Barbour remained active in the 1930s. The drop-off in the 1940s can be attributed to World War II, although Chapman Grant was active at this time. The 1950s mark the initiation of studies by Albert Schwartz and Ernest Williams in the region. Interest in Antillean herpetology blossomed in the 1960s with Orlando Garrido, James Lazell, A. Stanley Rand, Rodolfo Ruibal, Schwartz, Richard Thomas, and Williams as major contributors. By the 1970s, an ecological focus began to share the spotlight with alpha taxonomy through the contributions of Thomas Schoener, Judy Stamps, and others. Research activity has continued to increase in the region to the present day, with major emphases on conservation, ecology, and phylogenetics, with biologists who live on the islands as well as researchers from North America and Europe making valuable contributions. The number of publications that has appeared since Schwartz and Henderson (1991) rivals the number published during the previous 250 years.

genera or species groups. We remain indebted to him for that vision. We must also acknowledge the contributions of S. Blair Hedges and Daniel S. Townsend to S&H. Both contributed then unpublished observations to that book. However, because of the manner in which information was cited, we rarely were able to determine who contributed which observations in a particular account. Therefore, their previously unpublished observations (as well as those of Schwartz and Henderson) are now cited as "Schwartz and Henderson (1991)."

We sought to include data from all pertinent references published through the end of 2008 plus those accepted for publication in 2009 to which we had access. Despite our best efforts to be thorough, and even with the help of many colleagues who provided references or information, the fact that the topic is so broad and the resources so widely scattered almost guarantees that we missed some information. We purposely omitted some publications if they merely reiterated what appeared in an earlier publication, often choosing to credit the original source—although we sometimes included both original citations and subsequent summaries of data (e.g., Coy Otero and Lorenzo Hernández, 1982, who provided a compendium of what was then known about the parasites of Cuban amphibians and reptiles, and Stuart et al., 2008, who summarized data for amphibians threatened with extinction). Regardless, we undoubtedly were not always successful in finding, including, or citing all pertinent information. We apologize in advance to those researchers whose work we may have overlooked, the inclusion of which would have enhanced a particular species account and thus benefited the users of this book. Despite our probable omissions, we extracted information from more than 2,600 publications (plus a large number of personal communications and unpublished reports), a number that stands in sharp contrast to S&H, in which 883 publications were cited, including many that referred only to species descriptions and/or illustrations (i.e., topics not covered herein).

In summary, this compendium of natural history data in the form of individual species accounts was intended to serve two important functions: (1) To bring our accumulated knowledge of the natural history of West Indian amphibians and reptiles up to date, and, (2) because of the relative lack of information in many accounts, to point researchers in the direction of investigations figuratively crying out for someone to seek answers. We welcome anyone with an interest in natural history to join us in addressing the apparently inexhaustible supply of questions generated by observations of amphibians and reptiles in a part of the world that we continue to believe is exciting, important, and worthy of our dedication.

Where We Are Today

We are not very sanguine . . . that enough new discoveries will be made to permit unequivocal mapping of the distribution of the snakes in the Bahamas as it was on that fateful morning of October 12, 1492, when Columbus made landfall at Watling's Island. This event was not immediately prejudicial to the snake population, . . . but it sent in motion a long chain of destructive factors. . . . We can only regret that no "Cinquecento" herpetologist had opportunity to wander through the then extensive stands of tropical hardwoods and to observe the large boas which, we believe, may then have fattened on the richer avifauna in the cool glades on many of the islands.
—M. G. Netting and C. J. Goin (1944)

The State of West Indian Herpetological Natural History

In 1991, S&H estimated that only about 5% of the West Indian herpetofauna had been the subject of research dedicated to providing new natural history knowledge for a particular species or group of species (i.e., not fortuitous observations such as finding a clutch of eggs or luckily observing an act of predation, or general descriptions of habitat). Dedicated research could include examining preserved material for parasites, prey remains, or reproductive information, or it could be based on pre-planned field projects on any number of topics (e.g., habitat quantification, mark-recapture, thermal biology, social behavior, foraging behavior).

In 1999, Henderson and Powell determined that 12% of the West Indian herpetofauna had by then received dedicated study of some aspect of natural history. However, the 18 years that have elapsed since the publication of S&H have produced as many publications containing germane natural history observations as those of the preceding 140 years (figure 3). Consequently, the fact that, based on our criteria, 42.7% of the Antillean herpetofauna now has received dedicated research may not be surprising. This substantial increase can, we believe, be attributed to two factors: (1) in general, more biologists not based in the West Indies are making the region a focus of their fieldwork, and (2) Cuba and Puerto Rico are home to many biologists actively involved in herpetological fieldwork. Much, however, remains unknown about even those taxa for which some aspects of natural history have been addressed and fortuitous, anecdotal observations can (and do) contribute valuable information, especially for that portion of the herpetofauna that has not been the subject of "dedicated research." With the exception of some species now extinct, at least some habitat information is available on virtually every species.

Although field and laboratory studies of anoline lizards have dominated the literature on West Indian natural history and ecology, we were nevertheless somewhat surprised to determine that 100 species (63.7% of 157 species) had been the subjects of dedicated research (table 1). Species that have been most extensively covered include *Anolis aeneus*, *A. cristatellus*, *A. cybotes*, *A. gundlachi*, *A. lineatopus*, and *A. sagrei*, representing species from across the region. A great deal of the work on *Anolis* has been conducted by U.S.-based biologists, but also by West Indian nationals in Cuba and Puerto Rico. Also not surprising was the fact that, of 11 species of *Cyclura*, pertinent information is missing only for the extinct *C. onchiopsis*.

Natural history studies of frogs lag behind other major

Table 1. Families, genera, and number of species of frogs and reptiles that occur in the West Indies, and the number of species in each genus that have been the subjects of research dedicated to some aspect of their natural history.

Family	Genus	Number of Species	Number of Species with Dedicated Natural History Research
AMPHIBIA			
ANURA		197	53 (26.9%)
Aromobatidae	*Allobates*	1	1 (100%)
Bufonidae	*Bufo*	13	7 (53.8%)
Eleutherodactylidae	*Eleutherodactylus*	161	37 (23.0%)
Hylidae	*Hyla*	2	0
	Hypsiboas	1	0
	Osteopilus	8	3 (37.5%)
	Scinax	2	0
Leptodactylidae	*Leptodactylus*	3	2 (66.7%)
Microhylidae	*Gastrophryne*	1	0
Ranidae	*Rana*	3	1 (33.3%)
Strabomantidae	*Pristimantis*	2	2 (100%)
REPTILIA			
SQUAMATA (lizards)		364	191 (52.5%)
Anguidae	*Celestus*	21	11 (52.4%)
	Diploglossus	5	2 (40.0%)
Gekkonidae	*Gekko*	1	0
	Hemidactylus	5	2 (40.0%)
Gymnophthalmidae	*Bachia*	1	0
	Gymophthalmus	2	2 (100%)
Iguanidae	*Cyclura*	11	10 (90.9%)
	Iguana	2	2 (100%)
Leiocephalidae	*Leiocephalus*	25	15 (60.0%)
Phyllodactylidae	*Phyllodactylus*	4	0
	Tarentola	3	2 (66.7%)
	Thecadactylus	1	1 (100%)
Polychrotidae	*Anolis*	157	100 (63.7%)
Scincidae	*Mabuya*	4	2 (50.0%)
Sphaerodactylidae	*Aristelliger*	7	2 (28.6%)
	Gonatodes	2	1 (50.0%)
	Sphaerodactylus	88	23 (26.1%)

Teiidae	*Ameiva*	22	14	(61.9%)
	Cnemidophorus	1	1	(100%)
	Kentropyx	1	0	
Xantusiidae	*Cricosaura*	1	1	(100%)
SQUAMATA (amphisbaenians)		17	6	(35.3%)
Amphisbaenidae	*Amphisbaena*	15	4	(26.7%)
Cadeidae	*Cadea*	2	2	(100%)
SQUAMATA (snakes)		148	56	(37.8%)
Boidae	*Boa*	2	1	(50.0%)
	Corallus	2	2	(100%)
	Epicrates	9	7	(77.8%)
Colubridae	*Chironius*	1	1	(100%)
	Mastigodryas	1	1	(100%)
	Pantherophis	1	0	
	Tantilla	1	0	
Dipsadidae	*Alsophis*	8	8	(100%)
	Arrhyton	8	1	(12.5%)
	Borikenophis	3	2	(66.7%)
	Caraiba	1	1	(100%)
	Clelia	2	0	
	Cubophis	5	2	(40.0%)
	Haitiophis	1	1	(100%)
	Hypsirhynchus	8	6	(75.0%)
	Ialtris	4	2	(50.0%)
	Liophis	4	2	(50.0%)
	Magliophis	2	2	(100%)
	Tretanorhinus	1	1	(100%)
	Uromacer	3	3	(100%)
Leptotyphlopidae	*Leptotyphlops*	9	0	
Natricidae	*Nerodia*	1	0	
Tropidophiidae	*Tropidophis*	27	8	(29.6%)
Typhlopidae	*Ramphotyphlops*	1	0	
	Typhlops	41	4	(9.7%)
Viperidae	*Bothrops*	2	2	(100%)
REPTILA: TESTUDINES		8	6	(75.0%)
Emydidae	*Trachemys*	5	4	(80.0%)
Pelomedusidae	*Pelusios*	1	1	(100%)
Testudinidae	*Chelonoidis*	2	1	(50.0%)
REPTILIA: CROCODILIA		3	3	(100%)
Alligatoridae	*Caiman*	1	1	(100%)
Crocodylidae	*Crocodilus*	2	2	(100%)
TOTAL		737	315	(42.8%)

taxonomic groups, with only 27% of 197 species receiving attention, and only 23% of the 161 species of *Eleutherodactylus* have been the subjects of some concentrated research. U.S.- and Puerto Rican-based biologists lead the way in work on *Eleutherodactylus* (especially *E. antillensis*, *E. cooki*, *E. coqui*, and *E. johnstonei*; work on Puerto Rican species has, in general, been covered exceptionally well), with some coming from Cuban nationals as well. Two species of *Bufo* (*B. marinus* and *B. peltocephalus*), as well as two species of *Osteopilus* (*O. brunneus* and *O. septentrionalis*) have received more than passing attention.

Of the three major West Indian radiations, the diminutive geckos in the genus *Sphaerodactylus* also are underrepresented in dedicated research, with only 26% of 87 species having received some attention. The small size and secretive habits of these geckos undoubtedly have prohibited more extensive fieldwork. Of the 148 species of snakes, fewer than 40% have received some attention, primarily by U.S.- and Puerto Rican-based researchers. Nine of 13 species of boids (particularly *Corallus grenadensis*, *Epicrates inornatus*, and *E. monensis*) and 27 of 43 species of alsophiines (*Alsophis*, *Arrhyton*, *Borikenophis*, *Cubophis*, *Hypsirhynchus*, *Ialtris*, *Magliophis*, and *Uromacer*), especially *B. portoricensis*, have been the focus of research. Although dietary information existed for 33 of 51 species of Greater Antillean macrostomatan snakes recognized in 1996 (Rodríguez-Robles and Greene, 1996), we found dedicated natural history information on only 4 of 41 species of *Typhlops* and none of the 9 species of *Leptotyphlops* known to occur throughout the region. As for *Sphaerodactylus*, their secretive habits undoubtedly discourage more extensive coverage. Cuba, Hispaniola, and Grand Cayman have been the sites of turtle and crocodilian fieldwork, with 75% of the turtles and all three species of crocodilians having been studied to some extent.

Despite our documentation of increased attention to the natural history/ecology of West Indian frogs and reptiles, the number of species receiving prolonged attention (i.e., field study spanning one year or more) still remains remarkably small. Among anurans, such data are available only for *Eleutherodactylus cooki*, *E. coqui*, and *E. johnstonei*; multi-year fieldwork with *Pristimantis euphronides* is currently ongoing. Lizards fare much better; several species of *Cyclura* have received extensive study, with *C. cychlura* the focus of an ongoing field study for almost 30 years. *Cyclura carinata*, *C. lewisi*, *C. nubila*, *C. pinguis*, *C. rileyi*, and *C. stejnegeri* also have been the subjects of long-term field studies. Despite the attention received by anoline lizards, only *Anolis aeneus*, *A. distichus*, *A. lineatopus*, *A. sagrei*, and *A. stratulus* get singled out in the context of long-term attention. Among the Teiidae, only *Ameiva exsul* and *A. pleii* meet these criteria. The snakes *Epicrates inornatus* and *E. monensis* stand out (and a current long-term study is focused on *E. angulifer*); *Corallus grenadensis* is borderline, with data available for multiple years, but limited with regard to specific individuals. Among colubrids, only *Borikenophis portoricensis* has been targeted for long-term fieldwork in the British Virgin Islands. *Crocodylus acutus* has received prolonged attention in Haiti and the Dominican Republic, as has *C. rhombifer* in Cuba.

Threats to the Herpetofauna

Since the arrival in the Antilles of Amerindians, frogs and reptiles have been both beneficiaries and victims of human activities. An *Anolis aeneus* undoubtedly perched on the first thatched hut erected on Grenada. On various islands, piles of discarded vegetable debris became homes for species of *Sphaerodactylus*, *Celestus*, and/or *Bachia*. On St. Vincent, broken pieces of pottery became shelters for *Ameiva ameiva*, and, as forests were cleared, those same *Ameiva* exploited the new open habitat for basking and foraging. Conversely, as forests on St. Vincent were cleared, the ranges of *Pristimantis shrevei* and *Chironius vincenti* contracted, but, with more edge habitat, that of the arboreal boa *Corallus cookii* probably expanded (S. B. Powell et al., 2007). With the arrival of Europeans, habitat transformations accelerated, and many alien plants and animals were introduced. The 18th and 19th centuries saw the clear-cutting of forests to make way for agriculture (especially sugar), reducing habitats available for forest-adapted species. Today, burgeoning human population growth coupled with rampant development (often tourism driven) on many islands does not bode well for many segments of the herpetofauna. For example, recent assessments (Young et al., 2004; Stuart et al., 2008) found that 84% of amphibian species are threatened, with 37% listed as Critically Endangered, 36% as Endangered, and 11% as Vulnerable. No other region of the world has such a high proportion of threatened amphibians, and even within the Neotropics, the Greater Antilles represent a geographic concentration of threatened species (Bolaños et al., 2008). Henderson and Powell (2001) estimated that human-introduced resources benefited at best 5–10% of the herpetofauna (almost entirely reptiles), whereas at least 50% (mostly amphibians) have been negatively affected. Habitat alterations, however, have had an impact to some degree on 100% of the West Indian herpetofauna.

Introduced Invasive Predators and Competitors

Over the past 500+ years (and likely long before that), alien species have entered the West Indian islands, almost always with human assistance, intentional or inadvertent, and, as a group, have had a devastating impact on the herpetofauna. Non-native carnivores consume amphibians and reptiles, large mammalian herbivores severely alter habitats and may compete with herbivorous lizards, and invasive amphibians and reptiles may compete with and/or consume native species or even compromise the gene pools of related endemics through hybridization.

The Small Indian Mongoose (*Herpestes javanicus*) was introduced initially into Jamaica in 1872 (Espeut, 1882) and subsequently became established on more than 20 additional islands, some as late as the early 20th century (Hoagland et al., 1989). *Herpestes* has been implicated in the extirpation or extinction of several species of squamate reptiles, most notably ground-dwelling lizards and snakes (Powell and Henderson, 2005a: table 2), but anguid (*Celestus, Diploglossus*) and scincid (*Mabuya*) lizards have likely been impacted by this voracious predator as well (Powell and Henderson, 2005a). Mongooses also are possible nest predators on American Crocodiles (*Crocodylus acutus*) on Hispaniola (Thorbjarnarson, 1988), and they are known to prey on frogs (*Eleutherodactylus* spp.) in forested areas in eastern Puerto Rico (Viella, 1998).

Introduced opossums (*Didelphis marsupialis*) and raccoons (*Procyon lotor*) may be implicated in the extirpation of *Leptodactylus fallax* on Martinique, where they hunt along streams and rivers, and raccoons are known predators of *Cyclura* in the Bahamas and likely predators of *Iguana delicatissima* and *I. iguana* in the Lesser Antilles (Breuil, 2009).

Domestic cats (*Felis domesticus*) prey on a variety of lizards, including *Anolis* (Rand, 1967a), *Cyclura* (Iverson, 1978), *Leiocephalus* (Iverson, 1978), and *Ameiva* (Daudin and de Silva, 2007). Snake populations also have been impacted by house cats. Schwartz and Henderson (1991) noted an instance of predation on *Borikenophis portoricen-*

Table 2. Presence or absence of species of *Ameiva, Alsophis, Liophis,* and the mongoose (*Herpestes javanicus*) on Lesser Antillean islands.

Island	Mongoose	*Ameiva*	*Alsophis*	*Liophis*
Sombrero	–	+	x	x
Anguilla	–	+	+	x
Scrub	–	+	+	x
Little Scrub	–	+	x	x
St.-Martin/St. Maarten	+	+	+*	x
St.-Barthélemy	–	+	+	x
Saba	–	x	+	x
St. Eustatius	–	+	+	x
St. Christopher	+	+*	–	x
Nevis	+	+*	–	x
Antigua	+	+*	–	x
Great Bird	–	+	+	x
Barbuda	–	+	x	x
Redonda	–	+	x	x
Montserrat	–	+	+	x
Guadeloupe	+	x	+*	+*
Terre-de-Haut	–	x	+	+
Terre-de-Bas	–	x	+	x
Marie-Galante	+	x	–	+*
Dominica	–	+	+	+
Martinique	+	x	–	x
Rocher de Diamant	–	x	x	+
St. Lucia	+	x	x	–
Maria Major	–	x	x	+
Barbados	+	x	x	–
St. Vincent	+	+*	x	x
Grenada	+	+*	x	x

Notes: "+" indicates presence and "–" indicates absence of the mongoose or absence of a species of *Ameiva, Alsophis,* and/or *Liophis* from an island where that species once occurred. An asterisk (*) indicates that, if the species still occurs on the island, it is very rare or restricted to isolated enclaves. "x" denotes the absence of any evidence of that taxon occurring on that island. Many small islands and cays not listed here may support *Ameiva* populations and have no mongooses. *Alsophis* also is known from Barbuda, but only as a fossil.

sis. Grant (1940a) observed a decline in *Epicrates subflavus* in the Portland Point area of Jamaica after the introduction of house cats, and, in 1991–1992, three *Epicrates monensis granti* were rescued from house cats on St. Thomas, U.S. Virgin Islands (Tolson and Henderson, 1993). García et al. (2001) found that 33% of the stomachs of feral cats on Isla Mona contained the remains of reptiles.

Dogs (*Canis familiaris*), in collaboration with cats, are responsible for the near-extirpation of *Cyclura carinata* from Pine Cay in the Caicos Islands (Iverson, 1978). Of an estimated adult population of nearly 5,500 iguanas (15,000 total population in June-July 1974), by mid-1976, no iguanas were flushed during five transect surveys, and evidence by spoor of only five iguanas was found in June 1976. Similarly, cats and dogs have been implicated in the extermination of *Tropidophis bucculentus* on Navassa (Powell, 1999a) and as predators of the large anuran *Leptodactylus fallax* (Faaborg and Arendt, 1985). Dogs have been observed as nest predators on *Crocodylus acutus* on Hispaniola (Thorbjarnarson, 1988).

Likely introduced inadvertently during early European settlement, rats (*Rattus norvegicus* and *R. rattus*) may be predators on eggs and young of reptiles, and may modify habitats by exploiting seeds and seedlings. Dodd and Ortiz (1983) suggested that predation by rats may account for the rarity of *Sphaerodactylus micropithecus* on Isla Monito. Rats also have been implicated in frequent attacks on racers (*Alsophis sajdaki*) prior to the rat eradication program on Great Bird Island (Daltry, 2007). Tolson and Henderson (2006) suggested that small species of snakes were most vulnerable, and Tolson (1988) found that *Epicrates monensis granti* was absent at all sites on the Puerto Rico Bank that did not have a rat predator to control rodent populations, thus rendering those sites unsuitable for the boas. When *E. m. granti* was reintroduced to islands where rats had been eradicated, boa populations flourished.

Grazers such as goats (*Capra hircus*), cattle (*Bos taurus*), sheep (*Ovis aries*), and donkeys (*Equus asinus*) can quickly eliminate palatable vegetation from an area, degrading habitats that once provided food and cover for native frogs and reptiles. On The Quill on St. Eustatius, we have observed the efficient manner in which goats eliminate ground vegetation and low understory in an otherwise beautiful stand of forest. *Alsophis rufiventris* nevertheless remained plentiful. Conversely, Tolson et al. (2007) failed to find evidence of *Epicrates m. monensis* in goat-browsed depression forests on Isla Mona. Cattle on Anegada (British Virgin Islands) compete directly and successfully with the only remaining natural population of *Cyclura pinguis* (Bradley and Gerber, 2005). Additionally, pigs (*Sus scrofa*) destroy native vegetation by rooting and also may consume snakes (Tolson and Henderson, 2006). Netting and Goin (1944) believed that the introduction of pigs to Stranger's Cay on the Little Bahama Bank was responsible for the extirpation of *Epicrates exsul*, and Hedges and Díaz (2009) noted the impact of pigs (killing metamorphs, disturbing adults, clouding the water) on a population of *Bufo longinasus* in Cuba.

Three West Indian species of anuran amphibians (*Bufo marinus*, *Osteopilus septentrionalis*, and *Eleutherodactylus johnstonei*) are widespread outside their native ranges. Although Lever (2003) considered all West Indian populations of *B. marinus* to be introduced, some may have arrived in the islands by natural means. This toad occurs in a wide range of habitats, from nearly intact rainforest to urban settings made up mostly of concrete (e.g., Mallery et al., 2007). The range of the Cuban Treefrog (*O. septentrionalis*) has increased in recent years, largely as a result of transport of ornamental plants and building materials from the southeastern United States to resort sites (Townsend et al., 2000); the frogs are hidden either in the plants themselves or in containers of plants and building materials. These large frogs prey on smaller anurans or out-compete them for food, especially in altered habitats around human habitations, including outbuildings, swimming pools, decorative gardens, and other sources of fresh water (including cisterns for drinking water). *Eleutherodactylus johnstonei*, like *O. septentrionalis*, has been introduced to many islands via accidental transport (Kaiser, 1992). Once established, these frogs quickly become ubiquitous at low to moderate elevations, and populations seem to explode in a variety of habitats, but especially in disturbed situations. However, these frogs also occur in relatively undisturbed habitats, and may have the potential to out-compete endemic species of, for example, *Pristimantis shrevei* on St. Vincent (Powell and Henderson, 2007) and *P. euphronides* on Grenada (Henderson and Berg, 2006). On Grenada, *E. johnstonei* and *P. euphronides* once appeared to have essentially parapatric distributions based on elevational boundaries (Kaiser, 1992), but now the two species are syntopic in some forested sites at elevations over 500 m (Henderson and Berg, 2006). The large ranid frog, *Rana catesbeiana*, native to the United States, has been introduced to each of the Greater Antilles. Although it feeds largely on invertebrates, it is a voracious predator known to take fishes, smaller individuals of its own species, and smaller species of native frogs (i.e., *Leptodactylus albilabris* on Puerto Rico; Thomas and Joglar, 1996).

As with the frogs, a few reptiles have been remarkably successful in establishing a presence on islands where they are not native. The emydid turtle, *Trachemys scripta*, is native to the southeastern United States, but has been introduced to a number of West Indian islands on which it may compete for resources or hybridize with native congeners (Breuil, 2002b; Powell and Incháustegui, 2009). *Iguana iguana* raises similar concerns where it has been introduced to islands that harbor the West Indian endemic *I. delicatissima* (e.g., Breuil, 2002b). Again, both competition for resources and interbreeding may ensue. Two anoline species, *Anolis cristatellus* and *A. sagrei*, have been incredibly successful colonizers. The former is native to the Puerto Rico Bank, but has been in the Dominican Republic for decades (Fitch et al., 1989) and was recently documented on Dominica, where it may be displacing the island endemic *A. oculatus* in some habitats (Malhotra et al., 2007). *Anolis sagrei* is native to the Bahamas, Cuba (and associated islands), and Little Cayman. In the West Indies, it has been introduced (likely accidentally in most cases and often via populations in the southeastern United States, themselves introduced) to Jamaica (which may or may not have been human mediated), Grand Cayman, Swan Island, St. Vincent, and Grenada (as well as to locations far removed from the West Indies; Lever, 2003). On Grand Cayman, *A. sagrei* has caused behavioral changes and shifts in habitat use by the endemic *A. conspersus* (Losos et al., 1993). Like *Osteopilus septentrionalis*, it was probably introduced to some islands with building materials or plants. On Grenada and St. Vincent, it appears to be restricted to severely altered habitats in urban and suburban situations, and the displacement of endemic anoles currently seems unlikely (Greene et al., 2002; Treglia et al., 2008). The widely established *Ramphotyphlops braminus* may compete with endemic typhlopids or leptotyphlopids (e.g., with *Leptotyphlops carlae* on Barbados; Hedges, 2008c). The colubrid snake *Pantherophis guttatus* (formerly *Elaphe guttata*) is native to the eastern United States and bred by the tens of thousands in captivity for the pet trade. It is a proficient climber, and its diet consists largely of endotherms, at least in the species' native range. Individuals have been documented on several West Indian Islands, but, to the best of our knowledge, reproducing populations are known so far only from Grand Cayman (Seidel and Franz, 1994) and St. Thomas in the U.S. Virgin Islands (Perry et al., 2003a; Platenberg and Boulon, 2006). Should its distribution on St. Thomas expand, it could have a "severe detrimental impact on native wildlife" (Platenberg and Boulon, 2006) every bit as devastating as that of the introduction of the Brown Tree Snake (*Boiga irregularis*) on the native wildlife of Guam and possibly that of Hawaii (e.g., Kraus and Cravalho, 2001; Rodda and Savidge, 2007).

An introduced catfish (*Claria gariepinus*) is a potential predator of eggs and tadpoles of water-breeding frogs (*Bufo*, *Osteopilus*) in Cuba, where it now occurs throughout the island in most rivers and subterranean waters (Hedges and Díaz, 2009).

Although many habitats in the West Indies are dominated by non-native plants, and invasive species of plants may negatively impact frogs and reptiles, no data currently document such effects (see, however, comments on the potential effects of Coralita, *Antigonon* sp., on St. Eustatius; Powell et al., 2005; Powell, 2006). Nevertheless, as noted by Wilson et al. (2006), invasive plants effectively alter habitats and prey bases in much the same way as introduced herbivores.

Habitat Loss and Alteration

Habitat destruction is the major cause of species extinction (Tilman et al., 2002). Recent research (Gallant et al., 2007) has demonstrated that many regions of the world that support the richest assemblages of amphibians also are undergoing the highest rates of landscape modification. That probably applies to reptiles as well. The West Indies have a long history of habitat alterations, beginning with the first human inhabitants about 4,000 years ago. Some modifications to habitats probably occurred with the arrival of the earliest colonizers; for example, Archaic hunter-gatherers might have introduced exotic food crops and engaged in some low-level food production (Newsom and Wing, 2004). Between 500 and 250 BC, habitat alterations likely accelerated with the arrival of people in the Lesser Antilles and Puerto Rico from the Orinoco drainage and the river systems of South America's northeastern coast. These invaders made ceramics and grew food crops. They brought with them dogs and agoutis (*Dasyprocta agouti*) and domesticated plants from the mainland that included manioc (*Manihot esculenta*), the primary food crop (Wilson, 1997). Ceramic Age cultivation of corn (*Zea mays*), sweet potato (*Ipomoea batatas*), and manioc (among other cultivated food crops) certainly would have required extensive expanses of cleared ground (although corn may not have been as widely cultivated in the West Indies as on the Neotropical mainland, being overshadowed by the primary system of root cropping; Newsom and Wing, 2004; but see Lane et al., 2008). Habitat destruction increased exponentially with the arrival of European settlers in the late 15th century. Forest giants were felled for housing and boat building, and vast expanses of forest were cleared for agriculture, most notably for sugarcane and later for bananas. In addition, use of woody plants for fuel (wood or charcoal) occurred whenever and wherever people settled, even ephemerally—and that exploitation continues to this day. The landscapes of many islands became dissected by plantations, creating a patchwork of habitat islands that, to this day, dictates intra-island distributions and inhibits gene flow among anuran and reptilian populations.

Myers et al. (2000) and M. L. Smith et al. (2005) considered the West Indies a biodiversity hotspot; however, only 11.3% of the original vegetation remains in the region (Meyers et al., 2000), and a more realistic estimate of intact vegetation might be 10% (Hedges, 2006). Although numbers may be misleading, the percentage of primary forest remaining on specific islands ranges from virtually none in Haiti (on Hispaniola) to about 36% on Dominica (FAO, 2005). Despite the exploitation by native frogs, lizards, and snakes of introduced coconut palms, banana, cacao, and coffee trees, and human edifices and associated debris (table 3), relatively few species have benefited from human-influenced resources (Henderson and Powell, 2001), although edge-loving anoles may be far more abundant now than when West Indian forests were intact (e.g., G. Peters, 1970). Introduced orchard and forest trees (e.g., mahogany) may simulate natural vegetation patterns, but complex habitats are used more frequently by more species, and habitats associated with clear-cuts are rarely used (Henderson and Powell, 2001). On a more positive note, satellite image-based mapping on several islands in the Lesser Antilles indicated that cultivated land areas declined 60–100% from about 1945 to 2000, and forest cover increased 50–950%, a trend likely to continue where sugarcane cultivation previously dominated (Helmer et al., 2008).

Caribbean islands are popular tourist destinations, and tourism is important to the economies of many island nations. Tourism also places pressure on the islands to provide housing and vacation-related activities for visitors. Expansive resorts and golf courses impact a variety of habitats, but especially those in coastal areas. Some islands promote and feature nature-related tourism (e.g., St. Eustatius, Dominica), but such "nature" tours often are nothing more than a short ride in a minibus and an abbreviated hike in a national park or reserve—nevertheless, they do provide incentives for conserving the remaining natural areas.

Accidental Killing

For terrestrial (nonmarine) wildlife, the major threat from accidental killing likely occurs on roads. Anurans often are victims of automobiles on rainy nights, and reptiles, especially larger species, are frequently killed while crossing or perhaps thermoregulating on roads. Some of the reptiles we (or colleagues) have observed dead or alive on roads in the West Indies include turtles (*Trachemys stejnegeri*), lizards (*Iguana* spp., *Cyclura* spp., *Ameiva* spp., *Leiocephalus* spp., *Anolis* spp.), and a variety of snakes, including arboreal species (e.g., *Boa nebulosa*, *Corallus grenadensis*, *Epicrates striatus*, *U. oxyrhynchus*). As resident and transient human populations increase on various islands, so too will paved roads, automobile traffic, and wildlife mortality.

Intentional Killing

Exploitation of West Indian frogs and reptiles has a long history (e.g., Powell, 2003a). On Montserrat, as many as 50 individuals of the critically endangered frog *Leptodactylus fallax* were killed daily for consumption by local people and tourists (Daltry and Gray, 1999). Large iguanid lizards and their eggs have been exploited for food, likely for more than 2,000 years. Ample evidence suggests that species of *Cyclura* and *Iguana* were important for Amerindian subsistence; *Cyclura* has been reported from archeological sites on Jamaica, the Turks and Caicos, and the Bahamas (Wing, 1989; Newsom and Wing, 2004); similarly, *Iguana* remains have been taken at sites on the Virgin Islands, St. Eustatius, St. Christopher, Antigua, Marie-Galante, St. Lucia, and Grenada (Wing, 1989; Newsom and Wing, 2004), although those from the Puerto Rico Bank (i.e., the Virgin Islands) probably represent *Cyclura pinguis* (Wing, 2001 lists those remains as "Iguanidae"). Today, *Iguana* is still harvested for human consumption in the Lesser Antilles, and, although all species of *Cyclura* are protected, subsistence hunting still occurs (Powell and Incháustegui, 2009).

Several species of snakes are exploited for so-called medicinal purposes. On Dominica, *Boa nebulosa* is killed for fat processed as "snake oil" (Malhotra et al., 2007), and the Cuban *Epicrates angulifer* is hunted for its meat, fat (for curing hemorrhoids and respiratory problems), and hide (for belts and shoes; Rivalta et al., 2003a). During a seven-year span on Martinique, more than 50,000 *Bothrops lanceolatus* were killed for bounties (Pinchon, 1967). Snakes also are routinely killed for no reason other than for being snakes (e.g., Viña Dávila et al., 1984; Thomas et al., 2007). Certainly residents of West Indian islands (with the notable exceptions of Martinique and St. Lucia) must be aware that their resident snakes are not a threat, especially since no one on those islands has ever died from snakebite envenomation.

Freshwater turtles (*Trachemys stejnegeri*) are (or were) routinely hunted for food on Puerto Rico (K. P. Schmidt, 1928), Hispaniola (K. P. Schmidt, 1921; Seidel and Incháustegui, 1984; Powell, 2003a), and Marie-Galante (Breuil, 2002b). Likewise, tortoises (*Chelonoidis carbonaria*), as a popular source of protein, are raised and maintained in captivity for that purpose. On some islands, pens of tortoises in the yards of residential areas are not uncommon (although many involve "pets" rather than dinner).

Crocodilians (*Crocodylus acutus* and *C. rhombifer*) have

a long history of exploitation for a variety of purposes. In the Dominican Republic, *C. acutus* is hunted for meat and for fat for the local pharmaceutical industry (Powell, 2003a); it also is killed indiscriminately for "sport." At Etang Saumâtre in Haiti and at Lago Enriquillo in the Dominican Republic, locals kill male *C. acutus* to collect their penises for use as an aphrodisiac (Schubert and Santana, 1996). In Cuba, *C. rhombifer* is farmed for its meat and hides (e.g., Rivalta et al., 2003a).

Although certain species of West Indian frogs and reptiles have been exploited as food for many hundreds of years, no one before (or since) has expressed so great an appreciation for the gustatory pleasures of eating a frog or a turtle as has Thomas Barbour. In reference to the large frog *Leptodactylus fallax*, Barbour (1930a) mused that they are "hunted for good reason, for their delicious flesh. To expect that this should not be so, would be to expect too much of mere human beings." About the freshwater turtle, *Trachemys decussata*, he on the one hand (Barbour and Ramsden, 1919) warned that, if precautions were not soon taken, the species would be exterminated; conversely, he (Barbour, 1934) observed "how extremely pleasant a pastime eating tortoises [= *Trachemys*] really is." See also Henderson and Powell (2004b) for additional observations by Barbour regarding the dining pleasures provided by the West Indian herpetofauna.

Table 3. Percentages of West Indian frogs, lizards (excluding amphisbaenians), and snakes that are known to associate, at least occasionally, with one or more of 13 habitat variables resulting from human impact.

	Frogs	Lizards	Snakes
Variable	(197)	(364)	(148)
Edificarian[a]	7.6	30.8	14.2
Roads, road cuts, airport runways	6.1	9.3	3.4
Human debris[b]	11.7	32.4	20.9
Posts[c]	0.0	19.0	0.0
Rock piles and roadside berms	1.5	11.3	6.1
Mine shafts, tunnels, quarries	0.5	<0.1	0.7
Non-native trees[d]	21.8	39.6	28.4
Cultivated fields[e]	7.1	12.9	5.4
Pastures and clearings	7.1	13.5	14.2
Gardens[f]	6.6	16.2	10.8
Potted plants	1.0	<0.1	1.4
Artificial bodies of water[g]	6.6	0.0	1.4
Eats human food or scraps, *Rattus/Mus*, other introduced prey or forage, or is fed by humans	0.0	5.8	14.9

Notes: These are undoubtedly very low estimates. Almost every species found in the West Indies is affected to some degree by human activity, and a large number must exploit human-related resources; however, we list only those that have been explicitly recorded in the literature or we have personally witnessed. Nevertheless, the relatively high percentages in many categories speak eloquently to the ecological tolerances of many species (especially lizards). Numbers of species are listed in parentheses below each category.

a. e.g., buildings, ruins, rock walls, drains.
b. e.g., garbage dumps, *Cocos* trash, compost piles, boards and slats.
c. e.g., fences, dock pilings, telephone poles, railroad ties.
d. e.g., coconut, banana, coffee, cacao, mango, citrus, nutmeg, breadfruit, guava, papaya, *Casuarina*.
e. e.g., sugarcane, corn, pineapple.
f. e.g., yams, strawberries, ornamentals.
g. e.g., cisterns, troughs, pools, road ruts, rice fields.

Restricted Population Sizes and Ranges

Many West Indian endemics are geographically limited to either a single small island or a small area on a larger island (i.e., one of the Greater Antilles) and are, as a consequence, vulnerable to stochastic events, including inadvertent or intentional alterations of already limited habitats. Examples of the former include the melanistic species of *Ameiva* on Redonda (*A. atrata*), Sombrero (*A. corvina*), and Little Scrub Island off Anguilla (*A. corax*), other species of *Ameiva* (*A. alboguttata* on Isla Mona and *A. desechensis* on Desecheo), the extant Navassa Island endemics (*Anolis longiceps, Celestus badius, Sphaerodactylus becki*, and *Aristelliger cochranae*), and several species of *Alsophis* and *Liophis*. *Alsophis antiguae* is extirpated from Antigua, and the closely related *A. sajdaki* until recently was restricted to Great Bird Island (a 9.9 ha Antiguan satellite); it since has been introduced to Rabbit Island (2 ha) and Green Island (43 ha; Daltry, 2007). *Liophis cursor*, now extirpated from Martinique, occurs only on Rocher de Diamant (0.06 km^2), and *L. ornatus*, formerly on St. Lucia, today is found only on Maria Major (about 10 ha). The distribution of the teiid lizard *Cnemidophorus vanzoi* also was restricted to the Maria Islands (Major and Minor), with a total potential range of 10–11 ha, and a translocated population now on Praslin Island (1.1 ha; Dickenson et al., 2001). Other examples include *Allobates chalcopis* on Martinique, *Gonatodes daudini* on Union Island, *Mabuya macleani* on Carrot Rock, *Diploglossus montisserrati* on Montserrat, and *Typhlops annae* on St.-Barthélemy. Representatives of genera that have major radiations in the Greater Antilles (*Eleutherodactylus, Sphaerodactylus, Anolis, Tropidophis, Typhlops*) often are known from only a single collecting locality (sometimes from only a single specimen), or several localities in close proximity, often in specialized habitat, or from a small island satellite to the major island (especially Cuba and Hispaniola). Although ranges of some of these species may eventually be found to be broader, others are truly ecologically limited—and consequently vulnerable to the negative effects of human activities.

Climate Change

The earth is undergoing human-mediated climate changes that are adversely affecting biodiversity on a global scale (e.g., Vinnikov et al., 1999). One subtle but insidious response to global warming is that species with temperature-related sex determination (e.g., species of *Trachemys* and crocodilians) could develop skewed sex ratios. Other long- and short-term effects of altered temperature regimes have been studied in Puerto Rico. Burrowes et al. (2004) analyzed temperature and precipitation data between 1970 and 2000. The analysis indicated a significant warming trend and an association between years with extended periods of drought and the decline of frogs. Furthermore, the 1970s and 1990s were significantly drier than average, and were characterized by substantial amphibian extirpations and declines. Burrowes et al. (2004) suggested a possible synergistic interaction between drought and the pathological effect of the chytrid fungus on frog populations (the "climate-linked epidemic" hypothesis of Pounds and Crump, 1994). Similarly, Stewart (1995) observed a drop in numbers of *Eleutherodactylus coqui* at El Verde, Puerto Rico, that was correlated with an increased number of consecutive days without at least 3 mm of rain. Juveniles cannot survive extensive droughts, and extended dry periods may be lethal to adults that are inhibited from feeding in order to avoid desiccation (Stewart, 1995).

Natural Disasters

The threat of hurricanes is an annual event in the Caribbean, and they can have substantial effects on vegetation, forest structure, and biotic communities (e.g., Walker et al., 1991). Because the islands have been subjected to hurricanes for thousands of years, however, they probably have caused few extinctions or even extirpations, although Breuil (2002b) implicated hurricanes in the extinction of *Ameiva major*. On the other hand, because of the intermittent nature and unpredictability of hurricanes, native species probably have not evolved behaviors adaptive to surviving hurricanes. Based on data collected between January 1987 and October 1990, Woolbright (1991) found that the impact of Hurricane Hugo in September 1989 on frog populations in Puerto Rico depended on frog species and age class. Survivorship estimates of adult *Eleutherodactylus coqui* for the period including the storm were within the normal range from previous years, but juveniles suffered severe reductions. By October 1990, adult population density had increased fourfold over prehurricane numbers (although adults were smaller), and juvenile numbers were increasing but had not reached pre-Hugo levels. Woolbright speculated that the increase in numbers might have been due to an increase in retreat sites and a decrease in invertebrate predators. Auditory censuses indicated density changes ranging from a 14% increase in previously common *E. hedricki* to an 83% decrease in the already rare *E. richmondi* (Woolbright, 1991). Population densities of *E. coqui* since Hurricane Hugo have been influenced by succession (Woolbright, 1996). Woolbright (1997) further discussed the longer-term impact of Hurricane Hugo, noting that population declines, not apparent immediately after Hugo, were observed during the span of years (1990–1994) following the hurricane. *Eleutherodactylus portoricensis* also was found to have been negatively impacted by Hugo, and Joglar and Burrowes (1991) suggested that it might not have been capable of adapting to storm-mediated changes in its habitat.

The impact of Hurricane Hugo on Puerto Rican anoline lizards was perhaps even more dramatic. *Anolis stratulus*, a canopy species, was abundant near ground level one month post-Hugo, as was the ground-to-canopy *A. evermanni*; *A. gundlachi*, typically an inhabitant of moist, shaded sites in the forest interior, responded to increased light penetration and decreased humidity by using perch sites in protected locations closer to ground level than pre-Hugo (Reagan, 1991). One year post-Hugo, *A. stratulus* began to reinvade the forest canopy.

At higher elevations in the Sierra de Baoruco on Hispaniola, Sifers et al. (2001) found that anoles were largely restricted to edges created along roads into otherwise densely shaded cloud forest. From those edges, anoles rapidly invaded an adjacent clear-cut, from which they were largely excluded after several years of secondary growth. That sequence probably simulates effects of hurricanes during which edge-dependent species thrive temporarily, then retreat when forests reinvade a relatively short-term clearing. Conversely, shade-dependent eleutherodactylid frogs were as abundant in the clearings immediately after the clear-cut as in surrounding forest. Presumably, however, the intense insolation and resultant dry substrates would force them to vacate such sites, only to reinvade them after recovery by forests that have evolved in areas subjected to periodic hurricanes.

After Hurricane Georges struck Puerto Rico in September 1998, Wunderle et al. (2004) found that boas (*Epicrates inornatus*) became more visible and moved more frequently. High hurricane winds stripped leaves, vines, and branches from trees, rendering boas more visible and limiting access to arboreal sites. Post hurricane, boas shifted from heavily damaged to less damaged sites in response to changes in foliage cover and prey availability. On Isla Mona, canopy destruction by Hurricane Georges forced large *Epicrates monensis* to the ground to forage, as smaller branches in the understory could not support the mass of large boas. As a result, significantly more injuries (tail amputations, lacerations, punctures, missing eyes) to adult boas were recorded post-Georges (Tolson et al., 2007).

On Grenada, short-term effects of Hurricane Ivan (September 2004) were documented by Henderson and Berg (2005a). At that time (November 2004), the frog *Pristimantis euphronides* was encountered at prehurricane levels at Grand Etang despite the absence of crown vegetation providing shaded daytime retreats. However, surveys at the same site in December 2007 (RWH) and March 2008 (C. S. Berg, pers. comm.) indicated a substantive decrease in the numbers of *P. euphronides*. Treeboas (*Corallus grenadensis*) at Mt. William were observed foraging in the crowns of trees lying on the ground, while others foraged in leafless trees and shrubs. Data collected during later visits strongly indicated that boas had moved away from the severely damaged site, presumably to others in which vegetation structure was more similar to pre-Ivan stature. What we do not know is how many animals, especially arboreal species, perish during a hurricane; the numbers could be substantial—and effects on populations already small for various other reasons could be catastrophic.

Perhaps the most dramatic evidence of the potential impact of hurricanes on island herpetofaunas comes from the area around Great Exuma in the Bahamas. Immediately after Hurricane Lili hit on 19 October 1996, Spiller et al. (1998) revisited study sites on islands that had been censused just before Lili hit (9–17 October). On islands with some protection from the hurricane, the mean number of individual lizards (*Anolis* spp.) per island was 34% lower than prehurricane numbers. On exposed islands, all anoles were exterminated and, one year later, no anoles were found on those islands. Rather than wind velocity, however, these "catastrophic extinctions" were likely caused by a storm surge of nearly 5.0 m. Spiller et al. concluded that (1) larger organisms are more resistant to the immediate impact of moderate disturbance; (2) extinction risk is related to population size when disturbance is moderate, but not when it is catastrophic; (3) after a catastrophic disturbance, the recovery rate among different types of organisms is related to dispersal ability; and (4) the absence of lizards from small low-lying islands may be the result of their poor dispersal abilities coupled with a hurricane-related surge that completely inundates an island.

Hurricanes also have the potential for adding species to an island's fauna. During the 1995 hurricane season, at least 14 *Iguana iguana* appeared on the eastern beaches of Anguilla, arriving on a mat of logs and uprooted trees (some more than 9.0 m long); except for a few escaped "pets" elsewhere on the island, the species had not previously occurred on Anguilla. According to Censky et al. (1998), the lizards likely originated on Guadeloupe, and were transported to Anguilla along the paths of hurricanes Luis and Marilyn. Prototypical *Anolis longiceps* may have arrived on Navassa via hurricanes (Powell, 1999a), and Censky and Powell (2001) suggested a similar scenario to explain the presence of *Ameiva corax* on Little Scrub Island. Both Navassa and Little Scrub are in opposite directions from the presumed islands of origin than would have been predicted by over-water dispersal along prevailing winds and currents (i.e., other than during hurricanes).

Volcanic eruptions on West Indian islands, although less frequent than hurricanes, can be more devastating, as each eruption has the potential of killing virtually everything in its path. The 1979 eruption of La Soufrière on St. Vincent was considered less extensive than that of 1902 (Fiske and Sigurdsson, 1982), and the endemic frog *Pristimantis shrevei* was collected near the summit in 1990 (Kaiser et al. 1994b). Kaiser and Henderson (1994) reported that ground vegetation was beginning to cover the bare volcanic debris in 1992, and *Eleutherodactylus johnstonei* was found syntopically with *P. shrevei* at that time. Since 1995, the Soufrière Hills Volcano on Montserrat has been vacillating in levels of activity, some requiring evacuation of human residents from portions of the island. In 1995, dead and dying juvenile *Leptodactylus fallax* were found in shallow rock pools affected by ash-laden run-off (Kaiser and Day, 1995). By 1998, at least 10% of the habitat of *L. fallax* had been destroyed by pyroclastic flows, and the species' entire habitat had been affected by acid rain and volcanic ash fallout (Daltry and Gray, 1999).

Alien Parasites

The fungal parasite *Batrachochytrium dendrobatidis* has been implicated in amphibian extinctions and population declines in the Neotropics, including the West Indies. Chytridiomycosis has so far been documented on Cuba (Díaz et al., 2007a), Puerto Rico (based on frogs collected in 1976; Burrowes et al., 2004, 2008), Hispaniola (Joglar and Burrowes, 2005; Joglar et al., 2007), and Dominica (Daltry, 2007; Zoological Society of London, 2008; Daniells et al., 2008). Reports from additional islands are probably inevitable. *Eleutherodactylus coqui*, a West Indian endemic, apparently has introduced the fungus to Hawaii (Beard and O'Neill, 2005). Burrowes et al. (2004) provided a smorgasbord-like list of diseases found in Puerto Rican frogs collected in 2000.

Environmental Pollution

Although we have no documented evidence demonstrating the impact of pesticides on any West Indian herpetofauna, anecdotal accounts suggest that they have had a deleterious and sometimes substantial effect on some species of Antillean frogs and reptiles. Sugarcane dominated island economies until the late 19th century. Today, banana cultivation is the principal wide-scale cash crop, and plantations cover many thousands of hectares of land in the Greater and Lesser Antilles, including formerly forested hillsides. This industry has introduced the subsidized use of chemical fertilizers and extensive applications of pesticides and herbicides. As noted by Powell and Henderson (2007), although these chemicals may not have directly affected the herpetofauna, they surely had a negative impact on the arthropodan prey of many frogs and lizards—and, for all we know, a "direct" impact may have been felt as anurans absorbed chemicals through their skin (which may in turn impact frog-eating snakes). Burrowing reptiles (e.g., *Amphisbaena*, *Typhlops*) may be similarly vulnerable, in that they may spend their entire lives on and in the soil of agricultural areas. Short of finding animals dead in the field, we may never be able to assess the full impact of pesticide pollution.

Depleted Prey Bases

An estimated 77–78% of the original rodent fauna of the West Indies is now extinct; more specifically, 93% are extinct on Hispaniola, and 100% in the Lesser Antilles (Morgan and Woods, 1986; Woods, 1989). Based on field research by Peter J. Tolson in Cuba, predation by *Epicrates angulifer* on native hutias (*Capromys pilorides*) is very common (i.e., every adult boa examined had hutia hair in its feces). Hutias (and other rodents) on other islands (e.g., Bahamas, Hispaniola, Jamaica, Puerto Rico) were undoubtedly important prey of large boids (Tolson and Henderson, 2006). Similarly, prior to the arrival of Europeans and the introductions of *Mus* and *Rattus*, Antillean populations of *Corallus* likely fed on now-extinct populations of murid rodents (probably *Oryzomys*; Pregill et al., 1994; Henderson et al., 1996b).

Despite having suffered dramatic declines in numbers on many islands, species of *Cyclura* remain important prey in the diets of *Epicrates* and alsophiine snakes (e.g., Levering and Perry, 2003; Knapp and Owens, 2004; Tolson et al., 2007). The long-term impact of these declines on snake populations is unknown. According to Tolson and Henderson (2006, based on fieldwork by Tolson), the loss of *C. pinguis* on most of the Puerto Rico Bank probably has been somewhat mitigated by the introduction of *Iguana iguana*, upon which *Epicrates* frequently preys.

Overpopulation

The West Indies already includes several of the most densely populated countries in the world (e.g., Barbados), and the human population on some Lesser Antillean islands was expected to double between mid-1994 and 2025 (Jenkins and Harcourt, 1996b). The population on Cuba was predicted to increase from 11.1 million in mid-1994 to 12.9 million in 2025 (Jenkins and Harcourt, 1996a), that of the Dominican Republic from 7.8 million to 11.4 million, and Haiti from 7.0 million to 13.1 million (Harcourt and Ottenwalder, 1996). In 2007, that in Cuba already had reached 11,268,000, Haiti 9,598,000, the Dominican Republic 9,760,000, Puerto Rico 3,991,000, and Jamaica 2,714,000 (United Nations, 2007).

The growth of the human population on Antillean islands (as elsewhere) aggravates many of the factors already negatively impacting the herpetofauna. More people need more housing, more housing requires more development, and more development means that more habitats will be transformed to accommodate those additional people. Development also may require that building materials be imported from other islands or, more likely, from the United States. Those materials may harbor frogs, lizards, or snakes that are alien to the West Indies, thereby introducing species that may eat, compete with, hybridize with, or infect with equally alien parasites and pathogens species native to that particular island. Such habitat alterations render species already stressed by habitat contraction even more vulnerable to extirpation or extinction.

Commercial Exploitation

Legal commercial exploitation of West Indian frogs and reptiles appears to be of minimal concern, although "species with restricted ranges [and] high levels of endemism (e.g., small island species) . . . could be detrimentally affected by even a small number of individuals being removed from the wild" (Schlaepfer et al., 2005). Most island governments do not allow commercial collection and exportation of native wildlife. However, the boids (*Corallus*, *Epicrates*, *Boa*) do have commercial appeal. Species of *Epicrates* have been coveted in the live animal trade, and many have been removed illegally from their respective island ranges (Tolson and Henderson, 1993, 2006). Some species of *Epicrates* are bred in captivity and can be purchased legally. Among the lizards, *Cyclura* spp. and *Iguana delicatissima* have the most cachet; as with Antillean *Epicrates*, some species of *Cyclura* are bred in captivity and can be purchased legally. However, with so many populations in decline, the human predisposition to place exaggerated value on rarity fuels disproportionate exploitation of rare species, rendering them even rarer and thus more desirable, ultimately leading them into an extinction vortex (Courchamp et al., 2006).

A Current Overview of Natural History

Although lacking the detail of the 737 species accounts that follow, this very brief summary of the natural history of West Indian amphibians and reptiles is intended to provide a snapshot of the broad disparities in distribution and the diversities of habitats used by the herpetofauna, body sizes and population densities of its members, types of food consumed, and, in turn, the predators and parasites that feed on West Indian frogs and reptiles.

Amphibians

Distribution

Species of West Indian frogs exhibit distributions that may encompass tens of thousands of square kilometers on Cuba or Hispaniola and with an altitudinal distribution approaching 2,000 m (e.g., *Eleutherodactylus abbotti*), whereas others are known from a single locality on one of those same two islands. For example, on Cuba, Hedges (1999) determined that the average area of species distributions was 110,922 km^2, but that five species were known from only their type localities. On Hispaniola, the average area of distribution was 9,597 km^2, and six species were known from only their type localities (Hedges, 1999). In the Lesser Antilles, the average distribution area falls to 1,029 km^2, but 50% of the anuran fauna have distributions no greater than 170 km^2 (about 3% of the total area of the Lesser Antilles; Hedges, 1999). Whereas one species may be restricted to a single site on one island, other species have multi-island distributions (e.g., *Eleutherodactylus johnstonei* and *Osteopilus septentrionalis* each occur on more than 20 islands) and occupy a wide range of habitats.

Habitats

West Indian anurans occur in primary and secondary rainforest, cloud forest, elfin forest, pinewoods, upland deciduous forest, sclerophyll forest, cactus-acacia forest, cactus savanna, mesic grasslands, pastures, mangroves, limestone karst formations, palustrine herbaceous wetlands, plantations and groves that produce bananas, cacao, coffee, and coconuts, sugarcane and flooded rice fields, and coastal areas; along railroad and trolley embankments; in urban situations that include residential and hotel gardens and parks, abandoned edifices, and those still in use; on and along paved and unpaved roads; and in urban alleys and street gutters. Within these general habitats, frogs may be found under and on leaf litter, rocks, palm fronds, livestock excrement, and logs; in forest canopy to 15–20 m above ground level, in tree hollows, on branches, trunks, and leaf surfaces; in or on terrestrial and arboreal bromeliads; in boulder-strewn torrents, lakes, lagoons, streams, rivers, ditches, and puddles; in snail shells; in burrows that were self-excavated, or in those of other species (e.g., tarantulas, Burrowing Owls); on cliff faces, in caves, and in limestone crevices, sinks, and fractures; in abandoned mineshafts and dungeons; on and among boulder and rock jumbles; perched on a wide variety of broadleaf vegetation; in coconut and banana trash; under discarded boards and piles of lumber; on fence posts; in outhouses, shower rooms, and cisterns.

Activity

Depending on the species, frogs may be active by day or during the night, they may be crepuscular, or they may exhibit a combination of activity periods, often depending on weather conditions.

Densities

Few studies have addressed population densities, and all deal with species of *Eleutherodactylus*. Density estimates range from 100/ha for *E. richmondi* to 27,300/ha for *E. johnstonei* and to 45,000/ha (all age classes) for *E. coqui*.

Diet

All West Indian frogs are carnivorous; their diets include a broad taxonomic range of insects (but especially ants, coleopterans, lepidopterans, orthopterans, leafhoppers, hymenopterans, and termites—undoubtedly reflecting their availability in West Indian habitats), spiders, mites, millipedes, centipedes, scorpions, crustaceans, snails and slugs, earthworms, fishes, frogs (*Gastrophryne carolinensis, Eleutherodactylus coqui, E. planirostris, Osteopilus septentrionalis, Leptodactylus albilabris, Rana catesbeiana*), frog eggs (*E. portoricensis*), lizards (*Sphaerodactylus goniorhynchus, Anolis evermanni, Anolis gundlachi*), snakes (*Typhlops dominicanus, Alsophis manselli*), bats, and small rodents.

Parasites

At least 20 species of nematodes, 2 species of trematodes, and 5 species of ticks parasitize Antillean anurans.

Predators

Flies, ants, centipedes, millipedes, spiders, crickets, snails, conspecifics, and snakes (*Magliophis exiguus*) prey on anuran eggs; subadults and adults fall prey to spiders, scorpions, tarantulas, centipedes, land crabs, frogs (*Eleutherodactylus cochranae, E. coqui, E. pantoni, E. portoricensis, E. wightmanae, Osteopilus dominicensis, O. septentrionalis, Leptodactylus fallax, Rana catesbeiana*), lizards (*Thecadactylus rapicauda, Anolis cristatellus, A. equestris, A. gundlachi, A. marmoratus, A. porcatus, A. smallwoodi, Mabuya sloanii, Sphaerodactylus klauberi, Ameiva exsul*), snakes (*Epicrates striatus, Chironius vincenti, Mastigodryas bruesi, Alsophis antillensis, A. manselli, A. rijgersmaei, A. rufiventris, A. sanctonum, A. sibonius, Borikenophis portoricensis, Caraiba andreae, Cubophis cantherigerus, C. caymanus, C. vudii, Hypsirhynchus funereus, H. parvifrons, H. polylepis, Ialtris haetianus, I. dorsalis, Liophis cursor, L. juliae, Magliophis exiguus, M. stahli, Tretanorhinus variabilis, Uromacer catesbyi, Tropidophis celiae, T. curtus, T. haetianus, T. melanurus, T. parkeri*), a variety of birds (hawks, thrashers, woodpeckers, owls, cuckoos, todys, kingbirds, thrushes), mongooses, domestic cats and dogs, black rats, and humans.

Of 161 species of *Eleutherodactylus*, we have documentation of only 24 species being preyed upon by 39 species of frogs and reptiles. Based on our data, West Indian frogs and reptiles preyed upon a total of 32 species of West Indian anurans; in reality, we believe that number is much higher. Published information seems to indicate that *Eleutherodactylus coqui* and *Osteopilus septentrionalis* are consumed by a disproportionate number of species. These numbers no doubt reflect, at least in part, the great deal of fieldwork devoted to studying *E. coqui* and the amount of information accumulated regarding its predators, and, in the case of *O. septentrionalis*, to its broad geographic distribution (which continues to increase) and the number of different predators to which it is exposed.

Reproduction

Because species of *Eleutherodactylus* (161 of 197 species; 82.7%) comprise the bulk of the West Indian anuran fauna, reproduction via direct development (wherein miniatures of the adults hatch from eggs) is the predominant mode. A notable exception to this is *E. jasperi*, which is ovoviviparous. Species of *Eleutherodactylus* exhibit egg/nest guarding, with the male, the female, or both sexes sharing the responsibility. Within other anuran genera, a swimming larval (i.e., tadpole) stage is typical. Clutch sizes range from one in *E. iberia* to 45,000 in *Bufo marinus*.

Size

Anurans endemic to the West Indies range in size from 12.3 mm SVL (*Eleutherodactylus iberia*) to 210 mm SVL (*Leptodactylus fallax*). One species that may or may not be native to the region, *Bufo marinus*, attains a SVL of 230 mm, and the introduced *Rana catesbeiana* reaches 270 mm in Cuba.

Reptiles

Distribution

Like those of the frogs, the distributions of reptiles in the West Indies vary tremendously. Many are restricted to a single island where they may be known from a single locality or have a nearly islandwide distribution covering hundreds to thousands of square kilometers. Others occur on one or several island banks that may encompass several islands (e.g., *Ameiva auberi* on more than 50 islands in the Bahamas). A small handful of species are great colonizers and occur on dozens of islands (e.g., *Hemidactylus mabouia* and *Anolis sagrei*).

Habitats

Reptiles in the West Indies occur in rainforest, cloud forest, montane river valleys, broadleaf forest, dry scrub forest, thorn scrub woodlands, cactus savanna, deciduous forest, pine forest, Sea Grape stands, sandy and rocky beaches, limestone outcrops, coastal terraces and dunes, mangroves, xeric woods, oil palm and pineapple plantations, sugarcane fields, abandoned lots and garbage dumps; groves of bananas, coconuts, cacao, mangos, citrus, breadfruit, and coffee; abandoned sawmills and sugar mills; pastures; residential gardens, hotel grounds, and parks. Within these various habitats, reptiles can be found in leaf litter, under rocks and among boulders, in caves, on cliff faces, in rock crevices, and on cemetery walls; in ponds, rivers, streams, lagoons and swamps, both freshwater and brackish; on grasses, twigs, tree branches and trunks, and leaf surfaces from near ground level to 20 m or more in tree crowns; in terrestrial and arboreal bromeliads; in cacti, *Agave*, acacia, termitaria, tree hollows, burrows, drain pipes, and conch shells; under rotted logs and on stone walls; under banana, cacao, and coconut trash; in outbuildings, houses, hotel rooms, restaurants, and abandoned edifices.

Activity

Most Antillean lizards are diurnal, but many geckos (Gekkonidae, Phyllodactylidae, some Sphaerodactylidae) are nocturnal. Some diurnal lizards (*Anolis* spp.) are known to exploit the "night-light" niche, wherein they forage for insects attracted to artificial illumination at night. Whether or not the diminutive lizards of the genus *Sphaerodactylus* can be labeled diurnal or nocturnal is questionable; some exhibit nocturnal tendencies, others diurnal, and some are associated with leaf litter and, by day, may be active in a perpetual twilight. Compared with the lizards, more snakes are primarily nocturnal, especially among the boids, tropidophiids, the one natricid, and the viperids. Most of the dipsadids are diurnal, but species of *Arrhyton* appear to be nocturnal. The two native colubrids (*Chironius*, *Mastigodryas*) are diurnal. The secretive *Leptotyphlops* and *Typhlops* may be largely nocturnal, but we lack data for the vast majority of species. The tortoises (*Chelonoidis*) are diurnal, but the aquatic turtles (*Trachemys*, *Pelusios*) exhibit both diurnal and nocturnal activities, as do the crocodilians.

Densities

Most work on population densities has centered on lizards, with relatively little data available for snakes. Calculated densities range from fewer than 1.0/ha for the large iguanid *Cyclura cornuta* to an average of 52,800/ha for the diminutive *Sphaerodactylus macrolepis* on Guana Island (British Virgin Islands). A density of >100/ha has been documented for the boid snake *Epicrates monensis*.

Diets

The large iguanids in the genera *Cyclura* and *Iguana* are almost exclusively herbivorous, consuming a wide range of plant material, including leaves, vines, cacti, fruits, and flowers, along with some animal matter (insects, centipedes, snails, slugs, crabs, bird and mammal feces, birds and their eggs, carrion, conspecifics, human garbage). Although some Antillean lizards consume vegetable matter in addition to a variety of invertebrates and occasional vertebrates, most are primarily or exclusively insectivorous, consuming a broad taxonomic range of insects (ants and other hymenopterans, termites, flies, beetles, roaches, butterflies and moths, including larvae, orthopterans, hemipterans). Other invertebrate prey includes spiders, tarantulas, centipedes, millipedes, earthworms, and gastropods, and vertebrate material includes shed skin of lizards, frogs (*Eleutherodactylus coqui*, *E. johnstonei*, *Osteopilus septentrionalis*), other lizards (2 species of *Celestus*, 2 *Hemidactylus*, 3 *Cyclura*, 2 *Leiocephalus*, 32 *Anolis*, 7 *Sphaerodactylus*, 3 *Ameiva*), birds, and eggs of lizards and turtles.

Of 157 species of *Anolis*, the literature documented 36 species that prey on frogs and other reptiles (often congenerics and conspecifics). *Anolis cristatellus*, besides taking a wide taxonomic range of invertebrates, consumes frogs, geckos, other anoles, and has at least attempted to eat a snake.

Published information indicates that 60 species of West Indian snakes exploit a minimum of 137 species of frogs (*Bufo marinus*, *B. peltocephalus*, *Eleutherodactylus abbotti*, *E. alcoae*, *E. antillensis*, *E. armstrongi*, *E. atkinsi*, *E. blairhedgesi*, *E. coqui*, *E. dimidiatus*, *E. eunaster*, *E. glandulifer*, *E. glaphycompus*, *E. heminota*, *E. hypostenor*, *E. iberia*, *E. inoptatus*, *E. johnstonei*, *E. martinicensis*, *E. oxyrhyncus*, *E. planirostris*, *E. ruthae*, *E. sciagraphus*, *Osteopilus dominicensis*, *O. pulchrilineatus*, *O. septentrionalis*, *Leptodactylus albilabris*, *L. validus*), lizards (4 species of *Celestus*, 2 *Diploglossus*, 5 *Cyclura*, 1 *Hemidactylus*, 2 *Iguana*, 8 *Leiocephalus*, 47 *Anolis*, 14 *Sphaerodactylus*, 13 *Ameiva*), 2 species of *Amphisbaena*, other snakes (*Epicrates inornatus*, *Borikenophis portoricensis*, *Caraiba andreae*, *Cubophis vudii*, *Magliophis exiguus*, *Typhlops platycephalus*, *T. pusillus*, *Tropidophis melanurus*, *Bothrops caribbaeus*), and 1 turtle (*Eretmochelys imbricata*). Ten of 13 species of boid snakes eat other reptiles (mostly anoles), and 37 of 51 species of dipsadid snakes are known to consume frogs and other reptiles. Dipsadid species that are particularly catholic in prey choice are *Cubophis cantherigerus* (at least 14 species of frogs, lizards, snakes, and turtles), *Hypsirhynchus parvifrons* (24 species of frogs and lizards), and *Borikenophis portoricensis* (22 species of frogs, lizards, and snakes). Trophically narrow dipsadids include *Ialtris haetianus* (almost exclusively species of *Eleutherodactylus*) and the arboreal *Uromacer oxyrhynchus* (exclusively lizards, and almost all *Anolis*). As with the frogs, we feel that the number of reptiles known to exploit frog and reptilian prey is much lower than their actual number in nature, as is the number of species of frogs and reptiles that fall prey to reptiles. The diets of *Leptotyphlops* and *Typhlops* are restricted to small invertebrates.

Parasites

At least 61 species of nematodes, a variety of trematodes (e.g. *Urotrema* spp.), protozoan blood (*Leishmania*) and malarial (*Plasmodium* spp.) parasites, tapeworms, ticks, and mites infect members of the reptilian fauna.

Predators

In the West Indies, lizards fall prey to spiders, scorpions, land crabs, barracuda, frogs (*Eleutherodactylus coqui*, *Osteopilus dominicensis*, *O. septentrionalis*, *Leptodactylus fallax*), other lizards (2 species of *Celestus*, 2 *Hemidactylus*, 3 *Cyclura*, 9 *Leiocephalus*, 35 *Anolis*, 1 *Aristelliger*, 9 *Ameiva*, and 1 *Amphisbaena*), and snakes (*Boa nebulosa*, 2 *Corallus*, 7 *Epicrates*, 31 alsophiines [*Alsophis*, *Arrhyton*, *Borikenophis*, *Cubophis*, *Hypsirhynchus*, *Ialtris*, *Magliophis*, *Uromacer*], 2 *Liophis*, 9 *Tropidophis*). Snakes are preyed upon by frogs (*Leptodactylus fallax*), at least attempted by a lizard (*Anolis cristatellus*), and other snakes (*Epicrates angulifer*, *E. inornatus*, *Arrhyton taeniatum*, *Borikenophis portoricensis*, *Clelia errabunda*, *Cubophis cantherigerus*, *C. vudii*, *Ialtris agyrtes*, and *Magliophis stahli*). In addition, at least 35 species of birds prey on at least 59 species of reptiles (Powell and Henderson, 2008a). Further trophic exploitation is perpetrated by domestic and feral cats and dogs, raccoons, pigs, Cuban solenodons, rats, mongooses, and humans. Considering the dozens of potential invertebrate and vertebrate predators, as well as dozens of potential parasites, members of the West Indian herpetofauna clearly are under continual threat of predation in one form or another.

Reproduction

With the exception of the anguid genera *Celestus* and *Diploglossus* and the scincid genus *Mabuya*, Antillean lizards are egg layers. All members of the species-rich genera *Anolis* and *Sphaerodactylus* and diminutive snakes in the genus *Leptotyphlops* typically lay single-egg clutches, whereas members of other egg-laying genera exhibit considerable variability in clutch size, often reflecting the size of the female (i.e., larger females produce larger clutches). Boid, natricid, and tropidophiid snakes are viviparous, the rest are oviparous, as are the turtles and crocodilians.

Size

West Indian lizards range in size from the tiny *Sphaerodactylus ariasae* (17.9 mm SVL) to the massive *Cyclura nubila* (745 mm SVL). Snakes range in size from a total length of 104 mm in *Leptotyphlops carlae* to about 4.0 m SVL in *Epicrates angulifer*. The crocodilian *Crocodylus rhombifer* may exceed 4.0 m in total length.

Future Emphases

Despite overdevelopment, a burgeoning human population, mongooses, tourists, habitat-destroying goats, and the occasional hurricane or volcanic eruption, we cannot imagine a more rewarding, intriguing, or enjoyable place to do fieldwork than the West Indies, and we are not alone in thinking that (Henderson and Powell, 2003). Whatever preconceived image one may have of the region, it is likely to some degree accurate and to a greater degree false. Yes, spectacular palm-fringed beaches are gently lapped by water in dazzling hues of blue and green; quaint, pastel-colored houses are surrounded by vines of bougainvillea in wondrous shades of blue, orange, and purple, perhaps with a radiantly red Flamboyant Tree (*Delonix regia*) providing some shade; and, yes, pasty-white tourists from cruise ships wander through port towns buying souvenirs hawked by sidewalk venders. The West Indies are, however, more than just desirable tourist destinations. They are back roads through tropical forests and small villages where the alluring smell of mealtime fires makes one slow down to enjoy the aroma. Stately rows of Royal Palms (*Roystonea* sp.) where great plantation houses once stood and the ruins of long-abandoned sugar mills remind one of an era and economy that impacted the lives of humans and herpetofauna—but kilometer after kilometer of bananas suggest that things have perhaps not changed all that much. As a further reminder of challenges facing the Antilles, one can drive all night on some islands, through rainforest, long expanses of agriculture, and even concrete-dominated towns, and never be out of earshot of *Eleutherodactylus johnstonei*, a species that likely should not be on many of the islands on which it sings. Finally, the West Indian experience is the people. Those on many islands are friendly, glad to see you, and often amazingly willing to help you if help is needed. We have had folks file out of church on a rainy night in Grenada and help change a flat tire, and, when gas stations were closed on a Sunday morning on Dominica, a friendly woman sent her children searching for the station owner so we could get much-needed fuel. On a late afternoon in St. David, Grenada, two elderly women, dressed in their finest, spent a half-hour looking at a road map, trying to help RWH get to a locality with a name they had never heard. Bystanders lifted a vehicle out of a roadside ditch on St. Vincent, and commiserated with RP over the heavy traffic, narrow road, and crazy driver who had forced him into the ditch.

Aside from the setting, the history, and the question whether the total number of species of frogs and reptiles is 737, 800, or 1,000, the West Indian herpetofauna remains attractive for future research for multiple reasons. It is an amazingly rich fauna with spectacular radiations of several genera. It harbors species with localized distributions in specialized situations and others that are ubiquitous across many habitats (and sometimes on multiple islands). It includes a surprising number of species that occur at phenomenally high densities, and others that are alarmingly scarce. Perhaps most importantly, and despite the progress made over the past 15–20 years, we still know surprisingly little about most species. The hundreds of islands that support hundreds of species are discrete living laboratories ideal for projects ranging from autecologies to simple or complex community studies. Although no technically pristine habitats remain (Hedges, 2006, 2008b), and those that come the closest are essentially restricted to tiny, largely inaccessible islets, opportunities still exist for studying frogs and lizards in magnificent forests and, perhaps more importantly, in habitats that have been variously altered by human activity, be it subtly or flagrantly (Henderson and Powell, 1999, 2001). As humans continue to "remake" West Indian landscapes (e.g., Rios-López and Aide, 2007), monitoring the responses of native and introduced herpetofauna to altered habitats becomes increasingly critical (e.g., Glor et al., 2001; Germano et al., 2003; Mallery et al., 2007; Powell et al., 1996b; S. B. Powell et al., 2007; Ackley et al., 2009), especially since dramatically transformed areas often are those from which native species disappear and introduced species of frogs, lizards, or snakes may thrive.

Adding impetus to the need for natural history research on West Indian species is the sad fact that, of the species covered in this book, at least 22 are presumably "extinct"—and have disappeared in historical times. They are the frogs *Eleutherodactylus eneidae*, *E. jasperi*, *E. karlschmidti*; the lizards *Ameiva cineracea*, *A. major*, *Celestus anelpistus*, *C. occiduus*, *Cyclura onchiopsis*, *Leiocephalus cuneus*, *L. eremitus*, *L. herminieri*, *L. onaneyi*, *Sphaerodactylus gilvitorques*, *Tarentola albertschwartzi*; and the snakes *Alsophis antiguae*, *A. antillensis*, *Borikenophis sanctaecrucis*, *Clelia errabunda*, *Hypsirhynchus ater*, *H. melanichnus*, *Liophis perfuscus*, and *Tropidophis bucculentus*. Additional species that have not been observed in the field in several decades include frogs (e.g., *Bufo fluviaticus*, *Eleutherodactylus orcutti*, *E. semipalmatus*; Hedges, 1993, 1999; Hedges and Díaz, 2009), lizards (*Anolis roosevelti*; Dodd, 1994), and snakes (*Clelia clelia*; Henderson, 2004). Still others with multi-island distributions have suffered extirpations on one or more islands (e.g., *Alsophis rufiventris* on St. Christopher and Nevis, although these snakes remain abundant on Saba and St. Eustatius; Maley et al., 2006 and references therein), and *Liophis* on several islands (Powell and Henderson, 2005a). Finally, many species with formerly extensive ranges now are restricted to mere fragments of their historical distributions. Examples include *Ameiva ameiva* on Grenada and St. Vincent (Germano et al., 2003; Mallery et al., 2007), and *Cyclura cornuta* and *Crocodylus acutus* on Hispaniola (Powell et al., 2000; Powell and Incháustegui, 2009).

In many instances, the reasons for extinctions, extirpations, and declines are obvious and well documented (and have been listed above), but for others, the causes are unknown or at best speculative. If any hope exists that appropriate conservation measures and management decisions can be implemented (and enforced), the relevant national authorities and international agencies need data that can come only from dedicated research on both affected species and those that are thriving, if only to fully understand the criteria that relegate populations to one category or the other.

Consequently, the promotion of responsible scientific research is necessary and should be encouraged (e.g., Hedges and Thomas, 1991b; Powell, 2007a). Restrictions should be streamlined and limited to steps necessary to confirm the legitimacy and identities of scientists seeking to remove animals from the wild. Numbers of specimens to be collected should be determined by circumstances unique to each situation and the needs of the investigation rather than by an arbitrary limit established by bureaucrats, even those with the best of intentions. Restricting the responsible collection of animals that are threatened or potentially threatened in the wild can easily result in lack of sufficient knowledge to develop effective management programs or, when the very existence of entire populations is threatened, protocols for rearing those species in captivity. Ironically, a lack of information is frequently cited as justification for prohibiting even responsible scientific collection, which has never been implicated directly in the extinction of any species. Ultimately, restrictions that preclude accurate scientific assessments are shortsighted and inevitably place more populations and communities at risk of extirpation or extinction.

Part I Amphibia

Section 1 Anura

1 Family Aromobatidae

Allobates chalcopis Kaiser, Coloma, and Gray, 1994

Distribution: Southeastern slope of Montagne Pelee, Martinique; elevation ~500 m.

Habitat: Montane rainforest consisting of a few tall trees (to 30 m), shrub thickets, palm brakes, ferns, and stands of bamboo, by day under large leaves and rocks in a ravine (Kaiser et al., 1994a); grassy areas near lava flows (Hedges et al., 2004c; Stuart et al., 2008).

Activity: Crepuscular, shy of artificial light (Kaiser et al., 1994a). **Behavior:** Secretive, retreats under dry leaves and rocks when approached, also rapid succession of jumps (sometimes into water) with rapid changes of direction (Kaiser et al., 1994a). **Parasites:** Nematode *Cosmocerca longispicula* (Moravec and Kaiser, 1994). **Reproduction:** Calling peaks at dawn and especially at dusk, but some during the day (Kaiser et al., 1994a); 6 eggs (likely 2 clutches of 3 eggs each) on a decaying palm leaf hidden among decomposing foliage on a steep slope on 29 August, no adult in attendance, but males calling within 3.0 m of the nest site, eggs hatched 6 days later (Kaiser and Altig, 1994); adults presumably transport larvae to water as in other species in the genus (Hedges et al., 2004c; Stuart et al., 2008). **Size:** SVL to 18.4 mm (Kaiser et al., 1994a).

Figure 4. *Allobates chalcopis* (Aromobatidae) is the only representative of the family in the West Indies. Photograph by Hinrich Kaiser.

Conservation Status: "Vulnerable" on the IUCN Red List (Hedges et al., 2004c).

Remarks: This species was formerly placed in the genus *Colostethus*.

2 Family Bufonidae

Figure 5. Puerto Rican *Bufo lemur* (Bufonidae) breeds almost exclusively in temporary ponds. Photograph by Alejandro J. Sánchez Muñoz.

Bufo cataulaciceps Schwartz, 1959

Distribution: Western Cuba, Isla de la Juventud; elevational range, sea level to ~70 m.

Habitat: Pinewoods and forests, beneath pine logs and rocks on Isla de la Juventud, may spend much time below ground during nonreproductive times of year (Rivalta et al., 2003b); often in open areas (Garrido and Schwartz, 1968b).

Activity: Largely nocturnal, but juveniles often active by day (Díaz and Cádiz, 2008). **Behavior:** Males are very cautious (Rivalta et al., 2003b); may become immobile when handled, seemingly dead (Díaz and Cádiz, 2008). **Defense:** Exudes venomous skin secretions from various points on dorsum, not just from parotoid glands (Díaz and Cádiz, 2008). **Reproduction:** Explosive breeder after heavy rains (Hedges and Diaz, 2004a; Stuart et al., 2008); males call intensively after heavy rains (Díaz and Cádiz, 2008); calling (4 July–18 August) from margins of temporary pools, from flooded pastures, shallow water (7–8 cm), moist earth, or gravel at pool's edge, males very shy, amplexus axillary, eggs in still water in clumps rather than strings (Schwartz, 1959a; Schwartz and Henderson, 1991); males in August form small choruses at dusk in flooded or poorly drained terrain in savannas and call from dense herbaceous vegetation in which they are difficult to locate (Alonso Bosch and Rodríguez Gómez, 2003c). **Size:** SVL to 30 mm in males, to 27 mm in females (Schwartz, 1959a).

Conservation Status: "Vulnerable" (Vales et al., 1998); "endangered" on the IUCN Red List (Hedges and Diaz, 2004a).

Remarks: Recent work by Frost et al. (2006) resurrected *Peltophryne* as the genus for this species.

Bufo empusus Cope, 1862

Distribution: Cuba (including Isla de la Juventud, Archipiélago de Camagüey); elevational range, sea level to ~70 m.

Habitat: In more or less xeric and in mesic areas with friable soil, occurs in perfectly cylindrical burrows (carefully rimmed with smoothly patted clay), 15–25 cm deep, narrow at the bottom where toad rests, often clustered into large "colonies" (Barbour, 1914, 1945; Schwartz and Henderson, 1991); frogs and their burrows relatively common in lawns at the Botanical Garden in La Habana (G. Peters, 1974); may spend several months in burrows during periods of drought (Estrada, 1993b).

Activity: Adults primarily nocturnal, juveniles often active by day, seek refuge in burrows by day and during the dry season (Díaz and Cádiz, 2008). **Behavior:** Practices phragmosis, inflates body to resist removal from burrow (Barbour, 1945; Valdés de la Osa and Ruíz García, 1979; Rivalta et al., 2003b). **Parasites:** Nematodes *Aplectana cubana*, *Oswaldocruzia lenteixeirai*, and *Batracholandros bassi* (Baruš, 1973; Coy Otero and Lorenzo Hernández, 1982). **Predators:** Barn Owl (*Tyto alba furcata*; Arredondo Antúnez and Chirino Flores, 2002). **Reproduction:** Explosive breeder after heavy rains (Hedges and Diaz, 2004b; Stuart et al., 2008); males call May–August, often in enormous numbers, from edges (and in deeper water) of large rain pools, flooded ditches, and flooded road ruts after very heavy rains (Ruiz García, 1987); also call from cylindrical refuges in mud, under a cover of water of 10–15 cm (Garrido et al., 1986); may be stimulated to call by noise (and water) from large sprinklers used in irrigation systems, amplexus axillary (Ruíz García, 1987); call in September from exposed sites at the edge of water (Alonso Bosch and Rodríguez Gómez, 2003c). **Size:** SVL to 76 mm in males, to 74 mm in females (Rivalta et al., 2003b); to 90 mm (Díaz and Cádiz, 2008).

Conservation Status: "Vulnerable" on the IUCN Red List (Hedges and Diaz, 2004b).

Remarks: Recent work by Frost et al. (2006) resurrected *Peltophryne* as the genus for this species.

Bufo florentinoi Moreno and Rivalta, 2007

Distribution: Cuba (Península de Zapata); elevational range, near sea level.

Habitat: Semideciduous sclerophyll forest on dogtooth limestone substrates, diurnal refuges include roots of plants, holes in dogtooth limestone, and small depressions that temporarily accumulate water (Moreno and Rivalta, 2007); also on patios of houses where humid conditions persist (Díaz and Cádiz, 2008).

Abundance: Commonly encountered during rainy periods, but rarely seen when conditions are dry (Díaz and Cádiz, 2008). **Activity:** Nocturnal (Moreno and Rivalta, 2007). **Behavior:** Increases its body volume with air immediately upon being touched (Moreno and Rivalta, 2007). **Reproduction:** At least two types of calls at night from exposed limestone rocks or from crevices between rocks (Alonso Bosch et al., 2007); males call near small depressions that accumulate water and are suitable for reproduction (Díaz and Cádiz, 2008). **Size:** SVL in males to 112 mm, in females to 133 mm (Moreno and Rivalta, 2007).

Conservation Status: "Critically endangered" on the IUCN Red List (Rivalta G., 2008).

Bufo fluviaticus Schwartz, 1972

Distribution: Hispaniola, Cibao Valley near the type locality; elevational range 150–175 m.

Habitat: Occurs along rivers and creeks in areas that are xeric (*Acacia*-cactus) but have local broadleaf gallery forests (Schwartz, 1972).

Reproduction: Calling males in October, males call from very shallow (3.8 cm) running water in riffles, perched on wet (but not submerged) rocks (to 15 cm in diameter), 22 males in shallows adjacent to a small gravel bar in area 3.0 × 0.9 m (Schwartz, 1972). **Size:** Male SVL to 44.3 mm, females unknown (Schwartz, 1972).

Conservation Status: "Critically endangered" (Powell et al., 2000), "critically endangered" on the IUCN Red List, has not been encountered since its original discovery (Hedges et al., 2004k; Stuart et al., 2008).

Additional Pertinent Literature: Powell and Pregill (1991).

Remarks: Recent work by Frost et al. (2006) resurrected *Peltophryne* as the genus for this species.

Bufo fractus Schwartz, 1972

Distribution: Hispaniola, eastern Dominican Republic in the vicinity of Higüey; elevation <500 m.

Habitat: Occurs in mesic situations, a juvenile on grass of a hotel lawn at night (Schwartz, 1972); recorded only from degraded habitat, unknown whether it occurs in primary forest outside the breeding season (Hedges et al., 2004l; Stuart et al., 2008).

Reproduction: Breeding 13 June, 28 August, 3 September, choruses may be dense and deafening, extending from shortly after dark until well after dawn, calling sites include flooded pastures, a deep flooded pit at an old construction site, and along a slowly flowing stream in freshly plowed pasture (Schwartz, 1972); eggs laid and larvae develop in temporary pools (Hedges et al., 2004l; Stuart et al., 2008). **Size:** Male SVL to 66.8 mm, female SVL to 79.5 mm (Schwartz, 1972).

Conservation Status: "Endangered" on the IUCN Red List, not encountered during extensive surveys in 2003 (Hedges et al., 2004l; Stuart et al., 2008).

Additional Pertinent Literature: Powell (1992a).

Remarks: Recent work by Frost et al. (2006) resurrected *Peltophryne* as the genus for this species.

Bufo fustiger Schwartz, 1960

Distribution: Western Cuba; elevational range, sea level to ~390 m.

Habitat: Wet secondary forest with many clearings and some plantations (G. Peters, 1974); broadleaf forest, along stream banks in mesic situations, under rocks and palm trunks or in cavities in mogotes by day (Ruíz García, 1987); may occur in disturbed situations as long as the freshwater body remains undisturbed (Hedges and Diaz, 2004c); coastal thickets, agricultural areas, montane and submontane forests, by day under rocks, logs, inside cavities between rocks, burrows of freshwater crabs (*Epilobocerca* sp.; Díaz and Cádiz, 2008).

Abundance: Common (Díaz and Cádiz, 2008). **Activity:** Nocturnal (Díaz and Cádiz, 2008). **Diet:** Stomach with 40 cc of insects, beetles, and large slugs (Ruíz García, 1987). **Longevity:** In captivity, 20 years (Curator of Reptiles, Chicago Zoological Park in Slavens and Slavens, 2003). **Reproduction:** Explosive breeder in some areas, eggs laid and tadpoles develop in still water (Hedges and Diaz, 2004c); males calling 27 December at creekside (Schwartz, 1960b); form loud nocturnal chorusing aggregations in May along shorelines of montane rivers and bodies of water under gallery forests, males call while positioned among roots of riparian vegetation at water's edge (Alonso Bosch and Rodríguez Gómez, 2003c); amplexus axillary, eggs in double strings, ~2 mm in diameter (Ruíz García, 1987). **Size:** SVL to 198 mm in females (Schwartz, 1960b). **Vocalization:** Diphasic advertisement call (Alonso Bosch and Rodríguez Gómez, 2003c).

Conservation Status: On the IUCN Red List as a species of "least concern" (Hedges and Diaz, 2004c).

Remarks: Recent work by Frost et al. (2006) resurrected *Peltophryne* as the genus for this species. This species may represent a pattern variant of *B. peltocephalus* (Díaz and Abreu Guerra, 2005).

Bufo guentheri Cochran, 1941

Distribution: Hispaniola, lowlands of northern Haiti through the Cibao Valley, Plaine de Cul de Sac (Haiti)-Valle de Neiba and into the San Juan Valley (Dominican Republic); elevational range 40 m below sea level to 107 m.

Habitat: Occurs in very dry and very wet situations, including cactus savanna (Mertens, 1939); diurnally under rocks, under a beam and log at edge of or in *Cocos* plantation, in slat pile adjacent to lake (Schwartz, 1972; Schwartz and Henderson, 1991); occasionally near water with heavy concentrations of sulphur (Hedges et al., 2004m; Stuart et al., 2008); in grass on hotel grounds in early evening (RP).

Activity: Primarily nocturnal, juvenile active in leaf litter by day (RP). **Diet:** A sample of 1,109 prey items from 204 toads included 636 ants in 59 stomachs, 136 beetles in 53 stomachs, 127 termites in 2 stomachs, 128 immature beetles in 13 stomachs, 22 leafhoppers in 1 stomach, 15 spiders in 7 stomachs; beetles and ants accounted for the greatest volume of food (Parsons, 1995). **Parasites:** Nematode *Skrjabinoptera* sp. (Parsons et al., 1995). **Reproduction:** Males call from rain-flooded low areas (0.5–3.0 m deep), often extensive, flooded rice field, small roadside pool, cactus wood with muddy area flooded by leaky water main, cattle pond, flooded borrow pit in thorn scrub, large flooded desert pond, flooded ditches in desert, gravel bar in slack river water, mud bar in small stream, above deep pool in river, standing water at gas station, muddy edge of creek backwater 92 m from ocean, shore near or adjacent to water, often in herbaceous cover, also while floating in water, often with forelimbs propped on supporting branches or other plant material, in amplexus, male grasps female dorsal to forelimbs, not behind them, peak breeding season July-September, but may breed throughout the year (Schwartz, 1972; Schwartz and Henderson, 1991); large assemblage depositing eggs in shallow pool in stream bed on 15 April (Lynn, 1958); gravid female on 17 March (Mertens, 1939); metamorphs 20 May and 10 June (K. P. Schmidt, 1921); breeds throughout the year in some areas (Hedges et al., 2004m; Stuart et al., 2008). **Size:** SVL in females to 101 mm (Schwartz and Henderson, 1991).

Conservation Status: "Vulnerable" on the IUCN Red List (Hedges et al., 2004m).

Additional Pertinent Literature: Powell (1992a).

Remarks: Recent work by Frost et al. (2006) resurrected *Peltophryne* as the genus for this species.

Bufo gundlachi Ruibal, 1959

Distribution: Cuba (including Isla de la Juventud, Archipiélago de Sabana-Camagüey); elevational range, sea level to ~70 m.

Habitat: Xeric (especially) and mesic grasslands, diurnally retreats in burrows made near rocks that are well embedded and under grass, burrows L-shaped with 2.5 cm vertical, 3–4 cm horizontal portions (Ruíz García, 1987); savannas, pastures, cultivated open areas, swamps (Díaz and Cádiz, 2008).

Activity: Nocturnal (Estrada, 1993b); juveniles may be active by day (Díaz and Cádiz, 2008); retreats into burrows near rocks by day (Hedges and Díaz, 2004d; Stuart et al., 2008); aestivates in burrows during the dry season (Díaz, 2006). **Behavior:** Uses inflation of body to aid in burrowing, also inflates itself when handled (Valdés de la Osa and Ruíz García, 1979). **Reproduction:** Occurs May–October (Díaz and Cádiz, 2008); "explosive breeders during the rains" (Díaz, 2006); males call (10 June–27 August) while floating in water (0.15–0.5 m deep) of large flooded rain puddles, marshes, pastures, borrow pits, ditches, often in large numbers (Ruíz García, 1987; Schwartz and Henderson, 1991); males form large choruses in September, calling in open areas while floating in temporary pools in flooded lowlands, often surrounded or covered by grass, never totally exposed (Alonso Bosch and Rodríguez Gómez, 2003c); eggs 1.5 mm in diameter (Rivalta et al., 2003b); eggs laid in still water (Hedges and Díaz, 2004d; Stuart et al., 2008). **Size:** SVL in females to 36 mm (Schwartz and Henderson, 1991).

Conservation Status: "Vulnerable" on the IUCN Red List (Hedges and Díaz, 2004d).

Remarks: Recent work by Frost et al. (2006) resurrected *Peltophryne* as the genus for this species.

Bufo lemur Cope, 1868

Distribution: Puerto Rico, and the U.S. and British Virgin Islands; elevational range, sea level to ~50 m.

Habitat: Semiarid, limestone karst formations with well-drained soil in lowlands (Rivero et al., 1980); by day under limestone boulders in moist gorge behind bathhouses at Coamo Springs Hotel (K. P. Schmidt, 1928); also under bark of dead mangroves, under stones and boulders in dry gully with low grasses and scattered large trees (Rivero et al., 1980).

Defense: Relies on immobility and cryptic coloration (K. P. Schmidt, 1928); when approached, flattens body and legs and partially stretches legs to sides (Rivero et al., 1980). **Diet:** Includes ants, beetles, millipedes (K. P. Schmidt, 1928), and roaches (*Periplaneta americana*; Schwartz and Henderson, 1991). **Longevity:** In captivity, 10 years, 8 months (K. Murphy in Slavens and Slavens, 2003). **Parasites:** Heavily parasitized by the fluke *Mesocoelium* sp., cercariae in the gall bladder, nematodes in large intestine, and larval pentastomids in lungs (García Díaz, 1967). **Reproduction:** Breeding known to occur March–July, 4:1 sex ratio favoring males, up to 1,000 males at one site, males emerge first, females follow, breed, then leave ponds during one night, toads may travel up to 4 linear km from breeding pond (Moreno and Canals in Miller, 1985); exhibits axillary amplexus, known to breed in cattle watering troughs, pair in amplexus in cement reservoir ~10 × 6.7 m, eggs in "rosary" strings, female (SVL 95.6 mm) produced 15,761 eggs (Rivero et al., 1980); metamorphosis 35–62 days (Paine, 1984) and 20–30 days (Miller, 1985) in captivity, 25 days in nature (Moreno and Canals in Miller, 1985); eggs on 7 May, metamorphosed 22–26 July, SVL of metamorphs 7 mm (Rivero et al., 1980). **Size:** SVL in males to ~85 mm, in females to 120 mm (Rivero, 1998).

Conservation Status: "Critically endangered" on the IUCN Red List (Hedges et al., 2004nn; Angulo, 2008a); considered "threatened" by the U.S. Fish & Wildlife Service (2008); northern and southern populations are genetically distinct (CBSG, 2005), and the northern population is considered extirpated (Johnson, 1999); "endangered" (Moreno, 1991d); almost certainly extinct on Virgin Gorda (W. E. Rainey in Johnson, 1988); habitat loss and competition with *Bufo marinus* are major causes of population declines, one population occupies an ephemeral pool precariously close to the ocean in the Guanica Commonwealth Forest, and the site serves as a parking lot during the busy summer season (Barber, 2008); despite a single observation on St. John (U.S. Virgin Islands) in 1978 (Norton, 1997, 1998), the species is considered extirpated throughout the U.S. Virgin Islands (Platenberg and Boulon, 2006); thought to have been extinct from 1931 to 1974 until rediscovered by J. A. Rivero, recovery programs that emphasize captive breeding rather than field studies and habitat protection have not been effective (Joglar, 2005a); some reintroductions of captive-bred tadpoles appear to have been successful in that some adults subsequently have been found (Zippel, 2009; R. Thomas, in litt., 22.IV.2008).

Remarks: Recent work by Frost et al. (2006) resurrected *Peltophryne* as the genus for this species.

Bufo longinasus Stejneger, 1905

Distribution: Cuba, where four subspecies occur in western Pinar del Río Province (*B. l. longinasus*), Meseta de Cajálbana (*B. l. cajalbanensis*), Sierra de Trinidad (*B. l. dunni*), and the Sierra del Guaso (*B. l. ramsdeni*); elevational range 100–820 m.

Habitat: Under rocks and in litter in and along a mountain stream in February, submerged frogs (mostly males) among many tadpoles would intermittently rise to the surface to breathe (*B. l. dunni*; G. Peters, 1974); pinewoods (*B. l. longinasus*, *B. l. cajalbanensis*) and mesic broadleaf forest (*B. l. dunni*, *B. l. ramsdeni*), by day, under rocks, logs, or under bark 1.5 m above ground (Ruiz García, 1987); at night, *B. l. dunni* to 1.5 m in shrubs, emergent aquatics (*Cyperus alternifolius*), and mossy saplings (Schwartz and Henderson, 1991).

Activity: Nocturnal, but males call by day and night (Ruíz García, 1987; Alonso Bosch and Rodríguez Gómez, 2003c); often diurnally active (Díaz and Cádiz, 2006). **Behavior:** Inflates body when threatened (Valdés de la Osa and Ruíz García, 1979). **Diet:** Includes ants, centipedes, and small beetles (Ruíz García, 1987); primarily ants (Díaz and Cádiz, 2006). **Disease:** Chytrid fungus, *Batrachochytrium dendrobatidis* (Díaz et al., 2007; Díaz and Cádiz, 2008). **Parasites:** Nematodes *Aplectana* sp. and *Oswaldocruzia lenteixeirai* (Baruš, 1973; Coy Otero and Lorenzo Hernández, 1982). **Reproduction:** Males call throughout year, but especially June–August (Schwartz and Henderson, 1991); call day and night while floating in water (pools or slow-flowing creeks) and from banks of brooks and rivulets, amplexus axillary, eggs deposited in double strings or clumps of 20–30 on leaves or branches of submerged vegetation (Ruíz García, 1987); in September, male *B. l. dunni* called from exposed sites over a slow-flowing mountain stream, choosing logs, stones, and even branches of trees at heights to 1 m above ground or water (Alonso Bosch and Rodríguez Gómez, 2003c); activity seasonal, April–November, males call from water or shore, from perches on branches, leaves, and roots, females remain hidden in leaf litter and move to water only to mate, amplexus occurs as soon as the female arrives and eggs are quickly released, egg deposition principally around sunken roots or vegetative debris, rarely around rocks or on bare substrate, pairing often interrupted to move to new site, at each of which 20–90 eggs are deposited, breeding air temperatures 26–31 °C, water temperatures 25–26 °C (Díaz and Cádiz, 2006); eggs laid and larvae develop in streams (Hedges and Díaz, 2004e; Stuart et al., 2008). **Size:** SVL to 36 mm (Schwartz and Henderson, 1991); larvae to 19 mm total length (Díaz and Cádiz, 2006).

Conservation Status: "Endangered" on the IUCN Red List (Hedges and Díaz, 2004e); *B. l. longinasus* considered "vulnerable" (Rodríguez Gómez and Alonso Bosch, 2000); individuals of *B. l. ramsdeni* have not been seen since the subspecies was described by Barbour (1914); captive breeding program attempted without success (Valdés de la Osa and Ruíz García, 1980a), but basic parameters have been established since (Díaz and Cádiz, 2006).

Remarks: Recent work by Frost et al. (2006) resurrected *Peltophryne* as the genus for this species; subspecies may represent full species (Valdés de la Osa and Ruíz García, 1980a).

Bufo marinus Linnaeus, 1758

Distribution: Mexico, Central America, South America, Trinidad & Tobago, Little Tobago; introduced widely (e.g., Australia, Hawaiian Islands); likely introduced (as opposed to natural overwater dispersal from mainland or oceanic islands, e.g., Trinidad & Tobago) into the West Indies, currently known from Grand Cayman, Hispaniola, Jamaica, Puerto Rico, U.S. and British Virgin Islands, Anguilla (?), Antigua, St. Christopher, Nevis, Montserrat, Guadeloupe, Marie-Galante, Martinique, St. Lucia, St. Vincent, Carriacou, Grenada, Barbados; introduction into Cuba failed (Honegger, 1970; Lever, 2003).

Habitat: In a wide variety of situations, but usually in altered habitats (agricultural, urban), diurnally under fallen trees, loose boards, matted *Cocos* fronds (Schwartz and Henderson, 1991); on Grand Cayman, under a pile of brush (Burton and Echternacht, 2003); on Hispaniola, widespread, common in agricultural areas (Lever, 2003), but also occurs in coastal areas (Mertens, 1939); under driftwood in dry riverbed by day (RP); on Jamaica, at night on fields and lawns, by day beneath fallen trees, loose boards, and matted *Cocos* fronds (Lynn, 1940); at edge of Montego Bay and on flats of the Cabarita River (Neill, 1958); on roads at night (Schwartz and Fowler, 1973); on St. Lucia, "can be seen on any night doing its best to get killed on the road" (Long, 1974); on Cabarita Island (Jamaica), in garden area or adjacent forest (Crombie et al., 1983); forages at night on manicured hotel grounds, pairs in amplexus in garden pools (Hilgenhof, 1998b); on Puerto Rico, widespread, but common around houses and hotels in coastal regions (Hedges, in prep.); breeds at low elevations but found along roads and trails in the mountains (Drewry, 1970b); in coastal wetlands, with abundance increasing with salinity (Rios-López, 2008a); subtropical dry forest with understory of the cycad *Zamia* and the broadleaf *Antherium* (Carpenter and Gillingham, 1987); in a bat cave (Rodríguez and Reagan, 1984); in the U.S. Virgin Islands, in forests, grasslands, coastal scrub, beaches, agricultural pastures, and urban areas (Platenberg and Boulon, 2006); on Antigua, in most gardens and in pastures (Lever, 2003); in leaf litter in remaining patches of forest (RP); on Montserrat, often associated with human-modified habitats (urban and rural) such as roads, gardens, and agricultural areas (RWH); on Guadeloupe, in sugarcane fields, gardens, pastures, banana plantations, ravines, beaches, and culverts and under bridges (Breuil, 2002b); on Martinique, in cultivated areas (primarily *Musa* groves) and littoral dry forest (Breuil, 2009); congregate in large numbers around watering systems on hotel grounds (Bach and Schönecker, 2008); hotel lawns, around lighted areas at night (RWH); on St. Vincent, in agricultural areas, roadside ditches, gardens, vacant lots, abundant on the grounds of the botanical garden (Mallery et al., 2007); on Grenada, in cane fields and other agricultural situations, around livestock, residential areas (e.g., gardens, lawns), watering troughs, mudflats, roadside ditches (RWH); on Barbados, in nurseries (Lever, 2003).

Activity: Nocturnal (Breuil, 2002b); observed active at 0400 h on Cabarita Island, Jamaica (Crombie et al., 1983). **Activity Centers:** On Puerto Rico, males 389.9 ± 275.8 m^2, and females 166.5 ± 77.5 m^2 (Carpenter and Gillingham, 1987). **Competition:** May interfere with tadpoles of *Leptodactylus albilabris*, with which it shares ephemeral ponds for breeding (Platenberg, 2007). **Defense:** Secretions (cardiotoxic, vasoconstricting, convulsant, hallucinogenic) from enlarged parotoid glands may kill or sicken animals that bite the toad (Rivero, 1998). **Diet:** Includes beetles, ants, spiders, and millipedes (Jamaica; Lynn, 1940); on Puerto Rico, scarabeid beetles (*Parachalepus barbatus*) and May Beetles (*Phyllophaga portoricensis*) comprise bulk of prey items, but millipedes, scorpions, weevils, roaches, and other beetles are consumed (Wolcott, 1937); on St. Lucia, "will eat almost anything," cocoa beans, cigarette butts, twigs, leaves, insects, stones, and millipedes found in stomachs (Long, 1974); on Guadeloupe (where it will hunt at night on terraces for light-attracted insects), includes beetles (50%), ants (66%), centipedes (40%), and roaches (25%; Breuil, 2002b); tadpoles feed on vegetable matter until metamorphosis (Rivero, 1998). **Mortality:** Despite high temperature tolerances by both tadpoles and adults, dehydration appears to be the major cause of mortality (Malhotra and Thorpe, 1999). **Parasites:** Nematode *Foleyella brachyoptera* (Dominican Republic; Alfau, 1980); harbors ticks *Amblyomma rotundum* (Jamaica; Newstead, 1909; Morel, 1967); *A. dissimile* on Puerto Rico (although presumably introduced with toads and not found since; Maldonado Capriles and Medina Gaud, 1977; Guglielmone et al., 2003); Hispaniola (Jakowska, 1972), St. Lucia (Robinson, 1926; Morel, 1967; Lever, 2003), and Guadeloupe (Audebaud and Courmes, 1961; Morel, 1967; Breuil, 2002b); infested with the trematode *Mesocoelium monas* on Puerto Rico (Acholonu, 1970) and Jamaica (Mettrick and Dunkley, 1968); pathogenic *Salmonella* on Puerto Rico (Burrowes and Joglar, 2005). **Predators:** In captivity, *Epicrates inornatus* (Thomas and Gaa Kessler, 1996); Red-tailed Hawks (*Buteo jamaicensis*; Santana and Temple, 1988) and Cattle Egrets (*Bubulcus ibis*; Gassett et al., 2000); Giant Centipede (*Scolopendra alternans*) at El Verde, Puerto Rico (Carpenter and Gillingham, 1984; Waide and Reagan, 1996; Joglar, 2005a); River Crab (*Epilobocera sinuatifrons*; Stewart and Woolbright, 1996; Joglar, 2005a). **Reproduction:** Males call from almost any standing water, including rivers and creeks with low flow gradient (Schwartz and Henderson, 1991); on Puerto Rico, called from shortly after dark and stopped suddenly shortly before daylight in November on a rainy night near semipermanent water (Grant, 1932d); in coastal wetland on Puerto Rico, males calling on 8 and 22 August and 20 October (Rios-López, 2008a); on Jamaica, calling from pond no more than 6.0 m from the ocean (Neill, 1958); amplexus axillary (Rivero, 1998); males (155.5 and 120.0 mm SVL) displaced amplectant males (105.0 and 106.5 mm SVL, respectively) in Puerto Rico (Vargas-Salinas, 2005); in Puerto Rico, in small (~100 m^2) pond, breeding events last one night after heavy rains, especially at onset of wet seasons (March–May, August), males abundant (23–52 individuals) from 1900 h, females present 2000–2330 h with most in amplexus, breeding event sex ratio male biased (11.5:3.25), males alert to any movement or vibration, respond vigorously and attempt to clasp source (including tree trunks, turtles, dead female *Rana catesbeiana*, and birds on water surface), females arriving at pond during scramble clasped as soon as entered water with up to 5 males struggling for possession, in larger (>100 m^2) pond in Puerto Rico, males (to 9) form choruses, calling from mounds of leaves and branches, separated from one another by >2.0 m, male interaction reduced in choruses, more discriminating in what was clasped (Vargas-Salinas, 2007); on St. Lucia, eggs in long gelatinous strings deposited in still or slow-flowing water (Long, 1974); eggs (to 35,000) in long "rosary" strings, hatch after 36 h (Rivero, 1998) to 4 days (Schwartz and Henderson, 1991); may lay eggs in small, temporary accumulations of water (e.g., cavities in fallen trees), especially in the wet season, but may breed during the dry season in more permanent bodies of water (Malhotra and Thorpe, 1999); tadpoles with high heat tolerance, presumably an adaptation for breeding in temporary pools (Heatwole et al., 1968); metamorphs 12–14 mm (Rivero et al., 1980); metamorphs in coastal wetland on 18 December (Rios-López, 2008a); on Guadeloupe, tadpoles first half of July, metamorphs end of July (Breuil, 2002b). **Salinity Tolerance:** In laboratory experiments, tadpole survival to metamorphosis decreased from 67.0% at 1 ppt to 38.2% at 8 ppt, 4 of 19 metamorphs at 8 ppt failed to develop one forelimb (Rios-López, 2008a). **Size:** SVL to 230 mm (Rivero, 1998). **Water Hole Fidelity:** In Puerto Rico, marked toads returned to water holes on a regular and consistent basis depending on rainfall patterns, average time between visits to water holes 6.7 days (range 3–17 days), and moved 15–165 m from activity center to water hole (Carpenter and Gillingham, 1987).

Conservation Status: On the IUCN Red List as a species of "least concern" (Solís et al., 2004a, 2008a).

Additional Pertinent Literature: Easteal (1986).

Remarks: Recent work by Frost et al. (2006) placed this species in the genus *Chaunus* and that by Chaparro et al. (2007) assigned it to the genus *Rhinella*.

Bufo peltocephalus Tschudi, 1838

Distribution: Cuba (including Isla de la Juventud, Archipiélago de Sabana-Camagüey, cays of the north coast of Holguín Province), not known from western Cuba; elevational range, sea level to ~800 m.

Habitat: In many situations, mesic or xeric, including near ocean, but also broadleaf forest, cultivated fields, and along stream banks, diurnally under rocks, mouths of tile drains, and under palm trunks (Schwartz and Henderson, 1991); in burrows of Burrowing Owls (*Athene cunicularia*; Valdés de la Osa, 1978) and those of freshwater crabs (*Epilobocera*; Díaz and Cádiz, 2008); on patios and around houses in rural areas (Garrido et al., 1986; Rivalta et al., 2003b); between slabs of a hut (Garrido, 1980b); on grounds of hotel in Santa Lucia (Kwet, 1995); in the mouth of a cave system (Edwards, 1997); in holes formed by dogtooth limestone

(Rivalta et al., 2004); on lawns after dark at the U.S. Naval Base at Guantanamo (RP).

Abundance: Common and frequently seen throughout the year (Díaz and Cádiz, 2008). **Activity:** Nocturnal, with two activity peaks 2200–2300 h and 0100–0300 h (Sampedro Marín et al., 1982b). **Behavior:** Inflates body to prevent ingestion (Valdés de la Osa and Ruíz García, 1979). **Diet:** Movement-oriented predator (Rivalta et al., 2003b); diet includes insects (hymenopterans, 84% ants, beetles, roaches, lepidopterans, hemipterans, earwigs, orthopterans, flies), millipedes, spiders and other arachnids, centipedes, crustaceans, snails, earthworms; average weight of stomachs 7.93 g in males ($N = 36$), 8.36 in females ($N = 12$), average number of prey items 40.3 in males, 56.0 in females (Sampedro Marín et al., 1982). **Hybridization:** Presumed hybrids with *B. taladai* (Schwartz, 1960b). **Longevity:** In captivity, 16 years, 7 months (T. Loomaaed in Slavens and Slavens, 2003). **Parasites:** Nematodes *Abbreviata baracoa*, *Aplectana hamatospicula*, *Aplectana* sp., *Oswaldocruzia lenteixeirai*, and *Rhabdias elegans* (Baruš, 1973; Coy Otero and Lorenzo Hernández, 1982); cestode *Batrachotaenia bufonis* and trematode *Mesocoelium monas* (Pérez Vigueras, 1942; Coy Otero and Lorenzo Hernández, 1982); ticks *Amblyomma cruciferum* (Schulze, 1941; Morel, 1967; Guglielmone et al., 2003) and *A. rotundatum* (Pérez Vigueras, 1934a; Morel, 1967); *A. dissimile* (V. Černý, 1966, 1969; Guglielmone et al., 2003); ticks (Ruiz García, 1987). **Reproduction:** Can be an explosive breeder in some areas, eggs laid and tadpoles develop in still water (Hedges and Díaz, 2004f; Díaz and Cádiz, 2008); males call 6 April–25 December from flooded ditches and slow running streams, amplexus axillary, pair in amplexus 6 April (Sampedro Marín and Berovides Álvarez, 1982); dense choruses along edges of shallow, slowly running streams in May, not in a montane region and not covered by forest (Alonso Bosch and Rodríguez Gómez, 2003c); on Cayo Coco, reproduction depends on presence of ponds and lagoons (Estrada, 1993b). **Size:** SVL to 163 mm in females (Schwartz and Henderson, 1991).

Conservation Status: On the IUCN Red List as a species of "least concern" (Hedges and Díaz, 2004f).

Remarks: Recent work by Frost et al. (2006) resurrected *Peltophryne* as the genus for this species.

Bufo taladai Schwartz, 1960

Distribution: Cuba, disjunct populations in central, east central, and eastern Cuba; elevational range, sea level to ~850 m.

Habitat: Occurs in mesic broadleaf forest, banana and cacao groves, cultivated areas in montane foothills, under rocks by day (Schwartz, 1960b); in burrows of freshwater crabs (*Epilobocera* sp.; Díaz and Cádiz, 2008); associated with spring-fed stream in plantations, in litter, calling from rocks, or foraging on bare ground, burrows in stream banks are horizontal or sloping slightly upward, tadpoles in pools and puddles in and along the stream (G. Peters, 1974); larvae occur in small streams with a rapid current and in ponds at sides of rivers and between rocks on pond bank, at depths of 2–17 cm at the bottom of ponds in muddy substrate or around rocks with green algae, water temperature 20–21 °C (Diaz et al., 2000).

Abundance: Common (Díaz and Cádiz, 2008), "very common in suitable habitat" (Hedges and Díaz, 2004g; Stuart et al., 2008). **Activity:** Even at temperatures of 16.0–19.5 °C, toads active at dusk (G. Peters, 1974); nocturnal (Diaz et al., 2000). **Foraging and Diet:** Two individuals sat next to fallen oranges catching fruitflies (G. Peters, 1974); diet of a 103.9 mm SVL female included 16 mites, 1 scorpion (*Centruroides anchorellus*), 1 harvestman, 17 ants (including *Solenopsis* sp.), 5 orthopterans, 1 beetle, 1 hemipteran, and 1 centipede (Fong Grillo and Garcés G., 2002b); by 2.0 cm in length, larvae cannibalize other larvae, also eat small fish (*Girardinus* sp.; Rivalta et al., 2003b). **Hybridization:** Presumed hybrids with *B. peltocephalus* (Schwartz, 1960b). **Parasites:** Nematodes *Aplectana hamatospicula*, *Oswaldocruzia lenteixeirai*, *Rhabdias elegans* (Baruš, 1973; Coy Otero and Lorenzo Hernández, 1982), and larval *Porrocaecum* sp. (Baruš, 1972, 1973; Coy Otero and Lorenzo Hernández, 1982); cestode *Batrachotaenia bufonis* (Freze and Rysavy, 1976). **Reproduction:** Males call (9 July–19 August) from shallow creeks and flooded ditches, amplexus axillary (Schwartz, 1960b); males calling in June among herbaceous vegetation in a flooded area and in February entirely exposed along the shore of a montane river (Alonso Bosch and Rodríguez Gómez, 2003c); females deposit gelatinous strings of eggs at depths of 5–10 cm between roots along streams with relatively rapid currents and around ponds with circulating water (Rivalta et al., 2003b); eggs laid in permanent water (Hedges and Díaz, 2004g; Stuart et al., 2008). **Size:** SVL to 147 mm in females (Schwartz, 1960b); metamorph 11.1 mm (G. Peters, 1974).

Conservation Status: "Vulnerable" on the IUCN Red List (Hedges and Díaz, 2004g; Stuart et al., 2008).

Remarks: Recent work by Frost et al. (2006) resurrected *Peltophryne* as the genus for this species.

Figure 6. Like most species in the family Eleutherodactylidae, Haitian *Eleutherodactylus bakeri* (*left*) lays terrestrial eggs that are attended by the male. This male was calling while guarding three clutches of eggs from three different females on the inside of a bromeliad leaf. Female-biased sexual size dimorphism is common in these frogs, but more extreme than most species in *E. oxyrhyncus* (*right*), also from southern Haiti. Photographs by S. Blair Hedges.

Eleutherodactylus abbotti Cochran, 1923

Distribution: Hispaniola, islandwide, but absent from very dry areas; elevational range, sea level to > 1,800 m.

Habitat: Mesic wooded or forested situations (including pine forests), often under litter, logs, and trash by day, once under fallen communal stick nest of Palm Chat (*Dulus dominicus*; Schwartz and Henderson, 1991); in urban areas (Hedges et al., 2004rrr); under surface debris in recent clear-cut in the Sierra de Baoruco (RP).

Diet: Based on a sample of 18 frogs, includes ants in 11 of 18 stomachs, orthopterans (10/18), other hymenopterans (6/18), beetles (immature; 5/18), flies (5/18), and spiders (5/18), the most frequently consumed prey items were ants (23), orthopterans (21), and immature coleopterans (20), ants and orthopterans accounted for the greatest prey volume (A. K. Howard et al., 1997). **Parasites:** Larval nematodes *Terranova caballeroi*, with frogs presumably serving as intermediate hosts and frog-eating *Hypsirhynchus parvifrons* the final host, also a juvenile (possibly an oxyuroid) and two immature nematodes, apparently spiruroids (Powell et al., 1998a). **Predators:** *Ialtris haetianus* (Henderson and Schwartz, 1986); *Hypsirhynchus parvifrons* (Henderson et al., 1987a). **Reproduction:** Calls from ground, near ground, low bushes, and shrubs to 1.8 m above ground, often calls during the day in great numbers (Schwartz and Henderson, 1991); clutch size 11–23 (Hedges, in prep.); lays eggs among leaves above ground or on the ground at the base of reeds, hatchlings use "egg teeth" to rip first the egg's inner and then the outer capsule (Noble, 1926); breeds by direct development (Hedges et al., 2004rrr). **Size:** SVL to 25 mm (Schwartz and Henderson, 1991).

Conservation Status: Of "least concern" on the IUCN Red List (Hedges et al., 2004rrr).

Additional Pertinent Literature: Schwartz (1977c).

Eleutherodactylus acmonis Schwartz, 1960

Distribution: Eastern Cuba, Yunque de Baracoa in northern Guantánamo Province; elevational range 500–1,150 m; the extension of the range into the Sierra Maestra (Schwartz and Henderson, 1991) is probably in error (Díaz and Cádiz, 2008).

Habitat: On wet ground in a moist ravine in hardwood forest and on a shrub 0.6 m above ground (Schwartz, 1960a; Schwartz and Henderson, 1991); female on a rocky path in late afternoon on a rainy day (G. Peters, 1974); in leaf litter in humid forests, especially under piles of palm leaves, seeks refuge in cracks and crevices of rocks (Díaz and Cádiz, 2008); on the ground and in rocky crevices (Hedges and Díaz, 2004h; Stuart et al., 2008).

Abundance: Relatively common in its restricted range (Díaz and Cádiz, 2008). **Reproduction:** Eggs laid on the ground, breeds by direct development (Hedges and Díaz, 2004h; Stuart et al., 2008). **Size:** SVL in males to 21.2 mm, in females to 24.5 mm (Schwartz, 1960a); to 28 mm (Díaz and Cádiz, 2008).

Conservation Status: "Vulnerable" (Vales et al., 1998); "endangered" on the IUCN Red List (Hedges and Díaz, 2004h).

Eleutherodactylus adelus Díaz, Cádiz, and Hedges, 2003

Distribution: Western Cuba, pine forests of Alturas de Pizarras del Sur; elevational range ~130–170 m.

Habitat: *Pinus tropicalis* forest with herbaceous vegetation of *Lycopodiella* sp., ferns (*Pteris* sp.) and grass (*Eleocharis* sp.), soil acidic and sandstone derived (Díaz et al., 2003); on the ground, not recorded from outside forest habitats (Hedges and Díaz, 2004i; Stuart et al., 2008); during the dry season under leaf litter (that provides a humid microhabitat) at the edges of streams that cross the pine habitat (Díaz and Cádiz, 2008).

Abundance: Moderately common but very cryptic (Hedges and Díaz, 2004i; Stuart et al., 2008). **Reproduction:** Males call during early morning, less frequently during the night (Díaz and Cádiz, 2008); a clutch with three developing eggs (4.7–5.3 mm) directly on ground attached to another just-hatched clutch of four, hatchlings 3.7–3.8 mm SVL (Díaz et al., 2003). **Size:** Mean male SVL 11.4 mm, mean female SVL 14.5 mm (Diaz et al., 2003).

Conservation Status: "Endangered" on the IUCN Red List (Hedges and Díaz, 2004i).

Eleutherodactylus albipes Barbour and Shreve, 1937

Distribution: Cuba, higher elevations in the Sierra Maestra; elevational range 1,200–1,974 m.

Habitat: Upland wet forest (Alonso Bosch and Rodríguez Gómez, 2003b); occurs in leaf litter and under rocks and logs in rainforest and cloud forest, juveniles in mosses (Díaz and Cádiz, 2008); not recorded from disturbed forest (Hedges and Díaz, 2004j; Stuart et al., 2008).

Abundance: Uncommon (Hedges and Díaz, 2004j; Stuart et al., 2008). **Behavior:** Movements markedly slow and deliberate (Díaz and Cádiz, 2008). **Reproduction:** Call unknown (Díaz and Cádiz, 2008); female (31.25 mm) beneath a stone in a small cavity "guarding" a clutch of 42 eggs (mean diameter 4.3 mm) with embryos at development stage 11 or 12 in June (Alonso Bosch et al., 2005); eggs laid on the ground, breeds by direct development (Hedges and Díaz, 2004j; Stuart et al., 2008). **Size:** SVL to 33 mm (Schwartz and Henderson, 1991); SVL to 31.1 mm (Díaz and Cádiz, 2008).

Conservation Status: "Vulnerable" (Vales et al., 1998); "critically endangered" on the IUCN Red List (Hedges and Díaz, 2004j).

Eleutherodactylus alcoae Schwartz, 1971

Distribution: Hispaniola, lower Barahona Peninsula, including Isla Beata (Dominican Republic); elevational range, sea level to >600 m.

Habitat: By day, in fractures and crevices of limestone and in moist caves on the Peninsula de Barahona, in semixeric forest on slopes of the Sierra de Baoruco, in more mesic situations at Los Patos (eastern extreme of Sierra de Baoruco; Schwartz and Henderson, 1991); under rocks near cattle tank by day (RP).

Abundance: Very common in suitable habitat (Hedges et al., 2004jj; Stuart et al., 2008). **Predators:** *Ialtris haetianus* (Henderson and Schwartz, 1986); *Hypsirhynchus parvifrons* (Henderson et al., 1987a). **Reproduction:** Males call from ground, limestone cavities, shrubs, bushes, saplings, and vine tangles to heights of 1.8 m, homosexual pair in amplexus on exposed rock (Schwartz and Henderson, 1991); eggs laid on the ground, breeds by direct development (Hedges et al., 2004jj; Stuart et al., 2008). **Size:** SVL in females to 45.0 mm (Schwartz and Henderson, 1991).

Conservation Status: "Endangered" on the IUCN Red List (Hedges et al., 2004jj).

Additional Pertinent Literature: Schwartz (1977d).

Eleutherodactylus alticola Lynn, 1937

Distribution: Eastern Jamaica, Blue Mountains; elevational range, ~1,650–2,250 m.

Habitat: Under rocks along mountain trails or in open woods (Lynn, 1937); not tolerant of habitat disturbance (Hedges et al., 2004aaa; Stuart et al., 2008).

Reproduction: Eggs often attached to roots or sticks or to surfaces of small stones lying beneath larger rocks, and adhere to such objects and each other (Lynn, 1937); egg clutches in company of male or female (Lynn, 1940); 8–41 adherent eggs/clutch (in July), require at least 21 days for development (Lynn and Dent, 1943). **Size:** SVL to 26 mm (Schwartz and Henderson, 1991).

Conservation Status: "Critically endangered" on the IUCN Red List, last collected in the 1980s (Hedges et al., 2004aaa; Stuart et al., 2008).

Eleutherodactylus amadeus Hedges, Thomas, and Franz, 1987

Distribution: Hispaniola, Massif de la Hotte (Haiti); elevational range 1,000–2,340 m.

Habitat: Closed-canopy forest, including recently disturbed areas where it may be "hanging on" rather than "persisting" (Hedges et al., 2004sss; Stuart et al., 2008); diurnally under rocks and logs, nocturnally to 50 cm above ground in herbaceous vegetation (Hedges et al., 1987).

Abundance: Very common within its restricted range (Hedges et al., 2004sss; Stuart et al., 2008). **Reproduction:** Males call from low herbaceous vegetation in late afternoon and early evening, eggs (11–27) on ground under objects, in association with 1–3 adults, eggs 3.4–5.1 mm diameter, hatchlings 4.2–4.4 mm SVL (Hedges et al., 1987); clutch size 10 (Hedge, in prep.); breeds by direct development (Hedges et al., 2004sss; Stuart et al., 2008). **Size:** SVL to 25 mm (Hedges et al., 1987).

Conservation Status: "Critically endangered" on the IUCN Red List (Hedges et al., 2004sss).

Additional Pertinent Literature: Hedges and Powell (1998a).

Eleutherodactylus amplinympha Kaiser, Green, and Schmid, 1994

Distribution: Dominica, higher elevations; elevational range 300–1,200 m.

Habitat: Occurs in the transition zone between montane rainforest (few trees, shrub thickets, palm brakes, moist moss mats, and ferns) and elfin woodland (Kaiser and Henderson, 1994; Kaiser et al., 1994b); arboreal habitat generalist with preference for somewhat open higher-elevation rainforests, has been found in disturbed habitats along access roads supporting hydroelectric projects (Hedges and Powell, 2004a; Stuart et al., 2008).

Parasites: Plerocercoid cysts, *Porrocaecum* larvae, and polymorphid cystacanths (Goldberg et al., 1998d). **Reproduction:** Mass of 13 eggs on thin mat of ground moss in January attended by male (SVL 25 mm), 7 females (SVL 25.0–46.5 mm) with mean of 29.3 eggs (Kaiser et al., 1994b); eggs in bromeliad (Malhotra and Thorpe, 1999); clutch size 37 (Hedges, in prep.); develops directly (Hedges and Powell, 2004a; Stuart et al., 2008). **Size:** SVL in males to 26.4 mm, in females to 49.7 mm (Kaiser et al., 1994b); females to >50 mm SVL (Malhotra and Thorpe, 1999).

Conservation Status: "Endangered" on the IUCN Red List (Hedges and Powell, 2004a).

Eleutherodactylus andrewsi Lynn, 1937

Distribution: Jamaica, Blue Mountains and John Crow Mountains; elevational range ~545–1,970 m.

Habitat: Occurs in deep woods (Lynn, 1937); on streamside rocks, beneath trash, litter, and rocks on ground, calling males on ground under raised rocks on road cut and from other more or less concealed sites (Schwartz and Fowler, 1973); can tolerate only very temporary disturbances (Hedges et al., 2004bbb; Stuart et al., 2008).

Diet: Includes ants (Schwartz and Henderson, 1991). **Reproduction:** Ovarian clutch of 21 eggs (Schwartz and Henderson, 1991); clutch size 28 (Hedges, in prep.); males call from concealed terrestrial sites, eggs laid on ground, breeds by direct development (Hedges et al., 2004bbb; Stuart et al., 2008). **Size:** SVL to 26 mm (Schwartz and Henderson, 1991).

Conservation Status: "Endangered" on the IUCN Red List (Hedges et al., 2004bbb).

Eleutherodactylus antillensis Reinhardt and Lütken, 1862

Distribution: Puerto Rico, Vieques, Isla Culebra, U.S. and British Virgin Islands; introduced (?) on St. Croix (Seaman, 1961); elevational range, sea level to ~1,220 m.

Habitat: In Puerto Rico, at heights of 0.5–1.8 m in open, dry areas and dry forest (Stewart and Woolbright, 1996); in cities and other urban situations (Joglar, 1998); in meadows and along road cuts at any altitude on low shrubs or broadleaved plants (Drewry, 1970b); by day under grass roots, loose tree bark, rocks, logs, trash, and may use bromeliads

(*Tillandsia*) as diurnal retreats (Rivero, 1998); in tarantula burrow (Schwartz and Henderson, 1991); along railroad and trolley embankments north of Hotel Eureka in Santurce (K. P. Schmidt, 1928); on Vieques, under flat stones in nearly dry bed of small creek (Stejneger, 1904); on St. John (U.S. Virgin Islands), under banana sheaths ("sometimes in groups of eight or more"), occasionally in axils of bromeliads (Grant, 1932g); on St. Croix, in Christiansted gardens and along Canaan gut and Salt River (Seaman, 1961); bromeliads may be important habitat components on some islands in the British Virgin Islands (Ovaska et al., 2000); on Guana (British Virgin Islands), important habitat features include bromeliads and other plants that hold moisture in their axils (for diurnal retreats), and a layer of leaf litter for oviposition (Ovaska, 2005).

Abundance: Very common, even in disturbed habitats (Hedges et al., 2004yy; Angulo, 2008b). **Activity:** Initiated shortly after nightfall and terminated shortly before dawn (van Berkum et al., 1982). **Biomass:** On Guana, for adults, low estimate of 180–242 g/ha (in 2001) to a high estimate of 629–845 g/ha (in 1999; Ovaska, 2005). **Disease:** *Batrachochytrium dendrobatidis* fungal infections (Burrowes et al., 2008). **Foraging and Diet:** Forages from grasses, forbs, and shrubs (Stewart and Woolbright, 1996); average observed feeding rate 1.0 prey/h, average feeding rate based on stomach analysis 0.8 prey/h, average nonfeeding rate 1.0 move/h, diet on Puerto Rico includes ants (11.7% of total individuals in sample), flies (15.8%), beetles (14.8%), leafhoppers (32.8%), on St. John, ants (70.1%), flies (3.9%), beetles (6.0%), leafhoppers (3.9%; Jones, 1982). **Physiology:** Rehydrates more rapidly than *E. coqui* from high elevations (van Berkum et al., 1982). **Population Density:** 400 adults/ha (Stewart and Woolbright, 1996); monthly means (February 2001–February 2002) 33.3–177.8/ha depending on habitat at several sites (Rios-López and Aide, 2007); on Guana, estimated adult densities ranged from low (in 2001) of 82–110/ha to a high (in 1999) of 286–384/ha (Ovaska, 2005). **Predators:** Cattle Egrets (*Bubulcus ibis*) on St. Croix (Gassett et al., 2000); *Borikenophis portoricensis* (Rodríguez-Robles and Leal, 1993a; Henderson and Sajdak, 1996; Joglar, 2005a). **Reproduction:** On Puerto Rico, reproduces year-round (Joglar, 1998); males call from low bushes, branches (Rivero, 1998), and barbed-wire fencing, as well as from high in trees (Schwartz and Henderson, 1991); in the British Virgin Islands, call from terrestrial and arboreal bromeliads (14% on Guana, 4% on Virgin Gorda) and *Agave* (24% on Guana), 74% perched in vegetation <2.5 m high while calling, 26% at heights >2.5 m, 60–84% were on leaves or branches of trees and shrubs, depending on island (Ovaska et al., 2000); males produce advertisement calls, courtship calls, and aggressive calls, possibly in defense of calling territory and/or advertisement of occupancy of a territory (Ovaska and Caldbeck, 1997b); in captivity, when gravid females approached males to within 0–10 cm, males would switch from loud advertisement calls to much softer courtship calls consisting of 1–10 quickly repeated single notes (Michael, 1996); on Guana, courtship observations lasted 20–130 min, female would approach male on elevated vegetation, male would descend and change call from advertisement to courtship, process repeated until pair reached the ground, eventually both would move under leaf litter, two nests 450 cm and 480 cm from original calling sites of two males, egg masses 2–3 cm under leaf litter (Ovaska and Caldbeck, 1997a); clutch size 12–26 in Puerto Rico, gravid females with 6–29 eggs (Joglar, 1998); on Guana, clutch size 25–42, eggs 4.0–4.5 mm in diameter within 24 h of oviposition (Ovaska, 2005); male in attendance with 24-egg clutch on Puerto Rico (Ovaska and Estrada, 2003). **Size:** SVL in males to 28.1 mm (Joglar, 1998), in females to 43.2 mm (Ovaska et al., 2000).

Conservation Status: On the IUCN Red List as a species of "least concern" (Hedges et al., 2004yy; Angulo, 2008b), although numbers are declining "wherever *Osteopilus septentrionalis* spreads" (Perry and Gerber, 2006); "stable" (Moreno, 1991d).

Eleutherodactylus aporostegus Schwartz, 1965

Distribution: Hispaniola, Tiburon Peninsula (Haiti); elevational range, sea level to ~900 m.

Habitat: In subterranean chambers on grassy hillside adjacent to a slow-flowing stream (Schwartz, 1965e).

Size: SVL in males to 47.8 mm (Schwartz, 1965e).

Conservation Status: As until recently this species was considered a subspecies of *E. ruthae*, which is considered "endangered" (Hedges et al., 2004mm), we assume that this species is, at best, also endangered.

Eleutherodactylus apostates Schwartz, 1973

Distribution: Hispaniola, Massif de la Hotte (Haiti); elevational range ~335–1,650 m.

Habitat: Closed-canopy forest (Hedges et al., 2004ttt; Stuart et al., 2008); wooded ravine bordering a small creek (Schwartz, 1973b); under rocks, males call from muddy stream bank near water (Schwartz and Henderson, 1991).

Abundance: Once common in "good habitat," which probably no longer exists (Hedges et al., 2004ttt; Stuart et al., 2008). **Size:** Male SVL to 22.7 mm; female SVL to 43.2 mm (Schwartz, 1973b).

Conservation Status: "Critically endangered" on the IUCN Red List, last recorded in 1991, although no surveys have occurred since (Hedges et al., 2004ttt; Stuart et al., 2008).

Additional Pertinent Literature: Schwartz (1977e).

Eleutherodactylus armstrongi Noble and Hassler, 1933

Distribution: Hispaniola, Sierra de Baoruco (Dominican Republic) and Montagne Noire and Massif de la Selle (Haiti); elevational range ~150–1,700 m.

Habitat: Upland pinewoods and forest (but absent from that habitat in some areas), also in coffee groves and lowland and upland broadleaf forests, diurnal retreats include living bromeliads, immatures and some adults under rocks, logs, boards, and wood chips by day in mesic forest, also in ends of hollow stems, bases of banana leaves and certain mountain palms, and occasionally among damp leaves and stones on ground (Noble and Hassler, 1933); under surface debris in recent clear-cut (RP); occurs in degraded habitat (Hedges et al., 2004n; Stuart et al., 2008).

Abundance: Common in suitable habitat (Hedges et al., 2004n; Stuart et al., 2008). **Diet:** Based on a sample of 32 frogs, the most frequently consumed prey were ants (192), adult beetles (25), beetle larvae (18), and orthopterans (15), in terms of prey volume ants comprised the greatest volume (0.248 ml), followed by adult beetles (0.068 ml), mites and scorpions (0.055 ml), and flies (0.053 ml; A. K. Howard et al., 1997). **Parasites:** Larval nematodes *Terranova caballeroi*, with frogs presumably serving as intermediate hosts and frog-eating *Hypsirhynchus parvifrons* the final host, also an adult male *Oswaldocruzia lenteixeirai*, an adult female trichostrongyloid (but not *Oswaldocruzia*), and four additional unidentified nematodes (Powell et al., 1998a). **Reproduction:** Males call from shrubs (1.5–4.6 m), forest canopy, and *Pinus* trunks (Schwartz, 1978a); clutch size 31–49 (Hedges, in prep.); eggs laid on the ground under rocks or in bromeliads (Hedges et al., 2004n; Stuart et al., 2008). **Size:** SVL to 31 mm in males, to 43 mm in females (Schwartz, 1978a).

Conservation Status: "Endangered" on the IUCN Red List (Hedges et al., 2004n).

Additional Pertinent Literature: Schwartz (1978a).

Eleutherodactylus atkinsi Dunn, 1925

Distribution: Cuba, two currently recognized subspecies occur throughout much of Cuba, Isla de la Juventud, and possibly the Archipiélago de Sabana-Camagüey (*E. a. atkinsi*), and extreme eastern Cuba (*E. a. estradai*; replacement name for *E. a. orientalis*; Lynch, 1991); elevational range, sea level to >1,200 m.

Habitat: In various ecosystems, forested areas, but also open sites, such as savannas and grasslands, and cities, where it is less frequently encountered than other human commensals (Díaz and Cádiz, 2008); in mesic areas, including scrub, mesic woods and forests, groves of coffee, cacao, and bananas, rocky cave mouth, cornfield edge, small caves, and in a pasture (Schwartz and Henderson, 1991); also dry upland forest (Dunn, 1925); in houses in Santiago de Cuba (Alayo Dalmau, 1951); on a forest path on a sunny afternoon (K. Uhlenhaut in G. Peters, 1974); in a cave as far as 50 m from the opening (M. Stubbe in G. Peters, 1974); by day in dense grass and herbaceous growth, where disturbed while looking for *Anolis cyanopleurus* (G. Peters, 1974); by day under rocks under trash and fallen palm thatch, often under debris near human habitations (Schwartz and Henderson, 1991); in pastures under rocks and livestock excrement (Estrada, 1994); enters human habitations, under bed in hotel (Kwet, 1995); one of only two species (with *E. ionthus*) in a coffee plantation and one of only two in a managed timber forest (with *E. dimidiatus*) in eastern Cuba (Fong Grillo, 1999); often in gardens and yards (Kunz, 2005).

Abundance: Common (Díaz and Cádiz, 2008). **Activity:** Nocturnal (Estrada, 1993b). **Reproduction:** Males call from the ground to heights of 1.5 m above ground on fallen palm fronds, shrubs, and trees, 94 spherical eggs (likely laid 2 July) with mean diameter 3.5 mm in small depression in moist soil, on 8 July mean diameter 4.5 mm, 11 July 6–7 mm, 16 July begin to hatch, froglet SVL 4.65 mm, 42 eggs under tin cans on 12 July, hatched 16–28 July (Novo Rodríguez et al., 1987); males call on warm, humid, still nights from the ground or from leaves, branches, and trunks at heights of 0.5–3.0 m (Díaz and Cádiz, 2008); female attends nest (Alonso Bosch and Rodríguez Gómez, 2003b); develops directly (Hedges and Díaz, 2004k). **Size:** SVL to 43 mm in females (Schwartz and Henderson, 1991).

Conservation Status: On the IUCN Red List as being of "least concern" (Hedges and Díaz, 2004k).

Eleutherodactylus audanti Cochran, 1934

Distribution: Hispaniola, "South Island," Massif de la Hotte and Massif de la Selle (Haiti) and Sierra de Baoruco (Dominican Republic); elevational range ~800–2,500 m.

Habitat: Closed-canopy forest and along forest edges (Hedges et al., 2004o; Stuart et al., 2008); in highland forest edges, grasslands, and pinewoods (Shreve and Williams, 1963; Schwartz and Henderson, 1991); under rocks and pieces of debris in mesic situations by day (Schwartz, 1979c) and in bromeliads (Schwartz and Henderson, 1991); also under trunks and branches of fallen trees, in and around moist decaying wood under large logs among otherwise dry ground litter, and in crevices in roadside banks (Shreve and Williams, 1963); under surface debris in recent clear-cut (RP).

Abundance: Moderately common in suitable habitat (Hedges et al., 2004o; Stuart et al., 2008). **Reproduction:** Clutch size 13–17 (Hedges, in prep.); eggs laid under rocks, breeds by direct development (Hedges et al., 2004o; Stuart et al., 2008). **Size:** SVL in males to 22 mm, in females to 25 mm (Schwartz, 1979c).

Conservation Status: On the IUCN Red List as "vulnerable" (Hedges et al., 2004o).

Additional Pertinent Literature: Schwartz (1979c).

Remarks: Some of the natural history information presented for this species might refer to *E. melatrigonum* and/or *E. notidodes.*

Eleutherodactylus auriculatoides Noble, 1923

Distribution: Hispaniola, Cordillera Central (Dominican Republic); elevational range ~790–1,890 m.

Habitat: Arboreal in montane closed forest (Hedges et al., 2004p; Stuart et al., 2008); occurs in broadleaf forest, diurnally in bromeliads (Schwartz, 1980a; Schwartz and Henderson, 1991).

Abundance: "Relatively uncommon" even in suitable habitat (Hedges et al., 2004p; Stuart et al., 2008). **Reproduction:** Males call from shrubs, ferns, tree ferns, and saplings (0.3–3.5 m above ground), and occasionally on herbs and aroids (i.e., almost any horizontal surface including tops of fence posts), and most often from high in the forest canopy, male apparently guards eggs in bromeliads (Schwartz, 1980a; Schwartz and Henderson, 1991); clutch size 9 (Hedges, in prep.). **Size:** SVL in males to 27 mm, in females to 33 mm (Schwartz 1980a).

Conservation Status: On the IUCN Red List as "endangered" (Hedges et al., 2004p).

Additional Pertinent Literature: Schwartz (1980a).

Eleutherodactylus auriculatus Cope, 1863

Distribution: Cuba, islandwide at low to moderate elevations, also Isla de la Juventud; elevational range, sea level to >1,300 m.

Habitat: Arboreal in mesic primary and secondary forests, can tolerate shade-grown coffee plantations, but relies on tall trees, so not tolerant of disturbed areas (Hedges and Díaz, 2004l); in mesic woods or forest, at times within dense vine tangles and at times near streams (Schwartz and Henderson, 1991); one individual with *Sphaerodactylus bromeliarum* in a bromeliad that fell from 10 m above the ground (G. Peters and Schwartz, 1972); in thick grass, another in a bromeliad with *E. planirostris* (G. Peters, 1974); in soil, leaf litter, under rocks and logs, in herbaceous vegetation, around shrubs, trees, and epiphytes (Estrada, 1994; Díaz and Cádiz, 2008).

Abundance: Common, but more frequently heard than seen (Díaz and Cádiz, 2008). **Activity:** Nocturnal (Garrido and Schwartz, 1968b). **Reproduction:** Males call in large choruses when raining, during humid mornings and afternoons, and through portions of the night during the rainy season, at heights 1–2 m above ground level from small branches and leaf axils (Estrada, 1994); males call from 0.3–2.1 m high in shrubs, on tree trunks and branches, vines, leaves and dead stumps (Schwartz and Henderson, 1991); males may call (telegraphic click) up to 800 times/min from among vines and shrubs (Rodríguez Gómez and Alonso Bosch, 2003a); also call from under leaves on ground (Garrido and Schwartz, 1968b); call mainly at sunset (Díaz et al., 2005a); call at heights of 0.5–3.0 m from herbaceous growth and leaves, branches, and trunks of trees, either exposed or hidden in foliage, rarely from the ground, usually at dusk and dawn, occasionally by day in dense vegetation and only when very humid, females deposit eggs near calling males, which revert to a shorter and less intense call, with the vocal sac only partially distended, males emit soft calls to "guide" females to hidden sites in the litter, where eggs are deposited (Díaz and Cádiz, 2008); eggs 20/clutch in bromeliad (Ruiz García, 1987); clutch size 11 (Hedges, in prep.); breeds by direct development (Hedges and Díaz, 2004l). **Size:** SVL in males to 19 mm, to 24 mm in females (Rodríguez Gómez and Alonso Bosch, 2003a).

Conservation Status: On the IUCN Red List as being of "least concern" (Hedges and Díaz, 2004l).

Eleutherodactylus bakeri Cochran, 1935

Distribution: Hispaniola, Tiburon Peninsula (Haiti); elevational range ~890–2,325 m.

Habitat: Arboreal species dependent on closed-canopy forest, where it retreats into bromeliads by day (Hedges et al., 2004uuu; Stuart et al., 2008); bromeliadicolous and saxicolous, in bromeliads and vegetation at heights of 1.5 m at night, also on a rock face with trickling water (Schwartz, 1980b).

Abundance: Once common in suitable habitat that no longer exists (Hedges et al., 2004uuu; Stuart et al., 2008). **Reproduction:** Male attends nest (Townsend, 1996); may "guard" multiple clutches from different females (S. B. Hedges, in litt., 09.V.2008; figure 6); eggs laid in bromeliads, development direct (Hedges et al., 2004uuu; Stuart et al., 2008); clutch size 14–26 (Hedges, in prep.). **Size:** SVL in males to 37 mm, to 35 mm in females (Schwartz, 1980b).

Conservation Status: "Critically endangered" on the IUCN Red List (Hedges et al., 2004uuu).

Additional Pertinent Literature: Schwartz (1980b).

Eleutherodactylus barlagnei Lynch, 1965

Distribution: Guadeloupe (Basse-Terre); elevational range, sea level to ~1,400 m.

Habitat: Aquatic, in disturbed habitats only when trees occur along streams (Hedges et al., 2004d; Stuart et al., 2008); in humid forest, as well as in elfin woodlands on upper reaches of Soufrière, where it occurs in boulder-strewn mountain torrents (Breuil, 2002b); by day under rocks embedded in coarse sand on small bars in stream (Kaiser and Henderson, 1994); perches at night on boulders (0.3–1.8 m diameter and often moss-covered) in stream (with as many as 6 frogs/boulder), exhibits predilection for wet vertical substrates (Schwartz, 1967b); also under leaves and between tree roots (in company with *Eleutherodactylus johnstonei*, *E. martinicensis*, and *Sphaerodactylus*), near the Caribbean coast, in rivers on small dams comprising dead leaves and small rocks, prefers moss and *Sphagnum* substrates, as well as bases of bromeliads (*Guzmania*, *Pitcairnia*) on Soufrière (Breuil, 2002b).

Activity: Nocturnal (RWH). **Reproduction:** Males call from rocks in slack water along stream edges (not from rocks or boulders in rushing stream or from banks or rocks on banks; Schwartz, 1967b); clutch of approximately 10 eggs in a deep, narrow crack in the center of a rock wall 15 m high at about 900 m elevation (Kaiser and Henderson, 1994); eggs on vegetation or in rock crevices, development direct (Hedges et al., 2004d; Stuart et al., 2008); clutch size 7–18 (Hedges, in prep.). **Size:** SVL in males to 23 mm, in females to 32 mm (Schwartz, 1967b).

Conservation Status: "Endangered" on the IUCN Red List (Hedges et al., 2004d); may be susceptible to upstream pollution (Kaiser and Henderson, 1994); populations appeared to be declining in the 1990s (Breuil, 2002b).

Additional Pertinent Literature: Kaiser et al. (2004).

Eleutherodactylus bartonsmithi Schwartz, 1960

Distribution: Eastern Cuba; elevational range ~30–215 m.

Habitat: In virgin hardwood forest (Schwartz, 1960a); on a grassy path and under rocks (G. Peters, 1974); arboreal, also in coffee plantations, but not in open habitats (Hedges and Díaz, 2004m; Stuart et al., 2008); in semideciduous forests on limestone, diurnal retreats in rocky soil (Díaz and Cádiz, 2008).

Abundance: Common, frequently encountered when calling, but rarely found at other times (Díaz and Cádiz, 2008). **Activity:** Nocturnal (Díaz and Cádiz, 2008). **Reproduction:** Males calling before, during, and after an early evening rain from branches and leaves of trees and trunks of saplings (Schwartz, 1960a); males call from leaves and stems at heights of 0.5–3.0 m (Díaz and Cádiz, 2008); breeds by direct development (Hedges and Díaz, 2004m; Stuart et al., 2008). **Size:** Male SVL to 23.4 mm, female SVL to 26.2 mm (G. Peters, 1974).

Conservation Status: "Vulnerable" (Vales et al., 1998); "critically endangered" on the IUCN Red List (Hedges and Díaz, 2004m).

Eleutherodactylus blairhedgesi Estrada, Díaz, and Rodríguez, 1997

Distribution: Northwestern Cuba; elevational range around sea level.

Habitat: Limestone cliff faces 3–7 m from surf along the coast, by day in crevices and holes of a cliff, male on leaf of *Coccoloba uvifera* 25 cm above ground, 3.0 m from the beach, and 8.0 m from the surf (Estrada et al., 1997); some retreats occasionally inundated by waves, especially during the high surf associated with cold fronts (Díaz and Cádiz, 2008).

Abundance: Very common within its restricted range (Hedges and Díaz, 2004n; Stuart et al., 2008). **Activity:** Observed active 1830–1930 h until 0600–0700 h (Estrada et al., 1997); nocturnal, immediately after dark, leaves diurnal retreats and disperses throughout the coastal vegetation, later returning to the same refuges (Díaz and Cádiz, 2008). **Behavior:** Escapes by jumping large distances, movements easily heard in dry leaves (Díaz and Cádiz, 2008). **Predators:** Include *Tropidophis* sp. (Estrada et al., 1997). **Reproduction:** Vocal activity sporadic in early evening, one male calling at 0420 h, 5.0 m from the surf (Estrada et al., 1997); males call sporadically, with peaks at dawn and dusk, call from the ground and from leaves of Sea Grapes (*Coccoloba uvifera*) and other plants (Díaz and Cádiz, 2008); eggs laid on the ground, breeds by direct development (Hedges and Díaz, 2004n; Stuart et al., 2008). **Size:** SVL to at least 31.0 mm (Estrada et al., 1997); to 36 mm (Díaz and Cádiz, 2008). **Thermal Biology:** Body temperatures of three individuals were 26.4, 27.2, and 28.6 °C at corresponding air temperatures of 26.0, 29.0, and 29.2 °C and substrate temperatures of 26.0, 28.0, and 28.4 °C (Estrada et al., 1997).

Conservation Status: "Critically endangered" on the IUCN Red List (Hedges and Díaz, 2004n).

Eleutherodactylus bothroboans Schwartz, 1965

Distribution: Hispaniola, Cordillera Central (Dominican Republic) at elevations ~640 m.

Habitat: Occurs in open pine forest (Schwartz, 1965e).

Reproduction: Males call from underground cavities (Schwartz, 1965e). **Size:** SVL in males to 48.8 mm (Schwartz, 1965e).

Conservation Status: Until recently this species was considered a subspecies of *E. ruthae*, which is listed as "endangered" on the IUCN Red List (Hedges et al., 2004mm); we assume this species is, at best, also endangered.

Eleutherodactylus bresslerae Schwartz, 1960

Distribution: Eastern Cuba, Cuchillas de Baracoa; elevational range ~30–220 m.

Habitat: Rocks and cliffs in virgin coastal broadleaf forest at night (Schwartz, 1960a); coastal limestone forests (Díaz and Cádiz, 2008).

Abundance: Moderately common in suitable habitat (Hedges and Díaz, 2004o; Stuart et al., 2008). **Activity:** Nocturnal (Díaz and Cádiz, 2008). **Reproduction:** Males call from low vegetation (to heights of 1 m) and from the ground (Díaz and Cádiz, 2008); breeds by direct development (Hedges and Díaz, 2004o; Stuart et al., 2008). **Size:** SVL in males to 30.4 mm, in females to 45.8 mm (Schwartz, 1960a); SVL to 46 mm (Díaz and Cádiz, 2008).

Conservation Status: "Vulnerable" (Vales et al., 1998); "critically endangered" on the IUCN Red List (Hedges and Díaz, 2004o).

Eleutherodactylus brevirostris Shreve, 1936

Distribution: Hispaniola, Massif de la Hotte (Haiti); elevational range ~575–2,375 m.

Habitat: Under rocks and logs in a recent clearing and along a path in scrubby cloud forest (Schwartz, 1973b); whether the species can persist outside forested habitats is unknown (Hedges et al., 2004vvv; Stuart et al., 2008).

Abundance: Formerly common in good habitat, which no longer exists (Hedges et al., 2004vvv; Stuart et al., 2008). **Reproduction:** Calls from cavities in the ground under leaf litter and on low (15 cm) vegetation (Schwartz and Henderson, 1991); eggs laid on the ground, breeds by direct development (Hedges et al., 2004vvv; Stuart et al., 2008). **Size:** SVL to 28 mm (Schwartz and Henderson, 1991).

Conservation Status: "Critically endangered" on the IUCN Red List, not recorded since 1991, but no more recent surveys have occurred (Hedges et al., 2004vvv; Stuart et al., 2008).

Additional Pertinent Literature: Schwartz (1980d).

Eleutherodactylus brittoni Schmidt, 1920

Distribution: Puerto Rico; elevational range, sea level to ~1,080 m.

Habitat: Occurs in open, sunny, meadows (Drewry and Rand, 1983); pastures and young sugarcane fields (Rivero, 1998); also on roadside shrubs (K. P. Schmidt, 1920, 1928); in meadows and along road cuts at any altitude, usually on grasses (Drewry, 1970b); coastal grass-dominated wetlands (Rios-López, 2008c).

Activity: Nocturnal (Joglar, 1998). **Disease:** *Batrachochytrium dendrobatidis* fungal infections (Burrowes et al., 2008). **Foraging and Diet:** Females and juveniles eat earlier in the night than males, and calling males do not eat or, at best, eat less than females and juveniles, the diet includes collembolans, dipterans, hemipterans, hymenopterans, lepidopterans, and spiders (Lavigne and Drewry, 1970). **Reproduction:** Males call from low perches (0–1.5 m) on grass stems (Drewry and Rand, 1983) and herbaceous plants (Schwartz and Henderson, 1991); initiate calling before nightfall and stop shortly after dawn, calling peaks at dusk and dawn (Drewry, 1970a; Drewry and Rand, 1983; Joglar, 1998); preserved females with 2–5 eggs with diameters 1.76–2.3 mm (Joglar, 1998); eggs laid on the ground, development direct (Hedges et al., 2004oo; Rios-López, 2008c). **Size:** SVL in males to 16.7 mm, in females to 17.1 mm (Joglar, 1998).

Conservation Status: On the IUCN Red List as being of "least concern" (Hedges et al., 2004oo; Rios-López, 2008c); "few or no individuals located in recent surveys" (J. W. Wiley in Johnson, 1988); "stable" (Moreno, 1991d).

Eleutherodactylus caribe Hedges and Thomas, 1992

Distribution: Hispaniola, Tiburon Peninsula (Haiti); elevational range around sea level.

Habitat: Coastal mangrove (*Rhizophora mangle*) swamp with standing water (Hedges and Thomas, 1992a).

Reproduction: Some males calling 1–2 cm above waterline to 2.0 m high on mangrove leaves, but usually ~1.0 m above water (Hedges and Thomas, 1992a); clutch size 11 (Hedges, in prep.). **Size:** Males SVL to 18.1 mm, female SVL to 20.1 mm (Hedges and Thomas, 1992a).

Conservation Status: "Critically endangered" on the IUCN Red List, a distribution apparently limited to coastal mangroves suggests that the species' future might be tenuous (Hedges and Thomas, 2004a; Stuart et al., 2008).

Eleutherodactylus casparii Dunn, 1926

Distribution: Cuba, Sierra de Trinidad; elevational range, sea level to ~830 m (300–800 m; Díaz and Cádiz, 2008).

Habitat: Occupies a wide variety of mesic situations, including natural forest and some anthropogenic habitats such as plantations, as well as introduced vegetation, tolerates disturbance but requires large trees (Hedges and Díaz, 2004p; Stuart et al., 2008); in leaf litter and herbaceous vegetation, under rocks and other objects (Estrada, 1994).

Abundance: "Very common" (Hedges and Díaz, 2004p; Stuart et al., 2008). **Reproduction:** Males call at dusk, rarely during the day, from the ground and herbaceous growth or from refuges between rocks, females vocalize sporadically (Díaz and Cádiz, 2008); eggs laid on the ground, breeds by direct development (Hedges and Díaz, 2004p; Stuart et al., 2008). **Size:** SVL in males to at least 20.0 mm, in females to 25.3 mm (Schwartz, 1960a); SVL to 27 mm (Díaz and Cádiz, 2008).

Conservation Status: "Endangered" on the IUCN Red List (Hedges and Díaz, 2004p).

Additional Pertinent Literature: Schwartz (1974a).

Eleutherodactylus cavernicola Lynn, 1954

Distribution: Jamaica, Portland Ridge and Hellshire Hills; elevational range 10–15 m.

Habitat: Inhabitant of caves in moderately well-forested areas (Schwartz and Henderson, 1991); exposed on moist earth and calling from solution cavities in limestone near (within 9.0 m) the entrance of a small cave (Schwartz and Fowler, 1973).

Reproduction: Eggs laid among rocks, breeds by direct development (Hedges et al., 2004ccc; Stuart et al., 2008). **Size:** SVL in males to 38.0 mm, to 41 mm in females (Hedges, in prep.).

Conservation Status: "Critically endangered" on the IUCN Red List, very rare, last seen in the 1990s, but no surveys have been conducted since (Hedges et al., 2004ccc; Stuart et al., 2008).

Eleutherodactylus chlorophenax Schwartz, 1976

Distribution: Hispaniola, Massif de la Hotte (Haiti); elevational range ~990–1,290 m.

Habitat: Wet limestone forests and ravine pinelands, in solution pit caves to 97 m below the surface, hopping about or hiding under rocks and organic debris at sinkhole bottom, or wedging into nooks and crannies on vertical walls of a sink, also on the surface under a rock in a dry stream bed, and from a notch in the side of a boulder 2.0 m above a floodplain (Franz and Cordier, 1986); one male on a branch above a small deep sinkhole (Schwartz and Henderson, 1991).

Reproduction: Eggs laid on the ground, breeds by direct development (Hedges et al., 2004www; Stuart et al., 2008); clutch size 29 (Hedges, in prep.). **Size:** SVL in males to 59 mm (Schwartz and Henderson, 1991).

Conservation Status: "Critically endangered" on the IUCN Red List (Hedges et al., 2004www).

Additional Pertinent Literature: Cunningham et al. (1998).

Eleutherodactylus cochranae Grant, 1932

Distribution: Puerto Rico, Isla Vieques, Isla Culebra, U.S. and British Virgin Islands; elevational range, sea level to 521 m.

Habitat: Arboreal in xeric to mesic forests, also in anthropogenic habitats (Hedges et al., 2004zz); disturbed and natural grass-dominated wetlands (Rios-López, 2008d); wooded areas that are usually xeric, occurs in bromeliads (*Tillandsia*; Rivero, 1998); in axils of arboreal bromeliads, sometimes as many as 30 or more individuals in one plant (Grant, 1932e); on St. John, exclusively in axils of bromeliads (Grant, 1932g); in moist meadows, marshes, and scrubland below 100 m (Drewry, 1970b); in marshes, meadows, and on lawns (Drewry and Rand, 1983); uses bromeliads and coconut husk piles as diurnal retreats (Schwartz and Henderson, 1991).

Activity: Nocturnal (Joglar, 1998). **Behavior:** Based on captive observations, both sexes give retreat defense calls (Michael, 1995). **Foraging and Diet:** Sit-and-wait predator, diet in Puerto Rico includes leafhoppers (62.7% of total individual prey items), ants (13.4%), flies (13.0%), mites (3.5%), spiders (3.5%), on St. John (U.S. Virgin Islands), diet includes spiders (44.1%), ants (12.0%), orthopterans (11.1%), isopods (10.2%), beetles (5.7%), leafhoppers (3.9%), on average consume 1.5 prey/h (based on stomach contents), based on observations, average feeding rate 2.2 prey/h and average movement rate 1.4 move/h (Jones, 1982). **Reproduction:** Likely breeds year-round in Puerto Rico (Joglar, 1998); males call from heights of 0–2 m on grass or weeds and to 8.0 m in dry forest from before dusk to after dawn, with weak peaks at dusk and dawn (Drewry and Rand, 1983); on Culebra, call from bromeliads (*Tillandsia*) high in trees in xerophytic forest (Rivero and Joglar, 1979); on Tortola (British Virgin Islands), male calling from a crack in a small dead tree at 1.7 m above ground, female approached male and both entered crack in tree, male continued to call intermittently (Ovaska and Caldbeck, 1997a); in Puerto Rico, gravid females in March, September, October, clutches of 2–8 eggs in bromeliads (Joglar, 1998); in captivity, showed strong satellite male behavior, males stay with eggs until hatching and then for as long as 5 days post-hatching, males do not eat hatchlings, but females do, 5–10 (mean 8) eggs/clutch, 16–20 (mean 18) days to hatch, egg size 4–6 mm (mean 5 mm) at hatching, 7–8 day intervals between clutches common when adequate food is provided (Michael, 1995); in captivity, when gravid females approached males to within 0–10 cm, males would switch from loud advertisement calls to much softer courtship calls consisting of 1–10 quickly repeated single notes (Michael, 1996). **Size:** Male SVL to 20.5 mm, female SVL to 24.1 mm (Joglar, 1998).

Conservation Status: On the IUCN Red List as being of "least concern" (Hedges et al., 2004zz; Rios-López, 2008d); "stable" (Moreno, 1991d).

Eleutherodactylus cooki Grant, 1932

Distribution: Puerto Rico, Sierra de Panduras; elevational range ~90–300 m.

Habitat: On walls and in crevices of perpendicular rock faces, in deep cracks under husks of exfoliating granite or on "damp earth where earth and boulder meet" (Grant, 1932c); cavernicolous, in caves and cavities formed by enormous granitoid boulders, sometimes under canopy cover and sometimes in open situations (Joglar, 1998); boulder clusters near rivers and in humid forests (Hedges et al., 2004pp; Stuart et al., 2008).

Activity: Nocturnal, but active by day inside caves (peak in-cave activity 1300–1800 h; Burrowes, 2000) that do not receive illumination (Joglar et al., 1996); leaves caves a short time before dark and returns a little before dawn (Joglar, 1998). **Behavior:** A poor swimmer, does not enter water voluntarily, if placed in water, escapes as quickly as possible, movements in water contrast to its agility on boulders (Grant, 1932c). **Diet:** Includes insects (beetles, leafhoppers, flies, ants and other hymenopterans, lepidopterans) as well as spiders, diplopods, and chilopods (Joglar, 1998). **Egg Mortality:** Eggs eaten by male conspecifics, by snails (*Caracolus caracola* and *Polydontes acutangula*), and by insects (phorid flies), dehydration also cause of mortality (Burrowes, 2000, 2005). **Parasites:** An "infestation of small red ticks on the legs and sides of these frogs" (Grant, 1932c); tick *Ornithodoros talaje* (Maldonado Capriles and Medina Gaud, 1977; Goldberg et al., 1996d; Guglielmone et al., 2003). **Predators:** Conspecifics and snails (*Caracolus caracola*) on eggs (Burrowes, 2000, 2005; Joglar, 2005a). **Reproduction:** "An eyed egg is 6 mm. in diameter; uneyed, from 4 mm. up," laid "in a single layer in a clump of about 25 on the perpendicular face of a damp rough boulder," at each clutch, "a frog was sitting next to the eggs, facing away from them" (Grant, 1932c); reproductive activity correlated with precipitation and air temperature, males do not call outside of caves (Joglar et al., 1996); amplexus consists of male placing head over female's pelvic region with arms wrapped around outer part of her hind legs (supported on a rock), this position maintained for 12 h (1500–0300 h) before oviposition begins (Burrowes, 2000, 2005); by day in caves, males call, brood eggs, attend egg clutches without brooding, and call while attending clutches (Joglar et al., 1996); most calling between dusk and midnight (Drewry and Rand, 1983); average clutch size 17.35 ± 5.67 eggs (Burrowes, 2000, 2005); egg diameter 4.4–6.3 mm, eggs deposited in rock depressions (49.2%), rock crevices (23.0%), on flat rock surfaces (19.7%), or other microhabitats (8.1%), males attend nests to incubate and defend the eggs (85% of 58 clutches, the 15% unattended may have been only temporarily abandoned), may attend more than one clutch (22.4% double clutches, 5.2% triple clutches), and cover clutch(es) with portions of bodies (Joglar et al., 1996); males may guard multiple clutches from as many as four different females at a time, mean developmental time from oviposition to hatching was 28.66 ± 2.56 days (Burrowes, 2000, 2005); hatching success 61.5%–68.4% (Joglar et al., 1996), but 85% in nests with multiple clutches (Burrowes, 2000, 2005); hatchlings remain together as a group for several days, then disperse (Joglar et al., 1996). **Size:** Male SVL to 47.4 mm (Joglar et al., 1996), female SVL to 54.0 mm, hatchling SVL >8 mm (Grant, 1932c).

Conservation Status: "Endangered" on the IUCN Red List (Hedges et al., 2004pp); "vulnerable" (Rios-López, 2008e); "decreasing in some areas, urgent need of survey" (Moreno, 1991d); listed as a "threatened" species by U.S. Fish & Wildlife Service (2008); "few or no individuals located in recent surveys" (J. W. Wiley in Johnson, 1988).

Eleutherodactylus coqui Thomas, 1966

Distribution: Puerto Rico, where it occurs at sea level to 1,338 m; introduced on Culebra, Vieques, St. Thomas, St. John, and St. Croix, U.S. Virgin Islands, and in Hawaii (E. W. Campbell, 1996 in Kraus et al., 1999; McKeown, 1998); introductions into the Dominican Republic (Joglar and Rios-López, 1998) appear to have failed (Hedges et al., 2006).

Habitat: Wet forests, from ground level to 18.5 m above ground (Stewart and Woolbright, 1996); in bromeliads, palm axils, curled leaves, tree holes, holes in cutbanks, under rocks, trunks, or trash (Rivero, 1998); also in disturbed areas, including residential sites, around buildings and houses (Joglar, 1998); frequents human artifacts (Stewart and Woolbright, 1996); dead leaves of *Cecropia peltata* and *Prestoea montana* provide important daytime retreats at El Verde (Townsend, 1989); generally use habitat components in proportion to their availability, but adults use a much greater variety of components than juveniles, adults and juveniles have positive associations with shrubs and negative associations with grasses, vines, and ferns (exceptions include *Philodendron angustatum* and *Danea nodosa*, both with broad leaf structures for support), strong association with *Cecropia schreberiana* and *Heliconia carabea*), adults positively select for objects near the ground (leaf litter, fallen branches, roots, soil), juveniles prefer leaves and avoid leaf litter (perhaps because of predation pressures; Beard et al., 2003b); agricultural lands including plantations and arable land (Hedges et al., 2006; Rios-López, 2008f); habit of burying itself in potted plants apparently has facilitated dispersal of frogs introduced in the U.S. Virgin Islands (Platenberg, 2007).

Activity: Nocturnal, emerges from diurnal retreats at dusk and returns to retreats at dawn (Drewry and Rand, 1983; Woolbright, 1985); will forage at night lights (M. M. Stewart in Henderson and Powell, 2001; Perry et al., 2008). **Behavior:** Both males and females give aggressive calls while defending retreat sites, most aggressive calls heard between 0545–0630 and 1800–1845 h (Stewart and Rand, 1992); in staged contests, 67% of residents responded aggressively when diurnal retreats were entered by a conspecific, residents won 93% of contests (Stewart and Rand, 1991). **Biomass:** At El Verde, 3.26 kg dry mass/ha (Stewart and Woolbright, 1996). **Desiccation:** Reduces rate of evaporative water loss by adjusting postures and activities in response to changing conditions of water availability during nocturnal activity periods (Pough et al., 1983). **Ecosystem Processes:** At two spatial scales (1.0 m^3 and at 20 × 20 m), frogs reduced aerial invertebrates and leaf herbivory, but had no effect on litter invertebrates, at smaller scale frogs increased foliage production rates (measured as the number of new leaves and new leaf area produced) by 80% and decomposition rates by 20%, likely the result of elimination and excretion rather than controlling prey (Beard et al., 2003a). **Foraging and Diet:** Sit-and-wait forager (Woolbright and Stewart, 1987); emerges at dusk and sits on foliage surfaces, trunks, or limbs until dawn, nonbreeding females and some males ascend to canopy to forage (weather permitting), once reaching their perches, move little or not at all throughout the night, at dawn (most at 0545–0600 h), they parachute (live frogs drop more slowly than a dead frog) to ground level (Stewart, 1985); prey capture movements usually short distances (<5.0 cm) and account for about 50% of total number of moves during a night (Woolbright, 1985); average adult foraging rate 3.2 prey items/night, and adults may harvest >10,000 insects/ha of forest/night, pre-adults (17,305/ha) average 6 prey items, suggesting they could consume 104,000 insects/ha/night, all age classes combined may harvest 114,000 prey/night/ha, stomach contents of 173 adults yielded prey from >100 species representing 60 families, 28 orders, 9 classes, and 5 phyla (Stewart and Woolbright, 1996); diet includes snails, spiders, centipedes, insects, and frog eggs, predominant prey includes (by frequency of occurrence) ants (*Paratrechina* spp., 11.0% of total number; *Camponotus ustus*, 6.5%; *Paratrechina microps*, 5.4%; other ants, 5.4%), orthopterans (gryllids *Cyrtoxipha gundlachi*, 3.1%; other gryllids 3.1%), and *Eleutherodactylus* eggs (4.9%); by percentage of total volume, predominant prey were orthopterans (gryllids *Gryllus assimilus*, 23.5%; *Amphiacusta caraibea*, 11.6%; *Orocharis* sp., 10.4%) and roaches (*Pseudosymploce personata*, 5.6%; other blattellids, 7.4%; Stewart and Woolbright, 1996); vertebrate prey includes *Eleutherodactylus coqui*, *Anolis gundlachi* (Stewart and Woolbright, 1996), and *A. evermanni* (Leal and Thomas, 1992); males do not forage while calling (usually before midnight), and consumed only 18% of their night's food by 2400 h, on annual basis, males may lose 16% of their potential food intake due to calling activity (Woolbright and Stewart, 1987). **Disease:** Chytrid (*Batrachochytrium dendrobatidis*) infections (Joglar, 2005a; Burrowes et al., 2008). **Growth Rates:** Adult females grow faster than adult males at all body sizes, male growth stopped at about 38 mm SVL and females continued to grow to SVL >50 mm, growth rates of both sexes declined with increasing body size (Woolbright, 1989); hatchlings require about 316 days to reach 28 mm SVL, the minimum size of reproductive males (Stewart and Woolbright, 1996). **Homing:** Male homing success high (19 of 20 successful) at 20 m, moderate at 40 m (13 of 20), and low at 100 m (5 of 20), success rate not SVL-dependent, elapsed time from displacement to return to home territory was 1–22 days with return times greater for longer displacements, no relationship between homing success and compass direction displaced (Gosner and Woolbright, 1995). **Hurricane Impact:** Hurricane Hugo hit in September 1989, large posthurricane drop in the number of juveniles (likely due to severe drought that followed Hugo), but a sixfold increase in the number of adults, likely due to an increase in retreat sites, decrease in invertebrate predators, warmer temperatures, and increased insect abundance (Woolbright, 1991). **Immobilization:** The nettle *Urera baccifera* has the potential for immobilizing (i.e., making temporarily unresponsive) small frogs, after removal from leaf, adult frog unresponsive for 3–4 min (Ten Eyck and Kruger, 2002). **Mortality Rates:** About 94% of adults do not survive one year, a few live at least 6 years (Stewart and Woolbright, 1996). **Movements:** Move more frequently when vegetation is wet, total nightly movements average 3.0–4.5 m, with most occurring soon after dusk and shortly before dawn, males unlikely to move or forage while calling (Woolbright, 1985); all size classes forage higher above ground when conditions are wet (Stewart and Woolbright, 1996). **Parasites:** Nematode *Oswaldocruzia lenteixeirai* (Moravec and Kaiser, 1995). **Population Size and Density:** At El Verde, 3,264 adults/ha and 20,570/ha for all age classes, maximum density (El Verde) 24,800/ha, maximum density at 30 m above sea level 45,000/ha, and 2,500/ha at 750 m asl (Stewart and Woolbright, 1996); density dependent on number of available nest and retreat sites (Stewart and Pough, 1983), long periods without rain, and major hurricanes (Stewart and Woolbright, 1996); in central mountain range of Puerto Rico, adult wet season density 8–25/100 m^2, dry season density 3–19/100 m^2, with density of froglets and juveniles also greater during the wet season (Fogarty and Vilella, 2002); at El Yunque, populations in elfin forest declined during the mid-1990s but recovered by the end of the decade, in other habitats, populations have declined consistently (Burrowes and Joglar, 2005). **Population Structure:** Composed of juveniles (SVL 6–17 mm), subadults (18–23 mm), and adults (24–55 mm; Stewart and Woolbright, 1996); between 1987–1995, major factor influencing population size was Hurricane Hugo (1989), which deposited much of the forest canopy onto the forest floor, followed by succession, treefalls similar to hurricanes on a local scale also can impact numbers (Woolbright, 1996). **Predators:** Minimally, 39 predators on eggs, pre-adults, and adults (Stewart and Woolbright, 1996); spiders, *Stasnia portoricensis* (Formanowicz et al., 1981; Reagan, 1996b; Stewart and Woolbright, 1996; Armas, 2001; Joglar, 2005a), *Avicularia laeta* (Stewart and Woolbright, 1996; Joglar, 2005a), Ctenizidae (Stewart and Woolbright, 1996; Joglar, 2005a), *Olios* sp. and *Oligoctenus ottleyi* (Formanowicz et al., 1981; Armas, 2001; Joglar, 2005a), and a ground spider (Corinnidae; Daza et al., 2008); an amblypygid, *Phrynus longipes* (Formanowicz et al., 1981; Reagan, 1996b; Stewart and Woolbright, 1996; Armas, 2001; Joglar, 2005a); scorpions, *Tityus obtusus* (Stewart and Woolbright, 1996; Villanueva-Rivera et al., 2000; Armas, 2001; Joglar, 2005a; L. O. Nieves, in litt., 13.II.2008) and *Scolopendra alternans* (Formanowicz et al., 1981; Joglar, 2005a); frogs, *E. coqui* and *E. wightmanae* (Rivero and Seguí Crespo, 1992; Stewart and Woolbright, 1996; Joglar, 1998, 2005a); lizards, *Ameiva exsul* (Rivero and Seguí Crespo, 1992; Stewart and Woolbright, 1996; Joglar, 2005a), *Anolis gundlachi* (Stewart and Rand, 1991; Stewart and Woolbright, 1996; Joglar, 2005a), *A. cristatellus* (Stewart and Woolbright, 1996; Joglar, 1998, 2005a), *Sphaerodactylus klauberi* (Stewart and Woolbright, 1996; Thomas and Gaa Kessler, 1996; Joglar, 2005a), and *Mabuya sloanii* (Rivero and Seguí Crespo, 1992; Joglar, 1998, 2005a); snakes, *Borikenophis portoricensis* (Woolbright, 1989; Schwartz and Henderson, 1991; Rodríguez-Robles and Leal, 1993a, 1993b; Stewart and Woolbright, 1996; Joglar, 2005a), *Magliophis exiguus* (Schwartz and Henderson, 1991, eggs; Stewart and Woolbright, 1996; Thomas and Gaa Kessler, 1996, in captivity; Joglar, 2005a), and *Epicrates inornatus* (Reagan, 1984, in captivity; Stewart and Woolbright, 1996; Joglar, 2005a); birds, Puerto Rican Screech Owl (*Otus nudipes*), Pearly-eyed Thrasher (*Margarops fuscatus*), Red-legged Thrush (*Turdus plumbeus*; Woolbright, 1989; Stewart and Woolbright, 1996; Joglar, 2005a), Red-tailed Hawk (*Buteo jamaicensis*; Santana and Temple, 1988; Stewart and Woolbright, 1996; Joglar, 2005a), Broad-winged Hawk (*B. platypterus brunnescens*), Puerto Rican Lizard Cuckoo (*Saurothera vieilloti*; Stewart and Woolbright, 1996; Joglar, 2005a); mammals, mongooses (*Herpestes javanicus*), rats (*Rattus rattus*), and cats (*Felis catus*; Stewart and Woolbright, 1996; Joglar, 2005a); *E. coqui* composed up to 40% of prey brought to nestling Screech Owls (*O. nudipes*; Snyder et al., 1987); up to 42% of the diet of nestling Red-tailed Hawks (*B. jamaicensis*; Santana and Temple, 1988); the most important nest predators are conspecifics (Townsend et al., 1981, 1984; Stewart and Pough, 1983; Stewart and Woolbright, 1996; Joglar, 2005a); other nest predators include phorid flies (*Megaselia scalaris*; Villa and Townsend, 1983; Joglar, 2005a); ants (*Paratrechina* sp. and *Iridomyrmex melleus*; Stewart and Woolbright, 1996; Joglar, 2005a); crickets (*Amphiacusta caraibea*; Stewart and Woolbright, 1996; Joglar, 2005a); spiders (*Stasina portoricensis*; Joglar, 1998, 2005a); snails, *Subulina* sp. (Reyes Campos, 1971; Joglar, 2005a), *Polydontes acutangula* (Stewart and Woolbright, 1996; Joglar, 2005a); unidentified centipedes and millipedes (*Liomus obscurus*; Stewart and Woolbright, 1996; Joglar, 2005a). **Reproduction:** At El Verde, calling and oviposition occurred every month, although activity slowed during the dry season and was highest during the wet season, male chorus density associated with temperature and rainfall (Townsend and Stewart, 1994); male calling sites (usually elevated, >25 cm high; median = 78.0 cm; 92.6% of 297 observations) on open surfaces (84.9%) or shallow depressions lacking close cover, leaf surfaces (19.5% of 297 observations), tree trunks and branches (25.6%), axils of Sierra Palm (*Prestoea montana*) leaves (33.0%; Townsend, 1989); males call dusk to dawn, but most calling occurs between dusk and midnight (Drewry and Rand, 1983; Stewart and Rand, 1992); most intense calling shortly after dusk, decreasing throughout the night, may use 1–4 different calling sites, all within a 2.0-m diameter (Woolbright, 1985); calling males exhibit site fidelity, using the same site for as many as 10 consecutive nights (Reyes Campos, 1971); calling males seem to space themselves by mutual avoidance (Stewart and Bishop, 1994); in playback experiments, males engaged in competitive calling (increased call rate in response to stimulus calls; Benedix and Narins, 1999); females demonstrate preference for relatively high call rates (Lopez, 1996); female breeding asynchronous, with at least one female mating during 63% of all wet season nights (Townsend and Stewart, 1994); 83% of observed matings prior to 2130 h (Lopez, 1996); gravid females approach and make contact with calling males and are led to nest sites by males, preliminary amplexus ensues with males on females' backs (no clasping), after 5.5–7.5 h, females initiate reverse hind leg clasps (males never clasps females) that lasts 8.5–12.0 h, ovulation and insemination then occur (Townsend and Stewart, 1986a); fertilization is internal (Townsend et al., 1981); oviposition occurs early (0730–1240 h) on the following day and requires 2–4 h, females leave nests at dusk and males remain sitting on clutches (Townsend and Stewart, 1986a); clutch deposition associated with temperature and rainfall (Townsend and Stewart, 1994); nest sites usually on ground (<25 cm high; median = 7.0 cm; 68.7% of 134 nests) with substrates that include curls of dead, dehiscent leaves of *Cecropia peltata* (41% of 134 nests), rolled petioles of dead dehiscent Sierra Palm fronds (20.2%), and curled dead leaves or leaflets of other species (18.6%), and 84% in completely enclosed sites (Townsend, 1989); clutch size 3–45, (3.5–5.5 mm egg diameter; Hedges, in prep.); clutch size is correlated with female body size (Townsend and Stewart, 1994); parental (male) care includes egg brooding and nest defense against conspecific egg cannibals, defense behavior includes aggressive calling, biting, sustained biting, wrestling, blocking against intruders, care lasts from oviposition to hatching (17–26 days), and sometimes a few days post-hatching (Townsend et al., 1984); the development period is temperature dependent (Townsend and Stewart, 1986b); during pre-hatching development, males present in nests 97.4% of time during day and 75.8% of time at night (Townsend et al., 1984); brooding males have higher frequencies than nonbrooding males of empty stomachs, lose significant amounts of body mass, and miss opportunities to mate (Townsend, 1986); when parental care is absent, risks of desiccation and cannibalism are greater (Taigen et al., 1984); mean interclutch interval is ~8 weeks during the wet season (Townsend and Stewart, 1994); in captivity, shows strong satellite male behavior, males rehydrate eggs and vigorously defend them from other frogs, insects, or inanimate objects, 12–30 (mean 23) eggs/clutch, 15–33 (mean 22) days to hatch, egg size 4–6 mm (mean 5 mm) at hatching, 7–8 day intervals between clutches common when adequate food is provided (Michael, 1995); in captivity, when gravid females approach males to within 0–10 cm, males switch from loud advertisement calls to much softer courtship calls consisting of 1–10 quickly repeated single notes (Michael, 1996). **Sex Ratio (M:F):** 1:1 (Stewart, 1985); 2.5 ± 0.5:1 (Woolbright et al., 2006). **Size:** Male SVL to 50.3 mm, female SVL to 63 mm (Joglar, 1998), but varies with elevation (larger at higher elevations; Narins and Smith, 1986); also gradient of decreasing body size from east to west across Puerto Rico (Fogarty and Vilella, 2002). **Vocalization:** First note of two-note call involved in male-male interactions, whereas second note directed to females, when the recipient of the call can be identified, only the one appropriate note is expressed (Narins and Capranica, 1976).

Conservation Status: On the IUCN Red List as being of "least concern" (Hedges et al., 2006; Rios-López, 2008f); upland populations may be "near threatened" (Burrowes et al., 2004); "stable" (Moreno, 1991d).

Remarks: Because the two species are easily confused (Joglar, 1998) and some publications did not distinguish between them (e.g., Cintrón, 1970; Drewry, 1970a, 1970b), some data pertaining to *E. coqui* may be included in the account on *E. portoricensis*.

Eleutherodactylus corona Hedges and Thomas, 1992

Distribution: Hispaniola, Massif de la Hotte (Haiti) at elevations ~1,120 m.

Habitat: Arboreal, by day in bromeliads 1–3 m above ground on standing dead trees in remnant cloud forest (Hedges and Thomas, 1992b); recorded only from forest edges, but these probably are not suitable habitat (Hedges and Thomas, 2004b; Stuart et al., 2008).

Reproduction: At night, calling males, adult females, and a large (presumably communal) cluster of 50 eggs in different stages of development in a large orchid growing on a tree, eggs began hatching on 9–10 June, two females with masses of 0.49 g and 0.42 g each contained 3 enlarged ova (3.3 mm diameter) with a total mass of 0.075 g and 0.060 g, respectively (Hedges and Thomas, 1992b); eggs laid in bromeliads or orchids, which are required for reproduction, development direct (Hedges and Thomas, 2004b; Stuart et al., 2008); clutch size 3 (Hedges, in prep.). **Size:** Male SVL to 18.4 mm SVL, female SVL to 19.1 mm (Hedges and Thomas, 1992b).

Conservation Status: "Critically endangered" on the IUCN Red List (Hedges and Thomas, 2004b); as a forest inhabitant in Haiti, the species' future is uncertain.

Eleutherodactylus counouspeus Schwartz, 1964

Distribution: Hispaniola, distal Tiburon Peninsula, Massif de la Hotte and Monts Cartaches (Haiti); elevational range ~300–760 m.

Habitat: In caves, limestone sinks, crevices (often deep) in surface limestone, and horizontal branches along rock fences (Schwartz and Henderson, 1991); beyond the lighted zone at the bottom of a deep sink, active and accurate jumpers "quite at home" in total darkness (Franz and Cordier, 1986); caves and crevices in closed, humid forests (Hedges et al., 2004xxx; Stuart et al., 2008).

Abundance: Moderately common (Hedges et al., 2004xxx; Stuart et al., 2008). **Reproduction:** Calling males on rocks, in cavities in cave walls, or on horizontal branches 0.6–1.0 m from rock faces, males call at 0900–1000 h from the twilight zone to well back (30 m) in caves (Schwartz and Henderson, 1991); eggs laid in caves, development direct (Hedges et al., 2004xxx; Stuart et al., 2008); clutch size 35 (Hedges, in prep.). **Size:** SVL to 57 mm in females (Schwartz and Henderson, 1991).

Conservation Status: "Endangered" on the IUCN Red List (Hedges et al., 2004xxx).

Additional Pertinent Literature: Schwartz (1980e).

Eleutherodactylus cubanus Barbour, 1942

Distribution: Eastern Cuba, Sierra Maestra; elevational range 800–1,830 m.

Habitat: By day in cloud forest leaf litter (Hedges et al., 1995); closed, humid forests at high elevations (Hedges and Díaz, 2004q; Stuart et al., 2008); also in pine (*Pinus maestrensis*) litter (Díaz and Cádiz, 2008).

Abundance: Common in suitable habitat (Hedges and Díaz, 2004q; Stuart et al., 2008); common but difficult to locate (Díaz and Cádiz, 2008). **Behavior:** Moves slowly and deliberately, but jumps abruptly when disturbed (Díaz and Cádiz, 2008). **Reproduction:** Males call during the day and during the first few hours after sunset (Díaz et al., 2005a); female with a single egg (Hedges et al., 1995); breeds by direct development (Hedges and Díaz, 2004q; Stuart et al., 2008). **Size:** Female SVL to 13 mm (Schwartz and Henderson, 1991); SVL to 15 mm (Díaz and Cádiz, 2008).

Conservation Status: "Vulnerable" (Vales et al., 1998); "critically endangered" on the IUCN Red List (Hedges and Díaz, 2004q).

Eleutherodactylus cundalli Dunn, 1926

Distribution: Jamaica; elevational range, sea level to ~635 m.

Habitat: Wet limestone forest and dry scrub forest; on stone walls (Lynn, 1940); under rocks and logs and in bromeliads by day (Hedges, in prep.); at night, on high shrubs and along a limestone cliff and its talus, at Windsor Cave on rocks, stalactites, debris, and walls within the twilight zone, on cliffs and plants to heights of 2.4 m, on rocky jumbles on forested hillsides, in arboreal and terrestrial bromeliads, at the petiole base of *Musa* leaves at 1.5 m, from a hollow bamboo stem at 1.2 m, on the wall of a wet culvert, on bare rocks in a road cut, along a small mountain stream, and in piles of *Cocos* trash on mud behind a low, sandy beach ridge (Schwartz and Fowler, 1973); in Windsor Great Cave, males occurred from entrance to 20 m into the main passage, 30 m into the flood passage, and 87 m into the "glade" passage (Diesel et al., 1995); rocks and caves in closed-canopy moist forests, does not occur in open habitats (Hedges et al., 2004ddd; Stuart et al., 2008); on Cabarita Island, on a dead palm frond at forest edge (Crombie et al., 1983).

Abundance: "Moderately common" (Hedges et al., 2004ddd; Stuart et al., 2008). **Activity:** Nocturnal (Stewart, 1977). **Foraging and Diet:** Forages on the forest floor, on exposed leaves, vines, and large shrubs, elevated crevices in banks and rock walls (Stewart and Martin, 1980); diet includes ants (63% of diet by frequency of occurrence in a sample of 1,006 prey items), arachnids (mites, spiders; 10%), termites (5%), springtails (5%), beetles (7%), leafhoppers, snails, crickets, roaches, and thrips combined (6%; Stewart, 1977). **Parasites:** Infested with small roundworms embedded in the skin on sides of body (Lynn, 1940). **Reproduction:** Males usually call from limestone rocks or in caves, exhibit high calling site fidelity, usually within 1.0 m of original site over a period of several weeks (Diesel et al., 1995); amplexus observed in March (Diesel et al., 1995) and late June (Lynn, 1940); minimum SVL of females at reproduction 27 mm (Stewart, 1977); in interior of Windsor Great Cave, egg clutches deposited 7–87 m from cave entrance, clutch size 33–76 (Diesel et al., 1995); attended by female (Stewart, 1977); translocated females returned to clutches, hatching of an entire clutch required at least 24 h, hatchlings (30–72 in number; 6.0 mm SVL) transported on back, flanks, and head of female at cave sites (Diesel et al., 1995); eggs laid on the ground, development direct (Hedges et al., 2004ddd; Stuart et al., 2008). **Size:** Male SVL to 39 mm, female SVL to 47 mm (Hedges, in prep.).

Conservation Status: "Vulnerable" on the IUCN Red List (Hedges et al., 2004ddd).

Eleutherodactylus cuneatus Cope, 1862

Distribution: Eastern Cuba; elevational range, sea level to ~1,970 m.

Habitat: Semiaquatic, along streams on rocks, on the ground, and juveniles to 0.3 m above the ground on leaves (Schwartz and Henderson, 1991); occasionally far from water, but generally riparian, even in disturbed habitats such as agricultural lands or urban areas with streams (Hedges and Díaz, 2004r); most individuals ~20 cm from water, many at edges of streams, adults usually under cover objects (rocks, palm debris, trash) on the ground, in depressions in earthen banks, or below the ground, juveniles sometimes on leaves of shrubs to 30 cm above the ground (Fong Grillo and Hero, 2006); along streams or in wet leaf litter or moist sites in rocky outcrops, also in caves (Díaz and Cádiz, 2008).

Abundance: Very common (Díaz and Cádiz, 2008). **Behavior:** Readily responds to recorded calls by approaching the source (Díaz and Cádiz, 2008). **Movements:** Immigration to a streamside site high, average four-month residence, only two individuals present for the entire year of the survey, individuals moved as much as 82 m, some as much as 4 m in five days (Fong Grillo and Hero, 2006). **Population Density:** Numbers of adults decreased in August and remained low until February, then increased until July, numbers of juveniles peaked March–April and remained low during the rest of the year (Fong Grillo and Hero, 2006). **Population Structure:** Adults composed the greatest proportion of samples (Fong Grillo and Hero, 2006). **Reproduction:** Males call from among rocks, lianas, stumps, and other cover, from banks of mountain streams and river margins, clutch of 43 eggs under a rock on the bank of a mountain road in July at 1,300+ m (Estrada and Hedges, 1998); males call sporadically all night long (Díaz et al., 2005a); reproductive activity highly seasonal, calling begins in March, peaks in June-July, and extends through September, females with eggs in March, June, July, and August (Fong Grillo and Hero, 2006). **Size:** Male SVL to 35.0 mm, female SVL to 58.6 mm (Estrada and Hedges, 1998); SVL to 60.3 mm (Díaz and Cádiz, 2008).

Conservation Status: "Vulnerable" for *E. sierramaestrae* (Vales et al., 1998), which is now in the synonymy of *E. cuneatus* (Estrada and Hedges, 1998); on the IUCN Red List as being of "least concern" (Hedges and Díaz, 2004r).

Remarks: Frogs referred to *Eleutherodactylus cuneatus* prior to the publication of Estrada and Hedges (1998) are now referred to *E. riparius*; frogs previously referred to *E. sierramaestrae* are now referred to *E. cuneatus* (this account).

Eleutherodactylus darlingtoni Cochran, 1935

Distribution: Hispaniola, Massif de la Selle (Haiti); elevational range ~1,720–2,200 m.

Habitat: Under rocks and logs and in and around limestone sinkholes in pine (*Pinus occidentalis*) forest by day at high elevations (1,720–2,200 m; Hedges, 1992); has not been recorded from disturbed habitats (Hedges et al., 2004yyy; Stuart et al., 2008).

Reproduction: Eggs laid on the ground, breeds by direct development (Hedges et al., 2004yyy; Stuart et al., 2008). **Size:** Male SVL to 25.3 mm, female SVL to 39.6 mm (Hedges, 1992).

Conservation Status: "Critically endangered" on the IUCN Red List, last observed in the 1980s, but no surveys have been conducted since (Hedges et al., 2004yyy; Stuart et al., 2008).

Additional Pertinent Literature: Schwartz (1982a).

Eleutherodactylus dimidiatus Cope, 1862

Distribution: Cuba, where two subspecies are currently recognized in central and eastern Cuba (*E. d. dimidiatus*) and in western Cuba (*E. d. amelasma*); rare in lowlands, more abundant in montane forests; elevational range, sea level to ~1,375 m.

Habitat: Mesic forests (sometimes in pinewoods), under surface rocks, logs, palm debris, and other cover objects by day (G. Peters, 1974; Ruiz García, 1987); also at cave entrances and sinks (Schwartz and Henderson, 1991); one of only two species (with *E. atkinsi*) in a managed timber forest in eastern Cuba (Fong Grillo, 1999); occurs in leaf litter in various types of forest (Díaz and Cádiz, 2008).

Abundance: Moderately common in suitable habitat (Hedges and Díaz, 2004s; Stuart et al., 2008). **Parasites:** Nematodes *Aplectana cubana*, *Aplectana* sp. (Baruš, 1972; Coy Otero and Lorenzo Hernández, 1982; Coy Otero and Ventosa, 1984); *Oswaldocruzia lenteixeirai*, and larval *Porrocaecum* sp. (Baruš, 1972; Coy Otero and Lorenzo Hernández, 1982); *Aplectana* sp. (Baruš, 1972, 1973; Coy Otero and Lorenzo Hernández, 1982). **Reproduction:** Males call from the ground and from vegetation to heights of 1.2 m (Díaz and Cádiz, 2008); nest (on 21 March) with 24 eggs (3.3–4.4 mm in diameter) under the fallen trunk of *Carapa guianensis* in contact with the ground, in a small depression, partially covered by a female (SVL 32.4 mm), 10 days later, mean egg diameter 5.5 mm, embryos well developed, occasionally twisting, eggs hatched at 17 days postcollection, mean froglet SVL 5.8 mm (Estrada, 1987c); female attends nest (Alonso Bosch and Rodríguez Gómez, 2003b). **Size:** SVL to 45 mm in females (Schwartz and Henderson, 1991).

Conservation Status: "Near threatened" on the IUCN Red List (Hedges and Díaz, 2004s).

Eleutherodactylus diplasius Schwartz, 1973

Distribution: Hispaniola, Tiburon Peninsula, Massif de la Hotte and Monts Cartaches (Haiti); elevational range, sea level to >1,200 m.

Habitat and Natural History: Unknown.

Size: Male SVL to 35.5 mm, females SVL to 39.4 mm (Schwartz, 1977f).

Conservation Status: Until recently, this species was considered a subspecies of *E. wetmorei*, which is considered "vulnerable" on the IUCN Red List (Hedges et al., 2004gg); we assume this species is, at best, also vulnerable.

Eleutherodactylus dolomedes Hedges and Thomas, 1992

Distribution: Hispaniola, Massif de la Hotte (Haiti); elevation ~1,120 m.

Habitat: On leaves of small trees (to 3.0 m) and on top of the horizontal trunk of a large fallen tree at ~2.0 m above the ground (Hedges and Thomas, 1992b); arboreal, recorded from forest edges, but these probably are not ideal habitat (Hedges and Thomas, 2004c; Stuart et al., 2008).

Reproduction: Calling during light rain and strong wind (Hedges and Thomas, 1992b); eggs laid on the ground, breeds by direct development (Hedges and Thomas, 2004c; Stuart et al., 2008). **Size:** Male SVL to 21.6 mm, females unknown (Hedges and Thomas, 1992b).

Conservation Status: "Critically endangered" on the IUCN Red List (Hedges and Thomas, 2004c).

Eleutherodactylus eileenae Dunn, 1926

Distribution: Central and western Cuba; elevational range, sea level to ~830 m.

Habitat: Normally terrestrial, although males call from arboreal sites, not recorded outside forested habitats (Hedges and Díaz, 2004t; Stuart et al., 2008); in mesic forested areas under rocks and palm trash and in rock crevices (Schwartz and Henderson, 1991); in complex mogote vegetation, semideciduous coastal forests, gallery forests, and secondary forests, diurnal retreats include leaf litter, cracks in rocks, fallen logs, and bromeliads, but always close to the ground (Díaz and Cádiz, 2008).

Abundance: Common (Díaz and Cádiz, 2008). **Activity:** Nocturnal, leaves diurnal refuges to call (males) and search for food (Díaz and Cádiz, 2008). **Diet:** Hymenopterans (primarily ants) and arachnids, but also beetles, orthopterans, leafhoppers, termites, and dipterans (Alonso Bosch et al., 2001). **Parasites:** Nematodes *Aplectana* sp. (Baruš, 1972, 1973; Coy Otero and Lorenzo Hernández, 1982). **Reproduction:** Males form large choruses during rainy afternoons and humid nights during the rainy season (Estrada, 1994); males leave soil retreats at dusk and move vertically to arboreal calling sites (Alonso Bosch et al., 2001); arboreal sites 0.6–2.4 m above the ground, usually from a horizontal position on the leaves of trees (especially *Syzigium jambos*; Rodríguez Gómez and Alonso Bosch, 2003a); vocal activity occurs until dawn, but with peaks at dusk and dawn, between peaks males forage but eat few prey items to minimize weight (Alonso Bosch et al., 2001); males call from leaves, branches, tree trunks, and rocks, usually 1–2 m high, but occasionally to heights of 4–5 m, usually call during crepuscular periods, but sometimes by day when weather is favorable, some vocalizing males attending eggs in litter (Díaz and Cádiz, 2008); eggs laid on the ground, breeds by direct development (Hedges and Díaz, 2004t; Stuart et al., 2008); clutch size 37 (Hedges, in prep.). **Size:** Male SVL to 26 mm (Dunn, 1926a), female SVL to 33 mm (Schwartz and Henderson, 1991). **Vocalization:** In addition to the advertisement call, males produce two types of aggressive calls (Alonso Bosch and Rodríguez Gómez, 2001).

Conservation Status: "Near threatened" on the IUCN Red List (Hedges and Díaz, 2004t).

Eleutherodactylus emiliae Dunn, 1926

Distribution: Cuba, Sierra de Trinidad; elevational range ~350–850 m.

Habitat: Wet forests with arborescent ferns, under rocks, decayed tree trunks, and in leaf litter (Alonso Bosch and Rodríguez Gómez, 2003b); terrestrial in closed-canopy, humid forest (Hedges and Díaz, 2004u; Stuart et al., 2008).

Abundance: "Uncommon" (Hedges and Díaz, 2004u; Stuart et al., 2008). **Behavior:** Walks, only occasionally executes small jumps (Ruiz García, 1987). **Reproduction:** Vocalization has not been described (Díaz and Cádiz, 2008); eggs laid on the ground, breeds by direct development (Hedges and Díaz, 2004u; Stuart et al., 2008). **Size:** SVL to 27 mm (Schwartz and Henderson, 1991).

Conservation Status: "Vulnerable" (Vales et al., 1998); "endangered" on the IUCN Red List (Hedges and Díaz, 2004u); appears to be declining within the last decade (Díaz and Cádiz, 2008).

Eleutherodactylus eneidae Rivero, 1959

Distribution: Puerto Rico, interior uplands; elevational range ~300–1,150 m.

Habitat: Largely arboreal in rainforest and elfin woodland (Hedges and Díaz, 2009); also in cloud forest (Rivero, 1959); at night on forest floor or on leaves or trunks of bushes, palms, and tree ferns to 1.5–1.8 m above the ground (Rivero, 1959); most (>65%) inside petioles, some (~35%) under petioles, none in bromeliads (Cintrón, 1970); in moss on stones, tree trunks, and earthen banks, in low brush (Drewry, 1970b); by day in crevices, holes in mud, under rocks, logs, and trash, prefers mud banks along trails and roads (Rivero, 1998); terrestrial in extremely humid closed-canopy forest (Hedges and Joglar, 2004a; Hedges and Rios-López, 2008a; Stuart et al., 2008).

Activity: Nocturnal (Rivero, 1959); calling sporadic before 2200 h or later, peaking ~0200 h or later, calling activity maximal in summer, but apparently influenced most intensely by humidity, and is quite high on wet winter nights (Drewry, 1970b). **Competition:** Concentrates calling activity during periods of lowest activity of other species, apparently avoiding acoustic competition (Drewry, 1970b). **Diet:** Stomach contents included beetles and ants (most abundant), termites, ichneumonoid wasps, beetle larvae, and mosquitoes, mean food item size ~0.26 cm (Cintrón, 1970); diet in males includes insects (orthopterans, leafhoppers, hymenopterans dipterans, psocopterans) and arachnids (spiders; Lavigne and Drewry, 1970). **Reproduction:** Males frequently call from daytime hiding places, but are timid at this time and will not call when illuminated with a flashlight, but are "fearless" when calling at night from elevated perches (Drewry, 1970b); call beginning at dusk at Toro Negro, where calling peaks earlier than at El Yunque (where the peak is reached after midnight [Rivero, 1998]); males call from the ground, shrubs, cutbanks, and tree trunks (to 1.5 m; Schwartz and Henderson, 1991); gravid females 26.6–28.9 mm SVL (Joglar, 1998); clutch size 8–22, eggs 2.3–4.9 mm diameter (Hedges, in prep.); breeds by direct development (Hedges and Joglar, 2004a; Hedges and Rios-López, 2008a; Stuart et al., 2008). **Size:** Male SVL to 24.8 mm (Joglar, 1998), female SVL to 30.0 mm (Schwartz and Henderson, 1991).

Conservation Status: "Critically endangered" on the IUCN Red List, the species is possibly extinct (Hedges and Joglar, 2004a; Hedges and Rios-López, 2008a; Stuart et al., 2008); "few or no individuals located in recent surveys" (J. W. Wiley in Johnson, 1988); despite intensive searches, it has not been seen since 1990 (Burrowes and Joglar, 1991; Joglar and Burrowes, 1996; Burrowes et al., 2004; Joglar, 2005a; Joglar et al., 2007; Hedges and Díaz, 2009); "endangered" (Moreno, 1991d).

Eleutherodactylus erythroproctus Schwartz, 1960

Distribution: Cuba, Sierra de Camarones and Pan de Matanzas.

Habitat: Occurs in broadleaf forest among rock slides and talus slopes (Schwartz, 1960a).

Size: Male SVL to 15.5 mm, female SVL to 20.0 mm (Schwartz, 1960a).

Conservation Status: Until recently, this species was considered a subspecies of *E. zugi*, which is considered "endangered" on the IUCN Red List (Hedges and Díaz, 2004ccc); we assume this species is, at best, also endangered.

Eleutherodactylus etheridgei Schwartz, 1958

Distribution: Cuba, extreme southeastern lowlands; elevational range, sea level to ~150 m.

Habitat: Moist areas in otherwise xeric habitat, by day under a plank buried in mud near a temporary drainage ditch also under small rocks (exposed to direct sunlight most of day) on an abandoned road (Schwartz, 1958b); in rocky terrain, including small elevations of volcanic origin known as *monitongos* (Díaz and Cádiz, 2008); has not been recorded outside forest habitat (Hedges and Díaz, 2004v; Stuart et al., 2008).

Abundance: Moderately common in suitable habitat (Hedges and Díaz, 2004v; Stuart et al., 2008). **Activity:** Abandons cracks and crevices at nightfall to move about rocks and xeric vegetation, especially common during the rainy season (Díaz and Cádiz, 2008). **Reproduction:** Forms intense choruses during the rainy season at dusk and during the first hours of darkness, males call from the ground, from cavities in rocks, from roots of trees, and from low vegetation (Díaz and Cádiz, 2008); clutch size 15–22 (Hedges, in prep.); breeds by direct development (Hedges and Díaz, 2004v; Stuart et al., 2008). **Size:** Male SVL to 18.1 mm, female SVL to 19.0 mm (Schwartz, 1958b); SVL to 27.5 mm (Díaz and Cádiz, 2008).

Conservation Status: "Vulnerable" (Vales et al., 1998); "endangered" on the IUCN Red List (Hedges and Díaz, 2004v).

Eleutherodactylus eunaster Schwartz, 1973

Distribution: Hispaniola, Massif de la Hotte (Haiti); elevational range ~575–1,300 m.

Habitat: Arboreal, occurs in mesic hardwood closed-canopy forests, has not been recorded from disturbed forest (Hedges et al., 2004zzz; Stuart et al., 2008).

Predators: *Ialtris haetianus* (Henderson and Schwartz, 1986). **Reproduction:** Males call from concealed sites on shrubs and thorn bushes 0.2–0.9 m above ground (Schwartz and Henderson, 1991); eggs laid on the ground, breeds by direct development (Hedges et al., 2004zzz; Stuart et al., 2008). **Size:** Male SVL to 19.7 mm, female SVL to 21.4 mm (Schwartz, 1973b).

Conservation Status: "Critically endangered" on the IUCN Red List, last recorded in 1991, but no surveys have been conducted since then (Hedges et al., 2004zzz; Stuart et al., 2008).

Additional Pertinent Literature: Schwartz (1982b).

Eleutherodactylus flavescens Noble, 1923

Distribution: Hispaniola, eastern Dominican Republic; elevational range, sea level to ~900 m.

Habitat: Leaf litter in mesic primary and secondary forests and occasionally in mangroves, but does not adapt well to degraded forests (Hedges et al., 2004q; Stuart et al., 2008); occurs in wooded situations, including mangroves very close to the high tide mark (Mertens, 1939); occasionally adjacent to or on vertical rock faces in mountain streams and in coastal situations in close proximity to the ocean (Schwartz, 1982c).

Reproduction: Males call from branches, stems, and leaves of trees, shrubs, and bushes (to at least 6.0 m above the ground; Schwartz and Henderson, 1991); in coastal situations in close proximity to the ocean, males call from limestone boulders, low cliff faces, and vegetation (Schwartz, 1982c); lays eggs among leaves above ground or on the ground at the base of reeds (Noble, 1926); development direct (Hedges et al., 2004q; Stuart et al., 2008). **Size:** Male SVL to 31 mm, female SVL to 41 mm (Schwartz, 1982c).

Conservation Status: "Near threatened" on the IUCN Red List, some coastal populations near La Romana may be extirpated (Hedges et al., 2004q; Stuart et al., 2008).

Additional Pertinent Literature: Schwartz (1982c).

Eleutherodactylus fowleri Schwartz, 1973

Distribution: Hispaniola, Massif de la Selle (Haiti); elevational range ~1,045–1,300 m.

Habitat: Upland wooded situations including coffee groves and cutover hardwoods near human habitation, open meadows, occurs in bromeliads to 15 m above the ground (Schwartz, 1973b; Schwartz and Henderson, 1991); in deciduous cloud forest, in bromeliads by day (Schwartz, 1973b); high-elevation mesic forests, in bromeliads in tall trees (Hedges et al., 2004r; Stuart et al., 2008).

Reproduction: Males call from sites to 7.6 m above the ground (and higher on bromeliad leaves), lowest calling male at 1.8 m in a folded leaf of a bromeliad, 12 embryonated eggs (diameter 7.5–8.5 mm) on 25 August, two hatched 10 September, hatchling SVL 6.6–7.1 mm (Schwartz, 1973b); development direct (Hedges et al., 2004r; Stuart et al., 2008). **Size:** Male SVL to 31.1 mm, female SVL to 33.0 mm SVL (Schwartz, 1973b).

Conservation Status: "Critically endangered" on the IUCN Red List (Hedges et al., 2004r).

Additional Pertinent Literature: Schwartz (1982d).

Eleutherodactylus furcyensis Shreve and Williams, 1963

Distribution: Hispaniola, Massif de la Selle (Haiti and near the border with the Dominican Republic); elevational range ~800–2,100 m.

Habitat: Upland deciduous forest, under logs and rocks among pines (Schwartz, 1983b); on bare earthen road cuts, a burned clearing, and a freshly tilled cornfield (Schwartz and Henderson, 1991); upland mesic pine forests, recorded from recently disturbed forest, but that probably is not suitable habitat (Hedges et al., 2004y; Stuart et al., 2008).

Abundance: Once moderately common in suitable habitat (Hedges et al., 2004y; Stuart et al., 2008). **Reproduction:** Males call from among dead leaves and other forest debris on the ground (Schwartz, 1983b); clutch size 32 (Hedges, in prep.); eggs laid on the ground, development direct (Hedges et al., 2004y; Stuart et al., 2008). **Size:** Male SVL to 28 mm, female SVL to 37 mm (Schwartz, 1983b).

Conservation Status: "Critically endangered" on the IUCN Red List, last recorded in 1990, but no surveys have been conducted since (Hedges et al., 2004y; Stuart et al., 2008).

Additional Pertinent Literature: Schwartz (1983b).

Eleutherodactylus fuscus Lynn and Dent, 1943

Distribution: Western Jamaica, very localized and disjunct; elevational range ~120–680 m.

Habitat: Terrestrial in wet closed-canopy limestone forest (Hedges et al., 2004eee; Stuart et al., 2008); in wooded situations or clearings in woods, along a road shoulder and along a path through a hillside clearing, in banana plantations under banana trash and under a banana shelter, on red earth mounds heaped around yam or sweet potato plants adjacent to *Musa*, and among grass, shrubs, ferns, and trees on a cleared hillside (one frog at 0.18 m above the ground, the rest at ground level; Schwartz and Fowler, 1973).

Abundance: Locally abundant within its restricted and fragmented distribution (Hedges et al., 2004eee; Stuart et al., 2008). **Reproduction:** Males call from the ground in semi-open areas (Hedges, in prep.) and from exposed rocks along roadsides (Schwartz and Fowler, 1973); eggs laid on the ground, breeds by direct development (Hedges et al., 2004eee; Stuart et al., 2008). **Size:** Male SVL to 33.7 mm, female SVL to 37.3 mm (Schwartz and Fowler, 1973).

Conservation Status: "Critically endangered" on the IUCN Red List, not recorded for decades in the eastern portion of its range (Hedges et al., 2004eee; Stuart et al., 2008).

Eleutherodactylus glamyrus Estrada and Hedges, 1997

Distribution: Eastern Cuba, Sierra Maestra; elevational range ~800–1,975 m.

Habitat: Arboreal in closed mesic cloud forests (Hedges and Díaz, 2004w; Stuart et al., 2008); in rainforest and cloud forest on herbaceous growth and shrubs at heights of 0.3–3.0 m, hides by day in the dense growth of tree ferns, in epiphytic bromeliads or those growing on rocks, or in leaf litter (Díaz and Cádiz, 2008); in vegetation along trails in forests at the upper elevations of Pico Turquino (Estrada and Hedges, 1997c); on vegetation at night, in leaf litter by day (Díaz et al., 2005a).

Abundance: Common in suitable habitat (Hedges and Díaz, 2004w; Stuart et al., 2008). **Activity:** Nocturnal (Díaz and Cádiz, 2008). **Reproduction:** At night, males call from horizontal surfaces of leaves and ferns at 0.5–2.0 m above the ground, calling by day is from more concealed sites, females in leaf litter and in a hole in the ground beneath a stone with a clutch of 6 eggs, a pair in amplexus in leaf litter by day (August) and the female later produced 11 eggs (Estrada and Hedges, 1997c); males call sporadically all night long (Díaz et al., 2005a); call at low (15–18 °C) temperatures at 100% relative humidity while perched horizontally on leaves and fronds of tree ferns, mating and egg deposition in leaf litter (Díaz and Cádiz, 2008); eggs laid on the ground, breeds by direct development (Hedges and Díaz, 2004w; Stuart et al., 2008). **Size:** Male SVL to 20.0 mm, female SVL to 24.1 mm SVL (Estrada and Hedges, 1997c).

Conservation Status: "Endangered" on the IUCN Red List (Hedges and Díaz, 2004w).

Eleutherodactylus glandulifer Cochran, 1935

Distribution: Hispaniola, extreme western Tiburon Peninsula, Massif de la Hotte and vicinity (Haiti); elevational range ~300–1,885 m.

Habitat: Terrestrial in streamside habitats in closed-canopy forests (Hedges and Thomas, 2004d; Stuart et al., 2008); under logs in a recent clearing and under rocks along a path through dwarf (scrubby) cloud forest (Schwartz, 1983c); in pine forests and in cavities on vertical walls of the shafts in sinkhole caves, as well as under debris at the bottom of sinkholes (to depths of 27 m below the surface; Franz and Cordier, 1986).

Parasites: 99 of 104 specimens with light to heavy infestations of encysted nematodes on thighs and near the vent (and occasionally elsewhere on the body; Schwartz, 1983b). **Predators:** *Ialtris haetianus* (Henderson and Schwartz, 1986). **Reproduction:** May call in late afternoon (1645 h) from under rocks, males attend eggs (Townsend, 1996); eggs laid on the ground, breeds by direct development (Hedges and Thomas, 2004d; Stuart et al., 2008). **Size:** Male SVL to 36 mm, female SVL to 53 mm (Schwartz, 1983c).

Conservation Status: "Critically endangered" on the IUCN Red List (Hedges and Thomas, 2004d).

Additional Pertinent Literature: Schwartz (1983c).

Eleutherodactylus glanduliferoides Shreve, 1936

Distribution: Hispaniola, Massif de la Selle (Haiti); elevational range 1,515–2,120 m.

Habitat: On the ground or on grass stems to 0.2 m above the ground (Schwartz and Henderson, 1991); high-elevation moist forests (Hedges et al., 2004aaaa).

Reproduction: Males call from low perches (0.2 m; Schwartz and Henderson, 1991); eggs laid on the ground, breeds by direct development (Hedges et al., 2004aaaa); clutch size 5 (Hedges, in prep). **Size:** Female SVL to 16 mm (Schwartz and Henderson, 1991).

Conservation Status: "Critically endangered" on the IUCN Red List, known only from a few completely deforested localities on mountains above densely populated Port-au-Prince, the species is "possibly extinct," last seen in 1985, but few surveys have been conducted since (Hedges et al., 2004aaaa; Hedges, 2008b; Stuart et al., 2008).

Additional Pertinent Literature: Schwartz (1983d).

Eleutherodactylus glaphycompus Schwartz, 1973

Distribution: Hispaniola, Massif de la Hotte (Haiti); elevational range ~575–1,480 m.

Habitat: Rock crevices (erosion fissures between layers in limestone outcroppings and fragment jumbles) at night, one at night on vegetation (Schwartz, 1973b); in moist forest (Hedges and Thomas, 2004e; Stuart et al., 2008).

Abundance: Moderately common in suitable habitat (Hedges and Thomas, 2004e; Stuart et al., 2008). **Predators:** *Ialtris haetianus* (Henderson and Schwartz, 1986). **Reproduction:** Males call day and night (Schwartz, 1973b); call from 0.6–0.9 m above ground (one from a *Musa* leaf; Schwartz and Henderson, 1991); eggs laid on the ground, breeds by direct development (Hedges and Thomas, 2004e; Stuart et al., 2008); clutch size 27 (Hedges, in prep.). **Size:** Male SVL to 19.3 mm, female SVL to 28.7 mm (Schwartz, 1973b).

Conservation Status: "Endangered" on the IUCN Red List (Hedges and Thomas, 2004e).

Additional Pertinent Literature: Schwartz (1983e).

Eleutherodactylus glaucoreius Schwartz and Fowler, 1973

Distribution: Eastern Jamaica; elevational range, sea level to ~1,650 m.

Habitat: Wet limestone forest, rainforest, and elfin woodland (Hedges, in prep.); in xeric forest (on the ground and on rocks) and *Musa* groves, on leaves of low shrubs in rainforest, in and on a rock wall, on shrubs and trees to heights of 2.1 m, on dogtooth limestone cliffs, on *Musa* stumps, and on the bank of a mountain stream (Schwartz and Fowler, 1973); requires rainforest, but can occur in secondary forest (Hedges et al., 2004fff; Stuart et al., 2008).

Reproduction: Males call from the ground, rocks, or low shrubs, clutch size 20–37 (Hedges, in prep.); female attends eggs (Townsend, 1996); eggs laid on the ground, breeds by direct development (Hedges et al., 2004fff; Stuart et al., 2008). **Size:** Male SVL to 27.7 mm, female SVL to 31.6 mm (Schwartz and Fowler, 1973).

Conservation Status: "Near threatened" on the IUCN Red List (Hedges et al., 2004fff).

Eleutherodactylus goini Schwartz, 1960

Distribution: Cuba, Sierra de los Órganos and Sierra del Rosario; elevational range, sea level to >700 m.

Habitat: Rocky situations in broadleaf mesic forest (Schwartz, 1960a); in various forest formations, including those associated with mogotes, in leaf litter, river valleys, anthropogenic situations, rocky areas, and cave entrances (Díaz and Cádiz, 2008); leaf litter in rocky habitats, not recorded outside of forest habitats (Hedges and Díaz, 2004x; Stuart et al., 2008).

Abundance: Common and abundant (Hedges and Díaz, 2004x; Stuart et al., 2008). **Activity:** Nocturnal (Díaz and Cádiz, 2008). **Behavior:** Flees with long jumps when disturbed (Díaz and Cádiz, 2008). **Reproduction:** Males call from the ground or low vegetation to heights of 1 m (Díaz and Cádiz, 2008); clutch size 27–29 (Hedges, in prep.); breeds by direct development (Hedges and Díaz, 2004x; Stuart et al., 2008). **Size:** Male SVL to 34.3 mm, female SVL to 36.6 mm (Schwartz, 1960a).

Conservation Status: "Vulnerable" on the IUCN Red List (Hedges and Díaz, 2004x).

Additional Pertinent Literature: Schwartz (1974a).

Eleutherodactylus gossei Dunn, 1926

Distribution: Jamaica, two currently recognized subspecies are widespread on the island (*E. g. gossei*) and occur in extreme eastern Jamaica (*E. g. oligaulax*); elevational range, sea level to >1,500 m; introduced on Bermuda.

Habitat: Terrestrial in wet limestone forest, rainforest, and cloud forest, as well as moist rural gardens and formerly forested areas, agricultural areas might not be suitable due to pollution from agrichemicals (Hedges et al., 2004ggg); mesic habitats often surrounded by xeric situations, *E. g. gossei* under a rotting log in pasture, on rocks of a road cut, under banana trash, on a path through woods near a clearing, and on shrubs and in grass in a *Musa* grove, under *Cocos* trash in a grassy grove, on mud behind a beach ridge, under roadside rocks and trash in woods, and in loggers' chips at the base of a stump, *E. g. oligaulax* in an overgrown *Musa* grove, on the ground and on leaves of shrubs at heights to 1.5 m (Schwartz and Fowler, 1973).

Abundance: Common in suitable habitat (Hedges et al., 2004ggg). **Activity:** Nocturnal, dusk to dawn (Stewart, 1977). **Foraging and Diet:** Forages in banks and rock walls, among grass, low forbs, and twigs on the forest floor (Stewart and Martin, 1980); diet includes ants (72% by frequency of occurrence in sample of 1,167 prey items), arachnids (6%), termites (6%), springtails (5%), beetles (4%), leafhoppers + snails + crickets and roaches + thrips (4%; Stewart, 1977). **Reproduction:** Mature females 26–33 mm SVL (Schwartz and Henderson, 1991); males call from the ground among dead leaves and rocks, and from elevated perches to heights of 1.5 m in a viny tangle, 1.0–1.2 m in shrubs, and 1.5 m on a tree limb (Schwartz and Fowler, 1973); calling heights of 24 males 0–1.5 m (mean 11.8 cm; Stewart, 1977); nests with 49 and "upwards of seventy eggs" with female in attendance (Dunn, 1926b); clutches of 31–86 (Hedges, in prep.); yolked eggs/female 25–87 (mean 47.0; Schwartz and Henderson, 1991); eggs laid on the ground, development direct (Hedges et al., 2004ggg). **Size:** Male SVL to 28 mm, female SVL to 34 mm (Schwartz and Fowler, 1973).

Conservation Status: On the IUCN Red List as being of "least concern" (Hedges et al., 2004ggg).

Eleutherodactylus grabhami Dunn, 1926

Distribution: Central and western Jamaica; elevational range ~150–670 m.

Habitat: In woods, often under overhanging rocks (Dunn, 1926b); at night on a rock wall bordering a wooded hill (Lynn, 1940); in mesic woodlands, on rocks of a road cut, on terrestrial bromeliads, on a rock wall between an open road and an exposed pasture, on the ground and on leaves of low shrubs along a forest path, and associated with dog-tooth limestone on a steep hillside (Schwartz and Fowler, 1973); associated with terrestrial and arboreal bromeliads, intolerant of any disturbance to its habitat (Hedges et al., 2004hhh; Stuart et al., 2008).

Reproduction: Males call from rocks and low shrubs along a trail leading through cleared woodland, also from leaves of low herbaceous vegetation (Schwartz and Fowler, 1973); eggs laid on the ground, breeds by direct development (Hedges et al., 2004hhh; Stuart et al., 2008). **Size:** Male SVL to 21.1 mm, female SVL to 29.0 mm (Schwartz and Fowler, 1973).

Conservation Status: "Endangered" on the IUCN Red List, only a few records in the past decade, may no longer occur throughout much of its historic range (Hedges et al., 2004hhh; Stuart et al., 2008).

Eleutherodactylus grahami Schwartz, 1979

Distribution: Hispaniola, northwestern Haiti; elevational range ~20–330 m.

Habitat: In an area of limestone ridges with boulder and rock jumbles on a red earth substrate, may use rock crevices and caves as daytime retreats (Schwartz, 1979a).

Abundance: Relatively common in suitable habitats (Hedges and Thomas, 2004f; Stuart et al., 2008). **Reproduction:** Calling males on the wall at a cave entrance and on rocks next to a water pipe (Schwartz and Henderson, 1991); eggs laid on the ground, breeds by direct development (Hedges and Thomas, 2004f; Stuart et al., 2008); clutch size 23–34 (Hedges, in prep.). **Size:** SVL to 25 mm, based on one male specimen (Schwartz, 1979a).

Conservation Status: "Endangered" on the IUCN Red List (Hedges and Thomas, 2004f).

Eleutherodactylus greyi Dunn, 1926

Distribution: Central Cuba; elevational range, sea level to ~900 m.

Habitat: Rocky areas in mesic forest, on rock fences, rock walls, under large rocks in cave entrances, in caves, on rocky slopes near streams and rivers (Schwartz and Henderson, 1991; Estrada, 1994); also in leaf litter, in cavities and holes in the ground or in root tangles, adults and juveniles in dense bamboo (*Bambusa* sp.) stands, in stacks of cut sugarcane, adults often exposed on the ground, rocks, or tree trunks to heights >1 m (Díaz and Cádiz, 2008); restricted to closed-canopy forests (Hedges and Díaz, 2004y; Stuart et al., 2008).

Abundance: Common in suitable habitat but patchily distributed (Hedges and Díaz, 2004y; Díaz and Cádiz, 2008; Stuart et al., 2008). **Activity:** Nocturnal (Díaz and Cádiz, 2008). **Behavior:** Agile but not shy (Schwartz and Henderson, 1991). **Size:** Female SVL to 64 mm (Schwartz and Henderson, 1991).

Conservation Status: "Endangered" on the IUCN Red List (Hedges and Díaz, 2004y).

Eleutherodactylus griphus Crombie, 1986

Distribution: Jamaica, Cockpit Country; elevational range ~250–640 m.

Habitat: In pockets of leaf litter in wet limestone forest, in undisturbed forest around tree falls, at forest edge along a road and, after dark, along rock walls and cliffs (Crombie, 1986); restricted to pristine closed-canopy forests and intolerant of disturbances (Hedges, 2004a; Stuart et al., 2008).

Diet: Includes ants, spiders, cockroaches, and lepidopteran larvae (Crombie, 1986). **Reproduction:** Smallest reproductive females 15.5–15.7 mm SVL, smallest male with enlarged testes 9.4 mm SVL (Crombie, 1986); eggs laid on the ground, breeds by direct development (Hedges, 2004a; Stuart et al., 2008). **Size:** Male SVL to 14.2 mm, female SVL to 18.4 mm (Crombie, 1986).

Conservation Status: "Critically endangered" on the IUCN Red List (Hedges, 2004a).

Eleutherodactylus gryllus Schmidt, 1920

Distribution: Puerto Rico, interior uplands; elevational range ~300–1,180 m; now known to occur only in cloud forest at elevations >700 m (Hedges and Joglar, 2004b; Hedges and Rios-López, 2008b; Stuart et al., 2008).

Habitat: In rainforest, cloud forest, and elfin woodland (Hedges, in prep.); only about 5% in bromeliads (Cintrón, 1970); in meadows and along road cuts in bromeliads or broad-leaved plants (Drewry, 1970b); prefers edge situations (Rivero, 1998); in banana groves (Drewry and Rand, 1983); by day in bromeliads and under moss on rocks (K. P. Schmidt, 1920, 1928); not recorded outside forest habitats (Hedges and Joglar, 2004b; Hedges and Rios-López, 2008b; Stuart et al., 2008).

Activity: Nocturnal (K. P. Schmidt, 1920, 1928), although calling may start during the afternoon (Joglar, 1998). **Diet:** Juveniles take mites, hemipterans, dipterans, and hymenopterans (Lavigne and Drewry, 1970). **Population Size and Density:** At El Yunque, the population in elfin forest declined consistently during the 1990s (Burrowes and Joglar, 2005). **Reproduction:** Males start calling in the afternoon (~1600 h; Joglar, 1998); call from leaves and branches, usually at 1–4 m above the ground, with peaks just before dusk and just after dawn (Drewry and Rand, 1983); may call from high in the canopy (Rivero, 1998; Schwartz and Henderson, 1991); males observed attending 4–5 eggs in axils of bromeliads (G. E. Drewry in Joglar, 1998); a gravid female (SVL 18.6 mm) in April had 5 eggs (diameter 1.6–2.0 mm), juveniles in November-December were 6.2 mm and 7.1 mm SVL (Joglar, 1998); development direct (Hedges and Joglar, 2004b; Hedges and Rios-López, 2008b; Stuart et al., 2008). **Size:** Male SVL to 19.1 mm, female SVL to 19.5 mm (Joglar, 1998).

Conservation Status: "Endangered" on the IUCN Red List (Hedges and Joglar, 2004b; Hedges and Rios-López, 2008b); "few or no individuals located in recent surveys" (J. W. Wiley in Johnson, 1988); "decreasing in some areas, need of status survey" (Moreno, 1991d).

Eleutherodactylus guanahacabibes Estrada and Novo Rodríguez, 1985

Distribution: Cuba, Península de Guanahacabibes (extreme western tip of island); elevational range, sea level to ~20 m.

Habitat: Rocky areas (caverns, sinks, "dogtooth" limestone) with high humidity and reduced illumination, clivicolous or saxicolous (Estrada and Novo Rodríguez, 1985a); also in coastal formations, leaf litter of small swamps, and rustic human houses (Díaz and Estrada, 2000; Díaz and Cádiz, 2008).

Abundance: Moderately common in suitable habitat (Hedges and Díaz, 2004z; Stuart et al., 2008). **Activity:** Predominantly nocturnal (Díaz and Cádiz, 2008). **Reproduction:** Males call by day from cave interiors (Alonso Bosch and Rodríguez Gómez, 2003a); also call from the ground, rocky limestone surfaces, fallen trunks, and leaves of small palms (*Coccothrinax argentata*), male choruses heard just after sunset and throughout the night after heavy rains, closest distance between two calling males was <1.0 m in rocky situations, females vocalize when gravid, possibly to declare their state and receptivity to males (Díaz and Estrada, 2000); males call from depressions in rocks and from low vegetation (Díaz and Cádiz, 2008); eggs laid on the ground, breeds by direct development (Hedges and Díaz, 2004z; Stuart et al., 2008); clutch size 8 (Hedges, in prep). **Size:** Female SVL to 23 mm (Estrada and Novo Rodríguez, 1985a); SVL to 28 mm (Díaz and Cádiz, 2008).

Conservation Status: "Vulnerable" (Vales et al., 1998); "endangered" on the IUCN Red List (Hedges and Díaz, 2004z).

Eleutherodactylus guantanamera Hedges, Estrada, and Thomas, 1992

Distribution: Southeastern Cuba; elevational range ~60–1,150 m.

Habitat: Arboreal in mesic upland areas, some individuals in bromeliads by day (Hedges et al., 1992; Hedges and Díaz, 2004aa; Stuart et al., 2008); in bromeliads 3–12 m high in rainforest, semideciduous forest, pine forest, gallery forest, and various secondary forest formations, occasionally in introduced plants (e.g., *Pandanus*; Díaz and Cádiz, 2008).

Abundance: Moderately common in suitable habitat (Hedges and Díaz, 2004aa; Stuart et al., 2008); common, but easier to hear than see (Díaz and Cádiz, 2008). **Activity:** Nocturnal (Díaz and Cádiz, 2008). **Reproduction:** Males call from leaves of small trees and a banana leaf at 3.0 m, but most call from bromeliads high in trees, large females with 5 and 9 large (3 mm) ova (Hedges et al., 1992); males "drive" females to egg-deposition sites in bromeliads using calls much less intense than the advertisement call, pairs have been detected by vocalization inside holes in tree trunks, eggs usually attended by males (Díaz and Cádiz, 2008); eggs laid in bromeliads, breeds by direct development (Hedges and Díaz, 2004aa; Stuart et al., 2008). **Size:** Male SVL to 32.8 mm, female SVL to 37.3 mm (Hedges et al., 1992).

Conservation Status: "Vulnerable" (Vales et al., 1998); "vulnerable" on the IUCN Red List (Hedges and Díaz, 2004aa).

Eleutherodactylus gundlachi Schmidt, 1920

Distribution: Southeastern Cuba, Sierra Maestra, Sierra del Cobre, Sierra de la Gran Piedra, and to the east; elevational range ~200–1,970 m.

Habitat: Deep hardwood forest, under stones, rocks, and debris along brooks by day, at night on the ground, in and about ruins, and in rock piles (Schwartz and Henderson, 1991); in leaf litter of various forest formations, including those affected by human activities (Díaz and Cádiz, 2008); terrestrial in closed mesic forests (Hedges and Díaz, 2004bb; Stuart et al., 2008).

Abundance: Common in suitable habitat (Hedges and Díaz, 2004bb; Stuart et al., 2008). **Activity:** Nocturnal, although some calling occurs by day (Díaz and Cádiz, 2008). **Reproduction:** Males call from rock piles (Schwartz and Henderson, 1991); call mainly around dusk (Díaz et al., 2005a); call from the ground, small plants, root masses, and piles of rocks (Díaz and Cádiz, 2008); call from the ground, breeds by direct development (Hedges and Díaz, 2004bb; Stuart et al., 2008). **Size:** Female SVL to 23 mm (Schwartz and Henderson, 1991); SVL to 25 mm (Díaz and Cádiz, 2008).

Conservation Status: "Endangered" on the IUCN Red List (Hedges and Díaz, 2004bb).

Eleutherodactylus haitianus Barbour, 1942

Distribution: Hispaniola, Cordillera Central (Dominican Republic); elevational range ~1,545–2,455 m (but may range higher).

Habitat: Pine forests, under rocks, logs, and sheets of moss (Schwartz and Henderson, 1991); terrestrial, not recorded outside forested areas (Hedges et al., 2004z; Stuart et al., 2008).

Abundance: Can be locally common, but is patchily distributed (Hedges et al., 2004z; Stuart et al., 2008). **Reproduction:** Males call from leaves of herbaceous growth, ferns, and low bushes at heights of 3.2–30.5 cm (Schwartz and Henderson, 1991); egg clutches under moss mats (Schwartz, 1983f); clutch size 6–9 (Hedges, in prep.); eggs laid on the ground, development direct (Hedges et al., 2004z; Stuart et al., 2008). **Size:** Male SVL to 17.0 mm, female SVL to 21.0 mm (Schwartz, 1983f).

Conservation Status: "Endangered" on the IUCN Red List (Hedges et al., 2004z).

Additional Pertinent Literature: Schwartz (1983f).

Eleutherodactylus hedricki Rivero, 1963

Distribution: Puerto Rico, interior uplands; elevational range ~455–1,150 m.

Habitat: Entirely arboreal (Joglar, 2005a); occurs in rainforest and cloud forest (Hedges, in prep.); in or near a tree hole at "almost any height above ground" (Drewry, 1970b); by day in holes and cracks in trees (Rivero, 1963).

Activity: Nocturnal (Rivero, 1963); call strongly at sunset, peak within 2 h after sunset, individuals drop out progressively until activity either ceases or reaches a minimum earlier than either *E. portoricensis* or *E. wightmanae* on all but the driest nights, often fading out by midnight, resuming with a smaller peak at daybreak and then fading out, little or no nighttime calling between early December and late February (Drewry, 1970b). **Behavior:** Males may have an ability to home based on cues other than visual (Rivero, 1963). **Disease:** Chytridomycosis suspected of causing observed population declines (Hedges et al., 2004qq; Angulo, 2008c). **Population Density:** At El Verde, maximum of 200/ha (Stewart and Woolbright, 1996). **Reproduction:** Males call March/April–December from holes and cracks in tree trunks (including *Cecropia*) and branches (0–18.5 m above ground; Stewart and Woolbright, 1996); calling starts and peaks prior to dusk and tapers off gradually until just past midnight (Drewry and Rand, 1983); sometimes calls during day (Rivero, 1998); female follows male into a cavity, spends the day and deposits eggs, 14–32 eggs (diameter 4.0–4.5 mm) attached by gelatinous threads to wood encircling the pith cavity, possibly multiple clutches at one nest site (Rivero, 1963); females enter tree holes of males in early evening, amplexus and oviposition occurs the next morning, and females leave the second evening, males may associate with as many as three clutches at different stages of development (Drewry, 1970b); male attends nest (Rivero, 1998); may take 22 days for eggs to hatch, juveniles 3.5–8.0 mm SVL (Rivero, 1963); development direct (Hedges et al., 2004qq; Angulo, 2008c; Stuart et al., 2008). **Size:** Male SVL to 35.0 mm, female SVL to 34.6 mm (Joglar, 1998). **Territoriality:** Males highly territorial, pushing and biting at another frog experimentally released in their tree holes (Drewry, 1970b).

Conservation Status: "Endangered" on the IUCN Red List (Hedges et al., 2004qq; Angulo, 2008c); "few or no individuals located in recent surveys" (J. W. Wiley in Johnson, 1988); "decreasing in some areas, need of status survey" (Moreno, 1991d).

Eleutherodactylus heminota Shreve and Williams, 1963

Distribution: Hispaniola, Massif de la Selle and east on the Tiburon Peninsula (Haiti), an isolated record in the Sierra de Baoruco (Dominican Republic); elevational range, sea level to ~1,700 m.

Habitat: Occurs in bromeliads and the axils of banana plants (Schwartz and Henderson, 1991); requires closed-canopy forests (Hedges et al., 2004kk; Stuart et al., 2008).

Abundance: Although moderately common in suitable habitat, rarely encountered in much of its range (Hedges et al., 2004kk; Stuart et al., 2008). **Predators:** *Ialtris haetianus* (Henderson and Schwartz, 1986). **Reproduction:** Males call from bromeliads (to 4.0 m above the ground) and from *Musa*, eggs deposited in bromeliads (Schwartz and Henderson, 1991); development direct (Hedges et al., 2004kk; Stuart et al., 2008). **Size:** Male SVL to 21 mm (Schwartz, 1983g), female SVL to 30 mm (Shreve and Williams, 1963).

Conservation Status: "Endangered" on the IUCN Red List (Hedges et al., 2004kk).

Additional Pertinent Literature: Schwartz (1983g).

Eleutherodactylus hypostenor Schwartz, 1965

Distribution: Hispaniola, "South Island," Sierra de Baoruco (Dominican Republic), Massif de la Selle and Massif de la Hotte (Haiti); elevational range, sea level to ~1,060 m.

Habitat: Specialized burrowing species in closed mesic broadleaf forests, as well as shade-grown coffee and cacao plantations (Hedges et al., 2004ll; Stuart et al., 2008); in mesic broadleaf forests as well as coffee groves, in earthen cavities in open situations among coffee trees, adjacent to a roadside log (Schwartz, 1965e); occasionally on open ground or on low branches of shade trees (Schwartz and Henderson, 1991).

Abundance: Locally uncommon and patchily distributed (Hedges et al., 2004ll; Stuart et al., 2008). **Predators:** *Hypsirhynchus parvifrons* (Henderson et al., 1987a). **Reproduction:** Males call from shallow underground chambers 1–3 cm below the surface in loose soil (Hedges and Thomas, 1987); individuals on the ground or on vegetation (to 1.0 m) are noncalling adult males (Schwartz and Henderson, 1991); eggs laid underground (Hedges et al., 2004ll; Stuart et al., 2008). **Size:** Male SVL to 54.4 mm (Schwartz, 1965e), females unknown (Schwartz, 1984).

Conservation Status: "Endangered" on the IUCN Red List (Hedges et al., 2004ll).

Additional Pertinent Literature: Schwartz (1984).

Eleutherodactylus iberia Estrada and Hedges, 1996

Distribution: Eastern Cuba, Cuchillas de Moa; elevational range, sea level to ~850 m.

Habitat: In secondary hardwood forest under leaf litter and among roots of ferns by day and night (Estrada and

Hedges, 1996b); in secondary vegetation and crops along a road (Fong Grillo et al., 2005a); closed rainforest on poorly drained soils, with maintenance of local humidity critically important (Hedges and Díaz, 2004cc; Stuart et al., 2008); under palm fronds in open areas near forests, in leaf litter, under a dense cover of ferns (*Pteris* sp.; Díaz and Cádiz, 2008).

Abundance: Common in suitable habitat (Hedges and Díaz, 2004cc; Stuart et al., 2008); common, but often difficult to detect (Díaz and Cádiz, 2008). **Activity:** Largely nocturnal (Díaz and Cádiz, 2008). **Predators:** *Caraiba andreae* (Rodríguez Gómez and Alonso Bosch, 2003b). **Reproduction:** Males call by day and night but active and moving about in leaf litter only at night, female within a few centimeters of a single egg (Estrada and Hedges, 1996b); breeds by direct development (Hedges and Díaz, 2004cc; Stuart et al., 2008). **Size:** Male SVL to 11.5 mm, female SVL to 12.3 mm (Rodríguez Gómez and Alonso Bosch, 2003b).

Conservation Status: "Critically endangered" on the IUCN Red List (Hedges and Díaz, 2004cc); "vulnerable" (Rodríguez Gómez and Alonso Bosch, 2000).

Eleutherodactylus inoptatus Barbour, 1914

Distribution: Hispaniola, islandwide but disjunct and absent from xeric regions; elevational range, sea level to ~1,700 m.

Habitat: Mesic wooded areas, occasional subadults resting (in water retention position) on leaves or trunks of small saplings or herbaceous plants during the day (Schwartz and Henderson, 1991); on vine-covered trees (Noble, 1923b); sensitive to severe habitat destruction, but can persist in plantations such as coffee and banana as long as trees and shade are available (Hedges and Thomas, 2004g); jumping across paved road at night (RP).

Predators: *Hypsirhynchus parvifrons* (Henderson et al., 1987a). **Reproduction:** Calling males on bare ground to at least 6.0 m in trees, occasionally call from "moderate" concealment (cavities in rocks or sites partly covered by vegetation; Hedges and Thomas, 1987), or from a hole in a trunk (Schwartz and Henderson, 1991); clutch size 33–40 (Hedges, in prep.); eggs (20–34/clutch, July–October) under logs and rocks and in leaf litter (Schwartz and Henderson, 1991); male attends nest (Townsend, 1996); hatchlings have very small "egg teeth" and hatching occurs only after eggs are soaked in water (Noble, 1926); breeds by direct development (Hedges and Thomas, 2004g). **Size:** Female SVL to 88.0 mm (Hedges and Thomas, 1987).

Conservation Status: On the IUCN Red List as a species of "least concern" (Hedges and Thomas, 2004g).

Eleutherodactylus intermedius Barbour and Shreve, 1937

Distribution: Southeastern Cuba, Sierra Maestra, Sierra del Cobre, and mountains to the east; elevational range ~450–1,970 m.

Habitat: In leaf litter in forest adjacent to an old coffee plantation (Estrada, 1992); occurs in leaf litter and under logs and rocks in rainforest and cloud forest (Díaz and Cádiz, 2008); known only from moist closed forest (Hedges and Díaz, 2004dd; Stuart et al., 2008).

Abundance: Uncommon even in suitable habitat (Hedges and Díaz, 2004dd; Stuart et al., 2008); common (Díaz and Cádiz, 2008). **Activity:** Crepuscular to nocturnal, although male choruses may extend into the morning (Díaz and Cádiz, 2008). **Reproduction:** Both males and females call from the ground (Díaz and Cádiz, 2008); male calling from inside a hole 1 cm in diameter in a small gully along a natural ravine surrounded by dense forest, female on a nest in a chamber 3 cm deep and 2 cm in diameter, five yellowish eggs with diameters 3.0–3.8 mm (mean 3.2 ± 0.2 mm; Estrada, 1992); clutch size 6 (Hedges, in prep.); females guard nests (Alonso Bosch and Rodríguez Gómez, 2003b); breeds by direct development (Hedges and Díaz, 2004dd; Stuart et al., 2008). **Size:** Male SVL to 16.4 mm, female SVL to 16.3 mm (Estrada, 1992); female SVL to 15 mm (Alonso Bosch and Rodríguez Gómez, 2003b); SVL to 20 mm (Díaz and Cádiz, 2008).

Conservation Status: "Endangered" on the IUCN Red List (Hedges and Díaz, 2004dd).

Eleutherodactylus ionthus Schwartz, 1960

Distribution: Southeastern Cuba; elevational range, sea level to 1,230 m.

Habitat: In bromeliads (Rodríguez Gómez and Alonso Bosch, 2003a); one of only two species (with *E. atkinsi*) in a coffee plantation in eastern Cuba (Fong Grillo, 1999); arboreal in closed mesic forest (Hedges and Díaz, 2004ee; Stuart et al., 2008); in bromeliads from ground level to heights of 4 m in rainforest, cloud forest, and mesic pine forest (Díaz and Cádiz, 2008).

Abundance: Moderately common in suitable habitat (Hedges and Díaz, 2004ee; Stuart et al., 2008); common but more frequently heard than seen (Díaz and Cádiz, 2008). **Activity:** Nocturnal (Díaz and Cádiz, 2008). **Reproduction:** Occasionally heard calling by day (Díaz et al., 2005a); males call, often throughout the night, from bromeliads and leaves in the canopy (Díaz and Cádiz, 2008); breeds by direct development (Hedges and Díaz, 2004ee; Stuart et al., 2008). **Size:** Male SVL to 30.2 mm, female SVL to 40.0 mm (Schwartz, 1960a).

Conservation Status: "Endangered" on the IUCN Red List (Hedges and Díaz, 2004ee).

Eleutherodactylus jamaicensis Barbour, 1910

Distribution: Jamaica; elevational range ~120–1,290 m.

Habitat: Lowland and montane old-growth forests (Hedges et al., 2004iii; Stuart et al., 2008); in dense forest, pastures with isolated clumps of trees, and trees along open road margins (Schwartz and Fowler, 1973); in terrestrial bromeliads on limestone substrates (Schwartz and Fowler, 1973); arboreal bromeliads (Dunn, 1926b; Lynn, 1940); some bromeliads as low as 1.2 m, others high in trees, one frog in a petiole base of a *Heliconia* plant (Schwartz and Fowler, 1973); under coconut husks with land crabs in a supratidal situation near Montego Bay (Grant, 1939).

Abundance: Uncommon (Hedges et al., 2004iii; Stuart et al., 2008). **Activity:** Observed active on tips of bromeliad leaves at night (Lynn, 1940). **Defense:** When handled, exudes bright blue slime from sides and thighs (Dunn, 1926b; Schwartz and Fowler, 1973); presses dorsoventrally flattened body tightly between bromeliad leaves (P. Vogel, 2000b). **Reproduction:** Males call from bromeliads (Schwartz and Fowler, 1973); lay eggs (adherent to each other and to bromeliad leaves) in bromeliads above water level in batches of ~12 (Dunn, 1926b); six "batches" of eggs observed in late August–early September (Lynn, 1940); clutch of 28 from Quick Step on 8 August (Lynn and Dent, 1943); 12–28 eggs deposited immediately above water level in bromeliads (P. Vogel, 2000b). **Size:** Male SVL to 22.0 mm, female SVL to 29.0 mm (Schwartz and Fowler, 1973).

Conservation Status: "Endangered" on the IUCN Red List (Hedges et al., 2004iii).

Eleutherodactylus jasperi Drewry and Jones, 1976

Distribution: Puerto Rico, Sierra de Cayey; elevational range ~650–850 m.

Habitat: Open, "somewhat xeric" forest with dense bromeliad growth, frogs in water-filled bromeliads (*Vriesia*, *Hohenbergia*, and *Guzmania*) in trees, on the ground, and growing on vertical surfaces of cliffs, often in large groups (Drewry and Jones, 1976); prefers bromeliads devoid of leaf litter and other organic debris, bromeliads used as daytime retreats (Joglar, 1998).

Population Size: From 1973–1975, estimated between 1,000 and 2,000 individuals (C. Díaz in Joglar, 1998). **Reproduction:** Ovoviviparous, gravid females April–August, females may reproduce more than once annually (Drewry and Jones, 1976); 3–6 young (1–6 ova), intra-oviductal development dependent on yolk throughout development, 6.2–7.3 mm at birth, 33 days post-amplexus (Wake, 1978); 1–3 young (Hedges and Joglar, 2004c; Angulo, 2008d; Stuart et al., 2008). **Size:** SVL in males to 21.4 mm (Joglar, 1998), in females to 21.5 mm (Wake, 1978).

Conservation Status: "Critically endangered" on the IUCN Red List (Hedges and Joglar, 2004c; Angulo, 2008d); a U.S. Fish and Wildlife "threatened" species (U.S. Fish and Wildlife Service, 2007); the type locality has been cleared for house construction (Moreno, 1991a); this species is possibly extinct (Hedges and Joglar, 2004c; Angulo, 2008d; Stuart et al., 2008); despite intensive searches, it has not been observed since 1981 (J. W. Wiley in Johnson, 1988; Moreno, 1991c; Burrowes and Joglar, 1991; Joglar and Burrowes, 1996; Burrowes et al., 2004; Hedges and Joglar, 2004c; Joglar, 2005a; Joglar et al., 2007; Angulo, 2008d; Hedges and Díaz, 2009; Stuart et al., 2008); "endangered" (Moreno, 1991d).

Eleutherodactylus jaumei Estrada and Alonso Bosch, 1997

Distribution: Southeastern Cuba; elevational range ~150–950 m.

Habitat: Closed mesic forest (Hedges and Díaz, 2004ff; Stuart et al., 2008); secondary forest (at 150–200 m elevation) converted to fruit and coffee plantation, adults in leaf litter between tree roots at 0900 h, adults and juveniles under leaf litter and rocks 1900–2000 h (Estrada and Alonso Bosch, 1997).

Abundance: Common in suitable habitat (Hedges and Díaz, 2004ff; Stuart et al., 2008). **Activity:** Diurnal (Díaz and Cádiz, 2008). **Behavior:** In dry season, congregates in leaf litter and rock piles in arroyos (Díaz and Cádiz, 2008). **Reproduction:** Female with 2 oviductal eggs (Estrada and Alonso Bosch, 1997); breeds by direct development (Hedges and Díaz, 2004ff; Stuart et al., 2008). **Size:** Male SVL to 11.6 mm, female SVL to 12.7 mm (Estrada and Alonso Bosch, 1997); SVL to 13 mm (Díaz and Cádiz, 2008).

Conservation Status: "Critically endangered" on the IUCN Red List (Hedges and Díaz, 2004ff).

Eleutherodactylus johnstonei Barbour, 1914

Distribution: Widely distributed in the Lesser Antilles, Anguilla (may be extirpated), St.-Barthélemy, St.-Martin/St. Maarten, Saba, St. Eustatius, St. Christopher, Nevis, Barbuda (presumably extirpated; Malhotra and Thorpe, 1999), Antigua, Montserrat, Guadeloupe, Terre-de-Haut (Les Saintes), La Désirade, Marie-Galante, St. Lucia, Martinique, St. Vincent, the Grenadines (Bequia, Mustique, Canouan, Union Island, Petit St. Vincent, Carriacou), Grenada, Barbados (many populations introduced); on Dominica, restricted to narrow strip of Caribbean coastland, may occur only in disturbed situations (Malhotra and Thorpe, 1999); because the species has not been found during recent surveys (e.g., Quick, 2001; Alexander, 2007), the Dominican population is assumed to have been extirpated and the species has been removed from that island's list of known herpetofauna (A. James, pers. comm.; Ackley et al., 2009); elevational range, sea level to at least 1,300 m; outside of the Lesser Antilles, introduced to Tortola (British Virgin Islands; A. J. Sánchez Muñoz, in litt., 10.VIII.2008), Jamaica, Bermuda, Trinidad, Curaçao, and the Neotropical mainland; which Lesser Antillean populations of this species are naturally occurring and which were introduced with human agency is uncertain; Late Quaternary fossil evidence from noncultural sites suggests that the species is native to several of the Leeward Islands (Antigua, Barbuda; Pregill et al., 1994), and that it has been introduced elsewhere in the Lesser Antilles; populations outside the Lesser Antilles are certainly introduced.

Habitat: Very adaptive, mainly in disturbed habitats, including mowed fields, yards, gardens, sugarcane fields, towns, houses, and plantations, when in forests, usually in clearings and gaps (Hedges et al., 2004g, 2008b); in Jamaica, in disturbed situations from sea level to about 550 m, sugarcane fields, and on banana, ginger, coffee, and bamboo plants, also on trees and shrubs to 1.5 m above ground, on roads at night, and on bromeliads (Schwartz and Fowler, 1973); readily enters homes (Long, 1974); around dwellings, roadsides, forest margins, agricultural areas, river banks, and occasionally in bromeliads (Stewart, 1977); on Anguilla, in watered yards and gardens in The Valley (Hodge et al., 2003), but possibly extirpated (Lever, 2003); on St.-Martin/St. Maarten, at higher elevations and in well-watered gardens at lower elevations, readily takes advantage of suitable artificial conditions and may be quite common around and in human habitations, gardens, and plant nurseries, on Saba, abundant in generally moist habitats at higher elevations, in leaf litter, under rocks and logs, in bromeliads, and in association with natural and human debris, often edificarian, in showers, on St. Eustatius, abundant only on the Quill (especially in the crater), but also in moist pockets in gardens elsewhere on the island (Powell et al., 2005); also in sinks and water tanks on Saba (RP); on St. Christopher, under rocks in a coastal *Terminalia* grove (Schwartz, 1967b); on Antigua, in forest remnants, particularly along streams, under rocks by day, also in watered yards and gardens, less abundant in drier areas, often edificarian (RP); on Marie-Galante, in gardens, sugarcane fields, cutover sugarcane (Henderson et al., 1992); on Dominica, in residential gardens, roadside and beachside vegetation, and around soap factory sheds (Kaiser and Wagenseil, 1995); on Martinique, among dry breadfruit leaves bordering a dusty road through a sugarcane field (Schwartz, 1967b); in dry and wet forests, often in bromeliads (Breuil, 2009); on St. Vincent, "ubiquitous," urban (cement ditches, houses, gardens), agricultural, and more or less natural areas, both xeric and mesic (Mallery et al., 2007); in dense forest (even at high elevations still dominated by *Pristimantis shrevei*), streambeds, leaf litter, and urban alleys (Diaz-Lameiro et al., 2007); on Mustique, trees and bushes at night in residential garden (Henderson et al., 1992); by day under bark, stones, in damp corners of houses, bathrooms, and other "wet" areas (de Silva and Wilson, 2006); on Grenada, "ubiquitous," in urban (homes, cement ditches, small parks, roadside ditches), agricultural, and more or less natural areas at elevations below 600 m (Germano et al., 2003); on Barbados, in a forested gully (with a riverbed of coral-limestone rubble that was dry most of the year) with sparse ground vegetation and *Hippomane mancinella* overstory creating a nearly closed canopy, also in flowerbeds (with dense ground vegetation that included seasonally abundant herbaceous plants and bromeliads) and roadside hedges, potential diurnal retreats included cracks and cavities under and within rocks, as well as bromeliads and the bases of *Sancevieria* spp. and *Hymenocallis caribaea* (Ovaska, 1991a); also in snail (*Strophocheilus oblongus*) shells (Schwartz, 1967b).

Activity: Nocturnal (Germano et al., 2003; RWH); in Barbados, leave diurnal retreats (bromeliads) at 1830–1840 h, abandon bromeliads entirely after 2000 h (Ovaska, 1991a); on Saba, will forage at night lights (Perry, 2006). **Behavior:** "Great jumpers and can easily clear a distance of forty times their length" (Long, 1974); in captivity, when threatened, either adpress themselves tightly against the substrate or jump quickly (Reubold, 1983). **Diurnal Retreats:** Ground-level cover objects used by 1–9 frogs on Barbados (Ovaska, 1991a). **Foraging and Diet:** Feeds largely on insects (Long, 1974); in captivity, observes prey carefully and often for extended periods before jumping and capturing it (Reubold, 1983); sit-and-wait predator, on Jamaica, of 2,121 prey items (3.95/frog), 44% ants, 19% arachnids (mites, spiders), 11% leafhoppers, and 10% springtails (Stewart, 1977); on Barbados, during 60-min observations, made strikes at prey at a mean rate of 7.3 ± 6.90/h from 2.3 ± 2.05 foraging sites (Ovaska, 1991b); on Barbados, the diet included ants (in 79.6% of 113 stomachs), springtails (36.3%), leafhoppers (35.4%), adult dipterans (23.9%), earwigs (22.1%), and mites (20.4%), ants composed 51.4% of the total number of prey items (N = 939), individual frogs had 1–74 prey items, mean number/stomach 7.6 ± 8.49 (Ovaska, 1991b). **Growth:** On Barbados, most growth occurred during the wet season (Ovaska, 1991a). **Movements:** On Barbados, mean nightly movements of males (distance between two farthest sightings) 1.09–1.10 m, for females 1.72–1.99 m, after 150–220 days, mean distances from initial points of capture for adult males (at two sites) 1.67–2.96 m (range 0–6.1 m), for females 1.11–2.05 m (0.5–4.6 m), after 240–300 days for males at one site, mean 1.59 m (0.5–5.4 m), and for females 1.88 m (0.5–3.6 m; Ovaska, 1992). **Parasites:** Polymorphid cystacanths on Grenada (Goldberg et al., 1998d). **Population Density:** On Barbados, 0.32–2.73/m^2, depending on habitat (Ovaska, 1991a). **Predators:** Lizards *Thecadactylus rapicauda* (Henderson and Berg, 2005b); snakes *Alsophis antillensis, A. rufiventris* (Henderson and Sajdak, 1996); *Mastigodryas bruesi* (RWH). **Reproduction:** Calling in May (1987) on Anguilla (Censky, 1989); on Barbados, reproduction seasonal and peaks July–August (early part of wet season), number of calling males highest July–August, males start calling at 19–20 mm SVL (Ovaska, 1991a); on Barbados, all calling sites of a male within a small area (monthly means 0.2–0.8 m^2), sizes of calling ranges for males were 0–1.49 m^2 in June–July, and overlap between adjacent calling ranges was 0–100% in June–July, males defend calling sites against intruding males via displays and fighting that can involve physical contact, besides calling from territories, other mating tactics included satellite behavior and opportunistically interfering in courtships of others (Ovaska and Hunte, 1992); on Jamaica, males call from perches 12–229 cm (mean 107.6 cm) above the ground, 25% of 53 females gravid in April, and 1–49 yolked eggs/female (Stewart, 1977); both sexes attend nest (Townsend, 1996); eggs laid on the ground, breeds by direct development (Hedges et al., 2004g, 2008b). **Size:** Male SVL to 25 mm, female SVL to 35 mm (Kaiser and Hardy, 1994a).

Conservation Status: On the IUCN Red List as a species of "least concern" (Hedges et al., 2004g, 2008b).

Additional Pertinent Literature: Kaiser and Hardy (1994a).

Eleutherodactylus juanariveroi Rios-López and Thomas, 2007

Distribution: Puerto Rico, northern coastal areas; elevation ~17 m.

Habitat: Palustrine herbaceous wetland (seasonally flooded with fresh water) in subtropical moist forest zone, 0.4–1.2 m high in herbaceous vegetation that includes ferns (*Blechnum serrulatum, Thelypteris interrupta*), flat sedges (*Cyperus* sp.), spike rushes (*Eleocharis* sp.), and vines and grasses (Rios-López and Thomas, 2007).

Activity: Nocturnal (1900–2200 h), although calling may start earlier (Rios-López and Thomas, 2007). **Reproduction:** Males call (starting at 1630 h and decreasing significantly before midnight) from vegetation 0.6–1.2 m above ground level, clutch size 1–5, deposited on leaf axils (21 clutches) or leaf surfaces (3 clutches) of Bull-tongue Arrowhead (*Sagittaria lancifolia*; Rios-López and Thomas, 2007); thought to breed year-round, but frequency of calling males higher in May–November than dry months of January–March, embryos hatch in ~19 days (Rios-López, 2008g). **Size:** Male SVL to 16.0 mm, female SVL to 17.3 mm (Rios-López and Thomas, 2007).

Conservation Status: Probably threatened based on potential development at the type locality (Rios-López and Thomas, 2007); "critically endangered" on the IUCN Red List (Rios-López, 2008g).

Eleutherodactylus jugans Cochran, 1937

Distribution: Hispaniola, Massif de la Selle (Haiti and Dominican Republic); elevational range ~1,240–2,150 m.

Habitat: Under rocks and other surface debris in pine and broadleaf forest (Schwartz and Henderson, 1991); terrestrial, does not tolerate disturbed habitats (Hedges et al., 2004aa; Stuart et al., 2008).

Abundance: Moderately common in suitable habitat, which is rapidly disappearing (Hedges et al., 2004aa; Stuart et al., 2008). **Reproduction:** Eggs laid on the ground, breeds by direct development (Hedges et al., 2004aa; Stuart et al., 2008). **Size:** SVL to 33 mm in females (Cochran, 1941).

Conservation Status: "Critically endangered" on the IUCN Red List (Hedges et al., 2004aa).

Eleutherodactylus junori Dunn, 1926

Distribution: Central Jamaica; elevational range ~600–835 m.

Habitat: In banana plantations, females on ground in a *Musa* grove and an adjacent pasture (Schwartz and Fowler, 1973).

Abundance: Extremely rare, seen or heard on only a few occasions since described in 1926 (Hedges, 2004b; Stuart et al., 2008). **Reproduction:** Males call from superficial cavities in limestone rock and from exposed earth in a *Musa* grove, also from a crack in a cutbank (Schwartz and Fowler, 1973); call from holes in rocks (Dunn, 1926b); presumably breeds by direct development (Hedges, 2004b; Stuart et al., 2008). **Size:** Male SVL to 24.0 mm, female SVL to 27.7 mm (Schwartz and Fowler, 1973).

Conservation Status: "Critically endangered" on the IUCN Red List (Hedges, 2004b); threatened by habitat destruction, and females may not be able to locate calling males because of the much louder calls of introduced *E. johnstonei* (S. B. Hedges in Johnson, 1988).

Eleutherodactylus karlschmidti Grant, 1931

Distribution: Puerto Rico, eastern mountains and associated lowlands, isolated record from the western mountains; elevational range ~45–630 m.

Habitat: Aquatic, inhabiting boulder-strewn torrents in rainforest (Rivero, 1998); near waterfalls and the swifter parts of streams at all altitudes in the mountains (Drewry, 1970b); female in a large hollow log over a stream (Schwartz and Henderson, 1991).

Activity: Largely nocturnal (Drewry and Rand, 1983). **Behavior:** Retreats under rocks or into crevices (Rivero et al., 1963). **Diet:** Includes terrestrial (67.6% of 482 ingested "particles") and aquatic (32.4%) prey, 91% insects (44.5% dipterans, ~15% hymenopterans, ~15% trichopterans, 9.1% hemipterans, 7.1% coleopterans), 5.3% arachnids, 4.1% acarines, 1.5% crustaceans, and 0.82% mollusks (Rivero et al., 1963). **Disease:** Chytrid (*Batrachochytrium dendrobatidis*) infections (Hedges et al., 2004rr, 2008c; Joglar, 2005a; Burrowes et al., 2008; Stuart et al., 2008). **Reproduction:** Males call from before dusk to before dawn, with a major peak from after dusk to midnight (Drewry and Rand, 1983); call from boulders, water's edge, and under waterfalls (Schwartz and Henderson, 1991); gravid female (77.4 mm SVL) in September with 113 eggs (2.3–3.1 mm diameter), juveniles (20.3–28.3 mm SVL) in May, June, and September (Joglar, 1998); breeds by direct development (Hedges et al., 2004rr, 2008c; Stuart et al., 2008). **Size:** Male SVL to 58.6 mm, female SVL to 78.5 mm (Joglar, 1998).

Conservation Status: "Critically endangered" on the IUCN Red List, this species is possibly extinct (Hedges et al., 2004rr, 2008c; Stuart et al., 2008); despite intensive searches, it has not been observed since 1976 (J. W. Wiley in Johnson, 1988; Burrowes and Joglar, 1991; Burrowes, 2004; Hedges et al., 2004rr, 2008c; Joglar, 2005a; Joglar et al., 2007; Stuart et al., 2008; Hedges and Díaz, 2009); "endangered" (Moreno, 1991d).

Eleutherodactylus klinikowskii Schwartz, 1959

Distribution: Western Cuba, Sierra de los Órganos and Sierra del Rosario; elevational range ~70–200 m.

Habitat: Rocky areas (cliff faces, talus slopes, mouths of large caves, interiors of smaller caves), occasionally on the leaves of trees and shrubs (to 1.5 m) close to cliff faces (Schwartz, 1959d); in terrestrial bromeliads by day (Alonso Bosch and Rodríguez Gómez, 2003a); in leaf litter that accumulates at cave entrances, in the dry season under rocks on the banks of small riverbeds near mogotes (Díaz and Cádiz, 2008); never recorded outside forest habitats (Hedges and Díaz, 2004gg; Stuart et al., 2008).

Abundance: Moderately common in suitable habitat (Hedges and Díaz, 2004gg; Stuart et al., 2008). **Activity:** Largely nocturnal (Díaz and Cádiz, 2008); active by day in cave interiors, where frogs seek refuge in carpets of dead leaves (Schwartz, 1959d). **Behavior:** Shy at night and very wary by day (Schwartz, 1959d). **Reproduction:** Males call from solution holes, crevices, and other irregularities in rock faces, may call by day after heavy rains (Schwartz, 1959d); males occasionally call from vegetation at heights of 0.5–1.5 m above the ground (Díaz and Cádiz, 2008); eggs laid on the ground, breeds by direct development (Hedges and Díaz, 2004gg; Stuart et al., 2008); clutch size 8 (Hedges, in prep.). **Size:** SVL in males to 24.2 mm, in females to 26.7 mm (Schwartz, 1959d); SVL to 27 mm (Díaz and Cádiz, 2008).

Conservation Status: "Vulnerable" (Vales et al., 1998); "endangered" on the IUCN Red List (Hedges and Díaz, 2004gg).

Eleutherodactylus lamprotes Schwartz, 1973

Distribution: Hispaniola, Massif de la Hotte (Haiti); elevational range ~820–1,455 m.

Habitat: In a bromeliad in a cloud forest–coffee grove situation (Schwartz, 1973b); forested limestone hill, floodplain of the Riviere du Sud, in large, water-filled bromeliads at 975–1,200 m elevation (Franz and Cordier, 1986); not recorded from disturbed forests (Hedges and Thomas, 2004h; Stuart et al., 2008).

Abundance: Moderately common in suitable habitat (Hedges and Thomas, 2004h; Stuart et al., 2008). **Reproduction:** Male calling from a leaf at 1.2 m (Schwartz and Henderson, 1991); eggs laid in bromeliads, breeds by direct development (Hedges and Thomas, 2004h; Stuart et al., 2008). **Size:** Male SVL to at least 28.7 mm (Schwartz, 1973b).

Conservation Status: "Critically endangered" on the IUCN Red List (Hedges and Thomas, 2004h).

Eleutherodactylus leberi Schwartz, 1965

Distribution: Southeastern Cuba, northern foothills of the Sierra Maestra and north of Guantánamo; elevational range ~390–465 m.

Habitat: Mesic hardwood forest in river valleys, takes refuge by day between rocks (Rodríguez Gómez and Alonso Bosch, 2003a); mesic forests, including wet limestone forests (Hedges and Díaz, 2004hh; Stuart et al., 2008); secondary forests in karst areas, also in plantations (e.g., coffee, bananas), diurnal refuges in bromeliads and rock crevices (Díaz and Cádiz, 2008).

Abundance: Moderately common in suitable habitat (Hedges and Díaz, 2004hh; Stuart et al., 2008); common, but more frequently heard than seen (Díaz and Cádiz, 2008). **Reproduction:** Males call from piles of large rocks (to 2.0 m in diameter) overgrown with aroids, *Musa*, and climbing cacti (Schwartz, 1965d); males ascend vegetation to heights >4 m at dusk and call from large leaves and branches, call throughout much of the night (Díaz and Cádiz, 2008); breeds by direct development (Hedges and Díaz, 2004hh; Stuart et al., 2008). **Size:** Male SVL to 32.4 mm (Schwartz, 1965d); SVL to 34 mm (Díaz and Cádiz, 2008).

Conservation Status: "Vulnerable" (Vales et al., 1998); "endangered" on the IUCN Red List (Hedges and Díaz, 2004hh).

Eleutherodactylus lentus Cope, 1862

Distribution: U.S. and British Virgin Islands, possibly introduced in the latter; elevational range, sea level to ~10 m.

Habitat: Ground dwelling in more or less wooded, semixeric situations, under surface debris, bricks, and stones (Schwartz and Henderson, 1991)

Behavior: Thought to be mute (e.g., Schwartz and Henderson, 1991), but found calling on Jost Van Dyke, British Virgin Islands (G. Perry, pers. comm.). **Foraging and Diet:** Average observed feeding rate 2.6 prey/h, average movement rate 2.4 moves/h (Jones, 1982). **Reproduction:** Eggs laid on the ground, breeds by direct development (Hedges and Thomas, 2004i; Stuart et al., 2008). **Size:** SVL to 38 mm (Schwartz and Henderson, 1991).

Conservation Status: "Endangered" on the IUCN Red List (Hedges and Thomas, 2004i); "stable" (Moreno, 1991d).

Eleutherodactylus leoncei Shreve and Williams, 1963

Distribution: Hispaniola, Sierra de Baoruco (Dominican Republic) and Massif de la Selle (Haiti); elevational range ~1,180–2,300 m.

Habitat: Under rocks and logs in upland pine (*Pinus occidentalis*) and hardwood forests by day, one crossing a road in pine forest at night (Hedges, 1992); under stones, on roadside banks, and in low vegetation (Schwartz and Henderson, 1991); intolerant of disturbed habitats (Hedges et al., 2004bb; Stuart et al., 2008).

Reproduction: Males call from the ground, eggs laid on the ground, breeds by direct development (Hedges et al., 2004bb; Stuart et al., 2008); clutch size 32 (Hedges, in prep.). **Size:** Male SVL to 26.1 mm, female SVL to 42 mm (Hedges, 1992).

Conservation Status: "Critically endangered" on the IUCN Red List (Hedges et al., 2004bb).

Eleutherodactylus limbatus Cope, 1862

Distribution: Cuba, widely distributed but disjunct; elevational range, sea level to 1,150 m.

Habitat: Mesic forests, by day under stones and damp leaf litter on limestone substrates (Barbour, 1914; Valdes, 1989); in rocky patches of ground covered with tangled vegetation and taller trees, may congregate under rocks (10 under one rock) in dry streambeds during the dry season, 100 collected in one day, also under a rotten fallen bromeliad (Dunn, 1926c); under loose stones in open forest (R. Piechocki in G. Peters, 1974); also under logs and, in drier situations, under *Agave* leaves (Valdes, 1989).

Abundance: Common in suitable habitat, but particularly susceptible to habitat disturbances (Hedges and Díaz, 2004ii; Stuart et al., 2008). **Activity:** Nocturnal, becomes active at dusk (Schwartz and Henderson, 1991). **Behavior:** When disturbed, may quickly "disappear" (Barbour, 1914); moves slowly and deliberately, but jumps quickly when disturbed (Díaz and Cádiz, 2008). **Foraging and Diet:** Diet includes ants and small beetles (Ruiz García, 1987); will forage in leaf litter during the morning (Rodríguez Gómez and Alonso Bosch, 2003b). **Reproduction:** Males may call by day (Díaz and Cádiz, 2008); call from about 0.3 m above ground level on plants or rocks (Dunn, 1926c); commonly heard calling from the ground (Hedges and Díaz, 2004ii; Stuart et al., 2008); males compete vocally and physically, sometimes wrestling with each other (Rodríguez Gómez and Alonso Bosch, 2003b); clutch size 1, deposited beneath ground litter (Dunn, 1926c; Schwartz and Henderson, 1991); egg on 4 August (Dunn, 1926c). **Size:** Male SVL to 12.2 mm (Rodríguez Gómez and Alonso Bosch, 2003b); to 17 mm (Schwartz and Henderson, 1991); SVL to 13 mm (Díaz and Cádiz, 2008).

Conservation Status: "Vulnerable" on the IUCN Red List (Hedges and Díaz, 2004ii).

Eleutherodactylus limbensis Lynn, 1958

Distribution: Hispaniola, Massif du Nord (Haiti); elevational range, near sea level to >1,000 m.

Habitat: The moist bank of the Rivière du Limbé (Lynn, 1958).

Size: Male SVL to 35 mm, female SVL to 43 mm (Schwartz, 1970c).

Conservation Status: Until recently, this species was considered a subspecies of *E. schmidti*, which is considered "critically endangered" (Hedges et al., 2004u); we assume this species is, at best, also critically endangered.

Eleutherodactylus locustus Schmidt, 1920

Distribution: Puerto Rico, interior uplands; elevational range ~270–1,050 m.

Habitat: Upland mesic broadleaf forest, where it seems to prefer openings in forest (Schwartz, 1969a); mossy forest meadows and forest margins (Drewry and Rand, 1983); along trails and roads (Rivero, 1998); <5% in bromeliads, under but not in petioles (Cintrón, 1970); in dense but well-lighted thickets and fern banks at elevations above 450 m (Drewry, 1970b); diurnally under leaf litter, trunks, roots, and trash on ground (Rivero, 1998).

Activity: Largely nocturnal (Drewry and Rand, 1983). **Diet:** Adult diet includes entognathans (springtails), insects (orthopterans, coleopterans, psocopterans, dipterans, hymenopterans, lepidopterans) and arachnids (spiders), juvenile diets include entognathans and insects (coleopterans, dipterans, hymenopterans; Lavigne and Drewry, 1970). **Disease:** *Batrachochytrium dendrobatidis* fungal infections (Hedges et al., 2004ss; Burrowes et al., 2008; Hedges and Rios-López, 2008c; Stuart et al., 2008). **Parasites:** Nematode *Aplectana* sp. (Rivero, 1998). **Population Size:** Considered very low, individual population sizes ranging from 20 to >50 (but <100; Joglar, 1998). **Reproduction:** Males call from leaf surface (K. P. Schmidt, 1928); from vines and bromeliads (Hedges, in prep.); also from low bushes, grasses, and ferns (from ground level to 1.5 m above the ground; Drewry and Rand, 1983); may ascend to 4.5–6.0 m along mossy trunks (Rivero, 1998); call from before dusk to after dawn (Drewry and Rand, 1983); gravid females (21.8–28.9 mm SVL) with 6–10 eggs (2.0–2.8 mm diameter), gravid females in March and October at Bosque de Carite (Joglar, 1998); egg clutches in axils of bromeliads (Joglar, 2005a); eggs laid on the ground, breeds by direct development (Hedges et al., 2004ss; Hedges and Rios-López, 2008c; Stuart et al., 2008). **Size:** Male SVL to 22.7 mm, female SVL to 23.5 mm (Joglar, 1998).

Conservation Status: "Critically endangered" on the IUCN Red List, some populations extirpated (Hedges et al., 2004ss; Hedges and Rios-López, 2008c; Stuart et al., 2008); populations have not recolonized some areas at El Yunque where they were abundant in the past (Burrowes and Joglar, 2005; Joglar et al., 2007); "decreasing in some areas" (Moreno, 1991d).

Eleutherodactylus lucioi Schwartz, 1980

Distribution: Hispaniola, extreme northwestern Haiti; elevation ~100 m.

Habitat: In a ravine on a rock 0.7 m above a stream (1–2 m wide, shallow and rock-strewn) and projecting from the bank in remnant riparian forest (Schwartz, 1980c).

Size: SVL to 24 mm, based on one specimen (Schwartz, 1980c).

Conservation Status: "Critically endangered" on the IUCN Red List, not recorded since first described (Hedges and Thomas, 2004j; Stuart et al., 2008).

Eleutherodactylus luteolus Gosse, 1851

Distribution: Western Jamaica; elevational range, sea level to ~680 m.

Habitat: Wet limestone forest, dry scrub forest, and coastal areas, although associated with closed-canopy forest, can tolerate some habitat degradation (Hedges et al., 2004jjj; Stuart et al., 2008); in moist *Coccoloba* litter, in a terrestrial bromeliad (Schwartz and Fowler, 1973); on the leaves of bushes at heights of 0.3–1.0 m (Lynn and Dent, 1942); in palm trash within 15 m of the beach (Goin, 1953); by day beneath stones (Lynn, 1940).

Reproduction: Males call from the ground, rocks, or low vegetation, eggs in June, females with "ripe ovarian eggs" 18 June–28 July (Lynn, 1940); eggs laid in depressions in damp ground under stones or logs, clutch size 30–35 (Barbour, 1910); may be associated with bromeliads, breeds by direct development (Hedges et al., 2004jjj; Stuart et al., 2008); clutch size 34–37 (Hedges, in prep.). **Size:** Male SVL to 17 mm, female SVL to 25 mm (Hedges, in prep.).

Conservation Status: "Endangered" on the IUCN Red List (Hedges et al., 2004jjj).

Eleutherodactylus maestrensis Díaz, Cádiz, and Navarro, 2005

Distribution: Eastern Cuba, Sierra Maestra; elevational range ~900–1,640 m.

Habitat: Terrestrial in leaf litter in mountain rainforests and patches of pine (*Pinus maestrensis*), most individuals under leaf litter, one under a streamside stone (Díaz et al., 2005a; Díaz and Cádiz, 2008).

Abundance: A rare species (Díaz and Cádiz, 2008). **Behavior:** Moves slowly and deliberately, but jumps quickly when disturbed (Díaz and Cádiz, 2008). **Diet:** Cockroaches (*Epilampra* spp.), isopods, and firefly larvae (Díaz et al., 2005a). **Reproduction:** Unaware of vocalization (Díaz and Cádiz, 2008); inguinal amplexus (Díaz et al., 2005a); presumed to breed by direct development (Stuart, 2006). **Size:** Male SVL to 19.1 mm, female SVL to 32.4 mm (Díaz et al., 2005a).

Conservation Status: On the IUCN Red List as "data deficient" (Stuart, 2006); the species' recent discovery and apparently restricted distribution suggest that protective status may be warranted.

Eleutherodactylus mariposa Hedges, Estrada, and Thomas, 1992

Distribution: Eastern Cuba, Meseta del Guaso; elevation ~720 m.

Habitat: Arboreal in mesic closed-canopy forest (Hedges and Díaz, 2004jj; Stuart et al., 2008); thick, secondary growth adjacent to a coffee grove (Hedges et al., 1992); semideciduous forests in karst regions, diurnal retreats in rock fissures (Díaz and Cádiz, 2008).

Abundance: Moderately common at only known location (Hedges and Díaz, 2004jj; Stuart et al., 2008); difficult to observe, but frequently heard when weather is favorable (Díaz and Cádiz, 2008). **Reproduction:** Males call from horizontal perches 2–3 m high on vines and small trees growing on jagged limestone rock and from vertical positions on a large tree trunk and from within folds of a dead leaf at 3.0 m above ground (Hedges et al., 1992); calling begins at dusk, with males calling from leaves and branches of trees and bushes at heights of 60 cm to >4 m (Díaz and Cádiz, 2008); breeds by direct development (Hedges and Díaz, 2004jj; Stuart et al., 2008). **Size:** Male SVL to 36.1 mm, females unknown (Hedges et al., 1992).

Conservation Status: "Vulnerable" (Vales et al., 1998); "critically endangered" on the IUCN Red List (Hedges and Díaz, 2004jj).

Eleutherodactylus martinicensis Tschudi, 1838

Distribution: St.-Barthélemy, Antigua, Guadeloupean Archipelago (Guadeloupe, Les Saintes, La Désirade, Iles de Petite Terre, Marie-Galante), Dominica, Martinique; populations on St.-Martin/St. Maarten (?), St.-Barthélemy probably introduced, those on at least parts of Guadeloupe and Dominica possibly introduced; elevational range, sea level to at least 1,250 m.

Habitat: By day in the petiole bases of banana leaves, in bromeliads (to 11 m above the ground), under loggers' chips, palm trash, cacao husk piles, among fallen bricks and rocks of ruins (Schwartz, 1967b); on Antigua, beneath stones and coconut trash on the beach (Lynn, 1957); on Guadeloupe, mesic and xeric forests, savannas, in bromeliads, also in disturbed and artificial situations, including banana and coconut groves, sugarcane fields, orchards, gardens, inside houses (washrooms, kitchens), and greenhouses, by day in corners where damp, in *Cocos* trash piles, under rocks and logs, in banana leaves and in flower pots, high on Soufrière, associated with plants *Clusia* sp. and *Cyathea* sp. (Breuil, 2002b); on Terre-de-Haut (Les Saintes), under rocks in a pasture near a freshwater pond (Schwartz, 1967b); on Marie-Galante, in dry scrub under a rock (Schwartz, 1967b); on Dominica, in rainforest, dry woodlands, cacao and banana groves, in rock pile in stand of *Coccoloba*, and on large rocks and boulders in montane streams, at night, retreated into a hole in moss- and liverwort-covered roadside cutbank (Schwartz, 1967b); abundance of frogs not related to distance from water (Martin, 2001); on Dominica, in both natural and extensively altered habitats (Ackley et al., 2009).

Abundance: Very common, especially in undisturbed habitats (Hedges et al., 2004e; Stuart et al., 2008). **Activity:** Nocturnal (RWH); in rainforest at top of Pic Napoléon (Soufrière, Guadeloupe) with ambient temperature 14.5 °C (Schwartz and Henderson, 1991); activity increased significantly with increased relative humidity (Wolman, 2002). **Behavior:** Observed escaping into a crab hole in mud adjacent to the ocean (Schwartz and Henderson, 1991). **Foraging and Diet:** Hunts at night on walls near lights that attract insects (Breuil, 2002b); readily accepted ants in captivity, but refused termites (Bayless, 2005). **Longevity:** In captivity, 7 years and 2 months (Breuil, 2002b). **Parasites:** *Porrocaecum* larvae and oligacanthorhynchid cystacanths on Dominica and Guadeloupe, plerocercoid cysts on Dominica only (Goldberg et al., 1998d); chytridiomycosis has been documented (Daniells et al., 2008). **Predators:** Centipedes (*Scolopendra*) on Marie-Galante (Breuil, 2002b); snakes (*Alsophis sibonius* on Dominica and *A. sanctonum* on Terre-de-Bas; Henderson and Sajdak, 1996); Guadeloupe Woodpecker (*Melanerpes herminieri*; Villard, 1999). **Reproduction:** Males call (often with head angled toward the ground) from small mounds of dirt, leaves, branches (rarely >2.0 m above the ground), will call by day in wet forest, will call all night (Breuil, 2002b); also call from herbaceous growth, moss, wet earth, mud, shrubs, arums, cacao and banana trunks and low leaves, *Heliconia*, and ferns (usually within 0.6 m of the ground), choruses can be "quite loud" along stream courses in Dominica (Schwartz, 1967b); on Dominica, eggs and hatchlings among dead leaves of a large bromeliad (17 March), clutch of 29 eggs under a piece of palm fiber in a *Cocos* grove on wet earth on 13 March, hatched 21 March, another clutch of 40 eggs in a rolled palm petiole on the rainforest floor on 13 March that hatched 23 March (Schwartz and Henderson, 1991). **Size:** Male SVL to 32 mm, female SVL to 47 mm (Kaiser and Hardy, 1994b).

Conservation Status: "Near threatened" on the IUCN Red List (Hedges et al., 2004e).

Additional Pertinent Literature: Kaiser and Hardy (1994b).

Eleutherodactylus melacara Hedges, Estrada, and Thomas, 1992

Distribution: Eastern Cuba, Sierra Maestra; elevational range ~845–1,975 m.

Habitat: Arboreal in closed moist forest (Hedges and Díaz, 2004kk; Stuart et al., 2008); cloud forest (some undisturbed, others partially cutover) on Pico Turquino, by day in bromeliads 2–4 m above the ground (Hedges et al., 1992); in bromeliads to heights >12 m in rainforest and cloud forest, juveniles in low forest vegetation (Díaz and Cádiz, 2008).

Abundance: Moderately common in suitable habitat (Hedges and Díaz, 2004kk; Stuart et al., 2008). **Activity:** Nocturnal (Díaz and Cádiz, 2008). **Reproduction:** Large eggs and hatchlings in bromeliads, clutch sizes likely small (5–10), calling from bromeliads 4.0+ m high (Hedges et al., 1992); calls sporadically all night long, occasionally heard calling by day (Díaz et al., 2005a); calls from bromeliads or adjacent leaves, eggs in bromeliads (Díaz and Cádiz, 2008); breeds by direct development (Hedges and Díaz, 2004kk; Stuart et al., 2008). **Size:** Male SVL to 28.7 mm, female SVL to 35.6 mm (Hedges et al., 1992); to ~36 mm (Díaz and Cádiz, 2008).

Conservation Status: "Endangered" on the IUCN Red List (Hedges and Díaz, 2004kk).

Eleutherodactylus melatrigonum Schwartz, 1966

Distribution: Hispaniola, Cordillera Central (Dominican Republic); elevation ~1,525 m.

Habitat and Natural History: Unknown.

Size: Male SVL to 20.3 mm (Schwartz, 1966).

Conservation Status: Until recently this species was considered a subspecies of *E. audanti*, which is considered "vulnerable" (Hedges et al., 2004o); we assume this species is, at best, also vulnerable.

Additional Pertinent Literature: Schwartz (1979c).

Eleutherodactylus michaelschmidi Díaz, Cádiz, and Navarro, 2007

Distribution: Southeastern Cuba, northern slope of the Sierra Maestra; elevational range ~160–410 m.

Habitat: Limestone mountains and premontane valleys covered by secondary vegetation, remnants of deciduous forest, and coffee plantations where on rocky surfaces, in cave entrances, and on fallen tree trunks (Díaz et al., 2007b; Díaz and Cádiz, 2008).

Abundance: Common (Díaz and Cádiz, 2008). **Activity:** Nocturnal, active 2100–0100 h (Díaz et al., 2007b). **Diet:** Includes ants (*Camponotus* sp., *Odontomachus* sp.), cockroaches, crickets, beetles (including the carabid *Aspidoglossa mexicana*), an opilionid, and acari (Díaz et al.,

2007b). **Reproduction:** Males call from rocks, less frequently from fallen logs and other objects (Díaz and Cádiz, 2008). **Size:** Male SVL to 44.0 mm, female SVL to 48.1 mm (Díaz et al., 2007b).

Conservation Status: "Endangered" on the IUCN Red List (Díaz, 2008a).

Eleutherodactylus minutus Noble, 1923

Distribution: Hispaniola, Cordillera Central (Dominican Republic); elevational range ~880–2,300 m (and possibly higher).

Habitat: Mesic upland broadleaf forest and, occasionally, pine forest (Schwartz and Henderson, 1991).

Abundance: Moderately common in suitable habitat (Hedges et al., 2004s; Stuart et al., 2008). **Reproduction:** Males call from herbaceous plants and terrestrial bromeliads, not more than 1.0 m above the ground (Schwartz and Henderson, 1991); eggs laid on the ground, breeds by direct development (Hedges et al., 2004s; Stuart et al., 2008). **Size:** SVL to 19 mm (Schwartz and Henderson, 1991).

Conservation Status: "Endangered" on the IUCN Red List (Hedges et al., 2004s).

Eleutherodactylus monensis Meerwarth, 1901

Distribution: Isla Mona; elevational range, sea level to ~10 m.

Habitat: Dry scrub forest (Hedges et al., 2004tt; Angulo, 2008e; Stuart et al., 2008); by day and night on walls of shallow caves containing water (Rivero, 1998); in sinkholes, under galvanized sheets for covering water reservoirs (Schwartz and Henderson, 1991); by day and night in and under leaf litter, occasionally in bromeliads and on *Ipomoea pescaprae* (Joglar, 1998).

Abundance: Abundant in its restricted range (Hedges et al., 20004tt; Angulo, 2008e; Stuart et al., 2008). **Reproduction:** Males call from rocks (Joglar, 1998); amplexus in June (Rivero, 1998); gravid females (SVL 37.8–40.3 mm) in July and October (Joglar, 1998); eggs laid on the ground, breeds by direct development (Hedges et al., 20004tt; Angulo, 2008e; Stuart et al., 2008). **Size:** Male SVL to 33.1, female SVL to 49.4 (Joglar, 1998).

Conservation Status: "Vulnerable" on the IUCN Red List (Hedges et al., 20004tt; Angulo, 2008e); "stable" (Moreno, 1991d).

Eleutherodactylus montanus Schmidt, 1919

Distribution: Hispaniola, Cordillera Central (Dominican Republic); elevational range ~1,270–2,425 m (and possibly higher).

Habitat: Broadleaf and pine forests, primarily in fern understory and cutover broadleaf forest, diurnal retreats under logging debris, around old home sites, under rocks and logs (Schwartz and Henderson, 1991); closed-canopy forest and forest remnants (Hedges et al., 2004t; Stuart et al., 2008).

Abundance: Common in suitable habitats at high elevations (Hedges et al., 2004t; Stuart et al., 2008). **Reproduction:** Males call from tall (0.65–2.4 m) herbaceous and arboreal vegetation, including ferns (*Pteris*), occasionally from the ground, males guard eggs (Schwartz and Henderson, 1991); eggs laid on the ground, breeds by direct development (Hedges et al., 2004t; Stuart et al., 2008); clutch size 1 (Hedges, in prep.). **Size:** Male SVL to 44 mm, female SVL to 43 mm (Schwartz and Henderson, 1991).

Conservation Status: "Endangered" on the IUCN Red List (Hedges et al., 2004t).

Eleutherodactylus nortoni Schwartz, 1976

Distribution: Hispaniola, "South Island," Massif de la Hotte and Massif de la Selle (Haiti) and extreme western Sierra de Baoruco (Dominican Republic); elevational range ~575–1,515 m.

Habitat: Deciduous forest, forest remnants, coffee groves, overgrown ravines (Schwartz, 1976c); upland broadleaf forest (Schwartz and Henderson, 1991); common in sinkholes to 97 m below the surface, among rocks and organic debris on the sink bottom, also on vertical walls of sinks (Franz and Cordier, 1986); not recorded from disturbed habitats (Hedges et al., 2004cc; Stuart et al., 2008).

Reproduction: Males call (night and day) from 1.5–3.1 m above the ground on vines, tree branches (including *Coffea*), and shrubs, also from open and closed cavities in limestone (Schwartz, 1976c); eggs laid on the ground, breeds by direct development (Hedges et al., 2004cc; Stuart et al., 2008); clutch size 19–24 (Hedges, in prep.). **Size:** Male SVL to 65.7 mm (Schwartz, 1976c).

Conservation Status: "Critically endangered" on the IUCN Red List, last seen in the Dominican Republic in 1998, but no subsequent surveys have been conducted (Hedges et al., 2004cc; Stuart et al., 2008).

Eleutherodactylus notidodes Schwartz, 1966

Distribution: Hispaniola, Sierra de Neiba (Dominican Republic); elevational range 1,340–1,830 m.

Habitat: Under rocks (Schwartz, 1979c).

Reproduction: Eight egg clutches accompanied by one or more adults under rocks in a roadbed and along the roadside on 21 July in the Sierra de Neiba after a heavy rain, some eggs embryonated and hatched upon handling, others showed no signs of development (Schwartz, 1979c); 22 eggs accompanied by one adult under a rock on 11 August, hatched 20 August (Schwartz and Henderson, 1991); clutch size 28–34 (Hedges, in prep.). **Size:** SVL in males to 21.9 mm (Schwartz, 1966).

Conservation Status: Until recently this species was considered a subspecies of *E. audanti*, which is considered "vulnerable" (Hedges et al., 2004o); we assume this species is, at best, also vulnerable.

Additional Pertinent Literature: Schwartz (1979c).

Eleutherodactylus nubicola Dunn, 1926

Distribution: Jamaica, Blue Mountains; elevational range ~1,060–1,880 m.

Habitat: Terrestrial in dense closed-canopy deciduous, cloud, and elfin forests, intolerant of disturbed habitats (Hedges et al., 2004kkk; Stuart et al., 2008); by day under rocks along mountain trails (Lynn, 1940); on the ground, on monocot leaves (0.65–1.0 m above ground), under leaves adjacent to a drainage ditch, 10 individuals (along with 22 *E. pantoni*, 10 *E. cundalli*, and one *E. gossei*) in a roadside pile (~3.5 × 2.5 × 1.2 m) of short wet logs (Schwartz and Fowler, 1973).

Abundance: Locally common in suitable habitat (Hedges et al., 2004kkk; Stuart et al., 2008). **Activity:** Nocturnal (Lynn, 1940). **Diet:** Includes (at Hardwar Gap) ants (54.3% of total prey items), hemipteran nymphs (15.7%), beetles (7.4%), and arachnids (5.7%; Goin and Cooper, 1950). **Parasites:** Ticks in axilla or groin or on hind legs (Lynn, 1940). **Reproduction:** Males call from atop rocks, trailside banks, and mouths of burrows in banks (Lynn, 1940); also from bases of grass clumps, shallow hollows under roots, depressions beneath sparse litter, eggs deposited in small hollows beneath rocks (Goin and Cooper, 1950); nests contain 26–75 eggs (Hedges, in prep.); egg diameter 4.0 mm, development requires one month (Lynn, 1940); 12 clutches from 1 July to 2 August (Lynn and Dent, 1943); females attend nests (Lynn, 1940; Townsend, 1996); hatchling 6.0 mm SVL (Goin and Cooper, 1950). **Size:** Male SVL to 35 mm (Lynn, 1940), female SVL to 49 mm (Hedges, in prep.).

Conservation Status: "Endangered" on the IUCN Red List (Hedges et al., 2004kkk).

Eleutherodactylus olibrus Schwartz, 1958

Distribution: Western Cuba, Sierra de los Órganos.

Habitat: Arboreal (Schwartz, 1958d); seeks refuge in and calls from Traveler's Palms (*Ravenala madagascariensis*), also in ornamental terrestrial bromeliads (Díaz and Cádiz, 2008).

Abundance: Common, but easier to hear than to see (Díaz and Cádiz, 2008). **Reproduction:** Clutch size 9 (Hedges, in prep.). **Size:** Male SVL to 28.2 mm (Schwartz, 1958d); to 31 mm SVL (Díaz and Cádiz, 2008).

Conservation Status: Until recently this species was considered a subspecies of *E. varians*, which is considered "vulnerable" (Hedges and Díaz, 2004zz); we assume this species is, at best, also vulnerable.

Eleutherodactylus orcutti Dunn, 1928

Distribution: Eastern Jamaica; elevational range ~225–1,215 m.

Habitat: Riparian (Hedges and Díaz, 2009); streams in mesic, wooded areas, on boulders and rocks in small mountain streams (Schwartz and Fowler, 1973); on leaves, stems, and rocks within 1.0 m of and directly over water, by day under rocks and logs (Schwartz and Henderson, 1991); sits in spray from waterfalls (Lynn, 1940).

Behavior: Escapes by jumping into stream and swimming under aquatic plants or detritus dams (Schwartz and Henderson, 1991). **Reproduction:** Males call from rocks at the base of a small waterfall, from floating debris in pools (Lynn, 1940); also from seepage in forest (Schwartz and Fowler, 1973); females with large ovarian eggs 18 June–27 July (Lynn, 1940); possibly live-bearing, but unconfirmed (Hedges, 2004c; Stuart et al., 2008). **Size:** Male SVL to 33.7 mm, female SVL to 38.4 mm (Schwartz and Fowler, 1973).

Conservation Status: "Critically endangered" on the IUCN Red List, possibly extinct, since not observed since 1985 despite extensive efforts (Hedges, 2004c; Stuart et al., 2008; Hedges and Díaz, 2009).

Eleutherodactylus orientalis Barbour and Shreve, 1937

Distribution: Southeastern Cuba, known only from the Yunque de Baracoa; elevation ~300 m.

Habitat: Undisturbed moist forest (Hedges and Díaz, 2004ll; Stuart et al., 2008); in leaf litter in forests and in coffee and cacao groves that have not been thinned (Rodríguez Gómez and Alonso Bosch, 2003b).

Abundance: Moderately common in suitable habitat (Hedges and Díaz, 2004ll; Stuart et al., 2008). **Activity:** Diurnal (Díaz and Cádiz, 2008). **Reproduction:** Breeds by direct development (Hedges and Díaz, 2004ll; Stuart et al., 2008). **Size:** SVL to about 13 mm (Barbour and Shreve, 1937); SVL to 17 mm (Díaz and Cádiz, 2008).

Conservation Status: "Critically endangered" on the IUCN Red List (Hedges and Díaz, 2004ll).

Eleutherodactylus oxyrhyncus Duméril and Bibron, 1841

Distribution: Hispaniola, Massif de la Hotte and Massif de al Selle (Haiti); elevational range ~335–1,210 m.

Habitat: Under rocks in upland disturbed broadleaf forest (Schwartz and Henderson, 1991); in upland primary forest, although collected at forest edge, this is unlikely to be suitable habitat (Hedges and Thomas, 2004k; Stuart et al., 2008).

Abundance: Moderately common in suitable habitat, which has largely disappeared (Hedges and Thomas, 2004k; Stuart et al., 2008). **Predators:** *Ialtris haetianus* (Henderson and Schwartz, 1986). **Reproduction:** Eggs laid on the ground, breeds by direct development (Hedges and Thomas, 2004k; Stuart et al., 2008). **Size:** Male SVL to 28 mm, female SVL to 55 mm (Schwartz and Henderson, 1991).

Conservation Status: "Critically endangered" on the IUCN Red List (Hedges and Thomas, 2004k).

Eleutherodactylus pantoni Dunn, 1926

Distribution: Jamaica, two currently recognized subspecies occur in central Jamaica (*E. p. pantoni*) and western Jamaica (*E. p. amiantus*); elevational range, sea level to ~1,640 m.

Habitat: Forests and more open situations (Dunn, 1926b); beneath stones and in holes in the ground, with some "small burrows occupied night after night, presumably by the same individual" (Lynn, 1940); *E. p. pantoni* on a mossy bank and mossy rocks in a stream, on an open road cut, and under rocks in an open field (Schwartz and Fowler, 1973); *E. p. amiantus* in a terrestrial bromeliad, on a cutbank and in a pasture, on rocks and the ground in cleared woodland, intergrades between *E. p. pantoni* and *E. p. amiantus* on the forest floor, in terrestrial bromeliads, on cliff faces, on the ground in a banana grove, and in a pasture (Schwartz and Fowler, 1973); can tolerate moderate habitat alterations (Hedges et al., 2004lll; Stuart et al., 2008).

Diet: Includes beetles, flies, and *Sphaerodactylus goniorhynchus* (Lynn, 1940); also ants (55.3% of all prey items), winged ants (16.5%), beetles (6.8%), and arachnids (5.8%; Schwartz and Henderson, 1991). **Reproduction:** Males call from the ground or holes in limestone rocks (Dunn, 1926b); *E. p. pantoni* calling from 0.6 m high on a small tree stub, a wet roadside stack of logs, along a rainforest trail, and on road cuts, *E. p. amiantus* calls from limestone rock surfaces, intergrades between *E. p. pantoni* and *E. p. amiantus* call from small depressions in limestone rocks, under wet, fallen bark at the entrance to Windsor Great Cave (Schwartz and Fowler, 1973); gravid females 18 June–29 July, nests with 104 nonadherent eggs on 20 June (Lynn, 1940); 78 and 86 eggs (Lynn and Dent, 1943); eggs laid on the ground, breeds by direct development (Hedges et al., 2004lll; Stuart et al., 2008); clutch size 51 (Hedges, in prep.); females attend nests (Lynn, 1940). **Size:** Male SVL to 38.0 mm, female SVL to 43.6 mm (Schwartz and Fowler, 1973).

Conservation Status: "Near threatened" on the IUCN Red List (Hedges et al., 2004lll).

Eleutherodactylus parabates Schwartz, 1964

Distribution: Hispaniola, Sierra de Neiba (Dominican Republic); elevational range ~1,450–1,870 m.

Habitat: Under rocks and wood adjacent to a road through dense rainforest (hardwoods and *Cubanola dominguensis*), in the vicinity were low areas supporting small marshes with cattails (Schwartz, 1964b); also in low (1.0–2.0 m) bromeliads (Schwartz and Henderson, 1991).

Abundance: Moderately common in suitable habitat, although this habitat is declining rapidly (Hedges et al., 2004hh; Stuart et al., 2008). **Reproduction:** Males call from arboreal sites 0.3–1.0 m above ground (Schwartz and Henderson, 1991); eggs laid on the ground, breeds by direct development (Hedges et al., 2004hh; Stuart et al., 2008); clutch size 30–37 (Hedges, in prep.). **Size:** SVL to 24.1 mm (Schwartz, 1964b).

Conservation Status: "Critically endangered" on the IUCN Red List (Hedges et al., 2004hh).

Eleutherodactylus paralius Schwartz, 1976

Distribution: Hispaniola, southeastern Dominican Republic.

Habitat: Deep, rocky, well-forested mesic ravine that cuts through cultivated sugarcane fields, under trash pile behind exposed beach, along paths at night in grassy riverside pasture and on rocks adjacent to a gravel road, in *Cocos* trash piles at the edge of a coastal palm grove, beneath large rock in a moist pile of dead grass near the ocean, in stands of *Acacia* woods, often associated with limestone regions, at night on bare ground (Schwartz, 1976b); in a nearly dry fountain in Santo Domingo, in a swimming pool near La Romana (RP).

Reproduction: Males call from caves, large rocks, herbaceous growth, and shrubs (*Zamia* sp.) to ~1.0 m above ground, call erratic (Schwartz, 1976b). **Size:** Male SVL to 29.5 mm, female SVL to 40.4 mm (Schwartz, 1976b).

Conservation Status: Until recently this species was considered a subspecies of *E. weinlandi*, which is considered a species of "least concern" (Hedges et al., 2004ii); we assume this species also is of least concern.

Eleutherodactylus parapelates Hedges and Thomas, 1987

Distribution: Hispaniola, Massif de la Hotte (Haiti); elevational range ~950–1,050 m.

Habitat: Fossorial species in closed forest (Hedges et al., 2004bbbb; Stuart et al., 2008); upland broadleaf forest, forest remnants, and coffee groves, one nonvocal male on herbaceous vegetation 1.0 m above ground along a road cut (Hedges and Thomas, 1987).

Abundance: Moderately common in original habitat, which has largely disappeared (Hedges et al., 2004bbbb; Stuart et al., 2008). **Reproduction:** Males call from shallow, underground chambers, most of which have no evident external opening, one male called from open cavity beneath a mat of dead vegetation, all calling (November) in early morning (0100–0500 h), most in a relatively small area with the distance between individuals about 1.0 m (Hedges and Thomas, 1987); eggs laid underground, breeds by direct development (Hedges et al., 2004bbbb; Stuart et al., 2008). **Size:** SVL in males to 52 mm (Hedges and Thomas, 1987).

Conservation Status: "Critically endangered" on the IUCN Red List (Hedges et al., 2004bbbb).

Additional Pertinent Literature: Hedges and Powell (1998b).

Eleutherodactylus patriciae Schwartz, 1964

Distribution: Hispaniola, Cordillera Central (Dominican Republic); elevational range ~2,000–3,050 m.

Habitat: Upland pine forest, under logs, rocks, boards, sod, and human debris (Schwartz and Henderson, 1991); terrestrial in closed forest and forest remnants, not recorded from disturbed habitats (Hedges et al., 2004dd; Stuart et al., 2008).

Abundance: Moderately common in suitable habitat (Hedges et al., 2004dd; Stuart et al., 2008). **Disease:** Chytridiomycosis (Joglar and Burrowes, 2005; Joglar et al., 2007; Hedges et al., 2004dd; Stuart et al., 2008). **Predator:** Scorpions (*Tityus quisqueyanus*; Armas and Abud, 1992; Armas, 2001). **Reproduction:** Males call from rocks, ferns, and bushes to 1.0 m above ground, pair in "amplexus" (25 July, 1600 h) in a small cavity in the ground with the male still calling (male's legs on female's shoulders and extending over arms), eggs 18–24/clutch (25 July) under rocks and sod (Schwartz and Henderson, 1991); breeds by direct development (Hedges et al., 2004dd; Stuart et al., 2008). **Size:** Male SVL to 29 mm, female SVL to 35 mm (Schwartz, 1964b).

Conservation Status: "Endangered" on the IUCN Red List (Hedges et al., 2004dd).

Eleutherodactylus paulsoni Schwartz, 1964

Distribution: Hispaniola, Tiburon Peninsula (Haiti); elevational range, sea level to ~750 m.

Habitat: Terrestrial in closed forest (Hedges and Thomas, 2004l; Stuart et al., 2008); in leaf litter under large trees along the bank of an intermittent river in thorn scrub, at a cave entrance, and on the ground among dead leaves in grass about 2.0 m high (Schwartz, 1964d).

Reproduction: Eggs laid on the ground, breeds by direct development (Hedges and Thomas, 2004l; Stuart et al., 2008); clutch size 11 (Hedges, in prep.). **Size:** SVL to 26 mm (Schwartz, 1964d).

Conservation Status: "Critically endangered" on the IUCN Red List, common in the 1960s, encountered at only two locations since, last seen in 1991, but no subsequent surveys have been conducted (Hedges and Thomas, 2004l; Stuart et al., 2008).

Eleutherodactylus pentasyringos Schwartz and Fowler, 1973

Distribution: Eastern Jamaica; elevational range, sea level to ~1,275 m.

Habitat: Wet limestone forest, rainforest, cloud forest, under rocks and logs by day (Hedges, 2004d; Stuart et al., 2008); overgrown *Musa* grove, on the ground and on and under rocks adjacent to a stream (Schwartz and Fowler, 1973).

Abundance: Fairly common and more abundant than some other species of frogs in the Blue Mountains (Hedges, 2004d; Stuart et al., 2008). **Reproduction:** Males call from palm axils and on the base of tree fern at 0.5 m above the ground (Schwartz and Fowler, 1973); 78–104 eggs attended by a female (Lynn and Dent, 1943); eggs laid on the ground, breeds by direct development (Hedges, 2004d; Stuart et al., 2008). **Size:** Male SVL to 32.3 mm, female SVL to 39.3 mm (Schwartz and Fowler, 1973).

Conservation Status: "Vulnerable" on the IUCN Red List (Hedges, 2004d).

Eleutherodactylus pezopetrus Schwartz, 1960

Distribution: Southeastern Cuba; elevational range ~100–720 m.

Habitat: On and about rocks at the base of a cliff in closed deciduous woods, at the entrance of a small cave by day (Schwartz, 1960a); petricolous in limestone forests (Díaz and Cádiz, 2008).

Abundance: Rare (Díaz and Cádiz, 2008). **Activity:** Nocturnal (Díaz and Cádiz, 2008). **Behavior:** "Agile" (Schwartz, 1960a). **Reproduction:** Males call from rocks, leaves, and branches to heights of 1.5 m, begin calling at dusk and continue sporadically through the night when conditions are favorable (Díaz and Cádiz, 2008); breeding by direct development (Hedges and Díaz, 2004mm; Stuart et al., 2008). **Size:** Male SVL to 41.8 mm, female SVL to 49.3 mm (Schwartz, 1960a); SVL to 50 mm (Díaz and Cádiz, 2008).

Conservation Status: "Vulnerable" (Vales et al., 1998); "critically endangered" on the IUCN Red List, only about 20 animals have been collected (Hedges and Díaz, 2004mm; Stuart et al., 2008).

Eleutherodactylus pictissimus Cochran, 1935

Distribution: Hispaniola, mostly the "South Island," but extending into adjacent lowlands of the "North Island," where three currently recognized subspecies occur (*E. p. pictissimus*, *E. p. apantheatus*, and *E. p. eremus*); elevational range, sea level (or below in the Plaine de Cul de Sac/Valle de Neiba) to ~1,750 m.

Habitat: Terrestrial in dry scrub forest, mesic broadleaf forest, and secondary forests (Hedges et al., 2004h; Stuart et al., 2008); *E. p. pictissimus* occurs in a wide range of habitats (shaded ravines, grassy pastures, in a cave 46 m from the entrance, *Acacia* woods), including those subject to saltwater flooding, in moist *Cocos* trash (husks and fronds), rock piles, and under isolated rocks, *E. p. apantheatus* under moist palm fronds, moist *Musa* trash in a banana plantation (in an otherwise desertlike habitat), beachside palm trash, and in coconut trash in shaded to open coffee groves, by day under rocks in a moist and shaded ravine bed, at night on rocks along a streambed, at 15 cm above the ground on a fern, and in *Acacia* woods, *E. p. eremus* in huge piles of palm trash in a *Cocos* grove with large shade trees and from under palm fronds on mud at a mangrove edge (Schwartz, 1976b).

Abundance: Moderately common in suitable habitat (Hedges et al., 2004h; Stuart et al., 2008). **Activity:** Active in caves by day, active at night in *Acacia* woods and semimesic woods (Schwartz, 1976b). **Reproduction:** Amplexus axillary (Schwartz and Henderson, 1991); eggs laid on the ground, breeds by direct development (Hedges et al., 2004h; Stuart et al., 2008). **Size:** SVL in males to 34 mm, to 43 mm in females (Schwartz, 1976b).

Conservation Status: "Vulnerable" on the IUCN Red List (Hedges et al., 2004h).

Eleutherodactylus pinarensis Dunn, 1926.

Distribution: Western Cuba, including Isla de la Juventud; elevational range, sea level to ~380 m.

Habitat: Primarily saxicolous in mesic wooded areas such as caves, a moist abandoned mineshaft, and on rocks, near cliffs, and in sinkholes (Garrido and Schwartz, 1968b; Ruiz Garcia, 1987); not encountered outside forested areas (Hedges and Díaz, 2004nn; Stuart et al., 2008).

Abundance: Uncommon, seen only in suitable habitat June–October (Hedges and Díaz, 2004nn; Díaz and Cádiz, 2008; Stuart et al., 2008). **Activity:** Active dusk to dawn (Díaz and Cádiz, 2008). **Movements:** Capable of jumping large distances (Díaz and Cádiz, 2008). **Reproduction:** Males call from the ground, rocks, leaves, and branches to heights of 1.5 m, call sporadically through the night, with periods of silence lasting 15–30 min (Díaz and Cádiz, 2008); breeds by direct development (Hedges and Díaz, 2004nn;

Stuart et al., 2008). **Size:** SVL to 62 mm (Schwartz and Henderson, 1991); SVL to 70 mm (Díaz and Cádiz, 2008).

Conservation Status: "Endangered" on the IUCN Red List (Hedges and Díaz, 2004nn).

Eleutherodactylus pinchoni Schwartz, 1967

Distribution: Guadeloupe (Basse-Terre); elevational range, sea level to ~1,470 m.

Habitat: In elfin woodlands on slopes of Soufrière (Kaiser and Henderson, 1994); in rainforest, forest edge, *Theobroma* groves, in forest leaf litter, piles of wet cacao husks, under wet *Musa* trash, under palm husks and fronds at the edge of a forest clearing, under logs and rocks in deep forest shade, may leave forested areas at night to forage in artificially cleared areas, often hopping about leaf litter before noon in rainforest or cacao groves (Schwartz, 1967b); also in moss and peat moss, in crevices of rocky cliffs, and at bases of trees among roots (Breuil, 2002b); occurs in secondary forests but not outside forest habitats (Hedges et al., 2004f; Stuart et al., 2008).

Activity: Largely nocturnal (Breuil, 2002b). **Reproduction:** Males call from ground or low (15 cm) herbaceous growth, including from within a raspberry thicket, at Matouba at 1000 h calling from deep grass (<1.0 m) near abandoned farmhouses (where much trash and debris provide cover; Schwartz, 1967b); eggs laid on the ground and in bromeliads (Hedges et al., 2004f; Stuart et al., 2008); clutch size 5–8 (Hedges, in prep). **Size:** Male SVL to 16 mm, female SVL to 20 mm (Schwartz, 1967b).

Conservation Status: "Endangered" on the IUCN Red List, probably affected adversely by competition with introduced *E. johnstonei* and volcanic activity (Hedges et al., 2004f; Stuart et al., 2008).

Additional Pertinent Literature: Kaiser et al. (2003).

Eleutherodactylus pituinus Schwartz, 1964

Distribution: Hispaniola, Cordillera Central (Dominican Republic); elevational range ~1,210–1,770 m.

Habitat: Upland pinewoods and forest (Schwartz, 1964b); has not been recorded outside forested habitats (Hedges et al., 2004ee; Stuart et al., 2008).

Disease: Chytridiomycosis (Joglar and Burrowes, 2005; Joglar et al., 2007; Hedges et al., 2004ee; Stuart et al., 2008). **Reproduction:** Males call from pine-clad hillsides, edges of thickets or clusters of dense shrubs, one called from an exposed rock on the steep face of a road cut 7.6 m above the road (Schwartz, 1964b); eggs laid on the ground, breeds by direct development (Hedges et al., 2004ee; Stuart et al., 2008); clutch size 19 (Hedges, in prep.). **Size:** SVL to 28.9 mm (Schwartz, 1964b).

Conservation Status: "Endangered" on the IUCN Red List (Hedges et al., 2004ee).

Eleutherodactylus planirostris Cope, 1862

Distribution: Bahama Islands, Cayman Islands, Cuba (three subspecies recognized); elevational range, sea level to ~725 m; introduced to Jamaica, Caicos Islands (?), Grenada (?), southeastern United States, Mexico, and the Hawaiian Islands.

Habitat: Exceedingly adaptable, may be in a diversity of habitats, including anthropogenic situations, often in piles of rocks or stacks of bricks and other similar situations (Díaz and Cádiz, 2008); mostly terrestrial in mesic and xeric habitats, including forests, caves, beaches, nurseries, gardens, and urban areas, in arboreal bromeliads on Grand Cayman (Hedges et al., 2004b); in the Bahama Islands, shaded situations, well-watered nurseries, under flower pots, large cans, and other human debris, as well as in tropical hammocks, beach *Cocos* trash, under rocks in burned-over area in coppice, under coral fragments in *Coccoloba-Bursera-Terminalia* vegetation near shore (Schwartz and Henderson, 1991); in Cuba, in diverse habitats, takes refuge under rocks, lumber, rubbish (Garrido and Schwartz, 1968b); near the opening of a cave, in a bromeliad with *E. auriculatus* (G. Peters, 1974); in fissures of "dogtooth" limestone (Estrada, 1993b); also in parks and gardens, as well as patios (Alonso Bosch and Rodríguez Gómez, 2003b; Kunz, 2005); in the Cayman Islands, on Grand Cayman, occasionally in bromeliads, in gardens and around buildings where humus is adequate, under rocks, wood, and other surface debris, under rocks in sinkholes, cave entrances, and woodland limestone ridge at Old Man Bay, on Cayman Brac, around a hotel complex (Seidel and Franz, 1994); on Jamaica, on roadside grass at a height of 0.15 m (Lynn, 1937); under dead banana leaves (Goin and Cooper, 1950); under rocks in an open pasture, under banana trash, in an arboreal bromeliad 2.0 m above the ground, in grass at a pasture's edge, under coastal palm trash, in a xeric palmetto pasture near the coast, under moist rotten wood, on the grounds of a small hotel, from earthen bank of a culvert in a field, on the ground in a *Musa* grove, under *Cocos* trash in a grassy coconut grove, and in debris and drift (masses of dead cane, grass, coconut fronds, shrubs, and bushes) caught against a bridge across the Hectors River (Schwartz and Fowler, 1973); also in terrestrial bromeliads in xeric, open woods over karst (Stewart, 1977); daytime retreat in coconut husk piles (Stewart and Martin, 1980).

Abundance: Very common (Díaz and Cádiz, 2008). **Activity:** Primarily nocturnal, but may extend activity into the early morning, especially during the rainy season (Díaz and Cádiz, 2008). **Diet:** On Jamaica, forages on low grass of lawns, pastures, and roadsides, very close to the ground (Stewart, 1977), of 3,509 prey items (8.44/frog), 63% were ants and 18% arachnids (mites, spiders; Stewart, 1977); on Cuba, diet includes ants (*Odontomachus* sp., *Prenolepis* sp.; Goin, 1947). **Parasites:** In Cuba, nematodes Spiruridae gen. sp. and *Oswaldocruzia lenteixeirai* (Baruš, 1972; Coy Otero and Lorenzo Hernández, 1982). **Predators:** *Osteopilus septentrionalis* (Meshaka, 1996a); *Cubophis caymanus* on Grand Cayman and *C. vudii* in the Bahamas (Henderson and Sajdak, 1996); *C. cantherigerus* on Cuba (Alonso Bosch and Rodríguez Gómez, 2003b). **Reproduction:** Males call from on or near (0.5 m) the ground, often on leaves and debris (Schwartz and Henderson, 1991); call from the ground or from perches <1 m above the ground, eggs laid on the ground, often in a small depression excavated by the female (Díaz and Cádiz, 2008); on Jamaica, 69% females gravid in April, 93% in June (Stewart, 1977); clutch size 11–19 (Hedges, in prep.); on Cuba, eggs hatched 10–13 days post-collection (Dunn, 1926d); on Grand Cayman, eggs stuck to the underside of a stone at the bottom of a rock pile in August (Grant, 1940c); also under rocks in July in a cave near Old Man Bay (Seidel and Franz, 1994); apparently no parental care of eggs (Schwartz and Henderson, 1991); breeds by direct development (Hedges et al., 2004b). **Size:** Male SVL to 34 mm, female SVL to 36 mm (Schwartz, 1974a); SVL to 27.3 mm on Cuba (Díaz and Cádiz, 2008).

Conservation Status: On the IUCN Red List as a species of "least concern" (Hedges et al., 2004b).

Additional Pertinent Literature: Schwartz (1974a).

Eleutherodactylus poolei Cochran, 1938

Distribution: Hispaniola, Massif du Nord (Haiti); elevational range ~550–650 m.

Habitat: Moist dungeons of the Citadelle Laferrière and the vicinity of the Citadelle (Schwartz and Henderson, 1991).

Reproduction: Males call from the ground (moist cannon embrasure) and from leaves of small herbaceous plants 0.3–0.9 m above the ground on a rocky hillside (Schwartz and Henderson, 1991); eggs laid on the ground, breeds by direct development (Hedges and Thomas, 2004m; Stuart et al., 2008); clutch size 24–31 (Hedges, in prep.). **Size:** SVL to 34 mm in females (Schwartz and Henderson, 1991).

Conservation Status: "Critically endangered" on the IUCN Red List (Hedges and Thomas, 2004m).

Eleutherodactylus portoricensis Schmidt, 1927

Distribution: Puerto Rico, interior uplands; elevational range ~270–1,180 m.

Habitat: Mesic upland broadleaf forest, diurnally under forest debris and leaf litter (Schwartz and Henderson, 1991); mean perch heights and diameters at two upland forest sites 97.5 ± 7.32 cm, 115.8 ± 7.92 cm, 131.1 ± 12.19 cm, 131.1 ± 20.73 cm (heights) and 6.1 ± 0.81 cm, 6.4 ± 1.32 cm, 6.6 ± 1.93 cm (diameters) for males, 82.3 ± 9.14 cm and 3.6 ± 1.07 cm for females, 76.2 ± 11.58 cm and 1.8 ± 0.91 cm for "undetermined" individuals (Turner and Gist, 1970); most (>75%) in bromeliads, some (>20%) inside petioles, few (<5%) under petioles (Cintrón, 1970); under bark, in curled leaves, and on palm-leaf axils or foraging at night throughout the shrub layer to heights of ~4 m (Drewry, 1970b); most frequently on leaves (50%), bromeliads (18%), shrubs (14%), trees (9%), and the ground (9%; Joglar, 1998); adults will exploit bird nests (especially those of *Coereba flaveola* at heights of 1.0–8.0 m) as retreats (Pérez-Rivera and Nadal, 1996); largely terrestrial, not recorded from outside forested habitats (Hedges et al., 2004uu; Angulo, 2008f; Stuart et al., 2008).

Activity: Nocturnal (K. P. Schmidt, 1928); calling strongly at sunset, peaks within 2 h after sunset, individuals drop out progressively until activity either ceases or reaches a minimum an hour or so before daybreak, resuming with a smaller peak at daybreak and then fading out, however, calling ceases shortly after sundown when temperatures drop below 19 °C or relative humidity below 85% (Drewry, 1970b). **Behavior:** Both males and females produce a "low growling sound" during territorial disputes (Drewry, 1970b). **Diet:** Stomach contents included ants (most abundant), beetles, ichneumonoid wasps, and centipedes, mean food item size ~0.5 cm (Cintrón, 1970); beetles (23.9% by frequency of occurrence) and hymenopterans (ants; 28.2%) were most frequently consumed, but beetles (34.3%) and orthopterans (31.9%) composed greatest volume of prey, roaches (12.1%) and spiders (10.6%) also composed significant portions of prey volume (Stewart and Woolbright, 1996). **Disease:** *Batrachochytrium dendrobatidis* fungal infections (Hedges et al., 2004uu; Angulo, 2008f; Burrowes et al., 2008; Stuart et al., 2008). **Movements:** Distances between successive recaptures 0–39.3 m for males and 1.5–30.3 m for females (Turner and Gist, 1970). **Population Size and Density:** Almost 1,000 males/ha (Turner and Gist, 1970); at El Yunque, population in elfin forest declined during the mid-1990s but recovered by the end of the decade, in other habitats, populations have declined consistently (Burrowes and Joglar, 2005). **Predators:** Parental males may eat their own eggs if disturbed early during brooding (Stewart and Woolbright, 1996; Joglar, 2005a). **Reproduction:** Males call from the same kind of shelter used by day (Drewry, 1970b); call from dusk to after dawn with a major peak near midnight, use bushes and tree trunks as calling sites (Drewry and Rand, 1983); calling sites 0.56–1.7 m above the ground on leaves (50%), trees (42%), and shrubs (8%), egg clutches (16–20/clutch, diameter 4.0–7.3 mm) in April–June and September, gravid females (32.2–42.9 mm SVL) in February, May, and July, 80% of nests in bromeliads (Joglar, 1998); males attend nests (Townsend, 1996), never associated with more than one clutch (Drewry, 1970b); may use bird nests, especially those of *Coereba flaveola*, as nest sites (Pérez-Rivera and Nadal, 1996); eggs laid in bromeliads, breeds by direct development (Hedges et al., 2004uu; Angulo, 2008f; Stuart et al., 2008). **Size:** Male SVL to 40.2 mm, female SVL to 42.9 mm (Joglar, 1998); hatchling SVL 6 mm (Grant, 1932c).

Conservation Status: "Endangered" on the IUCN Red List (Hedges et al., 2004uu; Angulo, 2008f); "decreasing in some areas" (Moreno, 1991d).

Remarks: Because the two species are easily confused (Joglar, 1998) and some publications did not distinguish between them (e.g., Cintrón, 1970; Drewry, 1970a, 1970b), some data pertaining to *E. coqui* may be included in this account.

Eleutherodactylus principalis Estrada and Hedges, 1997

Distribution: Eastern Cuba, area between the Cuchillas de Moa and Cuchillas de Toa; elevational range ~300–1,000 m.

Habitat: Mesic closed forests, female in leaf litter (Estrada and Hedges, 1997a); in rainforest and mesic pine forests, diurnal retreats on the ground (Díaz and Cádiz, 2008).

Abundance: Moderately common in suitable habitat (Hedges and Díaz, 2004oo; Stuart et al., 2008); common, but most easily detected when males are calling (Díaz and Cádiz, 2008). **Reproduction:** Males call from horizontal perches 0.8–2.0 m above the ground, usually between 2400 h and 0600 h (Estrada and Hedges, 1997a); call preferentially from leaves at heights of 1–2 m (Díaz and Cádiz, 2008); breeds by direct development (Hedges and Díaz, 2004oo; Stuart et al., 2008). **Size:** Male SVL to 19.2 mm, female SVL to 22.9 mm ($N = 1$; Estrada and Hedges, 1997a).

Conservation Status: "Endangered" on the IUCN Red List (Hedges and Díaz, 2004oo).

Eleutherodactylus probolaeus Schwartz, 1965

Distribution: Hispaniola, extreme eastern Dominican Republic; elevational range, sea level to ~60 m.

Habitat: Semimesic broadleaf forest with an understory of *Bryophyllum* and *Zamia* on a limestone base, difficult to locate in dense understory (Schwartz, 1976b).

Abundance: Moderately common in suitable habitat (Hedges et al., 2004i; Stuart et al., 2008). **Reproduction:** Males call from leaves of *Zamia* and *Bryophyllum*, shrub limbs, and from a pile of dead, leafless slash, 0.3–1.8 m above ground (Schwartz, 1976b); eggs laid on the ground, breeds by direct development (Hedges et al., 2004i; Stuart et al., 2008). **Size:** SVL to 30 mm (Schwartz, 1976b).

Conservation Status: "Endangered" on the IUCN Red List (Hedges et al., 2004i).

Eleutherodactylus rhodesi Schwartz, 1980

Distribution: Hispaniola, northwestern Haiti; elevation ~30 m.

Habitat: Presumably in mesic cultivated situations, under trash in a *Musa* grove, juvenile under dry leaves on a slope adjacent to a dry drainage ditch (Schwartz, 1980c); terrestrial in leaf litter and under rocks (Hedges and Thomas, 2004n).

Abundance: Moderately common in suitable habitat (Hedges and Thomas, 2004n; Stuart et al., 2008). **Reproduction:** Eggs laid on the ground, breeds by direct development (Hedges and Thomas, 2004n; Stuart et al., 2008); clutch size 17 (Hedges, in prep.). **Size:** Male SVL to 26.1 mm, female SVL to 29.8 mm (Schwartz, 1980c).

Conservation Status: "Critically endangered" on the IUCN Red List (Hedges and Thomas, 2004n).

Eleutherodactylus richmondi Stejneger, 1904

Distribution: Puerto Rico, islandwide, but disjunct; elevational range ~40–1,150 m.

Habitat: Terrestrial in wet limestone forest, rainforest, cloud forest, and elfin woodland (Hedges et al., 2004vv; Angulo, 2008g; Stuart et al., 2008); under stones and leaves on damp ground at El Yunque (Stejneger, 1904); under stones and palm leaves on a trail or on damp ground (K. P. Schmidt, 1928); most (>75%) under petioles, some (~25%)

inside petioles, none in bromeliads (Cintrón, 1970); in burrows, stone crevices, and decaying wood (Drewry, 1970b); rarely on leaves above ground level, by day under logs, roots, trash, and rocks (Rivero, 1998); females usually on the forest floor, males call from vegetation to heights of 1.5 m (Joglar, 1998).

Activity: Largely nocturnal (Drewry and Rand, 1983); calling not concentrated, continues at about the same level throughout the night (Drewry, 1970b). **Biomass:** 26,500 mg dry/ha at El Verde (Stewart and Woolbright, 1996). **Disease:** *Batrachochytrium dendrobatidis* fungal infections (Hedges et al., 2004vv; Angulo, 2008g; Burrowes et al., 2008; Stuart et al., 2008). **Foraging and Diet:** Stomach contents included beetles (most abundant), ants, ichneumonoid wasps, termites, earthworms, and flies, mean food item size ~0.22 and ~0.26 cm at two locations (Cintrón, 1970); ambush forager, stomachs at El Verde yielded prey representing 5 classes and 8 orders, with arachnids most important in terms of prey volume (61%) and second (23.1%) by frequency of occurrence to hymenopterans (30.8%), centipedes/millipedes composed 15.1% by volume, and lepidopteran larvae 13.4% by volume (Stewart and Woolbright, 1996). **Population Density:** At El Verde, maximum of 100/ha (Stewart and Woolbright, 1996). **Reproduction:** Males frequently call from daytime hiding places, but are timid at this time and will not call when illuminated with a flashlight, are "fearless" when they call at night from elevated perches (Drewry, 1970b); call from just before dusk to just past dawn, slight peak around midnight, from crevices, sometimes climb plants 2–600 cm (Drewry and Rand, 1983); at Maricao call from leaf litter on the ground to heights of 0.53 m, at El Yunque from horizontal leaf surfaces 0.9–1.5 m high, clutches in February, March, and June, and gravid females in August, gravid females (SVL 39.6–42.3 mm) had 20–28 eggs 2.5–3.5 mm diameter (Joglar, 1998); clutch in January in a rotten log with a female in attendance (H. Heatwole in Drewry, 1970b); January coincides with maximal calling activity (Drewry, 1970b); one clutch in a rotten log with an adult (sex undetermined) in attendance, another in June in a partially rotten log, but unattended (Rivero, 1998); two clutches in January in a fallen tree-fern log with one female in attendance, one clutch with 29 eggs in an advanced stage of development (embryos visible and with adult coloration), eggs 5.4–6.3 mm (mean 5.76 mm) in diameter, hatchling SVL 7.5–8.0 mm (mean 7.7 mm), second clutch more recently deposited, 29 eggs, diameters 5.0–5.5 mm (mean 5.26 mm; Medina Díaz and Joglar, 2008). **Size:** Male SVL to 29.9 mm, female SVL to 43.5 mm (Joglar, 1998).

Conservation Status: "Critically endangered" on the IUCN Red List (Hedges et al., 2004vv; Angulo, 2008g); populations have not recolonized some areas at El Yunque where they were abundant in the past (Burrowes and Joglar, 2005; Joglar et al., 2007); "decreasing in some areas" (Moreno, 1991d).

Eleutherodactylus ricordii Duméril and Bibron, 1841

Distribution: Cuba, southeastern Cuba, Sierra Maestra, Sierra de la Gran Piedra, Cuchillas de Toa; elevational range ~30–1,150 m.

Habitat: Mesic woods and forest, at night on moist stairs, ground about ruins, on rocky hill slopes, along a creek, and on roads (Schwartz and Henderson, 1991; Hedges and Díaz, 2004pp); in mesic forests of all types, in leaf litter, under rocks and fallen logs, between roots, but also takes refuge in rocky outcrops (Díaz and Cádiz, 2008).

Abundance: Moderately common in suitable habitat (Hedges and Díaz, 2004pp; Stuart et al., 2008). **Activity:** Nocturnal, but males occasionally call during the day (Díaz and Cádiz, 2008). **Behavior:** Moves using "irregular" jumps (Díaz and Cádiz, 2008). **Reproduction:** Breeds by direct development (Hedges and Díaz, 2004pp). **Size:** SVL to 40 mm (Schwartz and Henderson, 1991).

Conservation Status: "Vulnerable" on the IUCN Red List (Hedges and Díaz, 2004pp).

Eleutherodactylus riparius Estrada and Hedges, 1998

Distribution: Cuba, except extreme eastern areas, also Isla de la Juventud, Cayo Coco, and Archipiélago de Sabana-Camagüey; elevational range, sea level to ~830 m.

Habitat: Semiaquatic, spring runs, rivers, lakes, and flooded rice fields in very moist areas under rocks, stones, and logs (Schwartz and Henderson, 1991); occurs along river banks and in forests (Alayo Dalmau, 1951; Estrada, 1993b, 1994); occurs along banks of streams, rivers, and lagoons in primary and secondary semideciduous forest, pine forest, dry forest, plantations, and coffee groves with an annual rainfall of 1,300–2,000 mm, beneath small rocks and trunks, in herbaceous vegetation, in fluvial caverns, under rocks in rivulets, and in litter, at night around streamside rocks and vegetation (Estrada and Hedges, 1998); also in temporarily inundated flats, may move some distance from water into patios and gardens of rural homes during the rainy season (Díaz and Cádiz, 2008); in forests and disturbed situations as long as streams or swamps are present (Hedges and Díaz, 2004qq).

Abundance: Common (Díaz and Cádiz, 2008). **Activity:** Primarily nocturnal, but occasionally active by day (Díaz and Cádiz, 2008). **Behavior:** When threatened, jumps into water and hides at the bottom (Dunn, 1926d; Hedges et al., 1995; Estrada and Hedges, 1998). **Diet:** Stomach of a large female contained two caterpillars (G. Peters, 1974); a subadult female (SVL 24 mm) consumed a juvenile, the stomach of the juvenile contained homopterans (Valdés de la Osa and Ruíz García, 1980b); at times cannibalistic (Ruiz García, 1987). **Parasites:** Nematodes *Aplectana cubana*, *Aplectana* sp., *Oswaldocruzia lenteixeirai*, and larval *Abbreviata* sp. (Baruš, 1972; Coy Otero and Lorenzo Hernández, 1982) and larval *Porrocaecum* sp. (Baruš, 1972, 1973; Coy Otero and Lorenzo Hernández, 1982); numerous spirurid nematodes in stomach and free in body cavity (G. Hartwich in G. Peters, 1974). **Reproduction:** Males call from the ground near streams, in mesic woods, and occasionally in pinewoods, 27–42 eggs under stones at water's edge or among dirt and gravel on 14–16 July, began hatching 21–22 July (Dunn, 1926d); choruses burst suddenly into sound every 15–30 min (Díaz and Cádiz, 2008); left and right oviducts of a large female contained 26 and 20 yolked eggs 1.5–3.0 mm in diameter (G. Peters, 1974); clutch size 51–86 (Hedges, in prep.); breeds by direct development (Hedges and Díaz, 2004qq). **Size:** Male SVL to 28.2 mm, female SVL to 41.6 mm (Estrada and Hedges, 1998); SVL to 41.8 mm (Díaz and Cádiz, 2008).

Conservation Status: On the IUCN Red List as species of "least concern" (Hedges and Díaz, 2004qq).

Remarks: Frogs referred to *Eleutherodactylus cuneatus* prior to the publication of Estrada and Hedges (1998) are now referred to *E. riparius* (this account).

Eleutherodactylus rivularis Díaz, Estrada, and Hedges, 2001

Distribution: Southeastern Cuba, Sierra Maestra; elevational range ~80–240 m.

Habitat: Montane streams with grassy margins in disturbed situations, primary vegetation along the streams is *Cyperus* sp. (Díaz et al., 2001); along sandy or pebbled shores of rivers in open montane areas, may venture some distance from the river in drains and ditches near human settlements, such as Santo Domingo along the Río Yara (Díaz and Cádiz, 2008).

Abundance: Moderately common in suitable habitat (Hedges and Díaz, 2004rr; Stuart et al., 2008; Díaz and Cádiz, 2008). **Activity:** Nocturnal (Díaz and Cádiz, 2008). **Diet:** Includes spiders (Lycosidae, Tetragnathidae), ants, beetles (Tenebrionidae, Staphylinidae), and flies (Díaz et al., 2001). **Reproduction:** Beginning at dusk (after 1800 h), males (1–4 m apart) call from river beaches of sand and pebbles, and from emergent rocks, most females with enlarged follicles or oviductal eggs in December and February, clutch of 42 eggs in excavated hole (22.3 × 20.8 mm and 14.3 mm deep) under a piece of cloth and moss sheaths on 23 February 4.0 m from the edge of the Río Jibacoa, nest temperature 25 °C at 1115 h, egg diameter 4.2–4.6 mm, eggs hatched throughout the day, froglets with SVL 5.0–5.6 mm (Díaz et al., 2001); calling decreases gradually after midnight (Díaz and Cádiz, 2008); breeds by direct development (Hedges and Díaz, 2004rr; Stuart et al., 2008). **Size:** Male SVL to 27.5 mm, female SVL to 31.3 mm (Díaz et al., 2001).

Conservation Status: "Critically endangered" on the IUCN Red List (Hedges and Díaz, 2004rr).

Eleutherodactylus rogersi Goin, 1955

Distribution: Great Bahama Bank (except South Bimini Island and New Providence), Cat Island, San Salvador Island, and Little San Salvador Island.

Habitat and Natural History: Unknown.

Size: SVL of female holotype 31.8 mm (Goin, 1955).

Conservation Status: Until recently, this species was considered a subspecies of *E. planirostris*, which is considered a species of "least concern" (Hedges et al., 2004b); on the IUCN Red List as "data deficient" (Angulo, 2008h).

Additional Pertinent Literature: Schwartz (1974a).

Eleutherodactylus ronaldi Schwartz, 1960

Distribution: Southeastern Cuba, Sierra Maestra, Sierra de Boniato, and northern Guantánamo Province; elevational range ~210–1,060 m.

Habitat: Lowland and upland mesic broadleaf forest, occasionally in pine forest (Schwartz, 1960a; Schwartz and Henderson, 1991); various types of woodland and shrubland formations, including clearings and disturbed habitats, diurnal retreats on the ground among rocks and in tree cavities and under bark (Díaz and Cádiz, 2008); primary and secondary wet forest (Hedges and Díaz, 2004ss; Stuart et al., 2008).

Abundance: Common in suitable habitat (Hedges and Díaz, 2004ss; Stuart et al., 2008). **Activity:** Crepuscular and nocturnal (Díaz and Cádiz, 2008). **Reproduction:** Males usually call from low (3.0 m) perches in shrubs or small trees with dense foliage in pastures or other grassy areas (Schwartz, 1960a; Schwartz and Henderson, 1991); call from herbaceous growth and bushes, choruses often in dense reeds (*Gynerium saccharoides*) along river banks, at heights of 0.5–4.0 m (Díaz and Cádiz, 2008); breeds by direct development (Hedges and Díaz, 2004ss; Stuart et al., 2008). **Size:** Male SVL to 27.9 mm (Schwartz, 1960a), female SVL to 28.0 mm (Schwartz and Henderson, 1991); SVL to 30.1 mm (Díaz and Cádiz, 2008).

Conservation Status: "Vulnerable" on the IUCN Red List (Hedges and Díaz, 2004ss).

Eleutherodactylus rucillensis Cochran, 1939

Distribution: Hispaniola, Cordillera Central (Dominican Republic); elevational range 1,200–1,770 m.

Habitat: On stream banks, edges of marshes and ponds, and wet rocks in a cutover mountain ravine (Schwartz, 1970c).

Size: Male SVL to 44.4 mm, female SVL to 58.0 mm (Schwartz, 1970c).

Conservation Status: Until recently this species was considered a subspecies of *E. schmidti*, which is considered "critically endangered" (Hedges et al., 2004u); we assume this species is, at best, also critically endangered.

Eleutherodactylus rufifemoralis Noble and Hassler, 1933

Distribution: Hispaniola, Sierra de Baoruco (Dominican Republic); elevational range ~725–1,370 m.

Habitat: Upland mesic broadleaf forest, under a stick pile and banana trash in a banana-coffee grove (Schwartz and Henderson, 1991); among damp leaves and stones on ground, and at the base of a banana leaf (Noble and Hassler, 1933); terrestrial cloud forest species in wet broadleaf and pine forests (Hedges et al., 2004ff; Stuart et al., 2008); in surface debris in a recently clear-cut area (RP).

Reproduction: Eggs laid on the ground, breeds by direct development (Hedges et al., 2004ff; Stuart et al., 2008); clutch size 4–7 (Hedges, in prep.). **Size:** SVL to 20 mm in females (Ruder et al., 1995).

Conservation Status: "Critically endangered" on the IUCN Red List (Hedges et al., 2004ff).

Additional Pertinent Literature: Ruder et al. (1995).

Eleutherodactylus ruthae Noble, 1923

Distribution: Hispaniola, eastern Dominican Republic; elevational range, sea level to ~900 m.

Habitat: Wooded (xeric or pine) areas, as well as broadleaf mesic forest (usually open), also in a hillside pasture and a corn field (Schwartz, 1965e; Hedges et al., 2004 mm; Stuart et al., 2008).

Predators: *Hypsirhynchus parvifrons* (Henderson et al., 1987a). **Reproduction:** Males usually call from (apparently) self-constructed underground chambers, chamber completely roofed and smoothed inside, with no visible evidence of chamber on the surface, males may occasionally call from above ground (e.g., from a rock pile), eggs deposited in the chamber and develop therein, hatchlings presumably escape the chamber when rains soften the ground (Schwartz, 1965e); development direct (Hedges et al., 2004 mm; Stuart et al., 2008). **Size:** Male SVL to 50.0 mm, female SVL to 52.3 mm (Schwartz, 1965e).

Conservation Status: "Endangered" on the IUCN Red List (Hedges et al., 2004 mm).

Eleutherodactylus schmidti Noble, 1923

Distribution: Hispaniola, Cordillera Central (Dominican Republic); elevational range, sea level to ~1,750 m.

Habitat: Aquatic, associated with springs and spring runs, streams, river edges, and ponds, diurnally under rocks and stones near water (Mertens, 1939); mesic closed-canopy rainforest (Hedges et al., 2004u; Stuart et al., 2008).

Reproduction: Males call from banks of streams and rivers, often in undergrowth and dead leaves as far as 4.0 m from water (Schwartz and Henderson, 1991); eggs laid on the ground, breeds by direct development (Hedges et al., 2004u; Stuart et al., 2008). **Size:** SVL in females to 58 mm (Schwartz and Henderson, 1991).

Conservation Status: "Critically endangered" on the IUCN Red List, not recorded since the 1980s despite extensive surveys in 1998 and 2000 (Hedges et al., 2004u; Stuart et al., 2008).

Eleutherodactylus schwartzi Thomas, 1966

Distribution: Puerto Rico Bank, U.S. and British Virgin Islands; elevational range, sea level to ~225 m.

Habitat: Semimesic woods and forests, often associated with terrestrial bromeliads (Schwartz and Henderson, 1991); dry scrub forest, not recorded outside forested areas (Hedges and Thomas, 2004o; Stuart et al., 2008).

Reproduction: Based on observations from Tortola, Virgin Gorda, and Guana combined, 89% of calling males perched in vegetation <2.5 m, 11% >2.5 m, calling males on Tortola perched on trees or shrubs (75%) and herbaceous vegetation (25%), males on Virgin Gorda called from bromeliads (84%), trees and shrubs (13%), and herbaceous vegetation (2%; Ovaska et al., 2000); on Great Dog Island (British Virgin Islands), five nests in ground bromeliads (*Hohenbergia antillana*) 25–40 cm above the ground on 10 October, one clutch in an arboreal bromeliad (*Tillandsia* sp.) at at height of 1.8 m, four nests contained single clutches, one contained two clutches, males attended each nest, the male in the nest with two clutches switched from clutch to clutch, clutch size 5–13, egg diameter 4.4–8.8 mm (Ovaska et al., 1998); development direct (Hedges and Thomas, 2004o; Stuart et al., 2008). **Size:** Male SVL to 23.1 mm, female SVL to 33.7 mm (Thomas, 1966a).

Conservation Status: "Endangered" on the IUCN Red List (Hedges and Thomas, 2004o); populations in the U.S. Virgin Islands may be extirpated (Philbosian and Yntema, 1977; Platenberg and Boulon, 2006; Hedges and Thomas, 2004o; Stuart et al., 2008); "decreasing in some areas" (Moreno, 1991d).

Eleutherodactylus sciagraphus Schwartz, 1973

Distribution: Hispaniola, Massif de la Hotte (Haiti); elevational range ~1,060–1,080 m.

Habitat: Closed mesic upland forests (Hedges and Thomas, 2004p; Stuart et al., 2008); under rocks (Schwartz and Henderson, 1991).

Predators: *Ialtris haetianus* (Henderson and Schwartz, 1986). **Reproduction:** Eggs laid on the ground, breeds by direct development (Hedges and Thomas, 2004p; Stuart et al., 2008). **Size:** Male SVL to 16.7 mm, female SVL to 20.9 mm (Schwartz, 1973b).

Conservation Status: "Critically endangered" on the IUCN Red List, encountered only once during the past two decades (Hedges and Thomas, 2004p; Stuart et al., 2008).

Eleutherodactylus semipalmatus Shreve, 1936

Distribution: Hispaniola, Massif de la Hotte and Massif de la Selle (Haiti); elevational range ~300–1,700 m.

Habitat: Uplands with grasses and shrubs (Schwartz and Henderson, 1991); riparian or aquatic in mesic hardwood forests, not recorded from disturbed forest habitats (Hedges and Thomas, 2004q; Stuart et al., 2008); associated with streams (Hedges and Díaz, 2009).

Reproduction: Males call from grasses and shrubs to ~0.9 m above ground (Schwartz and Henderson, 1991); eggs laid on the ground, breeds by direct development (Hedges and Thomas, 2004q; Stuart et al., 2008). **Size:** SVL in males to at least 28 mm (Shreve, 1936).

Conservation Status: "Critically endangered" on the IUCN Red List (Hedges and Thomas, 2004q); possibly extinct, since not observed since 1985 despite extensive efforts (Hedges and Thomas, 2004q; Stuart et al., 2008; Hedges and Díaz, 2009).

Eleutherodactylus simulans Díaz and Fong Grillo, 2001

Distribution: Eastern Cuba; elevational range <200 m.

Habitat: Terrestrial in rainforest, semideciduous forest, evergreen forest, coastal vegetation, secondary growth, and anthropogenic situations (Díaz and Cádiz, 2008); margins of a montane stream with dominant vegetation comprising *Syzigium jambos*, *Ficus* sp., *Carapa guianensis*, *Inga* sp., *Piper aduncum*, ferns (*Blechnum occidentale*, *Thelyptheris* sp.), and mosses, also in a residential garden, by day under a streamside log, at night in leaf litter, in holes of streamside talus, and in grass clumps (Díaz and Fong Grillo, 2001).

Abundance: Uncommon even in suitable habitat (Díaz, 2004; Stuart et al., 2008); common (Díaz and Cádiz, 2008). **Activity:** Active 2200–2400 h (Díaz and Fong Grillo, 2001); nocturnal, but males may call by day during heavy rain (Díaz and Cádiz, 2008). **Diet:** Includes spiders (*Corinna* sp.) and ants (*Wasmannia auropunctata* and *Odontomachus* sp.; Díaz and Fong Grillo, 2001). **Reproduction:** Males call from holes in streamside talus or within clumps of grass (Díaz and Fong Grillo, 2001); development direct (Díaz, 2004; Stuart et al., 2008). **Size:** Male SVL to 18.7 mm, female SVL to 23.9 mm (Díaz and Fong Grillo, 2001).

Conservation Status: "Endangered" on the IUCN Red List (Díaz, 2004).

Eleutherodactylus sisyphodemus Crombie, 1977

Distribution: Jamaica, Cockpit Country, known only from the vicinity of the type locality; elevation ~450 m.

Habitat: By day in litter in wet limestone forest on limestone ridges (Crombie, 1977).

Diet: Includes ants and winged hymenopterans (Crombie, 1977). **Reproduction:** Female with 6 ova (0.5–1.0 mm diameter) in left oviduct and 3 smaller ova in right oviduct (Crombie, 1977); eggs laid on the ground, breeds by direct development (Hedges et al., 2004mmm; Stuart et al., 2008); clutch size 10 (Hedges, in prep.). **Size:** Male SVL to 13.8 mm, female SVL to 17.9 mm (Crombie, 1977).

Conservation Status: "Critically endangered" on the IUCN Red List (Hedges et al., 2004mmm).

Eleutherodactylus sommeri Schwartz, 1977

Distribution: Hispaniola, Massif du Nord (Haiti) east into the western Cordillera Central (Dominican Republic); elevational range ~700–1,040 m.

Habitat: Remnant deciduous forest and coffee and banana groves (Schwartz, 1977f).

Size: Male SVL to 35.2 mm, female SVL to 40.7 mm (Schwartz, 1977f).

Conservation Status: Until recently this species was considered a subspecies of *E. wetmorei*, which is considered "vulnerable" (Hedges et al., 2004gg); we assume this species is, at best, also vulnerable.

Eleutherodactylus staurometopon Schwartz, 1960

Distribution: Cuba, Isla de la Juventud.

Habitat: Broadleaf forest bordering streams through pinelands (Schwartz, 1960a).

Reproduction: Males call from epiphytes in tall palms (Schwartz, 1960a). **Size:** Male SVL to 28.1 mm (Schwartz, 1960a).

Conservation Status: Until recently this species was considered a subspecies of *E. varians*, which is considered "vulnerable" (Hedges and Díaz, 2004zz); we assume this species is, at best, also vulnerable.

Eleutherodactylus symingtoni Schwartz, 1957

Distribution: Western Cuba, scattered localities in the Sierra de los Órganos, Sierra del Rosario, Sierra de Camarones; elevational range 70–200 m.

Habitat: Mesic broadleaf forest, associated with rocks and talus (i.e., rocks and soil, not rock faces), also darkened areas of caves (Schwartz and Henderson, 1991); moist closed-canopy forests, not recorded from outside forested areas (Hedges and Díaz, 2004tt; Stuart et al., 2008).

Activity: Nocturnal (Díaz and Cádiz, 2008). **Behavior:** Seeks refuge among rocks and crevices (Schwartz and Henderson, 1991); males rapidly approach recorded calls (Díaz and Cádiz, 2008). **Diet:** Includes roaches (*Eurycotis*), orthopterans, hemipterans (Cicadidae), and gastropods (Alonso Bosch and Rodríguez Gómez, 2003b). **Disease:** Population declines within suitable habitat are suggestive of chytridiomycosis (S. B. Hedges and L. M. Díaz in Stuart et al., 2008). **Reproduction:** Males call from dusk to dawn, sometimes during the day, from the ground, rock walls of ancient ruins, and from vegetation to heights of 1 m, calling males are widely spaced, whether attributable to low population densities or to territoriality is unknown (Díaz and Cádiz, 2008); breeds by direct development (Hedges and Díaz, 2004tt; Stuart et al., 2008). **Size:** Male SVL to 68 mm, female SVL to 69 mm (Schwartz and Henderson, 1991).

Conservation Status: "Vulnerable" (Vales et al., 1998); "critically endangered" (Rodríguez Gómez and Alonso Bosch, 2000); "critically endangered" on the IUCN Red List (Hedges and Díaz, 2004tt).

Eleutherodactylus tetajulia Estrada and Hedges, 1996

Distribution: Eastern Cuba, Cuchillas de Moa; elevational range 300–600 m (600–800 m; Díaz and Cádiz, 2008).

Habitat: Secondary hardwood forest in areas of high rainfall (>1,600 mm/yr) and humidity (Estrada and Hedges, 1996a); in rainforest leaf litter (Díaz and Cádiz, 2008); terrestrial in closed mesic forest (Hedges and Díaz, 2004uu; Stuart et al., 2008).

Abundance: Uncommon (Hedges and Díaz, 2004uu; Stuart et al., 2008); at two localities, 10 individuals collected within 1.0 m^2 of leaf litter by day (Díaz and Cádiz, 2008). **Behavior:** Moves slowly and deliberately (Díaz and Cádiz, 2008). **Reproduction:** Calling male and gravid female (with 6 eggs likely from a different female) together in a hole formed by decayed center of a tree fern ~10 cm above the ground, males calling from concealed locations on the forest floor by day and night (Estrada and Hedges, 1996a); nests in hollow logs (Díaz and Cádiz, 2008); breeds by direct development (Hedges and Díaz, 2004uu; Stuart et al., 2008); males may guard the nest (Alonso Bosch and Rodríguez Gómez, 2003a). **Size:** Male SVL to 12.3 mm, female SVL to 14.0 mm (Estrada and Hedges, 1996a); SVL to 15 mm (Díaz and Cádiz, 2008).

Conservation Status: "Critically endangered" on the IUCN Red List (Hedges and Díaz, 2004uu); "vulnerable" (Rodríguez Gómez and Alonso Bosch, 2000).

Eleutherodactylus thomasi Schwartz, 1959

Distribution: Cuba, where three currently recognized subspecies occur, *E. t. thomasi* in the Sierra de Cubitas and Sierra de Najasa, *E. t. trinidadensis* along the southern coast adjacent to the foothills of the Sierra de Trinidad, and *E. t. zayasi* in the Pan de Matanzas; elevational range, sea level to 390 m.

Habitat: Rocky situations in broadleaf forest, either mesic or semixeric, under and on large rocks, talus, at cave entrances to deep within caves on the floor, walls, and formations, also on much eroded and pocketed limestone and under trash, juveniles on shrubs to 1.0 m above ground, and artificial rock steps after a heavy rain (Schwartz, 1959b); on large rocks and associated with caves in both xeric and mesic broadleaf forests, not recorded outside forested habitats (Hedges and Díaz, 2004vv; Stuart et al., 2008).

Activity: Nocturnal, although males may call by day from within caves (Díaz and Cádiz, 2008). **Behavior:** Seeks refuge in interstices of talus (Schwartz, 1960a); escapes by jumping large distances (Díaz and Cádiz, 2008). **Reproduction:** Males call from the ground or from branches and leaves to heights of 1.5 m (Díaz and Cádiz, 2008); breeds by direct development (Hedges and Díaz, 2004vv; Stuart et al., 2008). **Size:** Male SVL to 31.8 mm, female SVL to 43.4 mm (Schwartz, 1959b).

Conservation Status: "Endangered" on the IUCN Red List (Hedges and Díaz, 2004vv).

Eleutherodactylus thorectes Hedges, 1988

Distribution: Hispaniola, Morne Macaya, Massif de la Hotte (Haiti); elevational range 1,700–2,340 m.

Habitat: Ground dwelling in closed pinewoods (*Pinus occidentalis*) with woody shrubs, tree ferns, bromeliads, and climbing bamboo, in rotten logs, sphagnum, under pine litter and decaying leaves, (Hedges, 1988); somewhat arboreal in closed montane pine and cloud forest (Hedges et al., 2004cccc; Stuart et al., 2008).

Reproduction: Gravid females in January, February, June, and September (i.e., reproduction probably continuous), clutch size 3 (Hedges, 1988); eggs laid on the ground, breeds by direct development (Hedges et al., 2004cccc; Stuart et al., 2008). **Size:** SVL in females to at least 15.1 mm (Hedges, 1988).

Conservation Status: "Critically endangered" on the IUCN Red List, last recorded in 1991, but no recent surveys have been conducted (Hedges et al., 2004cccc; Stuart et al., 2008).

Additional Pertinent Literature: Hedges and Powell (1998c).

Eleutherodactylus toa Estrada and Hedges, 1991

Distribution: Eastern Cuba; elevational range 195–900 m.

Habitat: Riparian or aquatic, in and near streams in mesic closed forest (Hedges and Díaz, 2004ww; Stuart et al., 2008); active or under rocks and trunks along roadside banks in pine forest (*Pinus cubensis*) and rainforest by day (Estrada and Hedges, 1991); clearings in forests or along roads and paths, often on slopes of gullies and channels, but not infrequently on riverine shores, by day under rocks, in leaf litter, in holes in the ground or in rocks (Díaz and Cádiz, 2008).

Abundance: Moderately common in narrow suitable habitat (Hedges and Díaz, 2004ww; Díaz and Cádiz, 2008; Stuart et al., 2008). **Activity:** Primarily nocturnal, but a few individuals may be active by day (Estrada and Hedges, 1991; Díaz and Cádiz, 2008). **Behavior:** Movements very rapid, characterized by sudden jumps of considerable distance (Díaz and Cádiz, 2008). **Reproduction:** Males call from the ground (Díaz and Cádiz, 2008); breeds by direct development (Hedges and Díaz, 2004ww; Stuart et al., 2008). **Size:** Male SVL to 22.3 mm, female SVL to 32.5 mm (Estrada and Hedges, 1991).

Conservation Status: "Vulnerable" (Vales et al., 1998); "endangered" on the IUCN Red List (Hedges and Díaz, 2004ww).

Eleutherodactylus tonyi Estrada and Hedges, 1997

Distribution: Eastern Cuba, Meseta de Cabo Cruz; elevational range, sea level to 70 m.

Habitat: In caves in a terraced limestone platform overlain by semideciduous forest and dry coastal forest, by day on the ground and under limestone rocks to 50 m inside caves and near entrances, caves with high humidity, areas with frogs usually moist (Estrada and Hedges, 1997b); associated with rocks, leaf litter, and caves in lowland closed-canopy forests, not recorded outside forested habitats (Hedges and Díaz, 2004xx; Stuart et al., 2008).

Abundance: Moderately common in caves and other appropriate habitat (Hedges and Díaz, 2004xx; Stuart et al., 2008). **Activity:** Nocturnal, but active by day inside caves (Díaz and Cádiz, 2008). **Reproduction:** Males call by day from exposed and concealed sites in caves (Estrada and Hedges, 1997b); call from the ground or from bushes to heights of 1 m (Díaz and Cádiz, 2008); breeds by direct development (Hedges and Díaz, 2004xx; Stuart et al., 2008). **Size:** Male SVL to 21.5 mm, female SVL to 33.7 mm (Estrada and Hedges, 1997b).

Conservation Status: "Critically endangered" on the IUCN Red List (Hedges and Díaz, 2004xx).

Eleutherodactylus turquinensis Barbour and Shreve, 1937

Distribution: Eastern Cuba, Sierra Maestra; elevational range ~450–1,830 m.

Habitat: Aquatic (Barbour and Shreve, 1937); undisturbed cloud forest as well as a deforested valley with small secondary growth of bushes, frogs cling to wet rocks (by day and night) underwater or in the splash zone of small mountain streams (0.1–2.0 m wide), females sit on rocks surrounding a small, sheltered pool in a stream, males, females, and juveniles at one site exposed on rocks in streams by day and night (Hedges et al., 1995); streams often cold (11–20 °C), daytime retreats under rocks or in holes in streamside slopes (Díaz and Cádiz, 2008); restricted to high-altitude montane streams in closed cloud forest (Hedges and Díaz, 2004yy; Stuart et al., 2008).

Abundance: Common in suitable habitat, although availability of habitat is limited (Hedges and Díaz, 2004yy; Stuart et al., 2008). **Activity:** Primarily nocturnal, but especially juveniles may be active during the day along river margins, even in full sun (Díaz and Cádiz, 2008). **Behavior:** When frightened may remain submerged for several minutes (Díaz and Cádiz, 2008). **Parasites:** Nematode *Physalopteroides valdesi* (Coy Otero and Baruš, 1980; Coy Otero and Lorenzo Hernández, 1982). **Reproduction:** Males call at night from crevices between rocks on stream banks, others call from concealed locations on the ground next to streams, a large (45.1 mm SVL) female with 24 enlarged eggs (Hedges et al., 1995); eggs laid on land, breeds by direct development (Hedges and Díaz, 2004yy; Stuart et al., 2008). **Size:** Females to 53 mm SVL (Schwartz and Henderson, 1991); SVL to 54.5 mm (Díaz and Cádiz, 2008).

Conservation Status: "Vulnerable" (Vales et al., 1998); "critically endangered" on the IUCN Red List (Hedges and Díaz, 2004yy).

Eleutherodactylus tychathrous Schwartz, 1965

Distribution: Hispaniola, Sierra de Neiba (Dominican Republic); elevation ~800 m.

Habitat: Among rocks on a wooded floodplain of a small river, males call from burrows (Schwartz, 1965e).

Size: SVL in single male 57.8 mm (Schwartz, 1965e).

Conservation Status: Until recently this species was considered a subspecies of *E. ruthae*, which is considered "endangered" (Hedges et al., 2004mm); we assume this species is, at best, also endangered.

Eleutherodactylus unicolor Stejneger, 1904

Distribution: Puerto Rico, Sierra de Luquillo; elevational range ~660–1,040 m.

Habitat: Montane forests (elfin forest), in subterranean burrows in a root plexus and under stones (Drewry and Rand, 1983; Snyder et al., 1987); in burrows in the mossy forest floor at very high elevations (Drewry, 1970b); also under moss (Rivero, 1998); lives underground in elfin forests, not recorded outside forested habitats (Hedges et al., 2004ww; Rios-López, 2008h; Stuart et al., 2008).

Disease: *Batrachochytrium dendrobatidis* fungal infections (Burrowes et al., 2008). **Population Size and Density:** At El Yunque, population in elfin forest declined slightly during the 1990s (Burrowes and Joglar, 2005); currently believed to be stable (Hedges et al., 2004ww; Rios-López, 2008h; Stuart et al., 2008). **Reproduction:** Males call from 2 to 7 cm below litter surface (Drewry and Rand, 1983); from the root cavity of tree ferns, call dusk to dawn even at "winter" temperatures of 16–17 °C (Schwartz and Henderson, 1991). **Size:** Male SVL to 15.5 mm, female SVL to 17.3 mm (Joglar, 1998). **Thermal Biology:** Inactive at temperatures >19 °C (Joglar, 1998).

Conservation Status: "Vulnerable" on the IUCN Red List (Rios-López, 2008h); "critically endangered" on the IUCN Red List (Hedges et al., 2004ww); "stable" (Moreno, 1991d).

Eleutherodactylus varians Gundlach and Peters, 1864

Distribution: Central Cuba; elevational range, sea level to 845 m.

Habitat: Lowland mesic forest, by day in leaf bases of bromeliads (Schwartz, 1969a); high in trees (G. Peters, 1974); arboreal, sometimes in bromeliads, but also outside bromeliads in mesic forest, not recorded outside forested habitats (Hedges and Díaz, 2004zz; Stuart et al., 2008); in various forest formations (evergreen and deciduous forests, rainforest, mogote vegetation, secondary forests, and, more rarely, pine forests), diurnal retreats and reproductive sites usually in bromeliads or palm axils (Díaz and Cádiz, 2008).

Abundance: Moderately common in suitable habitat (Hedges and Díaz, 2004zz; Stuart et al., 2008). **Activity:** Nocturnal (Díaz and Cádiz, 2008). **Reproduction:** Males call from branches (often covered with dense stands of epiphytes) and leaves, but rarely from tree trunks, 1.0–4.5 m above ground, probably also from much higher in the canopy, clutch size to 30 (Ruiz García, 1987); males call from leaves, branches, trunks, palm fronds, and bromeliads to heights >3 m, call throughout the night, occasionally during the day (Díaz and Cádiz, 2008); eggs in nests 30 September to 5 October, nests in axils of bromeliads (*Tillandsia*), 10 ± 2.3 (6–13) eggs/nest, nests 1.5 ± 0.8 m (0.6–3.0 m) high, on trees with mean trunk diameter 5.9 ± 11.3 cm (2–40 cm), mean temperature of egg clutch 26.1 ± 1.1 °C at mean air (26.6 ± 1.2 °C) and substrate (26.2 ± 1.4 °C) temperatures, incubation after collection 5–20 days, mean diameter of eggs 4.4–6.6 mm, nests guarded by males (3 of 5), female (1), or both (1), mean hatchling SVL 5.8–6.5 mm (Estrada, 1990); eggs deposited within moist leaf-bases of bromeliads, clutch of 12 well-embryonated eggs (agglutinated to each other) on 6 July in the base of a dead bromeliad with much plant debris (thus protected from desiccation), hatched 0900 h 14 July–1230 h 16 July (Schwartz and Henderson, 1991); breeds by direct development (Hedges and Díaz, 2004zz; Stuart et al., 2008). **Size:** Female SVL to 40 mm (Schwartz and Henderson, 1991); to 29.3 mm SVL (Díaz and Cádiz, 2008).

Conservation Status: "Vulnerable" on the IUCN Red List (Hedges and Díaz, 2004zz).

Eleutherodactylus varleyi Dunn, 1925

Distribution: Cuba, islandwide, but disjunct, also on Isla de la Juventud; elevational range, sea level to 900 m.

Habitat: In pastures, along roadsides, on lawns, in coffee groves, by day under trash, leaf, and palm frond piles (Schwartz, 1958b); along stream (G. Peters, 1974); especially associated with *Zoysia* (Poaceae) in gardens and around patios (Ruiz García, 1987; Kwet, 1995; Kunz, 2005); mesic secondary and primary forests and grasslands, agricultural areas and other disturbed habitats (Hedges and Díaz, 2004aaa); when in forests, associated with leaf litter and herbaceous vegetation, also in urban parks and gardens (Díaz and Cádiz, 2008).

Abundance: Very common in suitable habitats (Hedges and Díaz, 2004aaa). **Reproduction:** Males call from grass blades and herbaceous leaves 2.0–50.0 cm above ground (Schwartz, 1958b); females attend nests (Alonso Bosch and Rodríguez Gómez, 2003b); nest with 4 eggs in a coffee grove (24 March), egg size 3.6–3.8 mm, female several centimeters from the nest, hatched 29 March, mean froglet SVL 3.7 mm, after 30 days, 4.8 mm (Estrada, 1987c); clutch size 9 (Hedges, in prep.); eggs laid on the ground, breeds by direct development (Hedges and Díaz, 2004aaa). **Size:** Male SVL to 12.3 mm, female SVL to 15.0 mm (Schwartz, 1958b); SVL to 20 mm (Díaz and Cádiz, 2008).

Conservation Status: On the IUCN Red List as a species of "least concern" (Hedges and Díaz, 2004aaa).

Eleutherodactylus ventrilineatus Shreve, 1936

Distribution: Hispaniola, Massif de la Hotte (Haiti); elevational range 1,700–2,340 m.

Habitat: Open sites in montane pine and cloud forests, most abundant in areas that have been burned or are under cultivation (Franz and Cordier, 1986); under logs and in leaf litter under tree ferns (Schwartz and Henderson, 1991); terrestrial in open sites in montane pine and cloud forests (Hedges and Thomas, 2004r; Stuart et al., 2008).

Reproduction: Females attend nests (Townsend, 1996); eggs laid on the ground, breeds by direct development (Hedges and Thomas, 2004r; Stuart et al., 2008). **Size:** Female SVL to 31 mm (Schwartz and Henderson, 1991).

Conservation Status: "Critically endangered" on the IUCN Red List, last recorded in 1991, but no subsequent surveys have been conducted (Hedges and Thomas, 2004r; Stuart et al., 2008).

Eleutherodactylus warreni Schwartz, 1976

Distribution: Hispaniola, Île de la Tortue (Haiti); elevation ~400 m.

Habitat: Terrestrial on rocks and in leaf litter in xeric hardwood forests (Hedges and Thomas, 2004s; Stuart et al., 2008); around rocks and in leaf litter (Schwartz and Henderson, 1991).

Reproduction: Eggs laid on the ground, breeds by direct development (Hedges and Thomas, 2004s; Stuart et al., 2008). **Size:** SVL to 33 mm (female; Schwartz and Henderson, 1991).

Conservation Status: "Critically endangered" on the IUCN Red List, not collected since first described (Hedges and Thomas, 2004s; Stuart et al., 2008).

Eleutherodactylus weinlandi Barbour, 1914

Distribution: Hispaniola, where two currently recognized subspecies occur, *E. w. weinlandi* in central Haiti and the east central Dominican Republic, and *E. w. chersonesodes* in eastern Hispaniola (Dominican Republic); elevational range, sea level to ~800 m.

Habitat: Lowland mesic situations (Schwartz and Henderson, 1991); in coastal situations (Mertens, 1939); by day under *Cocos* trash piles within 3.1 m of a mangrove edge (*E. w. chersonesodes*), in coffee groves (*E. w. chersonesodes*), in cacao groves in piles of cacao husks (*E. w. weinlandi*), under logs and rocks, and in piles of rotting logs in edificarian situation at an abandoned sugar mill (*E. w. weinlandi*; Schwartz, 1976b); terrestrial in a variety of mesic habitats, including plantations, woods, gardens, ravines, and trash piles in urban areas (Hedges et al., 2004ii).

Reproduction: Males call from caves, rocks, or low vegetation, eggs are laid on the ground, breeds by direct development (Hedges et al., 2004ii); clutch size 14 (Hedges, in prep.). **Size:** Female SVL to 40 mm (Schwartz, 1976b).

Conservation Status: On the IUCN Red List as a species of "least concern" (Hedges et al., 2004ii).

Eleutherodactylus wetmorei Cochran, 1932

Distribution: Hispaniola; two currently recognized subspecies occur in the Massif de la Hotte in Haiti (*E. w. wetmorei*) and the Massif de la Selle and Morne l'Hôpital in Haiti and into the Dominican Republic (*E. w. ceramerus*); elevational range, sea level to ~1,330 m.

Habitat: Broadleaf forest in both uplands and lowlands, arboreal and bromeliadicolous, also in axils of banana leaves (Schwartz, 1968a; Hedges et al., 2004gg; Stuart et al., 2008).

Abundance: Moderately common in suitable habitat (Hedges et al., 2004gg; Stuart et al., 2008). **Reproduction:** Males call from low vegetation (*Sansevieria* leaf at 0.3 m), coffee tree at 3.0 m to forest canopy (to 12 m above the ground), clutch of 10 eggs in an arboreal bromeliad on 18 July (Schwartz, 1968a); breeds by direct development (Hedges et al., 2004gg; Stuart et al., 2008); clutch size 37 (Hedges, in prep.). **Size:** Male SVL to 33.4, female SVL to 38.3 mm (Schwartz, 1977f).

Conservation Status: "Vulnerable" on the IUCN Red List (Hedges et al., 2004gg).

Eleutherodactylus wightmanae Schmidt, 1920

Distribution: Puerto Rico, interior uplands; elevational range ~150–1,200 m.

Habitat: In forest litter layer (Drewry and Rand, 1983; Stewart and Woolbright, 1996); by day under rocks, rotting tree trunks, and trash on the forest floor (Rivero, 1998); not in bromeliads nor in association with petioles (Cintrón, 1970); in leaf litter and decaying wood (Drewry, 1970b); terrestrial, not recorded outside mesic upland forest habitats (Hedges and Joglar, 2004d; Angulo, 2008i; Stuart et al., 2008).

Activity: Nocturnal (Joglar, 1998); begin calling strongly at sunset, peak within 2 h after sunset, individuals drop out progressively until activity either ceases or reaches a minimum an hour or so before daybreak, resuming with a smaller peak at daybreak and then fading out (Drewry, 1970b). **Behavior:** Retreats under leaf litter or under a root at least disturbance (Rivero, 1998). **Biomass:** 38,300 mg dry/ha at El Verde (Stewart and Woolbright, 1996). **Disease:** *Batrachochytrium dendrobatidis* fungal infections (Burrowes et al., 2008; Hedges and Joglar, 2004d; Angulo, 2008i; Stuart et al., 2008). **Foraging and Diet:** Sit-and-wait predator, based on 26 adults (mean number of prey 3.0) from El Verde Field Station, diet includes pseudoscorpions, spiders, harvestmen, mites, millipedes, springtails, roaches, beetles, flies, hemipterans, hymenopterans, lepidopteran larvae, and a small *E. coqui* (Stewart and Woolbright, 1996); also at El Verde, orthopterans, barklice, beetles, hemipterans, hymenopterans, flies, lepidopterans, and termites, along with spiders, mites, and millipedes (Lavigne and Drewry, 1970). **Population Density:** May be the most abundant frog in the Luquillo Forest, "with individuals in almost any square meter of leaf litter" (Drewry, 1970b); density at El Verde 400/ha (Stewart and Woolbright, 1996). **Reproduction:** Males frequently call from daytime hiding places, but are timid at this time and will not call when illuminated with a flashlight, but are "fearless" when calling at night from elevated perches (Drewry, 1970b); call dusk to dawn with strong peaks at those times (Drewry and Rand, 1983); call from the ground or from very low (30–45 cm) vegetation (K. P. Schmidt, 1928; Drewry and Rand, 1983; Schwartz and Henderson, 1991); clutch sizes in preserved gravid females (20.2–23.0 mm SVL) collected in March, June, and August were 4–6, egg diameter 1.7–2.9 mm (Joglar, 1998); adult possibly guarding clutch of 4 eggs in folded *Cecropia* leaf on the forest floor in April, clutch of 5 eggs (4.31–4.52 mm, $N = 3$) in folded *Cecropia* leaf on forest floor in September (Joglar et al., 2005a); males attend nests (Townsend, 1996);

breeds by direct development (Hedges and Joglar, 2004d; Angulo, 2008i; Stuart et al., 2008). **Size:** Male SVL to 21.0, female SVL to 22.5 mm (Joglar, 1998).

Conservation Status: "Endangered" on the IUCN Red List (Hedges and Joglar, 2004d; Angulo, 2008i); populations have not recolonized some areas at El Yunque where they were abundant in the past (Burrowes and Joglar, 2005; Joglar et al., 2007); "stable" (Moreno, 1991d).

Eleutherodactylus zeus Schwartz, 1958

Distribution: Western Cuba, Sierra de los Órganos and Sierra del Rosario; elevational range ~70–300 m.

Habitat: Mesic broadleaf forest on a limestone base, saxicolous or clivicolous, occasionally on the ground or decaying logs, not under surface objects by day, presumably uses cliff crevices as retreats (Schwartz, 1958a); also in shelter of rock crevices and in a cave interior near the entrance (Estrada et al., 1986); always associated with limestone, not recorded outside forest habitats (Hedges and Díaz, 2004bbb; Stuart et al., 2008).

Abundance: Can be locally common, but is uncommon in many localities (Hedges and Díaz, 2004bbb; Stuart et al., 2008). **Activity:** Crepuscular to nocturnal, may be active in early morning or even midday in caves where light levels are comparable to light levels at dusk outside of the cave (Estrada et al., 1986). **Behavior:** Lethargic at night, but can climb and jump with agility (Estrada et al., 1986). **Diet:** Includes roaches (*Eurycotis*) that inhabit cliff crevices (Schwartz, 1958a). **Parasites:** Ticks (Goldberg et al., 1996d). **Predators:** Arachnid *Phrynus* is a possible predator (Schwartz, 1958a). **Reproduction:** Breeds by direct development (Hedges and Díaz, 2004bbb; Stuart et al., 2008). **Size:** Male SVL to 63.7 mm, female SVL to 83 mm (Schwartz, 1958a); to 123.2 mm (Díaz and Cádiz, 2008).

Conservation Status: "Vulnerable" (Vales et al., 1998); "endangered" on the IUCN Red List (Hedges and Díaz, 2004bbb).

Eleutherodactylus zugi Schwartz, 1958

Distribution: Cuba, Sierra del Rosario; elevational range 150–390 m.

Habitat: Mesic broadleaf forests, by night in the deep forested gorge of the Río Manantiales on spray-wetted leaves of shrubs at heights of 5–120 cm, and on wet rock surfaces to 1.2 m above ground level, under palm trash and other debris by day (Schwartz, 1958c); between and under rocks, under logs, and in leaf litter (Díaz and Cádiz, 2008); terrestrial, on rocks in the Sierra de Camarones, not recorded outside forested habitats (Hedges and Díaz, 2004ccc; Stuart et al., 2008).

Abundance: Rare, encountered only once in several decades of fieldwork (S. B. Hedges in Hedges and Díaz, 2004ccc and in Stuart et al., 2008); common (Díaz and Cádiz, 2008). **Parasites:** Nematodes *Aplectana cubana* and *Oswaldocruzia lenteixeirai* (Baruš, 1972; Coy Otero and Lorenzo Hernández, 1982). **Predators:** Arachnids *Phrynus* sp. and *Ctenus vernalis* (Novo Rodríguez et al., 1985; Armas, 2001); adults taken by spiders (*Ctenus vernalis*; Alonso Bosch and Rodríguez Gómez, 2003b). **Reproduction:** Males call from near or on the ground (Schwartz and Henderson, 1991); call by day or night when weather is suitable, from the ground or leaves of small plants or fallen palm (*Roystonea regia*) fronds to heights that rarely exceed 0.5 m (Díaz and Cádiz, 2008); eggs laid on the ground, breeds by direct development (Hedges and Díaz, 2004ccc; Stuart et al., 2008). **Size:** Male SVL to 16.8 mm (Schwartz, 1958c), female SVL to 20.0 mm (Schwartz, 1960a).

Conservation Status: "Endangered" on the IUCN Red List (Hedges and Díaz, 2004ccc).

4 Family Hylidae

Figure 7. *Osteopilus septentrionalis* (Hylidae) is native to the West Indies, but has been introduced widely onto islands where it does not occur naturally. This individual was found on Saba, probably an introduction from St. Maarten, where the population also is introduced. Photograph by John Magor.

Hyla cinerea Schneider, 1799

Distribution: Southeastern United States, introduced in Puerto Rico.

Habitat: In the United States, inhabits mesic situations, including woods, most often those adjacent to permanent ponds or pools or pools with emergent aquatic vegetation (Wright, 1932).

Natural History: Essentially nothing is known about the Puerto Rican population.

Size: SVL to 57.5 mm in females (Duellman and Schwartz, 1958).

Conservation Status: The species has not been seen recently on Puerto Rico (Rivero, 1998); on the IUCN Red List as a species of "least concern" (Hammerson and Hedges, 2004c); as an alien species in the West Indies, the conservation status of *H. cinerea* is not a concern, except in the context of the impact it may have on native species.

Hyla squirella Bosc in Daudin, 1800

Distribution: Eastern North America, introduced on the Little Bahama Bank.

Habitat: In the United States, in a wide variety of habitats, utilizes almost any small body of water (Duellman and Schwartz, 1958); on Stranger Cay (Bahamas), on reeds in brackish-water marsh (Barbour, 1904).

Natural History: Essentially nothing is known about the Bahamian population. **Size:** SVL to 43.0 mm (Duellman and Schwartz, 1958).

Conservation Status: On the IUCN Red List as a species of "least concern" (Hammerson and Hedges, 2004d); as an alien species in the West Indies, the conservation status of *H. squirella* is not a concern, except in the context of the impact it may have on native species.

Hypsiboas heilprini Noble, 1923

Distribution: Hispaniola, islandwide, but disjunct, principally but not exclusively in uplands; elevational range, sea level to ~1,850 m.

Habitat: Mesic broadleaf forest, where associated with rapidly flowing or torrential streams (Schwartz and Henderson, 1991); stream-breeding amphibian found only in fast-flowing mountain streams in mesic broadleaf forests (Hedges et al., 2004v; Stuart et al., 2008).

Reproduction: Males call from surfaces or within cover of crevices of large rocks, boulders, and nearly vertical rock faces over which water is sheeting, also from shrub leaves 0.65 m above a stream, from branch tip over pool at foot of a 6.0 m waterfall, vine tangles ~9.1 m above a stream, monocot leaves at water level on a river bank, and on an aroid leaf above a muddy area, a mating pair in <2.5 cm of water (1 July), clutch of eggs in the wall of a streamside cave on 14 July (Schwartz and Henderson, 1991). **Size:** Male SVL to 54.3 mm, female SVL to 52.0 mm (Trueb and Tyler, 1974).

Conservation Status: Considered "vulnerable" on the IUCN Red List, mountain stream habitat is particularly sensitive, the species is absent from a number of streams in Haiti (Hedges et al., 2004v; Stuart et al., 2008).

Remarks: Until recently, this species had been placed in the genus *Hyla* (Faivovich et al., 2005).

Osteopilus brunneus Gosse, 1851

Distribution: Jamaica; elevational range, sea level to >1,500 m.

Habitat: Open woods (Dunn, 1926b); wet limestone forest (Thompson, 1996); rainforest and cloud forest (Hedges, in prep.); frequently associated with arboreal and terrestrial bromeliads (*Aechmea paniculigera* and *Hohenbergia* spp.) by day and night (Gosse, 1851; Barbour, 1910; Lannoo et al., 1987; Thompson, 1996); also on leaves of shrubs and trunks of trees (to 2.1 m above the ground), on the ground, on rocks beside a river, on the concrete railing of a bridge, on paved and gravel roads at night, in a coffee tree adjacent to a rocky cliff, and on the side of a palm 1.8 m above the ground (Schwartz and Fowler, 1973); in open woods, parkland, and montane forests, usually associated with bromeliads, mainly arboreal, can tolerate some degree of habitat disturbance, can live in good secondary forest if bromeliads are available (Hedges et al., 2004nnn).

Activity: Nocturnal (Lannoo et al., 1987). **Reproduction:** Bromeliads serve as calling sites (bouts of chorusing intermittent, lasting 30 sec to 2 min, once or twice/h) and for egg deposition and larval development, 3–36 tadpoles/bromeliad (Lannoo et al., 1987); although different species of bromeliads chosen as nest sites had different physical qualities (width, central axil diameter, amount of leaf litter), depth and volume of water, and exposure to sun were similar, *Hohenbergia* bromeliads used as nest sites versus those not used differed in width (nest sites smaller), axil diameter (smaller), water depth (shallower), water volume (greater), and pH (lower; Thompson, 1996); eggs late May–early September (Lynn, 1940); mother feeds tadpoles fertilized and unfertilized eggs (Lannoo et al., 1987), depending on larval development (fertilized eggs early in development, unfertilized eggs later; Thompson, 1996); tadpoles may take as many as 180 eggs (P. Vogel, 2000b). **Size:** Male SVL to 50.6 mm, female SVL to 75.8 mm (Schwartz and Fowler, 1973); tadpole SVL ~2 mm at hatching and ~15 mm at metamorphosis (P. Vogel, 2000b).

Conservation Status: On the IUCN Red List as a species of "least concern" (Hedges et al., 2004nnn).

Osteopilus crucialis Harlan, 1826

Distribution: Jamaica, current distribution centered in the Cockpit Country; elevational range, sea level to ~1,200 m.

Habitat: Wet limestone forest and rainforest (Hedges, in prep.); also on wet paved road in deciduous forest (Schwartz and Fowler, 1973); often in hollow trees with openings at heights of 1.2–4.0 m, but frequently calls from greater heights (Dunn, 1926b; Garrick et al., 1985); also encountered in bromeliads (*Hohenbergia urbiana*) attached to limestone outcrops and on tree trunks (Garrick et al., 1985); in mesic broadleaf woods and forests on tree trunks and in bromeliads, requires large dead trees, not found in substantially altered habitats, but can be found in regenerating forests (Hedges et al., 2004ooo; Stuart et al., 2008); male on a rock in the middle of a trail on two consecutive nights (A. J. Sánchez Muñoz, in litt., 17.III.2008).

Abundance: Rarely encountered or heard calling except in pockets of suitable habitat (Hedges et al., 2004ooo; Stuart et al., 2008). **Activity:** Nocturnal (Dunn, 1926b). **Behavior:** May practice phragmosis (Dunn, 1926b). **Defense:** Skin secretions irritating to eyes of mammals (Hedges, in prep.). **Reproduction:** Males call from tree hollows, sometimes by day (Dunn, 1926b); from bromeliads (Garrick et al., 1985); call from lichen-covered tree trunks and rock walls (Sánchez Muñoz, 2008); tadpoles (and presumably eggs) in bromeliads (Dunn, 1926b). **Size:** Male SVL to 92.6 mm, female SVL to 122 mm (Trueb and Tyler, 1974).

Conservation Status: "Endangered" on the IUCN Red List (Hedges et al., 2004ooo).

Osteopilus dominicensis Tschudi, 1838

Distribution: Hispaniola, islandwide, except in very xeric regions; elevational range, sea level to ~2,000 m; introduced in the Turks and Caicos Islands.

Habitat: Occurs in a variety of habitats, including scrub savanna, pine forest, and human-altered habitats (Mertens,

1939); mangrove edge on Isla Saona (RWH); primarily (but not exclusively) arboreal (Schwartz and Henderson, 1991); on the concrete walls of a spring at Turgeau, Haiti (Lynn, 1958); in an open cistern of an outdoor restaurant in Haiti (Olson, 1995); in artificial pools and fountains in Santo Domingo, Dominican Republic (Powell and Henderson, 2008c).

Activity: Largely nocturnal, juveniles are active by day (RWH and RP). **Behavior:** Apparently practices phragmosis (Williams et al., 1963); metamorphs exhibit a clumped distribution on Isla Saona (Schwartz and Henderson, 1991); an adult male (>65 mm SVL) parachuted from 11.8 m above the ground and landed 6.0 m from the base of the building from which it jumped (Bowersox et al., 1994a). **Defense:** Irritating skin secretions (Lynn, 1958). **Diet:** On Hispaniola, beetles (followed by orthopterans and adult lepidopterans) were the most frequently consumed prey and accounted for the greatest volume of prey (followed by orthopterans and leafhoppers), but orthopterans occurred in the most stomachs, followed closely by coleopterans, *Eleutherodactylus* sp. and *Anolis* sp. also were consumed (Duer et al., 1992). **Parasites:** Nematodes in stomachs (Duer et al., 1992). **Predators:** Barn Owl (*Tyto alba*; Wetmore and Swales, 1931); *Hypsirhynchus parvifrons* (Henderson et al., 1987a); *Uromacer catesbyi* (Henderson et al., 1987c); in Haiti, *Rana catesbeiana* (Neils and Bugbee, 2007). **Reproduction:** Opportunistic breeder in any temporary or permanent body of standing water (Hedges et al., 2004j); males call from the ground, bushes, and saplings, as well as from roofs and residential yards, breeding in March–April, in Santo Domingo, chorus during heavy rain on 21 February lasted all night (Mertens, 1939); amplexus on 3 September (Schwartz and Henderson, 1991); breeding in open cistern of outdoor restaurant (Olson, 1995); lays eggs in standing pools of rain water, sometimes stagnant (Noble, 1927), swimming pools, ditches, pools in slow-running creeks or streams (Schwartz and Henderson, 1991); also in roof-mounted cisterns, tadpoles in puddles of rust-stained water (Mertens, 1939); in puddles in tire ruts in dirt roads and in fountains in January, March, and June (RP); larger tadpoles with algae on their backs (Mertens, 1939). **Size:** Male SVL to 66 mm, female SVL to 99 mm (Schwartz and Henderson, 1991).

Conservation Status: On the IUCN Red List as a species of "least concern" (Hedges et al., 2004j).

Osteopilus marianae Dunn, 1926

Distribution: Jamaica, uplands; elevational range ~120–880 m.

Habitat: In bromeliads (Dunn, 1926b; Schwartz and Fowler, 1973); by day in wide-leaved bromeliad with spiny edges, in wet limestone forest, rainforest, pine forest (Hedges, 1987); also in deciduous forest (Schwartz and Fowler, 1973); in "thick woods"(Dunn, 1926b); restricted to old-growth forest, not tolerant of disturbances (Hedges et al., 2004ppp; Stuart et al., 2008).

Miscellany: No vocal sac, but vocalizes by means of vibrations (Hedges et al., 2004ppp; Stuart et al., 2008). **Reproduction:** Males heard calling in January and September from the narrow-leaved bromeliad *Tillandsia* at 3–4 m above the ground (Hedges, 1987); eggs and tadpoles in bromeliads (P. Vogel, 2000b; Hedges, in prep.). **Size:** Male SVL to 31.0 mm, female SVL to 34.0 mm (Schwartz and Fowler, 1973).

Conservation Status: "Endangered" on the IUCN Red List (Hedges et al., 2004ppp).

Osteopilus pulchrilineatus Cope, 1869

Distribution: Hispaniola; islandwide, but disjunct; elevational range, sea level to ~1,100 m.

Habitat: Broadleaf forests, their remnants, and human-disturbed situations (Schwartz and Henderson, 1991); mesic broadleaf forests, records from 40 years ago in anthropogenic habitats such as rice fields, but recent records are all from forests (Hedges et al., 2004w; Stuart et al., 2008).

Predation: *Uromacer catesbyi* (Henderson et al., 1987c). **Reproduction:** Males call from leaves 0.6–2.4 m above the ground (Lynn, 1958); call in flooded situations (pools, rice fields, pastures) after heavy rains, in backwaters left by receding rivers after floods, from vegetation near the ground, amplexus axillary, pair on a small stick 0.3 m above the water on 24 August (Schwartz and Henderson, 1991); eggs laid and larvae develop in still water (Hedges et al., 2004w; Stuart et al., 2008). **Size:** Male SVL to 39.5 mm (Trueb and Tyler, 1974), female SVL to 43.0 mm (Lynn, 1958).

Conservation Status: "Endangered" on the IUCN Red List, many historic populations appear to have been extirpated, not found in the Cordillera Central during extensive searches in 1998–2000, last recorded from Haiti in 1985 (Hedges et al., 2004w; Stuart et al., 2008).

Osteopilus septentrionalis Duméril and Bibron, 1841

Distribution: Bahama Islands (Little and Great Bahama banks, Long Island, Cat Island, Conception Island, Rum Cay,

San Salvador, Crooked-Acklins Bank, Great Inagua Island), Cuba (Isla de la Juventud, Archipiélago de los Canarreos, Archipiélago de Sabana-Camagüey, cays off the northern coast of Holguín Province, Cayos de San Felipe), and the Cayman Islands (Grand Cayman, Cayman Brac, Little Cayman); introduced to Puerto Rico, U.S. and British Virgin Islands, Anguilla, St.-Martin/St. Maarten, St.-Barthélemy, Antigua, Nevis, and Saba (apparently no breeding population); elevational range, sea level to ~1,110 m.

Habitat: Generally in mesic habitats but sometimes in xeric situations, lives in all kinds of disturbed habitats, including towns and houses, also present in forests, mangroves, and coastal areas, can tolerate brackish water, on the ground and tree trunks (Hedges et al., 2004a); frequently a human commensal, associated with cisterns, drainage ditches, basements, axils of banana trees and palms, flower pots, ornamental plantings (Roming, 1993); in the Bahamas, around human habitations, but also in pine forest (Schwartz and Henderson, 1991); on New Providence, in citrus trees and sisal plants by day (Barbour, 1904); on Cuba, in essentially all terrestrial ecosystems (Díaz and Cádiz, 2008); in cultivated plantations (including *Musa* groves) and cisterns (Barbour and Ramsden, 1919; Estrada, 1993a); after a heavy rain, "hundreds of giant treefrogs appeared from beneath the roofing tiles . . . almost every tile seemed to have a treefrog under it" (Neill and Allen, 1957); by day in cracks and crevices of building walls, in cisterns (barrels), and on rafters, forage at night on lawns (Z. Vogel, 1965d); under a tarpaulin covering a power boat, in cisterns and even toilets on Antigua (Lindsay and Cooper, 2008); diurnal retreats in the foliage of trees, hedges, and bushes, axils of banana plants, tree ferns, bromeliads, and palms, in rock crevices, rock piles, and caves, also dark recesses in buildings, often more than one individual at a suitable site, with 5–12 smaller individuals in a single banana axil, 4 large frogs in a rotten palm log with 2 *Tarentola americana* and 4 large *Leiocephalus carinatus*, 8 frogs in the rotten core of a thorn palm (*Bactris cubensis*) standing in a puddle of water, 2 individuals in the trunk of a tree fern (*Alsophila aquilina*) on the edge of a spring, forage at night, often on the ground, especially near water (G. Peters, 1974); may spend the day secreted in the base of *Musa* leaves (Ruiz García, 1987); also in leaves of the palm *Thrinax radiata* (Estrada, 1993a); in mangroves, mountain rainforest (to 1,000 m), and urban situations (Rodríguez Gómez and Alonso Bosch, 2003a; Kunz, 2005); will enter houses and may seek shower stalls and bathrooms for humidity, also discharge tanks of toilets (Ruiz García, 1987); also in weakly chlorinated swimming pools, canals, puddles, and drainage ditches (Herrmann, 2006); in third-floor rooms of a hotel (Krüger, 2008); also in high corners or on beams of rooms in abandoned houses, and in spherical nests of grassquits (*Tiaris*; Schwartz and Henderson, 1991); by day in hollow trunks and in bromeliads (Rodríguez Gómez and Alonso Bosch, 2003a); between leaves and bark of palms (Garrido et al., 1986); may seek diurnal refuge in burrows of *Bufo empusus* (Schwartz and Henderson, 1991); by day, between sugarcane leaves (Kwet, 1995); in the Cayman Islands, in areas of human activity (vicinity of buildings, cisterns, and rain barrels), on roads, in *Typha* marshes, flooded pastures (Seidel and Franz, 1994); on Puerto Rico, in ponds and ditches (Hedges, in prep.); in the U.S. Virgin Islands, mesic and xeric situations, in cisterns, ponds, ditches, and ravines with pooled water (Platenberg and Boulon, 2006); in the British Virgin Islands, human-altered situations, including cisterns and a plant nursery (Owen et al., 2006; Perry et al., 2006); on Anguilla, in residential areas (gardens, outbuildings, cisterns), on hotel grounds, and in an abandoned well (Townsend et al., 2000); on St.-Martin/St. Maarten, in scrub situations with some standing water (RWH); in wall crevices, gutters, open canals, behind window shutters, on mango and *Ficus* trees, and in axils of palm, coconut, and banana trees (Breuil, 2002b; RP); very adaptable, even to situations dramatically altered by human activity, in almost any area of high humidity and near freshwater, which is necessary for breeding, may utilize cisterns, drains, ornamental ponds, and swimming pools (Powell et al., 2005); on St.-Barthélemy, along roof beams, in cisterns, under planks, and in banana trees (Breuil, 2002b); on Saba, waifs perched on a house (Powell, 2007b; J. Magor; pers. comm.).

Activity: Nocturnal, at least during the dry season (G. Peters, 1974); largely nocturnal (Roming, 1993; Townsend et al., 2000); hides by day in natural or artificial crevices, axils of leaves (especially some palms), or hangs from undersides of large leaves (Powell et al., 2005). **Behavior:** "When caught [by a snake] the frog . . . lifts up its voice in an imitation of a kitten in trouble" (English, 1912); "scream when attacked by a predator and may respond accordingly when handled by humans" (Powell et al., 2005). **Competition:** May interfere with tadpoles of *Leptodactylus albilabris*, with which it shares ephemeral ponds for breeding (Platenberg, 2007). **Defense:** Skin secretions may be irritating (Z. Vogel, 1965d; MacLean, 1982). **Disease:** May be a vector for pathogens (Hedges et al., 2004a). **Foraging and Diet:** On St.-Martin/St. Maarten, hunts on mango, fig, and coconut trees, on walls, and around insect-attracting garden lights, will change foraging site about every 0.5 h (Breuil, 2002b); will forage around night lights on Anguilla (Henderson and Powell, 2001) and in the British Virgin Islands (Perry et al., 2008); on New Providence (Bahamas), prey includes arachnids, flies (Chironomidae), lepidopteran larvae, orthopterans, and roaches (Meshaka, 1996a); on Cuba, forages around insect-attracting lights (Amorín Ponce et al., 2003); conspecifics (G. Peters, 1974); also on Cuba, prey includes orthopterans (roaches), beetles (Elateridae), isopods (Oniscidae), arachnids, and the pulmonate gastropods (Polygeridae and Subulinidae), as well as conspecifics and *Anolis sagrei*, on Grand Cayman, primarily roaches, also *Eleutherodactylus planirostris* and *Gastrophryne carolinensis* (Meshaka, 1996a); on St.-Martin/St. Maarten, roaches, beetles, ants, on St.-Barthélemy, roaches (Breuil, 2002b); on Puerto Rico, adults and clutches of *Eleutherodactylus coqui* (Joglar, 2005a); will consume lizards with lengths exceeding that of the frog itself (Platenberg, 2007). **Longevity:** In captivity, 12 years, 11 months (Bowler, 1977). **Parasites:** In Cuba, trematodes Digenea gen. sp. (metacercariae; Baruš and Moravec, 1967a) and *Mesocoelium monas* (Pérez Vigueras, 1942; Coy Otero and Lorenzo Hernández, 1982); nematodes *Batracholandros bassi* (Walton, 1940; Baruš and Moravec, 1967a; Baruš, 1973; Coy Otero and Lorenzo Hernández, 1982; Coy Otero and Ventosa, 1984; Goldberg et al., 1994a); larval *Eustrongylides* sp. (Walton, 1940; Goldberg et al., 1994a); *Hammerschmidtiella diesingi* (Coy Otero et al., 1980; Coy Otero and Lorenzo Hernández, 1982; a parasite of roaches probably ingested adventitiously with prey; C. R. Bursey, in litt., 10.XII.2007); larval oxyurids, *Aplectana* sp., *A. hamatospicula*, *Physalopteroides valdesi*, *Foleyellides brachyoptera*, and larval *Contracaecum* sp., *Porrocaecum* sp., and *Abbreviata* sp. (Coy Otero and Ventosa, 1984; Goldberg et al., 1994a); *Oswaldocruzia lenteixeirai* (Pérez Vigueras, 1938; Baruš and Moravec, 1967a; Coy Otero and Lorenzo Hernández, 1982; Coy Otero and Ventosa, 1984; Goldberg et al., 1994a); *Parapharyngodon osteopili* (Adamson, 1981; Goldberg et al., 1994a); in the Bahamas (San Salvador Island), nematodes *Oswaldocruzia lenteixeirai*, *Parapharyngodon osteopili*, female strongylid, and a larval oxyurid (Goldberg et al., 1994a); *Physalopteroides bahamensis* (Bursey and Goldberg, 1994; Goldberg et al., 1994a); sporozoan *Lankesterella* sp. (Zajicek and Mauri Méndez, 1969); in the Cayman Islands, Bahamas, and Cuba, nematode *Skrjabinoptera scelopori* (Meshaka, 1996c). **Predators:** Conspecifics (Meshaka, 1996a); *Anolis porcatus* (on larvae; L. Moreno in Rodriguez Schettino, 1999b); snakes *Epicrates striatus* in the Exuma Islands, Bahamas (Franz et al., 1993); *Cubophis cantherigerus* on Cuba (Novo Rodríguez and Arazoza Rodríguez, 1986; Henderson and Sajdak, 1996); *C. vudii* on New Providence, Eleuthera, Andros, Cat Island, and Crooked Island (Henderson and Sajdak, 1996); West Indian Red-bellied Woodpecker (*Melanerpes superciliaris*) on Cuba (Estrada, 1993a); a Giant Waterbug (*Lethocerus americanus*) feeding on a dead adult male *O. septentrionalis*, planarians eating eggs (Díaz, 1999); Cattle Egrets (*Bubulcus ibis*) on St. Croix (Gassett et al., 2000); on Cayman Islands, West Indian Red-bellied Woodpecker (*M. s. caymanensis*), Loggerhead Kingbird (*Tyrannus caudifasciatus caymanensis*), and Greater Antillean Grackle (*Quiscalus niger*; Johnston, 1975); Barn Owl (*Tyto alba furcata*) on Cuba (Arredondo Antúnez and Chirino Flores, 2002). **Reproduction:** Males call from vegetation near pooled rainwater, eggs laid in still water, including pools, marshes, flooded pastures, and ditches (Hedges et al., 2004a); on Grand Cayman, "within a few hours [eggs] are tadpoles and within 4 or 5 weeks these tadpoles are frogs" (English, 1912); reproduction begins during the dry winter season, tadpoles as early as 14 November in deep pools of a stream, on 9 February in a spring-fed roadside ditch, metamorphs in grass and herbaceous growth, males actively calling on rainy nights in February, some individuals calling every night in early December, sometimes even during the day (G. Peters, 1974); males call from leaves, branches, trunks of saplings, vertical walls adjacent to pools of rainwater, pinewoods, mixed pine and hardwood, and flooded areas in *Terminalia* stands (Schwartz and Henderson, 1991); call from artificial pools (Díaz et al., 2005b); in northwestern Puerto Rico, breeding is explosive and events last only one night, mating is random and non-assortative by size (mature females 57–107 mm SVL versus 41.5–67 mm for mature males), larger females lay significantly more eggs than smaller females (Vargas-Salinas, 2006); on Cuba, eggs deposited in rainwater in ornamental jars in a garden (Neill and Allen, 1957); on St.-Martin/St. Maarten, call late July–mid-August, tadpoles in late July in wells, cisterns, small rivers, large metal containers at abandoned sugar factory (Breuil, 2002b); on Anguilla, males call in June from branches of mangroves over water (RP); reproduction associated with the rainy season (May–October), although males may call throughout the year, females lay clutches of ~130 eggs in scattered masses, eggs capable of surviving exposure to saline water, eggs may hatch in as few as 2 days, tadpoles grow rapidly (Powell et al., 2005); on Cuba, males call primarily at night, especially during rain, during peak reproductive periods, calling can extend into the morning, breeding is seasonal (mostly April–November, but throughout the year in the mountains, amplexed pairs float in water, deposit eggs on the bottom or attached to vegetation in ponds, streams, rivers, and artificial bodies of water, tadpoles primarily in still or slowly moving water (Díaz and Cádiz, 2008); eggs deposited in freshwater situations, both permanent and temporary (Estrada, 1993b); on Cayman Brac, call from trees near a cistern, amplexus (axillary) in the cistern (Seidel and Franz, 1994); on Grand Cayman, choruses in *Typha* marshes, flooded pastures, and roadside ditches with Black Mangrove (*Avicennia nitida*) continued from dusk until about 2400 h, males appear at dusk, with females arriving 1–2 h later (Seidel and Franz, 1994); on St.-Barthélemy, eggs in pools, in garden basins, and cisterns, tadpoles in cisterns at the end of July (Breuil, 2002b); females may lay up to 2,000 eggs in 2–3 clutches during a breeding season (in captivity), eggs float on the surface, hatching in 24 h at temperatures 22–26 °C, tadpoles active at lengths 10–12 mm, maximum larval size 50–70 mm, metamorph SVL 18–22 mm (Roming, 1993); in captivity, amplexus lasted to at least 20 h, ~2,000 eggs 1.0–1.5 mm in diameter, gelatinous envelope ~3 mm in diameter, some eggs no older than 24 h began hatching (but some eggs did not hatch for 45 days), tadpoles ~7 mm long, fore- and hindlimbs at 40–45 mm, the tail resorbed 2–5 days later, metamorphs 12–15 mm, larger frogs cannibalized smaller individuals, some males began calling at 6 months of age (Königstedt and Königstedt, 1982); in captivity, ~2,000 eggs 1.0–1.5 mm in diameter, hatching begins within 24 h, tadpoles begin metamorphosis after ~40 days at ~45 mm total length (Herrmann, 2006). **Size:** Female SVL to 165.0 mm (Meshaka, 1996b); males only about half as large (Z. Vogel, 1965d). **Survivorship:** In northwestern Puerto Rico, 0.72 in males (95% confidence interval 0.66–0.78) and 0.91 (0.88–0.93) in females (Vargas-Salinas, 2006). **Thermoregulation:** Adult females perch in direct sunlight on tree trunks in forested areas (Seidel and Franz, 1994).

Conservation Status: On the IUCN Red List as a species of "least concern" (Hedges et al., 2004a); introduced populations compete with and prey on native species (Hedges et al., 2004a); populations of *Eleutherodactylus martinicensis* on St.-Barthélemy might have declined due to predation by *O. septentrionalis* and on St.-Martin/St. Maarten, *E. johnstonei* may be in decline for the same reason (Breuil, 2002b); on Puerto Rico, the presence of *O. septentrionalis* affects native frogs by (1) displacing them from diurnal refuges, (2) reducing calling frequencies, (3) causing them to capture fewer prey, (4) reducing attendance of clutches, (5) reducing hatching success, and (6) preying on both adults and clutches (N. Ríos, M. Cardona, and R. Joglar in Burrowes and Joglar, 2005).

Additional Pertinent Literature: Duellman and Crombie (1970).

Osteopilus vastus Cope, 1871

Distribution: Hispaniola, disjunct islandwide distribution; elevational range, sea level to ~1,700 m.

Habitat: Occurs in broadleaf forest, most often in gallery forest along creeks and streams (Mertens, 1939; Schwartz and Henderson, 1991; Hedges et al., 2004x; Stuart et al., 2008).

Activity: Nocturnal, occasionally exposed in direct bright sunlight (1000–1100 h) in water conservation posture (Mertens, 1939); during dry weather hides high in the tops of trees, then descends at night to court, in the afternoon during the rainy season in a drizzling mist on low branches not far from streams (Williams et al., 1963). **Defense:** Skin secretions may be dangerous to some animals (other frogs, dogs), and may cause skin irritation in humans (Noble, 1923a; Williams et al., 1963). **Disease:** Chytridiomycosis (Joglar and Burrowes, 2005; Joglar et al., 2007); dead animals and animals with deformities (one individual with an eye on its back) suggestive of chytridiomycosis (Hedges et al., 2004x; Stuart et al., 2008). **Longevity:** In captivity, 6 years, 7 months (Curator of Reptiles, Chicago Zoological Park in Slavens and Slavens, 2003). **Reproduction:** Males call from tree limbs 1.8–9.0 m above running water, eggs small, 2.5 mm diameter (Mertens, 1939); eggs in small basins in gravel and stone at edges of pools in a mountain torrent (Noble, 1927); tadpoles and metamorphs in pools and slow-flowing segments of a lowland stream in a coffee plantation in the Sierra de Baoruco in June (RP). **Size:** Male SVL to 108.8 mm, female SVL to 141.9 mm (Trueb and Tyler, 1974).

Conservation Status: "Endangered" on the IUCN Red List, few records in recent decades other than in the Cordillera Central and Santo Domingo (Hedges et al., 2004x; Stuart et al., 2008).

Additional Pertinent Literature: Henderson and Henderson (2004).

Osteopilus wilderi Dunn, 1925

Distribution: Jamaica; elevational range ~120–880 m.

Habitat: Open woods (Dunn, 1926b); wet limestone forest and rainforest (Hedges, in prep.); in arboreal and terrestrial bromeliads (Lynn and Dent, 1943; Schwartz and Fowler, 1973); can occur in secondary growth, as long as bromeliads are present (Hedges et al., 2004qqq; Stuart et al., 2008).

Reproduction: Males call from bromeliads (Sánchez Muñoz, 2008); eggs and tadpoles in bromeliads (P. Vogel, 2000b; Hedges et al., 2004qqq; Stuart et al., 2008). **Size:** Male SVL to 27.9 mm, female SVL to 29.2 mm (Schwartz and Fowler, 1973).

Conservation Status: "Endangered" on the IUCN Red List (Hedges et al., 2004qqq).

Scinax ruber Daudin, 1802

Distribution: Eastern Panama and much of the Guianas and the Amazon Basin of South America; introduced to Puerto Rico, St. Lucia, and Martinique; reports of the species on St.-Martin/St. Maarten are in error (Powell et al., 1992).

Habitat: On Puerto Rico, swimming or floating in ponds or ditches (Rivero, 1998); on St. Lucia, climbing on the wall of a hotel dining room at 1900 h (Schwartz and Henderson, 1991); on Martinique, in pastures, meadows, orchards, banana groves, gardens, and edificarian situations (Breuil, 2009); at night on bushes and shrubs (Breuil and Ibéné, 2008).

Activity: Hides by day in construction materials, hunts at night on walls and trees (Breuil, 2009). **Reproduction:** Reproductively active all year on Puerto Rico (Rios-López, 1999); on Martinique, reproduces in tanks, livestock water troughs, barrels, abandoned appliances, ponds, and other small bodies of water (Breuil and Ibéné, 2008; Breuil, 2009). **Size:** Male SVL to 41.2 mm, female SVL to 43.8 mm (Duellman and Wiens, 1993).

Conservation Status: On the IUCN Red List as a species of "least concern" (Solís et al., 2004b, 2008b); as an introduced species in the West Indies, the primary concern is its potential impact on native species.

Scinax cf. *x-signatus* Spix, 1824

Distribution: Venezuela to Surinam, eastern, southern, and southeastern Brazil; introduced to Guadeloupe (Basse-Terre and Grande-Terre), Marie-Galante, and La Désirade.

Habitat: Freshwater lakes, brackish lagoons with mangroves, edges of forest swamps, and pastures devoid of trees (Breuil and Ibéné, 2008).

Activity: Nocturnal (Breuil and Ibéné, 2008). **Reproduction:** In various situations, including freshwater lakes, brackish lagoons with mangroves, edges of forest swamps, and pastures devoid of trees (Breuil and Ibéné, 2008). **Size:** Maximum SVL 55 mm (Breuil and Ibéné, 2008).

Conservation Status: This species was first discovered on Grand-Terre in 2003, probably was introduced with ornamental vegetation from the Neotropical mainland, and it has rapidly extended its range on Grande-Terre (Breuil and Ibéné, 2008); on the IUCN Red List as a species of "least concern" (Trefaut Rodrigues et al., 2004); as an introduced species in the West Indies, the primary concern is its potential impact on native species.

5 Family Leptodactylidae

Figure 8. *Leptodactylus validus* (Leptodactylidae) presumably is native to St. Vincent and Grenada, but lack of genetic divergence with populations on Trinidad and Tobago and mainland South America is suggestive of a recent introduction. Photograph by Robert Powell.

Leptodactylus albilabris Günther, 1859

Distribution: Puerto Rico, Isla Vieques, Isla Culebra, U.S. and British Virgin Islands, Hispaniola; elevational range, sea level to ~1,030 m.

Habitat: Semiaquatic (Schwartz and Henderson, 1991); in streams, marshes, and irrigated fields, as well as drainage ditches and gutters in forests, open country, and urban areas (Hedges et al., 2004xx; Rios-López, 2008b; Stuart et al., 2008); not arboreal, rarely encountered far from water (Rivero, 1998); in Puerto Rico, in streams and marshy areas (Stejneger, 1904); "especially abundant" in moist meadows and irrigated sugarcane fields (Bhajan, 1966); also in towns and villages, in drains and gutters ("regardless of filth" if muddy water accumulates) as well as moist meadows, irrigated sugarcane fields, coffee plantations (concealed beneath logs and stones), and beneath loose stones on a bath house terrace (K. P. Schmidt, 1920, 1928); breeds in coastal wetlands, but abundance declines as salinity increases (Rios-López, 2008a); on St. John, "plentiful" under rocks in muddy ponds along dry stream beds, also froth nests containing eggs and larvae (Grant, 1932g); in the Virgin Islands, "common in stream beds and around water holes" (Seaman, 1961); in marshy spots where the forest is not dense (Drewry, 1970b); in British Virgin Islands, in area of the Slob (Anegada), in wet areas under dense shrubs on coral-limestone substrate covered by leaf litter and humus, and in muddy ditches around the airport on Beef Island (Ovaska et al., 2000); on Puerto Rico, in leaf litter under trees (especially *Bucida buceras*) presumably foraging for insects (A. J. Sánchez Muñoz, in litt., 17.III.2008); abundant in gutters in Road Town, Tortola (J. Lazell, in litt., 21.IV.2008).

Abundance: Abundant in suitable habitat (Hedges et al., 2004xx; Stuart et al., 2008); highest May–November, with peak August–November, lowest December–March, to 168.1 individuals/ha in coastal wetlands (Rios-López, 2008a). **Activity:** Largely nocturnal (RP). **Competition:** Tadpoles may be negatively affected by those of introduced *Osteopilus septentrionalis* and *Bufo marinus* in the U.S. Virgin Islands, where the species share ephemeral ponds for breeding (Platenberg, 2007). **Diet:** On Puerto Rico, includes land snails (in 4 of 25 stomachs), spiders (2), ants (2), beetles (2), bugs (2), flies (2), moth (1), lepidopteran larva (1), cockroach (1; K. P. Schmidt, 1920, 1928); larvae scavenged on dead *Anolis* (Lebron et al., 1995). **Disease:** *Batrachochytrium* *dendrobatidis* fungal infections (Burrowes et al., 2008). **Larval Development:** In the laboratory, Puerto Rican tadpoles from a montane population maintained at 20 °C (approximating the environmental temperature) metamorphosed within 111 days, whereas those from a lowland population maintained at 30 °C (approximating the environmental temperature) metamorphosed within 25 days, at 30 °C the montane larvae transformed in 30.5 days, and the lowland larvae required 252 days at 20 °C for metamorphosis (Bhajan, 1966). **Population Density:** Maximum density in suitable habitat 10,830 individuals of all sizes per hectare (Joglar et al., 2005b); 13.1 adults/ha in inland forested habitats (Rios-López and Aide, 2007); to 168.1 adults/ha in coastal wetlands (Rios-López, 2008a). **Predators:** On Puerto Rico, Greater Antillean Grackle (*Quiscalus niger*; Wetmore, 1916); *Rana catesbeiana* (Thomas and Joglar, 1996); *Borikenophis portoricensis* in captivity (N. Ríos and R. Joglar in Joglar, 2005a); Red-legged Thrush (*Turdus plumbeus*; Rolle, 1963; Joglar, 2005a); Cattle Egrets (*Bubulcus ibis*) on St. Croix (Gassett et al., 2000). **Reproduction:** Adapted to "breeding" in temporary ponds (Heatwole et al., 1968); males call (diurnally and nocturnally) from beneath rock edges or sticks or flat on the ground (K. P. Schmidt, 1928); from within a well on Anegada (British Virgin Islands; Carey, 1972); from submerged or emergent clumps of aquatics or matted shoreline grasses, spring seepages, rain pools, wet but cracked mud in a forest (Schwartz and Henderson, 1991); on Puerto Rico, eggs in January (El Yunque) and March (Río Piedras), on 7 April eggs and all stages of tadpoles in little streams behind the town of Utuado (Bhajan, 1966); in the British Virgin Islands, call from inside tufts of grass and from small cavities in mud banks close to water's edge (Beef Island), on Jost Van Dyke in a riverbed (Old Hill) and a marshy site in Great Harbor, and on Anegada from burrows of land crabs (Ovaska et al., 2000); from roadside gutters in Road Town (Tortola) in March (J. Lazell, in litt., 21.IV.2008); on Puerto Rico, nests typically subspherical smooth-walled cavities hollowed out of soft earth, about 3.0 cm deep and 6.0 cm at greatest diameter, under a stone beside a stream, under fallen palm fronds and palm debris in a swampy area, and within 1–2 m of a small pond, where nests were covered with vegetable debris, stones, or domes of earth (Dent, 1956); on Puerto Rico, a nest (foam "ball" 7.0 cm in diameter with central canal) under a large flat stone in a small shallow stream on 7 April (Stejneger, 1904); on 1 October, a nest (8 cm diameter) under a rotten log beside a pool of standing water with 150–200 eggs (2.5–3.0 mm diameter) uniformly distributed through the foam and with no central hollow, on 27 August, a 3.0-cm deep excavation (6–7 cm in diameter) filled with white foam with 75–100 tadpoles (K. P. Schmidt, 1920, 1928); tadpoles develop within nests (rarely) or in temporary pools or channels after being washed away by water runoff during heavy rains (Rivero, 1998); tadpoles tolerate water temperatures to 40–42 °C (Heatwole et al., 1968); in Puerto Rico, reproduction apparently not seasonal (Dent, 1956), with months during which juveniles are abundant varying from year to year (Joglar et al., 2005b); eggs hatched on the fourth day after deposition, metamorphosis completed in 73 days in captivity, but frogs were nutrition deprived and 35 days is likely more accurate (Dent, 1956). **Salinity Tolerance:** In laboratory experiments, survival to metamorphosis decreased from 41.1% at 2 ppt to 1.7% at 8 ppt, growth in length of tadpoles in 8 ppt not affected, but mass was significantly less than in controls (Rios-López, 2008a). **Size:** Male SVL to 43 mm (Hedges, in prep.); female SVL to 56 mm, smallest juveniles 11.2 mm SVL (Joglar et al., 2005b).

Conservation Status: On the IUCN Red List as a species of "least concern" (Hedges et al., 2004xx; Rios-López, 2008b); "stable" (Moreno, 1991d; Joglar et al., 2005b); populations in the Dominican Republic considered "endangered" (Hedges et al., 2004xx; Stuart et al., 2008).

Leptodactylus fallax Müller, 1926

Distribution: Montserrat and Dominica, apparently extirpated on Guadeloupe, Martinique, and St. Christopher, might once have occurred on St. Lucia and Antigua, unsuccessful attempts to introduce it to Jamaica and Puerto Rico; elevational range, sea level to ~400 m on Dominica and to ~1,000 m on Montserrat.

Habitat: Optimal habitat appears to be undisturbed primary forest with permanent waterflow, with which frogs are closely associated in the dry season, but from which they disperse widely during the rainy season (G. García, 2008); mesic habitats in woods, scrub, plantation hillsides, and a wet palm grove in a river valley (Lescure, 1979); on Dominica, in mountainous terrain with abandoned tree crops and gardens in various stages of succession to forest, and second growth moist evergreen forest with a few deciduous species (Davis et al., 2000); on Dominica, also in fruit and vegetable plantations (Kaiser and Henderson, 1994); on Montserrat, distribution highly localized in the northern part of the island (Kaiser and Henderson, 1994); appears to prefer habitats with high vegetation (Ajayi, 2001); dense secondary vegetation, plantations (Dominica only), ravines, and flooded forest, terrestrial, hiding in burrows during the day in moist forest, apparently associated with certain soil types that allow the digging of nests (Fa et al., 2004; Stuart et al., 2008).

Activity: Ground dwelling, nocturnal (Lescure, 1979); primarily nocturnal, spends days in holes, burrows, or moist leaf litter, leaving retreats at dusk to forage or search for mates during the breeding season (G. García, 2008). **Behavior:** Males compete for territories, resulting battles involve elevation on hindlimbs in an effort to force the competitor to the ground (G. García, 2008); peptide skin secretions produced only by males apparently promote aggressive behavior (King et al., 2005). **Disease:** Chytridiomycosis on Dominica (A. James, pers. comm.; Fa et al., 2004; G. García, 2008; Stuart et al., 2008; Zoological Society of London, 2008). **Foraging and Diet:** Sit-and-wait forager, diet (Dominica) includes orthopterans (Gryllidae, especially the cricket *Amphiacusta*), beetles (especially scarabeids), harvestmen (*Vonones*), gastropods, millipedes, centipedes (including *Scolopendra*), and decapod crustaceans, vertebrates include *Eleutherodactylus, Anolis, Typhlops dominicanus, Boa nebulosa*, bats, and a small rodent (Lescure, 1979; Brooks, 1982); on Montserrat, adult female (175 mm SVL, 510 g) captured a female *Alsophis manselli* (650 mm total length, 75 g), could not accommodate the entire snake, so the tail protruded from the mouth (Buley, 2003). **Longevity:** To 12 years of age, mature in 2 years, pair in captivity bred at 22 months of age (G. García, 2008). **Population Size and Density:** On Montserrat, density estimates along 200-m trail segments of 252 ± 75.7 frogs at Fairy Walk, 39 ± 6.2 at Weetwater Ghaut, and 7.5 ± 3.0 at Cat Ghaut (G. García, 2008); sampling showed considerable seasonal variation, with highest numbers during the dry season (January–March), when frogs are most closely associated with streams (R. P. Young, 2008 in G. García, 2008). **Population Structure:** Drop in mean SVL of sampled frogs in recent years suggestive of reductions in fecundity or juvenile survivorship on Montserrat (Daltry and Gray, 1999). **Predation:** Heavily hunted for food by humans (Krintler, 1986; Kaiser and Henderson, 1994; Malhotra and Thorpe, 1999; Kaiser and Day, 1995); as many as 50 frogs killed per day until recently (Daltry and Gray, 1999); falls prey to cats and dogs (Faaborg and Arendt, 1985); predation by mongooses (*Herpestes javanicus*) may have been responsible for the species' extirpation on islands where it no longer occurs, 5–27% of frogs captured during the dry season in 2005 on Montserrat had rat bites (G. García, 2008). **Reproduction:** On Dominica, males call on rainy nights (rarely during the day) at least mid-February–September (Lescure, 1979; Davis et al., 2000), and possibly all year, males (spaced ~20–100 m apart) maintained the same calling sites over a 4–5-month span (Davis et al., 2000); call primarily April–September on Dominica (Malhotra and Thorpe, 1999); call most intensely in May and June on Montserrat (G. García, 2008); call has territorial significance (Lescure, 1979); calling males spaced 6–12 m apart (Ajayi, 2001); calls increase in frequency from beginning to end, females sometimes initiate calls or respond to male calls, advertisement calls of males did not overlap, and all-night callers took turns on different nights, shorter-term callers made a duet or trio, with each male's call alternating antiphonally (Davis et al., 2000); 2,024 eggs deposited in a foam nest at the bottom of a burrow (0.5 m long, 0.2–0.3 m deep; Lescure, 1979); on 13 April, a male calling from a cavity under a patio at 2300 h was soon joined by a female as the male called ~120 times/min, the female joined his calls every fourth note in a synchronous manner, a foam nest on 22 April, 15–20 larvae on 25 April (no yolk sac present on 24-day-old larvae), on 7 June, froglets jumped back into the foam nest when disturbed (Davis et al., 2000); based on captive observations of frogs from Montserrat, males fought (wrestling while on hind legs) and the dominant male occupied a nesting burrow, females were enticed into the burrow with a trilling call and a terrestrial foam nest was produced after 9–14 h, males and females stayed close to nests and defended them aggressively during larval development (42–57 days), females fed larvae (26–43/nest) unfertilized eggs (10–13 provisioning events supplied a total of 10,000–25,000 eggs), trophic eggs only food source for developing larvae (Gibson and Buley, 2004); in captivity, produce only one clutch/year, 25–43 metamorphs/nest (G. García, 2008). **Size:** To 210 mm SVL (Daltry and Gray, 1999); male SVL to 158.7 mm, female SVL to 167.2 mm (Heyer, 1979); adult mass to >900 g, tadpoles to 150 mm, metamorph SVL ~30 mm, mass 1–3 g (G. García, 2008).

Conservation Status: On the IUCN Red List as "critically endangered" (Fa et al., 2004); sold openly in markets on Dominica (Krintler, 1986); monitoring of commercial take recommended on Montserrat (Johnson, 1988); on Montserrat, dead and dying juveniles in shallow rock pools affected by ash-laden runoff (Kaiser and Day, 1995); at least 10% of the species' habitat on Montserrat destroyed by pyroclastic flows and all of its habitat affected by acid rain (with pH levels as low as 2.0) and volcanic ash fallout, range reduced to <17 km^2, nevertheless, frogs still "patchily abundant" in the moist forest of the Centre Hills (Daltry and Gray, 1999; G. García, 2008); feral pigs (*Sus scrofa*) implicated in habitat destruction on Montserrat (G. García, 2008); population declines from 1999 to 2005 on Montserrat not nearly as dramatic as those on Dominica, but some areas more vulnerable to human activity than those sampled (R. P. Young, 2008 in G. García, 2008); chytridiomycosis was first recognized as a threat in December 2002 on Dominica, between 2002 and 2004, the disease is thought to have reduced populations by 60% (S. McIntyre, 2003 and C. Magin, 2004 in G. García, 2008) or even 70% (Zoological Society of London, 2008); populations on Montserrat appear free of chytridiomycosis (G. García et al., 2007).

Additional Pertinent Literature: Kaiser (1994).

Leptodactylus validus Garman, 1887

Distribution: St. Vincent, Grenadines (Bequia), Grenada, Trinidad & Tobago, and northern South America; elevational range, sea level to ~730 m; the species may have been introduced into the West Indies by human activity (Yanek et al., 2006).

Habitat: Terrestrial in forest edges often close to stream banks, shaded gullies, and caves, also in meadows and at roadsides in parks and rural yards, adapts well to altered habitats (Hardy et al., 2004); mesic habitats, ground dwelling, in shallow water-filled roadside ditches (St. Vincent and Grenada), under exposed board in concrete enclosure under catchment spigot (Schwartz and Henderson, 1991); in rain-saturated pasture adjacent to a wooded area, and in water-filled tire ruts on muddy roads (Grenada; Henderson and Berg, 2005a); in leaf litter at the Botanic Garden on St. Vincent (RWH and RP).

Activity: Largely nocturnal (Schwartz and Henderson, 1991). **Predation:** *Mastigodryas bruesi* (Schwartz and Henderson, 1991). **Reproduction:** Breeds in waterbodies such as streams and drainage ditches, eggs laid in foam nests in temporary ponds and ditches, and the larvae develop in water (Hardy et al., 2004); a calling site on Grenada under an overhang under a bank at water's edge after a heavy morning rain (Schwartz and Henderson, 1991); in shallow, slow-moving streams in the Botanic Garden on St. Vincent on rainy nights (RP). **Size:** Male SVL to 42.9 mm, female SVL to 51.5 mm (Treglia, 2006).

Conservation Status: On the IUCN Red List as a species of "least concern" (Hardy et al., 2004).

6 Family Microhylidae

Figure 9. Introduced *Gastrophryne carolinensis* (Microhylidae) is firmly established on Grand Cayman, where it occurs in cultivated areas over most of the island. This male was part of a breeding aggregation in a flooded field in Georgetown. Photograph by Arthur C. (Sandy) Echternacht.

Gastrophryne carolinensis Holbrook, 1836

Distribution: Southeastern United States; introduced on the Little and Grand Bahama banks, and on Grand Cayman, Cayman Islands.

Habitat: On Grand Cayman, occurs in cultivated areas over most of the island (Seidel and Franz, 1994).

Predation: *Osteopilus septentrionalis* (Meshaka, 1996a).
Size: Snout to urostyle length 22–32 mm (Seidel and Franz, 1994).

Conservation Status: On the IUCN Red List as a species of "least concern" (Hammerson and Hedges, 2004a); as an introduced species in the West Indies, our concern is for the impact these frogs may have on native species; they probably arrived in George Town (Grand Cayman) in the root systems of exotic plants from nurseries in the Miami, Florida, area (Seidel and Franz, 1994).

Additional Pertinent Literature: Nelson (1972).

7 Family Ranidae

Figure 10. *Rana catesbeiana* (Ranidae), introduced onto the Greater Antilles for exploitation as a delicacy, has become ubiquitous in suitable habitat on all four islands. This individual was captured in a hallway of the science building at the University of Puerto Rico at Humacao. Photograph by Luis O. Nieves.

Rana catesbeiana Shaw, 1802

Distribution: North America east of the 100th meridian into northern Mexico; introduced widely as a potential food source; introduced into the West Indies on Cuba (including Isla de la Juventud), Puerto Rico, Hispaniola, and Jamaica.

Habitat: On Cuba, in bodies of water of various sizes, along shores, often in cavities (Sampedro et al., 2003); in rivers, ponds, lakes, and other bodies of water, may move away from water when conditions are sufficiently moist, along shores, often hidden in vegetation (Díaz and Cádiz, 2008); on Hispaniola, ponds and slow-moving streams (Powell et al., 1999); locally abundant on lawns near bodies of water in a resort/golf complex (RP); on Jamaica, in a herbaceous swamp (Mahon and Aiken, 1977); on Puerto Rico, in large bodies of water such as lakes, ponds, and bogs, but also in sluggish streams, fountains, and tanks (Rivero, 1998).

Activity: On Cuba, active by day and night, although primarily nocturnal (Sampedro et al., 2003). **Behavior:** Jumps into water from the shore at the least sign of danger, often while emitting a characteristic alarm call (Díaz and Cádiz, 2008). **Foraging and Diet:** One captured individual on Cuba contained the bones of a bat (Z. Vogel, 1965d); at two sites in Cuba, beetles (most important at both sites), orthopterans, roaches, hymenopterans, earwigs, dragon- and damselflies, flies, hemipterans, neuropterans, lepidopterans, spiders, millipedes, annelids, gastropods, crustaceans, fishes (second in importance at one site, not taken at all at the other), amphibians, and a mammal (Sampedro Marín et al., 1985); at the Ciénaga de Zapata (Cuba), juveniles foraged nocturnally farther from water than adults, diets of adults (in decreasing order of importance) included beetles, orthopterans, hemipterans, roaches, earwigs, spiders, gastropods, fishes, lepidopterans, crustaceans, juvenile conspecifics, millipedes, hymenopterans, and flies, diets of juveniles (in decreasing order of importance) included beetles, orthopterans, spiders, earwigs, crustaceans, millipedes, hemipterans, flies, fishes, gastropods, lepidopterans, and hymenopterans (Montañez Huguez et al., 1996); on Cuba, the diet includes insects (especially beetles), arachnids, millipedes, snails, annelids, and fish, adults will cannibalize smaller individuals (Sampedro Marín et al., 2003); on Jamaica, takes largely aquatic prey (81 of 102 prey items) comprising predominantly hemipterans (41 of 102 prey items), snails (20), and fishes (11; Mahon and Aiken, 1977);

on Puerto Rico, includes roaches, crickets, termites, borers, weevils, cutworms, fire ants, and conspecifics (Perez, 1951); also on Puerto Rico, preys on the endemic anuran *Leptodactylus albilabris* (Thomas and Joglar, 1996); in Haiti, on *Osteopilus dominicensis* (Neils and Bugbee, 2007). **Predation:** On Puerto Rico, Indian Mongoose (*Herpestes javanicus*; Wolcott, 1953); on Cuba, *Tretanorhinus variabilis* (Neil, 1965) and Barn Owls (*Tyto alba furcata*; Arredondo Antúnez and Chirino Flores, 2002). **Reproduction:** In Cuba, eggs in December–May, metamorphosis requires 6 months (Sampedro et al., 2003); in Puerto Rico, breed May–July, males aggressively territorial (Rivero, 1998); in La Habana (Cuba), in an artificial pond, metamorphosis variable, but begins in June when environmental variables (water temperature, sedimentation) stabilize, relative abundance of larvae peaks in January and April when water temperature is at its lowest, a significant negative correlation exists between water temperature and monthly abundance of larvae (Sampedro Marín et al., 1992). **Size:** SVL to 270 mm on Cuba (Sampedro Marín et al., 2003).

Conservation Status: On the IUCN Red List as a species of "least concern" (Santos-Barrera et al., 2006); as an alien species in the West Indies, the conservation status of *R. catesbeiana* is not a concern, except in the context of the impact it may have on native species.

Remarks: Recent work by Frost et al. (2006) placed this species in the genus *Lithobates.*

Rana grylio Stejneger, 1901

Distribution: Southeastern United States, but introduced on Great Bahama Bank (Andros and New Providence islands) and on Puerto Rico.

Habitat: In the Bahamas, occurs in mangroves (Schwartz and Henderson, 1991); on Puerto Rico, in a flooded channel where females, subadults, and juveniles were partially submerged in open water surrounded by vegetation (Rios-López and Joglar, 1999).

Reproduction: Males call from large clumps of *Typha dominguensis* (not openly exposed) in March, August, and September–November (Rios-López and Joglar, 1999). **Size:** Female SVL to 162 mm in the United States (Conant and Collins, 1998).

Conservation Status: On the IUCN Red List as a species of "least concern" (Hammerson et al., 2008); as an alien species in the West Indies, the conservation status of *R. grylio* is not a concern, except in the context of the impact it may have on native species.

Additional Pertinent Literature: Altig and Lohoefener (1982).

Remarks: Recent work by Frost et al. (2006) places this species in the genus *Lithobates.*

Rana sphenocephala Cope, 1889

Distribution: Eastern United States, but introduced on the Little Bahama Bank.

Habitat: In the United States, occurs in shallow freshwater habitats, occasionally entering brackish water (Duellman and Schwartz, 1958).

Size: SVL to 127 mm (Conant and Collins, 1998).

Conservation Status: On the IUCN Red List as a species of "least concern" (Hammerson and Hedges, 2004b); as an alien species in the West Indies, the conservation status of *R. sphenocephala* is not a concern, except in the context of the impact it may have on native species.

Remarks: Recent work by Frost et al. (2006) places this species in the genus *Lithobates.*

8 Family Strabomantidae

Figure 11. *Pristimantis shrevei* (Strabomantidae) is one of only two West Indian endemics in this newly erected family that has a broad distribution on the Neotropical mainland. Photograph by Robert Powell.

Pristimantis euphronides Schwartz, 1967

Distribution: Grenada, uplands; elevational range ~300–840 m.

Habitat: Rainforests and forest edges and montane meadows surrounded by agriculture, on the ground and on vegetation (Hedges and Powell, 2004b; Stuart et al., 2008); wet, forested uplands (Henderson and Berg, 2006); also in grass and low herbaceous growth in a small montane meadow surrounded by forest and nutmeg, noncalling individuals on bromeliads growing low on trees, under a plank at Grand Etang (Schwartz, 1967b); on low (0.3 m) to moderately high (1.8 m) perches, usually ferns or leaves of broadleaf vegetation, along forest trails at Grand Etang (RWH and C. S. Berg, unpublished data).

Abundance: Moderately common in suitable habitat (Hedges and Powell, 2004b; Stuart et al., 2008). **Activity:** Nocturnal, but active on the ground in a cacao grove during the late afternoon (Schwartz, 1967b). **Encounter Rates:** 2.2–30.7/person-hour of searching depending on site, time of year, and prevalence of *Eleutherodactylus johnstonei* in 2004–2006 (Henderson and Berg, 2006). **Reproduction:** Males call from 1.0–1.3 m on leaves and trunks of trees to 1.0 m in diameter in rainforest (Schwartz, 1967b); eggs laid on the ground, breeds by direct development (Hedges and Powell, 2004b; Stuart et al., 2008). **Size:** Male SVL to 24.3 mm, female SVL to 39 mm (Schwartz, 1967b).

Conservation Status: "Endangered" on the IUCN Red List (Hedges and Powell, 2004b); range continues to contract as a consequence of habitat alterations and competition with introduced *Eleutherodactylus johnstonei* (Henderson, 2008).

Additional Pertinent Literature: Sander et al. (2003).

Remarks: Until recently (Heinicke et al., 2007), this species was assigned to the genus *Eleutherodactylus*.

Pristimantis shrevei Schwartz, 1967

Distribution: St. Vincent, uplands; elevational range ~275–925 m.

Habitat: Arboreal in rainforests, forest edges, and montane meadows surrounded by agriculture, on the ground and on vegetation (Hedges and Powell, 2004c; Stuart et al., 2008); in and at the edges of upland (>300 m) rainforests (Schwartz, 1967b; Diaz-Lameiro et al., 2007).

Parasites: Plerocercoid cysts *Rhabdias elegans*, polymorphid cystacanths (Goldberg et al., 1998d); nematodes *Oswaldocruzia brevispicula* and *Strongyloides* sp., acanthocephalan *Centroryhnchus* sp. (Moravec and Kaiser, 1995). **Predators:** *Chironius vincenti* (RWH). **Reproduction:** Males call from dusk well into the night (Treglia, 2006); males call from leaves of trees and shrubs, whereas females are on the ground or the leaves of herbaceous plants (Schwartz, 1967b); eggs laid on the ground, breeds by direct development (Hedges and Powell, 2004c; Stuart et al., 2008). **Size:** Male SVL to 28.0 mm, female SVL to 40.1 mm (Treglia, 2006).

Conservation Status: "Endangered" on the IUCN Red List (Hedges and Powell, 2004c); principal threats include deforestation causing further range contraction and fragmentation and possible susceptibility to the chytrid fungus (Powell and Henderson, 2007).

Other Pertinent Literature: Díaz-Lameiro et al. (2008).

Remarks: Until recently (Heinicke et al., 2007), this species was assigned to the genus *Eleutherodactylus*.

Part II Reptilia

Section 1 Squamata (lizards)

9 Family Anguidae

Figure 12. *Celestus badius* (Anguidae), like many of its congeners, is a semifossorial denizen of thick leaf litter, through which it "swims," surfacing only occasionally. Photograph by Robert Powell.

Celestus agasepsoides
Thomas, 1971

Distribution: Hispaniola, disjunct range with populations in lowlands associated with the Sierra Martín García and lower elevations of that range and in the lower Península de Barahona; elevational range, sea level to 460 m.

Habitat: Semifossorial to fossorial, under rocks in shade or diffuse sunlight in dry forest (L. R. White et al., 1992a); under rocks and other surface debris in xeric woods and open xeric hillsides with *Haitiella* (Arecaceae) woods (Thomas, 1971; Schwartz and Henderson, 1991).

Diet: Stomach (N = 41, 6 of which were empty) contents included adult beetles (N = 14, total volume = 0.060 mm^3, in 10 stomachs), larval beetles (14, 0.095, 9), beetle pupae (3, 0.040, 3), crickets (9, 0.108, 9), roaches (2, 0.005, 2), fly (1, 0.003, 1), ants (3, 0.003, 1), centipedes (2, 0.030, 2), plus shed skin and some unidentified plant matter presumed to have been ingested adventitiously (L. R. White et al., 1992a). **Predator:** *Ameiva chrysolaema* (Sproston et al., 1999). **Size:** Adult SVL to 71 mm (Schwartz and Henderson, 1991).

Conservation Status: Not assessed.

Additional Pertinent Literature: White and Powell (1996).

Celestus anelpistus
Schwartz, Graham, and Duval, 1979

Distribution: Hispaniola (Dominican Republic), San Cristóbal Province, known only from the type locality.

Habitat: Mesophilic (Schwartz and Henderson, 1991); along a creek (no longer extant) in cleared lowland broadleaf forest, associated with the root systems of Jabilla (*Hura crepitans*), some with trunk diameters approaching 1.0 m, and branches laden with bromeliads (*Tillandsia*), a possible sighting on a talus slope with broadleaf forest about 5 km SW of the type locality (Schwartz et al., 1979).

Behavior: Most active at night in captivity (Schwartz et al., 1979). **Reproduction:** Live bearing, two wild-caught females gave birth to a total of 42 young between 16 July and 3 August (Schwartz et al., 1979). **Size:** Male SVL to 285 mm (Schwartz et al., 1979), maximum female SVL unknown.

Conservation Status: "Critically endangered" (McGinnity and Powell, 2004a), primarily because of habitat destruction; the species was included in lists of threatened Hispaniolan species by Aquino (1983) and Powell et al. (2000); possibly extinct (Powell et al., 2000; Powell and Henderson, 2003b; McGinnity and Powell, 2004a); a recent report of a giant anguid made by a botanist from the National Botanical Garden of Santo Domingo in the Jarabacoa region of the Cordillera Central (S. J. Incháustegui and M. Hernández in McGinnity and Powell, 2004a) might have been *C. anelpistus*, offering hope that the species might still be extant.

Additional Pertinent Literature: Henderson (1988a).

Celestus badius Cope, 1868

Distribution: Navassa Island.

Habitat: Grassy areas around former human habitation, rock faces and concrete buttresses in the old pipeline trench, leaf litter in densely shaded forest (Powell, 1999a).

Activity: Diurnal, active from dawn, but most frequently encountered after 0930–1000 h until well into twilight (Powell, 1999a). **Behavior:** "Swims" through leaf litter, occasionally "surfacing" to scan surroundings (Powell, 1999a). **Diet:** Large male with a partially swallowed earthworm, presumably take litter-dwelling arthropods opportunistically (Powell, 1999a). **Reproduction:** Many females gravid and juveniles abundant in April (Thomas, 1966c); juveniles abundant in late July–early August (Powell, 1999a). **Size:** Male SVL to 104 mm (Powell, 1999c). **Thermoregulation:** Apparently basks, but cloacal temperatures of two adults in litter at 1130 and 1500 h were 31.4 and 32.3 °C, temperatures differed little from that of the litter, suggesting a willingness to thermoconform when appropriate ambient temperatures are available (Powell, 1999a).

Conservation Status: Although the species is confined to an area of 5.2 km^2, it is ubiquitous and abundant there (Thomas, 1966c; Powell, 1999a) and probably not in need of conservation efforts beyond those provided by the relative inaccessibility of Navassa Island and the island's inclusion in the U.S. National Wildlife Refuge system.

Additional Pertinent Literature: Powell (1999c).

Celestus barbouri Grant, 1940

Distribution: Jamaica, northern and central regions; elevational range >600 m.

Habitat: Mesophilic (Schwartz and Henderson, 1991); around the margin of a cultivated sink in an open area at the northern edge of the Cockpit Country (Schwartz and Henderson, 1991); in coconut trash in dry scrub forest (Underwood, 1959a; in trash piles (Grant, 1940a).

Size: Male SVL to 106 mm, female SVL to 94 mm (Schwartz and Henderson, 1991).

Conservation Status: Unknown.

Celestus costatus Cope, 1861

Distribution: Hispaniola, widely distributed on the main island (*C. c. costatus, C. c. chalcorhabdus, C. c. leionotus, C. c. melanchrous, C. c. neiba, C. c. oreistes, C. c. psychonothes*); elevational range, sea level to >2,300 m; also known from Île-de-la-Tortue (*C. c. emys*), Île-à-Vache (*C. c. nesobous*), Isla Saona (*C. c. saonae*), Isla Catalinita (*C. c. aenetergum*), and the Cayos Siete Hermanos (Cayo Tororú).

Habitat: Largely mesophilic, associated with ground debris (logs, stones, human trash, piles of wood, in leaf litter, especially in cacao groves), most abundant in broadleaf forest, including coffee and cacao plantations (Schwartz and Henderson, 1991); also in pine forest and under limestone rocks at disturbed sites, such as pastures and newly dug gardens (Franz and Cordier, 1986); in an active oil palm plantations (Glor et al., 2001); common in pastures and gardens of mountain plateaus in southern Haiti (Franz, 2003); occasionally on trunks of trees (RP) or in bromeliads as high as 6 m above the ground (Schwartz and Henderson, 1991).

Activity: On Isla Catalinita, actively moving through leaf litter in an *Ameiva*-like fashion between 0900 h and 1130 h (Schwartz and Jacobs, 1989).

Diet: Apparently omnivorous, on Cayo Tororú, one individual had eaten two roaches (combined volume = 0.020 mm^3) and melon scraps, another individual ate melon pulp (Burns et al., 1992); includes *Sphaerodactylus armstrongi* (Schwartz and Henderson, 1991). **Longevity:** In captivity, 12 years, 4 months and still alive at last report in 1998 (K. Wright in Slavens and Slavens, 2003). **Parasites:** Gastric nematodes (*Physalopteroides* sp.; Burns et al., 1992). **Predation:** *Celestus warreni* (Incháustegui et al., 1985); *Hypsirhynchus parvifrons* (Henderson et al., 1987a; J. W. Smith et al., 1994). **Reproduction:** Live-bearing, 3–9 embryos per clutch (N = 12; Greer, 1967). **Size:** Male SVL to 127 mm, female SVL to 116 mm (Schwartz and Henderson, 1991).

Conservation Status: Not assessed.

Celestus crusculus Garman, 1888

Distribution: Jamaica, where two currently recognized subspecies (*C. c. crusculus, C. c. cundalli*) are widespread on the main island; elevational range, sea level to >1,200 m.

Habitat: Largely mesophilic, but also on scrubby hillsides (Schwartz and Henderson, 1991); under surface debris, such as piles of coconut husks (Garman, 1888); also under *Coccoloba* leaves, rocks in the entrance to Portland Cave, under a termitarium, rock piles, and a terrestrial bromeliad

(Grant, 1940a; Schwartz, 1971a; Schwartz and Henderson, 1991; Crother, 2003).

Abundance: In pitfall traps in the Hellshire Hills, 48 of 51 *Celestus* were *C. crusculus* (other 3 were *C. duquesneyi*; Wilson and Vogel, 2000). **Diet:** Includes insects (Grant, 1940a). **Reproduction:** Live bearing, litter size 2–5 in females 56–74 mm SVL (*N* = 7; Greer, 1967); litter size 1–3 (Hedges, in prep.). **Size:** Male SVL to 90 mm, female SVL to 85 mm (Schwartz and Henderson, 1991).

Conservation Status: Not assessed.

Celestus curtissi Grant, 1951

Distribution: Hispaniola, where four currently recognized subspecies occur from the Plaine de Cul de Sac, north to St.-Marc, east into Valle de Neiba and Île-de-la-Gonâve (*C. c. curtissi*); Barahona Peninsula (*C. c. aporus*); northwestern Haiti, including Île-de-Tortue (*C. c. diastatus*); southeastern coastal areas in the Dominican Republic (*C. c. hylonomus*); islas Catalina and Saona; elevational distribution, sea level to ~550 m.

Habitat: Largely xerophilic, under surface debris, rocks, logs, in slat piles and piles of dry palm fronds (Schwartz and Henderson, 1991); under railroad ties (Ober, 1970).

Reproduction: Live bearing, litter size 2–4 (*N* = 4; Greer, 1967); 5 young born to a female (SVL = 85 mm) on 15–25 September, SVL 30–32 mm (Ober, 1970). **Size:** Male SVL to 86 mm, female SVL to 85 mm (Ober, 1970; Schwartz and Henderson, 1991).

Conservation Status: Not assessed.

Celestus darlingtoni Cochran, 1939

Distribution: Hispaniola, Cordillera Central (Dominican Republic); elevational distribution 1,600–2,500 m.

Habitat: Mesophilic (Schwartz and Henderson, 1991); inhabits high, well-drained, pine-clad slopes with sparse ground cover and little shade, under rocks and pieces of pine bark, among the ruins of buildings, and in piles of weathered boards and lumbering slash at old sawmill sites (on moist sawdust piles), all in relatively moist situations (often at edges of bogs) in pine forest (Schwartz and Incháustegui, 1976; Schwartz and Henderson, 1991).

Activity: Lethargic on cold mornings (Schwartz and Incháustegui, 1976). **Reproduction:** Litter size 2 (Hedges, in prep.). **Size:** Male SVL to 85 mm, female SVL to 70 mm (Schwartz and Henderson, 1991).

Conservation Status: Unknown.

Celestus duquesneyi Grant, 1940

Distribution: Jamaica, Portland Ridge and Hellshire Hills in southern Jamaica.

Habitat: Leaf litter in dry forest (Grant, 1940a).

Size: Female SVL to 111 mm and 21.1 g, males unknown (Wilson and Vogel, 2000).

Conservation Status: Apparently very rare and local, no specimens had been collected since the 1930s until the species was rediscovered in the Hellshire Hills in 1997 (Wilson and Vogel, 2000; Wilson, 2003); status listed as "data deficient" on the IUCN Red List (Gibson, 1996d); currently available information suggests that this species would qualify for critically endangered status according to IUCN criteria (IUCN, 2001); Johnson (1988) recommended surveys to evaluate the status of this species.

Celestus fowleri Schwartz, 1971

Distribution: Jamaica, known only from the vicinity of the type locality in the Cockpit Country, Trelawny Parish; elevation ~160 m.

Habitat: Mesophilic (Schwartz and Henderson, 1991); in deciduous forest, in bromeliads to at least 2.5 m above the ground (Schwartz, 1971a); in epiphytic bromeliads (R. Diesel in P. Vogel, 2000b).

Activity: Presumably diurnal (Schwartz, 1971a). **Size:** Adult female SVL to 105 mm, males unknown (Schwartz and Henderson, 1991).

Conservation Status: Status listed as "data deficient" on the IUCN Red List (Gibson, 1996e); the presumably restricted distribution and dependence on forests that are becoming increasingly fragmented suggest that the species may be endangered.

Celestus haetianus Cochran, 1927

Distribution: Hispaniola, where three currently recognized subspecies occur in the Massif de la Selle (*C. h. haetianus*, *C. h. surda*) and the Sierra de Baoruco (*C. h. mylica*); elevational range ~790–2,700 m.

Habitat: Mostly mesophilic, in upland broadleaf forests, but occasionally in drier habitats in pine forests, associated with rotten logs or under rocks and other surface debris, in disturbed areas in both pine and cloud forests, extremely abundant in abandoned sawmills with old boards and slats (Schwartz and Henderson, 1991); *C. h. haetianus* under rocks in disturbed areas in both pine and cloud forests (Franz and Cordier, 1986).

Diet: Presumably opportunistic, stomach contents of 27 individuals from broadleaf forests in the Sierra de Baoruco included adult and larval beetles, roaches, flies, ants and other hymenopterans, lepidopteran larvae, crickets, spiders, centipedes, snails, earthworms, lizard skin, and plant material assumed to have been adventitiously ingested (Cisek et al., 1990). **Predation:** *Hypsirhynchus parvifrons* (Henderson et al., 1987a). **Reproduction:** Live bearing, litter size 3–8 (Greer, 1967; Hedges, in prep.); spermatogenesis throughout the year, females show vitellogenesis in January and are gravid March–August, fully developed embryos 22 August (Guillette et al., 1988). **Size:** Male SVL to 98 mm, female SVL to 89 mm (Schwartz and Henderson, 1991).

Conservation Status: Not assessed, but upland forests with which the species is associated are being destroyed at an increasingly rapid pace (Hedges, 1999).

Celestus hewardii Gray, 1845

Distribution: Jamaica, widely distributed; elevational range, sea level to nearly 1,000 m.

Habitat: Mesophilic (Schwartz and Henderson, 1991); in rocky talus around the edge of a doline pasture, in stone walls, under trash piles (Grant, 1940a; Schwartz and Henderson, 1991).

Activity: Emerges shortly after sunrise to bask, often exposing only the anterior portion of its body, apparently diurnal, running across rock walls later in the day (Grant, 1940a). **Diet:** Stomachs contained flat sluglike animals known locally as "cow's tongues" (Grant, 1940a). **Size:** Adult SVL to 132 mm (Schwartz and Henderson, 1991).

Conservation Status: Not assessed, but Johnson (1988) recommended surveys to evaluate the status of this species.

Celestus macrotus Thomas and Hedges, 1989

Distribution: Hispaniola, Massif de la Selle (Haiti); elevation ~2,020 m.

Habitat: Mesophilic (Schwartz and Henderson, 1991); moist and mossy areas in high-elevation pine forest (*Pinus caribaea*), under rocks and rotten logs with *C. haitianus* (Thomas and Hedges, 1989).

Size: Female SVL to 60 mm, males unknown (Thomas and Hedges, 1989).

Conservation Status: Unknown.

Celestus maculatus Garman, 1888

Distribution: Cayman Islands, Little Cayman Island and Cayman Brac.

Habitat: Under trash, especially *Coccoloba* leaves, and under stone piles (Grant, 1940c); on Cayman Brac, under rocks at the base of a rock wall in Sea Grape (*Coccoloba*) stands near the beach, under rocks in heavily shaded woodlands near the airport, on Little Cayman, under trash and *Cocos* debris associated with abandoned houses (Seidel and Franz, 1994).

Behavior: Disappeared into a network of holes in coral rock when uncovered (Seidel and Franz, 1994). **Diet:** Includes insects (Grant, 1940c). **Reproduction:** On Cayman Brac, female (58 mm SVL) with two large yolked follicles in January (Seidel and Franz, 1994); litter size three (Hedges, in prep.). **Size:** SVL to 92 mm (Grant, 1940c).

Conservation Status: Not assessed.

Celestus marcanoi Schwartz and Incháustegui, 1976

Distribution: Hispaniola, Cordillera Central (Dominican Republic); elevational range 1,500–1,800 m.

Habitat: Mesophilic (Schwartz and Henderson, 1991); in montane river valleys where *Danthonia domingensis* (Poaceae) dominant, surrounded by pine forest, soil locally dry, covered with large rocks (Schwartz and Incháustegui, 1976).

Activity: Lethargic (i.e., "almost immobile") early in the day (frost usual at 0 °C in December–January when the type series was collected), more active as the day warms (Schwartz and Incháustegui, 1976). **Density:** One lizard per ten rocks suggests abundance (Schwartz and Incháustegui, 1976). **Size:** Male SVL to 78 mm, female SVL to 71 mm (Schwartz and Incháustegui, 1976).

Conservation Status: Not assessed.

Celestus microblepharis Underwood, 1959

Distribution: Jamaica, known only from the type locality at Bascobel, St. Mary Parish; elevation near sea level.

Habitat: Coastal scrub backed by limestone hills covered by scrub forest, under *Cocos* trash (Underwood, 1959).

Size: Female SVL to 87 mm, males unknown (Schwartz and Henderson, 1991).

Conservation Status: Listed on the IUCN Red List as "data deficient" (Gibson, 1996g); lack of information suggests rarity or possible extinction; Johnson (1988) recommended surveys to evaluate the status of this species.

Celestus molesworthi Grant, 1940

Distribution: Jamaica, along the northern shore.

Habitat: Mesophilic (Schwartz and Henderson, 1991); under piles of coconut husks or other rubbish (Grant, 1940a).

Behavior: Does not use legs when trying to escape, instead wriggling with astonishing speed (Grant, 1940a). **Reproduction:** Neonate SVL 26 mm, tail 30 mm (Grant, 1940a). **Size:** Male SVL to 85 mm (average 74 mm), average adult female SVL 69 mm (Grant, 1940a).

Conservation Status: Not assessed.

Celestus occiduus Shaw, 1802

Distribution: Jamaica, distribution unknown.

Habitat: Said to frequent woody and marshy districts (Shaw, 1802); reported to have lived in swamps (Schwartz, 1970a).

Diet: Reported to have eaten fish and fruit (Schwartz, 1970a). **Size:** Male SVL to 305 mm, female SVL to 256 mm (Schwartz and Henderson, 1991).

Conservation Status: Presumably extinct (Cousens, 1956; Schwartz, 1970a; Johnson, 1988), possibly attributable to predation by introduced mongooses (Grant, 1940b); listed as "extinct" on the IUCN Red List (World Conservation Monitoring Centre, 1996e).

Celestus sepsoides Gray, 1852

Distribution: Hispaniola, islandwide, but absent from large areas (Schwartz and Henderson, 1991); predominantly in the northeastern Dominican Republic and the Tiburón Peninsula of Haiti, also on Île-de-la-Gonâve and Île Grande Cayemite; elevational range, sea level to 790 m (possibly as high as 950 m).

Habitat: Mesophilic (xerophilic only on Grand Cayemite, where anthropomorphic influences may have rendered previously mesic habitat xerophytic; Thomas, 1971; Schwartz and Henderson, 1991); under rocks, logs, trash piles near the beach, under termitaria fragments in a mixed coffee/cacao plantation, under soggy cacao husks, wood chips, and other surface objects, especially large, rotten logs in broadleaf forest, never encountered in open (Schwartz and Henderson, 1991); also in abandoned gardens and abandoned cacao (Glor et al., 2001).

Reproduction: Live bearing, litter size 1 (Greer, 1967); litter size 2 (Hedges, in prep.). **Size:** Male SVL to 78 mm, female SVL to 72 mm (Schwartz and Henderson, 1991).

Conservation Status: Not assessed.

Celestus stenurus Cope, 1863

Distribution: Hispaniola, widely distributed across the main island (*C. s. stenurus*, *C. s. alloeides*, *C. s. rugosus*, *C. s. weinlandi*), on Île-à-Vache (*C. s. stenurus*), Île Grande Cayemite (unassigned subspecifically), Île-à-Cabrit (*C. s. weinlandi*), Cayos Siete Hermanos (*C. s. rugosus*; Cayo Monte Grande, records for Cayo Tororú in Burns et al.,1992 are in error); elevational range, sea level to ~1,160 m.

Habitat: Mesophilic (Schwartz and Henderson, 1991); although seemingly prefers drier sites than *C. costatus* when the two species occur together, also in locally moist or shaded conditions in otherwise xeric areas, under surface debris, railroad ties, palm fronds, cacao husk piles, leaf litter, in a fallen, rotten cactus trunk, particularly abundant in piles of rocks along a stream channel in association with dense bracken fern glades, common in pastures and gardens of mountain plateaus in southern Haiti, occasionally encountered on the surface in leaf litter in heavily shaded cacao groves (Franz and Cordier, 1986; Schwartz and Henderson, 1991; Franz, 2003); also in active oil palm plantations, abandoned gardens, abandoned cacao groves, and associated with mogotes (Glor et al., 2001).

Activity: Diurnal (Schwartz and Henderson, 1991). **Reproduction:** Live bearing, 6 young born 4–6 June, neonate SVL 36–37 mm, on 20 June 37–39 mm, average gain 0.12 mm/day, on 21 July 39–41 mm, average gain 0.06 mm/day in 31 days, another female gave birth to 5 young on 12–15 May, SVL at 6 days 36–37 mm, 32 days later SVL 38–41 mm, average increase of 0.10 mm/day, on 21 July 42 mm, average daily gain 0.06 mm/day, growth rapid for 2 weeks, then tapers off (Ober, 1968). **Size:** Male SVL to 172 mm, female SVL to 143 mm (Schwartz and Henderson, 1991).

Conservation Status: Not assessed.

Celestus warreni Schwartz, 1970

Distribution: Hispaniola, northern and northwestern Haiti, Île-de-la-Tortue, northern Dominican Republic.

Habitat: Haitian populations are almost invariably in mesic areas, once locally abundant in lowland forests, frequently associated with root systems of large trees (Incháustegui et al., 1985); on Île-de-la-Tortue, in banana groves (Schwartz and Henderson, 1991); Dominican populations formerly assigned to *C. carraui* in dry forest in a rocky arid area and in a cacao plantation (Incháustegui et al., 1985).

Activity: Outside burrows during day and on asphalt roads at night (Schwartz and Henderson, 1991). **Behavior:** Agonistic behavior between males in captivity (Lawler and Norris, 1979). **Diet:** Apparently an opportunistic forager, stomach contents of 36 individuals included highly toxic millipedes (Rhinocricidae; $N = 12$), cicadas (*Ciceroprocta* sp.; 7), centipedes (*Scolopendra alternans*; 5), slugs (Gastropoda; 5), a tarantula (*Scopelobates*?), various other arthropods (spiders, tailless whip scorpions, earwigs, bugs, beetle adults and larvae, lepidopteran larvae), two lizards (*Anolis* sp. and *Celestus costatus*), and bird feathers (probably *Quiscalus niger*), although no other bird parts were present (Incháustegui et al., 1985); hatchlings tested with surface chemicals performed more tongue flicks toward cricket stimuli than to lettuce, cologne, or water (and significantly more toward crickets than water), indicating that prey chemical discrimination occurs, and that *C. warreni* hatchlings are unlikely to consume leafy plant material (Cooper and Bradley, 2009). **Longevity:** In captivity, 7 years, 6 months (H. Cohen in Slavens and Slavens, 2003). **Reproduction:** Live-bearing, 10 gravid females in August had 8–27 young (SVL 32–47 mm), litter size positively correlated with female SVL (Incháustegui et al., 1985); maximum litter size 34 (Lawler and Norris, 1979). **Size:** Male SVL to 283 mm, female SVL to 260 mm (Incháustegui et al., 1985; Schwartz and Henderson, 1991). **Tail Autotomy:** 20 of 24 specimens with broken tails (Incháustegui et al., 1985).

Conservation Status: "Critically endangered" on the IUCN Red List (McGinnity and Powell, 2004b), primarily due to habitat destruction, predation by introduced predators, and persecution by local residents (Powell et al., 2000; Powell and Henderson, 2003b; McGinnity and Powell, 2004b); included in a list of threatened Hispaniolan species by Aquino (1983; as *C. carraui*).

Additional Pertinent Literature: Henderson (1988b, 1988c).

Diploglossus delasagra Cocteau, 1838 or 1839

Distribution: Western Cuba and Isla de la Juventud.

Habitat: Mesophilic (Schwartz and Henderson, 1991); under surface objects (rocks, logs) in or near broadleaf forest and forest edges, once in a formal garden (introduced?), occasionally uncovered during plowing (Schwartz and Henderson, 1991); in a pile of boards in an orchard on the Península de Zapata (Garrido, 1980b); in leaf litter on Cayo Coco (Estrada, 1993b).

Abundance: Uncommon (Schwartz and Henderson, 1991). **Parasites:** Nematode *Aplectana rysavy* (Coy Otero and Baruš, 1979b; Coy Otero and Lorenzo Hernández, 1982). **Predation:** *Cubophis cantherigerus* (Henderson and Sajdak, 1996); Cuban Solenodon, *Solenodon cubanus* (Abreu and de la Cruz, 1988). **Reproduction:** Two oviductal eggs with white shells and small whitish embryos (Greer, 1967); litter size five (Hedges, in prep.). **Size:** Male SVL to 121 mm, female SVL to 115 mm (Schwartz and Henderson, 1991).

Conservation Status: Considered to be at "lower risk" of extinction (Rodríguez Schettino and Chamizo Lara, 1998a).

Diploglossus garridoi Thomas and Hedges, 1998

Distribution: Cuba, known only from the vicinity of the type locality in the Sierra Maestra in southeastern Cuba.

Habitat: Mesophilic (Schwartz and Henderson, 1991); beneath a pile of cut vegetation debris near a stream in a generally open area in surroundings that are a mixture of predominately pine forest with some disturbed broadleaf vegetation (Thomas and Hedges, 1998a).

Reproduction: Gravid female on 30 June with 4 shelled eggs (Thomas and Hedges, 1998a). **Size:** Female SVL 103 mm (Thomas and Hedges, 1998a).

Conservation Status: Unknown.

Diploglossus montisserrati Underwood, 1964

Distribution: Montserrat, known only from vicinity of the type locality at Woodlands Spring; elevation ~180 m.

Habitat: Mesophilic (Schwartz and Henderson, 1991); ground dwelling in woodlands near streams and springs (Underwood, 1964); semifossorial, with individuals under a rock and in the roots of a tree (Ogrodowczyk et al., 2006).

Activity: Diurnal (Underwood, 1964) or nocturnal (Ogrodowczyk et al., 2006). **Behavior:** Very sensitive to disturbance, an individual on the forest floor ceased moving in response to any noise or movement of the observer (Ogrodowczyk et al., 2006). **Size:** Male SVL to 180 mm (only known specimen), female size unknown (Underwood, 1964); an unsexed specimen had SVL 180 mm, tail length 230 mm, and body mass 170 g (Ogrodowczyk et al., 2006).

Conservation Status: "Critically endangered" on the IUCN Red List (Day, 1996c); few sightings and no additional specimens had been taken since description of the holotype (Stevens and Waldman, 2001) until 2006 in the Woodland Springs area (Anonymous, 2006a; Ogrodowczyk et al., 2006); effects of recent volcanic activity on moist woodland habitats seemingly preferred by these lizards is unknown (Powell and Henderson, 2005a).

Diploglossus nigropunctatus Barbour and Shreve, 1937

Distribution: Eastern Cuba, along the northern and western coasts.

Habitat: In a field near the edge of woods (Barbour and Shreve, 1937); subterranean, in association with ant nests (Martinez Reyes and Moreno, 2003).

Reproduction: Holotype (adult female) curled around and among 4 eggs under a chunk of rotten wood lying loosely on ground in a field near the edge of woods (Barbour and Shreve, 1937); in June near Guantánamo, male with 5 eggs (9 × 17 mm) under a stone (Barbour and Ramsden, 1919). **Size:** Male SVL to 121 mm, female SVL to 115 mm (Schwartz and Henderson, 1991).

Conservation Status: Unknown.

Diploglossus pleii Cope, 1861

Distribution: Puerto Rico, widely distributed on the main island; elevational range <50–670 m.

Habitat: Mesophilic (Schwartz and Henderson, 1991); semifossorial in moist, wooded habitats deep in lower layers of thick drifts of leaf litter and other decomposing vegetation (Thomas and Gaa Kessler, 1996); in leaf litter, under logs in coffee groves, and in sugarcane (Rivero, 1998); also in mesic wooded situations including borders of pastures and a sugarcane field, under banana trash, in a mesic ravine on a limestone ridge, and in a dry, powdery log in a dry coffee grove (Schwartz and Henderson, 1991).

Behavior: Bites without hesitation when captured, when uncovered and grasped, wriggles violently and often autotomizes tail, difficult to hold because of smoothness and frantic behavior, hides in leaf litter, leaving elongate tail exposed in a possible effort to deceive predators (Rivero, 1998). **Diet:** Based on 36 individuals with food in their stomachs (and a total of 71 prey items), 23.9% contained stemmiulid millipedes, 19.7% with adult beetles, 18.3% with earwigs, 9.8% gastropods (slugs), 7.0% with termites, 5.6% with spiders, and 2.8% with centipedes (*Scolopendra*), plant matter probably adventitiously ingested (Thomas and Gaa Kessler, 1996). **Parasites:** Trematodes (*Mesocoelium monas*; Hoffman, 1935; Cofresí-Sala, 1964). **Predator:** *Epicrates inornatus* (A. J. Sánchez Muñoz, in litt., 17.III.2008). **Reproduction:** Live bearing, 2–4 yolky oviductal eggs per clutch ($N = 4$; Greer, 1967); oviductal egg 11 × 18 mm (K. P. Schmidt, 1920). **Size:** Male SVL to 125 mm, female SVL to 105 mm (Schwartz and Henderson, 1991).

Conservation Status: "Stable" (Moreno, 1991d).

10 Family Gekkonidae

Figure 13. *Hemidactylus mabouia* (Gekkonidae) functions largely as a human commensal, often exploiting insect prey attracted to artificial lights associated with buildings. Photograph by Robert Powell.

Gekko gecko Linnaeus, 1758

Distribution: Native to southeastern Asia; introduced to Japan, the Philippines, Madagascar, Samoa, the Hawaiian Islands, the United States (Florida), and onto Martinique and Guadeloupe (Grande-Terre) in the West Indies.

Habitat: On Martinique, edificarian, associated with human dwellings and the electricity meter boxes on buildings (Henderson et al., 1993).

Diet: Includes bats (*Molussus molussus* and *Tadarida brasiliensis*) and "other nocturnal lizards" (Breuil, 2009). **Size:** To 185 mm SVL (outside West Indies; Taylor, 1963).

Conservation Status: Introduced to Martinique in the early 1970s (Henderson et al., 1993) and, as an introduced species, its impact on native species is of concern as it has the potential to be a predator on smaller lizards (e.g., *Hemidactylus, Anolis*) and frogs (e.g., *Eleutherodactylus*).

Hemidactylus angulatus Hallowell, 1854

Distribution: In the West Indies, Cuba (including Archipiélago de Sabana-Camagüey), Hispaniola (including Isla Saona), Puerto Rico, and Isla Mona; elsewhere in western Africa, the probable source of West Indian populations, presumably via the slave trade (Weiss and Hedges, 2007); whether the populations in Colombia formerly known as *H. leightoni* are conspecific is unknown.

Habitat: Primarily edificarian, often in and on hotels, occasionally in rocks at dockside, under tin at an old mill site, in lumber piles, in palm trash piles, at times under human debris, including old, exposed cardboard in halophytic *Salicornia* flats (Schwartz and Henderson, 1991); 101 individuals all on stone or masonry (Grant, 1932d); edificarian (Alayo Dalmau, 1951); often exploiting insects attracted to lights at night, but also associated with natural and human debris in open areas (Collette, 1961, as *H. mabouia*; Rivero, 1998; Powell and Parmerlee, 1993b); in a hotel in Barahona, Dominican Republic, perch heights ranged from 0.1–5.0 m (mean 3.7 ± 1.2 m), 86 of 107 observations on or within 0.2 m of the ceiling, only 10 were within 1 m of the ground (Lynxwiler et al., 1991); in Cuba, spends day behind objects in houses, in crevices in walls or frames of doors and windows (Martínez Reyes and Hernández Marrero, 2003); in the Dominican Republic, foraging at night on unlit concrete walls in Santo Domingo and concrete fence posts in Barahona (RP).

Activity: Nocturnal, beginning at dusk (Lynxwiler et al., 1991); by day under cover such as wall hangings, rocks, or trash (Powell and Parmerlee, 1993b). **Behavior:** Squeaks when handled (Grant, 1932b). **Foraging and Diet:** On Hispaniola, adult flies (largely Chironomidae and Culicidae; 45.6% of all food items, 17.3% of total food volume, found in 28.2% of stomachs examined), moths (24.8%, 33.1%, 29.6%), beetles (6.7%, 5.5%, 11.3%), roaches (2.7%, 14.2%, 5.6%), spiders (4.0%, 12.6%, 5.6%), leafhoppers (10.1%, 1.6%, 5.6%), orthopterans (2.0%, 5.5%, 4.2%), hymenopterans (1.3%, 6.3%, 1.4%), fly larvae (1.3%, 0.8%, 2.8%), isopods (1.3%, 1.6%, 1.4%; Powell et al., 1990c); geckos flattened themselves against substrate when prey (moths) came into sight, crept slowly to within 12 cm, paused, then lunged (Lynxwiler et al., 1991); attempted infanticide, forages around night lights (Bowersox et al., 1994a). **Home Range and Territoriality:** No evidence of territoriality, home ranges 0.06–125.7 m^2 (mean 18.9 ± 24.5 m^2; Lynxwiler et al., 1991). **Movements:** In a hotel, 30 of 62 resighted animals moved less than 5 m and only one moved >50 m (Lynxwiler et al., 1991). **Parasites:** In Cuba, nematodes *Cyrtosomum longicaudatum* (Coy Otero and Baruš, 1973a, 1979b; Coy Otero and Lorenzo Hernández, 1982) and *Spauligodon antillarum* (Baruš and Coy Otero, 1974; Coy Otero and Baruš, 1979b; Coy Otero and Lorenzo Hernández, 1982); on Hispaniola, nematode *Trichospirura teixeirai* (Powell et al., 1990a); pentastomid *Raillietiella* spp. (Riley et al., 1991; Powell et al., 1993); protozoan *Eimeria helenlevineae* (Bui et al., 1992b). **Population Density:** 0.026 (Lincoln-Peterson estimate) and 0.021 (Schnabel estimate) per m^2 of sampled walls and ceilings in hotel in Barahona, Dominican Republic (Lynxwiler et al., 1991). **Reproduction:** In captivity, eggs hatched in 57–69 days (Teale, 1930); eggs tolerant of long exposure to salt water, deposited November–April, sometimes communally deposited (many eggs under exposed cardboard in *Salicornia* flats), two eggs (30 December) 10.8–11.7 × 9.0 mm (Schwartz and Henderson, 1991); spermatogenesis in 31 of 34 males from 9 different months (no samples available for February, April, October), female reproductive activity from 9 months, 1–16 nonyolked ovarian follicles <0.5–2 mm, 1–2 yolked ovarian follicles 2–10 mm in long axis, 1–2 shelled oviductal eggs 7–12 mm in long axis (Powell et al., 1990c). **Sex Ratio (M:F):** 81:79 (Powell et al., 1990c); 61:12 (Lynxwiler et al., 1991). **Size:** Male SVL to 71 mm (Lynxwiler et al., 1991); smallest adult male SVL 46 mm (Powell et al., 1990c); male mass to 7.0 g (Lynxwiler et al., 1991); female SVL to 68 mm (Powell et al., 1990c; Lynxwiler et al., 1991) and 4.7 g (Lynxwiler et al., 1991), smallest adult female SVL 50 mm (Powell et al., 1990c) and 49 mm (Lynxwiler et al., 1991). **Tail Autotomy:** During capture of animal, tail broke without being touched (Grant, 1932b); on Hispaniola, 79 of 133, rate not significantly greater in males, positive correlation between size and rate of tail loss in males (Powell et al., 1990c); on Hispaniola, 4 of 12 females, 19 of 61 males, 3 of 34 juveniles (total 26 of 107, 24.3%; Lynxwiler et al., 1991). **Thermal Biology:** On Hispaniola, cloacal temperatures 29.4–32.4 °C (mean 30.1 ± 0.9 °C), slightly higher than air temperatures and within 1.2–1.6 °C of substrate temperatures (Lynxwiler et al., 1991).

Conservation Status: "Stable" in Puerto Rico (Moreno, 1991d).

Additional Pertinent Literature: Powell and Maxey (1990).

Remarks: This species was formerly known as *Hemidactylus haitianus*.

Hemidactylus garnotii Duméril and Bibron, 1836

Distribution: Native to southeastern Asia and the Indo-Australian Archipelago; introduced to a number of Pacific islands (including the Hawaiian Islands), the United States (Florida), and, in the West Indies, the Bahama Islands (Little Bahamas Bank, Abaco, Man O'War Cay; Great Bahama Bank, New Providence).

Habitat: On walls of an old school (Meshaka, 1995).

Reproduction: Parthenogenetic, female (53.9 mm SVL) on 17 November yolking left and right follicles (Meshaka, 1995). **Size:** Adult SVL (in the United States) to 64 mm (Conant and Collins, 1998).

Conservation Status: Not assessed.

Hemidactylus mabouia Moreau de Jonnès, 1818

Distribution: Central Africa (where presumably endemic), eastern coast of South America from Uruguay to Guyana and along most of the length of the Río Amazonas in Brazil, Ecuador, and Peru; in the West Indies known from Grenada, St. Vincent, Barbados, St. Lucia, Martinique (and associated islets), Dominica, Îles des Saintes, Guadeloupe, Antigua, Montserrat, St. Christopher, St. Eustatius, Saba, St.-Martin/St. Maarten, Anguilla, Puerto Rico Bank (including the Virgin Islands, and islas Mona, Vieques, and Culebra), Hispaniola, Cuba, Turks and Caicos Islands, Grand Cayman Island, Bahama Islands, and many satellites of nearly all major islands; recently introduced into southern Florida, Atlantic islands (Cape Verde Islands, Madeira, São Thomé, Principe, Trinidade, Abrolhos Archipelago off Brazil), Macronesian islands, and islands in the Gulf of Guinea; populations on the Greater Antilles and those in the Turks and Caicos Islands, Bahama Islands, and on Grand Cayman presumably are introduced; individuals in Jamaica (Fläschendräger, 1999b) may represent a very recent introduction.

Habitat: Human commensal (Currat, 1980); primarily edificarian, in warehouses (at night), under planks, loose concrete slabs, loose rocks in open areas, in *Cocos* trash piles, under piles of coastal wrack, in rock piles (Schwartz and Henderson, 1991); usually near houses, also in rock piles and under bark (Malhotra and Thorpe, 1999); also associated with natural and human debris in open areas, buildings of greater complexity (overhanging roofs, more crevices, angles, and associated debris) harbored more geckos (K. G. Howard et al., 2001); on Vieques (Puerto Rico), on old stone buildings (Grant, 1932a); 92 individuals all in frame buildings (Grant, 1932d); on St. John (U.S. Virgin Islands), in frame buildings and under living banana sheaths (Grant, 1932g); more abundant in rural homes on Grenada (Germano et al., 2003); frequently utilizes night lights, but presence or absence of lights had no effect on gecko abundance on several buildings on Anguilla, perch heights (adults 0–3.5 m, mean 2.7 ± 0.3 m, $N = 75$; juveniles 0–9.0 m, mean 0.8 ± 0.1 m, $N = 70$) were significantly correlated with size, presumably because larger geckos higher on walls preyed on smaller geckos (K. G. Howard et al., 2001); in the Bahamas, on Long Island (Krysko and Borgia, 2005) and on Eleuthera (Krysko and Thomas, 2007), under bark of *Casuarina* trees by day and on building walls by night; on Isla Mona, under eaves of building by day (RP); on Martinique, hides in and under rocks in near-treeless habitats (Breuil, 2009).

Activity: Comes out to hunt in the late evening (Long, 1974); nocturnal, but also seen by day (Malhotra and Thorpe, 1999); on Anguilla, most activity 2000–0100 h, with larger geckos active longer into the night (K. G. Howard et al., 2001). **Behavior:** Vocalizes, hides in cracks or joints of boards (A. Serrano in Grant, 1932d); quickly retreats into crevices when approached, emits "sharp clicking sounds" (Long, 1974). **Competition:** Displaced at night lights by *Anolis gingivinus* on St. Maarten (Powell and Henderson, 1992). **Defense:** Tail autotomy (Currat, 1980). **Desiccation:** Standard rate of water loss 12.4 mg/g of initial body weight/h, vital limit (percent of initial body weight lost before death) 24.7, and survival time 17.5 h (Heatwole and Veron, 1977). **Foraging and Diet:** Insectivorous, frequently feeding on insects attracted to artificial lights (Currat, 1980); around night lights foraging for insects on Puerto Rico (R. Powell in Perry et al., 2008), Guana (British Virgin Islands; Perry et al., 2008), Anguilla (K. G. Howard et al., 2001), St. Maarten (Powell and Henderson, 1992), St. Lucia (Long, 1974), and Dominica (Daniells et al., 2008); approaches prey "slowly and stealthily like a cat, then suddenly pounce[s]" (Long, 1974). **Parasites:** In Cuba, larval nematode *Porrocaecum* sp. (Coy Otero and Baruš, 1979b; Coy Otero and Lorenzo Hernández, 1982); in Puerto Rico, pentastomid *Raillietiella hemidactyli* (Self and García Díaz, 1961); mite *Geckobia hemidactyli* (Martínez Rivera et al., 2003). **Population Density:** 0.04, 0.04, 0.21, 0.12/m^2 on four buildings on Anguilla (K. G. Howard et al., 2001). **Predation:** On Guana Island (British Virgin Islands), consumed by *Anolis cristatellus wileyae* (Owen and Perry, 2005); also on Guana, by a Pearly-eyed Thrasher (*Margarops fuscatus*) that killed the gecko by alternately bashing it on a concrete floor and pecking its head (J. Lazell, in litt., 21.X.2007). **Reproduction:** On Carriacou, eggs in rock crevices, tree hollows, holes in stone walls and house walls, door latches, behind books in a bookcase, evidence of communal nesting, eggs in June–July, 9.3–9.8 mm × 7.9–8.4 mm (mean 9.6 × 8.1 mm, $N = 7$), hatchling (16 July) SVL 21 mm, total length 40 mm (Pendlebury, 1972); 3 or 4 eggs together on St. Lucia (Long, 1974); eggs under rocks in forest on Hog Island, Grenada (Williamson et al., 2002). **Size:** Male SVL to 68 mm (Schwartz and Henderson, 1991), female SVL to 61 mm (K. G. Howard et al., 2001). **Thermal Biology:** On Anguilla, cloacal temperatures (14 adults, 5 juveniles) 26.2–28.7 °C (mean 27.2 ± 0.2 °C), differed from substrate temperatures by 0–1.7 °C (mean 0.2 ± 0.2 °C; K. G. Howard et al., 2001).

Conservation Status: "Stable" in Puerto Rico (Moreno, 1991d).

Additional Pertinent Literature: Powell et al. (1998b).

Hemidactylus palaichthus Kluge, 1969

Distribution: Dennery Island and the Maria Islands off the southeastern coast of St. Lucia; from outside the West Indies, northern South America, including Trinidad, Tobago, and Little Tobago.

Habitat: Saxicolous on Maria Islands and terrestrial on Maria Major, hides under rocks and in crevices in cliffs (Corke, 1987); on Maria Major, in small crevices, including on a rock close to the shoreline (T. R. Lewis, 2001); on Maria Minor, under rocks in the small area of woodland, as well as under rocks in grassy scrub (Buley et al., 1997).

Abundance: "The species appears to be abundant on both [of the Maria] islands" (Buley et al., 1997). **Activity:** Nocturnal (Corke, 1987). **Reproduction:** Eggs under stones in grassy areas of Maria Minor (Corke, 1987); eggs and juveniles in September on Maria Major (Buley et al., 1997). **Size:** Male SVL to 63 mm (Schwartz and Henderson, 1991), female SVL to at least 56 mm (Buley et al., 1997).

Conservation Status: Not assessed.

Additional Pertinent Literature: Powell (1990b).

Hemidactylus turcicus Linnaeus, 1758

Distribution: Cuba; from outside the West Indies, circum-Mediterranean south into Africa, Minorca, Socotra, Canary Islands; introduced into the Western Hemisphere on the Florida Keys and peninsular Florida, with scattered populations along the Gulf Coast to the Península de Yucatán.

Habitat: Populations in the Western Hemisphere are almost entirely edificarian (Schwartz and Henderson, 1991).

Activity: Presumably nocturnal (RP). **Parasites:** Nematodes *Spauligodon californiensis* (Baruš and Coy Otero, 1974; Coy Otero and Baruš, 1979b; Coy Otero and Lorenzo Hernández, 1982), *Skrjabinoptera phrynosoma*, and larval *Cyrnea* sp. (Coy Otero and Baruš, 1979b; Coy Otero and Lorenzo Hernández, 1982). **Size:** Adult SVL to 55 mm (Schwartz and Henderson, 1991).

Conservation Status: Not assessed.

Additional Pertinent Literature: McCoy (1970).

11 Family Gymnophthalmidae

Figure 14. *Gymnophthalmus underwoodi* (Gymnophthalmidae) is parthenogenetic; consequently, only a single individual is necessary to establish a colony. These semifossorial lizards have become established on a number of Lesser Antillean islands in recent years. Photograph by Robert Powell.

Bachia heteropus Lichtenstein, 1856

Distribution: The Grenada Bank, where known from Bequia, Battowia, Mustique, Canouan, Union, and Grenada (*B. h. alleni*); the species also occurs in Trinidad, Tobago, and northern Venezuela.

Habitat: On Grenada, in wet *Cocos* trash and large piles of wet mulch (RWH); also in decaying, soggy cacao trash and in wet piles of *Musa* stem cuttings (Schwartz and Henderson, 1991); under pots in a plant nursery (RWH, RP); in the Grenadines, in rotting logs and leaf litter and, during the dry season, seeks refuge in dry, cracked earth when disturbed (Daudin and de Silva, 2007); on Union Island, under rocks in a trail (RP).

Diet: Arthropods and their larvae (Malhotra and Thorpe, 1999). **Size:** SVL to 57 mm (Dixon, 1973).

Conservation Status: Not assessed; "locally common" in the Grenadines (Daudin and de Silva, 2007).

Gymnophthalmus pleii Bocourt 1881

Distribution: Guadeloupe, Dominica, Martinique (*G. p. pleii*), St. Lucia (*G. p. luetkeni*), Maria Major (*G. p. nesydrion*), and Maria Minor (Buley et al., 1997).

Habitat: On St. Lucia, frequently seen on the ground running through grass and leaves, on Maria Island (*G. p. nesydrion*), "seem to prefer dryer coastal areas," very common in litter on cultivated ground (Long, 1974); on Guadeloupe, in litter and grassy areas (Breuil, 2002b); on Dominica, in dry forest leaf litter where sun hits the forest floor (RWH, RP); on Martinique, in coastal litter, banana groves, gardens, near houses, bases of walls and ruins (on Îlet Chancel; Breuil, 2002b); also in xeric woodlands adjacent to ruins, on St. Lucia, in sand dunes (under *Terminalia* leaves and *Coccoloba*), *Musa* and *Theobroma* groves, ravines, and cultivated areas, on Maria Major, in northern grassland and central wooded areas among leaf litter and rocks (Schwartz and Henderson, 1991).

Activity: Diurnal, on Guadeloupe, active during hottest hours of the day (Breuil, 2002b). **Behavior:** When chased, will freeze beneath a stone or leaf, seeming to have disappeared (Long, 1974); seeks shelter in leaf litter and in cracks and crevices in the substrate (Breuil, 2002b). **Competitive Exclusion:** Apparently allopatric with *G. underwoodi* on Dominica (Daniells et al., 2008; Ackley et al., 2009). **Desiccation:** On Martinique, the rate of water loss 1.79 mg/g/h (Leclair, 1978). **Diet:** Insectivorous (Long, 1974). **Reproduction:** Bisexual, on Guadeloupe, hatchlings in April, July, and August (Breuil, 2002b). **Size:** Adult SVL to 48 mm (Schwartz and Henderson, 1991). **Thermal Biology:** On Guadeloupe, basks before starting to forage (Breuil, 2002b); considerably more heliophilic than sympatric *Sphaerodactylus* on Martinique (Bach and Schönecker, 2008).

Conservation Status: Not assessed; the arrival of *Gymnophthalmus underwoodi* may displace this species on Martinique (Breuil, 2009).

Gymnophthalmus underwoodi Grant, 1958

Distribution: Antigua (including Great Bird Island), Barbuda, St.-Martin/St. Maarten, Guadeloupean Archipelago, Dominica, Martinique, St. Vincent, the Grenadines (Bequia, Mustique, Union), Grenada, and Barbados; outside the West Indies, occurs in Trinidad and Tobago, and northeastern South America; at least some of the Lesser Antillean populations represent introductions.

Habitat: On Antigua, entering a building through the crack under a door (RP; Powell and Lindsay, 1999); on St.-Martin/St. Maarten, in leaf litter on sand 400–500 m from shore (van Buel and Powell, 2006); on Guadeloupe, in open sunny areas (pastures, gardens, *Musa* groves, lawns, behind beaches, and along a trail in humid forest), on Basse-Terre, in leaf litter where sunlight hits the ground (Breuil, 2002b); entering the front entrance of a café and under boards and trash near the beach (Schwartz and Henderson, 1991); on Marie-Galante, in dry forests, ravines, cracks in old coral, in sunny situations at edges of thickets, in sugarcane, and among ruins of sugar mills, on La Désirade, in *Croton* thickets with abundant litter (Breuil, 2002b); on Dominica, in dry forest leaf litter (RWH, RP; Daniells et al., 2008); on St. Vincent, in leaf litter in relatively dry open areas not occupied by *Sphaerodactylus vincenti* (Treglia, 2006); in the Grenadines, in grass and leaf litter primarily in sandy coastal areas, but also in higher forested situations (Daudin and de Silva, 2007); on Grenada, in leaf litter of dry forests at low elevations (Williamson et al., 2002); on Barbados, on roadside hillsides in thin *Swietenia* litter with some under rocks and cardboard in a hot, dry area with partial shade, also among dead leaves on sand among trees along the beach (Schwartz and Henderson, 1991).

Abundance: On Great Bird Island (Antigua), up to 5 within 1.0 m^2 (Daltry, 2007). **Activity:** Diurnal (Currat, 1980; RWH, RP). **Competitive Exclusion:** Apparently allopatric with *G. pleii* on Dominica (Daniells et al., 2008; Ackley et al., 2009). **Diet:** Small insects and larvae, especially ants and termites (Currat, 1980). **Predators:** Cats, wild birds, chickens (Breuil, 2002b). **Reproduction:** Unisexual, all female populations that reproduce via parthenogenesis (Williamson and Powell, 2004 and references therein); 4–6 oval (6 × 3 mm) eggs laid in June and July in litter or in soil to depths of 12–20 cm (Currat, 1980); based on captive animals that originated in Trinidad and Suriname, clutch size 1–4, mean egg size 9.4 × 6.5 mm, development time to 61+ days, hatchling SVL 16–19 mm (Hardy et al., 1989). **Size:** SVL to 43 mm (Hardy et al., 1989). **Thermal Biology:** On La Désirade, basks along trails and roadsides (Breuil, 2002b).

Conservation Status: Not assessed; appears to be an adept colonizer that can become extremely abundant in a very short time (Daltry, 2007).

Additional Pertinent Literature: Williamson and Powell (2004).

12 Family Iguanidae

Figure 15. Aggressive displays of competing male *Cyclura lewisi* (Iguanidae) may include gaping, lateral presentations, crest and dewlap erection, and elevation on all four limbs. Such displays can escalate into actual combat if neither participant submits. Photograph by John F. Binns (International Reptile Conservation Foundation).

Cyclura carinata Harlan, 1824

Distribution: Two historically recognized subspecies in the Turks and Caicos Islands (*C. c. carinata*) and Booby Cay (east of Mayaguana Island, Bahamas; *C. c. bartschi*), although subspecific recognition of *C. c. bartschi* is not warranted (Stephen, 2005, 2006; Bryan et al., 2007).

Habitat: Xerophilic (Schwartz and Henderson, 1991); locally abundant on smaller islands with herbaceous to shrub stage growth (Schwartz and Carey, 1977); on Booby Cay, iguanas on sandy and rocky beaches, rocky areas, sparsely and densely vegetated areas, and open areas, used sandy areas and holes between rocks for burrows, in bushes and trees to heights of 1.8 m (Wasilewski, 1998); most abundant in rocky coppice and sandy strand habitats, sandy areas are required for nesting (Gerber and Iverson, 2000).

Activity: Diurnal activity bimodal during warmer months, high midday temperatures result in reduced activity, winter activity reduced and primarily at midday, basks following emergence and prior to termination of day's activity, rest of day spent feeding, interacting with other lizards, and thermoregulating, which involves shuttling in sun/shade mosaic (Iverson, 1979). **Behavior:** Performs headbobs to declare territories, challenge males, and for sex recognition (Iverson, 1979); most headbob displays consist of a series of alternating single and double headbobs, each display usually beginning with 1–2 single headbobs or a fast "trill" of 6–15 bobs, followed by a set of 1–6 double headbobs, followed by single headbob or two, head nods and tail raises typical of other iguanas also occur (Martins and Lamont, 1998); aggressive displays include lateral orientation, gaping, gular expansion, arched back, crest elevation, inflated trunk, lateral compression, and single or double headbobs, aggression also may involve tail lashing and fighting (Carpenter, 1982); will take to water as an escape tactic, but only as a last resort (Iverson, 1979). **Diet:** Primarily herbivorous at all ages (95.6% of all food items and 94.8% of volume; Auffenberg, 1982); feeds both terrestrially and arboreally, 60% of macroplant species (including a few toxic forms) eaten, with plants taken falling into three categories: (1) Common staples of high caloric content, (2) common staples of low caloric value (roughage), (3) foods eaten frequently in spite of their uncommonness (plants of high preference levels), certain individual plants are repeatedly browsed whereas others of the same species are ignored, caloric content in the form of nonstructural carbohydrates is not an important factor in leaf and fruit choice, animal prey includes crabs (*Clibanarius* and *Cardisoma*), insects (termites, beetle larvae and adults, cicada nymph, dipteran larvae), a slug, and carrion (fish, lizard, bird, rodent, and canned meats, especially sardines), during the winter, restricted to items that are more difficult to digest (leaves), which may pass through the digestive tract almost intact (Iverson, 1982); cannibalism (Iverson, 1979; Auffenberg, 1982); one well-developed circular valve and 4 more distal semilunar valves partition the proximal colon ($N = 15$), presumably to increase digestive efficiency (Iverson, 1980); fruits of Turk's Cap Cactus (*Melocactus azureus*; Sánchez Muñoz, 2008). **Home Range and Territoriality:** Adult males are territorial throughout the year (Gerber and Iverson, 2000); average home range 980 m^2 in females, 1,260 m^2 in males, 1,590 m^2 in dominant males (Iverson, 1979). **Parasites:** Nematodes (*Alaeuris travassosi*, *Ozolaimus monhystera*, *Paralaeuris cyclurae*; Dosse, 1939); very large oxyurid nematode (*Cyrtosomum mega*) populations in colon, perhaps mutualistic rather than parasitic (Bowie and Franz, 1974); to >15,000 presumably commensal or even mutualistic nematodes in a healthy adult (Iverson, 1982). **Physiological Demands:** Glucocorticoids (used to evaluate heightened physiological demands under varying conditions) significantly greater in female *C. c. carinata* than in males, tended to be inversely related to body condition in females but not in males, in January, somewhat higher in the morning, affected by reproductive state and possibly by seasonal rainfall (MacDonald et al., 2007). **Population Density:** Estimated densities along flush transects on Pine Cay 1.7–36.8 adults/ha (Iverson, 1978); 90.3/ha in optimum habitat for juveniles, 31.1/ha for adults (Iverson, 1979); 558 iguanas in 14.5/ha on Booby Cay, Mayaguana (Conners et al., 2005, 2006). **Population Sizes:** Estimated 50,000 *C. c. carinata* and 1,000 *C. c. bartschi* (Alberts, 2004); estimated 30,000 *C. c. carinata* and 200–300 *C. c. bartschi* (Hudson and Alberts, 2004). **Predators:** Feral dogs and cats (Iverson, 1978, 1979; Mitchell et al., 2002; Mitchell, 2003; Gerber and Alberts, 2005, 2006). **Reproduction:** Cycles are synchronized with climatic cycles, courtship in early May, probably serial polygyny with monogyny among some males, eggs 2–9/clutch in early June, females defend burrows for several days to several weeks after nesting, but are not territorial at other times, hatching occurs after approximately 90 days, hatchlings average 79.8 mm SVL and 14.6 g, juvenile growth rate averaged 19.2 mm SVL/yr, males reach sexual maturity at approximately 220 mm SVL, 375–475 g, age about 7 years, females 185–200 mm, 200–300 g, 6–7 years (Iverson, 1979); individuals in recently established populations on French, Six Hills East, Bay, and Middle cays exhibit larger average adult sizes and reach maturity more rapidly (1.5–2.5 years versus 6–7 years) than on source cays, Big Ambergris and Little Water (Gerber and Alberts, 2005, 2006). **Sex Ratio:** 1:1 at all age classes (Iverson, 1979). **Size:** Adult male SVL to 360 mm, female SVL to 292 mm (Schwartz and Henderson, 1991). **Survivorship and Life Expectancy:** Survivorship positively correlated with body size, life expectancy 14.0 years (Iverson, 1979). **Tail Autotomy:** 32.6% of males and 28.6% of females on Pine Cay had broken tails (Iverson, 1979).

Conservation Status: The species is listed in CITES Appendix I (UNEP-WCMC, 2008); "critically endangered" on the IUCN Red List (Gerber, 1996a, 2004b; Gerber and West Indian Iguana Specialist Group, 2000); both *C. c. carinata* and *C. c. bartschi* as "threatened" on the U.S. List of Endangered and Threatened Wildlife and Plants (U.S. Fish & Wildlife Service, 2008) primarily due to predation by introduced mammals (dogs and cats), competition with and degradation of habitat by introduced herbivores (e.g., cattle and goats), habitat alteration by human activity, and exploitation by humans; a conservation and management plan exists (Burton and Bloxom, 2007a); domestic dogs and cats reduced the population (estimated 5,500 animals) on Pine Cay almost to extirpation in 3 years following construction of a hotel and tourist facility (Iverson, 1978).

Cyclura collei Gray, 1845

Distribution: Jamaica, Hellshire Hills (extirpated elsewhere on the island, including Goat and Little Goat islands).

Habitat: Xerophilic, tropical dry forest in low limestone hills along shores and in areas between the coast and interior uplands, in hollow trees in pastures, where congregated (Schwartz and Henderson, 1991); retreat site fidelity is strong, with hatchlings preferring dead hollow trees 6–15 cm in diameter with a northwesterly aspect (Van Veen and Wilson, 2004, 2005); 17 of 155 observations of headstarted iguanas released into the wild were in trees to heights of ~7 m (D. S. Lewis et al., 2008).

Behavior: Will climb, can climb a smooth fence post (C. B. Lewis, 1944); headbob displays consist of a single headbob or three single headbobs with pauses of 0.7–1.0 sec between them (Martins and Lamont, 1998). **Foraging and Diet:** Leaves, fruits, and flowers of a wide variety of plant species, supplemented occasionally by animal matter, including snails (P. Vogel, 2000a); of 25 types of plants consumed, 3–4 species of vines are most important (Van Veen and Wilson, 2004, 2005); foraging in tree canopies (D. S. Lewis et al., 2008). **Home Range and Territoriality:** 0.59–5.82 ha for 6 repatriated iguanas, home ranges overlapped considerably, probably reflecting recency of release (D. S. Lewis et al., 2008); injuries presumably from fighting in released iguanas are indicative of territorial behavior (R. van Veen in D. S. Lewis et al., 2008). **Movements:** Farthest distance traveled from release site ($N = 6$) 196–455 m (D. S. Lewis et al., 2008). **Population Size:** Estimated 150 (Alberts, 2004); estimated 100–200 (Hudson and Alberts, 2004). **Predators:** Hatchling consumed by *Epicrates subflavus* (Wilson and Van Veen, 2005, 2006); introduced predators include mongooses (*Herpestes javanicus*), rats, cats, dogs, and possibly pigs (C. B. Lewis, 1944; P. Vogel, 2000a; Wilson et al., 2004a, 2004b); dogs used for hunting pigs are able to take even adult iguanas (Woodley, 1980). **Reproduction:** Age at maturity 7–8 years in captivity (R. Hudson, pers. comm. to J. B. Iverson); nesting activities "apparently late March to early May" on Goat Island in the 1930s (C. B. Lewis, 1944); females begin digging trial holes long before egg laying, eggs deposited in nests in two known communal nesting areas in mid-June, with hatchlings emerging 85–87 days later, females aggressively guard nests up to 2 weeks after egg laying, using threat displays, biting, and chasing, clutch sizes average 17 eggs (16–20), hatching success ranges 0–100% and appears to be related to female body size (P. Vogel, 1994); in 2005, ovideposition ranged from 24 May to 20 June for 14 nests in two known communal nesting sites (Wilson and Van Veen, 2005, 2006). **Size:** Adult male SVL to 428 mm, female SVL to 378 mm (Schwartz and Henderson, 1991). **Survival:** Eight of 20 headstarted iguanas survived 24 months or more (Wilson et al., 2004a); all 6 repatriated iguanas survived at least 3.5 months (D. S. Lewis et al., 2008). **Weight Loss:** Over at least 3.5 months after release, 6 headstarted iguanas lost an average of 1.69 g/day (D. S. Lewis et al., 2008).

Conservation Status: The species is listed in CITES Appendix I (UNEP-WCMC, 2008); "critically endangered" in the IUCN Red List (Gibson, 1996b); "endangered" on the U.S. List of Endangered and Threatened Wildlife and Plants (U.S. Fish & Wildlife Service, 2008); principal causes include degradation of habitat by charcoal burners and predation by introduced mongooses (*Herpestes javanicus*), dogs, and cats (Woodley, 1980; P. Vogel, 1992; Alberts, 1993; P. Vogel et al., 1996; Wilson et al., 2004a, 2004b); ongoing threats include proposed development of the Hellshire Hills (Woodley, 1980; P. Vogel, 2000a).

Cyclura cornuta Bonnaterre, 1789

Distribution: Hispaniola, widespread, but disjunct distribution on the main island (probably once islandwide in xeric lowlands), also Isla Beata, Île-de-la-Petite Gonâve, Île-de-la-Tortue, Île Grande Cayemite, and possibly Île-de-la-Petite Cayemite and Isla Saona.

Habitat: Xerophilic to semimesophilic (Schwartz and Henderson, 1991); dry forest with sand, coral, or eroded limestone substrates in coastal terraces, main-island lowlands, and satellite islands, often using crevices as retreats or digging burrows in sand or fossil coral (Mertens, 1939; Carey, 1975; Noble, 1923c; Ottenwalder, 2000a); basks on roads (RP); roadkill recorded in 1939 (Mertens, 1939); readily climbs into trees to feed (Powell et al., 2002).

Behavior: May take to water as an escape tactic (Noble, 1923c); individual basking in full sun on an angled tree trunk 2 m above ground escaped into *Opuntia* thicket when disturbed (Mertens, 1939); head "rolls" and headbobs used in displays (Martins and Lamont, 1998); often become "tame" with feeding or regular exposure to humans (Banbury et al., 2000; Banbury and Ramos, 2005). **Diet:** Feeds on "harsh fruits and shore crabs," responded to bait composed of mashed sweet potato soaked in rum (Eyerdam, 1954); seeds of *Ziziphus rignoni* germinated more rapidly after being consumed and passed by iguanas (Hartley et al., 2000); dermatophagy (Weldon et al., 1993); a variable number of circular (3–4) and more distal semilunar valves (5–7) partition the proximal colon, presumably to increase digestive efficiency (Iverson, 1980). **Longevity:** In captivity, 22 years, 11 months and still alive at last report in 1990 (R. Noegel in Slavens and Slavens, 2003). **Parasites:** Include ticks *Amblyomma albopictum* (Carey, 1975; Wiewandt, 1977; Guglielmone et al., 2003), *A. cruciferum* (Neumann, 1901; Morel, 1967), and *Ornithodoros elongatus* (probably, found in container with an iguana; Kohls et al., 1965; Morel, 1967; Guglielmone et al., 2003); nematodes (*Atractis cruciata*; Linstow, 1902). **Population Size and Density:** 0.03 iguanas/ha east of Pedernales in 1985 (Incháustegui, 2003b); 0.70 iguanas/ha on Isla Cabritos (G. Santana in Incháustegui, 2003b); estimated 17,000 (Alberts, 2004); estimated 10,000–17,000 (Hudson and Alberts, 2004). **Population Structure:** Juveniles compose about one-third of the population on Isla Beata (Grupo Jaragua in Wiewandt and García, 2000). **Predators:** Humans (primarily for food), feral dogs and cats, mongooses (*Herpestes javanicus*), and pigs take adults, juveniles, and eggs (Ottenwalder, 2000a); feral dogs (Rupp et al., 2008). **Reproduction:** Age at maturity 3–4 years (Ottenwalder, 2000a); in captivity 3–4 years (Boylan, 1984, 1985); eggs warmed by sunlight in tunnels 0.4–0.6 m long (Noble, 1923c; Haast, 1969, in captivity); aggregate nesting on Isla Beata (Noble, 1923c); eggs in two clutches (in captivity at the San Diego Zoo) averaged 68.4 × 46.7 mm and 79.8 g ($N = 21$) and 67.6 × 48.6 mm and

83.0 g ($N = 17$), seven hatchlings measured 258–294 mm total length, 162–190 mm tail length, and 47.4–61.9 g body mass (Shaw, 1969); nesting mid-June through mid-July at four sites in the southwestern Dominican Republic, hatching early September through mid-October, 87.0–100.0% of nests hatched (0.9–1.6 nests/ha) at sites where *C. ricordii* also present, 567–1,851 nests/ha at two sites where *C. ricordii* not present (primarily rocky areas adjacent to "fondos" [= areas with well-drained soils] where *C. ricordii* nests are concentrated), lowest values in 2007, when feral dogs were active in the nesting areas (Rupp et al., 2008). **Size:** Male SVL to 463 mm, female SVL to 510 mm (Schwartz and Henderson, 1991).

Conservation Status: The species is listed in CITES Appendix I (UNEP-WCMC, 2008); "vulnerable" on the IUCN Red List (Day, 1996a; Ottenwalder, 1996a); recent information suggests that the status of the species may require reevaluation and upgrading to a higher level of protection (Ottenwalder and Powell, 2002); major threats include dry forest disturbances for charcoal production and agriculture (Ottenwalder, 2000a) and human exploitation (Powell, 2003a); this species was included in lists of threatened Hispaniolan species (Aquino, 1983; Powell et al., 2000).

Additional Pertinent Literature: Glor et al. (2000).

Cyclura cychlura Cuvier, 1829

Distribution: Bahama Islands, where three currently recognized subspecies occur on Andros Island (*C. c. cychlura*) and the northern (*C. c. inornata*) and central and southern Exuma Cays (*C. c. figginsi*); populations have been introduced onto several islands in the Exumas and off New Providence Island (Iverson, 2000; Knapp and Malone, 2003; Iverson et al., 2006a); some movement of individual *C. c. inornata* from cay to cay has been attributed to relocation by humans not engaged in conservation (e.g., tourists, fishermen; Hines and Iverson, 2006a, 2006b).

Habitat: Xerophilic to semimesophilic (Schwartz and Henderson, 1991); sandy beaches, xeric limestone devoid of vegetation, and areas of vegetation with or without sand or rock substrates, one preferred habitat is under the open canopy of the pine (*Pinus caribaea* var. *bahamensis*) barrens (*C. c. cychlura*; Buckner and Blair, 2000; Knapp and Owens, 2005; *C. c. figginsi*; Knapp, 2000b); in areas of rock and sand in the northern Exumas, including suboptimal areas of bare, honeycombed limestone (*C. c. inornata*; Schwartz and Carey, 1977; Iverson, 2000); along the strand and in the rocky interior (presumably site of dens) in the central Exumas (*C. c. figginsi*; Schwartz and Carey, 1977); limestone crevices and sand burrows are used as refugia (*C. c. figginsi*; Knapp, 2000b); occasionally enter the sea for short periods (Knapp, 1995).

Activity: Most feeding 0830–1400 h, iguanas sought deep cover in dense vegetation 1030–1500 h (Wilcox et al., 1973). **Behavior:** Lack of territorial or hierarchal behavior attributable to small island size and high population densities (*C. c. figginsi*; Carey, 1976; Knapp, 2000a, 2000b); but very aggressive in captivity (J. Schmidt, 2009); apparent dominance hierarchies in *C. c. inornata* develop during the nonbreeding season (Iverson, 2000); aggressive displays include face-offs and single or double headbobs, aggression may involve chasing and gaping in *C. c. cychlura*, displays include lateral presentation, lowered head, gular extension, crest erection, trunk inflation, gaping, and single or double headbobs in *C. c. figginsi* (Carpenter, 1982); displays in *C. c. inornata* consist of short bouts of mostly single but occasionally double headbobs with no obvious pattern to the choice of single or double headbobs (Martins and Lamont, 1998). **Diet:** Coprophagous on feces of Zenaida Doves (*Zenaida aurita*) and White-crowned Pigeons (*Patagioenas leucocephala*; Coenen, 1995); dermatophagy (Murphy, 1969; Weldon et al., 1993); forages on beach detritus, leaves and flowers of Ricewood (*Strumphia maritima*; Carey, 1976); herbivorous (45% arboreal, 55% terrestrial), preferred food is seasonal and includes flowers, fruits, young buds, and leaves of *Rachicallis americana*, *Reynosia septentrionalis*, *Strumpfia maritima*, *Jacquinia keyensis*, *Erithalis fruticosa*, *Coccoloba uvifera*, *Coccothrinax argentata*, *Eugenia axillaries*, *Suriana maritima*, and rotting fruits of *Casasia clusiifolia* (*C. c. figginsi*; C. Coenen, 1972 in Knapp, 1995; Wilcox et al., 1973; S. L. Windrow, 1977 in Hayes et al., 2004; Knapp, 1995, 2000b, 2001a); feeds on fruits, leaves, and flowers of most plants present on islands, including *Byrsonima lucida* and *Coccoloba uvifera* (*C. c. cychlura*; Knapp et al., 1999); opportunistically insectivorous (termites; *C. c. cychlura*; Auffenberg, 1982) or carnivorous, with crab (*Cardisoma* sp.) claws in feces (*C. c. cychlura*; Knapp et al., 1999; *C. c. inornata*; Iverson, 2000); takes table scraps and fresh produce from tourists, especially on Leaf Cay (*C. c. inornata*; Iverson, 2000; Iverson et al., 2004b, 2006a; Knapp and Hines, 2006a, 2006b), although most appears to be consumed by relatively few individuals (Iverson et al., 2004b); *C. c. inornata* facultatively takes Audubon's Shearwater (*Puffinus lherminieri*) nestlings and possibly those of White-crowned Pigeons (*Patagioenas leucocephala*) and Bahamas Mockingbirds (*Mimus gundlachii*; Hines et al., 2002a); 5 circular and 2–3 distal semilunar valves in *C. c. cychlura* ($N = 3$), 4 circular and 3 semilunar valves in *C. c. inornata* ($N = 1$), and modally 4 (2–6) circular and 3 (2–6) semilunar valves in *C. c. figginsi* ($N = 14$) partition the proximal colon, presumably to increase digestive efficiency (Iverson, 1980); drink rainwater from depressions in rocks (Wilcox et al., 1973). **Genetic Diversity:** Both source (Leaf Cay) and translocated (Alligator Cay) populations of *C. c. inornata* exhibited low average allelic diversity and heterozygosity for 8 microsatellite loci known to be polymorphic in *C. cychlura* (Knapp and Malone, 2003). **Growth:** Individual *C. c. figginsi* grew 8.33 mm SVL/yr over 3 years (Coenen, 1995); mean annual growth rate significantly faster in males (1.764 cm/yr) than in females (1.139 cm/yr; *C. c. inornata*; Iverson and Mamula, 1989); annual growth rates in an introduced population of *C. c. inornata* on Alligator Cay for both males and females (35.3 mm and 30.1 mm, respectively) were significantly greater than for the founder population (Knapp, 2001a); one second-year *C. c. figginsi* on Pasture Cay (Exuma Islands) had SVL = 13.7 cm and body mass = 118 g (Knapp et al., 2004b, 2005). **Home Range:** 30.58 ha for males and 5.63 ha for females, the largest reported for free-ranging *Cyclura* (Knapp and Owens, 2005). **Life History:** Survival of *C. c. inornata* higher on U Cay than on Leaf Cay (with more human activity), higher in females than males (which are "bolder"), and exhibited a declining trend as growth rates neared zero, presumably because these populations are near or at carrying capacity (Iverson et al., 2006a); of 16 juvenile *C. c. inornata* captured and marked at age 0.5 years, 9 were alive at age 12.5 years (typical age at maturity), an annualized rate of 95.3%, of 4 other juveniles marked at age 1.5 years, 3 were still alive 10 years later, an annual rate of 97.2% (highest rates recorded for juveniles of any lizard), life table demonstrates extreme longevity, very high survival throughout life, greatly delayed maturity (longer than for any studied lizard), extreme iteroparity, and low reproductive output, net reproductive rate ($R_0 = 3.4$) and cohort generation time (21 years) match the 3.2-fold population size increase estimated between 1982 and 2004 (Iverson, 2007a). **Longevity:** May exceed 20 years in nature (Iverson and Mamula, 1989); 36.7 years (adding estimated age at first capture) for *C. c. inornata* in nature (Iverson et al., 2004b); in captivity for *C. c. figginsi*, 23 years, 6 months and still alive at last report in 1990 (R. Noegel in Slavens and Slavens, 2003). **Mortality:** Death due to being wedged between tree trunks (Iverson et al., 2005); 3 individuals trapped in a 50-gallon plastic drum (Iverson, 2008). **Nest Sites:** On Guana Cay (Exuma Islands), major nesting area on beach characterized by fine, windblown sand with numerous small palms and tussocks of grass (C. Coenen, 1972 in Knapp, 1995). **Parasites:** Nematode *Atractis opeatura* (Leidy, 1890); oxyurid nematodes in 4 of 6 fecal samples and bacterial isolates in all but one of 17 animals (6 with *Corynebacterium* spp., 5 with coagulase-negative *Staphylococcus* spp. and *Salmonella* spp., and 4 with *Enterococcus* spp.) are probably commensals or part of the normal enteric microbial flora, respectively (*C. c. inornata*; James et al., 2006); ticks (*Amblyomma dissimile* on *C. c. cychlura* and *A. torrei* on *C. c. figginsi*; Durden and Knapp, 2005). **Population Density:** ~3.2/ha (*C. c. figginsi* on Guana Cay, Exuma Islands; Carey, 1976); <0.2 iguanas/ha on South Andros (Knapp and Owens, 2008a). **Population Size:** Population on Guana Cay (Exuma Islands), 64 individuals (Knapp, 1995); estimated total population sizes 3,500 *C. c. cychlura*, 1,500 *C. c. figginsi*, and 1,000 *C. c. inornata* (Alberts, 2004); estimated 2,500–5,000 *C. c. cychlura*, 1,000–2,000 *C. c. figginsi*, and 400–500 *C. c. figginsi* (Hudson and Alberts, 2004); *C. c. inornata* increased from ~150 in 1970 to "perhaps" 1,300 in 2006 (Iverson, 2007b). **Predation:** Hatchlings consumed by *Epicrates striatus fowleri* (Knapp and Owens, 2004; Knapp et al., 2004b, 2005) and by *Cubophis vudii* (Knapp et al., 2004b, 2005); apparently gape-limited *C. vudii* failed in predation attempts on iguanas (Knapp and Owens, 2008b); eaten by humans (e.g., Catesby, 1743; Barbour, 1904; Knapp, 2005; Knapp et al., 2004b, 2005); humans (for food), feral pigs (on nests), feral dogs and possibly cats on Andros Island (*C. c. cychlura*; Knapp et al., 1999; Buckner and Blair, 2000); humans (for illegal wildlife trade), dogs, and possibly rats (*C. c. figginsi*; Knapp, 2000b). **Reproduction:** Males "guard" females on Andros (*C. c. cychlura*), attendant male chasing an intruder to 150 m, males of equal size engage in displays, face-offs, and lateral head-pushing (Knapp and Owens, 2008a); copulation on Andros on 24 April (P. Hartman in Knapp and Owens, 2008a) and 25 April (A. Mitchell in Knapp and Owens, 2008a); nest on Guana Cay (*C. c. figginsi*) covered with damp sand on 6 June, entrance blocked by driftwood apparently placed by a nest-guarding female (Knapp, 1995); nests on Guana Cay ~61 cm long and 8–13 cm deep, with females actively defending incomplete tunnels but not nests after ovideposition (Knapp, 2000a); eggs laid in termite (primarily *Nasutitermes rippertii*) mounds on Andros Island and defended by females on tops of termitaria (*C. c. cychlura*; Knapp, 2004, 2005; Knapp and Owens, 2008a; Knapp et al., 1999, 2004b, 2005, 2006); two nests on Andros in coarse sand composed largely of gastropod (*Batillaria* sp.) shell fragments, both nests in termitaria and elsewhere often used repeatedly (6 of 12 in termitaria used again by same female), nesting began in early May through mid-June, mostly (46%) in open pine followed by dry evergreen shrubland (36%) and closed pine habitats (18%), mean canopy cover 72 ± 20%, eggs deposited at tunnel entrances and pushed into central chambers with head and forelimbs, nests plugged and covered with scattered soil or sand, nest temperatures (mean 32.6 ± 1.5 °C) higher than ambient (29.2 ±

4.8 °C), egg mortality due to crab (*Cardisoma guanhumi*) predation, desiccation, fungi, flooding, flaccid and died immediately, or ruptured by females, nests defended for at least 6 weeks against females approaching to within 10 m, at hatchling emergence 23% of females still in attendance (Knapp and Owens, 2008a); excavations of egg chambers revealed 3 eggs in each of 2 nests (Coenen, 1995); mating in *C. c. inornata* occurs in mid-May, with females migrating 30–173 m to potential nesting sites in mid- to late June, females often abandon initial excavations and average time from initiation of digging to completion of a nest is 6 days, females defended the nest during construction and for 3–4 weeks after egg deposition, burrows average 149 cm in length and terminal egg chambers usually angled off the main burrow, depth averages 28 cm and is inversely correlated with the shadiness of the site, indicating that depths correlate with temperature (mean 31.4 °C), only one in 3 females nest each year, with frequency increasing with body size, so the largest females nest yearly, females mature at 26–27 cm SVL, ~750 g body mass, and 12 years of age, clutch size 1–10 (mean 4.6) correlated with female body size and age, eggs average 66 × 35 mm and 49 g, with mass not correlated with female body size, but smaller females produced more elongate eggs, clutch mass averaged 16.5% of female body mass, hatching occurs in late September and early October after ~80–85 days incubation, hatchlings average 9.5 cm SVL and 33 g body mass, survivorship to emergence was 78.9% and inversely correlated with soil moisture (Iverson, 2000; Iverson et al., 2004a); female *C. c. inornata* on Leaf Cay actively defend nests, even against larger males, eggs oblong, about the width of a golf ball but twice as long, clutch accounted for nearly 20% of gravid female weight (Valiulis et al., 2003); *C. c. inornata* on Flat Rock Reef Cay did not actively defend nests, demonstrated higher clutch frequency (40–50% versus ~33% on Leaf and U cays), and grew faster (32 cm SVL in 10 years versus 32 cm SVL in 12 years), resulting in maturation of females in less than a decade versus 12 or more years on Leaf and U cays (Hines and Iverson, 2006a, 2006b); female *C. c. cychlura* on Andros Island initiated oviposition 5–6 weeks earlier, deposited significantly more eggs per clutch (9.8 ± 3.3, $N = 34$ versus 4.5 ± 1.7, $N = 73$), heavier total clutches (618.9 ± 230.0 g versus 223 ± 85 g, $N = 73$), and wider eggs (40.3 ± 2.8 mm versus 34.8 ± 1.8 mm, $N = 63$) than *C. c. inornata* in the northern Exuma Islands, egg mass (59.8 ± 4.9 g) was positively correlated with female body size (38.7 ± 4.3 cm SVL, $N = 32$; 3006 ± 1185 g, $N = 31$) on Andros, whereas egg length (67.5 ± 4.6 mm, $N = 63$) was inversely correlated with female body size (31.5 ± 3.0 cm SVL, $N = 70$; 1283 ± 418 g, $N = 57$) in the Exumas (egg mass in the Exumas = 49.1 ± 3.9 mm, $N = 63$; egg length on Andros = 70.7 ± 3.6, $N = 34$), but hatchling sizes (9.6 ± 0.4 cm SVL and 42.4 ± 4.9 g, $N = 37$ on Andros versus 9.5 ± 0.2 cm and 35.4 ± 3.3 g, $N = 12$ in the Exumas) were comparable (Knapp et al., 2004b, 2005, 2006); female *C. c. cychlura* with scoliosis and kyphosis nested in an active termite (*Nasutitermes rippertii*) mound, clutch with 7 eggs, mean length and width (67.3 mm and 37.2 mm), clutch size and egg length comparable to and egg width slightly smaller than those previously recorded, all eggs hatched and hatchlings normal (Owens and Knapp, 2007); 24 of 30 eggs hatched and emerged in 2006 on Flat Rock Reef Cay (*C. c. inornata*; Iverson, 2007c); in captivity, *C. c. figginsi* laid 5–8 eggs (some infertile), incubation 87–92 days at 30.2 °C (J. Schmidt, 2009). **Sex Ratio (M:F):** In *C. c. figginsi* on Guana Cay (Exuma Islands), 17:15 (Knapp, 1995); in *C. c. inornata* on Leaf and U cays changed from heavily male-biased to nearly 1:1 over 25 years, presumably attributable to the recovery of these populations from intense harvesting (particularly of females) and the recent removal of large males by either poachers or tourists (G. R. Smith and Iverson, 2006); increasingly female biased, presumably attributable to removal of large males by "mischievous human activities" (Iverson, 2008). **Size:** Male SVL to 65.5 cm and 133.9 cm total length and 7.8 kg (Knapp, 2002), female SVL to 57 cm and 115 cm total length (Iverson et al., 2006a). **Tail Autotomy:** 44.4% of male and 38.5% of female *C. c. cychlura* on Dissertation Cay, Andros, 14.1% of male and 15.7% of female *C. c. figginsi* on several Exuma cays, and 0% of male and 9.5% of female *C. c. inornata* on Alligator Cay (introduced) had broken tails (Knapp, 2000a); 186 of 1,839 sexable *C. c. inornata* (10.5% of males and 9.6% of females) had broken or regenerated tails, rates did not differ between sexes, but were higher on U Cay than on Leaf Cay (Iverson et al., 2004b).

Conservation Status: The species is listed in CITES Appendix I (UNEP-WCMC, 2008); "vulnerable" in the IUCN Red List (Knapp et al., 2004a); all three subspecies as "threatened" on the U.S. List of Endangered and Threatened Wildlife and Plants (U.S. Fish & Wildlife Service, 2008), although populations assigned to *C. c. cychlura* (Andros Island) are "endangered" (Knapp and Buckner, 2004a), those assigned to *C. c. figginsi* (southern Exuma Islands) are "critically endangered" (Knapp and Buckner, 2004b), and those assigned to *C. c. inornata* (northern Exuma Islands) are "endangered" (Blair and West Indian Iguana Specialist Group, 2000); populations on Leaf and U cays were nearly extirpated in the early 1900s because of heavy hunting pressure (for food), but increased to about 150 lizards in 1970, and number more than 500 in 2006 (Iverson et al., 2006a).

Additional Pertinent Literature: Iverson et al. (2006b).

Cyclura lewisi Grant, 1940

Distribution: Grand Cayman Island, once islandwide.

Habitat: Holes in limestone (C. B. Lewis, 1940); in sandy and rocky scrub (Schwartz and Carey, 1977); currently exploits a mosaic of natural and somewhat disturbed habitats in response to availability of thermoregulatory sites, food, substrates suitable for nesting, predation pressure, and human activity (Burton, 2000); captive-bred animals released into the wild "preferred" modified habitats over natural habitats and used artificial retreats more frequently than natural retreats (Goodman et al., 2005a).

Activity: Iguanas emerged earlier and retreated later in summer months than in the fall, most of the time inactive, usually in states of low or medium alertness, of the small percent of time males spent active (20.3 ± 2.3% summer, 19.6 ± 3.0% fall), they moved (8.3 ± 4.2%, 6.4 ± 2.5%), foraged (8.1 ± 4.8%, 10.2 ± 2.2%), and engaged in social situations (11.4 ± 21.1%, 1.9 ± 3.2%), much inactive time spent basking (29.9 ± 9.2%, 23.5 ± 4.4%), females active (18.3 ± 1.7%, 16.9 ± 4.4%), moving (6.1 ± 1.4%, 5.7 ± 2.4%), foraging (9.0 ± 2.5%, 8.3 ± 3.4%), engaged in social situations (20.7 ± 15.2%, 12.7 ± 20.3%), basking (23.5 ± 4.2%, 23.9 ± 9.3%), in general, iguanas spend little time in trees or bushes or inside retreats during the day (Goodman, 2007). **Behavior:** Produces an "S-roll" (the nose is raised and lowered as in a headbob, but is also moved in a horizontal plane during the lowering motion), although some individuals appear to produce standard headbobs in addition to the S-rolls, the twisting motion is not as dramatic as the full rolling motion produced by *C. cornuta* and *C. stejnegeri*, displays are series of ~5 S-rolls separated by pauses of ~0.7 sec, a pattern very similar to that of *C. nubila* from Cuba (Martins and Lamont, 1998); adult males show little long-term fidelity to retreats, but females appear to occupy small, fixed territories centered on 1–2 favored retreats, the number of suitable refugia may control female distribution, whereas male density may be regulated primarily by territorial interactions (Burton, 2000); hatchlings tongue-touched the ground as they moved away from nests, 3 of 5 climbed into trees within 20 min, suggesting greater arboreality than in adults (Goodman and Burton, 2005); extensive tongue-touching of retreats, substrates, and feces suggests the importance of chemosensation, supplemental feeding appeared to increase incidents of aggression toward humans, iguanas responded to increasing levels of human and vehicular disturbance (Goodman, 2007). **Diet:** Plums are preferred over all other fruit, mangoes are taken in season, and pawpaws are eaten by captive animals (C. B. Lewis, 1940); almost exclusively herbivorous, known to consume at least 45 species of plants in 24 families (~80% leaves and 20% fruits), particularly important are leaves of *Rhynchosia minima*, *Chamaecrista nictitans*, and *Stylosanthes hamata*, whole plants of *Spermacoce confusa*, and seedlings of *Waltheria indica*, all of which are weeds that thrive in disturbed open areas, iguanas gorge on fruits of *Picrodendron baccatum*, *Hippomane mancinella*, and *Myrcianthes fragrans* when available, these occur primarily in stands of primary vegetation, the diet is presumably very different than it was prior to human settlement, exceptions to strict herbivory include consumption of fungi, crabs, and cicadas, with the latter presumably scavenged, coprophagy and soil ingestion have been observed (Burton, 2000); slugs (*Veronicella* sp.; Goodman, 2003, 2007); primarily herbivorous, eating plants, the majority of which were noncultivated, but geophagy, coprophagy (including that of an agouti, *Dasyprocta punctata*), consumption of invertebrates (climbing into tree to feed on larval moths, *Pseudosphinx tetrio*) occurs, one iguana chased low-flying butterflies for ~5 m and attempted to catch them, consumes shed skin and debris (plastic spoon, plastic bags with food scraps, metal twist-tie, with efforts to climb into trash cans), supplemental feeding contributed little to the overall diet, but did result in aggregate feeding of females and juveniles, but never males (Goodman, 2007). **Home Range:** Released, head-started females had mean home ranges of 0.24 ha, released males 0.65 ha with considerable overlap, for a density of 4–6 iguanas/ha, attributable largely to an unnatural single age class in an area surrounded by unpopulated habitat (Burton, 2005a, 2005b, 2006); mean home ranges of males in summer (13.567 ± 14.426 ha using minimum convex polygon and 4.559 ± 3.517 ha using 95% contours for fixed kernel use distributions) were larger than in fall (1.600 ± 1.656 ha and 0.804 ± 1.102 ha) and much larger than that of females in summer (0.623 ± 0.312 ha and 0.209 ± 0.181 ha) and fall (0.856 ± 0.754 ha and 0.467 ± 0.486 ha), with males moving longer distances in summer than in fall and than females in either season (Goodman et al., 2005b). **Hybridization:** *Cyclura nubila caymanensis* will hybridize with *C. lewisi* if "brought together" (Davis, 2000; Burton, 2004c). **Longevity:** In captivity, at least 54 years, estimated ~69 years (Adams, 2004). **Mortality:** Seven iguanas "murdered" by humans at the Queen Elizabeth II Botanic Park (Binns, 2008a; S. Hicks, 2008). **Parasites:** Piroplasm *Sauroplasma* in erythrocytes of captive-reared juveniles (Alberts et al., 1998). **Population Density:** Density of head-started females 1.6–2.0/ha (Burton, 2005a). **Population Size:** Estimated 30 in nature (Alberts, 2004); estimated 100–175 total (Hudson and Alberts, 2004). **Predators:** Wild and domestic cats and dogs, and humans, who perceive iguanas as a threat to their crops (Burton, 2000, 2007); *Cubophis caymanus* preys on hatchlings (Goodman and Burton, 2005). **Reproduction:** May nest biannually, one female laid 9 eggs on 24 May and 7 more 53 days later, a second female laid 10 eggs on 10 June and 6 more 60 days later, eggs in first clutches larger than those in second clutches, and all eggs in first clutches viable, whereas 2 in each second clutch died early in incubation (Burton, 2004b); wild hatchlings (94.3 ± 7.4 mm SVL, *N* = 6) smaller than captive-bred hatchlings (100.8 ± 3.6 mm SVL, *N* = 17), possibly attributable to substrate moisture, temperature, or nutritional state of mother (Goodman and Burton, 2005); in captivity, eggs 66 × 45 mm, mean hatchling SVL 97 mm (Grant and Lemm, 2007). **Size:** Known male SVL to 550 mm, although larger individuals have been seen, the largest female SVL (425 mm) was recorded from a free-ranging individual in Queen Elizabeth II Botanic Park (F. J. Burton, pers. comm.).

Conservation Status: The species is listed in CITES Appendix I (UNEP-WCMC, 2008); "critically endangered" in the IUCH Red List (Burton, 2004a); "endangered" on the U.S. List of Endangered and Threatened Wildlife and Plants (U.S. Fish & Wildlife Service, 2008), primarily due to habitat destruction and alteration complicated by predation by feral mammals.

Cyclura nubila Gray, 1831

Distribution: Two currently recognized subspecies occur on Cuba (*C. n. nubila*), where widespread, but with a disjunct distribution on the main island (probably once islandwide in xeric lowlands), and on satellites Isla de la Juventud, Archipiélago de los Canarreos, Cayos de San Felipe, Jardin de la Reina, Archipiélago de Sabana-Camagüey, and Cayo Cinco Leguas, and on Little Cayman Island and Cayman Brac ("Sister Isles"; *C. n. caymanensis*); introduced on Isla Magueyes, Puerto Rico (*C. n. nubila*) and on Grand Cayman (*C. n. caymanensis*), although the latter probably has not established a breeding population (Burton, 2004c).

Habitat: Xerophilic to mesophilic (Schwartz and Henderson, 1991); on Cuba, primarily coastal, often where rocks provide retreats into which animals quickly escape when threatened (Gundlach, 1880); iguanas exploit a wide variety of spatial resources, possibly facilitated by the absence of competitors (Perera, 1984); rocky hills and precipices with crevices for hiding and ledges for basking (Barbour and Ramsden, 1919); may live in areas that completely lack water for drinking (Buide, 1951); seeks refuge in caves and "dogtooth" limestone (Garrido and Schwartz, 1968b; Rodríguez Schettino, 1999b); adult female in a hollow tree trunk (Garrido and Schwartz, 1969); common on beaches, especially those with exposed rocks, edges of the exposed central platform on Cayman Brac is particularly suitable for burrows in crevices and small caves (Cooper, 1958; Schwartz and Carey, 1977; Gerber, 2000a); coastal distributions dramatically affected by development (Estrada, 1993a); on grounds of a historic coastal fort (Krüger, 2008); on Cayo Largo, on the northern exposure of an eroded limestone hill, with three plant associations, grass on shallow soil, palm thickets, and hardwood thickets edged by Prickly Pear (*Opuntia*), in the latter usually associated with Prickly Pear (L. H. Saxe in Sutcliffe, 1952); on Cayo del Rosario, one juvenile in mangroves (*Rhizophora mangle*), other larger individuals on the beach or in adjacent vegetation (Berovides Álvarez, 1980); on Cayo Coco, may climb on branches of trees and bushes to heights of 2 m (Estrada, 1993b); basks on inclined trunks and branches of mangroves (C. B. Lewis, 1944); forages in coastal mangroves and an isolated population inhabits an inland karstic mountain area (Perera, 2000); juveniles on Cayman Brac around native houses (Carey, 1966); in piles of lumber and building materials, among yuccas, palmettos, cacti, and in ground cover of fallen leaves, especially *Bursera simaruba* (Schwartz and Henderson, 1991); predominant orientation of burrows (with 0.6–1.1 iguanas/refuge) southward on Cayo Cruz del Padre, also in mangroves, excellent climber, to 8 m above the ground on trees or rocks (Rodríguez Schettino, 1999b); burrows primarily in beach dunes on Cayo Sijú (Beovides-Casas and Mancina, 2006); on Cayo Largo del Sur, where both sandy and rocky soil are present, iguanas prefer the latter, possibly minimizing energy expended in burrow construction (Perera, 1985a); at Guantanamo Bay, iguanas about four times more abundant along coastlines than inland (Alberts, 2003); in mesophyllous evergreen and semideciduous forest, mangrove forest, coastal and subcoastal microphyllous forest, and semi-desert thorny scrubwoodland (Rodríguez Schettino et al., 1999); on Little Cayman, adults seem to be concentrated in disturbed areas, where frequent contacts with humans results in, among other things, high roadkill mortalities (Goetz, 2008).

Abundance: On cayos of the Archipiélago de Sabana-Camagüey, iguanas common when humans absent or at low density, but rare or absent when human activity is obvious, and "it is lamentable that an animal as typical and interesting as the iguana is disappearing due to the ego and absurdity of humans" (Garrido, 1973d). **Activity:** Diurnal, rarely seen on cloudy or rainy days during the cooler months, leaving refuges only when the sun is shining and only around midday, during the rainy season, activity is bimodal, from 0930–1200 h and 1530–1800 h (Rodríguez Schettino et al., 1999); most active during midday, when the sun is highest (Buide, 1951); active 0730–0830 h (temperature = 27 °C) on clear days, does not emerge from dens on rainy days or days when cloudy early, basks until 1000 h, then feeds (1000–1300 h), rests in shade before resuming activity after 1600 h, retires to dens by 1730 h, juveniles retreat just after 1200 h, almost always before 1400 h (on the Sister Isles; Grant, 1940c); active throughout most of the day, emerges on clear days when the temperature reaches ~27 °C, usually around 0830 h, with height of activity occurring between 1000 and 1300 h, rarely seen in the open after about 1430 h, resumes activity around 1600 h, not seen after 1730 h (Cayman Brac; Carey, 1966); on Little Cayman, begins to emerge ~1000 h, most active during the hottest part of the day (Townson, 1980); at Guantanamo Bay, vigorous bursts of social activity follow emergence, relatively peaceful migration to nearby areas of vegetation for midday feeding, ending in a second bout of social interactions in late afternoon prior to entering nighttime refuges (Alberts, 2003). **Behavior:** Initially reacts by inflating its body and "featuring a ferocious aspect," but quickly learns to take food offered by hand (Gundlach, 1880); very shy upon emergence from dens, but "tame" easily in captivity (on Sister Isles; Grant, 1940c); readily climbs smooth fence posts (C. B. Lewis, 1944); when disturbed, rarely covers the entire distance to refuge in one dash, instead uses short spurts of 14–18 m, during pauses "like a dog on the alert," stretching its head upward and bobbing, when cornered "opens its mouth in threatening fashion and gives voice to a disconcerting hoarse noise," readily climbs stone walls and tall trees, huddles together on cool nights (Buide, 1951); flees when encountered at substantial distances, retires to thickets in which they are concealed, sometimes seen in groups of 4–5 (on Cayo Largo, Cuba), elsewhere usually only a single individual at a time (L. H. Saxe in Sutcliffe, 1952); younger individuals will bipedal when fleeing, if captured, whip tails (Z. Vogel, 1965a); males responded to approaches by headbobbing and broadside challenge displays, interactions involved largest male in a population on Little Cayman chasing the largest female, which, in turn, chased other females, 9 hatchlings were solitary (Townson, 1980); losers in male-male agonistic encounters flee considerable distances (~7 m), stopping at intervals to look back at the threat, eventually taking refuge in a burrow in sand or under a rock, the encounter itself involves the peculiar pitching of the head, gaping, and circling the opponent, collectively comprising ritualized behaviors that circumvent actual fighting in many instances (Berovides Álvarez, 1980); produces an "S-roll" (the nose is raised and lowered as in a headbob, but is also moved in a horizontal plane during the lowering motion), although some individuals appear to produce standard headbobs in addition to the S-rolls, the twisting motion is not as dramatic as the full rolling motion produced by *C. cornuta* and *C. stejnegeri*, displays are series of ~5 S-rolls separated by pauses of ~0.7 sec, headbobs produced by animals in the Puerto Rican population lasted more than twice as long as those from Cuba or by *C. lewisi* (Martins and Lamont, 1998); agonistic male-male interactions consist of "lumbering" chases, occasionally with prolonged pushing, hissing, and biting matches, dominant males are larger, have higher testosterone levels and more scent glands (Alberts, 2003); dense aggregations in high anthropogenic usage areas exhibited more aggressive behavior and social interactions than individuals in low usage areas (*C. n. nubila* at the U.S. Naval Base at Guantanamo Bay; Lacy and Martins, 2003); aggressive displays include face-offs, lowered head, gaping, crest elevation, and headbobs or head shaking, aggression may involve biting, tail lashing, chasing, and fighting (Carpenter, 1982); will enter the sea to escape threats (Hardy, 1956); juveniles may enter water and hide among seaweeds (*Thalassia testudinum*; Rodríguez Schettino, 1999b); on Cayman Brac, iguanas cross large open areas when disturbed instead of running into a mangrove swamp (Carey, 1966); on Cayos San Felipe, seeks refuge in burrows when chased by dogs trained to catch jutias (*Capromys*), but fight tenaciously (attacking dogs and men) when cornered (Varona and Garrido, 1970). **Biomass:** Adult biomass ~11.7 kg/ha (Cayo Sijú; Beovides-Casas and Mancina, 2006). **Burrows:** Burrows under bushes, especially *Tournefortia gnaphaloides* and *Suriana maritima* (*C. n. caymanensis*; C. B. Lewis, 1940); ~1.2 m long and ~30 cm beneath surface (Grant, 1940c); may occupy crab burrows (Neill, 1958); an adult male resided in a small cave on Cayman Brac (Carey, 1966); on Little Cayman, lizard wedged in one corner of a small (10–30 cm high, <1 m long) chamber (Seidel and Franz, 1994); on Cayo Sesenta Arroba, many (7 of 19) burrows face southeast, on Cayo del Rosario, most (10 of 16) face south, mean width and height of burrows 15.94 ± 2.62 cm and 10.50 ± 1.34 cm, mean length and depth 147.64 ± 59.42 cm and 40.56 ± 15.89 cm, mean distance between burrows 15.04 ± 8.56 m, in September, burrow occupancy 60.0%, with two iguanas in only one burrow, in October, occupancy 79.6%, with 8 of 19 burrows housing two or more iguanas, with 8 iguanas in one burrow, burrow density on Cayo Sesenta Arroba 7.7/ha (Cubillas Hernández and Berovides Álvarez, 1991); on Cayo del Rosario, 13.7 burrows/ha (Berovides Álvarez, 1998); also on Cayo del Rosario, mean widths and heights of burrow entrances 21.58 ± 7.48 cm and 9.04 ± 2.76 cm, mean distance between burrows 18.37 ± 15.82 m, mean density of burrows 14.90 ± 9.70 burrows/ha (González Rossell et al., 2001); mean burrow entrance heights and widths 7.62 ± 1.61 cm and 13.33 ± 4.02 cm on Cayo Obispo and 6.05 ± 2.99 cm and 14.43 ± 5.59

cm and 5.89 ± 2.28 and 14.85 ± 5.14 cm, respectively, at two sites on Cayo Verde (Berovides Álvarez et al., 2003). **Competition:** On Cayo del Rosario, insects, anoles, and insectivorous birds are potential competitors with juvenile iguanas, and phytophagous insects may compete with adults (Berovides Álvarez, 1980); on Cayo Coco, competes for vegetation with introduced cattle (Estrada, 1993b). **Diet:** Feeds in cultivated areas on sweet potato vines and fallen papayas, Railroad Vine (*Ipomoea pescaprae*), *Cordia caymanensis*, *Ernodea littoralis*, "plums," and mangoes, eating flowers, fruits, and occasionally leaves (on Cayman Brac; C. B. Lewis, 1940); "omnivorous" (Buide, 1951); in supratidal situation, feeds on *Opuntia* and crabs (Proctor, 1950; L. H. Saxe in Sutcliffe, 1952); "phytophagous generalists," stomach contents of 4 individuals on Cayo del Rosario included *Ambrosia hispida* (2 stomachs), *Lantana involucrata* (4), *Strumphia maritima* (3), and *Erithalis fruticosa* (1; Berovides Álvarez, 1980); phytophagous, feeding on 25 different plants on Cayo Largo del Sur, taking fruits, flowers, and/or leaves primarily of *Avicenia*, *Canavalia*, *Capparis*, *Chrysobalanus*, *Conocarpus*, *Eragrostis*, *Laguncularia*, *Opuntia*, *Rachilcalis*, *Rizophora*, *Sporobolus*, *Strumpfia*, *Suriana*, and *Thrinax*, with seasonal changes reflecting abundance of preferred items, and opportunistically consuming animal matter, including crabs (*Cardisona guandhumi*; *C. n. nubila*; Perera, 1985b); "generalist phytophage with a certain degree of selectivity, and . . . an opportunistic omnivore," during the wet season, a notable preference for fruits of Coastal Thistle (*Strumphia maritima*) and Prickly Pear (*Opuntia dilenii*), when not available, preference for plants in mangrove forests and grasses, but eats at least 19 different plants plus occasional mollusks and beetles (Rodríguez Schettino, 1999b); on Cayo de Rosario, consumed leaves (24% of total stomach contents) and fruits (18%) more frequently than flowers (16%) and stems (6%), most frequently consumed plants were *Enthalis fruticosa*, *Lantana involucrata*, *Strumpfia maritima*, *Emodea littoralis*, *Ambrosia hispida*, *Suriana maritima*, *Chamaesyce camagüeyensis*, *Metopium toxiferum*, *Thrinax radiata*, and *Chrysobalanus icaco* (González Rossell et al., 2001); consumed decaying Zenaida Dove (*Zenaida macroura*), scavenged dead nestling White-winged Dove (*Z. asiatica*), and may take small birds (e.g., Cuban Grassquits, *Tiaris canora*) trapped in mistnets (Gerber et al., 2002); >40 species of plants are known to be consumed on Little Cayman (leaves of *Bauhinia divaricata*, *Capparis flexuosa*, *C. cynophallophora*, *Stylothanthes hamata*, *Ipomea pes-caprae*, *I. violacea*, and *Canavalia rosea* are taken year-round, flowers of *Tabebuia heterophylla*, *Ipomea* sp., and *Canavalia rosea*, and fruits of *Picrodendron baccatum*, *Hippomane mancinella*, *Cordia sebestena*, *Myrcianthes fragrans*, *Guapira discolor*, *Coccoloba uvifera*, and *Citharexylum fruticosum* are taken in great quantities when available, adults sometimes congregate in areas with fruiting *C. uvifera* and *H. mancinella* trees), and iguanas occasionally scavenge on carcasses (e.g., land crabs) or prey on slow-moving insects, such as lepidopteran larvae (Gerber, 2000a); favor red grapes, with which individuals can be lured into nets (Alberts, 2003); eight species of plants on Cayo Sijú, with fruits of *Chrysobalanum icaco* and leaves of *Batis maritima* most frequently represented, and remains of a crab (*Cardisoma guandhumi*) and hemipteran insects in scat samples (N = 30; Beovides-Casas and Mancina, 2006); cannibalism of juveniles on Isla Magueyes (Pérez-Buitrago et al., 2006); dermatophagy (Weldon et al., 1993); 4–5 circular and 2–3 distal semilunar valves partition the proximal colon (N = 2), presumably to increase digestive efficiency (Iverson, 1980). **Energetics:** Population on Isla Magueyes expended 4,800 kJ/ha/day (Christian et al., 1986). **Growth:** Juvenile growth ~100 mm SVL per year for the first 2 years (*C. n. caymanensis*; Gerber, 2000a). **Hybridization:** *Cyclura n. caymanensis* will hybridize with *C. lewisi* if "brought together" (Davis, 2000; Burton, 2004c). **Injuries:** Indicative of "aggressive habits," injuries to 3 adult animals were tip of tail, half of third toe missing, tip of tail, tips of third and fourth toes on one hindfoot, and tip of fifth toe on the other, more than a third of tail regenerated, right forefoot, tips of third, fourth, fifth toes on one hindfoot, and tips of first, second, fourth, and fifth toes and entire third toe of other hindfoot missing (L. H. Saxe in Sutcliffe, 1952). **Longevity:** In captivity for *C. n. nubila*, 24 years, 7 months and still alive at last report in 1994 (B. Christie in Slavens and Slavens, 2003); for *C. n. caymanensis*, 33 years and still alive at last report in 1990 (R. Noegel in Slavens and Slavens, 2003). **Mortality:** On Little Cayman (*C. n. caymanensis*), an estimated 50–60 adults are killed on roads annually (Goetz, 2009). **Movements:** Cuban fishermen report iguanas swimming from cay to cay, crossing narrows of >180 m by using the tail like a "sculling oar" (Buide, 1951). **Nest Destruction:** On Cayo Coco (Archipiélago de Sabana-Camagüey), introduced pigs destroy many nests (Estrada, 1993b). **Parasites:** Ticks *Amblyomma albopictum* (Neumann, 1899; Pérez Vigueras, 1934a, 1934b; Clifford and Kohls, 1962; V. Černý, 1966, 1969; Morel, 1967; Coy Otero, 1999; Guglielmone et al., 2003); *A. torrei* (Pérez Vigueras, 1934a; Whittick, 1939; Grant, 1940c; V. Černý, 1966, 1969; Morel, 1967; Coy Otero, 1999; Guglielmone et al., 2003); *Amblyomma rotundatum* (Pérez Vigueras, 1934a; Morel, 1967; Guglielmone et al., 2003); *Amblyomma* sp. on 97% of free-ranging juveniles at Guantanamo Bay, 5.8 ± 1.5 ticks per individual (Alberts et al., 1998); mites *Ornithodoros cyclurae* (de la Cruz, 1984; Coy Otero, 1999; Guglielmone et al., 2003) and *Cyclurobia javieri* (Coy Otero, 1999); feces contained *Balantidium* (ciliated protozoan) and nematode eggs, "heavy nematode infestations were found in the large bowel of all of the iguanas" (L. H. Saxe in Sutcliffe, 1952); intestinal nematodes *Atractis opeatura* (Pérez Vigueras, 1935b; Baruš and Coy Otero, 1969a; Baruš et al., 1969; Coy Otero, 1970; Coy Otero and Baruš, 1979b; Coy Otero and Lorenzo Hernández, 1982; Coy Otero et al., 1987; Coy Otero, 1999); *Oswaldofilaria brevicaudata*, *Paralaeuris cyclurae*, *Alaeuris travassosi* (Baruš and Coy Otero, 1969a; Baruš et al., 1969; Coy Otero, 1970; Coy Otero and Baruš, 1979b; Coy Otero and Lorenzo Hernández, 1982; Coy Otero et al., 1987; Coy Otero, 1999); *Oswaldocruzia lenteixeirai* (Coy Otero and Baruš, 1979b; Coy Otero and Lorenzo Hernández, 1982; Coy Otero, 1999); and *Ozolaimus monhystera* (Pérez Vigueras, 1936; Baruš and Coy Otero, 1969a; Baruš et al., 1969; Coy Otero, 1970; Coy Otero and Baruš, 1979b; Coy Otero and Lorenzo Hernández, 1982; Coy Otero et al., 1987; Coy Otero, 1999); oxyuroid nematodes in free-ranging juveniles at Guantanamo Bay (Alberts et al., 1998); 50% of large intestinal volume filled with nematodes (Rodríguez Schettino, 1999b); nematodes presumably commensal or even mutualistic (Iverson, 1982); piroplasm *Sauroplasma* in erythrocytes of 71% of 1993 captive hatchlings, but none in 1994, and 79% of free-ranging juveniles and >60% of cells of 26% of the same free-ranging lizards were infected with the haemogregarine *Hepatozoon* and hemococcidian *Schellackia* (Alberts et al., 1998). **Population Density:** Densities declining in many areas due to human activities and, in some areas, hunting, mean density 9.6 iguanas/ha on Cayo del Rosario (Rodríguez Schettino, 1999b); on Cayo del Rosario, 3 individuals/ha (Berovides Álvarez, 1980); 9.9 ± 0.65 individuals/ha (6.5–12.8 iguanas/ha in different months; González Rossell et al., 2001); densities on 11 Cuban cayos 9.3–30.0 individuals/ha, with highest densities (25.0, 30.0, and 30.0/ha) on rocky islets and the lowest density (9.3 iguanas/ha) on sandy Cayo del Rosario, where estimates based on numbers of burrows or distances between burrows were even lower (Berovides Álvarez, 1998); densities ranged 22.5–25.2/ha on Cayo Obispo and 10.5–18.2/ha on Cayo Verde in 1998–2000, and was 14.2/ha on Cayo Sotavento in 1998 (Berovides Álvarez et al., 2003); densities higher in wet than dry seasons and differed by habitat, with density in xerophytic coastal scrub (6.7 ± 6.3/ha) higher than in sand vegetation (3.6 ± 2.7/ha) and mangrove forests (2.9 ± 2.9/ha), with the most juveniles in the coastal scrub (Cayo Sijú; Beovides-Casas and Mancina, 2006); estimated 40 animals on Cayman Brac (*C. n. caymanensis*) in 1993 (Gerber, 1996b); ~1,800 iguanas on Little Cayman (Goetz, 2008); ~500 iguanas on 7.2-ha Isla Magueyes, Puerto Rico (García, 2005b, 2006b). **Population Size:** Estimated 40,000 *C. n. nubila* and 1,500 *C. n. caymanensis* (Alberts, 2004); estimated 40,000–60,000 *C. n. nubila* and 1,000 *C. n. caymanensis* (Hudson and Alberts, 2004); estimated 422 iguanas (50–75/ha) on Isla Magueyes, Puerto Rico (Álvarez et al., 2007); estimated 40 *C. n. caymanensis* on Cayman Brac (F. J. Burton in Goetz, 2008). **Predators:** On Little Cayman and Cayman Brac, dogs (C. B. Lewis, 1940); a very young iguana under artificial conditions panicked in the presence of *Cubophis fuscicauda*, suggesting that these snakes regularly prey on hatchling iguanas (Carey, 1966); potential predators of eggs on Cayo del Rosario include landcrabs, hermit crabs (*Coenobita clypeatus*), Greater Antillean Grackles (*Quiscalus niger*), gulls (*Larus atricilla*), and Royal Terns (*Sterna maxima*), with the three birds also potentially preying on juvenile iguanas (Berovides Álvarez, 1980); feral and domestic dogs (on adults and juveniles), cats (on juveniles), and pigs (on eggs; Estrada, 1993b); stuffed and sold as souvenirs to tourists in Cuba (M. Černý, 1992); humans are not a major factor because consumption of iguana meat or eggs is not common (*C. n. nubila*; Perera, 2000); in some areas, are hunted for food (Fong Grillo et al., 2005a); both iguanas and eggs were considered a delicacy on Cayman Brac (Carey, 1966); feral cats, domestic dogs, and possibly rats (*C. n. caymanensis*; Gerber, 2000a); feral cats and humans (in the form of road mortality) appear to be the major threats on Little Cayman (Goetz, 2008). **Reproduction:** Courtship probably begins in April and egg deposition from April to June, often in areas where iguanas do not normally occur, eggs 69.0–77.8 mm in length, clutch of 9 eggs 67.8–75.5 mm length (Rodríguez Schettino, 1999b); reach maturity at 2–3 years of age (Perera, 2000); on Little Cayman Island, courtship and mating occur in April and early May, females dig nest holes and deposit eggs between 1730 and 1800 h from May to early June in beach sand to within ~9 m of high tide mark, 2 females contained 8 and 20 eggs (C. B. Lewis, 1940); maturity at 983 g body mass on Isla Magueyes, courtship began in late March or early April, females apparently move to males during the mating season, once a female responds to a male, he restricts her movements to the immediate vicinity of the burrow for up to 7 weeks (post-copulatory guarding?), if she attempts to leave, he circles her while vigorously displaying, several burrows initiated before one completed, usually in sandy or grassy areas, often in sides of embankments, nests reused in consecutive years, burrows >1 m long and <50 cm deep, females extremely secretive and do not defend nest vigorously, eggs 2–14 (mean 5.6) with mean mass 46.4 g, incubation 70–78 days, mean SVL of hatchlings 107 mm, major source of egg mortality is ant predation (Christian, 1987); clutch of freshly laid eggs (in captivity) 82 × 42 mm, hatchling SVL 84 mm, 13.4 g (J. Schmidt, 1994); on Isla Magueyes, nest depths 26–70 cm, mean temperatures 30.5–33.5 °C, but nests dried at different rates after soaking rain, CO_2 concentrations related to depth and clutch size (Christian and Lawrence, 1991); hatchling size not related to soil water potential of nest, but hatchlings in dry nests had more yolk and less fat, indicating lower efficiency at converting yolk to fat (Christian et al., 1991); mean clutch size (8.6 ± 0.84) positively correlated with female SVL (300–465 mm; Alberts, 1995); clutch sizes 15–30 (Perera, 2000); for *C. n. caymanensis*, maturity at 2–3 years at 30.8 cm SVL, mating is polygynous, with large males with larger territories mating most frequently and small males moving from territory to territory attempting to court females and avoid detection by resident males, females frequently leave home ranges to mate or nest, lay 7–25 (mean 15) eggs, nest chamber 10–50 cm below the surface in interior patches of soil or coastal sandy areas, females guard nests from one day to several weeks, with extent depending on number of nest sites and other females in the vicinity, hatchlings emerge in early August to early September after incubation of 63–80 days (mean 72 days), median percentile of fertile eggs per clutch is 97.5 (35–100) and median percentile of hatchlings emerging from each nest is 91.5 (0–100), hatchlings average 107 mm SVL (97–121 mm) and 50 g (33–69 g), juveniles disperse immediately after emergence and often travel long distances before settling (to 5.2 km from the nest; Gerber, 2000a). **Sex Ratio:** Appears skewed toward females (*C. n. caymanensis*; Gerber, 2000a). **Size:** Male SVL to 745 mm, female SVL to 623 mm (Schwartz and Carey, 1977); on Cayo Sijú, mean adult male total length 1,000.8 ± 110.3 mm (710–1,119 mm) and body mass 4.1 ± 1.1 kg (2.0–5.5 kg; $N = 26$), mean adult female total length 824.3 ± 62.9 mm (690–910 mm) and body mass 1.7 ± 0.4 kg (1.1–2.4 kg; $N = 18$; Beovides-Casas and Mancina, 2006). **Survival:** Four of 45 headstarted iguanas sighted at least once within 24 months (Alberts et al., 2004). **Territoriality:** Males intensely territorial, occupy territories throughout the year, with largest males holding territories in the best habitats (*C. n. caymanensis*; Gerber, 2000a); dominant males actively defend territories that overlap with those of several females, lower-ranking males do not defend territories, but move between those of dominant males from which they are continually evicted, when dominant males relocated, previously subordinate males would defend identical territories, actively court females, and their testosterone levels rose dramatically (*C. n. nubila*; Alberts, 2003). **Thermal Biology:** Frequently (*C. n. caymanensis*) basks on inclined gnarled trunks and branches of mangroves (*Conocarpus*, *Rhizophora*; C. B. Lewis, 1944); body temperatures relatively stable (mean 38.6 ± 1.4 °C) from 0915 to 1800 h (on Isla Magueyes; Christian et al., 1986); heliothermic (Rodríguez Schettino, 1999b).

Conservation Status: The species is listed in CITES Appendix I (UNEP-WCMC, 2008); "vulnerable" in the IUCN Red List (Day, 1996b), as are Cuban populations (Alberts and Perera, 1996; L. Rodríguez Schettino and A. Chamizo Lara in Rodríguez Schettino, 1999a), although Cuban populations have been considered "endangered" (Rodríguez Schettino and Chamizo Lara, 1998a) and those on Little Cayman Island and Cayman Brac "critically endangered" (Gerber, 1996b); both subspecies are listed as "threatened" on the U.S. List of Endangered and Threatened Wildlife and Plants (U.S. Fish & Wildlife Service, 2008); primary threats are habitat alteration (coastal distributions dramatically affected by development; Estrada, 1993a) and degradation and predation by introduced mammals (dogs and cats).

Cyclura onchiopsis Cope, 1885

Distribution: Navassa Island.

Habitat: Presumably xerophilic, in tropical dry forest.

Natural History: Unknown. **Size:** Adult male SVL to 420 mm, female SVL to 378 mm (Schwartz and Carey, 1977).

Conservation Status: Almost certainly extinct (Powell, 1999a).

Additional Pertinent Literature: Powell (2000).

Cyclura pinguis Barbour, 1917

Distribution: Historically the entire Puerto Rico Bank, now only on Anegada Island (British Virgin Islands), reintroduced onto Guana, Necker, and Norman islands (Goodyear and Lazell, 1994; Lazell, 1995, 2000, 2002, 2005a; Binns, 2003; Bradley and Gerber, 2005).

Habitat: Xerophilic (Schwartz and Henderson, 1991); primarily in limestone areas, with retreats in natural cavities or beneath boulders (Carey, 1972, 1975; Mitchell, 2000a); dry tropical forest and scrub, seasonally in coastal strand, occasionally in salt marsh or salt pond habitats, on Anegada, closely tied to more porous limestone habitats, where it uses holes, caves, and other rocky shelters as retreats, but uses adjacent sandy areas for burrowing and nesting, fewer sandy areas on Guana and Necker, so iguanas more arboreal, females on Guana nest in *Coccoloba*-dominated beach strand (Mitchell, 2000b); can and will climb trees (Goodyear and Lazell, 1994; Lazell, 1997); most commonly encountered in sandy scrub or rocky woodland (Bradley and Gerber, 2005); will bask on rock walls and especially juveniles not adverse to foraging around buildings on Guana Island (RP).

Activity: Usually emerges from dens around 0700 h (mean ambient temperatures ~25 °C), moves to sunny areas and assumes a prostrate basking posture, after about 1 h (mean ambient temperatures ~27 °C), begins to search for food, stopping frequently to sample small bites of nearby plants, basking continues after feeding, but rarely in full sun (usually a sun-shade mosaic), during midafternoons (mean ambient temperatures ~30 °C), usually remains in full or partial shade, after ~1500 h, when temperatures drop below 30 °C, moves into more exposed sites, no iguanas seen after 1800 h (Carey, 1975). **Behavior:** Aggressive displays involve stereotypical vertical headbobs (Carpenter, 1982); juveniles engage in aggressive encounters with other juveniles, these involve headbobs and pushups and, on one occasion, biting, chases often followed by up to 10 min of displays (Perry et al., 2003b); adults on Anegada extremely shy, even basking adjacent to dense sheltering vegetation, in contrast, hatchlings are quite naïve (K. Bradley in Werning, 2009). **Diet:** Eats *Coccoloba* leaves, *Opuntia* fruit, and whole wild nutmegs (Grant, 1937b); primarily herbivorous, but takes some animal matter, juveniles probably insectivorous (Carey, 1972, 1975); bulk of diet consists of leaves and fruits, composed largely of forage not taken by introduced ungulates (30% *Croton discolor*, which contains compounds toxic to ungulates, and >55% fruits of *Byrsonima*, *Coccoloba*, *Eleaodendron*, and *Eugenia*, which livestock ignores), but animal matter (beetles, caterpillars, centipedes, roaches) forms about 1% of the natural diet (Mitchell, 1999, 2000a, 2000b); native plants taken on Guana include large quantities of *Centrosema virginiana*, flowers of *Tabebuia heterophylla*, seed pods of *Capparis cynophallophora*, and leaves of *Stigmophyllon emarginatum* and *Capparis flexuosa* (Mitchell, 2000b). **Growth:** Adulthood at 7–9 years and 350–400 mm SVL (Carey, 1975); estimated 50 mm increase in SVL/yr based on 5 recaptures and 47 captive-raised individuals, maturity at 4–7 years, 37.8 cm SVL (Gerber, 2000b). **Home Range:** Males 116.06–985.11 m^2 (*N* = 5), females 155.47–412.31 m^2, aside from food, available shelter is the most important factor for selecting a home range (Carey, 1972, 1975); recent population structure differs, with home ranges expanding 100-fold with a decline in population size, and male home ranges averaging 6.6 ha (2.2–12.3 ha) and those of females 4.2 ha (2.8–5.6 ha), home ranges broadly overlap and male centers of activity associated with home ranges of females and those of females with burrows (Mitchell, 1999, 2000a, 2000b). **Parasites:** Ticks (*Amblyomma antillorum*), with 20–30 (maximum 59) ticks per adult lizard, most frequently on the thorax and in the region of the femoral pores (Kohls, 1969; Carey, 1975; Keirans, 1985; Guglielmone et al., 2003); piroplasm *Sauroplasma* in erythrocytes of captive-reared juveniles (Alberts et al., 1998). **Population Density:** 2.03 iguanas/ha (Carey, 1975); 0.36/ha in 1991, with the lower density due to degraded forage attributable to competition with introduced mammalian herbivores (Mitchell, 2000a). **Population Size:** 200 wild adults and 44 repatriated iguanas on Anegada, with another 80 in the headstart facility, 130 on Guana, and 30 on Necker (Perry and Mitchell, 2003; Gerber, 2004a; Bradley and Gerber, 2005; Lazell, 2006; Perry et al., 2007a); estimated 400 (Alberts, 2004); estimated 200 (Hudson and Alberts, 2004). **Population Structure:** Adults (87.5% of population) outnumber juveniles (12.5%), adults appear to be monogamous, living in separate but closely adjacent burrows (Carey, 1972, 1975). **Predators:** Dogs (Grant, 1937b); feral dogs and cats (Mitchell, 2000a, 200b); humans (Wing, 2001; Lazell, 2005a); *Borikenophis portoricensis* and American Kestrels (*Falco sparvarius*) take juveniles on Guana Island (Levering and Perry, 2003); also on Guana, a 621-mm SVL, 88-g *Borikenophis portoricensis* consumed a hatchling iguana of 40.4 g, a mass representing 45.9% of the snake's pre-meal mass (RWH, RP). **Reproduction:** Cycle appears to be synchronized with two yearly rainfall maxima, adult females with 12–16 yolked ovarian follicles, loss of fecundity presumably due to competition with domestic livestock (Carey, 1972, 1975); mating season on Anegada May–June, females dig nests in sandy areas from late June through July in which they lay up to 20 eggs (average 13), hatchlings emerge in the fall (Gerber, 2000b, 2004a; Bradley and Gerber, 2005); on Guana, egg laying in June–July and hatchlings emerge in September–October (Perry et al., 2007a). **Sex Ratio:** Equal with suggested monogamy (Carey, 1975); 2:1 (M:F), with "pairs" no longer clearly definable (Mitchell, 2000a, 2000b). **Size:** Male SVL to 560 mm (Mitchell, 2000a), female SVL to 476 mm (Schwartz and Henderson, 1991); mean hatchling SVL on Guana 108 mm, iguanas of similar length on Guana weigh more than those on Anegada, suggesting that the latter are food-limited (Perry et al., 2007a). **Survival:** 86% survival of headstarted individuals released in 2003–2004 (Gerber et al., 2007). **Tail Autotomy:** 37.5% of males and 11.8% of females on Anegada had broken tails (Gerber, 2000b). **Thermal Biology:** Heliothermic, exhibits behavioral thermoregulation, cloacal temperatures of 16 active animals ranged 1.2–9.8 °C (mean 6.3 °C) above air temperatures at the time of capture (Carey, 1972).

Conservation Status: The species is listed in CITES Appendix I (UNEP-WCMC, 2008); "critically endangered" in the IUCN Red List (Mitchell, 1996); "endangered" on the U.S. List of Endangered and Threatened Wildlife and Plants (U.S. Fish & Wildlife Service, 2008); principal causes include habitat degradation by introduced mammalian herbivores (i.e., cattle, goats, burros) and predation by introduced mammalian predators (cats and dogs) on Anegada (Mitchell, 2000b; Bradley and Gerber, 2005); a species recovery plan exists (Burton and Bloxom, 2007b); reestablished populations elsewhere in the British Virgin Islands appear to be doing well or are too recent to be evaluated, with that on Guana thriving (Perry et al., 2007a); "threatened" (Moreno, 1991d).

Cyclura ricordii Duméril and Bibron, 1837

Distribution: Hispaniola, Valle de Neiba and the lower Península de Barahona (Dominican Republic).

Habitat: Xerophilic, in very arid situations, closely associated with thorn scrub woodlands (e.g., cactus scrub, *Acacia* scrub, and very bleak rock and sand situations) with seemingly optimal habitats characterized by alluvial powdery clay soils bordered by marine sediments and low forested hills with high plant diversity, burrows in flat, salt-encrusted playas on Isla Cabritos, prefers self-excavated burrows with entrances under dense thorny vegetation (especially *Cylindropuntia caribaea* on Isla Cabritos), shrubs, stumps, or exposed rocks, but will use hollow tree trunks and rock cavities where soil is not available, dens are absent from areas covered by rock or gravel along the southern shore of Lago Enriquillo, with concentrations in banks of dry creek beds and ravines cutting through marine sediments (Noble, 1923c; Schwartz and Henderson, 1991; Ottenwalder, 2000b; Ramer, 2004; Rupp et al., 2007).

Behavior: Aggressive displays involve stereotypical vertical headbobs (Carpenter, 1982); displays consist of long series of headbobs with progressively longer pauses between them, usually beginning with a series of ~7 headbobs with virtually no pauses, followed by an additional set of ~3 single headbobs with inter-bob pauses of ~0.2 sec, frequently followed by one long pause and then another series of ~2 headbobs (Martins and Lamont, 1998). **Diet:** Seeds of *Ziziphus rignoni* germinated more rapidly after being consumed and passed by iguanas (Hartley et al., 2000); 4 circular valves (only the first well developed) and 2 small semilunar valves partition the proximal colon (N = 1), presumably to increase digestive efficiency (Iverson, 1980). **Intestinal Microflora:** 17 bacterial isolates included *Salmonella* sp., *Escherichia coli*, *Brevibacterium* sp., *Enterobacter cloacae*, and *Entercoccus durans* (Maria et al., 2007). **Population Density:** Estimated 8/ha on Isla Cabritos in 1985, but only 1.33/ha in 2000 (Incháustegui, 2003a; G. Santana in Incháustegui, 2003b); based on numbers of burrows in a very small area near Cabo Rojo, a density of 31.5 iguanas/ha was calculated, but only two small areas in the entire region were inhabited by the species (Arias et al., 2004). **Population Size:** Estimated 1,300 (Alberts, 2004); estimated 2,000–4,000 (Hudson and Alberts, 2004). **Predators:** Humans (subsistence) and introduced carnivores (dogs, cats, mongooses; Ottenwalder, 2000b; Rupp et al., 2005); feral dogs (Rupp et al., 2008). **Reproduction:** Nesting sites in fine, sandy soils, eggs laid May–June (synchronized with the onset of the early rainy period), females lay 2–18 eggs (mean 11.1 ± 4.2) in egg chambers that are ~40 cm deep and have a stable temperature of 30–31 °C, incubation lasts 95–100 days, hatching is synchronized with the onset of the second rainy season in September and October, hatchlings average 87.4 mm SVL and 30 g, females reach maturity at 2–4 years of age (Ottenwalder, 2000b); nesting early March through mid-June, hatching mid-June through mid-September, 86.9–92.4% of nests hatched (9.2–11.7 nests/ha) at 4 sites in the southwestern Dominican Republic (almost exclusively in "fondos," areas with deep, well-drained soils), one nest with a record of 20 hatched eggs (Rupp et al., 2008); hatchling emerged on 2 September (Rupp et al., 2005); clutch size (in captivity at ZooDom, the national zoo in Santo Domingo) 18–24, hatchling SVL 82.4–91.0 mm (mean 87.4 mm, N = 11), total length 211.2–237.6 mm (mean 224.2 mm, N = 10), and body mass 25.5–35.0 g (mean 30.0 g, N = 11; Sanlley Castro and Duval, 1979). **Size:** Adult male SVL to 485 mm (4,200 g), female SVL to 450 mm (3,410 g; Maria et al., 2007).

Conservation Status: The species is listed in CITES Appendix I (UNEP-WCMC, 2008); "critically endangered" in the IUCN Red List (Ottenwalder, 1996b), primarily as a result of habitat alteration and degradation by human activities (agriculture and charcoal production), predation by introduced mammalian carnivores (dogs, cats, mongooses), competition with introduced mammalian herbivores (e.g., cattle, goats), and exploitation by humans, which is ongoing in at least some areas; a major part of the area inhabited by iguanas along the southern shore of Lago Enriquillo is not protected (Rupp et al., 2007); a captive breeding program exists at the Parque Zoológico Nacional in the Dominican Republic (Maria et al., 2007); this species was included in lists of threatened Hispaniolan species (Aquino, 1983; Powell et al., 2000).

Additional Pertinent Literature: Glor et al. (1998).

Cyclura rileyi Stejneger, 1903

Distribution: Bahama Islands, where three currently recognized subspecies occur on San Salvador Island and satellites (*C. r. rileyi*), southern Exuma Cays (*C. r. cristata*), and cays in the Bight of Acklins (*C. r. nuchalis*); Malone et al. (2000) questioned the appropriateness of subspecific recognition, based on a lack of haplotypic differences, but suggested that a final resolution should be determined by analysis of more rapidly diverging nuclear markers, such as microsatellites; distribution once more widespread on the main islands and presumably on cays currently without populations (Hayes, 2000a, 2000b; Hayes and Montanucci, 2000).

Habitat: Xerophilic to mesophilic (Schwartz and Henderson, 1991); exploits varied main island communities, plus coastal rock, sand strand and sea oats, and coastal coppice plant communities on offshore cays, and blacklands coppice and mangrove communities on cays within inland lakes, but most abundant in the vicinity of limestone outcrops and/or patches of Sea Grape (*Coccoloba uvifera*) and, on some cays, patches of Buttonwood (*Conocarpus erectus*), presumably require sandy areas for nest construction (*C. r. rileyi*; Hayes, 2000a); coastal rock habitats with *Strumpfia maritima* and Sea Grape, least abundant in dense Thatch Palm (*Thrinax morrisii*) forest (*C. r. cristata*; Hayes, 200b).

Behavior: *Cyclura r. rileyi* displays comprise series of up to 7 single and double headbobs, mixed, with no obvious pattern to the choice of single or double headbobs (Martins and Lamont, 1998); somewhat arboreal, climbing into mangroves in cormorant (*Phalacrocorax*) rookeries on San Salvador Island (*C. r. rileyi*; Paulson, 1966); males on cays in Acklins Bight (*C. r. nuchalis*) engage in jousting matches involving open-mouthed territorial displays, chase other males out of defended areas, and bear scars of bite marks (Hayes and Montanucci, 2000); aggressive displays involve headbobs or pushups (Carpenter, 1982); larger males guard females and smaller males may engage in forced copulation (Hayes et al., 2004); will take to water as an escape tactic when harassed (Hayes et al., 2004). **Commensalism:** Hermit crabs in all and *Sphaerodactylus corticola* in some iguana burrows (Gicca, 1980). **Diet:** Observed feeding only on fruits of Sea Grapes (*Coccoloba uvifera*; Gicca, 1980); on Green Cay (*C. r. rileyi*), consumes 7 species of plants, with Silver Sea Oxeye (*Borrichia arborescens*, flowers especially), Wild Thyme (*Rhachicallis americana*, both flowers and leaves), and *Opuntia stricta* (a preferred item despite scarcity at the time of the survey) most commonly represented in scat samples, nonplant items included a Purple Gallinule (*Porphyrula martinica*) and two unidentified songbirds (probably scavenged) and one Bridled Tern (*Sterna anaethetus*) chick (likely predated), conspecific hatchlings ($N = 2$), a dead land crab, a grasshopper (iguanas occasionally seen lunging for grasshoppers), a hermit crab, unidentified insect material (in 22% of scats), and sand and soil (Hayes et al., 2004); dermatophagy (Weldon et al., 1993); a single circular and 4 distal semilunar valves partition the proximal colon ($N = 2$), presumably to increase digestive efficiency (Iverson, 1980). **Home Range and Territoriality:** Adult males territorial throughout the year (Hayes, 2000a; Hayes and Montanucci, 2000); restricted activity to "what appeared to be rather small territories within the vegetation" (Gicca, 1980); mean 95% fixed kernel estimates of home range sizes for males and females, respectively, on Green Cay (*C. r. rileyi*) were 439 and 628 m^2 (S. Cyril, 2001 in Hayes et al., 2004); on North Cay (*C. r. nuchalis*), sizes were 2,047 m^2 for gravid females and 397 m^2 for non-gravid females (B. Thornton, 2000 in Hayes et al., 2004); on White Cay (*C. r. cristata*), 86% adaptive kernel estimates of home ranges of 5 males and 2 females were 2,656 m^2 (S. Fry, 2001 in Hayes et al., 2004); home range sizes reflect island sizes and possibly population densities (Hayes et al., 2004). **Injuries:** 16.0% of 75 *C. r. cristata*, 12.3% of 57, 14.9% of 87, and 20.4% of 53 *C. r. nuchalis* from Fish Cay, North Cay, and a translocated population, and 33.3% of 3, 34.5% of 35, 33.7% of 86, 13.3% of 15, 75.0% of 16, 21.1% of 19, and 51.6% of 31 *C. r. rileyi* from Gaulin, Goulding, Green, Guana, Low, Manhead, and Pigeon cays, respectively, had missing digits, 42.7% of 75 *C. r. cristata*, 24,6% of 57, 32,6% of 86, and 51.9% of 54 *C. r. nuchalis* from Fish Cay, North Cay, and a translocated population, and 0% of 3, 22.9% of 35, 22.6% of 84, 40.0% of 15, 50.0% of 16, 36.4% of 22, and 64.5% of 31 *C. r. rileyi* from Gaulin, Goulding, Green, Guana, Low, Manhead, and Pigeon cays, respectively, had injured tails, with toe loss increasing with age and tail injuries more common on rat-infested islands (Carter and Hayes, 2004). **Longevity:** In captivity, 9 years, 2 months (D. Blody in Slavens and Slavens, 2003). **Mortality:** Death due to being wedged between tree trunks (Iverson et al., 2005). **Parasites:** Ticks only on Guana Cay (*C. r. rileyi*; Hayes, 2000a). **Population Size and Density:** Estimated 500 *C. r. rileyi*, 200 *C. r. cristata*, and 13,000 *C. r. nuchalis* (Alberts, 2004); estimated 500–1,000 *C. r. rileyi*, 150–200 *C. r. cristata*, and 15,000 *C. r. nuchalis* (Hudson and Alberts, 2004); for *C. r. rileyi*, maximum population 35 on Green Cay (6 iguanas/ha), maximum population 28 on Man Head Cay (7.5/ha), maximum population 30 on Low Cay (2.5/ha), all probably at or near carrying capacity (Gicca, 1980); 300–400 *C. r. nuchalis* on Bush Hill Cay (Iverson, 2007c). **Population Structure:** Percent juveniles was 17% (White Cay; *C. r. cristata*), 1–10% (mean 6%, three cays; *C. r. nuchalis*), and 0–20% (mean 8%, six cays; *C. r. rileyi*; Hayes et al., 2004). **Predators:** "Other than man, the only likely natural predator[s] were sea gulls (capable of consuming young) and ospreys (capable of consuming both young and adult iguanas)" (Gicca, 1980); Ospreys (*Pandion haliaetus*) regularly prey on adult iguanas in Acklins Bight, a Barracuda (*Sphyraena barracuda*) attacked and consumed an adult iguana swimming across a mangrove channel (*C. r. nuchalis*; Hayes et al., 2004); humans (for illicit animal trade) and introduced mammals (raccoons, rats; Hayes, 2000a, 2000b). **Reproduction:** For *C. r. rileyi*, the smallest mature female was 21.5 cm SVL and 340 g, courtship and mating probably occur in May (copulation during the last week in May), nest excavation begins during the first week in July, with larger females nesting earlier and most nests beneath overhanging rocks, 18 of 22 females engaged in active nest defense and 13 remained near the nest and defended them for more than 5 days, with nest defense most common in areas where nests were closely placed, nest burrows 30–116 cm long (mean 53 cm, $N = 5$) and egg chambers 18–28 cm deep (mean 22 cm), burrows often change direction, one egg chamber apparently contained 2 clutches, clutch size 3–6 ($N = 5$) on Green Cay (one clutch on Low Cay with 10 eggs), reproductive investment averaged 21.7%, $N = 9$) and was associated with female size, mean egg size 53.4 × 29.5 mm and 27.7 g ($N = 21$) with egg mass negatively correlated with clutch size, 3 eggs at time of emergence 42–47 g, 3 recently hatched juveniles 21.5 g and 8.2–8.5 cm SVL, eggshell fragments suggest that some burrows are reused, ovideposition begins shortly after excavation and continues for up to 2 weeks, emergence of hatchlings in September or October after incubation times of 91–92 days (Hayes, 2000a; S. Cyril, 2001 in Hayes et al., 2004; Hayes et al., 2004); clutch with 5 eggs (*C. r. rileyi*; Stejneger, 1903); for *C. r. nuchalis*, mating system polygynous, possibly polygamous, smallest mature female 19.5 cm SVL and 260 g, copulation late May–early June, females may move up to 1 km to nest, with nest digging beginning during the last week in June, with larger females nesting later, nest density highest in beachfront sand (1,822/ha), intermediate in interior sandy habitat (578/ha), and lowest in rocky interior habitat (244/ha), ovideposition begins shortly after excavation and continues for up to 2 weeks, 5 of 14 females engaged in active nest defense, nest burrows 69–235 cm long (mean 129 cm, $N = 13$) and egg chambers 14–40 cm deep (mean 22 cm) with size correlated positively with female size, burrows often change direction and may contain multiple chambers, one nest contained clutches in 2 chambers, nest temperatures ranged 25–33 °C, clutch size 2–5 (mean 3.1, $N = 14$), positively correlated with female size, reproductive investment averaged 26.8%, $N = 11$) and also was associated with female size, mean egg size 55.4 × 30.2 mm ($N = 43$) and 27.1 g ($N = 41$), with egg mass negatively correlated with clutch size, eggshell fragments suggest some burrows are reused, emergence of hatchlings in September or October (B. Thornton, 2000 in Hayes et al., 2004; Hayes et al., 2004). **Size:** Adult male SVL to 306 mm, adult female SVL to 254 mm (Schwartz and Henderson, 1991); mean and maximum sizes vary considerably from cay to cay and are positively correlated with plant diversity, suggesting nutritional constraints on body size (*C. r. rileyi*; Schwartz and Carey, 1975; Hayes, 2000a).

Conservation Status: The species is listed in CITES Appendix I (UNEP-WCMC, 2008); "endangered" in the IUCN Red List (Carter and Hayes, 1996a), but the population on White (Sandy) Cay (*C. r. cristata*) is "critically endangered" (Carter and Hayes, 1996b), those on Fish and North Cay in Acklins Bight (*C. r. nuchalis*) are "endangered" (Carter et al., 1996a), and that on San Salvador (*C. r. rileyi*) is "critically endangered" (Carter et al., 1996b); *C. r. rileyi* is listed as "endangered" and *C. r. cristata* and *C. r. nuchalis* as "threatened" on the U.S. List of Endangered and Threatened Wildlife and Plants (U.S. Fish & Wildlife Service, 2008).

Cyclura stejnegeri Barbour and Noble, 1916

Distribution: Isla Mona, essentially islandwide.

Habitat: Xerophilic (Schwartz and Henderson, 1991); among rocks (Bowdish in Stejneger, 1904); along rocky coasts, inland regions of cliffs, caves, and rocky areas (K. P. Schmidt, 1928; Wiewandt, 1977, 1979; Wiewandt and García, 2000); generally terrestrial, but especially young animals capable of climbing rocks and inclined trees with "unexpected" agility (Rivero, 1998); abundance highest in "cliff habitat," lowest in the coastal plain, and intermediate on the plateau, low on the western side of the island where native vegetation has been displaced by plantations of *Casuarina equisetifolia* and *Swietenia mahagoni*, abundance not correlated with plant density (food plants present in all habitats) but positively correlated with the presence of sinkholes, which are used as retreats (Pérez-Buitrago and Sabat, 2000); hatchlings settled in territories spent 63% of time in trees or other elevated perches at heights 1.2–12.0 m above ground (Pérez-Buitrago and Sabat, 2007).

Activity: Dens (to ~4 m deep with terminal chamber ~0.5 m high) serve as nocturnal retreats and may be used during the day, basks on exposed "lookouts" primarily in the morning and late afternoon, feeding is not a daily requirement, about 95% of normal daily routine is rest, with or without basking (Wiewandt, 1977, 1979). **Behavior:** Retreats by running into a rock hole, but "ostrich-like," the tail is not always withdrawn (Grant, 1944b); usually shy, retreating into a den when disturbed, but can become "tame" when exposed regularly to human intrusions, may take to water as an escape tactic, males fight for mating territories, with fights serving as ritual tests of strength and involving face-to-face pushing matches with mouths gaped, usually lasting <15 min, and ending without violence when one or both contenders back slowly away, occasional battles last to 1 h and may involve biting and tail lashing (Wiewandt, 1977); aggressive displays include lateral orientation, gaping, dewlap extension, crest elevation, and headbobs or head rolls (Carpenter, 1982); aggressive behavior may be directed at humans (Tenlén, 1997a); produces "rolls" (one side of head is raised, when lowered, the opposite side is raised, often repeated in long series) as well as headbobs, rolls occur at the beginning of displays (usually) or at the end, almost always accompanied by 1–2 series of standard headbobs, consisting of up to 6 (mean 2.0 ± 0.42) up-and-down motions per series (Martins and Lamont, 1998). **Blindness:** Blind iguanas (Haneke, 1995); 9 of 28 animals captured in 1998–1999 were blind (cataracts), also observed subsequently (Hudson and Alberts, 2004; Pérez-Buitrago et al., 2007a). **Foraging and Diet:** Descends "occasionally to the tillable area to feed on the young corn and cotton" (K. P. Schmidt, 1928); consumes 71 of the 400 plants species known to occur on Mona, food includes fruits (especially *Opuntia* spp. and *Hippomane mancinella*), leaves, flowers, and some animal matter (especially seasonally abundant *Pseudosphinx tetrio* caterpillars and pupae), preferred plants are those low in cellulose, low in aromatic compounds, and easy to obtain, leaves (difficult to digest) may pass through the digestive tract almost intact, diets of juveniles like those of adults, except for the size of items eaten, known to eat mice in captivity (Wiewandt, 1977, 1979); may "graze" (moving from food item to food item), presumably selecting favored growth, or "gorge" (sit in one place and eat everything within reach; RP). **Growth:** Estimated 52.3 mm SVL during first year, but one field-captured juvenile grew 60 mm SVL in 0.7 years (85.7 mm/yr; Wiewandt, 1977); at 5 months, 3 individuals had gained no weight (average -1 ± 2.6 g) and SVL increased 0.4 ± 0.36 cm, fourth individual gained 119 g and increased SVL 4.11 cm (Pérez-Buitrago and Sabat, 2007); growth of 9 wild iguanas 4.1–10.5 mm/month (mean 6.3 ± 2.2 mm/month) and did not differ significantly from growth rates of headstarted iguanas, body condition (mass/SVL) declined after release to levels comparable with those of wild animals of comparable size and wild adults with SVL >400 mm (Pérez-Buitrago et al., 2008). **Home Range and Territoriality:** Males select territories with retreats attractive to females, females select retreats according to location and structure, favoring those with good drainage, ventilation, nearly horizontal resting places under low ceilings, and within ~1.7 m of surface, if females attempt to leave a male's territory, he pursues her and tries to "herd" her back (Wiewandt, 1977, 1979); mean home range size for 9 head-started iguanas 19.8 ha (2.4–22.2 ha; García, 2005a, 2006a); mean dispersal range of 8 radiotracked individuals (minimum convex polygon) 10.7 ha (2.6–22.2 ha; García et al., 2007); males exhibited uniform dispersion, whereas females were randomly distributed, mean male home range size 0.6 ± 0.2 ha (0.21–0.81 ha; minimum convex polygon) and 0.09 ± 0.05 ha (0.02–0.15 ha; 50% kernal algorithm), mean female home range size 0.28 ± 0.09 ha (0.13–0.34 ha, minimum convex polygon) and 0.07 ± 0.04 ha (0.03–0.13 ha; 50% kernal algorithm), male-female overlap extensive (mean 86.6%; 66.4–100.0%), male-male overlap minimal (mean 2.9%; 0–11%), suggesting that males are highly territorial and that home ranges correspond to territories, sizes of home ranges suggest that this may be partially responsible for low density if male carrying capacity is attained (Pérez-Buitrago et al., 2007b); mean juvenile home range size 297 m^2 (0–539 m^2; Pérez-Buitrago and Sabat, 2007). **Movements:** Headstarted iguanas moved 471–6,396 m (mean 2,844 ± 2,122 m), distances traveled between consecutive monitoring days 56–157 m (mean 76 ± 33 m; García et al., 2007); hatchlings dispersed 102–5,080 m from release site, moving on average ~250 m/day (Pérez-Buitrago and Sabat, 2007); another cohort of headstarted iguanas moved 441–1,249 m (mean 633 ± 147 m; Pérez-Buitrago et al., 2008). **Nesting:** Plantations of invasive Australian Pine (*Casuarina equisetifolia*) in nesting sites produce accumulations of leaf litter and organic matter that iguanas cannot penetrate (Wiewandt, 1977). **Parasites:** Ticks *Amblyomma cruciferum* (Bequaert, 1932; Morel, 1967) and *A. torrei* (Maldonado Capriles and Medina Gaud, 1977); ticks (*A. cruciferum*) and nematodes (Wiewandt, 1977); tick loads (*Amblyomma* sp.) positively correlated with SVL, ticks aquired quickly after release from captivity (one iguana acquired 7 ticks in 4 days), loads of recaptured headstarted iguanas (28.2 ± 25.4 ticks) comparable to those of wild iguanas (32.2 ± 22.0 ticks) of comparable size (Pérez-Buitrago et al., 2008). **Population Density:** 0.33 iguanas/ha (Wiewandt, 1977), but may have declined (Wiewandt and García, 2000); 0.21 lizards/ha (J. A. Moreno in Pérez-Buitrago and Sabat, 2000); 2.36 ± 3.3 iguanas/ha using line transects 16 m wide (Pérez-Buitrago and Sabat, 2000); in study area of relatively "natural" habitat 2.75 iguanas/ha (Pérez-Buitrago et al., 2007b). **Population Size:** 2,000–4,000 individuals (Wiewandt, 1977); 1,150 (J. A. Moreno in Pérez-Buitrago and Sabat, 2000); 5,093 ± 2,500 (Pérez-Buitrago and Sabat, 2000); estimated 1,500 (Alberts, 2004); estimated 1,500–2,000 (Hudson and Alberts, 2004). **Population Structure:** Juveniles only 5–10% of population (Wiewandt, 1977) or 2 of 118 observations (Wiewandt and García, 2000). **Predators:** Humans (historically), feral cats and pigs (Wiewandt, 1977; Moreno, 1995; Rivero, 1998; Wiewandt and García, 2000; García et al., 2001; Bendon, 2004). **Reproduction:** Mating occurs during an annual 2-week period, males lie in wait for females in male's territory 2–3 m from her retreat, with eyes fixed on burrow entrance, when she appears he displays with a series of toss-roll headbobs, dashes to her side with head lowered, pauses briefly for another series of head movements, grabs her wherever possible and restrains her while maneuvering for a better position, unwilling female may turn and bite or, more commonly, struggle into crevice to dislodge the male, this sequence may continue for days, with females retreating for progressively shorter periods after each episode, when she submits, she becomes limp, copulation lasts 1–2 min, females store sperm and eggs presumably may be "fathered" by several males, as females often copulate with more than one male, eggs are deposited about one month after mating (apparently cued by photoperiod), eggs in a tunnel ~1 m long, terminating in an egg chamber just large enough for the female to turn around, clutch sizes 5–19 soft-shelled eggs, females refill passages, but not the egg chamber, by kicking soil into the entrance hole and pushing it with forelimbs into the tunnel, then "ramming" it with her head until all vestiges of the hole are obliterated, hole preparation, egg laying, and covering requires about one day under ideal circumstances, nests are guarded by females against other females when they return to nest sites 1–10 days after deposition, defenses against intruders include head rolling, broadside challenges, huffing, biting, tail lashing, and kicking sand at the rival or into her hole, eggs 51 × 83 mm, 12 eggs deposited 19 June began to hatch 81 days later, 15 eggs deposited 20 June hatched 83–87 days later, hatchling body mass (N = 82) 37.6–67.5 g (mean 50.9 g), SVL 95.6–111.9 mm (mean 104.6 mm), total length 216.4–318.7 mm (mean 287.9 mm; N = 81), maturity at 6–7 years, 38.0 cm SVL (Wiewandt, 1977, 1979); maximum distance moved by a female to a nesting site 12.8 km, hatching success 55–86%, hatching season very short, only 3 weeks in October (Pérez-Buitrago et al., 2007b). **Size:** Male SVL to 533 mm, female SVL to 508 mm (Powell and Glor, 2000). **Survival:** Estimated 22% at end of first 20 weeks of life (Pérez-Buitrago and Sabat, 2007). **Tail Autotomy:** 50.0% of males and 14.3% of females had broken tails (Wiewandt, 1977).

Conservation Status: The species is listed in CITES Appendix I (UNEP-WCMC, 2008); "endangered" in the IUCN Red List (Garcia et al., 2000); "threatened" on the U.S. List of Endangered and Threatened Wildlife and Plants (U.S. Fish & Wildlife Service, 2008); principal causes include predation by feral cats, habitat degradation by feral goats, and possibly disease (e.g., Haneke, 1995; García et al., 2007); a species recovery plan exists (D. C. Díaz, 1984); "vulnerable" (Johnson, 1988); "endangered" (Moreno, 1991d); protected as "threatened" or "endangered" (Joglar, 2005b).

Remarks: Some authorities continue to recognize this taxon as a subspecies of *C. cornuta*.

Additional Pertinent Literature: Powell and Glor (2000).

Iguana delicatissima Laurenti, 1768

Distribution: Historically on Anguilla, St.-Martin/St. Maarten, Île Fourchue (including Îlet au Vent and Petite Islette), Les Îles Frégates, Île Chevreau (or Bonhomme), St.-Barthélemy, St. Eustatius, St. Christopher, Nevis, Antigua, Barbuda, Guadeloupe (Grande-Terre, Basse-Terre, and Iles de la Petite Terre), La Désirade, Les Îles des Saintes (Terre-de-Bas, Terre-de-Haut), Dominica, Martinique (including Îlet Chancel), and Marie-Galante; populations on St.-Martin/St. Maarten, Nevis, St. Christopher, Antigua, Barbuda, Les îles des Saintes, and Marie-Galante appear to have been extirpated (Pasachnik et al., 2006), and those on Martinique (Maillard and Breuil, 2007) and Guadeloupe (Basse-Terre, Grande-Terre, and some satellites, but not Petite Terre and La Désirade; Breuil et al., 2007) may be "lost" to hybridization with invasive *Iguana iguana*; an individual has been found on Puerto Rico (Rivero, 1998); elevational range, sea level to 400 m on Martinique (Breuil, 2002b).

Habitat: Essentially arboreal (to 30 m in humid forests), but varies by habitat and island, juveniles more likely to stay on or near the ground (Breuil, 2002b); various habitats ranging from xeric thornforest and cactus-dominated rocky outcrops to moister lowland forests at elevations from sea level to about 300 m, populations in moister areas may be infra-dispersed, whereas those in xeric areas or when associated with artificial habitats such as gardens are more likely to be clumped, arboreal or terrestrial (Lazell, 1973; Schwartz and Henderson, 1991; Tenlén, 1997b; Breuil, 2002b); on Dominica, seen exclusively in the vicinity of streams (Krintler, 1986); highly arboreal, may move from tree to tree by "leaping" between them, rarely on the ground, sleeps in crowns of trees (Malhotra and Thorpe, 1999); on St.-Barthélemy, runs across the streets (Wijffels, 1976); gathers at sites where water is available, even roadside ditches (Wijffels, 1997a); frequently in association with forests sporting abundant *Ipomea tiliacea*, on Guadeloupe, in ravines in moist lowland forests, often associated with banana trees overhanging streams, also on xeric cliffs characterized by succulents and in mangroves (Breuil, 2002b); on La Désirade, mostly restricted to dry areas and savannas, often characterized by cacti, succulents, and deciduous scrub on sandy or rocky substrates, uses rock crevices, logs, and burrows for refuges and sleeping sites, juveniles primarily in thornbushes at heights of 1–2 m (Schardt, 1998a; Breuil, 2002b); on Petite Terre, most abundant in wooded areas (Breuil, 2002b); on

Anguilla, largely restricted to thick scrub and cliffs along the northern coast, often arboreal, but will readily use low vegetation or rocks, burrows in loose soil or naturally occurring rock crevices and caves, not associated with human-modified habitats (Hodge et al., 2003); most abundant on St. Eustatius in apparently inhospitable scrub, often associated with White Cedars (*Tabebuia heterophylla*), where threats are minimal, uses trees, shrubs, rocks, and readily comes to ground, in gardens readily uses non-native plants for food and shelter, will seek refuge in exotic Razor Plants (*Pandanus* sp.), not uncommon on porches and roofs of buildings where fed or at least not molested, may sleep in trees, often on exposed branches, or in burrows (Powell et al., 2005); where habitat permits, sleeps almost exclusively high in trees (Breuil, 2002b).

Activity: Diurnal, sleeps in trees or rock crevices, associated with rock piles or cliff faces, much of day spent in thermoregulation or foraging (Lazell, 1973); main activity during early morning (~0830 h) and afternoon (~1500 h) on La Désirade (Schardt, 1998a); on Anguilla in June (N = 5), 86.0% of time not in burrows basking on rocks, 6.5% basking in trees, 0.2% feeding, 4.0% nonreproductive interactions, 2.0% courtship and mating, 1.8% roaming, and 0.4% responding to an observer (Pasachnik et al., 2002); some individuals may become active at dawn or remain active until dusk, but most activity occurs from ~0900 h to mid- to late afternoon, although some individuals may seek shelter during the heat of the day (Powell et al., 2005). **Behavior:** May bipedal and swim (Lazell, 1973); may interact with conspecifics at distances of 10 m, responds to disturbance by climbing higher into a tree or jumping to the ground (Schardt, 1998a); extremely alert while basking on rocks, flees by running across litter-covered ground or jumping to trees below the rocks (Tenlén, 1997b); quite capable of defending themselves by biting and scratching with their powerful claws (Wijffels, 1997a); may congregate on fruit-filled trees, but usually only one dominant male present (Breuil, 2002b); on St.-Barthélemy, iguanas easily tamed when fed, readily enter houses, sunbathe and copulate on a deck adjacent to a house (Breuil, 2003); exceedingly wary on Anguilla, relies on crypsis, but often drops to the ground from elevated perches when approached at considerable distances and flees quickly (Hodge et al., 2003); on St. Eustatius, readily tamed when fed, if disturbed may "dive" from arboreal or rocky perches, crashing to the ground, and moving considerable distances at high speed through thick, often thorn-encrusted vegetation, headbobbing used by both males and females, sometimes at substantial distances, but aggressive displays, almost always between males, involve lateral compression, push-ups, crest erection, and dewlap extension, rare fights involve biting and tail lashing (Powell et al., 2005); on Dominica, male "tasted" iguana scat and appeared to scent-trail another individual, subsequently squatted by elevating its hindlegs and tail, adpressing vent and thighs against the substrate, probably leaving scent marker immediately adjacent to the scat (J. S. Parmerlee, Jr., pers. comm.). **Competition:** On Terre-de-Haut, apparently displaced by *I. iguana* (Schardt, 1998b; Breuil, 2002b). **Foraging and Diet:** Food includes fruits of Prickly Pear (*Opuntia*), bird eggs, and carrion (Lazell, 1973); on La Désirade, eats leaves, flowers, and grasses, sometimes grazing alongside goats, a juvenile attempted to catch a cricket (Schardt, 1998a); diet includes 28 species of plants, especially *Tabebuia pallida*, *Hippomane mancinella*, *Clerodendron aculeatum*, *Capparis flexuosa*, *C. cynophallophora*, *Pisonia fragans*, *Guajacum officinale*, *Bursera simaruba*, *Conocarpus erecta* (Barré et al., 1997); may consume flowers and leaves of more than 50 species of trees and bushes in xeric coastal lowlands of Dominica, where iguanas may be important seed dispersal agents (Malhotra and Thorpe, 1999); forages mostly in the morning, frequently associated with *Ipomea tiliacea*, on which it avidly feeds, generalist vegetarian, diet varies by habitat and island, but may include introduced species (e.g., ornamentals), on St.-Barthélemy, feeds on leaves (especially during the dry season, and favoring young leaves) and fruits (during the rainy season) of *Melicoccus bijugatus*, *Cordia dentata*, *Leucaena leucocephala*, *Tabebuia pallida*, *Randia aculeata*, consumption of fruits and seeds may facilitate dissemination of these species, some food items are toxic to birds and mammals, drinks by licking leaves or opening mouth with head elevated during rains, may "taste" food before eating (Breuil, 2002b); on Anguilla, particularly fond of leaves, flowers, and fruits of Chink Bush (*Bourreria succulenta*) and flowers of White Cedar (*Tabebuia heterophylla*; Hodge et al., 2003); readily consumes ornamental plants, and appears to exhibit a particular fondness for *Hibiscus* sp. on Dominica (Daniells et al., 2008); on St. Eustatius, individuals may "gorge" (consume all vegetation within reach before moving) or "graze" (take only a few bites before moving), frequently feeding on Tamarind (*Tamarindus indica*) leaves (S. P. Fogarty, unpublished data); on Dominica, female ate dry dog food in a bowl (E. A. Daniells and N. J. Vélez Espinet, pers. comm.). **Home Range and Territoriality:** On Dominica, male territories during the breeding season 1,000 m^2, female territories 2,000 m^2, female ranges overlap those of other females and sometimes several males, suggesting a female dominance hierarchy (Day et al., 2000; Breuil, 2002b); males probably defend territories outside of the reproductive season (Breuil, 2002b). **Hurricane Effects:** May be adapted to habitats subjected to periodic hurricanes, since low metabolic requirements and large size allow survival during periods when habitats dramatically altered by storms, although population crashed after prolonged severe drought on Petite Terre (Breuil, 2002b). **Hybridization:** With *I. iguana* (Breuil, 1998, 2000a, 2000b, 2000c, 2000d, 2000e, 2001, 2002a, 2002b, 2003, 2004; Breuil and Sastre, 1994; Day and Thorpe, 1996; Powell, 2004e; Powell and Henderson, 2005a). **Longevity:** May live to 20 years of age (Malhotra and Thorpe, 1999); those living in proximity of humans probably live at least 15 years (Breuil, 2002b); in captivity, 4 years, 5 months (Snider and Bowler, 1992). **Miscellany:** Tropical Mockingbird (*Mimus gilvus*) harassed an adult male and a female at the base of a tree in which the bird's nest contained hatchlings by spreading its wings while perched on a rock in the vicinity of the iguanas and by grabbing and pecking at their tails (J. S. Parmerlee, Jr., pers. comm.). **Mortality:** Increasingly susceptible to road mortality (Malhotra and Thorpe, 1999); during movements to historic coastal nesting habitats, on Dominica 9 carcasses on 5-km stretch of road (Knapp, 2007a, 2007b); numerous eggs and hatchlings killed on seaside nesting slope on Dominica by Hurricane Dean in August 2007 (Knapp, 2007b); iguanas killed in floodwaters and hatchlings entombed by rock slides on sloped nesting area on Dominica after Hurricane Dean (Knapp and Valeri, 2008). **Movements:** Female 1,000 m from the original site of capture 2 years later, another female at original site of capture, a third female had moved 1,800 m into a mangrove stand (Breuil, 2000c); in forested areas, moves directly from tree to tree, rarely descending to the ground, readily enters water and may swim between islands (Breuil, 2002b). **Parasites:** On St.-Barthélemy and St. Eustatius, nematodes *Ozolaimus* (Leussink, 1958); on Dominica, ticks *Amblyomma* sp. (Lazell, 1973); *A. antillorum* (Kohls, 1969; Guglielmone et al., 2003); on La Désirade, oxyuroid nematodes (Schardt, 1998a). **Population Size and Density:** On St. Eustatius, ~300 animals in 1992 (M. Day and B. Leysner in Reichling, 2000), <300 in 1999 (Reichling, 2000), ~425 (275–650) in 2004 (Fogarty et al., 2004); in 1993 on Petite Terre, 4,000–6,000 adults, from 10,000 to ~5,000 after 1995 hurricanes, recovered to ~10,000 in 1999, 2,500–4,000 iguanas died due to drought in 2001, ~300 adults on Îlet Chancel (Breuil, 2002b; 2003); estimated 10,000 on Îlets de la Petite Terre "outside low population periods" (Lorvelec et al., 2002); greatest density on La Désirade, ~20–30 individuals, mostly adults, in an area of ~100 m^2 with exposed rock outcroppings along Baie Mahaut, one tree with 5 adults and one juvenile, juveniles generally scattered, unlike clumped distribution of adults (Schardt, 1998a); on Îlet Chancel, 600 individuals (Maillard and Breuil, 2007); density on Petite Terre about 10-fold that on Chancel, with the limiting factor appearing to be the availability of suitable nesting sites (Breuil, 2009). **Predation:** Humans, cats, and predatory birds, eggs eaten by mongooses (*Herpestes javanicus*) and humans (Currat, 1980); predators vary by island, include *Alsophis* spp., *Boa nebulosa*, Broad-winged Hawk (*Buteo platypterus*), Kestrels (*Falco sparvarius*), and possibly opossums (*Didelphis marsupialis*), *Ameiva pleii* on St.-Barthélemy often congregate around nests, hermit crabs and rats are other nest predators (Breuil, 2002b); on Dominica, *Ameiva fuscata* eats eggs (Day et al., 2000; Knapp, 2007b); *A. fuscata* eats embryos and hatchlings, Kestrel (*F. sparvarius*) with hatchling iguana (C. R. Knapp, pers. comm.); *Boa nebulosa* (Knapp et al., 2009); *B. nebulosa* probing iguana burrows while actively foraging on a hillside on Dominica (A. Mitchell, pers. comm.); humans illegally catch iguanas for food on Dominica (James, 2004); presumed predators include Kestrels (*F. sparvarius*), Red-tailed Hawks (*Buteo jamaicensis*), some seabirds, large racers (*Alsophis* sp.), and introduced dogs, cats, rats, mongooses, and humans (Powell et al., 2005). **Reproduction:** On Dominica, females lay eggs in burrows in sandy ground, migrating from higher ground to beaches (Malhotra and Thorpe, 1999); copulation in April–May (Knapp, 2007b); females gravid in July–August, eggs usually laid August–October, clutch of 22 eggs with lengths of about 35 mm in sand under Sea Grapes (*Coccoloba*) at a depth of 100 mm, on Antigua, female with 18 eggs with diameters 6–17 mm, on Île Fourchue, female with 17 eggs with average diameter about 25 mm on 12 July, on Dominica and Les Îles des Saintes hatching presumably December–February (Lazell, 1973); polygamous, with male-to-female ratio of 1:1 to 1:7 (Day et al., 2000) or even 1:12 on St.-Barthélemy (J.-C. Plassais in Breuil, 2002b); reproduction seasonal (tied to rainy season, when food for young is most abundant), sexually mature at 3 years, but males may be much older before mating because smaller individuals are unable to defend territories, defense usually without contact, displays involve lateral presentation, head-bobbing, undulation, gular and body inflation, and modest extensions of the dewlap, may rarely escalate into shoving matches and actual combat, sometimes leading to severe injuries, usually only the largest males mate, hold the female by her nape, but may release hold during copulation (Breuil, 2002b); eggs laid June–mid-August on Petite Terre and La Désirade (M. Breuil and B. Thiébot in Breuil, 2002b; Barré et al., 1997; Lorvelec et al., 2000); on St.-Barthélemy, eggs laid July–August (J.-C. Plaissais in Breuil, 2002b; Breuil, 2002b); less seasonally restricted in more mesic environments, on Dominica, gravid females February–August and egg laying 4 March–12 July, with second clutches likely (Day et al., 2000); to 18 eggs/clutch, deposition in sand on beaches (R. Breton, 1647 in Breuil, 2002b); on Guadeloupe, mate in March and lay on beaches in May, accompanied by males (J.-B. Du Tertre, 1667 in Breuil, 2002b); on Dominica, females may move 460–900 m to egg-deposition sites (Day et al., 2000); on St.-Barthélemy and Petite Terre, some females lay eggs within their home ranges, while others move considerable distances (Breuil, 2000c); nests in sand ~1 m deep ending in a chamber large enough to allow the female to turn around, in rocky areas, nest chambers less extensive (Breuil, 2002b); on Îlet Chancel, suitable nesting sites are restricted and females often excavate eggs laid by other females (Bach and Schönecker, 2008); on Martinique, clutch size 20–30 (Bouton, 1630 in Breuil, 2002b); on Guadeloupe, 13–25 (J.-B. Du Tertre, 1667 in Breuil, 2002b); on Dominica and Martinique, 15–30 (Anonyme De Carpentras, 1618–1620 in Breuil, 2002b); on Dominica, 8–18, strongly correlated with female size, average weight 25 g (Day et al., 2000); to 22 eggs with a length of 44 mm (Schardt, 1998a); freshly unearthed (by other females) eggs 17–22 g, 45 × 25 mm (Breuil, 2002b); incubation 3 months (Day et al., 2000); captive hatchling 20 g, SVL 77 mm (Reichling, 2000); hatchlings 17.6–18.9 g, SVL 80–83 mm (Gibson, 2001); egg and hatchling sizes vary by island and habitat (Breuil, 2002b); because clutch size is related to female size, the smaller size of iguanas on Chancel (compared to Petite Terre) limits recruitment into the population (Breuil, 2009). **Sex Ratio (M:F):** 1:1.3 (Maillard and Breuil, 2007); 1:1 to 1:7 (Day et al., 2000) or even 1:12 on St.-Barthélemy (J.-C. Plassais in Breuil, 2002b). **Size:** Male SVL to 434 mm, female SVL to 401 mm (Breuil, 2002b); iguanas on Chancel are smaller than those on Petite Terre (Breuil, 2009). **Swimming:** Male with body temperature 28 °C jumped into water 25–26 °C, swam on the surface for 80 sec, then dove for 16 sec at a depth of 2–3 m, when emerged, floated perpendicular to water's surface, before climbing onto a boat, within 2 min body temperature rose to 29 °C, 6 minutes later dove back into the water, held his breath for 28 min while immobile at 3 m at 26 °C, swam on the surface for 6 min (M. Breuil and B. Thiébot in Breuil, 2002b; Breuil, 2003). **Thermal Biology:** Position themselves to facilitate effective thermoregulation (Lazell, 1973); mean body temperatures on La Désirade 37.98 °C (36.53–38.77 °C) at sites with temperatures 30.5–31.0 °C at 0800 h, 41.0–42.2 °C at midday (but in shaded sites where iguanas are most frequently seen ~28–31 °C), ~34–35 °C at 1500 h, dropping to 30.3–31.0 °C in late afternoon and 28.5–29.0 °C at 1800 h, nighttime temperatures 24.5–25.0 °C (Schardt, 1998a); heliothermic, males warm at exposed arboreal sleeping sites, females often bask on the ground, when foraging often shuttle in and out of direct sun, highest temperature recorded for a male digesting food 39.4 °C, rest in shade during the hottest parts of the day (Breuil, 2002b); on St. Eustatius, maintains above ambient temperatures through the night even when sleeping on exposed perches (RP).

Conservation Status: Most populations remain extremely vulnerable to human exploitation for food, habitat alteration, introduced predators and competitors, and hybridization with introduced *Iguana iguana* (e.g., Breuil, 2002b, 2009; Powell, 2004e; Powell and Henderson, 2005a; Powell et al., 2005; Maillard and Breuil, 2007); on Chancel, an additional threat is posed by tourists who trample one of two principal communal nesting sites, although recent improvements (including fencing) have reduced the risk from tourists and sheep, also a proposal to relocate iguanas from Chancel to Îlet-à-Ramiers may provide another refuge from possible interbreeding with invasive *I. iguana* (Breuil, 2009); some populations (e.g., Dominica) are afforded protection under local laws (James, 2004); the species is listed on CITES Appendix I (UNEP-WCMC, 2008); on the IUCN Red List as "vulnerable" (Breuil and Day, 1996), but many individual populations meet criteria for "critically endangered" status (Breuil, 2002; Fogarty et al., 2004; Powell, 2004a; Powell and Henderson, 2005a); Powell (2004e) and Powell and Henderson (2005a) recommended a reevaluation and a probable upgrade to "endangered."

Additional Pertinent Literature: Pasachnik et al. (2006).

Iguana iguana Linnaeus, 1758

Distribution: Mainland from Mexico to Brazil, including numerous continental islands; introduced into southern Florida, Hawaii, and Israel; in the West Indies, on Grenada and the Grenadines, St. Vincent, Barbados, St. Lucia, Martinique, Marie-Galante, Les Îles des Saintes, Guadeloupe (Grande-Terre, Basse-Terre, and Iles de la Petite Terre), Montserrat, Antigua, Barbuda, Saba, St.-Martin/St. Maarten, Anguilla, St. Croix (may have been introduced by Caribs for food; Beatty in Grant, 1937b), Virgin Islands, Puerto Rico (including Culebra), Grand Cayman, and Swan Islands; elevational range, sea level to ~800 m; population on Barbados apparently extirpated (although some captive animals remain), those on Martinique, Marie-Galante, Les Îles des Saintes, Guadeloupe, and St. Croix possibly introduced or of mixed origin (e.g., the historic population on St. Croix presumably natural, but most currently present lizards probably are descendants of introduced individuals), and those on Antigua, Barbuda, St.-Martin/St. Maarten, Anguilla, Virgin Islands, Puerto Rico, and Grand Cayman are introduced, although at least some individuals on Anguilla arrived via natural over-water dispersal, almost certainly from Guadeloupe (Censky et al., 1998); populations on St. Lucia, Saba, and Montserrat (and possibly the historical population on St. Croix) may be distinct at the species level from those elsewhere (Malone and Davis, 2003; Powell, 2004e).

Habitat: Xeric habitats, where populations usually are clumped, and mesic habitats, where individuals are infradispersed, in *Acacia* thickets, mangrove edges, on a cemetery wall, in bushes, trees, or open rocky ground, on cliff faces, in rocky crevices, and around a water hole in xeric scrub (Îles des Saintes, Îlet à Cabrit; Lazell, 1973; Breuil, 2002b); largely arboreal on St. Lucia, but frequently ventures onto the ground (Long, 1974); congregated in dry forests and dispersed in moist areas on windward slopes, on the Îles des Saintes, concentrated along coastal cliffs, where essentially terrestrial, in scattered trees on savannas, and purely arboreal in the contiguous canopies of mesic forests, on Guadeloupe, in ravines and in urban areas, gardens (where they may pillage vegetables and ornamentals), and generally anywhere where trees occur, on interior cliffs of the Atlantic Coast where they compete with *I. delicatissima*, but also on banana trees in ravines (Breuil, 2002b); on Puerto Rico, on the ground and in trees, generally near water, into which they jump at the slightest disturbance (Rivero, 1998); abundant in urban areas and occasionally a problem when basking on airport runways (Engeman et al., 2005); on Terre-de-Haut, abundant along beaches, especially in association with Manchineel (*Hippomane mancinella*) or Sea Grape (*Coccoloba uvifera*) and in inland pockets, overwhelmingly in dry habitats, concentrations of individuals on piles of stones and trees around an artificial pond and along an overgrown, ~1.5 m-high fence, adults usually feed on the ground, juveniles primarily in bushes to heights of ~2 m, but also on the ground and on palms, may bask on coastal debris, sleeps at night on bushes, trees, and piles of rocks (Schardt, 1998b); on Saba, mostly arboreal in forested regions at all but the highest elevations, also exploits open areas with sparse vegetation where iguanas bask on rock outcroppings, on St.-Martin/St. Maarten, associated with human homes and gardens, although readily exploits newly established coastal mangrove stands, especially trees overhanging water (Powell et al., 2005); on Grand Cayman, frequently edificarian (Binns, 2008b).

Activity: Diurnal, juveniles sleep in low trees and bushes, adults probably in large trees, rock piles, rock crevices, or burrows (Lazell, 1973); first evident ~0830 h (Schardt, 1998b); on Îles des Saintes, emerge ~3 h after sunrise after spending nights in mangoes and other large trees (Breuil, 2002b); may be active from dawn to dusk, but most activity from mid-morning to late afternoon (Powell et al., 2005). **Behavior:** Told by residents that iguanas, when disturbed, would dive into the sea and cling to rocks (Grant, 1932h); leap into brackish lagoons and swim away on Water Island near St. Thomas (Grant, 1937b); may bipedal and swim, engages in headbobbing and dewlap shaking (Lazell, 1973); good swimmers that readily enter water (Long, 1974); when approached to ~20 m on Terre-de-Haut, most individuals climbed higher into trees, but one adult male jumped to the ground, ran into the sea, and emerged several minutes later ~30 m farther down the beach (Schardt, 1998b); when disturbed, color darkens, hiss and bite, use tail as a whip (Long, 1974); will climb to tops of trees, jump behind rocks, hide behind vegetation, or enter the sea, remaining submerged for some time, territorial defense displays involve headbobbing, gaping, hissing, and tail lashing (Breuil, 2002b); unless habituated to humans, quite wary, often relies on crypsis, but can move very quickly when alarmed, to escape climbs higher into trees or drops to the ground and moves considerable distances at high speed, on St. Maarten, may dive into water when disturbed (Powell et al., 2005). **Foraging and Diet:** Largely plant material (e.g., mangoes and Prickly Pear [*Opuntia*] fruit; Lazell, 1973), but also bird eggs (Arendt, 1987; Lazell, 1973); feeding on offal at the Grenada Zoo (Long, 1974); feeds primarily on leaves and flowers, but also on vegetation washed up on beaches (Schardt, 1998b); opportunistic vegetarian, essentially arboreal, but feeds on the ground as well as in trees, consumes a variety of vegetation (constrained by habitat) and animal matter, on savannas on Basse Terre and Îles des Saintes, is in direct competition with cattle and goats, and will graze beneath the feet of mammalian herbivores, also feeds on many cultivated plants, scavenges (including human garbage), and eats poultry feces, likes fallen mangoes (Breuil, 2002b); occasionally takes chicks (P. Perron in Breuil, 2002b); leaves, flowers, and fruits, readily adapts to non-native plants, can become pests in gardens, where they show a particular fondness for *Hibiscus* flowers, may occasionally feed on bird eggs or carrion (Powell et al., 2005); "feast on handouts and ornamental plants" in the U.S. Virgin Islands (Platenberg, 2007); iguanas from Montserrat and St. Croix had 4 valves partitioning the proximal colon, whereas mainland animals had 5, 6 (modal), or 7 valves, presumably to facilitate digestion (Iverson, 1982). **Mortality:** Especially high in late April–early May when females migrate to egg-laying sites (Breuil, 2002b). **Parasites:** Intestines and liver infected with nematodes (Hill, 1954), apparently *Ozolaimus* (Leussink, 1958); specimen from Bequia infected with ticks (*Aponomma*; Lazell, 1973); nematode *Alaeuris vogelsangi* in an iguana from Puerto Rico, probably introduced with the host from Venezuela (Dyer et al., 1999). **Predators:** Humans, cats, and predatory birds, eggs eaten by mongooses (*Herpestes javanicus*) and humans (Currat, 1980); *Borikenophis portoricensis* on Water Island,

U.S. Virgin Islands (Grant, 1932h); on Carriacou (Grenadines), *Corallus grenadensis* (Pendlebury, 1974); have a "reputation as excellent eating and their flesh is reputed to taste like chicken" (Long, 1974); dogs (on St. Lucia; Bendon, 2003) and probably mongooses (Morton, 2007); other than humans and domestic animals, Broad-winged Hawks (*Buteo platypterus*) are the main predators (Malhotra and Thorpe, 1999); presumed predators include Kestrels (*Falco sparvarius*), Red-tailed Hawks (*Buteo jamaicensis*), some seabirds, large racers (*Alsophis* spp.), and introduced dogs, cats, rats, mongooses, and humans (Powell et al., 2005); on Grand Cayman, a Great Egret (*Ardea alba*) captured a juvenile iguana (T. Ebanks, in litt., 15.XII.2008); on Grand Cayman, *Cubophis caymanus* (T. Ebanks, in litt., 23.II.2009). **Reproduction:** Females mature at ~3 years, later in males who must defend territories, males highly territorial February–March, females migrate to egg-laying sites in late April–early May, female (890 g) with 17 eggs 38–45.5 × 22–28.2 mm, 14–24 g, largest eggs in distal oviducts, incubation 90–92 days, hatchlings July–early August, corresponding to onset of the rainy season (Breuil, 2002b); eggs laid in warm, well-drained sites on beaches, in old charcoal pits, and gardens (Breuil, 2000a, 2000c); 17 eggs hatched in 14 weeks (Underwood, 1962; Long, 1974); presumably lays eggs December–February, on Montserrat, female 275 mm SVL with 16 eggs (42.0–46.8 × 26.6–28.4 mm) on 24 March (Lazell, 1973); on Puerto Rico (although all data may not apply to Puerto Rican populations), reproductive season November–December, males establish territories from which other males are actively expelled, courtship similar to the aggressive displays between males, copulation depends on female's response, may last 15 min, but males rarely copulate more than once/day, eggs (17–68/clutch) require high incubation temperatures, nests in clearings or on beaches, may be aggregated or dispersed, apparently depending on type of soil, nests may be used for several years, may be 1 m deep and connected by tunnels to 20 m in length with several openings to the surface, egg deposition may be communal (Rivero, 1998); in Puerto Rico, clutch sizes 12–90 (A. González and V. Ríos in Joglar, 2005b); egg deposition on Terre-de-Haut at end of dry season (April–May), favored nesting sites along beaches among clumps of vegetation, large numbers of hatchlings in October (Schardt, 1998b); may move considerable distances to two known nesting beaches on St. Lucia, 3 females moved 80 m, 1.57 km, and 1.61 km from nesting sites back to home ranges, 5 nests on St. Lucia with 17–21 eggs (mean 19.5), hatchlings on nesting beaches mid-May–mid-August (Graham and Morton, 2004, 2005); average clutch size 23 (14–28) in 14 nests with 63% hatching success (Morton, 2007). **Size:** Adult SVL to 500 mm (Schwartz and Henderson, 1991); recent hatchlings (October on Terre-de-Haut) 82–95 mm SVL (mean 88 mm) and 13–22 g (mean 19.4 g; Schardt, 1998b). **Thermal Biology:** Position themselves to facilitate effective thermoregulation (Lazell, 1973); in habitats frequented by iguanas temperatures of 34.3 °C between 0930 and 1030 h, peaking at 41.8 °C at 1215–1330 h, decreasing to 35.7 °C at 1500 h and to a minimum of 24.0 °C at night (Schardt, 1998b).

Conservation Status: Listed in CITES Appendix II (UNEP-WCMC, 2008), but export quotas exist for many countries, primarily for live animals (pet trade) or products (leather goods and meat), no distinction is made for native versus introduced or for continental versus insular populations; hunting is usually prohibited, but enforcement of laws is lax at best, the species is considered a game animal in Grenada and St. Vincent, with a regulated hunting season and bag limits (on Grenada), which are routinely ignored by local hunters (Powell, 2004e; Powell and Henderson, 2005a, 2007); population on St. Croix "greatly reduced in numbers (probably as much by the activity of man as by that of the mongoose)" (Seaman and Randall, 1962); presumably introduced populations in the U.S. Virgin Islands are afforded special protection under USVI Code Title 12 (Platenberg, 2007).

Remarks: Native West Indian populations currently assigned to *Iguana iguana* may represent more than one species (e.g., Stephen, 2007).

13 Family Leiocephalidae

Figure 16. An adult male *Leiocephalus lunatus* (Leiocephalidae) exhibiting defensive posture. Photograph by Eladio Fernández.

Leiocephalus altavelensis Noble and Hassler, 1933

Distribution: Hispaniola, Isla Alto Velo (Dominican Republic).

Habitat: Xerophilic, among rocks, leaves, and grass along the coast and to the top of the highest hill in the island's center (Noble and Hassler, 1933).

Reproduction: Clutch size 1 (Hedges, in prep.). **Size:** Male SVL to 71 mm, female SVL to 63 mm (Schwartz, 1967d).

Conservation Status: Not assessed.

Remarks: The population on Isla Alto Velo may not have diverged sufficiently from *Leiocephalus barahonensis* to warrant recognition as a distinct species (M. E. Gifford, in litt., 05.V.2008).

Leiocephalus barahonensis Schmidt, 1921

Distribution: Hispaniola, where four currently recognized subspecies occur from extreme southeastern Haiti, east across the Barahona Peninsula, north along the eastern coast of the peninsula, and west along the northern slopes of the Sierra de Baoruco to near the Dominican-Haitian border, also on Cayo Pisje (off the eastern coast of the Barahona Peninsula) and Isla Beata; elevational range, sea level to ~620 m.

Habitat: Xerophilic to mesophilic (Schwartz and Henderson, 1991); in coastal zones and scrub and cactus savannas, on rock piles in association with *Celestus costatus* (Mertens, 1939); in *Acacia*-cactus associations, but often in less rigorous habitats such as those along the edges of mesic or xeric forests, in *Cocos* groves, open areas on limestone plateau (Isla Beata), areas of exposed limestone and small arborescent cacti, shaded hammock woods, and mangrove flats

(Schwartz, 1967d; Schwartz and Henderson, 1991); on Cayo Pisaje, all over the island, especially on the beach, foraging among human debris (Yeska et al., 2000); on the Barahona Peninsula, associated closely with rock piles, whereas *Anolis longitibialis* is almost exclusively on rock outcrops or trees and *Ameiva chrysolaema* on bare ground (RP).

Activity: Active throughout the day at one site on the Barahona Peninsula, but lizards retreat under rocks at times when sky is overcast at another site (Micco et al., 1997). **Behavior:** Curls tails (Pregill, 1992); males performed wave-like headbob/pushups in sets of 2 following movement to a new perch, elevation effected by neck flexion and forelimb extension, with head moved in circles and extended during the low part of the display and retracted at the apex, no tail curling observed (Micco et al., 1997); "highly agitated" individuals follow headbobs with a wavelike raising and lowering of body and tail (R. A. Sosa in Nelson et al., 2001); mean approach (242.8 ± 26.75 cm) and flight (45.6 ± 10.78–99.0 ± 35.32 cm) distances were greater in more open habitats (Gifford et al., 2008). **Foraging and Diet:** Stomachs of animals from 2 sites on the Barahona Peninsula contained spiders, beetles (adults, most important at one site, and larvae), earwigs, roaches, flies (adults and larvae), hemipterans, ants (most important at the second site) and other hymenopterans, termites, lepidopterans (adults and larvae), scorpionflies, orthopterans, unidentified insect pupae, centipedes, millipedes, isopods, snails, a juvenile conspecific, flowers, fruits and seeds, other plant material, shed skin, and grit, mean prey size not correlated with SVL (Micco et al., 1997). **Home Range and Territoriality:** Highly territorial (Mertens, 1939). **Longevity:** In captivity, 4 years, 8 months (J. Grizzell in Snider and Bowler, 1992). **Parasites:** Nematode *Skrjabinoptera leiocephalorum* (Greve and Powell, 1989; Powell et al., 1990b); apicomplexan *Eimeria leiocephali* (Daszak and Ball, 1991; but see Huntington and Cisper, 1994). **Performance:** Mean sprint speeds (180.6 ± 8.74–209.7 ± 5.73 cm/sec) were greater in more open habitats (Gifford et al., 2008). **Reproduction:** Female with 4 enlarged follicles 27 August (Schwartz and Henderson, 1991); proportion of gravid females at 2 sites on the Barahona Peninsula in June, 8 of 21 and 7 of 12, ratios of juveniles to adults 23:29 and 8:36, suggesting different peaks of reproductive activity, clutch sizes invariably 2 at both sites, fat bodies minimal in males and females (Micco et al., 1997); mean clutch size 2.06 ± 0.54, mean egg length 15.05 ± 1.22 mm, mean egg volume 112.50 ± 13.64 mm^3, with egg volume positively correlated with female SVL (Gifford and Powell, 2007). **Sex Ratio (M:F):** Adults 24:12 and 8:21 at two sites on the Barahona Peninsula (Micco et al., 1997).

Size: Male SVL to 84 mm (Micco et al., 1997), female SVL to 64 mm (Schwartz, 1967d); smallest mature male 54 mm SVL, smallest mature female 46 mm SVL (Micco et al., 1997); estimated female SVL at maturity 47.8 mm (Gifford and Powell, 2007). **Tail Autotomy:** Tail-break frequencies (0.59–0.69) did not differ by sex or habitat (Gifford et al., 2008). **Thermal Biology:** Mean body and substrate temperatures vary by site, 36.9 ± 1.4 °C and 35.2 ± 2.6 °C versus 35.3 ± 1.5 °C and 32.3 ± 0.4 °C, with lizards at the former site basking more frequently (24 of 33) than those at the latter (12 of 31), although suitable basking sites were abundant at both locations (Micco et al., 1997).

Conservation Status: Not assessed.

Additional Pertinent Literature: Powell (1999d).

Leiocephalus carinatus Gray, 1827

Distribution: Thirteen currently recognized subspecies occur in the Bahama Islands (Great and Little Bahama banks, Cat Island, Little San Salvador Island), Cuba (mostly coastal lowlands, widespread but disjunct, also Isla de la Juventud and other satellites), Cayman Islands (Grand Cayman, Little Cayman, Cayman Brac), Swan Islands; elevational distribution, sea level to ~365 m; introduced into southeastern Florida.

Habitat: Xerophilic (Schwartz and Henderson, 1991); on Andros and Great Abaco islands (Bahamas), about old refuse sites and on logs and debris at lumbering sites in interior pine forest (Schwartz and Henderson, 1991); on Bimini, along beaches or in rocky areas, in and around shrubby growth back from beaches, around human habitations on ground or rock walls, shunning densely shaded areas, such as those where Australian "Pine" (*Casuarina*) had been planted (Bogert, 1982); on Grand Cayman, "on stony ground near the sea" (English, 1912); on open rocky ground, preferably near a beach, sometimes on roads and stone walls, spends much time sunning on tops of rocks (Grant, 1940c); on Little Cayman Island and Cayman Brac, apparently requires open beach habitat, using driftwood, rocks, and other elevated perches, in association with broken coral rocks and beach wrack, in *Coccoloba* thickets and yards of residents, on rock walls around cemetery and residences (Seidel and Franz, 1994); in Cuba, in "dog-tooth" limestone areas with sand interstices, cacti, mogotes (Pinar del Río Province), windrows of broken coral heaped in mounds on the beach, under *Coccoloba*, along sea cliffs, in abandoned rock quarries, on driftwood, in old rock walls, under limestone slabs, on stacks of lumber adjacent to rocky outcroppings near the shore (Schwartz and Ogren, 1956; Schwartz, 1959c, 1959e; Buide, 1966; Schwartz and Garrido, 1967; Garrido, 1980b; Ševčik, 1989; Schwartz and Henderson, 1991; Estrada, 1993b; Rodríguez Schettino, 1999b); along busy roadsides in urban areas (Z. Vogel, 1965a); of 5 sampled sites, only in dry, open areas near the coast, where it is abundant (Viña Dávila et al., 1984); in yards where it basks on concrete blocks (Sexto, 1989); mean height above ground on the Península de Guanahacabibes 0.8 m (Rodríguez Schettino, 1999b); also in the forest, where one individual was basking 15 m high on the trunk of a large tree (*Ceiba* sp.), on decorative rock piles and even on paved areas of La Habana's "malecon," on elevated perches near burrows, on an abandoned plantation near a beach, burrows under bushes and clump-grasses, some to 40 cm deep (Petzold et al., 1970); may forage so close to ocean as to be sprayed by breaking waves (Schwartz, 1959c); in crowns of low, rough-barked trees along the coast (Ševčik, 1989); in mogote complex of vegetation, coastal and subcoastal microphyllous forest, semidesert thorny scrubwoodland, and urban zones (Rodríguez Schettino et al., 1999); often associated with humans (Kunz, 2005; Krüger, 2008).

Activity: Individuals on the bluff behind East End, Grand Cayman, usually seen only in late afternoon (Grant, 1940c); completely diurnal, presumably from just after sunrise (on fair, warm days) until just after sunset, forages or basks on exposed rocks and outcroppings during the intense heat of the day but seeks shelter in shaded spots (Schwartz, 1959c); lizards encountered as early as 0830 h and as late as 1640 h, probably active before and after that period (Bogert, 1982); retreats into burrows ~1800 h (Petzold et al., 1970); greatest number of individuals active during hours of intense heat, adults often remain buried in sand or in burrows during the dry season (Rodríguez Schettino, 1999b); during the summer, often remains active during rain (Sexto, 1989); if a refuge is uncovered 0800–1000 h, multiple lizards may be crowded together (L. Moreno in Rodríguez Schettino, 1999b). **Behavior:** Easily tamed (Grant, 1940c; Sexto, 1989); very wary, retreats hastily with tail in "watch-spring" curl over back (Cuban name "Perrito" [= little dog]), exposes yellow to white underside, when inside a burrow, may stop in opening and watch intruder, juveniles at times easily taken (under rocks) partially buried in sand (Schwartz, 1959c); tail curling occurs in juveniles on the second day after hatching (Petzold, 1962) or immediately after hatching, when curled horizontally as well as vertically (Petzold et al., 1970); although often in close association with humans, much more "shy" than anoles, may congregate (usually one large male, several females and subadults) on elevated perches in late afternoon, male often in a ~30 cm opening of a 1 m long burrow under a rock wall, when threatened, escaped through a smaller (~10 cm) opening, in drainage pipes in walls in coastal spray zone (Petzold et al., 1970); males from Bimini (in captivity) wave entire tail slowly while pausing during retreat, tail rolling and curling important in often very severe struggles for social position that also involve bobbing, strutting, and inflating sides of necks (Evans, 1953); males run short distances (3.0–7.6 m) before stopping, rising on forelimbs, and watching intruder alertly (Schwartz and Henderson, 1991); young individuals hide under cover, when surprised, they remain motionless while partially burying themselves in sand (Rodríguez Schettino, 1999b); in the Exuma Islands, male-male aggression involves headbobbing and symmetrical tail curling, resident charging and chasing intruder from area, also charged Bahamas Mockingbird (*Mimus gundlachii*; Knapp, 2001b); aggressive interactions with Bahamas Mockingbirds increase in frequency after bird eggs hatch (Hines et al., 2002b); in addition to social displays, tail curling may deter pursuit and deflect predatory strikes from body to tail, which can be autotomized, curling more frequent when predator approaching than when moving away, of greater intensity and frequency when approached by a predator than when moving spontaneously (Cooper, 2001); flight-initiation distance, distance fled, and likelihood of entering a refuge increased as distance to refuge increased, mean onset of escape behavior 2.6 ± 0.9 m for lizards that entered refuges and 3.9 ± 0.9 m for those that did not, distance fled 0.6 ± 0.2 m and 1.0 ± 0.2 m, tail displays almost always initiated simultaneously with onset of flight and apparently does not compromise escape speed (Cooper, 2007); hatchlings fled by climbing into trees (Krüger, 2008). **Foraging and Diet:** Omnivorous, food includes flowers of Railroad Vine (*Ipomoea pescaprae*), insects, and lizards (Grant, 1940c); *Anolis luteosignifer* (A. S. Rutty in Grant, 1940c); smaller lizards, especially juvenile *A. sagrei* (Z. Vogel, 1965a); extent of herbivory generally higher on smaller islands, food includes buds, flowers, seeds, small fruits, anoline lizards (*A. sagrei*), thrips, termites, dragonflies, orthopterans, hemipterans, beetles (adults and larvae), neuropterans, lepidopterans (adults and larvae), flies (adults and larvae), roaches, ants and other hymenopterans, millipedes, centipedes, spiders, pseudoscorpions, amphipods, isopods, decapods, and mollusks, will lap water from ice cubes and quickly learns to solicit cubes, also feeds on food scraps dropped by diners in a hotel restaurant (Schoener et al., 1982); for *L. c. cayensis* on 4 cays off the southern coast of Camagüey Province, decapods, spiders, pseudoscorpions, mites, centipedes (*Scolopendromorpha*), ants (most important) and other hymenopterans, flies, orthopterans, beetles (adults and larvae), roaches, earwigs, hemipterans, termites, caterpillars, and conspecifics (Armas, 1987); known to be cannibalistic in "semi-captivity" in an enclosure at a zoo (Martínez Reyes and Rodríguez Schettino, 1987); a facultative omnivore, eats arthropods, with ants most commonly taken, one sample contained 16 types of arthropods, a conspecific, and vegetable matter, especially fruits, another sample contained 95.9% ants and 3.4% beetles, another 73.8% ants, 13.1% vegetable matter, and the remainder assorted arthropods, ate "cooked hens' eggs," at the zoo, eats "sweetmeats and remnants of food offered other species, at the same time capturing the insects they find" (Rodríguez Schettino, 1999b); arthropods (ants, most important, other hymenopterans, beetles, hemipterans, lepidopterans, orthopterans, spiders, and mites) and plant material, with consumption of the latter highest in August–November and the consumption of animal matter highest in January–May (Fong Grillo and Garcés G., 2002a); in the Bahamas, feeds on bread and chips and begs for food (Valiulis et al., 2003); scavenges, including dead fish washed ashore on the beach (J. Binns in Powell, 2004d); ingests sap of Bay Cedar (*Suriana maritima*) on Alligator Cay, Exumas, Bahamas (G. R. Smith et al., 2008); *Anolis sagrei* (Vanhooydonck et al., 2009). **Home Range and Territoriality:** Males fight for territories, accounting for the many mutilated tails (Grant, 1940c); as many as 4 individuals in 1 m^2 of rock surface (Schwartz, 1959c); males fiercely territorial, but home ranges of several females may overlap that of a single male (Powell, 2004d). **Longevity:** In captivity, 10 years, 10 months (Curator of Reptiles, Chicago Zoological Park, in Snider and Bowler, 1992). **Parasites:** In Cuba, nematodes *Physaloptera squamatae*, larval *Abbreviata* sp., larval *Porrocaecum* sp. (Baruš and Coy Otero, 1969a; Coy Otero, 1970; Coy Otero and Baruš, 1979b; Coy Otero and Lorenzo Hernández, 1982; Coy Otero, 1999); *Cyrtosomum longicaudatum* (Baruš and Coy Otero, 1969a; Coy Otero, 1970; Coy Otero and Baruš, 1973a, 1979b; Coy Otero and Lorenzo Hernández, 1982; Coy Otero, 1999); *C. scelopori*, *Oswaldocruzia lenteixeirai*, *Parapharyngodon cubensis*, *Skrjabinoptera phrynosoma*, larval *Cyrnea* sp., and Nematoda gen. sp. (Coy Otero and Baruš, 1979b; Coy Otero and Lorenzo Hernández, 1982; Coy Otero, 1999); *Atractis scelopori*, *P. squamatae*, *Spauligodon cubensis*, *O. lenteixeirai*, and larval *Porrocaecum* sp. on Grand Cayman Island (Goldberg et al., 1995); ticks (*Amblyomma albopictum* and *A. torrei*) in Cuba (Coy Otero, 1999; Guglielmone et al., 2003); *A. dissimile* on Andros Island (Durden and Knapp, 2005). **Population Size and Density:** 700 lizards/ha (Schoener et al., 1982); at the National Zoological Park in La Habana, 238.1 lizards/ha, encounter rates at different localities 1.9 individuals/hr in June, 1.8/hr in cliffs, and 0.6/hr in coastal shrubwoodlands in June, 1–3/hr at 3 other localities (Rodríguez Schettino, 1999b). **Predation:** *Cubophis*, birds, presumably rats, dogs, and cats (Schwartz and Henderson, 1991); cats in urban areas (Sexto, 1989). **Reproduction:** Copulation involves a male sitting on a female with jaws at the female's nape, lets himself slide back (male's vent slightly protruded), male curves underside of venter laterally without releasing hold on the female, female bends to one side and approximates her vent to that of the male, hemipenis (dark red in color, ~5 mm long) introduced for a few seconds, both tails slightly uncurled, pair separates and rests calmly next to each other (Petzold et al., 1970); young appear in early August (Grant, 1940c); in field, major period of hatching apparently mid-July, but gravid female on 27 July, eggs cream colored, leathery shelled, 25.1 × 11.5 mm on 27 June (Schwartz, 1959c); in captivity (from wild-caught female), shriveled eggs (14–16 × 9–11 mm), one hatched in 91–101 days, hatchling SVL 35 mm, 2 later eggs by same female 16.3 × 10.5 mm and 17.5 × 11.8 mm (Petzold, 1962); in captivity, 6–7 eggs/clutch, deposited 29 March–28 April, 19–21 × 12–13 mm at deposition, 25–26 × 16–19 mm at hatching, incubation 56–74 days (at 27 °C), hatchling SVL 32–36 mm (Petzold et al., 1970); young found throughout the year, reproduction continuous, increasing in some months, preserved eggs measured 19.8 × 10.9 mm, eggs buried in earth or sand, in July, egg measuring 25.1 × 11.5 mm in oviduct of female with SVL = 102 mm (Rodríguez Schettino, 1999b); clutches to 9 oviductal eggs of large size (M. Martínez Reyes in Rodríguez Schettino, 1999b); eggs buried adjacent to active termite (*Nasutitermes rippertii*) mound in front of a tunnel excavated and filled by *Cyclura cychlura*, eggs 25.2 × 12.2 mm and 24.9 × 14.2 mm (Owens and Knapp, 2004). **Size:** Male SVL to 133.2 mm, female SVL to 115.8 mm (Rodríguez Schettino, 1999b). **Thermal Biology:** Mean body temperature maintained within relatively narrow limits, even if ambient temperatures are as much as 12.5 °C below "acceptable" body temperatures determined in the laboratory, mean cloacal and air temperatures on Bimini in September–October 34.5 ± 0.28 °C and 30.8 °C, in January 34.5 ± 0.20 °C and 25.3 °C (Bogert, 1982); two individuals during the wet season on the Península de Guanahacabibes 33.2 and 35.6 °C at air and substrate temperatures of 29.4 and 31.4 °C (Rodríguez Schettino, 1999b).

Conservation Status: Not assessed, probably of least concern, although the presence of exotic predators (cats) affects the species' presence and population sizes on small islands in the Bahamas (Dodd and Franz, 1996).

Leiocephalus cubensis Gray, 1840

Distribution: Cuba, where five currently recognized subspecies occur essentially islandwide and on satellites (Isla de la Juventud, Cayos de San Felipe, Archipiélago de los Canarreos, Jardines de la Reina, Archipiélago de Sabana-Camagüey); elevational range, sea level to at least 365 m.

Habitat: Mesophilic to xerophilic (Schwartz and Henderson, 1991); habitats variable, including hot coastal beaches, *Coccoloba* stands, sugarcane and sisal fields, along grassy roadways, in scrub growth, dense broadleaf forest, under rocks along beaches, palm jungle, in stone walls at a deserted house, among herbaceous plants and trees, but usually does not enter forests (Schwartz, 1959e; Garrido, 1980b; Schwartz and Henderson, 1991; Estrada, 1993a; Rodríguez Schettino, 1999b; Martínez Reyes and Moreno García, 2003); few individuals in dense forest (Krüger, 2008); not as commonly encountered on beaches as *L. carinatus*, but frequently in lowlands paralleling the coast, often associated with *Ameiva auberi* (Buide, 1966); on the Cayos de San Felipe, in the interface between beach habitat and low vegetation (Varona and Garrido, 1970); four large individuals in a rotten palm log with two *Tarentola americana* and four *Osteopilus septentrionalis* (G. Peters, 1974); edificarian situations, including patios, gardens, empty lots, and abandoned buildings and building sites (Collette, 1961; Martínez Reyes and Fernández García, 1994; Rodríguez Schettino, 1999b; Kunz, 2005); in an urban area (La Habana), 41.0% on the ground, 27.1% on sidewalks or other human-built structures, 24.2% on rocks, and 6.8% in leaf litter (Martínez Reyes and Fernández García, 1994); seeks refuge in caves or under rocks and logs (Estrada, 1994; Martínez Reyes and Moreno García, 2003); in coastal and subcoastal microphyllous forest, cultivated areas and pastures, and urban zones (Rodríguez Schettino et al., 1999).

Activity: Diurnal, spends many hours basking, but leaves refuges only on relatively hot days, emerging only every 3–4 days during the dry season (Rodríguez Schettino et al., 1999); active as early as 0700 h (Schwartz, 1959e). **Behavior:** Aposematic tongue extension (Mertens, 1946); wary and difficult to approach, but not as easily frightened as *L. carinatus*, only curls distal third of tail when fleeing, not like "watch-spring" curl of *L. carinatus* (Schwartz, 1959e); large male chased a medium-sized *Rattus* and an adult male *Anolis sagrei* from its retreat, carrying the latter in its jaws (Fernández Milera, 1984). **Foraging and Diet:** Omnivorous, but mostly insects, especially ants (79.3% during the dry season and 73.4% during the wet season, primarily *Solinopsis geminata* in males and *Paratrechina anthracina* in females during the dry season, also substantial numbers of *Odontomachus ruginodes* and *Wasmannia auropunctata* during the wet season, when selectivity was lower) and larvae (11.1% during the dry season and 7.4% during the wet season), but also beetles, flies, hemipterans, roaches, earwigs, silverfish, spiders, isopods, gastropods, an annelid, and 4 instances of saurophagy, a neonatal *Anolis sagrei* and 3 conspecifics, in addition, considerable plant material, mostly flowers in males and seeds in females (Martínez Reyes and Fernández García, 1994); facultative omnivore that eats mainly insects, but will also take remnants of human food, ants made up 74.6% of one sample and 15.8% of stomachs contained vegetable matter (mostly seeds), 4 instances of saurophagy (including *Anolis sagrei*), three of them cannibalism (Rodríguez Schettino, 1999b); male ate a dead *Anolis scriptus* (Iverson and Smith, 2006). **Parasites:** Nematodes *Cyrtosomum longicaudatum*, *Physaloptera squamatae*, *Skrjabinoptera phrynosoma* (Coy Otero and Baruš, 1979b; Coy Otero and Lorenzo Hernández, 1982; Coy Otero, 1999); *Oswaldocruzia lenteixeirai*, larval *Abbreviata* sp., larval *Porrocaecum* sp. (Baruš and Coy Otero, 1969a; Coy Otero, 1970; Coy Otero and Baruš, 1979b; Coy Otero and Lorenzo Hernández, 1982; Coy Otero, 1999); *C. scelopori* (Baruš and Coy Otero, 1969a; Coy Otero, 1970; Coy Otero and Baruš, 1973a, 1979b; Coy Otero and Lorenzo Hernández, 1982; Coy Otero, 1999); *Parapharyngodon cubensis* (Baruš and Coy Otero, 1969a, 1969b; Coy Otero, 1970; Coy Otero and Baruš, 1979b; Coy Otero and Lorenzo Hernández, 1982; Coy Otero, 1999); and *Trichospirura teixeirai* (Baruš and Coy Otero, 1969a; Coy Otero, 1970; Coy Otero and Baruš, 1979a, 1979b; Coy Otero and Lorenzo Hernández, 1982; Coy Otero, 1999); ticks (*Amblyomma torrei*) on Cuba (Coy Otero, 1999). **Population Size and Density:** Encounter rate at Playa Larga, Matanzas Province 24 lizards/hr (Rodríguez Schettino, 1999b). **Reproduction:** Most reproductive activity in April–July, when juveniles and subadults are rare or not observed, females gravid, egg lengths 14.4–18.1 mm in April, 13.5–16.8 mm in July (Martínez Reyes, 1994); females begin courtship with quick headbobs to which the male responds in like fashion, male then pursues female, positions himself, bites her nape until she stops moving, moves his body in wavelike fashion from front to back, placing his tail perpendicular to and below the female, and initiates copulation, separation rapid, after which the female drags her cloaca on the ground, females with 1–4 ovarian follicles, ovulation when follicles 11.5–13.5 mm, 66.6% female with 2 eggs, 33.3% with three, oviductal eggs 13.5–18.0 mm, 2 preserved eggs originally laid in August in a flowerpot 18.0 × 10.9 mm and 16.3 × 9.4 mm, one egg immediately before eclosion in September 19.0 ×

13.0 mm (Rodríguez Schettino, 1999b). **Size:** Male SVL to 121 mm (Schwartz, 1959e), female SVL to 106.3 mm (Rodríguez Schettino, 1999b). **Thermal Biology:** Heliothermic (Rodríguez Schettino, 1999b).

Conservation Status: Entire species not assessed, but *L. c. pambasileus* and *L. c. minor* are "vulnerable" (L. Rodríguez Schettino and A. Chamizo Lara in Rodríguez Schettino, 1999a).

Leiocephalus cuneus Etheridge, 1964

Distribution: Fossil remains known from Anguilla, Barbuda, Antigua, and Guadeloupe.

Habitat: Probably coastal (Breuil, 2002b).

Natural History: Agile, bold, active, takes food scraps, on the ground, where they eat the eggs of other lizards and of turtles (Du Tertre, 1667 in Breuil, 2002b).

Conservation Status: Not assessed, but almost certainly extinct (Pregill, 1992; Breuil, 2002b).

Leiocephalus endomychus Schwartz, 1967

Distribution: Hispaniola, Plateau Central (Haiti).

Habitat: Hillside above a lake, on the rocky wall of a roadcut, in a shaded cavity in a wall 3.5 m above a road (Schwartz, 1967).

Size: Male SVL to 69 mm, female to 59 mm (Schwartz, 1967d).

Conservation Status: Not assessed.

Leiocephalus eremitus Cope, 1868

Distribution: Navassa Island.

Habitat and Natural History: Unknown.

Size: Female SVL to 64 mm (only known specimen; Schwartz, 1967d).

Conservation Status: Listed as "extinct" in the IUCN Red List (World Conservation Monitoring Centre, 1996c).

Additional Pertinent Literature: Powell (1999e).

Leiocephalus greenwayi Barbour and Shreve, 1935

Distribution: Bahama Islands, East Plana Cay.

Habitat: Xerophilic (Schwartz and Henderson, 1991); abundant in rocky and sandy areas, more common in interior scrub than on beaches (Clough and Fulk, 1971; Schwartz and Henderson, 1991).

Activity: Diurnal (Clough and Fulk, 1971). **Foraging and Diet:** Food includes anoline lizards (*Anolis brunneus*), hemipterans, beetles (adults and larvae), lepidopterans (adults and larvae), flies, roaches, ants and other hymenopterans, spiders, and decapods (Schoener et al., 1982). **Size:** Male SVL to 75 mm, female SVL to 71 mm (Schwartz and Henderson, 1991).

Conservation Status: Not assessed.

Leiocephalus herminieri Duméril and Bibron, 1837

Distribution: Martinique (Pregill, 1992; Breuil, 2003, 2009).

Habitat and Natural History: Unknown.

Size: Male SVL to 126 mm (Schwartz and Henderson, 1991), female SVL to 139 mm (Boulenger, 1885).

Conservation Status: Not collected since 1837 (Honegger, 1981); "extinct" in the IUCN Red List (World Conservation Monitoring Centre, 1996d).

Leiocephalus inaguae Cochran, 1931

Distribution: Bahama Islands, Great Inagua Island.

Habitat: Xerophilic (Schwartz and Henderson, 1991); entire coastal area and extending far inland, where it basks on rocks, logs, or other vantage points, in holes among rocks, under boards, in hollow decayed palms, and even in empty conch shells (Noble and Klingel, 1932); in open and exposed areas (not woods), in and about building materials, large rocks, especially near the shoreline in association with driftwood (Schwartz, 1967e).

Activity: Usually appears at 0800–1000 h (earliest at 0745 h at temperature of 25.5 °C), later when windy or cloudy (Noble and Klingel, 1932). **Behavior:** Aggression "is largely bluff," involving lateral presentation and compression (exposing ventrolateral pattern), elevation on all four limbs, flatten throats to "form feeble fans or dewlaps" (Noble and Bradley, 1933). **Foraging and Diet:** Will attack and devour other lizards, including *Ameiva maynardii* (Noble and Klingel, 1932); food includes buds, flowers, seeds, small fruits, thrips, termites, webspinners, orthopterans, hemipterans, beetles (adults and larvae), neuropterans, lepidopteran larvae, flies (adults and larvae), ants and other hymenopterans, and spiders (Schoener et al., 1982). **Movements:** Relocated lizards did not remain at sites of release, one had moved ~400 m in 2 h, another the same distance in 2 weeks, but none returned to original sites of capture (Noble and Klingel, 1932). **Reproduction:** Egg laying presumably in autumn, preserved eggs measured 24 × 11 mm, 22 × 10 mm, and 19 × 10 mm (Noble and Klingel, 1932); courtship (longest 5 min) and coitus (longest 45 sec) very quick, with few preliminaries (one male rubbed his cloaca on the substrate), male seizes female by the back of her neck (may be repeated), may bob while holding the female, nape hold continues through copulation, female elevated hindquarters and swayed, presumably facilitating intromission, a homosexual copulation observed (Noble and Bradley, 1933). **Size:** Male SVL to 90 mm, female SVL to 74 mm (Schwartz, 1967e).

Conservation Status: Not assessed.

Leiocephalus loxogrammus Cope, 1887

Distribution: Bahama Islands, where two currently recognized subspecies occur on San Salvador Island (*L. l. parnelli*) and on Rum Cay and Conception Island (*L. l. loxogrammus*).

Habitat: Xerophilic to semimesophilic (Schwartz and Henderson, 1991); occurs in sandy areas behind and near beaches, in coastal dune vegetation, under rocks and among ruins in low coppice, occasionally in scrub and shrubby growth, rarely in palm groves or other well-shaded situations (Schwartz and Henderson, 1991).

Foraging and Diet: Food includes buds, flowers, seeds, small fruits, thrips, earwigs, orthopterans, hemipterans, beetles (adults and larvae), neuropterans, lepidopterans (adults and larvae), flies, roaches, ants and other hymenopterans, spiders, and isopods (Schoener et al., 1982). **Size:** Male SVL to 90 mm, female SVL to 70 mm (Schwartz and Henderson, 1991).

Conservation Status: Not assessed.

Leiocephalus lunatus Cochran, 1934

Distribution: Hispaniola, where six currently recognized subspecies occur along the southeastern coast of the Dominican Republic and on islas Saona, Catalina, and Catalinita.

Habitat: Xerophilic to occasionally semimesophilic (Schwartz and Henderson, 1991); in coastal zones, on limestone formations, mostly adjacent to the coastline, on beaches with coral rubble, *Yucca*, and *Coccoloba* (Mertens, 1939); only from the edge of the sea to adjacent stands of *Coccoloba uvifera*, where lizards hide in driftwood and an empty coconut shell (Böker, 1939); primarily a beach lizard, often restricted to a narrow zone along the ocean front, in semiarid cultivated land at the foot of limestone bluffs on and among large concrete slabs and trash, *Cocos* groves, garbage dumps, shaded *Coccoloba* stands, wet rocky woods (one specimen), grassy savannas, on and about rocks on a semimesic, forested limestone ridge, semiarid scrub grading into mangrove flats and adjacent to shaded xeric woods on Isla Saona (Schwartz, 1967d; Schwartz and Henderson, 1991).

Activity: May not become active until midafternoon, probably avoids the heat of the day (Schwartz, 1967d). **Behavior:** Holds tail in "watch-spring" curl over back when running, very conspicuous, reveals lizards that are otherwise very well camouflaged by dorsal color and pattern, seeks shelter in *Cocos* shells (Mertens, 1939); tail curling used in antagonistic male-male interactions (M. E. Gifford, pers. comm.). **Home Range and Territoriality:** Highly territorial (Mertens, 1939). **Predation:** *Uromacer catesbyi* (Henderson et al., 1987c). **Reproduction:** Mating on 6 March (Mertens, 1939); mean clutch size 1.60 ± 0.52, egg length 13.45 mm, egg volume 92.45 mm^3 (Gifford and Powell, 2007). **Size:** Male SVL to 67 mm, female SVL to 60 mm (Schwartz, 1967d); estimated female SVL at maturity 46.5 mm (Gifford and Powell, 2007). **Tail Autotomy:** 8 of 14 adults with regenerated tails, not significantly related to size or to sex (RP).

Conservation Status: Not assessed.

Leiocephalus macropus Cope, 1863

Distribution: Cuba, where eleven currently recognized subspecies occur islandwide, but in disjunct areas that include the extreme eastern coast and adjacent lower and interior montane slopes, northern Holguín Province, northern Camagüey coast, interior Ciego de Ávila Province, northern Matanzas coast and interior, Sierra de Escambray, Península de Guanahacabibes, Sierra del Rosario, Cayo Damas of the southeastern coast, and Isla de la Juventud.

Habitat: Mesophilic, occasionally in xeric areas within otherwise mesic situations (Schwartz and Henderson, 1991); in broadleaf forests, wooded and rocky areas, on the rocky floor of broadleaf woods, in open *Eugenia* woods behind a beach, in scrub growth and among *Cocos* palms, on bare "dog-tooth" limestone with some shade, open forest, meadows, and pastures, *Coccoloba* forests back of beaches, on moist mud of mangrove forest, among driftwood and brush along a river bottom, on rocks in an open road cut with scattered shrubs and grassy tufts, in an abandoned cornfield between old roots and cornstalks (Hardy, 1958a; Garrido, 1979; Schwartz and Garrido, 1967; Schwartz and

Henderson, 1991; Rodríguez Schettino, 1999b); adjacent to a small stream, in rainforest and humid and shaded forests, suns on rocks and logs, capable of climbing to ~1 m on erect trunks (Garrido, 1979); on the Guanahacabibes Peninsula, in more shaded habitats than sympatric *L. carinatus* and *L. stictigaster* (Garrido and Schwartz, 1968b); female under a bush at night, presumably had not retired to its "customary nocturnal retreat" and was sleeping in the open (Schwartz and Garrido, 1967); all of 5 sampled sites in dry, open areas near the coast, where lizards are abundant (Viña Dávila et al., 1984); in the Biosphere Reserve "Sierra del Rosario," associated with prominent rock outcrops in a ravine in evergreen forest, preferentially on rocks but also on trunks and branches of bushes (Martínez Reyes, 1995); sandy or stony ground covered by leaf litter is the most favored habitat, at night or on cool mornings, under chunks of coral, rocks, or logs, but only in areas shaded by day (Rodríguez Schettino, 1999b); in mesophyllous evergreen and semideciduous forest, coastal and subcoastal microphyllous forest, and semidesert thorny scrubwoodland (Rodríguez Schettino et al., 1999).

Activity: Diurnal (Rodríguez Schettino et al., 1999); bimodal, with a major peak at 1000–1100 h and a minor peak at 1300–1400 h (Sampedro Marín et al., 1979; Berovides Álvarez and Sampedro Marín, 1980); most active 1000–1100 h, takes refuge during intensely hot times of day, active only after 1000 h during the wet season when mean air and substrate temperatures 29.0 °C and 27.4 °C (Rodríguez Schettino, 1999b). **Behavior:** Flees rapidly, using crevices in rocks as refuges or burying itself under leaf litter or in sand or loose soil (Rodríguez Schettino, 1999b). **Competition:** Perch (insolation) overlap with sympatric *L. raviceps* 0.639, activity overlap 0.936, dietary overlap 0.966 (Berovides Álvarez and Sampedro Marín, 1980). **Foraging and Diet:** At two localities in Guantánamo Province, plants constituted 82.4% of material ingested by males, 57.7% of that ingested by females, animal prey included ants and other hymenopterans (89.5% of all prey), beetles, lepidopterans, termites, hemipterans, roach, mite, scorpion, spider, centipedes, and snail (Sampedro Marín et al., 1979); in August in Guantánamo Province, 90.00% ants, 2.31% caterpillars, 4.62% beetles, and 3.08% other arthropodan orders (Berovides Álvarez and Sampedro Marín, 1980); omnivorous, forages in leaf litter, feeds mainly on cockroaches, insect larvae, moths and nocturnal butterflies sheltered under dead leaves, a sample contained large amounts of vegetable material (mainly flowers), along with arthropods (mostly ants, beetles, lepidopterans, hemipterans, termites, and cockroaches, but also mites, scorpions, spiders, centipedes) and mollusks (Rodríguez Schettino, 1999b); *Anolis jubar* (Fong Grillo and del Castillo, 2002; Martínez Reyes and Moreno García, 2003). **Parasites:** Ticks *Amblyomma torrei* (V. Černý, 1969; Coy Otero, 1999; Guglielmone et al., 2003); nematodes *Cyrtosomum longicaudatum*, *Physaloptera squamatae* (Coy Otero and Baruš, 1979b; Coy Otero and Lorenzo Hernández, 1982; Coy Otero, 1999); *Oswaldocruzia lenteixeirai*, larval *Abbreviata* sp., larval *Porrocaecum* sp. (Baruš and Coy Otero, 1969a; Coy Otero, 1970; Coy Otero and Baruš, 1979b; Coy Otero and Lorenzo Hernández, 1982; Coy Otero, 1999); and *Parapharyngodon cubensis* (Coy Otero and Baruš, 1973b, 1979b; Coy Otero and Lorenzo Hernández, 1982; Coy Otero, 1999). **Reproduction:** Gravid females in February, March, and May, but not September, ovarian follicles 3.1–10.2 mm long, oviductal eggs 12.0–20.3 mm, all females had 2 oviductal eggs (M. Martínez Reyes in Rodríguez Schettino, 1999b); mean clutch size 1.75 ± 0.25 (range 1–2; G. R. Smith and Nickel, 2002); clutch size 2 (Hedges, in prep.). **Size:** Male SVL to 95 mm (Schwartz and Henderson, 1991), female SVL to 75 mm (Schwartz and Garrido, 1967). **Thermal Biology:** In shade 90.5% of time (Sampedro Marín et al., 1979); in full sun 9.52% of time, in shade 71.43%, in partial shade 19.05% (Berovides Álvarez and Sampedro Marín, 1980).

Conservation Status: Entire species not assessed, but *L. m. hyacinthurus* considered "endangered" and *L. m. lenticulatus*, *L. r. hoplites*, *L. m. felinoi*, and *L. m. torrei* "vulnerable" (L. Rodríguez Schettino and A. Chamizo Lara in Rodríguez Schettino, 1999a).

Leiocephalus melanochlorus Cope, 1863

Distribution: Hispaniola, where two currently recognized subspecies occur on the Tiburon Peninsula and on Île-à-Vache (Haiti); elevational range, sea level to ~1,650 m.

Habitat: Mesophilic to semixerophilic (Schwartz and Henderson, 1991); in open areas (including beaches, field margins, mangrove edges, margins of broadleaf woods), upland pine forest (Schwartz and Henderson, 1991); shade loving, may climb to ~1.8 m above the ground on trees, shrubs, or posts (Schwartz, 1965c); common at drier sites on the plateau, particularly in areas modified by agriculture, along paths, and in pastures (may have moved into forested areas along deforested corridors to newly created gardens), also among bracken ferns and scrubby pines along a floodplain (Franz and Cordier, 1986).

Predation: *Uromacer catesbyi*, *U. frenatus* (Henderson et al., 1987c). **Reproduction:** Clutch size 4 (Hedges, in prep). **Size:** Male SVL to 130 mm, female SVL to 102 mm (Schwartz, 1965c).

Conservation Status: Not assessed.

Leiocephalus onaneyi Garrido, 1973

Distribution: Cuba, from the vicinity of the type locality on top of Loma de Mocambo, Guantánamo Province.

Habitat: Xerophilic (Schwartz and Henderson, 1991); in karst topography ("dogtooth limestone") in the rain shadow of the Sierra de Baracoa, with annual average rainfall <900 mm and natural vegetation with a predominance of spiny bushes with small leaves, only on the top of the hill, not on slopes, which have different vegetation (Garrido, 1973a); in semidesert thorny scrubwoodland (Rodríguez Schettino et al., 1999).

Activity: Diurnal (Rodríguez Schettino et al., 1999). **Size:** Male SVL to 68 mm, female SVL to 72 mm (only three specimens known; Garrido, 1973a).

Conservation Status: Considered "critically endangered" (Rodríguez Schettino and Chamizo Lara, 1998a) or possibly extinct (L. Rodríguez Schettino and A. Chamizo Lara in Rodríguez Schettino, 1999a).

Leiocephalus personatus Cope, 1863

Distribution: Hispaniola, where 12 currently recognized subspecies occur on the Tiburon Peninsula (including Grosse Caye) of Haiti, northern Haiti (into the Massif du Nord), east into the Dominican Republic, where essentially countrywide, except at very high elevations, the Valle de Neiba, and the Barahona Peninsula, also on the Cayos Siete Hermanos and Isla Cabras off Monte Cristi; elevational distribution, sea level to ~625 m.

Habitat: Mesophilic, if encountered in xeric areas, usually in the most shaded situations (Schwartz and Henderson, 1991); in coastal zones, scrub and cactus savannas, and pine forests (basking in morning sun on stumps 1 m high), along a garden fence in a city near the coast in dense herbaceous growth, basking on boards, tree roots, and chunks of coral, on coastal rocks and on beaches under *Cocos* debris, in pineapple plantations (Mertens, 1939); in the *Coccoloba* zone on beaches (H. Böker in Mertens, 1939); in cities in abandoned lots and on sidewalks, in *Cocos* groves, on rocks near river valleys, along beach dunes and among sea-edge driftwood, in pineapple fields, on a rocky bluff near mangrove flats, in *Coccoloba* stands, guava (*Psidium*) scrub, cactus woods, *Opuntia* thickets, sisal fields, *Acacia* and *Euphorbia lactea* woods, coffee plantation, pinewoods, broadleaf woods, grassy fields, and under rocks in an upland pasture (Schwartz, 1967d; Schwartz and Henderson, 1991; Powell, 2002a; Gifford et al., 2008); thrives in Cayes (Haiti) in yards of small farms, with the concentration probably a reflection of disturbed habitats (Olson, 1995); basks on roadside berms (Powell, 2003b); on Cayo Monte Chico (Cayos Siete Hermanos), in all available habitats except beaches (Burns et al., 1992).

Activity: Diurnal, with an activity period much longer than that of *Ameiva chrysolaema* (Mertens, 1939); often not active until the heat of the day has passed, rarely seen active in upland areas due to inclement cool weather, not out on cloudy days in some areas (Schwartz, 1967d). **Behavior:** Aggression "is largely bluff," involving lateral presentation and compression (exposing ventrolateral pattern), elevation on all 4 limbs, flattens throat to "form feeble fans or dewlaps" (Noble and Bradley, 1933); tail not tightly coiled, but may be waved as a possible predator deterrent (Mertens, 1939); aposematic tongue extension (Mertens, 1946); wary, quickly seeks shelter under rocks, logs, or flotsam, one in an old tennis shoe (Powell, 2002a); mean approach (115.1 ± 21.88–481.3 ± 94.42 cm) and flight distances (33.9 ± 6.76–162.5 ± 66.27 cm) in 4 populations were greater in more open habitats (Gifford et al., 2008); tail curling used in antagonistic male-male interactions (M. E. Gifford, pers. comm.). **Foraging and Diet:** Stomachs of 5 individuals from Cayo Monte Chico (Cayos Siete Hermanos) contained spiders, beetles (adults and larvae), a roach, flies, a hemipteran, ants (most important), lepidopterans (adult and larvae), some plant material, and grit (Burns et al., 1992). **Home Range and Territoriality:** Highly territorial (Mertens, 1939). **Longevity:** In captivity, 3 years, 4 months (J. Grizzell in Snider and Bowler, 1992). **Parasites:** Apicomplexan *Eimeria leiocephali* (Huntington and Cisper, 1994). **Performance:** Mean sprint speeds (152.8 ± 4.87–229.5 ± 7.49 cm/sec) were greater in populations occupying more open habitats (Gifford et al., 2008); polymorphic labial and tail color are honest signals of male performance (and presumably quality) but aspects of color variables that predict performance differ between populations in different habitats, more intense labial coloration significantly correlated with absolute and relative sprinting performance in a population occupying open habitat, low-intensity red-orange labials with high reflectance correlated with absolute and relative sprinting performance, and high-intensity, more yellowish tail color with less red/orange and low reflectance correlated with bite force in a population occupying a more densely vegetated habitat (M. E. Gifford, pers. comm.). **Predation:** *Hypsirhynchus parvifrons* (Henderson et al., 1987a); *Uromacer frenatus* (Henderson et al., 1987c). **Reproduction:** In captivity, courtship (longest 5 min) and coitus (longest 45 sec) very quick, with few preliminaries (one male rubbed his cloaca on the substrate), male seizes female by the back of her neck (may be repeated), may bob while holding the female, nape hold continues through copulation, homosexual courtship but not copulation observed (Noble and Bradley,

1933); mean clutch size 2.00 ± 0.59, mean egg length 15.49 ± 1.35 mm, mean egg volume 113.38 ± 12.54 mm^3 (Gifford and Powell, 2007). **Size:** Male SVL to 86 mm, female SVL to 63 mm (Schwartz, 1967d); estimated female SVL at maturity 42.5 mm (Gifford and Powell, 2007). **Tail Autotomy:** Tail-break frequencies (0.50–0.68) did not differ by sex or habitat (Gifford et al., 2008).

Conservation Status: Not assessed.

Leiocephalus pratensis Cochran, 1928

Distribution: Hispaniola, where two currently recognized subspecies occur in northern Haiti and on Île-à-Cabrit.

Habitat: Xerophilic to mesophilic (Schwartz and Henderson, 1991); grass and scrub along banks of small gullies (G. S. Miller in Cochran, 1928).

Behavior: Does not elevate or curl its tail while running (G. S. Miller in Cochran, 1928); seeks refuge in dead grass tussocks (R. Thomas in Schwartz, 1968e). **Size:** Male SVL to 64 mm, female SVL to 55 mm (Schwartz, 1968e).

Conservation Status: Unknown.

Leiocephalus psammodromus Barbour, 1916

Distribution: Turks and Caicos Islands, where six currently recognized subspecies occur throughout the archipelago.

Habitat: Xerophilic (Schwartz and Henderson, 1991); along rocky and sometimes steep coastal areas and in coastal scrub (Schwartz, 1967e); most often on litter (38.6% of observations) followed by trees (22.8%), bare sand (18.5%), and rocks (11.4%), more frequently in full shade (43.0%) than in sun/shade mosaic (38.5%) or full sun (18.5%; G. R. Smith, 1995).

Activity: Bimodal, with peaks in late morning (0900–1000 h) and afternoon (~1600 h), few active lizards at midday (G. R. Smith, 1994). **Foraging and Diet:** Male ate a dead *Anolis scriptus* (Iverson and Smith, 2006); feeds on fruits of Turk's Cap Cactus (*Melocactus azureus*; J. Burgess in Sánchez Muñoz, 2008); opportunistic conspecific necrophagy, an adult female feeding on the carcass of a road-killed adult male (Reynolds, 2009). **Predation:** Cats (Iverson, 1978). **Reproduction:** Seasonal with egg laying beginning late in the dry season (late April) and extending well into the wet season (September), fat body mass decreases in both sexes with the onset of reproduction, hatching late June–December, with some individuals probably producing multiple clutches, mean clutch size 1.71 ± 0.08 (range 1–3), mean preserved egg size 19.9 ± 0.91 × 9.2 ± 0.57 mm, neither clutch nor egg size related to female size, one-egg clutch buried in 5 cm sand, incubated in lab for 54 days after discovery, hatchling SVL = 33.1 mm (Smith and Iverson, 1993). **Size:** Male SVL to 110 mm, female SVL to 99 mm (G. R. Smith, 1992); males mature at 75–80 mm SVL (smallest 76.6 mm SVL) and females at 65–70 mm SVL (smallest 65.5 mm SVL) at about 1.5–2.0 years of age (G. R. Smith and Iverson, 1993). **Thermal Biology:** Most observations in early morning and late afternoon were in sun/shade mosaics, few in full sun during midday (G. R. Smith, 1994).

Conservation Status: Not assessed.

Leiocephalus punctatus Cochran, 1931

Distribution: Bahama Islands, Crooked-Acklins Bank, Samana Cay.

Habitat: Xerophilic (Schwartz and Henderson, 1991); 14% and 36% of lizards in arboreal situations on Crooked Island, 58% arboreal on Acklin's Island (Schoener, 1975).

Behavior: Larger lizards respond mostly to individuals in the same size class, smaller lizards more to larger individuals (Schoener, 1975). **Foraging and Diet:** Extent of herbivory generally higher on smaller islands, food includes buds, flowers, seeds, small fruits, anoline lizards (*Anolis brunneus*), earwigs, orthopterans, hemipterans, beetles (adults and larvae), neuropterans, lepidopterans (adults and larvae), flies (adults and larvae), roaches, ants and other hymenopterans, centipedes, spiders, pseudoscorpions, amphipods, isopods, and decapods, will readily accept bread, peanut butter, jelly, crackers, and cheese (Schoener et al., 1982). **Size:** Male SVL to 80 mm, female SVL to 61 mm (Schwartz and Henderson, 1991).

Conservation Status: Not assessed.

Leiocephalus raviceps Cope, 1863

Distribution: Cuba, where 5 currently recognized subspecies occupy disjunct areas along the southeastern coast and adjacent lower montane slopes in Pinar del Río, Matanzas, Villa Clara, and Holguín provinces.

Habitat: Xerophilic (Schwartz and Henderson, 1991); often very local, apparently restricted to areas with loose sand, into which lizards burrow when pursued, a typical plant association includes *Opuntia* on open and very dry hillsides, on a beach with *Ipomoea pescaprae* and scattered grass tufts, runs about on loose shale fragments on dry and sunny road cuts, juveniles sporadically in rocky areas (Schwartz, 1960c; Buide, 1966; Schwartz and Garrido, 1968a; Schwartz and Henderson, 1991; Rodríguez Schettino, 1999b); enters wooded areas only when extremely hard pressed, uses sandy construction areas for dispersal (Buide, 1966); in Pinar del Río Province (*L. r. jaumei*), inhabits only the "bleakest" portion of a gray sandy area (4 × 11 km) surrounded by yellow clays, inhabits an area in Holquín Province heavily disturbed by the removal of sand and has been "forced" to live in roadside ditches that are completely covered with grass (Schwartz and Garrido, 1968a); digs burrows in sand or seeks retreats under rocks, logs, and debris in human-modified habitats, 45.2% in sun, rest in shade (Rodríguez Schettino, 1999b); in coastal and subcoastal microphyllous forest and semidesert thorny scrubwoodland (Rodríguez Schettino et al., 1999).

Activity: In March, begins to emerge from burrows under fallen palm logs at 1000 h, most active ~1100 h and remains near burrows or fallen trunks above burrow mouths (Schwartz and Garrido, 1968a); bimodal, with a major peak at 1000–1100 h and a minor peak at 1300–1400 h (Sampedro Marín et al., 1979; Berovides Álvarez and Sampedro Marín, 1980); diurnal, emerges after 1000 h and retreats to shelters at sunset, usually active during the hottest parts of the day, more individuals out at 1000–1100 h and 1300–1400 h than at other times in August (Rodríguez Schettino et al., 1999). **Behavior:** Seeks refuge in land crab burrows as well as actively burrowing in sand (Schwartz, 1960c); does not carry its tail over its back or to one side (Buide, 1966); when disturbed, runs with great speed to burrow mouth, remaining 5–7.5 cm outside in an "attitude of observation" (Schwartz and Garrido, 1968a). **Competition:** Perch (insolation) overlap with sympatric *L. macropus* 0.639, activity overlap 0.936, dietary overlap 0.966 (Berovides Álvarez and Sampedro Marín, 1980). **Foraging and Diet:** At two localities in Guantánamo Province, plants constituted 38.9% of material ingested by males, 15.4% of that ingested by females, animal prey included ants and other hymenopterans (85.8% of all prey), beetles, lepidopterans, hemipterans, orthopterans, scorpions, spiders, and a snail (Sampedro Marín et al., 1979); in August in Guantánamo Province, 87.21% ants, 1.83% caterpillars, 7.99% beetles, and 2.97% other arthropodan orders (Berovides Álvarez and Sampedro Marín, 1980); omnivorous, consumes large amounts of vegetable matter in addition to arthropods (ants, beetles, lepidopterans, hemipterans, scorpions, spiders) and mollusks (Rodríguez Schettino et al., 1999). **Parasites:** Nematodes *Cyrtosomum longicaudatum*, *Physaloptera squamatae*, *Skrjabinoptera phrynosoma*, and larval *Porrocaecum* sp. (Coy Otero and Baruš, 1979b; Coy Otero and Lorenzo Hernández, 1982; Coy Otero, 1999). **Reproduction:** Eggs ready for deposition on 6 March, no juveniles seen, pair in copulation (Schwartz and Garrido, 1968a); egg laid in June 18.0 × 8.0 mm (Rodríguez Schettino et al., 1999); mean clutch size 1.67 ± 0.21 (range 1–2; G. R. Smith and Nickel, 2002). **Size:** Male SVL to 74.1 mm, female SVL to 62.7 mm (Rodríguez Schettino, 1999b). **Thermal Biology:** In shade 58.8% of time (Sampedro Marín et al., 1979); in full sun 41.18% of time, in shade 35.29%, in partial shade 23.53% (Berovides Álvarez and Sampedro Marín, 1980); heliothermic (Rodríguez Schettino, 1999b).

Conservation Status: Entire species not assessed, but *L. r. klinikowskii* is considered "endangered" (Rodríguez Schettino and Chamizo Lara, 1998a) or "critically endangered" and *L. r. delavarai* and *L. r. jaumei* "vulnerable" (L. Rodríguez Schettino and A. Chamizo Lara in Rodríguez Schettino, 1999a).

Leiocephalus rhutidira Schwartz, 1979

Distribution: Hispaniola, northwestern Haiti in the vicinity of the type locality in the Département de l'Artibonite.

Habitat: Presumably xerophilic (Schwartz and Henderson, 1991); at bases of hills in cactus-*Acacia*-*Agave* associations (Schwartz, 1979f).

Size: Male SVL to 66 mm, female SVL to 56 mm (Schwartz, 1979f).

Conservation Status: Not assessed.

Leiocephalus schreibersii Gravenhorst, 1837

Distribution: Hispaniola, where *L. s. schreibersii* occurs in extreme northwestern Haiti south along the Golfe de la Gonâve, to the northern shore of the base of the Tiburon Peninsula, east through the Plaine de Cul de Sac, Valle de Neiba, and into the Llanos de Azua, also in the Valle de Cibao in the extreme northwestern Dominican Republic and the Cayos Siete Hermanos, *L. s. nesomorus* occurs on Île de la Tortue.

Habitat: Xerophilic (Schwartz and Henderson, 1991); in coastal zones and cactus savannas, in brush and on piles of boards, in gardens and around shacks (Mertens, 1939); often very abundant in cactus-*Acacia* situations and along margins of sisal plantations and other (often irrigated) cultivations in xeric regions, in urban habitats on bleak lawns and about rock piles, abandoned walls, and concrete blocks and slabs, on Île de la Tortue, in more mesic and wooded areas (Schwartz, 1968e); adults of both sexes and juveniles use the same microhabitats and are closely associated (Jenssen et al., 1989); uses moderately sized rocks for surveillance and basking sites and rock crevices for nocturnal retreats and diurnal refugia from predators (Marcellini and Jenssen, 1989); on the Cayos Siete Hermanos, in all available habitats including deeply shaded woody clumps (*Acacia*) to exposed areas in *Opuntia* scrub, forays into open areas (beaches, human clearings) were of short duration (Burns et al., 1992); in and around a hotel and on adjacent walls in Barahona, refuges included burrows under logs or concrete (sometimes shared with crabs), drain pipes, and other human debris, lizards on the nearby beach used crab burrows in sand (Schreiber et al., 1993); often in more open, sparsely vegetated situations than sympatric *L. semilineatus* (Nelson et al., 2001).

Activity: Diurnal, with an activity period much longer than that of *Ameiva chrysolaema*, active until shortly before sunset (Mertens, 1939); "basked in the sun from sunrise to sunset" (Regal, 1978); few lizards active 0700–0759 h (first hour after sunrise), most emerged 0800–0959 h, begin preparation to enter burrows at 1700 h, by 1800 h, most individuals in burrows (Marcellini and Jenssen, 1989); lizards in a hotel courtyard active 0900–1600 h, occasionally as early as 0826 h and as late as 1630 h, lizards on a nearby beach emerged and retreated into burrows ~1 h earlier (Schreiber et al., 1993). **Behavior:** Agonistic displays involve lateral compression, elevation on all 4 limbs, protruded tongue, while rocking back and forth, before attempting to seize the rival male in the neck region (Noble and Teale, 1930); in captivity, head-nodding by one individual triggered similar behavior in others, aggression "is largely bluff," involving lateral presentation and compression (exposing ventrolateral pattern), elevation on all 4 limbs, flattens throat to "form feeble fans or dewlaps" (Noble and Bradley, 1933); tail elevated when running but not tightly curled and waved as in *L. lunatus* and *L. personatus*, can dig quickly into loose soil (Mertens, 1939); aposematic tongue extension (Mertens, 1946); movements (9.6/h) consist of short, rapid dashes (always 2 sec in duration) to catch insects or chase conspecifics, immobile (but alert) >99% of day, much less active than associated *Ameiva chrysolaema* (Regal, 1978); mean number of moves/min and distance/move for males 0.3 ± 0.06 and 0.7 ± 0.10 m, for females 0.3 ± 0.09 and 0.6 ± 0.05 m, juveniles occasionally perched on adult females, most agonistic male-male encounters along edges of home ranges, usually limited to ritualized aggressive signals and only once led to biting (Jenssen et al., 1989); proficient at avoiding capture (Marcellini and Jenssen, 1989; Schreiber et al., 1993); learn avoidance behavior quickly, when faced with a novel predator and a conditioned stimulus (human with a pole and noose), 80% of lizards uncatchable by the second trial, all by the third trial, with 75% establishing flight distances >10 m, females become unapproachable more quickly than males (Marcellini and Jenssen, 1991); male agonistic displays toward other males or humans begin with a lateral "face off" 0.5–1.0 m apart and involve 2 types of headbobs, one a large-amplitude, slow, alternating downward and upward motion of the head, the second with very rapid, low amplitude oscillations, displays begin with heads held high, 3–6 large bobs followed by 3–8 small bobs in rapid succession, 2 additional large bobs, and a variable number of small bobs, fights always initiated by the resident involved charges and collisions, once causing the recipient to roll across the ground, adult-juvenile aggression involved adults chasing juveniles (Schreiber et al., 1993); on Cayo Muertos (Cayos Siete Hermanos), climbs into *Acacia* to heights of 2 m, when disturbed, would "fall" to lower branches or to the ground, "arboreal" activity on other cays restricted to heights of 20 cm (never >40 cm; Burns et al., 1992); primary and secondary response distances 385 ± 105 cm and 150 cm, more difficult to approach closely than sympatric *L. semilineatus* and runs farther to find shelter (Nelson et al., 2001); displays may include (often in combinations) large headbobs (usually agonistic), small headbobs, "quick" headbobs (usually in the context of courtship), throat inflation, lateral compression, tongue protrusion, elevated tail, tail wagging, and stooping or bowing (Ackermann, 2006); rarely if ever "curls" its tail (Ackermann, 2006; M. E. Gifford, pers. comm.; RP). **Foraging and Diet:** Eats insects and plant material (Mertens, 1939); makes short dashes for prey, so quick that they are difficult to time (Regal, 1978); sit-and-wait forager, uses elevated perches to survey surroundings, 4 incidents of active foraging involved series of quick movements, scanning the immediate area, and rooting in sand or loose soil, when feeding on ants concentrated around a dead beetle or roach, lizard ate the dead insect after eating the ants, stomach contents of 17 individuals contained spiders and other arachnids, beetles, earwigs, roaches, flies, hemipterans, ants (present in greatest numbers and in the most stomachs), lepidopterans, orthopterans (present in greatest volume), shed skin, and plant material, mostly of marginal nutritional value (twigs and pieces of leaves), suggesting that herbivory may be facultative or seasonal (Schreiber et al., 1993); two-thirds of adults in a population near Monte Cristi readily attacked and consumed *A. distichus*, juveniles cannibalized, but not in areas of home range overlap with noncannibal males and females, cannibalistic females (usually feeding on insects) receive a large energy "boost" from ingesting young lizards (Jenssen et al., 1989); stomachs of 22 individuals from the Cayos Siete Hermanos contained spiders and other arachnids, beetles (adults and larvae), cockroaches, flies, hemipterans, ants (most important) and other hymenopterans, termites, lepidopterans (adult and larvae), neuropterans, orthopterans, isopods, lizard skin, nematodes, some plant material, and grit (Burns et al., 1992); stomach contents of lizards from the northern shore of the Bahía de Neiba included spiders, beetles (adults and larvae), roaches, flies (adults and larvae), hemipterans, ants (most important), other hymenopterans, lepidopterans (adults and larvae), grasshoppers, an arthopodan egg case, flowers, fruits and seeds, and lizard skin (Nelson et al., 2001). **Home Range and Territoriality:** Highly territorial (Mertens, 1939); mean home range for adult males 78.5 ± 6.0 m^2, for adult females 42.5 ± 5.0 m^2, and juveniles 6.3 ± 0.7 m^2, with those of males overlapping with a mean of 1.5 other males (mean overlap 35%), 1.5 females (35%), and 2.0 juveniles (8%), those of females with 1.6 males (88%), 0.9 females (22%), and 1.7 juveniles (11%), and those of juveniles with 1.1 males (100%), 1.0 females (96%), and 1.2 juveniles (20%), two cannibalistic males with large home range perimeters were nonterritorial (Jenssen et al., 1989); mean male home range size in a disturbed habitat in Barahona 74.8 ± 10.9 m^2, mean female home range 30.0 ± 20.1 m^2, 2 juvenile home ranges 3 m^2 and 7 m^2, adult home range sizes not correlated with SVL, each contained one favored refuge and one or more secondary refuges, no adult home ranges overlapped, although many in peripheral contact, juvenile home ranges overlapped with those of adult males and one with that of an adult female, but neither completely within an adult home range, males actively defended territories with agonistic displays and occasional fights, introduction of a larger male into an established territory resulted in displacement of the resident after one full day of interactions, the displaced resident subsequently displaced a smaller individual from its territory (Schreiber et al., 1993); uses displays in territorial defense, but fights are not infrequent (Ackermann, 2006). **Longevity:** In captivity, 5 years, 7 months (Aquarium/Reptile Curator, Milwaukee County Zoo, in Slavens and Slavens, 2003); 9–10 years in captivity (Ackermann, 2006). **Parasites:** Nematodes *Skrjabinoptera leiocephalorum* (Greve and Powell, 1989; Powell et al., 1990b) and *Physalopteroides* sp. (Burns et al., 1992); auricular myiasis by dipteran larvae (D. D. Smith et al., 1994). **Population Size and Density:** 142.9 lizards/ha in disturbed habitat (Schreiber et al., 1993); in presumably suboptimal habitat, 5.3 ± 1.0 in 8 plots (total area = 1,152 m^2) or ~46 lizards/ha (Nelson et al., 2001). **Predation:** *Uromacer oxyrhynchus* (Henderson et al., 1987c; RP); *Haitiophis anomalus*, *Hypsirhynchus ferox*, Hispaniolan Lizard Cuckoo (*Saurothera longirostris*), American Kestrel (*Falco sparvarius*; Marcellini and Jenssen, 1991). **Reproduction:** Male approaches female and bites her neck, sometimes rubbing his thigh on her back, he may or may not hold on after intromission, an attempted homosexual mating observed in captivity (Noble and Teale, 1930); in captivity, courtship (longest 5 min) and coitus (longest 45 sec) very quick, with few preliminaries (one male rubbed his cloaca on the substrate), male seizes a female by the back of her neck (may be repeated), may bob while holding the female, nape hold continues through copulation, male may thrust his pelvis during coitus, female elevated hindquarters and swayed, presumably facilitating intromission, a homosexual copulation observed (Noble and Bradley, 1933); in June, a male made a quick dash of ~0.5 m and grabbed a female by her tail, which broke, he lunged and secured a neck hold on the female, copulation lasted only "several seconds," the male performed several bouts of headbobbing after the female moved away (Jenssen and Marcellini, 1986); apparent courtship (but without copulation) involved a male approaching a female from behind, nudging her vent, retreating and bobbing once, the female moved and bobbed, the male arched his back and executed a series of extremely high-amplitude bobs (Schreiber et al., 1993); oviductal egg length 21 mm, in June, 2 females with enlarged oviducts, suggesting recent deposition of eggs (Nelson et al., 2001); nonresponsive females in captivity react to male overtures with bows and tail waving, freshly laid eggs ~5–7 mm long, hatchlings use "egg-teeth" to slit the shell (Ackermann, 2006); mean clutch size 2.47 ± 0.72, mean egg length 16.29 ± 1.31 mm, mean egg volume 117.99 ± 10.19 mm^3 (Gifford and Powell, 2007). **Size:** Male SVL to 107 mm (Schwartz, 1968e), female SVL to 83 mm (Burns et al., 1992); estimated female SVL at maturity 51.4 mm (Gifford and Powell, 2007). **Tail Autotomy:** Male grabbed a female by the tail prior to copulation, the tail broke and the male ate the wriggling segment after mating (Jenssen and Marcellini, 1986); 4 of 11 males and 9 of 36 females had regenerated tails, with larger individuals significantly more likely to have regenerated tails (RP). **Thermal Biology:** Precise thermoregulator, but functions as thermoconformer during the heat of the day, maintains a narrow range of body temperatures (mean 36.7 ± 0.17 °C) significantly above local air and site temperatures, moves frequently between sun and shade throughout the day, 80% basking frequency at burrow emergence (0900–0959 h), declines to 35% at 1200 h, oscillates 20–40% during the remainder of the afternoon, elevated and stable body temperatures allow lizards to remain active during the entire day (Marcellini and Jenssen, 1989); spends much of the day shuttling in and out of the sun, with periods in the sun longest in the morning and progressively shorter as the day progresses, spends much of the morning basking on elevated perches with an orientation at right angles to incident sunlight and bodies elevated on extended forelimbs, mean cloacal temperature 34.90 ± 2.75 °C at mean substrate temperature 35.95 ± 4.60 °C, lower body than substrate temperatures suggest thermal conformity (Schreiber et al., 1993); mean body temperatures <36 °C (mean 7.0 ± 1.5 °C above substrate temperatures), with females higher than males and juveniles, and body temperature farther from ambient than in sympatric *L. semilineatus* (Nelson et al., 2001).

Conservation Status: Not assessed.

Additional Pertinent Literature: Schreiber et al. (1995).

Leiocephalus semilineatus Dunn, 1920

Distribution: Hispaniola, Plaine de Cul de Sac (Haiti) east through the Valle de Neiba and into the Llanos de Azua, north into the Valle de San Juan (Dominican Republic); elevational distribution -20 m to >600 m.

Habitat: Xerophilic (Schwartz and Henderson, 1991); associated with gravelly or rocky areas, rocks used for sunning sites and as retreats on overcast days, also about garbage dumps, open, arid, flats, and virtually grassless and treeless savannas (Schwartz, 1968e; Schwartz and Henderson, 1991); throughout habitats associated with a dry riverbed (xeric mixed forest, *Acacia* woods and scrub, barren riverbed, islands with sparse herbaceous and shrubby plants and some clump grasses, sparsely vegetated slopes of the riverbed), but most abundant in open areas in the riverbed, when in other habitats, nearly always along edges (e.g., along the road) or in openings (a powerline right-of-way or gaps created by grazing goats), mean microhabitat temperatures 35.5 ± 3.0 °C (29.2–40.4 °C; Powell et al., 1996b); along roads and in sparsely vegetated rocky areas, but not in rock outcroppings receiving intense sunlight, perched on rocks (61.1%), usually the highest rock when in a pile, stumps, sticks, twigs, and debris (23.9%), dirt berms (10.4%), living vegetation (3.0%), or on the ground under brush (1.5%; Nelson et al., 2001).

Activity: Unimodal, begins at 0730 h, peaks about 1000 h, and declines throughout the remainder of the day until dark, but emergence and peak activity may occur later and retreat earlier on cloudy, overcast days (Nelson et al., 2001).

Behavior: About half of individuals under cover (rocks, logs, driftwood) during the heat of the day, remainder in stationary positions on or alongside cover, presumably basking or scanning the area for food, when disturbed, one individual retreated into a burrow occupied by a large theraphosid spider without any interaction (Powell et al., 1996b); frequently responds to being approached with headbobs and/or pushups, mean primary and secondary response distances 201 ± 41 cm and 102 ± 16 cm, with most animals moving to other perches or into cover (bunch grass, dead agaves, shrubs, cracks in rocks, and burrows), often perches on rocks under which the lizards took refuge, 75.1 ± 5.9% of time perching (presumably includes "sit-and-wait" foraging), 19.3 ± 5.9% interacting with conspecifics, 5.6 ± 2.5% moving, and 0.006 ± 0.005% feeding, with time of day not affecting frequencies of behaviors, although most observations during midday hours, intraspecific interactions usually adult male-to-male or male-female, juveniles and subadults usually ignored, even when in close proximity, males engage in headbobs and pushups and chasing or stealing perches toward other males, "highly agitated" individuals may follow headbobs with a wavelike raising and lowering of body and tail, male-female interactions involve headbobs, pushups, and occasional lateral wavelike movements by females (Nelson et al., 2001). **Foraging and Diet:** Uses a sit-and-wait foraging strategy (Powell et al., 1996b; Powell, 2004b); stomach contents from 2 sites include spiders, other arachnids, centipedes, beetles (adults and larvae), roaches, flies (adults and larvae), hemipterans, ants (most important), other hymenopterans, termites, lepidopterans (adults and larvae), earwigs, grasshoppers, a snail, fruits and seeds, other plant material, and lizard skin (Nelson et al., 2001). **Parasites:** Nematodes *Skrjabinoptera leiocephalorum* (Powell et al., 1990b). **Population Size and Density:** 25.0 ± 3.7 in eight plots (total area = 1,152 m^2) or ~217 lizards/ha (Nelson et al., 2001). **Reproduction:** Clutch size 1–2, in June, mean oviductal egg length 11.0 ± 3.2 mm, mean yolked ovarian follicle length 3.4 ± 2.57 mm, and mean testis length 2.3 ± 0.4 mm (Nelson et al., 2001); mean clutch size 1.78 ± 0.42, mean egg length 12.89 ± 1.37 mm, mean egg volume 82.84 ± 13.78 mm^3 (Gifford and Powell, 2007). **Size:** Male SVL to 53 mm, female SVL to 48 mm (Schwartz, 1968e); estimated female SVL at maturity 38 mm (Gifford and Powell, 2007). **Tail Autotomy:** 12 of 18 males and 6 of 6 females had regenerated tails, larger individuals more (but not significantly) likely to have regenerated tails (RP). **Thermal Biology:** Basks in fully insolated sites in the morning, but moves to shaded sites or under cover in the afternoon, mean body temperatures >36 °C (mean 2.90 ± 0.4 °C above substrate temperatures), with those of females slightly higher than those of males and juveniles, and body temperatures closer to ambient than in sympatric *L. schreibersii* (Nelson et al., 2001).

Conservation Status: Not assessed.

Additional Pertinent Literature: Nelson and Powell (2002).

Leiocephalus stictigaster Schwartz, 1959

Distribution: Cuba, where 12 currently recognized subspecies are widely distributed, but absent from large areas of the island, known from Pinar del Río, central Villa Clara, southern Sancti Spíritus, central and northeastern Camagüey, northern Holguín, northern Las Tunas, western Santiago de Cuba, and Granma provinces, also southeastern Santiago de Cuba, western Cienfuegos, and southern Matanzas provinces, Archipiélago de Sabana-Camagüey, and Isla de la Juventud.

Habitat: Xerophilic to mesophilic (Schwartz and Henderson, 1991); in sandy and rocky situations in scrub, along abandoned railways, at edges of sugarcane fields, in clearings in broadleaf forest, along edges of a lowland stream in dense riparian growth and associated pastures, in an oak (*Quercus*) grove, on gravelly soils in pinewoods along edges of lush, tropical arroyos, in dense grass and along sandy beaches, in halophytic growth, in stunted and scattered littoral growth, serpentine savannas, xeric areas on "dogtooth" limestone with low, scattered shrubs, among sisal plants, a lakeshore and adjacent pastures, dry clayey soils among fruit trees, in fencerows and in chinks and crannies of a stone fence in an edificarian situation, beneath rocks and leaves among grass and low vegetation (Schwartz, 1959e; Schwartz and Garrido, 1968a, 1968b; Garrido et al., 1986; Martínez Reyes et al., 1990; Schwartz and Henderson, 1991; Rodríguez Schettino, 1999b); on Cayo Santa María (Archipiélago de Sabana-Camagüey), on the sandy area of the beach and also in more wooded areas (Garrido, 1973d); on Cayo Coco (Archipiélago de Sabana-Camagüey), in all available vegetative associations (Estrada, 1993b); on Cayo Paredón Grande (Archipiélago de Sabana-Camagüey), in coastal scrub and lowlands (Estrada, 1998); always on the ground, takes refuge under cover or buries itself in sand, perches on rocks or fallen logs (Rodríguez Schettino, 1999b); in mesophyllous evergreen and semideciduous forest, pine forest, and coastal and subcoastal microphyllous forest (Rodríguez Schettino et al., 1999).

Activity: Diurnal, greatest number of active adults 1200–1300 h, most active juveniles 1000–1100 h during the wet season, after 1800 h, all individuals in nocturnal refuges (Rodríguez Schettino et al., 1999); inactive before 1000 h on cool mornings, abundant by 1100 h, when females and juveniles are first seen, on other days, often begins to emerge from burrows ~1030 h (Garrido and Schwartz, 1968b); peak activity for adults 1200–1300 h, for juveniles 1000–1100 h (Martínez Reyes et al., 1990). **Behavior:** Wary, quickly running to a refuge when disturbed (Rodríguez Schettino, 1999b); seeks shelter in crab burrows (Kwet, 1995). **Foraging and Diet:** Males and females feed primarily on ants and scolytid beetles, other food items include bees, other beetles, hemipterans, termites, flies, roaches, earwigs, larval and pupal insects, mites, spiders, crustaceans, fruits and flowers, and conspecifics (Martínez Reyes et al., 1990). **Parasites:** Ticks *Amblyomma albopictum* (Rodríguez Schettino, 1999b); nematodes *Cyrtosomum longicaudatum*, *C. scelopori*, *Oswaldocruzia lenteixeirai*, *Physaloptera squamatae*, and larval *Porrocaecum* sp. (Coy Otero and Baruš, 1979b; Coy Otero and Lorenzo Hernández, 1982; Coy Otero, 1999). **Population Size and Density:** On the Península de Guanahacabibes, populations denser on cliffs and in coastal scrub woodlands than in semideciduous forest on limestone substrates, relative abundance higher during the wet than in the dry season, encounter rates vary, 5–14 lizards/hr at different localities (Rodríguez Schettino, 1999b). **Predation:** Conspecifics (Martínez Reyes et al., 1990). **Reproduction:** Percent of individuals reproductively active highest in July (70% of October males had one testis with a diameter of 5 mm, 30% with 2, 75% of December males with one, 25% with 2, all April and July males had 2 testes with diameters of 5 mm, 82.3% of April females with active oviducts, 30% gravid, 100% of July females had active oviducts, 80% gravid), 66.7% of gravid females had a single oviductal egg, 33.3% had 2, mean April egg length 13.3 mm, mean July egg length 14.7 mm, number of ovarian eggs, oviductal eggs, and gonadal volume all positively correlated with female size, 3 gravid females in July laid 5 eggs in captivity, all hatched in August, eggs 15.0–17.0 × 7.0–8.3 mm, one neonate hatched after 53 days, 24.1 mm SVL (Martínez Reyes et al., 1990); one female contained 2 yolked follicles (G. R. Smith and Nickel, 2002). **Size:** Male SVL to 100 mm, female SVL to 80 mm (Schwartz and Garrido, 1968b); minimum sizes for reproductive activity 48.4 mm SVL in females, 60.0 mm SVL in males (Martínez Reyes et al., 1990). **Thermal Biology:** Mean body temperatures (34.3 ± 0.5 °C for males, 34.6 ± 0.6 °C for females) higher than those of air (31.3 ± 0.2 °C, 30.0 ± 0.5 °C) and substrate (33.5 ± 0.6 °C, 33.4 ± 0.3 °C), most individuals (57.3% of all individuals, 73.8% of adult females) in full sun (Martínez Reyes et al., 1990); heliothermic and thigmothermic, 57.7% of active individuals in full sun on the Península de Guanahacabibes, mean body temperature during the wet season 34.5 °C at mean air and substrate temperatures of 30.8 and 32.2 °C, once preferred temperature attained, lizards feed and engage in reproductive behavior in shaded areas (Rodríguez Schettino, 1999b).

Conservation Status: Entire species not assessed, but *L. s. lipomator* is considered "endangered" and *L. s. lucianus*, *L. s. parasphex*, *L. s. gibarensis*, *L. s. septentrionalis*, *L. s. astictus*, *L. s. naranjoi*, and *L. s. ophiplacodes* "vulnerable" (L. Rodríguez Schettino and A. Chamizo Lara in Rodríguez Schettino, 1999a).

Leiocephalus vinculum Cochran, 1928

Distribution: Hispaniola, Île de la Gonâve (Haiti).

Habitat: Xerophilic to mesophilic (Schwartz and Henderson, 1991); on rocky hillsides in xeric scrub with some large shade trees, on a hillside above a lake on a rock wall of a roadcut (Schwartz, 1967d).

Activity: Basking at 1300 h, but not seen between 1100 h and that time (Schwartz, 1967d). **Predation:** *Uromacer frenatus* (Henderson et al., 1987c). **Size:** Male SVL to 77 mm, female to 73 mm (Schwartz, 1967d).

Conservation Status: Not assessed.

14 Family Phyllodactylidae

Figure 17. Like many geckos, *Thecadactylus rapicauda* (Phyllodactylidae) regularly cleans the clear spectacles covering its eyes. Photograph by Mel José Rivera.

Phyllodactylus hispaniolae Schwartz, 1980

Distribution: Hispaniola, Valle de Neiba-Llanos de Azua and foothills of the Sierra Martín García (Dominican Republic).

Habitat: Xerophilic (Schwartz and Henderson, 1991); associated with limestone cliffs and boulders and exposed road cuts (Schwartz, 1979b).

Size: Male SVL to 56 mm, female SVL to 61 mm (Schwartz, 1979b).

Conservation Status: Unknown.

Additional Pertinent Literature: Bauer and Russell (2003, as *P. wirshingi hispaniolae*).

Phyllodactylus pulcher Gray 1830

Distribution: Barbados.

Habitat: Presumably arboreal (Dixon, 1962; Schwartz and Henderson, 1991); at least occasionally edificarian (Powell and Henderson, 2005a).

Size: Adult SVL to 62 mm (Schwartz and Henderson, 1991).

Conservation Status: Unknown; efforts to locate geckos in Barbados unsuccessful (Weiss and Hedges, 2007); known localities are limited and range has not been systematically investigated (Government of Barbados, 2002); considered "rare" (Ministry of Physical Development and Environment, Barbados, 2001); surveys should be encouraged (Powell and Henderson, 2005a), especially since *Phyllodactylus* elsewhere appear to be vulnerable to population declines when faced with competition from *Hemidactylus* (G. Köhler, pers. comm.).

Phyllodactylus sommeri Schwartz, 1980

Distribution: Hispaniola, northwestern Haiti.

Habitat: Xerophilic (Schwartz and Henderson, 1991); area of creviced and pock-marked limestone cliffs with various cacti, *Agave*, and *Acacia*, geckos on red earth substrate between 2 boulders (each about 1.0 m in diameter), also on the underside of a small rock in a rock pile, and on a rock face at 3.0 m above ground level (Schwartz, 1979b).

Behavior: Seeks refuge in solution holes and crevices (Schwartz, 1979b). **Reproduction:** Clutch size 2 (Hedges, in prep.). **Size:** Male SVL to 58 mm, female SVL unknown (Schwartz, 1979b).

Conservation Status: Unknown.

Additional Pertinent Literature: Bauer and Russell (2003, as *P. wirshingi sommeri*).

Phyllodactylus wirshingi Kerster and Smith 1955

Distribution: Southwestern Puerto Rico, including Isla Caja de Muertos.

Habitat: Xerophilic (Schwartz and Henderson, 1991); associated with limestone cliffs and, especially, boulders with crevices, but also by day in a pile of *Cocos* fronds on a sandy substrate, and on a paved road (Schwartz, 1979b); occasionally on roads at night (Rivero, 1998); under rocks, fallen logs, and other objects (Joglar, 2005b).

Size: Male SVL to 62 mm, female SVL to 66 mm (Schwartz, 1979b).

Conservation Status: "Stable" (Moreno, 1991d).

Additional Pertinent Literature: Bauer and Russell (2003).

Tarentola albertschwartzi Sprackland and Swinney, 1998

Distribution: Jamaica.

Habitat: Unknown.

Habitat and Natural History: Unknown.

Size: SVL of female (only known specimen) 137 mm (Sprackland and Swinney, 1998).

Conservation Status: Presumably extinct (Sprackland and Swinney, 1998).

Tarentola americana Gray, 1831

Distribution: Great Bahama Bank (*T. a. warreni*); Cuba, Isla de la Juventud, Archipiélago de Sabana-Camagüey (*T. a. americana*).

Habitat: Xerophilic (Schwartz and Henderson, 1991); in Cuba, associated with caves (Alayo Dalmau, 1951; Schwartz and Ogren, 1956; Hardy, 1957a; Z. Vogel, 1965b; Buide, 1966); rests in the twilight zone near entrances of caves by day and on cliffs (Schwartz, 1968b); in old erect dead trunks of mangroves (*Laguncularia*; Ruibal, 1957); in a storage shed, in a hollow Royal Palm (*Roystonea*; Schwartz, 1968b); in roofs of native shacks (Alayo, 1951; Buide, 1967; Viña Dávila et al., 1984); under a stone (Barbour, 1914); in and under dead dry cactus (Barbour and Ramsden, 1919); in a rotten palm log (G. Peters, 1974); in a live hollow tree in dense hardwood forest, occasionally on the ground associated with rough limestone rocks embedded in rocky soil (Schwartz, 1968b; Estrada, 1993a); under tree bark and in a hollow branch (Estrada, 1993b); in a hollow tree (Garrido and Schwartz, 1968b); under the bark of a palmetto (Garrido, 1973d); in a pile of rocks (Garrido et al., 1986); in the Bahamas, on Long Island, in abandoned stone houses and in *Cocos* trash, on Leaf Cay, around edges of limestone slabs in palm woods (Schwartz, 1968b); under large limestone slabs (Rabb and Hayden, 1957); on Waderick Wells (Exuma Islands), geckos "everywhere" in cave (Dodd, 2003).

Abundance: Three visits to a cave in Cuba yielded 60 specimens (Buide, 1966). **Foraging and Diet:** In Cuba, forages on cliffs, and, in the Bahamas, forages on walls and rafters in abandoned houses (Schwartz, 1968b); diet includes Black Widow Spiders (*Lactrodectus*; Schwartz and Henderson, 1991). **Parasites:** In Cuba, nematodes *Spauligodon cubensis* (Read and Amrein, 1953; Baruš and Coy Otero, 1969a; Coy Otero, 1970; Coy Otero and Baruš, 1979b; Coy Otero and Lorenzo Hernández, 1982), *Skrjabinodon* sp. (Baruš and Coy Otero, 1974; Coy Otero and Baruš, 1979b; Coy Otero and Lorenzo Hernández, 1982), *Alaeuris travassosi*, and larval *Porrocaecum* sp. (Coy Otero and Baruš, 1979b; Coy Otero and Lorenzo Hernández, 1982); mite *Geckobia tarentolae* (de la Cruz, 1973). **Predation:** Barn Owl (*Tyto alba furcata*; Arredondo Antúnez and Chirino Flores, 2002). **Reproduction:** In Cuba, eggs deposited in small hollows in roofs and lower walls of caves, or in small but extensive rocky cavities in cliff faces (Schwartz, 1968b); eggs in a cleft in the bark of a tree trunk and under a board, nest sites from ground level to 1.5 m above the ground (Novo Rodríguez, 1985a); clutch size 2, eggs adherent to tops and sides of cavities where deposited, nesting often communal (Schwartz, 1968b); 12–18 eggs/nest site (Novo Rodríguez, 1985b); in Cuba, eggs (11.0–12.5 × 12.5–15.6 mm) deposited March–September with hatching into October (Novo Rodríguez, 1985b); hatchling SVL 27.0–29.8 mm, 600–800 mg (Novo Rodríguez, 1985b). **Size:** Male SVL to 120 mm, female SVL to 111 mm (Schwartz and Henderson, 1991).

Conservation Status: Not assessed.

Tarentola crombiei Díaz and Hedges, 2008

Distribution: Cuba, southern coast of Granma, Santiago de Cuba, and Guantánamo provinces.

Habitat: Xerophilic, coastal xerophytic scrub on semidesertic lowlands and karstik marine terraces, by day in dry

Agave, at night on shrubs at heights of 0.5–3.0 m, perch diameters <5 cm, one female at night on the walls of a small volcanic elevation known as a "monitongo," co-occurs with *T. americana*, but not edificarian as is that species (Díaz and Hedges, 2008).

Behavior: Female rubbed her snout on *Tillandsia* sp. to assist in shedding (Díaz and Hedges, 2008). **Diet:** Roaches (Blaberinae and Blattidae), crickets, elaterid beetle, ants (*Camponotus* sp.), and very small stones in fecal samples (Díaz and Hedges, 2008). **Reproduction:** Clutch size 1, communal nests in dry clumps of *Agave*, eggs white, slightly fusiform, 11.0–13.3 × 8.2–9.9 mm, clutches in the same *Agave* as those of *T. americana*, but eggs of that species larger, usually depressed, round, and always in pairs (Díaz and Hedges, 2008). **Size:** Male SVL to 57.8 mm, female SVL to 56.6 mm, hatchling SVL 23.8–25.7 mm (Díaz and Hedges, 2008).

Conservation Status: Not assessed.

Thecadactylus rapicauda Houttuyn, 1782

Distribution: Widespread in the Neotropics from Mexico to the southernmost portions of the Amazon Basin (and Trinidad and Tobago); in the West Indies, on the U.S. Virgin Islands (St. Croix and possibly St. Thomas, probably introduced), British Virgin Islands (Necker Island), Anguilla, St.-Martin/St. Maarten, St.-Barthélemy, Saba, St. Eustatius, St. Christopher, Nevis, Barbuda, Antigua, Montserrat, Guadeloupe, Iles des Saintes, La Désirade, Iles de Petite Terre, Marie-Galante, Dominica, Martinique, St. Lucia (including Maria Major), St. Vincent, the Grenadines (Bequia, Mustique, Carriacou), and Grenada.

Habitat: Largely arboreal, but may forage underground (Lazell, 1995, 2006); generally prefers moderately humid environments, but may occur in dry forests or more mesic situations (Breuil, 2002b); on St. Croix, "normally inhabits old buildings and hollow trees" (Seaman, 1961); on Necker Island (British Virgin Islands), edificarian by day, by night in trees, in a bromeliad, underground, in cactus, and edificarian (Lazell, 1995, 2006); on Dominica, edificarian and arboreal (Krintler, 1986); shelters by day in crevices of trees and houses (Malhotra and Thorpe, 1999); actively foraging at night on the walls of an open-air restaurant (RP); on Saba, in an abandoned house, but generally less edificarian than *Hemidactylus mabouia* (Powell et al., 2005); on Anguilla, edificarian, on outside walls of a hotel, mean perch height 6.6 ± 1.2 m (0.3–9.1 m; K. G. Howard et al., 2001); in dry water tanks (RP); on St.-Martin/St. Maarten, sunning on a tree trunk in heavily altered forest (RWH); on Barbuda, primarily associated with human habitation (Malhotra and Thorpe, 1999); large numbers on scattered trees in dry savannas, probably in mesic forest on Pic du Paradis (St.-Martin), apparently showing a preference for dead trees with loose bark (often on the same trees with *Hemidactylus mabouia* and *Anolis gingivinus*), on a dead *Ficus* with introduced *Osteopilus septentrionalis*, edificarian, where may exploit the night-light niche, on St.-Barthélemy, edificarian (with *Hemidactylus*), under a bridge by day, in dry forests, under bark, rock crevices, by day in crevices on trunks of trees, on a Coconut Tree by day, on Guadeloupe, in coastal *Cocos*, houses, piles of rocks, in *Musa*, in humid forest, in bromeliads and under loose bark, in a *Philodendron* petiole by day 4 m above the ground, on Désirade, on vines in xerophytic forest, edificarian (with *Hemidactylus* and *Anolis* in a kitchen), on an abandoned refrigerator, in piles of debris (Breuil, 2002b); on the Iles des Saintes (Terre-de-Bas), in rock piles in mesic highlands, on Dominica, in bromeliads (Schwartz and Henderson, 1991); on a large boulder along a river in a heavily disturbed forest (RWH); on Dominica, in a crevice between rafters (J. S. Parmerlee, Jr., pers. comm.); on St. Lucia, in rock piles (Schwartz and Henderson, 1991); on Maria Major, in hollow trees (Corke, 1987); on St. Vincent, on tree trunks in rainforest (RWH) and dry upland forest by day (RP); on Martinique, in both dry and wet forests from sea level to >700 m asl (Breuil, 2009); on Grenada, on tree trunks in rainforest and in mixed agriculture (Germano et al., 2003; RWH); under *Cocos* fronds at the base of a tree, under the petiole base on an upright *Musa* stalk, on a *Musa* leaf in full sunlight at 1200 h (Schwartz and Henderson, 1991).

Activity: Nocturnal (Schwartz and Henderson, 1991); luciglomerant (congregate at a roost in daylight and disperse outward at night; Lazell, 1995, 2006); emerges from nocturnal retreats at dusk (Breuil, 2002b); will bask by day (Breuil, 2002b; Powell et al., 2005); most activity early in the evening before nighttime temperatures drop to their lowest points (Powell et al., 2005); generally active longer into the night than *Hemidactylus mabouia* and *Sphaerodactylus sputator* in the same habitat (K. G. Howard et al., 2001). **Activity Range:** On Necker Island, 133–1,496 m^2 (Lazell, 1995, 2006). **Behavior:** Vocal (West, 1793 in Grant, 1932d; Krintler, 1986; Breuil, 2002b; Vale, 2002); relies on crypsis and may be easily approached (Long, 1974); when threatened, hangs tightly to substrate, if approached too closely, will "squirrel" to the other side of a trunk or branch, or assumes a defensive posture involving the raising and lowering of its tail (Breuil, 2002b); territorial, with sometimes intense encounters resulting in broken tails, juveniles may wave tails to distract predators, adults elevate heads and tails and gape widely (exposing their blue tongues) when threatened, will bite when handled (Powell et al., 2005). **Biomass:** On Necker Island, 13.5 kg/ha (Lazell, 1995, 2006). **Escape Mechanism:** Regional integumentary loss (Bauer and Russell, 1992; Powell et al., 2005). **Foraging and Diet:** Insectivorous (Currat, 1980); on Necker Island (British Virgin Islands) apparently forage below ground as well as above, prey included crickets (Gryllidae; 12 of 54 prey items), roaches (Blattidae; 9/54), moths (Lepidoptera; 7/54), spiders (Araneida; 4/54), longhorn beetles (Cerambycidae; 4/54), and others, including a scorpion (Scorpionida), a walking stick (Phasmidae), and own shed skin (Lazell, 1995, 2006); forages on walls and adjacent trunks of Bougainvillea, equally effectively in total darkness and in light, 3 individuals foraged on one wall, but never simultaneously, eats various insects and other small arthropods, in captivity ate small vertebrates (*Hemidactylus*, *Anolis*, *Eleutherodactylus*), taking prey to 50 mm in length (Breuil, 2002b); on Grenada, consumed *Eleutherodactylus johnstonei* (Henderson and Berg, 2005); feeds primarily on nocturnal insects, with roaches most important, adults take small vertebrates, including conspecific juveniles (Powell et al., 2005); feeding on insects around artificial lights at night on Necker Island (British Virgin Islands; J. Lazell in Perry et al., 2008), Anguilla (R. Powell in Perry et al., 2008), and Dominica (Daniells et al., 2008; J. Lazell in Perry et al., 2008). **Interspecific Interactions:** An adult chased by an adult *Hemidactylus mabouia*, but the 2 species observed foraging side by side on other occasions (Breuil, 2002b). **Population Density:** On Necker Island, 500/ha (Lazell, 1995, 2006). **Reproduction:** Oviparous, eggs oval, ~1.5 cm long, shell calcareous, abandoned after laying under rocks or in crevices of trees (Currat, 1980); on Necker Island, eggs and juveniles (smallest with SVL of 38 mm) in October (Lazell, 1995, 2006); females usually with only one developing egg in one oviduct and one yolked follicle in the opposite ovary, eggs to 17.5 mm in length (Breuil, 2002b); apparently reproduce throughout the year, laying 2 hard-shelled eggs in cracks in rocks or walls or under bark or debris (Powell et al., 2005). **Size:** SVL in mainland males to 125 mm, females to 126 mm (Vitt and Zani, 1997). **Tail Autotomy:** Tail loss common, more likely a consequence of attempted predation than combat (male and female frequencies equal), but costly, since fat stored there facilitates survival during lean periods (Breuil, 2002b). **Thermal Biology:** Body temperatures (substrate temperatures) on Anguilla at night on a building ($N = 2$) 26.2 °C (25.5 °C) and 27.7 °C (26.5 °C; K. G. Howard et al., 2001); diurnal activity may increase during cooler weather (Breuil, 2002b).

Conservation Status: "Decreasing in some areas" in the U.S. Virgin Islands (Moreno, 1991d).

Additional Pertinent Literature: Russell and Bauer (2002).

15 Family Polychrotidae

Figure 18. Aggressive displays by male *Anolis allogus* (Polychrotidae) include extended dewlaps, raised nuchal and dorsal crests, and laterally curled tails. Photograph by Axel Fläschendräger.

Anolis acutus Hallowell, 1856

Distribution: U.S. Virgin Islands, throughout the St. Croix Bank.

Habitat: Generally xerophilic (Schwartz and Henderson, 1991); ubiquitous in dry forests where lizards usually perch on trees, bushes, posts, rocks, and other elevated sites from near the ground to about 3 m, with females and juveniles perching lower than males, but individuals may ascend to the tops of the highest trees (Lazell, 1972; Ruibal et al., 1972; Ruibal and Philibosian, 1974a, 1974b); mean perch height and diameter 2.39 ± 0.18 m and 13.2 ± 2.3 cm (Losos and de Queiroz, 1997a).

Activity: Diurnal, active from before dawn to after sunset with no peaks of activity or basking (Ruibal and Philibosian, 1974a, 1974b). **Behavior:** Escape behavior involves climbing up or down into cavities and crevices (Lazell, 1972); aggressive encounters usually of low intensity, ~1 min in duration, raising nuchal crest and moving toward an intruder, increase linearly as the number of lizards per tree increases (high-intensity encounters involve challenge displays and often last 10 min), males and females exhibit equivalent levels of aggression and aggression of one sex is independent of the other, but removal of an individual results in an immediate increase in aggression that subsides once the vacated territory is occupied (Ruibal and Philibosian, 1974a). **Ecomorphology:** Classified as a "generalist," most similar to "trunk-crown" (Losos and de Queiroz, 1997a); intermediate between "trunk-ground" and "trunk-crown" (Beuttell and Losos, 1999). **Foraging and Diet:** Most foraging in both sexes occurs on the ground (Lazell, 1972); total calories available from one full stomach 0.0595 kcal/g body weight (McManus and Nellis, 1973). **Habitat Partitioning:** Adults exclude juveniles from trees (Ruibal and Philibosian, 1974b). **Home Range and Territoriality:** Territories are small, mean distance to nearest neighbor 1.5 m in males, 0.7 m in females, 0.5 m for male-female pairs, territories less important in limiting density per tree than in most congeners, adult males home rapidly, with 50% returning when displaced 71–90 m (Ruibal and Philibosian, 1974b); mean territory size per lizard on large trees 1.2 m^2 for males and 0.5 m^2 for females (Philbosian, 1975). **Metabolism:** Lizards at 10 and 15 °C were noticeably sluggish and much darker than those at higher temperatures, metabolic rate (oxygen consumption) exponentially related to temperature, with the sharpest increase from 20 to 25 °C, the increase from 30 to 35 °C is negligible (McManus and Nellis, 1973). **Movements:** Migration rates low, adults very sedentary and, once territories are established, rarely move from the home tree, immigration into a tree devoid of residents is primarily by juveniles (Ruibal and Philibosian, 1974b). **Parasites:** Nematodes *Parapharyngodon cubensis* and larval *Ascarops* sp., acanthocephalan cystacanths *Centrorhynchus* sp. and *Oligacanthorhynchus* sp. (Goldberg et al., 1997b); nematode *Spauligodon anolis* (Goldberg et al., 1997b; Bursey and Goldberg, 1998). **Population Density and Size:** Densities to 0.43 lizards/m^2, 2,000–5,600 lizards/ha, biomass 12–23 kg/ha, single trees may have populations of 80 adults with a biomass of 300 g, egg and/or juvenile mortality may function to regulate population size (Ruibal and Philibosian, 1974b); mean number of lizards/tree 10.2 males and 7.2 females and males with SVL <50 mm, mean male-male distance 1.5 m, mean female-female distance 0.7 m, mean male-female distance 0.5 m (Philbosian, 1975). **Predation:** Cattle Egrets (*Bubulcus ibis*; Gassett et al., 2000). **Reproduction:** Eggs 10.5 × 7.0 mm and 11.0 × 7.6 mm, copulation in June and February (Lazell, 1972); ovigerous females present during most of the year (January–April with lowest reproductive potential, but ovigerous females may even then exceed 50% of the total number of adult females), maximum testicular sizes in June–October, female fat bodies rarely large, smallest during the dry season (December–April), no temporal pattern in males, maximum reproductive effort correlated with months of highest rainfall, with inverse true as well, females lay single eggs 10.5–11.4 × 7.0–8.6 mm in sandy or gravelly soils, egg mass 0.32 g 14 days after deposition, hatching occurs in about 30 days at 28 °C, hatchling SVL 10–21 mm, mass 0.21–0.30 g, average juvenile growth rate 0.07 mm/day (females 0.06 mm/day, males 0.11–0.14 mm/day), females mature in <1 year, males in about 1 year, minimum SVL sexually mature female 36 mm, sexually mature male 50 mm (Ruibal et al., 1972); reproductive activity higher in the more mesic western region of St. Croix, experimentally increased food levels increased reproductive effort in males but not in females, although fat levels increased in all food-augmented lizards (Rose, 1982). **Sex Ratio (M:F):** Apparently balanced, although a bias favors male captures (Lazell, 1972). **Size:** Male SVL to 67 mm, female SVL to 49 mm (Ruibal et al., 1972), smallest mature male SVL 48 mm, female 38 mm (Lazell, 1972); males reach maturity at 50 mm SVL, females at 36 mm SVL (Ruibal and Philibosian, 1974a). **Survival:** Mean survival over 12 months 12% for adult males, 37% for adult females (Ruibal and Philibosian, 1974b). **Thermal Biology:** Mean body temperature 30.28 °C (25.6–34.8 °C), differential between body and air temperatures from sunrise to noon (+0.66 °C) lower than those from noon to sunset (+1.37 °C), with the greatest differential from 1600 to 1900 h (+2.04 °C), when ambient temperatures fell more rapidly than body temperatures, upper level of thermal tolerance 39 °C (McManus and Nellis, 1973).

Conservation Status: "Stable" (Moreno, 1991d), but Platenberg and Boulon (2006) recommended monitoring.

Anolis aeneus Gray, 1840

Distribution: Grenada Bank, abundant and essentially ubiquitous on Grenada and the Grenadines; introduced on Trinidad and in Guyana.

Habitat: Xerophilic to mesophilic (Schwartz and Henderson, 1991); "ubiquitous" (Germano et al., 2003); on roadside trees, especially the shady sides, on fence posts, on stone walls (Underwood, 1959b); in *Acacia* and *Prosopis* scrub, xeric coastal scrub, cacao groves, *Cocos* trees and trash, on *Hippomane* adjacent to a salt pond (Petit St. Vincent), large trees on a lava hillside (Île de Caille), on and in buildings, gardens (fences, ornamental plants), and rainforest, less abundant in mangroves (Stamps, 1973, 1990b; Schwartz and Henderson, 1991; Germano et al., 2003; Harris et al., 2004; Powell and Henderson, 2008c); juveniles in clearings use "homesites" with many interconnecting twigs and plant stems and in which leaves are important as protection from rain and predators and because they collect dew and raindrops (Stamps, 1990b); near sea level on the dry leeward coast of Grenada, lizards perch from ground level to ~3 m above the ground (Lazell, 1972); males perch somewhat higher than females (Schoener and Gorman, 1968; Harris et al., 2004); females usually feed on tree trunks, fence posts, or wood piles, near but not on the ground (mean height 1.2 m; Stamps, 1973); mean perch heights and diameters for hatchlings, adult females, and adult males 14.36 ± 2.40 cm and 1.35 ± 0.15 cm, 50.59 ± 4.62 cm and 8.52 ± 6.30 cm, 169.7 ± 50.2 cm and 38.55 ± 12.60 cm, respectively, hatchlings perch on twigs or small plants, females on shrubs, rocks, or trees, and males on trees (Stamps, 1978); mean perch heights and diameters used by juveniles 19–21 mm SVL 35.2 ± 20.1 cm and 5.53 ± 4.31 mm, 22–24 mm SVL 38.4 ± 26.5 cm and 6.77 ± 4.27 mm, 25–27 mm SVL 40.1 ± 23.9 cm and 6.80 ± 4.52 mm, 28–31 mm SVL 38.7 ± 20.7 cm and 6.28 ± 4.47 mm (Stamps, 1983a); mean perch height 1.37 ± 0.06 m (Roughgarden et al., 1983a); on Grenada, mean perch height and diameter 1.87 ± 0.22 m and 16.7 ± 2.6 cm (Losos and de Queiroz, 1997a); on Hog Island (Grenada), in the absence of *A. richardii*, mean perch heights and diameters 128.2 ± 8.5 cm and 19.3 ± 1.8 cm on perches other than large *Cocos* palms and 327.0 ± 62.8 cm and 25.2 ± 1.4 cm on large

Cocos palms (Williamson et al., 2002); perch heights (110.9 ± 9.8–244.8 ± 27.7 cm) and diameters (4.5 ± 0.4–11.2 ± 3.2 cm) varied considerably in four different lowland habitats, and adult males perched higher (151.9 ± 9.5 cm) and on perches of larger diameter (7.8 ± 0.6 cm) than females and subadult males (125.8 ± 7.3 cm and 7.3 ± 0.6 cm) or juveniles (105.7 ± 18.2 cm and 4.6 ± 0.7 cm), overlap with sympatric *A. richardii* was high, although adult males of the latter species perched higher than any other size class of either species, most individuals were oriented vertically (facing up or down), except in mangroves, and the majority of individuals were in full or partial shade (Harris et al., 2004); at a lowland site on Grenada, mean perch heights 93.0 ± 48.6 cm for adult females and small males, 72.9 ± 19.9 cm for large adult males, significantly higher than *A. richardii*, with individuals often on cacti, rocks, and the ground, the latter less frequently than *A. richardii* and significantly more frequently at times when sympatric *Ameiva ameiva* were not active, most individuals were oriented vertically (Simmons et al., 2005); perches close to the ground at very low elevations (<30 m) and at intermediate heights at higher elevations, statistically indistinguishable from perches used by *A. richardii* (Buckley and Roughgarden, 2005b); sleeps tightly adpressed against the substrate (regardless of type) with eyes closed and hindlimbs extended, at mean height and diameter 1.2 ± 0.2 m and 3.1 ± 0.5 cm, respectively, most commonly on thornless plants (but commensurate with abundance of thorny plants), on leaves or branches at about equal frequencies, and usually facing away from the perch's point of origin, nearest neighbors did not differ by sex or species (syntopic with *A. richardii*, with which structural niche overlap was high, although *A. aeneus* used higher perches), a few individuals within 1 m of their previous perches on consecutive nights, but no evidence of strong sleep-site fidelity or evidence of perch selection that would minimize the predation risk by nocturnally foraging *Corallus grenadensis* aside from sleeping at the distal ends of branches (Poche et al., 2005).

Abundance: On Grenada, abundance is spatially variable at low elevations and remains present at low abundances at higher elevations (Buckley and Roughgarden, 2005a). **Activity:** Diurnal (Schwartz and Henderson, 1991); light showers do not interrupt activities, but lizards seek shelter during heavy rain (Stamps, 1976a); exploits the night-light niche (R. Powell in Perry et al., 2008). **Behavior:** Territorial display involves raising dorsal and nuchal crests, extending dewlap, raising and lowering the tail slowly, and headbobbing, spontaneous displays often accompany movement to a new position (Gorman, 1969); female social system ranges from territoriality in some habitats to dominance hierarchies (with more than 2 steps) in others, with female displays differing depending on social system, "multibob" in hierarchies (often associated with anxiety), "jerkbob" in territories, distance to nearest neighbor correlated with frequencies of either type of display (Stamps, 1973); 4 major types of behavioral patterns, "fan" (34% of all displays, given at lower temperatures, 25–27 °C), with forelimb elevation and dewlap extension, "fanbob" (62%, given at higher temperatures, >27 °C), similar to fan but with pushups ("signature bob"), "bob" (2%), identical to fanbob but without dewlap extension, "jerkbob" (2%), with series of abrupt bobs with or without dewlap extension, series of jerkbobs sometimes precedes a fanbob, sometimes accompanied by lateral compression of the body (but only in aggressive encounters), signature bobs both in aggression and courtship, gorged throats only in aggression, "multibobs" as an assertion display (Stamps and Barlow, 1973); males do not display and male-male aggression is reduced during nonbreeding periods (Stamps and Crews, 1976); female (but not juvenile) activity and aggression decrease with longer periods of no rainfall (Stamps, 1976a); ritualized combat may escalate into "jawlocking," with each individual attempting to dislodge the other by banging it against a perch, jawlocking in males often results in wounds, which may become septic, incapacitate the lizard, and can be fatal (Stamps, 1977a); hatchlings only a few days old have nonoverlapping ranges or are involved in dominance hierarchies (varying with microhabitat), aggression greatest when directed toward individuals slightly smaller than the aggressor, but juveniles are more aggressive than adults toward larger conspecifics, adults of both sexes and juveniles use all displays except courtship, all used "jerkbob" displays in indeterminate contexts, females and juveniles used "multibobs" almost as frequently, whereas males used "jerkfanbobs" at a high rate, duration of displays is longer in larger lizards (Stamps, 1978); in juveniles, increased food supply results in increased display rates, long-range effects were minimal, aggressive responses and home range sizes did not change, but lizards ignored added food during the wet season, ate it avidly during the dry season, when the immigration rate increased and remained high (Stamps and Tanaka, 1981b); adults ignore juveniles until they reach subadult sizes (Stamps, 1983a); juveniles often live in linear dominance hierarchies, with rank determined by size relative to others when the hierarchy is entered, rank gradually increases (Stamps, 1984b); juveniles "squirrel" around perches or dash into tangles of twigs and leaves in the presence of predators, exhibit "avid" interest in neighbors, probably in order to colonize a better territory when vacated by larger individuals moving back to forested areas (Stamps, 1990b); juveniles introduced into vacant patches of habitat in the presence of few competitors obtained more exclusive home ranges than did individuals settling at higher densities, when settling at high densities, juveniles that eventually attained high social status fought and chased their opponents more frequently than juveniles that were subordinate to other residents (Stamps and Krishnan, 1998). **Competition:** Juvenile *A. richardii* may aggressively dominate juvenile *A. aeneus*, possibly excluding them from shadier areas, juvenile *A. richardii* were the aggressors in 24 of 26 interspecific interactions (Stamps, 1983a). **Ecomorphology:** An ecomorphological "generalist" (Losos and de Queiroz, 1997a); equally distant from "trunk-ground," "trunk-crown," "twig," and "crown-giant" (Beuttell and Losos, 1999). **Foraging and Diet:** Males and females use head-down survey positions to search for prey (use of this position decreases significantly after feeding), foraging involves long periods of sitting motionless on elongated perches (may be indicative of a food-limited habitat) punctuated by short dashes to leaves or the ground to capture prey, before returning to favorite perches after catching prey (Stamps, 1977c); both sexes and even some *A. richardii* congregate to exploit seasonally abundant mulberry-like berries (Lazell, 1972); diet includes ants (greatest numbers in adult males and females and juveniles), other hymenopterans, flies, beetles (adults and larvae), hemipterans, lepidopterans (adults and larvae, latter in greatest volume in adults of both sexes and juveniles), booklice, thrips, termites, orthopterans, lacewings (adults and larvae), caddisflies, earwigs, springtails, spiders, pseudoscorpions, mites, centipedes, millipedes, isopods, other crustaceans, snails, annelids, and some plant matter, with prey sizes significantly smaller than those taken by *A. richardii*, with larger individuals taking larger prey (and containing fewer items) and subadult males taking larger prey than females of the same size, larger individuals take more plant matter, smaller lizards consume a wider variety of food, mean prey size of adult males 8.3 ± 5.7–11.1 ± 4.6 mm, females 9.1 ± 4.3–8.5 ± 4.9 mm, all size classes 8.2 ± 2.3–15.6 ± 10.0 mm and mean number of prey items/lizard 57.0 ± 17.8–44.3 ± 33.4 for males, 64.2 ± 29.9–61.5 ± 42.3 for females, 38.7 ± 27.5–71.3 ± 70.7 for all size classes (Schoener and Gorman, 1968); larger lizards took proportionately larger prey (Stamps, 1978); juveniles rarely catch prey on the ground (22 of 44 prey captures on twigs and branches, 20 on leaves, only 2 on the ground), choose prey to obtain a balanced diet (prefer a certain prey type even when it is relatively less abundant in the environment), lizards run or jump long distances to catch termites or caterpillars, but certain ants (*Crematogaster*) are rarely taken, despite being common in juvenile habitats, flies represented 17.1 ± 6.5% of gut samples (15.7 ± 8.8% of samples taken with sweep nets), hemipterans (mostly leafhoppers) 24.9 ± 10.7% (37.4 ± 13.6%), ants (other than *Crematogaster*) 25.0 ± 16.0% (4.72 ± 5.6%), hymenopterans (other than ants) 6.5 ± 3.6% (1.75 ± 0.45%), lepidopteran larvae 6.8 ± 4.3% (0.72 ± 1.25%), spiders 5.1 ± 3.3% (6.25 ± 4.97%), total for major taxa 85.4 ± 5.4% (66.5 ± 4.3%; Stamps et al., 1981); cannibalism (Stamps, 1983a); percent time moving 5.1 ± 0.8%, moves per minute 4.7 ± 0.8, percent feeding attacks while moving 0%, stomach contents included spiders, other arachnids, beetles (adult and larvae), flies, hemipterans, ants (which were the most important dietary item), lepidopterans (adults and larvae, which were the second most important dietary item, and presumably a reflection of exploiting seasonally abundant prey), a lacewing, roaches, shed lizard skin, and plant material, with flowers more important in females and fruits/seeds taken in about equal amounts by females and males (Simmons et al., 2005); nectarivory and possibly pollination of *Charianthus grenadensis* (Timmermann et al., 2008). **Growth:** Behavioral strategies for optimizing growth and survival may change ontogenetically, dry season growth rates in juveniles and adults are higher in watered sites (Stamps, 1977b); prey biomass and rainfall affect juvenile growth rates, even when food is abundant, growth rates low when drinking water curtailed, with larger individuals less affected by water deprivation than small lizards, wet season growth rate for juveniles with SVL 20–25 mm 0.163 ± 0.07 mm/day, dry season growth rate 0.090 ± 0.06 mm/day (Stamps and Tanaka, 1981a); growth rates highest for juveniles that traveled an average of about 400 cm/h, which is most likely when a juvenile has a relatively exclusive territory of ~0.5 m^2, more likely to be achieved by individuals with high social rank (Stamps and Eason, 1989). **Home Range and Territoriality:** Male home ranges may expand abruptly (or the overlap between home ranges may decrease) with the onset of the breeding season (0.40 m^2 and 2.8 m^2 at 2 sites in November, 0.3–0.4 m^2 and 0.4–12.2 m^2 in February–April, 22.8 m^2 and 7.9–14.0 m^2 in July–August, with a corresponding decrease in numbers of trees with more than one male), female home ranges remain relatively constant throughout the year, at a beach site, home ranges may include or be limited to the trunk, crown, and debris of a *Cocos* tree (Stamps and Crews, 1976); in general, females defend a food supply, males defend mates, during the breeding season, male territories overlap those of females and are determined by female spacing patterns, males spend most of their time (mean 89 ± 4%) in female territories, even though these are only 40 ± 23% of total male territories (Stamps, 1977a); home range sizes increase with food provisioning (Stamps and Tanaka, 1981b); at least 4 factors affect the value of territories for juveniles: food, perches, insolation and temperature, and proximity to potential predators (Stamps, 1983b, 1987a); mean female home range size 2.8 m^2 (Stamps, 1983c); growth rates negatively affected by the degree of territorial overlap (affected juveniles in all but the lowest status groups), with food depletion in the area of overlap the most likely reason, overlap is reduced by shifting rather than reducing sizes of territories, juveniles may tolerate overlapping territories because of a shortage of predator-safe homesites (Stamps, 1984a); small juveniles "float" around clearings (whereas larger juveniles cannot due to intense territoriality) and are more likely to find and hold an empty territory than larger juveniles, a strong negative relationship exists between size and aggression toward intruders, females more sedentary and aggressive than young males of about the same size (Stamps and Tollestrup, 1984); juveniles show preferences for homesites previously used by conspecifics (Stamps, 1987a); juveniles' preferences for homesites adjacent to established territorial residents lead to territorial aggregations independent of territorial quality, juveniles that arrive in empty territories require only about 6 h of residence to defend them successfully against later arrivals of comparable size (after several days, they can repel all intruders, regardless of size), juveniles previously resident in a homesite had a better chance of becoming settlers than did new arrivals (Stamps, 1987b, 1988a, 1988b); when large numbers of juveniles were introduced into a high-quality habitat patch, more settled in that patch (instead of emigrating), and net territory size (reflecting both territory size and overlap) was smaller (Stamps, 1990a); juvenile territories ~0.19 m^2 (Stamps, 1990b); settlers and nonsettlers (juveniles moving into previously settled habitats) did not differ in competitive ability (= size), prospective settlers did not avoid relatively full habitats nor did they show a preference for unsaturated habitats (Stamps, 1991); percent juveniles settled in territories about the same when introduced simultaneously and sequentially, sequential arrivals attacked at higher rates and lizards attacked at high rates were less likely to settle, but those attacked at lower rates initiated more chases and fights and settled at high rates when in sequential situation (Stamps, 1992); fights during first encounters among juveniles settling an area are most likely when lizards are approximately the same size, fights are more likely when encountering the first few opponents, chases more likely after meeting several unfamiliar animals (Stamps and Krishnan, 1994a); juveniles that won first encounters won subsequent encounters, reversals of dominance and co-equal status are rare, lizards engaging in fights during first encounters rarely engaged in subsequent interactions, those that chased during first encounters engaged in many subsequent interactions, animals that lost chases or fights are more likely to leave an area (Stamps and Krishnan, 1994b); neither winners nor losers of first encounters settled areas of encounters, dominants acquired space by repeatedly chasing subordinates, subordinates acquired space by persisting in the face of repeated attacks (Stamps and Krishnan, 1995); opponents whose last encounter was a fight were more likely to maintain mutually exclusive home ranges than those whose last encounter was a chase (Stamps and Krishnan, 1997). **Hybridization:** Hybridizes with *A. trinitatis* on Trinidad (Gorman, 2003, and references therein). **Movements:** Juveniles hatch in shady areas, migrate to clearings where they compete for foraging homesites with complex topographies and where predator efficiency is low, subadults migrate out of clearings and remain in shady areas throughout adult life (Stamps, 1983a, 1983b); hatchlings may move >150 m to a clearing (Stamps, 1990b). **Parasites:** Malarial parasites *Plasmodium azurophilum* (Staats and Schall, 1996b); nematodes *Parapharyngodon cubensis* and larval *Physocephalus* sp., acanthocephalan cystacanths *Centrorhynchus* sp. (Goldberg et al., 1997a). **Population Size and Density:** At 4 sites on Grenada, "rare" to 10.8 ± 2.6 lizards/100 m^2, hatchlings at one site 29.0 ± 8.6/100 m^2 (Roughgarden et al., 1983a); densities 0.6–3.5 lizards/m^2 in 4 garden sites, 0.9–1.3 lizards/m^2 in 3 scrub sites, and 1.9 lizards/m^2 in a beach site (Stamps, 1978); 1,163 females/ha (Stamps, 1983c); in 4 different habitats on Grenada, estimates were 2,700–9,360 and 2,010–9,240 lizards/ha using two different methods (Harris et al., 2004); in a study based on sleeping anoles, estimated population density was 1,125 lizards/ha, with more in a drier habitat than in an adjacent, more mesic habitat characterized by large trees, into which lizards may ascend to sleep (Poche et al., 2005). **Predation:** Ten species of large insectivorous birds eat unidentified Grenadian anoles (Wunderle, 1981); predators include larger conspecifics, *A. richardii*, and Carib Grackles (*Quiscalus lugubris*), predator avoidance may contribute to a shortage of living spaces (clearings are rare in natural habitats) and may lead to intense territoriality and aggression among the most vulnerable size classes (Stamps, 1983a); about 10% of juvenile time is spent avoiding predators, to which juveniles respond by adpressing close to a perch and freezing, "squirreling" to the opposite side of a perch, or running into cover (leaves or twigs in the center of homesites), the presence of *A. richardii* did not affect behavior, and lizards avoided predators only when they moved suddenly or approached (Stamps, 1983b); *Corallus grenadensis* (Henderson, 1993, 2002; Yorks et al., 2004); *C. grenadensis*, *Mastigodryas bruesi*, Carib Grackle (*Q. lugubris*), and Cattle Egret (*Bubulcus ibis*; Daudin and de Silva, 2007). **Reproduction:** Seasonal breeder, males court females using rapid, shallow headbobs ("courtship bob") interspersed with "fan" and "fanbob" (signature display), courtship results in copulation only when a female indicates receptivity by standing still and waiting for a male to bite the back of her neck, 1–60 sec after neck-bite, the male twists his tail around that of the female and inserts a hemipenis, mating of two types, lasting ~1 min (mean 1.06 ± 0.417 min) to 1 h (mean 69.7 ± 23.79 min), with only the latter apparently necessary for successful laying, receptivity occurs when one follicle of intermediate size (mean 5.2 ± 0.014 mm) is present in one ovary, female receptive periods occur at about 10-day intervals, both males and females wipe cloacae on substrata after mating (Stamps, 1975); maximal egg and sperm production occurs during the rainy season (May–November), with female fat body weight greatest during the dry season (Stamps and Crews, 1976); females retain shelled oviductal eggs during periods of drought, rainfall is a stimulus for females to dig holes for oviposition, but eggs are not laid unless soil moisture is adequate (Stamps, 1976b); males new to a territory court females more intensely than after several weeks of residence, subordinate males either do not court females or do so rarely (Stamps, 1977a); females lay and abandon single eggs every 10–14 days, burying them in their own territories, hatchlings emerge August–December (Stamps, 1990b); eggs 6.9–9.0 × 9.5–11.0 mm, hatchling SVL 20.5–21.8 mm on Trinidad (Quesnel and Kenny, 1959); hatchling total length ~45 mm (J. Schmidt, 1980a); in captivity, hatchling SVL 20–25 mm (Fläschendräger and Wijffels, 2009). **Sex Ratio (M:F):** Apparently balanced (Lazell, 1972); 1:1 (Stamps, 1983c). **Size:** Male SVL to 77 mm (to 80 mm in Guyana), female SVL to 55 mm, smallest mature male and female SVL 45 mm and 39 mm, respectively (Lazell, 1972); males mature at SVL 46–50 mm (D. Crews in Stamps, 1977a), females at ~39 mm (Stamps, 1977b); in the Grenadines, size is correlated with differences in insect productivity from island to island (Roughgarden and Fuentes, 1977). **Survival:** 6% of eggs in a study area appear as juveniles, once juveniles become established residents, mortality is very low, with 7 of 9 remaining until reaching adult size (Stamps, 1983a); most predators avoid fully insolated areas, hatchlings must leave forested areas for clearings, where survival is higher (Stamps, 1990b). **Thermal Biology:** Often absent in well-shaded forest, very abundant in sunny open areas, body temperatures 29.1–34.0 °C (most in 31–32 °C range), with 76% of temperatures 31.1 °C, with ambient air temperatures slightly lower (Schoener and Gorman, 1968); mean body temperature 30.3 ± 0.2 °C in coconut and mixed coconut-cacao habitat (Roughgarden et al., 1981); juveniles very rarely (5 min of 14.17 h of focal animal observations) in direct sun (Stamps, 1983a); on Grenada, mean microclimates used by *A. aeneus* and *A. richardii* are statistically indistinguishable at low-elevation (<300 m) sites, but *A. aeneus* perches in significantly warmer microclimates than *A. richardii* at high elevations (>300 m), *A. aeneus* maintains a constant difference between perch microclimate and mean available microclimate across the elevational gradient (Buckley and Roughgarden, 2005b).

Conservation Status: Not assessed.

Anolis agueroi Díaz, Navarro, and Garrido, 1998

Distribution: Cuba, known only from the vicinity of the type locality at the western end of the Sierra Maestra, Cabo Cruz, Granma Province.

Habitat: Evergreen microphyllous forest, an adult female on the trunk of *Allophylus cominia* (height 127 cm, diameter 10.4 cm), juveniles and an adult male in a Tamarind (*Lysiloma latisiliqua*) plantation at heights >10 m, except for the male, which was at a height of 4 m (Díaz et al., 1998b).

Activity: Diurnal (Rodríguez Schettino et al., 1999); all individuals captured in the morning, as early as 0715 h (Díaz et al., 1998b). **Diet:** Beetles, lepidopteran larvae, ants, slugs, vegetable matter, and rocks (Díaz et al., 1998b). **Ecomorphology:** "Twig-giant" (member of the *Chamaeleolis* clade), technically this is not an ecomorphological category as it evolved in only one clade (see definition in Williams, 1972). **Size:** Male SVL to 154 mm, female SVL to 158.1 mm (Díaz et al., 1998b).

Conservation Status: Unknown.

Anolis ahli Barbour, 1925

Distribution: Cuba, Sierra de Trinidad.

Habitat: Mesophilic (Schwartz and Henderson, 1991); generally in deeply shaded broadleaf forest (Ruibal, 1964; Ruibal and Williams, 1961b); in river valleys and arroyos (Estrada, 1994); occurs in less shaded situations along trails or roads, perches low on tree trunks (Ruibal and Williams, 1961b); in submontane rainforest and submontane mesophyllous evergreen forest (Rodríguez Schettino et al., 1999); on trunks of trees and bushes and on the ground (Estrada, 1994); males perch on trunks of trees at low heights, females more frequently on the ground, on rocks or fallen trunks, and in leaf litter (Rodríguez Schettino, 1999b); in a deep forested glen behind a hotel (Rodríguez Schettino, 2003); mean perch height and diameter 1.04 m and 8.5 cm (J. B. Losos, in litt., 21.IV.2008); sleeps primarily on leaves of low shrubs ~1 m above ground, but occasionally uses slender twigs (Schwartz and Henderson, 1991).

Activity: Diurnal (Rodríguez Schettino et al., 1999).

Behavior: Runs to the base of a tree or to the ground when approached (Ruibal and Williams, 1961b); sleeps with tail tightly curled or prehensily twisted around a leaf petiole or a minor surface irregularity (Hardy, 1958b). **Ecomorphology:** "Trunk-ground" (Williams, 1969). **Reproduction:** Mean duration of copulation 28.66 min (longest 31.02 min), separation of pair "normal" (Garrido, 1980a). **Size:** Male SVL to 61.7 mm, female SVL to 52.4 mm (Rodríguez Schettino, 1999b). **Thermal Biology:** Apparently nonheliothermic (Rodríguez Schettino, 1999b).

Conservation Status: Considered "vulnerable" (Rodríguez Schettino and Chamizo Lara, 1998a; L. Rodríguez Schettino and A. Chamizo Lara in Rodríguez Schettino, 1999a).

Anolis alayoni Estrada and Hedges, 1995

Distribution: Cuba, uplands of Holguín and Guantánamo provinces in eastern Cuba; elevational range 200–900 m.

Habitat: In forests ranging from very dry woods and dry vegetation associated with *Pinus cubensis* to secondary rainforest near streams, lizards perch on tree trunks and twigs at heights of 2–3 m and sleep on twigs several meters high (Estrada and Hedges, 1995); on vines and trunks at 3–4 m above the ground (Fläschendräger and Wijffels, 2009); in submontane mesophyllous evergreen forest, pine forest, and serpentine xeromorphous shrub-woodland (Rodríguez Schettino et al., 1999).

Activity: Diurnal (Rodríguez Schettino et al., 1999). **Behavior:** Often encountered in pairs (Fläschendräger and Wijffels, 2009). **Ecomorphology:** "Twig" (Estrada and Hedges, 1995). **Size:** Male SVL to 46.8 mm, mean female SVL 38.9 mm (Estrada and Hedges, 1995).

Conservation Status: Considered "vulnerable" (Rodríguez Schettino and Chamizo Lara, 1998a; L. Rodríguez Schettino and A. Chamizo Lara in Rodríguez Schettino, 1999a).

Anolis alfaroi Garrido and Hedges, 1992

Distribution: Cuba, known only from the type locality in Guantánamo Province (730 m), but probably widespread in the pine barrens that extend farther north and lie at the western end of the Cuchillas de Moa near the headwaters of the Río Toa (Garrido and Hedges, 1992).

Habitat: During the day, ~1 m high on bushes and grasses beneath pines (*Pinus cubensis*), sleeps at night on stems of ferns, grasses, and bushes, when sympatric with *A. cupeyalensis*, the latter perches closer to the ground (Garrido and Hedges, 1992); in pine forest (Rodríguez Schettino et al., 1999).

Activity: Diurnal (Rodríguez Schettino et al., 1999). **Behavior:** Escapes by jumping into cover or "squirreling" around stems (Garrido and Hedges, 1992). **Ecomorphology:** "Grass-bush" (Garrido and Hedges, 1992). **Size:** Male SVL to 36 mm, female SVL to 33 mm (Garrido and Hedges, 1992).

Conservation Status: Considered "vulnerable" (Rodríguez Schettino and Chamizo Lara, 1998a; L. Rodríguez Schettino and A. Chamizo Lara in Rodríguez Schettino, 1999a).

Anolis aliniger Mertens, 1939

Distribution: Hispaniola, widely distributed, with populations in the Cordillera Central, Sierra de Neiba, Cordillera Septentrional (Dominican Republic), and Massif du Nord, Montagne Noire, Massif de la Selle (Haiti); elevational range ~480–1,300 m.

Habitat: Mesophilic, in broadleaf forests or remnants in uplands, approaches the ground at forest edges, uses branches, upper trunk, and crown (Williams, 1965a; Rand and Williams, 1969); especially females on the ground among cultivated bromeliads, males on nearby trees or wooden supports for the cover over the bromeliad beds (Fläschendräger, 2001a, 2001b, 2001c; Fläschendräger and Wijffels, 2009); locally common in banana (*Musa*) plantations on leaves and banana bundles in the Cordillera Central (RP); mean perch heights and diameters 4.63 m and 18.58 cm (J. B. Losos in Calsbeek et al., 2006); 19.0% on leaves (J. B. Losos, in litt., 21.IV. 2008); sleeps 0.5–3 m above the ground on cut, dead branches in dense rainforest, on a dead and folded tree fern (*Cyathea*) frond, on a stick entangled in pendant vines, on leaves of low bushes, and on a vertical sapling stem (Schwartz and Henderson, 1991); may be closely associated with arboreal bromeliads (Fläschendräger and Wijffels, 2009).

Abundance: Where bromeliads were cultivated, ~30 animals in ~2 h (Fläschendräger and Wijffels, 2009). **Behavior:** When disturbed, moves upward, when displaying, entire body vertically compressed, showing much of venter, but the axillary color and pattern are not exposed, the dewlap is barely developed, the axillary color and pattern may be used as a flash pattern in agonistic behavior (Williams, 1965a); movement characterized as "crawler" (Moermond, 1981); moves per minute 1.19 (Johnson et al., 2008); moves per minute 1.19, percent time displaying 0.6, percent walking 85.7, percent jumping 3.3, percent running 11.0 (J. B. Losos, in litt., 21.IV.2008); vocal, in captivity, male sought refuge submerged in a water-filled bromeliad (Fläschendräger and Wijffels, 2009). **Ecomorphology:** "Trunk-crown" (Rand

and Williams, 1969); formally quantified (Beuttell and Losos, 1999). **Foraging:** Prey strikes either approach-pause or stationary (Moermond, 1981). **Reproduction:** Clutch size 1 (Hedges, in prep.); in captivity, copulation 28 sec, eggs ~5.5 × 10.5 mm, incubation 65–70 days at 24–26 °C (day) and ~20 °C (night), hatchling SVL ~20 mm (Fläschendräger and Wijffels, 2009). **Size:** Male SVL to 58 mm (Fläschendräger and Wijffels, 2009), female SVL to 57 mm (Schwartz and Henderson, 1991).

Conservation Status: Not assessed.

Additional Pertinent Literature: Powell (2002c).

Anolis allisoni Barbour, 1928

Distribution: Cuba, from La Habana Province east to Holguín, Granma, and Santiago de Cuba provinces; from outside the West Indies, on the Islas de la Bahía (Honduras) and cays and islands off the coast of Belize.

Habitat: Mesophilic (Schwartz and Henderson, 1991); savannas, coastal vegetation, common around human habitations, gardens, on fence posts, houses, *Cocos*, and *Roystonea*, usually high (>1.8 m) off the ground, 1–5 m (mean 1.7 m at Playa Caimito; Ruibal, 1961, 1964; Ruibal and Williams, 1961a; Buide, 1966; Schwartz and Henderson, 1991; Rodríguez Schettino, 1999b); on plastered walls of a newly constructed hotel in Granma Province (Dr. Schmidt in Bech, 1985); particularly abundant in ornamental *Scindapsus* (with leaves 70 cm long) in the immediate vicinity of houses (Vergner and Polák, 1990); in *Cocos*, *Coccoloba uvifera*, ornamental agaves, and between the slats of an outdoor table (Kwet, 1995); in Cuba, in mesophyllous evergreen and semideciduous forest and coastal and subcoastal microphyllous forest (Rodríguez Schettino et al., 1999); often in gardens (Kunz, 2005); mean perch height and diameter in Cuba 3.10 m and 12.1 cm (J. B. Losos, in litt., 21.IV.2008).

Activity: Diurnal, most individuals in May active 1200–1400 h (Rodríguez Schettino et al., 1999). **Behavior:** Male threat display is a stereotyped pattern of headbobs of about 3 sec duration, the dewlap is seldom extended in the field (but commonly extended in the laboratory), body laterally compressed, nuchal crest erected, and body raised in push-up (Ruibal, 1967); generally escapes upward, sometimes "squirrels," when on houses, hides under cornices (Rodríguez Schettino, 1999b); low index of arboreality corresponding to an association with edifices (M. A. Granda Martínez in Rodríguez Schettino, 1999b). **Ecomorphology:** "Trunk-crown" (Williams, 1969); categorized as a Cuban "tree anole" (G. Peters, 1970). **Foraging and Diet:** At Playa Caimito in May, stomach contents included ants (most important) and other hymenopterans, flies and mosquitoes, beetles, orthopterans, insect eggs, and spiders, ants also most important in a sample from Sancti Spíritus Province, followed by beetles and flies (Rodríguez Schettino, 1999b); nectarivory on palms (*Veitchia merillii*; Valido, 2006). **Home Range and Territoriality:** Usually only one male and several females on one tree (Kwet, 1995). **Hybridization:** Hybridizes with *A. porcatus* (Ruibal and Williams, 1961a; Glor et al., 2004). **Longevity:** In captivity, 1 year, 8 months (Snider and Bowler, 1992). **Parasites:** In Cuba, nematodes *Cyrtosomum scelopori* (Coy Otero and Baruš, 1973a, 1979b; Coy Otero and Lorenzo Hernández, 1982; Coy Otero, 1999); *Parapharyngodon cubensis* (Coy Otero and Baruš, 1973b, 1979b; Coy Otero and Lorenzo Hernández, 1982; Coy Otero, 1999); *Physaloptera squamatae* (Coy Otero, 1979; Coy Otero and Lorenzo Hernández, 1982; Coy Otero, 1999); *Oswaldocruzia lenteixeirai* and larval *Porrocaecum* sp. (Coy Otero and Baruš, 1979b; Coy Otero and Lorenzo Hernández, 1982; Coy Otero, 1999). **Population Size and Density:** Mean encounter rate at Playa Largo <1 individual/h (Rodríguez Schettino, 1999b). **Predation:** Barn Owl (*Tyto alba furcata*; I. Rodríguez and A. Hernández, 1994 in Arredondo Antúnez and Chirino Flores, 2002). **Reproduction:** Maximum reproductive activity during the rainy season, distinct drop in activity late November–February (Silva Lee, 1984); courtship characterized by bobbing, jiggling (very rapid and shallow headbobs accompanied by a jerky body motion and an advance that involves the alternate raising and trembling of each forelimb), and usually dewlapping, courtship display may or may not involve the standard bobbing pattern, but the pattern is always accompanied by dewlapping (Ruibal, 1967); motionless during copulation (Kwet, 1995); egg 11.5 × 6.5 mm, beige, and oval (Garrido, 1980a); on the grounds of the Museo Provincial in Camagüey, 31 (afternoon) to 46 juveniles (morning) on one small trunk (5 cm in diameter), communal egg deposition (as many as 221 eggs) in petioles of a Traveler's Palm (*Ravenala madagascariensis*), plaster white, oval eggs 12.3–12.7 × 4.5–6.2 mm (Garrido and Pareta, 1994); reproduces throughout the year, with increased egg production during the wet season (based on a larger percentage of gravid females), oviductal egg length 8.3–11.5 mm, eggs deposited in decayed wood or hollows of stone walls, eggs laid by captive females 12.6–18.0 × 5.8–8.0 mm (Rodríguez Schettino, 1999b); incubation in captivity ~42 days at 22–28 °C (Fläschendräger, 1986). **Size:** Male SVL to 91.3 mm, female SVL to 65.9 mm (Schoener, 1988). **Thermal Biology:** Mean body temperature in the morning 32.3 °C (28.2–36.2 °C) and afternoon 33.9 °C (31.0–37.5 °C; Ruibal, 1961); mean body temperature 33.5 °C (27.8–30.8 °C air temperature), 32.3 °C (30.6–32.5 °C), 33.6 °C (31.2–33.6 °C), 34.6 °C (33.0–35.0 °C; Ruibal and Philibosian, 1970); heliothermic, mean body temperature from two localities in Camagüey Province 33.06 °C at mean air temperature 31.6 °C (Rodríguez Schettino, 1999b).

Conservation Status: Not assessed.

Anolis allogus Barbour and Ramsden, 1919

Distribution: Cuba, islandwide, but distribution sporadic and often very local; elevational range, sea level to >1,800 m.

Habitat: Mesophilic (Schwartz and Henderson, 1991); nonheliothermic, confined to heavily shaded broadleaf forests, but also in mangroves, in and about deserted mine buildings with no arborescent growth, about the ruins of an upland plantation house, on rocks and fence posts along a well-shaded road, remains under rocks on the ground during overcast weather, males on small tree trunks within 2 m of ground, females often in leaf litter (Ruibal, 1961, 1964; Ruibal and Williams, 1961b; G. Peters and Schubert, 1968; Schwartz and Henderson, 1991; Rodríguez Schettino, 1999b); densely shaded sites in mesic forests, with a preference for localities near water, males on elevated perches (e.g., tree trunks and large rocks) to 1 m above the ground, females more likely to be on the ground or on moss-covered rocks in deep shade (Fläschendräger and Wijffels, 1996, 2009; Fläschendräger, 2002); in the Biosphere Reserve "Sierra del Rosario," most frequently in shaded interior forest, in draws with rocks and near bodies of water, preferentially on rocks, but also on the ground and trunks of trees and bushes (Martínez Reyes, 1995); in submontane rainforest, submontane mesophyllous evergreen forest, and mesophyllous evergreen and semideciduous forest (Rodríguez Schettino et al., 1999); mean male perch height 0.98 m, typically with head facing down, females and juveniles in leaf litter, on the ground, or on stones at mean height 0.61 m (Rodríguez Schettino, 1999b); mean perch height at Soroa 0.92 m, diameter 11.3 cm, behavioral diameter (mean diameter of all surfaces used during behavioral observations, means of individual means) 17.51 cm, 29.3% of observations on or within 1 m of rocks (J. B. Losos, in litt., 20.VIII.2007); mean perch height and diameter in Cuba west of Cienfuegos 1.28 m and 8.1 cm, in southeastern Cuba 0.88 m and 4.5 cm (J. B. Losos, in litt., 21.IV.2008); habitat light environment "full shade" (Fleishman, 2000); sleeps on leaves of shrubs (Schwartz and Henderson, 1991); females and juveniles may sleep under stones (Rodríguez Schettino, 1999b).

Activity: Diurnal (Rodríguez Schettino et al., 1999); first individuals observed after 0700 h, with activity increasing from 0730 h to a peak between 1230 and 1430 h (Rodríguez Schettino, 1999b). **Behavior:** Not shy, allows an intruder to approach within 30 cm before running down the trunk and onto the ground (Ruibal and Williams, 1961b); females are quicker to react to threats (Rodríguez Schettino, 1999b); mean approach distance 0.87 ± 0.05 m, usually runs to the ground, where cryptic color blends into complex microhabitat, first response "squirreling" (39 of 44) followed by running down (23 of 39) or jumping (5), running up (5), or staying in position (6) at about equal frequencies and first response jumping (5 of 44) followed by second responses jumping (3 of 5) and running down (Regalado, 1998); male threat display (about 12 sec) includes dewlap extension (either held extended or pulsed 8–12 times), tail tip curled laterally and wagged vertically and synchronously with headbobs, mouth partially open, large nuchal and dorsal crests raised (Ruibal, 1967); moves per minute 0.51 (Johnson et al., 2008); percent time displaying 3.7, percent walking 6.5, percent jumping 42.8, percent running 45.5 (J. B. Losos, in litt., 21.IV.2008). **Ecomorphology:** "Trunk-ground" (Williams, 1969). **Foraging and Diet:** Eats *Anolis rejectus* (Socarrás et al., 1988); diet includes cockroaches, beetles (most important in terms of numbers consumed), flies, hemipterans, ants, termites, lepidopteran larvae, orthopterans, barklice, crustaceans, millipedes, and spiders, males eat larger prey than females, but also take many small prey items (Silva Rodríguez and Estrada, 1994); uses a sit-and-wait strategy during the wet season, but forages actively in leaf litter and under cover during the dry season (Rodríguez Schettino, 1999b). **Home Range and Territoriality:** Males actively defend territories (Rodríguez Schettino, 1999b). **Parasites:** Cestode *Proteocephalus* sp. (Coy Otero, 1970; Coy Otero and Lorenzo Hernández, 1982; Coy Otero, 1999); nematodes *Cyrtosomum scelopori* (Baruš and Coy Otero, 1969a; Coy Otero, 1970; Coy Otero and Baruš, 1973a, 1979b; Coy Otero and Lorenzo Hernández, 1982; Coy Otero, 1999); *Oswaldocruzia lenteixeirai* (Baruš and Coy Otero, 1969a; Coy Otero, 1970; Coy Otero and Baruš, 1973a, 1979b; Coy Otero and Lorenzo Hernández, 1982; Coy Otero, 1999); larval *Abbreviata* sp. (Coy Otero, 1979; Coy Otero and Lorenzo Hernández, 1982; Coy Otero, 1999); *Parapharyngodon cubensis* (Baruš and Coy Otero, 1969a, 1969b; Coy Otero, 1970; Coy Otero and Baruš, 1979b; Coy Otero and Lorenzo Hernández, 1982; Coy Otero, 1999); *Physaloptera squamatae* and larval *Porrocaecum* sp. (Coy Otero and Baruš, 1979b; Coy Otero and Lorenzo Hernández, 1982; Coy Otero, 1999). **Reproduction:** Mean duration of copulation at 2 sites 19.04 min and 10.02 min (longest 37 and 32 min), separation of pair "normal," egg 8.5 × 6.1 mm, plaster white in color and oval (Garrido, 1980a); cyclical during the rainy season, with males with enlarged testes (>4 mm) and females with vitellogenic follicles February–July or with 1 or 2 oviductal eggs April–September (Silva Rodríguez and Estrada, 1984); clutch size 1 (Hedges, in prep.); in captivity, females lay eggs at intervals of 10–14 days, incubation 50–55 days at 23–29 °C, hatchling SVL ~19 mm (Fläschendräger and Wijffels, 2009). **Size:** Male SVL to 62.8 mm (Rodríguez Schettino, 1999b), female SVL to 49 mm (Schwartz, 1968f). **Thermal Biology:** Mean body temperature in the morning 28.7 °C (26.2–32.4 °C), afternoon 29.6 °C (26.2–33.5 °C), panting initiated at mean temperature of 30.3 °C (28.5–32.8 °C; Ruibal, 1961); nonheliothermic thermoconformer, mean body temperature at Camagüey 29.2 °C in July and August with mean air temperature 30.0 °C, in the Sierra del Rosario mean body and air temperatures 27.5 °C (Rodríguez Schettino, 1999b); mean body temperature 29.4 °C (30.6 °C air temperature), 27.7 °C (27.6–29.4 °C), 29.3 °C (30.4–31.0 °C; Ruibal and Philibosian, 1970); mean body temperature at Soroa 27.0 °C (J. B. Losos, in litt., 20.VIII.2007).

Conservation Status: Not assessed.

Anolis altavelensis Noble and Hassler, 1933

Distribution: Hispaniola, Isla Alto Velo (Dominican Republic).

Habitat: Xerophilic (Schwartz and Henderson, 1991); on rocks, trees, and bushes (Noble and Hassler, 1933).

Activity: From sunrise until long into dusk (Noble and Hassler, 1933). **Behavior:** Very wary and difficult to approach (Noble and Hassler, 1933). **Ecomorphology:** "Trunk?" (Losos, 2009). **Size:** Male SVL to 47 mm, female SVL to 44 mm (Schwartz and Henderson, 1991). **Thermal Biology:** Remains in shaded places during the heat of the day (Noble and Hassler, 1933).

Conservation Status: Unknown.

Anolis altitudinalis Garrido, 1985

Distribution: Cuba, picos Turquino, Cardero, and Sierra Maestra; elevational range >1,000–1,720 m (Garrido and Hedges, 2001).

Habitat: Mesophilic (Schwartz and Henderson, 1991); in moist montane forests in leaf litter (females), on the trunk of a tree ~4 m high (male), on the ground among roots adjacent to a small creek (gravid female), and among leaves of cultivated strawberry plants (Garrido, 1985).

Activity: Diurnal (Rodríguez Schettino et al., 1999). **Behavior:** Females on the ground try to escape by running, a male on a tree trunk did not try to escape (Garrido, 1985). **Ecomorphology:** "Trunk-crown" (Losos, 2009). **Reproduction:** Egg creamy white, oval, semielliptical, 10 × 5.7 mm (Garrido, 1985). **Size:** Male SVL to 52 mm, female SVL to 51 mm (Garrido, 1985). **Thermal Biology:** Apparently nonheliothermic (Rodríguez Schettino, 1999b).

Conservation Status: Considered "vulnerable" (L. Rodríguez Schettino and A. Chamizo Lara in Rodríguez Schettino, 1999a).

Anolis alumina Hertz, 1976

Distribution: Hispaniola, the Península de Barahona (Dominican Republic), including the southern slopes of the Sierra de Baoruco, west into extreme southeastern Haiti, and east along the coast to the northern uplands.

Habitat: Xerophilic (Schwartz and Henderson, 1991); in undergrowth of pine savanna, perching <60 cm high on vegetation, sleeps on living and dead desert shrubs, on dead *Acacia* in and along margins of cleared fields, on dead grass and twigs or near meadows, under rocks in hammock woods, and on a tree leaf 1.2 m high (Hertz, 1976).

Behavior: Drops to the ground to escape, scurrying into dense clumps of grass or between rocks (Hertz, 1976). **Ecomorphology:** "Grass-bush" (Hertz, 1976). **Reproduction:** Clutch size 1 (Hedges, in prep.). **Size:** Male SVL to 40 mm, female SVL to 37 mm (Schwartz and Henderson, 1991).

Conservation Status: Not assessed.

Anolis alutaceus Cope, 1861

Distribution: Cuba, islandwide, but local, and on Isla de la Juventud; elevational distribution to >1,000 m (Fläschendräger and Wijffels, 1996, 2009).

Habitat: Mesophilic (Schwartz and Henderson, 1991); largely terrestrial, in broadleaf forests that are moderately damp and have low light intensity, on small tree trunks, stems, twigs, grasses, and rocks, usually <50 cm above the ground, also in holes and at the base of walls in ruins, rock piles, and among sticks and leaves of forest floor, occasionally in weeds in open areas, but never in full sunlight (Collette, 1961; Ruibal, 1964; Garrido and Schwartz, 1968b; Garrido, 1980a; Schwartz and Henderson, 1991; Fläschendräger and Wijffels, 1996, 2009; Fläschendräger, 2002); in the Biosphere Reserve "Sierra del Rosario," primarily in secondary growth, associated with herbaceous and bushy strata, preferentially on herbaceous growth, but also on trunks and branches of bushes (Martínez Reyes, 1995); in submontane rainforest, submontane mesophyllous evergreen forest, mesophyllous evergreen and semideciduous forest, mogote complex of vegetation, coastal and subcoastal microphyllous forest, and cultivated areas and pastures (Rodríguez Schettino et al., 1999); on thin twigs and lianas, especially in shrubs (Estrada, 1994); usually below 1 m above the ground, sometimes in caves and in deep holes in dogtooth limestone (Rodríguez Schettino, 1999b); in dense herbaceous growth immediately adjacent to a waterfall, directly "above" *A. allogus*, which inhabits the moss-covered boulders and leaf litter below the thin perches used by *A. alutaceus*, also in dense herbaceous growth and on bushes, particularly common on leaves of tree ferns (*Cyathea* spp.), to heights of ~1 m (Fläschendräger, 1998; Fläschendräger and Wijffels, 2009); mean perch height at Soroa 1.56 m, diameter 1.2 cm, behavioral diameter (mean diameter of all surfaces used during behavioral observations, means of individual means) 7.42 cm, 10.4% of observations on or within 1 m of rocks (J. B. Losos, in litt., 20.VIII.2007); mean perch height and diameter 0.93 m and 3.1 cm, 73.0% on leaves (J. B. Losos, in litt., 21.IV.2008); sleeps on long leaves of vines and trees, grasses, and ferns up to 1.5 m above the ground (Schwartz and Henderson, 1991; Rodríguez Schettino, 1999b).

Activity: Diurnal (Rodríguez Schettino et al., 1999); most individuals active in early morning, retreats to deep shade by midday (Rodríguez Schettino, 1999b). **Behavior:** Escapes by retreating into holes in walls or by hopping to the ground and hiding under leaves (Collette, 1961); leaps instead of running or climbing, uses long tail for balancing while hopping (Ruibal, 1964; G. Peters, 1970); often immobile and difficult to see (Garrido and Schwartz, 1968b); after jumping, often relies on crypsis and remains immobile (Garrido, 1980a); "squirrels" to the other side of a stem or twig to hide or adroitly jumps into dense cover (Fläschendräger, 1998); male aggressive display includes lateral presentation, head movement with body laterally compressed, nodding, dewlap extension (dewlap remains extended), 6–7 vigorous flexions in two sequences, protruded tongue, and raised nuchal crest, display may last 23 sec, but usually only 13–14 sec (Garrido, 1980a); moves per minute 0.64 (Johnson et al., 2008); percent time displaying 0.7, percent walking 34.8, percent jumping 56.0, percent running 13.2 (J. B. Losos, in litt., 21.IV.2008). **Ecomorphology:** "Grass-bush" (Williams, 1961); categorized as a "grass anole" (G. Peters, 1970). **Foraging and Diet:** Ants (Barbour and Ramsden, 1919); approaches prey very slowly (Fläschendräger and Wijffels, 2009). **Home Range and Territoriality:** Apparently not highly territorial, with no defined home ranges, which explains apparent syntopy with other grass anoles (G. Peters and Schubert, 1968). **Parasites:** Nematodes *Skrjabinoptera phrynosoma* and larval *Porrocaecum* sp. (Coy Otero and Baruš, 1979b; Coy Otero and Lorenzo Hernández, 1982); *Spauligodon anolis* (Baruš and Coy Otero, 1974; Coy Otero and Baruš, 1979b; Coy Otero and Lorenzo Hernández, 1982; Coy Otero, 1999). **Population Density and Abundance:** In December, 17 individuals seen or captured in slightly over 2 h, 6 individuals collected in 1.5 h (Collette, 1961); in great numbers in dense vegetation (Mudde, 1999). **Predation:** *Tropidophis maculatus* (Collette, 1961). **Reproduction:** Clutch size 1 (Hedges, in prep.); single egg under decayed bark (Barbour and Ramsden, 1919); small oval eggs in bromeliads (Dunn, 1926d); eggs among leaves and under small rocks (B. B. Collette in Rand, 1967c); courtship elaborate and prolonged, including up-and-down neck movement, lateral movement, vibration and raising of head, nodding, and dewlap extension taking 6–12 sec, usually 9–10 sec depending on number of body flexions, female watches attentively and immobile, even when male bites her neck (if she is receptive), male mounts female and is immobile from 7–10 sec, female nods at beginning of copulation, mean duration of copulation 9.01 min (longest 23 min), separation "rapid," lays a single oblong egg under decayed bark or in palm trash, eggs cream colored, 7.8 × 5.0 mm, hatch 22 July (Garrido, 1980a); in captivity, egg immediately after deposition ~5 × 7 mm, shortly before hatching ~7 × 9 mm (Fläschendräger, 1998); reproduces throughout the year, incubation 40–45 (occasionally as many as 60) days at 23–28 °C, hatchling SVL ~15 mm (Fläschendräger and Wijffels, 2009). **Size:** Male SVL to 37.5 mm (Garrido, 1980a), female SVL to 37 mm (Schwartz and Henderson, 1991). **Thermal Biology:** Active at temperatures below those of other anoles (Collette, 1961); nonheliothermic (Rodríguez Schettino, 1999b); increasingly thermophilic at higher elevations (Fläschendräger and Wijffels, 1996, 2009; Fläschendräger, 2002); mean body temperature at Soroa 28.3 °C (J. B. Losos, in litt., 20.VIII.2007).

Conservation Status: Not assessed.

Anolis anfiloquioi Garrido, 1980

Distribution: Cuba, extreme eastern Cuba near Baracoa, Guantánamo Province; a specimen from Levisa, Nícaro may also pertain to this species (Schwartz and Henderson, 1991).

Habitat: Mesophilic (Schwartz and Henderson, 1991); in shaded overgrown areas under leafy tree at edge of forest, weedy and grassy areas, river margins in trees, bushes, and vines (Garrido, 1980a); in mesophyllous evergreen and semideciduous forest (Rodríguez Schettino et al., 1999); males perch on branches and leaves of small, sparse bushes and grasses at heights ~0.5 m, females and juveniles lower, often on the ground (Rodríguez Schettino, 1999b).

Activity: Diurnal (Rodríguez Schettino et al., 1999). **Behavior:** Temperament generally placid, makes no effort to escape, when pressed, jumps from low branches and hides in leaf litter, response to an intruder includes immobility, followed by lateral presentation, slight and low neck movement, extension of dewlap, tongue protrusion, and nodding (Garrido, 1980a). **Ecomorphology:** "Grass-bush" (Losos, 2009). **Population Density:** Densities low, at Pinares de Mayarí, only individuals encountered (Fläschendräger and Wijffels, 2009). **Size:** Male SVL to 40.5 mm (Rodríguez Schettino, 1999b), female SVL to 39 mm (Garrido, 1980a). **Thermal Biology:** Apparently nonheliothermic (Rodríguez Schettino, 1999b).

Conservation Status: Considered "vulnerable" (Rodríguez Schettino and Chamizo Lara, 1998a; L. Rodríguez Schettino and A. Chamizo Lara in Rodríguez Schettino, 1999a).

Anolis angusticeps Hallowell, 1856

Distribution: Cuba and Bahamas, where two currently recognized subspecies are widely distributed on Cuba and many satellites, including Isla de la Juventud (*A. a. angusticeps*), and on the Great Bahama Bank and Cat Island in the Bahamas (*A. a. oligaspis*).

Habitat: Mesophilic (Schwartz and Henderson, 1991); on small bushes and tree trunks 0.6–1.8 m above the ground in hardwoods and pine savanna, often in edge situations on small islands, on fence posts, rocks (especially those that are white or gray), walls of ruins, trunks of palms, *Coccoloba* and *Terminalia* along the coast, on wire of barbed-wire fence adjacent to woods, sloping tree trunk 2.4 m above ground, tops of *Bursera* and *Lysiloma* saplings, on tamarind sapling 1.5 m above ground, on *Roystonea* trunks, more arboreal on Cuba than in the Bahamas (Barbour and Ramsden, 1919; Barbour, 1930c; Collette, 1961; Ruibal, 1964; Buide, 1966; Hardy, 1967; Schoener, 1968; Schwartz and Thomas, 1969; Garrido, 1975d; Garrido et al., 1986; Schwartz and Henderson, 1991; Estrada, 1998; Rodríguez Schettino, 1999b); in the Biosphere Reserve "Sierra del Rosario" (Cuba), primarily in secondary vegetation, preferentially on branches of bushes but also on branches of trees (Martínez Reyes, 1995); forest edges, open stands of trees, especially those with abundant shrubs, and pines, usually on twigs, but occasionally on trunks of pines and other trees (Fläschendräger, 2002); to perch heights of ~2.5 m on small tree trunks and particularly on vertical narrow (4–8 cm) branches in forested areas and coastal scrub characterized by *Coccoloba uvifera* and *Terminalia* sp. (G. Peters and Schubert, 1968); in Cuba, in cloud forest, montane rainforest, submontane rainforest, submontane mesophyllous evergreen forest, mesophyllous evergreen and semideciduous forest, pine forest, coastal and subcoastal microphyllous forest, semidesert thorny scrubwoodland, serpentine xeromorphous shrub-woodland, and urban zones (Rodríguez Schettino et al., 1999); on Bimini, in the upper branches of light, gray-colored trees at heights of 1.8–7.6 m (Oliver, 1948); also on Bimini, nearly always in tree crowns on branches of small diameter (no individuals on the ground), those on elevated perches rarely on leaves, largest percentage of adult males (21%) on perches 1.7–2.6 m high and with diameters <2–4.5 cm, plurality of females (adults and subadults, 42%) on perches >2.6 m high and with diameters <2–4.5 cm, most juveniles (50%) on perches >2.6 m high and (75%) with diameters <1.6 cm (Schoener, 1968); on the Archipiélago de los Canarreos (Cuba), on small-diameter branches high in trees and on palms (*Thrinax radiata*; Estrada, 1993a); on branches or twigs in bushes, crowns, or jutting from fallen trees, none on trunks or on the ground, mean perch height and diameter for males 2.2 ± 0.3 m and 2.3 ± 0.5 cm, for females 1.8 ± 0.4 m and 0.7 ± 0.2 cm (Irschick and Losos, 1996); on Cayo Paredón Grande (Archipiélago de Sabana-Camagüey), exclusively on trunks of palms (*Coccothrinax littoralis*; Estrada, 1998); generally perches at heights >1 m, mean height on the Península de Guanahacabibes, 2.28 m, with adults on trunks and high branches, juveniles on grasses and small bushes (Rodríguez Schettino, 1999b); lizards most frequently on very small (<2 cm diameter) perches, with relatively high habitat breadth of 91.7 (Irschick and Losos, 1999); 62% of observations on Bimini at heights <2 m (Mattingly and Jayne, 2004); mean heights on Bimini 173 ± 9 cm (Mattingly and Jayne, 2005); mean perch height at Soroa 2.79 m, diameter 11.6 cm, behavioral diameter (mean diameter of all surfaces used during behavioral observations, means of individual means) 3.26 cm, no observations on or within 1 m of rocks (J. B. Losos, in litt., 20.VIII.2007); mean perch height and diameter in the Bahamas 1.84 m and 1.7 cm, 3.09 m and 0.9 cm in Cuba west of Cienfuegos, 2.10 m and 9.9 cm in southeastern Cuba (J. B. Losos, in litt., 21.IV.2008); in the Bahamas, sleeps lengthwise on small-diameter branches and twigs 1.2–3.6 m high in well-developed coppice, only on blackland or incipient blackland, on high twigs and branches of small diameter (Schwartz and Thomas, 1969); in Cuba, sleeps longitudinally on thin branches with the tail extended and on the branch, demonstrates no preference for type, size, or height of sleeping perches (Rodríguez Schettino, 1999b).

Activity: Diurnal (Rodríguez Schettino et al., 1999). **Behavior:** Often rests on trunks with head facing upward, remains motionless until approached very closely, relying on crypsis, but climbs quickly when pressed (Oliver, 1948; Collette, 1961; Schwartz and Thomas, 1969; Rodríguez Schettino, 1999b); "squirrels" quickly when faced with a threat (Silva Lee, 1984); in male-male encounters, sometimes sex-recognition does not occur and males initiate courtship displays, when recognition occurs, they immediately assume a lateral presentation with continuous head-bobbing, color darkens, and they compress their bodies laterally while elevating the anterior body (push-up) and fully extending the dewlap 8–13 times with tongue protruding, after a pause of a few seconds, the sequence may be repeated, albeit usually of shorter duration, sometimes without tongue protrusion or dewlap extension, but with moderated headbobs, the encounter terminates with one contender retreating or with an attack that involves biting (Garrido, 1991); lateral presentation during aggressive displays involves each individual leaning toward the other in a manner that presents the dorsum (Fläschendräger and Wijffels, 1996, 2009); displays infrequently, moves slowly out of sight when disturbed, males walk 89.3 ± 2.0%, run 1.9 ± 0.9%, jump 8.8 ± 1.7%, females walk 92.4 ± 3.7%, run 0.5 ± 0.4%, jump 7.1 ± 3.6% of locomotor movements (Irschick and Losos, 1996); 11.4% of distance traveled on trunk bases, 16.0% on trunks, 72.6% on branches, 8.8% of movements with jumps, mean movement distance 7.3 cm (median 3.9 cm) at speed 4.9 cm/sec (2.7 cm/sec), mean pause duration 32.8 sec (10.3 sec; Mattingly and Jayne, 2004); during an escape chooses broadest diameter 83% of time, longest length 40%, shallowest incline 52%, and straightest angle 73% (Mattingly and Jayne, 2005); neither absolute nor size-corrected dewlap size significantly related with bite force, 75% of combat trials involved dewlap extensions, 5% biting, and 100% headbobbing (Lailvaux and Irschick, 2007); moves per minute 1.19 on South Bimini and 2.49 in western Cuba (Johnson et al., 2008); in western Cuba, percent time displaying 5.5, percent walking 65.7, percent jumping 14.4, percent running 20.0 (J. B. Losos, in litt., 21.IV.2008). **Diet:** On Bimini, food items were ants, other hymenopterans, aphids (greatest numbers in adult males and females), other hemipterans, flies (greatest numbers and volumes in females and juveniles), thrips, beetles, barklice, lepidopterans (adults and larvae, greatest numbers and volume in subadult males), lacewings, spiders, mites, and plant matter (Schoener, 1968); crawls slowly about the vegetation, inspecting it carefully for food, most obtained from twigs, feeds most actively early in the day or on cloudy days (Schoener, 1979). **Ecomorphology:** Categorized as a Cuban "tree anole" (G. Peters, 1970); "twig" (Williams, 1972); formally quantified (Beuttell and Losos, 1999). **Growth:** On Bimini, males and females grow more rapidly than *A. sagrei*, suggesting less food pressure on arboreal species (Schoener, 1979). **Home Range:** On Bimini in *Terminalia* habitat, median home range size of males 9.02 m^2, of females 5.37 m^2, home ranges generally do not overlap, but that may be attributable to relative rarity rather than active defense (Schoener, 1979); home ranges of adult males (11.51 ± 9.19–38.20 ± 49.66 m^2 at 3 Bahamian sites) larger than those of females (4.96 ± 6.49–7.32 ± 4.42 m^2 at 3 sites) and subadult males (4.96 ± 1.89–8.26 ± 0.48 m^2 at 2 sites; Schoener and Schoener, 1982b). **Injuries:** Injuries to tails, front toes, and back toes about equal in males and females, with front toe rates presumably due to intraspecific aggression (*A. angusticeps* is less territorial than sympatric Bahamian anoles) and tail and hind toe rates presumably due to predation, tail-break rates unusually low compared with other Bahamian anoles examined (Schoener and Schoener, 1980b). **Movements:** Mean distance per movement 15.6 ± 1.3 cm at mean speed 10.5 ± 2.4 cm/sec, 0.72 ± 0.19 moves/min, 2.12 ± 0.35% time moving (Irschick, 2000b). **Performance:** Mean maximal speed 0.8 ± 0.10 m/sec, escape speed 0.8 ± 0.10 m/sec, feeding speed 0.4 ± 0.12 m/sec, undisturbed activity 0.3 ± 0.09 m/sec, undisturbed jumping distance 16.7 ± 3.18 cm (Irschick and Losos, 1998); average gross speed over a 5-stride interval on a 4.8-cm diameter perch 49 ± 6.1 cm/sec and 44 ± 5.5 cm/sec with 90° turning angle, stride-one speed 55 ± 8.0 cm/sec and 25 ± 4.0 cm/sec with 90° turning angle (Mattingly and Jayne, 2005). **Population Density and Abundance:** In December, 5 individuals taken in 70 min, 5 in 30 min, and only 2 in 2 weeks in a different year (Collette, 1961); male densities in various Bahamian habitats 0.019–0.154 individuals/m^2, female densities 0.011–0.095 individuals/m^2 (Schoener and Schoener, 1978b); on Bimini, in "short-blackland" habitat, 0.013 males/m^2, 0.018 females/m^2, in "forest-path" habitat, 0.030 males/m^2, 0.008 females/m^2, in "*Terminalia*" habitat, 0.108 males/m^2, 0.067 females/m^2 (Schoener, 1979); in various habitats at multiple sites on 2 Bahamian islands, male densities are 0.000–0.154 lizards/m^2, female densities 0.000–0.096 lizards/m^2, total densities 0.000–0.250 lizards/m^2 (Schoener and Schoener, 1980a, 1980b); in Cuban coastal scrub, encounter rate of 5.76 lizards/h, in ecotone between coastal scrub and dry semideciduous forest 3 individuals/h, on the Península de Guanahacabibes 0.3 lizards/h in coastal scrub and 0.4/h in semideciduous forest (Rodríguez Schettino, 1999b). **Predation:** *Tropidophis maculatus* (Collette, 1961). **Reproduction:** Reproduction is seasonal and correlated with the rainy season (Fläschendräger and Wijffels, 1996, 2009); mean duration of copulation 9 min (longest 23 min), separation of pair "normal," egg 10 × 5 mm, cream colored and semielliptical (Garrido, 1980a); upon sighting a female, the male approaches her, begins to headbob, gradually increasing the rate while simultaneously expanding his throat and extending his dewlap to about one-fourth of its maximal capacity, the female responds with headbobs but otherwise remains tranquil, the male then grasps her neck, gently shaking her 2–3 times, and then remains immobile for a few seconds before attempting intromission, mean duration of copulation 9 min (6–23 min), during which the male loosens his hold on the female's neck and often shakes his head once or twice, the female may also shake her head, the male brusquely terminates the encounter (Garrido, 1991); communal nest with 1,625 eggshells and 65 recently laid eggs, mean egg size 9.9 ± 0.11 × 7.0 ± 0.06 mm, mean neonatal SVL 17.4 ± 0.11 mm (Novo Rodríguez, 1985a); nests communally on Cayo Coco, eggs in September, 9.7 × 5.5 mm, 9.2 × 4.6 mm (at oviposition), increased in size to 10.1 × 7.1 mm, additional eggs 9.4 × 6.7 mm, 8.9 × 6.3 mm, 9.7 × 6.5 mm, 9.5 × 6.4 mm, eggs deposited at intervals of 11 and 20 days with measurements of 10.1 × 6.9 mm and 10.5 × 6.9 mm, hatched 22 days after deposition (Estrada, 1993b); in captivity, incubation 45–55 days at 23–28 °C (Fläschendräger and Wijffels, 2009). **Sex Ratio:** In various habitats at multiple sites on 2 Bahamian islands F/M 0.63–1.50 (excluding data with sample sizes of 3 for one or both sexes; Schoener and Schoener, 1980a); 0.96:1.04 M:F (Schoener and Schoener, 1980b; T. W. Schoener in Stamps, 1983c). **Size:** Male SVL to 53 mm, female SVL to 47 mm (Schwartz and Thomas, 1969); lizards in populations in western Cuba are smaller than those elsewhere on the island (G. Peters, 1970); mean SVL of 28 hatchlings 17.4 mm (Rodríguez Schettino, 1999b); hatchling SVL in captivity ~13.5 mm (Fläschendräger and Wijffels, 2009). **Survival:** Survival higher on smaller island (Bimini) than on larger island (Andros), presumably due to predation by a greater abundance and variety of birds on Andros (Schoener and Schoener, 1978a); annual survival in 4 habitats on two Bahamian islands 0.000–0.500 for males and 0.111–0.429 for females (Schoener and Schoener, 1980b); relatively constant age-specific mortality, percent survival after 12 months of all individuals on Andros and Bimini 21.4% (Schoener and Schoener, 1982a). **Thermal Biology:** Heliothermic in moderately moist area of low light intensity and cooler temperatures (Ruibal, 1964; Fläschendräger and Wijffels, 1996, 2009); mean body temperature in deciduous forest during the wet season 32.4 °C, higher than mean air and substrate temperatures (28.8 °C and 29.3 °C; Rodríguez Schettino, 1999b); mean body temperature at Soroa 29.2 °C (J. B. Losos, in litt., 20.VIII.2007).

Conservation Status: Not assessed.

Anolis argenteolus Cope, 1861

Distribution: Eastern Cuba.

Habitat: Mesophilic (Schwartz and Henderson, 1991); in shaded broadleaf forest, on large and small tree trunks, particularly the lower portions, also on limestone faces and rocky sides of a ravine, rocks, houses, on *Coccoloba* near the coast (Barbour and Ramsden, 1919; Barbour, 1930c; Alayo Dalmau, 1951; Schwartz and Ogren, 1956; Cooper, 1958; Ruibal, 1964; Schwartz and Henderson, 1991; Kwet, 1995; Vergner and Polák, 1996; Rodríguez Schettino, 1999b); when on tree trunks, usually on the shaded side (Fläschendräger and Wijffels, 1996, 2009); in shaded draws with higher humidity than surrounding areas, usually on cliffs, also on walls of houses and singly on trunks of trees (Fläschendräger and Wijffels, 2009); in mesophyllous evergreen and semideciduous forest, coastal and subcoastal microphyllous forest, semidesert thorny scrubwoodland, cultivated areas and pastures, and urban zones (Rodríguez Schettino et al., 1999); males perch higher than females, all individuals sleep high in crowns (Rodríguez Schettino, 1999b); coefficient of arboreality high (M. A. Granda Martínez in Rodríguez Schettino, 1999b); on *Coccoloba* at heights of 1–2 m (Rodríguez Schettino, 2003b); mean perch height and diameter 2.09 m and 23.1 cm (J. B. Losos, in litt., 21.IV.2008).

Activity: Diurnal (Rodríguez Schettino et al., 1999); seldom seen before midday, peak activity in the afternoon (G. Peters and Schubert, 1968); bimodal, with peaks in early morning and late afternoon (Fläschendräger and Wijffels, 1996, 2009; Rogner, 2002); active until well into dusk (Kwet, 1995); begins to leave sleeping sites after 0600 h, activity increases and then stabilizes ~1000 h, during the wet season, number of active individuals decreases from 1100 to 1400 h, then increases again (although fewer lizards are active than during the morning), in the dry season, number of individuals is stable until ~1500 h, diminishing thereafter, few lizards active after 1800 h and none after 1900 h (Rodríguez Schettino, 1999b). **Behavior:** Escapes by running into dense growth of vines and shrubbery (Schwartz and Ogren, 1956); when jumping to escape may turn completely around (M. Černý, 1992); flight via series of short, irregular jumps, during which lizards rapidly change direction

(Vergner and Polák, 1996); escapes rapidly into tree crowns or to the highest part of walls, occasionally jumping from branch to branch, advertisement displays rare, consist of headbobs and dewlap extensions, but no other threatening gestures (Rodríguez Schettino, 1999b); threat displays in captivity involve rhythmic tail waving (Baedermann, 1989); vocalizes (G. Peters, 1970; Rodríguez Schettino et al., 1999). **Competition:** Rarely syntopic with *A. lucius*, which has similar habitat requirements, but both species on trunks of trees adjacent to cliffs in Camagüey (Estrada, 1987a). **Diet:** Feeds mainly on ants (Rodríguez Schettino, 2003b); also takes other hymenopterans, flies, beetles, hemipterans, lepidopterans, orthopterans, roaches, insect larvae (mainly lepidopterans), spiders, and pseudoscorpions, males take larger prey than females (Rodríguez Schettino and Martínez Reyes, 1996; Rodríguez Schettino, 1999b). **Ecomorphology:** Undetermined; categorized as a Cuban "gecko anole" (G. Peters, 1970). **Home Range and Territoriality:** Defends territories only "slightly" or not at all, pairs and males may share a single tree or even a trunk without aggressive displays (Rodríguez Schettino, 1999b). **Miscellany:** "Windows" in lower eyelids presumed to act as "sunglasses," reducing light intensity when shade-dwelling lizards move into bright sun (Williams and Hecht, 1955); may serve to detect movement while sleeping (Vergner and Polák, 1996). **Parasites:** Nematode *Cyrtosomum scelopori* (Coy Otero and Baruš, 1979b; Coy Otero and Lorenzo Hernández, 1982; Coy Otero, 1999); trematode *Turquinia cubensis* (Coy Otero, 1989; Coy Otero, 1999). **Population Size and Density:** At La Mula, density 82–200 lizards/ha, biomass 0.19–0.40 kg/ha (Rodríguez Schettino, 1999b). **Population Structure:** Adults:immatures (subadults and juveniles) 7.7:1, large proportion of adults throughout the year (Rodríguez Schettino, 1999b). **Predation:** *Anolis imias* along the southern coast of Guantánamo Province (J. de la Cruz in Socarrás et al., 1988; Rodríguez Schettino, 1999b). **Reproduction:** Breeds year-round and lays eggs in communal nests (Rodríguez Schettino, 2003b); courtship with bobs, dewlap extensions, and short "runs" at the female, mean duration of copulation 1.18 min (longest 1.23 min), separation of pair "hasty," egg 12.5 × 5.2 mm, cream colored and elliptical (Garrido, 1980a); 40 eggshells in a communal nest in September (Estrada, 1988); oval eggs (in captivity) 10 × 5 mm (Zernikov, 1987); 8 × 5 mm (Rogner, 2002); 9.5 ± 0.7 × 5.1 ± 0.8 mm (Vergner and Polák, 1996); in captivity, eggs ~6 × 12 mm, females lay 4–6 (rarely as many as 10) eggs/year, incubation 55–70 days at 23–30 °C, hatchling SVL ~23 mm (Fläschendräger and Wijffels, 1996, 2009); most gravid females March–September, most with one egg, but during the peak of the reproductive season 10% of females with 2, approximately 8 days between eggs laid during that period (~20 days during other months), mean length of eggs 12.0 mm, eggs in holes of tree trunks and under "lapels" of limestone cliffs, size of fat bodies inversely related to reproductive activity (Rodríguez Schettino, 1999b); eggs laid in the "smallest clefts and crevices" (Lambrechts, 2005). **Sex Ratio (M:F):** 0.9:1 (Rodríguez Schettino, 1999b). **Size:** Male SVL to 59.8 mm, female SVL to 51.0 mm, smallest hatchling SVL 21.4 mm, maximum SVL of immature males 40.9 mm, females 35.9 mm (Rodríguez Schettino, 1999b); in captivity, hatchling SVL 18 mm (Zernikov, 1987), 17.21 ± 1.17 mm (Vergner and Polák, 1996). **Thermal Biology:** Nonheliothermic, mostly in shaded sites, mean body temperatures (27.0–30.0 °C) vary but are maintained above ambient temperatures (Rodríguez Schettino, 1999b).

Conservation Status: Unknown.

Anolis argillaceus Cope, 1862

Distribution: Extreme eastern Cuba, but not along the southeastern coast.

Habitat: Mesophilic (Schwartz and Henderson, 1991); broadleaf forest and other dense growth in both upland and lowland situations, but also in savannalike habitats, coffee plantations, and *Guazuma ulmifolia* stands, on trunks of relatively large trees and fence posts, clumps of grass along a road, coastal *Coccoloba* on seaside cliffs (Barbour and Ramsden, 1919; Schwartz and Ogren, 1956; Ruibal, 1964; Lando and Williams, 1969; Garrido, 1975b; Schwartz and Henderson, 1991); on trunks and large limbs of pines, avoids those without lichens, which appear to facilitate crypsis (G. Peters and Schubert, 1968); in mesophyllous evergreen and semideciduous forest, pine forest, coastal and subcoastal microphyllous forest, and serpentine xeromorphous shrubwoodland (Rodríguez Schettino et al., 1999); generally with head down on tree and bush trunks with rugose bark (Rodríguez Schettino, 1999b); mean perch height and diameter 2.30 m and 1.8 cm (J. B. Losos, in litt., 21.IV.2008).

Activity: Diurnal, males begin to leave shelters under bark when in direct sun (Rodríguez Schettino et al., 1999; Rodríguez Schettino, 1999b). **Behavior:** Moves slowly, when trying to escape, seeks shelter in vegetation or interstices of bark, does not climb (Rodríguez Schettino, 1999b). **Ecomorphology:** Categorized as a "bark anole" (G. Peters, 1970); "trunk" (Williams and Rand, 1977), but not tested morphologically. **Parasites:** Nematodes *Cyrtosomum scelopori* (Coy Otero, 1970; Coy Otero and Baruš, 1973a, 1979b; Coy Otero and Lorenzo Hernández, 1982; Coy Otero, 1999); *Physalopteroides valdesi* (Coy Otero and Baruš, 1980; Coy Otero and Lorenzo Hernández, 1982; Coy Otero, 1999);

and larval *Porrocaecum* (Coy Otero and Baruš, 1979b; Coy Otero and Lorenzo Hernández, 1982; Coy Otero, 1999). **Reproduction:** Mean duration of copulation 3.8 min (longest 5.25 min), separation of pair "normal" (Garrido, 1980a); courtship of 2 sequences, 7–9 bobs followed by a second series of 4 bobs, with both series ending in "trembling," dewlap totally extended during first sequence, but only partially during second, when situated above female, male grabs her neck and remains immobile for several seconds before intromission, copulation 2–5.25 min (mean 3.9 min) during which the male releases the neck bite at times, only to bite again, postcopulatory separation is brusque (Garrido, 1988); 10 eggs deposited communally in semidecayed pine (*Pinus cubensis*), some eggs in chamber with ants (*Odontomachus haematoda*), eggs 7.1–13.0 × 6.3–8.9 mm when collected (10 April), 9.4–10.4 × 6.9–7.5 mm (9 May), hatching irregular, first on 13 April, last after 34 days, mean neonatal SVL 16.8 ± 0.24 mm (Estrada, 1987b); majority of females captured at Cupeyal in January gravid, but not along the Río Toa near Baracoa (Rodríguez Schettino, 1999b). **Size:** Male SVL to 48.0 mm (Garrido, 1975d), female SVL to 44.8 mm (Rodríguez Schettino, 1999b). **Thermal Biology:** Apparently heliothermic, air temperature where lizards are present 31.8 °C, substrate 32.0 °C in an evergreen forest (Rodríguez Schettino, 1999b).

Conservation Status: Unknown.

Anolis armouri Cochran, 1934

Distribution: Hispaniola, Massif de la Hotte, Massif de la Selle (Haiti), and Sierra de Baoruco (Dominican Republic); elevational range 760 to >2,400 m.

Habitat: Xerophilic (Schwartz and Henderson, 1991); in pine forests, often in disturbed sites in rock piles adjacent to cleared garden plots (in Haiti), on boulders and cobbles of a streambed (Franz and Cordier, 1986; Schwartz, 1989); on trunks of trees as well as fallen branches and rock piles (RP).

Behavior: Not an adept climber, ascends only occasionally onto shrubs or low bushes (Schwartz, 1989), but commonly in situations comparable to those used by *A. cybotes* (e.g., males on trunks of large trees; RP). **Diet:** Hymenopterans (ants), orthopterans, and adult beetles were the most important prey items in a sample from the Sierra de Baoruco, with larval beetles, non-ant hymenopterans, adult and larval lepidopterans, and damselflies of lesser importance, dermatophagy, plant material and grit presumably ingested adventitiously (Lenart et al., 1994). **Ecomorphology:** "Trunk-ground," formally quantified (Glor et al., 2003). **Hybridization:** With *A. cybotes* in areas of ecological disturbance (Schwartz and Henderson, 1988; Schwartz, 1989). **Parasites:** Apicomplexa *Eimeria schwartzi* and *Isospora hendersoni* (Cisper et al., 1995); nematodes *Skrjabinoptera leiocephalorum* (Lenart et al., 1994; Goldberg et al., 1998b); *Atractis scelopori*, *Oswaldocruzia lenteixeirai*, *Spinicauda spinicauda*, and larval *Abbreviata* sp. and *Physocephalus* sp., acanthocephalan cystacanth *Centrorhynchus* sp. (Goldberg et al., 1996c, 1998b). **Predation:** An *Anolis* cf. *armouri* was the only non-anuran prey item of *Ialtris haetianus* (Henderson and Schwartz, 1986); *Hypsirhynchus parvifrons* (Henderson et al., 1987a). **Reproduction:** Stage of embryonic development before eggs are deposited is not advanced, possibly reflecting proximity to the sea, which would moderate climatic extremes (Huey, 1977); clutch size 1 (Hedges, in prep.). **Size:** Male SVL to 74 mm (Goldberg et al., 1996c), female SVL to 56 mm (Schwartz, 1989). **Thermal Biology:** When overcast, hides under surface objects (rocks, logs), during sunny days, basks on same objects (Schwartz, 1989).

Conservation Status: Not assessed.

Additional Pertinent Literature: Lenart et al. (1995).

Anolis bahorucoensis Noble and Hassler, 1933

Distribution: Hispaniola, where two currently recognized subspecies occur in the Sierra de Baoruco (Dominican Republic; *A. b. bahorucoensis*) and the Massif de la Selle (Haiti; *A. b. southerlandi*); elevational range ~45–1,400 m.

Habitat: Mesophilic (Schwartz and Henderson, 1991); shaded broadleaf forest, coffee plantations, gallery forest, and ravines, associated primarily with bushes and small trees, on ferns and low bushes along a stream in a coffee plantation, on pineapple plants, fence posts, and in leaf litter, increasingly abundant along edges at higher elevations (Noble and Hassler, 1933; Schwartz and Henderson, 1991; Lenart et al., 1997; Cast et al., 2000; Sifers et al., 2001; Fläschendräger and Wijffels, 2009); often along forest edges (Flores et al., 1994); abundance declines in increasingly altered sites (Lenart et al., 1997); mean perch heights and diameters along rainforest edge 75.9 ± 9.0 cm and 5.5 ± 1.1 cm, in rainforest interior 63.3 ± 8.3 cm and 4.8 ± 0.7 cm, with males perching higher (85.0 ± 8.8 cm) and on perches of greater diameter (6.0 ± 1.0 cm) than females (48.4 ± 6.8 cm and 3.4 ± 0.5 cm; Cast et al., 2001); mean perch heights and diameters of males along edges in cloud forest 86.8 ± 6.9 cm and 4.7 ± 1.1 cm, females 47.0 ± 6.9 cm and 1.3 ± 0.3 cm, both sexes combined at 4 sites were 56.0 ± 9.3–74.7 ± 16.6 cm and 1.6 ± 0.3–5.9 ± 4.0 cm (Sifers et al., 2001); mean perch heights and diameters 1.50 m and 2.68 cm (J. B. Losos in Calsbeek et al., 2006); sleeps on twigs, branches, and lianas in coffee and cacao plantations at heights of 0.3–2.4 m (Schwartz, 1978b); on horizontal leaves at an average height of 1.43 m (0.6–3.0 m; $N = 26$) in thick, shrubby vegetation (Fitch and Henderson, 1987).

Activity: Peak diurnal activity 1500–1700 h, when individuals forage in leaf litter (Fitch and Henderson, 1987). **Behavior:** Escapes by dropping to the ground and hopping or jumping from danger (Schwartz, 1978b); shy and elusive, flush at 2–3 m, minimal response, fleeing over leaf litter, is a series of rapid hops, escape tactics include hiding in leaf litter, running up slender stems and flattening against them with legs flexed, running up stems to tops of bushes and hiding in foliage or returning to the leaf litter, jumping from bush to bush, running along the upper- or underside of sticks into shelter (dead brush, vine tangles, dense ferns), at night, quiescent and not easily disturbed by lights, if disturbed, darts away, hopping from leaf to leaf in an erratic course, males display rarely (probably an anti-predator strategy) and at intervals of ~5 min, displays of small amplitude (no independent dewlap movement, torso immobile, head and neck move only ~2 mm in a vertical plane), a series of about 7 vibratory nods with regular pauses between series, display initiated by female with slow, deep bowing followed by 4–5 series of very rapid nods, sometimes seen in pairs, possibly suggestive of pair-bonding (Fitch and Henderson, 1987); capable of vocalization (Schwartz and Henderson, 1991; Fläschendräger and Wijffels, 2009); mean primary approach distance before reacting 124.0 ± 24.2 cm, mean secondary distance 72.4 ± 22.2 cm, most common (32%) primary responses "squirreling" and jumping to another substrate, most common (27%) secondary responses moving down, jumping to another substrate, moving to the ground (Schneider et al., 2000); displays include extremely rapid head "nods" and may include lateral movements of the head (Fläschendräger, 1990); displays by both sexes usually (83% for males, 100% for females) at distances of 40 cm and composed of head nods of shallow amplitude (2.5 ± 0.9 mm vertical head displacement) and extremely short duration (0.07–0.10 sec) with a cadence defined by "bursts" of multiple nods at rates of 11.6 ± 0.8 nods/sec, lasting 0.1–1.4 sec, containing 1–17 nods, and separated by pauses 0.09–1.10 sec, 3 stereotyped display patterns and one non-stereotyped behavior ("stepbobbing") modified by mouth-gape/tongue-out, lowered hyoid, raised nuchal and dorsal crests, and lateral compression of the body, unusual in that dewlap movement was unpredictable and did not involve pulsing or retraction, dewlap use not quantified because its small size precluded observation from many angles (Orrell and Jenssen, 1998); although generally unobtrusive, presumably fearing predation by *A. cybotes*, shares exposed perches with the latter at higher elevations, where presumably thermoregulation is more important than evading the notice of a prospective predator (Powell, 2003b); moves per minute 0.51 (Johnson et al., 2008). **Ecomorphology:** "Grass-bush" (Williams, 1963a); formally quantified (Beuttell and Losos, 1999). **Foraging and Diet:** Stomach contents of 20 individuals from rainforest contained spiders, other arachnids, beetles (adults and larvae), roaches, flies, hemipterans, hymenopterans (including ants, which were the most important dietary item), termites, lepidopterans (adults and larvae), orthopterans, and a snail (Cast et al., 2001); stomach contents of 26 individuals from cloud forest contained spiders, beetles (adults and larvae), roaches, flies (adults and larvae), hemipterans, hymenopterans (including ants, which were the most important dietary item), lepidopterans (adults and larvae), orthopterans, and shed skin (Sifers et al., 2001); taking butterflies (*Calisto*) in a large flowering roadside shrub (B. I. Crother, in litt., 26.II.2008). **Home Range:** Small home ranges, mean 2.3 m^2 in females, 1.5 m^2 in males (Fitch and Henderson, 1987). **Movements:** Occasionally within 1–2 m of nocturnal perches, but usually do not use the same sleeping sites, especially if disturbed, daytime recaptures first seen an average of 1.9 m (0–6 m) from a previous location, males (1.5 m average, $N = 11$) moved less than females (2.3 m average, $N = 13$; Fitch and Henderson, 1987). **Parasites:** Coccidian *Isospora reui* (Cisper et al., 1995); nematode *Oswaldocruzia lenteixeirai* and acanthocephalan cystacanth *Centrorhynchus* sp. (Goldberg et al., 1996c, 1998b). **Population Size and Density:** In rainforest interior 4,839 or 4,979 lizards/ha depending on method used, along forest edge 6,993 or 8,194 lizards/ha (Cast et al., 2000); at 4 cloud forest sites, 1,479–21,499 lizards/ha concentrated along edge habitats (Sifers et al., 2001). **Predation:** Intense predation pressure by *A. cybotes* and *A. coelestinus*, probably primarily on juveniles (Fitch and Henderson, 1987). **Reproduction:** In captivity, copulation lasts to 3 min, eggs at deposition 5 × 8 mm, possible temperature-dependent sex determination (Fläschendräger, 1990); female may lay 5–6 fertile eggs after separation from a male, hatchling SVL 20 mm (Bundt, 2007); clutch size 1, incubation 40–70 days at 19–28 °C, hatchling SVL ~18 mm (Fläschendräger and Wijffels, 2009); clutch size 1 (Hedges, in prep.). **Size:** Male SVL to 51 mm (Schwartz, 1978b), female SVL to 45 mm (Bundt, 2007). **Thermal Biology:** Mostly in shade, a few in sun-shade mosaic (Fitch and Henderson, 1987); mean body, substrate, and air temperatures for males were 24.4 ± 0.6 °C, 22.1 ± 0.4 °C, and 22.1 ± 0.5 °C, females 25.3 ± 0.6 °C, 22.5 ± 0.5 °C, and 22.4 ± 0.5 °C (Cast et al., 2000); mean male and female body temperatures at 4 sites in cloud forest 25.4 ± 0.5 °C and 26.4 ± 0.7 °C at mean substrate and air temperatures of 23.6 ± 0.4 °C and 23.5 ± 0.4 °C (Sifers et al., 2001).

Conservation Status: Not assessed; Powell and Incháustegui (2009) included the species among those dependent on increasingly vulnerable forest habitats.

Additional Pertinent Literature: Sowell et al. (1995).

Anolis baleatus Cope, 1864

Distribution: Hispaniola, where 10 currently recognized subspecies have a combined range covering most of the Dominican Republic, except areas occupied by *A. ricordii* and *A. barahonae* and the extreme northwestern portion of the country, also on Isla Saona and islets in the Bahía de Samaná; elevational range, sea level to ~1,220 m.

Habitat: Mesophilic to semi-xerophilic (Schwartz and Henderson, 1991); highly arboreal, crowns of tall trees (including mangoes and palms), in gallery forest, coffee and cacao plantations, dense broadleaf forest, often along streams, springside hardwoods in hilly region, and heavily shaded mangrove swamps on Isla Saona, but not in agricultural areas on the island (Mertens, 1939; A. S. Rand in Williams, 1962a; Schwartz, 1974c; R. I. Crombie in Schwartz, 1977a; Henderson and Sajdak, 1983; Schwartz and Henderson, 1991); in active cacao plantations, abandoned pastures, gardens, and cacao, also in forest and associated with mogotes (Glor et al., 2001); occasionally comes to ground (Mertens, 1939); in green enclaves in otherwise urban areas (Powell and Henderson, 2008c); a juvenile ~3 m high on a small branch and two adults from a large tree in an agricultural area (Rand and Williams, 1969); mean perch height and diameter 10.0 m and 14.0 cm (J. B. Losos, in litt., 21.IV.2008); sleeps on exposed branches with hindlimbs drawn up (Mertens, 1939); sleeps at heights of ~1 m (juveniles) to >10 m (adults) on pendant and semipendant vines, tree limbs, shrubs, and even *Acacia* (Schwartz, 1974c; R. I. Crombie in Schwartz, 1977a; Henderson and Sajdak, 1983).

Activity: Most commonly encountered in late afternoon (Mertens, 1939); may be active at dusk (Haberham, 2003). **Behavior:** When captured, vocalizes, extends dewlap, and tries to bite, adpresses itself to branch when threatened (Mertens, 1939); sleeps with tail dangling, awakens quickly and either wanders slowly upward or falls or jumps to the ground and scurries away (Schwartz, 1974c); exhibits "tongue-bunch" defensive/agonistic behavior (Schwenk and Mayer, 1991). **Ecomorphology:** "Crown-giant" (Rand and Williams, 1969). **Predation:** Ridgway's Hawk (*Buteo ridgwayi*; Wiley and Wiley, 1981). **Relative Abundance:** Least abundant (19 of 990 observations) of anoles in and around Los Haitises National Park, Dominican Republic (Glor et al., 2001). **Reproduction:** Copulation on 8 November on a tree trunk 1.2 m above the ground at 1225 h in a coffee plantation (Schwartz, 1974c); clutch size 1 (Hedges, in prep.). **Size:** Male SVL to 180 mm, female SVL to 148 mm (Schwartz, 1974c).

Conservation Status: Not assessed; the ongoing destruction of forests might have long-term negative consequences (Powell et al., 2000; Powell and Incháustegui, 2009).

Anolis baracoae Schwartz, 1964

Distribution: Extreme eastern Cuba, including Cuchillas de Toa and Sierra del Purial.

Habitat: Mesophilic (Schwartz and Henderson, 1991); broadleaf forest and riverine woods or forests in otherwise xeric areas (Schwartz and Garrido, 1972; Schwartz and Henderson, 1991); tolerant of human modified situations (Garrido, 1975a; Kunz, 2005); in submontane rainforest, submontane mesophyllous evergreen forest, and mesophyllous evergreen and semideciduous forest (Rodríguez Schettino et al., 1999); usually high in trees, even in the outskirts of Baracoa, although often along rivers, especially during the dry season, juveniles more likely to be in shrubs and small trees (Holanova and Hribal, 2005).

Activity: Diurnal, descends trunks during the heat of the day (Rodríguez Schettino et al., 1999); usually not active early in the morning or late afternoon or on cloudy and rainy days, descends from high in the trees 1000–1300 h (Holanova and Hribal, 2005). **Behavior:** Slow-moving and deliberate, escapes by "squirreling" and then running upward, aggressive displays include dewlap extension and gaping (Rodríguez Schettino, 1999b); quickly moves to the opposite side of the branch when frightened (Holanova and Hribal, 2005). **Ecomorphology:** "Crown-giant" (Williams, 1972). **Foraging and Diet:** Eats gastropods (*Polymita picta*) and hummingbirds (*Chlorostilbon ricordii*; Rodríguez Schettino, 1999b); small lizards, snails, and birds, eats fruits in captivity (Holanova and Hribal, 2005). **Parasites:** Cestode *Sparganum* sp. (Coy Otero and Baruš, 1979b; Coy Otero and Lorenzo Hernández, 1982); nematodes *Cyrtosomum scelopori*, *Oswaldofilaria brevicaudata*, and larval *Abbreviata* sp. (Baruš and Coy Otero, 1969a; Coy Otero, 1970; Coy Otero and Baruš, 1979b; Coy Otero and Lorenzo Hernández, 1982; Coy Otero, 1999); *Trichospirura teixeirai* (Coy Otero and Baruš, 1979a, 1979b; Coy Otero and Lorenzo Hernández, 1982; Coy Otero, 1999); *Cyrtosomum longicaudatum*, *Oswaldocruzia lenteixeirai*, *Physaloptera* *squamatae*, *Piratuba digiticauda*, and larval *Porrocaecum* sp. (Coy Otero and Baruš, 1979b; Coy Otero and Lorenzo Hernández, 1982; Coy Otero, 1999). **Predation:** Adults have few enemies except possibly predatory birds, juveniles may be eaten by adults or snakes in the genus *Tropidophis* (Holanova and Hribal, 2005). **Reproduction:** Reproductive activity extends March–October, in captivity, may mate 2–3 times per week, courtship involves headbobbing and dewlap extension, copulation on vertical branches facing down, females deposit 1 egg (very rarely 2, second egg apparently retained until deposition propitious), eggs 23–25 × 6–7 mm, increasing to 27 × 20 mm during incubation, total length of neonates 13 cm, females mature at 10–12 months (Holanova and Hribal, 2005); in captivity, incubation 74–81 days at 26 °C and nightly cooling (Reisinger, 2003). **Size:** Male SVL to 172 mm (Schwartz and Henderson, 1991), female SVL to 158 mm (Schwartz, 1964c). **Thermal Biology:** Presumably heliothermic (Rodríguez Schettino, 1999b); shuttled from sun to shade at heights of 4–7 m in *Cecropia* sp. (Fläschendräger and Wijffels, 2009).

Conservation Status: Considered "vulnerable" (Rodríguez Schettino and Chamizo Lara, 1998a; L. Rodríguez Schettino and A. Chamizo Lara in Rodríguez Schettino, 1999a).

Anolis barahonae Williams, 1962

Distribution: Hispaniola, where four currently recognized subspecies occur on the Península de Barahona and in the Sierra de Baoruco (Dominican Republic); elevational range, sea level to ~800 m.

Habitat: Mesophilic to semixerophilic (Schwartz and Henderson, 1991); highly arboreal in riverine and gallery broadleaf forests, coffee plantations, edge situations in relatively xeric lowland forest (Schwartz, 1974c; Lenart et al., 1997), apparently absent from highly altered sites (Lenart et al., 1997); mean perch height and diameter 8.25 m and 33.0 cm (J. B. Losos, in litt., 21.IV.2008); juveniles sleep on shrubs and low trees and adults on limbs, branches, and vines to ~4.5 m above the ground (Schwartz, 1974c).

Behavior: Escapes by climbing high into tree crowns (RP). **Diet:** No correlation between SVL and mean prey size (larger individuals continued to eat small items), stomachs contained beetles, earwigs, flies, hemipterans, hymenopterans (including ants), lepidopterans (adults and larvae, which were the most important food item, probably reflecting opportunistic exploitation of a seasonally abundant resource), orthopterans (second most important prey), roaches, snails, shed skin, leaves, fruits/seeds (third most important food item), miscellaneous plant material, and grit, probably adventitiously ingested and suggestive of individuals feeding on the ground (Bowersox et al., 1994b). **Ecomorphology:** "Crown-giant" (Rand and Williams, 1969); formally quantified (Irschick et al., 1997). **Parasites:** Nematodes *Urotrema wardi*, *Ascarops* sp., and *Atractis scelopori* (Goldberg et al., 1996c, 1998b). **Predation:** *Tropidophis haitianus* (Schwartz and Henderson, 1991). **Reproduction:** Sperm present in all individuals ($N = 51$) in June–September with SVL 123 mm, only one of 22 adult females contained 2 shelled ova, one contained 2 nonshelled ova (one apparently retained until deposition favorable), and 11 contained one shelled and one nonshelled ovum, no ovary contained more than one yolked follicle, but 13 females had yolked follicles in both ovaries, volume of coelomic fat positively correlated with SVL, but did not differ by sex or by ovigerous/non-ovigerous females (Allen et al., 1994). **Size:** Male SVL to 172 mm, female SVL to 162 mm (Allen et al., 1994); smallest mature male SVL 123 mm, smallest mature female SVL 125 mm (Bowersox et al., 1994).

Conservation Status: Not assessed; deforestation of broadleaf forests on which the species depends may have long-term negative consequences (Powell and Incháustegui, 2009).

Additional Pertinent Literature: Calderón et al. (1995).

Anolis barbatus Garrido, 1982

Distribution: Cuba, Sierra del Rosario; elevational distribution, sea level to ~500 m (Fläschendräger and Wijffels, 1996, 2009).

Habitat: Mesophilic (Schwartz and Henderson, 1991); in broadleaf forest on limestone hills, with a 15–25 m canopy and many understory trees and shrubs, generally in closed-canopy, shaded and humid situations with considerable vegetative detritus, often associated with tangles or bundles of vegetation (Garrido, 1982a; Rodríguez Schettino, 1999b; Leal and Losos, 2000); in the Biosphere Reserve "Sierra del Rosario," in secondary vegetation and evergreen forest, preferentially on branches of trees, but also on trunks of trees and bushes and branches of bushes (Martínez Reyes, 1995); in submontane rainforest, submontane mesophyllous evergreen forest, and mesophyllous evergreen and semideciduous forest (Rodríguez Schettino et al., 1999); perch diameters and heights for 7 individuals were 0.5–4.0 cm and 1.5–4.0 m (subadult male), 6.0–19.0 cm and 4.0 m (adult female), 4.0 cm and 3.0–3.5 m (adult male), 0.3–6.0 cm and 1.25–1.75 m (adult female), 0.3–6.0 cm and 0–7.5 m (adult female), 0.50–0.75 cm and 2.0 m (juvenile), and 1.0–3.5 cm and 7.0–12.0 m (adult male; Leal and Losos, 2000); adult male at ~3 m, adult female at ~1 m on a Sandbox Tree

(*Hura crepitans*), apparently a pair maintaining visual contact (Fläschendräger, 2002); mean perch height at Soroa 4.55 m, diameter 4.7 cm, behavioral diameter (mean diameter of all surfaces used during behavioral observations, means of individual means) 2.93 cm, no observations on or within 1 m of rocks (J. B. Losos, in litt., 20.VIII.2007); rests at night on branches narrower than its body (Garrido, 1982a); sleeps on horizontal branches or vines or on trunks (Rodríguez Schettino, 1999b); sleeps on small-diameter perches, especially leafless twigs (Leal and Losos, 2000).

Activity: Diurnal, focal animal observation conducted 0700–1500 h, spent 63.3% of time stationary, 31.2% moving, and 5.5% displaying (Leal and Losos, 2000). **Behavior:** Agonistic behavior similar to courtship except that, while positioned laterally, males "swell with pride," stretch head up and down, and fully expand dewlap and dorsal crest, repeated at intervals (Garrido, 1982a); in captivity, did not react to the presence of *Tropidophis melanurus* (Hottmar, 1991); apparently solitary (Fläschendräger and Wijffels, 1996, 2009); but a pair observed in close proximity (Fläschendräger and Wijffels, 2009); moves slowly and does not try to escape, aggressive male-male behavior similar to courtship, but headbobs repeated more frequently and dewlaps extended and dorsal crests erected, same behavior directed toward a snake (*Caraiba andreae*) introduced into the enclosure (Rodríguez Schettino, 1999b); "assertion displays" (not directed at another individual) consisted of dewlapping and headbobbing, with the former in two distinct ways, "basic dewlap displays" and "dewlap rolls," both given with head elevated and forelimbs extended, dewlap slowly extended, not pulsed repeatedly, instead slowly partially retracted and then fully extended in the course of the display or kept fully extended, in "dewlap rolls," dewlap fully extended while the head rotated around long axis, tilted first to one side and then to the other, with bouts lasting 37–307 sec, headbobbing by both males and females, but only males flexed forelimbs in synchrony with headbobs, modifiers included gradual color changes, with the body becoming lighter, and erection of dorsal and nuchal crests, movements usually (89%) by walking, either in a slow and deliberate fashion or while slowly rocking back and forth, as many as 20 times per complete cycle of limb movements, usually covering 10–30 cm, two instances of running up and down vines and one on the ground immediately after being released with body undulating laterally while elevated above the substrate, jumping from branch to branch occurred after turning body perpendicular to the perch before launching, distances short (10–20 cm, mean 11.8 ± 1.0 cm; Leal and Losos, 2000). **Ecomorphology:** "Twig-giant" (Hass et al., 1993), member of the *Chamaeleolis* clade); technically this is not an ecomorphological category as it evolved in only one clade (see definition in Williams, 1972). **Foraging and Diet:** In captivity, comes to ground to feed avidly on slugs (*Veronicella*; Garrido, 1982a); in one sample ($N = 5$), one insect, one millipede, some mollusks (*Zachrysia auricoma*), and abundant vegetable matter, especially fruits (M. Martínez Reyes in Rodríguez Schettino, 1999b); an active forager, a female descended slowly down a liana from an initial height of 7 m to about 30 cm above the the ground, stopping occasionally and assuming a "watch position," jumped to the ground, walked a short distance, and caught what appeared to be a snail, jumped to a stem about 10 cm high, manipulated the food for ~60 sec, at the end of which a crunching sound could be heard, after swallowing, moved and surveyed habitat for >47 min, then, from a perch ~50 cm, jumped to the ground, walked a short distance, and caught what appeared to be a slug, manipulated it for 98 sec, before walking and surveying habitat for >21 min before jumping back to the ground and catching another prey item, which may have been a beetle pupa, which was manipulated for 75 sec (Leal and Losos, 2000); large head size and massive jaw musculature ideally suited for molluscivory, juveniles can successfully hunt snails a few days after hatching (Herrel and Holanova, 2008). **Movements:** Individuals moved 0.5–11.5 m during focal animal observations (Leal and Losos, 2000). **Reproduction:** Courtship initiated by mutual inspection with female reacting first, slowly raising her head before letting it fall rapidly, giving the impression of an affirmative nod, action repeated twice with interval of several seconds, male responds in similar fashion but may alternate with lateral head movement, as though shaking his head, male lowers his head to the substrate, moves it laterally back and forth about 4 times, pauses for a few minutes, and repeats vertical head movements, male approaches the female while engaged in head movements until he mounts her from the rear (delaying for several minutes in that position) until his snout is under the posterior margin of her casque, repeats his head orientation movement, female rapidly changes her perch, leaving male to repeat the entire process at least 3 and as many as 5 times until she lowers her head and he slowly rubs his snout beneath her casque, activity continues at intervals, each repetition increasing in rhythm until male attempts to bite her neck or nuchal fold without sacrificing a firm grip, pair is immobile for about 50 sec before male insinuates his tail under hers and intromission is accomplished, pair is completely immobile during copulation, which can last to 4 h (mean 22.7 min, $N = 11$; Garrido, 1982a); egg length 22 mm (Silva Lee, 1984); in captivity, females bury single eggs (14 × 22 mm) at 5–8 day intervals, shell initially soft and sticky (soil particles adhere to it), but becomes hard and brittle, egg increased during incubation to 27 × 22 mm, hatchling total length 76–82 mm, mass 1.9–2.2 g (Hottmar, 1991); hatchling SVL in captivity 36 mm, SVL 43 mm after 38 days (Fläschendräger and Wijffels, 2009). **Size:** Male SVL to 170 mm, female SVL to 157 mm (Garrido, 1982a). **Thermal Biology:** Body temperatures 27.1 °C (at 1030 h), 27.3 °C (at 1000 h), 27.5 °C (at 1700 h), and 32.5 °C (at 1425 h), first 3 active in shade, the latter perching completely exposed in full sun (Leal and Losos, 2000); mean body temperature at Soroa 28.6 °C (J. B. Losos, in litt., 20.VIII.2007).

Conservation Status: Considered "endangered" (Rodríguez Schettino and Chamizo Lara in Rodríguez Schettino, 1999a); assessment based on IUCN Red List criteria (Ruiz Urquiola et al., 1998a).

Anolis barbouri Schmidt, 1919

Distribution: Hispaniola, widely distributed on the "South Island" (Haiti and Dominican Republic) and also in the Sierra de Neiba and Cordillera Central on the "North Island" (Dominican Republic); elevational range ~300–1,710 m.

Habitat: Mesophilic (Schwartz and Henderson, 1991); primarily terrestrial, 64 of 68 individuals captured on the ground (Flores et al., 1994); in densely shaded mesic broadleaf forest and coffee plantations with considerable leaf litter, in a mat of dead vines and twigs and cut branches on ground, on twigs of dead shrubs, piles of dead grass in an open area between coffee and sugarcane plantings, in a crevice of a large tree 1.2 m above the ground in a field actively being cut and weeded (may have been driven there by human activity), 1.3 m high on a stump, on trunks of trees <1 m high, on boulders in a recently cut ravine, in weeds adjacent to forest, on a path with dense undergrowth in coffee plantation, in *Musa-Cacao* cultivation in a ravinelike river valley, in a river valley with some natural vegetation, wet montane forest with dense climbing grass ("tibisí") and vines, in dead pine needles and piles of wood chips near the ground, among twigs of dead shrubs, with the apparent dependence on shade and litter affected by human activity (Schwartz and Incháustegui, 1980; Franz and Cordier, 1986; Flores et al., 1994; A. K. Howard et al., 1999a); may burrow (Flores et al., 1994); sleeps near the ground (hindlimbs flexed) on small vines or other vegetation (Schwartz and Incháustegui, 1980); in captivity, sleeps on twigs and leaves, occasionally on a bromeliad, hanging on solely with hindlimbs and tail (Fläschendräger, 2000).

Activity: Throughout the morning and into the early hours of the afternoon, "disappears" after 1400 h (Flores et al., 1994), but reported as late as 1600 h (SEA/DVS, 1992) and active in about equal numbers from 0700–1630 h (A. K. Howard et al., 1999a). **Behavior:** When pressed, hops from twig to twig or drops to the ground and "disappears" in ground cover (i.e., leaf litter) or low growth (Schwartz and Incháustegui, 1980; Flores et al., 1994; A. K. Howard et al., 1999a); behavior when undisturbed very placid and lethargic (Schwartz and Incháustegui, 1980); immobile for extended periods (to 15 min of observation; Flores et al., 1994); individuals can be aggressive when captured (Flores et al., 1994; A. K. Howard et al., 1999a); moved abruptly and became agitated at birdlike clicking sounds (Flores et al., 1994); capable of vocalization (Fläschendräger and Wijffels, 1996, 2009); jumping distance 40.1 ± 1.6 cm (80.6% of expected ability), when handled, became immobile and adopted a vertically compressed posture while extending the abdomen to give a flat and rounded profile, similar behavior observed in the field, may enhance crypticity or interfere with ingestion by a predator (Autumn and Losos, 1997); male-male approach behavior very unlike that of other anoles, with imprecise or absent lateral presentation the only postural modifier used, highly variable (2–92 sec) and nonstereotypical displays consist of slow, large-amplitude headrocking with a continuously extended dewlap often accompanied by head or eye shifts that may facilitate depth perception (Jenssen and Feely, 1991). **Ecomorphology:** Undetermined, far from all categories (Beuttell and Losos, 1999); extremely specialized for terrestrial habitats (Hass et al., 1993). **Foraging and Diet:** An extremely well-camouflaged sit-and-wait forager (Schwartz and Incháustegui, 1980; Fläschendräger, 2000); stomach contents include spiders, centipedes, millipedes, isopods, beetles (most important), beetle larvae, flies (adults and larvae), hemipterans, ants and other hymenopterans, lepidopterans (adults and larvae), neuropteran, orthopterans, and a thysanopteran, standardized dietary niche breadth indicative of a generalist, and dietary overlap higher with syntopic frogs (*Eleutherodactylus abbotti* and *E. armstrongi*) than sympatric anoles (*A. coelestinus*, *A. cybotes*, *A. distichus*), of the lizards, highest with *A. cybotes*, which, of the 3 anoles, is most likely to consume terrestrial prey (A. K. Howard et al., 1999a); are quick when grabbing prey, which they bite with considerable strength (Fläschendräger, 2000). **Parasites:** Immature nematodes *Terranova* (*caballeroi*?) for which *A. barbouri* may serve as an intermediate host, with *Hypsirhynchus parvifrons* a possible definitive host (A. K. Howard et al., 1999a). **Reproduction:** In captivity, breeds throughout the year, males "assault" females, grasp them by the nape, copulation (to 126 minutes) inevitably on the ground, bury 1 or occasionally 2 eggs (one presumably retained until deposition propitious) in self-excavated cavities, mean size 5.7 × 9.6 mm, after 12 days 8.0 × 11.7 mm, incubation ~52 days at 22–26 °C, hatchling SVL ~17 mm (Fläschendräger, 2000; Fläschendräger and Wijffels, 2009); clutch size 1 (Hedges, in prep.). **Size:** Male SVL to 44 mm (possibly 47 mm; K. P. Schmidt, 1919), female SVL to 55 mm (Schwartz and Incháustegui, 1980). **Thermal Biology:** Never seen basking, 67 of 70 individuals in shaded conditions, body temperatures (26.0 ± 1.9 °C) well above air temperatures (22.6 ± 1.5 °C), although linear regression suggests thermoconformity rather than regulation (Flores et al., 1994); of 13 lizards presented with a thermal gradient, 4 moved to the cool end and stayed there while 9 chose an intermediate position, mean body temperature of the 9, 25.2 ± 0.7 °C, of all 13, 24.1 ± 0.7 °C, mean critical thermal minimum 11.7 ± 0.3 °C, mean critical thermal maximum 34.9 ± 0.6 °C (Autumn and Losos, 1997).

Conservation Status: Not assessed; ongoing deforestation may cause problems for this species in the future (Powell and Incháustegui, 2009).

Additional Pertinent Literature: A. K. Howard et al. (1999b).

Remarks: This species was formerly assigned to the genus *Chamaelinorops.*

Anolis bartschi Cochran, 1928

Distribution: Western Cuba, Pinar del Río Province, lowlands and Sierra de los Órganos and Sierra del Rosario.

Habitat: Mesophilic (Schwartz and Henderson, 1991); on rock faces, cliffs, rock piles, small superficial cave walls, basks on vertical cliff faces, retreats into crevices when pressed, both sexes use naked, open, or covered rock faces, rarely venturing onto the ground or into trees (Cochran, 1928; Ruibal, 1964; G. Peters, 1970; Estrada and Novo Rodríguez, 1986a; Rodríguez Schettino, 1999b); in mogote complex of vegetation (Rodríguez Schettino et al., 1999); perch heights of males (258.1 ± 130.0 cm in wet season, 279.2 ± 102.0 cm in dry season) and females (206.5 ± 91.1 cm and 199.0 ± 91.8 cm) do not differ significantly (Estrada and Novo Rodríguez, 1986a); usually at heights of 1–3 m, as high as 6 m, sleeps in crevices, holes, and under rocks (Rodríguez Schettino, 1999b); mean perch height and diameter 2.97 m and 2.0 cm (J. B. Losos, in litt., 21.IV.2008); may sleep adpressed to the surface of cave walls with limbs widely spread (Vergner and Polák, 1996); coefficient of arboreality very low, typical of petricolous species (M. A. Granda Martínez in Rodríguez Schettino, 1999b).

Activity: Diurnal (Rodríguez Schettino et al., 1999); forages in late afternoon and into the evening (Z. Vogel, 1965a); usually leaves nocturnal retreats after 0900 h during the dry season and after 0700 h during the wet season, number of active lizards increases rapidly until 1000 h, diminishes thereafter until 1700 h in the wet season and 1400 h in the dry season, no lizards active after 1900 h in the wet season and 1700 h in the dry season (Rodríguez Schettino, 1999b); first encountered 0800–1000 h when they abandon nocturnal refuges, peak activity 1500–1700 h when sunlight is less direct (Holanova and Hribal, 2004). **Behavior:** Vocal when handled (G. Peters, 1970; Rodríguez Schettino et al., 1999); often "walk" with toes, especially on the hindlimbs, elevated (in effect, "walking on their heels"), may hang on vertical surfaces solely by their forelimbs, threat displays in captivity involve rhythmic tail waving and body inflation, which exposes bright flank colors (Baedermann, 1989); females quicker to escape than males, both move to rock crevices, but pause before entering (Rodríguez Schettino, 1999b); during male-male interactions, both individuals elevate on all 4 limbs, swell their throats, open mouths, and attack (Rodríguez Schettino, 1999b; Rodríguez Schettino, 2003b); juveniles often congregated near mouths of caves (Holanova and Hribal, 2004); head bobs very slow (Fläschendräger and Wijffels, 2009). **Body Condition:** Condition did not vary for either sex between wet and dry seasons (Rodríguez Schettino and Martínez Reyes, 1992). **Ecomorphology:** Undetermined; categorized as a Cuban "gecko anole" (G. Peters, 1970). **Foraging and Diet:** Feeds on insects, including ants (most important in both sexes and during both wet and dry seasons), other hymenopterans, beetles, flies and mosquitoes, lepidopterans, roaches, and various larvae, spiders, isopods, centipedes and millipedes, gastropods, shed skin (presumably their own), and flowers and fruits, larger males take larger prey than females (Rodríguez Schettino and Martínez Reyes, 1992; Rodríguez Schettino, 1999b; Rodríguez Schettino, 2003); will take smaller conspecifics (Holanova and Hribal, 2004); reacts strongly in captivity to the presence of other anoles, which in nature might be a considerable portion of its prey (Fläschendräger and Wijffels, 2009). **Home Range and Territoriality:** Not strongly territorial, individuals of both sexes often in proximity on cliffs and cave walls, but males may display actively in defense of shelters (Rodríguez Schettino, 1999b); home ranges 2–3 m^2 (Holanova and Hribal, 2004). **Parasites:** Nematodes *Cyrtosomum scelopori* (Baruš and Coy Otero, 1969a; Coy Otero, 1970; Coy Otero and Baruš, 1973a, 1979b; Coy Otero and Lorenzo Hernández, 1982; Coy Otero, 1999); *Oswaldocruzia lenteixeirai*, *Parapharyngodon cubensis*, *Skrjabinoptera phrynosoma*, and larval *Porrocaecum* sp. (Coy Otero and Baruš, 1979b; Coy Otero and Lorenzo Hernández, 1982; Coy Otero, 1999); sporozoan red blood cell parasite *Haemogregarina* sp. (Zajicek and Mauri

Méndez, 1969). **Population Size and Density:** Density 180–470 individuals/ha, biomass 1.01–3.08 kg/ha at Cueva del Indio (Rodríguez Schettino, 1999b); as many as 10 individuals in a few m^2 (Fläschendräger, 2002). **Population Structure:** Adults:immatures (subadults and juveniles) 9.9:1, with 77% and 100% of the population composed of adults in February and April, respectively (Rodríguez Schettino, 1999b). **Predation:** Natural predators include *A. luteogularis*, rarely *A. chamaeleonides*, snakes in the genus *Tropidophis* (especially *T. melanurus*), and occasionally *Cubophis cantherigerus* and *Epicrates angulifer* (Holanova and Hribal, 2004). **Reproduction:** During courtship, males inflate their throats (Holanova and Hribal, 2004); in captivity, courtship is of short duration, females not infrequently injured (even killed) by males biting during copulation (Baedermann, 1989); clutch size 1 (Hedges, in prep.); in captivity, egg immediately after deposition 7.9 × 4.7 mm (Wijffels, 1972); one (rarely 2, if one retained) eggs 9 × 13 mm, females lay 6–7 (maximum 12) eggs/year, incubation 60–70 days at 23–28 °C (Fläschendräger and Wijffels, 1996, 2009); females deposit eggs communally in crevices on the sides or walls of shallow caves, in the twilight zone or in crevices on cliff faces (Silva Rodríguez et al., 1982); communal nests with 34 and 14 eggs (or shells) in July and August, 6 other nests with shells of 10–20 eggs, eggs white and ovoid, mean size 15.5 ± 0.15 × 9.40 ± 0.65 mm, most females gravid and in stages III or IV in May and August, 100% of females in stage I in October (Estrada and Novo Rodríguez, 1985b); major reproductive activity in mid-March, with enlarged testes and high percentages (83–86%) of gravid females in April, May, and August, none in October–February, when body fat highest (Novo Rodríguez and Estrada, 1986); in captivity, eggs immediately after deposition 16.8 ± 1.3 × 9.2 ± 0.8 mm, becoming increasingly less oval during development (Vergner and Polák, 1996); eggs shortly before hatching 19.0 × 12.5 mm (Skála, 1989); first gravid females in February, in October females without developing ovarian follicles or oviductal eggs, 15% of ovigerous females with 2 eggs, mean lengths of eggs 15.5–18.0 mm, size of fat bodies inversely proportional to reproductive activity (Rodríguez Schettino, 1999b); breeds February–September (Rodríguez Schettino, 2003b). **Sex Ratio (M:F):** 0.9:1 (Rodríguez Schettino, 1999b). **Size:** Male SVL to 76 mm (Ruibal, 1964), female SVL to 63.6 mm (Rodríguez Schettino, 1999b); mean SVL of hatchlings 20.5 mm (Rodríguez Schettino, 1999b). **Thermal Biology:** Avoids direct sunlight (G. Peters, 1970); fully active at temperatures 18–20 °C (G. Peters and Schubert, 1968); nonheliothermic, body temperatures tied to variations in the temperatures of substrates, thermoecologically passive, but complemented by effective exploitation of microhabitat temperatures, mean male cloacal, substrate, and air temperatures 25.04 ± 1.86 °C, 24.40 ± 1.52 °C, and 24.6 ± 2.08 °C (dry season), 27.20 ± 1.80 °C, 27.10 ± 1.65 °C, and 27.60 ± 2.11 °C (wet season), mean female cloacal, substrate, and air temperatures 24.60 ± 1.71 °C, 23.7 ± 2.14 °C, and 24.5 ± 2.28 °C (dry season), 27.40 ± 1.61 °C, 27.01 ± 1.97 °C, and 27.50 ± 2.02 °C (wet season; Estrada and Novo Rodríguez, 1987).

Conservation Status: Considered at "lower risk" of extinction (Rodríguez Schettino and Chamizo Lara, 1998a; L. Rodríguez Schettino and A. Chamizo Lara in Rodríguez Schettino, 1999a).

Anolis bimaculatus Sparrman, 1784

Distribution: St. Christopher Bank (St. Eustatius, St. Christopher, Nevis); introduced on St. Maarten, but colony may have failed (Wijffels, 1997b; Powell et al., 2005); elevational range, sea level to ~600 m (Fläschendräger and Wijffels, 1996, 2009), but abundance declines with increasing elevation (Buckley and Roughgarden, 2005a).

Habitat: Xerophilic to mesophilic (Schwartz and Henderson, 1991); highly arboreal, in abandoned *Cocos* grove (Nevis), on fence posts and rock walls (St. Eustatius), adapts readily to human-altered conditions and readily enters homes and other buildings, perches on walls, fences, arbors, water tanks, woodpiles, and trash, as well as natural substrates such as shrubs, trees of all sizes, rocks, and cliff faces (Underwood, 1959b; Lazell, 1972; Wijffels, 1984; Schwartz and Henderson, 1991; Fläschendräger and Wijffels, 1996, 2009; Wijffels, 1997b; Powell et al., 2005); males may perch as high as 3.0–4.5 m above ground, but usually are at 0.9–1.8 m, females and juveniles lower (Lazell, 1972); mean perch height on St. Christopher 0.90 ± 0.09 m, on St. Eustatius 1.52 ± 0.12 m (Roughgarden et al., 1983a); adult males and females and juveniles may forage on the ground (Schwartz and Henderson, 1991; Powell et al., 2005); on St. Christopher, mean perch height and diameter 1.89 ± 0.27 m and 12.7 ± 3.3 cm (Losos and de Queiroz, 1997a); perch heights and diameters varied considerably at 2 sites on St. Eustatius (heights 80.5 ± 14.3 cm and 159.7 ± 20.6 cm, diameters 13.5 ± 6.9 cm and 23.1 ± 12.4 cm) with the differences reflecting availability of large tree trunks, males perched higher (162.7 ± 26.1 cm) and on perches of larger diameter (33.4 ± 16.3 cm) than females (96.3 ± 14.5 cm and 9.0 ± 2.6 cm), none were on the ground, most individuals perched facing down at one site and up at the other, and most were in fully shaded or partially shaded sites, perch heights were significantly higher than those used by sympatric *A. schwartzi* (Medina Díaz et al., 2005); sleeps on leaves or twigs at heights similar to those used by day, but also may sleep in rock walls, under rocks, or in crevices in cliffs or artificial situations (Powell et al., 2005).

Abundance: Often in large numbers around human habitations where suitable cover is available, such as the grounds of the Montpelier Plantation Inn on Nevis (Malhotra and Thorpe, 1999); on St. Christopher, greatest abundance at sea level, declining rapidly to zero at higher elevations (Buckley and Roughgarden, 2005a). **Activity:** An hour or two after dawn to dusk (Adolph and Roughgarden, 1983); but may exploit the "night-light niche" (Schwartz and Henderson, 1991; Fläschendräger and Wijffels, 1996, 2009; Powell et al., 2005; R. Powell in Perry et al., 2008). **Behavior:** Interacts frequently with *Anolis schwartzi*, with males displaying toward one another (Lazell, 1972); males move head from side to side just prior to attacks in male-male interactions (Fläschendräger and Wijffels, 1996, 2009); when disturbed, will "squirrel" to the other side of a trunk or branch, jumps to another branch or (rarely) to the ground, usually climbs upward into the canopy (Powell et al., 2005). **Ecomorphology:** Because of large size, less close to "trunk-crown" than many Lesser Antillean anoles (Losos and de Queiroz, 1997a; Beuttell and Losos, 1999). **Foraging and Diet:** 99.65% arthropods, 0.96% vertebrates, 1.19% mollusks, and 0.67% plant material, 0.90 overlap with *A. schwartzi* and 0.50 with *Ameiva erythrocephala* (Adolph and Roughgarden, 1983); when anoles (*A. bimaculatus* and *A. schwartzi*) excluded from enclosures, forest-floor arthropods (mainly dipterans) increased 2–3 fold, large web-building spiders increased 20–30 fold, in turn causing a 25% decrease in the abundance of insects between the forest floor and the canopy, exclusion of anoles had no obvious effect on the degree of herbivory (Pacala and Roughgarden, 1984); mean number of prey items per stomach 19.6 ± 20.5, mean arthropodan prey length 3.0 ± 1.9 mm, with 71.9% of prey 0–4 mm in length, and 27.2% 4–10 mm in length, prey in the 4–10 mm range taken more frequently than prey caught in traps (Rummel and Roughgarden, 1985); annelids (frequency 0.08), gastropods (0.08), centipedes (0.02), millipedes (0.02), mites (0.02), juvenile spider (0.10), beetles (adults, 1.48), beetles (larvae, 0.02), springtails (0.02), ants (8.52), other hymenopterans (2.12), flies (adults, 0.56), flies (larvae, 0.20), hemipterans (1.46), termites (0.54), isopods (0.04), lepidopterans (adults, 0.36), lepidopterans (larvae, 9.58), orthopterans (0.12), and thrips (0.02; J. D. Rummel in Goldwasser and Roughgarden, 1993; Roughgarden, 1995); various insects, smaller conspecifics, and *A. schwartzi* (Fläschendräger and Wijffels, 1996, 2009); more selective of larger prey than *A. schwartzi*, search from higher vantage points than *A. schwartzi*, closer, for example, to caterpillars on leaves, both species equally likely to take ants present at all heights (Roughgarden, 1995); insects and other small arthropods, smaller lizards (including conspecifics), laps nectar and eats flowers and fruits (Powell et al., 2005); catches fish in captivity (Fläschendräger and Wiffels, 2009). **Home Range and Territoriality:** Home range on St. Eustatius 3–4 m² (Roughgarden, 1995). **Longevity:** Maximum life span ~2–3 years, "typical" life span ~1 year (Roughgarden, 1995). **Parasites:** Nematode *Thelandros cubensis* on St. Eustatius, with montane populations more heavily infected than coastal animals, nematodes *Parapharyngodon cubensis* and acanthocephalan cystacanth *Centrorhynchus* sp. (Dobson et al., 1992); mean parasite species/host 0.66 in xeric habitat and 0.52 in mixed habitat on St. Eustatius (Dobson and Pacala, 1992); nematodes *Parapharyngodon cubensis* (Goldberg et al., 1996b); *Spauligodon anolis* (Dobson et al., 1992; Goldberg et al., 1996b; Bursey and Goldberg, 1998); on St. Christopher (but not St. Eustatius), malarial parasites *Plasmodium azurophilum* and *P. floridense* (Staats and Schall, 1996b). **Population Size and Density:** 2,220–3,000 lizards/ha on St. Eustatius (Roughgarden, 1995); estimates at 2 sites on St. Eustatius 993–1,854 and 1,042–1,667 lizards/ha using 2 different methods (Medina Díaz et al., 2005). **Population Structure:** Exceptionally large males may be solitary (are avoided even by females that see them as potential predators), may not be reproductively active (Fläschendräger and Wijffels, 1996, 2009). **Predation:** Pearly-eyed Thrasher (*Margarops fuscatus*), American Kestrel (*Falco sparverius*; McLaughlin and Roughgarden, 1989; Powell et al., 2005); presumably *Ameiva erythrocephala* and *Alsophis rufiventris* on St. Eustatius (Powell et al., 2005). **Reproduction:** Probably reproductively active throughout the year, eggs are buried in soil substrate (Fläschendräger and Wijffels, 1996, 2009); copulation in June (Powell et al., 2005); clutch size 1 (Hedges, in prep.); hatchling SVL in captivity ~26 mm (Fläschendräger and Wijffels, 2009). **Resource Partitioning:** Selected perch heights and growth rates not affected by the presence of *A. schwartzi* in enclosures (Pacala and Roughgarden, 1982, 1985; Roughgarden, 1995); in enclosures that limited perch heights to 1.5 m, growth, egg production, and prey selection were lowered in the presence of *A. schwartzi* (Rummel and Roughgarden, 1985; Roughgarden, 1995). **Size:** Male SVL to 125 mm (Fläschendräger and Wijffels, 1996, 2009), female SVL to 70 mm (Schwartz and Henderson, 1991), smallest mature male and female SVL 62 mm and 52 mm, respectively (Lazell, 1972); large size of individuals in the vicinity of human habitation may reflect an absence of predation and a willingness to exploit abundant food (Wijffels, 1997b). **Thermal Biology:** Mean body temperatures (St. Christopher) 32.9 ± 0.3 °C in coconut habitat

(Roughgarden et al., 1981); on St. Christopher, mean microclimates used by *A. bimaculatus* and *A. schwartzi* are statistically indistinguishable at low-elevation (<300 m) sites (Buckley and Roughgarden, 2005b).

Conservation Status: Not assessed.

Anolis birama Garrido, 1990

Distribution: Cuba, only from the vicinity of the type locality near the mouth of the Río Cauto.

Habitat: In riverine habitat with vegetation dominated by *Cailliea glomerata*, but also with *Guasuma tomentosa*, *Calycophyllum candidissimun*, and *Conocarpus erecta*, most individuals perching on *Cailliea glomerata* and *Guasuma tomentosa*, but presumably is not restricted to the vicinity of the river and may extend onto isolated hills within the marshy region (Garrido, 1990); in pine forest (Rodríguez Schettino et al., 1999).

Activity: Diurnal (Rodríguez Schettino et al., 1999). **Ecomorphology:** "Trunk-ground" (Losos, 2009). **Size:** Male SVL to 65 mm, female SVL to 43 mm (Garrido, 1990).

Conservation Status: Considered "endangered" (Rodríguez Schettino and Chamizo Lara, 1998a; L. Rodríguez Schettino and A. Chamizo Lara in Rodríguez Schettino, 1999a).

Anolis bremeri Barbour, 1914

Distribution: Cuba, where two subspecies occur in Pinar del Río Province in western Cuba (*A. b. bremeri*) and on Isla de la Juventud, north of Ciénaga de Lanier (*A. b. insulaepinorum*).

Habitat: Xerophilic (Schwartz and Henderson, 1991); on the thatched roof of a house (Barbour, 1930c); in palm woods (*Thrinax* and *Sabal*) and forests (Pinar del Río) and pine forests (Isla de la Juventud), frequently in areas disturbed by humans for cultivation and pastures (Garrido, 1972); in mesophyllous evergreen and semideciduous forest (Rodríguez Schettino et al., 1999); sandy savannas where palms abundant, lizards perching on trunks of palms, *Eucalyptus*, and bushes, males perch higher than females and juveniles, with the latter more abundant on fence posts, bushes, and the ground, whereas males are on palm trunks and into the leaves (Rodríguez Schettino, 1999b); 11.1% on leaves (J. B. Losos, in litt., 21.IV. 2008).

Activity: Diurnal (Rodríguez Schettino et al., 1999). **Behavior:** Displays with little variation in the number of headbobs, typically differentiated by distinct pauses and coordinated with dewlap extension (Regalado and Garrido, 1993); escapes by running into herbaceous growth and hiding among grasses and stones (Rodríguez Schettino, 1999b). **Ecomorphology:** "Trunk-ground" (Williams and Rand, 1977). **Hybridization:** With *A. sagrei* (Garrido, 1980a). **Parasites:** Nematodes *Cyrtosomum scelopori* (Coy Otero and Baruš, 1973a, 1979b; Coy Otero and Lorenzo Hernández, 1982; Coy Otero, 1999); *Parapharyngodon cubensis* (Baruš and Coy Otero, 1969a, 1969b; Coy Otero and Baruš, 1979b; Coy Otero and Lorenzo Hernández, 1982; Coy Otero, 1999); *Oswaldocruzia lenteixeirai*, *Physaloptera squamatae*, *Skrjabinoptera phrynosoma*, *Spauligodon cubensis*, larval *Porrocaecum* sp., and Nematoda gen. sp. (Coy Otero and Baruš, 1979b; Coy Otero and Lorenzo Hernández, 1982; Coy Otero, 1999). **Reproduction:** Mean duration of copulation 64.9 min (longest 127 min; *A. b. insulaepinorum*), separation of pair "normal," egg 10 × 5 mm, white, and oval (Garrido, 1980a); mean duration of precopulatory behaviors 70 sec, male holds the female through copulation but does not maintain display elements, copulation characterized by 3 abrupt thrusts, separation rapid, copulation with *A. sagrei* (but no offspring produced), egg 8.4 × 5.1 mm (Regalado and Garrido, 1993). **Size:** Male SVL to 72 mm (Garrido, 1972), female SVL to 52.3 mm (Rodríguez Schettino, 1999b).

Conservation Status: Not assessed.

Anolis breslini Schwartz, 1980

Distribution: Hispaniola, Presqu'île du Nord-Ouest (Haiti).

Habitat: Xerophilic (Schwartz and Henderson, 1991); very arid scrub forest and semidesert (Schwartz, 1980f; Glor et al., 2003).

Ecomorphology: "Trunk-ground" (Rand and Williams, 1969, by implication); formally quantified (Glor et al., 2003). **Size:** Male SVL to 60 mm, female SVL to 45 mm (Schwartz, 1980f).

Conservation Status: Unknown.

Anolis brevirostris Bocourt, 1870

Distribution: Hispaniola, where three currently recognized subspecies occur in the Valle de Neiba/Plaine de Cul de Sac, Île-à-Cabrit, and an isolated population on Cap de Léogâne (*A. b. brevirostris*), Valle de San Juan into the Llanos de Azua and south onto the northern slopes of the

Sierra de Baoruco (*A. b. deserticola*), and the southeastern coast of the Tiburon Peninsula east across the Barahona Peninsula and north along the coast to Barahona, also on Isla Beata and Cayo Pisaje (*A. b. wetmorei*).

Habitat: Xerophilic, but occasionally in semimesic to mesic forests (Schwartz and Henderson, 1991); prefers larger trees in woods and flourishes in the proximity of springs and in taller *Acacia* around villages, on fence posts in variegated sun-and-shade mosaics, on vines, shrubs, and cliff faces, on the ground under palm logs, stacks of palm fronds, coconut husks piles, and in *Cocos* trash (Webster and Burns, 1973; Arnold, 1980; Schwartz and Henderson, 1991); along the coast, in *Cocos*, *Coccoloba uvifera*, and *Terminalia* sp. (Fläschendräger and Wijffels, 2009); in a park, adults moved freely between all 3 of the broadleaf trees (*Aleurites molucana*, *Cassia siamea*, *Terminalia catappa*) at mean perch heights 2.1 ± 1.1 m (males) and 1.4 ± 0.9 m (females), with only 2 adults on the ground, only a few individuals, almost all juveniles, used palms and shrubs, only observation on a leaf was an introduced male chased by a resident (Moster et al., 1992); restricted to trunks of *Acacia* and mixed xeric scrub along a dry riverbed, mean microhabitat temperatures 29.9 ± 2.7 °C (27.0–35.4 °C), lowest of all sympatric lizards (Powell et al., 1996b); one female on a mangrove on Cayo Pisaje (Yeska et al., 2000); mean perch heights and diameters 1.03 m and 21.07 cm (J. B. Losos in Calsbeek et al., 2006); sleeps 1–6 m above the ground on vines, shrubs, *Ficus*, and especially on *Acacia* (Schwartz and Henderson, 1991).

Activity: Active from dawn to dusk (Moster et al., 1992); exploits the "night-light niche" (Bowersox et al., 1994; Perry et al., 2008). **Behavior:** Displays differ from those of sibling species *A. caudalis* and *A. websteri*, with the greatest difference between geographically adjacent *A. caudalis*, headbobs in core (type-A) display usually 6 (5–7), 6.94 ± 0.29 sec in duration, usually not followed by dewlap pulses, tail-lift posture not observed, two type-B displays, a long bob followed by a short bob and a long bob followed by 2 short bobs (Jenssen and Gladson, 1984); 91 of 156 observations (including 2 females headbobbing during copulation) involved "type-A" behavior (bobs often followed by dewlapping), 31 involved dewlap extension, 17 push-ups, 43 push-ups followed by dewlap extension, and 5 tail raising, 7 males engaged in "type-B" behavior (series of headbobs with simultaneous dewlap extension), twice combined with tail lashing ("type-C" behavior), most male-male interactions agonistic, often appearing to involve defense of a territory or a female, many included chases and biting, introduced females accepted, one male displaced, females sometimes assumed positions vacated by males, escape behavior primarily "squirreling" followed by climbing higher on the trunk or to higher branches or to those farther from the trunk, chased by *A. chlorocyanus* (Moster et al., 1992); mean moves per minute 1.12, percent time displaying 1.4, percent walking 2.8, percent jumping 0.3, percent running 96.9 (J. B. Losos, in litt., 21.IV.2008). **Ecomorphology:** "Trunk" (Williams, 1983); formally quantified (Beuttell and Losos, 1999). **Foraging and Diet:** 16 of 24 instances of prey capture involved a sit-and-wait strategy, other 8 instances involved "joint-stepping" (Moermond, 1981); most captures on trees, sometimes involved backward stepping or jumping, 3 attempts to catch flying insects were not successful, stomach contents included spiders, beetles, flies, hemipterans, hymenopterans (including ants, which were the most important food item), lepidopterans (adults and larvae), a dragonfly, plant material, and grit (Moster et al., 1992). **Home Range and Territoriality:** A branch or group of branches at approximately the same height (Moster et al., 1992). **Hybridization:** Hybridization possibly frequent with *A. distichus*, but hybrids probably infertile (Webster, 1977); extent of hybridization with *A. distichus* variable, no evidence at montane sites, but extensive at 2 lowland sites where interspecific differences in dewlap color were minimal or variable (Williams and Case, 1986). **Movements:** Of 43 animals marked, 19 were on 2 different trees, 5 on 3 different trees, and 2 on 4 different trees (Moster et al., 1992). **Parasites:** Nematode *Skrjabinoptera leiocephalorum* (Moster et al., 1992; Goldberg et al., 1998b); trematode *Urotrema scabridum*, nematodes *Atractis scelopori*, *Parapharyngodon cubensis*, and larval *Abbreviata* sp. (Goldberg et al., 1996c, 1998b). **Population Size and Density:** In an altered habitat, 3.6 lizards per tree and ~800 lizards/ha (Moster et al., 1992). **Reproduction:** Copulation 5 times in June, during which males engaged in type-A or B behavior and 2 females in type-A behavior, separation was abrupt, with females either remaining in place or (usually) running upward (Moster et al., 1992). **Sex Ratio (M:F):** 33:27 in a small population in altered habitat, 14:5 in a sample when the first lizards seen were captured, suggesting that males are more visible (Moster et al., 1992). **Size:** Male SVL to 53.1 mm (Schoener, 1988), female SVL to 46 mm (Moster et al., 1992). **Thermal Biology:** Did not bask in a shaded park, 81% of observations in shade, 19% in diffuse sunlight, none in full sun except when responding to human disturbances, mean body temperature 30.6 ± 3.4 °C at mean substrate and air temperatures of 28.7 ± 3.2 °C and 28.6 ± 1.5 °C (Moster et al., 1992); along a dry riverbed, microhabitat temperatures (mean 29.9 ± 2.7 °C, with a high of 35.4 °C) lower than those of *A. whitemani* (Powell et al., 1996b).

Conservation Status: Not assessed.

Additional Pertinent Literature: Moster et al. (1995).

Anolis brunneus Cope, 1894

Distribution: Bahama Islands, Crooked-Acklins Bank, West Plana Cay.

Habitat: Xerophilic, on *Musa*, buildings, large trees, and *Cocos* palms (Schwartz and Henderson, 1991); "about gardens and often coming on the verandas of houses," "most often seen on *Pandanus, Agave, Musa* and similar plants" (Barbour, 1930c); perches lower on Acklins Island, where *A. brunneus* occurs alone, than on Crooked Island, in the presence of *A. sagrei*, where *A. brunneus* also is less frequently encountered on leaves (Schoener, 1975, 1976); on Acklins Island, mean perch height and diameter 1.68 ± 0.22 m and 2.3 ± 0.5 cm (Losos and de Queiroz, 1997a); sleeps on trees and bushes (Schwartz and Henderson, 1991).

Ecomorphology: "Trunk-crown" (Losos et al., 1994; Losos, 2009); descendant of trunk-crown stock, but closest to a "generalist" (Losos and de Queiroz, 1997a). **Diet:** "Ants, small spiders, and other insects" (Barbour, 1930c). **Predation:** *Leiocephalus greenwayi* and *L. punctatus* (Schoener et al., 1982); *Cubophis vudii* (Henderson and Sajdak, 1996). **Size:** Male SVL to 76 mm (Ruibal, 1964), female SVL to 70 mm (Schwartz and Henderson, 1991).

Remarks: This species was frequently referred to as *A. carolinensis* in the older literature.

Anolis carolinensis Voight, 1832

Distribution: In the West Indies, known only from an introduced population on Anguilla and a few individuals (one voucher specimen) on Grand Cayman Island; native to the southeastern United States; introduced widely in the Pacific on Hawaii (Oahu and Coconut Island, Molokai, Hawaii, Lanai, Kauii, Maui), Ogasawara (Bonin) Islands (one island), Northern Marianas (Guam, Tinian, Saipan), Republic of Palau (four islands), Ryuku Islands (Okinawa), and probably other islands in the region.

Habitat: West Indian populations are known to occur only in association with imported ornamental vegetation on the grounds of resort hotels, usually high on trunks or even in crowns of trees on Anguilla (Eaton et al., 2001; Powell, 2002b; Hodge et al., 2003; Powell and Henderson, 2008c).

Behavior: Shy, will creep to other side of trunk or branch or climb up into the canopy (Hodge et al., 2003). **Ecomorphology:** "Trunk-crown" (Beuttell and Losos, 1999). **Size:** In the southeastern United States, male SVL to 75 mm (Conant and Collins, 1998), female SVL to 69 mm (R. M. Goodman, in litt., 10.VIII.2007).

Conservation Status: Listed as "least concern" on the IUCN Red List (Frost and Hammerson, 2007).

Anolis caudalis Cochran, 1932

Distribution: Hispaniola (Haiti), along the coast of the Golfe de la Gonâve, two isolated populations on the northern shore of the Tiburon Peninsula, Île de la Gonâve, Île de la Petite Gonâve, and Île a Cabrit.

Habitat: Xerophilic (Schwartz and Henderson, 1991); in *Acacia* scrub and xeric woodlands, prefers high perches (>2 m) on large trees, in a semicultivated lowland site in Haiti, in June (wet season), mean male perch height and diameter 2.9 ± 0.27 m and 18.2 ± 3.01 cm, female 1.5 ± 0.18 m and 19.4 ± 3.89 cm, juvenile 0.2 ± 0.06 m and 1.8 ± 1.09 cm, in November (dry season), mean male perch height and diameter 2.9 ± 0.25 m and 19.8 ± 2.59 cm, female 2.3 ± 0.22 m and 18.8 ± 2.60 cm (Jenssen et al., 1988).

Behavior: Displays differ from sibling species *A. websteri* and *A. brevirostris*, headbobs in core (type-A) display usually 6 (5–7), 5.75 ± 0.15 sec in duration, with 2.15 ± 0.20 dewlap pulses, tail-lift posture usually observed, single type-B display of alternating long and short bobs (Jenssen and Gladson, 1984). **Ecomorphology:** "Trunk" (Williams, 1983). **Population Size and Density:** Population sizes very stable across seasons (Jenssen et al., 1988). **Predation:** *Hypsirhynchus parvifrons* on La Gonâve (Franz and Gicca, 1982) and elsewhere on Hispaniola (Henderson et al., 1987a); *Uromacer oxyrhynchus* (Henderson et al., 1987c). **Reproduction:** No evidence of pre-mating isolating mechanisms in sympatry with *A. websteri*, males mated indiscriminately with females of either species, females were selective, but mated infrequently (0.04 copulations/day), mean duration of conspecific intromission 13.4 ± 1.34 min, decreasing to 3.5 ± 0.33 min when with *A. websteri* females (Jenssen, 1996). **Sex Ratio (M:F):** 1.2:1 across seasons (Jenssen et al., 1988). **Size:** Male SVL to 48 mm, female SVL to 44 mm (Arnold, 1980).

Conservation Status: Not assessed.

Anolis centralis Peters, 1970

Distribution: Cuba, northeastern Camagüey Province to western Holguín Province, also apparently in the Río Cauto basin and perhaps on the Península de Banes.

Habitat: Xerophilic (Schwartz and Henderson, 1991); on fences, trees, in shrubby areas, and on spiny bushes and neighboring xerophytes (Garrido, 1975b); an individual on the rocky bed of a deeply shaded stream (G. Peters, 1970); in coastal and subcoastal microphyllous forest, semidesert thorny scrubwoodland, and serpentine xeromorphous shrub-woodland (Rodríguez Schettino et al., 1999); on Cayo Coco (Archipiélago de Sabana-Camaguey), especially along the ecotone between dry coastal scrub and microphyllous evergreen forest (Estrada, 1993b); mean perch height and diameter 1.54 m and 5.4 cm (J. B. Losos, in litt., 21.IV.2008).

Activity: Diurnal (Rodríguez Schettino et al., 1999). **Behavior:** Aggressive displays of males ($N = 2$) include mouth opening, lateral presentation, darkening of color, lateral compression, raising nuchal and dorsal crests, a rapid nod with throat inflated, and gradual approach toward other male before sudden elevation on all limbs with dewlap completely extended, entire process lasts approximately 12 sec (Garrido, 1988); to escape, moves slowly, stops, then runs again (Rodríguez Schettino, 1999b). **Ecomorphology:** Undetermined; characterized as a "bark anole" (G. Peters, 1970). **Parasites:** Nematode *Cyrtosomum scelopori* (Coy Otero and Baruš, 1979b; Coy Otero and Lorenzo Hernández, 1982; Coy Otero, 1999; note that the original association of parasite and host did not provide collecting localities, consequently, the hosts could be *A. litoralis*, which was elevated from subspecific status subsequent to the publication of Coy Otero and Baruš, 1979b). **Reproduction:** Mean duration of copulation 2.7 min (longest 7.33 min), separation of pair "hasty," egg (April) 7.4 × 4.8 mm, plaster white and oval (Garrido, 1980a); courtship display (~14.5 sec) is of shorter duration and less elaborate than that of *A. litoralis*, males raise head backward in a sequence of 4–5 nods, preceded by one bob, the process continuing until the female acquiesces, dewlap only partially extended, after intromission, male remains immobile (mean 14.6 sec) while he bites her neck, mean copulatory time 2.7 min, during which female nods occasionally, male leaves female hastily or crawls over her (Garrido, 1988). **Size:** Male SVL to 42.7 mm (Navarro, Pacheco et al., 2001). **Thermal Biology:** Apparently nonheliothermic (Rodríguez Schettino, 1999b).

Conservation Status: Not assessed.

Anolis chamaeleonides Duméril and Bibron, 1837

Distribution: Cuba, widely distributed across the island, but local, also on Isla de la Juventud; elevational range, sea level to ~850 m.

Habitat: Mesophilic (Schwartz and Henderson, 1991); humid broadleaf and shaded forests, moderately common in coffee plantations, very occasionally on fence posts, perches head-down on tree trunks, sleeps fully exposed on vines or limbs 1.5 to >3.5 m above the ground (Ruibal, 1964; Garrido and Schwartz, 1968a; Schwartz and Henderson, 1991); occasionally in open areas (in a yard, on a road, in an area around a swimming pool) but always in the vicinity of contiguous forest, mean perch height 4.2 m above the ground (Rodríguez Schettino, 1999b); in submontane rainforest, submontane mesophyllous evergreen forest, mesophyllous evergreen and semideciduous forest, coastal and subcoastal microphyllous forest, and cultivated areas and pastures (Rodríguez Schettino et al., 1999); rarely descends to the ground except when females bury eggs in soil (Fong Grillo and Garcés G., 2001).

Activity: Diurnal (Rodríguez Schettino et al., 1999). **Behavior:** Easy to catch (Gundlach, 1880); in captivity, "very sluggish, even when freshly captured, and could be left on a laboratory table without much danger of its wandering far away; it would often remain in the same spot for hours or even days without changing its position" (Wilson, 1957); when disturbed, moves slowly to the opposite side of a perch (Rodríguez Schettino, 1999b); in captivity, hatchlings are active and aggressive, resist handling and try to bite (Fong Grillo and Garcés G., 2001). **Ecomorphology:** Categorized as a Cuban "giant anole" (G. Peters, 1970); "twig-giant" (Hass et al., 1993), member of the *Chamaeleolis* clade; technically this is not an ecomorphological category as it evolved in only one clade (see definition in Williams, 1972). **Foraging and Diet:** In captivity, moves deliberatively toward prey until about 5 cm away, smaller insects are taken with a sudden forward and downward flick of the tongue, larger prey usually seized with the mouth, although tongue typically extruded first and plays an important part in manipulating prey once taken (Wilson, 1957); insects, millipedes, and other lizards, approaches prey slowly, opens mouth slightly, moves tongue into position in front of mouth, shoots tongue at prey animal, which sticks, allowing prey to be drawn into mouth, chewed for a few seconds or minutes and then swallowed (Fong Grillo and Garcés G., 2001); an abundance of snails on rocky soils in the Sierra de Cubitas draws lizards to lower heights during the day (Díaz, 2006). **Movements:** Very "chameleon-like," slow movements and independent movement of the eyes (Ruibal, 1964). **Parasites:** Nematodes *Cyrtosomum longicaudatum* (Coy Otero and Baruš, 1973a, 1979b; Coy Otero and Lorenzo Hernández, 1982; Coy Otero, 1999) and *Alaeuris travassosi* (Pérez Vigueras, 1938; Coy Otero and Baruš, 1979b; Coy Otero and Lorenzo Hernández, 1982; Coy Otero, 1999). **Predation:** In captivity, reacted to a 60-cm *Tropidophis melanurus* with a lateral presentation and body constriction, elevation of nuchal crest, and gaping (Hottmar, 1991). **Reproduction:** Copulation in April (L. F. de Armas in Rodríguez Schettino, 1999b); female laid (May) a white egg in a collecting sack and then another in a depression in the soil of a terrarium, first (28 × 15 mm) did not develop, the second (28 × 15 mm at time of deposition) measured 28 × 22 mm and 7.8 g after 59 days of incubation, hatching after 65 days of incubation (July), neonate SVL 44.6 mm, weight 2.5 g (Estrada, 1994); white, oval eggs soft and sticky when laid, but become hard and rough shortly after deposition, 22 and 24 mm in length, hatched in captivity after 49 and 52 days, hatchlings measured 76 and 79 mm (Fong Grillo and Garcés G., 2001). **Size:** Male SVL to 177 mm, female SVL to 172 mm, a female with an open umbilicus was SVL 46 mm (Garrido and Schwartz, 1968a). **Thermal Biology:** Apparently nonheliothermic, never observed basking (Rodríguez Schettino, 1999b).

Conservation Status: Considered "vulnerable" (L. Rodríguez Schettino and A. Chamizo Lara in Rodríguez Schettino, 1999a).

Anolis chlorocyanus Duméril and Bibron, 1837

Distribution: Hispaniola, where two currently recognized subspecies occur in the south-central Dominican Republic (*A. c. cyanostictus*) and essentially everywhere else on the "North Island," other than very xeric areas, although it may be in oases and locally in the Massif de la Selle and south onto the Barahona Peninsula into the city of Barahona, Île de la Gonâve, Île de la Tortue, and Isla Saona (*A. c. chlorocyanus*); elevational range, sea level to >1,220 m; introduced in Florida.

Habitat: Mesophilic (Schwartz and Henderson, 1991); on tree trunks (including *Cocos*), large leaves, fence posts, generally prefers moister areas, but occasionally on bushes and trees in mesic microhabitats in dry scrub forest, common in human-modified habitats, including on and in houses (Mertens, 1939); on leaves and trunks in broadleaf forest, but also in edificarian situations and in gardens, *Cocos*, *Citrus-Musa*, and *Theobroma* plantings, and other more-or-less shaded situations, also on a bridge abutment, cutbanks, fence posts, and roadside *Acacia* (Rand, 1962b; Williams, 1965a; Schwartz and Henderson, 1991; Fläschendräger and Wijffels, 1996, 2009; Powell and Henderson, 2008c); in active oil palm and cacao plantations, abandoned pastures, and gardens, as well as forests and associated with mogotes (Glor et al., 2001); on tree trunks and fence posts, but also ranges high in trees and out among smaller branches, may occasionally descend to the ground to reach isolated trees (Rand, 1962b); in the La Romana area, where *A. cristatellus* is established, *A. chlorocyanus* favors trees with smooth, slick surfaces, presumably because its larger toepads provide superior clinging ability (Fitch et al., 1989); mean perch height and diameter 3.57 m and 10.62 cm (J. B. Losos in Calsbeek et al., 2006); 25.9% on leaves (J. B. Losos, in litt., 21.IV. 2008); sleeps from <1 m to >4.5 m above the ground on leaves and horizontal or vertical pendant vines (Schwartz and Henderson, 1991).

Abundance: Uncommon, 77 of 990 observations of anoles in and around Los Haitises National Park, Dominican Republic (Glor et al., 2001). **Activity:** Diurnal (Rand, 1962b). **Behavior:** Sleeps with hindlimbs extended (Mertens, 1939); vocal "alarm call" (Mertens, 1946); when approached, immediately begins to climb up, usually slowly and deliberately until out of reach, if it cannot go higher, moves into foliage to hide (Rand, 1962b); rests with head and body close to the substrate with neck bent little or not at all, sleeps with hindlimbs extended along the tail, moves slowly and deliberately, sure climber, spends very little time on the ground, moving from tree to tree in one quick process (down tree, rapidly across ground, jumping onto second tree), makes horizontal jumps to >25 cm (Rand, 1962b); capable of vocalization (T. A. Jenssen in Milton and Jenssen, 1979); may enter water as an escape strategy, swimming by lateral undulations of body and tail with tightly adpressed limbs, may submerge and cling to vegetation for at least 1.5 min (Powell and Parmerlee, 1993a); agonistic display similar to that of *A. coelestinus* (but more rapid and with shorter intervals between bobbing sequences), consists of lateral presentation, body compressed, small dorsal and nuchal crests erected, dewlap fully extended, not pumped (no tail, mouth, tongue, or pawing movements, and no push-ups), head held high until bobbing display begins, may try to reach a perch above the competitor, at times with biting, terminates with retreat of one individual (Garcea and Gorman, 1968; Bels, 1984, 1986); in La Romana, fought with large male *A. cristatellus* (Fitch et al., 1989); in Santo Domingo, potentially agonistic male-male encounter with *A. porcatus* (Powell et al., 1990e); in La Romana, one large male with the entire head of another in its mouth, presumably a territorial dispute gone awry, after escape the "loser" ascended an adjacent tree to the same height as the other individual, dewlapped repeatedly before ascending into the canopy (Powell and Parmerlee, 1991); in Barahona, chased *A. brevirostris* (Moster et al., 1992); moves per minute 1.17 (Johnson et al., 2008); percent time displaying 3.8, percent walking 70.2, percent jumping 8.2, percent running 21.6 (J. B. Losos, in litt., 21.IV.2008). **Competition:** Introduced *A. porcatus* may be excluding native *A. chlorocyanus* from modified habitats in Santo Domingo (Powell et al., 1990e). **Ecomorphology:** "Trunk-crown" (Rand and Williams, 1969); formally quantified (Beuttell and Losos, 1999). **Foraging and Diet:** Rarely comes to ground to feed, getting most food on leaves and limbs or trunk (Rand, 1962b); 6 of 9 stomachs contained flies, mean prey size 0.8 ± 0.9 cm (Fitch et al., 1989); near La Romana, an adult male (SVL ~65 mm) with the entire head of a female *A. cristatellus* (SVL ~40 mm) in its mouth, but it escaped during an attempt at repositioning (Powell and Parmerlee, 1991); cannibalism (Bowersox et al., 1994). **Home Range and Territoriality:** Highly territorial, only one male and one or more females in a single tree (Mertens, 1939; Rand, 1962b). **Hybridiza-**

tion: With *A. coelestinus* (R. Garcia et al., 1994). **Longevity:** In captivity, 2 years, 1 month and still alive at last report in 1992 (J. Grizzell in Snider and Bowler, 1992). **Parasites:** Cestode *Oochoristica maccoyi*, trematode *Urotrema wardi*, nematodes *Atractis scelopori*, *Parapharyngodon cubensis*, and larval *Ascarops* sp. and *Physocephalus* sp. (Goldberg et al., 1996c, 1998b). **Population Size and Density:** 0.25 person-hours to collect each individual at a mesic site in San Cristóbal Province (Cusumano et al., 1991). **Predation:** Ridgway's Hawk (*Buteo ridgwayi*; Wiley and Wiley, 1981); *Hypsirhynchus parvifrons* on La Gonâve (Franz and Gicca, 1982), on Hispaniola (Henderson et al., 1987a); *Uromacer catesbyi*, *U. frenatus*, *U. oxyrhynchus* (Henderson et al., 1987c); adult conspecifics (Bowersox et al., 1994). **Reproduction:** Clutch size 1 (Hedges, in prep.); egg (11.5 × 7.5 mm) in a pile of debris at the base of a *Cocos* palm hatched 30 October, egg collected 22 July hatched around 20 August (Schwartz and Henderson, 1991). **Size:** Male SVL to 79.5 mm, female SVL to 60.9 mm (Schoener, 1988).

Conservation Status: Not assessed.

Additional Pertinent Literature: Ramos and Powell (2001a).

Anolis christophei Williams, 1960

Distribution: Hispaniola, Massif du Nord (Haiti), Cordillera Central and Cordillera Septentrional (Dominican Republic), also in the eastern Dominican Republic; elevational range ~75 m to nearly 1,300 m.

Habitat: Mesophilic (Schwartz and Henderson, 1991); in moist, shaded woods, particularly stream-associated gallery forests and mesic ravines, on trunks 1.2–1.8 m above the ground, fallen logs, fence posts in riparian wood, on small trees 0.6–6.1 m above the ground (Thomas and Schwartz, 1967; Rand and Williams, 1969); also on large rocks and small branches, frequently in the spray zone of mountain streams (Schirra, 1993); on moss-covered and plant-overgrown rocks and among exposed roots along deeply shaded eroded streambanks (Fläschendräger and Wijffels, 2009); mean perch height and diameter 1.96 m and 27.2 cm, do not use leaves (J. B. Losos, in litt., 21.IV.2008); sleeps, usually with hindlimbs flexed and across leaflets or leaves (instead of aligned with the stems), in streamside plants, bushes, tree ferns, ferns in coffee and cacao plantations, on leaves (not stems) of vines (Thomas and Schwartz, 1967).

Abundance: Abundant in appropriate habitat, 3–4 individuals of all ages encountered in 10-m stretches (Fläschendräger and Wijffels, 2009). **Activity:** Diurnal (Rand and Williams, 1969). **Behavior:** Movement characterized as "runner" (Moermond, 1981); presumably due to effective crypsis, escape behavior weakly developed, but individuals of both sexes readily take to water when disturbed (Schirra, 1993); escape into deep cracks and crevices along an eroded streambank, readily dive into water, male hatchlings extend dewlaps in captivity (Fläschendräger and Wijffels, 2009); mean moves per minute 0.64, percent time displaying 1.8, percent walking 15.1, percent jumping 1.8, percent running 83.1 (J. B. Losos, in litt., 21.IV.2008). **Ecomorphology:** "Trunk" (Rand and Williams, 1969); formally quantified and determined to be morphologically distant from all other ecomorphs (Beuttell and Losos, 1999); undetermined (Losos, 2009). **Foraging and Diet:** Prey strikes either approach-pause or stationary (Moermond, 1981). **Home Range and Territoriality:** Males and females strongly territorial, apparently with relatively large home ranges, distances between adult males to 8 m (Schirra, 1993). **Predation:** *Hypsirhynchus parvifrons* (Henderson et al., 1987a). **Reproduction:** In captivity, males approach females while headbobbing and extending dewlaps, females remain in place or allow approach after a brief pursuit, males grasp females by their napes, copulation lasts 3–8 min, single eggs deposited in excavated holes, hatchling SVL ~20 mm (Schirra, 1993); clutch size 1 (Hedges, in prep.); eggs at deposition 5 × 8 mm, incubation 36–43 days at 22–25 °C, hatchling SVL ~19 mm (Fläschendräger and Wijffels, 1996, 2009). **Size:** Male SVL to 49 mm, female SVL to 45 mm (Thomas and Schwartz, 1967). **Thermal Biology:** Basking behavior not observed (Rand and Williams, 1969).

Conservation Status: Not assessed.

Anolis chrysops Lazell, 1964

Distribution: Terre-de-Haut and Terre-de-Bas, Les Iles de la Petite Terre, east of Guadeloupe.

Habitat: "Arid lowlands typical of islands too low in elevation to stop significant rainfall" (Lazell, 1972); in low, wind-blown vegetation (Heselhaus and Schmidt, 1990); primarily in forested areas and on dry rock walls (Breuil, 2002b).

Reproduction: Seasonal, copulation primarily in July–August (Breuil, 2002b). **Size:** Male SVL to 73 mm (Lazell, 1964a), female SVL to 48 mm, smallest mature male and female SVL 62 mm and 46 mm, respectively, but data undoubtedly incomplete (Lazell, 1972).

Conservation Status: Unknown.

Anolis clivicola Barbour and Shreve, 1935

Distribution: Cuba, Sierra Maestra and Sierra del Cobre; elevational range ~1,350–2,000 m.

Habitat: Mesophilic (Schwartz and Henderson, 1991); ferns and grasses 1 m above ground, on a supporting post of a hut, where an individual sought shelter under the roof from incessant rain, prefers areas of high humidity and shade (Garrido, 1980a); in cloud forest and montane rainforest (Rodríguez Schettino et al., 1999); on trunks of trees, bushes, and tree ferns, males perch higher (to 6 m) than females and juveniles, which are on the ground, on roots, or in the herbaceous layer of vegetation (Rodríguez Schettino, 1999b).

Activity: Diurnal (Rodríguez Schettino et al., 1999); most individuals encountered during the warmest times of day (Rodríguez Schettino, 1999b). **Behavior:** Jumps from branch to branch or onto the ground to escape (Rodríguez Schettino, 1999b). **Ecomorphology:** "Grass-bush" (Williams, 1961). **Reproduction:** Copulation on 26 May (Garrido, 1980a); clutch size 1 (Hedges, in prep.). **Size:** Male SVL to 49.4 mm (Rodríguez Schettino, 1999b), female SVL to 45 mm (Schwartz and Henderson, 1991). **Thermal Biology:** Apparently nonheliothermic (Rodríguez Schettino, 1999b).

Conservation Status: Considered "vulnerable" (Rodríguez Schettino and Chamizo Lara, 1998a; L. Rodríguez Schettino and A. Chamizo Lara in Rodríguez Schettino, 1999a).

Anolis coelestinus Cope, 1862

Distribution: Hispaniola, where three currently recognized subspecies occur throughout most of the "South Island," locally into oases of the Plaine de Cul de Sac/Valle de Neiba (*A. c. coelestinus*), and on Île Grande Cayemite (*A. c. demissus*) and Île-a-Vache (*A. c. pecuarius*); elevational range, sea level to >1,700 m.

Habitat: Mesophilic to somewhat xerophilic (Schwartz and Henderson, 1991); in shaded situations and relatively cool localities, wooded areas in cool ravines, coffee plantations, and broadleaf forest, on trees and bushes, occasionally on fence posts, often exclusively arboreal on trees and saplings down to a diameter of 7.5 cm, commonly seen in Haiti in association with human structures (one lived on a lone papaya on a third floor patio for 3 months), on rocks in open areas, particularly where the forest had been cleared for gardens, also on boulders and cobbles in a streambed and on exposed slopes near a deep ravine (Mertens, 1939; Williams, 1965a; Schwartz, 1969b; Williams and Webster, 1974; Franz and Cordier, 1986; Olson, 1995; Lenart et al., 1997); 25.8% on leaves, 9.0% on grass blades, 0.7–7.6% on stems of varying diameters, and 40.5% on rocks at a Haitian site (Moermond, 1979a); 8.5% on leaves or grasses, 9.4% on perches <0.7 cm in diameter, 7.7% on perches 0.7–2.5 cm in diameter, 18.8% on perches >2.5 cm in diameter, and 6.8% on rocks at heights >90 cm, 16.2% on leaves or grasses, 8.5% on perches 0.7–2.5 cm in diameter, 3.4% on perches >2.5 cm in diameter, and 4.3% on rocks at heights 31–90 cm, 6.8% on leaves or grasses, 2.6% on perches <0.7 cm in diameter, 3.4% on perches 0.7–2.5 cm in diameter, and 3.4% on rocks or ground at heights <30 cm (Moermond, 1979b); appears tolerant of human-mediated habitat alterations, most abundant species in two disturbed upland habitats in the Sierra de Bahoruco (Lenart et al., 1997); mean perch heights and diameters along rainforest edge 191.7 ± 36.3 cm and 4.7 ± 1.8 cm, in rainforest interior 153.3 ± 89.5 cm and 10.1 ± 0.0 cm, with males perching higher (225.0 ± 25.0 cm) and on perches of greater diameter (6.1 ± 1.2 cm) than one female (125.0 cm and 20.0 cm; Cast et al., 2000); mean perch heights and diameters of males along edges in cloud forest 103.2 ± 10.4 cm and 10.4 ± 1.3 cm, females 97.8 ± 13.2 cm and 6.2 ± 1.2 cm, both sexes combined at four sites were 70.0 ± 15.4–141.8 ± 30.7 cm and 2.0 ± 1.0–10.6 ± 1.1 cm (Sifers et al., 2001); mean perch height and diameter 2.53 m and 20.92 cm (J. B. Losos in Calsbeek et al., 2006); in coastal *Coccoloba uvifera* (Fläschendräger and Wijffels, 2009).

Behavior: Percent jumps (fraction of total movements) by adults in crowns 26 and 28% at two Haitian sites, juveniles at one site primarily in bushes jump 47%, juveniles in grass-herb habitat jump more than crawl or run, individuals of all ages in scrub jump, crawl, or run in about equal frequency, adults in trees crawl more frequently than jump or run (Moermond, 1979a); at 2 different sites, percent crawls 34 and 52%, runs 40 and 20%, jumps 26 and 28%, percent moves to different surface 28% (adults) and 47% (juveniles; Moermond, 1979b); mean primary approach distance before reacting 110.6 ± 55.0 cm, mean secondary distance 50.5 ± 8.9 cm, most common (50%) primary response "squirreling," most common (33%) secondary response jumping to another substrate (Schneider et al., 2000); agonistic display similar to that of *A. chlorocyanus* (see that account), but slower and with longer intervals between bobbing sequences (Garcea and Gorman, 1968); vocal "alarm call" (Mertens, 1946); capable of vocalization (T. A. Jenssen in Milton and Jenssen, 1979); moves per minute 1.27 (Johnson et al., 2008). **Ecomorphology:** "Trunk-crown" (Rand and Williams, 1969); formally quantified (Beuttell and Losos, 1999). **Foraging and Diet:** Scanning from high, broad perch takes less time than scans from perches with limited overviews and much less than scans of the ground from elevated perches, resulting in relatively short waiting intervals and distance/move, mean 4.2 min between location changes, mean 3.1 min between location or posture changes, mean distance/move 22 cm, distances moved per prey capture longer than those of anoles surveying dense habitat clusters but shorter than those ranging from elevated perches to the ground, 56% of prey captures on the surface where the lizard was perched (Moermond, 1979b); preys on *A. bahorucoensis* (Fitch and Henderson, 1987); stomach contents of 25 individuals from 3 variously altered upland habitats in the Sierra de Bahoruco included spiders and another arachnid, orthopterans (the most important dietary item), beetles (adults and larvae), hemipterans, roaches, lepidopterans (adult and larvae), flies (adults and larva), ants, a snail, and plant matter (probably adventitiously ingested), dietary niche breadth was greater than that of *A. cybotes* in the 2 most altered habitats, but less in a more natural situation, and more than that of sympatric *A. distichus* in all habitats (Lenart et al., 1997); dietary niche overlap with *A. barbouri* is low (A. K. Howard et al., 1999a); stomach contents of 5 individuals from rainforest contained beetles (adults, the most important dietary item, and larvae), a fly, ants, a lepidopteran larva, orthopterans, shed skin, flowers, and presumably adventitiously consumed grit (Cast et al., 2000); stomach contents of 24 individuals from cloud forest contained spiders, beetles (adults and larvae), roaches, flies, hemipterans, hymenopterans (including ants, the most important dietary item), lepidopterans (adults and larvae), orthopterans, snails, shed skin, and fruits/seeds (Sifers et al., 2001); an arboreal cricket (Sánchez Muñoz, 2008). **Home Range and Territoriality:** Typically 1 male and 2 females together (Fläschendräger and Wijffels, 2009). **Hybridization:** With *A. chlorocyanus* (R. Garcia et al., 1994). **Parasites:** Larval mites (*Eutrombicula alfreddugesi*) more prevalent and at higher intensities in higher-elevation, moister situations (Zippel et al., 1996); trematode *Mesocoelium monas*, cestode *Oochoristica maccoyi*, nematodes *Atractis scelopori*, *Parapharyngodon cubensis*, and *Physalopteroides bahamensis* (Goldberg et al., 1996c, 1998b). **Population Size and Density:** At 3 cloud forest sites, 5,662–12,985 lizards/ha concentrated in edge habitats (Sifers, 2001). **Predation:** *Hypsirhynchus parvifrons* (Henderson et al., 1987a); *Epicrates striatus* (Henderson et al., 1987b); *Uromacer catesbyi*, *U. frenatus* (Henderson et al., 1987c); domestic chicken (*Gallus gallus*; Bowersox et al., 1994); *Ameiva chrysolaema* carrying a dead adult male, unknown whether killed or scavenged (RP). **Size:** Male SVL to 84 mm (Schwartz, 1969b), female SVL to 61.5 mm (Schoener, 1988). **Thermal Biology:** Mean male and female body temperatures at 4 sites in cloud forest 28.8 ± 0.9 °C and 28.8 ± 0.8 °C at mean substrate and air temperatures of 23.9 ± 0.3 °C and 24.0 ± 0.4 °C (Sifers et al., 2001).

Conservation Status: Not assessed.

Additional Pertinent Literature: Ramos and Powell (2001b).

Anolis confusus Estrada and Garrido, 1991

Distribution: Cuba, only from the vicinity of the type locality near Cabo Cruz and extending north to the vicinity of Las Coloradas, Granma Province.

Habitat: Xerophilic, in dry evergreen forest bordering the dry coastal and subcoastal vegetation in a band more or less parallel to the coast, where lizards perch on trunks of trees and shrubs ~2 m above the ground, with females more frequently on smaller trees or on the ground (Estrada and Garrido, 1991); in mesophyllous evergreen and semideciduous forest (Rodríguez Schettino et al., 1999).

Activity: Diurnal (Rodríguez Schettino et al., 1999). **Ecomorphology:** "Trunk-ground" (Estrada and Garrido, 1991, by implication). **Size:** Mean male SVL 53.0 mm, female SVL 48.0 mm (Rodríguez Schettino, 1999b).

Conservation Status: Considered "vulnerable" (Rodríguez Schettino and Chamizo Lara, 1998a; L. Rodríguez Schettino and A. Chamizo Lara in Rodríguez Schettino, 1999a).

Anolis conspersus Garman, 1887

Distribution: Grand Cayman Island, where two currently recognized subspecies occur islandwide and on Booby Cay in the North Sound.

Habitat: Mesophilic (Schwartz and Henderson, 1991); in more open areas, not thick forest, also on mangroves (especially on Booby Cay), palms, fence posts, an abandoned WW II watchtower, and coral heads tossed up into windrows (Grant, 1940c; Schwartz and Henderson, 1991); only sporadically in mangrove or buttonwood swamps, Sea Grape (*Coccoloba uvifera*) thickets, or areas of sparse windblown vegetation, adults at a variety of perch heights (adult males to >4 m) in both forest/scrub and open habitats (higher in the former, correlated with the abundance of taller trees and bushes), with few on the ground (0.5% of adult males), males perched higher than females, juveniles mostly on the ground (55.6%), subadults less frequently (Avery, 1988); show "ecological release" to the extent that they largely exclude other anoles from a relatively large island geographically close to a rich major source island (Williams, 1969; note, however, that *A. sagrei* has colonized Grand Cayman; see "Competition" below for more information); most adult males (32%) on perches 0.9–1.5 m high and diameters 7.6 cm, most adult females (32%) and subadult males (33%) on perches 0.9–1.5 m high with diameters 1.3–7.6 cm, most juveniles on perches 0.3–0.6 m high with diameters 1.3 cm (25%) or 1.3–7.6 cm (25%), 5 of 138 adult males, 17 of 239 adult females, 3 of 46 subadult males, and 14 of 51 juveniles were on the ground (Schoener, 1967); mean adult male and female perch heights 1.7 ± 1.12

m (2.2% on the ground) and 1.4 ± 0.86 m (2.2% on ground; Lister, 1976); males establish territories on trunks of large trees and other objects above the ground, females and juveniles are more secretive, basking on rocks, logs, and other ground debris (Seidel and Franz, 1994); males display from exposed perches 1–3 m above ground, large females on lower trunks and large buttress roots of *Ficus aurea* in association with territorial males, adults at a wide variety of heights with males higher than females, but few at ground level, juveniles on the ground much more frequently than adults, mean perch height and diameter at sites where *A. sagrei* does not occur 1.70 ± 0.17 m and 11.8 ± 1.2 cm (Losos and de Queiroz, 1997a); most frequently in the shade of broken canopy woodland, where UV-reflectance profile would be most effective (Macedonia et al., 2000).

Behavior: Subadult males (patterned like females) in territories of large males, dominant males retreat by ascending, whereas subdominant males, females, and juveniles dash to the ground and seek refuge under objects (Seidel and Franz, 1994); males often stationary with heads down, juveniles more often moving, when stationary, are horizontal or in head-up position, juveniles move to the ground in the presence of larger lizards, subsequently ascend on vertical structures and assume a head-up position (representing alertness, not food-seeking), intraspecific interactions frequent (Avery, 1988); display behavior includes headbobs in "volleys," also single-bob elements of brief duration repeated a variable number of times, dewlap pulses and headbobs tend to alternate (Macedonia and Stamps, 1994). **Competition:** In open habitats where *A. sagrei* has become established, perch heights have shifted upward, mean perch heights in open habitats on the western end of Grand Cayman in 1967 were 1.86 ± 0.06 m (males), 1.22 ± 0.03 m (female-sized lizards), and 0.85 ± 0.26 m (juveniles), in 1991 2.43 ± 0.19 m, 1.64 ± 0.13 m, and 0.63 ± 0.16 m, in closed habitat in 1967 1.95 ± 0.09 m, 1.25 ± 0.03 m, and 0.64 ± 0.16 m, in open habitats on eastern end in 1987 0.88 ± 0.07 m (males only), in 1991 1.38 ± 0.38 m, in closed habitat in 1987 1.43 ± 0.12 m, in 1991 1.44 ± 0.28 m, islandwide in 1966 1.49 ± 0.25 m (males), 0.96 ± 0.14 m (female-sized lizards), and 0.61 ± 0.31 m (juveniles), in 1991 2.08 ± 0.13 m, 1.27 ± 0.08 m, and 0.71 ± 0.11 m (Losos et al., 1993). **Diet:** Food items were ants (most important by numbers in adult males, adult females, subadult males, and juveniles and by volume in juveniles), other hymenopterans, flies, beetles (adults and larvae), barklice, termites, hemipterans, lepidopterans (adults and larvae), lacewings (adults and larvae), thrips, orthopterans, roaches (most important by volume in adult males, adult females, and subadult males), spiders, pseudoscorpions, centipedes, mites, isopods, snails, an anole (probably a conspecific), and plant matter, with larger lizards taking larger prey (Schoener, 1967); floral nectarivory (possibly extrafloral) in *Schomburgkia thomsoniana* var. *thomsoniana* (Orchidaceae; Echternacht and Gerber, 2000). **Ecomorphology:** "Trunk-crown" (Losos and de Queiroz, 1997a; Losos, 2009); intermediate between trunk-ground and trunk-crown (Beuttell and Losos, 1999). **Home Range and Territoriality:** Males establish territories in January–February (Seidel and Franz, 1994). **Parasites:** Protozoan malarial parasite *Plasmodium floridense* (Telford, 1975; Ayala, 1977, 1978); nematodes *Spauligodon anolis* (Bursey and Goldberg, 1995, 1998; Goldberg et al., 1995) and larval *Porrocaecum* sp. (Goldberg et al., 1995). **Predation:** *Cubophis caymanus* (Grant, 1940a; R. W. Henderson in Seidel and Franz, 1994; Henderson and Sajdak, 1996); Loggerhead Kingbird (*Tyrannus caudifasciatus caymanensis*), Greater Antillean Grackle (*Quiscalus niger*), Smooth-billed Ani (*Crotophaga ani*; Johnston, 1975); rarely seen fleeing from potential predators (Bananaquit, *Coereba flaveola*, La Sagra's Flycatcher, *Myiarchus sagrae*, or racer, *C. caymanus*; Avery, 1988); *Pantherophis guttatus* (Seidel and Franz, 1994). **Reproduction:** Reproduction distinctly seasonal, testicular weights declined during October, when fat body size doubled, most females were reproductive in September–October, although not all were ovigerous, but half were nonreproductive in December–January, fat bodies not evident except in January, indicating that egg production declines after the onset of the dry season (Licht and Gorman, 1970); males establish territories in January–February (Seidel and Franz, 1994). **Size:** Male SVL to 73 mm (Licht and Gorman, 1970), female SVL to 50 mm (Schwartz and Henderson, 1991); smallest sexually mature male <50 mm SVL and female 38 mm SVL (Licht and Gorman, 1970). **Tail Autotomy:** 11 of 60 territorial males and 2 of 35 females and juveniles had regenerated or missing tails (Seidel and Franz, 1994). **Thermal Biology:** Intermediate between nonbasking thermoconformers and extensively basking thermoregulators, may aestivate during periods of drought, 25.5% of lizards in trees and shrubs and 27.6% of lizards on the ground in direct sun, but rates are higher in the morning and evening, lower at midday (Avery, 1988).

Conservation Status: Not assessed, although potentially threatened by competitive interactions with introduced *A. sagrei*.

Anolis cooki Grant, 1931

Distribution: Southwestern Puerto Rico, Isla Caja de Muertos, and a small cay off Punta Águila.

Habitat: Xerophilic (Schwartz and Henderson, 1991); frequently on rocks, only sites where allopatric with *A. cristatellus* on xeric limestone ridges vegetated with cacti and thorny scrub, these may be suboptimal refugia from competition with *A. cristatellus* (Marcellini et al., 1985); mean adult male and female perch heights 0.6 ± 0.27 m (3.9% on the ground) and 0.5 ± 0.20 m (11.3% on ground; Lister, 1976); uses a less diverse set of perch heights and diameters during the dry winter season (Lister, 1981); where syntopic with *A. cristatellus*, on standing dead vegetation and small bushes rather than larger and more complex habitats, larger males control larger and more complex habitats, so large male *A. cooki* are most likely to encounter smaller size classes of *A. cristatellus* (Jenssen et al., 1984); 29.7% of lizards on rocks and the ground in January, 8.8% in August, mean perch height 0.8 ± 0.1 m in January, 0.9 ± 0.1 m in August, mean perch diameter 4.5 ± 0.5 cm in January, 7.7 ± 0.9 cm in August (Hertz, 1992a); in southwestern Puerto Rico, adult males perch higher (mean ~0.9 m) than subadult males and adult females or juveniles, but perch diameters did not vary significantly, overlap with sympatric *A. cristatellus* very high for all size classes (Genet, 2002); mean perch heights and diameters 1.49 m and 7.75 cm (J. B. Losos in Calsbeek et al., 2006); habitat light environment "no shade" (Fleishman, 2000).

Behavior: In intruder-release experiments with male *A. cristatellus*, *A. cooki* males very aggressive, with interactions similar in intensity to conspecific interactions, suggesting that *A. cooki* is under competitive pressure from *A. cristatellus*, agonistic interactions only 1/6 as intense as with *A. monensis* or *A. evermanni* (Ortiz and Jenssen, 1982; Jenssen et al., 1984); head-down posture in 10 of 13 individuals, moves/min 0.13 ± 0.06, percent time moving 0.28 ± 0.14% (Cooper, 2005); approach distances decrease as perch heights increase (all with perch heights ≤2.1 m fled, 5 of 7 with perch heights >2.1 m did not), approach distances greater when near a path than when farther away from a path, most "squirrel," some subsequently flee up, one moved to the ground and under a bush (Cooper, 2006a). **Competition:** In direct interference competition with *A. cristatellus*, with the latter dominant (Marcellini et al., 1985). **Desiccation:** Mean critical activity point (percent of original body weight at which the lizard could not right itself when turned on its back) 71.4 ± 3.54%, mean rate of water loss 0.14% body mass/h (Hillman and Gorman, 1977). **Ecomorphology:** "Trunk-ground" (Williams, 1972); formally quantified (Beuttell and Losos, 1999). **Population Size and Density:** 198.33 ± 43.88 and 107.83 ± 22.86 and 36.00 ± 5.37 lizards/ha at 3 sites where the species is sympatric with *A. cristatellus*, age distribution heavily skewed in favor of adult males and 78% of the population consisted of adults and subadult males, suggesting that recruitment was poor (Genet, 2002). **Reproduction:** Only 2 of 30 females reproductively active during the dry winter season, 15 of 30 females reproductively active in a wet-season sample (Lister, 1981); clutch size 1 (Hedges, in prep.). **Size:** Male SVL to 70 mm (Marcellini and Jenssen, 1983), female SVL to 59 mm (Schwartz and Henderson, 1991). **Thermal Biology:** Maintains highest average body temperatures of species in the *A. cristatellus* complex, mean body temperatures at 3 sites at various times of day varied from 26.5 ± 0.1 to 34.1 ± 0.3 °C at air temperatures of 26.2 ± 0.1 to 33.0 ± 0.3 °C (overall means for all sites and times were 32.0 ± 0.1 °C and 30.2 ± 0.1 °C, respectively), critical thermal maximum 38.9 ± 0.3 °C (Huey and Webster, 1976); in an open situation 32.7 ± 0.94 °C and 29.2 ± 0.77 °C (Lister, 1976); basks early and late in the day and avoids sunny perches at midday, sunny weather basking rates 0.64 in January and 0.54 in August, mean body/model temperatures 30.4 ± 0.8/34.4 ± 0.5 °C (sun), 31.0 ± 0.3/31.7 ± 0.2 °C (filtered sun), 31.3 ± 0.7/28.4 ± 0.2 °C (shade), 30.2 ± 0.4/26.9 ± 0.3 °C (overcast) in January and 31.7 ± 0.3/35.3 ± 0.5 °C (sun), 31.1 ± 0.5/33.9 ± 0.1 °C (filtered sun), 33.1 ± 0.2/31.6 ± 0.1 °C (shade) in August (Hertz, 1992a); moderately effective thermoregulator, more effectively in winter than summer, optimal body temperature for sprint speed 30.4 ± 1.8 °C (Hertz et al., 1993). **Vision:** In stark contrast with sympatric *A. cristatellus*, *A. cooki* dewlaps reflect little UV, which contrasts with UV-rich background radiance (Leal and Fleishman, 2002).

Conservation Status: "Threatened" (Moreno, 1991d), and Moreno (1991b) indicated that this species should be listed as "threatened" by the federal (U.S.) government, although the Commonwealth of Puerto Rico already lists it as such (Joglar, 2005b); the relatively restricted range, which is vulnerable to development, and possible competitive pressures from *A. cristatellus* suggest that close monitoring of the situation is warranted (Marcellini et al., 1985; Ortiz, 1990, 1991; Genet, 2002); the conservation status of populations varies greatly, demes from Punta Águila, Morrillos, and Bahía Ballena inhabit protected areas, are larger, genetically diverse, and seemingly stable, the population from Playa Santa showed a high level of genetic diversity, but occurs in an area intensively developed and its long-term survival is uncertain, no lizards were found on Caja de Muertos during recent (2006 and 2007) surveys (Rodríguez-Robles et al., 2008).

Additional Pertinent Literature: Jenssen (1990).

Anolis cristatellus Duméril and Bibron, 1837

Distribution: Puerto Rico Bank, where two currently recognized subspecies occur essentially islandwide on Puerto Rico, including satellite islands, and introduced in the La Romana area of the Dominican Republic, Dominica, St.-Martin/St. Maarten, southeastern Florida, and Costa Rica (*A. c. cristatellus*); also on islands east of Puerto Rico, Isla

Vieques, Isla Culebra, Isla Culebrita, and the U.S. and British Virgin Islands, but not the St. Croix Bank (*A. c. wileyae*); elevational range, sea level to >980 m.

Habitat: Mesophilic to somewhat xerophilic (Schwartz and Henderson, 1991); adapts well to altered habitats (Powell and Henderson, 2008c); on rocks near a river at Adjuntas (elevation ~440 m), but not in woodlands or coffee plantations (Stejneger, 1904); open fields and roadsides, frequently on fence posts (K. P. Schmidt, 1920, 1928); usually on perches of large to medium diameter close to the ground (Rand, 1964a); "invaded" a high-elevation site after irradiation promoted leaf fall and increased illumination (Turner and Gist, 1970); at an upland site on Puerto Rico, along a secondary road edge, most adult males on perches >12.5 cm in diameter at heights <0.9 m (33%) with no lizards on the ground or on rocks, most "small" individuals on elevated perches at heights <0.9 m with diameters 2.2–12.5 cm (52%) and 11 of 19 on the ground, in open forest, most adult males (20%) on perches <0.9 m high with diameters of >12.5 cm and 6 of 50 on the ground, most "small" individuals (23%) <0.9 m high on perches with diameters 6.4–12.5 cm and 12 of 125 on the ground and one on a rock (Schoener and Schoener, 1971b); at 2 relatively low-elevation sites at El Yunque (Puerto Rico), juveniles close to the ground and to 2 m high on perches of very small to moderate (<1–8 cm) diameters, females close to the ground and to 2–3 m high on perches of small to large (2–16 cm) diameters, males either on high (>3 m) perches or close to the ground, 78% in sunny situations, but more frequently in shade than sun at higher elevations, presumably due to those habitats being preoccupied by *A. gundlachi*, which is "dominant" in that habitat (García Moll, 1978); in the Virgin Islands, juveniles on the ground by day (MacLean, 1982); on Puerto Rico, mean perch height and diameter 1.2 ± 0.06 m and 13.3 ± 3.41 cm (Losos, 1990a); 6.7% of lizards on rocks and the ground in January, 0% in August, mean perch height 1.1 ± 0.1 m in January, 0.8 ± 0.1 m in August, mean perch diameter 10.0 ± 2.2 cm in January, 9.4 ± 1.5 cm in August (Hertz, 1992a); on Puerto Rico, mean perch heights and diameters 1.20 ± 0.72 m and 10.8 ± 13.5 cm for males and 0.89 ± 0.63 m and 16.9 ± 20.2 cm for females (Butler and Losos, 2002); on Guana (British Virgin Islands), 1–3 m with most individuals 1–2 m above ground (Lazell, 2005a) or to ~2 m (Perry, 2005); in southwestern Puerto Rico, larger males control larger and more complex habitats (Jenssen et al., 1984); also in southwestern Puerto Rico, adult males perch higher (mean >1.2 m) and on perches of greater diameter (mean >8 cm) than subadult males and adult females or juveniles, overlap of all size classes with sympatric *A. cooki* is very high (Genet, 2002); uses a less diverse set of perch heights and diameters during the dry winter season in dry habitats in southwestern Puerto Rico (Lister, 1981); uses "post-vantage" behavior to select habitat, with larger individuals choosing trees and smaller lizards grass (Kiester et al., 1975); in open forest and open fields, on fence posts (resting head-down, usually one individual per post), in deforested areas, on buildings, and on brush piles (Carey, 1972; MacLean, 1982; Schwartz and Henderson, 1991; Perry, 2005); not in elfin woods (Miranda-Castro et al., 2000); favors sunny areas, utilizing more sunny situations than most other Puerto Rican anoles (Schwartz and Henderson, 1991); on Guana, most individuals in moderately shaded situations (Lazell, 2005a); sleeps in crowns of *Cocos* trees (Schwartz and Henderson, 1991); on Puerto Rico, mean adult male and female perch heights 0.9 ± 0.41 m (5.5% on the ground) and 0.8 ± 0.41 m (7.8% on ground; Lister, 1976); mean male perch height and diameter 1.18 ± 0.11 m and 1.13 ± 0.40 m, mean female perch height and diameter 0.71 ± 0.09 m and 0.73 ± 0.33 m (Van-Hooydonck et al., 2007); habitat light environment "partial shade" (Fleishman, 2000); in the introduced population in the Dominican Republic, males perched higher on trunks than females, juveniles, and hatchlings, with smaller individuals mostly on exposed roots near the bases of trees and all hatchlings on the ground, favored more shaded sites than native *A. distichus* (Fitch et al., 1989); not on leaves (J. B. Losos, in litt., 21.IV.2008); in the Virgin Islands, males move to tops of trees to sleep, leaving lower perches to females and juveniles (MacLean, 1982); on Cayo Diablo (Puerto Rico), sleeps primarily in *Coccoloba* (N = 82 of 104 observations) and *Cassythia* (N = 13), where adult males slept at mean height of 1.70 ± 0.10 m on branches of mean diameter of 9.75 ± 2.47 mm, adult females at 1.58 ± 0.12 m and 5.48 ± 0.60 mm, and juveniles at 1.04 ± 0.12 m and 5.81 ± 0.94 mm (Chandler and Tolson, 1990); does not sleep on same plants used by day, usually sleeps on leaves of woody plants taller than 1 m, uses plants with broad leaves, narrow leaves, compound narrow or broad leaves in proportion to their abundance in the habitat, only lizards with SVL >40 mm sleep higher than 1 m, lizards with SVL 45–49 mm sleep higher than smaller lizards and those with SVL >50 mm sleep significantly higher than those in all other size classes (Clark and Gillingham, 1990); may sleep under rocks as well as on leaves and branches, when sympatric with *Epicrates monensis granti*, males tend to prefer branches that are thicker, higher, and more distant from the trunk than those used by females and subadults (Rivero, 1998); abundant on "islands" of vegetation in intensely developed areas like the airport in San Juan (RP).

Activity: Primarily diurnal, but may forage at night to exploit the "night-light" niche in Puerto Rico, the British Virgin Islands, and the Dominican Republic, requiring a shift in circadian rhythm (Garber, 1978; Schwartz and Henderson, 1991; Perry and Lazell, 2000; Henderson and Powell, 2001; Perry et al., 2008; RP); apparently using visual cues, moves from leaf litter or adjacent vegetation to sleeping sites about 15 min before sunset, remaining there until sunrise (Clark and Gillingham, 1990); some individuals active all day long, but activity peaks during late morning and late afternoon (Lazell, 2005a); on *Conocarpus erectus* on a beach on St. Thomas (U.S. Virgin Islands), male displaying to a female at 2230 h under a full moon (A. J. Sánchez Muñoz, in litt., 17.III.2008). **Behavior:** "Extremely pugnacious . . . [males] nod their heads up and down three or four times, at the same time expanding their frills; their eyes glisten with rage, and after waving their tails from side to side for a few seconds, as if to gather energy, they dart at each other furiously, rolling over and over, and holding firmly with their teeth . . . the conflict generally ends in one of the combatants losing his tail, which is often devoured by the victor" (Austen, 1867); sleeps with eyes closed, chins pressed to leaf surface, front limbs forward, and hindlimbs extended, 72% oriented toward base of leaf, 54% oblique head up, 43% horizontal, 3% oblique head down, mean nocturnal recapture radius 0.45 ± 0.24 m, indicative of sleep-site fidelity, 14% of recaptures at the previous site, 21% within 10 cm of the previous site, 98% within 1 m of the previous site for 5 consecutive nights (Clark and Gillingham, 1990); on Puerto Rico, mean display rate (proportion of time spent displaying) 0.043 ± 0.008, percent walks 27.1 ± 5.9, percent runs 58.3 ± 6.3, percent jumps 14.6 ± 2.0, moves/min 1.10 ± 0.11, distance jumped in nature 45.7 ± 6.1 cm (Losos, 1990a); in intruder-release experiments with male *A. cooki*, *A. cristatellus* males were very aggressive, with interactions similar in intensity to conspecific interactions, suggesting that *A. cristatellus* may exert competitive pressure on *A. cooki*, agonistic interactions only 1/6 as intense as with *A. gundlachi* or *A. evermanni* (Ortiz and Jenssen, 1982; Jenssen et al., 1984); male threat display (mean 20 sec) includes raising head high by stretching neck, extending dewlap 4–10 times (about 2 sec between pulses), raising nuchal crest, and tail wags (Ruibal, 1967); demonstrates territorial aggression toward *A. gundlachi*, but less intense than toward conspecifics (Hess and Losos, 1991); male sought refuge submerged in a brackish pond, clinging to sticks for 4 min (Heatwole and Torres, 1963); less cryptic adult males had greater approach distances than females and juveniles or *A. stratulus* and did not frequent trees on which they were most cryptic, most common escape behavior was "squirreling" and then ascending, except when on fence posts or on trees with dense vegetation around the base, in which case lizards would descend (Heatwole, 1968a); mean approach distances 150 cm (males), 110 cm (females, juveniles), usually flee upward (Carey, 1972); as "pursuit-deterrent signals" against *Borikenophis portoricensis*, lizards used flight, immobility, body thrashing, biting, dewlapping, tongue-bunch, headbobbing, lateral body compression, lateral presentation, push-ups, crouching, and tail autotomy, with dewlapping, tail lashing, tongue-bunch, headbobbing, lateral body compression, lateral presentation, and push-ups also used for social interactions, rendering them multipurpose behaviors, subduction time by snakes increased linearly with the time the lizard spent biting the snake (Leal and Rodríguez-Robles, 1995); lizards display actively in response to a predator model (and fled rarely), suggesting that the displays function as predator-deterrent signals (Leal and Rodríguez-Robles, 1997a); intensity of signals (push-ups) is positively correlated with condition (= endurance), indicating that signals are "honest" in that they allow a predator or prospective mates or rivals to assess quality (Leal, 1999); uses head-down posture, moves/min 0.15 ± 0.04, percent time moving 0.28 ± 0.07% (Cooper, 2005); size of tail crest does not affect outcome of male-male interactions, sizes of home ranges, or number of females associated with a male, but is important in identifying an individual as an adult male (subadults, normally ignored, with artificial crests were immediately attacked by territorial males; Perry, 2005); approach distances decrease as perch heights increase (all with perch heights ≤ 1.9 m fled, none with perch heights >1.9 m did), approach distances greater when exposed to view, "squirrel," some subsequently flee up (Cooper, 2006a); resident on a small palm challenged by slightly larger interloper on Guana (British Virgin Islands), led to a 2-hr long series of challenges involving lateral presentation, elevation of nuchal and dorsal crests, body inflation, elevated tail, and gaping jaws, with both lizards circling the other in an apparent attempt to establish and hold the higher position, leading to jaw-locking and efforts to throw the other off the perch, in which the interloper succeeded on 3 occasions, after which the resident quickly reengaged, eventually the larger interloper "won," with the former resident displaced to the distal end of a palm frond, from which he disappeared shortly thereafter, the "winner" copulated with a resident female within minutes after the battle was terminated (Powell, 2006a); both absolute and size-corrected dewlap size are significantly related with bite force, 41% of combat trials involved dewlap extensions, 73% biting, and 64% headbobbing (Lailvaux and Irschick, 2007); display speed 14.7–31.5 mm/sec, strong correlation with maximum speed of moving vegetation, suggesting that males experiencing high levels of visual background noise produce high-speed displays (Ord et al., 2007); male approach distance, flight distance, and final distance 1.32 ± 0.27 m, 0.46 ± 0.10 m, and 1.63 ± 0.26 m, female approach distance, flight distance, and final distance 1.07 ± 0.17 m, 0.33 ± 0.05 m, and 1.27 ± 0.18 m (VanHooydonck et al., 2007); moves per minute 0.41 and 1.10 (Johnson et al., 2008). **Colonization:** Five juveniles in association with ornamental plants transported from Beef to Guana Island (Perry et al., 2006). **Competition:** In the introduced Dominican population, occurs on the same trees as native *A. distichus* despite similar height preferences and food habits (primarily ants), but differs in the average size of prey, native *A. chlorocyanus* similar in size, but confined to trees with smooth or slippery bark, which are avoided by *A. cristatellus*, has apparently excluded native *A. cybotes* from areas occupied by *A. cristatellus*, primarily by preying on hatchlings and outreproduction (Fitch et al., 1989); is displacing native *A. oculatus* where established in Dominica (Malhotra et al., 2007; Eales et al., 2008); in Dominica, males respond aggressively toward male *A. oculatus* introduced into their territories (RP). **Desiccation:** Mean critical activity points (percent of original body weight at which the lizard could not right itself when turned on its back) 71.6 ± 3.31% and 68.1 ± 6.44%, mean rates of water loss 0.19 and 0.18% body mass/h (Hillman and Gorman, 1977); on Puerto Rico, mean water loss rate 0.083 ± 0.006 percent initial body weight/h, considerably lower than those for high-elevation *A. gundlachi* (Hertz et al., 1979); in the British Virgin Islands, mean water-loss rates (1 ± 0.3–10 ± 4.7 mg/g/h) higher and mean cutaneous resistance to water loss (29 ± 14.6–199 ± 73.4 sec/cm) lower in more mesic habitats (Dmi'el et al., 1997; Perry et al., 1999, 2000) and were significantly altered (1.7 ± 0.5–3.8 ± 1.4 mg/g/h and 100 ± 34–229 ± 62 sec/cm) during a major drought, with effects on animals on usually mesic islands the most dramatic (Perry et al., 1999); size of crests incurs no costs in regard to water-loss rates (Perry, 2005). **Ecomorphology:** "Trunk-ground" (Rand and Williams, 1969); formally quantified (Beuttell and Losos, 1999). **Foraging and Diet:** In captivity, "any anolis that felt inclined to feed would crouch down and gradually steal towards a fly that happened to have settled . . . having crept up to within a short distance, the anolis would seize the insect by a sudden spring, and discuss it at leisure" (Austen, 1867); sit-and-wait forager (Perry, 1999); attacks launched by immobile lizards that detected prey while scanning visually (Cooper, 2005); readily accepted supplemental food (mealworms) at all times of day during the summer breeding season, adult males exhibited significant increases in liver and fat body mass after one week of supplements, suggesting that the "normal" lack of fat bodies during the breeding season reflects insufficient food to allow energy storage (Licht, 1974); beetles (adults and larvae), cockroaches, mantid, orthopterans, ants, lepidopteran larvae, hemipterans, flies, spiders, 2 anoles, and "vegetable matter," mostly brightly colored seeds (K. P. Schmidt, 1920, 1928); in Puerto Rico, snails, earthworms, isopods, millipedes, pseudoscorpions, spiders, cockroaches, earwigs, orthopterans, thrips, lacewings, ants and other hymenopterans, beetles (adults and immature), lepidopterans (adults and larvae), flies (adults and larvae), hemipterans, berries, including those of *Cordia corymbosa*, *Volkameria aculeate*, and *Solanum seaforthianum*, conspecifics, and cast skin, presumably their own (Wolcott, 1923); eats *Eleutherodactylus* sp. and is "fond of the juicy pulp of the mango, and can be easily baited with this" (Danforth, 1925); in the Virgin Islands, insects ranging from ants to large roaches, small fruits, small frogs and other lizards, including the young of its own species (MacLean, 1982); quickly learns to modify foraging behavior when rewarded with food (Shafir and Roughgarden, 1994); food in the Dominican Republic includes beetles (especially weevil *Diaprepes*), orthopterans (cockroaches, grasshoppers, crickets, mantids), ants (most abundantly taken items), caterpillars, a large cicada, flies, spiders, brightly colored seeds, *Sphaerodactylus*, and two conspecifics, mean prey length and volume 0.90 ± 0.14 cm and 0.81 ± 0.23 cm^3 (males) and 0.68 ± 0.05 cm and 0.30 ± 0.05 cm^3 (females; Fitch et al., 1989); takes *Sphaerodactylus* spp., attempted to eat a juvenile *Borikenophis portoricensis* (Thomas and Gaa Kessler, 1996); stomach contents in the British Virgin Islands includes ants (most important by numbers and volume), beetles, crickets, leafhoppers, whip scorpions, cactus fruit and other plant material, and pebbles (Lazell, 2005a); in the British Virgin Islands, eats fruits of *Melocactus intortus* (Cactaceae) and *Rivina humila* (Phytolaccaceae) and bright yellow flowers of *Tecoma stans* (Lazell and Perry, 1997; Lazell and Mitchell, 1998); on Necker Island (British Virgin Islands), eats berries of *Lantana involucrata* (Lazell, 2006); consumed *Hemidactylus mabouia* in the British Virgin Islands (Owen and Perry, 2005); fruitflies on fallen mangoes and mango pulp on Dominica (Daniells et al., 2008); eats *Anolis krugi* (Sánchez Muñoz, 2008); adult male ate a *Typhlops hypomethes* (Colón Archilla, 2009); individuals quickly congregated around a broken termite gallery on Guana Island, British Virgin Islands (RP); subadult male on Guana Island grabbed a juvenile snail (*Zachrysia provisoria*), not native to the Virgin Islands (native to Cuba), immediately spit it out and spent ~10 min licking its lips and gaping (M. E. Gifford, pers. comm.); on St. John (U.S. Virgin Islands), ate a paper wasp (*Myschocyttarus* sp.), on Puerto Rico, attempted to eat Monarch Butterfly (*Danaus plexippus*) caterpillars on Giant Milkweed (*Calotropis procera*) bushes, but spit out the caterpillar within a few minutes, female lizard lapping cane sugar from a Bananaquit (*Coereba flaveola*) feeder grabbed by the tail and hurled from feeder by a bird, male lizards feed on scraps of Sara

Lee pound cake, boiled white rice, and moist dog food (A. J. Sánchez Muñoz, in litt., 17.III.2008). **Home Range:** Mean territory size per lizard on large trees 19.0 m^2 for males and 1.5 m^2 for females (Philbosian, 1975); females occupy smaller home ranges within those of males, individual ranges 1–34 m^2 (Rivero, 1998); in the Dominican Republic, limited to one part of a tree (but overlapped with those of other individuals and of *Anolis distichus*), mean home range sizes for adult males 3.4 ± 0.3 m^2, subadult males 4.2 ± 2.3 m^2, adult females 4.4 m^2, juvenile 2.4 m^2, and hatchlings 2.5 ± 0.7 m^2 (Fitch et al., 1989); in the British Virgin Islands, male home ranges average ~20 m^2 (Perry, 2005). **Homing:** May return to the same perch or immediate vicinity to sleep for many consecutive nights (Rivero, 1998); 9 of 13 males and 7 of 12 females returned to home territories within an average of 3 days when relocated 11–62 m (2–26 territory diameters) to presumably unfamiliar release sites (Jenssen, 2002). **Interclass Interactions:** On Guana Island (British Virgin Islands), Green-throated Caribs (*Eulampis holosericeus*) chase anoles away from flowers on which they feed (Boal, 2008a). **Movements:** Mean number of moves per minute 0.36 ± 0.53, percent time moving 0.83 ± 1.35 (Perry, 1999). **Parasites:** Trematodes *Allopharynx riopedrensis* (Acholonu, 1976; Coy Otero, 1989); *Mesocoelium monas* (highest infection rate of Puerto Rican anoles, 38.6%; Cofresí-Sala, 1964; Acholonu, 1970, 1976); from Puerto Rico, nematodes *Spauligodon anolis* (Chitwood, 1934; Acholonu, 1976; Bursey and Goldberg, 1998); *Pharyngodon travassosi*, and acanthocephalans *Lueheia inscripta*, *Centrorhynchus* sp. "A," *Centrorhynchus* sp. "B" (Acholonu, 1976); from 7 islands in the British Virgin Islands, nematodes *Parapharyngodon cubensis*, larval *Porrocaecum* sp., *Spauligodon anolis*, *Trichospirura teixeirai*, *Physaloptera* sp., larval acanthocephalans *Centrorhynchus* sp. and Oligacanthorhynchidae, cestode *Oochoristica maccoyi*, and trematodes *Rhabdias* sp. and *Mesocoelium monas* (Goldberg et al., 1998a); on Hispaniola, nematodes *Parapharyngodon cubensis*, *Skrjabinoptera leiocephalorum*, larval *Ascarops* sp. and *Physocephalus* sp., acanthocephalan cystacanths *Centrorhynchus* sp. (Goldberg et al., 1998b); protozoan malarial parasite *Plasmodium azurophilum* (Schall and Vogt, 1993). **Performance:** On Puerto Rico, mean maximum sprint speed (fastest time over an interval of 0.25 m) 0.116 ± 0.002 sec, jump 84.9 ± 1.43 cm, cling 2.837 ± 0.335 Neutons (Losos, 1990a); clinging ability 1.14 ± 0.32 Neutons, divided by mass 0.14, with the relatively low value characteristic of trunk-ground ecomorphs (Elstrott and Irschick, 2004); males with relatively high endurance display more frequently in the field, are more likely to win fights staged in captivity, and tend to control larger territories (Perry et al., 2004; Perry, 2005); peak velocity 1.81 ± 0.24 m/sec, peak acceleration 38.1 ± 7.1 m/sec^2 (Toro et al., 2004); acceleration 32.66 ± 2.36 m/sec, sprint speed 1.24 ± 0.06 m/sec (Vanhooydonck et al., 2006a). **Population Size and Density:** Mean number of lizards/tree 1.0 male and 2.6 females, mean male-male distance 6.8 m (only 2% of "pairs" in same tree), mean female-female distance 2.1 m, mean male-female distance 1.4 m (Philbosian, 1975); 4,000–7,200 lizards/ha on Guana (British Virgin Islands; Lazell, 1991); in *Leucaena* plots on Guana with little litter 1,000 lizards/ha, biomass 0.69 kg/ha, in *Coccoloba* plots with ample litter 1,100 lizards/ha, biomass 1.68 kg/ha (Rodda et al., 2001); at 3 sites in southwestern Puerto Rico, where the species is sympatric with *A. cooki*, estimates were 262.00 ± 13.86, 452.33 ± 69.50, and 296.00 ± 70.18 lizards/ha, with nearly half the population composed of juveniles (Genet, 2002). **Predation:** *Ameiva exsul* (K. P. Schmidt, 1920); Indian Mongoose (*Herpestes javanicus*) in Puerto Rico (Wolcott, 1953); Pearly-eyed Thrasher (*Margarrops fuscatus*; Lu, 2009); Red-tailed Hawks (*Buteo jamaicensis*; Santana and Temple, 1988); *Epicrates monensis granti* (Tolson, 1988; Chandler and Tolson, 1990); on Puerto Rico, Caja de Muertes, and British Virgin Islands, *Borikenophis portoricensis* (Rodríguez-Robles and Thomas, 1992; Rodríguez-Robles and Leal, 1993a; Leal and Rodríguez-Robles, 1995; Henderson and Sajdak, 1996; Perry, 2005; Sánchez Muñoz, 2008); *Magliophis exiguus* (Thomas and Gaa Kessler, 1996); *Anolis cuvieri* (A. J. Sánchez Muñoz, in litt., 17.III. 2008). **Reproduction:** Clutch size 1 (Hedges, in prep.); eggs "two or three in number" (suggesting communal deposition), about 10 × 6 mm, uniformly oval, surface white and striate, frequently under edges of logs or stones, or in debris about the base of banana plants (K. P. Schmidt, 1920, 1928); semicommunal egg deposition in a coffee plantation, 11 eggs under one log, 4 under another, and 1 under a third, eggs not associated with each other, also a single egg under a rock pile (Rand, 1967c); maximal male reproductive activity March–August, some sperm production throughout the year, but testicular weights decrease during winter months (but less than in most other species of anoles), size of abdominal fat bodies inversely correlated with testis weight (Licht and Gorman, 1975); all females nonreproductive during the dry winter season in xeric habitat in southwestern Puerto Rico, 14 of 30 females reproductively active in a wet-season sample (Lister, 1981); deposits a single egg at a time, but 6 of 40 females in the Dominican Republic had eggs in both oviducts, largest oviductal eggs 11 × 7 mm (Fitch et al., 1989); females may produce eggs every 2 weeks during the rainy season, which may last 8 months on Dominica (Eales et al., 2008); 52% of females that had produced

2 or more progeny had mated with multiple males (M. A. Johnson, R. Kirby, C. C. Fresquez, and S. Wang in Eales et al., 2008). **Size:** Male SVL to 77 mm (Schall and Vogt, 1993), female SVL to 73 mm (Schwartz and Henderson, 1991); in the British Virgin Islands, adult males from more mesic islands are smaller than those from drier islands (Perry et al., 2000). **Source of Introduced West Indian Populations:** Genetic distances (based on electrophoretic data) were greater between the population in La Romana (Dominican Republic) and those in Mayagüez (Puerto Rico) than between the population in La Romana and that in Río Piedras (Puerto Rico; Zani et al., 1993); mtDNA haplotypes suggest that the Dominican population originated in the Ponce/Guanica area of Puerto Rico (Kolbe et al., 2007b). **Tail Autotomy:** 30% of males ($N = 70$) and 21% of females ($N = 57$) had intact tails in the Dominican Republic (Fitch et al., 1989); 31 of 71 males (44%) and 12 of 33 females (36%) in another sample from the Dominican Republic had regenerated tails, rate not significantly related to size or sex (RP). **Thermal Biology:** Body temperatures maintained slightly higher above ambient (mean 1.2 °C, –1.6–4.0 °C) than *A. stratulus* (mean 0.7 °C, –0.7–2.8 °C), not attributable to size, individuals exposed to cold temperatures (11.5 ± 0.5 °C) were capable of locomotor movements, but did not attempt to feed, *A. cristatellus* survived at a lower rate than high-elevation *A. evermanni* and *A. krugi* and at a rate comparable to low-elevation *A. stratulus*, with smaller individuals within the same species succumbing first, heat stress monitored under rising temperatures was noted after onset in low-elevation *A. stratulus* and high-elevation (~550–640 m) *A. evermanni* and *A. gundlachi*, preferred higher environmental temperature in the laboratory than in the field (Heatwole et al., 1969); mean body temperatures in the field vary by site and time of day at 8 sites from 23.3 ± 0.1 to 33.4 ± 0.3 °C at air temperatures ranging from 22.9 ± 0.2 to 31.7 ±0.3 °C (overall means for all sites 29.0 ± 0.2 °C and 27.4 ± 0.1 °C, respectively), critical thermal maxima at 2 sites were 38.1 ± 0.4 °C and 38.9 ± 0.3 °C (Huey and Webster, 1976); in an open situation on Puerto Rico 30.6 ± 1.44 °C and 28.9 ± 1.38 °C (Lister, 1976); lizards died or lost weight under conditions simulating mean minimum January temperatures of the eastern Puerto Rican rainforest, apparently excluded from upland rainforest habitats by thermal requirements that cannot be met consistently in such habitats (Gorman and Hillman, 1977); mean critical thermal maximum at elevation of 5 m 40.7 ± 0.1 °C, considerably higher than that of *A. gundlachi* at higher elevations (Hertz et al., 1979); regulates temperature behaviorally in open habitats but passively tolerates lower and more variable temperatures in forest (where basking sites are few), basks frequently in open habitats, rarely in closed forest, due to high "cost" in latter, lizards in open areas should be able to move or run faster than lizards in the forest in the morning, but ambient temperatures in the open are high at midday and body temperatures are above thermal preferences, whereas forest lizards then may have a performance advantage, temporal and spatial variation in physiological performance may be dynamic (Huey, 1983); mean minimum and maximum body temperatures at an elevation of 5 m on Puerto Rico 24.3 ± 0.1–30.4 ± 0.2 °C at maximum and minimum air temperatures of 24.2 ± 0.1 °C and 28.3 ± 0.2 °C (Hertz, 1983); mean body temperature in the Dominican Republic 29.1 °C (24.6–32.6 °C, $N = 84$), averaged 1.6 °C above air temperatures (Fitch et al., 1989); in sunny weather, 92.2% of lizards basking at a high elevation site in January, 83.5% in August, 66.7% at a low elevation site in January, 55.9% in August, mean body temperatures in the sun, filtered sun, shade, and overcast at a high elevation site in January 26.2 ± 0.7 °C, 24.1 ± 0.7 °C, 21.5 ± 1.2 °C, 20.1 ± 0.4 °C, in August 29.6 ± 0.4 °C, 27.2 ± 0.4 °C, 26.3 ± 0.9 °C, 25.1 ± 0.6 °C, at a low elevation site in January 28.7 ± 0.6 °C, 26.8 ± 0.5 °C, 22.8 ± 0.6 °C, 26.5 ± 0.7 °C, in August 31.0 ± 0.6 °C, 29.9 ± 0.3 °C, 28.1 ± 0.4 °C, 28.2 ± 0.3 °C (Hertz, 1992a, 1992b); behaviorly thermoregulates effectively in some habitats but not in others, varying by site and sometimes even within sites from being a moderately effective thermoregulator to a thermoconformer, thermoregulating more effectively throughout the year at upland sites and during the summer in lowland desert, avoids activity at midday at lowland sites and early and late in the day at montane sites, optimal body temperature for sprint speed 27.3 ± 1.8 °C (Hertz et al., 1993); uses more brightly illuminated and warmer habitats than those used by *A. gundlachi*, lizards selected perches that were brighter than randomly selected sites in the morning and afternoon, but not at midday, perch sites were warmer than randomly selected sites only in the morning, environmental temperatures (which are highly correlated with body temperatures in the field) significantly correlated with light intensity, lizards selected microhabitats in the laboratory that varied inversely with experimental temperatures, suggesting that lizards use light intensity to identify thermally suitable microhabitats (Hertz et al., 1994); size of crest provides no advantages or costs in thermoregulation (Perry, 2005). **Vision:** Dewlaps, but not rest of body, highly UV reflective, most effective in UV-rich environment, UV-bright spot associated with open mouth, part of typical anoline threat display (Fleishman et al., 1993); dewlaps highly visible in typical habitat (Persons et al., 1999); in a lowland populations sympatric with *A. cooki*, *A. cristatellus* dewlaps strongly reflect UV, which contrasts

with background radiance (Leal and Fleishman, 2002); reflectance and transmittance of dewlaps varied so that dewlaps in xeric habitats were darker and those from mesic habitats were brighter (Leal and Fleishman, 2004).

Conservation Status: Both *A. c. cristatellus* and *A. c. wileyai* "stable" (Moreno, 1991d).

Anolis cupeyalensis Peters, 1970

Distribution: Cuba, Sierra de Nipe and Sierra del Cristal.

Habitat: Mesophilic (Schwartz and Henderson, 1991); near the deforested peak among grasses, ferns, and pine seedlings (G. Peters, 1970); in areas of tall grass bordering a small creek, populations apparently very local and disjunct (Garrido, 1975c); in mesophyllous evergreen and semideciduous forest (Rodríguez Schettino et al., 1999); perches on stems and leaves of tall grasses and crawler ferns in shaded places near the ground (Rodríguez Schettino, 1999b).

Abundance: Very high population densities in suitable habitat (Fläschendräger and Wijffels, 2009). **Activity:** Diurnal (Rodríguez Schettino et al., 1999); most individuals active during the early hours of the day and in the afternoon, but may be active throughout the day on cool rainy days (Rodríguez Schettino, 1999b); bimodal, active at 0930 h at an elevation of 660 m, retreat into dense vegetation (clumps of grass and under fern fronds) at midday (Fläschendräger and Wijffels, 2009). **Behavior:** Generally immobile, escapes by jumping among leaves (Rodríguez Schettino, 1999b). **Ecomorphology:** Categorized as a "grass anole" (G. Peters, 1970); "grass-bush" (Losos, 2009). **Foraging and Diet:** Captive individuals show a strong preference for spiders of the families Araneidae, Salticidae, and Thomisidae over insects (O. H. Garrido in Alayón García, 1976). **Size:** Male SVL to 33.5 mm (G. Peters, 1970), female SVL to 32 mm (Garrido, 1975c). **Thermal Biology:** Apparently nonheliothermic (Rodríguez Schettino, 1999b).

Conservation Status: Considered "vulnerable" (Rodríguez Schettino and Chamizo Lara, 1998a; L. Rodríguez Schettino and A. Chamizo Lara in Rodríguez Schettino, 1999a).

Anolis cuvieri Merrem, 1820

Distribution: Puerto Rico, islandwide but local, possibly also Isla Vieques; elevational range, sea level to 1,192 m.

Habitat: Mesophilic (Schwartz and Henderson, 1991); in coffee plantations (usually on large shade trees) above 2.4 m, most common in secondary forests (abandoned coffee plantations) 30–50 years old, in various upland habitats including elfin forest (K. P. Schmidt, 1928; Pérez-Rivera, 1985; Schwartz and Henderson, 1991; Rivero, 1998; Miranda-Castro et al., 2000); brown juveniles are primarily terrestrial, whereas green adults are primarily arboreal (Rand and Andrews, 1975; Gorman, 1977; Pérez-Rivera, 1985; Reagan, 1996a); one juvenile at El Yunque (elevation 450 m) 2 m high on a perch with diameter 2 cm (García Moll, 1978); adult females come to ground to lay eggs, 11 of 18 individuals were females (2 unsexed juveniles, 5 adult males), all seen at heights of 5 m at El Verde (Reagan, 1996a); associated with forests, especially in karst habitats (Rios-López and Puente-Colón, 2007); primarily in the forest canopy or subcanopy (Reagan, 1996a); in San Juan, on tall trees along a road behind a modern hotel (G. C. Gorman in Wijffels, 1975 and Fläschendräger and Wijffels, 1996, 2009); on trunks (to 80 cm diameter), twigs, leaves, and a boulder on the ground, other than the boulder, perch heights 3–14 m (mean 6.5 m; Dial et al., 1994); on palm (*Prestoea montana*) trunk 3 m high, relocated via telemetry at heights of 6 m, 15 m, and 8 times in crowns 15–25 m above the ground—without ever seeing the lizard despite intense searching (Losos et al., 1990); mean perch heights and diameters 2.84 ± 1.69 m and 10.8 ± 4.6 cm for males and 2.77 ± 0.66 m and 9.8 ± 0.6 cm for females (Butler and Losos, 2002); mean perch height and diameter 3.25 m and 8.50 cm (J. B. Losos in Calsbeek et al., 2006); apparently sleeps near ground level (R. Thomas in Reagan, 1996a).

Activity: Forages principally 0900–1100 h and 1430–1730 h (Pérez-Rivera, 1985). **Behavior:** Lies closely adpressed to branches and trunks, moves slowly to the other side when pressed, bites tenaciously (Pérez-Rivera, 1985); 3 of 5 individuals responded to a model predator (*Borikenophis portoricensis*) by "pursuit-deterrent signaling" that included lateral orientation, dewlapping, push-ups, flight (by moving 30 cm to a higher perch), or flattening the body against the substrate, push-ups only after the model had reached the base of the perch, another watched the putative predator, but did not signal (Leal and Rodríguez-Robles, 1997b); can bipedal on pavement or level ground (A. J. Sánchez Muñoz, in litt., 17.III.2008). **Ecomorphology:** "Crown-giant" (Rand and Williams, 1969); formally quantified (Beuttell and Losos, 1999). **Foraging and Diet:** Both adults and juveniles forage on the ground as well as arboreally, while foraging, traverses ground, branches, trunks very carefully, never approaching other individuals to within 1 m, moving 1–3 m where observes surroundings for 3–6 min, can locate prey (20 mm in size) at about 3 m distance, when prey located, remains quiet but follows prey with its eyes without moving its head, then slowly moves toward prey and, when <1 m away, rushes to it and traps it in jaws, waits for rapidly moving prey (lepidopterans, wasps, small birds) until prey comes within 25–40 cm and then attacks, if prey small, swallows immediately, if large (earthworms), rubs against substrate until in pieces, food includes insects (lepidopteran adults and larvae, roaches, beetles, hemipterans, wasps, and others that were not identified), isopods, millipedes, earthworms, slugs, *A. pulchellus*, and a hummingbird (*Chlorostilbon maugaeus*; Pérez-Rivera, 1985); one individual had eaten 4 large snails (Wolcott, 1923); of 14 stomachs with food, all contained invertebrates (mostly orthopterans, only one ant), 3 contained anoles (one an *A. krugi*), and one contained a small fruit (Rand and Andrews, 1975); phasmid, hemipteran, and anoline skin ("doubtless its own"), possibly berries and fruits, including coffee berries (K. P. Schmidt, 1920, 1928); seeds and a variety of insects, including beetles, a hemipteran, ants, and a mite (Losos et al., 1990); fecal matter contained an *Anolis* jaw and bone fragments, snail opercula and shell fragments, and fibrous vegetative material (Dial et al., 1994); *Anolis stratulus* (possibly *A. evermanni*; Dial and Roughgarden, 1995); at El Verde, ants, flies, lepidopteran larvae (most important prey item by volume), beetles, orthopterans, phasmids, spiders, snails (most important prey by numbers and second most important by volume), lizard skin and a tail (*A. gundlachi*), plant matter, and unidentified material (R. Garrison and D. P. Reagan in Reagan, 1995, 1996a); *Anolis cristatellus* (A. J. Sánchez Muñoz, in litt., 17.III. 2008). **Home Range and Territoriality:** Home ranges 275–326 m² in males and 68 m² in females (Losos et al., 1990; Dial et al., 1994); may occupy the same tree (or 2–3 contiguous trees) for several weeks or months (Rivero, 1998). **Longevity:** In captivity, 9 years, 3 months (K. Murphy in Slavens and Slavens, 2003). **Movements:** Mean distance moved by an adult male 18.1 ± 7.5 m (Losos et al., 1990). **Parasites:** "Mature distome fluke" (Wolcott, 1923); trematode *Mesocoelium monas*, with lowest infection rate (12.7%) of Puerto Rican anoles (Cofresí-Sala, 1964). **Performance:** Mean maximum sprint speed (fastest time over an interval of 0.25 m) 0.114 ± 0.006 sec, jump 73.5 ± 4.20 cm, cling 14.857 ± 0.485 Neutons (Losos, 1990a); acceleration 21.12 ± 2.18 m/sec, sprint speed 0.89 ± 0.07 m/sec (Vanhooydonck et al., 2006a). **Population Density:** Maximal density at El Verde, 30 lizards/ha (Dial et al., 1994); 2 individuals (1:1) in 14 Tabonuco (*Dacryodes excelsa*) crowns = 1.33 lizards/ha (Dial et al., 1994); mean density in Sabana Seca was ~39 lizards/ha (0–133/ha) during cooler, drier months (December–April) and increased to ~74 lizards/ha (0–200/ha) in the warmer, wetter months (May–November; Rios-López and Puente-Colón, 2007). **Predation:** Juvenile coloration may serve as a signal to adults that it is "non-food" (Rand and Andrews, 1975); Red-tailed Hawk (*Buteo jamaicensis*; Santana and Temple, 1988); Broad-winged Hawk (*B. platypterus*), mongoose (*Herpestes javanicus*; Schwartz and Henderson, 1991); *Epicrates inornatus* (Wiley, 2003); an egg, probably *A. cuvieri*, eaten by *Magliophis stahli* (Puente-Rolón, 2001); at El Verde, Lizard Cuckoos (*Saurothera vieilloti*), Pearly-eyed Thrashers (*Margarops fuscatus*), Red-tailed Hawks (*Buteo jamaicensis*), and *Borikenophis portoricensis* regularly feed on anoles (Reagan and Waide, 1996); adult lizard entered a Puerto Rican Parrot (*Amazona vittata*) nest, emerged within 3 min and began to bask, female parrot emerged in 5 min, hopped to the vine, "crunched the lizard's neck in her bill," and flew downhill out of sight, whether she consumed any part of the lizard is unknown (Snyder et al., 1987). **Reproduction:** In captivity, copulation 20–90 min, eggs 23 × 12 mm at deposition, 30 × 16 mm at hatching, incubation ~95 days at 20–25 °C (Appler, 1963); copulation for 35 min, egg white, soft shelled, 21 × 11 mm, glued to side of branch in captivity, pair-bonding has been suggested based on observations of sleeping pairs in nature, pairs only during months (July and October) when juveniles (mean SVL 41 mm, $N = 7$) were present, the months May–October may represent the breeding season in Sabana Seca (Rios-López and Puente-Colón, 2007). **Size:** Male SVL to 140 mm (Reagan, 1996a), female SVL to 119.4 mm (Schoener and Schoener, 1971b); female SVL in captivity 124 mm (Appler, 1963).

Conservation Status: "Stable" (Moreno, 1991d), but association with large trees may be cause for concern in some areas.

Anolis cyanopleurus Cope, 1861

Distribution: Cuba, where two currently recognized subspecies occur in mountainous areas of extreme eastern Cuba, from eastern Holquín Province through much of Guantánamo Province to Punta Maisí, but absent from the southern coast of the province; elevational range 300–600 m.

Habitat: Mesophilic (Schwartz and Henderson, 1991); in grass in wooded situations, sleeps on sugarcane and other monocot leaves to 60 cm above ground (Ruibal, 1964); grasses along watercourses, weeds along a fence in a shaded pineapple plantation (Garrido, 1975c); in mesophyllous evergreen and semideciduous forest (Rodríguez Schettino et al., 1999).

Activity: Diurnal (Rodríguez Schettino et al., 1999); active in early morning and late afternoon, may be locally abundant under very unfavorable weather conditions (rainy, cool, windy) during even the "hottest" part of day (Garrido, 1975c). **Behavior:** Generally immobile, jumps from perch to perch to escape (Rodríguez Schettino, 1999b). **Ecomorphology:** "Grass-bush" (Williams, 1961); categorized as a "grass anole" (G. Peters, 1970). **Home Range and Territoriality:** Apparently not highly territorial, with no defined home ranges, which explains apparent syntopy with other grass anoles (G. Peters and Schubert, 1968). **Parasites:** Nematode *Cyrtosomum scelopori* (Coy Otero and Baruš, 1979b; Coy Otero and Lorenzo Hernández, 1982; Coy Otero, 1999). **Reproduction:** Mean duration of copulation 22.1 min, longest 33 min (Garrido, 1980a); gravid female in June (Rodríguez Schettino, 1999b). **Size:** Male SVL to 43 mm (Garrido, 1975c), female SVL to 36.2 mm (Rodríguez Schettino, 1999b). **Thermal Biology:** Apparently nonheliothermic (Rodríguez Schettino, 1999b).

Conservation Status: Entire species not assessed, but *A. c. orientalis* is considered "vulnerable" (L. Rodríguez Schettino and A. Chamizo Lara in Rodríguez Schettino, 1999a).

Anolis cybotes Cope, 1862

Distribution: Hispaniola, where three currently recognized subspecies occur essentially islandwide (*A. c. cybotes*), except for very xeric areas, where the species is restricted to oases, and the western tip of the Tiburon Peninsula, where *A. haitianus* occurs, also Île-à-Vache, Île de la Tortue, Île Grande Cayemite, Île à Cabrit, Île de la Gonâve (*A. c. doris*), Isla Saona (*A. c. ravifaux*), where apparently uncommon, Isla Catalina, Isla Catalinita (*A. c. ravifaux*?), Isla Cabras, Cayos Siete Hermanos (Cayo Monte Grande); introduced into southeastern Florida; elevational range, sea level to ~1,830 m.

Habitat: Mesophilic (Schwartz and Henderson, 1991) to somewhat xerophilic (RP); at broadleaf forest edges and edges of dense thickets, xeric coastal forest, highly developed agricultural areas, edificarian situations, and other disturbed habitats, uncommon in dense forests, common on trees (including palms), fence posts, on rocks in a deep ravine, rock piles, and artificial perches (including on and in houses), frequency in shaded habitats decreases with elevation (Mertens, 1939; Rand, 1962b; Williams and Webster, 1974; Hertz and Huey, 1981; Schwartz and Henderson, 1982, 1991; Jenssen et al., 1988; Fobes et al., 1992; Olson, 1995; Lenart et al., 1997; Cast et al., 2000; Sifers et al., 2001; Powell and Henderson, 2008c); on a dense mat of water hyacinths (*Eichhornia* sp.) on a backwater of a river (Mertens, 1939); in active pastures, oil palm and cacao plantations, abandoned pastures and gardens, and mogote vegetation (Glor et al., 2001); *A. c. ravifaux* is saxicolous (Schwartz and Henderson, 1982; Gifford et al., 2002; Glor et al., 2003); exposed tree trunks and fence posts <3 m from the ground, especially smaller individuals frequently on the ground, avoids small twigs and foliage during the day, does not appear to be an adept climber (Rand, 1962b); 7.8% on leaves, 2.6% on stems ~3 mm in diameter, and 88.9% on rocks at a Haitian site (Moermond, 1979a); 0.8% on leaves or grasses and 6.2% on perches >2.5 cm in diameter at heights >90 cm, 5.5% on perches <0.7 cm in diameter, 0.8% on perches 0.7–2.5 cm in diameter, 0.8% on perches >2.5 cm in diameter, and 35.9% on rocks at heights 31–90 cm, 1.6% on leaves or grasses, 1.6% on perches <0.7 cm in diameter, 2.3% on perches 0.7–2.5 cm in diameter, 0.8% on perches >2.5 cm in diameter, and 43.7% on rocks or ground at heights <30 cm (Moermond, 1979b); tolerant of human-mediated habitat alterations in upland habitats in the Sierra de Bahoruco (Lenart et al., 1997); males perch on large trees at heights <2 m, whereas females perch on all available vegetative structure below males, and juveniles frequently are on the ground in most habitats (Hertz and Huey, 1981; Fitch et al., 1989); in a semicultivated lowland site in Haiti, in June (wet season), mean male perch height and diameter 1.7 ± 0.17 m and 13.7 ± 2.82 cm, female 1.2 ± 0.15 m and 8.3 ± 1.81 cm, juvenile 0.5 ± 0.11 m and 2.9 ± 0.48 cm, respectively, in November (dry season), mean male perch height and diameter 1.7 ± 0.13 m and 11.9 ± 1.64 cm, female 1.0 ± 0.09 m and 6.7 ± 1.02 cm, juvenile 0.5 ± 0.30 m and 3.1 ± 1.44 cm, respectively (Jenssen et al., 1988); mean perch heights in a disturbed habitat at sea level 1.29 ± 1.16 m for males and 0.86 ± 0.84 m in females (Fobes et al., 1992); mean perch heights and diameters along rainforest edge 71.4 ± 10.4 cm and 10.4 ± 2.2 cm, in rainforest interior 19.4 ± 6.9 cm and 3.9 ± 0.9 cm, with males perching higher (61.5 ± 14.1 cm) and on perches of greater diameter (13.3 ± 3.6 cm) than females (50.7 ± 11.0 cm and 6.0 ± 1.7 cm; Cast et al., 2000); mean perch heights and diameters of males along edges in cloud forest 68.0 ± 8.0 cm and 7.9 ± 1.4 cm, females 44.3 ± 6.4 cm and 5.5 ± 1.0 cm, both sexes combined at 3 sites were 42.6 ± 5.6–66.4 ± 9.7 cm and 4.7 ± 1.6–8.1 ± 1.5 cm (Sifers et al., 2001); not an adept climber (Fitch et al., 1989); more at home on ground, especially in areas where ground-dwelling lizards (*Leiocephalus*, *Ameiva*) are absent or when they are inactive (Fobes et al., 1992; Moster et al., 1992); does not actively forage on ground, prey apparently sighted from elevated perches (Schwartz and Henderson, 1991); mean perch height and diameter 1.43 m and 24.52 cm (J. B. Losos in Calsbeek et al., 2006); not on leaves (J. B. Losos, in litt., 21.IV.2008); sleeps on leaves and small branches with hindlimbs partly flexed, and apparently moves to higher perches than those used during the day in habitats with large trees (Rand, 1962b; Fobes et al., 1992).

Abundance: Moderately abundant, 155 of 990 observations of anoles in and around Los Haitises National Park, Dominican Republic (Glor et al., 2001). **Activity:** No evidence of activity restricted by weather conditions (Hertz and

Huey, 1981); basically diurnal (Rand, 1962b), but exploits the "night-light niche" (Henderson and Powell, 2001; Perry et al., 2008). **Behavior:** Sleeps with hindlimbs drawn up, individuals from high-elevation populations more shy than those from low elevations (Mertens, 1939); vocal "alarm call" (Mertens, 1946); allows a very close approach, then "squirrels," if pressed, runs down, frequently to the ground, where it might remain or move quickly to another tree (Rand, 1962b); rests facing down, anterior part of body elevated and neck bent (sometimes well over 90°) so the head is parallel to the ground, spends relatively long periods in one spot, but changes of position occur rapidly (Rand, 1962b); capable of vocalization (T. A. Jenssen in Milton and Jenssen, 1979); percent jumps (fraction of total movements) by adults on lower tree trunks 21% at a Haitian site, juveniles on ground jump 48%, on rocks at a second site jump 9%, whether in trees or on rocks, run more frequently than jump, crawl rarely in trees (Moermond, 1979a); at 2 different sites, percent crawls 0 and 3%, runs 89 and 77%, jumps 9 and 21%, percent moves to different surface 21% (adults) and 48% (juveniles; Moermond, 1979b); orientation up, down, or horizontal did not differ significantly in a disturbed habitat, escape behavior includes "squirreling," moving to an adjacent branch, jumping from the ground to a tree and ascending, hiding under a water pump, and retreating under the loose bark of palm trees, never responding at distances >3 m, responses of males to introduced males involved dewlapping and headbobbing (7–13 repetitions) by one or both individuals, followed by an advancing attack by the resident that involved lunging and biting, throat inflation and lateral compression occurred in 1 of 3 incidents, which lasted a few seconds to >1 min and terminated with the intruder escaping, all 3 displaced individuals returned to their home ranges within 24–48 h (Fobes et al., 1992); mean primary approach distance before reacting 151.6 ± 24.1 cm, mean secondary distance 56.3 ± 5.9 cm, most common (28%) primary response jumping to another substrate, most common (39%) secondary response moving to a lower perch (Schneider et al., 2000); has only one species-specific signature display, which seems geographically stable, at least in Haiti, sometimes preceded by a somewhat variable head roll or a "head dip," which also was seen as lizards assumed new perches and may enhance visual discrimination (Jenssen, 1983); where sympatric with *A. marcanoi*, aggressive display less elaborate and ritualized than that of *A. marcanoi*, comparable to low-level dewlap display of *A. marcanoi* (see that account), although occasionally involving elevation on all 4 limbs, dewlap extended or pulsed, all modified by nuchal and dorsal crest erection and lightening of body color, especially in dominant males, headbobs with dewlap extended or pulsed dewlap with head moved only slightly, in addition, lateral orientation and compression and gular inflation increased apparent lizard size, quickly escalated to include biting and chasing (Losos, 1985a); moves per minute 0.48 and 0.89 (Johnson et al., 2008); percent time displaying 5.4, percent walking 26.9, percent jumping 16.6, percent running 56.5 (J. B. Losos, in litt., 21.IV.2008). **Competition:** Largely excluded from urban areas around La Romana where *A. cristatellus* has been introduced (Fitch et al., 1989). **Desiccation:** Mean critical activity point (percent of original body weight at which the lizard could not right itself when turned on its back) 78.8 ± 8.04%, mean rate of water loss 0.18% body mass/h (Hillman and Gorman, 1977); vital limit (percent of initial body weight at death) 64.3 ± 0.9%, rate of weight loss (percent of initial body weight lost per hour) 0.114 ± 0.009%, survival time 340.4 ± 31.7 h (Hertz, 1980a); rates of fractional weight loss (percent of initial body weight lost/h) increased and vital limits and survival times decreased with increasing elevation (all were correlated with initial body weight), suggesting that frequent basking behavior at high elevations may induce selection for water conservation that is nearly as effective as that in lowland populations (Hertz, 1980b). **Ecomorphology:** "Trunk-ground" (Rand and Williams, 1969); formally quantified (Beuttell and Losos, 1999). **Foraging and Diet:** Uses "watch-position" to maintain a careful scrutiny of a large area, obtains at least part of its food from the ground, eats dragonflies and cockroaches (Rand, 1962b); scanning the ground from an elevated perch takes longer than scans from other perches, increasing waiting intervals and distance/move, mean 6.3 min between location changes, mean 4.5 min between location or posture changes, mean distance/move 31 cm, distance moved per prey capture variable but relatively long, 79% of prey captures on the same surface as that on which the lizard is perched (Moermond, 1979b); *Sphaerodactylus* sp. (RP in Schwartz and Henderson, 1991); conspecific juveniles, also preys on *A. bahorucoensis* and *A. coelestinus* (Fitch and Henderson, 1987); 3 of 5 stomachs in the La Romana area contained ants, mean prey size 0.76 ± 0.24 cm and 0.53 ± 0.22 cm^3 (Fitch et al., 1989); stomachs of 3 lizards from the Siete Hermanos (identified as *A. whitemani*) contained spiders, beetles, a hemipteran, ants (the most important prey item), lepidopteran larvae, a dragonfly, and an orthopteran (Burns et al., 1992); in disturbed habitat, one individual actively foraging on the ground was eating prey items too small to identify, an adult lunged for a bee caught in a spider web, and another was on a leaf with a bee in its mouth, stomach contents included spiders, other arachnids, beetles (adults and a larva), roaches (greatest volume), flies (adults and larvae), hemipterans, hymenopterans (including ants, most important by numbers), a termite, lepidopterans (adults and larvae), a lacewing, a dragonfly, orthopterans (in the most stomachs), an isopod, shed skin, plant material, and grit, with the latter presumably adventitiously ingested (Fobes et al., 1992); stomach contents of 34 individuals from 3 variously altered upland habitats in the Sierra de Bahoruco included spiders and other arachnids, orthopterans, termites, beetles (adults and larvae), which were the most important prey items, roaches, lepidopterans (adult and larva), fly larvae, hymenopterans (including ants), centipedes, an earthworm, a *Sphaerodactylus* sp., shed skin, and plant matter (probably adventitiously ingested), dietary niche breadth was greater than that of *A. coelestinus* in the least altered habitat, but less in the more heavily altered situations, and more than that of sympatric *A. distichus* in all habitats (Lenart et al., 1997); dietary niche overlap with *A. barbouri* lower than between *A. barbouri* and litter-dwelling frogs (*Eleutherodactylus*) but higher than with other sympatric anoles, attributed to *A. cybotes* being more likely to take terrestrial prey (A. K. Howard et al., 1999a); eats *Anolis distichus* (T. A. Jenssen in Gerber, 1999); stomach contents of 17 individuals from rainforest in June contained spiders, another arachnid, beetles, flies, hemipterans, hymenopterans (mostly ants), lepidopterans (adults and larvae), roaches, crickets (the most important dietary item), a millipede, and an isopod (Cast et al., 2000); stomach contents of 28 individuals from cloud forest in June contained beetles (adults and larvae), springtails, roaches, stick insects, fly larvae, hemipterans, ants, lepidopterans (adults and larvae, the most important dietary item), lacewings, orthopterans, a millipede, snails, leeches, fruits/seeds, shed skin, and grit, probably ingested adventitiously (Sifers et al., 2001); consumes conspecific juveniles (Fitch, 2003). **Home Range and Territoriality:** Highly territorial, only one male and one or more females on any one tree, any intruder of either sex chased away (Mertens, 1939; Rand, 1962b); male two-dimensional home ranges (mean 14.36 ± 10.48 m^2) in a disturbed habitat contained 1–7 bushes or trees (mean 3.3 ± 2.4), 4 of which were used by 2 individuals and one large Croton bush (*Codiaeum* sp.) by 4 males, and were significantly larger than female home ranges (mean 3.97 ± 4.57 m^2) that contained 1–4 (mean 2.5 ± 1.3) bushes or trees, none of which were used by more than one individual (Fobes et al., 1992). **Injuries:** 17 of 115 (14.8%) of males with lesions on jaws attributed to intraspecific aggression (Bowersox et al., 1994). **Movements:** Mean moves per minute 0.07 (Moermond, 1979b). **Parasites:** Protozoan malarial parasites *Plasmodium azurophilum* and *P. floridense* (Telford, 1975; Ayala, 1977, 1978); nematodes *Skrjabinoptera leiocephalorum* (Fobes et al., 1992; Goldberg et al., 1998b); Apicomplexa *Isospora hendersoni* (Cisper et al., 1995); larval mites (*Eutrombicula alfreddugesi*) more prevalent and at higher intensities in higher-elevation, moister situations (Zippel et al., 1996). **Population Size and Density:** In a highly disturbed courtyard at sea level, 367 lizards/ha (Fobes et al., 1992); 136–144/ha (Schell et al., 1993b); in rainforest interior 3,139 or 3,715 lizards/ha depending on method used, along forest edge 833 lizards/ha (Cast et al., 2000); at 3 cloud forest sites (only one juvenile was in the forest interior), 2,034–8,972 lizards/ha concentrated in edge habitats (Sifers, 2001). **Predation:** *Hypsirhynchus parvifrons* on La Gonâve, (Franz and Gicca, 1982) and on Hispaniola (Henderson et al., 1987a); *Epicrates fordii, E. gracilis, E. striatus* (Henderson et al., 1987b); *Uromacer catesbyi, U. frenatus, U. oxyrhynchus* (Henderson et al., 1987c); domestic chicken (*Gallus gallus*; Bowersox et al., 1994); *E. striatus* (Fernández, 2007). **Reproduction:** Clutch size 1 (Hedges, in prep.); buries eggs in the ground (Rand, 1962b, 1967c); reproduction distinctly seasonal, but amplitude can vary by location or from year to year, testicular weights decline during November-December, although spermatogenic and interstitial cell activity is not uniformly depressed, peaks in January–February and remains high through the summer, fat bodies present throughout the year, but greatest in November as testes regress, females show a decline in reproductive activity in November–January when fat bodies are greatly enlarged (Licht and Gorman, 1970); stage of embryonic development before eggs are deposited is positively correlated with elevation (Huey, 1977); 23 of 28 females contained ova and/or follicles in June, 3 contained 2 ova (suggesting an ability to retain eggs until deposition propitious), copulation observed, but no high frequency of association between males and females (Fobes et al., 1992); in captivity, eggs ~6 × 9 mm, incubation 30–40 days at 23–28 °C (Rogner, 2000). **Sex Ratio (M:F):** 1:1 in November, 2:3 in June in Haiti, presumably because females were more closely associated with males in June (breeding season), resulting in multiple females per male (Jenssen et al., 1988); 18:27 and 27:16 in two samples from Barahona, Dominican Republic (Fobes et al., 1992). **Size:** Male SVL to 81.1 mm (Schoener, 1988), female SVL to 66 mm (Schwartz, 1989); smallest sexually mature male 52 mm SVL and female 45 mm SVL (Licht and Gorman, 1970); body weight greatest at higher elevations (Hertz, 1980b). **Tail Autotomy:** In a disturbed habitat, 6 of 18 adult males and 10 of 27 adult females had regenerated tails, frequencies did not differ significantly, nor was the frequency of autotomy correlated with SVL (Fobes et al., 1992). **Thermal Biology:** Experimental voluntary maximal temperature 34.3 ± 0.07 °C, critical thermal maximum 38.4 ± 0.13 °C (Hertz, 1980a); basks more frequently at higher elevations, with lowland populations essentially thermal conformers, body temperatures higher (mean 31.6 °C) at lower elevations (air temperatures 25.8–32.6 °C) than at higher elevations (mean 29.9 °C) at air temperatures of 20.8–30.2 °C (Hertz and

Huey, 1981); in a disturbed habitat at sea level, most individuals (94%) the shade, even during early morning hours, mean body and substrate temperatures of males (30.4 ± 1.3 °C and 28.7 ± 1.6 °C) did not differ significantly with those of females (30.8 ± 1.6 °C and 29.3 ± 1.7 °C; Fobes et al., 1992); mean body, substrate, and air temperatures for males 25.1 ± 1.2 °C, 22.8 ± 5.8 °C, and 22.8 ± 0.8 °C, females 28.7 ± 1.1 °C, 25.2 ± 1.0 °C, and 24.8 ± 0.9 °C (Cast et al., 2000); mean male and female body temperatures at 4 sites in cloud forest 29.0 ± 0.3 °C and 29.8 ± 0.7 °C at mean substrate and air temperatures of 23.5 ± 0.3 °C and 23.5 ± 0.4 °C (Sifers et al., 2001).

Conservation Status: Not assessed.

Additional Pertinent Literature: Fobes et al. (1993).

Anolis darlingtoni Cochran, 1935

Distribution: Hispaniola, Massif de la Hotte (Haiti); elevational range ~1,360–1,500 m.

Habitat: In 1984, in a disturbed patch of forest surrounded by cleared areas and with small trees, Sierra Palms, tree ferns, and herbaceous growth, but few large broadleaf trees (Thomas and Hedges, 1991); adults on large branches about 1 m above the ground, juveniles on tips of dead branches of living trees about 3–6 m above ground, sleeps on twigs (Schwartz and Henderson, 1991); juveniles sleeping vertically on tips of dead tree fern branches 1–4 m above the ground, 3 adults sleeping horizontally on limbs of small trees ~1 m above the ground (Thomas and Hedges, 1991).

Ecomorphology: "Twig" (Williams, 1983); formally quantified (Beuttell and Losos, 1999). **Size:** Male SVL to 74 mm, adult females unknown (Thomas and Hedges, 1991).

Conservation Status: Unknown, association with rapidly disappearing Haitian forests is cause for concern.

Anolis delafuentei Garrido, 1982

Distribution: Cuba, Sierra de Trinidad; elevation ~700 m.

Habitat: Mesophilic (Schwartz and Henderson, 1991); in a humid and shaded ravine with remnants of dense broadleaf forest alternating with calcareous rocks (Garrido, 1982b); in submontane mesophyllous evergreen forest (Rodríguez Schettino et al., 1999).

Activity: Diurnal (Rodríguez Schettino et al., 1999). **Ecomorphology:** "Trunk-ground" (Losos, 2009). **Size:** Male SVL 61 mm (only holotype known; Garrido, 1982b).

Conservation Status: Considered "critically endangered" (Rodríguez Schettino and Chamizo Lara, 1998a; L. Rodríguez Schettino and A. Chamizo Lara in Rodríguez Schettino, 1999a).

Anolis desechensis Heatwole, 1976

Distribution: Isla Desecheo (Puerto Rico).

Habitat: Xerophilic to semi-xerophilic (R. Thomas, in litt., 22.IV.2008); general ecology and structural habitat essentially like those of *A. cristatellus* (Heatwole, 1976); occurs throughout the island, but appears to be most common (with greatest range in sizes, i.e., age groups) in canopy forest areas and their margins (Evans et al., 1991); most common near the shore at the vegetation line, "almost every pile of rocks had an adult male and several females associated with it," males in a conspicuous location, females usually in crevices between rocks, also common in deciduous woodlands, but uncommon in thorny cactus-scrub on upper slopes and ridge tops, uses low perches in early morning and late afternoon, sleeps on slender branches (Meier and Noble, 1991).

Activity: Forages actively from sunrise to sunset (Evans et al., 1991); becomes active at dawn (Meier and Noble, 1991). **Behavior:** Usually head-downward on lower trunks of *Bursera simaruba*, females rarely seen in exposed situations, except when mating, appears to avoid perches on or near the ground during periods of high *Ameiva* activity (Meier and Noble, 1991). **Diet:** Berries (Gorman and Stamm, 1975); flies, grasshoppers, moths, anole egg (Meier and Noble, 1991). **Ecomorphology:** "Trunk-ground" (Gorman and Stamm, 1975; Losos, 2009); descended from trunk-ground ecomorphs, but possesses a more generalized trunk-ground morphology (Losos and de Queiroz, 1997a). **Predation:** Introduced Rhesus Monkeys (*Macaca mulatta*; Evans, 1989; Evans et al., 1991); *Borikenophis variegatus* (Meier and Noble, 1991). **Size:** Male SVL 57 mm, female SVL to 45 mm (Schwartz and Henderson, 1991).

Conservation Status: "Common but restricted, protect habitat" (Moreno, 1991d); some protected status may be warranted (Moreno, 1991b).

Remarks: *Anolis desechensis* may not be distinct at the species level from *A. cristatellus* from southwestern Puerto Rico (Rodríguez-Robles et al., 2007).

Anolis desiradei Lazell, 1964

Distribution: La Désirade, east of Guadeloupe.

Habitat: In "arid lowlands typical of islands too low in elevation to stop significant rainfall" (Lazell, 1972); fence posts in cultivated and wild cotton fields (Schwartz and Henderson, 1991); on tree trunks, fence posts, rocks, buildings (Breuil, 2002b).

Biomass: 0.97 g/m^2 (Wright, 1981). **Ecomorphology:** Undetermined. **Natural History:** Unknown. **Size:** Male SVL to 80 mm (Lazell, 1964a), female SVL to 53 mm, smallest mature male and female SVL 51 mm and 44 mm, respectively (Lazell, 1972).

Conservation Status: Unknown.

Anolis distichus Cope, 1861

Distribution: Hispaniola and the Bahama Islands; 17 currently recognized subspecies occur on the Great Bahama Bank, Cat Island, Little San Salvador Island, introduced on the Little Bahama Bank (Grand Bahama Island), islandwide on Hispaniola, Île de la Tortue, Île-à-Vache, Île Grande Cayemite, Île Petite Cayemite, Isla Saona, Isla Catalina, Isla Cabras, Cayos Siete Hermanos (Cayos Monte Chico and Tercero); several subspecies introduced in southeastern Florida; elevational range, below sea level to at least 1,830 m.

Habitat: Mesophilic to somewhat xerophilic (Schwartz, 1968d; Schwartz and Henderson, 1991); "eurytopic," in coastal zones with *Ipomoea* and *Coccoloba*, scrub savanna, pine forests, and human-modified habitats, including on and in houses (Mertens, 1939); on trees (especially *Cocos*, *Lysiloma*, *Terminalia*), fence posts, in active (sometimes very open) and abandoned agricultural areas and at forest edges, in oases or even shaded sites in xeric regions (otherwise often inhabited by members of the *A. brevirostris* complex), rainforests, pine, and cloud forests at higher elevations, shaded and mesic forests, coffee and cacao plantations, and associations of orchard trees (mango, breadfruit) and Royal Palms (*Roystonea*), coppice, occasionally in pine forests, in the Bahamas, often on isolated pale- or gray-barked trees (*Ficus*, *Coccoloba*, *Terminalia*), on *Cocos* and *Coccothrinax*, on limestone walls and street cutbanks, in ruins, on trunks of pines and other trees in drier areas, in the ecotone where forest was being cut, on isolated trees in an agricultural area, on large boulders, along a stream channel, exposed slopes bordering a deep ravine, occasionally on houses and other buildings, under rocks (Mertens, 1939; Oliver, 1948; Rand, 1962b; Schoener, 1968; Schwartz, 1968d; Williams and Webster, 1974; Franz and Cordier, 1986; Schwartz and Henderson, 1991; Olson, 1995; Lenart et al., 1997; Cast et al., 2000; Glor et al., 2001; Sifers et al., 2001; Powell and Henderson, 2008c); "this is the lizard so well known to tourists as the one which swarms over the trunk of the giant ceiba tree in the Court House Square at Nassau" (Barbour, 1930c); in the Exumas, on a roof beam (Dodd, 2003); on Bimini, usually at heights below 1.8 m (Oliver, 1948); 3.0–4.5 m high on trunks, rarely on the ground or in smaller branches of bushes or trees, when more than one individual is in a small tree, they are of opposite sexes with the male closer to the ground than the female, on one large tree, a large male near the base, a female above him, a slightly smaller female above her, and a juvenile at about 2.1 m highest of all, with stratification maintained despite considerable movement by individuals (Rand, 1962b); on Bimini, nearly always on trunks of fairly large trees (only 2 of 329 observation on the ground), those on elevated perches always on trunks or branches (never on leaves), most adults (19%) on perches 0.9–1.6 m high and with diameters 5–13 cm, most juveniles (21%) on perches 1.7–2.6 m high and with diameters 5–13 cm (Schoener, 1968); 4.4% on leaves, 2.0% on stems 12 mm in diameter, 5.6% on stems 12–18 mm in diameter, 13.8% on stems 18–24 mm in diameter, and 73.2% on rocks (Moermond, 1979a); 1.1% on perches <0.7 cm in diameter, 9.0% on perches 0.7–2.5 cm in diameter, 16.8% on perches >2.5 cm in diameter, and 20.8% on rocks at heights >90 cm, 0.6% on perches <0.7 cm in diameter, 4.5% on perches 0.7–2.5 cm in diameter, 6.7% on perches >2.5 cm in diameter, and 17.4% on rocks at heights 31–90 cm, 1.1% on leaves or grasses, 0.6% on perches 0.7–2.5 cm in diameter, 6.7% on perches >2.5 cm in diameter, and 14.6% on rocks or ground at heights <30 cm (Moermond, 1979b); appears tolerant of human-mediated habitat alterations, abundance approximately equal in 3 variously disturbed upland habitats in the Sierra de Bahoruco (Lenart et al., 1997); forages on the ground more frequently at locations where ground-dwelling lizards (*Ameiva*, *Leiocephalus*) are absent or rare (Moster et al., 1992); on Exuma Island, mean adult male and female perch heights 1.8 ± 0.90 m (0.7% on the ground) and 1.3 ± 0.71 m, none on the ground (Lister, 1976); mean perch heights for males 97.84 ± 0.61 cm, females 68.82 ± 0.32 cm, immature individuals 69.49 ± 0.43 cm (Fitch et al., 1989); mean perch heights and diameters along rainforest edge 160.5 ± 11.8 cm and 27.8 ± 2.1 cm, in rainforest interior 148.9 ± 19.9 cm and 29.6 ± 6.3 cm, with males perching higher (179.7 ± 12.3 cm) and on perches of greater diameter (30.3 ± 2.9 cm) than females (148.5 ± 17.7 cm and 26.0 ± 3.8 cm; Cast et al., 2000); mean perch heights and diameters of males along edges in cloud forest 121.5 ± 23.0 cm and 15.5 ± 2.9 cm, females 129.0 ± 49.4 cm and 17.5 ± 4.2 cm, both sexes combined at 2 sites were 100.3 ± 11.7–205.0 ± 49.4 cm (one individual at a third site perched at 30 cm) and 17.0 ± 2.2–26.8 ± 9.7 cm (one individual at a third site was on a perch with a diameter of 4 cm; Sifers et al., 2001); 90% of observations on Bimini at heights <2 m

(Mattingly and Jayne, 2004); mean height on Bimini 177 ± 11 cm (Mattingly and Jayne, 2005); mean perch height and diameter 2.57 m and 18.33 cm on Hispaniola (J. B. Losos in Calsbeek et al., 2006); green color of individuals at higher elevations blends well with trunks covered by epiphytes, lichens, and mosses (Fläschendräger and Wijffels, 2009); 8.3% on leaves (J. B. Losos, in litt., 21.IV.2008); sleeps with hindlimbs partially flexed (Mertens, 1939) on leaves and twigs of small shrubs within 1 m of the ground and on small limbs 1.5–3.5 m above ground in *Sabal*, groves of large trees, and forests, also on *Lysiloma* pods (Schwartz, 1968d; Schwartz and Henderson, 1991).

Abundance: By far the most abundant (681 of 990 observations) of anoles in and around Los Haitises National Park, Dominican Republic (Glor et al., 2001). **Activity:** Diurnal, seen in the sun more frequently in early morning and appears to feed more actively in the morning (Rand, 1962b); exploits the "night-light niche" (RP in Perry et al., 2008). **Behavior:** Sleeps with hindlimbs drawn up, moves to the other side of a tree when approached and then down to its base to hide, very shy (Mertens, 1939); alert and active, remains in position until an intruder is very close, if on the ground, runs to a nearby tree or shrub, relies on agility to escape capture (Oliver, 1948); individuals maintaining vertical stratification on a large tree react immediately to another lizard approaching too closely, male by headbobbing and dewlap extension, female by charging the intruder, in all cases, the intruder retreats immediately, around the tree and up, rests facing either up or down or angled across tree with head and at least anterior body raised and neck bent so that head is parallel to the substrate, restless, seldom remains in one place more than a few minutes, then quickly moves a few centimeters and rests again, responds to an intruder by "squirreling" and moving up or down at the same time, only when hard pressed will an individual leave a tree, will hide under bark or in a hole in a tree (Rand, 1962b); percent jumps (fraction of total movements) on tree trunks 14 and 37% at two Haitian sites, on rocks at the second site jump 16%, in scrub, trees, or on rocks, run more frequently than jump, rarely crawl (Moermond, 1979a); at 2 different sites, percent crawls 3 and 0%, runs 79 and 86%, jumps 16 and 14%, percent moves to different surface 11% for adults and 30% for juveniles (Moermond, 1979b); movement characterized as "runner" (Moermond, 1981); mean primary approach distance before reacting 83.5 ± 14.7 cm, mean secondary distance 56.5 ± 15.4 cm, most common primary (45%) and secondary (55%) response "squirreling" (Schneider et al., 2000); 33.5% of distance traveled on trunk bases, 31.6% on trunks, 34.9% on branches, 78.7% of movements with jumps, mean movement distance 4.4 cm (median 2.8 cm) at speed 32.8 cm/sec (31.5 cm/sec), mean pause duration 18.0 sec (2.5 sec; Mattingly and Jayne, 2004); during escape chooses broadest diameter 94%, longest length 38%, shallowest incline 46%, and straightest angle 90% (Mattingly and Jayne, 2005); 6 display types, some with unusual characteristics, largest repertoire of any studied anole (Jenssen, 1983); defends territories against *A. bahorucoensis* (Fitch and Henderson, 1987); shares space with introduced *A. cristatellus* in La Romana area, moves more frequently and rapidly than *A. cristatellus*, and may avoid predation by speed and agility (Fitch et al., 1989); neither absolute nor size-corrected dewlap size significantly related with bite force, 91% of combat trials involved dewlap extensions, 23% biting, and 91% headbobbing (Lailvaux and Irschick, 2007); moves per minute 1.21 (Johnson et al., 2008); percent time displaying 4.0, percent walking 28.0, percent jumping 2.9, percent running 69.2 (J. B. Losos, in litt., 21.IV.2008). **Desiccation:** Mean critical activity point (percent of original body weight at which the lizard could not right itself when turned on its back) 70.3 ± 2.52%, mean rate of water loss 0.43% body mass/h (Hillman and Gorman, 1977). **Ecomorphology:** "Trunk" (Rand and Williams, 1969); formally quantified (Beuttell and Losos, 1999). **Foraging and Diet:** Feeds by snapping tiny insects off the surface of bark, but capable of distinguishing (and rejecting) potential prey items (Rand, 1962b); scanning from broad perch takes less time than scans from perches with limited overviews and much less than scans of the ground from elevated perches, resulting in relatively short waiting intervals and distances/move, mean 2.9 min between location changes, mean 2.5 min between location or posture changes, mean distance/move 27 cm, distance moved per prey capture longer than those of anoles surveying dense habitat clusters but shorter than those ranging from elevated perches to the ground, 85% of prey captures on same surface as that on which the lizard was perched (Moermond, 1979b); prey strikes either approach-pause or stationary (Moermond, 1981); on Bimini, feeds most commonly on aggregated prey (ants), searches actively and then sits passively in front of trail and "gobbles" the ants up as they pass by, most obtained on trunks and branches (but may come to ground to eat ants), feeds most actively early in the day or on cloudy days, frequently also in late afternoon and at dusk (Schoener, 1979); on Bimini, food items were ants (greatest numbers and volumes in adults, subadults, and juveniles of both sexes), other hymenopterans, aphids, other hemipterans, flies, thrips, beetles (adults and larvae), barklice, lepidopterans (adults and larvae), orthopterans, lacewings, termites, spiders, mites, and plant matter, with larger individuals taking larger food items (Schoener, 1968); on Exuma Island, stomachs contained ants (59.3%), other hymenopterans, flies, beetles (adults and larvae), termites, leafhoppers (29.7%), other hemipterans, lepidopteran, earwigs, spiders, pseudoscorpions, and mites (Lister, 1976); in the Dominican Republic, an adult pursued a hatchling *A. cristatellus*, stomachs contained ants (most common, 82.6% of 46 stomachs), flies, and beetles, with mean prey size of males 0.57 ± 0.09 cm and 0.24 ± 0.07 cm^3 and females 0.27 ± 0.07 cm and 0.10 ± 0.02 cm^3 (Fitch et al., 1989); stomachs of 3 lizards from the Cayos Siete Hermanos contained a spider, beetles, ants (most important prey item), and an orthopteran (Burns et al., 1992); stomachs of 15 lizards from Santo Domingo (lowland) contained spiders, beetles, roaches, flies, hemipterans, hymenopterans (including ants, which were the most important prey type), a termite pupa, lepidopterans, orthopterans, an isopod, plant material, and grit, 15 lizards from Monteanada Nueva (elevation ~1,100 m) contained spiders, beetles (adults and larvae), roaches, flies (adults and a larva) hemipterans, hymenopterans (including ants, which were the most important prey type), lepidopterans (adults and larvae), orthopterans, lizard skin, reptilian eggshell, plant material and grit, although ants the most important item in both samples, calculated dietary niche breadths much higher for the upland population (Cullen and Powell, 1994); stomach contents of 25 individuals from 3 variously altered upland habitats in the Sierra de Bahoruco included spiders, orthopterans, termites, beetles (adults and larvae), aphids, hemipterans, lepidopterans (adult and larva), flies (adult and larvae), hymenopterans (including ants, which were the most important prey item in all 3 plots), and plant matter (probably adventitiously ingested), dietary niche breadth less than those of sympatric *A. coelestinus* and *A. cybotes*, reflecting the large proportion of ants in the diet (Lenart et al., 1997); dietary niche overlap with *A. barbouri* low (A. K. Howard et al., 1999a); stomach contents of 29 individuals from rainforest contained miscellaneous arachnids (no spiders), a springtail, beetles (adults and larvae), flies, hemipterans, hymenopterans (mostly ants, which were the most important dietary item), termites, lepidopteran larvae, a roach, orthopterans, a snail, shed skin, and fruits/seeds (Cast et al., 2000); stomach contents of 4 individuals from cloud forest contained roaches, leafhoppers, ants (which were the most important dietary item), a lepidopteran, lacewings, a leech, and miscellaneous plant matter (not fruits or seeds), which was probably ingested adventitiously (Sifers et al., 2001). **Growth:** On Bimini, males grow more rapidly than *A. sagrei*, suggesting less food pressure on arboreal species (Schoener, 1979). **Home Range and Territoriality:** Highly territorial (Mertens, 1939); territories smaller than those of *A. chlorocyanus* and *A. cybotes*, in areas of high density, more than one male may defend a territory in the same large tree (Rand, 1962b); on Bimini in "short-blackland" habitat, median home range size of males 9.93 m^2, of females 7.44 m^2, in *Terminalia* habitat, of males 9.45 m^2, of females 8.20 m^2, male home ranges "fairly exclusive," but can overlap substantially, sometimes in areas of intensive feeding, in *Terminalia* habitat, 4 largest males had territories arranged radially about the largest trunk, females can occur in "nested" ranges, sometimes almost perfectly contained within ranges of large males, subadult males can have very large home ranges that overlap with many individuals, displays are frequent, but aggression rare (Schoener, 1979); home ranges of adult males (7.86 ± 5.17–30.66 ± 15.59 m^2 at 7 Bahamian sites) larger than those of females (5.88 ± 4.77–26.43 ± 13.29 m^2 at 3 sites) and subadult males (4.38 ± 6.14–7.23 ± 9.00 m^2 at 3 sites; Schoener and Schoener, 1982b); home ranges included parts of trees and overlapped with those of other conspecifics and with those of introduced *A. cristatellus* (Fitch et al., 1989). **Hybridization:** Hybridization possibly frequent with *A. brevirostris*, but hybrids probably infertile (Webster, 1977); extent of hybridization with *A. brevirostris* variable, no evidence at montane sites, but extensive at 2 lowland sites where interspecific differences in dewlap color were minimal or variable (Williams and Case, 1986). **Injuries:** Injuries to tails, front toes, and back toes higher in males than in females, with front toe rates highest (presumably due to intraspecific aggression) and tail and hind toe rates (presumably due to predation) lowest (Schoener and Schoener, 1980b). **Parasites:** Protozoan malarial parasite *Plasmodium floridense* (Telford, 1975; Ayala, 1977, 1978); in the Dominican Republic, Apicomplexa *Isospora guarocuyai* (Bui et al., 1992a); on San Salvador, nematodes *Atractis scelopori*, *Parapharyngodon cubensis*, *Physalopteroides bahamensis*, larval *Physaloptera* sp., and adult female *Strongyloides* sp. (Goldberg et al., 1994b); *Skrjabinoptera leiocephalorum* (Lenart et al., 1994; Goldberg et al., 1998b). **Performance:** Clinging ability 0.94 ± 0.18 Neutons, divided by mass 0.39, with the very high value apparently characteristic of small trunk ecomorphs (Elstrott and Irschick, 2004); average gross speed over an interval of 5 strides on a 4.8-cm diameter perch 98 ± 8.8 cm/sec and 49 ± 4.5 cm/sec with 90° turning angle, stride-one speed 101 ± 7.0 cm/sec and 19 ± 1.0 cm/sec with 90° turning angle (Mattingly and Jayne, 2005); acceleration 30.53 ± 2.42 m/sec, sprint speed 1.05 ± 0.06 m/sec (Vanhooydonck et al., 2006a). **Population Size and Density:** Male densities in various Bahamian habitats 0.064–0.298 individuals/m^2, female densities 0.043–0.248 individuals/m^2 (Schoener and Schoener, 1978b); on Bimini, in "short-blackland" habitat, 0.052 males/m^2, 0.066 females/m^2, in "forest-path" habitat, 0.045 males/m^2, 0.030 females/m^2, in *Terminalia* habitat, 0.209 males/m^2, 0.174 females/m^2 (Schoener, 1979); in various habitats at multiple sites on 2 Bahamian islands, male densities 0.000–0.299 lizards/m^2, female densities 0.000–0.249 lizards/m^2, total densities 0.000–0.548 lizards/m^2 (Schoener and Schoener, 1980a, 1980b); in rainforest interior on Hispaniola 1,701 or 1,694 lizards/ha depending on method used, along forest edge 3,853 or 1,764 lizards/ha (Cast et al., 2000); at two cloud forest sites, 257–444 lizards/ha concentrated along edge habitats, by far the least abundant of congeners at these sites (Sifers, 2001). **Predation:** *Uromacer oxyrhynchus* (in collecting bag) and *Tropidophis haetianus* (Mertens, 1939); *Hypsirhynchus parvifrons* (Henderson et al., 1987a); *Epicrates gracilis* (Henderson et al., 1987b); *Uromacer catesbyi*, *U. frenatus*, *U. oxyrhynchus* (Henderson et al., 1987c); *Anolis cybotes* (T. A. Jenssen in Gerber, 1999); Hispaniolan Lizard Cuckoo (*Coccyzus longirostris*; R. E. Glor, pers. comm.). **Reproduction:** Copulation 1 March (Mertens, 1939); copulation 6 July, 3 copulations early to mid-afternoon on tree trunks 1.2–1.8 m above ground, in one instance oriented diagonally, male on higher side above female, head on her neck, but not biting (Rand, 1962b); clutch size 1 (Hedges, in prep.); in captivity, females bury eggs, egg size at deposition 6 × 9 mm, incubation 40–50 days at 23–28 °C, hatchling SVL ~19 mm (Fläschendräger and Wijffels, 1996, 2009). **Sex Ratio:** In various habitats at multiple sites on two Bahamian islands F/M 0.63–1.40:1 (excluding data with sample sizes of 3 for one or both sexes; Schoener and Schoener, 1980a); 1.11:0.89 M:F (Schoener and Schoener, 1980b; T. W. Schoener in Stamps, 1983c). **Size:** Male SVL to 58.4 mm, female SVL to 50.2 mm (Schoener, 1988). **Survival:** Survival higher on smaller island (Bimini) than on larger island (Andros), presumably due to predation by a greater abundance and variety of birds on Andros (Schoener and Schoener, 1978a); annual survival in 4 habitats on 2 Bahamian islands 0.000–0.571 for males and 0.000–0.400 for females (Schoener and Schoener, 1980b). **Thermal Biology:** In an open situation on Exuma Island 31.0 ± 2.10 °C and 28.3 ± 0.85 °C (Lister, 1976); along a rainforest edge on Hispaniola, mean body, substrate, and air temperatures for males 27.1 ± 0.7 °C, 23.9 ± 0.5 °C, and 22.3 ± 0.5 °C, females 26.4 ± 1.1 °C, 24.1 ± 0.8 °C, and 23.1 ± 0.6 °C (Cast et al., 2000); mean male and female body temperatures at 3 sites in cloud forest 31.5 ± 0.1 °C and 26.8 °C (N = 1) at mean substrate and air temperatures of 23.5 ± 0.4 °C and 23.1 ± 0.3 °C (Sifers et al., 2001).

Conservation Status: Not assessed.

Additional Pertinent Literature: Schwartz (1971c).

Anolis dolichocephalus Williams, 1963

Distribution: Hispaniola, where three currently recognized subspecies occur on the western Tiburon Peninsula in Haiti, on Monts Cartaches, Massif de la Hotte, and adjacent lowlands; elevational range, sea level to >1,000 m.

Habitat: Mesophilic (Schwartz and Henderson, 1991); can persist in extremely disturbed situations, in broadleaf forest or remnants, banana groves, coffee plantations, shaded, shrubby situations, on leaf blades of bracken ferns among scattered pines in a river floodplain, 0.3–1.2 m high on small branches and vertical stems, less frequently on rocks in an overgrown limestone boulder pile (Williams, 1963a; Schwartz, 1978b; Franz and Cordier, 1986; Schwartz and Henderson, 1991); "snugly snake" along midribs of bracken ferns (Franz, 2003); sleeps on small branches or vines (in banana grove, coffee plantation, and other cultivated but shrubby or arborescent situations), on ferns and herbaceous vegetation in a road cut (Schwartz, 1978b).

Behavior: Escapes by dropping to the ground and hopping or jumping from danger (Schwartz, 1978b). **Ecomorphology:** "Grass-bush" (Williams, 1963a). **Size:** Male SVL to 52 mm, female SVL to 43 mm (Schwartz, 1978b).

Conservation Status: Not assessed.

Anolis equestris Merrem, 1820

Distribution: Cuba, where 11 currently recognized subspecies occur throughout most of the island (absent in the extreme east and west, local along the southern coast), Archipiélago de Sabana-Camagüey, and cays north of Matanzas; introduced into southeastern Florida and the upper Florida Keys and Hawaii (Oahu); elevational distribution to nearly 1,000 m.

Habitat: Generally mesophilic, at times somewhat xerophilic, often edificarian (Schwartz and Henderson, 1991); associated with large trees (including mangoes and palms) in or near broadleaf forest or woods, can be abundant in mangroves, at times in open areas (fields, savannas, agricultural areas) and areas sometimes substantially altered by human activity (including fills along coastlines), occasionally in sinkhole caves, often perches head-down on trunks rarely less than 3 m high (Gundlach, 1880; Schwartz and Ogren, 1956; Collette, 1961; Ruibal, 1964; Buide, 1966; Garrido, 1975a; Schwartz and Henderson, 1991; Estrada, 1993b; Rodríguez Schettino, 1999b; Garrido et al., 2001); "in orchards of mangos and other dense fruit trees" (Barbour, 1930c); in submontane rainforest, submontane mesophyllous evergreen forest, mesophyllous evergreen and semideciduous forest, pine forest, mangrove forest, coastal and subcoastal microphyllous forest, cultivated areas and pastures, and urban zones (Rodríguez Schettino et al., 1999); on Cayo Cinco Leguas (*A. e. cincoleguas*), occupies all arboreal habitats available, including mangroves and coastal scrub (Garrido, 1981); on large trees and especially palms (Breustedt, 1991); often associated with yards and gardens (Kunz, 2005); adults favor large trees, but juveniles often in bushes (Fläschendräger and Wijffels, 1996, 2009; Anonymous, 2006b); mean perch height and diameter 4.03 m and 15.3 cm (J. B. Losos, in litt., 21.IV.2008); one individual sleeping in a crevice near a cave entrance (Schwartz and Ogren, 1956).

Activity: Diurnal, basks in tree crowns during the morning, descends around noon, possibly to avoid intense sun during the heat of the day (Z. Vogel, 1965a; Rodríguez Schettino et al., 1999); Cubans sometimes speak of these lizards as descending to "kiss the ground" at midday (J. B. Losos, in litt., 21.IV. 2008). **Behavior:** Aggressive, wary, capable of inflicting a painful bite (Barbour and Ramsden, 1919; Ruibal, 1964; Buide, 1985); when threatened, turns laterally to attacker (Red-legged Thrush, *Turdus plumbeus*), opens mouth, hisses, while slowly maneuvering for position, "quick to freeze" when approached and hard to see among leaves, will often slowly "squirrel" around a trunk, keeping just out of sight (Ruibal, 1964); both sexes "display," always in green color phase, aggressive displays associated with general cadence, number of headbobs, location of large and small headbobs, usually ending in lateral head movements, possibly arising from ritualized mouth wiping, social and defensive displays can be distinguished by the long duration of dewlap extension and gaping in defensive behavior (Font and Kramer, 1989). **Ecomorphology:** "Crown-giant" (Rand and Williams, 1969); formally quantified (Beuttell and Losos, 1999); categorized as a Cuban "giant anole" (G. Peters, 1970). **Foraging and Diet:** An opportunistic omnivore (Rodríguez Schettino, 1999b); stalks prey from distance of several meters and rushes 1 m (can be enticed to descend to the foot of a tree by crickets at a distance of 5 m), food includes fruit, tree frogs (*Osteopilus septentrionalis*), and birds (Gundlach, 1880; Barbour and Ramsden, 1919; Ruibal, 1964); *Anolis porcatus* (Alayo Dalmau, 1951; O. H. Garrido, pers. comm.); other anoles (Collette, 1961; Z. Vogel, 1965a); in captivity, otherwise lethargic individuals quickly react to and grab smaller anoles (Breustedt, 1991); Cuban Emerald Hummingbird (*Chlorostilbon ricordii*; Ševčik, 1990); during feeding, the mechanical resistance of food items influences the number of cyclic movements (mechanical reduction), decrease of which when fruit is eaten may be caused by juice production, swallowing affected by the efficiency of the hyoid-lingual complex (Bels and Baltus, 1988). **Home Range and Territoriality:** Males establish territories in the canopy, fights are frequent when males meet on the same tree, aggressive displays involve dewlap extension, open mouths, lateral orientation, headbobs, push-ups (sometimes on all 4 limbs), and tail elevation (Rodríguez Schettino, 1999b). **Longevity:** In captivity, to 10–15 years (Breustedt, 1991); 16 years, 6 months and still alive at last report in 1992 (H. Cohen in Slavens and Slavens, 2003). **Parasites:** Trematode *Neodiplostomum centuri* (Macko and Rysavy, 1968; Coy Otero and Lorenzo Hernández, 1982); cestode *Sparganum* sp. (Baruš and Coy Otero, 1969a; Coy Otero, 1970; Coy Otero and Lorenzo Hernández, 1982; Coy Otero, 1999); nematodes *Physaloptera squamatae, Oswaldocruzia lenteixeirai, Cyrtosomum longicaudatum* (Baruš and Coy Otero, 1968, 1969a; Coy Otero, 1970; Coy Otero and Baruš, 1979b; Coy Otero and Lorenzo Hernández, 1982; Coy Otero, 1999); *C. scelopori* (Baruš and Coy Otero, 1969a; Coy Otero, 1970; Coy Otero and Baruš, 1973a, 1979b; Coy Otero and Lorenzo Hernández, 1982; Coy Otero, 1999); *O. brevicaudata, Piratuba digiticauda* (Baruš and Coy Otero, 1968; Coy Otero and Baruš, 1979b; Coy Otero and Lorenzo Hernández, 1982; Coy Otero, 1999); sporozoan red blood cell parasite *Haemogregarina* sp. (Zajicek and Mauri Méndez, 1969). **Performance:** Clinging ability 4.78 ± 0.91 Neutons, divided by mass 0.09, with the low value characteristic of crown-giant ecomorphs (Elstrott and Irschick, 2004); acceleration 13.84 ± 1.60 m/sec, sprint speed 1.29 ± 0.10 m/sec (Vanhooydonck et al., 2006a). **Population Size and Density:** On Cayo Santa María (*A. e. potior*), 0.38 individuals/ha for a total population estimate of 165 lizards, on Cayo Francés, total population estimated at ~80 individuals, on other small cayos, population estimates of 18, 12, 5, and 3 are likely too small for long-term viability, with small populations reflecting small island size and limited habitat (Fernández Méndez and Manso Valdés, 1998). **Predation:** Most important predator on *A. e. potior* is the Cuban Lizard Cuckoo (*Saurothera merlini santamaria*; Fernández Méndez and Manso Valdés, 1998); American Kestrel (*Falco sparvarius*; Ševčik, 1990; Rodríguez Schettino, 1999b); Red-legged Thrush (*Turdus plumbeus*) took a juvenile lizard (Rodríguez Schettino, 1999b); Lizard Cuckoo (*Saurothera merlini*) suffocated trying to swallow a subadult lizard (M. S. Buide in Rodríguez Schettino, 1999b). **Reproduction:** Egg (*A. e. potior*) 23.5 × 13.3 mm, cream colored, and elliptical (Garrido, 1980b); females descend to lay eggs (23 × 12.8 mm), repeating the process every 48–72 hours in July–August, the hottest months, hatchlings rapidly double their size, so by the next winter months they have accumulated sufficient energy reserves to pass the coldest months, during which lizards feed infrequently (Fernández Méndez and Manso Valdés, 1998); courtship and mating high in trees, oviposition on the ground in holes near the bases of trees, hatching presumably toward the end of the wet season (July–August), when juveniles more abundant (Rodríguez Schettino, 1999b); in captivity, one female laid 12 eggs in one year, incubation 60–92 days at 18–30 °C (Bech, 1986); eggs 18 × 25 mm, hatchling total length 8–10 cm (Breustedt, 1991); hatchlings 14 cm total length (Anonymous, 2006b). **Size:** Male SVL to 188 mm, female SVL to 170 mm (Schwartz and Garrido, 1972). **Thermal Biology:** Heliothermic (Rodríguez Schettino, 1999b).

Conservation Status: Entire species not assessed, but *A. e. potior* and *A. e. brujensis* (Cayo Las Brujas) are considered "critically endangered" (Rodríguez Schettino and Chamizo Lara, 1998a); *A. e. potior* considered "critically endangered," *A. e. juraguensis* "endangered," and *A. e. buidei*, *A. e. verrepmemsos*, and *A. e. cincoleguas* "vulnerable" (L. Rodríguez Schettino and A. Chamizo Lara in Rodríguez Schettino, 1999a); assessment using IUCN Red List criteria (Fernández Méndez, 1998b; Fernández Méndez et al., 1998); development of a hotel on Cayo Santa María (*A. e. potior*) eliminated the habitat for 65 lizards, more than a third of estimated population, elimination of habitat on Cayo Brujas (*A. e. brujensis*) drastically reduced the numbers of lizards (Fernández Méndez and Manso Valdés, 1998).

Anolis ernestwilliamsi Lazell, 1983

Distribution: British Virgin Islands, Peter Island and Carrot Rock (island area 1.2 ha), possibly extinct on Peter Island (Perry, 2005).

Habitat: Xerophilic (Schwartz and Henderson, 1991); saxicolous, primarily on rocks under the canopy (*Coccoloba*) and on trunks of trees in the shady, dimly lit zone, somewhat protected from the full force of trade winds (Lazell, 1983).

Desiccation: Mean water-loss rate (1 ± 0.4 mg/g/h) and cutaneous resistance to water loss (168 ± 71.3 sec/cm) comparable to that of *A. cristatellus* in similarly arid habitat (Dmi'el et al., 1997); during a year of severe drought, mean water loss rate (2 ± 0.5 mg/g/h) slightly higher, but no change in mean cutaneous resistance to water loss (167 ± 56 sec/cm), which increased dramatically in *A. cristatellus* from more mesic islands (Perry et al., 1999). **Ecomorphology:** "Trunk-ground" (Losos, 2009), but not tested morphometrically (Brandley and de Queiroz, 2004). **Movements:** Mean moves per minute 0.60 ± 0.69, percent time moving 1.21 ± 1.48 (Perry, 1999). **Population Size and Density:** 2,000–3,000 probably present at a density of 2–4 individuals/100 m^2 (Lazell, 1983). **Sex Ratio (M:F):** Balanced (Lazell, 1983). **Size:** Male SVL to 82 mm, female SVL to 60 mm (Lazell, 1983).

Conservation Status: "Common but restricted, protect habitat" (Moreno, 1991d); the extremely restricted distribution and resultant vulnerability to stochastic events suggest that the species should be considered "critically endangered" according to IUCN Red List standards (IUCN, 2001); Carrot Rock should be set aside as a sanctuary or preserve (Lazell, 1983; Moreno, 1991b).

Anolis etheridgei Williams, 1962

Distribution: Hispaniola, Cordillera Central (Dominican Republic); elevational range 550–2,440 m (possibly higher).

Habitat: Mesophilic (Schwartz and Henderson, 1991); in rainforest, gallery forest, on bushes or small trees adjacent to paths, very rarely on walls and thatched roofs of rural houses, on slender perches (branches, twigs, fern stems) close to the ground (Thomas and Schwartz, 1967; Rand and Williams, 1969); a male in shade 1 m above the ground at 23 °C at 1300 h (Fläschendräger and Wijffels, 2009); sleeps with hindlimbs flexed within 3 m of the ground on tree ferns, shrubs, herbs, and ferns, draped over leaves and not aligned along branches or twigs, not on vines, long dangling tail reveals presence (Thomas and Schwartz, 1967).

Behavior: Movement characterized as "jumper" (Moermond, 1981). **Ecomorphology:** "Grass-bush" (Rand and Williams, 1969), but does not have the long limbs normally associated with that ecomorph; formally quantified (Beuttell and Losos, 1999); undetermined (Losos, 2009). **Foraging and Diet:** Prey strikes either jump, approach-pause, or stationary (Moermond, 1981). **Parasites:** Nematodes Cosmocerdidae gen. sp., *Porrocaecum* sp., larval Acuariidae gen. sp. and *Physocephalus* sp., acanthocephalan cystacanths *Centrorhynchus* sp. (Goldberg et al., 1998b). **Population Size and Density:** 35 sleeping individuals during 2 hours of collecting (Rand and Williams, 1969). **Reproduction:** Clutch size 1 (Hedges, in prep.). **Size:** Male and female SVL to 43 mm (Schwartz and Henderson, 1991). **Thermal Biology:** No evidence of basking (Rand and Williams, 1969).

Conservation Status: Not assessed.

Anolis eugenegrahami Schwartz, 1978

Distribution: Hispaniola, only from the vicinity of the type locality at Roche Parfait, 9 km NE Plaisance, Département du Nord, Haiti, at 215 m.

Habitat: Mesophilic (Schwartz and Henderson, 1991); aquatic, in boulder-strewn mountain stream flowing through gallery forest, never outside stream margins, sleeps exposed (no effort at concealment) on rock faces or in shallow depressions in large boulders, often on broken boulder edges, occasionally on hanging vegetation 15 mm above water and 30 mm from the bank and on a shrub growing out of a boulder (Schwartz, 1978c).

Behavior: Rests on rocks and fallen logs, takes refuge under wet boulders and behind a waterfall, does not hesitate to enter water at night to escape (Schwartz, 1978c). **Ecomorphology:** "Aquatic" (Williams, 1983), which is not an ecomorphological category; fits morphologically into the "trunk-crown" category (Leal et al., 2002); undetermined (Losos, 2009). **Parasites:** Nematodes *Atractis scelopori*, *Oswaldocruzia lenteixeirai*, *Parapharyngodon cubensis*, and larval *Ascarops* sp., acanthocephalan cystacanths *Centrorhynchus* sp. (Goldberg et al., 1998b). **Reproduction:** Clutch size 1 (Hedges, in prep.). **Size:** Male SVL to 72 mm, female SVL to 61 mm (Schwartz, 1978c).

Conservation Status: Unknown, but association with stream in threatened Haitian gallery forests suggests that some formal protected status is warranted.

Additional Pertinent Literature: Powell and Neland (1998).

Anolis evermanni Stejnegeri, 1904

Distribution: Puerto Rico, uplands; elevational range 20–1,205 m.

Habitat: Mesophilic (Schwartz and Henderson, 1991); closely associated with the "coffee belt" (Stejneger, 1904); "more often on larger trees, and at a considerable height from the ground" (K. P. Schmidt, 1928); at 3 sites at El Yunque, juveniles 1–3 m high on perches with diameters <1 cm, females mainly 1–3+ m high on perches of small to large (2–16+ cm) diameter, at the highest site (850 m), 3 males at 2 m and >3 m in palm canopies, at another site with fewer large trees, males low and at heights >3 m on perches of moderate to large (8–16+ cm) diameters, at a lower site (250 m), males much like those at the highest site, 83% in the sun at high elevation, 66–93% at intermediate elevations (García Moll, 1978); in deep, wet forests, males perch higher and on thicker perches than any other anole, favors thick stems of bamboo and upper green portions of palms, also along roadsides, on fence posts adjacent to a patch of woods, on shrubs and low trees in a coffee plantation, on streamside boulders (Losos, 1990b; Schwartz and Henderson, 1991; Rivero, 1998); also common on, in, and around buildings in heavily forested areas (Fleishman et al., 1997); mean perch height and diameter 3.4 ± 0.87 m and 35.3 ± 5.69 cm (Losos, 1990a); usually on large-diameter perches from 0.9–3.0 m (Rand, 1964a); regularly uses the canopy at heights >8 m at El Verde, although most (21%) on perches 1 m high and usually (42%) on perches with diameters 10 cm (Reagan, 1992, 1996a); at 2 upland forest sites, mean perch height and diameters 131.1 ± 11.58 cm and 10.7 ± 1.88 cm for large (SVL >50 mm) males, 134.1 ± 21.95 cm and 8.9 ± 2.08 cm for small (SVL = 50 mm) males, 106.7 ± 11.58 cm and 9.9 ± 1.83 cm for females, 118.9 ± 13.41 cm and 6.9 ± 1.12 cm for "undetermined" individuals (Turner and Gist, 1970); at an upland site along a secondary road edge, largest percentage of adult males (22%) on perches >12.5 cm in diameter at heights of 3.0–6.0 m, plurality of "small" individuals (46%) at heights <0.9 m on perches with diameters >12.5 cm, in open forest, largest percentage of adult males (19%) on perches 3.0–6.0 m high with diameters of 6.4–12.5 cm with 10 of 21 individuals on the ground or rocks, plurality of "small" individuals (16%) <0.9 m high on perches with diameters >12.5 cm and 2 of 49 lizards on the ground, in forest with cleared understory, largest percentage of adult males (33%) on perches 3.0–6.0 m high and with diameters 6.4–12.5 cm and no individuals on the ground, plurality of "small" individuals on elevated perches (36%) on perches with diameters >12.5 cm at heights <0.9 m and 7 of 14 lizards on the ground, in forest interior, largest percentage of adult males (15%) on perches 3.0–6.0 m high with diameters >12.5 cm and no lizards on the ground, plurality of "small" individuals" (17%) on perches 1.5–3.0 m high with diameters 2.2–5.7 cm, with 3 of 36 lizards on the ground, at El Verde, largest percentage of adult males (38%) on perches 3.0–6.0 m high usually (45%) with diameters >12.5 cm and 3 of 31 lizards on the ground, plurality of "small" individuals (32%) on perches 3.0–6.0 m high with varying diameters and 6 of 69 lizards on the ground and 3 on rocks (Schoener and Schoener, 1971b); at El Verde in July, mean perch heights, diameters, and insolation for adult males and females 198.6 ± 111.0 cm, 17.6 ± 11.6 cm, 714 ± 341 lux and 184.5 ± 104.3 cm, 17.3 ± 9.8 cm, 631 ± 408 lux, in January, 290.7 ± 189.7 cm, 17.1 ± 8.4 cm, 527 ± 406 lux and 229.1 ± 129.6 cm, 18.9 ± 9.8 cm, 568 ± 413 lux, with both males and females using a less diverse set of perch heights and diameters during the dry winter season (Lister, 1981); most frequently on very small (<2 cm diameter) or very large (>8 cm) perches, with intermediate habitat breadth of 60.3 (Irschick and Losos, 1999); mean perch heights and diameters 2.82 ± 4.28 m and 22.8 ± 45.5 cm for males and 1.99 ± 2.39 m and 14.9 ± 19.4 cm for females (Butler and Losos, 2002); mean male perch height and diameter 1.17 ± 0.12 m and 1.10 ± 0.29 m, mean female perch height and diameter 1.31 ± 0.23 m and 0.23 ± 0.05 m (VanHooydonck et al., 2007); mean perch height and diameter 3.35 m and 35.3 cm, 27.0% on leaves (J. B. Losos, in litt., 21.IV.2008); habitat light environment "full shade" (Fleishman, 2000).

Behavior: Male threat display includes push-ups on all 4 limbs, eversion of tongue, rapid extension and folding of dewlap (Ruibal, 1967); mean approach distance ~100 cm (Heatwole, 1968a); mean display rate (proportion of time spent displaying) 0.056 ± 0.010, percent walks 53.6 ± 4.3, percent runs 31.9 ± 4.6, percent jumps 14.6 ± 2.0, moves/min 1.27 ± 0.12, distance jumped in nature 39.3 ± 3.9 cm (Losos, 1990a); frequency of displays related to frequency of "non-walk" movements (e.g., running, jumping), displays more frequently when on trees than when on boulders, jumps proportionately more frequently when on boulders, rates of jumping and walking about the same on boulders and trees, but runs more frequently on trees, in general, moves more frequently when on trees (Losos, 1990b); head-down posture in 14 of 19 individuals, moves/min 0.40 ± 0.12, percent time moving (PTM) 0.81 ± 0.25%, PTM increases (not significantly) with increase in perch height (Cooper, 2005); approach distance decreases as perch height increases (all at perch heights of ≤ 2 m fled, none at perch heights >2 m did), flee up, "squirrel," or both (Cooper, 2006a); both absolute and size-corrected dewlap size significantly related with bite force, 83% of combat trials involved dewlap extensions, 42% biting, and 67% head-bobbing (Lailvaux and Irschick, 2007); male approach distance, flight distance, and final distance 1.25 ± 0.19 m, 0.44 ± 0.08 m, and 1.48 ± 0.20 m, female approach distance, flight distance, and final distance 1.11 ± 0.19 m, 0.30 ± 0.08 m, and 1.25 ± 0.18 m (VanHooydonck et al., 2007); moves per minute 1.27 (Johnson et al., 2008). **Biomass:** At El Verde, 2.944 kg/ha (Lister, 1981), 6.2 kg/ha (Reagan, 1996a), with the latter based on population size estimates that included the canopy. **Competition:** In experimental plots, the abundance of adult and juvenile *A. evermanni* increased in the absence of *A. gundlachi*, but no shift in structural habitat ("ecological release") was evident, could be the result of interspecific competition or intraguild predation (Leal et al., 1998). **Desiccation:** Mean critical activity point (percent of original body weight at which the lizard could not right itself when turned on its back) 74.8 ± 2.04%, mean rate of water loss 0.32% body mass/h (Hillman and Gorman, 1977). **Ecomorphology:** "Trunk-crown" (Williams, 1972); formally quantified (Beuttell and Losos, 1999). **Foraging and Diet:** Attack launched by an immobile lizard that detected prey while scanning visually (Cooper, 2005); 20 stomachs contained beetles, wasps, ants, caterpillars, spiders, anole skin ("doubtless its own"), and a juvenile conspecific (K. P. Schmidt, 1920, 1928); beetles (most important, adults and larvae, 41.4%), hemipterans (37.6%), wasps (<1%), ants (5.4%), lepidopterans (adults and larvae, 11%), flies (<1%), winged termite, springtails, and psocid (<1%), and spiders and a snail (1%), eats cast skin, cannibalism of juveniles (Wolcott, 1923); at El Verde in July, mean prey length and number of prey items/stomach were 2.7 mm and 25.8 items/stomach, in January, 2.6 mm and 27.1 items/stomach, food items were ants, other hymenopterans, hemipterans, flies, beetles (adults, greatest percent biomass in both wet, July, and dry, January, seasons, and larvae), lepidopterans (adults and larvae), orthopterans, spiders, annelids, and seeds/fruits (Lister, 1981); eats *A. stratulus* (Reagan, 1992); at El Verde, estimated 34.6 prey items/lizard led to an estimated daily consumption rate of 5.19 × 104 prey items/ha, feeds on small *A. gundlachi* and *A. stratulus* (Reagan, 1996a); at 2 sites at El Verde (one a streamside situation where lizards fed on stream-associated insects), stomachs contained hymenopterans (including ants, which were most important in numbers in males and females at both sites), flies, lepidopterans, beetles, leafhoppers (most important by volume in males at the forest site), other hemipterans (most important by volume in females at the forest site), orthopterans, roaches, spiders, shed skin, rocks and plant matter (most important by volume in males at the stream site), and others (R. Garrison and D. P. Reagan in Reagan, 1996a); also at El Verde, *A. gundlachi*, *A. stratulus*, tarantulas (*Avicularia* sp.), and crab spiders (*Stasina*; Reagan et al., 1996). **Home Range:** Individuals included 1–3 large branches in each home range (Dial et al., 1994). **Life Expectancy:** 0.6 year (Lister, 1981). **Movements:** Movements between successive recaptures at 2 sites in upland forest 1.5–39.6 m for large (SVL >50 mm) males, 2.1–7.7 m for small (SVL = 50 mm) males, 1.5–31.8 m for females, 3.3–6.7 m for "undetermined" individuals (Turner and Gist, 1970); mean distance per movement 18.5 ± 1.6 cm at mean speed 35.0 ± 2.4 cm/sec, 0.59 ± 0.15 moves/min, 0.69 ± 0.26% time moving (Irschick, 2000b). **Parasites:** Trematode *Mesocoelium monas* (Cofresí-Sala, 1964); nematodes *Spauligodon anolis* and *Pharyngodon travassosi* (Acholonu, 1976); protozoan malarial parasite *Plasmodium azurophilum* (Schall and Vogt, 1993). **Performance:** Mean maximum sprint speed (fastest time over an interval of 0.25 m) 0.137 ± 0.003 sec, jump 67.2 ± 1.70 cm, cling 3.148 ± 0.223 Neutons (Losos, 1990a); clinging ability 1.97 ± 0.18 Neutons, divided by mass 0.32, with the relatively high value apparently characteristic of trunk-crown ecomorphs (Elstrott and Irschick, 2004); mean maximal speed 1.2 ± 0.03 m/sec, escape speed 1.1 ± 0.08 m/sec, feeding speed 1.1 ± 0.10 m/sec, undisturbed activity 0.7 ± 0.11 m/sec, mean maximal jumping distance 67.0 ± 1.70 cm, feeding jumping distance 35.0 ± 4.79 cm, undisturbed jumping distance 42.1 ± 4.21 cm (Irschick and Losos, 1998); peak velocity 1.92 ± 0.08 m/sec, peak acceleration 44.9 ± 0.6 m/sec^2 (Toro et al., 2004); acceleration 29.96 ± 2.12 m/sec, sprint speed 1.13 ± 0.05 m/sec (Vanhooydonck et al., 2006a). **Population Size and Density:** At El Verde in July, numbers and biomass for adult males and females were ~973/ha (~3,383 g/ha) and ~1,015/ha (~2,368 g/ha), in January, for adult males, adult females, and juveniles ~338/ha (~1,522 g/ha), ~423/ha (~803 g/ha), and ~761/ha (~550 g/ha; Lister, 1981); also at El Verde, 1.0–17.3 individuals per Tabonuco (*Dacryodes excelsa*) crown (mean 7.2 ± 4.1 individuals/crown) = 0.14–0.90 individuals/m^2 (mean 0.30 ± 0.26 individuals/m^2 = 2,308 individuals/ha; Dial et al., 1994); based on relative abundance 1,500 individuals/ha (Reagan, 1996a). **Population Structure:** Adult:juvenile ratio 67:33% at El Verde (Turner and Gist, 1970). **Predation:** *Eleutherodactylus coqui* (Leal and Thomas, 1992); *Borikenophis portoricensis* (Rodríguez-Robles and Leal, 1993a); Red-tailed Hawks (*Buteo jamaicensis*; Santana and Temple, 1988); *Epicrates inornatus* (Wiley, 2003); at El Verde, *A. stratulus*, tarantulas (*Avicularia* sp.), and presumably whip scorpions (*Phrynus longipes*), forming "food loops" (Reagan, 1996b; Reagan et al., 1996); also, Lizard Cuckoos (*Saurothera vieilloti*), Pearly-eyed Thrashers (*Margarops fuscatus*), Red-tailed Hawks (*Buteo jamaicensis*), and *Borikenophis portoricensis* regularly feed on anoles (Reagan and Waide, 1996). **Relative Abundance:** 3–16% of wet season and 25–36% of dry season sightings of this species, *A. gundlachi*, and *A. stratulus* at El Verde (Reagan, 1992, 1996a). **Reproduction:** Maximal male reproductive activity March–August, testicular weights decrease during winter months, size of abdominal fat bodies inversely correlated with testis weight (Licht and Gorman, 1975); 35 of 36 females in a wet-season sample and 38 of 46 in a dry-season sample from relatively aseasonal mesic habitat at El Verde were reproductively active, 24 of 25 females in a wet-season sample but only 3 of 40 in a dry-season sample from strongly seasonal xeric habitat in southwestern Puerto Rico were reproductively active (Lister, 1981); clutch size 1 (Hedges, in prep.); egg in February white, ovoid, 11.2 × 6.0 mm, hatching 42 days after discovery, hatchling SVL 20.0 mm (Estrada, 2002). **Size:** Male SVL to 70 mm (Schwartz and Henderson, 1991), female SVL to 56 mm (Schall and Vogt, 1993). **Thermal Biology:** At low daytime temperatures (correlated with cloudiness or deep shade at higher elevations), body temperatures suggest that incoming radiation is insufficient to warm the body above ambient, individuals exposed to cold temperatures (11.5 ± 0.5 °C) were capable of locomotor movements, but did not attempt to feed, individuals at an elevation ~580 m survived at a rate intermediate between high-elevation *A. krugi* (with higher rate) and low-elevation *A. cristatellus* and *A. stratulus*, with smaller individuals within the same species succumbing first, heat stress monitored under rising temperatures was noted before onset in *A. cristatellus* and *A. stratulus* and comparable to high-elevation (~550–640 m) *A. gundlachi* (Heatwole et al., 1969); mean minimum and maximum body temperatures at elevation of 480 m on Puerto Rico 23.2 ± 0.2 °C and 25.7 ± 0.2 °C at maximum and minimum air temperatures of 22.7 ± 0.1 °C and 24.5 ± 0.1 °C, at 675 m 22.2 ± 0.1 °C and 25.1 ± 0.2 °C at air temperatures of 21.9 ± 0.1 °C and 24.2 ± 0.2 °C (Hertz, 1983). **Vision:** Neither dewlap nor rest of body highly UV reflective, which would be most effective in a UV-rich environment, but UV-bright spot is associated with an open mouth, part of the typical anoline threat display (Fleishman et al., 1993).

Conservation Status: "Stable" (Moreno, 1991d).

Anolis extremus Garman, 1887

Distribution: Barbados, essentially islandwide; introduced on St. Lucia, Bermuda, Trinidad, and in Caracas, Venezuela.

Habitat: Mesophilic (Schwartz and Henderson, 1991); in open situations, in yards, and by the creek in Bridgetown (Underwood, 1959b); "utterly ubiquitous and exceedingly abundant all over Barbados" (Lazell, 1972); on St. Lucia, exhibits enclave (or pocket) distribution (on *Cocos* trees, fence posts, hedge shrubbery; Gorman, 1976); on St. Lucia, essentially a human commensal, being particularly abundant in Castries and vicinity (Heselhaus and Schmidt, 1990); also on St. Lucia, mean perch height and diameter 1.60 ± 0.29 m and 9.0 ± 1.5 cm (Losos and de Queiroz, 1997a).

Activity: On St. Lucia, most active at 1000–1200 h and 1400–1600 h (J. Schmidt, 1979). **Behavior:** On St. Lucia, very shy and difficult to capture (J. Schmidt, 1979). **Competition:** On St. Lucia, interspecifically territorial with *A. luciae* and appears to be displacing the native species in some habitat pockets (Gorman, 1976). **Diet:** Predominantly insects, but also fruits and nectar (Fläschendräger and Wijffels, 2009). **Ecomorphology:** Closest to "trunk-crown" (Losos and de Queiroz, 1997a; Beuttell and Losos, 1999). **Home Range and Territoriality:** On St. Lucia, never more than one adult male per tree (J. Schmidt, 1979). **Hybridization:** No evidence of hybridization with *A. luciae* on St. Lucia (Giannasi et al., 1997). **Parasites:** Nematodes *Parapharyngodon cubensis*, *Trichospirura teixeirai*, and larval *Ascarops* sp. (Goldberg et al., 1997a). **Sex Ratio (M:F):** Apparently balanced (Lazell, 1972). **Size:** Male SVL to 83 mm, female

SVL to 60 mm smallest mature male and female SVL 46 mm and 42 mm, respectively (Lazell, 1972).

Conservation Status: Not assessed; however, the report of the species' near extinction on Barbados (Barbour, 1930c) was premature (Lazell, 1972).

Anolis fairchildi Barbour and Shreve, 1935

Distribution: Bahama Islands, Cay Sal Bank.

Habitat: Xerophilic, on palms in the interior (Schwartz and Henderson, 1991).

Ecomorphology: "Trunk-crown" (Losos, 2009). **Size:** Male SVL to 76 mm, female SVL unknown (Buden and Schwartz, 1969).

Conservation Status: Unknown.

Anolis ferreus Cope, 1864

Distribution: Marie-Galante, islandwide; elevational range, sea level to 204 m.

Habitat: Mesophilic to xerophilic (Schwartz and Henderson, 1991); "arid lowlands typical of islands too low in elevation to stop significant rainfall" (Lazell, 1972); on "tall mango trees—which are about the only tall trees there are, since this island is so completely cleared for sugar cane" (Barbour, 1930c); concentrated in "islands" of tall trees, but especially juveniles associated with hedges and lower ornamental plantings (Heselhaus and Schmidt, 1990); largely restricted to remaining stands of forest, mango stands, and coastal vegetation characterized by *Coccoloba uvifera* and *Hippomane mancinella* (Fläschendräger and Wijffels, 1996, 2009); in a coastal *Hippomane* grove, on large trees and palms about rural homes, in mesic forest at the mouth of a cave, on a tree at the edge of a sinkhole (Schwartz and Henderson, 1991); mean perch height and diameter 2.40 ± 0.27 m and 13.0 ± 3.1 cm (Losos and de Queiroz, 1997a); dry forest, very abundant behind beaches in *Hippomane*, *Cocos*, flamboyant, and mango, also in trees along edges and near bodies of water, along mangrove borders, less common in moist forest, on limestone cliffs, also in edificarian situations, especially on deserted buildings and ruins (Breuil, 2002b).

Activity: Active primarily during the middle of the day, retreats to high perches toward late afternoon (Malhotra and Thorpe, 1999). **Diet:** Lizards with larger jaws exploit larger prey items (Roughgarden, 1974). **Ecomorphology:** Closer to "trunk-crown" (differs primarily in size) than any other category or to the putative "generalist" (Losos and de Queiroz, 1997a); less close to "trunk-crown" than many Lesser Antillean anoles (Beuttell and Losos, 1999). **Parasites:** Nematodes *Parapharyngodon cubensis* (Dobson et al., 1992) and *Spauligodon anolis* (Dobson et al., 1992; Bursey and Goldberg, 1998); mean parasite species/host 0.94 (Dobson and Pacala, 1992). **Reproduction:** In captivity, females bury eggs in moist substrates (Fläschendräger and Wijffels, 1996, 2009); reproductive activity peaks July–September, males hold females by the head during copulation (Breuil, 2002b). **Size:** Male SVL to 120 mm (Fläschendräger and Wijffels, 1996, 2009), female SVL to 66 mm (Dobson et al., 1992), smallest mature male and female SVL 63 mm and 50 mm, respectively (Lazell, 1972).

Conservation Status: Not assessed, but protected by French legislation (Lorvelec et al., 2007).

Anolis fowleri Schwartz, 1973

Distribution: Hispaniola, Cordillera Central (Dominican Republic); elevational range 1,585–1,770 m.

Habitat: Mesophilic (Schwartz and Henderson, 1991); in deciduous forest, possibly in the canopy, although none observed except along roads, sleeps on tree branches 1.4–3.1 m above the ground, on vertical stems of shrubs, vine tangles, and (juveniles) draped across leaves of a shrub and on a shrub 60 cm high in a thicket (Schwartz, 1973c); sleeps on ferns near the ground and on a fern frond ~2 m high (Glor, 2003).

Ecomorphology: "Twig" (Williams, 1983); formally quantified and determined to be morphologically distant from all ecomorph categories (Beuttell and Losos, 1999); undetermined (Losos, 2009). **Size:** Male SVL to 77 mm, female SVL to 75 mm (Schwartz, 1973c).

Conservation Status: Unknown.

Anolis fugitivus Garrido, 1975

Distribution: Eastern Cuba, in the vicinity of the type locality 2 km S Aserio de Nuevo Mundo, Moa, Guantánamo Province and Nibujón, probably also in the Cuchillas de Toa.

Habitat: Mesophilic (Schwartz and Henderson, 1991); inhabits weedy and grassy areas, especially near rivers and creeks, associated with grasses of the genus *Arthostylidium*, especially in cleared areas that might provide avenues for expansion into similar habitats (Garrido, 1975c); in dense mats of saw grasses (Garrido, 2003); in submontane rainforest (Rodríguez Schettino et al., 1999).

Activity: Diurnal (Rodríguez Schettino et al., 1999); most individuals active during early hours of the day and in the afternoon, but may be active throughout the day in cloudy or rainy conditions (Rodríguez Schettino, 1999b). **Behavior:** Well-camouflaged, difficult to capture except when cloudy or drizzling (Garrido, 1975c); disappears into tangled mats of grass (Garrido, 2003). **Ecomorphology:** "Grass-bush" (Losos, 2009). **Population Size and Density:** Abundant in the vicinity of the type locality (Garrido, 1975c). **Reproduction:** Apparently a short reproductive cycle, females in January, February, and June were not gravid (Rodríguez Schettino, 1999b). **Size:** Male SVL to 36.2 mm, female SVL to 33.3 mm (Rodríguez Schettino, 1999b). **Thermal Biology:** Apparently nonheliothermic (Rodríguez Schettino, 1999b).

Conservation Status: Considered "critically endangered" (Rodríguez Schettino and Chamizo Lara, 1998a; L. Rodríguez Schettino and A. Chamizo Lara in Rodríguez Schettino, 1999a).

Anolis garmani Stejneger, 1899

Distribution: Jamaica, essentially islandwide; elevational range, sea level to >1,200 m; introduced into southeastern Florida and on Grand Cayman Island; however, breeding on Grand Cayman has not been confirmed, and these lizards are not expanding their range, if they remain present at all (F. J. Burton, pers. comm.).

Habitat: Mesophilic (Schwartz and Henderson, 1991); strongly arboreal (Fläschendräger and Wijffels, 1996, 2009); may move considerable distances by moving from tree to tree, jumping short distances if crowns are not contiguous, not edificarian, but often on verandas as well as "forest foliage" (Panton, 1928); at lower elevations associated only with large trees or bamboo clumps, fairly abundant in montane forest at moderate elevation, will descend to the ground to feed (Grant, 1940a; Underwood and Williams, 1959; Stettler, 1978); most common in bushy or forested areas around Kingston, occasionally in gardens and on vine tangles and once on a telephone pole, never on isolated trees in open pastures, not in edificarian situations or on stones or walls, the largest percentage of individuals (44.4%) on perches >3 m high and most (66.6%) on large-diameter (>7.6 cm) perches (Rand, 1967b); high in large trees with abundant epiphytes, generally descends to the ground only to deposit eggs (Krintler, 1985); on large trees at heights around 7 m (Hilgenhof, 1998a); remains common in relict woods and sometimes occurs along roads (Williams, 1999); mean perch height and diameter 3.5 ± 0.35 m and 33.2 ± 4.59 cm (Losos, 1990a); most (71.56%) individuals on tree trunks, 11.63% on ground, none at heights <30 cm, 7.00% at 30–59 cm, 12.79% 60–89 cm, 11.63% 90–119 cm, 18.60% 120–149 cm, 12.79% 150–179 cm, 12.79% 180–209 cm, 7.00% 210–239 cm, 2.33% 240–269 cm, 1.16% 270–299 cm, 2.33% 300–329 cm, none at heights = 330 cm, 5.26% on rocks, 6.32% on perches with diameters <2.5 cm, 15.79% 2.5–5.0 cm, 30.53% 5.1–7.5 cm, 15.79% 7.6–10.0 cm, 7.37% 10.1–12.5 cm, 1.05% 12.6–15.0 cm, 10.53% 15.1–17.5 cm, 7.37 on perches with diameters >17.6 cm, perches did not differ by sex or size class (G. L. Powell and Russell, 1992); mean perch heights and diameters 3.55 ± 1.95 m and 13.0 ± 8.8 cm for males and 2.60 ± 1.98 m and 17.6 ± 14.1 cm for females (Butler and Losos, 2002); 11.1% on leaves (J. B. Losos, in litt., 21.IV.2008); habitat light environment "full shade" (Fleishman, 2000); dewlap UV-reflectance compatible with visibility in typical habitat (Macedonia et al., 2000); one individual slept in a bamboo pipe under the roof of a veranda (Lachner, 1987).

Activity: Females spend 62% of their time feeding, whereas males spend only 20%, instead males spend 56% of their time engaged in social interactions (43% in male-male interactions and 13% dewlapping), 23% of male and 30% of female time unclassified (motionless, often in the sun, possibly basking or watching for conspecifics or predators), both sexes perch lower during late morning and early afternoon than at other times, often in sun with higher body temperatures (Salvato, 1977). **Behavior:** Can inflict a painful bite (Panton, 1928; Underwood and Williams, 1959); escaped from bird (*Tyrannus caudifasciatus*) attack by "squirreling" (Panton, 1928); if disturbed, descends tree to hide in rocks (Grant, 1940a); except two homosexual copulations, males never seen in close proximity unless displaying aggressively or fighting, fighting males attempt to bite each other on the jaw, neck, or body, adult females never seen in close proximity unless fighting or when one was clearly hiding from the other (Trivers, 1976); capable of vocalization (Panton, 1928; T. A. Jenssen in Milton and Jenssen, 1979); male display behavior was much more similar to that of *A. grahami* in western Jamaica than those from central and eastern Jamaica in that it involved quick up-and-down movements without plateaus (Jenssen, 1981); territorial defense readily escalates to fighting, which can prove fatal for the loser (Krintler, 1985); mean display rate (proportion of time spent displaying) 0.054 ± 0.013, percent walks 45.1 ± 6.9, percent runs 45.4 ± 7.5, percent jumps 9.5 ± 2.0, moves/min 0.81 ± 0.08, distance jumped in nature 50.8 ± 6.9 cm (Losos, 1990a); moves per minute 0.81 (Johnson et al., 2008). **Ecomorphology:** "Crown-giant" (Rand and Williams, 1969); formally quantified (Beuttell and Losos, 1999). **Foraging and Diet:** Eats adult male *A. lineatopus* and (apparently) vegetable matter (Gosse, 1850); frequently descends to the ground to feed on insects (Panton, 1928; Grant, 1940a); diet includes insects and other small invertebrates, *Lantana* berries, "young" chochos and tender shoots of the Chocho Vine (*Sechium edule*), and possibly cocoa pods (*Theobroma cacao*), also will scavenge scraps of food from plates, including bits of pineapple and ripe banana, but also meat gravy and sweet syrups "licked up with extreme gusto," particular fondness for mangoes, consumes shed skin (Panton, 1928); diet includes insects and other lizards (Underwood and Williams, 1959); 38.5% of stomachs contained fruit, 1.0 fruit/individual (Herrel et al., 2004); in captivity, known to eat poison-dart frogs (Baum, 2008). **Growth:** Males grow faster at all sizes than females (Trivers, 1972); rates decline with larger size in both sexes, rates greatest in summer and fall (Trivers, 1976). **Home Range and Territory:** Two mature males never occupy the same tree (Panton, 1928); large males defend correspondingly larger territories (may occupy several contiguous trees), smaller males (<104 mm SVL) disperse if in a large male's territory (Trivers, 1976); often only one pair in one tree crown (Krintler, 1985). **Injuries:** Adult males with half a dewlap and one hindlimb missing, adult female with useless hindlimb, all apparently in "excellent health" (Panton, 1928). **Longevity:** To 10 years (Baum, 2008). **Movements:** Often in the same tree day after day (Grant, 1940a). **Parasites:** Protozoan malarial parasite *Plasmodium floridense* (Telford, 1975; Ayala, 1977, 1978); acanthocephalan *Centrorhynchus* (*spinosus*?) and nematode *Cyrtosomum longicaudatum* (Bundy et al., 1987). **Performance:** Mean maximum sprint speed (fastest time over an interval of 0.25 m) 0.101 ± 0.003 sec, jump 97.0 ± 2.73 cm, cling 13.484 ± 1.728 Neutons (Losos, 1990a); clinging ability 3.87 ± 0.26 Neutons, divided by mass 0.11, with the low value characteristic of crown-giant ecomorphs (Elstrott and Irschick, 2004); acceleration 23.73 ± 3.45 m/sec, sprint speed 1.66 ± 0.10 m/sec (Vanhooydonck et al., 2006a). **Predation:** Humans, "once esteemed a delicacy by some," also rats (*Rattus rattus*; Panton, 1928); Smooth-billed Ani (*Crotophaga ani*; C. B. Taylor, 1920 in Panton, 1928); Jamaican Lizard Cuckoo (*Saurothera vetula*; Danforth, 1928); Barn Owl (*Tyto alba*; McFarlane and Garrett, 1989); Greater Antillean Grackles (*Quiscalus niger*) are capable of taking all but adult males (Graves, 2006). **Reproduction:** Mature at 12–15 months of age, probably breeds throughout the year, may store sperm as long as 3 months in captivity (Baum, 2008); female descended a tree at midday, dug a small hole with her snout, laid an egg, covered the hole with her snout, ascended the tree, another egg found nearby, white egg 15 × 11 mm with leathery shell, rough to the touch, incubation (indoors) 2 mo, 12 days, 21 h, possible communal egg deposition, with ~100 shells in the decaying core of a fallen *Ficus* and >40 together on the ground (Panton, 1928); in both sexes, reproductive success increases with size, but success of the largest males significantly greater than gains by females (Trivers, 1972, 1976); females with one mature egg or none, large individuals often with 1–2 eggs in earlier stages of development, females appear to copulate once per egg, pairs may copulate 4 times in a week if copulations disturbed, if not, female will go at least one month before copulating again, 90% of copulations by territorial males in their own territories, copulations occur in highly conspicuous situations, pair almost invariably facing down on exposed tree trunks, copulation time 25 min ($N = 1$), pair reluctant to disengage, homosexual copulation observed twice, once unusually in foliage rather than in the open (Trivers, 1976); reproductively active throughout the year, frequently 2 eggs (suggesting ability to retain eggs until deposition propitious), size ~12 mm, but double in size during development, deposited in litter or in mulch accumulated in branch axils, eggs covered after deposition, incubation 60–75 days at 24–26 °C, hatchling SVL 27 mm, shed within 1 h of hatching, egg tooth lost at that time (Krintler, 1985); in captivity, eggs 8 × 12 mm at deposition (Fläschendräger and Wijffels, 1996, 2009). **Sex Ratio (M:F):** 0.80:1 (Trivers, 1976). **Size:** Male SVL to 131 mm (Schwartz and Henderson, 1991), female SVL to 110.7 mm (G. L. Powell and Russell, 1992); smallest male copulating 87 mm SVL, smallest female 70 mm (Trivers, 1976); hatchling SVL 27 mm (Krintler, 1977). **Survival:** Female survival (to 90 mm SVL) higher than that of males (to 110 mm SVL), but probably not at larger sizes (Trivers, 1976). **Tail Autotomy:** Broken tail of an adult male grew ~19 mm in 6 weeks and another ~19 mm in a further 6 weeks, after breaking a second time at the same point, regeneration at the same rate and with the same final result (Panton, 1928). **Thermal Biology:** Avoids direct sunlight (Fläschendräger and Wijffels, 1996, 2009; Baum, 2008).

Conservation Status: Not assessed.

Additional Pertinent Literature: Russell and Bauer (1991a).

Anolis garridoi Díaz, Estrada, and Moreno, 1996

Distribution: Cuba, only from the vicinity of the type locality in the Sierra de Trinidad; elevation ~700 m.

Habitat: In upland hardwood forests, 4 of 10 animals on twigs, 2 on branches, 2 on leaves, one on the trunk of a shrub, and one in leaf litter at perch heights and diameters of 0.3–3.0 m and 0.7–4.0 cm, respectively, with males (1.78 ± 0.42 m, 2.56 ± 0.61 cm) higher and on perches of larger diameter than females (1.50 ± 0.16 m, 2.00 ± 1.59 cm), lizards in shade during morning and midday hours (Díaz et al., 1996); in submontane mesophyllous evergreen forest (Rodríguez Schettino et al., 1999).

Activity: Diurnal (Rodríguez Schettino et al., 1999). **Behavior:** Responds to disturbance by moving slowly to the opposite side of the twig, movement along twigs is very slow, using "semi-prehensile" tail to hold the twig when moving from one perch to another or when catching prey, sometimes touching tongue to the surface of the twig, 2 females observed for 42.5 and 23.5 min, walked or crept 19 and 10 times, usually just a step or two, did not run or ran 4 times (may have been disturbed by the observer), and jumped 3 times and once, male aggressive display involves raising the dorsal and nuchal crests, lateral presentation (perpendicular to the axis of the perch), lateral compression, and headbobbing while the temporal blotch turned darker, amplitude of headbobs and push-ups increased and, during the last headbob, the dewlap was extended, resident moved toward the intruder and bit him, the intruder fled while vibrating his tail-tip (Díaz et al., 1996). **Ecomorphology:** "Twig" (Díaz et al., 1996). **Foraging and Diet:** Sit-and-wait predator, spends most time motionless, eats insects (Díaz et al., 1996). **Miscellany:** Moves eyes independently of each other (Díaz et al., 1996). **Size:** Mean male SVL 39.5 ± 1.07 mm, female SVL 36.5 ± 0.03 mm (Díaz et al., 1996); male SVL to 41.8 mm, female SVL to 36.8 mm (Rodríguez Schettino, 1999b).

Conservation Status: Considered "vulnerable" (Rodríguez Schettino and Chamizo Lara, 1998a; L. Rodríguez Schettino and A. Chamizo Lara in Rodríguez Schettino, 1999a).

Anolis gingivinus Cope, 1864

Distribution: Anguilla and Sombrero island banks, "occurs throughout the Anguilla Bank, on every rock and cay that supports more than herb-stage vegetation, essentially islandwide on Anguilla, St.-Martin/St. Maarten, and St.-Barthélemy, and on many satellite islands, "on forty or more separate islands" (Lazell, 1972); on trees and palms, rocks, walls and buildings, along the coast, may be on rocks close enough to the sea that lizards are sprayed by waves (Wijffels, 1964); intentionally introduced onto Anguillita as part of an experiment in resource use (Roughgarden, 1995); ubiquitous, except in some areas that have been completely denuded of vegetation, common in garden shrubs and piles of rocks, on rocky cliffs and a few large trees in windswept areas, edificarian, abundant on rock walls, readily uses human debris (Breuil, 2002b); elevational range, sea level to 424 m (Pic du Paradis on St.-Martin/St. Maarten).

Habitat: Mesophilic to xerophilic (Schwartz and Henderson, 1991); primarily arboreal, but extremely adaptive, in brush and scrub, on walls, on houses, and in gardens, along the coast, almost entirely saxicolous (Fläschendräger and Wijffels, 1996, 2009); ubiquitous and sometimes superabundant, in coastal scrub, overgrown fields, *Cocos* groves, xeric woods, on bushes, shrubs, rocks and boulders, fence posts, *Opuntia*, rock walls, buildings, arbors, trees, including cultivated orchard species, but rarely high in crowns, frequently forages on the ground (F. W. King in Williams, 1962c; Schwartz and Henderson, 1991; Eaton et al., 2002; Shew et al., 2002b; Hodge et al., 2003; Powell et al., 2005; Powell and Henderson, 2008c); on Anguilla, "almost every large rock not surrounded by weeds had an anole sunning itself on it. The rocks seem to provide the anole with a good view of the surrounding area for they all darted behind the rock no matter what side they were approached from" (F. W. King in Williams, 1962c); on Sombrero, on the ground and in stone heaps, often beneath rocks (Lazell, 1964c); on rocks along the coast where they venture within the range of ocean spray, often function as human commensals, very abundant in developed areas (Stettler, 1978); mean perch height on St.-Martin/St. Maarten 1.46 ± 0.06 m (Roughgarden et al., 1983a); perch higher in the presence of *A. pogus*, mean 1.46 ± 0.06 m where the 2 species co-occur, mean 0.88 ± 0.03 m where solitary (Roughgarden et al., 1984); where solitary on St.-Martin/St. Maarten, 13% on the ground, 27% at heights 0–0.46 m, 16% at 0.46–0.9 m, 30% at 0.9–1.8 m, 13% >1.8 m, where co-occur with *A. pogus*, 5% on the ground, 12% at heights 0–0.46 m, 12% at 0.46–0.9 m, 33% at 0.9–1.8 m, 40% >1.8 m (Roughgarden, 1995); on St.-Martin/St. Maarten, mean perch height and diameter 2.23 ± 0.24 m and 9.7 ± 1.5 cm (Losos and de Queiroz, 1997a); mean perch heights and diameters at 6 sites on Anguilla for males 62.5 ± 3.9 to 140.7 ± 27.8 cm and 1.8 ± 0.3 to 20.1 ± 2.7 cm, respectively, and for females 55.9 ± 3.6 to 130.9 ± 32.8 cm and 1.1 ± 0.1 to 18.5 ± 4.6 cm, respectively, with male but not female SVL positively correlated with perch height and males perching significantly higher only at locations with substantial vertical structure (Eaton et al., 2002); at 2 sites on St.-Martin/St. Maarten males perched at mean heights 0.9 ± 0.1 m and 1.6 m, females 0.6 ± 0.0 m and 1.1 ± 0.3 m, juveniles at one site 0.1 ± 0.0 m (Pereira et al., 2002); sleeps on leaves and branches of bushes and trees, with males higher than females, often near daytime centers of activity on Anguilla (Shew et al., 2002b); may use artificial retreats on St.-Martin/St. Maarten (Powell et al., 2005).

Activity: Basically diurnal, but exploits the "night-light niche" on St. Maarten and Anguilla (Powell and Henderson, 1992; Breuil, 2002b; Hodge et al., 2003; Powell et al.,

2005; Perry et al., 2008); noticeably less active later in the day (Malhotra and Thorpe, 1999); individuals awakened at night feed and lap water from leaves (Shew et al., 2002b); peak activity often early and late (Breuil, 2002b). **Behavior:** On Sombrero, very skittish (Lazell, 1964c); generally calm and not as easily disturbed as other related species on nearby islands, both sexes react to conspecifics of the same sex with gular inflation, nuchal crest erection, dewlap extension (males only), and a lightening of body color, one individual usually retreats without fighting (Fläschendräger and Wijffels, 1996, 2009); on Anguilla, males especially were aggressive toward a noose (Eaton et al., 2002); often show sleep-site fidelity for short periods of time, females more so than males, and females more frequently sharing bushes or trees, with 3–4 on the same plant, when disturbed, either remained on the perch or fled to a nearby perch (Shew et al., 2002b); shy, responds to approach by "squirreling" or retreating farther into a crevice, if threatened, jumps to another perch or even to the ground, only rarely climbs upward (Hodge et al., 2003; Powell et al., 2005); approach distance (with a 12 × 12 cm square) varied considerably (0–130 cm, mean 14.1 ± 1.38 cm) and did not vary by perch height, diameter, substrate temperature, or orientation on perch, but was affected slightly by color, with "blue" eliciting an earlier response than other colors, perch height did not affect the type of response ("squirreling," running down or up, jumping), but anoles were more likely to jump from narrower perches and "squirrel" or run up on larger-diameter perches, orientation also affected responses, with anoles most likely to run up or down based on their initial orientation and more likely to jump when horizontal, substrate temperatures did not affect responses, but lizards in a sun/shade mosaic were more likely to jump than those in full shade or sun, responses did not vary by sex (Larimer et al., 2006). **Competition:** See *A. pogus* account (Roughgarden et al., 1984; Pacala and Roughgarden, 1985; Roughgarden, 1995; Pereira et al., 2002). **Desiccation:** Water loss rate per unit surface area lower than that of *A. pogus* (Roughgarden et al., 1983b). **Ecomorphology:** Qualifies as a "generalist" (Losos and de Queiroz, 1997a); closest to "trunk-ground" (Beuttell and Losos, 1999); undetermined (Losos, 2009). **Foraging and Diet:** Captures and eats juvenile and adult *A. pogus* (Roughgarden et al., 1983b; J. McLaughlin in Goldwasser and Roughgarden, 1993; Breuil, 2002b); readily recognizes even motionless prey, slow-moving prey (e.g., caterpillars) are pursued "at leisure," efforts to catch flying insects involve a gradual approach followed by a rapid leap (Fläschendräger and Wijffels, 1996, 2009); likelihood of females taking prey are less at greater distances, and distances at which prey would be taken are greater for larger prey and in situations with low prey abundance (Shafir and Roughgarden, 1998); stomach contents contained spiders, beetles, flies (adult and larvae), hemipterans, ants (most important) and other hymenopterans, lepidopteran adults and larvae, mecopterans, shed skin, and seeds, with the diversity of the diet substantially lower on Scrub Island than on Anguilla, no correlation between prey size and SVL for males or females, standardized dietary niche breadth based on importance values 0.26, standardized dietary overlap between males and females 81.7 (Eaton et al., 2002); ants, termites, spiders, centipedes, scorpions, *Sphaerodactylus parvus* (Breuil, 2002b); small insects and spiders, will lap nectar and eat flowers and fruits (Powell et al., 2005). **Growth:** 0.2 mm/day for juveniles, declining as lizards grow larger, females essentially stop growing at 50 mm SVL, males at 63 mm SVL (Roughgarden, 1995). **Home Range:** Male home ranges at 2 sites on St.-Martin/St. Maarten encompassed 1–3 female home ranges, with overlaps averaging 81 ± 2% and 30 ± 6%, sizes for males 1.7 ± 1.2 m^2 and 5.8 m^2, females 0.8 ± 0.2 m^2 and 3.6 ± 0.9 m^2, juveniles 2.2 ± 1.0 m^2, male-female home range size ratio 2.6:1 (Pereira et al., 2002). **Life Expectancy:** 0.62 year (on St.-Martin/St. Maarten; McLaughlin and Roughgarden, 1989). **Parasites:** 25% of adults infected with malaria (*Plasmodium azurophilum*, highest in relatively mesic upland sites where *A. pogus* is abundant), indicating less resistance to infection and greater vulnerability to effects than *A. gingivinus* (Schall, 1992); on Anguilla and St.-Bartélemy, only *P. floridense*, on St.-Martin/St. Maarten *P. floridense* and *P. azurophilium* (Staats and Schall, 1996b; Perkins, 2001); trematodes *Mecocoelium* sp. and *Allogyptus crenshawi*, nematode *Parapharyngodon cubensis* (Dobson et al., 1992); nematode *Spauligodon anolis* (Dobson et al., 1992; Goldberg et al., 1997a; Bursey and Goldberg, 1998); mean parasite species/host 0.22 on Anguilla, 0.84 on St.-Martin/St. Maarten (Dobson and Pacala, 1992); cestode *Oochoristica maccoyi* (Bursey and Goldberg, 1996); trematode *Mesoceolium monas*, cestode *Oochoristica maccoyi*, nematodes *Parapharyngodon cubensis*, *Trichospirura teixeirai*, and larval *Abbreviata* sp., *Ascarops* sp., and *Physocephalus* sp., acanthocephalan cystacanths *Centrorhynchus* sp. (Goldberg et al., 1997a). **Population Size and Density:** Densities at 8 sites on St.-Martin/St. Maarten 19.5 ± 3.4–118.6 ± 9.4/100 m^2 (Peterson estimate) and 21.9 ± 1.8–125.5 ± 5.6/100 m^2 (3-census independence-model estimate; Heckel and Roughgarden, 1979); mean density on St.-Martin/St. Maarten 7.6 ± 6.2–129.8 ± 5.5/100 m^2 (density varies by season and habitat) and 28.4–44.0/100 m^2 on Anguillita (Roughgarden et al., 1983b); density much higher in the absence of *A. pogus* on St.-Martin/St. Maarten, 50.1 ± 2.2, 43.2 ± 3.4, 42.4 ± 1.9, and 6.4 ± 0.5 lizards/100 m^2 when alone, 8.4 ± 2.6, 10.2 ± 2.0, and 19.7 ± 1.4 lizards/100 m^2 when sympatric with *A. pogus*, densities over a 4-year period at 3 sites ranged from 32.3 ± 4.2 to 89.8 ± 6.6 lizards/100 m² (no *A. pogus*), 50.1 ± 2.2–129.8 ± 5.5 lizards/100 m² (few *A. pogus*), and 7.6 ± 6.2–39.8 ± 4.7 lizards/100 m² (*A. pogus* abundant; Roughgarden et al., 1984); over 7 years at a lowland site on St.-Martin/St. Maarten 49.3 ± 2.3–129.8 ± 5.5 lizards/100 m², at an upland site 6.4 ± 1.4–39.8 ± 4.7 lizards/100 m² (Roughgarden, 1986); to 67.7/100 m² on Anguilla (McLaughlin and Roughgarden, 1989); at 6 sites on Anguilla, 21.5 ± 0.2 to 81.3 ± 0.1 lizards/144 m² (Eaton et al., 2002). **Predation:** American Kestrel (*Falco sparverius*) on Anguilla (68.8% of diet by frequency of occurrence, 73.9% by biomass); Pearly-eyed Thrasher (*Margarops fuscatus*) on St.-Martin/St. Maarten to mean 57.3% by biomass (McLaughlin and Roughgarden, 1989); thrashers sometimes selectively hunt anoles (Hodge et al., 2003; Powell et al., 2005); on Anguilla and St.-Barthélemy, *Alsophis rijgersmaei* (Henderson and Sajdak, 1996; Breuil, 2002b); *Ameiva pleii* (Hodge et al., 2003; Powell et al., 2005). **Reproduction:** Copulation in June (RP); copulation (1–7 min) follows an "energetic" courtship, males generally bite female napes, eggs at deposition 6 × 8 mm, shortly before hatching 10 × 16 mm (Fläschendräger and Wijffels, 1996, 2009); courtship in early morning, male displayed and rapidly approached female and tried to mount repeatedly, duration of copulation 130 sec with vigorous pelvic thrusts followed by quiescence, copulation in July–August (earliest probably in June, none in April), incubation 30–40 days (Breuil, 2002b); home range overlaps consistent with polygynous social structure, 3 copulations of one pair at one site and 2 copulations with 2 different females at a second site (Pereira et al., 2002). **Resource Partitioning:** Chose higher perches and grew more slowly in the presence of *A. pogus* in enclosures (Pacala and Roughgarden, 1982; Roughgarden et al., 1984); in experimental introductions on Anguillita, established population restricted *A. pogus* to peripheral sites (Roughgarden et al., 1984). **Sex Ratio (M:F):** 213:93, differed significantly from 1:1 (Eaton et al., 2002). **Size:** Male SVL to 72 mm (Lazell, 1972), female SVL to 54 mm (Dobson et al., 1992; Eaton et al., 2002); minimum male size at maturity 42 mm, female 41 mm (Lazell, 1972; Eaton et al., 2002). **Thermal Biology:** Heat tolerant, higher panting temperature than *A. pogus* (Roughgarden et al., 1983b, 1984); mean body temperature 29.1 ± 0.6 °C (Roughgarden et al., 1984); not an effective thermoregulator, with body temperatures averaging 30.1 ± 0.1 °C at mean substrate and air temperatures 28.9 ± 0.1 °C and 28.8 ± 0.1 °C, with a negative correlation between perch height and body and ambient temperatures on Anguilla, but not on Scrub Island, where less vertical structure was present and the compression of the vertical dimension subjected lizards to similar thermal regimes (Eaton et al., 2002).

Conservation Status: Not assessed, but protected by French legislation in St.-Martin and St.-Barthélemy (Lorvelec et al., 2007).

Anolis grahami Gray, 1845

Distribution: Jamaica, two currently recognized subspecies, nominate form essentially islandwide except extreme east, where *A. g. aquarum* occurs; introduced on Bermuda.

Habitat: Mesophilic (Schwartz and Henderson, 1991); usually arboreal, common in open, sunny situations, rarely in dense forest (Fläschendräger and Wijffels, 1996, 2009); in grazed pastures, well-watered shady gardens, xeric coastal growth, xeric forest, edificarian, commonly enters houses, climbs walls, and hides behind pictures, abundant on trees, walls, fence posts, limestone rocks, stones, and in brush heaps, small individuals often in hedges (Gosse, 1851; Rand, 1967b; Schoener and Schoener, 1971a; Schwartz and Henderson, 1991); often human commensals, on trees in proximity to human habitations and on buildings (Hilgenhof, 1998a); plurality (40.7%) of large adults on perches >3 m high and most (66.2%) on large-diameter (>7.6 cm) perches, plurality of smaller individuals (37.5%) also on perches >3 m high and most (55.0%) on large-diameter (>7.6 cm) perches (Rand, 1967b); mean perch diameter 17.7 ± 3.94 cm (Losos and Sinervo, 1989); on a wide variety of perch diameters, with an intermediate habitat breadth of 66.1 (Irschick and Losos, 1999); rarely on the ground, usually >1.5 m above the ground on perches >1.3 cm in diameter, maximum height above the ground ~7.5 m for large individuals and ~9 m for small lizards (Schoener and Schoener, 1971a); perches higher (Schoener, 1975, 1976) and on perches of slightly greater diameter when sympatric with *A. sagrei* than when alone on Jamaica (Schoener, 1975); mean adult male and female perch heights 1.7 ± 0.83 m (0.9% on the ground) and 1.0 ± 0.77 m (4.2% on the ground; Lister, 1976); mean perch height and diameter 2.5 ± 0.29 m and 17.7 ± 3.94 cm (Losos, 1990a); most individuals on stone walls (53.90%) or tree trunks (20.96%), 16.26% on the ground, 5.69% at heights <30 cm, 13.82% at 30–59 cm, 26.02% 60–89 cm, 17.89% 90–119 cm, 9.77% 120–149 cm, 3.25% 150–179 cm, 4.07% 180–209 cm, 1.63% 210–239 cm, none at 240–269 cm, 1.63% 270–299 cm, none at heights of 300 cm, 78.03% on rocks, 3.03% on perches with diameters <2.5 cm, 3.79% 2.5–5.0 cm, 9.85% 5.1–7.5 cm, 2.27% 7.6–10.0 cm, 1.52% 10.1–12.5 cm, none 12.6–15.0 cm, 1.52% 15.1–17.5 cm, 0.75% on perches with diameters >17.6 cm, perches did not differ by sex or size class (G. L. Powell and Russell, 1992); mean perch heights and diameters 2.04 ± 1.77 m and 10.2 ± 9.5 cm for males and 1.47 ± 1.10 m and 7.0

± 7.2 cm for females (Butler and Losos, 2002); mean male perch height and diameter 1.73 ± 0.19 m and 0.81 ± 0.22 m, mean female perch height and diameter 1.35 ± 0.17 m and 0.38 ± 0.07 m (VanHooydonck et al., 2007); 17.2% on leaves (J. B. Losos, in litt., 21.IV.2008); habitat light environment "partial shade" (Fleishman, 2000); dewlap UV-reflectance compatible with visibility in typical habitat (Macedonia et al., 2000); moves from perching on trunks and branches by day to using mostly leaves at night, male and female sleeping sites did not differ significantly, but both perched at inclinations more parallel to the ground and females perched higher and on perches of smaller diameter at night than by day (Singhal et al., 2007).

Activity: Diurnal (Schoener, 1970); earliest activity at Discovery Bay 0606 h, latest activity 1754 h, male displays peak at dawn and dusk, presumably to advertise territorial activity before and after a period of nocturnal inactivity (Ord, 2008). **Behavior:** Although no structure shown to be a sound-producing mechanism, vocalizes (Grant, 1940a; Underwood and Williams, 1959; Myers, 1971; T. A. Jenssen in Milton and Jenssen, 1979); during fights, produces a squeaklike sound in lunging attacks and multiple squeaks during mouth-holds, handheld lizards produce 2 kinds of growls, a chirp, and 2 kinds of squeals, respond only slightly to a playback of recorded fight-related and environmental sounds, suggesting that they are not responsive to sounds generally and that vocalizations have limited intraspecific significance, vocalizations may have some antipredator function, and may represent a preadaptation for social communication, greater reliance on visual than auditory information (Milton and Jenssen, 1979; Rothblum et al., 1979); mean display rate (proportion of time spent displaying) 0.097 ± 0.020, percent walks 31.5 ± 4.7, percent runs 49.3 ± 4.3, percent jumps 19.2 ± 2.0, moves/min 1.74 ± 0.26, distance jumped in nature 41.0 ± 4.6 cm (Losos, 1990a); when presented with videotaped displays of conspecifics and 3 congenerics (*A. carolinensis*, *A. conspersus*, *A. sagrei*), males respond sooner to conspecifics, pulse dewlaps and headbob for longer periods in response to conspecific displays, and incorporate modifiers (e.g., sagittal compression, raised crest, gorged throat) when responding to conspecifics (Macedonia and Stamps, 1994); tends to jump rather than run on perches of small diameter (Losos and Irschick, 1996); display behavior of males from western Jamaica differs considerably from those in central and eastern Jamaica in that both stereotypical unsolicited signature displays and responses to territorial invaders involve a quick up-and-down movement with no pauses at the tops of the bob amplitudes, whereas those of central and eastern populations exhibited distinct differences in cadence and bobs with clear plateaus (Jenssen, 1981); when sleeping perches are disturbed, lizards drop to the ground (Singhal et al., 2007); male approach distance, flight distance, and final distance 0.62 ± 0.12 m, 0.32 ± 0.04 m, and 0.93 ± 0.11 m, female approach distance, flight distance, and final distance 0.58 ± 0.14 m, 0.32 ± 0.04 m, and 0.75 ± 0.13 m (VanHooydonck et al., 2007); moves per minute 0.92 and 1.74 (Johnson et al., 2008). **Diet:** Eats *A. lineatopis* (Rand, 1967a); catholic, stomach contents were ants (82.0%), other hymenopterans, flies, beetles (adults and larvae), leafhoppers, lepidopterans (adults and larvae), earwigs, orthopterans, spiders, pseudoscorpions, centipedes, and mites (Lister, 1976); on Bermuda, predominantly hemipterans (e.g., *Aphis*), hymenopterans (especially ants, *Iridomyrmex, Pheidole, Brachymyrmex*), but includes other anoles (hatchling *A. valencienni*) and shed skin (Hicks and Trivers, 1983); presumed nectarivory (anole licking a flower; K. B. Sandved in Losos and de Queiroz, 1997b); 19.1% of stomachs contained fruit, 1.25 fruits/individual (Herrel et al., 2004). **Ecomorphology:** "Trunk-crown" (Williams, 1972); formally quantified (Beuttell and Losos, 1999). **Home Range and Territoriality:** Females defend territories against other lizards of similar size, regardless of species, males appear to be more tolerant of other species (Rand, 1967b); dewlap size (secondary sexual characteristic) correlated with bite force (related to fighting capacity) in this territorial species (Vanhooydonck et al., 2005). **Hybridization:** With *A. lineatopus neckeri*, hybrids probably infertile (Jenssen, 1977). **Movements:** Mean distance per movement 16.9 ± 1.4 cm at mean speed 22.9 ± 1.9 cm/sec, 0.76 ± 0.17 moves/min, 1.57 ± 0.24% time moving (Irschick, 2000b). **Parasites:** Protozoan malarial parasites *Plasmodium azurophilum* and *P. floridense* (Telford, 1975; Ayala, 1977, 1978); acanthocephalan *Centrorhynchus* sp., nematodes *Cyrtosomum scelopori* and *Thelandros* sp. (Bundy et al., 1987). **Performance:** Sprint speed declines on smaller perches (Losos and Sinervo, 1989); mean maximum sprint speed (fastest time over an interval of 0.25 m) 0.129 ± 0.005 sec, jump 74.3 ± 2.40 cm, cling 2.344 ± 0.200 Neutons (Losos, 1990a); mean maximal speed 1.5 ± 0.12 m/sec, escape speed 1.3 ± 0.07 m/sec, feeding speed 1.2 ± 0.07 m/sec, undisturbed activity 0.5 ± 0.05 m/sec, mean maximal jumping distance 74.3 ± 2.40 cm, feeding jumping distance 21.3 ± 4.33 cm, undisturbed jumping distance 19.0 ± 3.40 cm (Irschick and Losos, 1998); sprint speed reduced when turning, 1.21 ± 0.06 m/sec with no turn, 1.20 ± 0.06 m/sec over 30° turn (reduction to 99% of sprint speed without turns), 0.96 ± 0.08 m/sec over 90° turn (79%), about half of individuals jumped to traverse 90° turns, but potential benefits of jumping offset by pauses when preparing to jump or recovering balance after jumping (Higham et al., 2001); peak velocity 1.90 ± 0.19 m/sec, peak acceleration 40.6 ± 5.2 m/sec^2 (Toro et al., 2004); acceleration 28.81 ± 2.59 m/sec, sprint speed 1.21 ± 0.08 m/sec (Vanhooydonck et al., 2006a). **Predation:** Greater Antillean Grackle (*Quiscalus niger*; Graves, 2006). **Reproduction:** Egg under a log (Rand, 1967c); some females active in all months, but pronounced seasonal fluctuations are evident, with peak egg production corresponding to the rainy season (May–November), when nearly all females ovigerous, nearly half with 2 oviductal eggs, and essentially no fat bodies, female body fat weight highest during dry months, when reproductive activity drops, testicular weights decrease during wet months, remain small for 4–5 months, and began increasing rapidly in February, although spermatogenesis and androgen production continues throughout the year in most males, fat body weights are inversely proportional to testicular weights (Licht and Gorman, 1970); in captivity, incubation 38–45 days at 23–28 °C, hatchling SVL ~17 mm (Fläschendräger and Wijffels, 2009). **Size:** Male SVL to 75 mm (Rand, 1967b), female SVL to 64 mm (Schwartz and Henderson, 1991); smallest sexually mature male 49 mm SVL and female 38 mm SVL (Licht and Gorman, 1970). **Thermal Biology:** Maintains temperatures higher than ambient and warmer than those of *A. lineatopus* and *A. opalinus* (Rand, 1967b); in an intermediately insolated situation 31.2 ± 1.55 °C and 28.9 ± 1.02 °C (Lister, 1976). **Vision:** Moderate critical flicker-fusion frequency corresponded to association with semi-open habitats (Jenssen and Swenson, 1974).

Conservation Status: Not assessed.

Additional Pertinent Literature: Russell and Bauer (1991b).

Anolis griseus Garman, 1887

Distribution: St. Vincent, essentially islandwide.

Habitat: Mesophilic (Schwartz and Henderson, 1991); in heavily disturbed areas (Kingstown) to rainforest, *Musa* plantations (including in banana sheds), coastal *Cocos*, and nutmeg and cacao plantations, rare where vegetation is sparse (Treglia, 2006); arboreal, but will go to ground, prefers large trees, males perch 1.8–7.6 m above the ground, females lower (Lazell, 1972); largely in crowns of large trees (Heselhaus and Schmidt, 1990); mean perch height 1.25 ± 0.06 m (Roughgarden et al., 1983a); mean perch height and diameter 2.65 ± 0.30 m and 20.9 ± 3.2 cm (Losos and de Queiroz, 1997a); perch heights and diameters varied widely depending on available structural habitat at 5 sites on St. Vincent, with males perching higher (mean 212.2 ± 20.4 cm, 0–650 cm, 1 of 51 on the ground) and on perches of larger diameters (22.3 ± 3.5 cm, 1–90 cm) than females and subadult males (136.9 ± 13.0 cm, 0–350 cm, 6 of 56; 13.0 ± 1.8 cm, 1–80 cm) and juveniles (87.6 ± 16.7 cm, 0–400 cm, 10 of 33; 14.9 ± 3.6 cm, 1–65 cm), with most individuals of all size groups perching with heads down (106 versus 37 horizontal and 52 with heads up) and most in filtered light (108 versus none in full sun and 89 in full shade; Hite et al., 2008a); at a highly disturbed construction site, mean perch heights of males 176.2 ± 21.9 cm, females 145.0 ± 5.0 cm, used "natural" perches almost to exclusion of "artificial" perches, not obviously affected by the presence of introduced *A. sagrei* (Treglia et al., 2008).

Activity: Diurnal (RWH, RP). **Behavior:** Very shy, quick to flee at when approached, high-intensity display consists of dewlap extension, rapid tail lashes, headbobs, and dewlap retraction, displays similar to those of *A. luciae* (Gorman, 1968); approach distance often >15 m (J. Schmidt, 1980a). **Diet:** In Kingstown, descends to the ground to eat small, red, mulberry-like berries (Lazell, 1972); ate the decapitated head of an *A. trinitatis* (Gorman, 1968); readily takes *A. trinitatis* (J. Schmidt, 1980a). **Ecomorphology:** Close to "crown-giant" (Losos and de Queiroz, 1997a; Beuttell and Losos, 1999). **Parasites:** Nematodes *Parapharyngodon cubensis* and larval *Ascarops* sp. and *Physaloptera* sp. (Goldberg et al., 1997a). **Population Size and Density:** Rare in woods, in forest 55 ± 17/100 m^2 (Roughgarden et al., 1983); varied widely in 8 habitats (0–5,208 lizards/ha and 0–6,430/ha using 2 different methods, 9,618/ha at one site where only one method yielded an estimate (Hite et al., 2008a). **Predation:** *Corallus cookii* (Henderson, 1993, 2002). **Reproduction:** Some reproductive activity throughout the year, but a decline in the frequency of ovigerous females March–April (although all with yolked ovarian follicles), a gradual increase during the next 4 months, only 7 of 112 females had 2 oviductal eggs, female fat body weight peak in September/October and December/January, at which times all females were reproductive, male testicular weight relatively constant throughout the year, fat body weight peaks in August/September (Licht and Gorman, 1970); hatchling total length ~75 mm (J. Schmidt, 1980a); copulation in June (RP). **Sex Ratio (M:F):** Apparently balanced (Lazell, 1972). **Size:** Male SVL to 136 mm (Lazell, 1972), female SVL to 88 mm (Licht and Gorman, 1970); smallest sexually mature male SVL 63 mm (Licht and Gorman, 1970) and female SVL 57 mm (Lazell, 1972).

Conservation Status: Not assessed.

Anolis guafe Estrada and Garrido, 1991

Distribution: Cuba, Granma Province, Cabo Cruz to the vicinity of Punta Monje (~25 km west), never more than 5 km inland; elevational range, sea level to ~275 m.

Habitat: Xerophilic, coastal and subcoastal xeric scrub and penetrating inland into dry evergreen forests on the higher levels of the marine platforms, the substrate of which is largely "dog-tooth" limestone matched by lizards' cryptic coloration, males perch at 1 m on trunks of shrubs (Estrada and Garrido, 1991); in mesophyllous evergreen and semi-deciduous forest and coastal and subcoastal microphyllous forest (Rodríguez Schettino et al., 1999).

Activity: Diurnal (Rodríguez Schettino et al., 1999). **Behavior:** Male display with head and tail elevated and dewlap extended (Estrada and Garrido, 1991). **Ecomorphology:** "Trunk-ground" (Losos, 2009). **Size:** Mean male SVL 48.8 mm, female SVL 40.0 mm (Rodríguez Schettino, 1999b).

Conservation Status: Considered "vulnerable" (Rodríguez Schettino and Chamizo Lara, 1998a; L. Rodríguez Schettino and A. Chamizo Lara in Rodríguez Schettino, 1999a).

Anolis guamuhaya Garrido, Pérez-Beato, and Moreno, 1991

Distribution: Cuba, apparently confined to higher elevations of the mountainous region of Guamuhaya in the southern part of Sancti-Spíritus Province; elevational range, sea level to ~900 m.

Habitat: Mesophilic, cool and humid forest characterized by trees with trunk diameters 10–20 cm and 8 m in height, one male on a trunk 10 cm in diameter at a height of 2 m above the ground (Garrido et al., 1991; Rodríguez Schettino, 1999b); in submontane mesophyllous evergreen forest (Rodríguez Schettino et al., 1999).

Activity: Diurnal (Rodríguez Schettino et al., 1999). **Behavior:** Moves lethargically, when disturbed, climbs up and onto a branch (Garrido et al., 1991). **Ecomorphology:** "Twig-giant" (Garrido et al., 1991, by implication), member of the *Chamaeleolis* clade; technically this is not an ecomorphological category as it evolved in only one clade (see definition in Williams, 1972). **Foraging and Diet:** In captivity, catches small insects by extending the tongue to a distance of ~5 cm (Rodríguez Schettino, 1999b). **Size:** Holotype (adult male) SVL 162 mm, apparently adults of both sexes reach this approximate size, adult females unknown (Garrido et al., 1991).

Conservation Status: Considered "endangered" (Rodríguez Schettino and Chamizo Lara in Rodríguez Schettino, 1999a); assessment based on IUCN Red List criteria (Ruiz Urquiola et al., 1998b).

Anolis guazuma Garrido, 1984

Distribution: Cuba, Pico Turquino, Sierra Maestra; elevation ~600 m.

Habitat: Mesophilic (Schwartz and Henderson, 1991); in submontane mesophyllous evergreen forest (Rodríguez Schettino et al., 1999); only on peripheral branches and bare twigs in the crowns of *Guazuma tomentosa*, the twigs and branchlets of which the lizards mimic with their lack of rapid motility and their dorsal color and pattern, although sympatric with *A. angusticeps*, the latter is more likely to be on trunks or larger branches than *A. guazuma*, which occupies the crown (Garrido, 1983); generally perches 3–4 m above the ground toward the peripheral ends of twigs (Rodríguez Schettino, 1999b); mean perch height and diameter 2.30 m and 0.9 cm (J. B. Losos, in litt., 21.IV.2008).

Activity: Diurnal (Rodríguez Schettino et al., 1999). **Behavior:** Flatten themselves against twigs and remain immobile, relying on crypsis to avoid predation, in one male-male interaction, neither did push-ups, but one moved the head into a position in which the postocular spot was prominent and elevated the anterior body while the other nodded its head and opened its mouth, dewlaps were partially extended (Garrido, 1983); moves slowly, when pressed, jumps to ground and hides among grasses and beneath dead leaves (Rodríguez Schettino, 1999b). **Ecomorphology:** "Twig" (Garrido, 1983). **Size:** Male SVL to 48.5 mm (Rodríguez Schettino, 1999b), female SVL to 41 mm (Garrido, 1983). **Thermal Biology:** Presumably heliothermic (Rodríguez Schettino, 1999b).

Conservation Status: Considered "vulnerable" (Rodríguez Schettino and Chamizo Lara, 1998a; Rodríguez Schettino and Chamizo Lara in Rodríguez Schettino, 1999a).

Anolis gundlachi Peters, 1876

Distribution: Puerto Rico, uplands; elevational range ~240–1,200 m.

Habitat: Mesophilic (Schwartz and Henderson, 1991); in heavily shaded forest interiors more frequently than in sunny situations, most abundant in the "coffee belt" and higher forested areas, but reaches the coastal plain, usually on trees or other vegetation, but also on streamside boulders (Stejneger, 1904; Fowler, 1918; K. P. Schmidt, 1920, 1928; Losos, 1990b; Schwartz and Henderson, 1991); usually on perches of large to medium diameter close to the ground (Rand, 1964a); mean perch heights and diameters at 3 sites in irradiated upland forest 128.0 ± 7.92 cm and 8.6 ± 0.66 cm, 137.2 ± 17.07 cm and 7.6 ± 1.52 cm, 115.8 ± 22.56 cm and 7.6 ± 2.03 cm for large (SVL >50 mm) males, 82.3 ± 7.32 cm and 6.4 ± 0.76 cm, 67.1 ± 14.02 cm and 6.4 ± 1.52 cm, 70.1 ± 28.04 cm and 5.1 ± 1.83 cm for small (SVL 50 mm) males, 79.2 ± 5.49 cm and 5.8 ± 0.46 cm, 79.2 ± 18.90 cm and 8.1 ± 2.49 cm, 79.2 ± 20.73 cm and 6.9 ± 1.93 cm for females, 70.1 ± 5.49 cm and 4.6 ± 0.61 cm, 73.2 ± 11.58 cm and 5.1 ± 1.17 cm, 64.0 ± 24.99 cm and 4.1 ± 1.42 cm for "undetermined" individuals, with 7.0% of large males, 8.8% of small males, 14.9% of females, and 16.1% of "undetermined" on the ground (Turner and Gist, 1970); at 3 sites at El Yunque, juveniles close to the ground or to 2 m high on perches of small to moderate (1–8 cm) diameter, females mainly 1–2 m high on perches of small to large (2–16 cm) diameter, males primarily 1–3 m high on perches of moderate to large (8–16+ cm) diameter, ~80% of all individuals in deep shade (García Moll, 1978); mean perch diameter 33.5 ± 5.35 cm (Losos and Sinervo, 1989); mean perch height and diameter 1.3 ± 0.08 m and 33.5 ± 5.35 cm (Losos, 1990a); most frequently on large (>8 cm diameter) perches, with low habitat breadth of 45.2 (Irschick and Losos, 1999); adult males generally on conspicuous perches (e.g., large tree trunks, fallen logs, large branches), whereas juveniles are generally on narrow surfaces in bushes or at the bases of large trees (Irschick et al., 2000); in an upland forest with cleared understory, plurality of adult males (25%) on perches 3.0–6.0 m high with diameters 0.3–2.2 cm, largest number of "small" individuals (28%) on perches <0.9 m high with diameters 2.2–5.7 cm, with no individuals on rocks or the ground, in forest interior, most adult males (45%) on perches <0.9 m high with varying diameters and one of 70 lizards on the ground, most "small" individuals (75%) on perches <0.9 m high usually (64%) with diameters 0.3–5.4 cm and 8 of 173 lizards on the ground and one on a rock, at El Verde, largest percentage of adult males (44%) on perches 3.0–6.0 m high usually (32%) with diameters >12.5 cm and 2 of 205 lizards on the ground and one on a rock, plurality of "small" individuals (39%) on perches <0.9 m high with varying diameters and 4 of 89 lizards on the ground and 3 on rocks (Schoener and Schoener, 1971b); at El Verde in July, mean perch heights, diameters, and insolation for adult males and females were 182.5 ± 72.6 cm, 13.1 ± 8.3 cm, 730 ± 495 lux and 155.2 ± 67.1 cm, 12.6 ± 6.9 cm, 615 ± 322 lux, in January, 205.3 ± 93.3 cm, 14.3 ± 8.7 cm, 524 ± 334 lux and 144.9 ± 76.6 cm, 11.8 ± 8.8 cm, 541 ± 390 lux, with males but not females using a less diverse set of perch heights and diameters during the dry winter season (Lister, 1981); most (77%) individuals below 2 m, none above 5 m, a plurality (35%) on perches 20–30 cm in diameter, but all sizes used (Reagan, 1992, 1996a); mean perch heights and diameters 1.50 ± 0.78 m and 14.9 ± 15.8 cm for males and 1.13 ± 0.76 m and 8.3 ± 9.6 cm for females (Butler and Losos, 2002); males and females selected woody vegetation over nonwoody plants and Sierra Palms as perching sites (sexes did not differ in frequency), both sexes also selected vegetation with larger diameters than was randomly available, with males (10.6 ± 10.3 cm) selecting wider perches than females (8.7 ± 9.5 cm), perch heights did not differ significantly between males (1.62 ± 0.91 m) and females (1.49 ± 0.79 m), wider, woody perches presumably render lizards less conspicuous, provide them with the ability to scan more of their surroundings, and enhance locomotor functions (Rodríguez-Robles et al., 2005); mean perch heights 1.63 ± 0.50–3.53 ± 1.50 m and diameters 8.74 ± 6.50–14.45 ± 8.80 cm at 4 sites of varying elevations, with the variation in habitat use attributable to suitable microhabitat availability (Johnson et al., 2006); mean male perch height and diameter 1.29 ± 0.09 m and 0.56 ± 0.12 m, mean female perch height and diameter 0.98 ± 0.13 m and 0.79 ± 0.24 m (VanHooydonck et al., 2007); 11.4% on leaves (J. B. Losos, in litt., 21.IV. 2008); habitat light environment "full shade" (Fleishman, 2000); does not sleep on the same plants as those used by day, usually sleeps on leaves of woody plants taller than 1 m, uses plants with broad leaves, narrow leaves, compound narrow or broad leaves essentially in proportion to their abundance in the habitat, lizards with SVL >50 mm sleep significantly higher than those in all other size classes (Clark and Gillingham, 1990); sleeps on ground (in a small cavity in an earthen cutbank) to >2 m above the ground on leaves of shrubs and bushes (Schwartz and Henderson, 1991).

Activity: Apparently using visual cues, moves from leaf litter or adjacent vegetation to sleeping sites about 15 min before sunset, remaining there until sunrise (Clark and Gillingham, 1990). **Behavior:** Male threat display (24 sec) includes dewlap extension (stays extended during the entire response), body pushed up, tip of tongue everted, nuchal and dorsal crests raised (Ruibal, 1967); sleeps with eyes closed, chins pressed to leaf surface, front limbs forward, and hindlimbs extended, 72% oriented toward the base of the leaf, 40% oblique head up, 54% horizontal, 6% oblique head down, mean nocturnal recapture radius 0.19 ± 0.19 m, indicative of sleep-site fidelity, 36% of recaptures at the previous site, 55% within 10 cm, >96% within 1 m for 5 consecutive nights (Clark and Gillingham, 1990); mean display rate (proportion of time spent displaying) 0.055 ± 0.010, percent walks 33.9 ± 5.2, percent runs 42.9 ± 4.2, percent jumps 23.1 ± 2.0, moves/min 0.73 ± 0.08, distance jumped in nature 36.4 ± 4.0 cm (Losos, 1990a); frequency of displays related to frequency of "non-walk" movements (e.g., running, jumping), displays more frequently on trees than when on boulders, jumps proportionately more frequently when on boulders, rate of walking higher on trees than on boulders, but runs and jumps less frequently on trees, in general, moves less frequently when on trees (Losos, 1990b); tends to jump rather than run on perches of small diameter (Losos and Irschick, 1996); adult males tend to walk more and jump less than juveniles, but run at about the same frequency (Irschick et al., 2000); head-down posture in 23 of 30 individuals, moves/min 0.08 ± 0.03, percent time moving 0.13 ± 0.05% (Cooper, 2005); approach distances decrease as perch heights increase (all with perch heights of ≤ 2.1 m fled, 5 of 7 with perch heights >2.1 m did), approach distances greater at lower temperatures and on nonvertical perches, most "squirrel," some subsequently flee up (Cooper, 2006a); both absolute and size-corrected dewlap size significantly related with bite force, 17% of combat trials involved dewlap extensions, 28% biting, and 17% headbobbing (Lailvaux and Irschick, 2007); display speed 14.0–37.2 mm/sec, strongly correlated with maximum speed of moving vegetation, suggesting that males experiencing high levels of visual background noise produce high-speed displays (Ord et al., 2007); as the distance to receivers increased and viewing conditions deteriorated as a result of low light or high visual "noise" (windblown plants and rapidly shifting sunspots), males produced more broadcast displays that began with "alert signals" (a series of exaggerated 4-legged pushups), high-speed introductory movements greatly reduced the time required for receivers to respond to a robotic broadcast display under poor light conditions, lizards oriented more quickly when the display was preceded by either the species-typical alert (4-legged pushups) or a novel alert (a series of rapid dewlap extensions) than when the display lacked any alert (Ord and Stamps, 2008); male approach distance, flight distance, and final distance 1.51 ± 0.14 m, 0.57 ± 0.07 m, and 1.77 ± 0.15 m, female approach distance, flight distance, and final distance 1.53 ± 0.15 m, 0.35 ± 0.06 m, and 1.63 ± 0.15 m (VanHooydonck et al., 2007); moves per minute 0.37 and 0.73 (Johnson et al., 2008). **Biomass:** At El Verde, 11.8 kg/ha (Lister, 1981); 10.8 kg/ha (Reagan, 1996a). **Competition:** In experimental plots, the abundance of adult and juvenile *A. evermanni* increased in the absence of *A. gundlachi*, but no shift in structural habitat ("ecological release") was evident, could be the result of interspecific competition or intraguild predation (Leal et al., 1998). **Condition:** Body condition of males with intact tails over a 3-year period did not vary if infected with malaria (*Plasmodium* spp.) nor was condition affected by cumulative rainfall or low temperatures 1–3 months before sampling, but cumulative rainfall and low temperatures 4–6 months prior to sampling were significantly correlated with body condition, with condition highest after several months of warm, wet weather (Schall and Pearson, 2000). **Desiccation:** Standard rate of water loss for juveniles 3.8 mg/g of initial body weight/h, vital limit (percent of initial body weight lost before death) 24.5, and survival time 60 h (J. Molinari-Teron, 1969 in Heatwole and Veron, 1977); mean critical activity point (percent of original body weight at which the lizard could not right itself when turned on its back) 80.0 ± 3.85%, mean rate of water loss 0.40% body mass/h (Hillman and Gorman, 1977); rates of evaporative water loss exhibit same pattern, 0.154 ± 0.012% initial body weight/h (270 m elevation), 0.204 ± 0.026% initial body weight/h (480 m), 0.220 ± 0.015% initial body weight/h (675 m), and 0.229 ± 0.036% initial body weight/h (850 m), survival times vary by altitude, but vital limits of desiccation do not (Hertz et al., 1979). **Ecomorphology:** "Trunk-ground" (Williams, 1972); formally quantified (Beuttell and Losos, 1999). **Foraging and Diet:** Snails, spiders, mites, winged termites, dragonflies, ants (most important, 22.7%), beetles (adults and larvae), lepidopterans (adults and larvae), flies, hemipterans, and 2 large seeds (Wolcott, 1923); at El Verde in July, mean prey length and number of prey items/stomach 4.3 mm and 7.6 items/stomach, in January, 4.5 mm and 5.7 items/stomach, food items were ants, other hymenopterans, leafhoppers, flies, beetles (adults and larvae), lepidopterans (adults and larvae), orthopterans (greatest percent biomass in both wet, July, and dry, January, seasons), spiders, annelids, and seeds/fruits (Lister, 1981); *A. stratulus* (Reagan, 1992); at El Verde, estimates of 25.4 prey items/lizard led to an estimated daily consumption rate of 50,800 prey items/ha (Reagan, 1996a); stomachs contained ants (most important in numbers in males and females), flies, lepidopteran larvae, caddisflies, beetles, hemipterans, thrips, orthopterans (most important by volume in females), roaches, earthworms (most important by volume in males), *Eleutherodactylus* sp., and others (R. Garrison and D. P. Reagan in Reagan, 1996a); also at El Verde, *Eleutherodactylus coqui* and crab spiders (Stasina; Reagan et al., 1996); *Anolis krugi* (A. J. Sánchez Muñoz in Horstink, 2004); *Sphaerodactylus klauberi* (J. A. Román, in litt., 13.II.2008). **Growth and Longevity:** Reaches maturity in ~1 year and lives to 3 years; life expectancy 0.6 year (Lister, 1981). **Home Range:** 350 m^2 for males, 250 m^2 for females (Turner and Gist, 1970). **Metabolic Compensation for Altitude:** Despite lack of altitudinal compensation by thermoregulation, does not compensate metabolically, metabolic rates same for animals at moderate (350–400 m) and high (850–900 m) elevations, no change in metabolism after exposure to lower than usual temperatures (15 °C), but evidence of thermal stress after exposure to higher than usual temperatures (30 °C; Rogowitz, 1996). **Movements:** Distances moved between successive recaptures at 2 sites in upland forest 0–44.6 m for large (SVL >50 mm) males, 0–36.6 m for small (SVL 50 mm) males, 0–36.3 m for females, and 0–39.3 m for "undetermined" individuals (Turner and Gist, 1970); mean distance per movement 21.8 ± 2.2 cm at mean speed 37.6 ± 2.1 cm/sec, 0.37 ± 0.11 moves/min, 0.50 ± 0.18% time moving

(Irschick, 2000b). **Parasites:** Trematode *Mesocoelium monas* (Cofresí-Sala, 1964); protozoan malarial parasites *Plasmodium azurophilum* and *P. floridense* (Schall and Vogt, 1993). **Performance:** Sprint speed declines on smaller perches (Losos and Sinervo, 1898); mean maximum sprint speed (fastest time over an interval of 0.25 m) 0.116 ± 0.002 sec, jump 81.4 ± 1.89 cm, cling 2.3.5 ± 0.228 Neutons (Losos, 1990a); mean maximal speed 1.4 ± 0.16 m/sec, escape speed 1.3 ± 0.08 m/sec, feeding speed 1.1 ± 0.08 m/sec, undisturbed activity 0.5 ± 0.05 m/sec, mean maximal jumping distance 81.4 ± 1.89 cm, feeding jumping distance 31.4 ± 3.39 cm, undisturbed jumping distance 40.6 ± 3.52 cm (Irschick and Losos, 1998); peak velocity 1.81 ± 0.26 m/sec, peak acceleration 39.4 ± 6.3 m/sec^2 (Toro et al., 2004); acceleration 36.76 ± 2.77 m/sec, sprint speed 1.12 ± 0.08 m/sec (Vanhooydonck et al., 2006a). **Population Size and Density:** ~2,000 individuals/ha at El Verde (Turner and Gist, 1970); at El Verde in July, numbers and biomass for adult males and females were ~1,691/ha (~8,118 g/ha) and ~1,776/ha (~4,228 g/ha), in January, for adult males, adult females, and juveniles ~1,522/ha (~7,991 g/ha), ~1,395/ha (~2,664 g/ha), and ~1,057/ha (~888 g/ha; Lister, 1981); 1,976/ha (Reagan, 1992); as high as 800 individuals/ha (Rivero, 1998). **Population Structure:** Adult:juvenile ratio 66:34 at El Verde (Turner and Gist, 1970). **Predation:** Red-tailed Hawks (*Buteo jamaicensis*; Santana and Temple, 1988); spider *Oligoctenous ottleyi* (2.28 g) consuming a lizard with SVL 17 mm and 1.59 g (Clark and Gillingham, 1990); *Magliophis exiguus* (G. Rogowitz in Thomas and Gaa Kessler, 1996); *Epicrates inornatus* (Wiley, 2003); at El Verde, *Eleutherodactylus coqui* (Reagan, 1996b; Reagan et al., 1996; Stewart and Woolbright, 1996); also *A. evermanni*, tarantulas (*Avicularia* sp.), and whip scorpions (*Phrynus longipes*), forming "food loops" (Reagan, 1996b; Reagan et al., 1996); Lizard Cuckoos (*Saurothera vieilloti*), Pearly-eyed Thrashers (*Margarops fuscatus*), Red-tailed Hawks (*Buteo jamaicensis*), and *Borikenophis portoricensis* regularly feed on anoles (Reagan and Waide, 1996). **Relative Abundance:** 84–90% of wet season and 57–71% of dry season sightings of this species, *A. evermanni*, and *A. stratulus* at El Verde (Reagan, 1992, 1996a). **Reproduction:** Clutch size 1 (Hedges, in prep.); egg (5 × 11 mm) taken 17 February, hatched 18 March (Stejneger, 1904); maximal male reproductive activity March–August, testicular weights decrease during winter months, size of abdominal fat bodies is inversely correlated with testis weight (Licht and Gorman, 1975); all (N = 40) females in a wet-season sample and 17 of 40 in a dry-season sample from relatively aseasonal mesic habitat at El Verde were reproductively active, 24 of 25 females in a wet-season sample but only 1 of 30 in a dry-season sample from strongly seasonal xeric habitat in southwestern Puerto Rico were reproductively active (Lister, 1981). **Sex Ratio (M:F):** 57:55 at El Verde (Turner and Gist, 1970). **Size:** Male SVL to 76 mm, female SVL to 51 mm (Schall and Vogt, 1993). **Thermal Biology:** Heat stress monitored under rising temperatures was noted before onset in *A. cristatellus* and *A. stratulus* and comparable to high-elevation (~550–640 m) *A. evermanni* (Heatwole et al., 1969); lowest average body temperature of *A. cristatellus* group, mean body and air temperatures at 6 sites and at various times of day were 19.6 ± 0.2–29.2 ± 0.2 °C and 19.4 ± 0.1–28.9 ± 0.1 °C, critical thermal maximum 35.7 ± 0.5 °C, much less heat resistant than *A. cristatellus* (Huey and Webster, 1976); lizards survived and half gained weight under conditions simulating mean minimum January temperatures of the eastern Puerto Rican rainforest (conditions under which *A. cristatellus* died or lost weight), indicating tolerance of extreme thermal conditions in upland rainforest habitats (Gorman and Hillman, 1977); critical thermal maxima vary with elevation (all substantially lower than those of lowland *A. cristatellus*), 36.0 ± 0.2 °C (270 m elevation), 34.6 ± 0.3 °C (480 m), 33.5 ± 0.2 °C (675 m), and 33.0 ± 0.2 °C (850 m; Hertz et al., 1979); mean minimum and maximum body temperatures at an elevation of 270 m on Puerto Rico 24.0 ± 0.1 °C and 27.1 ± 0.1 °C at maximum and minimum air temperatures of 23.7 ± 0.1 °C and 26.5 ± 0.1 °C (overall mean body temperature 25.6 ± 0.1 °C at mean air temperature 25.1 ± 0.1 °C), at 480 m 22.3 ± 0.1 °C and 25.1 ± 0.1 °C at air temperatures of 22.1 ± 0.1 °C and 24.3 ± 0.1 °C (overall 24.1 ± 0.1 °C at 23.6 ± 0.1 °C), at 675 m 21.7 ± 0.1 °C and 24.5 ± 0.2 °C at air temperatures of 21.6 ± 0.1 °C and 23.9 ± 0.2 °C (overall 23.2 ± 0.1 °C at 22.8 ± 0.1 °C), at 850 m 20.8 ± 0.1 °C and 22.6 ± 0.2 °C at air temperatures of 20.6 ± 0.1 °C and 22.2 ± 0.1 °C (overall 21.8 ± 0.1 °C at 21.4 ± 0.1 °C), with opportunities for basking minimal or absent at all elevations (Hertz, 1981, 1983); in sunny weather, 25.0% of lizards basking at a high elevation site in January, 30.6% in August, 39.7% at a low elevation site in January, 26.0% in August, mean body temperatures in sun, filtered sun, shade, and overcast at a high elevation site in January none, 19.8 ± 0.9 °C, 19.1 ± 0.4 °C, 19.4 ± 0.2 °C, in August 20.9 °C (N = 1), 22.8 ± 0.3 °C, 21.9 ± 0.2 °C, 22.2 ± 0.2 °C, at a low elevation site in January 24.4 ± 0.3 °C, 24.4 ± 0.3 °C, 23.5 ± 0.2 °C, 24.8 ± 0.2 °C, in August 24.2 ± 2.1 °C, 26.3 ± 0.3 °C, 25.4 ± 0.2 °C, 26.0 ± 0.1 °C (Hertz, 1992b); prefers and is active at relatively low body temperatures, rarely basks, and is active throughout the day, essentially thermoconforms in shaded habitats, optimal body temperature for sprint speed 28.6 ± 2.5 °C (Hertz et al., 1993); uses less brightly illuminated and cooler habitats than those used by *A. cristatellus*, lizards occupy perches at random with respect to both light intensity and operative temperature at all times of day, environmental temperatures (which are highly correlated with body temperatures in the field) were significantly correlated with light intensity only at midday in habitat used by *A. gundlachi*, lizards selected microhabitats in the laboratory that were independent of experimental temperatures (Hertz et al., 1994). **Vision:** Neither the dewlap nor the rest of the body is highly UV reflective, which would be most effective in a UV-rich environment, but a UV-bright spot is associated with an open mouth, part of a typical anoline threat display (Fleishman et al., 1993); dewlaps are highly visible in typical habitat (Persons et al., 1999).

Conservation Status: "Stable" (Moreno, 1991d).

Anolis haetianus Garman, 1887

Distribution: Hispaniola, extreme western tip of the Tiburon Peninsula (Haiti); elevational range, sea level to 1,037 m.

Habitat: Largely mesophilic, but probably occurs in some relatively xeric situations (R. Thomas, in litt., 22.IV.2008).

Ecomorphology: "Trunk-ground" (Rand and Williams, 1969, by implication; Losos, 2009). **Size:** Male SVL to 75 mm, female SVL to 60 mm (Schwartz and Henderson, 1982).

Conservation Status: Unknown.

Anolis hendersoni Cochran, 1923

Distribution: Hispaniola, where two currently recognized subspecies occur in the eastern Massif de la Hotte, the Massif de la Selle (including the northern frontal ranges, Morne l'Hôpital, Montagne Noire), and associated lowlands (Haiti); elevational range, sea level to ~1,750 m.

Habitat: Mesophilic, but less so than related *A. bahorucoensis* and *A. dolichocephalus* (Schwartz and Henderson, 1991); can persist in extremely disturbed situations, in broadleaf forest or remnants and along forest edges, also on fence rows (*Yucca* hedges) on exposed, completely deforested mountainsides, in deciduous woods along creeks, in pinewoods and forests, along margins of a small stream in a cornfield, often in more or less unnatural situations such as road cuts (Schwartz, 1978b); 6.3% on leaves, 2.9% on grass blades, 9.2% on stems 3 mm in diameter, 55.9% on stems 3–6 mm in diameter, 16.0% on stems 6–12 mm in diameter, 3.3% on stems 12–18 mm in diameter, and 1.9% on stems 18–24 mm in diameter (Moermond, 1979a); 0.8% on leaves or grasses, 3.0% on perches <0.7 cm in diameter, 6.1% on perches 0.7–2.5 cm in diameter, and 1.5% on perches >2.5 cm in diameter at heights >90 cm, 5.3% on leaves or grasses, 30.5% on perches <0.7 cm in diameter, 19.1% on perches 0.7–2.5 cm in diameter, and 3.0% on perches >2.5 cm in diameter at heights 31–90 cm, 9.2% on leaves or grasses, and 16.8% on perches <0.7 cm in diameter at heights <30 cm (Moermond, 1979b); sleeps on shrubs and bushes 0.3–2.4 m above the ground aligned along stems and branches or lianas, not draped across leaves (Williams and Webster, 1974; Schwartz, 1978b; Schwartz and Henderson, 1991).

Behavior: Escapes by dropping to the ground and hopping or jumping from danger (Schwartz, 1978b); capable of vocalization (T. A. Jenssen in Milton and Jenssen, 1979); percent jumps (fraction of total movements) by adults in bushes at a Haitian site 46%, juveniles primarily in grass jump 43%, jumps or runs in about equal frequency, crawls less frequently (Moermond, 1979a); percent crawls 15%, runs 37%, jumps 44% (Moermond, 1979b). **Ecomorphology:** "Grass-bush" (Williams, 1963a). **Foraging and Diet:** Scanning from a high, broad perch takes longer than scans from perches with limited overviews (but less than scans of the ground from elevated perches), resulting in moderate waiting intervals and distances/move, mean 6.0 min between location changes, mean 4.3 min between location or posture changes, mean distance/move 27 cm, distance moved per prey capture longer than those of anoles surveying dense habitat clusters but shorter than those ranging from elevated perches to the ground, 43% of prey captures on the same surface as that on which the lizard is perched (Moermond, 1979b). **Parasites:** Trematode *Mesocoelium monas* and nematode Cosmocercidae gen. sp. (Goldberg et al., 1998b). **Predation:** *Hypsirhynchus parvifrons* (Henderson et al., 1987a). **Size:** Male SVL to 49 mm, female SVL to 42 mm (Schwartz, 1978b).

Conservation Status: Unknown.

Anolis homolechis Cope, 1864

Distribution: Cuba, where two currently recognized subspecies are essentially islandwide, but missing from some hot, dry areas, where they often are replaced by *A. jubar*; also on Isla de la Juventud, Archipiélago de los Canarreos, Cayos de San Felipe, and cays north of Cárdenas; elevational distribution, sea level to ~1,830 m.

Habitat: Mesophilic (Schwartz and Henderson, 1991); in filtered sunlight of forest margins and within more open and sparse woods, including palm-pine savannas, mogote broadleaf forest, and microphyllous semideciduous forest, but the species has adapted to disturbed situations such as gardens, farms, and pastures, where it is relatively common on fence posts and thatched huts (Schwartz and Ogren, 1956; Ruibal, 1961, 1964; Ruibal and Williams, 1961b; Buide, 1966; Garrido and Schwartz, 1968b; G. Peters and Schubert, 1968; Schwartz, 1968f; Lando and Williams, 1969; Schwartz and Henderson, 1991; Estrada, 1993a, 1994; Rodríguez Schettino and de Queiroz, 2002a; Kunz, 2005); less common than many congeners in towns, usually perches at heights of 1–2 m (Fläschendräger and Wijffels, 1996, 2009); at all sites surveyed in the Biosphere Reserve "Sierra del Rosario," occupying diverse habitats, but preferentially in sunny areas on trunks of trees, but also on trunks of bushes and on the ground (Martínez Reyes, 1995); "scrubby woods and especially . . . wooded ravines with streams" (Barbour, 1930c); in thornforest, especially near watercourses, and at higher elevations where it replaces the otherwise ubiquitous *A. sagrei* (Fläschendräger and Wijffels, 2009); in cloud forest, montane rainforest, submontane rainforest, submontane mesophyllous evergreen forest, mesophyllous evergreen and semideciduous forest, pine forest, mogote complex of vegetation, swamp forest, coastal and subcoastal microphyllous forest, semidesert thorny scrubwoodland, serpentine xeromorphous shrub-woodland, cultivated areas and pastures, and urban zones (Rodríguez Schettino et al., 1999); males perch head-down on small tree trunks a meter or less above the ground (or 0.9–2.4 m at Guantanamo Naval Base), often with tails curled laterally, females are more terrestrial, usually on the ground or on perches closer to the ground than males (Barbour and Ramsden, 1919; Ruibal and Williams, 1961b; Lando and Williams, 1969); where sympatric with *A. sagrei*, 58.34% of lizards on trees, 8.33% on bushes, and 33.33% on the ground (Berovides Álvarez and Sampedro Marín, 1980); also where sympatric with *A. sagrei*, less frequently on the ground and in bushes and more frequently in trees than expected (differences not significant), more commonly on trees (14 of 24) than on the ground (8) or bushes (2), mean perch heights and diameters at 3 localities 77.82 ± 9.08 cm and 7.52 ± 1.23 cm (morning), 46.38 ± 3.76 cm and 5.37 ± 1.26 cm (afternoon), and 62.58 ± 5.67 cm and 6.48 ± 0.89 cm (all day; Sampedro Marín et al., 1982a); on the Archipiélago de los Canarreos (Cuba), 19 of 25 males on trees (6 on the ground), 13 of 16 females on trees (3 on the ground), most individuals of both sexes on trunks, mean height and diameter for males 72.4 ± 9.2 cm and 10.8 ± 2.7 cm, for females 37.9 ± 6.6 cm and 11.0 ± 2.2 cm (Estrada, 1993a); in all kinds of forests, but largely in less densely shaded areas or along edges, in some plantations (*Coffea arabica* and *Persea americana*), males perch at mean heights <1.5 m, usually with heads facing down, tails may be prehensile, occasionally on the ground, fallen logs, or rocks, females and juveniles mainly on the ground, fallen logs, roots, or rocks and on trunks of trees and bushes generally at heights below those used by males, but no differences between sexes at some localities in La Habana Province, where mean heights varied by season and habitat (0.51 m, 0.68 m, 0.76 m, 1.29 m), along forest edge in the Sierra del Rosario mean height 0.93 m, in mogote complex of vegetation both sexes at mean height of 1.11 m, in semideciduous forest on limestone substrate on the Península de Guanahacabibes (mean height 0.67 m) and in the Jardín Botánico Nacional (0.66 m) regardless of sex or season, but at the latter site in a different year males (1.13 m) perched higher than females (0.84 m), possibly attributable to increased vegetation height providing greater variety of suitable perches, similar situation along the northeastern coast of La Habana Province, where mean height 0.84 m in coastal scrub and 1.52 m for males and 0.91 m for females in ecotone with taller trees (e.g., *Bursera simaruba*), numbers of individuals decrease with increases in elevation (Rodríguez Schettino, 1999b); 9 of 10 males (mean height 0.35 m; other on the ground) and 2 females on trunks of trees (both 0.2 m above the ground) in shade or partial shade, head downward (Rodríguez Schettino and de Queiroz, 2002a); mean perch height at Soroa 1.34 m, diameter 11.8 cm, behavioral diameter (mean diameter of all surfaces used during behavioral observations, means of individual means) 16.60 cm, 26.0% of observations on or within 1 m of rocks (J. B. Losos, in litt., 20.VIII.2007); basks on lower steps of staircases on hotel grounds (Rogner, 2002); habitat light environment "partial shade" (Fleishman, 2000); mean perch height and diameter in Cuba west of Cienfuegos 1.31 m and 22.5 cm, in southeastern Cuba 0.75 m and 3.0 cm (J. B. Losos, in litt., 21.IV.2008).

Activity: Diurnal (Rodríguez Schettino et al., 1999); bimodal, with peaks at 1000–1100 h and 1300–1400 h (Berovides Álvarez and Sampedro Marín, 1980); some individuals active shortly after dawn and a majority active at midday, in La Habana Province bimodal activity with peaks at 0900 and 1300 h, in the Sierra del Rosario, only females bimodal, the single activity peak for males from 1230–1430 h (Rodríguez Schettino, 1999b). **Behavior:** Tails carried rather sharply curled laterally (Barbour, 1920); generally rests head downward, keeps the tail curled when resting, but "twitch[es] and wriggle[s] the distal third nervously" when disturbed (Barbour, 1930c); laterally curled tail waving may be aposematic (Mertens, 1946); escape behavior involves "squirreling" (both sexes) and climbing higher into a tree (males) or dropping to the ground (females) and hiding or running to another tree (Lando and Williams, 1969); mean approach distance 2.10 ± 0.07 m, always prefers running up rather than down, first responses "squirreling" (174 of 249) followed by running up (78 of 174), jumping (64), running down (21), or staying in position (11), first response jumping (46 of 249) followed by second responses "squirreling" (25 of 46), and running up, jumping, or staying in place at about equal frequencies, first response running (23 of 249) followed by jumping (18 of 23), "squirreling" (3), or running up, first response running up (6 of 249) followed by "squirreling" (4), running up, or jumping (Regalado, 1998); males run into herbaceous growth or jump to another tree before hiding among stones or dead leaves, females and juveniles hide in leaf litter (Rodríguez Schettino, 1999b); male threat display initiated by pawing with the forefoot nearest the aggressor, push-up with large dewlap extended, tail whipping (total elapsed time 6 sec with 4 push-ups and 4 tail lashes), dewlap extended during the entire display or slightly pulsed 2–3 times, nuchal and dorsal crests erected (Ruibal, 1967); a subadult male defended his territory in a Rose Apple Tree (*Eugenia jambos*) by displaying with an extended dewlap, headbobs, and tail waves, as well as rotations of one forelimb in a fashion much like that when shed skin is removed from the head, after a pause, the resident repeated the display while assuming a lateral presentation, followed by an attack that included jaw locking and biting the body before the intruder fled (Ruiz García, 1975); moves per minute 0.67 (Johnson et al., 2008); mean percent time displaying 8.0, percent walking, 45.3, percent jumping 23.8, percent running 43.7 (do not add up to 100% because information from 2 data sets was combined; J. B. Losos, in litt., 21.IV.2008). **Competition:** Restriction to forest habitats in central Cuba may be correlated with the presence of *A. allisoni* around human habitations (Ruibal and Williams, 1961b); in Cuba, where sympatric with *A. sagrei*, perch heights (mean 68 cm for *A. homolechis* and 58 cm for *A. sagrei*) differed significantly, and *A. homolechis* had a higher proportion of ants in its diet, but types of perches used and times of activity differed only slightly, structural niche overlap 0.833, activity niche overlap 0.862, dietary niche overlap 0.581 (Berovides Álvarez and Sampedro Marín, 1980). **Ecomorphology:** "Trunk-ground" (Williams, 1969). **Foraging and Diet:** Responds quickly when a termite nest is opened (Barbour, 1930c); in captivity (at Guantanamo Naval Base) eats wild-caught roaches, moths, caterpillars, flies, millers, crickets and grasshoppers, and mealworms (Lando and Williams, 1969); captive individuals show a strong preference for spiders of the families Araneidae, Salticidae, and Thomisidae over insects (O. H. Garrido in Alayón García, 1976); a small adult female ate a large spider (*Ctenus variabilis*), apparently after actively searching for such prey, which is normally nocturnal and reclusive by day (Alayón García, 1976); on Cuba, ants (68.27%), caterpillars (14.42%), beetles (7.69%), spiders (9.62%; Berovides Álvarez and Sampedro Marín, 1980); at 3 sites, 32%, 10%, and 5% of prey items were spiders (versus insects), insect prey included ants and other hymenopterans (63%, 76%, 93%), lepidopterans (mostly caterpillars, 29%, 16%, 6%), and beetles (8%, 8%, 1%; Sampedro Marín et al., 1982a); includes cockroaches, beetles, flies, hemipterans, ants (most important in terms of numbers consumed), termites, lepidopteran larvae, orthopterans, barklice, thrips, crustaceans, and spiders (Silva Rodríguez and Estrada, 1994); feeds on small arthropods (mostly insects) detected from elevated perches and rapidly captured, this sit-and-wait strategy most common during the wet season, during the dry season it is supplemented by actively foraging in leaf litter and under cover, feeds mostly on ants (57.4–75.5% of prey taken), but also on lepidopterans, beetles, termites, insect larvae, leafhoppers, spiders, and conspecific juveniles, prey size generally larger for males, but they also consume many small prey items (Rodríguez Schettino, 1999b). **Parasites:** Nematodes *Cyrtosomum scelopori*, *Physaloptera squamatae*, *Skrjabinoptera phrynosoma*, larval *Abbreviata* sp., Nematoda gen. sp. (Coy Otero and Baruš, 1979b; Coy Otero and Lorenzo Hernández, 1982; Coy Otero, 1999); *Oswaldocruzia lenteixeirai*, *Parathelandros* sp., larval *Porrocaecum* sp. (Baruš and Coy Otero, 1969a; Coy Otero, 1970; Coy Otero and Baruš, 1979b; Coy Otero and Lorenzo Hernández, 1982; Coy Otero, 1999); *Parapharyngodon cubensis* (Baruš and Coy Otero, 1969a, 1969b; Coy Otero, 1970; Coy Otero and Baruš, 1979b; Coy Otero and Lorenzo Hernández, 1982; Coy Otero, 1999); and *Trichospirura teixeirai* (Coy Otero and Baruš, 1979a, 1979b; Coy Otero and Lorenzo Hernández, 1982; Coy Otero, 1999). **Population Size and Density:** At the Parque Zoológico Nacional in May, 1,166 lizards/ha, and in June, 2.068 kg/ha (via mark-recapture), using direct observations densities varied, 30.0–713.2 individuals/ha (Rodríguez Schettino, 1999b). **Predation:** *Anolis porcatus* (J. Novo in Socarrás et al., 1988; Rodríguez Schettino, 1999b); Lizard Cuckoo (*Saurothera merlini*) at a cave entrance where anoles were attracted to abundant cockroaches (*Periplaneta americana*; Rodríguez Schettino, 1999b); stalked by *Anolis luteogularis* at Soroa (P. E. Hertz in J. B. Losos, in litt., 21.IV.2008). **Reproduction:** In captivity, female responds to male display with headnods and tail elevation, after which male bites her neck and intromission occurs, copulation 4–12 min, egg ~6.5 × 10 mm, incubation ~35 days at 25–30 °C (to 17 °C at night), hatchling SVL 16–17 mm (Szidat, 1969); clutch size 1 (Hedges, in prep.); 2–3 eggs (suggesting communal deposition) in hollow recesses or under loose bark at the base of a tree (Lando and Williams, 1969); mean duration of copulation 4.4 min (longest 8 min), separation of pair "normal," egg 9.0 × 5.2 mm, plaster white and elliptical (Garrido, 1980a); cyclical during the rainy season, with males with enlarged testes (>4 mm) and females with vitellogenic follicles February–July, but activity of short duration, females bear 1–2 oviductal eggs April–September, after which activity declines (Silva Rodríguez and Estrada, 1984); egg (12.5 × 9.5 mm) taken from bromeliad 7 July, hatched 17 July (Schwartz and Henderson, 1991). **Sex Ratio (M:F):** 0.68:1 (Rodríguez Schettino, 1999b). **Size:** Male SVL to 70 mm (Garrido, 1973c), female SVL to 55.8 mm (Rodríguez Schettino, 1999b). **Thermal Biology:** Mean body temperature in the morning 31.2 °C (26.2–36.6 °C), in the afternoon 31.9 °C (29.4–35.0 °C), mean temperature at which panting is initiated 36.2 °C (32.8–39.4 °C; Ruibal, 1961); mean body temperature 32.1 °C (27.8–30.8 °C air temperature), 30.4 °C (30.0–31.5 °C), 30.6 °C (28.5–28.6 °C), 30.2 °C (29.0–29.8 °C), 32.8 °C (32.4–33.0 °C), 31.0 °C (29.2–32.0 °C; Ruibal and Philibosian, 1970); apparently heliothermic, but the percentage of individuals in full sun varies by locality, body temperatures (mean 31.8 °C) higher than air temperatures (mean 30.8 °C) in Camagüey Province in July–August, also in the Sierra del Rosario (32.1 °C and 29.9 °C), but did not vary significantly in the Sierra de Casas (27.9 °C and 27.2 °C), air temperatures at other sites occupied by these lizards varied from 29.4 to 33.8 °C, depending on habitat and season (Rodríguez Schettino, 1999b); body and air temperatures of 2 females 32.0 °C and 31.0 °C, mean body and air temperatures of 10 males 30.6 °C and 29.9 °C, air and body temperatures highly correlated and increased during the course of a morning (Rodríguez Schettino and de Queiroz, 2002a); mean body temperature at Soroa 29.0 °C (J. B. Losos, in litt., 20. VIII.2007).

Conservation Status: Entire species not assessed, but *A. h. turquinensis* is considered "vulnerable" (L. Rodríguez Schettino and A. Chamizo Lara in Rodríguez Schettino, 1999a).

Anolis imias Ruibal and Williams, 1961

Distribution: Cuba, extreme eastern part of the island.

Habitat: Unknown; Garrido and Jaume (1984), in emending the type locality, noted the occurrence of this species on highly xeric coastal cliffs, but P. J. Darlington (collector of the holotype) stated that it had been taken in the mesic (?) Sierra del Purial; in semidesert thorny scrubwoodland (Rodríguez Schettino et al., 1999); among rocky outcrops where vegetation is xerophytic and thorny bushes prevail, under slabs of limestone and in rock fissures, males perch on trunks of bushes and on rocks to heights of 2 m (Rodríguez Schettino, 1999b).

Activity: Diurnal (Rodríguez Schettino et al., 1999); becomes active when rocks are no longer in direct sunlight, individuals do not emerge until 1600 h (Rodríguez Schettino, 1999b); spends much of the day hidden in rock crevices and caves, active in the very early morning and evening (M. Černý, 1992). **Behavior:** Escapes by darting into rock crevices and fissures, males very aggressive, displays consist of dewlap extension and active attacks (Rodríguez Schettino, 1999b). **Ecomorphology:** "Trunk-ground" (Williams and Rand, 1977). **Foraging and Diet:** Possible cannibalism, males attack females (L. Rodríguez in Martínez Reyes and Rodríguez Schettino, 1987; F. González Bermúdez in Rodríguez Schettino, 1999b); eats *Anolis argenteolus* along the southern coast of Guantánamo Province (J. de la Cruz in Socarrás et al., 1988; Rodríguez Schettino, 1999b). **Size:** Male SVL to 65 mm (Ruibal and Williams, 1961b), female SVL to 46.5 mm (Rodríguez Schettino, 1999b). **Thermal Biology:** Apparently nonheliothermic (Rodríguez Schettino, 1999b).

Conservation Status: Considered "vulnerable" (Rodríguez Schettino and Chamizo Lara, 1998a; L. Rodríguez Schettino and A. Chamizo Lara in Rodríguez Schettino, 1999a).

Anolis incredulus Garrido and Moreno, 1998

Distribution: Cuba, only from the type locality on Pico Turquino, Sierra Maestra.

Habitat: Unknown.

Activity: Diurnal (Rodríguez Schettino et al., 1999). **Ecomorphology:** "Trunk-crown" (Losos, 2009). **Reproduction:** Gravid female in July (Garrido and Moreno, 1998). **Size:** Female SVL to 34 mm, males unknown (Garrido and Moreno, 1998).

Conservation Status: Unknown.

Anolis inexpectatus Garrido and Estrada, 1989

Distribution: Cuba, extreme eastern uplands.

Habitat: Xerophilic to mesophilic (Schwartz and Henderson, 1991); in undergrowth (ground level to 2 m) in pine (*Pinus cubensis*) forest, with bushes, small pines, vines, ferns, and herbaceous plants, in rainforest where it uses the same structural niche as in pine forest, also at interfaces of pine and rainforests, roadsides, and cleared areas, females often on ferns and lower vegetation, mean perch heights

1.5 m and perch diameters 1–2 cm (Garrido and Estrada, 1989); in submontane rainforest, submontane mesophyllous evergreen forest, and pine forest (Rodríguez Schettino et al., 1999).

Activity: Diurnal (Rodríguez Schettino et al., 1999). **Behavior:** Locomotor modality jumping (Garrido and Estrada, 1989). **Ecomorphology:** "Grass-bush" (Losos, 2009). **Size:** Male SVL to 36.5 mm, female SVL to 35.0 mm (Garrido and Estrada, 1989). **Thermal Biology:** Apparently nonheliothermic (Rodríguez Schettino, 1999b).

Conservation Status: Considered "vulnerable" (Rodríguez Schettino and Chamizo Lara, 1998a; L. Rodríguez Schettino and A. Chamizo Lara in Rodríguez Schettino, 1999a).

Anolis insolitus Williams and Rand, 1969

Distribution: Hispaniola, Cordillera Central (Dominican Republic); elevational range 1,050–1,770 m.

Habitat: Mesophilic (Schwartz and Henderson, 1991); on the branch of a small tree 2.4 m above the ground, often on twigs, especially those with a diameter about same as that of the body, prefers "half-shade" (Rand and Williams, 1969); mean perch height and diameter 3.36 m and 0.36 cm (J. B. Losos in Calsbeek et al., 2006); sleeps completely exposed on tips of twigs, vines, and branchlets, occasionally on green, leafy shrubs, 0.6–7.6 m high, sheltered during heavy and continuous rain under vine mats or curtains, 2 on top of each other on a pendant vine (Schwartz, 1974b); on bush or tree leaves or on slender, nearly horizontal twigs of bushes 1.2–3.7 m above ground (Williams and Rand, 1969).

Activity: Diurnal (Schwartz and Henderson, 1991). **Behavior:** When not pressed, movements are a slow, deliberate walk or series of rapid hops, movement characterized as "crawler" (Williams and Rand, 1969; Moermond, 1981); when captured, opens mouth and extends dewlap, escapes by jumping from branch to branch, jumps willingly and accurately and over distances to 30 cm with no loss of height (or even with the second perch above the first) or "squirrels" with body adpressed to substrate, may go completely beneath a horizontal perch, tail may be used as a "sliding hook," with the tip sufficiently flexible to grip small branches, tail frequently carried in an upright curve, vent close to the substrate, or tail carried straight (Williams and Rand, 1969); moves per minute 0.42 (Johnson et al., 2008); percent time displaying 0, percent walking 81.3, percent jumping 12.5, percent running 6.3 (J. B. Losos, in litt., 21.IV.2008). **Ecomorphology:** "Twig" (Williams and Rand, 1969); formally quantified (Beuttell and Losos, 1999). **Foraging and Diet:** Prey strikes either approach-pause or stalk (Moermond, 1981. **Reproduction:** Clutch size 1 (Hedges, in prep.). **Size:** Male SVL to 47 mm, female SVL to 44 mm (Schwartz and Henderson, 1991).

Conservation Status: Not assessed.

Anolis isolepis Cope, 1861

Distribution: Cuba, mountainous regions of eastern Cuba (Holguín, Guantánamo, Granma, and Santiago de Cuba provinces), except the Sierra Maestra.

Habitat: Mesophilic (Schwartz and Henderson, 1991); deep broadleaf forests, 2 individuals at about 2.4 m above the ground in leaves of the lowest canopy (Barbour and Ramsden, 1919; Ruibal, 1964); in cloud forest, montane rainforest, and submontane rainforest (Rodríguez Schettino et al., 1999); to 4 m above the ground (Rodríguez Schettino et al., 2003); walking on the tops of leaves in the canopy, mean perch height and diameter 7.13 m and 0.3 cm, all perching on leaves (J. B. Losos, in litt., 21.IV.2008).

Activity: Diurnal (Rodríguez Schettino et al., 1999). **Behavior:** Vocal, emits high-pitched squeaks when captured, aggressive, bites repeatedly (Ruibal, 1964); climbs to escape (Garrido, 1985). **Ecomorphology:** "Twig" (Williams, 1972), but morphologically trunk-crown (Losos, 2009). **Reproduction:** Egg 10.0 × 5.7 mm, cream or white, and oval (Garrido, 1980a). **Size:** Both male and female SVL to 40 mm (Garrido, 1985). **Thermal Biology:** Apparently nonheliothermic (Rodríguez Schettino, 1999b).

Conservation Status: Barbour and Ramsden (1919) described it as a "rare denizen of the deep mountain forests" and Ruibal (1964) said it was "probably the rarest of the Cuban anoles"; considered "vulnerable" (Rodríguez Schettino and Chamizo Lara, 1998a; L. Rodríguez Schettino and A. Chamizo Lara in Rodríguez Schettino, 1999a).

Anolis juangundlachi Garrido, 1975

Distribution: Cuba, only from the vicinity of the type locality in northern Matanzas Province.

Habitat: Mesophilic (Schwartz and Henderson, 1991); in grasses near watercourses (Garrido, 1975c); in mesophyllous evergreen and semideciduous forest (Rodríguez Schettino et al., 1999); grasslands in gallery forest, where perches on stems and leaves of grasses in shaded places near the ground (Rodríguez Schettino, 1999b).

Activity: Diurnal (Rodríguez Schettino et al., 1999); majority of active individuals at midday (Rodríguez Schettino, 1999b). **Behavior:** Shy, difficult to capture (Garrido, 1975c); generally immobile, jumps to escape and hides under leaves (Rodríguez Schettino, 1999b). **Ecomorphology:** "Grass-bush" (Losos, 2009). **Reproduction:** Mean duration of copulation 16.7 min (longest 22 min), separation of pair "hasty" (Garrido, 1980a). **Size:** Male SVL to 35.8 mm, female SVL to 34.3 mm (Rodríguez Schettino, 1999b). **Thermal Biology:** Apparently nonheliothermic (Rodríguez Schettino, 1999b).

Conservation Status: Considered "critically endangered" (Rodríguez Schettino and Chamizo Lara, 1998a; L. Rodríguez Schettino and A. Chamizo Lara in Rodríguez Schettino, 1999a); Hurricane Lili in 1996 dramatically altered the habitat of this species (Rodríguez Schettino, 1999b).

Anolis jubar Schwartz, 1968

Distribution: Cuba, where nine currently recognized subspecies occupy the lowlands and satellite islands along the northeastern and southeastern coasts, Isla Turiguanó, Archipiélago de Sabana-Camagüey (probably on cayos Romano, Guajaba, and Coco), and Los Ballenatos in the Bahía de Nuevitas.

Habitat: Xerophilic to mesophilic (Schwartz and Henderson, 1991); in both xeric and mesic forests, coastal woods, and scrub—virtually any wooded situation with some shade (Schwartz, 1968f; Schwartz and Henderson, 1991; Estrada, 1993b); in mesophyllous evergreen and semideciduous forest, coastal and subcoastal microphyllous forest, and semidesert thorny scrubwoodland (Rodríguez Schettino et al., 1999); absent from deltas, alluvial fans, coasts with barrier reefs, and mangroves, not edificarian (Ruibal and Williams, 1961b); especially abundant in microphyllous forest on Cayo Coco, on trunks of trees and bushes 0–2 m, with females and juveniles lower (<1 m) and primarily on bushes, in general on the ground, fallen logs, and stumps, males seek shelter in hollow branches and under bark of dead trees, females, subadults, and juveniles may share shelters of adult males, but generally seek refuges even closer to the ground (Estrada and Garrido, 1990); on Cayo Paredón Grande, Archipiélago de Sabana-Camagüey, on trunks of palms (*Coccothrinax littoralis*) 1–2 m above the ground or horizontally ~2 m above the ground on Black Olive Trees (*Bucida* sp.; Estrada, 1998); in coastal scrub with an abundance of Sea Grapes (*Coccoloba uvifera*) and along forest edges, openings, or ecotones in uplands, males perch on trunks of trees and bushes, usually at heights <2 m, sometimes on the ground, on fallen logs and branches, or in leaf litter with juveniles, mean height in a coastal area in Guantánamo Province 0.88 m (maximum 2.6 m for only one male), usually perches with head facing down (Rodríguez Schettino, 1999b); among *Coccoloba* trees on the ground or to heights of ~1 m (Rodríguez Schettino, 2003b); mean perch height and diameter 0.99 m and 10.3 cm (J. B. Losos, in litt., 21.IV.2008).

Activity: Diurnal (Rodríguez Schettino et al., 1999); number of individuals did not vary during the day (0800–1600 h) in July in the Sierra de Cubitas or on the southern coast of Guantánamo Province, but lizards at La Mula in July first appear at 1100 h and are active only until 1300 h (Rodríguez Schettino, 1999b). **Behavior:** Male threat display includes very rapid dewlap extension (remains extended during the entire display, ~8 sec) and slow retraction, rapid tail wagging (about 10 times), body raised in several push-ups, but headbobbing rare, nuchal and dorsal crests present (Ruibal and Williams, 1961b); vocal when handled (Rodríguez Schettino et al., 1999); escapes by running down and hiding among herbaceous growth or in leaf litter, occasional males climb and then jump to the ground or another tree (Rodríguez Schettino, 1999b). **Ecomorphology:** "Trunk-ground" (Williams and Rand, 1977). **Foraging and Diet:** Forages on the ground (Rodríguez Schettino, 1999b). **Home Range and Territoriality:** Male territories small, often use the same perch (Rodríguez Schettino, 1999b). **Parasites:** Nematodes *Cyrtosomum longicaudatum* (Coy Otero, 1979; Coy Otero and Lorenzo Hernández, 1982; Coy Otero, 1999); *C. scelopori*, *Parapharyngodon cubensis*, and larval *Porrocaecum* sp. (Coy Otero and Baruš, 1979b; Coy Otero and Lorenzo Hernández, 1982; Coy Otero, 1999); larval *Abbreviata* sp. (Coy Otero et al., 1987). **Population Size and Density:** Populations at La Mula declined steadily, apparently in response to human activity in the area, density at Cayo Coco 213–377 individuals/ha in evergreen microphyllous forest (Rodríguez Schettino, 1999b). **Predation:** *Leiocephalus macropus* (Fong Grillo and del Castillo, 2002; Martínez Reyes and Moreno García, 2003). **Reproduction:** Mean duration of copulation (*A. j. albertschwartzi*) 3.09 min (longest 8.3 min), (*A. j. balaenarum*) 5.06 min (12.15 min), (*A. j. cuneus*) 5.04 min (8.04 min), (*A. j. jubar*) 3.06 min (8 min), separation of pairs "normal" (Garrido, 1980a); clutch size 1 (Hedges, in prep.); egg white, deposited singly in moist soil under rocks (Estrada, 1993b). **Size:** Male SVL to 62 mm (Schwartz and Henderson, 1991), female SVL to 53.2 mm (Rodríguez Schettino, 1999b). **Thermal Biology:** Mean body temperature in the morning 30.8 °C (26.8–36.2 °C), in the afternoon 32.5 °C (29.8–35.4 °C; Ruibal, 1961); much more frequently in shade (72.7% of observations) than basking (27.3%) in coastal areas, but body temperatures of 11 males (mean 32.7 °C) higher than mean air temperature (31.6 °C), in the Sierra de Cubitas, mean body temperature 31.7 °C (Rodríguez Schettino, 1999b).

Conservation Status: "Of concern" (Buide et al., 1974); *A. j. balaenarum* and *A. j. santamariae* are considered "endangered" and *A. j. albertschwartzi*, *A. j. maisiensis*, and *A. j. cocoensis* "vulnerable" (L. Rodríguez Schettino and A. Chamizo Lara in Rodríguez Schettino, 1999a).

Anolis kahouannensis Lazell, 1964

Distribution: Îlet-à-Kahouanne, Guadeloupe Passage northwest of Guadeloupe.

Habitat: On *Hippomane* (Schwartz and Henderson, 1991); most abundant in *Hippomane* groves, in association with *Thespesia populnea*, and in mangroves, rare in xerophytic vegetation and at higher elevations, where restricted to a few large trees (Breuil, 2002b).

Ecomorphology: Undetermined. **Reproduction:** Copulation in August, many hatchlings in August (Breuil, 2002b). **Size:** Male SVL to 76 mm (Lazell, 1964a), female SVL to 48 mm, smallest mature male and female SVL 57 mm and 47 mm, although data may be incomplete (Lazell, 1972).

Conservation Status: Unknown.

Anolis koopmani Rand, 1961

Distribution: Hispaniola, southern slopes of Massif de la Hotte on the Tiburon Peninsula (Haiti); elevational range ~240–750 m.

Habitat: Mesophilic (Schwartz and Henderson, 1991); at the edge of a coffee plantation at dusk among small twigs about 1 m above the ground and on the ground in damp leaf litter, juvenile in low herbaceous vegetation about 15 cm above the ground (Rand, 1961); 36.3% on leaves, 15.8% on grass blades, 10.9% on stems 3 mm in diameter, 22.0% on stems 3–6 mm in diameter, and 5.6% on stems 6–12 mm in diameter (Moermond, 1979a); 9.9% on leaves or grasses and 0.6% on perches 0.7–2.5 cm in diameter at heights 31–90 cm, 55.9% on leaves or grasses, 27.6% on perches <0.7 cm in diameter, and 5.9% on perches 0.7–2.5 cm in diameter at heights <30 cm (Moermond, 1979b).

Behavior: Percent jumps (fraction of total movements) in grass and herbaceous growth 59%, jumps more frequently than runs, rarely crawls (Moermond, 1979a); percent crawls 6%, runs 35%, jumps 59% (Moermond, 1979b). **Ecomorphology:** "Grass-bush" (Williams and Webster, 1974).

Foraging and Diet: Scanning densely clustered habitats takes more time than scans of open clusters or broad surfaces but less time than scans of the ground from elevated perches, resulting in relatively short waiting intervals and distances/move, mean 3.9 min between location changes, mean 3.4 min between location or posture changes, mean distance/move 12 cm, distance moved per prey capture shorter than those of anoles surveying open habitat clusters or broad surfaces and much shorter than those ranging from elevated perches to the ground, 56% of prey captures on the same surface as that on which the lizard is perched (Moermond, 1979b). **Size:** Male SVL to 42 mm, female SVL to 35 mm (Williams and Webster, 1974).

Conservation Status: Unknown.

Anolis krugi Peters, 1876

Distribution: Puerto Rico; elevational range, sea level to ~1,205 m.

Habitat: Mesophilic (Schwartz and Henderson, 1991); shady, but open situations such as low bushes, vines, and ferns along roads and trails, in denser shade at lower elevations (Rivero, 1998); at edges of coffee plantations, occasionally in open forest or even forest interiors, usually on perches of small diameter at heights below 0.9 m, on ferns and vines in a moist, dark gorge, generally in cooler, shadier sites than *A. pulchellus* where the 2 species occur in sympatry (K. P. Schmidt, 1920, 1928; Rand, 1964a); in coffee groves, sweet potato patches, mountain meadows, and pastures (Wolcott, 1923); "invaded" a high elevation site after irradiation promoted leaf fall and increased illumination (Turner and Gist, 1970); at 2 upland sites, most adult males (51%) on perches 3.0–6.0 m high usually (51%) with diameters 0.3–2.2 cm and no lizards on the ground or rocks, most "small" individuals (67%) on perches <0.9 m high on leaves, with 4 of 83 lizards on the ground (Schoener and Schoener, 1971b); not infrequently on the ground (A. J. Sánchez Muñoz in Horstink, 2004); at El Yunque, juveniles in understory ground vegetation and to 1 m high on saplings and seedlings with diameters <1 cm, females and males perch higher and on perches of small to moderate (1–8 cm) diameters, males using perches of larger diameter than females, individuals in all age classes mostly in shaded areas near the ground (García Moll, 1978); mean perch height and diameter 0.6 ± 0.08 m and 5.7 ± 1.74 cm (Losos, 1990a); mean perch heights and diameters 0.91 ± 0.63 m and 6.2 ± 10.1 cm for males and 0.67 ± 0.56 m and 3.9 ± 9.6 cm for females (Butler and Losos, 2002); uses perches at heights of 0.64 ± 0.83–0.96 ± 0.40 m and with diameters of 2.29 ± 2.20–5.14 ± 8.30 cm at 4 sites of varying elevations on Puerto Rico, with variation reflecting an optimal microhabitat range selected regardless of availability (Johnson et al., 2006); mean male perch height and diameter 0.63 ± 0.12 m and 1.08 ± 0.30 m, mean female perch height and diameter 0.30 ± 0.12 m and 1.33 ± 0.84 m (VanHooydonck et al., 2007); males occupy higher and thicker perches than smaller individuals, juveniles on leaves, sleeps preferentially in areas of thick plant growth (mean height 94 cm), on low shrubs and ferns and on tangles of Spanish Moss (*Tillandsia*) 1.2–2.7 m above the ground, higher than *A. pulchellus* where the 2 species are sympatric, although the difference in bush habitat not significant (Goto and Osborne, 1989); habitat light environment "partial shade" (Fleishman, 2000); 32.1% on leaves (J. B. Losos, in litt., 21.IV.2008).

Behavior: Males elevate nuchal and dorsal crests when fighting (K. P. Schmidt, 1928); mean display rate (proportion of time spent displaying) 0.040 ± 0.008, percent walks 49.6 ± 5.3, percent runs 25.8 ± 6.0, percent jumps 24.7 ± 3.0, moves/min 0.84 ± 0.10, distance jumped in nature 28.2 ± 3.2 cm (Losos, 1990a); uses head-down posture when on vertical surfaces, moves/min 0.70 ± 0.19, percent time moving (PTM) 1.60 ± 0.41%, PTM decreases with increases in perch height (Cooper, 2005); approach distances increase as perch heights increase, most escape by running away on the ground, but sometimes climb bushes or other vegetation (Cooper, 2006a); male approach distance, flight distance, and final distance 1.66 ± 0.14 m, 0.54 ± 0.07 m, and 2.02 ± 0.14 m, female approach distance, flight distance, and final distance 1.33 ± 0.27 m, 0.42 ± 0.07 m, and 1.69 ± 0.33 m (VanHooydonck et al., 2007); moves per minute 0.50 and 0.84 (Johnson et al., 2008). **Desiccation:** Standard rate of water loss 3.1 mg/g of initial body weight/h, vital limit (percent of initial body weight lost before death) 19, and survival time 57 h (J. Molinari-Teron, 1969 in Heatwole and Veron, 1977); mean critical activity point (percent of original body weight at which the lizard could not right itself when turned on its back) 72.7 ± 4.46%, mean rate of water loss 0.41% body mass/h (Hillman and Gorman, 1977). **Diet:** Includes snails, sowbugs, spiders, ticks, earwigs, crickets, springtails, ants and other hymenopterans, beetles, thrips, flies (adults and larvae), hemipterans, and lepidopteran adults and larvae (most important, 33%; Wolcott, 1923); uses sit-and-wait foraging strategy (Cooper, 2005). **Ecomorphology:** "Grass-bush" (Williams, 1961); formally quantified (Beuttell and Losos, 1999). **Home Range and Territoriality:** Males vigorously defend territories (K. P. Schmidt, 1928). **Injuries:** "Not uncommon to find specimens with injured dewlaps and often with injured mouths, probably due to their habit of locking jaws" (K. P. Schmidt, 1928); female with an injured snout (presumably by a "rival") condemned to death by starvation (Sánchez Muñoz, 2008). **Parasites:** Trematode *Mesocoelium monas* (Cofresí-Sala, 1964); protozoan malarial parasite *Plasmodium azurophilum* (Telford, 1975; Ayala, 1977, 1978; Schall and Vogt, 1993); parasitic fly (*Philornis* sp.) maggot (A. J. Sánchez Muñoz, in litt., 17.III.2008). **Performance:** Mean maximum sprint speed (fastest time over an interval of 0.25 m) 0.140 ± 0.003 sec, jump 67.8 ± 1.07 cm, cling 2.162 ± 0.122 Neutons (Losos, 1990a); acceleration 34.87 ± 1.80 m/sec, sprint speed 1.11 ± 0.05 m/sec (Vanhooydonck et al., 2006a). **Predation:** *Anolis cuvieri* (Rand and Andrews, 1975); predators include *Borikenophis portoricensis*, *Magliophis exiguus*, *Anolis cristatellus*, *A. gundlachi*, and *A. cuvieri* (A. J. Sánchez Muñoz in Horstink, 2004), but observations of *M. exiguus* and *A. cuvieri* preying on *A. krugi* occurred in captivity (A. J. Sánchez Muñoz, in litt., 17.III.2008). **Reproduction:** Reproduction tied to the rainy season (Rivero, 1998); maximal male reproductive activity March–August, testicular weights decrease during winter months, size of abdominal fat bodies is inversely correlated with testis weight (Licht and Gorman, 1975); clutch size 1 (Hedges, in prep.); in captivity, egg 6.7 × 5.5 mm, incubation ~49 days at 20–26 °C, hatchling SVL 16 mm (Horstink, 2004). **Size:** Male SVL to 53 mm, female SVL to 44 mm (Schall and Vogt, 1993). **Thermal Biology:** At low daytime temperatures (correlated with cloudiness or deep shade at higher elevations), body temperatures suggest that incoming radiation is insufficient to warm the body above ambient, individuals exposed to cold temperatures (11.5 ± 0.5 °C) were capable of locomotor movements, but did not attempt to feed, individuals at an elevation ~580 m survived at a rate above *A. evermanni* at the same elevation and low-elevation *A. cristatellus* and *A. stratulus*, with smaller individuals within the same species succumbing first (Heatwole et al., 1969); mean critical thermal maxima 36.0 ± 0.2 °C (200 m elevation), 35.6 ± 0.1 °C (350 m), 35.2 ± 0.2 °C (500 m), and 34.9 ± 0.2 °C (650 m), significantly more heat resistant than sympatric *A. pulchellus* (Hertz, 1979b). **Vision:** Dewlaps, but not the rest of the body, are highly UV reflective, most effective in a UV-rich environment, a UV-bright spot is associated with the open mouth, part of a typical anoline threat display (Fleishman et al., 1993); dewlaps highly visible in typical habitat (Persons et al., 1999).

Conservation Status: "Stable" (Moreno, 1991d).

Anolis leachii Duméril and Bibron, 1837

Distribution: Antigua Bank (Antigua and Barbuda); introduced on Bermuda.

Habitat: Xerophilic to mesophilic (Schwartz and Henderson, 1991); highly arboreal, in xeric scrub, in open situations, on telephone poles adjacent to a pasture, on trees along an avenue on Antigua, edificarian, seeks shelter under shutters and eaves (Underwood, 1959b; Lazell, 1972; Schwartz and Henderson, 1991; Powell and Henderson, 2008c); adult males perch 1–5 m above the ground (sometimes higher), not known to forage on the ground, but females and juveniles may (Lazell, 1972); mean perch height on Barbuda 1.77 ± 0.06 m (Roughgarden et al., 1983a); like *A. bimaculatus* on the St. Christopher Bank, very capable of living as a human commensal (Wijffels, 1984); on Antigua, mean perch height and diameter 2.01 ± 0.21 m and 7.3 ± 1.3 cm (Losos and de Queiroz, 1997a); in shady thermal microhabitats 73% of time, mean perch height 122 ± 2.43 cm, uses higher perches during midday periods, most frequently used perches are trunks of trees, branches, and agaves (Kolbe et al., 2008); presumably sleeps on leaves and branches, but also under shutters and eaves (RP).

Activity: Active 0500–1900 h, peak at 0900 h (Kolbe et al., 2008); foraging for insects attracted to lights at 2000 h on Antigua (Schwartz and Henderson, 1991; Perry et al., 2008; RP). **Competition:** Aggressive interactions with *Anolis wattsi* usually will result in displacement of *A. wattsi* (Kolbe et al., 2008). **Diet:** Basically insectivorous, but eats *Opuntia* fruits and flowers on Great Bird Island (Lazell, 1972); on Barbuda, feeds on crumbs placed on a plate on the ground (Lachner, 1987); on Great Bird Island, attracted to canned tuna and ate bread (Schwartz and Henderson, 1991; Henderson and Powell, 1999); preys on *A. wattsi* (J. J. Kolbe, O. Davis in Kolbe et al., 2008). **Ecomorphology:** Because of large size, less close to "trunk-crown" than many Lesser Antillean anoles (Losos and de Queiroz, 1997a; Beuttell and Losos, 1999). **Home Range and Territoriality:** Extremely territorial in captivity (Fläschendräger and Wijffels, 2009). **Longevity:** In captivity, 7 years (A. Odum in Slavens and Slavens, 2003). **Parasites:** Cestode *Mesocoelium monas*, nematodes *Paraphyaryngodon cubensis*, *Spinicauda spinicauda*, *Trichospirura* sp., Physalopteridae gen. sp., and acanthocephalan cystacanths *Centrorhynchus* sp. (Dobson et al., 1992); nematode *Spauligodon anolis* (Dobson et al., 1992; Bursey and Goldberg, 1998); mean parasite species/host 1.14 in forests and 0.72 in xeric habitat on Antigua (Dobson and Pacala, 1992); cestode *Oochoristica anolis*, nematodes *Parapharyngodon cubensis* and larval *Abbreviata* sp. and *Physaloptera* sp. (Goldberg et al., 1996b). **Population Size and Density:** Density on Barbuda (several sites) 1.2 ± 0.2–5.9 ± 1.1 individuals/100 m^2 (Roughgarden et al., 1983a). **Predation:** *Alsophis antiguae* and *A. sajdaki* (Schwartz and Henderson, 1991; Henderson and Sajdak, 1996; Daltry et al., 2001). **Reproduction:** Females bury eggs in captivity, hatchling SVL ~23 mm (Fläschendräger and Wijffels, 1996, 2009). **Size:** Male SVL to 123 mm, female SVL to 70 mm (Lazell, 1972); in captivity, female SVL to 85 mm (Fläschendräger and Wijffels, 1996, 2009); smallest mature male and female SVL 68 mm and 52 mm, respectively (Lazell, 1972).

Conservation Status: Not assessed; probably "not currently compromised" (Daltry, 2007).

Anolis lineatopus Gray, 1840

Distribution: Jamaica, where four currently recognized subspecies occur essentially islandwide.

Habitat: Xerophilic to mesophilic (Schwartz and Henderson, 1991); in habitats ranging from open, dry areas to dark forests, in shady gardens, secondary growth, fencerows through grazed pastures, mangrove edges, rainforest, xeric coastal woods, berms, on porches, verandas, in houses, in a small cave, on trees, fence posts, stone walls, large and small rocks, ruined and occupied houses, hedges, bushes, and arboreal bromeliads (Underwood and Williams, 1959; Rand, 1967a, 1967b; Schoener and Schoener, 1971a; P. Vogel, 1984; Schwartz and Henderson, 1991); for cover uses dense vegetation, grasses, herbaceous plants or ferns that grow around a perch, dead sticks, large dead leaves, piles of rocks (uses sides of vertical perches and tops of slanting or horizontal perches), most perches within 1.8 m of the ground, with males higher and on larger diameter perches (e.g., tree trunks and fence posts) than females and juveniles, which are usually in bushes, hedges, and brush piles close to the ground, usually on vertical perches or the upper sides of slanted or horizontal perches with heads down (Rand, 1967a, 1967b); prefers drier and less shady situations when in sympatry with *A. reconditus* (Lazell, 1966); plurality (36.4%) of large adults on perches 0.9–1.5 m high and frequently (45.5%) on large-diameter (>7.6 cm) perches, most smaller individuals (60.3%) on perches <0.9 m high and plurality (44.9%) on large-diameter (>7.6 cm) perches (Rand, 1967b); lowest perch heights of abundant Jamaican anoles (comparable to those of *A. sagrei*), elevated perches from <0.9 to >6.0 m, males usually on perches of intermediate (2.2–5.7 cm) diameter and females and juveniles lower and on perches of smaller diameter (Schoener and Schoener, 1971a); at Discovery Bay, mean adult male and female perch heights 1.0 ± 0.46 m (7.7% on the ground) and 0.8 ± 0.32 m (19.8% on the ground), at Savanna-la-Mar, mean adult male and female perch heights 1.2 ± 0.52 m (2.3% on the ground) and 0.8 ± 0.52 m (5.5% on the ground; Lister, 1976); 76% perch <1.8 m above ground, 83% on perches

>1.3 cm in diameter, 26% adult males <1 m above ground, 60% of females and juveniles <1 m above ground (Jenssen, 1973); mean perch diameter 26.0 ± 4.78 cm (Losos and Sinervo, 1989; Losos, 1990a); mean perch height 1.0 ± 0.07 m (Losos, 1990a); lizards most frequently on small (2–4 cm diameter) perches, with intermediate habitat breadth of 64.8 (Irschick and Losos, 1999); adult males generally on conspicuous perches (e.g., large tree trunks, fallen logs, large branches), whereas juveniles generally on narrow surfaces in bushes or at the bases of large trees (Irschick et al., 2000); mean perch heights and diameters 0.87 ± 0.60 m and 6.2 ± 5.9 cm for males and 0.56 ± 0.44 m and 5.8 ± 5.9 cm for females (Butler and Losos, 2002); mean male perch height and diameter 0.71 ± 0.13 m and 0.61 ± 0.28 m, mean female perch height and diameter 0.80 ± 0.11 m and 0.49 ± 0.18 m (VanHooydonck et al., 2007); 13.0% on leaves (J. B. Losos, in litt., 21.IV.2008); habitat light environment "partial shade" (Fleishman, 2000); dewlap UV-reflectance compatible with visibility in typical habitat (Macedonia et al., 2000); sleeping sites may or may not be within an individual's home range and may be shared with other lizards with activity ranges that are exclusive by day, include exposed positions 0.3–3.0 m above ground on distal ends of branches, leaves, twigs, grass blades, larger individuals higher than smaller lizards, may use the same sleeping perch night after night, sleeping sites distinctly different from daytime perches (Rand, 1967a); moves from perching on trunks and branches by day to using mostly leaves at night, male and female sleeping sites did not differ significantly, but both perched higher, on perches of smaller diameter, and at inclinations more parallel to the ground at night than by day, some fidelity to sleeping sites (sleeping in a limited portion of the diurnal home range), males more spatially dispersed at night (females equally dispersed by day and night), most individuals sleep near the boundaries of diurnal home ranges (Singhal et al., 2007).

Activity: Diurnal, but will forage around lights at night, leave sleeping perches before dawn (0617 h, 19 °C), but remain concealed until the sun is up (0716 h), when they move to basking sites, although on the same perch for extended periods, displays frequently, rarely go 15 min without shifting position, one adult male shifted 12 times in one 4-min period, males patrol activity areas, sunset at 1744 h, rarely settles down to sleep (1811 h) until too dark to see (Rand, 1967a); earliest activity at Discovery Bay 0546 h, latest activity 1745 h, at Sun Valley Plantation 0643 h and 1707 h (although the latter might be compromised by curtailed sampling periods), male displays peak at dawn and dusk, presumably to advertise territorial activity before and after a period of nocturnal inactivity (Ord, 2008). **Behavior:** Allows a closer approach when moving slowly, also allows a closer approach by a stick than a head or body, remains motionless until darting around the tree and running up a short distance, flight distance inversely correlated with body temperature, compensating for reduced activity at lower-than-preferred temperatures at which anoles often remain in the open (Rand, 1964b); first reaction to an approaching threat is to remain still, sometimes flattening against the perch, if pressed, will "squirrel" and run either up (adult males) or down, even onto the ground (smaller individuals), when captured frequently defecates and shows most of the agonistic display, nuchal and dorsal crest erection, tail lashing, gaping with tongue protruding, dewlapping, but not flashed as when displaying, sleeps with limbs flexed or extended, slow to leave sleeping perches when disturbed, doing so only when shaken violently (Rand, 1967a); threat display includes nuchal and dorsal crest erection, dewlapping, and headbobbing (Rand, 1967a; Ruibal, 1967); displays in the following categories: (1) back jumping, (2) step bobbing, (3) dewlapping, (4) orientation, (5) posture, and (6) jaw fencing (blood may be drawn), adult males spend more time displaying than in any other activity except sleeping and watching, one male observed for 11 h gave 181 displays and had 2 conflicts with other males, copulated twice, and courted unsuccessfully several times, most agonistic behavior directed toward conspecifics of similar size (much less frequently toward larger or smaller individuals), most aggressive behavior occurs in a lizard's activity area and on favored perches, 85% of 182 disputes won by the larger lizard, some interspecific agonistic behavior with *A. garmani*, *A. grahami*, and *A. opalinus* (Rand, 1967a); mean display rate (proportion of time spent displaying) 0.031 ± 0.007, percent walks 15.3 ± 3.4, percent runs 57.5 ± 4.6, percent jumps 27.3 ± 3.0, moves/min 0.82 ± 0.09, distance jumped in nature 35.8 ± 3.6 cm (Losos, 1990a); tends to jump rather than run on perches of small diameter (Losos and Irschick, 1996); adult males tend to walk more and jump less than juveniles, but ran at about the same frequency (Irschick et al., 2000); flight initiation distance much shorter when a putative predator approaches slowly, even when the approach speed is initially fast, than when approach is fast, even when initially slow (Cooper, 2006b); 48% of combat trials involve dewlap extensions, 29% biting, and 67% headbobbing (Lailvaux and Irschick, 2007); when sleeping perch disturbed, lizards drop to the ground (Singhal et al., 2007); male approach distance, flight distance, and final distance 0.82 ± 0.19 m, 0.52 ± 0.11 m, and 1.14 ± 0.21 m, female approach distance, flight distance, and final distance 0.89 ± 0.11 m, 0.33 ± 0.03 m, and 1.09 ± 0.12 m (VanHooydonck et al., 2007); moves per minute 0.65 and 0.82 (Johnson et al., 2008). **Competition:** In the Kingston area, may force *A. opalinus* to select higher perches when sharing trees (Jenssen, 1973). **Ecomorphology:** "Trunk-ground" (Rand, 1967a); formally quantified (Beuttell and Losos, 1999). **Foraging and Diet:** Moves to the ground from elevated perches to feed (Grant, 1940a); when naïve males in natural habitat were presented with green, blue, or brown morphs of Sphinx Moth (Sphingidae) caterpillars, all of which forage on twigs, but only the latter rests cryptically on brown stems, 12 of 21 green caterpillars and only 1 of 20 brown individuals were attacked, suggesting that the lizards' foraging strategies exert strong selective pressure favoring brown morphs (Curio, 1970); waits on an elevated perch, may take food without moving, but usually moves to the ground or nearby foliage, with the largest individuals moving farthest (large adult males as much as 1.2–1.5 m), approach is a quick run, often broken by one or more pauses, and usually a pause just before prey is reached, during pauses, individuals cock their heads and look at the prey with one eye (presumably for better definition), with eyes moving independently of each other, then shifts back to binocular vision (presumably for better evaluation of direction and distance), may leap to catch an insect, small prey frequently eaten where caught, larger prey usually carried back to a perch, most prey items <5 mm regardless of lizard size, but larger lizards eat more large items, diet includes spiders, millipedes, orthopterans, roaches, earwigs, hemipterans, lepidopterans (adults and larvae), beetles (adults only), and hymenopterans (largely ants), occasionally dragonflies, flies, snails, small frogs, conspecifics, bits of vegetation, and shed skin, with large adult males taking larger prey and having fewer food items in stomachs, with smaller items mostly ants, in contrast, smaller individuals took a variety of small prey, predominantly not ants, a young male descended to the ground to eat a berry that had broken open and was moist inside (Rand, 1967a); stomach contents were ants (31.4%), flies (13.6%), beetles, termites, hemipterans, lepidopteran larvae, orthopterans, spiders, pseudoscorpions, centipedes, mites, isopods, gastropods, an anole, and plant matter (Lister, 1976); *A. valencieni* (Hicks and Trivers, 1983); naïve hatchlings cannot distinguish between stinging ants (*Solenopsis geminata*) and stingless ants (*Pheidole fallax*), explaining a general reluctance to take ants, lizards rarely attacked *S. geminata* when offered alone, but consumed considerable numbers of *P. fallax* when offered alone, older juveniles also confuse harmful and harmless species of ants, feeding on them according to their prior experience (P. Vogel and von Brockhusen-Holzer, 1984); prey-naïve hatchlings may reject or accept ants, rejection decreased with hatchling size, increased with ant size and abundance of stinging ants (*Solenopsis geminata*), those that initially accepted ants developed an aversion to most prey types and grew more slowly than those that rejected ants (P. Vogel, 1996); 4.3% of stomachs contained fruit, 1.2 fruits/individual (Herrel et al., 2004); mites, spiders, beetles, thrips, dermapterans, roaches, millipedes, flies (adults and larvae), lacewings, hemipterans, hymenopterans (with ants most important in numbers), isopods, termites, lepidopterans (adults and larvae, most important overall), mollusks, dragonflies, orthopterans, phasmids, aphids, other small arthropods, one juvenile anole (unidentified), lizard skin, fruit, seeds, and a stone (adventitious?), prey not taken in accordance with availability, with larvae disproportionately represented in stomach samples, mean prey length, width, mass, and numbers for males 7.26 ± 5.62 mm, 2.77 ± 1.48 mm, 0.02 ± 0.04 g, and 1.41 ± 1.15, females 4.20 ± 4.27 mm, 1.89 ± 0.91 mm, 0.05 ± 0.22 g, and 1.52 ± 0.99, juveniles 7.49 ± 18.23 mm, 1.46 ± 0.78 mm, 0.01 ± 0.02 g, and 1.95 ± 1.52, respectively, with males taking significantly "harder" prey correlated with much greater bite force and head size (Herrel et al., 2006); individuals get water from licking drops of dew or rain from leaves (Rand, 1967a). **Home Range and Territoriality:** Males actively defend territories, but females and juveniles do not (Grant, 1940a); lizards may use the same perch day after day, using only one or a few perches as centers for activity, adult male activity range usually 3–10 m^2 (maximally about 20 m^2), extends vertically 1–3 m, adult female range usually 1–2 m^2 (maximally 3–4 m^2), extends vertically 0.3–1.5 m, juvenile activity ranges smaller than those of adults, increase with size, in general, no 2 lizards of comparable size have widely overlapping ranges (regardless of size or sex), but individuals of different sizes may have widely overlapping ranges, only 2 of 30 trees with 2 or more males, only 1 of 21 trees with 2 or more females (Rand, 1967a); females defend territories against other lizards of similar size, regardless of species, males appear more tolerant of other species (Rand, 1967b); dewlap size (secondary sexual characteristic) is correlated with bite force (related to fighting capacity) in this territorial species (Vanhooydonck et al., 2005); mean diurnal home range size 4.7 m^2, nocturnal home range size 0.99 m^2 (Singhal et al., 2007). **Hybridization:** With *Anolis grahami*, hybrids probably infertile (Jenssen, 1977). **Movements:** Young males approaching adult size may be displaced from juvenile home ranges by adult males, movements 15–24 m and probably more to settle in unoccupied, subpreferable sites high in trees or on very small isolated perches (Rand, 1967a); mean distance per movement 18.3 ± 2.0 cm at mean speed 48.8 ± 4.5 cm/sec, 0.38 ± 0.10 moves/min, 0.48 ± 0.13% time moving (Irschick, 2000b). **Parasites:** Protozoan malarial parasites *Plasmodium azurophilum* and *P. floridense* (Telford, 1975; Ayala, 1977, 1978); in *A. l. lineatopus*, trematode *Mesocoelium monas*, acanthocephalan *Centrorhynchus spinosus*, nematodes *Cyrtosomum scelopori*, *Thelandros* sp. (P. Vogel and Bundy, 1987; Bundy et al., 1987), an oxyuroid, and in *A. l. neckeri*, nematodes *Rhabdias* sp. (Bundy et al., 1987) and

Spauligodon anolis (Bundy et al., 1987; Bursey and Goldberg, 1998). **Performance:** Sprint speeds decline on smaller perches (Losos and Sinervo, 1989); mean maximum sprint speed (fastest time over an interval of 0.25 m) 0.123 ± 0.002 sec, jump 75.7 ± 2.53 cm, cling 1.876 ± 0.244 Neutons (Losos, 1990a); clinging ability 0.97 ± 0.09 Neutons, divided by mass 0.16, with the relatively low value characteristic of trunk-ground ecomorphs (Elstrott and Irschick, 2004); mean maximal speed 1.7 ± 0.05 m/sec, escape speed 1.4 ± 0.12 m/sec, feeding speed 0.9 ± 0.08 m/sec, undisturbed activity 0.8 ± 0.12 m/sec, mean maximal jumping distance 75.7 ± 2.53 cm, feeding jumping distance 14.0 ± 3.61 cm, undisturbed jumping distance 28.1 ± 4.21 cm (Irschick and Losos, 1998); mean speed on broad and narrow perches 1.74 ± 0.22 m/sec and 0.84 ± 0.16 m/sec for adult males, 1.54 ± 0.22 m/sec and 0.73 ± 0.12 m/sec for adult females, and 1.26 ± 0.19 m/sec and 0.68 ± 0.07 m/sec for juveniles (Macrini and Irschick, 1998); mean absolute escape speed higher than feeding speed, which was higher than undisturbed speed for males, females, and juveniles (Irschick and Losos, 1998; Irschick, 2000a); sprint speed reduced when turning, 1.46 ± 0.06 m/sec with no turn, 1.16 ± 0.06 m/sec over 30° turn (reduction to 79% of sprint speed without turns), 0.73 ± 0.04 m/sec over 90° turn (50%), about half of individuals jumped to traverse 90° turns, but potential benefits of jumping offset by pauses when preparing to jump or recovering balance after jumping (Higham et al., 2001); peak velocity 1.90 ± 0.28 m/sec, peak acceleration 40.5 ± 4.7 m/sec^2 (Toro et al., 2004); acceleration 43.66 ± 2.53 m/sec, sprint speed 1.26 ± 0.05 m/sec (Vanhooydonck et al., 2006a). **Predation:** *Anolis garmani* (Gosse, 1850); domestic cats (maybe most important in area around Kingston), dogs, chickens, snakes (*Hypsirhynchus callilaemus*), presumably insectivorous and carnivorous birds, other anoles (rarely conspecifics), including *A. garmani*, *A. grahami*, *A. valencienni* (Rand, 1967a); ants (*Crematogaster brevispinosa*), few *A. lineatopus* eggs, juveniles, or adults in areas densely populated by these ants (P. Vogel, 1983); *Tropidophis jamaicensis* (Wilson and Vogel, 2000; B. S. Wilson, in litt., 28.IV.2006); Greater Antillean Grackles (*Quiscalus niger*) capable of taking all size classes (Graves, 2006). **Reproduction:** Clutch size 1 (Hedges, in prep.); eggs under logs and stones, "local boys . . . reported finding some . . . in the leaf litter" (Rand, 1967c); courtship display distinct from threat display in that it included with dewlapping a rapid series of headbobs (Ruibal, 1967); relatively little display behavior associated with consummated courtship ("very casual"), copulation probably occurs throughout the year, evidence of pair bonding (Schwartz and Henderson, 1991); larger females lay larger eggs than smaller females, sex ratio of hatchlings remains constant throughout year and does not differ from unity, hatchling SVL 17–22 mm (mean 19.5 mm), eggs laid under logs, edges of rocks, or in leaf litter, egg-laying rate in the wet season 7.7 days, hatchling weight varied with season, 199.9 mg in wet season and 215.0 mg in dry season, hatchling recruitment and juvenile growth highly predictable by rainfall (arthropod abundance higher in the wet season than in the dry season), growth (mean juvenile growth rate 0.19 mm/day in the wet season, 0.12 mm/day in the dry season) influenced by sex and site (P. Vogel, 1984); starvation not considered a mortality factor (Rand, 1967a); some females active in all months, but pronounced seasonal fluctuations are evident, with peak egg production corresponding to the rainy season (June–October), when nearly all females ovigerous and with essentially no fat bodies, female body fat weight highest during the dry months, when reproductive activity drops, testicular weights decrease during the wet months, remain small for 4–5 months, and begin increasing in February, although spermatogenesis and androgen production continue throughout the year in most males, fat body weights are inversely proportional to testicular weights (Licht and Gorman, 1970); in captivity, incubation 35–40 days at 30 °C (Bech, 1979); one female laid 18 eggs (12 fertile) in 6 months, incubation 55–60 days at 20–25 °C (U. Manthey, 1984 in Fläschendräger and Wijffels, 2009); hatchling SVL 25 mm (Fläschendräger and Wijffels, 2009). **Size:** Male SVL to 71 mm, female SVL to 51 mm, smallest sexually mature male 45 mm SVL and female 41 mm SVL (Licht and Gorman, 1970); sexually mature at SVL 35–39 mm (P. Vogel, 1984). **Thermal Biology:** "Preferred" body temperatures 28–31 °C, always close to and slightly above ambient temperatures, uses behavioral thermoregulation, apparently more careful to avoid extremely hot conditions than cool temperatures, color changes may play a role in thermoregulation, in early morning, when sunning, the body is darker than during the heat of the day (Rand, 1967a); maintains temperatures higher than ambient and intermediate between those of *A. grahami* (warmer) and *A. opalinus* (Rand, 1967b); in a closed situation 28.3 ± 1.02 °C and 27.6 ± 0.92 °C (Lister, 1976). **Vision:** In *A. l. neckeri*, relatively low critical flicker-fusion frequency corresponds to an association with shady habitats (Jenssen and Swenson, 1974).

Conservation Status: Not assessed.

Anolis litoralis Garrido, 1975

Distribution: Cuba, coastal lowlands in Santiago de Cuba, Guantánamo, and Holguín provinces, but also onto the slopes of adjacent uplands.

Habitat: Xerophilic (Schwartz and Henderson, 1991); on fences, trees, and spiny bushes and neighboring xerophytes in shrubby areas (Garrido, 1975b); in hot, dry coastal scrub with succulent/thorn vegetation, one male on a gravel beach with no elevated vegetation, only ground-hugging *Ipomoea pes-caprae* (Fläschendräger and Wijffels, 1996, 2009); often in thorn-free depressions between "ribs" of columnar cacti (*Pilosocereus, Ritterocereus*; Albinsky in Fläschendräger and Wijffels, 1996, 2009); in coastal and subcoastal microphyllous forest, semidesert thorny scrub-woodland, and serpentine xeromorphous shrub-woodland (Rodríguez Schettino et al., 1999); spends most of the day high in trees, descending only in early morning and late evening (Rogner, 2002); on thin branches and twigs not occupied by *A. sagrei* or *A. homolechis*, usually on low perches by day only in early morning or late afternoon (Fläschendräger and Wijffels, 2009).

Activity: Diurnal (Rodríguez Schettino et al., 1999). **Behavior:** Aggressive display of males ($N = 2$) includes lateral compression, raising nuchal and dorsal crests, 5 bobs with dewlap completely extended, entire process repeated several times (Garrido, 1988); to escape, moves slowly, stops, then runs again (Rodríguez Schettino, 1999b); when threatened, quickly climbs trunks of trees and moves out onto small branches, almost always encountered in pairs (pair-bonding?), in captivity, pairs often sleep on the same twig while maintaining body contact (Rogner, 2002); when a resident male was displaced from its territory, a smaller male occupied the territory within 3 days (Fläschendräger and Wijffels, 2009). **Ecomorphology:** Undetermined. **Foraging and Diet:** Gathers insects on branches of Sea Grapes (*Coccoloba uvifera*; Rodríguez Schettino, 1999b). **Parasites:** Nematode *Cyrtosomum scelopori* (Coy Otero and Baruš, 1979b; Coy Otero and Lorenzo Hernández, 1982; Coy Otero, 1999; note that the original association of parasite and host did not provide collecting localities, consequently, the hosts could be *A. centralis*; *A. litoralis* was elevated from subspecific status subsequent to the publication of Coy Otero and Baruš, 1979b). **Reproduction:** Mean duration of copulation 7.4 min (longest 41 min), separation of pair "rapid," eggs (April) oval, matte white or plaster white, 6.6–7.4 × 4.5–5.0 mm (Garrido, 1980a); courtship display (~28.0 sec) of longer duration and more elaborate than that of *A. centralis*, males raise head backward in a sequence of 6–7 nods, followed by 2–3 bobs, the process continuing until female acquiesces, dewlap fully extended during the initial sequence but only partially extended in the subsequent sequence, after intromission, male remains immobile (mean 33.5 sec) but releases his hold on her neck, mean copulatory time 7.4 min, during which female nods occasionally, copulation lasts 3–17 min (mean 5.39 min), male leaves female hastily or crawls over her (Garrido, 1988); in captivity, one female laid 7 eggs in one year (Rogner, 2007). **Size:** Male SVL to 48.1 mm (Navarro Pacheco et al., 2001); females somewhat smaller (Fläschendräger and Wijffels, 1996, 2009). **Thermal Biology:** Presumably heliothermic (Fläschendräger and Wijffels, 1996, 2009); apparently non-heliothermic, but body temperature of one male 33.4 °C at air temperature of 31.6 °C (Rodríguez Schettino, 1999b).

Conservation Status: Unknown.

Anolis lividus Garman, 1887

Distribution: Montserrat, essentially islandwide, sometimes exceedingly abundant; difficult to locate at elevations over ~760 m (Lazell, 1972).

Habitat: Mesophilic (Schwartz and Henderson, 1991); largely arboreal, with both adult males and females at comparable heights (~0.6–2.5 m) above the ground, but juveniles lower, readily exploit walls (Fläschendräger and Wijffels, 1996, 2009); but both sexes forage on the ground (Lazell, 1972); abundant in shady places in Plymouth and open situations in the cemetery (Underwood, 1959b); in mangroves, *Acacia*, and on walls (Schwartz and Henderson, 1991); males perch 0.6–2.4 m above ground on vertical substrates with >60% of adult males with SVL >56 mm on perches >1 m, 70% of males and females with SVL 41–55 mm on perches <1 m, 90% of individuals with SVL <40 mm on perches <50 cm above the ground (Brooks, 1981); mean perch height and diameter 2.01 ± 0.21 m and 7.3 ± 1.3 cm (Losos and de Queiroz, 1997a); perch heights are not significantly higher as elevation increases (Buckley and Roughgarden, 2005b).

Abundance: Abundance declines in an almost linear fashion with increasing elevation (Buckley and Roughgarden, 2005a). **Activity:** Diurnal (Schwartz and Henderson, 1991). **Diet:** Both sexes exploit similar trophic niches, mean number/lizard and size of prey items 19–20 and 2.9–3.4 mm, respectively, prey length in both sexes positively correlated with SVL, >88% of all prey items in all size classes <5 mm (Brooks, 1981); feeds on ants, adult female ate a dry, fragmented eggshell, probably of a Bananaquit (*Coereba flaveola*; Stevenson, 1991). **Ecomorphology:** Closest to "trunk-crown" (Losos and de Queiroz, 1997a; Beuttell and Losos, 1999). **Home Range and Territoriality:** Both sexes are aggressively territorial, distance between individuals usually 5–7 m (Fläschendräger and Wijffels, 1996, 2009). **Miscellany:** Males show eyeshine (Schwartz and Henderson, 1991). **Parasites:** Cestode *Allogyptus crenshawi*, nematodes

Paraphyaryngodon cubensis, Physalopteridae gen. sp., and acanthocephalan cystacanths *Centrorhynchus* sp. (Dobson et al., 1992); nematode *Spauligodon anolis* (Dobson et al., 1992; Bursey and Goldberg, 1998); mean parasite species/host 1.38 in forests and 0.51 in xeric habitat (Dobson and Pacala, 1992); malarial parasites *Plasmodium floridense* (Staats and Schall, 1996b). **Predation:** *Alsophis manselli* (Henderson and Sajdak, 1996). **Reproduction:** Clutch size 1 (Hedges, in prep.); hatchling SVL ~21 mm in captivity, two juveniles in a greenhouse in June (Fläschendräger and Wijffels, 2009). **Sex Ratio (M:F):** Apparently balanced (Lazell, 1972). **Size:** Male SVL to 72 mm (Brooks, 1981), female SVL to 53 mm (Schwartz and Henderson, 1991), smallest mature male and female SVL 44 mm and 40 mm, respectively (Lazell, 1972).

Conservation Status: Not assessed; effects of recent volcanic activity unknown.

Anolis longiceps Schmidt, 1919

Distribution: Navassa Island, islandwide.

Habitat: Xerophilic (Schwartz and Henderson, 1991); on the ground, rock faces, in open cactus scrub, large males usually on trunks and smaller branches of *Ficus* (Thomas, 1966c); uses almost every available habitat, in addition to *Ficus*, associated with a variety of trees (including *Metopium brownei*), foraging in leaf litter (especially females and juveniles), in skirts of dead, pendant fronds of fan palms (*Thrinax morrisii*), and in and on ruins at Lulu Bay, rare only in densely forested areas (although every opening was exploited) and absent only in completely open areas without vertical structure, perches to 3.5 m above the ground with males generally higher than females, but found at all heights, including the ground, before 1000 h, 27 of 36 individuals (including a pair in copula) on vegetation, 33 in full or partial sun, from 1000–1730, 42 individuals about equally abundant on vegetation and rock faces, with a few (8) on the ground or concrete posts, 38 in full shade, after 1730 h, 17 of 27 lizards on rock faces, 23 in full shade (Powell, 1999a); sleeps on palm fronds 1.8 m above the ground (Thomas, 1966c); 7 lizards sleeping on *Ficus* or *Metopium* leaves 1–3 m above the ground, one sleeping in a rock crevice (Powell, 1999a).

Activity: From just before dawn until full dark, although less frequently seen during early hours on cloudy days, primarily basking to 1000 h (on various perches, but most frequently on horizontal palm fronds and lower branches of medium-sized and large trees, toward evening, most evident on rock faces or boulders (Powell, 1999a). **Behavior:** Lizards on low perches (including lower branches of small palms and trees) moved away when approached (only one of 35 "dove" into leaf litter), anoles on the ground did "dive" into litter or "snaked" away into dense grass, those on trunks or other vertical structure would initially "squirrel" before moving up or down at about equal frequency, on rock faces, escape was almost invariably up, although an occasional individual would seek shelter in a crevice (Powell, 1999a). **Diet:** Ants (most frequently), hemipterans, and caterpillars (Powell, 1999a). **Ecomorphology:** "Trunk-crown" (Williams, 1969). **Reproduction:** Egg 7 × 12 mm on 26–28 April (Thomas, 1966c). **Size:** Male SVL to 83 mm (Thomas, 1966), female SVL to 76 mm (Schwartz and Henderson, 1991). **Thermal Biology:** Mean body temperature (32.2 ± 0.5 °C) significantly above substrate temperature (28.6 ± 0.3 °C), with male and female temperatures not differing significantly (Powell, 1999a).

Conservation Status: Not assessed, but the limited extent of the range suggests a vulnerability to stochastic events (Powell, 1999a).

Additional Pertinent Literature: Powell (1999b).

Anolis longitibialis Noble, 1923

Distribution: Hispaniola, where two currently recognized subspecies occur on the Península de Barahona (*A. l. specuum*) and Isla Beata (*A. l. longitibialis*).

Habitat: Xerophilic (Schwartz and Henderson, 1991); largely saxicolous in desert scrub on the main island (Hertz and Huey, 1981; Gifford et al., 2002); on rocks 56.4%, on trees 28.2%, and on posts, vines, stumps, or deadfall 15.4% of time (*A. l. specuum*; Gifford et al., 2002); 13 of 17 on rock faces, 3 in crevices, one on a tree (J. B. Losos in Gifford et al., 2002); on shrubs and cliff faces on Isla Beata (Schwartz and Henderson, 1991); mean perch heights and diameters 44.1 ± 6.0 cm and 17.5 ± 3.0 cm, respectively, and did not differ significantly by sex (Gifford et al., 2002); mean perch height and diameter 1.33 m and 9.0 cm (J. B. Losos, in litt., 21.IV.2008).

Diet: Stomachs of 45 individuals (32 males, 13 females) in June contained spiders and another arachnid, beetles (adults and larvae), roaches, a fly larva, hemipterans, hymenopterans (including ants), lepidopterans (an adult and larvae, which were the most important dietary item, suggestive of exploitation of seasonally abundant resources), a lacewing, orthopterans, millipedes, unidentified arthropods, snails, shed skin, flowers and other plant material, and grit, presumably ingested adventitiously (Gifford et al., 2002); stomachs of 8 lizards (4 males, 4 females) in January contained spiders (most important item), a beetle larva, flies, ants, a termite, a roach, crickets, an isopod, unidentified arthropod parts, a snail, miscellaneous plant material, and fruits/seeds, with the absence of lepidopteran larvae supporting the assumption that their importance in June was a seasonal phenomenon (Fontenot et al., 2003). **Ecomorphology:** "Trunk-ground" (Williams, 1983); formally quantified (Beuttell and Losos, 1999). **Habitat Partitioning:** On the Barahona Peninsula, evidence of habitat partitioning with *Ameiva chrysolaema* (almost entirely on soil substrates) and *Leiocephalus barahonensis* (usually on piles of small rocks; Gifford et al., 2002). **Reproduction:** Males in June with enlarged testes, 8 of 13 females ovigerous (5 with 2 eggs, but only 1 with 2 shelled eggs, indicative of ability to retain eggs) and the other 5 contained yolked ovarian follicles, egg size (8.3–11.4 × 4.9–6.9 mm) not correlated with female SVL, absence of juveniles suggests that June is the very beginning of the reproductive season in this xeric habitat (Gifford et al., 2002). **Size:** Male SVL to 72 mm, female SVL to 59 mm (Schwartz, 1979e). **Thermal Biology:** Most conspicuous on overcast days, rarely perches in full sun, body temperatures 30.2–34.3 °C (mean 32.2 °C) at air temperatures 29.2–33.8 °C (mean 31.1 °C; Hertz and Huey, 1981); on the Barahona Peninsula, in shade 77.1% and in partial shade 14.3% of time, mean male body, air, and substrate temperatures 30.5 ± 2.0 °C, 27.4 ± 0.4 °C, and 27.9 ± 0.4 °C, and those for females 32.2 ± 0.2 °C, 30.3 ± 0.7 °C, and 30.3 ± 0.8 °C, respectively, differences in body and substrate temperatures not significant, but females on perches where air temperatures are significantly higher than those of males (Gifford et al., 2002).

Conservation Status: Not assessed.

Additional Pertinent Literature: Gifford and Powell (2002a).

Anolis loysianus Duméril and Bibron, 1837

Distribution: Cuba, islandwide but local; elevational range to 1,120 m.

Habitat: Mesophilic (Schwartz and Henderson, 1991); on tree trunks, especially those with gray bark, on *Mangifera indica* in open, savannalike clearings, in areas formerly forested but cleared for agriculture, and on gray limestone on a stream bank (Barbour and Ramsden, 1919; Barbour, 1930c; Ruibal, 1964; Olexa, 1972; Schwartz and Henderson, 1991); at higher elevations, most abundant in valleys low on tree trunks and roots (M. Černý, 1992); in the Biosphere Reserve "Sierra del Rosario," primarily in the interior of evergreen forest on branches of trees (Martínez Reyes, 1995); in mesophyllous evergreen and semideciduous forest (Rodríguez Schettino et al., 1999); frequently on trees with bark and color that promote crypsis (*Bombyx* sp., *Lonchocarpus* sp., *Roystonea regia*, and *Guazuma tomentosa*), mean perch height on the Península de Guanahacabibes 1.55 m (Rodríguez Schettino, 1999b); on a trunk of a *Eucalyptus* tree (Fläschendräger and Wijffels, 2009); mean perch height at Soroa 3.07 m, diameter 22.4 cm, behavioral diameter (mean diameter of all surfaces used during behavioral observations, means of individual means) 16.20 cm, no observations on or within 1 m of rocks (J. B. Losos, in litt., 20.VIII.2007); mean perch height and diameter 0.79 m and 24.5 cm, none on leaves (J. B. Losos, in litt., 21.IV.2008).

Activity: Diurnal (Rodríguez Schettino et al., 1999). **Behavior:** Relatively inactive on trunks of trees where they are almost invisible (Olexa, 1972); relies heavily on crypsis when threatened (Fläschendräger and Wijffels, 1996, 2009); moves slowly except when pursued, takes shelter in fissures of trunks, escapes upward or jumps to another tree (Rodríguez Schettino, 1999b); moves per minute 1.83 (Johnson et al., 2008); mean percent time displaying 6.5, percent walking, 52.0, percent jumping 5.5, percent running 40.9 (J. B. Losos, in litt., 21.IV.2008). **Ecomorphology:** Categorized as a "bark anole" (G. Peters, 1970); "trunk" (Williams and Rand, 1977). **Foraging and Diet:** Approaches prey very deliberately (Fläschendräger and Wijffels, 1996, 2009); stomachs of 3 females contained ants (10 items in 2 stomachs), beetles (adults and a larvae, 7 items in 2 stomachs, including the largest items in the smallest individual, 5 and 6 mm in length), a spider, and a pseudoscorpion (Rodríguez Schettino, 2001). **Parasites:** Nematodes *Cyrtosomum scelopori* and *Oswaldocruzia lenteixeirai* (Coy Otero and Baruš, 1979b; Coy Otero and Lorenzo Hernández, 1982; Coy Otero, 1999). **Reproduction:** Copulation in September (L. F. de Armas in Rodríguez Schettino, 1999b); in captivity, hatchling SVL 17 mm (Olexa, 1972); fertile eggs laid by wild-caught female after 3 months in captivity, incubation 50–65 days at 23–26 °C, hatchling SVL ~15 mm (Fläschendräger, 2007); fertile eggs to 6 month after contact with a male (Fläschendräger and Wijffels, 2009). **Size:** Male SVL to 47.2 mm, female SVL to 38.2 mm (Rodríguez Schettino, 1999b). **Thermal Biology:** Apparently nonheliothermic (Rodríguez Schettino, 1999b); mean body temperature at Soroa 29.0 °C (J. B. Losos, in litt., 20.VIII.2007).

Conservation Status: Considered to be at "lower risk" of extinction (L. Rodríguez Schettino and A. Chamizo Lara in Rodríguez Schettino, 1999a).

Anolis luciae Garman, 1887

Distribution: St. Lucia and Pigeon Island; "throughout St. Lucia and on its coastal cays that support more than herb-stage vegetation" (Lazell, 1972).

Habitat: Mesophilic (Schwartz and Henderson, 1991); in lowland woods, coastal *Terminalia*, rainforest, river bottom forest, *Cocos* plantations, scrub woods along a beach, *Musa* groves (including on and in banana shacks), on *Ficus* growing about old ruins (Pigeon Island), in bushes, trees, on walls of houses, and on the ground (Schwartz and Henderson, 1991; Fläschendräger and Wijffels, 1996, 2009; Powell and Henderson, 2008c); abundant in lowlands and especially in cultivated areas, relatively rare in rainforest (Long, 1974); particularly abundant in human-modified habitats, such as hotel grounds and gardens, but also on walls and on and in buildings (J. Schmidt, 1979); on Maria Major in all wooded areas, in crevices on sheltered west-facing cliffs, forages on vertical rock faces (Corke, 1987); in some areas, so abundant that every tree and shrub supports lizards, human commensals, abundant in developed areas and ubiquitous in the ornamental gardens of hotels (Heselhaus and Schmidt, 1990); males perch 0.9–3.7 m above the ground, females and juveniles lower (Lazell, 1972); mean perch height and diameter 1.96 ± 0.23 m and 12.6 ± 1.7 cm (Losos and de Queiroz, 1997a); perch heights increase with increasing elevation (Buckley and Roughgarden, 2005b).

Abundance: Abundance variable at low and moderate elevations and declines with increasing elevation (Buckley and Roughgarden, 2005a). **Activity:** Activity peaks at 1000–1600 h (J. Schmidt, 1979). **Behavior:** On Pigeon Island, tends to be more "social," even males coexisting peacefully (J. Schmidt, 1979); more approachable on Maria Major than on the adjacent main island (Corke, 1987); resident responds to an intruder with lateral presentation, elevation on all 4 limbs, dewlap extended, tail raised and slowly lowered, headbob sequence long in duration (longer than any other species in the *A. roquet* group) and divided into series by pauses, dewlap may be partially retracted at the end of a series (Gorman, 1968). **Competition:** No less common than elsewhere within the area where *A. extremus* (from Barbados) has become established, "despite the apparent niche overlap" (Lazell, 1972); interspecifically territorial with introduced *A. extremus* and appears to be displaced by the invasive species in some habitat pockets (Gorman, 1976). **Ecomorphology:** Closest to "trunk-crown" (Losos and de Queiroz, 1997a; Beuttell and Losos, 1999). **Foraging and Diet:** When fed by humans, some lizards "developed a passion for Ribena" (a fruit drink) and regularly returned to the source (Long, 1974); female competing with Lesser Antillean Bullfinches (*Loxigilla noctis*) for pieces of raw coconut and yams at a bird feeder in the Soufriere Botanical Garden (A. J. Sánchez Muñoz, in litt., 27.VII.2008). **Hybridization:** No evidence of hybridization with introduced *A. extremus* (Giannasi et al., 1997). **Parasites:** Cestode *Oochoristica maccoyi*, nematodes *Parapharyngodon cubensis* and larval *Porrocaecum* sp. (Goldberg et al., 1997a). **Population Size and Density:** To 20 lizards on some large trees on Maria Major (Corke, 1987); not unusual to find an adult male, several subadult males, and as many as 10 females on one large tree (J. Schmidt, 1979; Heselhaus and Schmidt, 1990). **Reproduction:** Clutch size 1 (Hedges, in prep.); in captivity, copulation 70–90 sec, egg 8 × 11 mm buried in moist soil, incubation 78–84 days at 23–28 °C, hatchling SVL ~23 mm (Fläschendräger, 1988; Fläschendräger and Wijffels, 2009). **Sex Ratio (M:F):** Apparently balanced (Lazell, 1972). **Size:** Male SVL to 91 mm, female SVL to 63 mm, smallest mature male and female SVL 50 mm and 48 mm, respectively (Lazell, 1972). **Thermal Biology:** Heliothermic (Fläschendräger and Wijffels, 1996, 2009).

Conservation Status: Not assessed.

Anolis lucius Duméril and Bibron, 1837

Distribution: Cuba, from La Habana Province east to Holquín, Granma, and Santiago de Cuba provinces, reported from Pinar del Río Province, but record doubtful; introduced (?) on Archipiélago de los Canarreos (Garrido and Jaume, 1984; Estrada, 1993a).

Habitat: Mesophilic (Schwartz and Henderson, 1991); in entrances to caves (Gundlach, 1880); most frequently on limestone cliffs and in mouths of caves, also on very large tree trunks such as *Ficus* and *Roystonea*, prefers shade with a substrate of large, intricate vertical surfaces, on rocks at heights of 1–2 m, together with *Tarentola americana* in the hollow of large *Melicocca bijuga* (Barbour and Ramsden, 1919; Schwartz and Ogren, 1956; Allen and Neill, 1957; Ruibal, 1961, 1964; Rodríguez Schettino and Valderrama Puente, 1986; Marcellini and Rodríguez Schettino, 1987; Valderrama Puente and Rodríguez Schettino, 1988); on walls of a bat cave at depths of 22–30 m (H. M. Smith and Willis, 1955); in mesophyllous evergreen and semideciduous forest, mogote complex of vegetation, and coastal and subcoastal microphyllous forest (Rodríguez Schettino et al., 1999); perch heights from 0 (rarely) to 15 m (usually 1–2 m), males (1.72 ± 0.27 m, maximum 5.10 m) perch significantly higher than females (1.37 ± 0.10 m, maximum 3.75 m), perch diameters vary widely (0.01–8.00 m) because of the large main trunks of *Ficus* (Rodríguez Schettino and Valderrama Puente, 1986; Marcellini and Rodríguez Schettino, 1987); mean perch heights of males in dry (0.91 ± 0.17 m) and wet (1.22 ± 0.19 m) seasons, females (0.68 ± 0.19 m and 0.77 ± 0.10 m), and juveniles (0.58 ± 0.20 m and 0.79 ± 0.12 m; Valderrama Puente and Rodríguez Schettino,

1988); low coefficient of arboreality typical of petricolous and clivicolous species (M. A. Granda Martínez in Rodríguez Schettino, 1999b); mean perch height and diameter 1.56 m and 51.3 cm (J. B. Losos, in litt., 21.IV.2008); may perch on ceilings of caves, mean perch heights generally 1–2 m, except in coastal hardwood along the northern coast of La Habana Province, where vegetation is "reclined" and mean height 0.75 m, and at Playa Jibacoa, where vegetation low and mean height 0.79 m, sleeps in fissures and under "lapels" of rock or in interstices of tree trunks (Rodríguez Schettino, 1999b); sleeps on leaves of low shrubs (Schwartz and Henderson, 1991).

Activity: Diurnal (Rodríguez Schettino et al., 1999); seldom seen before midday, peak activity in afternoon (G. Peters and Schubert, 1968); active in early morning and late afternoon (G. Peters, 1970); spaced throughout the day (Marcellini and Rodríguez Schettino, 1987); bimodal in early morning and late afternoon at temperatures 26–29 °C (Fläschendräger and Wijffels, 1996, 2009); leaves nocturnal shelters after 0800 h in the wet season and after 0900 h in the dry season, activity peaks ~1000 h, diminishes slowly during the wet season and suddenly during the dry season, during the wet season activity increases again in the afternoon (not observed during the dry season), activity ceases by 1700 h in the wet season and 1500 h in the dry season (Rodríguez Schettino, 1999b). **Behavior:** Extremely wary, with "gecko-like" actions, "pugnacious" toward conspecifics (H. M. Smith and Willis, 1955); very shy, in caves, responds to threats by running quickly over the walls, slipping into crevices, or flattening themselves against rock walls, lizards outside caves sought refuge in fissures in rocks and beneath tangles of vines (Allen and Neill, 1957); often perches perpendicular to a twig, escapes with hops rather than by running, often changing direction, frequently moving onto very small twigs (Vergner and Polák, 1996); males defend territories on trees, females move rather freely and do not defend areas against other females, male threat display (duration 16 sec) includes forelimb extension, dewlap extension (during entire display), tongue (engorged with blood) protrusion, a series of shallow headbobs (no push-ups), nuchal crest raised (Ruibal, 1967); loudly vocal when handled, but even when not pressed (Barbour and Ramsden, 1916, 1919; Schwartz and Ogren, 1956; Allen and Neill, 1957; Ruibal, 1964; G. Peters, 1970; Vergner and Polák, 1996); vocalization may be associated with agonistic conspecific encounters (Baedermann, 1989; Alonso and Rodríguez, 2004). **Competition:** Rarely syntopic with *A. argenteolus*, which has similar habitat requirements, but both species occur on trunks of trees adjacent to cliffs in Camagüey (Estrada, 1987a). **Ecomorphology:** Undetermined; categorized as a "gecko anole" (G. Peters, 1970). **Foraging and Diet:** Moves quickly while hunting (Rodríguez Schettino, 1999b); consumes mainly ants, but takes other hymenopterans, beetles, lepidopterans, flies, roaches, termites, insect larvae, spiders, and isopods, occasionally saurophagous and cannibalistic, may occasionally eat fruits, males take somewhat larger prey than females (Rodríguez Schettino and Martínez Reyes, 1994; Rodríguez Schettino, 1999b). **Home Range and Territoriality:** Males defend portions of trees (*Ficus*) and females move rather freely and did not defend territories (Marcellini and Rodríguez Schettino, 1987). **Miscellany:** "Windows" in lower eyelids presumed to act as "sunglasses," reducing light intensity when shade-dwelling lizards move into bright sun (Williams and Hecht, 1955); may serve to detect movement while sleeping (Vergner and Polák, 1996). **Parasites:** Nematodes *Cyrtosomum scelopori*, *Oswaldocruzia lenteixeirai*, *Parapharyngodon cubensis*, *Physaloptera squamatae*, *Skrjabinoptera phrynosoma*, larval *Porrocaecum* sp., and Nematoda gen. sp. (Coy Otero and Baruš, 1979b; Coy Otero and Lorenzo Hernández, 1982; Coy Otero, 1999); protozoan blood parasite *Leishmania* (H. Jiménez in Rodríguez Schettino, 1999b). **Population Size and Density:** Gregarious, encounter rate in coastal scrub 11.5 lizards/h, density at Playa Jibacoa 428–1,404 individuals/ha, at the Jardín Botánico de Cienfuegos 930 individuals/ha (Rodríguez Schettino, 1999b); >1,000/ha (B. Král in Vergner and Polák, 1996). **Population Structure:** Adults:subadults and juveniles 1.4:1, with a greater percentage of adults April–July and subadults and juveniles September–March (Rodríguez Schettino, 1999b). **Predation:** Barn Owls (*Tyto alba*) and giant anoles (Rodríguez Schettino, 1999b). **Reproduction:** May be polygynous (Marcellini and Rodríguez Schettino, 1987); during courtship, males display actively with dewlap extensions (Holanova and Hribal, 2004); mean duration of copulation 3.95 min (longest 5.3 min), separation of pair "normal," egg 15.3 mm (Garrido, 1980a); copulation (24 August) in the shade on a cliff face 30 cm above ground at 1400 h, male astride female, grasping her behind forelimbs, his tail twisted below hers, which in turn was twisted to the right (Hardy, 1957a); eggs in numbers in crevices and on undersides of overhanging rocks of cliffs, eggs stick together and to rocks (Dunn, 1926d); eggs stuck to rocks singly or in pairs "after the manner of certain species of Geckos" (Barbour, 1930c); eggs (>100) are not brittle and are deposited communally, attached to roofs of crevices, hollows in rocks, in the twilight zone of cave entrances, in hollows of host trees parasitized by Strangler Figs (*Ficus aurea*), and in crevices of contorted growth of *Ficus*, zero of 7 females in December had eggs, 9 of 11 in September had eggs, 3 of 4 in June had eggs, mean length of posterior oviductal eggs 14.4 mm, mean weight

0.3 g, 7 eggs 14–16 mm long (mean 14.1 mm) in June, one hatched 27 June, another 15 July, hatchlings 60.5 mm in total length (Hardy, 1957a); reproductive activity begins in March (when first gravid females appear), extends through April–July, decreasing in August and ceasing in October (when last gravid females observed), suggesting a peak in reproductive activity during the rainy season, mean size of oviductal eggs 14.4 ± 0.20 mm and 0.30 ± 0.01 g (Valderrama Puente and Rodríguez Schettino, 1988); in captivity, single eggs laid at 14–16-day intervals, yellowish eggs 15 × 8 mm, incubation 60–70 days at 24–29 °C, total length of hatchlings 75 mm (Baedermann, 1989); in captivity, eggs at mean 19.9-day intervals, 14.9 ± 0.96 × 8.1 ± 0.67 mm and 0.49 ± 0.09 g (11.4% of female mass) at deposition, becoming more spherical and increasing by 71.6% (width), 33.6% (length), and 44.7% (mass), mean hatchling SVL 25.1 ± 1.4 mm (Vergner and Polák, 1996); 7 eggs 14.7–16.0 mm × 7.6–9.9 mm on 17 July, two hatched 28 July (Schwartz and Henderson, 1991); active spermatogenesis February–July, peak May–July, fat body mass diminishes May–July and peaks August–December, increased photoperiod, environmental temperatures, and relative humidity appear to increase testicular activity (Domínguez Díaz et al., 2006c). **Sex Ratio (M:F):** 77:117 (Marcellini and Rodríguez Schettino, 1987). **Size:** Male SVL to 69.1 mm (Domínguez Díaz et al., 2006c), female SVL to 56 mm (Marcellini and Rodríguez Schettino, 1987); minimum size at sexual maturity 50 mm for males (Valderrama Puente and Rodríguez Schettino, 1988) or 52 mm and age of ~7 months (Domínguez Díaz et al., 2006c), 42 mm for females (Hardy, 1957a). **Thermal Biology:** Mean body temperature in the morning 28.7 °C (24.8–31.5 °C), in the afternoon 29.9 °C (27.1–32.4 °C; Ruibal, 1961); mean body temperature 29.5 °C (29.2–30.8 °C air temperature), 29.7 °C (30.5–31.0 °C), 27.6 °C (27.6–30.2 °C; Ruibal and Philibosian, 1970); nonheliothermic, mean body temperature higher during the rainy season (30.1 °C) than during the dry season (28.5 °C), mean air temperature selected during the rainy season (27.7 °C) higher than that selected during the dry season (26.5 °C; Rodríguez Schettino and Valderrama Puente, 1986); most frequently (62%) in full shade, mean body temperature 29 °C, with upward shift between morning and afternoon, mean cloacal temperature 29.97 ± 1.71 °C at mean ambient temperature of 30.03 ± 1.57 °C did not differ significantly between males and females, suggesting thermal conformism (Marcellini and Rodríguez Schettino, 1987); mean body and air temperatures higher during the wet (30.4 ± 0.25 °C and 29.8 ± 0.19 °C) than during the dry (25.5 ± 2.06 °C and 22.7 ± 1.80 °C) season (Valderrama Puente and Rodríguez Schettino, 1988).

Conservation Status: Not assessed.

Anolis luteogularis Noble and Hassler, 1935

Distribution: Cuba; where nine currently recognized subspecies occur in western and central portions of Cuba, and on Isla de la Juventud, Archipiélago de los Canarreos, and the Cayos de San Felipe; elevational range, sea level to ~700 m.

Habitat: Mesophilic to somewhat xerophilic (Schwartz and Henderson, 1991); on Cayo Francés, almost invariably on palmettos (*Pseudophoenix* sp.) or the low characteristic vegetation of other cayos (Garrido, 1975a); on trunks and large branches, often on *Roystonea* trunks in edificarian situations (Garrido, 1980b); also on mangroves (*Laguncularia racemosa*; Garrido, 1973b); in the Biosphere Reserve "Sierra del Rosario," in secondary vegetation at the boundaries of forests and pastures primarily on branches but also on trunks of trees (Martínez Reyes, 1995); in microphyllous evergreen forest on the Archipiélago de los Canarreos (Estrada, 1993a); in addition to forests, also in a hole in dead bark on a tree 1 m above the ground, in mangroves, in a mogote complex of vegetation, and in a cave 1 m deep (Rodríguez Schettino, 1999b); in submontane rainforest, submontane mesophyllous evergreen forest, mesophyllous evergreen and semideciduous forest, pine forest, mogote complex of vegetation, swamp forest, mangrove forest, coastal and subcoastal microphyllous forest, cultivated areas and pastures, and urban zones (Rodríguez Schettino et al., 1999); in crowns of large trees (*Tectona* sp.), palms, and even large bushes, primarily along forest edges, near human habitations, and on larger trees in parks and gardens (Fläschendräger, 2002); often abundant in yards and gardens (Kunz, 2005); mean perch height at Soroa 5.86 m, diameter 12.8 cm, behavioral diameter (mean diameter of all surfaces used during behavioral observations, means of individual means) 11.11 cm, 3.3% of observations on or within 1 m of rocks (J. B. Losos, in litt., 20.VIII.2007); mean perch height and diameter 5.63 m and 11.3 cm, none on leaves (J. B. Losos, in litt., 21.IV.2008).

Activity: Diurnal, descends from crowns of trees during the heat of the day (Rodríguez Schettino et al., 1999); active at artificial lights at night (J. Losos in Perry et al., 2008). **Behavior:** Agile, rapidly runs and jumps in tree crowns, when pursued, flees upward and can jump to another tree, tends to "squirrel" on trunks (Rodríguez Schettino, 1999b); moves per minute 0.62 (Johnson et al., 2008); mean percent time displaying 2.1, percent walking 72.1, percent jumping 14.0, percent running 14.6 (J. B. Losos, in litt., 21.IV.2008). **Ecomorphology:** "Crown-giant" (Williams, 1972). **Foraging and Diet:** Eats insects, millipedes, molluscs (*Sachrysia auricoma*), and vegetable matter, mainly fruits (M. Martínez Reyes in Rodríguez Schettino, 1999b); stalked an *Anolis homolechis* at Soroa (P. E. Hertz in J. B. Losos, in litt., 21.IV.2008). **Parasites:** Nematodes *Cyrtosomum longicaudatum*, *C. scelopori*, *Oswaldocruzia lenteixeirai*, *Parapharyngodon cubensis*, *Skrjabinoptera phrynosoma*, *Piratuba* sp., larval *Abbreviata* sp., larval *Cyrnea* sp., and larval *Porrocaecum* sp. (Coy Otero and Baruš, 1979b; Coy Otero and Lorenzo Hernández, 1982; Coy Otero, 1999); ticks *Amblyomma torrei* (Rodríguez Schettino, 1999b; Coy Otero, 1999); sporozoan red blood cell parasite *Haemogregarina* sp. (Zajicek and Mauri Méndez, 1969). **Population Density:** Only 1–2 individuals per tree, encounter rate <1/h (Rodríguez Schettino, 1999b). **Predation:** Lizard Cuckoo (*Saurothera merlini*) hunting lizards attracted to a cave mouth by an abundance of cockroaches (*Periplaneta americana*; González Grau et al., 1989). **Size:** Male SVL to 191 mm, female SVL to 176 mm (Schwartz and Garrido, 1972). **Thermal Biology:** Heliothermic (Rodríguez Schettino, 1999b); mean body temperature at Soroa 29.4 °C (J. B. Losos, in litt., 20.VIII.2007).

Conservation Status: Of concern (Buide et al., 1974); *A. l. sanfelipensis* and *A. l. coctilis* are considered "endangered" and *A. l. jaumei*, *A. l. calceus*, *A. l. delacruzi*, and *A. l. nivevultus* "vulnerable" (L. Rodríguez Schettino and A. Chamizo Lara in Rodríguez Schettino, 1999a).

Remarks: *Anolis l. calceus* should be considered a geographic race of *A. equestris* rather than of *A. luteogularis* (Díaz and Abreu Guerra, 2005).

Anolis luteosignifer Garman, 1887

Distribution: Cayman Islands, Cayman Brac.

Habitat: Wooded karst regions and more open coastal areas, spends considerable time on the ground and in limestone crevices, also on trees and fence posts from which it darts to the ground for prey (Grant, 1940c; Seidel and Franz, 1994); mean perch height and diameter 0.96 ± 0.08 m and 9.7 ± 0.9 cm (Losos and de Queiroz, 1997a).

Behavior: Shy, difficult to approach (Grant, 1940c). **Ecomorphology:** "Trunk-ground" (Losos and de Queiroz, 1997a). **Predation:** *Leiocephalus carinatus* (A. S. Rutty in Grant, 1940c); *Cubophis fuscicauda* (Grant, 1940c). **Size:** Male SVL to 57 mm, female SVL to 37 mm (Grant, 1940c).

Conservation Status: Unknown.

Anolis macilentus Garrido and Hedges, 1992

Distribution: Cuba, only from the type locality on the Meseta del Guasco (650 m), Guantánamo Province.

Habitat: In compact brushy vegetation (ferns, bushes, and scattered trees) above a steep slope bordering the Río Pai, most lizards 0.5–2.0 m high on twigs and stems of high grasses, one male on the bark of a tree (Garrido and Hedges, 1992); in submontane mesophyllous evergreen forest (Rodríguez Schettino et al., 1999).

Activity: Diurnal (Rodríguez Schettino et al., 1999); most active at midmorning (Garrido and Hedges, 1992). **Behavior:** When encountered, remained still for a moment and then tried to escape by dashing down among lower grasses and bushes and remaining motionless and cryptic (Garrido and Hedges, 1992). **Ecomorphology:** "Grass-bush" (Garrido and Hedges, 1992). **Size:** Male SVL to 41 mm, female SVL to 36 mm (Garrido and Hedges, 1992). **Thermal Biology:** Apparently nonheliothermic (Rodríguez Schettino, 1999b).

Conservation Status: Considered "vulnerable" (Rodríguez Schettino and Chamizo Lara, 1998a; L. Rodríguez Schettino and A. Chamizo Lara in Rodríguez Schettino, 1999a).

Anolis marcanoi Williams, 1975

Distribution: Hispaniola, southern slopes of the Cordillera Central and associated lowlands (Dominican Republic); elevational range ~450–1,800 m.

Habitat: Xerophilic to somewhat mesophilic (Schwartz and Henderson, 1991); on fence posts or in *Cocos* groves, rocky, very open hillsides, locally syntopic with *A. cybotes*, but usually in enclaves surrounded by *A. cybotes* at lower elevations, with the reverse true at higher elevations (Williams, 1975; Schwartz, 1989); mean perch height and diameter 1.00 m and 11.6 cm (J. B. Losos, in litt., 21.IV.2008).

Behavior: Aggressive behavior progressively and hierarchically structured in contrast to that of sympatric *A. cybotes*, low-level dewlap display (initial response), slight elevation of head and anterior body, and dewlap extended with several headbobs, intermediate dewlap display involves either complete forelimb extension or elevation on all 4 limbs, snout pointed high, tail lifted (straight or slightly arched, occasionally curled over body), dewlap extended completely, full-elevation display involves limbs extended and body fully elevated, tail usually rigid and extended or held higher than body, snout very high and dewlap off the substrate, all modified by nuchal and dorsal crest erection and head-spot darkening, modifiers more common at higher levels of aggression and rarer or of shorter duration in subordinate males, series of 2–10 headbobs usually preceded dewlap extension but might accompany it, push-ups by fore-, hind-, or all limbs sometimes accompanied displays (especially those of greater intensity), in addition, lateral orientation and compression and gular inflation increased apparent lizard size, escalated to biting only when both individuals persisted at high aggression levels, chasing rare, usually terminated by flight or quiescence (assumption of submissive posture) of one lizard (Losos, 1985a); intraspecific aggression high, but interspecific aggression against *A. cybotes* low, aggression level intermediate against *A. cybotes* with dewlaps modified to look like those of *A. marcanoi*, aggression levels against conspecifics with dewlaps modified to look like those of *A. cybotes* low, lizards obviously recognized conspecifics (Losos, 1985b); males spent more time displaying synchronously with videotaped conspecific displays than to those of *A. cybotes*, suggesting that species recognition is mediated by display pattern and/or morphological clues, but conspecific displays did not elicit a greater overall level of aggression, probably attributable to a lack of interaction with the subject (Macedonia et al., 1994). **Desiccation:** Vital limit (percent of initial body weight at death) 64.3 ± 0.6%, rate of weight loss (percent of initial body weight/h) 0.122 ± 0.006%, survival time 299.5 ± 16.4 h (Hertz, 1980a). **Ecomorphology:** "Trunk-ground" (Williams, 1983); formally quantified (Beuttell and Losos, 1999). **Size:** Male SVL to 65 mm (Losos, 1985a), female SVL to 49 mm (Schwartz, 1989). **Thermal Biology:** Experimental voluntary maximal temperature 35.7 ± 0.18 °C, critical thermal maximum 40.5 ± 0.18 °C (Hertz, 1980a).

Conservation Status: Unknown.

Anolis marmoratus Duméril and Bibron, 1837

Distribution: Guadeloupe and some satellites, where six currently recognized subspecies occur; elevational range, sea level to ~900 m (Fläschendräger and Wijffels, 1996, 2009).

Habitat: Xerophilic to mesophilic (Schwartz and Henderson, 1991); in arid regions to montane rainforest, in canefields, *Musa* groves, coastal *Coccoloba*, *Terminalia*, and *Cocos*, coffee, vanilla, and *Theobroma* groves, mangroves (Îlet Fortune), generally perches on bushes and trees, but also under rocks (at dusk), and on rock walls, fence posts, buildings, and forest shacks (Schwartz and Henderson, 1991; Powell and Henderson, 2008c); at one site, large adult males and females inhabit palms with intermediate temperatures, small adult males and females of all sizes occur in cool mango trees, and only juveniles occupy warm banana trees (Huey and Webster, 1975); *A. m. marmoratus* especially often associated with banana trees, *A. m. girafus* with palm trunks, *A. m. speciosus* in towns on lampposts and walls (J. Schmidt, 1981); *A. m. marmoratus* in *Musa* plantations, especially subadults and juveniles, adults common on fence posts with associated lower vegetation and on larger trees in developed areas, *A. m. alliaceus* abundant in crowns of large (to 40 m) trees festooned with epiphytes and in dense *Yucca* growth used as fences around pastures, *A. m. caryi* preferentially associated with large trees and palm groves, *A. m. setosus* most abundant where it functions as a human commensal and in relatively open coastal woodlands, *A. m. speciosus* concentrated in artificially mesic situations such as gardens and parks (Heselhaus and Schmidt, 1990); lowland populations in very diverse habitats, usually at perch heights of 1–3 m, upland rainforest denizens primarily crown dwellers (Fläschendräger and Wijffels, 1996, 2009); large males conspicuous on tree trunks and fence posts 0.9–3.0 m above the ground, both sexes forage on the ground, mean perch height and diameter 1.67 ± 0.13 m and 9.4 ± 1.4 cm (Losos and de Queiroz, 1997a); sleeps sprawled on leaves (Heselhaus and Schmidt, 1990); requires elevated vertical perches, usually tree trunks, including mangroves and ornamental and orchard species, may use buildings (under roofs of which lizards seek shelter at night), walls, pylons, signposts, fences, but not sugarcane, sleeps at ends of branches, under bark, or (juveniles especially) under cover on the ground (Breuil, 2002b).

Activity: Diurnal, but exploits the "night-light niche" (Powell and Henderson, 1992; Schwartz and Henderson, 1991; Henderson and Powell, 2001; Perry et al., 2008); activity peaks early and late in the day, usually immobile in the shade during midday, but will readily exploit prey that comes into range (Breuil, 2002b). **Behavior:** Rainforest-dwelling *A. m. alliaceus* is especially shy, whereas *A. m. girafus* is easily captured by hand, although individuals will "squirrel" to the other side of the trunk, they rarely ascend to escape (J. Schmidt, 1981). **Biomass:** 0.40 g/m^2 for Guadeloupe (Wright, 1981). **Desiccation:** Rates of evaporative water loss varied significantly between populations and were negatively correlated with habitat aridity (Hillman et al., 1979). **Ecomorphology:** Closest to "trunk-crown" (Losos and de Queiroz, 1997a; Beuttell and Losos, 1999). **Foraging and Diet:** Use a sit-and-wait strategy, employing elevated perches to see prey primarily on the ground, juveniles forage mostly on the ground, opportunistic, eat ants in large quantities, other hymenopterans, fruit flies (especially in orchards), caterpillars and adult lepidopterans, beetles, *Sphaerodactylus*, *Eleutherodactylus*, rarely cannibalistic (especially when dominant males are defending territories), lap nectar and take juicy fruits, occasionally take bread offered by people (Breuil, 2002b). **Home Range and Territoriality:** Aggressively territorial (Fläschendräger and Wijffels, 1996, 2009). **Parasites:** Nematodes *Oswaldocruzia dorsarmata*, *O. jeanbarti*, *O. marechali*, and *O. mauleoni*

(Ben Slimane et al., 1995); nematodes *Oswaldocruzia marechali, Parapharyngodon cubensis* (Goldberg et al., 1997a), and *Spauligodon anolis* (Goldberg et al., 1997a; Bursey and Goldberg, 1998). **Population Size and Density:** Density of *A. m. alliaceus* relatively low in upland rainforests, in contrast densities very high in *Yucca* thickets around pastures, the comparatively low densities of *A. m. setosus* and *A. m. speciosus* in many natural habitats may be attributed to the xeric nature of those habitats (Heselhaus and Schmidt, 1990). **Predation:** *Alsophis antillensis* (Henderson and Sajdak, 1996); Guadeloupe Woodpecker (*Melanerpes herminieri*; Breuil, 2002b). **Reproduction:** Seasonal (June–August), first young appear in end of June/early July, copulation most frequently on vertical surfaces, male usually bites female's head or nape, 2–4 eggs (suggesting communal depostion) in moist soil (sometimes in flowerpots, which may facilitate dispersal of individuals and foster homogenization of otherwise distinctive populations), incubation probably ~6 weeks (Breuil, 2002b); probably breeds throughout the year, females bury eggs in captivity, incubation 45–55 days at 23–28 °C, a female can lay as many as 30 eggs/year (Fläschendräger and Wijffels, 2009). **Sex Ratio:** May vary by habitat, at one site, 37.6% of lizards on palms were males, 52.3% on mango trees, and 79.4% on banana trees (Huey and Webster, 1975). **Size:** Male SVL to 85 mm, female SVL to 57 mm, smallest mature male and female SVL 45 mm and 40 mm respectively (Lazell, 1972). **Thermal Biology:** Mean body temperature from 29 samples (N = 638) representing 9 localities 29.75 ± 0.08 °C with corresponding mean air temperature of 28.5 ± 0.08 °C, differences in body temperatures by location and time of day and the correlation between body and air temperatures suggest tolerance of a broad range of air and body temperature despite no conspicuous thermoregulatory behavior, basks only occasionally, and habitat discrimination at a lowland locality suggests thermal preferences despite being generally passive (Huey and Webster, 1975).

Conservation Status: Not assessed, but protected by French legislation (Lorvelec et al., 2007).

Anolis marron Arnold, 1980

Distribution: Hispaniola, southeastern coast of the Tiburon Peninsula (Haiti).

Habitat: Xerophilic, on *Coccoloba* along the coast (Schwartz and Henderson, 1991).

Ecomorphology: "Trunk" (Williams, 1983). **Size:** Male SVL to 50 mm, female SVL to 42 mm (Arnold, 1980).

Conservation Status: Unknown.

Anolis maynardii Garman, 1888

Distribution: Little Cayman Island, introduced on Cayman Brac.

Habitat: Mesophilic to somewhat xerophilic (Schwartz and Henderson, 1991; RP); arboreal (especially on *Cocos*), males to 4.5 m, females to 1.2 m above the ground on a variety of perches, including roadside scrub, cacti, and *Agave*, but occasionally on the ground, under old boards in grass (Grant, 1940c; Schwartz and Henderson, 1991; RP); large trees and abandoned buildings (Seidel and Franz, 1994); mean perch height and diameter 2.52 ± 0.28 m and 11.7 ± 1.7 cm (Losos and de Queiroz, 1997a).

Behavior: Faces upward on tree trunks, escapes upward on trees or buildings, stealthy in movement, apparently "stupid" (Grant, 1940c); escapes to upper portions of trees or buildings when pursued (Seidel and Franz, 1994). **Ecomorphology:** "Trunk-crown" (Williams, 1969). **Parasites:** Nematodes *Physocephalus* sp., *Gongylonema* sp., *Parapharyngodon cubensis, Atractis scelopori*, and a larval spirurid (Goldberg and Bursey, 1996b). **Reproduction:** Clutch size 1 (Hedges, in prep.). **Size:** Male SVL to 76 mm (Seidel and Franz, 1994), female SVL to 49.1 mm (Schoener, 1988).

Conservation Status: Not assessed.

Anolis mestrei Barbour and Ramsden, 1916

Distribution: Cuba, Pinar del Río and extreme western La Habana provinces.

Habitat: Mesophilic (Schwartz and Henderson, 1991); largely saxicolous, especially in shady portions of forest, more frequently on rocks and boulders (often along streams) and even cliff faces than on tree trunks, although these are not shunned (Ruibal, 1964; Ruibal and Williams, 1961b; Schwartz, 1968f; Schwartz and Henderson, 1991); in the Biosphere Reserve "Sierra del Rosario," in rock outcrops in gallery forest, near the river, and in the tourist center, preferentially on rocks but also on the ground (Martínez Reyes, 1995); found in mesophyllous evergreen and semideciduous forest and mogote complex of vegetation (Rodríguez Schettino et al., 1999); males on limestone rocks and walls, females and juveniles more frequently on the ground, on stones, or sometimes in leaf litter, occasional individuals on trunks of trees and bushes (Rodríguez Schettino, 1999b); on rocks 79% of time, on trunks adjacent to cliffs (21%), only 2 of 15 in the sun (Rodríguez Schettino and Chamizo Lara, 2001); mean perch height at Soroa 1.28 m, diameter 12.7 cm, behavioral diameter (mean diameter of all surfaces used during behavioral observations, means of individual means) 11.94 cm, 57.9% of observations on or within 1 m of rocks (J. B. Losos, in litt., 20.VIII.2007); habitat light environment "full shade" (Fleishman, 2000); mean perch height and diameter 1.74 m and 11.4 cm (J. B. Losos, in litt., 21.IV.2008); sleeps on large leaves, horizontal stems and twigs to ~1 m above ground, tail in a watch-spring curled laterally or prehensily (Hardy, 1958b).

Activity: Diurnal (Rodríguez Schettino et al., 1999); only females and juveniles active during early morning, adult males evident around midday (Ruibal and Williams, 1961b). **Behavior:** Escapes by running to the ground and hiding or by entering crevasses in limestone (Ruibal and Williams, 1961b); aggressive displays of males include elevation on all 4 limbs, raising nuchal and dorsal crests, elevation of tail, headbobs, and dewlap extension (Rodríguez Schettino, 1999b); moves per minute 0.59 (Johnson et al., 2008); mean percent time displaying 7.0, percent walking, 45.2, percent jumping 28.4, percent running 26.5 (J. B. Losos, in litt., 21.IV.2008). **Ecomorphology:** "Trunk-ground" (Williams and Rand, 1977). **Movements:** Tail prehensile (Hardy, 1958b). **Parasites:** Sporozoan red blood cell parasite *Haemogregarina* sp. (Zajicek and Mauri Méndez, 1969); nematodes *Cyrtosomum scelopori* (Coy Otero and Baruš, 1973a, 1979b; Coy Otero and Lorenzo Hernández, 1982; Coy Otero, 1999) and *Skrjabinoptera phrynosoma* (Coy Otero and Baruš, 1979b; Coy Otero and Lorenzo Hernández, 1982; Coy Otero, 1999). **Reproduction:** Mean duration of copulation 2.01 min (longest 12 min), separation of pair "normal," egg 13.5 × 9.5 mm (double yolked), cream colored and oval (Garrido, 1980a); clutch size 1 (Hedges, in prep.); reproductive activity appears to increase during the wet season, 2 empty eggshells in April, one in a hole in a rock, the other on soil at the base of the same rock (Rodríguez Schettino, 1999b). **Size:** Male SVL to 56.5 mm, female SVL to 48.5 mm (Rodríguez Schettino, 1999b). **Thermal Biology:** Nonheliothermic, in April on rocks at mean air temperature 26.0 °C (Rodríguez Schettino, 1999b); mean body and air temperature for males 28.2 ± 0.4 °C and 28.1 ± 0.4 °C, for females 28.0 ± 0.2 °C and 28.2 ± 0.3 °C, body temperatures did not differ significantly from air temperatures (thermally passive), nor did temperatures of males and females differ (Rodríguez Schettino and Chamizo Lara, 2001); mean body temperature at Soroa 27.3 °C (J. B. Losos, in litt., 20.VIII.2007).

Conservation Status: Not assessed.

Anolis monensis Stejnegeri, 1904

Distribution: Isla Mona and Isla Monito.

Habitat: Xerophilic (Schwartz and Henderson, 1991); more arboreal than closely related *A. cristatellus* from Puerto Rico, but uses perches generally comparable to those used by *A. cristatellus*, common on beaches as well as the central plateau, especially in *Casuarina* forest, perch diameter provides no evidence of infraspecific niche partitioning, but spatial separation evident in perch heights, with 60% of females at heights 0–0.6 m and 75% of males 0.9 m above the ground (Weaver et al., 1961; Gorman and Stamm, 1975); mean adult male and female perch heights 1.5 ± 0.75 m (3.0% on the ground) and 1.1 ± 0.69 m (9.0% on the ground; Lister, 1976); in open pasture land (R. Thomas in Rivero, 1998); on tree trunks and branches in all habitats on the island (Tenlén and Wijffels, 1999); 49 of 71 individuals on trees, 16 on bushes, 6 on the ground, mean perch height of adult males 235.0 ± 33.8 cm, adult females 88.5 ± 16.3 cm, juveniles 50.0 ± 26.5 cm (T. A. Jenssen, unpublished data); 85 of 115 observations of individuals on trees or trunk analogues (e.g., fence posts, large-diameter deadfall), 17 on bushes or small-diameter deadfall, 7 on the ground, 4 on rock walls, and 2 on leaves, mean perch height of adult males 90.0 ± 7.4 cm, adult females 54.9 ± 5.7 cm, juveniles 25.8 ± 5.6 cm, 28 of 56 adult males and 30 of 47 adult females facing down, all juveniles horizontal on deadfall and debris on or near the ground (RP); sleeps on leaves, with larger individuals sleeping higher (Weaver et al., 1961; Gorman and Stamm, 1975).

Behavior: Usual response to an intruder is flight, but, when agonistic encounters occur, escalate to fighting with minimal displays (Gorman and Stamm, 1975); 14 of 32 lizards escaped in the direction they were facing, 4 changed direction, and 14 "squirreled" around their perches (RP); male threat displays consist of a pattern of 5 (occasionally 6) bobs with which dewlap extension could not be predictably coordinated and variously modified by lateral presentation, lateral compression, protruded eyes, elevated nuchal crest, gaped mouth, protruded tongue, enhanced body color, and elevated tail, in contrast, displays directed to females involved lateral presentations, 2- and 4-legged push-ups, and dewlap extension, but lacked other modifiers of male-to-male displays (T. A. Jenssen and P. R. Ortiz, unpublished data); free-ranging lizards reluctant to display, even when confronted by a tethered intruder (T. A. Jenssen and RP, unpublished data). **Biomass:** 0.74 g/m^2 (Wright, 1981). **Diet:** Primarily insectivorous, but food includes red berries of *Rivina humilis* (Gorman and Stamm, 1975). **Ecomorphology:** Descended from "trunk-ground" stock, qualifies as a "generalist," but is most like a "trunk-ground" anole (Losos and de Queiroz, 1997a); intermediate between "trunk-ground" and "trunk-crown" (Beuttell and Losos, 1999); "trunk-ground" (Losos, 2009). **Predation:** *Ameiva alboguttata* (K. P. Schmidt, 1920); *Epicrates monensis* (K. P. Schmidt, 1926; Tolson et al., 2007); *Borikenophis variegatus* (Rodríguez-Robles and Leal, 1993a); American

Kestrels (*Falco sparvarius*; Barnes, 1946); feral cats (Garcia et al., 2001). **Reproduction:** Most females gravid in mid-October (Gorman and Stamm, 1975). **Size:** Male SVL to 60 mm (Gorman and Stamm, 1975), female SVL to 57 mm (Schwartz and Henderson, 1991), although the latter seems disproportionately large—largest females cited elsewhere are 44 mm SVL (Gorman and Stamm, 1975) and 46 mm SVL (T. A. Jenssen, unpublished data); smallest sexually mature female SVL = 38 mm (Gorman and Stamm, 1975). **Thermal Biology:** Probably thermally passive, cloacal temperatures 29.0–33.4 °C (almost 50% 31.0–31.4 °C), slightly higher than air temperatures (Gorman and Stamm, 1975); in an open and closed situation 31.8 ± 1.43 °C and 28.9 ± 1.10 °C, 30.6 ± 1.69 °C and 28.3 ± 0.97 °C (Lister, 1976).

Conservation Status: "Stable" (Moreno, 1991d).

Remarks: Populations on Monito form a monophyletic group nested within *A. monensis* from Mona and may be diverging (Rodríguez-Robles et al., 2007); "subtle morphological and behavioral differences" have been documented (R. Thomas and M. Leal in Rodríguez-Robles et al., 2007).

Anolis monticola Shreve, 1936

Distribution: Hispaniola, Monts Cartaches and Massif de la Hotte (Haiti); elevational range ~400–1,400 m.

Habitat: Mesophilic (Schwartz and Henderson, 1991); very agile, primarily saxicolous or at least associated with well-shaded limestone boulder jumbles, rocky cliffs, limestone rocks and boulders in coffee plantations, also on the ground, in the lighted portion of a deep solution pit, among rocks on a floodplain, base of a hill with slabs and limestone fragments, small, overgrown pile of boulders, in brush in a deep ravine, neighboring garden areas, but absent from drier patches on adjacent hillsides (Williams, 1962b; Thomas and Schwartz, 1967; Williams and Webster, 1974; Franz and Cordier, 1986; Schwartz and Henderson, 1991); 2.4% on perches >2.5 cm in diameter at heights >90 cm, 2.4% on leaves or grasses, 8.4% on perches <0.7 cm in diameter, 13.2% on perches 0.7–2.5 cm in diameter, and 1.2% on perches >2.5 cm in diameter at heights 31–90 cm, 4.9% on leaves or grasses, 14.4% on perches <0.7 cm in diameter, 6.0% on perches 0.7–2.5 cm in diameter, 4.8% on perches >2.5 cm in diameter, and 42.2% on rocks or the ground at heights <30 cm (Moermond, 1979b); displays from leaves of low shrubs overhanging a stream, into which they would fall when pursued (Franz, 2003); sleeps 0.3–1.2 m above the ground on *Coffea* leaves and leaves of a shrub (Weaver et al., 1961; Schwartz and Henderson, 1991).

Behavior: Capable of vocalization (Schwartz and Henderson, 1991); percent jumps (fraction of total movements) in rocks and boulders at a Haitian site 57 (Moermond, 1979a); percent crawls 11, runs 32, jumps 57 (Moermond, 1979b); will swim to escape (Franz and Cordier, 1986). **Ecomorphology:** Undetermined. **Foraging and Diet:** Scanning the ground from an elevated perch takes longer than scans from other perches, increasing waiting intervals and distances/move, mean 11.5 min between location changes, mean 8.8 min between location or posture changes, mean distance/move 23 cm, distance moved per prey capture variable but relatively long, 58% of prey captures on same surface as that on which the lizard was perched (Moermond, 1979b). **Parasites:** Cestode *Oochoristica maccoyi*, nematodes Acuariidae gen. sp., *Parapharyngodon cubensis*, larval *Physocephalus* sp., *Rhabdias* sp., *Trichospirura teixeirai*, acanthocephalan cystacanths *Centrorhynchus* sp. (Goldberg et al., 1998b). **Size:** Male SVL to 55 mm, female SVL to 39 mm (Thomas and Schwartz, 1967).

Conservation Status: Unknown.

Anolis noblei Barbour and Shreve, 1935

Distribution: Cuba, where two currently recognized subspecies occur in the Sierra de Nipe (*A. n. noblei*), Sierra Maestra, and Sierra de la Gran Piedra (*A. n. galeifer*); elevational range to ~600 m.

Habitat: Submontane rainforest, submontane mesophyllous evergreen forest, and mesophyllous evergreen and semideciduous forest (Rodríguez Schettino et al., 1999); on tree trunks and in crowns (Rodríguez Schettino, 1999b); at heights of 5–7 m in broadleaf trees, at another site, often in *Swietenia* 3–10 m above the ground (Fläschendräger and Wijffels, 2009); often associated with humans (Kunz, 2005); mean perch height and diameter 6.00 m and 9.3 cm (J. B. Losos, in litt., 21.IV.2008).

Activity: Diurnal (Rodríguez Schettino et al., 1999). **Behavior:** Escape tactics include "squirreling" and then moving upward (Rodríguez Schettino, 1999b); move to the other side of the trunk or branch at the least disturbance, an invader was chased down a tree and across the ground by a resident (Fläschendräger and Wijffels, 2009). **Ecomorphology:** "Crown-giant" (Williams, 1972). **Size:** Male SVL to 190 mm (Schwartz and Garrido, 1972), female SVL to 172.1 mm (Rodríguez Schettino, 1999b). **Thermal Biology:** Apparently heliothermic (Rodríguez Schettino, 1999b).

Conservation Status: Unknown.

Anolis nubilis Garman, 1887

Distribution: Redonda, islandwide.

Habitat: Mesophilic (Schwartz and Henderson, 1991); in shade of large rocks near the ground, but climbs as high as possible (~4.5 m) on the only trees on Redonda (*Casuarina*), uncommon on lava rocks and boulders near shore, on small bushes and shrub-like trees in the one ghut (= gulley) on the island (Lazell, 1972; Gorman and Stamm, 1975).

Abundance: "Not abundant" (Lazell, 1972). **Behavior:** In captivity, an adult male responded to a mirror image with lightened coloration, elevation on all 4 limbs, headbobs, and dewlap extended halfway (Fläschendräger and Wijffels, 1996, 2009). **Ecomorphology:** Undetermined. **Foraging and Diet:** In captivity, eats a variety of small arthropods, was quick and agile in capturing prey (Fläschendräger and Wijffels, 1996, 2009). **Sex Ratio:** Apparently balanced (Lazell, 1972). **Size:** Male SVL to 81 mm, female SVL to 52 mm, smallest mature male and female SVL 53 mm and 50 mm (but data may be incomplete; Lazell, 1972).

Conservation Status: Unknown (Daltry, 2007).

Anolis occultus Williams and Rivero, 1965

Distribution: Puerto Rico, islandwide, but local; elevational range 65–1,325 m.

Habitat: Mesophilic (Schwartz and Henderson, 1991); probably a canopy dweller, descending to lower levels at breaks in the canopy, where it occupies peripheral vegetation, including bushes, on ferns, upper surface of a leaf on a broadleaf green plant on the ground, on outer leaves of bromeliads (*Werauhia sintenisii*) on the forest floor, at the entrance of a cavity in a rotten buttress 1.8 m above the ground (Thomas, 1965a; Williams and Rivero, 1965; Webster, 1969; Schwartz and Henderson, 1991; Rivero, 1998); common in tabonuco forest to elevations of 980 m (Miranda-Castro et al., 2000); note that mean perch heights and diameters in Calsbeek et al. (2006) are actually data from Losos (1990a) for *A. valencienni* (see that account); sleeps with head facing the distal portions of perches on dead leafless vines and twigs 6 mm in diameter at forest edge, to 4.5 m above the ground (and probably higher), occasionally on upper surfaces of broad, stiff leaves, sleeping sites offer maximum early morning insolation (Thomas, 1965a; Webster, 1969); crawls on twigs and perched on a bromeliad (A. J. Sánchez Muñoz, in litt., 06.V.2008).

Activity: Wakes at 0630–0700 h (Webster, 1969); most animals active by day during rainy weather (A. J. Sánchez Muñoz, in litt., 06.V.2008). **Behavior:** Movements "slow" (Williams and Rivero, 1965); movements slow and desultory during the day, sleeps with forelimbs flexed, hindlimbs extended, tail hooked or curled about the substrate, held in place by the 5th toe on either side, when disturbed, holds tightly to the substrate, showing great tolerance for shaking and disturbance, if tolerance exceeded, will release hold, dropping toward the ground before catching and ascending a lower object, where it assumes a sleeping posture, eyes chameleonlike, with some independent movement, squeaks when handled (Thomas, 1965a); progresses using a slow crawl, hopping, or leaping, vocalizes (Webster, 1969); male and female frequently sleeping near each other, suggestive of pair-bonding (Gorman, 1980), but not observed by others (J. B. Losos, in litt., 21.IV.2008); when disturbed, a lizard moved toward the base of a twig in an attempt to hide among leaves instead of flattening itself against the branch (A. J. Sánchez Muñoz, in litt., 06.V.2008). **Ecomorphology:** "Twig" (Williams, 1972); formally quantified (Beuttell and Losos, 1999). **Performance:** Mean jump 27.7 ± 1.46 cm, cling 0.834 ± 0.159 Neutons (Losos, 1990a); acceleration 11.79 ± 3.43 m/sec, sprint speed 0.25 ± 0.01 m/sec (Vanhooydonck et al., 2006a). **Reproduction:** Clutch size 1 (Hedges, in prep.). **Size:** Male SVL to 42 mm (Williams and Rivero, 1965), female SVL to 40 mm (Schwartz and Henderson, 1991).

Conservation Status: "Stable" (Moreno, 1991d).

Anolis oculatus Cope, 1879

Distribution: Dominica, where four subspecies have been recognized (Lazell, 1962), although these are now generally considered to be ecotypes (Malhotra and Thorpe, 1991a, 1991b; Thorpe et al., 2004); the species occurs essentially islandwide except for the highest elevations; elevational range, sea level to >900 m; found at 10 of 11 variously altered sites, absent only where displaced by invasive *A. cristatellus* (Ackley et al., 2009).

Habitat: Xerophilic to mesophilic (Schwartz and Henderson, 1991); in essentially all vegetational types, except elfin woodland, in dry scrub woodlands, lime groves, *Cocos*, *Theobroma*, and *Musa* plantations, rainforests (primary and disturbed), *Coccoloba* forests, occupying a wide range of structural niches, but usually on vertical perches <2 m high (trees, bushes, fence posts, walls, boulders, road cuts, buildings), on the ground at a sandy stream-edge, will forage on the ground, occasionally seeks shelter under rocks, between rocks, or in root masses (Lazell, 1962; Ruibal and Philibosian, 1970; Schwartz and Henderson, 1991; Malhotra and Thorpe, 1997a, 1997b); *A. o. oculatus* on stone walls, trees with open foliage, dead stumps, and on houses, *A. o. cabritensis* primarily on *Cocos* trunks, *A. o. winstoni* abundant in *Musa* plantations, *A. o. montanus* at Trafalgar Falls on a large rock in midstream (Krintler, 1986); frequently in association with streams at higher elevations and along the mesic windward coast (Heselhaus and Schmidt, 1990); lowland populations arboreal in dry forests of the leeward coast and wet formations on the windward side, also plantations, gardens, walls, and houses, montane populations increasingly saxicolous (Fläschendräger and Wijffels, 1996, 2009); larger individuals feed mainly above the ground and those at a rainforest site perched higher than in dry scrub woodland, whereas small individuals (mainly juveniles) feed primarily (on ants) on the ground (percent prey captures by small lizards on ground 92.3, 37.5, and 49.0 at a dry scrub site and 2 rainforest sites, respectively, whereas corresponding percentages for large individuals were 25.0, 28.6, and 50.0), mean perch heights 0.33 m, 1.14 m, and 1.08 m for small individuals at a dry scrub site and 2 rainforest sites, respectively, 0.55 m, 1.29 m, and 0.64 m for individuals of medium size, and 0.73 m, 1.76 m, and 1.75 m for large individuals (Bullock et al., 1993); most anoles in primary rainforest on the ground, whereas most in trees in lowland dry forest (Valentine, 2002); plurality of males and majority of females on elevated perches with diameters 2–5 cm (Ferrara, 2003); sleeps on tips of branches while clinging to leaves (Malhotra and Thorpe, 1999).

Abundance: Anoles most abundant under closed or semi-closed canopies in dry habitats at lower elevations at temperatures of 34–37 °C (Tarbox, 2005). **Activity:** In xeric woodlands (Cabrits), active throughout the day, but activity (especially feeding) greatest at dawn and dusk (Malhotra and Thorpe, 1999); exploits the night-light niche (Powell and Henderson, 2008d). **Behavior:** Display behavior involves pumping dewlap in rhythmic manner, even when no recipient is apparent, in high-intensity displays, dewlap pumping is an integral part of a complex display involving headbobbing and tail movements (Gorman, 1968); territorial battles usually won by the resident male (Malhotra and Thorpe, 1999); an adult male displayed toward a male *Iguana delicatissima* that had displaced it from its perch (L. A. White, pers. comm.); *A. o. montanus* on boulder piles rapidly flee to the other side of the rock (Krintler and Dommaschk, 1980); *A. o. cabritensis* on *Cocos* trunks quickly flee upward into the crowns when threatened (Krintler and Dommaschk, 1980; Krintler, 1986). **Competition:** Rare or absent in areas where invasive *A. cristatellus* has become established (Malhotra et al., 2007; Eales et al., 2008); males respond aggressively toward male *A. cristatellus* introduced into their territories (Daniells et al., 2008). **Ecomorphology:** Undetermined. **Foraging and Diet:** Mean prey length 3.4 mm, mean prey volume/lizard 9.7 mm^3, mean number of prey items/lizard 29.5 (Andrews, 1976); sit-and-wait forager, 82% successful in 6.7 prey capture attempts/h, mean prey distance 19 cm and 85% of prey captures on the ground, numerous small insects, with ants (44–57% of all prey items), flies, and termites most important (by frequency and volume), and lepidopteran adults and larvae, orthopterans, cockroaches, hymenopterans (other than ants), hemipterans, beetles (adults), holometabolous insect larvae (other than lepidopterans), spiders, isopods, and miscellaneous items (mostly barklice, springtails, and mites) in lesser quantities, with larger lizards taking larger prey (but nearly all prey items <5 mm) and 32–62 prey items per individual (Andrews, 1979); ants, springtails, barklice, beetles, flies, spiders, neuropterans, isopods, millipedes, orthopterans, holometabolous larvae, termites, earthworms, true bugs, and other small arthropods, with ants most important in numbers and volume at a dry site, but large, soft-bodied prey more important in volume for large lizards in rainforest, increases in proportions of soft-bodied prey occurred at both sites during the wet season, prey capture success was 60.3%, 68.8%, and 78.6% for small individuals at a dry scrub site and 2 rainforest sites, respectively, 83.9%, 50.0%, and 100.0% for individuals of medium size, and 100.0%, 87.5%, and 100.0% for large individuals (Bullock et al., 1993); cockroaches, frequently forages on the ground, bulk of diet in xeric woodlands (Cabrits) composed of ants and termites, in rainforest sites, eats larger and more soft-bodied food than elsewhere (Malhotra and Thorpe, 1999); readily accepted grasshoppers in captivity, killed crickets and ate them later, refused a large katydid, earthworm, centipede, millipede, and beetles (Bayless, 2005); adult male on a perch 1.8 m high on a wall descended to the ground, ran 3.2 m to a small shrub, ascended to 30 cm and ate a small arthropod, assumed a head-down feeding position for 5 min, then returned to the original perch on the wall (RP). **Growth:** With low food 0.10 ± 0.06 mm/day to 0.17 ± 0.06 mm/day with high food (*A. o. winstoni*), females expected to reach maturity in 5–9 months (Andrews, 1976). **Life Expectancy:** Males 0.51 year, females 1.24 years (Andrews, 1979). **Home Range and Territoriality:** One adult male *A. o. winstoni* with 2–3 females on a banana tree (Krintler, 1986); both males and females defend territories, but those of females are smaller, with several overlapping with one male territory (Malhotra and Thorpe, 1999). **Movements:** Mean moves per minute 0.20 (Andrews, 1979); position shifts per 10 min were 1.2, 3.6, and 2.2 for small individuals at a dry scrub site and 2 rainforest sites, respectively, 2.3, 1.1, and 0.5 for individuals of medium size, and 0.9, 0.8, and 1.1 for large individuals, numbers of prey captures per 10 minutes were 1.3, 0.3, and 0.3 (small), 0.4, 0.01, and 0 (medium), and 0.5, 0.2, and 0.2 (large), and corresponding distances moved per 10 min were 0.61 m, 0.91

m, 0.05 m (small), 0.77 m, 0.44 m, and 0.18 m (medium), and 0.74 m, 1.05 m, and 1.29 m (large; Bullock et al., 1993). **Parasites:** Malarial parasites *Plasmodium azurophilum* (Staats and Schall, 1996b); nematodes *Oswaldocruzia lenteixeirai, Parapharyngodon cubensis, Spinicauda spinicauda*, larval *Ascarops* sp. and *Physaloptera* sp., trematode *Mesocoelium monas*, and juvenile acanthocephalan *Centrorhynchus* sp. (Goldberg and Bursey, 1996a); trematode *Mesocoelium monas*, nematodes *Oswaldocruzia marechali, Parapharyngodon cubensis, Trichospirura teixeirai*, and larval *Physaloptera* sp., and acanthocephalan cystacanth *Centrorhynchus* sp. (Goldberg et al., 1997a); nematode *Spauligodon anolis* (Goldberg and Bursey, 1996a; Goldberg et al., 1997a; Bursey and Goldberg, 1998). **Performance:** Anoles with larger bodies and shorter tails were faster (13.7 ± 1.5 versus 28.4 ± 3.2 sec/2 m) on a vertical track mimicking large trunks in primary rainforest, whereas anoles with longer tails performed comparably (8.9 ± 1.4 versus 8.1 ± 0.9 sec/2 m) on an inclined track (Loughridge, 2006). **Population Size and Density:** 48–58 lizards/100 m^2 and biomass of 169–179 g/100 m^2 (*A. o. winstoni*) with variation based on 2 different estimates of juvenile numbers (Andrews, 1979); mean density estimates at 10 coastal woodland sites ranged from 44.9–128.3 to 676.1–2,147.6 lizards/ha and from 0 to 146.1–849.6 lizards/ha in 8 rainforest sites, mean biomass estimates for the coastal woodland sites ranged from 0.46 ± 0.14 to 11.22 kg/ha and those for rainforest sites from 0 to 0.97 ± 0.38 kg/ha (Bullock and Evans, 1990); 33.5 anoles/ha (biomass 168.3 g/ha) in primary rainforest, 33.5 anoles/ha (118.3 g/ha) in secondary rainforest, 846.8 anoles/ha (4,232.8 g/ha) in dry forest (Morrison et al., 2002); extrapolation from biomass estimates suggests a breeding population of 200 million adults and, as females may produce an egg every 2 weeks, annual turnover may be one billion individuals (Thorpe et al., 2008b). **Predation:** Mangrove Cuckoos (*Coccyzus minor*) and Pearly-eyed Thrashers (*Margarops fuscatus*; Wunderle, 1981; Malhotra and Thorpe, 1997a); *Liophis juliae* (Henderson and Bourgeois, 1993); *Alsophis sibonius* (Henderson and Sajdak, 1996; Malhotra and Thorpe, 1999); Gray Kingbird (*Tyrannus dominicensis*; Malhotra and Thorpe, 1999); Brown Trembler (*Cinclocerthia ruficauda*; RP, RWH). **Reproduction:** Breeds throughout the year, but the number of gravid females highest in May and lowest in January–March, mean clutch size 1.1 ± 0.0 (1–2), no pattern to presence of 2 eggs at the same time, very rarely with 2 eggs in the same oviduct, mean egg length 10.9 ± 0.1 mm, mean yolked follicle number 1.2 ± 0.0 (1–3), mean follicle length 4.3 ± 0.2 mm, ovulation probably occurs when follicles 8–9 mm in length (Somma and Brooks, 1976); 63% of females with oviductal eggs in August (*A. o. winstoni*; Andrews, 1979); mean number of days between eggs 8.6 (*A. o. cabritensis* in captivity) for 3 females over 106 days of observation (Andrews and Rand, 1974); in captivity, female *A. o. montanus* defend oviposition sites (Welzel, 1981); reproductively active throughout the year, but peaks toward the end of the dry season in more seasonal areas of the island, eggs laid under rocks, leaves, bark, etc. (Malhotra and Thorpe, 1999); in captivity, eggs laid at intervals of 8–14 days, incubation 36–47 days at 23–28 °C (Fläschendräger and Wijffels, 2009). **Sex Ratio (M:F):** 0.65:1.35 (Andrews, 1979). **Size:** Male SVL to 96 mm (*A. o. montanus*; Lazell, 1962), female SVL to 64 mm (*A. o. montanus* and *A. o. winstoni*), smallest mature male and female SVL 44 mm and 42 mm, respectively (both *A. o. oculatus*; Lazell, 1972); at Rosalie, males are significantly larger in a deeply shaded and cool windbreak than those in warmer and more open coconut groves, large gradients in average lizard sizes over <100 m, with spatial size patterns strongly correlated with patches of insect abundance, but SVL strongly correlated with elevation, which may not reflect prey abundance (Malhotra and Thorpe, 1997a); at low and moderate elevations, lizard SVL and elevation are inversely related (Raetzsch, 2004); 50% of males mature at 37 mm SVL, 50% of females mature at 43 mm SVL (Somma and Brooks, 1976). **Survival:** 32.6% of males and 62.9% of females survived after 7 months, 25.0% of males and 53.5% of females after 11 months (all adults with SVL = 48 mm; Andrews, 1979). **Thermal Biology:** Rarely thermoregulates, except at coolest (= highest) localities, eurythermic, probably passive to a broad range of temperatures, mean body temperatures for *A. o. winstoni* 31.0 ± 0.19 °C (28.8–29.8 °C air temperature), 29.0 ± 0.12 °C (26.8–28.8 °C), 27.8 ± 0.09 °C (27.4–28.0 °C), for *A. o. oculatus* 26.9 ± 0.04 °C (26.2–26.8 °C), 29.8 ± 0.08 °C (28.6–29.8 °C), for *A. o. oculatus* at 350 m elevation 23.7 ± 0.15 °C (22.0–26.6 °C), for *A. o. montanus* at 850 m 23.8 ± 0.30 °C (20.4–22.7 °C; Ruibal and Philibosian, 1970); not heliothermic (*A. o. oculatus*), mean body temperature of adults 28.1 ± 1.7 °C at mean air temperature 27.8 ± 1.4 (females) and 28.3 ± 1.8 °C at mean air temperature 27.5 ± 2.0 °C (males), critical thermal maxima 33.0–36.4 °C (Brooks, 1968); mean body and air temperatures of free-ranging "North Caribbean" lizards 28.44 °C (25.5–31.4 °C) and 27.54 °C (25.1–30.5 °C), "South Caribbean" lizards 30.05 °C (27.7–31.8 °C) and 28.79 °C (26.0–30.4 °C), "montane" lizards 23.93 °C (22.3–26.7 °C) and 22.93 °C (21.8–25.0 °C), and "Atlantic" lizards 28.67 °C (27.0–31.1 °C) and 27.02 °C (25.4–28.1 °C), but mean temperatures of lizards from different regions did not vary significantly when contained in enclosures, suggesting that the apparent thermal preferences are attributable solely to the availability of thermal microclimates in different habitats (Malhotra and Thorpe, 1993); all but 3 of 50 anoles with body temperatures above ambient, but body and ambient temperatures are strongly correlated and chosen microhabitat temperatures did not differ significantly from that of a random nearby microhabitat, suggesting that ambient temperature is more important than microhabitat selection in determing body temperature (Girdler, 2004).

Conservation Status: Not assessed; the species remains abundant in most parts of the island, but the extent of genetic variation is not protected in established national parks and the area where the displacement by *A. cristatellus* is occurring affects nearly the entire range of the South Caribbean ecotype (Malhotra et al., 2007).

Additional Pertinent Literature: Malhotra and Thorpe (1992).

Anolis olssoni Schmidt, 1919

Distribution: Hispaniola, where eight currently recognized subspecies occur in disjunct ranges, also on Île de la Gonâve and Île à Cabrit; elevational range, sea level to ~700 m.

Habitat: Xerophilic (Schwartz and Henderson, 1991); in thornscrub, *Acacia* and cactus desert, lowland scrub, coastal areas (including clumps of grass overhanging water), woodland areas (on Île de la Gonâve), montane pine forests (Schwartz and Henderson, 1991; RP); at potentially high densities on grass stems and slender twigs of small bushes and herbaceous plants, sometimes on fence posts (mostly large males facing downward and neck bent so head almost horizontal) and strands of barbed wire (Williams and Rand, 1961; Schwartz, 1981; Schwartz and Henderson, 1991; J. W. Smith et al., 1994; Queral et al., 1995); restricted to clumps of grasses along the edge of a dry riverbed, mean microhabitat temperatures 31.6 ± 2.7 °C (27.0–35.4 °C; Powell et al., 1996b); perch heights 0.2–1.5 m (mean 0.6 ± 0.3 m), generally sprawled horizontally resting one or two limbs on leaves or blades of grass, with other limbs and tail extending to adjacent perches, with 48% observation on grass blades or stems, 21% on shrubs or herbaceous plants, 20% on leaves, and 11% on grass peduncles, none on the ground or in trees (J. W. Smith et al., 1994); mean perch height and diameter 0.46 m and 1.5 cm, 28.0% on leaves (J. B. Losos, in litt., 21.IV.2008); sleeps with hindlimbs fully extended (Williams and Rand, 1961) on branches and twigs of desert plants, guava bushes, grasses, low shrubs, in bunches of dead pine needles on a fallen branch on the ground (Schwartz, 1981; Schwartz and Henderson, 1991); conforms to the substrate, typically a blade of grass or plant stem, with heads facing up (J. W. Smith et al., 1994).

Activity: Most active in the morning (76%; Hertz, 1979a); begins moving 0500–0545 h and active throughout the day, but uses lower perches during midday while moving through dense vegetation (J. W. Smith et al., 1994). **Behavior:** Sleeps adpressed on grasses with hindlimbs extended, adpresses body tightly against grass blades when threatened (Mertens, 1939); escapes by jumping into grass cover, when chased into a rock pile, stays close to the surface (Williams and Rand, 1961); male-male interactions included "variety-1 behaviors" (3 sec to nearly 2 min in duration) with low-amplitude head and dewlap movements involving discrete high-frequency (2–4 bobs/sec), short-duration (0.05–0.08 sec/bob), low-amplitude (less than half of head-depth) head movements similar to "jiggling," with some altering the angle of the head relative to the body and others elevating the head while it remains parallel to the body axis, "variety-2 behaviors" (23 sec to 10 min) with a variety of actions in which the only common parameter was relatively high amplitude head and dewlap movements, with head/body movements including bobs, bows, push-ups (2–7/sec and <0.1–0.3 sec duration), and head raising singly or in combinations, with amplitudes 1–2 head-depths, and partial or full dewlap extensions (2.4/sec and 0.3–0.8 sec duration), and "variety-3 behaviors" (2–5 sec) consisting of sudden and simultaneous change in head and dewlap position to their greatest amplitudes, with the dewlap extended or pulsed, occasionally on hindlimbs only, with ventilation increasing dramatically (from 48 to 78/min), modifiers included lateral orientation and compression, eye protrusion, erection of nuchal and dorsal crests, gaping with tongue protrusion, changes in color (body became lighter and the middle third of the tail became very dark), and tail elevation or wagging, all physical attacks preceded by one or both males assuming "variety-3" postures (Queral et al., 1995); moves per minute 0.56 and 1.22 (Johnson et al., 2008); percent time displaying 5.2, percent walking 22.4, percent jumping 37.6, percent running 40.0 (J. B. Losos, in litt., 21.IV.2008). **Ecomorphology:** "Grass-bush" (Williams and Rand, 1961); formally quantified (Beuttell and Losos, 1999). **Foraging and Diet:** Contents of 28 stomachs contained spiders and another arachnid, beetles, hemipterans (with leafhoppers most important in numbers and numbers of stomachs), hymenopterans (including ants), lepidopterans (adults and larvae, which constituted the greatest volume of all prey items), a scorpionfly, a dragonfly, orthopterans, unidentified arthropodan material, plant material, and grit (presumably adventitiously ingested), male and female diets did not differ significantly, larger individuals took larger prey, one individual caught a moth in flight (J. W. Smith et al., 1994). **Home Range and Territoriality:** Territorial (Mertens, 1939); mean male home range size 25.4 ± 28.5 m^2, female 28.3 ± 75.6 m^2, with home range size not correlated with SVL, 13 of 20 male home ranges overlapped with those of at least one other male, 15 of 17 female home ranges overlapped with those of at least one other female, and 13 overlapped with at least one male, apparently nonterritorial, but males will respond aggressively to other males in close proximity in a sight-limited habitat (J. W. Smith et al., 1994; Queral et al., 1995). **Movements:** Mean distances moved between observations by adult males 3.7 ± 2.5 m, adult females 4.3 ± 3.9 m (J. W. Smith et al., 1994). **Parasites:** Apicomplexa *Eimeria avilae* (Cisper et al., 1995); trematode *Urotrema scabridum* and nematode *Parapharyngodon cubensis* (Goldberg et al., 1998b); larval mites (*Eutrombicula alfreddugesi*) present in higher-elevation, moister situations, but absent from a lowland population (Zippel et al., 1996). **Population Size and Density:** Near Barahona, 608 or 650/ha, depending on method used (J. W. Smith et al., 1994). **Population Structure:** Adult:juvenile ratio 86:26 in June (J. W. Smith et al., 1994). **Predation:** *Uromacer frenatus*, *U. oxyrhynchus* (Henderson et al., 1987c). **Reproduction:** Male-female interactions variable, with males beginning displays 7.5 cm to 1.8 m from a female, actively pursuing her, displays and pursuits characterized by frequent pauses of varying durations, displays including headbobs, push-ups, and 4 different dewlap modalities, fully extended, partially extended, double-pulsed, and prolonged (to 3 min) extension, 13 of 38 females with enlarged ovarian follicles or eggs, 2 with 2 eggs, suggesting an ability to retain eggs until deposition is propitious (J. W. Smith et al., 1994). **Sex Ratio (M:F):** Adults 48:38 (J. W. Smith et al., 1994). **Size:** Male SVL to 53 mm, female SVL to 47 mm, smallest mature female 39 mm (J. W. Smith et al., 1994). **Thermal Biology:** 80% active in sunny weather, thermally passive despite living in open habitats, mean body temperature on sunny days 33.2 ± 0.3 °C at mean air temperature 30.1 ± 0.3 °C, on cloudy days 30.6 ± 1.0 °C at mean air temperature 28.3 ± 1.1 °C (Hertz, 1979a); near Barahona, in shade 56% and partial shade 38% of time, not observed basking or overtly thermoregulating in any way and are probably thermally passive, mean body temperature 32.3 ± 1.3 °C at mean substrate temperature 30.5 ± 1.3 °C, at night mean body temperature 29.6 ± 1.0 °C at mean substrate temperature 27.5 ± 1.2 °C (J. W. Smith et al., 1992); along a dry riverbed, microhabitat temperatures (mean 31.6 ± 3.6 °C, with a high of 35.4 °C) comparable to those of *A. brevirostris*, but lower than those of *A. whitemani* (Powell et al., 1996b).

Conservation Status: Not assessed.

Additional Pertinent Literature: J. W. Smith et al. (1995).

Anolis opalinus Gosse, 1850

Distribution: Jamaica, where widely distributed, but records concentrated in some areas and sparse in others; elevational range, sea level to >1,500 m.

Habitat: Mesophilic (Schwartz and Henderson, 1991); in and about houses (Gosse, 1851); in bushy and forested areas, xeric woods, a *Musa* grove, along a road in forest, under bark on tree stumps, usually on trees and bushes (Grant, 1940a; Rand, 1967b; Schoener and Schoener, 1971a; Schwartz and Henderson, 1991); almost always on trees or bushes, sometimes on hanging vines, rarely on rocks, logs, or stumps, even young individuals not close to the ground, plurality (39.2%) of large adults on perches 1.8–3.0 m high, but many (36.9%) on perches >3 m high, most (58.7%) on large-diameter (>7.6 cm) perches, most smaller individuals (39.0%) on perches 1.8–3.0 m high, but many (36.0%) on perches 0.9–1.8 m high, most (48.4%) on large-diameter (>7.6 cm) perches (Rand, 1967b); strongest association with shaded habitats of common Jamaican anoles, perch heights from <0.9 m to >6 m, mostly on perches of intermediate size (6–13 cm), with males perching higher than females and juveniles (Schoener and Schoener, 1971a); near Mandeville, nearly all lizards perched at heights <2 m, around Kingston, 63% perched above 1.5 m and 94% on perches with diameters >1.2 cm, but lizards may have selected higher perches in the presence of *A. lineatopus* (Jenssen, 1973); mean adult male and female perch heights 1.6 ± 0.90 m (1.3% on the ground) and 1.5 ± 0.77 m (0.0% on the ground; Lister, 1976); mean perch height and diameter 1.4 ± 0.09 m and 24.2 ± 6.58 cm (Losos, 1990a); over half (54.86%) of all individuals on tree trunks, 17.09% on the ground, 5.13% at heights <30 cm, 9.40% at 30–59 cm, 27.35% 60–89 cm, 16.24% 90–119 cm, 7.69% 120–149 cm, 5.13% 150–179 cm, 8.55% 180–209 cm, 2.56% 210–239 cm, none higher, 28.83% on rocks, 6.31% on perches with diameters <2.5 cm, 9.01% 2.5–5.0 cm, 20.72% 5.1–7.5 cm, 6.31% 7.6–10.0 cm, 6.31% 10.1–12.5 cm, 8.11% 12.6–15.0 cm, 4.50% 15.1–17.5 cm, 9.91 on perches with diameters >17.6 cm, perches did not differ by sex or size class, except size classes used significantly different perch types and diameters (G. L. Powell and Russell, 1992); mean perch heights and diameters 1.84 ± 0.99 m and 9.5 ± 7.8 cm for males and 1.62 ± 1.30 m and 9.0 ± 11.2 cm for females (Butler and Losos, 2002); 13.6% on leaves (J. B. Losos, in litt., 21.IV.2008); habitat light environment "partial shade" (Fleishman, 2000); dewlap UV-reflectance compatible with visibility in typical habitat (Macedonia et al., 2000).

Activity: Diurnal (Schoener, 1970); earliest activity at Hardwar Gap 0710 h, latest activity 1657 h (possibly compromised by curtailed sampling periods), male displays peak at dawn and dusk, presumably to advertise territorial activity before and after a period of nocturnal inactivity (Ord, 2008). **Behavior:** Squeaks during aggressive interactions (Grant, 1940a; Underwood and Williams, 1959; Myers, 1971; T. A. Jenssen in Milton and Jenssen, 1979); mean display rate (proportion of time spent displaying) 0.078 ± 0.013, percent walks 19.8 ± 3.6, percent runs 69.6 ± 3.8, percent jumps 10.6 ± 2.0, moves/min 1.74 ± 0.19, distance jumped in nature 30.3 ± 4.7 cm (Losos, 1990a); lacks even one discrete display pattern (i.e., weak display stereotypy), with headbobs 4–11 and dewlap pulses 0–8, but most displays (75%) included an initial double-headbob series, a middle-bob series of successively quickening cadence, and a final-bob series of successively slowing cadence (Jenssen, 1979b); exhibit static display modifiers (optional body postures) that include erected crest, gorged throat, orbed eyes, opened mouth, protruded tongue, cocked head, and lateral compression, dynamic modifiers (optional display movements) include circumduction, rearing, head roll, dewlap pulsing, step-bobbing, rapid headbobs, and limb amplitude (Jenssen, 1979a); moves per minute 1.74 (Johnson et al., 2008). **Diet:** Stomach contents were ants (highest percentage), other hymenopterans, flies, beetles (adults and larvae), barklice, leafhoppers, lepidopteran larvae, lacewings, earwigs, orthopterans, spiders, pseudoscorpions, centipedes, mites, and isopods (Lister, 1976); mostly soft-bodied prey (except ants), by frequency ants compose 70% of food (46% of ants *Wasmannia auropunctata*), but only 12% by volume, lepidopteran larvae and orthopterans compose 46% of the diet by volume, rarely take other anoles (*A. valencienni* hatchlings), little evidence of dietary niche partitioning, ingest same volume of food in wet and dry seasons, but fewer, larger prey taken during the wet season (Floyd and Jenssen, 1983, 1984); volume of stomach contents not significantly different in wet and dry seasons, juveniles 14.5 ± 3.08 versus 23.5 ± 5.85 mm^3/g body weight, subadult males 23.3 ± 5.35 versus 21.0 ± 4.56 mm^3/g, all ages 15.9 ± 1.51 versus 20.1 ± 2.06 mm^3/g (Jenssen and Andrews, 1984). **Ecomorphology:** "Trunk-crown" (Williams, 1972); formally quantified (Beuttell and Losos, 1999); acts as a trunk-ground species in the absence of *A. lineatopus* (Jenssen, 1973). **Growth:** Growth faster during the wet than in the dry season for adult males (rate 0.011 versus 0.006 mm/day), adult females (rate 0.011 versus 0.008 mm/day), and juveniles (males 0.14 ± 0.010 versus 0.08 ± 0.014 mm/day, females 0.10 ± 0.012 versus 0.07 ± 0.017 mm/day; Jenssen and Andrews, 1984). **Parasites:** Protozoan malarial parasite *Plasmodium floridense* (Telford, 1975; Ayala, 1977, 1978); cestode *Oochoristica* sp. (Bundy et al., 1987). **Performance:** Mean maximum sprint speed (fastest time over an interval of 0.25 m) 0.142 ± 0.003 sec, jump 55.5 ± 1.62 cm, cling 1.275 ± 0.165 Neutons (Losos, 1990a). **Predation:** Greater Antillean Grackles (*Quiscalus niger*) capable of taking all size classes (Graves, 2006). **Reproduction:** In a high-elevation (690 m) population, males gametogenic, copulating, and advertising for mates throughout the year, but morphometric sexual structure size peaks in April–July and is at a minimum October–November, shifting the testicular cycle 2 months in advance of the wet season, percent gravid females varies from 42% (December) to 100% (August), with reproductive variables reaching maxima in June–October, occurrence of females with 2 oviductal eggs in April–October, larger females produced larger but not more eggs than smaller females, seasonal reproductive peak activity reached greatest levels several months after males (Jenssen and Nuñez, 1994); clutch size 1 (Hedges, in prep); in captivity, females lay eggs at intervals of 7–10 days, incubation ~36 days at 24–28 °C (Fläschendräger and Wijffels, 2009). **Size:** Male SVL to 54.8 mm, female SVL to 52.1 mm (G. P. Powell and Russell, 1992); juvenile SVL <37 mm (Jenssen and Andrews, 1984). **Thermal Biology:** Maintains temperatures higher than ambient and lower than those of *A. grahami* (warmer) and *A. lineatopus* (Rand, 1967b); shade loving at low elevations, seeks more open habitats at higher elevations (Schoener and Schoener, 1971a).

Conservation Status: Not assessed.

Additional Pertinent Literature: Russell and Bauer (1991c).

Anolis ophiolepis Cope, 1861

Distribution: Cuba, widely distributed, but local, also Isla de la Juventud.

Habitat: More or less xerophilic (Schwartz and Henderson, 1991); truly terrestrial, in pastures and savannas, may climb to about 45 cm, but rarely does so, sleeps on blades of grass or leaves of small low bushes and on salt marsh vegetation (Barbour and Ramsden, 1919; Neill, 1958; Ruibal, 1964; Schwartz and Henderson, 1991); in mesophyllous evergreen and semideciduous forest, cultivated areas and pastures, and urban zones (Rodríguez Schettino et al., 1999); open savannas, pastures, gardens, uncultivated terrain, marshes, and sugarcane fields, on stems of grasses and small bushes, seeks shelter on the ground among dead leaves or in piles of recently cut grass (Rodríguez Schettino, 1999b); rocky, grass-covered slopes paralleling roadside ditches (Fläschendräger, 2002); mean perch height and diameter 0.25 m and 0.8 cm (J. B. Losos, in litt., 21.IV.2008).

Activity: Diurnal (Rodríguez Schettino et al., 1999); active only when the sun is shining and usually only after 1000 h (Fläschendräger, 2002). **Behavior:** Runs when pressed and takes refuge in grass tussocks (Ruibal, 1964); agonistic male-male encounters involve rapid and complete extension of the dewlap and rapid, continuous headbobs with lateral presentation and lateral compression (Garrido, 1991); hides between the lowest part of the herbaceous stratum and leaf litter or in holes in the ground (Rodríguez Schettino, 1999b); seeks shelter in dense vegetation or under limestone rocks, in captivity, both males and females are intensely territorial (Fläschendräger, 2002). **Diet:** Primarily ants (Barbour and Ramsden, 1919). **Ecomorphology:** "Grass-bush" (Williams, 1961). **Parasites:** Nematodes *Cyrtosomum scelopori* and larval *Porrocaecum* sp. (Coy Otero and Baruš, 1979b; Coy Otero and Lorenzo Hernández, 1982; Coy Otero, 1999). **Reproduction:** Mean duration of copulation 4.31 min (longest 9 min), separation of pair "hasty," egg 7.6 × 4.1 mm, cream colored and elliptical (Garrido, 1980a); clutch size 1 (Hedges, in prep.); in captivity, egg 5.0 × 7.2 mm at deposition, 7.8 × 11.0 mm at hatching, incubation ~40–50 days at 25–28 °C, hatchling SVL ~14 mm (Fläschendräger, 2002; Fläschendräger and Wijffels, 2009); courtship involves extension of the dewlap while headbobbing at an increasing rate, until it is almost "vibrating" at the end, this is repeated 3–5 times (duration 7–10 sec, mean 8.5 sec), before mounting the female and biting her nape, remaining immobile for ~40 sec before attempting intromission, mean duration of copulation 4 min (2–9 min), during which the male continues to hold the female's neck while moving his head laterally, initially quick and rhythmically, but diminishing in intensity until completely immobile before dismounting (Garrido, 1991). **Size:** Male SVL to 39.8 mm, female SVL to 39.5 mm (Rodríguez Schettino, 1999b). **Thermal Biology:** Heliothermic, usually in filtered sun or shade low on grasses during the warmer wet season, exposed in sunny situations during the dry season (Rodríguez Schettino, 1999b).

Conservation Status: Not assessed.

Anolis oporinus Garrido and Hedges, 2001

Distribution: Cuba, only from the vicinity of the type locality on the northern slopes of the Sierra Maestra; elevation ~465 m.

Habitat: Limestone mogotes with coffee plantations and some pasture, the type specimen was collected in vegetation along a road (Garrido and Hedges, 2001).

Activity: Diurnal (Rodríguez Schettino et al., 1999). **Ecomorphology:** "Trunk-crown" (Losos, 2009). **Size:** SVL of only known specimen (adult female) 47 mm, males unknown (Garrido and Hedges, 2001).

Conservation Status: Unknown.

Anolis paternus Hardy, 1967

Distribution: Cuba, where two subspecies occur in extreme southwestern Cuba (*A. p. pinarensis*) and on Isla de la Juventud (*A. p. paternus*).

Habitat: Xerophilic (Schwartz and Henderson, 1991); on fence posts adjacent to a sandy savanna with scattered trees or large bushes (*Byrsonima*), on palmettos in sandy savanna, trunks of palms (*Roystonea*), and on grass in meadowland (Barbour and Ramsden, 1919; Ruibal, 1964; Hardy, 1967; Garrido, 1975d; Schwartz and Henderson, 1991; Rodríguez Schettino, 1999b); in cultivated areas and pastures, and urban zones (Rodríguez Schettino et al., 1999).

Activity: Diurnal (Rodríguez Schettino et al., 1999). **Behavior:** Male-male encounters involve lateral compression, push-ups, head elevation, and dewlap extension repeated 3–4 times (sometimes as many as 8 times), neither nuchal nor dorsal crests are erected, nor is the tongue protruded as in *A. angusticeps*, an encounter ends when one of the contenders assumes the posture of a nonreceptive female (Garrido, 1991); moves slowly, staying very close to the substrate, if pressed, tends to move upward, when on fence posts or grasses, retreats to and hides in herbaceous vegetation (Rodríguez Schettino, 1999b). **Ecomorphology:** "Twig" (Langerhans et al., 2006). **Reproduction:** Mean duration of copulation 2 min (longest 4 min), separation of pair "normal" (Garrido, 1980a); courtship begins with the male headbobbing rapidly, inflating his throat and extending his dewlap, usually repeated twice, but sometimes only once and occasionally 3 times, before approaching the female and grasping her by the neck, once positioned, he pauses for a second before attempting intromission, if the female is not receptive, she lashes her tail several times and flees, mean duration of copulation is 2.5 min (2–4 min), during which the male waves or vibrates his tail rhythmically, stopping several seconds before termination, which is sometimes gradual and other times abrupt (Garrido, 1991). **Size:** Male SVL to 50 mm (Garrido, 1975d), female SVL to 47.8 mm (Rodríguez Schettino, 1999b). **Thermal Biology:** Apparently heliothermic, basks frequently and is active mainly during the heat of the day (Rodríguez Schettino, 1999b).

Conservation Status: Not assessed.

Anolis pigmaequestris Garrido, 1975

Distribution: Cuba, Archipiélago de Sabana-Camagüey.

Habitat: Xerophilic (Schwartz and Henderson, 1991); on trunks of palmettos (*Pseudophoenix*), bushes, and generally shaded areas, with males favoring palmettos and females preferring shrubs (Garrido, 1975a); in coastal and subcoastal microphyllous forest (Rodríguez Schettino et al., 1999).

Activity: Diurnal (Rodríguez Schettino et al., 1999). **Behavior:** Females remain motionless and escape by jumping, males escape upward (Rodríguez Schettino, 1999b). **Ecomorphology:** "Crown-giant" (Williams, 1972). **Reproduction:** Egg laid by a female captured in October 23.5 × 13.3 mm (Rodríguez Schettino, 1999b). **Size:** Male SVL to 140 mm, female SVL to 121 mm (Garrido, 1975a). **Thermal Biology:** Presumably heliothermic based on apparent preference for sunny perches (Rodríguez Schettino, 1999b).

Conservation Status: Considered "critically endangered" (Rodríguez Schettino and Chamizo Lara, 1998a; Rodríguez Schettino and Chamizo Lara in Rodríguez Schettino, 1999a); assessment using IUCN Red List criteria (Fernández Méndez, 1998a).

Anolis placidus Hedges and Thomas, 1989

Distribution: Hispaniola, Sierra de Neiba (Dominican Republic); elevational range 1,710–1,900 m (Hedges and Thomas, 1989b).

Habitat: Mesophilic (Schwartz and Henderson, 1991); in mixed broadleaf cloud forest, sleeps 1–3 m above the ground on twigs of bushes or trees next to roads (Hedges and Thomas, 1989b).

Behavior: When pressed, remains tightly attached to the twig and must be pried loose, one individual was hopping across a road during a heavy rain (Hedges and Thomas, 1989b); escape behavior (in captivity) involves "squirreling" (Hedges and Thomas, 1989b). **Ecomorphology:** "Twig," with prehensile tail (Hedges and Thomas, 1989b). **Reproduction:** Clutch size 1 (Hedges, in prep.). **Size:** Male SVL to 45 mm, female SVL to 46 mm (Hedges and Thomas, 1989b).

Conservation Status: Unknown.

Anolis pogus Lazell, 1972

Distribution: St.-Martin/St. Maarten, essentially islandwide; known to have occurred on Anguilla (Lazell, 1972), where apparently extirpated; also may have occurred on St.-Barthélemy (Lazell, 1972; Schwartz and Henderson, 1991; Breuil, 2002b); failed experimental introductions on Anguillita (Roughgarden et al., 1983b, 1984).

Habitat: Xerophilic to mesophilic (Schwartz and Henderson, 1991); in rainforest and xeric scrub, often in wet, shady pockets, on trees, bushes, fence posts, rocks, stone walls, woodpiles, trash, and buildings, usually in the shade, but occasionally in open sunny areas (Lazell, 1972; Powell et al., 2005; Powell and Henderson, 2008c); largely absent from dry, open habitats (Fläschendräger and Wijffels, 1996, 2009), but taken in dry scrub with grasses, shrubs, xerophytic bromeliads, and cacti (Wijffels, 1962); mean perch height 0.79 ± 0.03 m (Roughgarden et al., 1983a, 1984); 15% on the ground, 25% at heights 0–0.46 m, 23% at 0.46–0.9 m, 28% at 0.9–1.8 m, 8% >1.8 m (Roughgarden, 1995); mean perch height and diameter 1.46 ± 0.21 m and 5.4 ± 0.9 cm (Losos and de Queiroz, 1997a); mean perch height of males at two sites 0.1 ± 0.0 and 0.1 ± 0.1 m, females 0.0 ± 0.0 and 0.1 ± 0.1 m, one juvenile was on the ground (Pereira et al., 2002); reported absence at xeric sea level sites attributed to competitive exclusion with *A. gingivinus* (Pacala and Roughgarden, 1982; Roughgarden et al., 1983b), but many populations occur in mesic pockets (often in highly altered habitats) at sea level (Powell et al., 1992).

Activity: Peak of feeding activity around dusk, often continuing until full dark (Malhotra and Thorpe, 1999). **Behavior:** Retreats into holes and crevices in rocks, tree trunks, under shutters of buildings, or into debris on the ground (Lazell, 1972); 2 males in proximity without interaction (Breuil, 2002b); shy, when disturbed will "squirrel" to the other side of a trunk or branch or retreat into a crevice, may jump to another perch or to the ground and hide in leaf litter or under cover, only rarely climbs upward (Powell et al., 2005). **Competition:** Presence caused *A. gingivinus* to select higher perches, grow more slowly in enclosures, contain smaller volumes of developing eggs per female, and have smaller volumes of insect prey (Pacala and Roughgarden, 1982, 1985; Roughgarden et al., 1984; Roughgarden, 1995); in experimental introductions on Anguillita, territories were peripheral to those of established undisturbed *A. gingivinus*, but occurred in centers of groves when *A. gingivinus* partially removed (Roughgarden et al., 1984; Roughgarden, 1995); mean overlap between each *A. gingivinus* and *A. pogus* was 94% and 18% at 2 sites on St.-Martin/St. Maarten, and 43% of agonistic interactions (many around food resources such as fruits with insects) and 26% of directed displays were interspecific, juvenile *A. gingivinus* consistently chased away by *A. pogus* males, *A. gingivinus* females won about half of the disputes with *A. pogus* males but most with *A. pogus* females, *A. gingivinus* males engaged in 4 interspecific interactions, one of which was won by a much smaller *A. pogus* male (Pereira et al., 2002); male *A. gingivinus* intolerant of *A. pogus* on perches of large diameter, but relatively oblivious to the presence of the smaller species on perches of smaller diameter or even in rock walls, degrees of tolerance may vary seasonally (Breuil, 2002b). **Desiccation:** Water loss rate per unit surface area higher than that of *A. gingivinus* (Roughgarden et al., 1983b). **Diet:** Gastropods (frequency 0.01), mites (0.07), isopods (0.035), juvenile spider (0.33), centipedes (0.01), beetles (adults, 0.70), beetles larvae (0.01), springtails (0.10), flies (adults, 2.01), flies (larvae, 0.07), hemipterans (1.70), ants (9.55), other hymenopterans (0.58), lepidopterans (adults, 0.32), lepidopterans (larvae, 1.18), orthopterans (0.11), thrips (0.12), and termites (0.07; J. D. Rummel in Goldwasser and Roughgarden, 1993); *Sphaerodactylus parvus* (Breuil, 2002b); small insects and spiders, will lap nectar and eat flowers and fruits (Powell et al., 2005); a captive population in a greenhouse ate snails (F. Molle in Fläschendräger and Wijffels, 2009). **Ecomorphology:** Close to "grass-bush" (Beuttell and Losos, 1999). **Home Range and Territoriality:** Home range ~1 m^2 (C. Bohlen in Roughgarden, 1995); each male home range associated with one female home range (mean overlap 91 ± 4% and 50 ± 9% at 2 sites), sizes 1.0 ± 0.1 and 3.1 ± 0.5 m^2 for males and 1.3 ± 0.5 and 2.6 ± 0.7 m^2 for females, one juvenile home range 2.6 m^2, size ratio of male to female home ranges 1.05, highest male-male home range overlap 3% and highest female-female overlap 9%, highest overlap between "non-paired" male and female 8%, low perch heights of males makes control of more than one female home range difficult (Pereira et al., 2002). **Life Expectancy:** 0.62 year (McLaughlin and Roughgarden, 1989). **Parasites:** Only 3% of adults infected with malaria (*Plasmodium azurophilum*), indicating greater resistance to infection and tolerance of effects than *A. gingivinus* (Schall, 1992; Staats and Schall, 1996b); nematode *Paraphyaryngodon cubensis* (Dobson et al., 1992); mean parasite species/host 0.44 on St.-Martin/St. Maarten (Dobson and Pacala, 1992). **Population Size and Density:** 26 individuals in an area of ~100 m^2 and 10 lizards from a 20-m stretch of stone wall and estimated that only 1 of every 3 captured (Lazell, 1972); 61.4 ± 7.0/100 m^2 (Peterson estimate) and 57.7 ± 2.4/100 m^2 (3-census independence-model estimate) and 30.0 ± 3.0/100 m^2 and 35.0 ± 3.4/100 m^2 (Peterson estimates only; Heckel and Roughgarden, 1979); 4.4 ± 1.8–56.8 ± 2.7 lizards/100 m^2 at 2 sites (Roughgarden et al., 1983b); at 3 sites where sympatric with *A. gingivinus* 10.5 ± 0.7, 17.7 ± 0.7, and 16.9 ± 0.8 lizards/100 m^2, over 4-year period densities at 2 sites 0–4.4 ± 1.8 and 10.5 ± 0.7–56.8 ± 2.7 lizards/100 m^2 (Roughgarden et al., 1984); over 7 years at a lowland site on St.-Martin/St. Maarten 0–4.4 ± 1.8 lizards/100 m^2, at an upland site 10.5 ± 0.7–56.8 ± 2.7 lizards/100 m^2 (Roughgarden, 1986, 1995); 1.0 individuals/m^2 of *A. pogus* and *A. gingivinus* at one site, 0.3 lizards of both species/m^2 at a second site (Pereira et al., 2002). **Predation:** Presumably American Kestrels (*Falco sparverius*), Pearly-eyed Thrashers (*Margarops fuscatus*; McLaughlin and Roughgarden, 1989; Roughgarden, 1995; Powell et al., 2005); *A. gingivinus* (Roughgarden et al., 1983b; Goldwasser and Roughgarden, 1993; Roughgarden, 1995); *Ameiva pleii* (Powell et al., 2005). **Reproduction:** High male-female home range overlap suggests monogamy, and all copulations ($N = 10$) at one site were between associated pairs, 3 of 4 at a second site between associated pairs, and all but 2 pairs copulated one or more times, with an average of 2 days between copulations, but 40% of intercopulation intervals were shorter (Pereira et al., 2002); reproduction largely associated with the onset of the rainy season in June–August (Breuil, 2002b). **Size:** Male SVL to 50 mm (Lazell, 1972), female SVL to 45 mm (Dobson et al., 1992), smallest mature male and female SVL 39 mm and 36 mm, respectively (Lazell, 1972). **Survival:** 6-month survivorship low, but population existed for 2 years after an experimental introduction on Anguillita (Roughgarden et al., 1984). **Thermal Biology:** Lower panting temperature than *A. gingivinus* (Roughgarden et al., 1983b, 1984); mean body temperature 28.3 ± 0.4 °C (Roughgarden et al., 1984).

Conservation Status: Not assessed, but protected by French legislation on St.-Martin (Lorvelec et al., 2007); absence from Anguilla and St.-Barthélemy suggests that populations may be vulnerable to certain human-mediated alterations of habitats.

Anolis poncensis Stejneger, 1904

Distribution: Puerto Rico, southern coast.

Habitat: Xerophilic (Schwartz and Henderson, 1991); on a hill of disintegrating coral limestone among scant grass and weeds (Stejneger, 1904); on arid cactus-covered hills (K. P. Schmidt, 1920, 1928); usually on perches of small diameter at heights below 0.9 m (Rand, 1964a); in exposed grassy areas, on fence posts, bushes, and Prickly Pear Cactus (*Opuntia*), where sympatric with *A. pulchellus*, less abundant in grassy areas than that species (Rivero, 1998); mean perch height and diameter 0.9 ± 0.16 m and 3.6 ± 0.36 cm (Losos, 1990a); mean perch heights and diameters 0.31 ± 0.14 m and 2.3 ± 1.5 cm for males and 0.30 ± 0.17 m and 3.4 ± 5.6 cm for females (Butler and Losos, 2002); 22.2% on leaves

(J. B. Losos, in litt., 21.IV.2008); sleeps on shrubs and dead branches of shrubs to 0.6 m above ground (Schwartz and Henderson, 1991).

Abundance: At times in dense "colonies" (Schwartz and Henderson, 1991). **Behavior:** Very agile (Stejneger, 1904); mean display rate (proportion of time spent displaying) 0.012 ± 0.004, percent walks 43.5 ± 7.2, percent runs 37.6 ± 5.4, percent jumps 18.9 ± 5.0, moves/min 0.58 ± 0.12, distance jumped in nature 32.0 ± 5.2 cm (Losos, 1990a); moves per minute 0.58 (Johnson et al., 2008). **Ecomorphology:** "Grass-bush" (Williams, 1961); formally quantified (Beuttell and Losos, 1999). **Parasites:** Trematode *Mesocoelium monas* (Cofresí-Sala, 1964). **Performance:** Mean maximum sprint speed (fastest time over an interval of 0.25 m) 0.142 ± 0.003 sec, jump 60.9 ± 1.36 cm, cling 1.312 ± 0.124 Neutons (Losos, 1990a). **Size:** Male SVL to 48 mm (Williams, 1983), female SVL to 41 mm (M. Leal in Jezkova et al., 2009).

Conservation Status: "Rare everywhere, protect habitat" (Moreno, 1991d); restriction to regions of Puerto Rico vulnerable to development suggests that habitat protection may be necessary, although the species remains locally common (Moreno, 1991b).

Anolis porcatus Gray, 1840

Distribution: Cuba, widely distributed and on Isla de la Juventud, Archipiélago de los Canarreos, Cayos de San Felipe, Archipiélago de Sabana-Camagüey, Archipiélago de los Colorados; introduced into the area around Santo Domingo, Dominican Republic (Hispaniola), southeastern Florida, and Aruba (G. van Buurt in Fläschendräger and Wijffels, 2009); no subspecies are usually recognized, although one (from Pinar del Río Province, Cuba) has been described (Pérez-Beato, 1997).

Habitat: Mesophilic (Schwartz and Henderson, 1991); often edificarian, common in gardens, on fences and fence posts (usually facing upward on upper part of post), in pastures on Algarrobas (*Samanea saman*) and *Roystonea*, *Cocos*, and *Ficus* at forest edges (but scarce in forest interiors), in banana groves, on palms (*Coccothrinax*) and *Coccoloba* in coastal thickets, on long leaves of *Agave* and *Fourcraea*, and in mangroves, when on trees, usually facing downward (Barbour and Ramsden, 1919; Schwartz and Ogren, 1956; Collette, 1961; Ruibal and Williams, 1961a; Ruibal, 1964; Lando and Williams, 1969; Heselhaus and Schmidt, 1990; Schwartz and Henderson, 1991; Rodríguez Schettino, 1999b; Powell and Henderson, 2008c); frequently on and in houses (Alayo Dalmau, 1951); on a curtain rod in a bungalow (Heselhaus, 1990); often a human commensal (Fläschendräger, 2002); on fences, hedges, palms, and roadside trees, even in La Habana, parks, and gardens, dependence on relatively open areas was probably reflected in a curtailed distribution prior to the arrival of humans (G. Peters, 1970); on fences, trunks of small trees, gardens, creeping plants, and on bushes 100–200 m from the nearest houses in Santa Damiana, Pinar de Río Province, where rare in forested zones but abundant in disturbed areas (Pérez-Beato, 1997); particularly abundant in crowns of trees and higher bushes in habitats modified by human activity (Vergner and Polák, 1990); in the Biosphere Reserve "Sierra del Rosario" (Cuba), associated primarily with tree trunks in open areas of secondary vegetation and pastures (Martínez Reyes, 1995); in submontane mesophyllous evergreen forest, mesophyllous evergreen and semideciduous forest, coastal and subcoastal microphyllous forest, semidesert thorny scrubwoodland, serpentine xeromorphous shrubwoodland, cultivated areas and pastures, and urban zones (Rodríguez Schettino et al., 1999); frequently in yards on ornamental vegetation (Kunz, 2005); juveniles use "bush-grass habitat" (Collette, 1961); on Cayo Cantilles, often on palmettos, where they are the only anoles, but also on forest trees (Garrido and Schwartz, 1969); on cayos Francés and Santa María, on trunks of palms and palmettos (Garrido, 1973d); on Cayo Guajaba (Archipiélago de Samana-Camagüey), uses high perches, most >3 m high (Garrido et al., 1986); on Cayo Inés de Soto, more likely to be on branches (than trunks) and on higher perches of smaller diameter than *A. sagrei* (Estrada and Novo Rodríguez, 1986b); on the Archipiélago de los Canarreos, coastal vegetation (especially in *Cocos nucifera*), microphyllous evergreen forest, and on cays without forest, uses palms and mangroves, where lizards use branches high in trees, lianas, and main trunks at substantial heights (Estrada, 1993a); often in habitats altered by human activity (Estrada, 1994); perches at heights of 1–5 m, with adults higher than juveniles, means 0.95 m at Jibacoa, 1.14 m for males at Guanajay, 1.85 m at cliffs on the Península de Guanahacabibes, where almost exclusively on palmettos, the tallest plants present, where syntopic with *A. sagrei*, *A. porcatus* perches higher and on perches of smaller diameter (Rodríguez Schettino, 1999b); mean perch height at Soroa 9.69 m, diameter 12.6 cm, behavioral diameter (mean diameter of all surfaces used during behavioral observations, means of individual means) 10.50 cm, no observations on or within 1 m of rocks (J. B. Losos, in litt., 20.VIII.2007); in the Dominican Republic, edificarian or associated with ornamental vegetation in parks and other green areas (Powell et al., 1990e); mean perch height and diameter in Cuba 2.30 m and 47.6 cm, in the Dominican Republic 1.74 m and 8.6 cm, all in Cuba on leaves (J. B. Losos, in litt., 21.IV.2008); in early morning, under bark and in tree crevices, one individual sleeping in the fold of a garden plant (Collette, 1961); sleeps with body extended over long leaves, small branches, or thin trunks, individuals in the early morning under tree bark and in crevices (Rodríguez Schettino, 1999b).

Activity: Diurnal (Rodríguez Schettino et al., 1999); most active in morning and afternoon, with a period of decreased activity at midday, when they take refuge in shaded places, like dead palm leaves at tops of trunks, in the axils of banana trees, or under window louvers (Collette, 1961; Lando and Williams, 1969); adult males bask on trunks of trees in the morning, but retreat into shady crowns later in the day (Vergner and Polák, 1990); activity peaks at 1030 and 1300 h, with peaks in mean weight of stomach contents at 1130 and 1500 h, in February at Guanajay, only one activity peak at 1230 h (Rodríguez Schettino, 1999b). **Behavior:** Aggregate (to 30 individuals) in winter under palm fronds or other cover, where they do not demonstrate territoriality (Collette, 1961); when on fence posts, escape upward, juveniles reach the top of a post and run along the fence to the next post (Collette, 1961); not shy, will readily challenge intruders (Lando and Williams, 1969); male threat display similar to that of *A. allisoni* (with dewlap extension and erection of dorsal and nuchal crests), but during the third ("high") headbob, head dropped rather than elevated, and dewlap extended (Ruibal, 1967); in Santo Domingo (Dominican Republic), potentially agonistic male-male encounters with native *A. chlorocyanus* and *A. cybotes* (Powell, 1990a); usually escapes upward, climbing to the tops of fence posts and then jumping to another perch (as far as 1 m), sometimes "squirrels" (L. Moreno in Rodríguez Schettino, 1999b); where arboreal tendencies low, associated with human buildings (M. A. Granda Martínez in Rodríguez Schettino, 1999b); moves per minute 1.52 (Johnson et al., 2008); mean percent time displaying 2.1, percent walking 59.0, percent jumping 23.0, percent running 21.0 (J. B. Losos, in litt., 21.IV.2008). **Competition:** Where sympatric with the ecologically similar *A. allisoni* in central Cuba, *A. porcatus* has been displaced into less ideal habitats (G. Peters and Schubert, 1968); the introduced population in Santo Domingo may be excluding native *A. chlorocyanus* from modified habitats (Powell, 1990a; Powell et al., 1990e). **Ecomorphology:** "Trunk-crown" (Williams, 1969); formally quantified (Beuttell and Losos, 1999); categorized as a Cuban "tree anole" (G. Peters, 1970). **Foraging and Diet:** Feeds avidly on guava fruit (Otero, 1950); includes dragonflies (Collette, 1961); in captivity at the Guantanamo Naval Base, took wild-caught moths, millers, flies, grasshoppers, mosquitoes, roaches, crickets, and mealworms in that order of preference (Lando and Williams, 1969); captive individuals show a strong preference for spiders of the families Araneidae, Salticidae, and Thomisidae over insects (O. H. Garrido in Alayón García, 1976); spiders (*Argiope trifasciata*; Armas and Alayón García, 1986); cannibalism, adult male taking a hatchling as it emerged from an egg (G. Alayón García in Socarrás et al., 1988); one report of a male swallowing a female *A. homolechis* in coastal scrub (J. Novo in Socarrás et al., 1988; Rodríguez Schettino, 1999b); adult male ate an adult arboreal scorpion (*Centruroides anchorellus*; Armas et al., 1999); facultatively omnivorous, but mainly insectivorous, bite guava fruits and Custard Apple (*Annona squamosa*), hymenopterans (mainly ants) are the most frequently taken prey, also eats flies and mosquitoes (including larvae), beetles, lepidopterans, orthopterans, insect eggs, and spiders, eats mainly ants, especially during the dry season, frequency of beetles taken increases in the wet season, also eats flies, lepidopterans, orthopterans, roaches, and spiders (Rodríguez Schettino, 1999b); hunts tree frog (*Osteopilus septentrionalis*) larvae in a receptacle for rainwater (L. Moreno in Rodríguez Schettino, 1999b); effectively captures flying insects (Fläschendräger, 2002); bottles of honey left open in houses (A. Fong Grillo in Valido, 2006). **Hybridization:** Hybridizes with A. *allisoni* (Ruibal and Williams, 1961a; Glor et al., 2004). **Parasites:** In Cuba, nematodes *Cyrtosomum scelopori* (Coy Otero and Baruš, 1979b; Coy Otero and Lorenzo Hernández, 1982; Coy Otero, 1999) and *Parapharyngodon cubensis* (Baruš and Coy Otero, 1969a, 1969b; Coy Otero, 1970; Coy Otero and Baruš, 1979b; Coy Otero and Lorenzo Hernández, 1982; Coy Otero, 1999); trematodes *Urotrema scabridum* and *U. wardi* (Coy Otero, 1970; Coy Otero and Lorenzo Hernández, 1982; Coy Otero, 1999). **Population Size and Density:** Frequently 5–6 adult males and at least as many females in the crown of each tree along streets in Guanabo, east of La Habana (Vergner and Polák, 1990); encounter rates in palmettos at the cliffs on the Península de Guanahacabibes, 12 lizards/h in the dry season and 4/h in the wet season, in coastal scrub at two sites, 1.9 and 4.0 lizards/h, in similar habitat at the Jardín Botánico Nacional, mean encounter rate over 9 months 23.6 individuals/h (Rodríguez Schettino, 1999b). **Predation:** *Anolis equestris* (Alayo Dalmau, 1951; O. H. Garrido, pers. comm.); orbweavers (*Argiope trifasciata*; Armas and Alayón García, 1987; Armas, 2001); scorpions (*Centruroides gracilis*; N. Morales, 1984 in Armas, 2001); hutias (*Capromys pilorides*) in captivity (Manójina et al., 1989); conspecifics (G. Alayón García in Socarrás et al., 1988); *Cubophis cantherigerus* (Henderson and Sajdak, 1996); snakes (*Tropidophis* and *Cubophis* eat *A. porcatus* in captivity), American Kestrels (*Falco sparvarius*), blackbirds (Icterinae) in urban zones, domestic cats (*Felis*), hutias and solenodons (*Solenodon cubanus*) readily eat anoles in captivity (Rodríguez Schettino, 1999b); introduced House Sparrow (*Passer domesticus*) took a juvenile anole, probably *A. porcatus* (Bello, 2000). **Reproduction:** Clutch size 1 (Hedges, in prep.); eggs in interstices between leaf bases of Coconut (*Cocos*) and Royal Palms (*Roystonea*), as many as 29 together (Dunn, 1926d); in eastern Cuba, mean duration of copulation 20.03 min (longest 59 min), separation of pair "normal," egg 15.5 × 9.0 mm, white or beige, and elliptical (Garrido, 1980a); reproduction essentially continuous, but with periods of peak activity (M. E. Rodríguez in Sanz Ochotorena, 2000); copulation in June in Santo Domingo, during which a larger male attempted to displace a slightly smaller male in copula, the encounter involved no overt agonistic behavior (e.g., biting), consisting instead of lateral pushing and shoving to establish or maintain position, all 3 lizards fled before the issue was resolved, the larger male and female were green and the smaller male was brown throughout the observation (Parmerlee et al., 1992); eggs ~8 × 12 mm (Schröter, 1986); egg production primarily February–August, when oviductal eggs 6.5–10.8 mm, testis length >3.0 mm in all months, and fat bodies small, number of hatchlings (SVL 21–31 mm) increases July–September (Rodríguez Schettino, 1999b); in La Habana, oviductal eggs first appear in April, termination of egg production (in December) in females is associated with a decrease in temperature, and gonadal quiescence in males is associated with a decrease in temperature and a shortened photoperiod (Sanz Ochotorena et al., 2000); spermatogenesis begins in November and peaks March–August, diminishes in September, and ceases in October, previtellogenic follicles November–February, vitellogenesis begins in March, oviductal eggs late April–October, male and female cycles are synchronized, indicating prenuptial spermatogenesis (Sanz Ochotorena et al., 2005). **Size:** Male SVL to 84 mm (Pérez-Beato, 1997), female SVL to 70 mm (Buden and Schwartz, 1969); smaller in central Cuba where sympatric with larger *A. allisoni* than in eastern or western Cuba (Schoener, 1977); lizards with SVL <45 mm are juveniles, mean SVL of 5 hatchlings 22.7 mm (Rodríguez Schettino, 1999b). **Source of the Introduced West Indian Population:** mtDNA haplotypes suggest that the Dominican population originated in the San José de Lajas area of Cuba (Kolbe et al., 2007b). **Thermal Biology:** Heliothermic, basks frequently, mean air temperatures 31.3 °C for males, 31.6 °C for females, and 34.0 °C for juveniles, most lizards in full or filtered sun, on cliffs on the Península de Guanahacabibes, mean body temperature 32.7 °C, higher than mean air and substrate temperatures (30.4 °C and 31.0 °C), in coastal shrub woodlands, air temperatures 27.4–38.4 °C, in the laboratory, panting began at 41.4 °C and thermal shock set in at 47.0 °C, maximum temperature tolerated 50.0 °C (Rodríguez Schettino, 1999b); mean body temperature at Soroa 31.9 °C (J. B. Losos, in litt., 20.VIII.2007).

Conservation Status: Not assessed.

Additional Pertinent Literature: Powell (1992c).

Remarks: Populations in eastern and western Cuba are distinctive and probably warrant status as distinct species (Glor et al., 2004).

Anolis porcus Cope, 1864

Distribution: Cuba, eastern provinces; elevational range, sea level to >600 m (almost certainly higher).

Habitat: Mesophilic (Schwartz and Henderson, 1991); in coffee plantations, dark and moist broadleaf forest with *Eugenia jambos*, on trunks of which lizards perch (Garrido and Schwartz, 1968a); in moist, shaded forests, often near rivers and streams, on trunks and branches of trees and bushes (Rodríguez Schettino, 1999b); in submontane rainforest, submontane mesophyllous evergreen forest, and mesophyllous evergreen and semideciduous forest (Rodríguez Schettino et al., 1999); in montane rainforest with mixture of pines, tree ferns, and broadleaf vegetation, often in tangles or bundles of vegetation, perch diameters and heights of 2 individuals 0.5–4.0 cm and 2.5–4.0 m (juvenile) and 2.0–8.0 cm and 2.0–4.5 m (adult; Leal and Losos, 2000); in pine groves (and no other habitats) in the Pico Mogote Ecological Reserve (Fong Grillo, 2006); mean perch height and diameter 8.67 m and 3.0 cm (J. B. Losos, in litt., 21.IV.2008).

Activity: Diurnal, focal animal observations conducted 0930–1700 h, mean percent time stationary 85.7% and moving 10.9%, most frequently by walking, using either normal or rocking movements (see *A. barbatus* account), 2 jumps from branch to branch were 15 and 20 cm (Leal and Losos, 2000). **Behavior:** Usually lethargic, moves slowly in captivity, remains in the same place for hours or even days without changing position, displays rarely (Wilson, 1957, 1994); escape tactics include "squirreling" and a rapid climb, at times hiding in terminal leaves of branches, if further disturbed, leaps to the ground (Garrido and Schwartz, 1968a); in captivity, did not react to the presence of *Tropidophis melanurus* (Hottmar, 1991); in captivity, actively moves between perches and jumps toward prey, displays frequently by expanding dewlap, opening mouth, and protruding tongue, when handled, extends dewlap and attempts to bite (Gorman et al., 1969); display repertoire simple, consisting solely of headbobs (Garrido, 1982a). **Ecomorphology:** Categorized as a Cuban "giant anole" (G.

Peters, 1970); "twig-giant" (Hass et al., 1993), member of the *Chamaeleolis* clade; technically this is not an ecomorphological category as it evolved in only one clade (see definition in Williams, 1972). **Foraging and Diet:** Refused fruit in captivity (Garrido and Schwartz, 1968a); catches prey by extruding tongue or by using both mouth and tongue (Rodríguez Schettino, 1999b); one individual initially perched 2 m high, moved ~2 m, stopping frequently and appearing to scan the area, jumped to a new perch and walked ~10 cm to capture what appeared to be a cockroach, moved another 10 cm into the shade, manipulated the food for nearly 10 min before swallowing, remained on the same perch actively scanning the area for >32 min (Leal and Losos, 2000); insects, millipedes, and other lizards (Fong Grillo and Garcés G., 2001); large head size and massive jaw musculature ideally suited for molluscivory (Herrel and Holanova, 2008). **Movements:** A juvenile moved 4.0 m and an adult 12.0 m during one focal observation (Leal and Losos, 2000). **Parasites:** Nematodes *Cyrtosomum longicaudatum*, *C. scelopori* (Baruš and Coy Otero, 1969a; Coy Otero, 1970, 1999; Coy Otero and Baruš, 1973a, 1979b; Coy Otero and Lorenzo Hernández, 1982); *Oswaldocruzia lenteixeirai*, *Piratuba digiticauda* (Baruš and Coy Otero, 1969a; Coy Otero, 1970, 1999; Coy Otero and Baruš, 1979b; Coy Otero and Lorenzo Hernández, 1982); *Parapharyngodon* sp. (Baruš and Coy Otero, 1969a, 1969b; Coy Otero, 1970; Coy Otero and Baruš, 1979b; Coy Otero and Lorenzo Hernández, 1982); *Trichospirura teixeirai* (Baruš and Coy Otero, 1969a; Coy Otero, 1970, 1999; Coy Otero and Baruš, 1979a, 1979b; Coy Otero and Lorenzo Hernández, 1982), and larval *Porrocaecum* sp. (Coy Otero and Baruš, 1979b; Coy Otero and Lorenzo Hernández, 1982; Coy Otero, 1999); trematode *Mesocoelium monas* (Coy Otero, 1970, 1999; Coy Otero and Lorenzo Hernández, 1982). **Reproduction:** Courtship similar to that of *A. barbatus*, but less complex, composed of stereotyped movements (Garrido, 1982a); eggs deposited on 4 September (in captivity), 2 females contained well-developed eggs on 18 September, egg 23 mm in length (Garrido and Schwartz, 1968a); egg 21.5 × 9.7 mm, gray, and oblong (Garrido, 1980a); female laid egg in September (Rodríguez Schettino, 1999b); in captivity, egg ~22 mm long, egg rolled in soil before covered, incubation 56–85 days at 22–28 °C (Karch, 2002). **Size:** Male SVL to 162 mm (Garrido and Schwartz, 1968a), female SVL to 171.2 mm (Rodríguez Schettino, 1999b). **Thermal Biology:** Apparently nonheliothermic, active only in shade (Rodríguez Schettino, 1999b).

Conservation Status: Considered "vulnerable" (Rodríguez Schettino and Chamizo Lara, 1998a; L. Rodríguez Schettino and A. Chamizo Lara in Rodríguez Schettino, 1999a).

Anolis pulchellus Duméril and Bibron, 1837

Distribution: Puerto Rico Bank, Puerto Rico and many satellite islands, Isla Vieques, Isla Culebra, and the U.S. and British Virgin Islands; elevational range, sea level to 945 m.

Habitat: Xerophilic to semimesophilic (Schwartz and Henderson, 1991); in open, exposed, grassy and successional situations (e.g., along roadsides and trails), but also in coastal *Coccoloba* (Heatwole et al., 1962; Schwartz and Henderson, 1991; Rivero, 1998; Miranda-Castro et al., 2000; Perry, 2005); on Puerto Rico, open fields (K. P. Schmidt, 1920, 1928); swamps, pastures, uncultivated fields, guinea-grass meadows (Wolcott, 1923); usually on perches of small diameter, often at extreme ends of dry branches, at heights <0.9 m, with females and juveniles especially in ground cover or on very low perches, generally in warmer, sunnier sites than *A. krugi* where the two species occur in sympatry (K. P. Schmidt, 1920, 1928; Rand, 1964a; Gorman and Harwood, 1977); on Anegada (British Virgin Islands), usually 1.25–2.00 m above the ground on branches 8 mm in diameter, but occasionally also on fence posts, with forelimbs flexed and hindlimbs extended (Carey, 1972); at 150 m in El Yunque, juveniles only on the ground, females on the ground and to heights of 2 m, males on the ground and to heights of 3 m, perch diameters used by adults of both sexes very small to moderate (<1–8 cm), 85% of individuals in sunny situations (García Moll, 1978); also on Puerto Rico, mean perch height and diameter 0.3 ± 0.03 m and 1.4 ± 0.29 cm (Losos, 1990a); on Puerto Rico, mean perch heights and diameters 0.51 ± 0.32 m and 4.5 ± 11.1 cm for males and 0.30 ± 0.20 m and 1.8 ± 4.0 cm for females (Butler and Losos, 2002); on Guana (British Virgin Islands), to 2.5 m above ground, but most individuals 0.5–1.5 m high in fully to partially insolated situations, no individuals at sites >70% shaded (Lazell, 2005a); mean male perch height and diameter 0.28 ± 0.06 m and 0.70 ± 0.32 m, mean female perch height and diameter 0.33 ± 0.08 m and 0.02 ± 0.01 m (VanHooydonck et al., 2007); 81.3% on leaves (J. B. Losos, in litt., 21.IV.2008); habitat light environment "no shade" (Fleishman, 2000); sleeps on grass blades or twigs (MacLean, 1982); sometimes sleeps in close proximity to other conspecifics (Gorman, 1980); mean height 0.65 m on twigs, dead shrub stems, and grass blades, with body and head adpressed to the perch, posture similar to daytime resting position, sleeps lower than *A. krugi* where the two species are sympatric, although the difference in bush habitat not significant (Gorman and Harwood, 1977; Goto and Osborne, 1989).

Activity: Basks in the morning, rarely seen 1100–1400 h, by 1500 h, moving and occupying elevated perches, apparently more active in the evening than in the morning, by 1900 h had settled on perches where they slept (Gorman and Harwood, 1977); on Guana (British Virgin Islands), most active in morning, tapering off at midday and through the afternoon and evening (Lazell, 2005a). **Behavior:** Escape distances for well-camouflaged lizards were 161 and 105 cm (Heatwole, 1968a); uses "post-vantage" behavior to select habitat (Kiester et al., 1975); on Puerto Rico, mean display rate (proportion of time spent displaying) 0.033 ± 0.008, percent walks 61.0 ± 4.7, percent runs 19.1 ± 3.8, percent jumps 20.0 ± 2.0, moves/min 0.99 ± 0.13, distance jumped in nature 19.3 ± 4.1 cm (Losos, 1990a); if disturbed, "squirrels" but maintains posture (Perry, 1999); voluntarily enters water as an escape tactic, swims (to >7.5 m) by undulations of body, at times stopping with the head raised, no part of the body penetrating the surface (Heatwole et al., 1962); lateral face-off, push-ups, and dewlapping used in "pursuit-deterrent signals" against *Borikenophis portoricensis* (Leal and Rodríguez-Robles, 1997b); uses head-down posture when on vertical surfaces, moves/min 0.81 ± 0.34, percent time moving (PTM) 1.35 ± 0.58%, PTM decreases (not significantly) with increases in perch height (Cooper, 2005); approach distances increase as perch heights increase, approach distances longer when exposed than when partially concealed, escapes by jumping to the ground and running away (Cooper, 2006a); male approach distance, flight distance, and final distance 1.47 ± 0.12 m, 0.30 ± 0.03 m, and 1.65 ± 0.09 m, female approach distance, flight distance, and final distance 1.15 ± 0.11 m, 0.23 ± 0.02 m, and 1.32 ± 0.11 m (VanHooydonck et al., 2007); moves per minute 0.99 (Johnson et al., 2008). **Desiccation:** Standard rate of water loss 2.9 mg/g of initial body weight/h, vital limit (percent of initial body weight lost before death) 20, and survival time 72 h (J. Molinari-Teron, 1969 in Heatwole and Veron, 1977); mean critical activity point (percent of original body weight at which the lizard could not right itself when turned on its back) 74.0 ± 4.24%, mean rate of water loss 0.40% body mass/h (Hillman and Gorman, 1977). **Diet:** Appears to be "strictly insectivorous" in the Virgin Islands (MacLean, 1982); takes snails, millipedes, centipedes, isopods, spiders, mites (total invertebrates other than insects ~8%), high proportion of ants (~20%, exceeded on Puerto Rico only by *A. stratulus*), other hymenopterans, orthopterans, cockroaches, earwigs, damsel flies, beetles (adults and larvae), lepidopterans (>20%, adults and larvae), flies, thrips, hemipterans, leafhoppers, aphids, and plant lice (Wolcott, 1923); anthers from grass flowers and seeds (Schwartz and Henderson, 1991); shows some "preference" for ants (Rivero, 1998); small invertebrates typically captured from ambush (Perry, 1999); in the British Virgin Islands, stomach contents include ants (most important in numbers and volume), crickets, hemipterans, flies, thrips, spiders, and other small arthropods (Lazell, 2005a); extrafloral nectarivory on *Leucaena leucocephala* (Leguminosae; Perry and Lazell, 2006); *Sphaerodactylus macrolepis* on Guana Island (J. B. Losos, in litt., 21.IV.2008). **Ecomorphology:** "Grass-bush" (Williams, 1961); formally quantified (Beuttell and Losos, 1999). **Home Range:** Relatively small (<20 m^2) and lizards show little vagility (Gorman and Harwood, 1977). **Movements:** Not too mobile, rarely moves away from areas of ~20 m^2 (Rivero, 1998); mean moves per minute 0.65 ± 0.74, percent time moving 1.78 ± 2.31 (Perry, 1999). **Parasites:** Trematode *Mesocoelium monas* (Cofresí-Sala, 1964); protozoan malarial parasite *Plasmodium floridense* (Telford, 1975; Ayala, 1977, 1978). **Performance:** On Puerto Rico, mean maximum sprint speed (fastest time over an interval of 0.25 m) 0.147 ± 0.007 sec, jump 57.9 ± 1.40 cm, cling 1.463 ± 0.147 Neutons (Losos, 1990a); clinging ability 0.51 ± 0.09 Neutons, divided by mass 0.29, with the relatively high value apparently characteristic of grass-bush ecomorphs (Elstrott and Irschick, 2004); peak velocity 1.50 ± 0.10 m/sec, peak acceleration 29.8 ± 5.1 m/sec^2 (Toro et al., 2004); acceleration 27.43 ± 1.98 m/sec, sprint speed 0.81 ± 0.05 m/sec (Vanhooydonck et al., 2006a). **Population Size and Density:** To 20,000–25,000/ha with biomass ~10 kg/ha (Gorman and Harwood, 1977); 1,000–1,800 lizards/ha on Guana (British Virgin Islands; Lazell, 1991); in *Leucàena* plots on Guana with little litter 50 lizards/ha, biomass 0.09 kg/ha, none in *Coccoloba* plots with ample litter (Rodda et al., 2001). **Predation:** *Magliophis stahli* (K. P. Schmidt, 1928); Indian Mongoose (*Herpestes javanicus*; Wolcott, 1953); *Anolis cuvieri* (Pérez-Rivera, 1985); on Puerto Rico, Culebra, and U.S. Virgin Islands, *Borikenophis portoricensis* (Rodríguez-Robles and Leal, 1993a; Henderson and Sajdak, 1996); *Magliophis exiguus* (Thomas and Gaa Kessler, 1996). **Reproduction:** Clutch size 1 (Hedges, in prep.); 2 eggs in a pile of coconut husks (Rand, 1967c); maximal male reproductive activity March–August, some sperm production throughout the year, but testicular weights decrease during winter months (but less than in other species of anoles), size of abdominal fat bodies is inversely correlated with testis weight (Licht and Gorman, 1975). **Size:** Male SVL to 51.3 mm (Schoener, 1988), female SVL to 45 mm (Lazell, 2005a). **Thermal Biology:** During early morning hours, lizards invariably select sunny perches and orient toward the sun, as temperatures increase, lizards shift to the shady sides of grass stems (Gorman and Harwood, 1977); mean critical thermal maxima at 5 m and 200 m elevation 36.1 ± 0.1 °C and 35.1 ± 0.3 °C, significantly less heat resistant than sympatric *A. krugi* (Hertz, 1979b); heliothermic (Fläschendräger and Wijffels, 2009). **Vision:** Parts of dewlaps, but not the rest of the body, are highly UV reflective, most effective in UV-rich environment, a UV-bright spot is associated with the open mouth, part of a typical anoline threat display (Fleishman et al., 1993); dewlaps highly visible in typical habitat (Persons et al., 1999).

Conservation Status: "Stable" (Moreno, 1991d).

Anolis pumilus Garrido, 1988

Distribution: Cuba, central and western Cuba, Archipiélago de Sabana-Camagüey, and Isla de la Juventud.

Habitat: Mesophilic (Schwartz and Henderson, 1991); apparently primarily (originally?) an inhabitant of woody situations with dead twigs and vine tangles, but favors direct sunlight, generally perches close to the ground at heights of <2 m (Collette, 1961); also in coastal vegetation with widely spaced bushes with drooping leaves, often spinose, on a spinose tree in an open field, seeks refuge in bushes and grasses in completely cutover areas, where apparently favors spinose plants of various sizes, in shrubby growth with dominant spinose plants on the Archipiélago de Sabana-Camagüey (Garrido, 1988); in coastal and subcoastal microphyllous forest, serpentine xeromorphous shrub-woodland, and urban zones (Rodríguez Schettino et al., 1999); on trunks and thin limbs, usually on spiny bushes (*Casearia aculeata* and *Acacia farnesiana*), at mean perch height 0.75 m and no higher than 2 m (Rodríguez Schettino, 1999b); mean perch height and diameter 1.89 m and 2.2 cm (J. B. Losos, in litt., 21.IV.2008).

Activity: Diurnal (Rodríguez Schettino et al., 1999); begins about 1000 h (Garrido, 1988); generally 2 activity peaks at 1030 and 1430 h, may take shelter during the heat of the day, active 1700–1900 h on the Península de Guanahacabibes during the wet season (Rodríguez Schettino, 1999b). **Behavior:** When disturbed in a vine tangle, runs back to cover instead of climbing or descending the trunk (Collette, 1961); uses holes in dogtooth limestone for shelter or to escape, aggressive displays include headbobs, dewlap extension, lateral orientation, elevation of nuchal crests, lateral compression of bodies, and elevation on all 4 limbs (Rodríguez Schettino, 1999b). **Ecomorphology:** Undetermined. **Foraging and Diet:** Stomach contents include hymenopterans (mostly ants), beetles, lepidopterans, flies, leafhoppers, and spiders (Rodríguez Schettino, 1999b). **Population Size and Density:** Encounter rates to 6 individuals/h (Rodríguez Schettino, 1999b). **Reproduction:** Long precopulatory display (mean 11.6 sec), male recognizes female, rapidly extends dewlap, makes short nod, repeated 4–5 times, if female receptive, she nods rapidly, if not she responds aggressively, courting male repeats above sequence and makes arrhythmic and spasmodic movements with limbs placed above her, then remains immobile 11–16 sec (mean 12.2 sec) and inserts hemipenis, copulation lasts 2.33–13.00 min (mean 4.88 min), postcopulatory separation brusque, male jumping backward or to side of female, eggs (March–April) oval and plaster white, 5.0–7.6 × 4.1–4.8 mm (Garrido, 1988). **Size:** Male SVL to 34.2 mm, female SVL to 39.2 mm (Rodríguez Schettino, 1999b). **Thermal Biology:** Apparently heliothermic, although a majority of individuals are in shadows of twigs and leaves at mean air temperature 28.8 °C (Rodríguez Schettino, 1999b).

Conservation Status: Considered to be at "lower risk" of extinction (Rodríguez Schettino and Chamizo Lara, 1998a; Rodríguez Schettino and Chamizo Lara in Rodríguez Schettino, 1999a).

Anolis quadriocellifer Barbour and Ramsden, 1919

Distribution: Cuba, Península de Guanahacabibes.

Habitat: Mesophilic (Schwartz and Henderson, 1991); on large trees in a clearing in broadleaf forest (Schwartz, 1968f); large males on trunks at heights of 0.9 m, females and juveniles lower, often on the ground or on rocks (Garrido and Schwartz, 1968b); in mesophyllous evergreen and semideciduous forest (Rodríguez Schettino et al., 1999); males on tree and bush trunks at mean heights <1 m above the ground, generally with heads facing down, also may perch on rocks that protrude only a few centimeters, females and juveniles are at lower heights on trunks and are more frequently on the ground, stones, roots, or in leaf litter (Rodríguez Schettino, 1999b).

Activity: Diurnal (Rodríguez Schettino et al., 1999); in coastal zones, often not seen until 1000 h, with females and juveniles emerging first, large adult males not until the sun is fully up (Garrido and Schwartz, 1968b); in forests and scrub, lizards active at 0700 h, regardless of sex or age, with increased activity during the day corresponding to rising temperatures, during the wet season, peak activity 1400–1600 h at sites where air temperatures are 27–31 °C, during the dry season, only isolated individuals are seen in sunny places, with activity decreasing dramatically when air temperatures <27 °C, no individuals are active after 1800 h (Rodríguez Schettino, 1999b). **Behavior:** Males assume a "typical" position with the body vertical and head elevated until nearly parallel with the ground (Garrido and Schwartz, 1968b); to escape, runs into the herbaceous stratum and hides among dead leaves, stones, and roots (Rodríguez Schettino, 1999b). **Ecomorphology:** "Trunk-ground" (Williams and Rand, 1977). **Foraging and Diet:** Feeds on arthropods, especially scolytid beetles and ants, but takes other hymenopterans, hemipterans, orthopterans, roaches, insect larvae, and spiders, males take larger food than females, but most prey items <3 mm in length (Rodríguez Schettino, 1999b). **Home Range and Territoriality:** Males (adults and juveniles) are fiercely territorial, displaying aggressively (headbobbing, dewlap extensions, limb extensions) and attacking, biting on the neck and head (Rodríguez Schettino, 1999b). **Parasites:** Nematodes *Cyrtosomum scelopori* (Coy Otero and Baruš, 1973a, 1979b; Coy Otero and Lorenzo Hernández, 1982; Coy Otero, 1999); *Oswaldocruzia lenteixeirai*, *Parapharyngodon cubensis*, and *Skrjabinoptera phrynosoma* (Coy Otero and Baruš, 1979b; Coy Otero and Lorenzo Hernández, 1982; Coy Otero, 1999). **Population Size and Density:** Encounter rates 1.0 and 4.4 individuals/h in cliffs and forests, 0.1–0.3 lizards/h during the dry season (Rodríguez Schettino, 1999b). **Predation:** *Cubophis cantherigerus* (Henderson and Sajdak, 1996). **Reproduction:** Apparently cyclic, juveniles and subadults only in October (Rodríguez Schettino, 1999b). **Size:** Male SVL to 55 mm (Schwartz, 1968f), female SVL to 48.5 mm (Rodríguez Schettino, 1999b). **Thermal Biology:** Heliothermic, basks on trunks and stones, often in filtered light, mean body temperatures during the wet season 30.9 °C at mean air and substrate temperatures of 28.8 and 28.7 °C (Rodríguez Schettino, 1999b).

Conservation Status: Considered "vulnerable" (Rodríguez Schettino and Chamizo Lara in Rodríguez Schettino, 1999a).

Anolis reconditus Underwood and Williams, 1959

Distribution: Jamaica, Blue Mountains; elevational range ~760–1,220 m.

Habitat: Mesophilic (Schwartz and Henderson, 1991); in montane mist forest to elfin woodland, also in a steep, dark, damp, rocky ravine, will hide under fallen logs, occasionally edificarian (about 12 anoles in a 6 × 6 m garage), perches on large trees (trunk diameters 10–45 cm) at heights of 0.9–4.9 m (usually at about 3.6 m), usually in vertical head-down position, often near a hole in the trunk, also under eaves of a garage (Underwood and Williams, 1959; Lazell, 1966; Hicks, 1973); prefers moister and shadier habitat than sympatric *A. lineatopus* (Lazell, 1966); dewlap UV-reflectance compatible with visibility in typical habitat (Macedonia et al., 2000); probably sleeps in the foliage in crowns of trees (Hicks, 1973).

Activity: Earliest observation 0750 h, active until "darkness was rapidly falling" (Hicks, 1973). **Behavior:** Wary and timid, escape behavior includes climbing higher on tree trunks, retreating into holes in tree trunks, crawling under objects, into rocks along a ravine edge, into a garage (Lazell, 1966); tends to escape upward, displays include headbobbing, 4-legged push-ups, front-legged push-ups, half push-up and pulled back, slightly cocked head preceding dewlap extension, and dewlapping combined with headbobbing, perches passively during heavy rain, making no attempt to seek shelter (Hicks, 1973). **Ecomorphology:** "Montane generalist" (Williams, 1983); formally quantified and determined to be closest to "trunk-ground" (Beuttell and Losos, 1999). **Foraging and Diet:** May employ both active and sit-and-wait foraging strategies, food includes flying insects (capable of picking them out of the air) and ants (Hicks, 1973). **Home Range:** May be "clumped" in small areas of optimal habitat (Lazell, 1966); large male with a home range shared by 2 or more females, encompassing 3 large trees and a few saplings, ground area about 45–70 m^2, one large male covered a stretch of ~25 m with many large trees in 2 h (Hicks, 1973). **Predation:** Greater Antillean Grackles (*Quiscalus niger*) capable of taking all size classes (Graves, 2006). **Reproduction:** Clutch size 1 (Hedges, in prep.); 2 females in February contained eggs, one with 2 eggs (suggesting an ability to retain eggs until deposition is propitious), ~10 mm and ~6 mm (Lazell, 1966); female captured in April continued to lay fertile eggs for 2 months, eggs immediately after deposition 8 × 15 mm, after 35 days 11 × 16 mm, with estimated incubation period under captive conditions ~45 days at 20–25 °C, hatchling SVL ~28 mm (Fläschendräger, 1999a). **Size:** Male SVL to 100 mm, female SVL to 84 mm (Hicks, 1973).

Conservation Status: Unknown.

Anolis rejectus Garrido and Schwartz, 1972

Distribution: Cuba, Cordillera de la Gran Piedra and Sierra de Boniato and nearby lowlands; elevational range 50–1,214 m.

Habitat: Open grassy hillsides with scattered trees and dense "waist-high" grass and low shrubs overgrowing a cemented drainage ditch (Schwartz and Ogren, 1956; Garrido and Schwartz, 1972); in submontane mesophyllous evergreen forest and pine forest (Rodríguez Schettino et al., 1999); on bushes and grasses, among tall grasses and sparse vegetation in deforested areas (Rodríguez Schettino, 1999b); in dense grass in dry pine forests (Fläschendräger, 1998; Fläschendräger and Wijffels, 2009); mean perch height and diameter 0.54 m and 0.4 cm (J. B. Losos, in litt., 21.IV.2008).

Activity: Diurnal (Rodríguez Schettino et al., 1999); most individuals active during the early morning hours and in the afternoon, but may be active throughout the day on cool rainy days (Rodríguez Schettino, 1999b). **Behavior:** Generally immobile, jumps and hides under leaves to escape (Rodríguez Schettino, 1999b); tight adpression against blades of grasses probably a predator deterrent against hunting Lizard Cuckoos (*Saurothera merlini*) that are seen frequently in habitat used by lizards, in captivity, territorial displays involve nuchal crest erection and 4-legged push-ups and may escalate into fighting (Fläschendräger, 1998). **Ecomorphology:** "Grass-bush," by implication (Garrido and Schwartz, 1972). **Predation:** *Anolis allogus* (Socarrás et al., 1988). **Reproduction:** In captivity, eggs ~8.0 × 10.7 mm, incubation 45–55 days at 20–24 °C, hatching SVL ~14 mm (Fläschendräger and Wijffels, 2009). **Size:** Male SVL to 37.0 mm (Schwartz and Henderson, 1991), female SVL to 36.7 mm (Rodríguez Schettino, 1999b). **Thermal Biology:** Apparently nonheliothermic (Rodríguez Schettino, 1999b); populations formerly assigned to *A. mimus* (see note below) heliothermic, adhere tightly to blades of grasses when basking, active even at 17 °C (Fläschendräger, 1998; Fläschendräger and Wijffels, 2009).

Conservation Status: Considered "endangered" (Rodríguez Schettino and Chamizo Lara in Rodríguez Schettino, 1999a), but that was before *A. mimus* was synonymized with *A. rejectus* (see note below), "*A. mimus*" was considered "vulnerable" (Rodríguez Schettino and Chamizo Lara, 1998a; Rodríguez Schettino and Chamizo Lara in Rodríguez Schettino, 1999a).

Remarks: *Anolis mimus* Schwartz and Thomas (1975) was synonymized with *A. rejectus* by Garrido and Hedges (2000).

Anolis richardii Duméril and Bibron, 1837

Distribution: Grenada Bank, Grenada (essentially island-wide) and the Grenadines, introduced on Tobago; elevational distribution, sea level to at least 870 m.

Habitat: Xerophilic to mesophilic (Schwartz and Henderson, 1991); ubiquitous, absent only from severely altered sites lacking vertical vegetative structure (Germano et al., 2003); although in such areas in the Grenadines and where introduced on Tobago (Lazell, 1972); abundant in roadside trees, especially on the shady sides (Underwood, 1959b); in well-shaded forest, *Cocos*, bamboo, and nutmeg groves, xeric forest above the coast, gardens and the arboretum, *Theobroma* and *Musa* groves, coastal *Hippomane*, mangroves, in ammunition tunnels and buildings, generally on large trees 1.5–2.5 m above the ground, frequently in the open on the shaded side of trees, occasionally on rocks and boulders along creeks and rivers (Schwartz and Henderson, 1991; Powell and Henderson, 2008c); adult males perch higher than smaller conspecifics and *A. aeneus* (Schoener and Gorman, 1968; Harris et al., 2004); mean perch height 1.37 ± 0.06 m (Roughgarden et al., 1983a); mean perch height and diameter 1.88 ± 0.16 m and 14.3 ± 1.8 cm (Losos and de Queiroz, 1997a); perch heights (129.5 ± 6.0–153.5 ± 14.5 cm) and diameters (7.1 ± 1.0–10.5 ± 1.0 cm) varied in 4 different lowland habitats, and adult males perched higher (183.1 ± 11.0 cm) and on perches of larger diameter (9.5 ± 0.9 cm) than females and subadult males (129.8 ± 6.4 cm and 8.7 ± 0.7 cm) or juveniles (105.8 ± 7.3 cm and 6.2 ± 0.9 cm), overlap with sympatric *A. aeneus* was high, although adult male *A. richardii* perched higher than any other size class of either species, most individuals were oriented vertically (facing up or down), with only juveniles assuming horizontal perches with any frequency, and very few individuals were in full sun (Harris et al., 2004); at a lowland site on Grenada, mean perch heights 21.2 ± 4.1 cm for juveniles, 30.2 ± 4.6 cm for adult females and small males, 57.9 ± 10.0 cm for large adult males, significantly lower than *A. aeneus*, with individuals often on rocks and the ground, the latter significantly more frequently than *A. aeneus* and significantly more frequently at times when sympatric *Ameiva ameiva* were not active, most large adult males were oriented vertically, whereas smaller size classes were most likely to be horizontal (Simmons et al., 2005); close to the ground at very low elevations (<30 m) and at intermediate heights at higher elevations, statistically indistinguishable from perches used by *A. aeneus* (Buckley and Roughgarden, 2005b); on palms, large broadleaf trees, bamboo, mangroves, and banana plantations, also in gardens, to heights of 7–8 m with heads facing down (Fläschendräger and Wijffels, 2009); sleeps tightly pressed against the substrate (regardless of type) with eyes closed and hindlimbs extended, at mean height and diameter of 1.0 ± 0.1 m and 3.0 ± 0.3 cm, respectively, most commonly on thornless plants (but commensurate with abundance of thorny plants), on leaves or branches at about equal frequencies, but large individuals (>80 mm SVL) much more frequently on branches, and usually facing the perch's point of origin, nearest neighbors did not differ by sex or species (syntopic with *A. aeneus*, with which structural niche overlap was high, although *A. aeneus* used somewhat higher perches), a few individuals within 1 m of the previous perch on consecutive nights, but no evidence of strong sleep-site fidelity or evidence of perch selection that would minimize predation risk by nocturnally foraging *Corallus grenadensis* aside from sleeping at the distal ends of branches (Poche et al., 2005).

Abundance: On Grenada, abundance gradually declines with increasing elevation (Buckley and Roughgarden,

2005a). **Activity:** Diurnal, but exploits the "night-light niche" (RWH in Perry and Fisher, 2006; Perry et al., 2008). **Behavior:** Not shy, displays of "low" (puffing or gorging of throat, dewlap extended, elevated on all 4 limbs, tail up, then down) to "high" (dewlap extended, vigorous bobbing, tail lashing, rising on all 4 limbs) intensities, displays with a single series of bobbing sequences, stereotyped in number of peaks and time in sequence (Gorman, 1968). **Competition:** Juvenile *A. richardii* may aggressively dominate juvenile *A. aeneus*, possibly excluding them from shadier areas, juvenile *A. richardii* were the aggressors in 24 of 26 interspecific interactions (Stamps, 1983a). **Ecomorphology:** Close to "crown-giant" (Beuttell and Losos, 1999). **Foraging and Diet:** Several individuals congregated with *A. aeneus* on the ground exploiting an abundance of mulberry-like berries (Lazell, 1972); includes ants (greatest numbers in adult and subadult males and adult females), other hymenopterans, flies, beetles (adults and larvae), hemipterans, lepidopterans (adults and larvae), booklice, thrips, termites, orthopterans, lacewings (adults and larvae), earwigs, spiders, pseudoscorpions, mites, centipedes, millipedes, crustaceans other than isopods, snails, annelids, anoles, vertebrate eggs, and some plant matter (greatest volume in adult and subadult males and adult females), with prey sizes significantly larger than those taken by *A. aeneus*, with larger individuals taking larger prey (and containing fewer items) and subadult males taking larger prey than females of the same size, larger individuals take more plant matter, smaller lizards consume a wider variety of food, mean prey size of adult males 9.8 ± 4.1–19.3 ± 9.3 mm, females 8.7 ± 2.1–14.0 ± 9.6 mm and mean number of prey items/lizard 24.5 ± 7.3–16.9 ± 16.1 for males, 20.7 ± 10.0–23.8 ± 13.4 for females, 12.3 ± 8.5–24.5 ± 7.3 for all size classes (Schoener and Gorman, 1968); *Anolis aeneus* subadult and juveniles (Stamps, 1983a); percent time moving 2.7 ± 0.4%, moves per minute 3.0 ± 0.4, percent feeding attacks while moving 0.0%, stomach contents included spiders, other arachnids, centipedes, a millipede, beetles (adult and larvae), springtails, earwigs, flies, hemipterans, hymenopterans (including ants, which were the second most important dietary item), lepidopterans (adults and larvae, which were the most important dietary item, and presumably a reflection of exploiting seasonally abundant prey), orthopterans (including crickets and roaches), shed lizard skin, and plant material, with flowers more important in females and fruits/seeds more important in males (Simmons et al., 2005); nectarivory and possibly pollination of *Marcgravia umbellata* (Timmermann et al., 2008). **Parasites:** Malarial parasites *Plasmodium azurophilum* (Staats and Schall, 1996b); acanthocephalan cystocanths *Centrorhynchus* sp. and nematodes *Parapharyngodon cubensis*, larval *Physocephalus* sp. (Goldberg et al., 1997a), and *Spauligodon anolis* (Goldberg et al., 1997a; Bursey and Goldberg, 1998). **Population Size and Density:** 0–18.0 ± 2.6 lizards/100 m^2 at 4 sites on Grenada (Roughgarden et al., 1983a); in 4 different lowland habitats, estimates were 830–12,390 or 12,750 lizards/ha using 2 different methods, with the low estimate in mangroves and the highest density in a dense streamside forest with large trees (Harris et al., 2004); in a study based on sleeping anoles, estimated population density was 2,354 lizards/ha, with more in drier habitat than in adjacent, more mesic habitat characterized by large trees, into which lizards may ascend out of view to sleep (Poche et al., 2005). **Predation:** Broad-winged Hawks (*Buteo platypterus*) were seen taking *A. richardii* and 9 other species of large insectivorous birds eat anoles (Wunderle, 1981); *Corallus grenadensis* (Henderson, 1993, 2002; Yorks et al., 2004). **Reproduction:** Hatchling SVL in captivity ~30 mm (Fläschendräger and Wijffels, 2009). **Sex Ratio:** Apparently balanced (Lazell, 1972). **Size:** Male SVL to 140 mm, female SVL to 77 mm, smallest mature male and female SVL 70 mm (or possibly 59 mm) and 56 mm, respectively (Lazell, 1972). **Thermal Biology:** Mean body temperature 28.4 ± 0.1 °C in coconut and mixed coconut-cacao habitat (Roughgarden et al., 1981); on Grenada, mean microclimates used by *A. aeneus* and *A. richardii* are statistically indistinguishable at low-elevation (<300 m) sites, but *A. aeneus* perches in significantly warmer microclimates than *A. richardii* at high elevations (>300 m), where *A. richardii* perches in relatively cooler microclimates (Buckley and Roughgarden, 2005b).

Conservation Status: Not assessed; the species has adapted well to human-mediated habitat alterations in many instances (Henderson and Berg, 2006; Daudin and de Silva, 2007).

Anolis ricordii Duméril and Bibron, 1837

Distribution: Hispaniola, throughout much of Haiti but only in the northwestern and south central Dominican Republic; elevational range, sea level to at least 1,068 m.

Habitat: Mesophilic (Schwartz and Henderson, 1991); on tree trunks in cultivated areas 1.5–4.6 m above ground, gallery forests in otherwise xeric areas, ravine woods, especially on trees and vines overhanging an intermittent streambed, coffee plantations (Schwartz, 1974c; Franz and Cordier, 1986; Schwartz and Henderson, 1991); mean perch height and diameter 10.33 m and 11.0 cm (J. B. Losos, in litt., 21.IV. 2008); sleeps 1.2–6.1 m above the ground in vine tangles, especially in dense mats or "curtains" under the canopy and in tree fern thickets, juveniles sleep lower than adults on horizontal vines (Williams, 1965b; Schwartz, 1974c).

Behavior: Vocal "alarm call" (Mertens, 1946); "squirrels" when pressed, moves to distal branches, may jump from tree to tree, and finally descend to the ground (A. S. Rand in Williams, 1962a; A. S. Rand and J. Lazell in Williams, 1962a); an individual moved through 3 trees before seeking refuge in dense grass on the ground (Schwartz, 1974c). **Ecomorphology:** "Crown-giant" (Rand and Williams, 1969). **Predation:** Barn Owls (*Tyto alba*; Wetmore and Swales, 1931; Etheridge, 1965). **Size:** Male SVL to 160 mm, female SVL to 153 mm (Schwartz, 1974c), although Etheridge (1965) extrapolated SVL to 190–192 mm from a suite of cranial elements and pelvises in pre-Columbian cave deposits.

Conservation Status: Not assessed; ongoing destruction of forests might have long-term negative consequences (Powell et al., 2000; Powell and Incháustegui, 2009).

Anolis rimarum Thomas and Schwartz, 1967

Distribution: Hispaniola, Chaîne de Marmelade (Haiti); elevation ~975 m.

Habitat: Mesophilic (Schwartz and Henderson, 1991); in a steep limestone jumble and talus with dense, natural second-growth woods, mostly small trees, brush, and vine tangles, as well as a few large trees, also on rocks, low twigs and branches, near the ground on trunks or exposed roots of trees (Thomas and Schwartz, 1967).

Behavior: When pressed, retreats with agility into rock crevices (Thomas and Schwartz, 1967). **Ecomorphology:** Undetermined. **Size:** Male SVL to 45 mm, female SVL to 40 mm (Thomas and Schwartz, 1967).

Conservation Status: Unknown.

Anolis roosevelti Grant, 1931

Distribution: Isla Culebra, Vieques, St. John (?), and Tortola (Puerto Rico Bank).

Habitat: Probably frequents high branches in *Bursera* and *Ficus* forest (Campbell and Dodd, 1982; MacLean, 1982).

Diet: Diet includes fruits of fig trees (Dumas in Campbell and Dodd, 1982). **Ecomorphology:** "Crown-giant" (Williams, 1972). **Size:** Male SVL to 162 mm, female SVL to 144 mm (Mayer, 1989).

Conservation Status: Unknown, possibly extinct (MacLean, 1982; Moreno, 1991b, 1991d; Thomas, 1999; Perry and Gerber, 2006), although a few individuals may survive on Culebra and a species-recovery plan exists (Campbell and Dodd, 1982); last specimen collected in 1932, and apparently reliable reports date to 1978 (Campbell and Dodd, 1982; Mayer, 1989); listed as "endangered" on the U.S. List of Endangered and Threatened Wildlife and Plants (U.S. Fish & Wildlife Service, 2008) and as "critically endangered" on the IUCN Red List (Conyers and Wingate, 1996); protected as "threatened" or "endangered" (Joglar, 2005b).

Additional Pertinent Literature: Dodd and Campbell (1982).

Anolis roquet Lacepède, 1788

Distribution: Martinique and satellites, where six currently recognized subspecies occur; introduced on Bermuda; elevational range, sea level to 1,370 m.

Habitat: Xerophilic to mesophilic (Schwartz and Henderson, 1991); ubiquitous and infradispersed (Lazell, 1972); often most abundant in human-modified habitats, effectively a human commensal, *A. r. summus* often associated with pigsties, which apparently attract insects, *A. r. zebrilus* uses large trees, ruins, and walls of houses as basking sites (M. Schmidt, 1993); largely arboreal, but exceedingly capable of adapting to various habitats, frequently associated with trees, bushes, cacti, bamboo, cliffs, walls, houses, gardens, and plantations (Fläschendräger and Wijffels, 1996, 2009); in dry rain shadows to wet lowlands and densely shaded portions of high forests, in pasture scrub, xeric woodlands, gardens, dead razor grass, *Ficus*, ruins, coastal *Hippomane*, *Musa* grove, on the ground, shrubs, trees, cacti, fence posts (including those in cold wind on the upper slope of Mt. Pelée), a bamboo fence row, road signs, and buildings (Schwartz and Henderson, 1991; Powell and Henderson, 2008c); on trees and rocks on the platform of Rocher de Diamant, where very abundant in 1963, males perch on rocks on the ground to ~3 m above the ground (Lazell, 1972); densest populations in the vicinity of human habitations, on tree trunks at heights of 1–3 m, *A. r. majolgris* associated with narrow branches from which they jump accurately to other perches over distances to 1.5 m, *A. r. summus* in rainforest at elevations >500 m (J. Schmidt, 1980b); *A. r. roquet* particularly abundant in parks, but also in hedges and vegetation along fences, females less obvious due to a preference for lower, denser undergrowth, *A. r. caracoli* in ruins and small stands of forest, where they appear to prefer shady perches, *A. r. majolgris* in dense branches of larger trees, especially abundant in locally mesic areas, *A. r. salinei* in low vegetation, but slightly more abundant in few available large trees, *A. r. summus* in rainforest trees covered with epiphytes, apparently most abundant in the vicinity of towns, *A. r. zebrilus* on large trees, ruins, and buildings (Heselhaus and Schmidt, 1990); *A. r. majolgris* particularly abundant on rock walls and adjacent dry forest, *A. r. summus* to heights of 1.5 m on deadfall, leaves, trunks, ferns, and bromeliads (Bach and Schönecker, 2008); perch heights at 20 m elevation 89.6 ± 4.3 cm, at 350 m 85.9 ± 3.9 cm, at 650 m 89.0 ± 4.3 cm, lizards at elevations of 350 m and 650 m use elevated perches in the sun more frequently than lizards at 20 m elevation (Hertz, 1983); mean perch height and diameter 1.81 ± 0.18 m and 9.4 ± 1.4 cm (Losos and de Queiroz, 1997a); both sexes forage on the ground (Lazell, 1972).

Activity: Diurnal (Schwartz and Henderson, 1991); at 20 m elevation, activity bimodal, at 350 m elevation, activity unimodal (peaking when sunshine most available), at 650 m elevation uniformly active over protracted day (Hertz, 1981, 1983); especially females forage actively in early morning and evening (M. Schmidt, 1993). **Behavior:** Display behavior includes dewlapping, bobbing, push-ups, tail wags, tail lashes, and tongue protrusion, challenge display behavior sequence includes extended dewlap, elevation on all 4 limbs, raising and lowering the tail extensively and slowly, lowering the body, giving a rapid headbob, and retracting the dewlap (Gorman, 1968); capable of vocalization (Myers, 1971); generally not particularly shy or wary, except the subspecies in the most xeric situations, *A. r. zebrilus* is "skittish and difficult to approach" and *A. r. caracoli* and *A. r. salinei* are "wild, shy, and difficult to noose" (Lazell, 1972); adult males highly aggressive, *A. r. caracoli* particularly difficult to capture, with approach distances of ~10 m (J. Schmidt, 1980b). **Desiccation:** Rate of water loss in juvenile *A. r. salinei* slow (1.71 mg/g/h), able to suffer weight loss to 38% (Leclair, 1978); rates of evaporative water loss varied significantly between populations and were negatively correlated with habitat aridity, rates of water loss dropped significantly after 6 weeks of acclimation to xeric conditions, although populational differences remained (Hillman et al., 1979); rates of fractional weight loss (percent of initial body weight lost per hour) increased, vital limits (percent of intial body weight at death) varied (generally decreasing), and survival times decreased by elevation (all were correlated with initial body weight), suggesting that frequent basking behavior at high elevations may induce selection for water conservation that is nearly as effective as that in lowland populations (Hertz, 1980b). **Ecomorphology:** Closest to "trunk-crown" (Losos and de Queiroz, 1997a; Beuttell and Losos, 1999). **Foraging and Diet:** Includes ants (greatest numbers and volumes in adult and subadult males and adult females), other hymenopterans, flies, beetles (adults and larvae), hemipterans, lepidopterans (adults and larvae), booklice, thrips, termites, orthopterans, lacewing larvae, springtails, earwigs, spiders, pseudoscorpions, mites, centipedes, millipedes, isopods, snails, annelids, conspecifics, and some plant matter, with larger individuals taking larger prey (and containing fewer items) and females taking larger prey than subadult males of the same size, larger individuals take more plant matter, smaller lizards consume a wider variety of food, mean prey size of males 4.8 ± 2.1–18.2 ± 6.7 mm, females 4.6 ± 1.5–9.6 ± 5.0 mm and mean number of prey items/lizard 141.0 ± 67.1–53.8 ± 62.6 for males, 96.8 ± 34.3–81.4 ± 22.9 for females, 12.3 ± 8.5–24.5 ± 7.3 for all size classes (Schoener and Gorman, 1968). **Home Range and Territoriality:** Usually 1 male to 3 females on a single tree, adult males do not tolerate the presence of even subadult males in their territories (J. Schmidt, 1980b); population densities of *A. r. roquet* may be so high in some areas (e.g., parks) that normal territorial boundaries break down and several adult males may occupy a single palm (Heselhaus and Schmidt, 1990). **Hurricane Effects:** Particularly tolerant of hurricanes, foraging on leafless trees within 2 days of Hurricane Iris in 1995, pattern variants out of range may have been transported with debris during storms (Wijffels, 2001). **Longevity:** In captivity, 6 years, 8 months and still alive at last report in 1992 (D. Grow in Slavens and Slavens, 2003). **Parasites:** Malarial parasite *Plasmodium azurophilum* (Ayala and Hertz, 1981; Staats and Schall, 1996b); nematodes *Parapharyngodon cubensis* and larval *Physocephalus* sp. (Goldberg et al., 1997a). **Population Size and Density:** Density of *A. r. caracoli* often very high, whereas that of *A. r. salinei* is much lower, probably attributable to the xeric habitats where the latter occurs (Heselhaus and Schmidt, 1990); to 12 individuals on a single "medium-sized" palm, >40 adult *A. r. summus* on a pigsty with an area of ~12 m² (M. Schmidt, 1993); >100 lizards/100 m² in banana plantations (Bach and Schönecker, 2008). **Predation:** *Liophis cursor* (Henderson and Bourgeois, 1993). **Reproduction:** Copulation in captivity usually in the morning, may bury 2 eggs in substrate (J. Schmidt, 1980b); in captivity, a single pair of *A. r. summus* produced 20 eggs in one year, hatchling SVL about 20 mm (Verstegen, 1988); one female produced 30 eggs in one year in captivity (Fläschendräger, 1986); 31 eggs in one year (U. Manthey, 1984 in Fläschendräger and Wijffels, 2009); incubation ~43 days at 23–28 °C, hatchling SVL in captivity ~24 mm (Fläschendräger and Wijffels, 2009). **Sex Ratio (M:F):** 1:2 or even 1:3 in many habitats (Heselhaus and Schmidt, 1990; M. Schmidt, 1993). **Size:** Male SVL to 86 mm, female SVL to 66 mm, smallest mature male and female SVL 49 mm and 46 mm, respectively (Lazell, 1972). **Thermal Biology:** Even *A. r. summus* (the subspecies with the highest altitudinal distribution) not notably heliophilic (Lazell, 1972); mean body and air temperatures are inversely correlated with higher elevation, where basking opportunities are limited (but basks more frequently) and lizards shift to more open habitats, at 20 m, elevate T_b above T_a only at cooler temperatures, at 350 m, uniformly elevate T_b above T_a over entire range of T_a, at 650 m, can bask and elevate T_b only in the middle of the day when temperatures are highest, mean T_b at 20 m 28.6 ± 0.1 °C (23.6–31.8 °C), mean T_b at 350 m 28.3 ± 0.2 °C (23.6–32.0 °C), mean T_b at 650 m 24.3 ± 0.1 °C (21.3–28.6 °C), mean critical thermal maxima 36.0 ± 0.1 °C at 20 m elevation, 35.9 ± 0.2 °C at 175 m, 35.9 ± 0.2 °C at 350 m, 35.5 ± 0.2 °C at 500 m, 35.2 ± 0.2 °C at 650 m (Hertz, 1981, 1983).

Conservation Status: Not assessed, but protected by French legislation (Lorvelec et al., 2007).

Anolis rubribarbus Barbour and Ramsden, 1919

Distribution: Cuba, extreme northeastern coast and adjacent interior; elevational range, sea level to >300 m.

Habitat: Mesophilic (Schwartz and Henderson, 1991); in broadleaf gallery forest along streams and rivers descending through pine forest, but not restricted to deep, shaded areas, males perch head-down about 1 m above the ground, females perch lower, are less exposed than males, and difficult to see against the grayish bark of some trees (Ruibal and Williams, 1961b); in closed-canopy forests associated with rivers, adult males on fallen tree trunks, females perch much lower and in less exposed situations, such as denser vegetation (Garrido, 1967); deeply shaded humid draws, on trunks and branches with heads down at heights of ~1 m (Fläschendräger and Wijffels, 2009); in submontane rainforest, submontane mesophyllous evergreen forest, mesophyllous evergreen and semideciduous forest, and pine forest (Rodríguez Schettino et al., 1999).

Activity: Diurnal (Rodríguez Schettino et al., 1999). **Behavior:** Shy, critical approach distance 2 m (Ruibal and Williams, 1961b; Garrido, 1967). **Ecomorphology:** "Trunk-ground" (Williams and Rand, 1977). **Parasites:** Nematodes *Cyrtosomum scelopori* and larval *Porrocaecum* sp. (Coy Otero and Baruš, 1979b; Coy Otero and Lorenzo Hernández, 1982; Coy Otero, 1999). **Size:** Male SVL to 65.9 mm, female SVL to 47.5 mm (Rodríguez Schettino, 1999b). **Thermal Biology:** Apparently nonheliothermic (Rodríguez Schettino, 1999b).

Conservation Status: The species no longer occurs at the type locality due to the destruction of forest (Ruibal and Williams, 1961b); considered "vulnerable" (Rodríguez Schettino and Chamizo Lara, 1998a; Rodríguez Schettino and Chamizo Lara in Rodríguez Schettino, 1999a).

Anolis ruibali Navarro Pacheco and Garrido, 2004

Distribution: Cuba, Meseta de Cabo Cruz.

Habitat: In xeric coastal scrub, along edges of evergreen forest, semideciduous forest, and secondary vegetation, where lizards commonly perch on *Acasia farnesiana*, *Dichrostachys glomerata*, other thorny plants and shrubs, and fence posts along forest edges and openings at heights <2 m (Navarro Pacheco and Garrido, 2004).

Activity: Diurnal (Rodríguez Schettino et al., 1999); most active in mid-morning (~1000 h), activity decreases at midday (Navarro Pacheco and Garrido, 2004). **Ecomorphology:** Undetermined. **Size:** SVL to 36.5 mm (sex not given; Navarro Pacheco and Garrido, 2004).

Conservation Status: Unknown.

Anolis rupinae Williams and Webster, 1974

Distribution: Hispaniola, Massif de la Hotte (Haiti).

Habitat: Mesophilic (Schwartz and Henderson, 1991); in a damp and shaded ravine with a rocky floor and few large trees, but slopes covered by thick growth of brush, on 2 July, sun reached the floor of the ravine at 0930 h (Williams and Webster, 1974).

Ecomorphology: Undetermined. **Size:** Male SVL to 56 mm, female SVL to 42 mm (Williams and Webster, 1974).

Conservation Status: Unknown; the association with dense Haitian forests suggests that some protected status may be warranted.

Anolis sabanus Garman, 1887

Distribution: Saba, essentially islandwide; elevational range, sea level to 870 m; a colony on St. Eustatius is no longer extant (Van den Heuvel in Wijffels, 1997c).

Habitat: Mesophilic (Schwartz and Henderson, 1991); islandwide, except in the driest areas, in bushes, low trees, on rocks, stone walls, or artificial situations such as buildings, walls, fences, arbors, water tanks, woodpiles, or trash (Wijffels, 1971; Lazell, 1972; Fläschendräger and Wijffels, 1996, 2009; Powell et al., 2005; Powell and Henderson, 2008c); less abundant at higher elevations, where more closely associated with trees (Wijffels, 1971); mostly on rock outcrops and roots at bases of trees, not as arboreal as other anoles (Petzold, 1981); preferentially on large tree trunks in moister uplands, on rocks at lower, drier sites (Heselhaus and Schmidt, 1990); highly saxicolous (Powell et al., 2005); males usually perch <1.5 m above the ground, females and juveniles usually <0.9 m above the ground (Lazell, 1972); mean perch height and diameter 1.88 ± 0.42 m and 10.4 ± 2.8 cm (Losos and de Queiroz, 1997a); majority of perches on trees or rocks, but often on buildings and walls in developed areas, males perched higher, only one female >2 m above ground and 7% of males 2–4 m high, 79% of females and 67% of males within 1 m of the ground, lizards equally abundant in shade, partial sun, and full sun (Staats et al., 1997); sleeps in and on buildings as well as vegetation, where sometimes tightly clumped (Schwartz and Henderson, 1991); sleeps on leaves and twigs at heights similar to those used by day, often favoring the very ends of slim branches, may also use crevices in natural or artificial situations (Powell et al., 2005).

Activity: Basically diurnal, but exploits the "night-light niche" (Schwartz and Henderson, 1991; Powell and Henderson, 1992; Henderson and Powell, 2001). **Behavior:** Adult males very wary, females less so, males aggressively territorial (Blok, 1971); retreats into cavities and crevices, may "squirrel" to other side of trunk or branch, jump to another perch or onto the ground, only rarely ascends to escape (Lazell, 1972; Powell et al., 2005); most individuals hold head off the substrate but do not extend forelimbs to the extent seen in other anoles (only 20% of females and 42% of males extended forelegs; Staats et al., 1997); displays from position pressed tightly against a vertical substrate with forelimbs more extended than in most anoles, lifts head and extends throat fan (Lazell, 1972); male-male territorial disputes involve little display behavior and rapidly escalate into combat, with biting and efforts to throw the other lizard off the perch (Powell et al., 2005). **Ecomorphology:** Qualifies as "trunk-crown" (Losos and de Queiroz, 1997a), but behaves more like "trunk-ground" (RP). **Foraging and Diet:** Small insects and spiders, will lap nectar and eat flowers and fruits (Powell et al., 2005). **Home Range and Territoriality:** Both sexes strongly territorial and aggressive (Fläschendräger and Wijffels, 2009). **Parasites:** Trematode *Mesocoelium* sp., nematodes *Paraphyaryngodon cubensis* (Dobson et al., 1992) and *Spauligodon anolis* (Dobson et al., 1992; Bursey and Goldberg, 1998); mean parasite species/host 0.82 (Dobson and Pacala, 1992); cestode *Mesocoelium monas* and nematodes *Parapharyngodon cubensis* and larval *Abbreviata* sp. (Goldberg et al., 1997a); malarial parasites *Plasmodium azurophilum* and *P. floridense* (Staats and Schall, 1996a, 1996b); parasitemia (parasite density in blood) low for three species of *Plasmodium*, infection altered blood cell composition and blood hemoglobin, but had no apparent effect on body temperatures, proportions of lizards with broken tails, perches used, foraging success, male-male interactions in experimental manipulations, body color, or symmetry in body color (Schall and Staats, 2002). **Predation:** *Alsophis rufiventris* (Henderson and Sajdak, 1996); presumably American Kestrels (*Falco sparverius*) and Pearly-eyed Thrashers (*Margarops fuscatus*; Powell et al., 2005). **Reproduction:** Copulation relatively "passive," duration 30–50 sec, in captivity, single eggs laid at 1–5-week intervals, with as many as 30 eggs in a reproductive season, females may mate immediately after laying, eggs in excavated holes that are covered after deposition, eggs 5 × 8 mm at laying, 8 × 13 mm prior to hatching, hatchling SVL 23 mm (Blok, 1971); in captivity, eggs laid at intervals of 5–7 days, incubation 35–45 days at 20–25 °C (Fläschendräger and Wijffels, 2009); copulation in June and October (RP). **Size:** Male SVL to 90 mm (Fläschendräger and Wijffels, 1996), female SVL to 55 mm (Staats et al., 1997); smallest mature male and female SVL 44 mm and 42 mm, respectively (Lazell, 1972); males may mature at SVL 40 mm (Fläschendräger and Wijffels, 1996, 2009). **Tail Autotomy:** More males than females had broken or regenerated tails, larger animals more likely to have injured tails (Staats et al., 1997); 35% of males and 31% of females had injured tails (Schall and Staats, 2002). **Thermal Biology:** Thermoregulating in midafternoon (February), with body flush to the substrate and limbs splayed (Schwartz and Henderson, 1991); mean body temperatures of males 29.2 °C, for females 29.1 °C (Staats et al., 1997).

Conservation Status: Not assessed.

Additional Pertinent Literature: M. A. Powell et al. (2006).

Anolis sagrei Cocteau in Duméril and Bibron, 1837

Distribution: Five currently recognized subspecies occur in the Bahama Islands (Cay Sal, Conception, Crooked-Acklins, Grand Bahama, Little Bahama, Rum Cay, and San Salvador island banks), Cuba and associated cays, Isla de La Juventud, and Little Cayman, including most cays with even rudimentary vegetation; introduced in Jamaica (which may or may not have been human-mediated), Grand Cayman, Swan Island, Grenada, St. Vincent, the Grenadines (Canouan), Barbados, the Atlantic Coast of Mexico to Belize, Islas de la Bahía, Aruba, Hawaii (Oahu and Coconut Island, Kauai), Taiwan, the southeastern United States, and California (Orange County).

Habitat: Cosmopolitan (Schwartz and Henderson, 1991); in woods, plantations, orchards, and gardens (Gundlach,

1880); on the ground and on fence posts (Stejneger, 1917); greatest ecological tolerance of species on Bimini, in greatest variety of habitats and abundant everywhere (Oliver, 1948); a variety of sunny habitats, including savannas, pine forests, secondary semideciduous forest, dry swamp forest (perching on mangrove prop roots), and in coastal regions, rarely in deep woods or forests (where restricted to openings), populations densest in areas of intermediate insolation, also abundant in edificarian situations such as yards and patios, where on buildings, the ground, brush, fallen logs and boards, fences, and piles of debris and rocks (Barbour and Ramsden, 1919; Grant, 1940a, 1940c; Oliver, 1948; Alayo Dalmau, 1951; Schwartz and Ogren, 1956; Ruibal, 1961, 1964; Buide, 1966; G. Peters and Schubert, 1968; Schoener, 1968; Williams, 1969; Schoener and Schoener, 1971a; Berovides Alvarez and Sampedro Marín, 1980; Seidel and Franz, 1994; Schwartz and Henderson, 1991; Estrada, 1998; Fläschendräger, 2002; Rodríguez Schettino and de Queiroz, 2002b; Powell and Henderson, 2008c); more arboreal on Cuba than on Jamaica (Williams, 1969); common in roadside vegetation, with females frequently on the ground (Heselhaus and Schmidt, 1990); also on sign posts and traffic islands in urban areas (Zeilfelder and Bartelt, 2006); in the Biosphere Reserve "Sierra del Rosario" (Cuba), mostly in areas modified by human activities, preferentially on the ground, but also on trunks of trees and bushes and on buildings (Martínez Reyes, 1995); in Cuba, in submontane rainforest, submontane mesophyllous evergreen forest, mesophyllous evergreen and semideciduous forest, pine forest, mogote complex of vegetation, swamp forest, mangrove forest, coastal and subcoastal microphyllous forest, semidesert thorny scrubwoodland, serpentine xeromorphous shrub-woodland, cultivated areas and pastures, and urban zones (Rodríguez Schettino et al., 1999); often associated with humans (Kunz, 2005); adult males typically with head facing down (Grant, 1940c); often singly on observation posts (e.g., tree trunks, rocks or boulders, tops of woodpiles, fence posts) from which they jump to the ground to feed, females and juveniles on the ground or slightly above the ground in weedy growth (Oliver, 1948); by far the most terrestrial anole on Bimini, 167 of 637 observations of this species were for lizards on the ground, those on elevated perches were more likely to be on branches than leaves, most adult males (26%) on perches 0.9–1.6 m high and with diameters 5–11 cm, most subadult males (26%), adult females (22%), and juveniles (60%) on the ground (Schoener, 1968); perches lower on Bimini, where *A. smaragdinus* is present, than on Cayman Brac, where *A. sagrei* historically occurred alone, perches lower and on perches of slightly smaller diameter when sympatric with *A. grahami* than when alone on Jamaica (Schoener, 1975); mean adult male, adult female, and subadult male (no subadults listed for some islands) perch heights on Jamaica 0.7 ± 0.30 m (8.1% on ground), 0.4 ± 0.15 m (16.5%), 0.4 ± 0.20 m (16.3%), on Exuma 0.9 ± 0.45 m (9.6%), 0.5 ± 0.25 m (25.1%), 0.5 ± 0.29 m (17.8%), on Little Cayman 1.4 ± 0.59 m (1.3%), 0.7 ± 0.37 m (3.4%), on Abaco 0.9 ± 0.51 m (8.7%), 0.8 ± 0.45 m (8.1%), 0.6 ± 0.35 m (11.3%), on Duck Cay 1.2 ± 0.63 m (5.2%), 1.0 ± 0.54 m (2.7%), on Cayman Brac 1.5 ± 0.76 m (1.3%), 1.0 ± 0.62 m (3.4%), and on Swan Island 1.5 ± 0.76 m (1.4%), 1.0 ± 0.59 m (6.0%), 0.7 ± 0.36 m (9.8%; Lister, 1976); on Abaco Island, females are polymorphic in pattern and morphs differ in perch-diameter occurrence (Schoener and Schoener, 1983b); females spend considerable time on the ground (4.3–32.1% of observations in various habitats, mean for all habitats 14.6%), with unstriped resident females increasing in frequency with increased availability of thick perches and increased diameter of perches selected, unstriped transient females decrease in frequency with increasing availability of thick perches, in habitat with greatest diversity of available perch sizes, unstriped individuals select perches of larger diameter than striped females (Schoener and Schoener, 1976); alternative female morphs did not differ significantly in perch diameters on Kidd Cay and Flamingo Bay Cay in the Bahamas (Calsbeek et al., 2008); on tree trunks, bushes, and wooden, metal (wire), and concrete fences and posts, to heights of 1.5 m above the ground, with females and juveniles lower than males in herbaceous strata, leaf litter, and on stones (Rodríguez Schettino, 1999b); where sympatric with *A. homolechis*, 41.67% of observations on trees, 13.89% on bushes, and 44.44% on the ground (Berovides Álvarez and Sampedro Marín, 1980); also where sympatric with *A. homolechis*, less frequently in trees and more frequently on the ground or in bushes than expected (differences not significant), mean perch heights and diameters 64.00 ± 7.99 cm and 3.18 ± 0.38 cm (morning), 49.86 ± 4.86 cm and 4.23 ± 0.58 cm (afternoon), and 57.81 ± 5.16 cm and 3.64 ± 0.35 cm (all day; Sampedro Marín et al., 1982a); mean perch height and diameter 0.61 m and 0.08 m (Sampedro et al., 1980); 0.7 m and 0.03 m (J. R. Llanes Echevarría in Rodríguez Schettino and de Queiroz, 2002b); 0.57 m and 0.22 m for 10 males (7 with heads facing up and 3 down; Rodríguez Schettino and de Queiroz, 2002b); on Cayo Inés de Soto, Cuba, 23 of 27 males on trees (18 on trunks, 5 on branches), 3 on bushes, 1 on the ground, 16 of 28 females on trees (13 on trunks), 4 on bushes, 8 on the ground, 16 of 25 subadults on trees (15 on trunks), 3 on bushes, 6 on the ground, 2 of 15 males at heights 0.5 m, 6 at 0.5–1.0 m, 3 at 1.0–1.5 m, 4 >1.5 m, 9 of 13 females at heights 0.5 m, 1 at 0.5–1.0 m, 2 at 1.0–1.5 m, 1 >1.5 m, 2 of 9 subadults at heights 0.5 m, 5 at 0.5–1.0 m, 1 at 1.0–1.5 m, 1 >1.5 m, 9 of 15 males, 8 of 13 females on perches with diameters 10.1–15.0 cm, 5 of 9 subadults on perches with diameters >15 cm, more likely to be on trunks, at lower heights, and on perches of larger diameter than *A. porcatus* (Estrada and Novo Rodríguez, 1986b); on Cayo Guajaba (Archipiélago de Samana-Camagüey), mostly in sandy coastal scrub and xeric interior regions on rocky ground (Garrido et al., 1986); mean perch height and diameter in Jamaica 0.4 ± 0.09 m and 15.4 ± 3.45 cm (Losos, 1990a); on the Archipiélago de los Canarreos, on palms and small trees, herbaceous vegetation, and on the ground (Estrada, 1993a); on Cayman Brac, mean perch height and diameter 0.96 ± 0.08 m and 9.7 ± 0.9 cm (Losos and de Queiroz, 1997a); most frequently on very small (<2 cm diameter) or very large (>8 cm) perches, with intermediate habitat breadth of 61.1 (Irschick and Losos, 1999); on Jamaica, mean perch heights and diameters 0.42 ± 0.35 m and 5.8 ± 7.6 cm for males and 0.29 ± 0.27 m and 8.0 ± 11.3 cm for females (Butler and Losos, 2002); structural habitat use indicative of resource partitioning among age classes (e.g., males tend to perch higher), but may be influenced by the presence of congeners (Estrada and Novo Rodríguez, 1986a) or predators, the introduction of *Leiocephalus carinatus* onto small Bahamian islands occupied by *A. sagrei* resulted in the latter shifting to higher perches of smaller diameter, which would increase selective pressures favoring longer hindlimbs (Losos et al., 2004, 2006); 95% of observations on Bimini at heights <2 m (Mattingly and Jayne, 2004); mean heights on Bimini 133 ± 10 cm (Mattingly and Jayne, 2005); mean perch height at Soroa 0.93 m, diameter 6.6 cm, behavioral diameter (mean diameter of all surfaces used during behavioral observations, means of individual means) 3.36 cm, 13.2% of observations on or within 1 m of rocks (J. B. Losos, in litt., 20.VIII.2007); mean perch height and diameter in Cuba west of Cienfuegos 0.63 m and 25.0 cm, in southeastern Cuba 0.75 m and 21.9 cm, 31.8% on leaves (J. B. Losos, in litt., 21.IV.2008); on Kidd Cay (near Great Exuma, Bahamas), mean perch height and diameter for males 0.71 ± 0.05 m and 8.34 ± 0.84 mm, for females 0.46 ± 0.05 m and 4.69 ± 0.81 mm (Calsbeek, 2008); use of narrow perches in scrub habitat changes over time, whereas that of broad perches remains stable, low-density, reduced intraspecific competition revealed significant selection on limb length driven by changes in perch diameter (Calsbeek, 2009); on Kidd Cay and Nightmare Cay, lizards increase use of vegetation in wet years (correlated with selection on limb length but not body size), but reduce arboreality in dry years (with relaxation of selective pressures; Calsbeek et al., 2009); habitat light environment "no shade" (Fleishman, 2000); dewlap UV-reflectance compatible with visibility in typical habitat (Macedonia et al., 2000); presumably sleeps under terrestrial objects (Schwartz and Henderson, 1991); on Grenada (Greene et al., 2002; Germano et al., 2003) and St. Vincent (Henderson and Powell, 2005; Treglia, 2006), apparently restricted to severely altered habitats, such as open fields or areas with little vegetation, in which native anoles are rare or absent; at a highly disturbed construction site on St. Vincent, mean perch heights of adult males 55.0 ± 9.6 cm, subadults and adult females 19.9 ± 5.2 cm, juveniles 5.3 ± 1.0 cm, used "artificial" perches (walls and foundation, miscellaneous construction materials and debris) at much greater frequencies than "natural" perches (Treglia et al., 2008).

Activity: Diurnal (Rodríguez Schettino et al., 1999); from shortly after dawn to dusk, ~0600–1800 h (Evans, 1938a); becomes active later in the day than most other anoles, with peak activity about midday, but active under rocks 0800–1000 h (Schwartz and Henderson, 1991); bimodal, with peaks at 1000–1100 h and 1400–1500 h (Berovides Álvarez and Sampedro Marín, 1980) or 0930–1100 h and 1300–1430 h (Fläschendräger and Wijffels, 1996, 2009); "partially active" under stones and other shelters 0830–1000 h, sometimes with activity peaking between 1130 h and 1530 h, sometimes with no fluctuations in daily activity (Rodríguez Schettino, 1999b); earliest activity at Discovery Bay 0540 h, latest activity 1635 h, male displays peak at dawn and dusk, presumably to advertise territorial activity before and after a period of nocturnal inactivity (Ord, 2008); exploits the "night-light niche" until at least 2230 h in the Bahamas (Schwartz and Henderson, 1991; Perry et al., 2008). **Behavior:** During territorial encounters involving dewlap extension, nuchal and dorsal crest erection, body elevation, lateral compression, gaping, tail lashing, and biting (only as a last resort), the resident of either sex becomes dark (almost black), whereas the invader retains the pale (olive) color, if the invader is a female, a resident female attacks, but a resident male stops her and courts the invader (Evans, 1938a); extended tail waving may be aposematic (Mertens, 1946); alert and active, remains in position until an intruder is very close, if on the ground, runs to a nearby tree or shrub, relies on agility to escape capture (Oliver, 1948); when pressed, individuals retreat quickly to ground and run away (Bels, 1983); when pursued, "disappears" among rocks and debris on the ground (Seidel and Franz, 1994); when escaping, runs into herbaceous growth, hides among leaves, stones, and grasses, sometimes when threatened, exhibits tonic immobility, slowly waving tail (to distract predator?), when on trunks perching with head facing down, climbs upward and turns with the head facing down again, when facing *A. allisoni*, some males seek to escape and others defend territories (Rodríguez Schettino, 1999b); mean approach distance 2.64 ± 0.14 m, individuals perched near the ground most likely to escape to the ground, whereas those at greater distances from the ground more likely to run up, first responses "squirreling" (168 of 201) followed by running up (87 of 168), running down (47), staying in position (24), or jumping (10), first response jumping (18 of 201) followed by second responses "squirreling" (8 of 18), jumping, running up, or staying in place, first responses running down (6 of 201), running up (5), or running (4) occurred less frequently (Regalado, 1998); 6.6% of distance traveled on trunk bases, 22.6% on trunks, 70.8% on branches, 14.9% of movements with jumps, mean movement distance 11.5 cm (median 5.3 cm) at speed 13.4 cm/sec (8.9 cm/sec), mean pause duration 67.3 sec (12.8 sec; Mattingly and Jayne, 2004); during an escape chooses the broadest diameter 95%, longest length 39%, shallowest incline 31%, and straightest angle 84% (Mattingly and Jayne, 2005); headbobbing is the primary method of territorial defense, females and juveniles invariably respond with headbobbing, whereas males respond by dewlap extension (Oliver, 1948); male threat displays are series of irregular headbobs, pushups, vertical tail wags, extended dewlaps (without regular sequence, but always a series of rapid extensions and foldings), and sometimes with erected nuchal and dorsal crests (Grant, 1940c; Ruibal, 1967); squeaks during aggressive interactions (Grant, 1940a); on Jamaica, mean display rate (proportion of time spent displaying) 10.0 ± 1.9, percent walks 18.5 ± 4.6, percent runs 55.8 ± 4.5, percent jumps 25.6 ± 3.0, moves/min 0.90 ± 0.13, distance jumped in nature 28.3 ± 2.6 cm (Losos, 1990a); tends to jump rather than run on perches of small diameter (Losos and Irschick, 1996); when experimentally placed on very small fragments of natural islands, in 1/3–1/2 of 54 trials, lizards leaped into water under their own volition, especially under sunny conditions (Schoener and Schoener, 1983a); can float for 1 h, 12 of 39 individuals still afloat after 24 h (A. Schoener and Schoener, 1984); wave tails erratically (in captivity), possible predator distraction (Cochran, 1989); males on elevated perches perform more displays and approached intruders more frequently when invasions occurred on the ground, but were quicker to attack intruders that entered territories on elevated perches (Calsbeek and Marnocha, 2006); absolute but not size-corrected dewlap size significantly related with bite force, 92% of combat trials involved dewlap extensions, 63% biting, and 100% headbobbing (Lailvaux and Irschick, 2007); moves per minute 0.65 (South Bimini), 0.57 (Cuba), 0.90 (Jamaica; Johnson et al., 2008); dewlap extensions may serve as predator deterrent against *Leiocephalus carinatus*, dewlaps are larger in both males and females on Bahamian islands with *L. carinatus* than on islands where the predator is absent (Vanhooydonck et al., 2009). **Competition:** In Cuba, where sympatric with *A. homolechis*, perch heights (mean 68 cm for *A. homolechis* and 58 cm for *A. sagrei*) differed significantly and *A. homolechis* had a higher proportion of ants in its diet, but types of perches used and times of activity differed only slightly, structural niche overlap 0.833, activity niche overlap 0.862, dietary niche overlap 0.581 (Berovides Álvarez and Sampedro Marín, 1980); on Grand Cayman, forced *A. conspersus* to use higher perches (Losos et al., 1993); on small Bahamian islands near Great Exuma, propagules of 5 lizards thrived (until hurricane surge submersed the islands), effects of *A. smaragdinus* were minimal, declining as populations of the latter declined (Losos and Spiller, 1999); at low density, reduced intraspecific competition revealed significant selection on limb length driven by changes in perch diameter (Calsbeek, 2009). **Desiccation:** Selection may favor individuals with fewer scales in hot and/or dry habitats (Calsbeek et al., 2006). **Ecomorphology:** "Trunk-ground" (Rand and Williams, 1969); formally quantified (Beuttell and Losos, 1999); *A. s. nelsoni* (Swan Islands) qualifies as a "generalist" (Losos and de Queiroz, 1997a). **Egg Survival:** Hatching success and incubation time did not differ among young (<10 days old) from eggs immersed in seawater for 3 or 6 h and those not immersed (Losos et al., 2003). **Foraging and Diet:** Responds quickly when a termite nest is opened (Barbour, 1930c); makes forays for insects in tall grass, but always ~6 m from a lookout (Evans, 1938a); on Bimini, frequently forages on the ground (Oliver, 1948); also on Bimini, food items were ants (greatest numbers in adult and subadult males), other hymenopterans, aphids, leafhoppers (greatest numbers and volume in juveniles), other hemipterans, flies (greatest numbers in females and greatest numbers and volumes in juveniles), thrips, beetles (adults and larvae), barklice, lepidopterans (adults, greatest volume in adult males, and larvae, greatest volume in subadult males), orthopterans, lacewings, termites, earwigs, spiders, mites, and plant matter, with larger individuals taking larger prey (Schoener, 1968); in captivity (on Cuba), eats a variety of insects, but avoids termites in the genera *Neotermes* and *Nasutitermes*, especially when fluttering their wings, dermatophagy, with one individual taking ~4 h to complete shedding (Ruiz García, 1977); "classical sit-and-wait predator," obtaining most of its food on the ground and a lesser amount from low vegetation, takes larger prey than sympatric congeners, resulting in longer food-handling times, actively feeds throughout the day, may sit motionless in the shade during midday heat, but will still "dart down at a prey if the opportunity arises" (Schoener, 1979); on Jamaica, feeds on larger prey than similarly sized individuals of other anoline species, resulting in longer prey-handling/swallowing times, stomachs contained ants (34.2%), other hymenopterans, flies (10.9%), beetles (adults and larvae, 11.1%), termites, hemipterans (mostly leafhoppers, 12.1%), lepidopterans (adults and larvae), orthopterans, spiders, mites, and isopods, on Exuma island, ants (31.8%), other hymenopterans, flies, beetles (adults and larvae, 10.0%), termites (25.5%), leafhoppers (12.1%), lepidopterans (adults and larvae), orthopterans, spiders, centipedes, and an anole, on Abaco Island, ants (11.2%), other hymenopterans, flies (10.6%), beetles, leafhoppers (65.5%), other hemipterans, lepidopterans (adults and larvae), lacewings, earwigs, orthopterans, spiders, pseudoscorpions, centipedes, mites, isopods, and gastropods, on Cayman Brac, ants (34.8%), other hymenopterans, flies, beetles (adults and larvae, 13.1%), termites, hemipterans, lepidopterans (adults and larvae), lacewings, orthopterans, spiders, pseudoscorpions, centipedes, mites, isopods, gastropods, an anole, and plant matter, on Swan Island, ants (51.2%), other hymenopterans, flies, beetles (adults and larvae), termites (12.3%), hemipterans, lepidopterans (adults and larvae), earwigs, orthopterans, spiders, mites, isopods, gastropods, an anole, and plant matter (Lister, 1976); especially males move from observation posts to make frequent forays to catch insects before returning to their post, feeds throughout the day, cannibalistic, at least in captivity (Cochran, 1989); on Cuba, spiders (11.84%), ants (26.32%), caterpillars (28.95%), beetles (13.16%), other arthropod orders (19.74%; Berovides Álvarez and Sampedro Marín, 1980); 12% of prey items were spiders (versus insects), insect prey includes ants and other hymenopterans (30%), lepidopterans (mostly caterpillars, 33%), beetles (15%), and others (22%; Sampedro Marín et al., 1982a); presumed predators of juvenile orbweavers (*Argiope trifasciata*; Armas and Alayón García, 1986); on small Bahamian islands, in experiments involving *A. sagrei*, *A. smaragdinus*, and 8 species of spiders, conclusions were (1) both spiders and lizards were generalists, and lizards may be both predators and competitors of spiders, (2) in the absence of lizards, spider populations may be limited by other factors to below the level at which competitive exclusion would occur, (3) in the absence of predators or other disturbances, spiders may coexist by partitioning resources, in plots where anoles had been removed, spider densities were 2.5 times as high as in control plots, thus spider survival, prey abundance, and prey consumption were all negatively affected by lizards (Schoener et al., 2002); lizard removal (92.6% *A. sagrei*, 7.4% *Ameiva auberi*, 0.5% *A. smaragdinus*) resulted in increased numbers and species of spiders in experimental enclosures, with the presence of lizards having a greater impact on the presence of rare than on abundant species (Spiller and Schoener, 1998); males take larger prey than females (Rodríguez Schettino, 1999b). **Growth:** Males generally grow faster than females at SVL >24 mm for Abaco and Bimini populations, rates of growth decline with size in both sexes (Schoener and Schoener, 1978b). **Home Range and Territoriality:** Nonresident males invariably defeated and chased from a perch, territory of individuals or pairs about 37 m^2 (Evans, 1938a); females strongly territorial (Evans, 1938b); both males and females defend territories, those of females contiguous but not overlapping (although night-time sleeping sites overlapped), territories of large males included territories of more than one female and a smaller male, those of females contained 1–5 subadults or juveniles, defoliation by hurricanes disrupts territories (Oliver, 1948); on Bimini in "short-blackland" habitat, median home range size of males 5.53 m^2, of females 2.94 m^2, in "forest-path" habitat, of males 3.23 m^2, of females 2.09 m^2, males defend territories from other males, whereas smaller individuals occupy areas largely exclusive of one another regardless of sex, overlap in males greatest when about 6 males extended their ranges into an area where a fruiting tree attracted a large numbers of insects, chases and fights were frequent, but most individuals persisted in attempts to exploit this temporary food supply (Schoener, 1979); has larger home ranges than more arboreal species, with those of adult males (2.84 ± 1.87–34.10 ± 24.78 m^2 at 14 Bahamian sites) larger than those of females (0.65 ± 0.43–6.25 ± 5.12 m^2 at 15 sites) and subadult males (1.40 ± 1.26–20.74 ± 12.23 m^2 at 10 sites), home range sizes increase with body size more in males than in females, within sex and site, home-range area unrelated to relative growth rate or survival, home range area in size classes inversely related to the density of the same or other classes, in adult females and subadult males, the relationship is strongest when using the density of all conspecifics, in adult males strongest when using only adult male density (Schoener and Schoener, 1982b); in Cuba, size is important in establishing dominance, with the largest males defending territories most effectively, perching at greater heights, usually first to invade territories of other individuals, and initiating more challenges, one male and several females, of which one is dominant, occupy a territory, juveniles and hatchlings may stay in a territory for several weeks before scattering to look for new territories (Rodríguez Schettino, 1999b); on 3 Bahamian islands, males expanded territory sizes when females were more widely scattered, male territories on Abaco 5.88–17.65 m^2, female territories 2.50–4.78 m^2, male territories on Bimini 9.38–34.10 m^2, female territories 1.57–6.25 m^2, male territories on Andros 2.84 m^2, female territories 0.65 m^2 (Stamps, 1999). **Hurricane Effects:** No naturally occurring populations on 5 small islands on the "protected" side of Great Exuma Island became extinct (Spiller et al., 1998); on 66 Bahamian islands near Great Abaco, island area is a better predictor of population sizes than altitude before Hurricane Floyd (1999), but altitude a better predictor immediately after the hurricane, after about one year, size again better predictor, by 19 months, 88% of islands formerly occupied ($N = 50$) supported populations, recovery fueled by overwater colonization and propagation from eggs that survived inundation, Hurricane Lili (1996) had lower wind speeds, but a higher storm surge, affecting higher islands and decimating vegetation, lack of hatchling propagation may reflect destruction of eggs or lack of eggs due to timing late in the reproductive season (Schoener et al., 2001). **Hybridization:** With *A. bremeri bremeri* (Garrido, 1980a). **Injuries:** Injuries to tails, front toes, and back toes higher in males than in females, with front toe rates highest (presumably due to intraspecific aggression) and tail and hind toe rates (presumably due to predation) lowest, rates increase with size (= age; Schoener and Schoener, 1980b). **Life Expectancy:** 1.0 year in males, 1.8 years in females (Schoener and Schoener, 1982a; but also see "Survival"). **Miscellany:** A large male *Leiocephalus cubensis*, apparently defending its retreat, carried an adult male *A. sagrei* away in its jaws (Fernández Milera, 1984). **Movements:** Mean distance per movement 18.9 ± 2.2 cm at mean speed 22.5 ± 3.4 cm/sec, 0.25 ± 0.05 moves/min, 0.64 ± 0.16% time moving (Irschick, 2000b). **Parasites:** In Cuba, sporozoan red blood cell parasite *Haemogregarina* sp. (Zajicek and Mauri Méndez, 1969); trematodes *Urotrema scabridum* (Coy Otero, 1970, 1999; Coy Otero and Lorenzo Hernández, 1982) and *U. wardi* (Pérez Vigueras, 1940; Coy Otero, 1970, 1999; Coy Otero and Lorenzo Hernández, 1982); nematodes *Atractis opeatura*, *Oswaldocruzia lenteixeirai*, *Ozolaimus monhystera*, *Parapharyngodon cubensis*, *Physaloptera squamatae*, larval *Abbreviata* sp., larval *Porrocaecum* sp., Nematoda gen. sp. (Coy Otero and Baruš, 1979b; Coy Otero and Lorenzo Hernández, 1982; Coy Otero, 1999); *Cyrtosomum scelopori* (Baruš and Coy Otero, 1969a; Coy Otero, 1970, 1999; Coy Otero and Baruš, 1973a, 1979b; Coy Otero and Lorenzo Hernández, 1982); *Skrjabinoptera phrynosoma* (Baruš and Coy Otero, 1969a; Coy Otero, 1970, 1999; Coy Otero and Baruš, 1979b; Coy Otero and Lorenzo Hernández, 1982); and *Trichospirura teixeirai* (Coy Otero and Baruš, 1979a, 1979b; Coy Otero and Lorenzo Hernández, 1982; Coy Otero, 1999); tick *Amblyomma torrei* (Rodríguez Schettino, 1999b; Coy Otero, 1999); mite *Eutrombicula alfreddugesi* (de la Cruz and Abreu, 1986; Coy Otero, 1999); on North Bimini, protozoan malarial parasite *Plasmodium floridense* (Telford, 1975; Ayala, 1977, 1978); in the Bahamas (Andros and San Salvador), trematode *Urotrema scabridum*, nematodes *Atractis scelopori*, *Oswaldocruzia lenteixeirai*, *Parapharyngodon cubensis*, *Physaloptera squamatae*, larval *Porrocaecum* sp., adult female *Strongyloides* sp., and a juvenile acanthocephalan *Centrorhynchus* sp. (Goldberg et al., 1994b); on Jamaica, trematode *Urotrema scabridum* (Urotrematidae), nematodes *Cyrtosomum scelopori* and *Thelandros* (*cubensis* ?), and acanthocephalan *Centrorhynchus* (*spinosus* ?; Bundy et al., 1987); on Grand Cayman, nematodes *Atractis scelopori*, *Physaloptera squamatae*, and larval *Abbreviata* sp. (Goldberg et al., 1995). **Performance:** On Jamaica, mean maximum sprint speed (fastest time over an interval of 0.25 m) 0.138 ± 0.004 sec, jump 61.6 ± 1.29 cm, cling 1.136 ± 0.132 Neutons (Losos, 1990a); mean maximal speed 1.8 ± 0.04 m/sec, escape speed 1.3 ± 0.08 m/sec, feeding speed 1.1 ± 0.08 m/sec, undisturbed activity 0.4 ± 0.08 m/sec, mean maximal jumping distance 61.6 ± 1.29 cm, undisturbed jumping distance 24.7 ± 3.20 cm (Irschick and Losos, 1998); average gross speed over a 5-stride interval on a 4.8-cm diameter perch 94 ± 9.4 cm/sec and 50 ± 4.0 cm/sec with 90° turning angle, stride-one speed 100 ± 12.0 cm/sec and 28 ± 3.0 cm/sec with 90° turning angle (Mattingly and Jayne, 2005); acceleration not dramatically affected but sprint speed reduced on perches of small diameter (VanHooydonck et al., 2006b); acceleration 30.15 ± 1.93 m/sec, sprint speed 1.23 ± 0.06 m/sec (Vanhooydonck et al., 2006a); selection favors long-limbed lizards that are fast on broad surfaces and that prefer broad substrates in nature and short-limbed lizards that are less sprint sensitive on narrow perches and that prefer narrow perches in nature (Calsbeek and Irschick, 2007; Calsbeek et al., 2007b); on Kidd Cay (near Great Exuma, Bahamas), mean sprint speed on narrow and broad perches for males 1.90 ± 0.43 m/sec and 2.43 ± 0.04 m/sec, for females 1.73 ± 0.38 m/sec and 2.02 ± 0.04 m/sec, stamina (running time until exhaustion for males 77.77 ± 1.62 sec, for females 63.83 ± 1.53 sec (Calsbeek, 2008); stamina was significantly dependent on body size (Calsbeek and Bonneaud, 2008). **Population Size and Density:** To 4,942/ha in optimal habitats (Lister, 1976); male densities in various Bahamian habitats 0.115–0.404 individuals/m^2, female densities 0.119–0.593 individuals/m^2 (Schoener and Schoener, 1978b); on Bimini, in "short-blackland" habitat, 0.122 males/m^2, 0.136 females/m^2, in "forest-path" habitat, 0.194 males/m^2, 0.194 females/m^2 (Schoener, 1979); may fluctuate from year to year on some islands, while remaining constant on others, in various habitats at multiple sites on 3 Bahamian islands, female densities 0.017–0.563 lizards/m^2, male densities 0.060–0.406 lizards/m^2, total densities 0.077–0.969 lizards/m^2 (Schoener and Schoener, 1980a, 1980b); on Cuba, encounter rates vary by habitat, 3.8 lizards/h in coastal scrub in northeastern La Habana Province, 8.4–83.3/h at the Jardín Botánico Nacional, 13.0–20.0/h on the Península de Zapata (Rodríguez Schettino, 1999b); female density on Abaco 0.18–1.14 individual/m^2, on Bimini 0.03–0.56/m^2, on Andros 0.37–0.70/m^2 (Stamps, 1999). **Predation:** Estimated efficiency of predation on females greater in open habitats in the Bahamas than in forests (Schoener and Schoener, 1980b); Zapata Wren (*Ferminia cerverai*; S. C. Brunner in Garrido, 1980b); may be cannibalistic (Schwartz and Henderson, 1991); on Little Cayman Island, *Tropidophis parkeri* (Grant, 1940c); on Cayman Brac, *Cubophis fuscicauda* (Grant, 1940c; R. W. Henderson in Seidel and Franz, 1994); on Cuba, orbweavers (*Argiope trifasciata*; Armas and Alayón García, 1987; Armas, 2001); *Osteopilus septentrionalis* (Meshaka, 1996a); *Cubophis cantherigerus* (Henderson and Sajdak, 1996); *Leiocephalus cubensis*, American Kestrel (*Falco sparvarius*), Cuban Blackbird (*Dives atroviolaceus*), Greater Antillean Grackle (*Quiscalus niger*; Rodríguez Schettino, 1999b); captive hutias (*Capromys pilorides*) readily consume anoles (Manójina et al., 1989); in the Bahamas, *Cubophis vudii* (Henderson and Sajdak, 1996); *Leiocephalus carinatus* (Schoener et al., 1982; Vanhooydonck et al., 2009); *Anolis smaragdinus* (J. B. Losos in Gerber, 1999); introduction of *L. carinatus* onto small Bahamian islands had an immediate effect on population density (lower) and perch heights (higher) of *A. sagrei* and longer term effects on perch diameter (smaller), percentage hatchlings (lower), body-size distribution (narrower), and body condition, especially of males (poorer), with behavioral changes persisting after density changes had leveled off, reversing the negative impact of *A. sagrei* on the number of species and densities of web spiders (Schoener et al., 2002, 2005); *Caraiba andreae* (Domínguez Díaz and Moreno García, 2003); Greater Antillean Grackles (*Quiscalus niger*) capable of taking all size classes on Jamaica (Graves, 2006); Bahamas Mockingbird (*Mimus gundlachii*) on Abaco (J. B. Losos, in litt., 23.IV.2008). **Reproduction:** In captivity, male seized female by neck, brought his tail under hers and inserted hemipenis, one copulation lasted ~5 min, during which the male's pelvis thrusted rhythmically (Noble and Bradley, 1933); females respond with headbobs to dewlap displays of territorial males or to a male's display directed to a female, females may headbob to attract male's attention (Evans, 1938b); mean duration of copulation 3.8 min (longest 12 min; *A. s. sagrei*), 2.6 min (3.12 min; *A. s. greyi*), 3.53 min (5.18 min; presumed hybrid *A. sagrei* × *A. b. bremeri*), separation of pairs "normal," egg 10 × 5 mm, cream colored and semielliptical (Garrido, 1980b); courtship (13 July) with male bobbing head and extending dewlap (but no push-ups), female responded with headbobbing (actions continued ~2 min), male moved to female's right side, crawled along her side until his head at the level of her neck, grasped back of her neck in his jaws, moved his body on top of hers while moving his tail under her tail to approximate vents (lapsed time to this position ~1 min), copulation began with vigorous pumping of male's posterior (30 sec), position retained ~3 min, then individuals separated, moved about 7.5 cm apart, and bobbed heads for ~1 min, copulation time ($N = 10$) 1.75 to 2 or 3 min (most frequent) or 4.75 min (Ruibal, 1967); duration of copulation 1.75 min (in a garden) to 4.25 min (in captivity), usually 2–3 min, male may "wipe" vent on substrate after copulation (Ruiz García, 1977); displays with considerable variation in number and amplitude of headbobs, which are differentiated by distinct pauses, dewlap extension only with the third bob and continued thereafter, mean duration of precopulatory behaviors 34 sec, male headbobs throughout precopulation, copulation mean duration 3.8 min and characterized by only slight movement, separation rapid (Regalado and Garrido, 1993); mated pair 13 August (Schwartz and Henderson, 1991); seasonal cycle in La Habana, with reproductive quiescence during the dry season, testicular weight lowest in October–November and greatest in April and July–August, females may become reproductive in February, but peak during June–August, no females reproductive November–January, body fat weight is inversely proportional to reproductive activity (Sanz Ochotorena and Uribe Aranzábal, 1999); in Jamaica, reproduction strongly seasonal, testicular weight maximal April–June, minimal November–January, ovigerous females most abundant April–July, declining to zero in November–February, with body fat weight inversely proportional to reproductive activity (Licht and Gorman, 1970); clutch size 1 (Hedges, in prep.); egg 8.7 × 5.8 mm hatched 28 July, egg deposited on 14 July hatched 29 August, hatchling SVL 16.0 mm, tail 28.0 mm, 4 hatchlings (12 and 25 June) with large umbilical scars SVL 24.5–29.0 mm, hatchling 3 August from egg taken 15 July (Schwartz and Henderson, 1991); reproductive activity cyclic, with notable increases in wet seasons, percent males with enlarged testes decreases in September, females with oviductal eggs only March–October, eggs usually deposited in humus or leaf litter or under stones (Rodríguez Schettino, 1999b); multiple paternity occurred in 80.7% of field-caught Bahamian females that produced more than one offspring, females store sperm for about 3 weeks before mating again, although females are capable of storing viable sperm for a mean of 55 days in the laboratory (Calsbeek et al., 2007a); in captivity, incubation 32–45 days at 22–28 °C, hatchling SVL ~15 mm (Fläschendräger, 1986; Fläschendräger and Wijffels, 2009). **Sex Ratio:** In various habitats at multiple sites on 3 Bahamian islands F/M 0.45–3.13 (excluding data with sample sizes of 3 for one or both sexes; Schoener and Schoener, 1980a); 0.74:1.26 M:F (Schoener and Schoener, 1980b). **Size:** Male SVL to 70 mm (Schwartz and Henderson, 1991), female SVL to 47.8 mm (Rodríguez Schettino, 1999b); smallest sexually mature male 35 mm SVL and female 34 mm SVL, from Jamaica, where these lizards appear smaller than elsewhere in their range (Licht and Gorman, 1970); in Cuba, smallest reproductive male 45 mm SVL, smallest reproductive female 36 mm SVL (Sanz Ochotorena and Uribe Aranzábal, 1999); on 3 Bahamian islands, male size distributions skewed in favor of large males, suggesting elevated growth or mortality rates for young adult males when compared to females of the same age (Stamps, 1999); selection favors large body size at high population densities (Calsbeek and Smith, 2007). **Source of Introduced West Indian populations:** Haplotype variation suggests that secondary introductions from south Florida are responsible for introduced populations in Grand Cayman (Kolbe et al., 2007a). **Survival:** Survival higher on smaller island (Bimini) than on larger island (Andros), presumably because of predation by a greater abundance and variety of birds on Andros (Schoener and Schoener, 1978a); annual survival in various habitats on 3 Bahamian islands 0.038–0.385 for males and 0.000–0.357 for females, negatively correlated with the estimated intensity of predation and lower on islands with more predatory bird species and in more open habitats (Schoener and Schoener, 1980b); survival greater in forested areas in the Bahamas than in more open habitats, males survived less well than females, possibly reflecting extreme territoriality in the species and associated visibility, relatively constant age-specific mortality, but on island with the fewest birds, large individuals of both sexes survived better, percent survival after 12 months of all individuals on Andros and Bimini in 2 different habitats 8.3 and 6.7% (Schoener and Schoener, 1982a); survival on small Bahamian islands was highest (80.0–91.4% over 6 months) with shortest vegetation (birds are more abundant on islands with higher vegetation), but shifted dramatically (6.3–15.2%) with the introduction of a terrestrial predator (*Leiocephalus carinatus*) in that survival was highest on islands with higher vegetation (Schoener et al., 2005); 44 of 133 (33%) males and 38 of 100 (38%) females recaptured after one year (Calsbeek and Smith, 2008); survival of alternative female morphs differed significantly as a function of social crowding, "bar" females survived better at low density and "diamond" and "diamond-bar" females survived better at higher density (Calsbeek et al., 2008). **Tail Autotomy:** In 26 days, male regenerated about 2.5 cm of tail (Ruiz García, 1977); tail break frequencies 0.32–0.55 for males and 0.09–0.48 for females on 7 Bahamian islands, highest for males and lowest for females on Grand Bahama (Vanhooydonck et al., 2009). **Thermal Biology:** Heliothermic, mean body temperature in the morning 32.2 °C (27.4–36.4 °C), in the afternoon 33.9 °C (28.3–36.1 °C; Ruibal, 1961); mean body temperature 32.8 °C (30.0–31.6 °C air temperature), 34.3 °C (34.0–35.0 °C), 31.8 °C (30.2–31.6 °C), 34.8 °C (33.0–34.0 °C; Ruibal and Philibosian, 1970); mean body and air temperatures in open, intermediate, and closed situations on Exuma Island 33.6 ± 1.50 °C and 28.7 ± 1.01 °C, 32.5 ± 1.85 °C and 29.1 ± 0.97 °C, 29.5 ± 1.46 °C and 27.9 ± 1.12 °C, in open and closed situations on Abaco Island 33.5 ± 2.21 °C and 27.5 ± 1.61 °C, 26.6 ± 1.94 °C and 25.5 ± 1.89 °C, in open and closed situations on Cayman Brac 33.7 ± 1.76 °C and 30.5 ± 0.87 °C, 32.4 ± 1.17 °C and 31.0 ± 0.76 °C, in a closed situation on Swan Island 32.1 ± 0.89 °C and 30.8 ± 0.96 °C, in an open situation at Discovery Bay, Jamaica 32.9 ± 0.83 °C and 29.4 ± 0.89 °C, and in an open situation at Savanna-la-Mar, Jamaica 33.7 ± 1.02 °C and 30.9 ± 0.96 °C (Lister, 1976); mean body temperature 33.1 °C at mean air temperature 31.9 °C in July and August, generally associated with habitats where air temperatures are high (Rodríguez Schettino, 1999b); mean body and air temperature for 7 males 32.1 °C and 31.0 °C, 6 of 10 males in shade, 4 in filtered sun during afternoon (Rodríguez Schettino and de Queiroz, 2002b); mean body temperature at Soroa 32.3 °C (J. B. Losos, in litt., 20.VIII.2007). **Vision:** Moderately critical flicker-fusion frequency corresponded with an association with semi-open habitats in Cuba, although not in Jamaica, where tested specimens were collected, suggesting a relatively recent colonization of Jamaica and a lack of adaptation to local conditions (Jenssen and Swenson, 1974).

Conservation Status: Entire species not assessed, but *A. s. greyi* is considered "vulnerable" (L. Rodríguez Schettino and A. Chamizo Lara in Rodríguez Schettino, 1999a).

Anolis schwartzi Lazell, 1972

Distribution: St. Christopher Bank, essentially islandwide on St. Eustatius, St. Christopher, and Nevis; elevational range, sea level to >900 m.

Habitat: Xerophilic to mesophilic (Schwartz and Henderson, 1991); in rainforest and xeric scrub, often most abundant in wet, shady pockets, on bushes, fence posts, rocks, stone walls, and buildings or other artificial situations, usually in shade, but occasionally in open sunny areas (Lazell, 1972; Fläschendräger and Wijffels, 1996, 2009; Powell et al., 2005); perches on trees and shrubs to 90 cm above the ground, females and juveniles more terrestrial, but all ages and sexes forage largely on the ground (Lazell, 1972); mean perch height on St. Christopher 0.12 ± 0.06 m, on St. Eustatius 0.03 ± 0.03 m (Roughgarden et al., 1983a); on St. Christopher, mean perch height and diameter 0.59 ± 0.05 m and 7.1 ± 2.9 cm (Losos and de Queiroz, 1997a); appears to favor rock piles or slides when available, when in short supply, those available dominated by large males, perch heights (23.5 ± 2.1–65.3 ± 11.3 cm) and diameters (6.7 ±

0.7–17.5 ± 3.0 cm) varied considerably by site, and adult males perched higher (46.3 ± 4.5 cm) and on perches of larger diameter (12.2 ± 1.3 cm) than females and subadult males (24.3 ± 1.9 cm and 10.3 ± 1.2 cm) or juveniles (18.3 ± 2.4 cm and 8.5 ± 2.5 cm), most individuals were on rocks or on the ground at sites where rocks were abundant, fewer lizards were on the ground at sites where the only available perches were trees and deadfall, vertical (up and down) and horizontal orientation at about equal frequencies, nearly all animals in full or partial shade, perch heights significantly lower than those of sympatric *A. bimaculatus* (Medina Díaz et al., 2005); perch height increases significantly with increasing elevation (and with decreasing abundance of *A. bimaculatus*) but is significantly lower than that of *A. bimaculatus* at all elevations where the latter is present (Buckley and Roughgarden, 2005b).

Abundance: On St. Christopher, increases in abundance from sea level to midelevations and declines at higher elevations (Buckley and Roughgarden, 2005a). **Activity:** An hour or two after dawn to dusk (Adolph and Roughgarden, 1983); exploits the "night-light niche" (Powell et al., 2005; Perry et al., 2008). **Behavior:** Retreats into holes and crevices in rocks, tree trunks, under shutters of buildings, or into debris on the ground (Powell et al., 2005); interacts frequently with *A. bimaculatus* (males display toward one another), *A. schwartzi* not always the "loser" (Lazell, 1972). **Competition:** Presence did not alter growth of, perch heights selected by, or volumes of developing eggs or prey contents in stomachs of *A. bimaculatus* in enclosures (Pacala and Roughgarden, 1982, 1985; Roughgarden, 1995); in enclosures that limited perch heights to 1.5 m, *A. schwartzi* used perches in hotter microclimates and were active at different times of day when *A. bimaculatus* was present, but growth, egg production, and prey selection were not affected (Rummel and Roughgarden, 1985; Roughgarden, 1995). **Ecomorphology:** Closest to "grass-bush" (Beuttell and Losos, 1999). **Foraging and Diet:** 97.18% arthropods, 0.13% mollusks, and 0.22% plant material, 0.90 overlap with *A. bimaculatus* and 0.55 with *Ameiva erythrocephala* (Adolph and Roughgarden, 1983); when anoles (*A. bimaculatus* and *A. schwartzi*) were excluded from enclosures, forest-floor arthropods (mainly dipterans) increased 2–3-fold, large web-building spiders increased 20–30-fold, in turn causing a 25% decrease in the abundance of insects between the forest floor and the canopy, exclusion of anoles had no effect on the extent of herbivory (Pacala and Roughgarden, 1984); gastropods, mites, spiders, isopods, scorpions, beetles (adults and larvae), springtails, flies (adults and larvae), hemipterans, ants and other hymenopterans, termites lepidopterans (adults and larvae), orthopterans, thrips, and other insects, less selective of larger prey than *A. bimaculatus*, searches from lower vantage points than *A. bimaculatus*, closer, for example, to fruitflies on the forest floor, both species equally likely to take ants found at all heights (Roughgarden, 1995); mean number of prey items per stomach 18.4 ± 12.7, mean arthropodan prey length 2.3 ± 1.5 mm, with 86.7% of prey 0–4 mm in length, and 12.7% 4–10 mm in length, prey in the 4–10 mm range taken more frequently than prey caught in traps (Rummel and Roughgarden, 1985). **Home Range and Territoriality:** Home range on St. Eustatius 1–2 m^2 (Roughgarden, 1995). **Parasites:** Nematodes *Paraphyaryngodon cubensis*, *Rhabdias* sp., Physalopteridae gen. sp., and acanthocephalan cystacanths *Centrorhynchus* sp. (Dobson et al., 1992); nematode *Spauligodon anolis* (Dobson et al., 1992; Bursey and Goldberg, 1998); mean parasite species/host 0.94 in forests and 0.36 in mixed habitat on St. Eustatius (Dobson and Pacala, 1992). **Population Size and Density:** At 5 sites on St. Christopher, 29.1 ± 1.9–76.3 ± 8.8 lizards/100 m^2 (Roughgarden et al., 1983); on St. Eustatius, 4,780–9,850 lizards/ha (Roughgarden, 1995); on St. Eustatius, density estimates at 6 sites were 653–6,354 and 729–6,812 lizards/ha using 2 different methods (Medina Díaz et al., 2005). **Predation:** *Alsophis rufiventris* (Henderson and Sajdak, 1996), including eggs (Heinz et al., 2005); *A. bimaculatus* (Fläschendräger and Wijffels, 1996, 2009); American Kestrel (*Falco sparverius*) and Pearly-eyed Thrasher (*Margarops fuscatus*; McLaughlin and Roughgarden, 1989; Powell et al., 2005); possibly *Ameiva erythrocephala* (Powell et al., 2005). **Reproduction:** Egg (on St. Christopher) 10.0 × 5.7 mm (Lazell, 1972); copulation in June (RP); lays eggs in leaf litter (Heinz et al., 2005). **Size:** Male SVL to 52 mm (Dobson et al., 1992), female SVL to 43 mm, smallest mature male and female SVL 34 mm and 37 mm, respectively (Lazell, 1972). **Thermal Biology:** On St. Christopher, mean body temperature 33.3 ± 0.4 °C in coconut habitat (Roughgarden et al., 1981); on St. Christopher, mean microclimates used by *A. bimaculatus* and *A. schwartzi* are statistically indistinguishable at low-elevation (<300 m) sites, the difference between perch microclimate and mean available microclimate is greater for *A. schwartzi* at elevations of 300 m, where it is essentially solitary, than at elevations <300 m (Buckley and Roughgarden, 2005b).

Conservation Status: Not assessed.

Additional Pertinent Literature: Hensley et al. (2006).

Anolis scriptus Garman, 1887

Distribution: Four currently recognized subspecies occur on the Bahama Islands and the Turks and Caicos Islands (Great Inagua, Little Inagua, Mayaguana, Plana Cays, Caicos, and Turks island banks and associated cays with any appreciable vegetation).

Habitat: Xerophilic (Schwartz and Henderson, 1991); on trees (*Conocarpus erecta*), palms, palmettos, major trunks in brush piles, stumps, stone walls, fences, buildings in more or less exposed locations, but rarely in thornscrub (Laska, 1970; Schwartz and Henderson, 1991); abundant foraging in roadside debris (Rabb and Hayden, 1957); shows "ecological release" by largely excluding other anoles from relatively large islands geographically close to a rich major source island (Williams, 1969); on Inagua and Sheep Cay, 59% of adult males perch 1.8–3.0 m above the ground, 68% of adult females 0.9–1.5 m above the ground, with perch diameters usually (70%) 1.3–7.6 cm, juveniles usually (74%) below 0.9 m, sometimes on the ground or in leaf litter (Laska, 1970); shift from terrestrial habitats to arboreal substrates early in the day (~0700 h) and back in the evening (~1800 h; G. R. Smith, 1994); most often on trees (61.3% of observations), and rocks (24.5%), rarely on walls (5.7%), bare sand (4.7%), or litter (1.9%), more frequently in full shade (50%) than in sun/shade mosaic (13.8%) or full sun (36.2%), use of terrestrial habitats decreased when *Leiocephalus psammodromus* was active (G. R. Smith, 1995); sleeps on leaves of shrubs, trees, and on debris 0.45–1.7 m above the ground (Schwartz and Henderson, 1991).

Activity: Bimodal with peaks after dawn and before dusk, few lizards active at midday (G. R. Smith, 1994). **Ecomorphology:** Descended from "trunk-ground" stock, classified as a "generalist" (Losos and de Queiroz, 1997a); "trunk-ground" (Brandley and de Queiroz, 2004; Losos, 2009). **Parasites:** Nematodes *Moaciria alvarengai*, *Parapharyngodon cubensis*, *Physaloptera squamatae*, larval *Abbreviata* sp. (Goldberg et al., 1996a), and *Spauligodon anolis* (Goldberg et al., 1996a; Bursey and Goldberg, 1998). **Predation:** *Tropidophis canus* (Noble and Klingel, 1932); Barn Owl (*Tyto alba*) on Mayaguana (Buden, 1974); feral cats (Iverson, 1978); *Tropidophis greenwayi* (Iverson, 1986); *Cubophis vudii* (Henderson and Sajdak, 1996). **Size:** Male SVL to 76 mm, female SVL to 64 mm (Schwartz and Henderson, 1991). **Thermal Biology:** 70% of observations in early morning and 75% in late afternoon were in full sun, primarily in full shade or sun/shade mosaics during the rest of the day (G. R. Smith, 1994).

Conservation Status: Not assessed.

Anolis semilineatus Cope, 1864

Distribution: Hispaniola, essentially islandwide, Île de la Tortue, Île Grande Cayemite, Île a Cabrit, Isla Catalina; elevational range −20–1,700 m.

Habitat: Generally mesophilic, but xerophilic to some extent (Schwartz and Henderson, 1991); in coastal zones and pine forests, generally in dry grassy areas but also in moister meadows and fields, cultivated areas (corn), and in mangroves, absent only in extremely xeric situations like thornscrub (Mertens, 1939); on grasses, but also on ferns, shrubs and bushes, corn, and mangroves, sometimes on fence posts (mostly large males facing downward with necks bent so heads are almost horizontal) and strands of barbed wire (Williams and Rand, 1961; Schwartz and Henderson, 1991); 39.3% on leaves, 6.1% on grass blades, 11.3% on stems 3 mm in diameter, 21.2% on stems 3–6 mm in diameter, 6.2% on stems 6–12 mm in diameter, and 10.0% on rocks (Moermond, 1979a); 2.0% on leaves, grasses, or perches <0.7 cm in diameter at heights 31–90 cm, 43.3% on leaves or grasses, 34.6% on perches <0.7 cm in diameter, 2.9% on perches 0.7–2.5 cm in diameter, and 17.3% on rocks at heights <30 cm (Moermond, 1979b); in dense herbaceous vegetation (mostly Rosaceae) in a clearing along a forest path at heights of ~1 m (Fläschendräger, 1998); in abandoned pastures, gardens, cacao, as well as forest and mogote vegetation (Glor et al., 2001); sleeps with hindlimbs fully extended 0.1–2.4 m above the ground on low shrubs, a tuft of pine needles on a fallen branch, an exposed leafless twig, low dead vines, *Zamia* (Cycadaceae) and *Bryophyllum* (Crassulaceae) leaves, grasses, ferns, and *Cocos* trash piles (Williams and Rand, 1961; Schwartz and Henderson, 1991).

Abundance: Uncommon, 58 of 990 observations of anoles in and around Los Haitises National Park, Dominican Republic (Glor et al., 2001). **Activity:** Most active in the morning (62.5%), but activity periods vary by elevation (Hertz, 1979a). **Behavior:** Sleeps adpressed on grasses with hindlimbs extended, escapes by jumping into grass cover (Mertens, 1939); when chased into a rock pile, stays close to the surface (Williams and Rand, 1961); jumps to avoid threats, then adpresses itself tightly to narrow stems or blades of grasses (Fläschendräger, 1998); percent jumps (fraction of total movements) 60% at both of 2 Haitian sites, jumps more frequently than runs or crawls (Moermond, 1979a); percent crawls 15%, runs 25%, jumps 60% (Moermond, 1979b). **Ecomorphology:** "Grass-bush" (Williams and Rand, 1961); formally quantified (Beuttell and Losos, 1999). **Foraging and Diet:** Scanning densely clustered habitats takes more time than scans of open clusters or broad surfaces but less time than scans of the ground from elevated perches, resulting in relatively short waiting intervals and distances/move, mean 4.1 min between location changes, mean 3.4 min between location or posture changes, mean distance/move 12 cm, distance moved per prey capture shorter than those of anoles surveying open habitat clusters or broad surfaces and much shorter than those ranging from elevated perches to the ground, 50% of prey captures on same surface as that on which lizards perched (Moermond, 1979b). **Longevity:** In captivity, 3 years, 2 months and still alive at last report in 1992 (J. Grizzell in Snider and Bowler, 1992). **Predation:** *Hypsirhynchus parvifrons* (Henderson et al., 1987a); *Uromacer catesbyi, U. frenatus, U. oxyrhynchus* (Henderson et al., 1987c). **Population Size and Density:** Six adult males in an area of 3–4 m^2 without evidence of territoriality (Mertens, 1939). **Reproduction:** Clutch size 1 (Hedges, in prep.); semicommunal egg deposition, in one small area, only 3 of 22 logs hid eggs, 1 with 12, 1 with 7, and 1 with 4 eggs, eggs scattered under logs with nests of large black stinging ants (Rand, 1967c); 2 eggs under rocks on 6 September hatched 13 September (Schwartz and Henderson, 1991); egg immediately after deposition ~5 × 7 mm, shortly before hatching ~7 × 10 mm, incubation 40–45 days at 23–28 °C (Fläschendräger, 1998). **Size:** Male SVL to 40 mm, female SVL to 36 mm (Schwartz and Henderson, 1991). **Thermal Biology:** 71.9% active in sunny weather, thermally passive despite living in open habitats, mean body temperature on sunny days 32.3 ± 0.4 °C at mean air temperature 30.0 ± 0.4 °C, on cloudy days 30.7 ± 0.8 °C at mean air temperature 28.5 ± 1.1 °C (Hertz, 1979a); heat resistance declines with increasing elevation, mean critical thermal maxima 39.6 ± 0.2 °C (5 m elevation), 38.2 ± 0.3 °C (535 m), and 37.1 ± 0.2 °C (1,190 m), variation much greater than in Puerto Rican *A. krugi* and *A. pulchellus* (Hertz, 1979b).

Conservation Status: Not assessed.

Anolis sheplani Schwartz, 1974

Distribution: Hispaniola, Sierra de Baoruco (Dominican Republic); elevational range 820 to >1,000 m.

Habitat: Mesophilic (Schwartz and Henderson, 1991); mesic deciduous forest modified by coffee and cacao plantations (Schwartz, 1974b; Schwartz and Henderson, 1991; Huyghe et al., 2007); at mean perch heights and diameters 2.24 ± 0.28 m and 4.07 ± 1.03 cm for males and 2.08 ± 0.17 m and 3.85 ± 0.74 cm, respectively (Huyghe et al., 2007); sleeps 0.9–4.3 m above the ground (probably higher) exposed on bare vines and twigs within and beneath dense curtains and mats of vines (Schwartz, 1974b; Schwartz and Henderson, 1991).

Behavior: Not easily disturbed, but may move unhurriedly if pressed at night, usual response to nocturnal annoyance is clutching the substrate more tightly without opening eyes (Schwartz, 1974b); when disturbed by day, slowly "squirrels" to the other side of a perch and holds on tightly, moves slowly, never jumps from one perch to another, "eager" to bite when captured, presumably assertive display of one recently released individual included elevated head, extended dewlap, erected nuchal crest, extended forelimbs, and extended tail, with only the tip curled (Huyghe et al., 2007); moves per minute 0.58 (Johnson et al., 2008). **Diet:** Beetles in all stomachs with contents and were the most important food item (Huyghe et al., 2007). **Ecomorphology:** "Twig" (Williams, 1983); formally quantified (Beuttell and Losos, 1999). **Performance:** Sprint speeds low (mean 3.54 ± 0.60 cm/sec for females, 3.97 ± 0.84 cm/sec for males), with lizards appearing to "walk" rather than run at even the highest recorded speeds, bite force positively correlated with body size and head length, height, and width (Huyghe et al., 2007). **Reproduction:** Clutch size 1 (Hedges, in prep.). **Size:** Male SVL to 41 mm, female SVL to 40 mm (Schwartz, 1974b).

Conservation Status: Not assessed.

Additional Pertinent Literature: Gifford and Powell (2002b).

Anolis shrevei Cochran, 1939

Distribution: Hispaniola, Cordillera Central (Dominican Republic); elevational range 1,550–2,500 m.

Habitat: Xerophilic (Schwartz and Henderson, 1991); in upland pine forests and savannas, where not uncommon, although rarely seen abroad and active because of frequent cool, rainy, and overcast days, under logs, rocks, and boards and debris at abandoned sawmill sites (Hertz and Huey, 1981; Schwartz, 1989).

Ecomorphology: "Trunk-ground," formally quantified (Glor et al., 2003). **Reproduction:** Stage of embryonic development before eggs are deposited is advanced in a population at 2,200 m (Huey, 1977); clutch size 1 (Hedges, in prep.). **Size:** Male SVL to 60 mm, female SVL to 50 mm (Schwartz, 1989). **Thermal Biology:** Postpones emergence and retreats under rocks or logs during cloudy periods, invariably basks during sunny weather, body temperatures 23.6–32.4 °C (mean 29.2 °C) at air temperatures 19.0–25.2 °C (mean 22.1 °C; Hertz and Huey, 1981).

Conservation Status: Unknown; effects of habitat destruction have not been assessed.

Anolis singularis Williams, 1965

Distribution: Hispaniola, Massif de la Hotte, Massif de la Selle, and Île de la Gonâve (Haiti), Sierra de Baoruco (Dominican Republic); elevational range 440–2,030 m.

Habitat: In bromeliads 3.0 m above the ground in pines (Schwartz and Henderson, 1991); high on trunks and into the canopy (Glor, 2003); perch height and diameter for one individual 3.00 m and 8.00 cm (J. B. Losos in Calsbeek et al., 2006).

Behavior: Very wary, at first disturbance scampers up the trunk into the canopy, may seek refuge in clusters of pine needles (Glor, 2003). **Ecomorphology:** "Trunk-crown" (Rand and Williams, 1969). **Predation:** Hispaniolan Trogon (*Priotelus roseigaster*; Fernández, 2007). **Reproduction:** Clutch size 1 (Hedges, in prep.). **Size:** SVL to 45 mm in both males and females (Schwartz and Henderson, 1991).

Conservation Status: Unknown.

Anolis smallwoodi Schwartz, 1964

Distribution: Three currently recognized subspecies occur in eastern Cuba; elevational range, sea level to 1,020 m.

Habitat: Mesophilic, but often in xeric areas (Schwartz and Henderson, 1991); on large trees, generally at heights ~4.5 m above ground, at times much lower (Lando and Williams, 1969); often tolerant of habitats modified by human activity (Garrido, 1975a; Kunz, 2005); in dry coastal habitats associated with small-leaved thorntrees at heights to 5 m (Fläschendräger and Wijffels, 1996, 2009); rare in dry coastal zones where few large trees remain (Fläschendräger and Wijffels, 2009); in submontane rainforest, submontane mesophyllous evergreen forest, mesophyllous evergreen and semideciduous forest, mangrove forest, and cultivated areas and pastures (Rodríguez Schettino et al., 1999).

Activity: Diurnal, descends from high in trees around midday (Lando and Williams, 1969). **Behavior:** Not docile or timid, 2 males on adjacent trees displayed toward one another, then descended to ground and fought, when disturbed, climbed a tree and jumped from tree to tree (Lando and Williams, 1969); "squirrels" and then climbs upward to escape, may jump to another tree, extends dewlap, opens the mouth, and extends and flexes limbs in aggressive displays, may resort to attacks, 2 males 10 m apart, displayed, descended to the ground, and attacked one another (Rodríguez Schettino, 1999b); often remains motionless in small-leaved foliage of scrubby trees where mottled light renders them cryptic, but, if approached closely, rapidly flees, usually upward (Rogner, 2002). **Ecomorphology:** "Crown-giant" (Williams, 1972). **Foraging and Diet:** In captivity (at Guantanamo Naval Base), ate flowers (*Hibiscus*), lettuce, apple slices, small pieces of bloody beef, raw chicken liver, other anoles, and House Mouse (*Mus*), but rejected mangoes, papayas, guavas, *Opuntia* fruits, and a dove egg, feeding very deliberate, each item scrutinized with one eye, then the other, food seized and shaken vigorously, crushed, then swallowed (Lando and Williams, 1969); *Osteopilus septentrionalis*, handled for 2 h prior to swallowing, during which the lizard shook the prey vigorously and hit it against the branch (Fong Grillo and Blanco Ojeda, 2002). **Home Range and Territoriality:** Have large home ranges, 2 individuals almost never together (Fläschendräger and Wijffels, 1996, 2009). **Longevity:** In captivity, 3 years, 11 months and still alive at last report in 1990 (Aquarium/Reptile Curator, Milwaukee County Zoo in Slavens and Slavens, 2003). **Parasites:** Larval nematodes *Porrocaecum* sp. (Coy Otero and Baruš, 1979b; Coy Otero and Lorenzo Hernández, 1982; Coy Otero, 1999). **Size:** Male SVL to 190 mm (Schwartz and Garrido, 1972), female SVL to 177 mm (Fong Grillo, 2002). **Thermal Biology:** Presumably heliothermic (Rodríguez Schettino, 1999b).

Conservation Status: Unknown.

Anolis smaragdinus Barbour and Shreve, 1935

Distribution: Two currently recognized subspecies occur in the Bahama Islands (Conception and Grand Bahama island banks); populations on Abaco and possibly Grand Bahama are introduced.

Habitat: Xerophilic (Schwartz and Henderson, 1991); on Bimini, in clusters of Australian Pine (*Casuarina*) needles to 3.0 m above the ground, on palms, agaves, and thick-leaved shrubs (Oliver, 1948); also on Bimini, most abundant in *Coccothrinax* scrub and incipient blackland, those on elevated perches (no individuals on ground) frequently (50 of 78 observations) on leaves, particularly large leaves of *Coccothrinax* or other palms, or straddling a cluster of smaller leaves, largest percentage of adult males (18%), females (adults and subadults, 42%), and juveniles (41%) on leaves at heights 1.0–1.6 m high (Schoener, 1968); on Exuma Island, mean adult male and female perch heights 2.2 ± 1.04 m (0.8% on ground) and 1.7 ± 0.86 m (0.0% on ground; Lister, 1976); mean perch height and diameter 1.8 m and 4.7 cm, respectively (Losos and Irschick, 1996); 81% of observations on Bimini at heights <2 m (Mattingly and Jayne, 2004); mean heights on Bimini 161 ± 14 cm (Mattingly and Jayne, 2005).

Activity: Diurnal, most active at midday on sunny days, disappearing during cold waves or on very cloudy days (Schwartz and Henderson, 1991). **Behavior:** Shy and retiring, begins to retreat as soon as an intruder is seen, "squirrels" to the other side of a limb and moves higher in a tree or shrub (to >9 m above the ground) as an escape tactic (Oliver, 1948); tends to jump rather than run on perches of small diameter (Losos and Irschick, 1996); 8.1% of distance traveled on trunk bases, 13.2% on trunks, 78.7% on branches, 11.0% of movements with jumps, mean movement distance 9.7 cm (median 5.0 cm) at speed 2.6 cm/sec (1.6 cm/sec), mean pause duration 30.3 sec (4.2 sec; Mattingly and Jayne, 2004); during escapes choose broadest diameter 90%, longest length 25%, shallowest incline 38%, and straightest angle 87% (Mattingly and Jayne, 2005); moves per minute 1.53 (Johnson et al., 2008). **Competition:** On small Bahamian islands near Great Exuma, many propagules of 5 lizards went extinct within 3 years, but not at a higher rate on islands where *A. sagrei* also was present, however, in the presence of *A. sagrei*, *A. smaragdinus* population densities were lower and the latter tended to perch higher relative to available vegetation, *A. smaragdinus* had little effect on *A. sagrei* over the 3 years of monitoring (Losos and Spiller, 1999). **Ecomorphology:** "Trunk-crown" (Williams, 1969). **Foraging and Diet:** On Bimini, food items were ants, other hymenopterans, aphids, leafhoppers (greatest numbers and volume in juveniles), other hemipterans, flies (greatest numbers and volumes in adults), thrips, beetles, barklice, lepidopterans (adults and larvae), orthopterans, lacewings, spiders, snails, and plant matter, with larger individuals tending to take larger prey (Schoener, 1968); on Exuma Island, stomach contents were hymenopterans (but no ants), flies, beetles (adults and larvae), leafhoppers (96.3%), other hemipterans, lepidopteran larvae, and mites (Lister, 1976); males sometimes "ambush" prey flying to the tips of twigs, both males and females crawl slowly about vegetation and search diligently for potential prey, most taken from leaves, feeds most actively at midday (Schoener, 1979); *Anolis sagrei* (J. B. Losos in Gerber, 1999). **Growth:** On Bimini, females grow more rapidly than *A. sagrei*, suggesting less food pressure on arboreal species (Schoener, 1979). **Home Range:** Males may fight viciously and in highly stylized fashion, but home ranges overlap extensively, perhaps because ranges are large (compared to *A. sagrei*) and in such dense foliage that they are less defensible, female ranges overlap to some extent with each other and with subadult males (Schoener, 1979); male home ranges large and overlapping, home ranges of adult males (10.92 ± 6.80–26.34 ± 34.56 m^2 at 2 Bahamian sites) larger than those of females (2.81 ± 3.32–29.08 ± 24.07 m^2; Schoener and Schoener, 1982b). **Injuries:** Injuries to tails, front toes, and back toes about equal in a small sample of males and females, with front toe rates presumably due to intraspecific aggression and tail and hind toe rates presumably due to predation, tail-break rates unusually low compared with other Bahamian anoles examined (Schoener and Schoener, 1980b). **Movements:** Lizards introduced onto a small islet dispersed across 17.7 m with no vegetation and 49.8 m between patches of vegetation normally used by the species, both the original and dispersed populations were exterminated by a hurricane (Losos and Spiller, 2005). **Parasites:** On North Bimini, protozoan malarial parasite *Plasmodium floridense* (Telford, 1975; Ayala, 1977, 1978). **Performance:** Average gross speed over a 5-stride interval on a 4.8-cm diameter perch 66 ± 12.3 cm/sec and 45 ± 3.6 cm/sec with a 90° turning angle, stride-one speed 57 ± 10.0 cm/sec and 34 ± 2.0 cm/sec with a 90° turning angle (Mattingly and Jayne, 2005). **Population Size and Density:** Male densities in various Bahamian habitats 0.025–0.074 individuals/m^2, female densities 0.018–0.038 individuals/m^2 (Schoener and Schoener, 1978b); on Bimini, in "short-blackland" habitat, 0.018 males/m^2, 0.027 females/m^2, in "forest-path" habitat, 0.052 males/m^2, 0.019 females/m^2, in "*Terminalia*" habitat, 0.049 males/m^2, 0.013 females/m^2 (Schoener, 1979); in various habitats at multiple sites on 2 Bahamian islands, female densities are 0–0.084 lizards/m^2, males densities 0–0.075 lizards/m^2, total densities 0–0.142 lizards/m^2 (Schoener and Schoener, 1980a, 1980b). **Sex Ratio:** In various habitats at multiple sites on 2 Bahamian islands F/M 0.71:1.50 (excluding data with sample sizes of 3 for one or both sexes; Schoener and Schoener, 1980a). **Size:** Male SVL to 64.1 mm, female SVL to 51.3 mm (Oliver, 1948). **Survival:** Annual survival in 4 habitats on 2 Bahamian islands 0 for males and 0–0.333 for females (Schoener and Schoener, 1980b); relatively constant age-specific mortality, percent survival after 12 months of all individuals on Andros and Bimini 0% (Schoener and Schoener, 1982a). **Thermal Biology:** In an open situation on Exuma Island, mean body and ambient temperatures 34.2 ± 1.85 °C and 28.6 ± 0.62 °C (Lister, 1976).

Conservation Status: Unknown.

Remarks: This species often is assigned to *A. carolinensis* in the older literature.

Anolis spectrum Peters, 1863

Distribution: Cuba, north central region.

Habitat: Mesophilic (Schwartz and Henderson, 1991); forests in lowland karst regions, primarily terrestrial, forages on the ground or no more than 0.6 m above the ground on small leafy or dry twigs, low shrubs, grass, and small bushes, seeks refuge in foliage of leafy trees (Ruibal, 1964; Garrido and Schwartz, 1972); in mesophyllous evergreen and semideciduous forest (Rodríguez Schettino et al., 1999); on small bushes in the herbaceous stratum of semideciduous and secondary forests, perch at heights <1 m, with some individuals in the leaf litter (Rodríguez Schettino, 1999b).

Activity: Diurnal (Rodríguez Schettino et al., 1999). **Behavior:** Moves largely by jumping, a female jumped from grass into the dry twigs of a small plant about 0.6 m high (Garrido and Schwartz, 1972); jumps and then remains motionless, difficult to see in dense grass (Garrido, 2003); some individuals escape by running (Rodríguez Schettino, 1999b). **Ecomorphology:** "Grass-bush" (Williams, 1961; Losos, 2009); categorized as a "grass anole" (G. Peters, 1970). **Home Range and Territoriality:** Apparently not highly territorial, with no defined home ranges, which explains apparent syntopy with other grass anoles (G. Peters and Schubert, 1968). **Reproduction:** Mean duration of copulation 11.7 min (longest 19 min), separation of pair "somewhat hasty," egg 6.5 × 4.1 mm, pure white and oval (Garrido, 1980a). **Size:** Male SVL to 42.1 mm (Rodríguez Schettino, 1999b), female SVL to 40 mm (Garrido and Schwartz, 1972). **Thermal Biology:** Apparently nonheliothermic (Rodríguez Schettino, 1999b).

Conservation Status: Considered "vulnerable" (Rodríguez Schettino and Chamizo Lara, 1998a).

Anolis strahmi Schwartz, 1979

Distribution: Hispaniola, where two currently recognized subspecies occur on the northern face of the Sierra de Baoruco (*A. s. strahmi*) and to the south of that range at lower elevations (*A. s. abditus*); elevational range, sea level to 900 m.

Habitat: Xerophilic (Schwartz and Henderson, 1991); largely saxicolous, inhabits crevices, crannies, and solution holes in exposed road cuts and cliff faces, sunning on these substrates and retreating when alarmed, foraging on vegetation close (1 m) to cliff faces, a juvenile in the basal rosette of dead *Agave* leaves on a cliff (F. G. Thompson in Schwartz, 1979e; Schwartz, 1979e; Glor, 2003; Glor et al., 2003).

Behavior: Not easily alarmed (relying on crypsis) and recovering quickly (10 min), escape tactics include jumping from a cliff face and remaining motionless in grassy and herbaceous ground cover (Schwartz and Henderson, 1991). **Diet:** Includes caterpillars (Schwartz and Henderson, 1991). **Ecomorphology:** "Trunk-ground" (Williams, 1983); formally quantified (Beuttell and Losos, 1999). **Predation:** *Uromacer catesbyi* actively foraging, and *Epicrates striatus* found in an area with abundant *A. strahmi* (Schwartz, 1979e). **Reproduction:** Clutch size 1 (Hedges, in prep.). **Size:** Male SVL to 79 mm, female SVL to 64 mm (Schwartz, 1979e).

Conservation Status: Unknown.

Anolis stratulus Cope, 1861

Distribution: Puerto Rico Bank, essentially islandwide on Puerto Rico, including some satellites, Isla Vieques, Isla Culebra, and throughout the U.S. and British Virgin Islands; elevational range, sea level to at least 945 m.

Habitat: Xerophilic to mesophilic (Schwartz and Henderson, 1991); more common on trees than fence posts (K. P. Schmidt, 1928); usually on large-diameter perches at 0.9–3.0 m (Rand, 1964a; Perry, 2005), although much more abundant at heights of 8–24 m in the canopy at El Verde, most on perches of small (10 cm) diameter (Reagan, 1992, 1996a); on St. John (U.S. Virgin Islands), from mangroves at sea level to hilltops (Grant, 1932g); on Puerto Rico, along a secondary road edge, most adult males on perches 6.4–12.5 cm at heights of 1.5–3.0 m (17%) or 3.0–6.0 m (17%), most "small" individuals at heights 1.5–3.0 m with diameters 2.2–5.7 cm (13%) or 6.4–12.5 cm (13%), in open forest, most adult males (22%) on perches 3.0–6.0 m high with diameters of 6.4–12.5 cm, most "small" individuals (19%) 3.0–6.0 m high on perches with diameters 2.2–5.7 cm, in forest with cleared understory, all adult males on perches 3 m high and with diameters 6.4 cm, most "small" individuals on perches with diameters 6.4 cm at heights <3 m (36%) or 3.0–6.0 m (18%), in forest interior, most adult males (32%) on perches 3.0–6.0 m high with diameters 6.4–12.5 cm, most "small" individuals" (20%) on perches <3 m high with diameters 2.2–5.7 cm, with none on rocks or the ground at any site (Schoener and Schoener, 1971b); also on Puerto Rico, mean adult male and female perch heights 3.5 ± 1.20 m (none on the ground) and 2.5 ± 1.08 m (none on the ground; Lister, 1976); at 3 sites at El Yunque, juveniles 1–2 m high on perches 2–16 cm in diameter at 2 sites and higher (>3 m) on perches of small to moderate (1–8 cm) diameters at the third site (with larger trees), adults of both sexes mainly 1–3+ m high on perches of small to large (2–16+ cm) diameters, in dense forest at higher elevations, occupied shaded to slightly unshaded microclimates (few sunny spots available), in varied situations at intermediate elevations, and mostly (66%) in sunny situations at low elevations (García Moll, 1978); at El Verde in July, mean perch heights, diameters, and insolation for adult males and females 278.2 ± 179.9 cm, 18.8 ± 10.8 cm, 865 ± 703 lux and 226.9 ± 116.5 cm, 16.4 ± 8.9 cm, 920 ± 805 lux, in January, 330.3 ± 195.8 cm, 14.8 ± 7.4 cm, 700 ± 376 lux and 283.9 ± 129.0 cm, 13.4 ± 5.8 cm, 668 ± 367 lux, with both males and females using a less diverse set of perch heights and diameters during the dry winter season (Lister, 1981); in the Virgin Islands, high in trees, although commonly seen on lower trunks of trees when raining (MacLean, 1982); on Puerto Rico, mean perch height and diameter 7.0 ± 1.38 m and 9.4 ± 1.91 cm (Losos, 1990a); in Tabonuco (*Dacryodes excelsa*) forest on Puerto Rico, largest percentage of lizards at heights >15 m (Reagan, 1995); on Puerto Rico, mean perch heights and diameters 10.49 ± 10.29 m and 14.2 ± 14.4 cm for males and 6.85 ± 9.22 m and 14.4 ± 25.5 cm for females (Butler and Losos, 2002); on Guana (British Virgin Islands), most frequently on tree trunks and found almost everywhere where these occur, except in the wettest forests, often at considerable heights on branches of large diameter, most frequently in the sun (less so in shade), also in trash on the ground and on grasses, especially in early morning (Perry, 2005); also on Guana, at perch heights to 8 m, but most individuals 2–4 m above the ground in partially to fully shaded situations (no individuals in <20% shade; Lazell, 2005a); mean male perch height and diameter 1.47 ± 0.14 m and 0.52 ± 0.16 m, mean female perch height and diameter 1.40 ± 0.20 m and 0.86 ± 0.34 m (VanHooydonck et al., 2007); 29.0% on leaves (J. B. Losos, in litt., 21.IV.2008); habitat light environment "partial shade" (Fleishman, 2000).

Activity: Greater number of movements/unit time during midday than in the morning or afternoon during the wet season, no such difference in the dry season, no significant differences in numbers of movements/unit time between males and females, but females more successful than males at foraging on a per movement basis (Reagan, 1986); on Guana (British Virgin Islands), peak activity in the morning, dropping at midday, and tapering off through the afternoon and evening (Lazell, 2005a); on Guana, adults show similar levels of activity throughout the day, whereas juveniles are less active during hot and dry midday periods, in general, the cooler the temperature, the more juveniles are active, presumably attributable to potential predation by *Ameiva exsul* (Nicholson et al., 2005). **Behavior:** When threatening, males protrude a "blood red" tongue, otherwise pink (Grant, 1932h); "tongue-protrude" behavior, used as a static modifier in assertive postures (Schwenk and Mayer, 1991); agile and aggressive, during aggressive displays, dark spots become evident, lateral flattening and orientation cause them to be quite "impressive" (Fläschendräger and Wijffels, 1996, 2009); cryptic *A. stratulus* selected trunks on which camouflage was most effective and exhibited shorter approach distances than less cryptic *A. cristatellus*, most common escape behavior is "squirreling" and then ascending (Heatwole, 1968a); on Puerto Rico, mean display rate (proportion of time spent displaying) 0.083 ± 0.017, percent walks 34.8 ± 6.0, percent runs 55.3 ± 6.2, percent jumps 9.9 ± 2.0, moves/min 1.84 ± 0.27, distance jumped in nature 31.3 ± 4.7 cm (Losos, 1990a); head-down posture in 18 of 28 individuals, moves/min 0.41 ± 0.07, percent time moving (PTM) 0.07 ± 0.11%, PTM decreases with increases in perch height (Cooper, 2005); mean approach distance 50 cm (Perry, 2005); perch height affects approach distance (21 of 23 with perch heights ≤ 2 m fled, but 2 of 3 with perch heights >2 m did not), flee up and "squirrel" (Cooper, 2006a); male approach distance, flight distance, and final distance 0.78 ± 0.13 m, 0.27 ± 0.03 m, and 0.92 ± 0.14 m, female approach distance, flight distance, and final distance 1.10 ± 0.30 m, 0.24 ± 0.04 m, and 1.24 ± 0.31 m (VanHooydonck et al., 2007); moves per minute 1.84 (Johnson et al., 2008). **Biomass:** At El Verde, 0.736 kg/ha (Lister, 1981), 55.7 kg/ha (Reagan, 1996a), with the latter based on population size estimates that included the canopy. **Desiccation:** Mean critical activity point (percent of original body weight at which the lizard could not right itself when turned on its back) 74.8 ± 5.89%, mean rate of water loss 0.20% body mass/h (Hillman and Gorman, 1977). **Ecomorphology:** "Trunk-crown" (Williams, 1972); formally quantified (Beuttell and Losos, 1999). **Foraging and Diet:** Sit-and-wait predator (Perry, 1999); attack launched by an immobile lizard that detected prey while scanning visually (Cooper, 2005); spiders, mites, springtails, ants (most important, >25%) and other hymenopterans, beetles (adults and larvae), cockroaches, orthopterans, earwigs (adults and pupa), lepidopterans (adults and larvae), flies, thrips, hemipterans, and berries (*Cordia polycephala*; Wolcott, 1923); ants (most frequently taken), beetles, cockroach, earwig, flies, spiders, and lizard skin ("doubtless its own"; K. P. Schmidt, 1920, 1928); "fond of the juicy pulp of the mango" (Danforth, 1925); eats ants and other small insects (MacLean, 1982); at El Verde in July, mean prey length and number of prey items/stomach 1.9 mm and 25.6 items/stomach, in January, 1.9 mm and 15.3 items/stomach, food items were ants (greatest percent biomass in both wet, July, and dry, January, seasons), other hymenopterans, hemipterans, flies, beetles (adults and larvae), lepidopterans (adults and larvae), orthopterans, spiders, and seeds/fruits (Lister, 1981); shows a "preference" for ants (Rivero, 1998); insectivorous (<10% of stomach contents noninsect material), mostly ants (60% of all prey items), hemipterans (chiefly leafhoppers) constituted 50% of food by volume, also took flies, lepidopterans, beetles, orthopterans, mites, spiders, and small amounts of plant material, mean prey size 2–3 mm (R. Garrison and D. Reagan in Reagan, 1995); in the British Virgin Islands, stomach contents included ants (most important in numbers and volume), beetles, crickets, spiders, and cactus fruit and other plant material (Lazell, 2005a); daily foraging rate 20.5 prey items/lizard during the wet season and 18.0 prey items/lizard during the dry season, average daily consumption rate 3.4 × 105 prey items/ha, mean 11.5 prey items/individual during the wet season (12.1 in males, 10.8 in females) and 8.7 prey items/individual during the dry season (6.9 in males, 10.4 in females), most prey captures during the wet season at El Verde at heights >15 m, more evenly distributed during the dry season, with most at 3 m, more moves (1,039 versus 647) in dry versus wet season, more frequent at midday than morning or afternoon, resulted in comparable numbers (27 versus 28) of prey captures (Reagan, 1986); also at El Verde, fewer prey items per stomach (8.7 versus 11.5), and lower consumption rates (17.4 versus 23.0 items/day), resulting in an estimated 3.4 × 105 prey items/ha/day (Reagan, 1996a); stomach contents were ants (most important in numbers in males and females), flies, lepidopterans (adults and larvae), beetles, leafhoppers (most important by volume in males and females), orthopterans (most important by volume in females), mites, spiders, shed skin, plant matter, and others (R. Garrison and D. P. Reagan in Reagan, 1996a); also at El Verde, take *A. evermanni* and presumably *A. gundlachi* (Reagan et al., 1996); on Guana, British Virgin Islands, adult male sipping nectar on shrub *Pedilanthus tithymaloides* (Perry and Lazell, 1997); lapping nectar from stipitate glands of *Inga versa* (Mimosoideae; Rios-López, 2004); licking a damaged mango still hanging from its stem on Puerto Rico (Sánchez Muñoz, 2008); lapping nectar of *Anthurium* sp. on St. John, U.S. Virgin Islands, and of orchids and banana (*Musa*) plants on Puerto Rico (A. J. Sánchez Muñoz, in litt., 17.III.2008); quickly congregate around a broken termite gallery on Guana, British Virgin Islands (G. Perry, pers. comm.). **Home Range:** Individuals included 1–3 large branches in each home range (Dial et al., 1994); adult and subadult home ranges 6.2 ± 1.2 m in diameter, with those of males larger than females, but only one of 26 home ranges calculated was large enough to encompass the full vertical extent of the 10–14-m canopy (Reagan, 1992, 1996a). **Hurricane Effects:** Hurricane Hugo (1989) restricted activity to within a few meters of the ground, one year later, *A. stratulus* had begun reinvading the canopy where branches remained and leaves had reappeared, but in areas of severe damage, populations remained low 4 years after the hurricane (Reagan, 1996a). **Interclass Interactions:** On Guana Island (British Virgin Islands), Antillean Crested Hummingbird (*Orthorhyncus cristatus*) attacked an anole apparently in order to protect its nest (Boal, 2008a). **Movements:** Mean moves per minute 0.94 ± 0.32 (Reagan, 1986); mean moves per minute 0.54 ± 0.49, percent time moving 1.09 ± 0.99 (Perry, 1999). **Parasites:** Trematode *Mesocoelium monas* (Cofresí-Sala, 1964); protozoan malarial parasite *Plasmodium azurophilum* (Schall and Vogt, 1993). **Performance:** On Puerto Rico, mean maximum sprint speed (fastest time over an interval of 0.25 m) 0.168 ± 0.007 sec, jump 55.1 ± 1.83 cm, cling 1.471 ± 0.119 Neutons (Losos, 1990a); acceleration 27.84 ± 1.27 m/sec, sprint speed 0.89 ± 0.04 m/sec (Vanhooydonck et al., 2006a). **Population Size and Density:** At El Verde in July and again in January, numbers and biomass for adult males and females were ~169/ha (~423 g/ha) and ~169/ha (~296 g/ha; Lister, 1981); estimated at 25,870 ± 7,005 (dry season) and 21,333 ± 6,638 (wet season) individuals/ha at El Verde (Reagan, 1986, 1992); 28,026 (95% confidence interval 23,487–35,329) and 24,145 (20,184–29,763) lizards/ha (Reagan, 1996a); also at El Verde, 0.0–33.1 individuals per Tabonuco (*Dacryodes excelsa*) crown (mean 7.3 ± 10.8 individuals/crown) = 0.00–0.99 individuals/m^2 (mean 0.21 ± 0.30 individuals/m^2 = 2,339 individuals/ha; Dial et al., 1994); 3,000–5,400 lizards/ha on Guana (British Virgin Islands; Lazell, 1991); in *Leucaena* plots on Guana with little litter 600 lizards/ha, biomass 0.63 kg/ha, in *Coccoloba* plots with ample litter 900 lizards/ha, biomass 1.00 kg/ha (Rodda et al., 2001). **Population Structure:** Adult:juvenile ratio 57:43% at El Verde (Reagan, 1996a). **Predation:** *Anolis evermanni, A. gundlachi* (Reagan, 1992); on Culebra, in the canopy on Puerto Rico, and in the British Virgin Islands, *Borikenophis portoricensis* (Rodríguez-Robles and Leal, 1993a; J. Gillingham in Reagan, 1995; Henderson and Sajdak, 1996); *Anolis cuvieri* (Dial and Roughgarden, 1995); whip scorpions (*Phrynus longipes*), tarantula (*Avicularia laeta*), and centipede (*Scolopendra alternans*; Reagan, 1995); at El Verde, *A. evermanni*, presumably *A. gundlachi*, and whip scorpions (*Phrynus longipes*), forming "food loops" (Reagan, 1996b; Reagan et al., 1996); Lizard Cuckoos (*Saurothera vieilloti*), Pearly-eyed Thrashers (*Margarops fuscatus*), Red-tailed Hawks (*Buteo jamaicensis*), and *Borikenophis portoricensis* regularly feed on anoles (Reagan and Waide, 1996). **Relative Abundance:** 0–7% of wet season and 4–17% of dry season sightings of this species, *A. evermanni*, and *A. gundlachi* at El Verde (Reagan, 1992, 1996a). **Reproduction:** Copulation in October, male crept close to female, gradually moving over her, tail bent under hers for intromission, remained perfectly still for 3–4 min, parted suddenly (R. E. Danforth in Danforth, 1925); maximal male reproductive activity March–August, some sperm production throughout the year in lowland populations, but testicular weights decrease during winter months with those in upland populations declining earlier and more rapidly, size of abdominal fat bodies is inversely correlated with testis weight (Licht and Gorman, 1975); female digging a nest cavity beneath ground litter during the wet season (Reagan, 1986); in captivity, incubation 40–50 days at 24–28 °C (Fläschendräger and Wijffels, 2009). **Sex Ratio (M:F):** Adult sex ratio 55:54 at El Verde (Reagan, 1996a). **Size:**

Male SVL to 61 mm (Schall and Vogt, 1993), female SVL to 50 mm (Lazell, 2005a). **Survivorship:** Marked individuals declined steadily from 81.5% to 27.9% of the population, mortality constant through life, mean population turnover time 526 days (Reagan, 1996a). **Thermal Biology:** Body temperatures maintained less above ambient (mean 0.7 °C, −0.7–2.8 °C) than *A. cristatellus* (mean 1.2 °C, −1.6–4.0 °C), not attributable to size, individuals exposed to cold temperatures (11.5 ± 0.5 °C) were capable of locomotor movements, but did not attempt to feed, lizards survived at a rate lower than those of high-elevation *A. krugi* and *A. evermanni* and comparable to that of *A. cristatellus*, with smaller individuals within the same species succumbing first, heat stress monitored under rising temperatures was noted before onset in *A. cristatellus* but after high-elevation (~550–640 m) *A. evermanni* and *A. gundlachi* (Heatwole et al., 1969).

Conservation Status: "Stable" (Moreno, 1991d).

Anolis terraealtae Lazell, 1964

Distribution: Two currently recognized subspecies occur on Îles des Saintes; elevational range, sea level to 309 m; introduced to French Guiana.

Habitat: In "arid lowlands typical of islands too low in elevation to stop significant rainfall," although *A. t. caryae* occurs on Terre-de-Bas, a low "snag island" capable of impeding sufficient cumulus to have a distinctly wetter climate than the other islands in the archipelago (Lazell, 1972); concentrated in stands of large trees and around buildings (Heselhaus and Schmidt, 1990); primarily in semideciduous forest, occupies all habitats except savannas (and may inhabit the few trees there), rare in xerophytic situations, often associated with human dwellings, gardens, walls, ornamental vegetation, juveniles in leaf litter, on Grand Îlet (*A. t. terraealtae*), one large *Ficus* occupied by 24 anoles, including 7 dominant males, on Terre-de-Bas (*A. t. caryae*) omnipresent, except in areas lacking trees and rocky areas at the northern end of the island (Breuil, 2002b).

Biomass: 1.79 g/m^2 on Terre-de-Bas and 1.40 g/m^2 on Terre-de-Haut (Wright, 1981). **Diet:** Eats *Sphaerodactylus* (Breuil, 2002b). **Ecomorphology:** Undetermined. **Population Density:** To 1 lizard/m^2 during the hatching season (Breuil, 2002b). **Predation:** Feral cats (*Felis*), birds American Kestrel (*Falco sparvarius*), Smooth-billed Ani (*Crotophaga ani*), primarily in open areas, in forests the principal predator is *Alsophis sanctonum* (Breuil, 2002b). **Reproduction:** Peaks July–August, hatchlings first apparent in July, most abundant in December (Breuil, 2002b). **Size:** Male SVL to 80 mm (Lazell, 1964a), female SVL to 55 mm, smallest mature male and female SVL 54 mm and 46 mm, respectively, but data are probably incomplete (Lazell, 1972).

Conservation Status: Not assessed, but protected by French legislation (Lorvelec et al., 2007).

Anolis terueli Navarro Pacheco, Fernández, and Garrido, 2001

Distribution: Cuba, from the coastal lowlands near Manzanillo (Granma Province) north and west to Camagüey Province, where it occurs both in lowland and adjacent upland habitats.

Habitat: On fence posts, on trunks of small trees, and in shrubs (*Dichrostachys glomerata*, *Gliricidia sepium*, *Cajanus indicus*) in otherwise open and often highly disturbed areas at heights 25–190 cm above the ground, one individual 25 cm high on the trunk of a Mallow (*Sida acutifolia*; Navarro Pacheco et al., 2001).

Activity: Diurnal (Rodríguez Schettino et al., 1999); most activity occurs in midmorning (~1030 h), when lizards move to the highest parts of the vegetation (Navarro Pacheco et al., 2001). **Ecomorphology:** Undetermined. **Size:** Male SVL to 40.0 mm, female SVL to 37.2 mm (Navarro Pacheco et al., 2001).

Conservation Status: Unknown.

Anolis toldo Fong Grillo and Garrido, 2000

Distribution: Cuba, only from the type locality on the uplands of Toldo, Moa, Holquín Province.

Habitat: Only known specimen on a branch at a height of 1.4 m in a small isolated patch of habitat in stratified sclerophilic lower montane forest (Fong Grillo and Garrido, 2000).

Activity: Diurnal (Rodríguez Schettino et al., 1999). **Behavior:** When handled, tries to bite constantly (Fong Grillo and Garrido, 2000; Fong Grillo, 2003). **Ecomorphology:** "Trunk-crown" (Losos, 2009). **Size:** Male SVL to 61.2 mm, female SVL unknown (Fong Grillo and Garrido, 2000).

Conservation Status: Unknown.

Anolis trinitatis Reinhardt and Lütken, 1862

Distribution: St. Vincent, essentially islandwide and many coastal cays, Chateaubelair Islet; introduced on Trinidad; elevational range, sea level to >900 m.

Habitat: Mesophilic (Schwartz and Henderson, 1991); on trees and fences in Kingstown along the waterfront (Underwood, 1959b); "ubiquitous" (Lazell, 1972; J. Schmidt, 1980a; Treglia, 2006); inhabits rainforest, *Cocos* plantations, *Theobroma* groves, pastures, lowland xeric forest, and *Musa* groves, edificarian, usually perches at heights <3 m on *Ficus*, tree trunks, shacks and buildings, ruins of an old mill, occasionally on the ground (Lazell, 1972; Schwartz and Henderson, 1991; Powell and Henderson, 2008b); usually at perch heights of 1–2 m on large tree trunks (J. Schmidt, 1980a); populations at lower elevations arboreal or on bushes, at higher elevations, more likely to associate with rocks (Fläschendräger and Wijffels, 1996, 2009); mean perch height 1.13 ± 0.06 m (Roughgarden et al., 1983a); mean perch height and diameter 1.65 ± 0.19 m and 26.6 ± 3.0 cm (Losos and de Queiroz, 1997a); may perch higher in the presence of invasive *A. sagrei* (Treglia, 2006; Treglia et al., 2008); perch heights varied widely depending on available structural habitat at 8 sites on St. Vincent, with males perching higher (mean 156.1 ± 5.4 cm, 0–660 cm, 32 of 494 on the ground) and on perches of larger diameters (20.3 ± 1.1 cm, 1–150 cm) than females and subadult males (105.9 ± 3.0 cm, 0–550 cm, 152 of 1,036; 15.9 ± 0.7 cm, 1–150 cm) and juveniles (126.3 ± 4.2 cm, 0–660 cm, 87 of 748; 17.2 ± 0.9 cm, 1–150 cm), with most individuals of all size groups perching with heads down (871 versus 444 horizontal and 481 with heads up) and most in filtered light (1,316 versus 60 in full sun and 417 in full shade; Hite et al., 2008a); at a highly disturbed construction site, mean perch heights of adult males 110.7 ± 6.2 cm, subadults and adult females 54.5 ± 3.2 cm, juveniles 57.9 ± 7.9 cm, used both "natural" and "artificial" perches, not obviously affected by presence of introduced *A. sagrei* (Treglia et al., 2008); sleeps high in crowns to which lizards move rapidly at dusk (Gorman and Stamm, 1975).

Activity: Emerges as early as 0530 h (Schwartz and Henderson, 1991); basically diurnal, but exploits the "night-light niche" on St. Vincent and Young Island (R. Powell in Perry et al., 2008). **Behavior:** Challenge display consists of very rapid tail lashes followed by rapid bobs with infrequent long pauses (Gorman, 1969); resident runs down and attacks an intruder by biting, locking jaws, breaks away and runs up, bobs rapidly, attacks again, backs away, extends dewlap halfway, rapid bob, attacks again, grabs jaws and flips to ground, but some encounters settled without biting, 2 males engaged in high-intensity displays, moved toward one another, butted heads, and one backed off (Gorman, 1968); pair sleeping on a leaf in contact with one another (C. S. Mallery, Jr., pers. comm.). **Desiccation:** Mean critical activity point (percent of original body weight at which the lizard could not right itself when turned on its back) 69.5 ± 3.27%, mean rate of water loss 0.19% body mass/h (Hillman and Gorman, 1977). **Ecomorphology:** Closest to "trunk-crown" (Losos and de Queiroz, 1997a; Beuttell and Losos, 1999). **Home Range and Territory:** 3-dimensional territory estimated at 1–2 m^3 (Gravelle and Simon, 1980). **Hybridization:** Hybridizes with *A. aeneus* on Trinidad (Gorman, 2003, and references therein). **Parasites:** Malarial parasites *Plasmodium azurophilum* (Staats and Schall, 1996b); larval nematode *Physocephalus* sp. (Goldberg et al., 1997a). **Population Size and Density:** At 2 localities on St. Vincent 5 ± 3 and 50 ± 10 lizards/100 m^2 (Roughgarden et al., 1983a); varied widely in 8 habitats (2,611–27,923 lizards/ha and 3,118–38,111/ha using 2 different methods; Hite et al., 2008a). **Predation:** Decapitated head eaten by *A. griseus* (Gorman, 1968); *A. griseus* readily takes these smaller anoles (J. Schmidt, 1980a); *Corallus cookii* (Henderson, 1993, 2002). **Reproduction:** Some reproductive activity throughout the year with only slight seasonal changes, declines in January–March (dry season) and November–December (wet season), only 8% of 295 females nonreproductive, all samples with ovigerous females, with proportions 40–85%, only 14 of 190 ovigerous females with 2 oviductal eggs, no cycle evident in female fat body weight, which usually was completely lacking (discernible only January–April, i.e., dry months), male testicular weight declines during wet winter months, with growth during January–February, when female reproductive activity is minimal, fat body weight peaks during times when testicular weight declines (driest months) and drops as testes grow (Licht and Gorman, 1970); copulation in June (Gravelle and Simon, 1980; RP); in captivity, eggs 6.5 × 9.0 mm, incubation 50–60 days at 25–28 °C, hatchling SVL ~18 mm (Fläschendräger and Wijffels, 1996, 2009); hatchling total length ~50 mm (J. Schmidt, 1980a). **Sex Ratio:** Apparently balanced (Lazell, 1972). **Size:** Male SVL to 74 mm, female SVL to 57 mm, smallest sexually mature male 45 mm SVL (Lazell, 1972) and female 41 mm SVL (Licht and Gorman, 1970). **Thermal Biology:** Erroneously reported to seek full sun throughout the day (J. Schmidt, 1980a); commonly in shaded microhabitats beneath tree crowns (Heselhaus and Schmidt, 1990; Fläschendräger and Wijffels, 1996, 2009; Hite et al., 2008a).

Conservation Status: Not assessed.

Anolis valencienni Duméril and Bibron, 1837

Distribution: Jamaica, islandwide; elevational range, sea level to ~1,220 m.

Habitat: Mesophilic (Schwartz and Henderson, 1991); in second growth forest, xeric woods, in open to heavily shaded situations on trees, bushes, and (rarely) fence posts, appears to favor trees with smooth, gray bark, occasionally in open residential areas and very dense secondary growth, never in mangroves or open pastures (Grant, 1940a; Rand, 1967b; Schoener and Schoener, 1971a; Hicks and Trivers, 1983); on perches at varying heights (17.3% <0.9 m, 30.4% 0.9–1.8 m, 26.1% 1.8–3.0 m, 26.0% >3.0 m; maximum to ~13 m), usually (65.2%) on perches of large diameter (>7.6 cm), although subjective impressions suggest a greater frequency on small-diameter perches, branches and twigs far from the trunk (Rand 1967b); uses perch diameters 0.3–2.2 cm (usually), but to 12.7 cm, and perch heights from <0.9 to >12.0 m above the ground (on average, greatest mean perch heights of common Jamaican anoles), with males perching higher than females and adults higher than juveniles (Schoener and Schoener, 1971a); forages on the ground to the outer leaves of branches to 13 m above the ground (Hicks and Trivers, 1983); mean perch diameter 6.0 ± 2.01 cm (Losos and Sinervo, 1989); mean perch height and diameter 2.3 ± 0.41 m and 6.0 ± 2.01 cm (Losos, 1990a; note that these same data are erroneously also attributed to *A. occultus* in Calsbeek et al., 2006); lizards most frequently on very small (<2 cm diameter) perches, but also on very large (>8 cm perches) with highest habitat breadth (102.4) of species examined (Irschick and Losos, 1999); mean perch heights and diameters 2.62 ± 1.66 m and 7.0 ± 21.5 cm for males and 2.18 ± 1.67 m and 2.3 ± 6.0 cm for females (Butler and Losos, 2002); mean male perch height and diameter 1.43 ± 0.31 m and 0.43 ± 0.15 m, mean female perch height and diameter 0.66 ± 0.51 m and 0.06 ± 0.01 m (VanHooydonck et al., 2007); 4.2% on leaves (J. B. Losos, in litt., 21.IV.2008); dewlap UV-reflectance compatible with visibility in typical habitat (Macedonia et al., 2000); perches primarily on branches by day and night, but uses perches of significantly smaller diameter at night, male and female sleeping sites did not differ significantly, but males perched lower and females higher than during the day, both sexes use perches closer to parallel with the ground at night (Singhal et al., 2007).

Activity: Diurnal (Schwartz and Henderson, 1991). **Behavior:** Tail carried rather sharply curled laterally (Barbour, 1920); deliberate and slow moving, relies on crypsis, spends much time adpressed to perch, rotates to side of branch opposite an observer (Grant, 1940a; Underwood and Williams, 1959; Hicks and Trivers, 1983); slowly sidles around perches to avoid detection, but will run to avoid capture (Losos and Sinervo, 1989); male threat display includes nuchal and dorsal crest erection, dewlapping, and tail wags (Ruibal, 1967); squeaks during aggressive interactions (Grant, 1940a; Milton and Jenssen, 1979); mean display rate (proportion of time spent displaying) 0.015 ± 0.007, percent walks 71.9 ± 4.9, percent runs 16.6 ± 4.1, percent jumps 11.4 ± 1.0, moves/min 1.55 ± 0.16, distance jumped in nature 28.8 ± 7.5 cm (Losos, 1990a); 65% of combat trials involve dewlap extensions, 0% biting, and 24% headbobbing (Lailvaux and Irschick, 2007); when sleeping perches are disturbed, lizards drop to the ground (Singhal et al., 2007); male approach distance, flight distance, and final distance 0.46 ± 0.15 m, 0.84 ± 0.34 m, and 0.87 ± 0.21 m, female approach distance, flight distance, and final distance 0.08 ± 0.03 m, 0.84 ± 0.32 m, and 0.44 ± 0.01 m (VanHooydonck et al., 2007); moves per minute 1.06 and 1.55 (Johnson et al., 2008). **Ecomorphology:** "Twig" (Williams and Rand, 1977); formally quantified (Beuttell and Losos, 1999). **Foraging and Diet:** Active forager on the ground, leaves, trunks, and limbs of trees, in holes in trunks, and in arboreal epiphytes (*Tillandsia*), searching for stationary, cryptic prey, which is usually within 8 cm of a lizard's head when first noticed (farthest distance 45 cm), 27 of 36 feeding attempts successful, occasionally preys on small conspecifics (Hicks and Trivers, 1983); moves slowly and appears to use an ambush foraging strategy, drinks from bromeliads in captivity (Hilgenhof and Hilgenhof, 2004); insects (Grant, 1940a); 29.2% of stomachs contained fruit, 1.57 fruits/individual (Herrel et al., 2004). **Home Range and Territoriality:** Adult females completely nonterritorial and usually nonaggressive, overlap with each other in daily movements, with as many as 40 occupying the same feeding space, males aggressive toward each other, but have broadly overlapping home ranges (Hicks and Trivers, 1983); dewlap size (secondary sexual characteristic) not correlated with bite force (related to fighting capacity) in this relatively nonterritorial species (Vanhooydonck et al., 2005); mean diurnal home range size 48.2 m^2, nocturnal home range size 37.2 m^2 (Singhal et al., 2007). **Miscellany:** Considered by residents to be the most poisonous lizard in Jamaica (Barbour, 1910). **Movements:** Females more vagile than males, mean distance per movement 14.9 ± 1.2 cm at mean speed 23.8 ± 4.4 cm/sec, 0.82 ± 0.24 moves/min, 7.15 ± 2.69% time moving (Irschick, 2000b). **Parasites:** Cestode *Oochoristica* sp., acanthocephalan *Centrorhynchus* sp., and nematodes *Thelandros* sp. (Bundy et al., 1987) and *Spauligodon anolis* (Bundy et al., 1987; Bursey and Goldberg, 1998). **Performance:** Sprint speed declines on smaller perches (Losos and Sinervo, 1989); mean maximum sprint speed (fastest time over an interval of 0.25 m) 0.153 ± 0.004 sec, jump 56.6 ± 1.52 cm, cling 3.102 ± 0.210 Neutons (Losos, 1990a); clinging ability 1.54 ± 0.26 Neutons, divided by mass 0.24, with the relatively high value apparently characteristic of twig ecomorphs (Elstrott and Irschick, 2004); mean maximal speed 1.2 ± 0.07 m/sec, escape speed 1.3 ± 0.16 m/sec, undisturbed activity 0.2 ± 0.08 m/sec, mean maximal jumping distance

56.6 ± 1.52 cm, undisturbed jumping distance 27.6 ± 6.12 cm (Irschick and Losos, 1998); sprint speed reduced when turning, 1.00 ± 0.06 m/sec with no turn, 0.74 ± 0.08 m/sec over a 30° turn (reduction to 74% of sprint speed without turns), 0.48 ± 0.01 m/sec over a 90° turn (48%), about half of individuals jumped to traverse 90° turns, but potential benefits of jumping offset by pauses when preparing to jump or recovering balance after jumping (Higham et al., 2001); peak velocity 1.70 ± 0.13 m/sec, peak acceleration 33.3 ± 3.6 m/sec^2 (Toro et al., 2004); distance jumping (e.g., escape from predators) involves a simple control strategy that can be rapidly applied and which incorporates more force during takeoffs to jump farther, jumping at different angles (e.g., movement in a complex 3-dimensional habitat) is more complex and involves a more time-consuming strategy (Toro et al., 2006); acceleration impeded but sprint speed not dramatically affected on perches of small diameter (VanHooydonck et al., 2006b); acceleration 21.38 ± 2.24 m/sec, sprint speed 0.94 ± 0.07 m/sec (Vanhooydonck et al., 2006a). **Predation:** Include conspecifics, *A. lineatopus*, *A. opalinus*, and *A. grahami* (Floyd and Jenssen, 1983; Hicks and Trivers, 1983); Greater Antillean Grackles (*Quiscalus niger*) capable of taking all size classes (Graves, 2006); *Epicrates subflavus* (Singhal et al., 2007). **Reproduction:** Clutch size 1 (Hedges, in prep.); in captivity, males approach females with headbobs and dewlap extension and grab females by the nape before intromission, copulation takes a few minutes, females can store sperm, may lay 10 eggs/year, white eggs ~7 × 11 mm, incubation 50–60 days at 27–32 °C (Hilgenhof and Hilgenhof, 2004); unsubstantiated reports of green-brown eggs with irregular surfaces and brown spots (van Straelen, 1928 in H. Eikamp, 1996 in Hilgenhof and Hilgenhof, 2004); eggs more cylindrical than those of many anoles, laid communally (2–30 eggs) in tree trunk holes from ground level to 6 m, about 100 eggs in the base of a cut tree (<0.3 m above ground) that had been filled with cement (eggs between cement and the periphery of the 3-m diameter stump; Grant, 1940a); egg laid 4 May hatched 54 days later, hatchling SVL 23 mm, tail 32 mm (Underwood and Williams, 1959); 25 eggs and 6 hatched eggshells in a tree cavity, 16 eggs in a hollow in rotten wood in the side of a large tree, eggs stuck to one another and to wood, 8 hatched eggs in an open internode of a large bamboo, 45 eggshells in a small hole in the vertical bank of a small stream (Rand, 1967c); breeds seasonally, females may copulate more than once per day during the summer, over a 6-week period, one female may copulate with 5 or more males, both males and females copulate more frequently with increased size, copulation may be aggressive, male bites female on nuchal skinfold, bites occur at intromission and are maintained throughout coitus (Hicks and Trivers, 1983); eggs hatched 5 July to 7 August (Schwartz and Henderson, 1991). **Sex Ratio (M:F):** At one site with large trees, 69:198, but biased because males are higher in trees and hard to see and catch, recapture data gave an estimate of 186:259, at a second site with fewer large trees, ratios were 35:44 in June, 92:97 in summer, and 24:31 in winter (Hicks and Trivers, 1983). **Size:** Male SVL to 80 mm (Rand, 1967b), female SVL to 65 mm (Schwartz and Henderson, 1991); smallest copulating male 54 mm SVL, smallest copulating female 50 mm SVL (Hicks and Trivers, 1983). **Thermal Biology:** Heliophilic (Fläschendräger and Wijffels, 1996, 2009; Hilgenhof and Hilgenhof, 2004). **Vision:** Moderate critical flicker-fusion frequency corresponded to an association with semi-open habitats (Jenssen and Swenson, 1974).

Conservation Status: Not assessed.

Anolis vanidicus
Garrido and Schwartz, 1972

Distribution: Cuba, Sierra de Trinidad and associated lowlands.

Habitat: Mesophilic (Schwartz and Henderson, 1991); associated with grassy areas (often adjacent to rubbish or litter piles used as retreats) within or adjacent to hardwood forest, sleeps on grasses, vines, and ferns within 1.5 m of the ground (Garrido and Schwartz, 1972); usually in the vicinity of streams (Ševčik, 1989); in submontane rainforest, submontane mesophyllous evergreen forest, and mesophyllous evergreen and semideciduous forest (Rodríguez Schettino et al., 1999); mean perch height and diameter 0.41 m and 0.2 cm (J. B. Losos, in litt., 21.IV.2008).

Activity: Diurnal (Rodríguez Schettino et al., 1999). **Behavior:** Escapes by running instead of jumping (Garrido, 2003). **Ecomorphology:** "Grass-bush," by implication (Garrido and Schwartz, 1972). **Reproduction:** Mean duration of copulation 14.3 min (longest 35 min), separation of pair "rapid," egg 8.2 × 4.3 mm, creamy yellow in color and oval (Garrido, 1980a). **Size:** Male SVL to 39 mm, female SVL to 37 mm (Garrido and Schwartz, 1972). **Thermal Biology:** Apparently nonheliothermic (Rodríguez Schettino, 1999b).

Conservation Status: Listed both as "vulnerable" and at "lower risk" of extinction (Rodríguez Schettino and Chamizo Lara in Rodríguez Schettino, 1999a); at "lower risk" of extinction (Rodríguez Schettino and Chamizo Lara, 1998a).

Anolis vermiculatus
Cocteau in Duméril and Bibron, 1837

Distribution: Cuba, Pinar del Río Province.

Habitat: Mesophilic (Schwartz and Henderson, 1991); an aquatic anole, on trunks and branches of trees (primarily *Eugenia jambos* (83.3%), but also *Roystonea regia*, *Calophyllum antillarum*, and *Hibiscus elatus*), bushes, and boulders along watercourses in broadleaf forest or pinewoods (Gundlach, 1880; Ruibal, 1964; Garrido, 1976b; González Bermúdez and Rodríguez Schettino, 1982; Rodríguez Schettino and Novo Rodríguez, 1985; Rodríguez Schettino et al., 1987; Rodríguez Schettino and Lizana, 1997; Rodríguez Schettino, 1999b, 2003); males primarily arboricolous, females and juveniles often on the ground and on rocks at water's edge (Fläschendräger, 2002); in the Biosphere Reserve "Sierra del Rosario," in gallery forest where associated with tree trunks immediately adjacent to rivers, but also on branches of trees and on the ground (Martínez Reyes, 1995); in mesophyllous evergreen and semideciduous forest (Rodríguez Schettino et al., 1999); males and females prefer shaded situations over those that are sunny or partially shaded (most individuals in shaded sites throughout the day), juveniles prefer direct sunlight, adult males perch higher (1.44 ± 0.12 m) than females (0.83 ± 0.27 m) or juveniles (0.30 ± 0.18 m), which are most frequently on the ground, juveniles with SVL greater than those of adult females behave like adult males in terms of perch heights, food, and selection of temperatures and degrees of insolation, posture either horizontal or vertical, with head either up or down (González Bermúdez and Rodríguez Schettino, 1982); adult males more arboreal than females and juveniles, frequently on large trunks to heights of 3 m (Silva Lee, 1984); males and females mostly on trunks (41 of 76 and 28 of 55, respectively), but also on rocks (16, 19), branches (13, 2), and the ground (6, 6), juveniles more frequently seen on rocks (7 of 22), trunks (6), branches (5), and the ground (4), mean perch heights and diameters for males (2.85 ± 0.23 m and 10.5 ± 1.29 cm) greater than those of females (1.06 ± 0.14 m, 7.7 ± 0.90 cm) and juveniles (0.98 ± 0.19 m, 6.2 ± 1.02 cm), with males and females more closely associated with water than juveniles (Rodríguez Schettino et al., 1987); most individuals at heights of 1–2 m above the ground or water, although some may range to 6 m high, mean distance from water for males 2.01 m, for females 0.93 m (Rodríguez Schettino, 1999b); coefficient of arboreality high (M. A. Granda Martínez in Rodríguez Schettino, 1999b); no more than 5 m from water (Sampedro Marín and Rodríguez Schettino, 2003); mean perch height at Soroa 0.86 m, diameter 12.4 cm, behavioral diameter (mean diameter of all surfaces used during behavioral observations, means of individual means) 6.68 cm, 58.5% of observations on or within 1 m of rocks (J. B. Losos, in litt., 20.VIII.2007); mean perch height and diameter 1.76 m and 15.4 cm (J. B. Losos, in litt., 21.IV.2008); sleeps, always head-down, on leaves, vines, or logs (Neill and Allen, 1957); sleeps on shrubs and rocks to 0.9 m above the ground and to 0.9 m from water's edge (Schwartz and Henderson, 1991); males sleep in high branches of trees, whereas females retreat to hollows in mud or under leaf litter (Rodríguez Schettino, 1999b).

Activity: Diurnal (Rodríguez Schettino and Lizana, 1997; Rodríguez Schettino et al., 1999); from 0830 to 1630 h, most individuals active at 1130 h, peak activity during the dry season 1100–1200 h (González Bermúdez and Rodríguez Schettino, 1982); bimodal activity with individuals most active 0800–1300 h and 1600–1700 h (Rodríguez Schettino and Novo Rodríguez, 1985); active from 0800–0900 h to 1600–1700 h, seeks shelter high in trees during heavy afternoon rains, sometimes more distant from water (Rodríguez Schettino et al., 1987); may begin to leave nocturnal sleeping sites after 0700 h during the wet season and 0800 h in the dry season, activity increases rapidly and peaks ~1100 h, decreasing thereafter, although a slight increase occurs ~1400 h during the wet season, activity ceases by 1800 h (Rodríguez Schettino and Lizana, 1997; Rodríguez Schettino, 1999b). **Behavior:** Escapes to water (Gundlach, 1880), especially when pursued or wounded (Stejneger, 1917); adult males readily jump from streamside vegetation and take refuge in water (to 0.75 m deep), often under rocks, stones, and roots, juveniles and females more likely to hide in vegetation or streamside holes (Neill and Allen, 1957; Garrido, 1976b; González Bermúdez and Rodríguez Schettino, 1982; Rodríguez Schettino and Lizana, 1997; Rodríguez Schettino, 1999b, 2003); adult males can run on water like *Basiliscus* and move on the stream bottom for as long as 30 min (Barbour and Ramsden, 1919; Ruibal, 1964; Buide, 1985); in captivity, can hold breaths for 50 min (Garrido, 1976b) or 25 min (Kaden, 1985); often "walks" with toes, especially on the hindlimbs, elevated (in effect, "walking on their heels"), may hang on vertical surfaces solely by their forelimbs (Baedermann, 1989); adult males run into crowns or jump into water to escape, aggressive displays (with elevation on all 4 limbs, nuchal crest erection, rudimentary dewlap extension, and open mouths) are common, if fights occur, males try to bite each other on the neck and overturn their opponents, do not cease if they fall into water while fighting (Rodríguez Schettino, 1999b); vocal when handled, uttering squeaks and mews (Neill and Allen, 1957; Ruibal, 1964; G. Peters, 1970; Vergner and Polák, 1996); very sensitive to sounds, escaping quickly when startled, when cornered, responds aggressively with lateral presentations and compression, "dewlap" extension, nuchal crest erection, headbobs, before leaping open-mouthed at the "enemy," simultaneously uttering a sharp squeak (Neill and Allen, 1957); "jiggles" head, mean moves per minute 0.88, mean percent time displaying 2.3, percent walking, 55.0, percent jumping 23.3, percent running 21.7 (J. B. Losos, in litt., 21.IV.2008). **Ecomorphology:** Categorized as a "gecko anole" (G. Peters, 1970); "aquatic," which is not an ecomorphological category, morphologically "crown-giant" (Leal et al., 2002); undetermined (Losos, 2009). **Foraging and Diet:** Insectivorous, roaches, craneflies, lacewings, beetles, and spiders (Neill and Allen, 1957); inclusion of plant material increases the variety of prey individuals are able to secure under given environmental conditions (Rodríguez Schettino and Novo Rodríguez, 1985); opportunistic feeder, food includes berries, fishes (*Girardinus* sp.), amphibians (*Eleutherodactylus*), reptiles (*Anolis* taken by one female), snails, insects (major food item, with ants and flies most frequently taken, but also beetles, lepidopterans, hemipterans, dragon- and damselflies, roaches, orthopterans, and various larvae), arachnids, crustaceans, with only males taking both terrestrial and aquatic prey (females and juveniles take only terrestrial prey), males take larger prey than females (Garrido, 1976b; González Bermúdez and Rodríguez Schettino, 1982; Rodríguez Schettino and Novo Rodríguez, 1985; Rodríguez Schettino and Lizana, 1997; Rodríguez Schettino, 1999b); effectively take insects (especially lepidopterans) floating on the stream surface (Fläschendräger, 2002); nestling doves (Buide, 1985). **Home Range and Territoriality:** Males highly territorial, each living on a different trunk, separated from other males by several meters (Rodríguez Schettino, 1999b). **Injuries:** Crescent-shaped scars attributable to biting only in males, most of which also had broken tails (Neill and Allen, 1957). **Parasites:** Nematodes *Cyrtosomum scelopori* (Coy Otero and Baruš, 1973a, 1979b; Coy Otero and Lorenzo Hernández, 1982; Coy Otero, 1999); *Parapharyngodon cubensis* and Nematoda gen. sp. (Coy Otero and Baruš, 1979b; Coy Otero and Lorenzo Hernández, 1982; Coy Otero, 1999); *Piratuba digiticauda* (Coy Otero and Rodríguez Schettino, 1982; Coy Otero, 1999). **Population Size and Density:** Encounter rates at Soroa 36–132 individuals/h, biomass 0.43–3.67 kg/ha (Rodríguez Schettino and Lizana, 1997; Rodríguez Schettino, 1999b). **Population Structure:** Adults:immatures (subadults and juveniles) 2.6:1, although the latter are evident during all months of the year (Rodríguez Schettino and Lizana, 1997; Rodríguez Schettino, 1999b). **Predation:** Probable predators include *Cubophis cantherigerus*, *Tretanorhinus variabilis* (Rodríguez Schettino, 1999b), *Caraiba andreae*, and *Tropidophis melanurus* (Fläschendräger, 2002). **Reproduction:** Eggs (9.0–11.2 × 15.0–16.8 mm) deposited communally in crevices of cliffs (M. Buide in Garrido, 1976b); oval egg 15.8 × 9.9 mm (Garrido, 1980a); reproduction cyclic, first gravid females in April, no oviductal eggs are present in September, during periods of peak reproductive activity, 30–60% of females contain 2 oviductal eggs, empty shells in sand on river banks, mean length of eggs deposited in captivity 19.5 mm, size of fat bodies is inversely related to reproductive activity (Rodríguez Schettino, 1999b); in captivity, incubation ~65 days at 20–28 °C (Kaden, 1985); courtship of short duration, females not infrequently injured (even killed) by males biting during copulation (Baedermann, 1989); eggs 16.0–16.5 × 11.0 mm, incubation 58–60 days at 28 °C (Vergner and Polák, 1996). **Sex Ratio (M:F):** 1.2:1 (Rodríguez Schettino and Lizana, 1997; Rodríguez Schettino, 1999b). **Size:** Male SVL to 135 mm (Vergner and Polák, 1996), female SVL to 84.9 mm (Rodríguez Schettino, 1999b); hatchling SVL 30 mm (Kaden, 1985). **Thermal Biology:** During the dry season, males (24.59 ± 0.32 °C in November, 25.36 ± 0.89 °C in February) and adult females (24.65 ± 0.32 °C, 25.60 ± 0.98 °C) prefer similar air temperatures (24.38 ± 0.25 °C and 24.25 ± 0.41 °C, 25.43 ± 0.72 °C and 24.60 ± 0.25 °C) and maintain cloacal temperatures similar to air temperatures, juveniles are heliothermic and maintain cloacal temperatures (27.53 ± 0.74 °C) above air temperatures (25.73 ± 0.33 °C), during rain, cloacal temperature is maintained higher than air temperature (González Bermúdez and Rodríguez Schettino, 1982); mean male (27.54 ± 0.49 °C), female (27.40 ± 0.62 °C), and juvenile (27.35 ± 0.39 °C) body temperatures higher than air temperatures (26.11 ± 0.23 °C, 25.64 ± 0.19 °C, and 26.15 ± 0.34 °C, respectively) during the rainy season (Rodríguez Schettino and Novo Rodríguez, 1985); mean male cloacal and air temperatures (25.7 ± 0.36 °C and 26.0 ± 0.37 °C) do not differ from those of females (26.3 ± 0.43 °C and 26.2 ± 0.36 °C), nor do body and air temperatures differ significantly (Rodríguez Schettino et al., 1987); active at body temperatures 24.7–27.9 °C, air temperatures 25.5–27.2 °C, and relative humidity 65–90% (Rodríguez Schettino and Lizana, 1997); mean body temperature at Soroa 29.0 °C (J. B. Losos, in litt., 20.VIII.2007).

Conservation Status: Considered to be at "lower risk" of extinction (Rodríguez Schettino and Chamizo Lara, 1998a, 1998b; Rodríguez Schettino and Chamizo Lara in Rodríguez Schettino, 1999a).

Anolis vescus Garrido and Hedges, 1992

Distribution: Cuba, only from the type locality on the southern face of the Sierra del Purial; elevation ~700 m.

Habitat: In compact brushy vegetation (ferns, bushes, and scattered trees) within 2 m of the ground in a relict pocket of forest adjacent to an orchard with bananas and crotus trees (Garrido and Hedges, 1992); in submontane mesophyllous evergreen forest (Rodríguez Schettino et al., 1999).

Ecomorphology: "Grass-bush" (Garrido and Hedges, 1992). **Size:** SVL of only known male 41 mm, female SVL to 37 mm (Garrido and Hedges, 1992).

Conservation Status: Considered "vulnerable" (Rodríguez Schettino and Chamizo Lara, 1998a; Rodríguez Schettino and Chamizo Lara in Rodríguez Schettino, 1999a).

Anolis wattsi Boulenger, 1894

Distribution: Two currently recognized subspecies occur essentially islandwide on Antigua (*A. w. wattsi*) and Barbuda (*A. w. forresti*), including many satellites ("on every coastal cay that supports more than herb stage vegetation," Lazell, 1972); *A. w. wattsi* introduced on St. Lucia (Castries Botanical Garden) and Trinidad, *A. w. forresti* introduced on St. Christopher; *A. w. forresti* considered a full species by some authorities (e.g., Roughgarden, 1995).

Habitat: Xerophilic to mesophilic (Schwartz and Henderson, 1991); on igneous rock outcrops (Barbour, 1930c); appears to prefer shady situations and is therefore probably somewhat localized in its occurrence (Underwood, 1959b); in rainforest and xeric scrub, often most abundant in wet, shady pockets (Schwartz and Henderson, 1991); mostly near the ground, often on rocks (Wijffels, 1960); above the tide line in *Coccoloba* (Antigua) and on trees (including palm trunks in a sinkhole in limestone highlands; F. W. King in Williams, 1962c); on Barbuda, also on bushes, fence posts, rocks, stone walls, and buildings, usually in shade, but occasionally in open sunny areas, penetrates mangroves only a few meters (Lazell, 1972; Schwartz and Henderson, 1991; Powell and Henderson, 2008c); *A. w. wattsi* often common in open areas, including coastal scrub, *A. w. foresti* even in desertlike habitats, both subspecies have benefited from the construction of rock walls (Fläschendräger and Wijffels, 1996, 2009); females and juveniles are more terrestrial than adult males, which perch 30–60 cm above the ground but forage on the ground (Lazell, 1972); on St. Lucia, in disturbed habitats, more closely associated with bushes and building walls than trees (J. Schmidt, 1979); mean perch height on Barbuda 0.61 ± 0.06 m (Roughgarden et al., 1983a); on Antigua, mean perch height and diameter 0.80 ± 0.07 m and 10.4 ± 2.6 cm (Losos and de Queiroz, 1997a); in shady thermal microhabitats 73% of time, substrates used most frequently are the ground, tree trunks, and twigs, mean perch height 24 ± 2.01 cm, perch heights vary by island (Antiguan satellites) and time of day, most often on the ground and mean perch height lowest on Rabbit Island at 6 cm (no *Ameiva griswoldi*), most often on tree trunks and highest on Red Head Island at 40 cm (no *Anolis leachii*), and intermediate on 3 other islands, uses higher perches during midday periods (to 70–110 cm) on islands with *A. griswoldi*, but not on islands with no *A. griswoldi*, most frequently on the ground and mean heights lowest during early morning (0500–0800 h) and evening (1600–1900 h), when *A. griswoldi* activity low or absent (Kolbe et al., 2008).

Activity: Reduced during hot midday hours (B. E. Smith and Baum, 2000); active 0500–1900 h, most active early (~0600 h) and evening (~1800 h), decreasing during periods when *Ameiva griswoldi* is most active (Kolbe et al., 2008). **Behavior:** Retreats into holes and crevices in rocks, tree trunks, under shutters of buildings, or into debris on the ground (Lazell, 1972); flees short distances when first approached, but resumes "normal" behavior within 5–7 min, after repeated surveys recovery time reduced to 2–3 min (B. E. Smith and Baum, 2000). **Competition:** Aggressive interactions with *Anolis leachii* usually will result in displacement of *A. wattsi* (Kolbe et al., 2008). **Ecomorphology:** Intermediate between "trunk-ground" (closest), "trunk-crown," and "grass-bush" (Beuttell and Losos, 1999). **Home Range:** Mean home range size 1.76–9.52 m^2 (Roughgarden, 1995). **Parasites:** Nematodes *Paraphyaryngodon cubensis*, Physalopteridae gen. sp., and acanthocephalan cystacanths *Centrorhynchus* sp. (Dobson et al., 1992); nematode *Spauligodon anolis* (Dobson et al., 1992; Goldberg et al., 1997a; Bursey and Goldberg, 1998); mean parasite species/host 0.93 in forests and 0.40 in xeric habitat on Antigua (Dobson and Pacala, 1992); nematodes *Paraphyaryngodon cubensis*, *Trichospirura teixeirai*, and larval *Abbreviata* sp. (Goldberg et al., 1997a). **Population Size and Density:** 2.7 ± 0.7–35.1 ± 3.5 lizards/100 m^2 at 4 sites on Barbuda (Roughgarden et al., 1983); mean number of lizards per point-count site over 2 years in 3 habitats on Great Bird Island and 3 other Antiguan satellite islands 0.23 ± 0.31–2.76 ± 1.20 (B. E. Smith and Baum, 2000; B. E. Smith et al., 2001), but a different method (mark-recapture) used on Maiden Island increased population size estimates by 2.38 (B. E. Smith et al., 2001); population size estimates for Green Island (43 ha) 73,100 ± 5,590 lizards, Great Bird Island (9.93 ha) 55,878 ± 8,263, Maiden Island (8.29 ha) 37,772 ± 5,720, Rabbit Island (2.13 ha) 18,659 ± 1,299, Red Head Island (0.93 ha) 4,371 ± 539 (B. E. Smith and Colbert, 2002); 7,143 lizards/ha in forested parts of Great Bird Island (Daltry, 2007). **Predation:** *Alsophis antiguae*/*A. sajdaki* (Schwartz and Henderson, 1991; Henderson and Sajdak, 1996; Daltry, 1999; Daltry et al.,

2001; B. E. Smith and Baum, 2000); possibly *Ameiva griswoldi*, which frequently chases *A. wattsi* (B. E. Smith and Baum, 2000; Kolbe et al., 2008); *Anolis leachii* (J. J. Kolbe, O. Davis in Kolbe et al., 2008). **Reproduction:** Clutch size 1 (Hedges, in prep.). **Size:** Male SVL to 58 mm (Lazell, 1972), female SVL to 50 mm (Dobson et al., 1992), with animals on Barbuda smaller (Lazell, 1972). **Thermal Biology:** Microhabitat use differs among Antiguan satellite islands and by time of day, lizards on islands without *Ameiva griswoldi* use sunny habitats more frequently than on other islands, and lizards on islands without *Anolis leachii* use shady habitats more frequently (Kolbe et al., 2008).

Conservation Status: Not assessed; the species is widely distributed on many islands, ecologically versatile, and locally abundant (Daltry, 2007).

Anolis websteri Arnold, 1980

Distribution: Hispaniola, northwestern Haiti, inland to Ennery and south to Montrouis.

Habitat: Xerophilic to semimesophilic, on trees in open woody situations, also on walls, trees, and bushes in very xeric locations (Schwartz and Henderson, 1991).

Behavior: Displays differ from sibling species *A. caudalis* and *A. brevirostris*, with the greatest difference between geographically adjacent *A. caudalis*, headbobs in core (type-A) display usually 6 (5–7), 7.28 ± 0.43 sec in duration, with 1.07 ± 0.04 dewlap pulses (if a second pulse, short and incomplete), tail-lift posture not observed, 3 type-B displays, a single headbob, series of double bobs, and alternating long and short bobs (Jenssen and Gladson, 1984). **Ecomorphology:** "Trunk" (Williams, 1983). **Reproduction:** No evidence of pre-mating isolating mechanisms in sympatry with *A. caudalis*, males and females mated indiscriminately with either species, coitus frequency for females (0.64 copulations/day), mean duration of intromission 1.2 ± 0.2 sec, analogous to a "cloacal kiss" (Jenssen, 1996). **Size:** Male SVL to 51 mm, female SVL to 47 mm (Arnold, 1980).

Conservation Status: Unknown.

Anolis whitemani Williams, 1963

Distribution: Hispaniola, Plaine de Cul de Sac (Haiti) and Valle de Neiba (Dominican Republic), also on adjacent xeric slopes of Montagnes de Trou-d'Eau and Sierra de Baoruco, east into the Llanos de Azua; a second subspecies (*A. w. lapidosus*) from the southeastern base of Presqu'île du Nord-Ouest probably represents a distinct species; elevational range, near sea level to ~458 m.

Habitat: Xerophilic (Williams, 1963b; Schwartz and Henderson, 1991); in open, hot, arid situations, such as *Acacia* woods, cactus stands, *Agave*, open areas of semidesert with at least some shrubby or arborescent vegetation, often on fence posts, may occur in xeric enclaves in otherwise more mesic areas, dry hillsides adjacent to a more mesic valley, at times syntopic with *A. cybotes* in disturbed areas (Williams, 1963b; Schwartz, 1980f; Powell et al., 1989b; Schwartz and Henderson, 1991); most abundant along sparsely vegetated edges of a dry riverbed in a xeric region, where assumed "watch position" on trunks of shrubs and small trees, mean microhabitat temperature 31.6 ± 3.6 °C (32.5–43.0 °C), warmest of sympatric anoles (Powell et al., 1996b); perched on a rock next to a road (Glor, 2003); sleeps to 3.5 m above the ground on *Acacia* and pendant vines (Schwartz, 1980f).

Behavior: Vocal when captured (Schwartz, 1980f). **Diet:** Includes ants (most frequent), leafhoppers, orthopterans, lepidopterans, and (less frequently) arachnids and chilopods, an *Anolis* tail in one individual and (adventitious?) plant matter (Carr et al., 1989). **Ecomorphology:** "Trunk-ground" (Rand and Williams, 1969); formally quantified (Glor et al., 2003). **Reproduction:** Appears to demonstrate a prolonged or continuous reproductive cycle, unusual in a xeric-adapted species, sperm in testes of males from all months except April and October–January and testicular size did not vary by month, ovigerous females in February–March and May–August (only months from which specimens available), shelled ova present bilaterally in 10 of 53 ovigerous females (suggesting ability to retain eggs until deposition propitious), a shelled and an unshelled egg in 8 females, mean length of shelled oviductal eggs 11.0 ± 1.4 mm, positive correlation between SVL and number of yolked follicles and/or oviductal eggs and between SVL and mean size of yolked follicles and/or oviductal eggs, weight of coelomic fat bodies is negatively correlated with the number of yolked follicles and/or oviductal eggs, with the least amount in females with ova (shelled or not), males contained more fat than females with ova, but less than females without eggs, juveniles had approximately the same amount of fat as males (Pflanz et al., 1991); clutch size 1 (Hedges, in prep). **Sex Ratio (M:F):** 106:72, but possibly skewed by the tendency to encounter and capture more males (Pflanz et al., 1991). **Size:** Male SVL to 67 mm (Schwartz, 1980f), female SVL to 54 mm, with the smallest mature male SVL 33 mm and female SVL 40 mm (Pflanz et al., 1991). **Tail Autotomy:** 17 of 98 individuals with regenerated tails, frequency higher in females than males (21 versus 15%), but not significantly so, nor was frequency significantly correlated with SVL, with the low rate attributable to low population density (24 person-hours to collect each individual) and consequently infrequent agonistic interactions (Cusumano et al., 1991). **Thermal Biology:** In and along a dry riverbed, exclusively in the shade, "squirreling" around trunks of shrubs and trees as the day progressed (RP); microhabitat temperatures warmer (mean 31.6 ± 2.7 °C, with high of 43.0 °C) than those of *A. brevirostris* and *A. olssoni* (Powell et al., 1996b).

Conservation Status: Not assessed.

Additional Pertinent Literature: Powell and Carr (1990).

16 Family Scincidae

Figure 19. West Indian skinks (Scincidae) may be far more diverse than the four currently recognized species. The populations in the Grenadines have traditionally been assigned to *Mabuya mabouya*. Photograph by Mark de Silva.

Mabuya lineolata Noble and Hassler, 1933

Distribution: Hispaniola, widespread but disjunct.

Habitat: Under rubbish in *Acacia* scrub, occasionally in mesic situations, such as broadleaf woods and *Musa* plantings (Schwartz and Henderson, 1991).

Size: SVL in males to 56 mm, in females to 54 mm (Schwartz and Henderson, 1991).

Conservation Status: Not assessed.

Mabuya mabouya Lacépède, 1788

Distribution: In the Lesser Antilles, on Guadeloupe, Les Saintes, La Désirade, Marie-Galante, Dominica, and St. Lucia; restricted to small mongoose-free satellite islands (Petite Terre, La Désirade, Îlet-a-Cachons) in the Guadeloupean Archipelago (Breuil, 2002b, 2009) and very rare or possibly extirpated on Martinique (Breuil, 2009); elevational range, sea level to >500 m; populations on St. Vincent, the Grenadines, and Grenada in the West Indies, and elsewhere in northerm South America (including Trinidad and Tobago) north to Panama are no longer associated with this species (Miralles, 2005).

Habitat: On La Désirade, in leaf litter, grass, and piles of vegetation in xeric situations, on Îles de la Petit Terre (Guadeloupean Archipelago), in forest with pear trees (Breuil, 2002b); on Dominica, in dry coastal woodlands, littoral woodland, and coconut plantations (Bullock and Evans, 1990); basically terrestrial but good climbers, may be seen in trees and on roofs of buildings (Malhotra and Thorpe, 1999); in cultivated areas where often under corrugated iron roofs of agricultural sheds (Malhotra et al., 2007); in disturbed, coastal scrub forest where observed on tree trunks, in leaf litter, on large boulders, and associated with ruins overgrown by *Ficus* (RP, RWH); in the Grenadines, on the ground in woody underbrush and between cacti, capable of climbing onto large pads of Prickly Pear (*Opuntia dillenii*) and dry tree trunks where it hides in holes, frequents areas with garbage and vegetable trash and often near human habitation (Daudin and de Silva, 2007); on Grenada, in litter in *Acacia* forest (RWH).

Activity: Diurnal (RWH); on Dominica, peak activity 1000–1400 h, rarely active before 1000 h or after 1600 h (Bullock and Evans, 1990). **Biomass:** On Dominica, mean values in dry scrub and littoral woodlands 0.01–0.63 kg/ha (maximum 1.9 kg/ha), in rainforest and elfin woodland, mean values 0.0–0.2 kg/ha (Bullock and Evans, 1990). **Foraging and Diet:** Active foragers feeding on arthropods, mainly insects and woodlice (Malhotra and Thorpe, 1999). **Population Densities:** On Dominica, mean densities 0–751/ha, with lower values in rainforest habitat (Bullock and Evans, 1990). **Reproduction:** Viviparous, on Dominica, gravid females with 3–5 embryos (mean 3.3 ± 0.7) in February, June, August, September, December (6 of 7 gravid females had embryos in both oviducts), number of follicles/female 1–7 (mean 3.3 ± 1.1), neonate SVL 29–30 mm (Somma and

Brooks, 1976). **Size:** Male SVL to 98 mm, female SVL to 106 mm (S. B. Hedges, in litt., 122.VI.2009). **Thermal Biology:** Mean body temperature (Dominica) of adult females 33.4 ± 2.6 °C at mean air temperature 27.9 ± 1.4 °C, mean body temperature of adult males 32.8 ± 1.7 °C at mean air temperature 28.7 ± 1.5 °C, mean body temperature significantly higher than mean air temperature, but average body temperature not significantly different between males and females (Brooks, 1968); on Dominica, retreats to burrows or under rocks in cool, wet weather (Bullock and Evans, 1990); heliothermic (Malhotra and Thorpe, 1999).

Conservation Status: Not assessed, but considered rare on the Guadeloupean Archipelago and on Martinique because of mongoose (*Herpestes javanicus*) predation (Breuil, 2002b, 2009); similar circumstances may affect *Mabuya* on other islands with mongooses.

Remarks: The taxonomic status of populations on St. Vincent and its satellites, the Grenadines (Bequia, Mustique, Canouan, Mayreau, Union, Carriacou), Grenada, and Barbados is in question (e.g., Miralles, 2005); the taxonomic status of West Indian skinks in general is uncertain, and populations currently assigned to *M. mabouya* throughout its range may comprise a number of species.

Mabuya macleani Mayer and Lazell, 2000

Distribution: Carrot Rock, British Virgin Islands.

Habitat: Associated with boulder piles in *Coccoloba uvifera* clumps in a draw, also in an area with few rocks, little vegetation, and soil substrate, total habitable area ~1.3 ha, one individual on a vine 3 cm above ground (Mayer and Lazell, 2000).

Population Size and Density: Estimated at 520 individuals or 400/ha (Mayer and Lazell, 2000). **Size:** SVL to 80.5 mm (sex not given; Mayer and Lazell, 2000).

Conservation Status: A range limited to a single island with only 1.3 ha of suitable habitat speaks strongly to the need for some formal protected status.

Mabuya sloanii Daudin, 1803

Distribution: Turks & Caicos Islands, Jamaica, Hispaniola, Isla Mona, Isla Monito, Puerto Rico including satellites, Isla Vieques, Isla Culebra, U.S. and British Virgin Islands, and Lesser Antilles (Redonda, Anguilla, St.-Barthélemy, St.-Martin/St. Maarten, Montserrat).

Habitat: Primarily xerophilic, but at times in more mesic situations (Schwartz and Henderson, 1991); on Puerto Rico, in hollow trees and occasionally in houses (W. Peters, 1876); on Mona, Culebra, and Buck Island (St. Thomas), hides in dense clumps of *Opuntia dillenii* (Grant, 1932d); on Jamaica, among leaf litter in forest on hard, porous limestone (Grant, 1940a); in Puerto Rico, at the bases of trees (K. P. Schmidt, 1928); in xeric woods and secondary scrub at edges of cultivated area (Thomas and Thomas, 1977); at bases of coconut palms, under rocks or rock fissures, under *Opuntia* clumps, in a knot hole of a fence post, in small trees 1–3 m above ground level, in a ground bromeliad (*Bromelia pinguin*), and inside a house, on Isla Mona, in coconut palm trash (usually one/pile; Rivero, 1998); in the U. S. Virgin Islands, on cays satellite to the main islands, in low dense vegetation on beaches and lower slopes of the cays, seeks shelter in grass, in brush litter, under rocks and other surface debris, in rocky fissures, and on branches of low shrubs (Platenberg and Boulon, 2006); on Anegada (British Virgin Islands), among rock piles in areas of forested limestone (Carey, 1972); on Guana (British Virgin Islands), on dry hillsides, often in association with cacti, on an *Agave* (RP); on Anguilla, on and in loosely constructed rock walls (Hodge et al., 2003); on St.-Barthélemy, in sunny situations on cacti, in tall grass, will enter houses (Breuil, 2002b).

Abundance: On St.-Barthélemy, relatively abundant (Breuil, 2002b). **Activity:** Diurnal, in Jamaica, "most frequently met in mid-afternoon" (Grant, 1940a). **Diet:** In the Lesser Antilles, omnivorous, eating fruits, insects, and other small animals (Currat, 1980); on Puerto Rico, includes cricket nymphs (Wolcott, 1923) and *Eleutherodactylus coqui* (Rivero and Seguí Crespo, 1992). **Predation:** Mongooses (*Herpestes javanicus*), predatory birds, and crabs (Currat, 1980). **Reproduction:** In Jamaica, all females contained young (Grant, 1940a). **Size:** Male SVL to 93 mm, female SVL to 99 mm (S. B. Hedges, in litt., 22VI.2009). **Thermal Biology:** Basks on human-constructed rock piles on Anegada (Carey, 1972); basks on piles of *Cocos* trash on Mona Island (Rivero, 1998); basks on rocks and emergent vegetation (including cacti) on Guana (RP).

Conservation Status: The species has been extirpated from some of the Virgin Islands, possibly attributable to the presence of the mongoose (*Herpestes javanicus*; Seaman, 1961; Carey, 1972; Platenberg and Boulon, 2006); Grant (1940a) suggested that the mongoose on Jamaica had reduced its numbers; on St.-Barthélemy, sometimes killed because of its snakelike appearance and movements (Breuil, 2002b); "common but restricted, protect habitat" (Moreno, 1991d).

Remarks: The taxonomy of Lesser Antillean *Mabuya* remains in a state of flux, and the names applied to those populations may change (e.g., Mayer and Lazell, 2000); the taxonomic status of West Indian skinks in general is uncertain, and populations currently assigned to *M. sloanii* throughout its range may comprise a number of species.

17 Family Sphaerodactylidae

Figure 20. In coastal habitats dominated by Sea Grape (*Coccoloba uvifera*) on Guana Island (British Virgin Islands), *Sphaerodactylus macrolepis* (Sphaerodactylidae) reaches population densities of 67,600/ha. Photograph by Robert Powell.

Aristelliger barbouri Noble and Klingel, 1932

Distribution: Bahamas, Great Inagua Island, Little Inagua Island, Sheep Cay.

Habitat: Xerophilic (Schwartz and Henderson, 1991); dry sheltered sites in decayed wood or under loose *Coccoloba*, *Cocos*, and *Rhizophora* bark, in a rotten *Laguncularia racemosa* limb, under a pile of palm fronds (Noble and Klingel, 1932).

Behavior: Very secretive, not seen active either by day or by night (Noble and Klingel, 1932). **Reproduction:** Minimal breeding size 35 mm SVL (Hecht, 1952); clutch size 1 (Noble and Klingel, 1932; Kluge, 1987); eggs (11.5–12.0 × 7.5–8.5 mm) tapered sharply at one end, usually deposited singly, sometimes communally (to $N = 21$) in small cavities in decayed wood deeper than bark, at times attached to bark, tightly fastened to the substrate and other eggs (when present), shell calcareous and very brittle, right ovary and oviduct function more frequently than the left, eggs on 7 January–14 February, incubation 75 days (under artificial conditions), hatchlings emerge by cracking and tearing shell despite presence of two premaxillary egg teeth, hatchling 18 mm SVL (Noble and Klingel, 1932). **Size:** Male SVL to 50 mm. (Hecht, 1952), female SVL to 46 mm (Schwartz and Henderson, 1991).

Conservation Status: Unknown.

Additional Pertinent Literature: Bauer and Russell (1993a, 1993b).

Aristelliger cochranae Grant, 1931

Distribution: Navassa Island.

Habitat: Xeric woodlands, associated with trees (adventitious roots of *Ficus*, pendant palm fronds of *Thrinax morrisii*, and *Metopium*; Thomas, 1966c; Powell, 1999a); also edificarian (especially in ruins at Lulu Bay) and on rock faces (Powell, 1999a).

Activity: Primarily nocturnal, but active by day in ruins, deep rock crevices, and under vegetation on rock faces (Powell, 1999a). **Diet:** Overlap between diet and prey availability in pendant palm fronds low (0.182; zero at species level) suggesting that lizards forage outside fronds, prey included spiders (importance value 0.197), pseudoscorpions (0.018), beetles (0.063), flies (0.019), ants (0.183), termites (0.049), moths (0.096), roaches (0.021), gecko eggs (0.271), and a hatchling (0.082; Gifford et al., 2000). **Reproduction:** Smallest female with eggs SVL 41.3 mm (Powell, 1999a); singly deposited eggs with brittle shells in pendant palm fronds often misshapen to conform to the leaves to which they were attached, 11.1 × 8.4 mm (Thomas, 1966c); 10.0–11.5 × 7.8–9.2 mm (*N* = 11), hatchling SVL 20.6 mm, total length 43.1 mm (Powell, 1999a). **Size:** Male SVL to 63 mm, female SVL to 53 mm (Schwartz and Henderson, 1991). **Thermoregulation:** Cloacal temperatures of geckos (*N* = 4) in ruins at 0939–0945 h 29.0–29.8 °C, of one individual on *Metopium* trunk at 1125 h 32.3 °C, and of geckos in pendant palm fronds (*N* = 8) at 1800–1830 h 30.1–32.3 °C, all significantly higher than ambient (Powell, 1999a).

Conservation Status: Although the species is confined to an area of 5.2 km², it is there ubiquitous and abundant (Powell, 1999a) and probably not in need of conservation efforts beyond those provided by the relative inaccessibility of Navassa Island and the island's inclusion in the U.S. National Wildlife Refuge system.

Additional Pertinent Literature: Bauer and Russell (1993a), Lynxwiler and Parmerlee (1993).

Aristelliger expectatus Cochran, 1933

Distribution: Hispaniola (Haiti and the southwestern Dominican Republic), widespread but localized, Île-de-la-Gonâve, Île-à-Cabrit, Isla Beata, Isla Alto Velo, Cayo Pisaje.

Habitat: Xerophilic, in rotten tree trunks and under palm bark, rocks, stones, and palm trash on the beach (Schwartz and Henderson, 1991); on walls, on a concrete barrier, or in the thatch of an abandoned shack (Lynxwiler et al., 1991; Schwartz and Henderson, 1991).

Reproduction: Minimal breeding size 35 mm SVL (Hecht, 1952); clutch size one (Hedges, in prep.); egg taken 9 August hatched 19 August, egg taken 29 July in a dead *Agave* on a cliff face hatched 27 August (Schwartz and Henderson, 1991). **Size:** Male SVL to 55 mm (Thomas, 1966c), female SVL unknown.

Conservation Status: Not assessed.

Additional Pertinent Literature: Bauer and Russell (1993a), Lynxwiler and Parmerlee (1993).

Aristelliger hechti Schwartz and Crombie, 1975

Distribution: North and South Caicos Islands, Six Hills Cays, and Little Ambergris Cay; possibly introduced in northern Hispaniola (Burns et al., 1992; Powell and Parmerlee, 1992).

Habitat: Xerophilic (Schwartz and Henderson, 1991); under rocks, in palm trash, in low scrub, *Batis maritima*, scattered living and decaying remains of *Cocos*, in a crumbling rock fence and a ruined building (Schwartz and Crombie, 1975).

Escape Strategy: Skin very fragile (Greene, 1988; Bauer and Russell, 1992; Bauer et al., 1992). **Size:** Male SVL to 90 mm, female SVL to 75 mm (Schwartz and Crombie, 1975).

Conservation Status: Unknown.

Additional Pertinent Literature: Bauer and Russell (1993a, 1993c).

Aristelliger lar Cope, 1862

Distribution: Hispaniola, widespread but localized, also Isla Catalina, Cayos Siete Hermanos (Monte Chico and Monte Grande), Cayo Levantado in Bahía de Samaná; populations in the Cayos Siete Hermanos possibly *A. hechti* (Burns et al., 1992; Powell and Parmerlee, 1992).

Habitat: Largely mesophilic, associated with large trees (including the bases of *Cocos* fronds), broadleaf woods in mesic ravines, limestone and karst areas, in irregularities in the substrate, crevices, and interfaces between rocks and tree roots (e.g., *Ficus*; Schwartz and Henderson, 1991); on Cayo Monte Chico, on deeply shaded trunks of small trees and large shrubs (Burns et al., 1992); in the Valle de Neiba north of Barahona, under rocks in xeric open scrub, often edificarian in villages along the western shore of the Península de Barahona (RP).

Activity: Primarily nocturnal, but occasionally diurnal (Thomas, 1966c; Burns et al., 1992; Powell and Parmerlee, 1992). **Diet:** Apparently omnivorous, on Cayo Monte Chico, individuals (*N* = 5) consumed spiders (*N* = 2, volume = 0.005 mm³, present in 2 stomachs), beetles (7, 0.070, 2), roaches (2, 0.025, 2), true bugs (2, 0.050, 1), crickets (5, 0.103, 3), ghost crab (*Ocypode* sp.; 1, 0.110, 1), and a flower (Burns et al., 1992); feces include spherical seeds (~6 mm diameter) and red pulp, also many tiny seeds (like strawberry achenes), and elongate (19 mm), many-seeded white fruit (Schwartz and Henderson, 1991). **Escape Strategy:** Skin fragile (Cochran, 1941); tails break readily (Hassler, 1930). **Longevity:** In captivity, 1 year, 11 months (Snider and Bowler, 1992). **Reproduction:** Minimal breeding size 100 mm SVL (Hecht, 1952); clutch size one (Hedges, in prep.); eggs very brittle, attached to sides of trees in the open, in crevices or under loose wood on dead limbs or in rotten hollows (Hassler, 1930); large, flattened oval eggs (ca. 20 mm) taken in the buttress cavity of a *Ficus* approximately 45 cm above the ground and adherent to the side of the crevice (Schwartz and Henderson, 1991). **Size:** Adult male

SVL to 135 mm, female SVL to 111 mm (Schwartz and Henderson, 1991).

Conservation Status: Not assessed.

Additional Pertinent Literature: Bauer and Russell (1993a, 1993d).

Aristelliger praesignis Hallowell, 1857

Distribution: Two currently recognized subspecies occur on Jamaica (mostly at elevations below 650 m) and the Cayman Islands (Grand Cayman, Little Cayman, Cayman Brac, Owen Island; *A. p. praesignis*), and on the Swan Islands (Swan and Little Swan islands; *A. p. nelsoni*).

Habitat: Mostly mesophilic (Schwartz and Henderson, 1991); in homes on Grand Cayman (English, 1912); forested areas on Swan Islands and homes on Swan Island (Barbour, 1914); on Cayman Brac, in houses, in crevices, or beneath palm thatch (Garman, 1888); generally arboreal, especially in rotten trees, but also on *Ficus* to nearly 7 m above the ground, in rock piles, old buildings and thatched dwellings, under large rocks, on a cliff overhang, in *Cocos* trash, in a terrestrial bromeliad, under a palm trunk pile adjacent to the ocean, under bark, on a stone wall, and under a termitarium >2 m above the ground (Grant, 1940c; Schwartz and Henderson, 1991); on Jamaica, in rafters of an old buildings, in wall crevices of ruins of abandoned sugar factories, and in hollow trees (Barbour, 1910); on Cabarita Island (Jamaica), by day under bark of dead trees and under a limestone slab in an open area (Crombie et al., 1983); male and 2 females on *Morinda citrifolia* on hotel grounds in eastern Jamaica (Hilgenhof, 1998b); by day between dry leaves of bromeliads (P. Vogel, 2000b).

Activity: Crepuscular (Barbour, 1914; Schwartz and Henderson, 1991) or nocturnal (Garman, 1888) or partially diurnal (Grant, 1940c); on Cabarita Island, emerges from tree holes at dusk and ascends toward the tops of trees (Crombie et al., 1983; may forage around artificial lights at night (Grant, 1940a, 1940b); eye size suggests combined diurnal/nocturnal activity (Werner, 1969). **Behavior:** Extremely placid, easily becomes "tame" (Grant, 1940a, 1940b); produces a "croak" like a rattle (English, 1912) or resembling the sound made by vibrating one's tongue against the roof of the mouth (Schwartz and Henderson, 1991); calls from dusk to dawn in choruses (Grant, 1940a, 1940b); voice functions in sexual competition (Hecht, 1952). **Diet:** "A most useful eater of cockroaches indoors" (English, 1912); insects (Grant, 1940c; Schwartz and Henderson, 1991); "seems to be fond of sugar" (Garman, 1888). **Escape Strategy:** Skin fragile (Barbour, 1910, 1921; Grant, 1940b, 1940c; Cochran, 1941; Bauer et al., 1989, 1992; Bauer and Russell, 1992); produces an odor like the musk of a garter snake (*Thamnophis* sp.; Garman, 1888). **Predation:** Barn Owl (*Tyto alba*; Johnston, 1974, 1975). **Reproduction:** Minimal breeding size 45 mm SVL (Hecht, 1952); eggs laid singly, often in communal nests in tree holes, hollow branches, crevices in the bark of trees (Barbour, 1910; Grant, 1940a); attached to fallen *Cocos* fronds (Grant, 1940c); shells very brittle, adhere to the substrate and other eggs, 13.5–14.5 × 11.0–11.5 mm (Grant, 1940a); eggs laid as frequently as twice per year, incubation 3 months (Hecht, 1952); female in January with a large yolked follicle (7.5 mm diameter) on the left side, a smaller follicle (2.5 mm) on the right side, but males in January did not demonstrate reproductive activity (Seidel and Franz, 1994). **Size:** Male SVL to 85 mm, female SVL to 65 mm (Schwartz and Henderson, 1991).

Conservation Status: Not assessed.

Additional Pertinent Literature: Bauer and Russell (1993a, 1993e).

Aristelliger reyesi Díaz and Hedges, 2009

Distribution: Cuba, Península de Hicacos (western Cuba).

Habitat: On limestone outcrops mostly covered by microphyllous forest, cliffs, cave entrances, rocky shelters, sinkholes, rustic wooden houses roofed with palm fronds, tree cacti (*Dendrocereus* sp.), and fig (*Ficus* sp.) root tangles at heights of 1–3 m above the ground, most individuals on limestone surfaces close to or partially hidden in small holes that serve as refugia (Díaz and Hedges, 2009).

Activity: Individuals active by day and night (Díaz and Hedges, 2009). **Behavior:** Two males fighting at midday on *Ficus* roots, skin fragile and easily damaged, some males vocal when caught (Díaz and Hedges, 2009). **Foraging and Diet:** Feces contained ants (including *Camponotus* sp.), adult neuropterans, beetles, shed skin, and stones (Díaz and Hedges, 2009). **Reproduction:** Females with single eggs readily visible through the body wall, eggs (10.6 × 8.4 mm and 10.8 × 9.1 mm) firmly attached to inner surfaces of bamboo tubes provided as shelters in terraria, absence of juveniles in early June suggests that the breeding season is just beginning at that time (Díaz and Hedges, 2009). **Size:** Male SVL to 63.6 mm, female SVL to 55.1 mm (Díaz and Hedges, 2009).

Conservation Status: Not assessed; the "fragile environment" of the Península de Hicacos is threatened by rapid development related to tourism (Díaz and Hedges, 2009).

Gonatodes albogularis Duméril and Bibron, 1836

Distribution: In the West Indies, Cuba, Hispaniola (Haiti, including Île-de-la-Gonâve and Île-à-Cabrit), Jamaica, and Grand Cayman Island; elevational range, sea level to ~915 m; elsewhere on mainland Central and South America, including numerous continental islands; introduced into southern Florida; note that the Cuban populations are assigned to *G. a. fuscus*, whereas other West Indian populations are collectively considered to be an endemic subspecies, *G. a. notatus* (Reinhardt and Lütken, 1862); the population on Grand Cayman (Williams, 1964; Schwartz and Thomas, 1975) may have been a recent introduction and apparently is extirpated (R. I. Crombie in Seidel and Franz, 1994; Röll, 2009); reports of the species in Martinique (Duméril and Bibron, 1836; Vanzolini and Williams, 1962) undoubtedly are in error.

Habitat: Mostly xerophilic, but occasionally mesophilic (Schwartz and Henderson, 1991); encountered solely in houses in Cuba (Gundlach, 1880; Alayon Dalmau, 1951) and Jamaica (Barbour, 1910); particularly abundant on wooden buildings (Vit, 1981); often edificarian and in human debris, but also in oceanside *Coccoloba* stands (Grant, 1940a; Collette, 1961; Schwartz and Henderson, 1991); in Cuba, frequently in old walls, using cracks for retreats, also active on buildings at night in La Habana (Z. Vogel, 1965b); in a variety of crevices, holes, and dense foliage that provide shelter, refuge, and food, also in fissures and crevices in walls in edifices (Martínez Reyes and Hernández Marrero, 2003); a human commensal, abundant in houses, gardens, parks, and plantations (Díaz, 2008b); frequently very abundant in Cuban hotels (Kunz, 2005); on Hispaniola, in open areas as well as more concealed situations (high walls and vegetation), including suitable habitat around human dwellings (Olson, 1995).

Activity: Diurnal (Schwartz and Henderson, 1991). **Behavior:** Climbs readily and scurries along trunks and branches when threatened, often hangs upside down beneath objects (Schwartz and Henderson, 1991); territorial in captivity, displays involve inflation of the throat and headbobs (Díaz, 2008b). **Longevity:** In captivity, to at least 9 years (Vit, 1981; Díaz, 2008b). **Parasites:** In Cuba, nematodes *Spauligodon antillarum* (Baruš and Coy Otero, 1974; Coy Otero and Baruš, 1979b; Coy Otero and Lorenzo Hernández, 1982); *Cyrtosomum scelopori* and larval *Abbreviata* sp. (Coy Otero and Baruš, 1979b; Coy Otero and Lorenzo Hernández, 1982). **Predator:** In Cuba, introduced House Sparrows (*Passer domesticus*; Bello, 2000). **Reproduction and Growth:** On Jamaica, eggs 8.0 × 6.5 mm, white and brittle, 10 eggs in one crevice in a house, suggestive of communal deposition (Grant, 1940a); in captivity, eggs (~9 × 8 mm) laid under cover, incubation 58–72 days at ~28 °C, hatchling total length ~30 mm (Vit, 1981); in captivity on Cuba, both males and females lick cloacal apertures after copulation, clutch size 1, eggs 7.6–8.6 × 6.2–7.6 mm, laid at intervals of 15–37 days (maximum 13/year), incubation usually 55–86 days (exceptionally 107–152 days) at ambient temperatures, hatchling SVL 17.2–18.7 mm, males reach SVL 25.4 mm at 1 month, 32.8 mm at 2 months, 34.3 mm at 3 months (when yellow head color develops), females reach SVL 22.3 mm at 1 month, 27.3 mm at 2 months, 33.3 mm at 3 months, and 39.0 mm at 6 months (Díaz, 2008b). **Size:** Male SVL to 40 mm (Schwartz and Henderson, 1991), female SVL to 39mm (Díaz, 2008b).

Conservation Status: Not assessed.

Remarks: Some authors have recognized *G. albogularis fuscus* as a full species (e.g., Röll, 2009).

Gonatodes daudini Powell and Henderson 2005

Distribution: Union Island, St. Vincent Grenadines.

Habitat: Dry forest, leaf litter and in association with rotting logs and loose soil, often interspersed with small stones and rocks (Powell and Henderson, 2005b).

Size: Adult male SVL 19.9–29.9 mm, female SVL unknown.

Conservation Status: The very limited distribution and the vulnerability of the forest habitat on Union Island speak to a need for implementation of conservation measures; the species certainly qualifies for red-listing as "critically endangered" according to IUCN criteria (IUCN, 2001).

Sphaerodactylus altavelensis Noble and Hassler, 1933

Distribution: Hispaniola, where four currently recognized subspecies occur, also on Isla Alto Velo (Dominican Republic) and Île à Cabrit (Haiti); elevational range −30–125 m .

Habitat: Xeric situations, occasionally mesic (Schwartz and Henderson, 1991); on Alto Velo (*S. a. altavelensis*), under loose rough chunks of limestone, under rocks, and in crevices in limestone outcrops (Noble and Hassler, 1933); individual questionably attributable to this subspecies on the Barahona Peninsula in roots of dry clump grasses (Powell et al., 1989c); *S. a. brevirostratus* in banana and coffee groves, among trash and rocks in a cave (Thomas and Schwartz, 1983a); *S. a. enriquilloensis* in wooded areas, banana groves, thornscrub, palm savanna, a garbage dump, and mesic oases in xeric scrub, in leaf litter, trash, and under rocks (Thomas and Schwartz, 1983a); *S. a. lucioi* in dead *Agave* rosettes and other objects on gravelly or rocky substrates (Thomas and Schwartz, 1983b); also under slats, human debris, and wet *Musa* leaves (Schwartz and Henderson, 1991).

Activity: Forages in leaf litter 1500–1730 h (Schwartz and Henderson, 1991). **Diet:** Isopterans consumed with greatest frequency, followed by arachnids, coleopterans, hymenopterans (ants), and lepidopteran larvae (Cunningham et al., 1993). **Predation:** *Hypsirhynchus parvifrons* (Henderson et al., 1987a). **Reproduction:** Eggs (2) 4.6–4.8 mm × 6.5–6.6 mm, also 3.5–4.0 mm × 5.0–6.0 mm on 19 June (Schwartz and Henderson, 1991); clutch size 1 (Hedges, in prep.). **Size:** Male SVL to 28 mm, female SVL to 29 mm (Schwartz and Henderson, 1991).

Conservation Status: Unknown.

Sphaerodactylus argivus Garman, 1888

Distribution: Cayman Islands, Grand Cayman and satellites (*S. a. lewisi*), Little Cayman and Owen Island (*S. a. bartschi*), and Cayman Brac (*S. a. argivus*).

Habitat: Xeric and mesic situations (Schwartz and Henderson, 1991); on Grand Cayman, under rocks at bottoms of sinkholes, in mouths of caves on a limestone ridge, and in bromeliads, usually associated with dry organic soils and large accumulations of terrestrial mollusk shells (Seidel and Franz, 1994); eggs in termite nests on Booby Cay (Grant, 1941); under trash piles on Little Cayman (Grant, 1941); under rocks, trash, and lumber around abandoned houses near the beach, along a nature trail, and in drier mangrove areas (Seidel and Franz, 1994); in leaf litter in *Coccoloba* stands (RP); on Cayman Brac, under rocks "below the bluff" (Grant, 1940c); also under coconut palms, lumber, and trash on the ground (Seidel and Franz, 1994).

Diet: Termites taken by *S. a. lewisi* in captivity (Grant, 1940c). **Predators:** *Cubophis fuscicauda* on Cayman Brac (Henderson and Sajdak, 1996); West Indian Red-bellied Woodpecker (*Centurus superciliaris caymanensis*) on Grand Cayman (Johnston, 1975). **Reproduction:** Clutch size 1 (Hedges, in prep.); eggs of *S. a. argivus* in June (Grant, 1940c); 2 eggs of *S. a. bartschi* in February in the remains of a complete skeleton of *Anolis maynardii* on a wooden shelf in an abandoned house, hatchling 14 mm total length (Seidel and Franz, 1994); eggs of *S. a. lewisi* in termite nests on Booby Cay in May (Grant, 1940c). **Size:** Male SVL to 28 mm, female SVL to 29 mm (Schwartz and Henderson, 1991).

Conservation Status: Not assessed.

Sphaerodactylus argus Gosse, 1850

Distribution: Two subspecies occur in the Bahama Islands (North Bimini Island and New Providence Island), Cuba, Jamaica, Isla Grande de Maiz (Nicaragua), and Key West (Florida; *S. a. argus*), and on Isla San Andres (*S. a. andresensis*).

Habitat: On North Bimini Island, on a ridge and along the beach in piles of shaded *Cocos* husks and trash (Schwartz and Henderson, 1991); on Cuba, in coastal areas, in leaf litter, under rocks, and around edifices (Chamizo Lara et al., 2003; Díaz and Abreu Guerra, 2005); in mangrove stands (Vergner, 1998); on Jamaica, in dry scrub forest, wet limestone forest, and rainforest, as well as open areas, beaches, pastures, hotel grounds, under palm fronds and *Cocos* husks (Hedges, in prep.); also under bark, trash, rocks, and occasionally in dry bromeliads high in trees (Grant, 1940a); under limestone rubble and palm trash on Cabarita Island (Crombie et al., 1983); by day on Jamaica, between dry leaves of bromeliads (P. Vogel, 2000b).

Activity: Primarily diurnal (Vergner, 1998). **Reproduction:** Clutch size 1 (Hedges, in prep.); in Jamaica, eggs "laid promiscuously under any object" and empty shells accumulate in such places from year to year (Grant, 1940a); on Cuba, eggs (7.9–8.1 × 5.7–5.8 mm) on 6 March (Schwartz and Henderson, 1991); on Jamaica, eggs hatched 3–5 September (Schwartz and Henderson, 1991); diameter of *S. a. argus* eggs 5.1–7.5 mm, annual egg production during the first year of life 4–12, incubation 53–80 days, SVL of male hatchlings 13–16 mm, female 15 mm, females mature at 28–31 mm SVL (Regalado, 2006). **Size:** SVL in males and females to 33 mm (Thomas, 1975).

Conservation Status: Not assessed; the Bahamian populations may well be a consequence of human mediation (Thomas, 1975).

Sphaerodactylus ariasae Hedges and Thomas, 2001

Distribution: Hispaniola, Isla Beata and the tip of the Barahona Peninsula (Dominican Republic).

Habitat: Xeric woods over a limestone substratum, in leaf litter around a sinkhole on Isla Beata, and in leaf litter in partially cutover (degraded) limestone forest near the coast on Cabo Beata (Hedges and Thomas, 2001).

Size: Male SVL to 17.7 mm, female SVL to 17.9 mm (Hedges and Thomas, 2001).

Conservation Status: Not assessed.

Sphaerodactylus armasi Schwartz and Garrido, 1974

Distribution: Cuba, southeastern coast.

Habitat: Xerophilic (Schwartz and Henderson, 1991); associated with *Agave* (Schwartz and Garrido, 1985).

Reproduction: Clutch size 1 (Hedges, in prep.). **Size:** Male SVL to 28 mm, female SVL to 30 mm (Schwartz and Garrido, 1985).

Conservation Status: Considered "vulnerable" (Rodríguez Schettino and Chamizo Lara, 1998a).

Sphaerodactylus armstrongi Noble and Hassler, 1933

Distribution: Hispaniola, where two currently recognized subspecies occur in uplands and along the eastern coast of the Barahona Peninsula (*S. a. armstrongi*) and the western Sierra de Bahoruco and Massif de la Selle (*S. a. hypsinephes*); elevational range, sea level to >1,800 m.

Habitat: Mesophilic (Schwartz and Henderson, 1991); under rocks in a damp streambed (*S. a. armstrongi*; Noble and Hassler, 1933); forested ravines (425 m), palm and deciduous forest (1,220 m), in a pile of thatching in broadleaf rainforest (1,765 m), a coffee grove, open fields with stacks of limestone (1,340 m), pinewoods (1,700 m), and small patches of shade created by *Clusia rosea* and *Mammea americana* (*S. a. hypsinephes*; Thomas and Schwartz, 1983a); also under rocks, logs, *Cocos* fronds and husks, and other surface debris (Schwartz and Henderson, 1991).

Predators: *Celestus costatus* (Schwartz and Henderson, 1991). **Size:** Male SVL to 30 mm, female SVL to 31 mm (Thomas and Schwartz, 1983a).

Conservation Status: Unknown.

Sphaerodactylus asterulus Schwartz and Graham, 1980

Distribution: Hispaniola, northwestern Haiti; elevational range, sea level to ~235 m.

Habitat: Xerophilic (Schwartz and Henderson, 1991); under dead basal *Agave* rosettes and in low rock piles (Schwartz and Graham, 1980).

Diet: Most common prey beetles, flies, termites, and orthopterans (Cunningham et al., 1993). **Reproduction:** Eggs 6.4–6.8 mm × 5.1–5.7 mm, hatchling 15 mm SVL (Schwartz and Graham, 1980). **Size:** Male and female SVL to 31 mm (Schwartz and Graham, 1980).

Conservation Status: Not assessed.

Sphaerodactylus beattyi Grant, 1937

Distribution: U.S. Virgin Islands (St. Croix and satellites) where two currently recognized subspecies occur on eastern St. Croix, Green Cay, and Buck Island (*S. b. beattyi*) and the southern coast of St. Croix (*S. b. seamani*).

Habitat: In cactus and desertlike terrain in areas where an accumulation of leaves provides some protection (Grant and Beatty, 1944); on St. Croix, in areas of dry forest (MacLean and Holt, 1979).

Desiccation: Water-loss rate 3.2 mg/g/h (Snyder, 1979). **Predation:** Eggs eaten by *Ameiva polops* (Meier et al., 1993). **Size:** Male SVL to 30 mm, female SVL to 29 mm (Schwartz and Henderson, 1991).

Conservation Status: *Sphaerodactylus b. beattyi* "stable," *S. b. seamani* "threatened" (Moreno, 1991d).

Sphaerodactylus becki Schmidt, 1919

Distribution: Navassa Island.

Habitat: Xerophilic (Schwartz and Henderson, 1991); under palm debris (R. Thomas in Powell, 1999a); under rocks and in leaf litter in second growth woods (Schwartz and Henderson, 1991); also in rock crevices, in cracks and crevices in walls of a railway trench, low on trunks of *Metopium*, in tangled roots and stems of *Ficus*, in ruins at Lulu Bay, and in an empty conch shell (Powell, 1999a).

Activity: Active day and night (but primarily at dawn and dusk), diurnal activity confined to deep shade (none seen fully exposed during daylight hours; Powell, 1999a). **Size:** Male SVL to 31 mm, female SVL to 29 mm (Schwartz and Henderson, 1991).

Conservation Status: Not assessed.

Additional Pertinent Literature: Powell (1999f).

Sphaerodactylus bromeliarum Peters and Schwartz, 1972

Distribution: Eastern Cuba, Cuchillas de Toa.

Habitat: Mesophilic (Schwartz and Henderson, 1991); in a bromeliad with *Eleutherodactylus auriculatus* that fell from 10 m above the ground (G. Peters and Schwartz, 1972; G. Peters, 1974).

Size: Female SVL to at least 24 mm (G. Peters and Schwartz, 1972).

Conservation Status: Considered "critically endangered" (Rodríguez Schettino and Chamizo Lara, 1998a).

Sphaerodactylus caicosensis Cochran, 1934

Distribution: Caicos Islands.

Habitat: Xerophilic, under debris in an urban area (Schwartz and Henderson, 1991).

Predators: *Tropidophis greenwayi* (Iverson, 1986). **Reproduction:** Clutch size 1 (Hedges, in prep.); egg on 11 January (5.7 × 7.7 mm) in a house (Schwartz, 1968c). **Size:** SVL to 32 mm in both sexes (Schwartz, 1968c).

Conservation Status: Not assessed.

Sphaerodactylus callocricus Schwartz, 1976

Distribution: Hispaniola, northeastern Dominican Republic and the Península de Samaná.

Habitat: Mesophilic (Schwartz and Henderson, 1991); under *Cocos* trash along a beach near limestone cliffs, under logs and rocks in forest (Schwartz, 1976a).

Reproduction: Clutch size 1 (Hedges, in prep.). **Size:** Male and female SVL to 28 mm (Schwartz, 1976a).

Conservation Status: Not assessed.

Sphaerodactylus celicara Garrido and Schwartz, 1982

Distribution: Eastern Cuba.

Habitat: Mesophilic (Schwartz and Henderson, 1991); to some extent edificarian (on walls and in the lobby of a hotel), also in a bromeliad and on a rock (Garrido and Schwartz, 1982).

Size: Male SVL to 32 mm, female SVL to 31 mm (Garrido and Schwartz, 1982).

Conservation Status: Considered "vulnerable" (Rodríguez Schettino and Chamizo Lara, 1998a).

Sphaerodactylus cinereus Wagler, 1830

Distribution: Hispaniola (Haiti), where two currently recognized subspecies occur in the Plaine de Cul de Sac (*S. c. cinereus*) and northern Haiti (*S. c. stejnegeri*).

Habitat: Xerophilic to occasionally mesophilic, under surface debris, commonly edificarian under wall slats of houses (Schwartz and Henderson, 1991).

Activity: Forages around artificial lights at night (J. Lazell in Perry et al., 2008). **Reproduction:** Clutch size 1 (Hedges, in prep.); 3 eggs under the bark of a tree 1.0 m above the ground on 7 July (Schwartz and Henderson, 1991). **Size:** SVL in males and females to 32 mm (Schwartz and Henderson, 1991).

Conservation Status: Not assessed.

Sphaerodactylus clenchi Shreve, 1968

Distribution: Hispaniola, where two currently recognized subspecies occur on the Samaná Peninsula (*S. c. clenchi*) and in the eastern Dominican Republic (*S. c. apocoptus*).

Habitat: Mesophilic (Schwartz and Henderson, 1991); on the Samana Peninsula, *S. c. clenchi* in palm trash in mesic *Cocos* groves, in decayed palm trash, at the inner edge of a mangrove border in a dense growth of herbaceous halophytes, under palm trash in an open grassy coconut grove, under palm fronds behind *Coccoloba* on a sandy beach, also in a cacao grove in upland broadleaf forest at 305 m in the Sierra de Samaná, *S. c. apocoptus* in palm trash in a coastal *Cocos* grove and on the trunk of a banana in an open coffee-banana grove (Schwartz, 1983a); also associated with mogotes (Glor et al., 2001).

Activity: Observed on a *Musa* trunk by day (Schwartz, 1983a); likely crepuscular (Leuck et al., 1990). **Behavior:** Lateral compression (often accompanied by back arching and leg extension) observed in 4% of male-male and male-female interactions, and 10% of female-female interactions, gular extension (~2/sec) in 8% of male-male, 18% of male-female, 30% of female-female interactions, tail waving in 14% of male-male, 20% of male-female, 56% of female-female interactions, push-ups in 74% of male-male, 58% of male-female, 4% of female-female interactions (Leuck et al., 1990). **Diet:** In sample of 39 lizards, most frequently consumed prey were spiders, beetles, hemipterans, and hymenopterans (ants; Cunningham et al., 1993). **Population Density:** >4,000 adults/ha (Cheng, 1983). **Predation:** Predation pressure apparently high based on the incidence of broken tails (Cheng, 1983). **Reproduction:** Eggs November–December (Schwartz and Henderson, 1991); 1 egg/clutch and 3–5 clutches/year (Cheng, 1983); incubation 54 days at 28–30 °C in laboratory (Teale, 1930). **Size:** Male SVL to 33 mm, female SVL to 32 mm (Schwartz, 1983a).

Conservation Status: Not assessed.

Sphaerodactylus cochranae Ruibal, 1946

Distribution: Hispaniola, southern shore of Samaná Bay (Dominican Republic).

Habitat: Mesophilic (Schwartz and Henderson, 1991); in a breakwater and a limestone wall (W. G. Hassler in Ruibal,

1946); among human refuse just within the mouth of a cave immediately south of the mouth of the Bahía de San Lorenzo (Thomas and Schwartz, 1983a); also associated with mogotes, almost exclusively in and around bromeliads (Glor et al., 2001).

Parasites: Small unidentified mites in the folds of skin below the eyes on the type specimen (Ruibal, 1946). **Size:** Male SVL to ~30 mm, female SVL to 28 mm (Thomas and Schwartz, 1983a).

Conservation Status: Not formally assessed, but considered at risk ("endangered"; Powell et al., 2000).

Additional Pertinent Literature: Powell and Henderson (2008b).

Sphaerodactylus copei Steindachner, 1869

Distribution: Hispaniola, where nine currently recognized subspecies occur in Haiti (*S. c. copei*, *S. c. astreptus*, *S. c. cataplexis*, *S. c. enochrus*, *S. c. pelates*, *S. c. picturatus*, *S. c. websteri*), including Île-à-Vache (*S. c. cataplexis*), Île de la Gonâve (*S. c. deuterus*), and Île Grande Cayemite (*S. c. polyommatus*); introduced to the Bahama Islands (New Providence and possibly Andros Island; *S. c. cataplexis*); elevational range, sea level to >900 m.

Habitat: Mesophilic to xerophilic (Schwartz and Henderson, 1991); most frequently associated with piles of rocks, in piles of limestone rocks in a moderately dense wooded area surrounded by arid scrub and rocky hillsides, in rocks on scrubby hillsides, in moist banana and coffee groves associated with rotting logs and ground debris, *S. c. polyommatus* in dry rocky terrain (Thomas, 1968a).

Predation: *Hypsirhynchus parvifrons* (Henderson et al., 1987a). **Reproduction:** Clutch size 1 (Hedges, in prep.). **Size:** Male SVL to 41 mm, female SVL to 40 mm (Schwartz and Henderson, 1991).

Conservation Status: Not assessed.

Sphaerodactylus corticola Garman, 1988

Distribution: Bahama Islands, where as many as six subspecies occur on the Crooked-Acklins Bank (*S. c. campter*), Rum Cay (*S. c. corticola*), San Salvador Cays (*S. c. soter*), East Plana Cay (*S. c. apporox*), Samana Cay (*S. c.* ssp.), and possibly Conception Island (*S. c.* ssp).

Habitat: Mesophilic to xerophilic, often around human settlements and associated with human debris (Schwartz and Henderson, 1991); on San Salvador (*S. c. soter*), under palm (*Thrinax*) leaves on the beach, under boards adjacent to ruins in low coppice, in piles of *Sabal* trash, under small rocks at the base of a large *Tamarindus* tree, on Man Head and Low cays under flat rocks among strand plants, on Green Cay in a pile of rocks in a *Coccoloba* thicket, on Acklin's Island (*S. c. campter*), in open and shaded situations, including *Cocos* and *Thrinax* trash in beach habitat, and in plantings of fruit trees (citrus, banana, sugar-apple), on East Plana Cay (*S. c. apporox*), under palm fronds and palm logs (Schwartz, 1968c); at night in open shrub vegetation (Clough and Fulk, 1971).

Commensalism: Occupy iguana (*Cyclura rileyi*) burrows (Gicca, 1980). **Reproduction:** Clutch size 1 (Hedges, in prep.); eggs (17 June–30 December) 7.5–8.4 × 5.9–6.3 mm and hatchlings 16–17 mm SVL (Schwartz, 1968c); egg taken 26 June hatched 30 June, egg taken 28 June hatched 10 July (Schwartz and Henderson, 1991). **Size:** Male SVL to 37 mm, female SVL to 39 mm (Schwartz, 1968c).

Conservation Status: Not assessed.

Sphaerodactylus cricoderus Thomas, Hedges, and Garrido, 1992

Distribution: Southeastern Cuba.

Habitat: Mesic woods, including coffee groves, in leaf litter on limestone rocks near the base of a small, well-shaded sinkhole (Thomas et al., 1992).

Size: SVL to 25 mm (Thomas et al., 1992).

Conservation Status: Considered "vulnerable" (Rodríguez Schettino and Chamizo Lara, 1998a).

Sphaerodactylus cryphius Thomas and Schwartz, 1977

Distribution: Hispaniola, Valle de Neiba and lower slopes of the Sierra de Bahoruco (Dominican Republic); elevational range, sea level to ~480 m.

Habitat: Xerophilic (Schwartz and Henderson, 1991); in a shaded oasis surrounded by very xeric conditions, in piles of thatch-palm fronds in scrub, and in cactus-*Acacia* scrub (Thomas and Schwartz, 1983a); under rocks and in dead *Agave* (Schwartz and Henderson, 1991).

Size: Male SVL to 21 mm, female SVL to 22 mm (Thomas and Schwartz, 1977).

Conservation Status: Not assessed.

Sphaerodactylus dacnicolor Barbour, 1910

Distribution: Northeastern Jamaica; elevational range, sea level to ~240 m.

Habitat: Mesophilic (Schwartz and Henderson, 1991); wet limestone forest, rainforest, and coastal areas with high rainfall (Hedges, in prep.); in leaf litter, piles of *Cocos* trash, and from bromeliads (Thomas, 1975).

Reproduction: Clutch size 1 (Hedges, in prep.); egg hatched on 5 September (Schwartz and Henderson, 1991). **Size:** SVL to 32 mm (Thomas, 1975).

Conservation Status: Not assessed.

Sphaerodactylus darlingtoni Shreve, 1968

Distribution: Hispaniola, where four currently recognized subspecies occur in the western Cordillera Septentrional (*S. d. darlingtoni*), Sierra Martín García (*S. d. bobilini*), Sierra de Neiba (*S. d. mekistus*), and the northeastern Dominican Republic (*S. d. noblei*).

Habitat: Mesophilic (Schwartz and Henderson, 1991); *S. d. darlingtoni* in a mixed coffee-cacao planting in deciduous montane forest (425 m), under rocks, palm trash, and logs, also in rock rubble at 1,220 m, *S. d. bobilini* in montane forest (600–850 m) under rocks and leaf litter, *S. d. mekistus* in montane forest (850 m) under a rock along a road cut, *S. d. noblei* in a coffee-cacao grove, under rocks and in rock rubble on a densely shaded forested slope of a ravine, under rocks in high-canopied forest, and under limestone rocks at forest edge (Thomas and Schwartz, 1983a); also associated with mogotes (Glor et al., 2001).

Activity: Active in leaf litter at 1600 h in coffee-cacao habitat (*S. d. noblei*; Thomas and Schwartz, 1983a). **Reproduction:** Eggs (*S d. mekistus*) 7.8–8.1 × 5.3–5.5 mm (hatched 22 September) under rocks along a roadside in montane rainforest (Thomas and Schwartz, 1983a). **Size:** Male SVL to 25 mm, female SVL to 29 mm (Thomas and Schwartz, 1983a).

Conservation Status: Not assessed.

Sphaerodactylus difficilis Barbour, 1914

Distribution: Hispaniola, where seven currently recognized subspecies are widely distributed primarily in the lowlands (but absent from central Haiti and the Tiburon Peninsula), also on Île de la Tortue; elevational range, sea level to at least 600 m.

Habitat: *Sphaerodactylus d. difficilis* in deep ground litter of cacao leaves, palm frond trash, in cacao groves, dead *Agave* in xeric scrub, the thatched roof of a kitchen of an occupied hut, and on the porch of a hotel, *S. d. lycauges* in urban/edificarian situations, including under trash in coffee groves, under fallen thatch roof, in semixeric coffee-cacao grove, and in scattered leaf litter on a xeric hillside, *S. d. euopter* "fairly common" in mesic habitat, *S. d. typhlopous* in coastal *Coccoloba* and *Terminalia*, in a hotel bungalow at Sosúa, under palm trash on a mesic hillside, under dry logs, under termitaria, in piles of dry wood, sticks, leaves, and palm slats in riverine woods adjacent to a *Musa* grove, in palm trash in a garbage dump, in *Acacia*-cactus lowlands, in a stack of old logs in the yard of an abandoned sugar mill, in coastal coconut trash, *S. d. typhlopous* under bark and woody bracket-fungus 1.8 m high on a leguminous tree in mesic riverine forest, at 3.7 m in a tree, in a hollow decayed tree limb, and from under bark of a standing dead tree in a shaded *Acacia* ravine, *S. d. peratus* in palm trash along a sandy beach, under trash in *Theobroma* and *Cocos* groves, *S. d. diolenius* in *Cocos* trash near shore, under fronds on mud adjacent to a mangrove swamp, in human dwellings, in fallen thatch of an abandoned hut, under palm fronds on sand flats and beach dunes, under a log in a limestone pasture, in rock rubble and human debris on the cave floor along the edge of a cave, and under a thatch pile adjacent to a dry rice field, *S. d. anthracomus* under palm trash in coastal groves, in a banana grove, under dry *Cocos* trash piles on sand behind mangroves, under palm trash in mesic woods (Schwartz, 1983a); in an active cacao plantation, abandoned gardens and cacao, and mogote vegetation (Glor et al., 2001); *S. d. diolenius* in palm debris associated with a rock wall in an urban area (RP).

Activity: In late afternoon, *S. d. difficilis* "running freely," *S. d. diolenius* "actively running about" in the morning in litter on floor of mesic cacao grove, *S. d. anthracomus* active in the morning in an abandoned schoolhouse (Schwartz, 1983a). **Foraging and Diet:** Based on a sample of 28 lizards, stomach contents included beetles, spiders, hymenopterans (ants), and termites, beetles in more stomachs than any other prey taxon, with spiders a close second (Cunningham et al., 1993); exploits the "night-light" niche (RP in Perry et al., 2008). **Predators:** *Hypsirhynchus parvifrons* (Henderson et al., 1987a; J. W. Smith et al., 1994). **Reproduction:** Eggs (*S. d. anthracomus*) 5.3–5.5 × 7.4–7.9 mm, hatchlings 14–15 mm SVL (Schwartz, 1983a); eggs taken 15 July hatched 9 August, 1 egg (5.0 × 7.0 mm) taken 24 July hatched 1–7 September, 4 eggs (4.2–4.8 × 6.2–6.7 mm) taken 24 July in a palm slat pile, egg taken 15 October hatched 3 November (Schwartz and Henderson, 1991). **Size:** SVL in males to 34 mm, in female to 33 mm (Schwartz, 1983a).

Conservation Status: Not assessed.

Sphaerodactylus dimorphicus Fong Grillo and Diaz, 2004

Distribution: Southeastern Cuba.

Habitat: Limestone with xerophytic vegetation, in dead and living *Agave*, under stones on the seashore, and around an abandoned structure (Fong Grillo and Díaz, 2004).

Size: Male SVL to 32.9 mm, female SVL to 33.4 (Fong Grillo and Díaz, 2004).

Conservation Status: Not assessed.

Sphaerodactylus docimus Schwartz and Garrido, 1985

Distribution: Cuba, southeastern coast.

Habitat and Natural History: Unknown.

Size: Female SVL to 30 mm (Schwartz and Garrido, 1985).

Conservation Status: Considered "vulnerable" (Rodríguez Schettino and Chamizo Lara, 1998a).

Sphaerodactylus elasmorhynchus Thomas, 1966

Distribution: Hispaniola, Massif de la Hotte (Haiti).

Habitat: Among roots of a standing (but rotten) tree on the side of ravine (overgrown with mesic vegetation) filled with a jumble of limestone rocks and boulders in an area of cutover countryside (Thomas, 1966e).

Size: Female SVL to 17 mm (Thomas, 1966e).

Conservation Status: Not assessed, but known only from a single specimen despite concerted efforts to find additional animals (R. Thomas, in litt., 22.IV.2008).

Sphaerodactylus elegans MacCleay, 1834

Distribution: Cuba, islandwide, but disjunct, also Isla de la Juventud, Archipiélago de Sabana-Camagüey, Archipiélago de los Canarreos (*S. e. elegans*), and Hispaniola, including Île Grande Cayemite and Île de la Gonâve (*S. e. punctatissimus*); introduced in the Florida Keys (*S. e. elegans*).

Habitat: Mesophilic to xerophilic (Schwartz and Henderson, 1991); predominantly edificarian, "rarely seen out of doors" (Barbour, 1921); in Cuba, often in villages and cities in houses and hotels (Garrido, 1973d; Schwartz and Garrido, 1985; Estrada, 1993a; M. Černý, 1994; Martínez Reyes, 1995; Holanova, 2003; Kunz, 2005; Díaz and Abreu Guerra, 2005); by day behind furniture, books, and pictures (often of "the Virgin"; Barbour and Ramsden, 1919); under tree bark (Collette, 1961); almost entirely arboreal, under loose bark of trees (Collette, 1962); also occasionally on tree trunks, under fiber on palm trunks, under palm fronds and trash adjacent to a cornfield near the ocean, in *Cocos* husks near a mangrove edge, locally common under palm slats of dwellings (Schwartz and Henderson, 1991); always in developed areas, often associated with buildings, root systems of Kapoc (*Ceiba pentandra*), frequently near arboreal termite nests, even with *Anolis sagrei* in palm axils (Vergner, 1998); frequently a human commensal, under rocks and deadfall or in leaf litter and artificial situations (under flower pots, behind wall hangings or books on shelves, even in drawers), any microhabitat that holds moisture (Röll, 2006); presumably transported to Cayo Francés (Cuba) in utensils and construction materials (Garrido, 1973d).

Activity: Forages at and after dusk on walls near lights, feeding on light-attracted insects, also active by day when overcast or in shade, foraging on house walls and interiors of houses and hotels (Schwartz and Garrido, 1985); "sallies forth when dusk comes" (Barbour and Ramsden, 1919); in Cuba, foraging at any time of day when in edificarian situations, but in natural situations active mainly in early morning and at dusk, when it forages on tree bark, in bark crevices, and on litter near trees (Regalado, 1997); diurnal, often becomes active when the sun comes out after rain, in captivity, primarily crepuscular (Vergner, 1998); human commensals often active throughout the day, more likely to be crepuscular in nature, basks only in the mornings (Röll, 2006). **Behavior:** Repertoire of 20 behaviors (body postures, head and tail movements), tongue-flicking critical for sex determination (detection of pheromones may be involved), males behave aggressively toward other males and courted females, females were passive with males, but were aggressive toward other females (Regalado, 1997); when an animal dropped to the ground, it burrowed into detritus, moved back to base of the tree, and climbed up again (Collette, 1962); very shy, does not readily flee when disturbed in refuges, but difficult to catch, often in male-female pairs, often aggressive in captivity, uses tail waving and readily bites (Vergner, 1998); never ventures far from refuges, especially juveniles, very shy, retreats into shelters at the first sign of danger (Holanova, 2003); frequently headbobs, but body elevation rare, instead uses an angled orientation with forelimbs drawn under the body, male and female behavior similar when interactions initiated (sex recognition is probably chemical), both sexes may be aggressive, battles in nature may last to 4 h before the "loser" retreats, will bite when handled (Röll, 2006). **Defense Mechanisms:** Tail autotomy, tails are very fragile (Vergner, 1998). **Foraging and Diet:** In captivity, approaches prey slowly while wagging its tail, grabs prey with jaws, chews, and licks lips after swallowing, will take fruit (Holanova, 2003). **Miscellany:** Eyes can move independently or can be used for binocular vision (Röll, 2006). **Parasites:** Nematodes *Parapharyngodon cubensis* (Coy Otero and Baruš, 1973b; Coy Otero and Baruš, 1979b; Coy Otero and Lorenzo Hernández, 1982) and

Spauligodon antillarum (Baruš and Coy Otero, 1974; Coy Otero and Baruš, 1979b; Coy Otero and Lorenzo Hernández, 1982). **Predation:** Scorpion *Rhopalurus junceus* (R. Regalado, 1984 in Armas, 2001); *Hypsirhynchus parvifrons* (Henderson et al., 1987a); *Uromacer frenatus* (Henderson et al., 1987c). **Reproduction:** Eggs laid communally (to 58 in one place) in trees, in abandoned termite nests, under bark, and in rotten ends of branches (Collette, 1962); egg on 2 August 8.0 × 5.9 mm (Schwartz and Henderson, 1991); lays eggs in crevices in rotten logs or under loose bark, in captivity, females can store sperm for at least one month, incubation 70–90 days (M. Černý, 1994); female in captivity laid eggs (~6 × 10 mm) 6 times/year at intervals of at least 4 weeks, incubation ~3 months at 25–32 °C (day) and 18–22 °C (night), hatchling total length ~20 mm (Klug, 1995); egg under bark, in captivity, male pursues the female, waving his tail, biting the tail, body, and neck of the female until establishing a hold on her nape just prior to intromission, copulation lasts 6–9 sec before the female shakes her tail and escapes, copulation often repeated several times, incubation 80–86 days at 27–29 °C, hatchlings ~30 mm total length and 0.11 g, female 1 month after hatching 20 mm SVL, 28 mm total length (Vergner, 1998); in captivity, courtship involves male approach, moving head from side to side, expanding the throat, and wagging the tail, female responds by elevating the side nearest the male, which then approaches the female's cloaca, he bites her neck, inserts a hemipenis, copulation lasts only a few seconds, after which the female flees, both lick their cloacae, females deposit eggs communally (in nature, 2 females and ~15 eggs together), do not bury eggs, one female laid 2 small eggs at one time, female may "protect" eggs from males or insects, eggs with calcified shells 8 × 6–7 mm, hatching in 70–90 days, hatchlings ~30 mm total length, mature in about 1 year (Holanova, 2003); in captivity, unresponsive females react to male overtures with tail waving and attempts to bite the male, males may continue courtship, female often "gives up" after 2–3 attempts, females may lay 2–3 fertile eggs after separation from a male, eggs 6 × 7–8 mm, 100–150 mg, hatch in 65–90 days at 25–29 °C, hatchlings 16–16.5 mm SVL, 90–130 mg, juvenile color pattern begins to fade at 3–4 weeks at SVL 22–24 mm (Röll, 2006). **Size:** SVL in males to 39 mm, in females to 37 mm (Schwartz and Garrido, 1985).

Conservation Status: Not assessed.

Sphaerodactylus elegantulus Barbour, 1917

Distribution: Antigua (and satellites) and Barbuda.

Habitat: Mesophilic, on Antigua, under stones and dead leaves beneath bushes, and under rock piles in a xeric "garden" (Schwartz and Henderson, 1991); gravid females on Barbuda in moist debris under palm fronds and rocks in Darby Cave (King, 1962).

Abundance: Common all over Barbuda (Malhotra and Thorpe, 1999). **Activity:** On a sidewalk at 2000 h at Dutchman's Bay, Antigua (Schwartz and Henderson, 1991). **Reproduction:** On Barbuda, gravid females in June–July, eggs (5.5 × 7.0 mm) in moist debris under palm fronds and rocks in Darby Cave (King, 1962). **Size:** Male SVL to 28 mm, female SVL to 29 mm (Schwartz and Henderson, 1991).

Conservation Status: Not assessed.

Sphaerodactylus epiurus Thomas and Hedges, 1993

Distribution: Hispaniola, Cordillera Oriental (Dominican Republic).

Habitat: Limestone hills, beneath limestone rocks (often in piles) in mesic, canopied situations (including entrances to caves), some on undersides of rocks (Thomas and Hedges, 1993).

Reproduction: Clutch size 1 (Hedges, in prep.). **Size:** SVL to 25 mm (Thomas and Hedges, 1993).

Conservation Status: Unknown.

Sphaerodactylus fantasticus Cuvier, 1836

Distribution: Central Lesser Antilles, where eight currently recognized subspecies occur on Montserrat (*S. f. ligniservulus*), Guadeloupe (*S. f. fantasticus*, *S. f. orescius*, *S. f. tartaropylorus*) and satellites (*S. f. karukera*), Marie-Galante (*S. f. anidrotus*), La Désirade (*S. f. hippomanes*), and Dominica (*S. f. fuga*).

Habitat: Mesophilic (Schwartz and Henderson, 1991); on Montserrat, on forested slopes and in *Cocos* trash (Schwartz and Henderson, 1991); on Guadeloupe, in humid forests, leaf litter in drier areas, often associated with manchineel and almond trees in coastal situations, along sugarcane fields, *Musa* groves, among mango and breadfruit trees, in *Acacia*, and in ravines, usually under logs, stones, piles of wet leaves (Breuil, 2002b); in garbage and *Cocos* trash, on La Désirade, in beachside coconut palm habitat, breadfruit, *Terminalia*, *Hippomane*, and *Coccoloba*, on Îlet Kahouanne, around mango trees and in a *Hippomane* grove (Schwartz and Henderson, 1991); on Dominica, in xeric woodlands (Malhotra and Thorpe, 1999); also in *Musa* groves, in and about old houses, in dry river ravines, in *Acacia* scrub, under logs, in the bottom of a sinkhole in a field, under breadfruit and *Terminalia* leaves, and on the edge of a cane field (Schwartz and Henderson, 1991).

Activity: At dusk under boards in a shack in forest (Schwartz and Henderson, 1991); on Guadeloupe, active by day and crepuscular (Breuil, 2002b). **Behavior:** Raises tail to expose contrasting underside as a possible predator deterrent (Breuil, 2002b). **Foraging and Diet:** *S. f. fantasticus* with small snail shells in stomach (King, 1962); may have binocular vision to aid foraging, forages for termites, ants, spiders, and springtails during cooler hours of the day (Breuil, 2002b). **Population Density:** 1.0/m^2 on Grand Terre (with 4–5 layers of leaf litter), 0.79/m^2 (0.27 g/m^2) in mangrove habitat and 0.20/m^2 in dry forest on Grand Îlet (Breuil, 2002b). **Predators:** On Guadeloupe, *Anolis marmoratus* saw a gecko from 2.0 m up in a tree, swallowed it head first (Breuil, 2002b). **Reproduction:** Clutch size 1 (Hedges, in prep.); egg in rotten wood on Guadeloupe 6.5 × 4.75 mm (Barbour, 1921); on Grande-Terre, 4 eggs in July with logs used as egg deposition sites (Breuil, 2002b); egg 4.9 × 7.4 mm (13 April) in dry fallen termitarium (Schwartz and Henderson, 1991). **Size:** Male SVL to 28 mm, female SVL to 29 mm (Schwartz and Henderson, 1991).

Conservation Status: Not assessed.

Sphaerodactylus gaigeae Grant, 1932

Distribution: Puerto Rico, including Vieques, Cayo Santiago, and Isla Piñeros.

Habitat: Mesophilic (Schwartz and Henderson, 1991); in dry scrub forest and wet upland forest under leaf litter, rocks, and logs in coffee plantations (Hedges, in prep.); on Isla Piñeros, under rocks in xeric situations (Rivero, 1998).

Desiccation: Water loss rates 80.60 ± 40.20 mg/g/h and 103.96 ± 4.88 mg/g/h (Nava, 2004). **Size:** SVL in males and females to 25 mm (Schwartz and Henderson, 1991).

Conservation Status: "Stable" (Moreno, 1991d).

Sphaerodactylus gilvitorques Cope, 1862

Distribution: Presumably Jamaica (see remarks in Schwartz and Henderson, 1988).

Habitat and Natural History: Unknown.

Size: SVL of only known female 27 mm (Schwartz and Henderson, 1991).

Conservation Status: Unknown, possibly extinct (S. B. Hedges, in litt., 18.IV.2008).

Sphaerodactylus goniorhynchus Cope, 1895

Distribution: Jamaica, widely distributed, but disjunct; elevational range, sea level to ~1,220 m.

Habitat: Mesophilic (Schwartz and Henderson, 1991); among roots of banana plants and under rocks (Barbour, 1910); in rotten logs in wooded country and heavy forests (Barbour, 1921); under trash (Grant, 1940a); in xeric coppice scrub, hammocklike woods associated with mangroves (Morant Point), coastal scrub, *Cocos* plantation in coconut trash, in bromeliads at night, *Coccoloba* litter, coral rubble, and under rocks (Schwartz and Henderson, 1991); under leaf litter in a dry wash (B. I. Crother, in litt., 26.II.2008).

Reproduction: Clutch size 1 (Hedges, in prep.); eggs in termitaria and under bark, hatchling 19 mm SVL (Barbour, 1921); egg (23 July) 3.5 × 4.5 mm, egg laid 1 August hatched 31 October, on 30 November, hatchling 13.5 mm SVL, 0.07 g (Schwartz and Henderson, 1991). **Size:** Male SVL to 32 mm, female SVL to 25 mm (Schwartz and Henderson, 1991).

Conservation Status: Not assessed.

Sphaerodactylus inaguae Noble and Klingel, 1932

Distribution: Bahama Islands, Great Inagua Island and Sheep Cay.

Habitat: Xerophilic (Schwartz and Henderson, 1991); widely distributed in heavily wooded coastal areas, in and about houses, beneath loose boards, shingles, and flat stones (Noble and Klingel, 1932).

Activity: Largely nocturnal, exits daytime "retreats at approximately six in the evening. After dark, many are seen running over the floors and walls of the native huts" (Noble and Klingel, 1932). **Reproduction:** Eggs deposited on undersides of tamarind shells, under flat stones, boards, and shingles, in hollow sisal stems and roof thatch, and in coconut fiber, egg on 20 January, eggs on 27 January and 16 February hatched on 14 April, a series of eggs on 16 February (and maintained at 28 °C starting 25 February) hatched 4 March, hatchling 14 mm SVL, egg dimensions 7.0 × 5.0 mm (Noble and Klingel, 1932). **Size:** Male SVL to 29 mm, female SVL to 28 mm (Schwartz and Henderson, 1991).

Conservation Status: Not assessed.

Sphaerodactylus intermedius Barbour and Ramsden, 1919

Distribution: Cuba, northwestern coast.

Habitat: Mesophilic (Schwartz and Henderson, 1991); in coastal or near-coastal habitats, in talus and rubble at the base of low limestone cliffs, from leaf and stone rubble at the base of a small tree on a cleared open beach (Schwartz and Garrido, 1985); under loose stones in a dry, open, scrubby, hillside pasture (Barbour, 1921).

Parasites: Nematode *Parapharyngodon sphaerodactyli* (Baruš and Coy Otero, 1974; Coy Otero and Baruš, 1979b; Coy Otero and Lorenzo Hernández, 1982). **Size:** Male SVL to 36 mm, female SVL to 35 mm (Schwartz and Garrido, 1985).

Conservation Status: Not assessed.

Sphaerodactylus kirbyi Lazell, 1994

Distribution: St. Vincent Grenadines, Bequia, Petit Nevis, Mustique, Mayreau.

Habitat: In leaf litter (Powell and Henderson, 2006; Daudin and de Silva, 2007).

Abundance: Common on Mustique in windward coastal habitat (Daudin and de Silva, 2007). **Size:** Male SVL to 23.6 mm, female SVL to 25 mm (Lazell, 1994).

Conservation Status: Not assessed.

Additional Pertinent Literature: Hite et al. (2008b).

Sphaerodactylus klauberi Grant, 1931

Distribution: Puerto Rico, uplands; elevational range ~50–1,220 m.

Habitat: Mesophilic (Schwartz and Henderson, 1991); interior upland moist habitats, primarily in leaf litter (Thomas and Gaa Kessler, 1996); in wet limestone forest, rainforest, cloud forest, and elfin forest where under rocks, logs, leaf litter, palm fronds, and *Cocos* husks (Hedges, in prep.).

Activity: Active in leaf litter by day (Thomas and Gaa Kessler, 1996). **Desiccation:** Standard rate of water loss 16.7 mg/g of initial body weight/h, vital limit (percent of initial body weight lost before death) 24.3, and survival time 15.5 h (Heatwole and Veron, 1977). **Diet:** In a sample of 157 prey items, mites 15.9%, spiders 15.3%, springtails 14.0%, isopods 13.4%, beetle larvae and adults 12.1%, flies 5.7%, lepidopteran larvae and an adult 5.1%, and an *Eleutherodactylus*, also termites, gastropods, harvestmen, millipedes, centipedes, orthopterans, barklice, silverfish, ants, trichopterans (Thomas and Gaa Kessler, 1996); *Eleutherodactylus coqui* (Stewart and Woolbright, 1996). **Population Density:** 10,000/ha, sometimes in high concentrations in thick litter piles (Thomas and Gaa Kessler, 1996). **Predation:** *Anolis gundlachi* (J. A. Román, in litt., 13.II.2008). **Reproduction:** Clutch size 1 (Hedges, in prep.); 2 eggs (17 March) 7.4–7.7 × 5.5–5.6 mm (Schwartz and Henderson, 1991). **Size:** Male SVL to 36 mm, female SVL to 37 mm (Schwartz and Henderson, 1991).

Conservation Status: "Stable" (Moreno, 1991d).

Sphaerodactylus ladae Thomas and Hedges, 1988

Distribution: Hispaniola, western base of Sierra Martín García (Dominican Republic).

Habitat: Xerophilic (Schwartz and Henderson, 1991); in "hilly and scrubby" area near a region of mixed agriculture (corn, coconuts; Thomas and Hedges, 1988); under rocks in dry scrub thornforest (RP).

Size: SVL to at least 27 mm (Thomas and Hedges, 1988).

Conservation Status: Unknown.

Sphaerodactylus lazelli Shreve, 1968

Distribution: Hispaniola, northern Haiti.

Habitat: Probably mesophilic (Schwartz and Henderson, 1991); under the bark of a tree in a shady gully (Shreve, 1968).

Size: Male SVL to at least 31 mm (Shreve, 1968).

Conservation Status: Unknown.

Sphaerodactylus leucaster Schwartz, 1973

Distribution: Hispaniola, Valle de Neiba to the Llanos de Azua and into the San Juan Valley (Dominican Republic); elevational range, sea level to ~370 m.

Habitat: Xerophilic (Schwartz and Henderson, 1991); in xeric and coastal woods, in an abandoned termitarium, under bark at the base of a large, standing *Acacia*, and in dead basal *Agave* rosettes (Schwartz and Graham, 1980).

Reproduction: Clutch size 1 (Hedges, in prep.). **Size:** Male SVL to 27 mm, female SVL to 31 mm (Schwartz and Graham, 1980).

Conservation Status: Not assessed.

Sphaerodactylus levinsi Heatwole, 1968

Distribution: Isla Desecheo.

Habitat: Xerophilic (Schwartz and Henderson, 1991); under stones and dead wood at edges of dry drainage gullies

(Heatwole, 1968b); most common under rocks in forests (with *Bursera simaruba*) on low slopes just above valley floors (Meier and Noble, 1990).

Population Density: Significantly higher in forested than in nonforested habitat, observed more frequently after rainfall (Meier and Noble, 1990). **Reproduction:** Egg (4.0 mm diameter) on 20 October partially embedded in soil under a rock with an adult (Meier and Noble, 1990). **Size:** SVL in males and females to 28 mm (Heatwole, 1968b).

Conservation Status: "Common but restricted, protect habitat" (Moreno, 1991d).

Sphaerodactylus macrolepis Günther, 1859

Distribution: Puerto Rico Bank, where nine currently recognized subspecies occur on Puerto Rico (*S. m. ateles*, *S. m. grandisquamis*), including Cayo Santiago, Cayo Algodones, Cayo Batata, Isla Piñeros (*S. m. guarionex*, *S. m. mimetes*, *S. m. phoberus*, *S. m. spanius*), and islands to the east (*S. m. macrolepis*), Isla Piñeros (*S. m. stibarus*), Isla Vieques and satellites (*S. m. inigoi*), and the U.S. and British Virgin Islands (*S. m. macrolepis*); elevational range, sea level to ~850 m.

Habitat: On Puerto Rico, *S. m. grandisquamis* in littoral situations under palm trash and other debris, in an abandoned shack in rolling topography, and in a patch of lowland forest with leaf-littered floor and much trash (Thomas and Schwartz, 1966); in houses (W. Peters, 1876); also in bathrooms and "semidark" rooms (Danforth, 1925); on Cayo Santiago, beneath coconut fronds and under hillside rocks, on Cayo Batata under loose rocks and under pieces of coral above the high-tide line on the beach (Heatwole et al., 1963); *S. m. stibarus* (Isla Piñeros) in *Coccoloba* leaves along the shore, in mango leaves, and in coconut trash in an abandoned grove, *S. m. phoberus* in mesic woods and old coconut plantations under moist piles of rotting coconut husks, tires, rocks, concrete slabs, and palm logs, *S. m. mimetes* in lowland coffee, in coastal palm groves, under logs and rocks on dry hills and plains, *S. m. ateles* in palm trash along the coast, under rocks on exposed hillsides, and in a wet, shaded ravine, *S. m. spanius* in forest clearings where trash is present, at the bases of banana plants in a high-canopied coffee grove, along edges of a temporarily disused picnic area (Thomas and Schwartz, 1966); on Puerto Rico, largely a leaf-litter denizen in some lowland areas, but in forested situations, in thick deposits of vegetative debris and aboveground situations, 2–10 m above the ground in bromeliads or under tree bark (Thomas and Gaa Kessler, 1996); on Isla Vieques and Cayo de Tierra, *S. m. inigoi* under rocks on an exposed, sunny hillside, and in mesic ravines, coastal *Coccoloba*, and palm trash (Thomas and Schwartz, 1966); on Anegada, under boards, pieces of tin, coconut husks, rocks, logs, moist leaf litter, and fallen palm fronds at bases of trees, usually on soil rather than limestone (Carey, 1972); on Tortola, in palm trash in a coastal *Cocos* grove, on Beef Island, in an abandoned coconut grove and open scrub (Thomas and Schwartz, 1966); on Guana, in *Leucaena-Acacia* stands and *Coccoloba* (Rodda et al., 2001); on Buck Island, in *Coccoloba* stands, on Virgin Gorda, in coastal coconut trash, on St. John, plentiful in ruins of plantation houses, on St. Croix, in coastal coconut groves, dry maritime woods, and upland woods (Thomas and Schwartz, 1966).

Activity: Diurnal in the Virgin Islands (Seaman, 1961); *S. m. spanius* active at dusk while crossing open tracts of lawn grass (Thomas and Schwartz, 1966); on Anegada, several *S. m. macrolepis* sitting (basking?) on a fallen palm frond by day (Carey, 1972). **Behavior:** Males will stand upright on hindlimbs to survey surroundings, but tail must be complete or regenerated (A. J. Sánchez Muñoz, in litt., 17.III.2008). **Desiccation:** Water-loss rate 4.2 mg/g/h on St. Croix (Snyder, 1975); 3.44 ± 0.44–4.95 ± 0.54 mg/g/h on Guana (Maclean, 1985); 70.83 ± 5.29–71.0 ± 34.2 mg/g/h on Puerto Rico (Nava, 2004). **Foraging and Diet:** In a sample of 146 prey items from northeastern Puerto Rico, 64.4% springtails, 10.3% isopods, 7.5% spiders, 4.1% flies, also ants, silverfish, termites, barklice, hemipterans, and mites (Thomas and Gaa Kessler, 1996); forages around artificial lights at night on Guana, British Virgin Islands (Perry and Lazell, 2000; Perry et al., 2008). **Population Density and Biomass:** On Guana (British Virgin Islands), in plots dominated by *Leucaena leucocephala* and *Acacia* sp., average density was 13,400/ha (biomass 3.60 kg/ha), in plots dominated by *Coccoloba uvifera*, average density was 52,800/ha (biomass 15.26 kg/ha), projected density in continuous *Coccoloba* litter 67,600/ha (Rodda et al., 2001). **Predators:** *Borikenophis portoricensis* on Culebra, Hans Lollick, and Guana (Henderson and Sajdak, 1996); on Guana, *Ameiva exsul* (RP) and *Anolis pulchellus* (K. de Queiroz and J. B. Losos, pers. comm.); Cattle Egret (*Bubulcus ibis*) on St. Croix (Gassett et al., 2000); Bridled Quail Dove (*Geotrygon mystacea*) on Guana (Boal, 2008b); Mangrove Cuckoo (*Coccyzus minor*) on Guana (M. E. Gifford, pers. comm.). **Reproduction:** Clutch size 1 (Hedges, in prep.); egg of *S. m. grandisquamis* 7.0 × 5.2 mm, hatchlings 14 mm SVL, 10 eggs of *S. m. phoberus* 6.7–7.2 × 5.0–5.8 mm, hatchlings 13–14 mm SVL, hatchling *S. m. macrolepis* on Tortola 12 mm SVL (Thomas and Schwartz, 1966); on Puerto Rico, eggs in bromeliads and under tree bark (Thomas and Gaa Kessler, 1996); eggs in a china cabinet at Mayagüez in May (Danforth, 1925). **Size:** Male SVL to 34 mm, female SVL to 35 mm (Schwartz and Henderson, 1991).

Conservation Status: All subspecies "stable," except *S. m. phoberus*, which is "threatened" (Moreno, 1991d).

Remarks: This species as currently recognized may represent a complex of species.

Sphaerodactylus mariguanae Cochran, 1934

Distribution: Bahama Islands, Mayaguana Island, Booby Cay, and Turks Islands, Grand Turk Island.

Habitat: Probably xerophilic, under surface debris (Schwartz and Henderson, 1991).

Reproduction: Eggs on 23–30 March, 7.2–7.5 mm × 5.7–5.9 mm (Schwartz and Henderson, 1991). **Size:** SVL in males and females to 41 mm (Schwartz and Henderson, 1991).

Conservation Status: Not assessed.

Sphaerodactylus microlepis Reinhardt and Lütken, 1862

Distribution: St. Lucia (*S. m. microlepis*) and Maria Major and the Vieux Fort Quarter of St. Lucia (*S. m. thomasi*); presence reported but not verified on Dominica (Malhotra and Thorpe, 1999).

Habitat: Primarily in leaf litter (Long, 1974); probably xerophilic (Schwartz and Henderson, 1991); on St. Lucia, in rainforest, under wet rocky rubble along paths and under dead leaves on mud, under rocks and large piles of palm trash in a *Cocos* grove and an adjacent pasture, in rich leaf cover of a cacao grove, in the wet remains (palm thatch sides and roof) of a collapsed shack on a moist shaded hillside and ravine, and under palm trash and rocks on a xeric (but shaded) hillside, at Anse de Sables (Vieux Fort), in leaves under *Coccoloba* trees immediately behind the strand, on Maria Major, under rocks in the shaded tree zone (Schwartz, 1964a).

Diet: In captivity, feeds readily on small flies and termites (Long, 1974). **Reproduction:** 5 eggs (5.5–5.8 × 6.7–7.4 mm) on 15 April on St. Lucia, hatched 15 May, 4 eggs (6.6–7.0 × 5.3–5.6 mm) on Maria Major on 3 April, hatched 15 May (Schwartz, 1964a). **Size:** SVL in males and females to 34 mm (Schwartz and Henderson, 1991).

Conservation Status: Not assessed.

Sphaerodactylus micropithecus Schwartz, 1977

Distribution: Isla Monito.

Habitat: Xerophilic (Schwartz and Henderson, 1991); under rocks with small amounts of soil at a temperature of 26 °C (air temperature 32 °C; Dodd and Ortiz, 1983).

Predators: Potentially *Rattus rattus* (Dodd and Ortiz, 1983). **Reproduction:** Likely March–November, egg 24 May, gravid females and juveniles 24–25 August (Dodd and Ortiz, 1983); gravid female (late August) with egg 6.25 × 9.2 mm (Dodd and Ortiz, 1984). **Size:** SVL to 36 mm (Dodd and Ortiz, 1983).

Conservation Status: On the IUCN Red List as "endangered" (Tolson, 1996c), but the listing is "out of date"; listed as "endangered" on the U.S. List of Endangered and Threatened Wildlife and Plants (Dodd and Ortiz, 1983; U.S. Fish & Wildlife Service, 2008); "endangered" (Moreno, 1991d); protected as "threatened" or "endangered" (Joglar, 2005b).

Additional Pertinent Literature: Dodd (1985).

Sphaerodactylus monensis Meerwarth, 1901

Distribution: Isla Mona.

Habitat: Xerophilic to semimesophilic (Schwartz and Henderson, 1991); among limestone boulders and from beneath pieces of coral (K. P. Schmidt, 1928); under trash (Grant, 1931b); under leaf litter and debris near the lighthouse (Heatwole, 1961); also in moderately mesic woods to xeric scrub in palm trash, among needles in open *Casuarina* plantings, and in scrub and scrubby woods on the limestone platform (Thomas and Schwartz, 1966).

Activity: Implication of nocturnal activity (K. P. Schmidt, 1928); active by day in leaf litter (RP). **Predators:** *Borikenophis variegatus* (M. Leal in Henderson and Sajdak, 1996); in stomachs of feral cats (García et al., 2001). **Size:** SVL in both sexes to 30 mm (Thomas and Schwartz, 1966).

Conservation Status: "Stable" (Moreno, 1991d).

Sphaerodactylus nicholsi Grant, 1931

Distribution: Puerto Rico, primarily coastal, but discontinuous.

Habitat: Xerophilic (Schwartz and Henderson, 1991); in piles of coconut trash on beaches, often in exposed areas (Rivero, 1998); in coconut groves under surface debris (Schwartz and Henderson, 1991); in southwestern Puerto Rico, in areas of dense shade under canopies of broad-leaved evergreen trees (*Pithecellobium dulce, Tamarindus indica, Bucida buceras, Hymenaea courbaril, Guaiacum officinale*) that produce large seeds or fruits (López-Ortiz and Lewis, 2004); gravid females at the bases of bushes or grass clumps (*Croton* sp., *Panicum maximum*; López-Ortiz and Lewis, 2002).

Activity: Walks about during the day over leaf-carpeted open deciduous forest floor (Rivero, 1998). **Behavior:** Repertoire of 20 behaviors (including copulation), males and females have similar repertoires, but used signaling behavior differently, headbob rates are indicative of male sex, especially during courtship, some behaviors (lifting of forelimb, stereotyped head lifting) used only by females, throat color may be used in sex recognition, but tongue flicking important in conspecific sex recognition (Regalado, 2003). **Commensalism:** 3–5 individuals often with pairs of *S. roosevelti* in tufts of grass 1.0 × 1–3 m in diameter (Boone, 2005). **Population Density:** In southwestern Puerto Rico, mean 4.1 ± 4.19 geckos/min (1–19) in good habitat (López-Ortiz and Lewis, 2004). **Predators:** *Borikenophis portoricensis* (Rodríguez-Robles and Leal, 1993a; Thomas and Gaa Kessler, 1996); likely *Ameiva exsul* based on *S. nicholsi* population crash after immigration of 5 *A. exsul* into a study plot (López-Ortiz and Lewis, 2004); *Anolis cristatellus* and *A. stratulus* (López-Ortiz and Lewis, 2004). **Reproduction:** In southwestern Puerto Rico (Cabo Rojo), gravid females 15.7–23.0 mm SVL, reproduction throughout most of the year, but most gravid females in July and most hatchlings September–February, monthly percentage of gravid females positively correlated with historical average monthly insolation, temperature, and day length, monthly percent of hatchlings negatively correlated with historical average monthly insolation, average monthly insolation of the year of the study (June 1997–May 1998), and average monthly day length (i.e., hatchlings most abundant when heat stress presumably least; López-Ortiz and Lewis, 2002); clutch size 1 (Hedges, in prep.). **Size:** Male SVL to 24 mm, female SVL to 25 mm (Schwartz and Henderson, 1991).

Conservation Status: "Stable" (Moreno, 1991d).

Sphaerodactylus nigropunctatus Gray, 1845

Distribution: Nine currently recognized subspecies occur in the Bahama Islands, Great Bahama Bank (*S. n. nigropunctatus*, *S. n. flavicauda*), Exuma Cays, Long Island, Little San Salvador Island, Green Cay (*S. n. gibbus*), Cat Island, Rum Cay (*S. n. decoratus*), Cay Sal Bank (*S. n. flavicauda*), Ragged Islands (*S. n. porrasi*), and Cuba (*S. n. alayoi*, *S. n. granti*, *S. n. lissodesmus*, *S. n. ocujal*, *S. n. strategus*), including Archipiélago de Sabana-Camagüey and Cayos los Ballenatos (*S. n. granti*).

Habitat: Xerophilic to mesophilic (Schwartz and Henderson, 1991); at the U.S. Naval Base at Guantanamo Bay, *S. n. strategus* under a piece of demolished brick masonry (Lando and Williams, 1969); on Mangrove Cay (Bahamas), *S. n. flavicauda* in chinks in the wall of a house, on Mangrove Cay (Andros Island), common around houses (Barbour, 1904); on Great Ragged Island, *S. n. porrasi* under rocks and other cover objects on the ground in the vicinity of human structures, on Little Ragged Island, *S. n. porrasi* on a rock wall and under rocks bordering a *Conocarpus* swamp (L. Porras, in litt., 17 July 2007); often in edificarian situations, although also in mesic broadleaf forest, more likely encountered in xeric woods under rocks, leaf litter, and human debris (Schwartz and Henderson, 1991); in Santiago de Cuba, may coexist under cover objects with *S. torrei* (Röll, 2006); under palm fronds and husk debris in a nursery at Guantanamo Bay (B. I. Crother, in litt., 26.II.2008).

Reproduction: Eggs (9) 17–28 June, 7.5–8.1 × 5.7–6.5 mm, hatched 24 June–4 July (Schwartz and Henderson, 1991); diameters of *S. n. granti* eggs 5.4–8.1 mm, annual egg production 3.0 in first year, 9.0 second year, incubation time 74–94 days (Regalado, 2006); *S. n. lissodesmus* laid 2 eggs sequentially and manipulated them with her hind limbs (L. M. Díaz in Regalado, 2006); male hatchling *S. n. granti* SVL 17.0 mm, female 20.0 mm, smallest mature female 28.0 mm SVL (Regalado, 2006). **Size:** SVL in males and females to 40 mm (Schwartz and Henderson, 1991).

Conservation Status: Not assessed.

Sphaerodactylus notatus Baird, 1858

Distribution: As many as five subspecies occur in the Bahama Islands, Great Bahama Bank, Cat Island, Little San Salvador Island (*S. n. amaurus*), Little Bahama Bank (*S. n. peltastes*), on Cuba, including Cayos los Ballenatos, apparently introduced on Morant Cays and Great Inagua Island, and probable subspecies on the Pedro Cays (*S. n. atactus*), on Isla de la Juventud and the Archipiélago de los Canarreos, Cayos de San Felipe (*S. n.* ssp.), and in the Swan Islands (*S. n. exsul*); also in southern Florida (Lazell, 1989).

Habitat: Xerophilic, at times mesophilic (Schwartz and Henderson, 1991); much of what is known about this species is based on observations in southern Florida (Duellman and Schwartz, 1958), we recommend caution in extrapolating this information to a West Indian context; in leaf litter, under rocks and logs in xeric (at times mesic) forest, sometimes edificarian, and associated with human debris and abandoned dwellings (Duellman and Schwartz, 1958); near Nassau (Bahamas) active on walls in limestone caves (Barbour, 1904); in Cuba, not edificarian, under dry leaf litter and rocks (Alayo Dalmau, 1951); only on the ground under rocks and in leaf litter (Collette, 1962); under stones and logs in a coppice near the coast and in a densely vegetated arroyo (Viña Dávila et al., 1984); under tree bark on the Archipiélago de los Canarreos (Estrada, 1993a); can function as a human commensal (M. Černý, 1994); under rocks, fallen trunks, and in leaf litter (Chamizo Lara et al., 2003).

Diet: In Florida, beetles, ants, caterpillars (Duellman and Schwartz, 1958). **Predators:** *Cubophis vudii* on Great Exuma (Henderson and Sajdak, 1996). **Reproduction:** In southern Florida, eggs 4.6–5.6 × 6.0–7.8 mm, often deposited communally (as many as 280) in rotten logs, bases of palm fronds, leaf debris, coconut husks (Duellman and Schwartz, 1958). **Size:** Male SVL to 33 mm, female SVL to 34 mm (Schwartz and Henderson, 1991).

Conservation Status: Listed as a species of "least concern" on the IUCN Red List (Hammerson, 2007a).

Additional Pertinent Literature: Schwartz (1970b).

Sphaerodactylus nycteropus Thomas and Schwartz, 1977

Distribution: Hispaniola, southern coast of the Tiburon Peninsula (Haiti).

Habitat: Xerophilic (Schwartz and Henderson, 1991); scrubby hillside woods around roots of small trees on the Morne Dubois Peninsula (Thomas and Schwartz, 1977).

Size: Male SVL to at least 21 mm (Thomas and Schwartz, 1977).

Conservation Status: Not assessed.

Sphaerodactylus ocoae Schwartz and Thomas, 1977

Distribution: Hispaniola, Sierra de Ocoa (Dominican Republic); elevational range ~150–215 m.

Habitat: Probably mesophilic (Schwartz and Henderson, 1991); palm trash in mesic situations and *Cocos* trash in a well-shaded grove, as both situations artificial, knowledge of original habitat is lacking (Schwartz and Thomas, 1977).

Size: Male SVL to 37 mm, female SVL to 36 mm (Schwartz and Thomas, 1977).

Conservation Status: Not assessed.

Sphaerodactylus oliveri Grant, 1944

Distribution: South central Cuba.

Habitat: Mesophilic (Schwartz and Henderson, 1991); type series under boards and trash of a dismantled cabin (Grant, 1944a); under surface debris, including rock piles (Estrada, 1993a); usually in or at the margins of more or less mesic broadleaf forest or woods (Schwartz and Henderson, 1991); in a mangrove stand (Vergner, 1998).

Activity: Primarily diurnal (Vergner, 1998). **Behavior:** Males use head and tail movements to "impress" one another, females compress and expand the ocelli on their shoulders when aroused or frightened (Vergner, 1998). **Predation:** Scorpion (*Rhopalurus garridoi*) a possible predator (Vergner, 1998). **Reproduction:** Communal nesting, ~20 eggs and several shells in rotten wood under a termite (*Nasutitermes costalis*) nest ~30 cm above the ground (Vergner, 1998); clutch size 1 (Hedges, in prep.). **Size:** Male SVL to 32 mm, female SVL to 34 mm (Schwartz and Henderson, 1991).

Conservation Status: Considered "vulnerable" (Rodríguez Schettino and Chamizo Lara, 1998a).

Sphaerodactylus omoglaux Thomas, 1982

Distribution: Hispaniola, southeastern Plaine de Cul de Sac and adjacent Massif de la Selle (Haiti).

Habitat: Xerophilic (Schwartz and Henderson, 1991); in leaf litter and beneath rocks within 200 m of the southern shore of Etang Saumâtre in a mixture of xerophytic scrub (*Acacia*, *Prosopis*, *Opuntia*, *Cephalocereus*) and more mesic woods (*Sabal* palms, *Swietenia*, *Mangifera*, *Tamarindus*, *Catalpa*), in shady situations with dense canopy, also in dead *Agave* rosettes in xerophytic scrub (Thomas, 1982).

Reproduction: Females with eggs in July–August (Thomas, 1982). **Size:** Male SVL to 18 mm, female SVL to 20 mm (Thomas, 1982).

Conservation Status: Not assessed.

Sphaerodactylus oxyrhinus Gosse, 1850

Distribution: Western Jamaica.

Habitat: Mesophilic (Schwartz and Henderson, 1991); holotype in a house (Gosse, 1850); typically in terrestrial (on limestone exposures in mesic deciduous woods) and arboreal bromeliads 0.6–2.4 m above the ground (Thomas, 1975; P. Vogel, 2000b).

Reproduction: Clutch size 1 (Hedges, in prep.). **Size:** Male SVL to 34 mm, female SVL to 33 mm (Schwartz and Henderson, 1991).

Conservation Status: Not assessed, but Johnson (1988) recommended surveys to evaluate the status of this species.

Sphaerodactylus parkeri Grant, 1939

Distribution: Southern Jamaica, elevational range, sea level to ~230 m.

Habitat: Mesophilic (Schwartz and Henderson, 1991); under trash in the company of similarly patterned crickets (Grant, 1940a); on xeric wooded hillsides, in *Cocos* groves and under *Cocos* trash, under leaf litter and rocks, in a rotten log (Schwartz and Henderson, 1991); at night, active on rock walls just inside a cave and well within the cave (Hedges, in prep.).

Reproduction: Clutch size 1 (Hedges, in prep.). **Size:** Male SVL to 33 mm, female SVL to 35 mm (Schwartz and Henderson, 1991).

Conservation Status: Not assessed.

Sphaerodactylus parthenopion Thomas, 1965

Distribution: British Virgin Islands (Virgin Gorda, Tortola, Prickly Pear Cay, and Moskito Island).

Habitat: Xerophilic (Schwartz and Henderson, 1991); on rocky hillsides covered with scrubby xeric woods, may be absent from purely littoral sites (Thomas and Schwartz, 1966).

Desiccation: Water loss rate 6.48 mg/g/h (MacLean, 1985). **Size:** Male SVL to 16 mm, female SVL to 18 mm (Schwartz and Henderson, 1991).

Conservation Status: "Rare everywhere, protect habitat" (Moreno, 1991d); the species' distribution may be curtailed by development (Perry and Gerber, 2006).

Sphaerodactylus parvus King, 1962

Distribution: Anguilla Bank, Anguilla, Scrub Island, St.-Barthélemy, St.-Martin/St. Maarten, Tintamarre Island, Dog Island.

Habitat: On Anguilla, beneath rocks in a cave (King, 1962); under rocky talus at base of trees in littoral scrub (Thomas and Schwartz, 1966); often under rocks (with area 36–4,524 mm^2, height 3–38 mm, N = 129) on a substrate of dried leaves, small rocks, and sticks, to 6 *S. parvus* under a single rock, sometimes shared with S. *sputator* (Nava et al., 2001); on St.-Barthélemy, behind mangroves (Breuil, 2002b); in piles of dead *Opuntia* in a pasture adjacent to mangroves (Thomas and Schwartz, 1966); on St.-Martin/St. Maarten, in leaf litter under rocks in a cave, under rocks from a stone wall (King, 1962); in leaves on a scrubby hillside near the ocean, and in rock piles with *S. sputator* (Thomas and Schwartz, 1966).

Abundance: Very common at higher elevations on St.-Martin/St. Maarten (Breuil, 2002b). **Activity:** On Anguilla, becomes active in the late afternoon and early evening with a peak at 1900–2100 h, but active in leaf litter until at least shortly past midnight, activity initiated when under-rock temperatures exceeded air temperature (Nava et al., 2001). **Movements:** Two observed foraging within 40–60 cm of the rocks where they were originally found, another 1.45 m from the original rock where it was seen 3 days previously, 4 of 12 subsequently located under the same rocks as where originally observed (Nava et al., 2001). **Population Density:** On Anguilla, 0.4–5.2/m^2 (4,000–52,000/ha), gecko density positively correlated with rock density (Nava et al., 2001). **Predators:** *Anolis* and *Ameiva* on Tintamarre, *Anolis gingivinus* and *A. pogus* on St.-Martin/St. Maarten (Breuil, 2002b); possible predators include *Alsophis rijgersmaei, Ameiva pleii, Anolis gingivinus, Hemidactylus mabouia, Thecadactylus rapicauda* (Hodge et al., 2003). **Reproduction:** Females with mature eggs June–July, on St.-Martin/St. Maarten, 1 egg in a cave 6 × 4.5 mm (King, 1962); reproduction from February to the end of July on St.-Martin/St. Maarten (Breuil, 2002b); based on shells under rocks, females may lay eggs communally (Hodge et al., 2003). **Size:** SVL to 24 mm (King, 1962).

Conservation Status: Not assessed.

Additional Pertinent Literature: Nava et al. (2002).

Sphaerodactylus perissodactylius Thomas and Hedges, 1988

Distribution: Hispaniola, western slopes of the Sierra Martín García (Dominican Republic); elevational range 400–500 m.

Habitat: Xerophilic (Schwartz and Henderson, 1991); in semixeric woodlands that grade into mesic woodlands on limestone substrates, under limestone rocks and rotten logs, and in leaf litter in a small shaded copse (Thomas and Hedges, 1988).

Size: Male SVL to 22 mm, female SVL to 23 mm (Thomas and Hedges, 1988).

Conservation Status: Not assessed.

Sphaerodactylus phyzacinus Thomas, 1964

Distribution: Îles des Saintes, Îlet à Cabrit, Terre-de-Bas, Terre-de-Haut, Grand Îlet.

Habitat: Mesophilic (Schwartz and Henderson, 1991); on Iles de la Petite Terre (Terre-de-Bas), under scattered and piled rocks and in *Coccoloba* litter (Schwartz and Henderson, 1991).

Biomass: On Grand Îlet, 0.27 g/m^2 (Breuil, 2002b). **Population Density:** On Grand Îlet, 2–3/10 m^2, similar at Crawen (Terre-de-Haut), in litter near mangroves on Grand Îlet, 0.79/m^2 (Breuil, 2002b). **Population Size:** On Grand Îlet, 50,000–100,000 (Breuil, 2002b). **Reproduction:** On Grand Îlet, 4 isolated eggs in litter (5 × 7 mm; Breuil, 2002b). **Size:** Male SVL to 25 mm, female SVL to 24 mm (Thomas, 1964).

Conservation Status: Not assessed.

Remarks: Until recently (Thorpe et al., 2008a), this species was considered a subspecies of *S. fantasticus*.

Sphaerodactylus pimienta Thomas, Hedges, and Garrido, 1998

Distribution: Cuba, Sierra Maestra.

Habitat: Appears to prefer 3-dimensional microhabitats in dead *Agave* on a vertical cliff face bordering a coffee plantation, under dead *Agave* on denuded mogotes (karst haystack hills), on areas of relief on limestone exposures, among limestone rocks at bases of limestone cliffs (Thomas et al., 1998). **Reproduction:** In captivity, diameter of eggs 6.6–9.6 mm, annual egg production during the first year of life 2–4, second year 4–7, incubation 53–125 days, hatchling male SVL 17–19 mm, female 18–19 mm, females produce the first oviductal egg at 32–33 mm SVL (Regalado, 2006). **Size:** Male SVL to 30 mm, female SVL to 36 mm (Thomas et al., 1998).

Conservation Status: Unknown.

Sphaerodactylus plummeri Thomas and Hedges, 1992

Distribution: Hispaniola, Barahona Peninsula (Dominican Republic).

Habitat: In xeric habitat on a large limestone escarpment, under dead *Agave* and other vegetative litter at bases of road cuts (Thomas and Hedges, 1992).

Size: SVL to 22 mm (Thomas and Hedges, 1992).

Conservation Status: Unknown.

Sphaerodactylus ramsdeni Ruibal, 1959

Distribution: Eastern Cuba, Sierra del Guaso and Sierra de la Gran Piedra.

Habitat: Mesophilic (Schwartz and Henderson, 1991); in a well-shaded (by saplings and shrubs) pile of moss-covered rocks (with diameters <20 cm) filling a small depression adjacent to ruins of a plantation house in broadleaf forested uplands, also under a rock on a limestone hill (Schwartz and Garrido, 1985).

Size: Male SVL to 29 mm, female SVL to 30 mm (Schwartz and Garrido, 1985).

Conservation Status: Considered "vulnerable" (Rodríguez Schettino and Chamizo Lara, 1998a).

Sphaerodactylus randi Shreve, 1968

Distribution: Hispaniola, Barahona Peninsula (Dominican Republic), where three currently recognized subspecies occur near Oviedo (*S. r. randi*), in the vicinity of Pedernales (*S. r. methorius*), and in the vicinity of Juancho and on Cayo Pisaje (*S. r. strahmi*).

Habitat: Xerophilic (Schwartz and Henderson, 1991); *S. r. randi* under fallen thatched roofs, boards, and other Hurricane Inez wreckage, *S. r. methorius* under stacks of boards, thatch piles, and human debris in *Acacia*, *S. r. strahmi* under a log on sand, under piles of *Cocos* fronds and husks on sand (Schwartz, 1977b).

Abundance: 39 *S. r. strahmi* in 5 variously sized (small and shallow, large and deep) piles of vegetative debris on sand (Schwartz, 1977b). **Reproduction:** Egg (*S. r. methorius*) on 3 July from an apparent communal nest site 1.4 m above the ground under the bark of a standing tree (hatched 9 July; hatchling SVL 15 mm), 2 eggs (7.5–7.6 × 5.0–5.1 mm) 30 December (Schwartz and Henderson, 1991). **Size:** Male SVL to 31 mm, female SVL to 32 mm (Schwartz, 1977b).

Conservation Status: Unknown.

Sphaerodactylus rhabdotus Schwartz, 1970

Distribution: Hispaniola, Valle de Neiba (Dominican Republic); elevational range, sea level to ~150 m.

Habitat: Xerophilic (Schwartz and Henderson, 1991); type series in fallen thatch (at the interface of thatch and a limestone base, and within accordion-like bases of palm fronds) of a construction workers' shack, also beneath a dry palm frond at an old, abandoned shack near a gypsum mine, and in an old pile of palm trunk cuttings in xeric scrub (Schwartz and Graham, 1980); under rocks and dead *Agave* (Schwartz and Henderson, 1991).

Size: Male SVL to 31 mm, female SVL to 32 mm (Schwartz and Graham, 1980).

Conservation Status: Not assessed.

Sphaerodactylus richardi Hedges and Garrido, 1993

Distribution: Central Cuba.

Habitat: In semixeric limestone scrub, also near the beach under thatch palm fronds, *Coccoloba* leaves, and limestone rocks (Hedges and Garrido, 1993); on the Zapata Peninsula, abundant among rocks along the coastline and in semideciduous forest on limestone (Díaz and Abreu Guerra, 2005).

Size: Male SVL to 31.1 mm, female SVL to 32.1 mm (Hedges and Garrido, 1993).

Conservation Status: Considered "vulnerable" (Rodríguez Schettino and Chamizo Lara, 1998a).

Sphaerodactylus richardsoni Gray, 1845

Distribution: Northern Jamaica, where two currently recognized subspecies occur in the vicinity of Montego Bay (*S. r. richardsoni*) and along the northern coast (*S. r. gossei*); elevational range, sea level to ~80 m.

Habitat: Mesophilic (Schwartz and Henderson, 1991); in dry scrub forest and wet limestone forest, *S. r. gossei* in piles of coconut husks (Grant, 1940a); also in coastal *Cocos* plantations, in a hotel bungalow (during midafternoon), under *Cocos* fronds, in piles of limestone rocks, and beneath and in rotten logs (Schwartz and Henderson, 1991).

Escape Mechanism: Regional integumentary loss (Bauer and Russell, 1992). **Size:** Male SVL to 40 mm, female SVL to 38 mm (Schwartz and Henderson, 1991).

Conservation Status: Not assessed, but Johnson (1988) recommended surveys to evaluate the status of this species.

Sphaerodactylus roosevelti Grant, 1931

Distribution: Southwestern Puerto Rico and Isla Caja de Muertos.

Habitat: Xerophilic (Schwartz and Henderson, 1991); in littoral situations in piles of *Cocos* palm trash, in *Coccoloba* litter, also in scrub or xeric woods short distances inland (Thomas and Schwartz, 1966; Boone, 2005); at night in cacti "several feet from the ground," usually under rocks, trunks, and dry *Agave* in Guánica forests (Rivero, 1998).

Behavior: Slow movements, will "freeze" among grass tussocks, perhaps relying on cryptic coloration for protection (Boone, 2005). **Commensalism:** Pairs, often accompanied by a subadult or a juvenile and as many as 3–5 *S. nicholsi*, in tufts of grass 1.0 × 1–3 m in diameter (Boone, 2005). **Escape Mechanism:** Skin is mechanically very weak, and may be used as an escape mechanism in response to arthropod predators (Bauer et al., 1992). **Reproduction:** Clutch size 1 (Hedges, in prep.); eggs in sand under palm trash 6.3–6.6 × 8.2 mm, hatched 21 May (Schwartz and Henderson, 1991); egg diameter 6.0–8.6 mm, annual egg production during the first year 5.0, second year 5.0, incubation 54–80 days, female SVL at hatching 17–22 mm, first oviductal egg at 32–33 mm SVL (Regalado, 2006); gravid females in April, egg dimensions 8.3–8.6 mm × 6.3–6.5 mm (Boone, 2005). **Size:** Male SVL to 39 mm, female SVL to 38 mm (Thomas and Schwartz, 1966).

Conservation Status: "Stable" (Moreno, 1991d).

Sphaerodactylus ruibali Grant, 1959

Distribution: Cuba, southeastern coast; elevational range, sea level to ~150 m.

Habitat: Xerophilic (Schwartz and Henderson, 1991); on grassy hills (Grant, 1959); under demolished brick masonry with many shells of hatched *Sphaerodactylus* eggs (Lando and Williams, 1969); under rocks and human debris, under scattered rocks with a shallow covering of dead leaves, under dead *Agave* (Schwartz and Henderson, 1991).

Reproduction: Clutch size 1 (Hedges, in prep.); based on the number of shells under demolished brick masonry, may nest communally (Lando and Williams, 1969). **Size:** Male SVL to 32 mm, female SVL to 30 mm (Schwartz and Garrido, 1985).

Conservation Status: Considered "vulnerable" (Rodríguez Schettino and Chamizo Lara, 1998a).

Sphaerodactylus sabanus Cochran, 1938

Distribution: Saba, St. Eustatius, St. Christopher, and Nevis.

Habitat: Mesophilic (Schwartz and Henderson, 1991); under bark fragments, rocks, logs, and *Coccoloba* leaves (Schwartz and Henderson, 1991); on Saba, in a wet shaded ravine (Schwartz and Henderson, 1991); under rocks and a board in a dry streambed (RP); on St. Eustatius, most abundant in deep, moist leaf litter (often *Coccoloba* litter), and less common in dry, dead *Agave* (Hensley et al., 2004); occasionally in association with human habitations and debris (Powell et al., 2005); on St. Christopher, under rocks, on Nevis, beneath coconut husks and fronds (King, 1962); in shaded *Coccoloba* and xeric scrub (Schwartz and Henderson, 1991).

Activity: Diurnally active in leaf litter and under cover (RP). **Behavior:** Exhibited 18 different behaviors depending on whether male-male, male-female, or female-female interactions (Wissmann et al., 2005). **Population Density:** On St. Eustatius, 727–1,427/ha in leaf litter; in *Agave*, 0.4 ± 0.1/plant, range 0–2 (Hensley et al., 2004). **Reproduction:** Females with mature shelled eggs in May–June (King, 1962); clutch size 1 (Hedges, in prep.). **Size:** Male SVL to 29 mm, female SVL to 28 mm (Schwartz and Henderson, 1991).

Conservation Status: Not assessed.

Additional Pertinent Literature: Hensley and Powell (2006).

Sphaerodactylus samanensis Cochran, 1932

Distribution: Hispaniola, southern shore of Samaná Bay and adjacent areas (Dominican Republic).

Habitat: Mesophilic (Schwartz and Henderson, 1991); in a crevice in a limestone wall of a cave behind a pile of rocks, shells, bones, and *Cocos* husks (Schwartz, 1976a); in an abandoned cacao plantation and associated with mogotes (Glor et al., 2001).

Size: Male SVL to 27 mm, female SVL to 29 mm (Schwartz and Henderson, 1991).

Conservation Status: Not assessed.

Sphaerodactylus savagei Shreve, 1968

Distribution: Hispaniola, where two subspecies occur in the southeastern Dominican Republic (*S. s. savagei*) and the extreme eastern coast plus islas Saona and Catalinita (*S. s. juanilloensis*).

Habitat: Xerophilic (Schwartz and Henderson, 1991); *S. s. savagei* in lumber trash on a hotel patio, in piles of papers at the abandoned office of a sugar mill in central La Romana, behind a picture in the corridor of a hotel, in palm trash on a beach, in semixeric woods, under piles of old *Cocos* trash in xeric scrub on sandy soil, under a rock in a dry pasture, *S. s. juanilloensis* in palm trash along beaches, under rocks and logs in semimesic woods, in piles of fallen bark, in dead tree stumps, among petiole bases of dead palmettos in dense dry woods, beneath boards adjacent to a shack in coastal scrub, on a living palm frond, and at 0.6 m above the ground on a cycad (*Zamia*) after heavy rain (Thomas and Schwartz, 1983a).

Predation: *S. s. savagei* eaten by *Tropidophis haetianus* (Thomas and Schwartz, 1983a). **Reproduction:** Clutch size 1 (Hedges, in prep.); eggs (4) 20 June, 6.9–7.6 × 4.7–5.3 mm (Schwartz and Henderson, 1991); in captivity, egg diameter 4.8–7.6 mm, annual egg production during the first year 7.0, incubation 74–98 days, SVL of female at hatching 14.0 mm (Regalado, 2006). **Size:** Male SVL to 33 mm, female SVL to 32 mm (Thomas and Schwartz, 1983a).

Conservation Status: Not assessed.

Sphaerodactylus scaber Barbour and Ramsden, 1919

Distribution: Cuba, central Cuba and the Archipiélago de Sabana-Camagüey.

Habitat: Mesophilic to somewhat xerophilic (Schwartz and Henderson, 1991); in deep woods under stones and decaying vegetation (Barbour and Ramsden, 1919); under surface debris (Schwartz and Henderson, 1991).

Size: Male SVL to 34 mm, female SVL to 32 mm (Schwartz and Henderson, 1991).

Conservation Status: Unknown.

Sphaerodactylus schuberti Thomas and Hedges, 1998

Distribution: Hispaniola, Sierra de Neiba.

Habitat: In and beneath dead *Agave* rosettes in xeric habitat (Thomas and Hedges, 1998c).

Size: SVL in males to 33 mm (Thomas and Hedges, 1998c).

Conservation Status: Unknown.

Sphaerodactylus schwartzi Thomas, Hedges, and Garrido, 1992

Distribution: Eastern Cuba.

Habitat: Xeric woods, in leaf litter, rotten wood, and fallen *Tillandsia* on limestone rocks and volcanic hills (Thomas et al., 1992).

Size: SVL to 20 mm (Thomas et al., 1992).

Conservation Status: Considered "vulnerable" (Rodríguez Schettino and Chamizo Lara, 1998a).

Sphaerodactylus semasiops Thomas, 1975

Distribution: Jamaica, Cockpit Country; elevational range ~335–550 m.

Habitat: Slightly xerophilic to mesophilic (Schwartz and Henderson, 1991); in rocky, well-drained situations, in bromeliads growing on trees, rock exposures, and the ground (Thomas, 1975); rarely outside bromeliads (P. Vogel, 2000b).

Defense Mechanism: Prominent ocelli may function to startle or frighten potential predators (Thomas, 1975). **Size:** Male SVL to 30 mm, female SVL to 32 mm (Schwartz and Henderson, 1991).

Conservation Status: Not assessed.

Sphaerodactylus shrevei Lazell, 1961

Distribution: Hispaniola, extreme northwestern Haiti.

Habitat: Xerophilic (Schwartz and Henderson, 1991); in a large rock pile 0.77 m deep, rocks varying from the size of a golf ball to almost that of a football, and thorn bushes sprouting within pile (Lazell, 1961).

Size: SVL in lone male 24 mm, female SVL to 30 mm (Schwartz and Henderson, 1991).

Conservation Status: Unknown.

Sphaerodactylus siboney Fong Grillo and Díaz, 2004

Distribution: Southeastern Cuba.

Habitat: Limestone with steplike terraces and xerophytic, low, thorny vegetation, in sandy coastal vegetation, under limestone rocks, in leaf litter, inside a dead *Agave underwoodi* and in the bromeliad *Tillandsia fasciculata* (Fong Grillo and Díaz, 2004).

Size: Male SVL to 31.3 mm, female SVL to 32.2 mm (Fong Grillo and Díaz, 2004).

Conservation Status: Unknown.

Sphaerodactylus sommeri Graham, 1981

Distribution: Hispaniola, northwestern Haiti; elevational range, sea level to ~680 m.

Habitat: Mesophilic to xerophilic (Schwartz and Henderson, 1991); in an oasis (surrounded by extensive desert) under leaf rubbish at the base of a fallen tree (Graham, 1981); also in a rock pile and dead *Agave* (Schwartz and Henderson, 1991).

Activity: Active at night on rocks (Schwartz and Henderson, 1991) **Size:** Male SVL to 30 mm, female SVL to 35 mm (Graham, 1981).

Conservation Status: Not assessed.

Sphaerodactylus sputator Sparrman, 1784

Distribution: Sombrero Island, Dog Island, Anguilla and satellites (Scrub and Little Scrub islands), St.-Martin/St. Maarten, St.-Barthélemy, Île Forchue, St. Eustatius, St. Christopher, Nevis; the population on Sombrero may represent a distinct species (Lazell, 1964c; Hodge et al., 2003).

Habitat: Mesophilic (Schwartz and Henderson, 1991); on Sombrero, beneath boards and stones near the lighthouse (Lazell, 1964c); on Anguilla, in rocky talus at the base of a tree (Schwartz and Henderson, 1991); on and around buildings and rock walls of a hotel complex (K. G. Howard et al., 2001); under rocks of various sizes on a hillside with mottled sunlight, sometimes in the company of *S. parvus* (RWH); on Little Scrub Island (Anguilla), under a rock in a grassy draw (RP); on St.-Martin/St. Maarten, in a pasture adjacent to a mangrove swamp and on a xeric scrubby hillside near the ocean (Schwartz and Henderson, 1991); in a humid ravine and under rocks along a trail (Breuil, 2002b); beneath rocks (King, 1962); on St. Eustatius, in leaf litter in forested areas, elsewhere in dead *Agave*, where the number of geckos is positively correlated with plant volume (mean 1.7 ± 0.2/agave, 0–5; Hensley et al., 2004); on St. Christopher, in a pile of coconut husks (King, 1962).

Activity: Crepuscular-nocturnal (Breuil, 2002b); reportedly abroad during rainstorms on Sombrero (Lazell, 1964c); active late afternoon and evening, peaking before midnight (K. G. Howard et al., 2001; Hodge et al., 2003); active around artifical lights at night on Anguilla (K. G. Howard et al., 2001; Perry et al., 2008). **Behavior:** Elevates tail when anxious, presumably to direct potential predators away from the body (Breuil, 2002b); exhibited 17 different behaviors depending on whether male-male, male-female, or female-female interactions (Wissmann et al., 2005). **Population Densities:** On St. Eustatius, on The Quill in leaf litter, 72–195/ha depending on litter depth and canopy structure, in *Agave* habitat, 801/ha (Hensley et al., 2004). **Predation:** Possible predators include birds, snakes (*Alsophis rijgersmaei* and *A. rufiventris*), larger lizards (*Ameiva pleii*, *A. erythrocephala*, *Anolis gingivinus*, *A. pogus*, *A. bimaculatus*, *A. schwartzi*, *Hemidactylus mabouia*, *Thecadactylus rapicauda*), and large predatory invertebrates (Hodge et al., 2003; Powell et al., 2005); when edificarian, remains low on walls where it quickly seeks refuge on the ground from larger geckos that generally forage higher (K. G. Howard et al., 2001). **Reproduction:** Gravid females February–June (King, 1962). **Size:** Male SVL to 35 mm, female SVL to 39 mm (Schwartz and Henderson, 1991). **Thermal Biology:** On St.-Martin/St. Maarten, under rocks in direct sunlight 0930–1300 h along a trail with temperatures under the rocks 27–33 °C (Breuil, 2002b); at night on Anguilla, 2 geckos had body and substrate temperatures 28.1 °C (K. G. Howard et al., 2001).

Conservation Status: Not assessed.

Sphaerodactylus storeyae Grant, 1944

Distribution: Cuba, Isla de la Juventud and associated cays.

Habitat and Natural History: Unknown.

Size: Male SVL to 32 mm, female SVL to 31 mm (Hedges and Garrido, 1993).

Conservation Status: Considered "vulnerable" (Rodríguez Schettino and Chamizo Lara, 1998a).

Sphaerodactylus streptophorus Thomas and Schwartz, 1977

Distribution: Hispaniola, where two currently recognized subspecies occur on the Barahona Peninsula west into the Massif de la Selle (*S. s. streptophorus*), and southern Haiti, including the base of the Tiburon Peninsula (*S. s. sphenophanes*); elevational range, sea level to ~425 m.

Habitat: Xerophilic to mesophilic (Schwartz and Henderson, 1991); *S. s. streptophorus* primarily in and under trash piles in *Acacia* scrub in xeric Barahona lowlands (Thomas and Schwartz, 1977); in dead *Agave*, grass clumps, and under rocks (Schwartz and Henderson, 1991); also known from mesic areas at intermediate elevations of the Baoruco and La Selle mountains, *S. s. sphenophanes* in mesic shaded situations (coffee, banana, and breadfruit groves), in leaf litter, under rocks, and in limestone jumbles, but also areas that are somewhat xeric, such as *Mangifera indica* leaf litter (Thomas and Schwartz, 1983a).

Reproduction: Clutch size 1 (Hedges, in prep.). **Size:** Male SVL to 24 mm, female SVL to 26 mm (Thomas and Schwartz, 1983a).

Conservation Status: Not assessed.

Sphaerodactylus thompsoni Schwartz and Franz, 1976

Distribution: Hispaniola, Barahona Peninsula and Isla Beata (Dominican Republic) and adjacent Haiti; elevational range, sea level to ~60 m.

Habitat: Xerophilic (Schwartz and Henderson, 1991); in areas of limestone ledges with shallow dry caves and xeric vegetation (*Agave*, *Acacia*, cacti), under rocks about 3.7 m above the base near the top of a small talus cone (loosely packed limestone rocks to 0.6 m in diameter and covered with dry leaves; Schwartz and Franz, 1976); in roots of bunch grasses (RP).

Size: Male SVL to 30 mm, female SVL to 33 mm (Schwartz and Henderson, 1991).

Conservation Status: Not assessed.

Sphaerodactylus torrei Barbour, 1914

Distribution: Southeastern Cuba, where two currently recognized subspecies (*S. t. torrei* and *S. t. spielmani*) occur; populations in buildings on El Toldo probably introduced and may not have become established (Fong Grillo et al., 2005b).

Habitat: Xerophilic (Schwartz and Henderson, 1991); *S. t. torrei* abundant in and around the city of Santiago de Cuba (Alayo Dalmau, 1951); also in buildings, trash piles, and stone walls, outside the city in arid coastal forest under natural debris, *S. t. spielmani* in open xeric forest 400 m inland under rotten logs, loose bark, and dead leaves at bases of trees, under rocks at the foot of a limestone escarpment behind a beach, in an abandoned pump house, about 2.0 m up on a large tree trunk (Schwartz and Garrido, 1985); under a dead *Agave* leaf (Schwartz and Henderson, 1991); under rocks, in leaf litter, often associated with dead *Agave* plants, under loose bark or in root tangles, to heights of ~2 m on branches or in orchids on large trees, can be a human commensal in cracks in walls or in piles of trash, in Santiago de Cuba, may coexist under cover objects with *S. nigropunctatus* (Röll, 2006).

Behavior: Sex recognition visual, males react to other males with head bobs, to females with lateral movement of the head, both sexes may react aggressively toward other individuals (Röll, 2006). **Parasites:** Nematode *Parapharyngodon sphaerodactyli* (Baruš and Coy Otero, 1974; Coy Otero and Baruš, 1979b; Coy Otero and Lorenzo Hernández, 1982). **Reproduction:** In captivity, eggs 7 × 9–10 mm, 170–180 mg, hatch in 70–90 days at 25–29 °C, hatchlings 18–19 mm SVL, 150–160 mg, juvenile color begins to fade at 3–4 months at SVL 26–28 mm, adult male color develops at 6–8 months (Röll, 2006) **Size:** Male SVL to 38 mm, female SVL to 39 mm (Schwartz and Henderson, 1991).

Conservation Status: Not assessed.

Sphaerodactylus townsendi Grant, 1931

Distribution: Puerto Rico Bank, eastern and southern Puerto Rico, Isla Caja de Muertos, Isla Platillo (= Morillito), islands east of Puerto Rico, Isla Vieques and satellites, cays of La Cordillera (Cayo Hicacos to Cayo Diablo), questionably from Isla Culebra, where the record was in error or the population has been extirpated (R. Thomas, in litt., 22.IV.2008).

Habitat: Xerophilic to semimesophilic (Schwartz and Henderson, 1991); in coastal palm groves, in *Coccoloba* litter, and leaf litter of xeric maritime woods, on Cayo de Tierra (off Vieques) under rocks and leaf litter beneath a large shade tree on the exposed, almost vegetation-free crown of the island (Thomas and Schwartz, 1966).

Desiccation: Standard rate of water loss 5.1 mg/g of initial body weight/h, vital limit (percent of initial body weight lost before death) 23.0, and survival time 71.7 h (Heatwole and Veron, 1977); water loss rates 42.76 ± 5.18–61.83 ± 28.77 mg/g/h depending on locality in Puerto Rico (Nava, 2004). **Predators:** *Borikenophis portoricensis* on Puerto Rico (Henderson and Sajdak, 1996). **Reproduction:** Eggs beneath bark and inside rotten wood of old tree trunks (Rivero, 1998); egg diameter 3.3–6.2, egg production during the first year 4–7, incubation duration 58–93 days, male SVL at hatching 12–13 mm, female 12–13 mm, first oviductal egg at 22–23 mm SVL (Regalado, 2006). **Size:** SVL in males and females to 28 mm (Thomas and Schwartz, 1966).

Conservation Status: "Stable" (Moreno, 1991d).

Sphaerodactylus underwoodi Schwartz, 1968

Distribution: Turks Islands.

Habitat: Xerophilic, primarily in edificarian situations (Schwartz and Henderson, 1991).

Reproduction: Clutch size 1 (Hedges, in prep.). **Size:** Male SVL to 30 mm, female SVL to 32 mm (Schwartz, 1968c).

Conservation Status: Not assessed.

Sphaerodactylus vincenti Boulenger, 1891

Distribution: Central and southern Lesser Antilles, Dominica (*S. v. monilifer*), Martinique (*S. v. adamas* on Rocher de Diamant, *S. v. festus*, *S. v. josephinae*, *S. v. pheristus*, *S. v. psammius*, *S. v. ronaldi*), St. Lucia (*S. v. diamesus*), and St. Vincent (*S. v. vincenti*).

Habitat: Xerophilic to mesophilic (Schwartz and Henderson, 1991); on Dominica, in xeric woodlands, cane fields, a bamboo grove, beaches, *Cocos* plantations, mangrove edges, a *Musa* grove, rainforest, under rocks, and in the outer dry leaves and central wet leaves of bromeliads, on Martinique, under leaves of breadfruit and bases of uprooted *Musa* stumps (Schwartz and Henderson, 1991); primarily in upland wet forest on Dominica (Malhotra and Thorpe, 1999); in bromeliads (*Glomeropitcairnia penduliflora*) at 1–3 m above ground level (Breuil, 2009); among rocks and in crevices in rock faces, into which lizards rapidly retreat when threatened (Krintler, 1986); on Rocher de Diamant (*S. v. adamas*), under *Cocos* trash, in an old wall under rocks on a platform (Schwartz and Henderson, 1991); under rotten logs and other debris that retain humidity (Leclair, 1978); on St. Lucia, under *Coccoloba* and *Terminalia* leaves (*S. v. diamesus*; Schwartz and Henderson, 1991); on St. Vincent (*S. v. vincenti*) in leaf litter of *Hibiscus elatus*, *Theobroma*, *Mangifera*, and *Cecropia* with litter depth and cover, shade, and soil temperatures critical factors (Steinberg et al., 2007)

Activity: Generally active in leaf litter by day on St. Vincent (Steinberg et al., 2007); crossing a trail in a rainforest clearing at 1100 h on St. Vincent (Schwartz and Henderson, 1991). **Behavior:** 23 types of behaviors observed, crouch, gular inflation, and anti-parallel-positioning more common in male-male interactions than in male-female interactions; tongue flick and rooting more common in female-female than male-male or male-female interactions, head-turn more common in female-female than male-female encounters, female-female more likely to be aggressive, but attacks nearly equal in male-male and female-female interactions, most consistent pattern regardless of pairing was stare, distance signal, approach or attack, contact, and withdrawal (Marcum et al., 2008). **Desiccation:** Water-loss rates in *S. v. psammius* 7.68 mg/g/h, death occurs at a loss of 23.9% of mass (Leclair, 1978); on St. Vincent (*S. v. vincenti*), lizards from 350 m asl 13.55 ± 0.79 mg/g/h, from 75 m asl 7.63 ± 0.95 mg/g/h (Steinberg et al., 2007). **Diet:** On St. Vincent, fly larvae, lepidopterans, and spiders most frequently consumed, prey items small (~1.0 mm^3), largest food item 29.3 mm^3, gravid female with 30 prey items, mostly dipteran larvae (Steinberg et al., 2007). **Movements:** Male *S. v. ronaldi* move longer distances than females (Leclair and Provencher, 1988). **Population Density:** At 8 sites on St. Vincent (*S. v. vincenti*), 625–56,255/ha, mean 2,562.5 ± 632.5/ha (Steinberg et al., 2007); on Martinique (*S. v. ronaldi*) 0.82/m^2 (Leclair and Provencher, 1988); in dry litter on south-facing slopes on Martinique, only 1–2/100 m^2 (Bach and Schönecker, 2008). **Reproduction:** On Martinique, (*S. v. ronaldi*) sexual maturity at 25–26 mm SVL in 19–20 months, newly mature females start laying eggs in September and keep producing thereafter, incubation ~60 days (Leclair and Provencher, 1988). **Sex Ratio:** In *S. v. ronaldi*, 1.6 females/male (Leclair and Provencher, 1988). **Size:** SVL in males and females to 40 mm (Schwartz, 1964a). **Tail Autotomy:** On Martinique, both sexes of *S. v. ronaldi* with high (62.3%) incidence of regenerated tails (Leclair and Provencher, 1988).

Conservation Status: Not assessed.

Additional Pertinent Literature: Steinberg et al. (2008).

Sphaerodactylus williamsi Thomas and Schwartz, 1983

Distribution: Hispaniola, northwestern Haiti.

Habitat: Probably mesophilic (Schwartz and Henderson, 1991); in a coastal oasis surrounded by *Agave*-cactus desert (Thomas and Schwartz, 1983b).

Size: Female SVL to at least 22 mm (Thomas and Schwartz, 1983b).

Conservation Status: Not assessed.

Sphaerodactylus zygaena Schwartz and Thomas, 1977

Distribution: Hispaniola, tip of Tiburon Peninsula (Haiti).

Habitat: Probably mesophilic (Schwartz and Henderson, 1991); near the sea in a mesic banana grove (Thomas and Schwartz, 1983a).

Reproduction: Clutch size 1 (Hedges, in prep.). **Size:** Male SVL to 28 mm, female SVL to 32 mm (Schwartz and Thomas, 1977).

Conservation Status: Not assessed.

Figure 21. Like many of its West Indian congeners, *Ameiva ameiva* (Teiidae) is a facultative omnivore. This individual is eating discarded fruit on a beach in Grenada. Photograph by Alejandro J. Sánchez Muñoz.

Ameiva alboguttata Boulenger, 1896

Distribution: Isla Mona.

Habitat: Chiefly on sandy terraces, occasionally on rocky tableland (K. P. Schmidt, 1928); on beaches and the limestone cap, at night under limestone boulders (Weaver et al., 1961; Heatwole and Torres, 1967).

Activity: Diurnal (Heatwole and Torres, 1967); surface activity largely restricted to midmorning through early afternoon, although some individuals reappear in the late afternoon (Powell, 2004c). **Foraging and Diet:** Active forager (Heatwole and Torres, 1967); 20 stomachs contained vegetable matter (chiefly red-coated seeds), beetles, crickets, land snails, and a juvenile *Anolis monensis* (K. P. Schmidt, 1920, 1928); subadult 4 m high on the trunk of a large (~60 cm diameter at 4 m), sloping tree trunk following a trail of ants, periodically taking prey until it disappeared in the crown at a height >6 m (Powell and Censky, 2002). **Predation:** *Borikenophis variegatus* (K. P. Schmidt, 1928; Heatwole and Torres, 1967). **Size:** Male SVL to 126.9 mm (Heatwole and Torres, 1967). **Thermal Biology:** Pauses in patches of direct sunlight to bask (Powell, 2004c).

Conservation Status: Not assessed.

Ameiva ameiva Linnaeus, 1758

Distribution: Two currently recognized subspecies occur in the West Indies, *A. a. tobagana* on the Grenada and St. Vincent island banks in the Lesser Antilles and *A. a. fulginosa* on the Swan Islands in the western Caribbean; elsewhere from Panama through tropical South America, including Trinidad and Tobago.

Habitat: Largely mesophilic, ubiquitous in sun-shade mosaics, along a lowland stream in a grassy floodplain, in cacao and *Musa* groves, in dry pasture (Grenada), among *Cocos* palms, in pastures, woods, cultivated fields of peas (Bequia; Schwartz and Henderson, 1991); abundant on the grounds of a hotel resort complex on Grenada (Simmons et al., 2005); in a cemetery and adjacent yards and gardens and on the grounds of an abandoned factory on St. Vincent (RP).

Activity: Diurnal, at 1030 h, in the shade of a Mango tree (Schwartz and Henderson 1991); on Grenada, first observations shortly before 0800 h, activity peaks 0900–1200 h (>75% of all observations), nearly stopping by 1400 h, a few individuals reappearing ~1600 h (Simmons et al., 2005). **Foraging and Diet:** Active forager (Schwartz and Henderson 1991); on Grenada, did not respond to bait (cat food, tuna) nor intent on scavenging food scraps, lizards sprint toward tethered anoles (*Anolis aeneus* and *A. richardii*) and jump into the air in an effort to catch them, percent time moving (PTM) 51.2 ± 3.9%, mean number of moves/min (MPM) 5.6 ± 0.5, and frequency of feeding attacks while mobile (PAM) 960 ± 2.0% for lizards with mean body temperatures 30.5 ± 0.2 °C, for 3 individuals with mean body temperatures 29.3 ± 0.2 °C under cool, cloudy, windy conditions when most lizards not active, PTM = 17.9 ± 12.3%, MPM = 2.1 ± 1.3, and PAM = 100%, in both instances, neither PTM, MPM, nor PAM differed by SVL (Simmons et al., 2005); gorges on discarded food at a campsite on Grenada (Sánchez Muñoz, 2008). **Home Range:** Mean adult male home range size 647.8 ± 252.2 m^2, adult female home range size 204.0 ± 54.5 m^2, with no correlation between home range size and SVL in either sex, all home ranges overlapped with those of 5 other individuals, with mean overlap between males 39.5 ± 2.8% and between females 45.1 ± 5.4% (Simmons et al., 2005). **Interfamilial Interactions:** Surface activity by *A. ameiva* curtailed time spent on the ground in both *Anolis aeneus* and *A. richardii* (Simmons et al., 2005). **Population Size and Density:** ~460 adults/ha, but high number may reflect an "activity center," extrapolated to an adult biomass of 15.2 kg/ha (Simmons et al., 2005). **Predation:** Presumably mongooses (*Herpestes javanicus*; Simmons et al., 2005; Treglia, 2006). **Reproduction:** Two copulations but no "guarding" behavior in June (Simmons et al., 2005). **Retreats:** Uses PVC drainage pipe (8.5 cm inside diameter) on the grounds of a guesthouse as a retreat from human activity and inclement weather (RWH). **Sex Ratio (M:F):** 13:8 in adults at a site on Grenada (Simmons et al., 2005). **Size:** Male SVL to 149 mm, female SVL to 141 mm (Schwartz and Henderson 1991). **Thermal Biology:** Mean cloacal temperatures (36.1 ± 0.3 °C, 32.5–39.0 °C) not significantly correlated with either air or substrate temperatures (Simmons et al., 2005).

Conservation Status: Not assessed, but restriction of the species to 2 (or possibly 3) isolated localities on St. Vincent (Corke, 1992; Treglia, 2006; Powell and Henderson, 2007) and to areas of intense human activity on Grenada (Germano et al., 2003; Henderson and Berg, 2006) and intense predation by domestic and feral cats in the Grenadines (Daudin and de Silva, 2007) suggest that monitoring and some protective status may be warranted for these insular populations (Powell and Henderson, 2005a, 2007); the status of the Swan Island populations is unknown.

Ameiva atrata Garman, 1888

Distribution: Redonda Island; elevational range, sea level to >210 m.

Habitat: Xerophilic, ground dwelling, on boulder-strewn beach and in *Opuntia*-covered uplands (Baskin and Williams, 1966).

Activity: Diurnal (Schwartz and Henderson, 1991). **Foraging and Diet:** Active forager (Schwartz and Henderson, 1991). **Size:** Female SVL to 104 mm (Barbour and Noble, 1915).

Conservation Status: Not assessed, the restricted range suggests that monitoring is warranted (Powell and Henderson, 2005a).

Ameiva auberi Cocteau, 1838

Distribution: Forty currently recognized subspecies (almost certainly a complex of species) occur on Cuba and satellites and in the Bahama Islands; elevational range, sea level to >300 m.

Habitat: Primarily in xeric and near-coastal areas in beach growth (*Coccoloba uvifera*, *Cocos* plantings, scrub, coastal woodlands, including mangroves), grassy areas, but at times in somewhat more mesic situations (cultivated areas, rather arid uplands, and along roads in otherwise relatively mesic forest), especially common on and behind beaches with interspersed rocks that serve as retreats, on sandy soils, often hard-packed and very hot, and on "dogtooth" limestone, at times common in and along margins of sugarcane fields (Cuba) or other inland disturbed areas (e.g., garbage dumps), but not in swampy areas (although may occupy surrounding *Acacia* stands), does not shun towns and even cities, especially favors parks and other more "natural" areas, occasionally common on stretches of cinders and ballast along railroad rights-of-way (Cuba), often under stones or rubbish such as fallen palm fronds and logs (Gundlach, 1880; Barbour and Ramsden, 1919; Buide, 1966; Schwartz, 1970d; Garrido, 1980b, Schwartz and Henderson, 1991; Estrada, 1993a); of 5 sampled sites, only in dry, open areas near the coast, where abundant (Viña Dávila et al., 1984); most abundant in areas with substantial leaf litter in which lizards seek refuge (Rogner, 2002); in coffee plantation with pineapple undergrowth (Krüger, 2008).

Behavior: Very shy and quick to hide (Barbour and Ramsden, 1919). **Burrows:** Under rocks, logs, and leaf litter (Estrada, 1993b). **Foraging and Diet:** Eats insects (Gundlach, 1880); seen "crawling about . . . leisurely picking up ants" (Barbour and Ramsden, 1919). **Parasites:** In Cuba, nematodes *Skrjabinoptera phrynosoma*, *Spauligodon cubensis*, larval *Abbreviata* sp. (Coy Otero and Baruš, 1979b; Coy Otero and Lorenzo Hernández, 1982); *Cyrtosomum scelopori* (Baruš and Coy Otero, 1969a; Coy Otero, 1970; Coy

Otero and Baruš, 1973a, 1979b; Coy Otero and Lorenzo Hernández, 1982); *Oswaldocruzia lenteixeirai*, *Physaloptera squamatae*, larval *Porrocaecum* sp. (Baruš and Coy Otero, 1969a; Coy Otero, 1970; Coy Otero and Baruš, 1979b; Coy Otero and Lorenzo Hernández, 1982); and *Parapharyngodon cubensis* (Baruš and Coy Otero, 1969a, 1969b; Coy Otero, 1970; Coy Otero and Lorenzo Hernández, 1982). **Predation:** American Kestrel (*Falco sparvarius*) and Cuban Pygmy Owl (*Glaucidium siju vittatum*; Schwartz, 1970d); domestic and feral cats (Dodd and Franz, 1996). **Reproduction:** One white, ovoid, smooth egg 11.8 × 15.5 mm taken 16 June under a rock in a depression 4 cm deep, soil temperature 29.4 °C, incubated in the laboratory at air temperatures 25.5–34.0 °C, 90–98% humidity, 13 days after collection egg 12.4 × 16.1 mm, hatchling 33.4 mm SVL, tail length 70.5 mm, total incubation period probably ~30 days (Estrada et al., 1987); on Cayo Coco, lays eggs under rocks (Estrada, 1993b). **Size:** Male SVL to 136 mm, female SVL to 115 mm (Schwartz, 1970d).

Conservation Status: Not assessed, but the presence of exotic predators (cats) affects the presence and population sizes on small islands (Dodd and Franz, 1996).

Ameiva chrysolaema Cope, 1868

Distribution: Sixteen currently recognized subspecies occur on Hispaniola and satellites; elevational range, −20 m to ~520 m.

Habitat: Xerophilic (Schwartz and Henderson, 1991); on sand and rocks along the coast, in scrub and cactus savannas, abundant in developed areas (Mertens, 1939); inhabits *Acacia* scrub and woods, xeric woodlands (especially without dense ground cover), mangrove borders, scrublands, open beaches with some cover, may retreat to forage during the heat of the day to less open and more shaded situations, occasionally in mesic areas about human habitations (Schwartz and Klinikowski, 1966; Schwartz, 1973a; Schwartz and Henderson, 1991); in a grove of trees, primarily Breadfruit (*Artocarpus altilis*), in a relatively mesic cultivated area (Schwartz, 1968g); uses 2 distinct types of burrows, one for shelter and one for egg deposition, as much as 2.75 m long (Klingel, 1929); in the Cayos Siete Hermanos, in open woody scrub, less common in low, dense scrub where *Leiocephalus* sp. are more abundant, on beaches only on Cayo Tercero, aggregates around fishing camps on Cayo Muertos, where discarded garbage appears to concentrate arthropodan prey (Burns et al., 1992); in all habitats associated with a dry riverbed (xeric mixed forest, *Acacia* woods and scrub, barren riverbed, islands with sparse herbaceous and shrubby plants and some clump grasses, sparse slopes of riverbed), most frequently in shaded situations, except when actively foraging amid trash, mean microhabitat temperatures 38.9 ± 3.8 °C (35.4–43.0 °C; Powell et al., 1996b); when a disturbed area in Barahona became overgrown with weeds, the population of *A. chrysolaema* declined dramatically (Powell et al., 1999); in xeric habitat on the Barahona Peninsula, *A. chrysolaema* uses sites with less tree cover, less ground cover, and closer to clearings than sympatric *A. taeniura*, with differences in tree cover significant (Sproston et al., 1999); on the Barahona Peninsula, almost always on soil substrates (*Anolis longitibialis* on rock outcrops and *Leiocephalus barahonensis* on rocks or piles of rocks; Gifford et al., 2002); amid boulders used to reinforce the coastline in Santo Domingo (Powell and Henderson, 2008c).

Activity: Diurnal, often inactive for 2–2.5 h after sunrise, first activity ~0900 h but peak activity not until hours later, activity drops off ~1400 h and ceases by 1600 h (Mertens, 1939); active only 4–5 h during the middle of the day (Regal, 1978); bimodal in June in altered habitat in Barahona, main activity peak begins late morning (1035 ± 45 min) and lasts through early afternoon (mean duration 241.7 ± 40.0 min), a smaller peak in activity ~1700 h, activity extended on overcast days (Schell et al., 1993a); unimodal in June at a xeric site on the Barahona Peninsula, with few lizards active before 1200 h and after 1600 h (Sproston et al., 1999). **Behavior:** Shelter burrows occupied by a single adult individual (Klingel, 1929); less skittish than *A. taeniura* (Mertens, 1939); vocal "alarm call" (Mertens, 1946); subadult male displaced a juvenile from a basking site (Schell et al., 1993a). **Foraging and Diet:** Intensive active forager, >70% of aboveground activity period devoted to foraging, except on windy day when largely immobile and basking, move one body length every 2–5 sec while foraging and constantly probing with their snouts (Regal, 1978); foraging behavior suggests an association of prey (insects) with yellow flowers, attracted to noise produced by potential prey (crickets) even when unseen (Klingel, 1929); in Barahona, focal animals eating a caterpillar, rat-tailed maggots (in garbage), a cockroach, earthworms, and mango peels, stomach contents ($N = 25$) were spiders, beetles (in most stomachs), roaches, flies, fly larvae (most important by volume), hemipterans, ants, other hymenopterans, termites (present in greatest numbers), lepidopteran larvae and adults, a dragonfly, orthopterans, caddisfly larva, crustaceans (isopod and decapod), *Anolis* egg, lizard tail (*Ameiva*), some plant material, and grit (Schell et al., 1993a); in the Cayos Siete Hermanos, stomach ($N = 3$) contents included adult beetles ($N = 4$, combined volume = 0.045 mm^3, present in 2 stomachs), immature beetles (3, 0.003 mm^3, 1), roaches (20, 0.380 mm^3, 3), caterpillar (1, 0.060 mm^3, 1), and plant material traceable to human refuse (Burns et al., 1992); on the Barahona Peninsula, stomach contents were roaches, beetles (adults and larvae), flies (adults and larvae), hemipterans, ants and other hymenopterans, termites (most important), lepidopterans (adults and larvae), biting lice, scorpionflies, antlions, orthopterans, spiders and other arachnids, centipedes, millipedes, snails, a lizard (*Celestus agasepsoides*), plant matter, shed skin, and grit (Sproston et al., 1999); in Barahona, adult *Anolis coelestinus*, unknown whether killed or scavenged (RP). **Home Range and Territoriality:** Nonterritorial, although individuals have favored burrows, will tolerate use by other lizards (Mertens, 1939); mean home range sizes 1,889.2 ± 1,691.5 m^2 (recapture radii) and 540.5 ± 352.4 m^2 (convex polygon), with those of males larger, but not significantly so, than those of females, and those of males overlapping with those of 0–4 other males (areas of overlap 0–100%), those of females overlapping with those of 0–6 other females (0–100%), and those of males overlapping with those of 1–6 females (18.0–95.1%; Schell et al., 1993a). **Parasites:** Gastric nematodes (*Cyrtosomum* sp.; Burns et al., 1992); pharyngeal myiasis by sarcophagid (Diptera) larvae (D. D. Smith et al., 1994); coccidian *Eimeria bautistae* (note that *Isospora jedlickai*, attributed to *A. chrysolaema*, is known to occur only in *A. taeniura*; Huntington et al., 1997); immature nematode *Terranova* sp. (probably *T. caballeroi*), with the lizard probably serving as a paratenic host (Sproston et al., 1997). **Population Size and Density:** 136 and 144/ha (using 2 different methods) in a partially confined and heavily altered habitat in Barahona (Schell et al., 1993a). **Predation:** Barn Owls (*Tyto alba*; Wetmore and Swales, 1931); American Kestrel (*Falco sparvarius*; Fernández, 2007); *Hypsirhynchus parvifrons* (Henderson et al., 1987a; RP); *H. ferox* (Henderson, 1984b); *Uromacer frenatus* (Henderson et al., 1987c). **Reproduction:** Egg deposition burrows communal, females leave after eggs laid, clutches ($N = 5$) deposited in a small accessory chamber at the back of the main burrow, then covered with packed earth, hatchlings apparently must dig out (Klingel, 1929); male often rubs his cloaca on the substrate before approaching a female and "poking" at and sometimes biting the back of her neck, rubbing his femoral pores on her, then throwing his tail under hers, and arching his body while inserting a hemipenis, female is passive throughout, a homosexual mating in captivity (Noble and Teale, 1930); in captivity, lizards mated during all months, male always initiates courtship, rubs his cloaca on the substrate before attempting copulation by lying on the back of a prospective partner, poking her neck or nipping the nape while the cloaca swells and intromission occurs, warm temperatures triggered reproductive activity, homosexual mating occurs (Noble and Bradley, 1933); mating in February in full sun (in captivity throughout the year; Mertens, 1939). **Sex Ratio (M:F):** 7:11 and 11:14 in two samples from Barahona (Schell et al., 1993a). **Size:** Male SVL to 160 mm, female SVL to 134 mm (Schwartz and Klinikowski, 1966). **Tail Autotomy:** 14 of 33 individuals with regenerated tails, significantly correlated with SVL but not with sex (RP). **Thermal Biology:** Basks in direct sunlight along edges of shaded areas, mean duration of basking episodes 94.2 ± 123.8 sec, mean cloacal temperatures 37.5 ± 1.9 °C significantly higher than air temperatures 31.5 ± 2.7 °C (Schell et al., 1993a); mean cloacal temperature 36.6 ± 0.4 °C at mean substrate temperature 34.1 ± 0.9 °C and 38.1 ± 0.5 °C at 33.6 ± 1.2 °C (Sproston et al., 1999).

Conservation Status: Not assessed.

Additional Pertinent Literature: Schell et al. (1993a), Sproston et al. (1998).

Remarks: *Ameiva leberi* is now considered a pattern variant of *A. chrysolaema ficta* (Sproston et al., 1999; Hower and Hedges, 2001; Gifford et al., 2004).

Ameiva cineracea Barbour and Noble, 1915

Distribution: Restricted in recent times to Grand Îlet off Petit-Bourg on the eastern coast of Basse-Terre, Guadeloupe, although fossil records suggest that it formerly ranged across Guadeloupe, La Désirade, Marie-Galante, and Les Saintes (Breuil, 2002b).

Habitat: Always on the ground (Du Tertre, 1677 in Breuil, 2002b).

Activity: Absent only during the hottest part of the day (Du Tertre, 1677 in Breuil, 2002b). **Diet:** Nibbles on the scraps left after cleaning fish, sometimes eats plants, if killed, others cannibalize remains (Du Tertre, 1677 in Breuil, 2002b); see also "predation" in the *A. fuscata* account. **Size:** Male SVL to 150 mm, female SVL unknown (Barbour and Noble, 1915).

Conservation Status: Listed as "extinct" in the IUCN Red List (World Conservation Monitoring Centre, 1996a); extinction occurred following a hurricane early in the 20th century (Honegger, 1981); introduced predators and consumption by Amerindians may have played roles in the extinction of this species (Breuil, 2002b).

Ameiva corax Censky and Paulson, 1992

Distribution: Little Scrub Island off Anguilla.

Habitat: Lizards are concentrated in vegetated areas (grasses, cacti, and some shrubs) on the protected southwestern side of the island and an area near the cliff on the east side, uses holes in limestone and excavated crevices under rocks as burrows (Hodge et al., 2003).

Activity: Some individuals active throughout the day, respond quickly to disturbance and possible foraging opportunities (A. M. White et al., 2002b); active earlier and later than *A. pleii*, with some individuals emerging at dawn and the last retreating after dusk (Hodge et al., 2003). **Behavior:** Do not fear humans and are very curious, approaching closely and nipping at proffered finger in an apparent effort to find food (Censky and Powell, 2001; Hodge et al., 2003). **Crypsis:** Selection appears to favor lighter colored lizards in periods after hurricanes when dark lichens are scoured off rocks and very dark lizards are more vulnerable to avian predation (Censky and Paulson, 1992; Censky and Powell, 2001; Hodge et al., 2003). **Foraging and Diet:** Ascends cacti (*Opuntia*) to heights approaching 2 m to feed on flowers and fruits (Powell and Censky, 2002); feeds seasonally on bird eggs (including those of Bridled Terns, *Sterna anaethetus melanoptera*), which they crack with their pointed snouts, on fish fry that fishermen bring to the island to use as bait, by scavenging food that birds drop or that blows in from the sea, and by exploiting food scraps, with the "obsession" for food leading to lizards "tasting" a wriggled finger (Censky and Powell, 2001; Hodge et al., 2003). **Parasites:** All individuals ($N = 12$) examined were infected with chigger mites, extent of areas covered by mites was not significantly correlated with indices of condition (Brennan et al., 2009). **Predation:** Presumably sea birds, including Frigatebirds (*Fregata magnificens*), Brown Boobies (*Sula leucogaster*), and Laughing Gulls (*Larus atricilla*; Hodge et al., 2003). **Sex Ratio (M:F):** 49:12 in a sample (A. M. White et al., 2002b), male bias possibly attributable to males spending more time active on the surface (E. J. Censky in A. M. White et al., 2002b). **Size:** Male SVL to 132 mm, female SVL to 96 mm (Censky and Paulson, 1992). **Thermal Biology:** Devotes less time and effort to thermoregulation than *A. pleii* (possibly diverting more time to foraging in a food-deficient habitat), mean cloacal temperatures (33.7 ± 0.5 °C) significantly higher than corresponding substrate temperatures (31.1 ± 0.5 °C), both strongly correlated with time of day, temperatures of males and females did not differ, in the laboratory, *A. corax* heated and cooled at rates comparable to those of nonmelanistic *A. pleii* (A. M. White et al., 2002b).

Conservation Status: Not assessed, but restriction to a single tiny island suggests that some protected status is warranted, although protective measures other than limiting human impact on Little Scrub Island are probably not necessary (Censky and Powell, 2001; Hodge et al., 2003); Powell and Henderson (2005a) recommended monitoring.

Additional Pertinent Literature: A. M. White et al. (2002a).

Ameiva corvina Cope, 1861

Distribution: Sombrero Island.

Habitat: Xerophilic (Schwartz and Henderson, 1991); ground dwelling, but will climb pit walls for bird eggs and basking sites (Lazell, 1964c); concentrated in the area near the lighthouse (prior to its abandonment), presumably to scavenge food, uses burrows in karst limestone (Hodge et al., 2003).

Activity: Diurnal (Schwartz and Henderson, 1991). **Behavior:** Somewhat skittish (Lazell, 1964c). **Foraging and Diet:** Includes eggs of ground-nesting birds, waits until a bird leaves its nest, moves in and smashes eggs, laps contents before the bird returns, large lizards can carry cracked eggs away from a nest (Lazell, 1964c); opportunistic, bird eggs, flowering plants, scraps from humans, food dropped by birds or blown in from sea, insects and other small invertebrates (Hodge et al., 2003). **Predation:** Larger individuals consume smaller conspecifics in traps (J. C. Daltry in B. E. Smith and Baum, 2000); presumably sea birds (see *A. corax* account; Hodge et al., 2003). **Size:** Male SVL to 133 mm, female SVL to 87 mm (Censky and Paulson, 1992).

Conservation Status: Not assessed, but restriction to a single small island suggests that some protected status is warranted; may be more vulnerable than *A. corax*, as population dropped dramatically after Hurricane Luis in 1995 and the island is dramatically altered by human activity, including excavated pits (for phosphate mining) that lizards use for burrows and in which many apparently drowned, as water cannot drain (Hodge et al., 2003); Powell and Henderson (2005a) recommended monitoring.

Additional Pertinent Literature: Shew et al. (2002a).

Ameiva desechensis Heatwole and Torres, 1967

Distribution: Isla Desecheo.

Habitat: Common in coastal areas, including shoreline margins, in habitats of maximum sunlight exposure, but frequently near some vegetational cover or shade, in cactus scrub and grassy areas, but apparently most abundant in areas with canopy forest, where the greatest range in sizes (i.e., age groups) is present (Evans et al., 1991).

Activity: Active at peak of midday temperatures, but forages under the forest canopy 0830–1700 h (Evans et al., 1991); most active during the heat of the day (Meier and Noble, 1991). **Foraging and Diet:** Digs in loose soil around fallen trunks of Almacigo (Gumbo Limbo Tree, *Bursera simaruba*), where lizards congregate and actively dig in a mixture of soil and wood detritus and even into decomposing trunks (as high as 1.8 m above the ground) to feed on insects and larvae (Evans et al., 1991). **Size:** Female SVL to 90 mm (Heatwole and Torres, 1967).

Conservation Status: Not assessed.

Ameiva dorsalis Gray, 1838

Distribution: Jamaica and Pigeon Island, widely distributed but disjunct in lowlands.

Habitat: Xerophilic to mesophilic (Schwartz and Henderson, 1991); terrestrial, abundant at seaside and on dunes (Alligator Pond), along coastal dunes among *Ipomoea* and *Coccoloba*, common in gardens in Kingstown, where it "alarms the womenfolk," and where it is considered a nuisance, pest control operators advertise lizard control as one of their services (Lambert, 1986).

Activity: Diurnal, emerges from burrows after the sun has warmed the ground (0800–0900 h), most active in the morning, retires during the heat of the day, reappears for a few hours in midafternoon (Grant, 1940a). **Foraging and Diet:** Active forager, omnivorous, one shot with another *A. dorsalis* in mouth (Grant, 1940a); "scratching like a bird among the sand" (Gosse, 1851). **Size:** Male SVL to 117 mm, female SVL to 85 mm (Schwartz and Henderson, 1991).

Conservation Status: Not assessed.

Ameiva erythrocephala Daudin, 1802

Distribution: St. Christopher Bank, St. Eustatius, St. Christopher, Nevis.

Habitat: Mesophilic (Schwartz and Henderson, 1991); ground dwelling, abundant and nearly ubiquitous in all but the most densely shaded habitats at higher elevations and very dry rocky areas with sparse scrub vegetation on St. Eustatius (Powell et al., 2005); on St. Christopher, abundant in low-lying uncultivated fields near Basse Terre (Barbour and Noble, 1915); appears to be most abundant in mesic enclaves, such as watered gardens at the airport on St. Eustatius (RP); populations on St. Christopher and Nevis are largely restricted to "urban" areas where encounters with mongooses are presumably less frequent (Barbour, 1930a; Westerman, 1953; Powell and Henderson, 2005a).

Activity: Diurnally active foragers (Powell et al., 2005), with a unimodal activity pattern extending from midmorning to midafternoon at one site on St. Eustatius (Kerr et al., 2005). **Behavior:** Nonterritorial, usually not aggressive toward other individuals, but adult-juvenile and male-male aggression occurs (Kerr et al., 2005). **Competition:** On St. Eustatius, competes for prey with anoles (0.50 dietary overlap) to a greater extent than do birds (Adolph and Roughgarden, 1983). **Foraging and Diet:** Forages on table-food scraps given to domestic animals (Kerr et al., 2005); active forager, presumably feeds on a variety of small arthropods and anoles (Powell et al., 2005). **Home Range:** Home range sizes of adult males, adult females, and juveniles did not differ significantly in a population inhabiting hotel grounds (St. Eustatius), but male-male and adult-juvenile home range overlaps were significantly smaller than male-female and female-female overlaps (Kerr et al., 2005). **Population Size and Density:** Extrapolated to 375/ha and 5.4 kg/ha from data collected on hotel grounds on St. Eustatius (Kerr et al., 2005). **Predation:** Known prey of *Alsophis rufiventris* (Henderson and Sajdak, 1996); *A. rufiventris* foraging in burrows presumably belonging to *A. erythrocephala* (Heinz et al., 2004); presumed prey of birds, mongooses (St. Christopher and Nevis), feral and domestic dogs and cats (Powell et al., 2005). **Reproduction:** Copulation followed by the female retreating into a burrow, while male "guarded" the burrow for a short period of time (Kerr et al., 2005). **Size:** Male SVL to 140 mm (Kerr et al., 2006), female SVL to 109 mm (Schwartz and Henderson, 1991); male to female sexual size dimorphism (SSD) index in a population on St. Eustatius was 1.37 (Kerr et al., 2005). **Tail Autotomy:** Broken tails more frequent in adult males than females and individuals with regenerated tails were larger than those with entire tails (Kerr et al., 2005).

Conservation Status: Not assessed, although Malhotra and Thorpe (1999) indicated that the species is "scarce" on St. Christopher and Nevis, no formal protected status exists at this time; the species is abundant on mongoose-free St. Eustatius (Kerr et al., 2005; Rojer, 1997b), but restricted to heavily trafficked "urban" areas on St. Christopher and Nevis (Barbour, 1930a, 1930b; Westerman, 1953; Powell et al., 2005; Powell, 2006b).

Additional Pertinent Literature: Kerr et al. (2006).

Ameiva exsul Cope, 1862

Distribution: Puerto Rico Bank, including Puerto Rico and nearly all of the British and U.S. Virgin Islands; elevational range, sea level to ~150 m; recently introduced on St. Croix, presumably with a shipment of car tires from either St. Thomas or Puerto Rico (W. Coles in Platenberg, 2007).

Habitat: Xerophilic (Schwartz and Henderson, 1991); thrives in disturbed habitats throughout the Puerto Rican lowlands (Rivero, 1998); on sandy or gravelly substrates, may follow river courses into the interior (Stejneger, 1904); most common on sandy soils in more or less arid open areas (e.g., beaches), but also in gardens, pastures, sugarcane plantations, vacant lots, roadsides, city parks and plazas, in and around human habitations, and along mangrove borders (Wolcott, 1923; K. P. Schmidt, 1928; Heatwole and Torres, 1967); on St. John, "near the buildings at Cruz Bay where the dogs keep the mongoose away," apparently nowhere else on the island (Grant, 1932g); essentially all terrestrial habitats on Anegada, but not on beaches (Carey, 1972a); seeks refuge in burrows under stones, other loose objects, ties of an abandoned tramway, concrete sidewalks, or directly into the ground (K. P. Schmidt, 1928; Grant, 1931a; Carey, 1972); associated with a bank on which lizards bask and into which they tunnel (A. R. Lewis, 1983); reaches peripheries of forests in open or edge situations, but does not occur under continuous canopy (Thomas and Gaa Kessler, 1996); in essentially all habitats on Guana Island (British Virgin Islands), but less abundant in dense forest and rare in open, grassy areas and mangroves (RP).

Activity: Strictly diurnal (Heatwole and Torres, 1967); active only at midday in the winter, but after April, active whenever the sun shines (Danforth, 1925); feeds at midday (K. P. Schmidt, 1928); apparently bimodal activity period in spring-fall and unimodal in winter, retires well before sunset (Grant, 1931a); only ~25% of lizards in an area active at any given time, generally active only during midday hours and sensitive to cloud cover (A. R. Lewis, 1983); activity peaks during midday, positively correlated with temperature and negatively correlated with humidity (Nicholson et al., 2005); occasional individuals active before 0800 h in October on Guana, activity peaks during late morning hours, and declines during the afternoon (RP). **Behavior:** Capable of bipedal locomotion (Heatwole and Torres, 1967; A. J. Sánchez Muñoz, in litt., 17.III.2008); dominance hierarchy size-based, but males do more nonforaging walking than females, >70% of time spent basking and "prowling," large animals spend more time feeding and less time digging than smaller individuals, digging and feeding significantly correlated in large lizards, independent in small individuals, males tend to interact agonistically with males more than with females (A. R. Lewis and Saliva, 1987); responds to an observer by fleeing or settling down to watch the observer (A. R. Lewis and Saliva, 1987); dominant (and sometimes subordinate) males "followed" females, at least one series of interactions never resulted in the displacement of a dominant male, which broke the tail of the second-ranking male in fights (A. R. Lewis et al., 1996); some males never "followed" females (large males more likely to "follow" than smaller males; "following" not extended "guarding" as in *A. pleii*; see that account for more information), but all females were "followed" on at least 3 days (A. R. Lewis et al., 2000); during the breeding season, males respond to female chemical cues in enclosures at significantly higher rates than to male or blank cues, females do not distinguish among male, female, or blank cues (Bofill and Lewis, 1999); 2 juveniles apparently resident on a barge transporting ornamental plants in the British Virgin Islands readily launched themselves into the sea, swam by undulating body and tail, returned to the boat, and climbed up its side (Perry et al., 2006); female (SVL 63.5 mm) responded to the proximity (~75 cm) of a *Borikenophis portoricensis* on Guana Island by headbobbing (M. E. Gifford, pers. comm.). **Desiccation:** Smaller individuals more vulnerable to water loss than larger animals (G. Perry et al. in Nicholson et al., 2005). **Foraging and Diet:** 20 stomachs contained many red-coated seeds, crickets, small crabs, lizard eggs, tail of *Anolis cristatellus* (K. P. Schmidt, 1920, 1928); feeds on lepidopteran larvae and pupae (most important, 34.5%), orthopterans (15%), hymenopterans (1%), beetles (adults and larvae, 12%), flies (adults and larvae, 8.7%), hemipterans (6.3%), sand fleas (4.3%), isopods (1%), spiders (1.3%), millipedes (0.1%), earthworms (3%), snails (2%), mushrooms (6.7%; Wolcott, 1923); also feeds on table scraps left exposed for pets, including onion trimmings and potato peelings, cactus fruits, bananas, apples, even "monkey-chow" on Cayo Santiago (Grant, 1931a; Heatwole et al., 1963; Heatwole and Torres, 1967); active forager, generalized omnivore, examines small objects and pokes "into clumps of grass, under leaves, etc., frequently extending [the] tongue in snake-like fashion" (Heatwole and Torres, 1967); either stomachs or fecal samples contained snails (at least 3 species), ants, earwigs, flies, crickets, cockroaches, beetles, caterpillars, insect eggs, spiders, mites, isopods, amphipods, earthworms, *Anolis* eggs, and plant matter, some food items (*Anolis* eggs) overconsumed, others (snails) underconsumed relative to availability, millipedes and termites, common in leaf litter, not consumed, depressed population numbers of land snails (A. R. Lewis, 1989); insects, snails, treefrogs, "working through fruit falls," probably eating both fruit and insects therein, leftover rice and beans, will jump into the air in attempts to catch low-flying bees (A. R. Lewis, 1986); in the British Virgin Islands (Guana), caught, dismembered, and consumed a ground tarantula (*Cyrtopholis bartholomaei*; Nieves, 2007); on Guana, large male actively foraging in leaf litter for large arthropods and *Sphaerodactylus macrolepis*, several of which were taken in the same area on consecutive mornings, another smaller adult jumping into the air in an effort to catch a moth disturbed when rooting in leaf litter (RP); crawls into "empty" spaghetti, ravioli, and meat cans to eat remaining bits (Schwartz and Henderson, 1991); may associate with introduced *Iguana iguana* in the U.S. Virgin Islands, where they "may have discovered the benefits of following around a messy eater" (Platenberg, 2007); ground tarantula (*Cyrtopholis portoricae*), insect larvae on drying dog feces (and fragments of feces), discarded remains of cheese pie and other human food, including cooked beans, boiled rice, and pieces of boiled squash (Pérez-Rivera and Molina-Opio, 2008); 2 individuals on Guana Island had eaten small red berries of *Cordia polycephala* (M. E. Gifford, pers. comm.). **Growth:** Growth rate negatively correlated with SVL (A. R. Lewis, 1983); populations with greater food supplies show higher absolute growth rates and large asymptotic sizes, males grow faster within populations than females and reach larger asymptotic sizes (A. R. Lewis, 1986). **Home Range and Territoriality:** Not territorial, home ranges of 3 males 725 m^2, 275 m^2, and 200 m^2, with the largest probably the most accurate as it was based on the most observations (A. R. Lewis, 1983); small adult female chased lizards (including a larger individual) away from a burrow she was using, average home range area of males (376.8 m^2) significantly larger than that of females (173.7 m^2), male and female home ranges overlapped those of males and females equally, with all home ranges showing some overlap (A. R. Lewis and Saliva, 1987). **Parasites:** Six parasitic worms, presumably nematodes (K. P. Schmidt, 1920); trematode *Mesocoelium monas* (Cofresí-Sala, 1964; Acholonu, 1976); cestode *Oochoristica* sp. and nematodes *Spauligodon anolis* and *Ozolaimus megatyphlon* (Acholonu, 1976); "scratch off" bright orange mites, with animals from "outer islands" more heavily infested than those on Puerto Rico (Grant, 1931a). **Population Size and Density:** 120 lizards/ha to 0.04 lizards/m^2 (400/ha) in a study area near Laguna Joyuda south of Mayagüez (A. R. Lewis, 1983); some coastal sites support average densities of 200 lizards/ha and may contain activity centers with substantially higher concentrations, at one site in disturbed habitat 260 lizards/ha (A. R. Lewis and Saliva, 1987); to 0.04 lizards/m^2 (A. R. Lewis et al., 1996); estimated 200 lizards/ha on Guana Island, British Virgin Islands (Rodda et al., 2001); estimated 34 lizards/ha in forest and 52/ha along White Bay beach on Guana (R. Powell in Lazell, 2005a). **Population Structure:** On Guana, adult:juvenile 1.24:1 in October (rainy season), period of sampling responsible for apparent high levels of recruitment (R. Powell in Lazell, 2005a). **Predation:** Greater Antillean Grackle (*Quiscalus niger*; Wetmore, 1916); *Borikenophis portoricensis* (Henderson and Sajdak, 1996; Joglar, 2005b); in captivity, *Epicrates inornatus* (M. Leal in Thomas and Gaa Kessler, 1996); *Buteo jamaicensis* (Santana and Temple, 1988); American Kestrels (*Falco sparvarius*), feral dogs and cats (A. R. Lewis and J. G. Rodríguez-Ramírez in A. R. Lewis, 1986); Indian Mongoose (*Herpestes javanicus*; A. J. Sánchez Muñoz, in litt., 17.III.2008). **Reproduction:** In captivity, courtship lasts 10 min and coitus 45 sec, during which the male holds the female's neck until intromission, when he bites her "squarely across the back," with the location of the bite probably attributable to the relatively small size of the male in this instance (Noble and Bradley, 1933); copulation occurs in the open, usually with the apparent cooperation of the female, but occasionally males may "run down" the female and overcome considerable resistance (A. R. Lewis et al., 1996); tail apparently necessary for copulation, males that lost tails (not yet regenerated) unsuccessful in attempts to copulate (F. MacKenzie in Heatwole and Torres, 1967); pink eggs (4–7 in clutch, 20–22 × 13–15.5 mm) deposited in humus slightly >100 mm below the surface (Wolcott, 1923); hatchling (5 February) with yolk-sac scar 36 mm SVL (Heatwole and Torres, 1967); high cost of reproduction, 3-egg clutch 15.8% body mass (J. G. Rodríguez-Ramírez in A. R. Lewis, 1986); seasonal in xeric southwestern Puerto Rico, activity more closely associated with day length than rainfall or temperature (less so elsewhere, female in November with 3 fully formed eggs in a wet area, copulation in January in Mayagüez), testes at maximal sizes June–August, females with eggs June–July, although individuals with enlarged ovarian follicles in all months except October–December, fat bodies reach maximal size in late fall and early winter, clutch sizes 1–3 (mean 2.00 ± 0.816), number of enlarged ovarian follicles 1–6 (mean 2.28 ± 1.320), probably produce 2 clutches of 1 egg per year, average investment per clutch 0.16 ± 0.025 of body mass (Rodríguez-Ramirez and Lewis, 1991); in experimental enclosures, at least 1 of 3, 2 of 2, and 3 of 3 clutches "fathered" by subordinate males (A. R. Lewis et al., 1996); largest of 2 males "fathered" a clutch of 3 eggs in an experimental enclosure (Tirado and Lewis, 1997); body size strongly correlated with male reproductive success, with largest males in experimental enclosures siring the most progeny (84% of clutches and 88% of eggs), although smaller males were not excluded from reproducing, multiple paternity in one clutch, clutch sizes average 3.36 ± 0.81 eggs, with clutch size not related to female SVL, some large females did not reproduce and some smaller females produced up to 3 clutches (A. R. Lewis et al., 2000); eggs laid in an excavated depression under a rock (Sánchez Muñoz, 2008). **Sex Ratio (M:F):** 23:32 at a site near Laguna Joyuda south of Mayagüez (A. R. Lewis, 1983); 38:42 and 33:42 in 2 Puerto Rican populations (A. R. Lewis, 1986). **Size:** Male SVL to 201 mm

(Heatwole and Torres, 1967), female SVL to 103 mm (A. R. Lewis, 1983); minimum size at maturity 69.0% mean adult size in males, 80.3% in females (Rodríguez-Ramirez and Lewis, 1991). **Survivorship:** 0.54 ± 0.33–1.42 ± 0.12 (Jolly estimates) and 0.27–1.94 (Manly-Parr estimates) at various times at a study site south of Mayagüez (A. R. Lewis, 1983). **Tail Autotomy:** Significantly more frequent in large males than in large females and in large animals of either sex than in smaller individuals (A. R. Lewis, 1986). **Thermal Biology:** Avoids cool mornings, basks for a few minutes in strips of sunlight when "hunting," quiescent during the heat of the day, seeking shady places (Grant, 1931a); mean body temperature at onset of daily activity 27.7 ± 7.45 °C, at cessation 34.3 ± 2.46 °C, lizards moved from sun to shade at mean body temperature 39.5 ± 1.42 °C and from shade to sun at mean temperature 32.5 ± 1.25 °C, heating rate in the first 5–10 min after moving into sun greater than cooling rate after a move to shade, lizards spend significantly more time in shade than sun (Rivera-Vélez and Lewis, 1994).

Conservation Status: "Stable" (Moreno, 1991d); however, K. P. Schmidt (1928), citing local inhabitants of Vieques, indicated that it had been extirpated there by mongooses (*Herpestes javanicus*); "mostly restricted to the towns by the mongoose" (Seaman, 1961).

Ameiva fuscata Garman, 1887

Distribution: Dominica, islandwide in lowlands.

Habitat: Mesophilic (Schwartz and Henderson, 1991); common just outside the town of Rouseau, "especially in the hills among the plantings of cocoa trees" (A. G. Ruthven in Barbour and Noble, 1915); ground dwelling, in cultivated areas, on berms, in *Musa* and *Cocos* plantations, in beach grasses, trees, and *Coccoloba*, coastal scrub, coastal cacao grove, a road edge in rainforest, cleared meadows, hillsides at sea level, dry creekbeds, along a river and a drainage ditch in a *Cocos* plantation (Schwartz and Henderson, 1991); coastal woodlands and associated cultivation (Bullock and Evans, 1990); may occur at "quite high altitudes in open cultivated areas" (Malhotra and Thorpe, 1999); 3 small (SVL 70–100 mm) individuals in trees to heights >1.5 m (Rudman et al., 2009).

Activity: Diurnal (Bullock and Evans, 1990); most frequently active during the hottest times of the day, may not be active at all in inclement weather (Malhotra and Thorpe, 1999). **Behavior:** Flight distance 5 m, rendering capture impossible (Krintler, 1986); bipedal (Malhotra and Thorpe, 1999); aggregation scavenging meat scraps on discarded beef bones, apparent a size hierarchy with the largest male driving away smaller males but seemingly tolerant of other smaller individuals (females?), some of which even crawled over him (L. A. White, pers. comm.; RWH). **Biomass:** In undisturbed coastal habitats, means 0.9 kg/ha and 26.0 kg/ha with a maximum of 43.8 kg/ha, in associated cultivation means lower (1.7–8.0 kg/ha), with a maximum of 20.5 kg/ha (Bullock and Evans, 1990). **Foraging and Diet:** Active forager (Bullock and Evans, 1990); takes fallen fruit, scavenges carrion, and eats invertebrates (Malhotra and Thorpe, 1999); eats eggs of *Iguana delicatissima* (Breuil, 2002b; Knapp, 2007b); eats hatchlings and embryonic *I. delicatissima* (C. R. Knapp, pers. comm.); hatchling Hawksbill Sea Turtles (*Eretmochelys imbricata*; Knapp and Prince, 2008); apparently stalking an *Anolis oculatus* in a tree, regularly eats the pulp of fallen mangoes, but did not eat the carcass of a dead *Ameiva* (Rudman et al., 2009). **Miscellany:** Tropical Mockingbird (*Mimus gilvus*) harassed an adult at the base of a tree in which a nest contained hatchlings by spreading its wings while perched on a rock in the vicinity of the lizard and by grabbing and pecking at its tail, lizard responded by turning toward the bird and "feigning" an attack (P. A. Turk, pers. comm.). **Population Densities:** 0–379/ha in coastal woodlands and adjacent cultivation, with highest densities in dry scrub woodland and coconut plantations (Bullock and Evans, 1990). **Predation:** American Kestrel (*Falco sparvarius*) taken in 1937 on Guadeloupe contained an *Ameiva*, which may have come from Dominica or might have been *A. cineracea* (G. K. Noble in Danforth, 1939). **Reproduction:** Not seasonal, oviductal eggs present March–July and December, but females with yolked follicles in every month, the number of gravid females peaks in April, >80% of males in every month sampled possessed sperm, clutch size 3–7 (mean 4.0 ± 0.8) positively correlated with female SVL (Somma and Brooks, 1976); clutch of 5 eggs excavated from an aggregate pile in September (A. James, in litt., 21.IX.2008). **Size:** Male SVL to 200 mm, female SVL to 154 mm (Schwartz and Henderson, 1991); 50% of males mature at 95 mm SVL, females at 111 mm SVL (Somma and Brooks, 1976). **Thermal Biology:** Heliophilic, mean adult female body temperature 35.9 ± 2.8 °C at mean air temperature 28.3 ± 1.5 °C and mean adult male body temperature 35.8 ± 2.9 °C at mean air temperature 28.3 ± 1.4 °C, critical thermal maximum 41.8–45.8 °C (Brooks, 1968); often basks along sides of roads, especially at higher elevations (Malhotra and Thorpe, 1999); effectively exploits pockets of sunlight in otherwise densely shaded areas (Daniells et al., 2008).

Conservation Status: Not assessed.

Additional Pertinent Literature: Malhotra and Thorpe (1995).

Ameiva griswoldi Barbour, 1916

Distribution: Antigua Bank, Antigua, Barbuda, and satellites.

Habitat: Mesophilic, ground dwelling (Schwartz and Henderson, 1991); digs shallow burrows in sandy soils in limestone areas on Antigua, restricting these lizards to the northern part of the island (B. E. Smith et al., 2002); in sunny areas 84% of time, mean perch height 0 ± 2.39 cm (Kolbe et al., 2008).

Activity: Diurnal (Schwartz and Henderson, 1991); retreats into self-excavated burrows during the heat of the day and at night (B. E. Smith et al., 2002); emerges as early as 0600 h, peak activity at 0900 h, less frequently active early in the morning and in the afternoon (Kolbe et al., 2008). **Behavior:** Very "tame" on Great Bird Island (Schwartz and Henderson, 1991); poor climber, falling even from gently sloping large tree trunks (B. E. Smith and Baum, 2000). **Foraging and Diet:** Active forager (Schwartz and Henderson, 1991); on Great Bird Island, crawled into a can of tuna (in spring water, not oil) to eat, also ate canned baked beans in tomato sauce (Schwartz and Henderson, 1991); on Great Bird Island, are "fed" by tourists and reach "enormous sizes," on Red Head Island, frequently chases and eats *Anolis wattsi* (B. E. Smith and Baum, 2000; Kolbe et al., 2008); on Red Head Island, feeds on eggs of Laughing Gulls (*Larus atricilla*), scraps of seafood around nests, probably on arthropods and other invertebrates attracted to detritus around the gull colony (B. E. Smith et al., 2002). **Interfamilial Interactions:** Often chases *Anolis wattsi* into bushes that *A. griswoldi* cannot climb (B. E. Smith and Baum, 2000). **Population Size and Density:** Green Island 5,496 (95% confidence interval 4,857–6,135), density 127.8 lizards/ha (113.0–142.7 lizards/ha), Red Head Island 435.1 (401.8–468.4), 483.4 lizards/ha (446.5–520.3 lizards/ha), Great Bird Island 259.3 (203.8–314.8), 26.3 lizards/ha (20.6–32.0 lizards/ha), Lobster Island 58.1 (46.7–69.5), 290.5 lizards/ha (233.5–347.5 lizards/ha), with the low density on Great Bird Island presumably attributable to predation by *Alsophis sajdaki* (B. E. Smith et al., 2002). **Predation:** *Alsophis sajdaki* (Daltry, 1999; Daltry et al., 2001; B. E. Smith et al., 2002); presumably Black Rats (*Rattus rattus*), at least on eggs, on Antigua, road mortality is a factor in developed areas (B. E. Smith et al., 2002); feral cats on Barbuda (Powell and Henderson, 2005a). **Size:** Male SVL to 124 mm, female SVL to 97 mm (Schwartz and Henderson, 1991).

Conservation Status: Not assessed; very localized on Antigua, still abundant on Barbuda (at least locally; B. E. Smith et al., 2002), despite extensive predation by feral cats (Powell and Henderson, 2005a), and on several Antiguan satellite islands, although apparently extirpated from others (B. E. Smith and Baum, 2000; B. E. Smith et al., 2001, 2002); Powell and Henderson (2005a) suggested monitoring.

Ameiva lineolata Duméril and Bibron, 1839

Distribution: Six currently recognized subspecies occur in disjunct populations on Hispaniola, including a recently discovered population in the extreme eastern Dominican Republic (M. E. Gifford, pers. comm.), and satellites (Islas Beata, Cabras, Catalina, and Île à Cabrit); elevational range −20 m to ~500 m.

Habitat: Xerophilic (Schwartz and Henderson, 1991); on sand and rock along coastlines and in cactus savannas (Mertens, 1939); in open *Acacia*-cactus woods and in slightly more mesic areas under cultivation, under a rock in xeric savanna at 1700 h and under boulders during early afternoon, under thatch at the edge of a pasture (Schwartz, 1965b; Schwartz and Henderson, 1991); in most habitats associated with a dry riverbed (*Acacia* woods and scrub, barren riverbed, islands with sparse herbaceous and shrubby plants and some clump grasses, sparse slopes of riverbed), but most abundant in open areas in the riverbed, mean microhabitat temperatures 34.5 ± 3.2 °C (29.8–43.0 °C; Powell et al., 1996b).

Behavior: Males may fight (Mertens, 1939); fewer than half of individuals are under cover during the heat of the day, the remainder is actively foraging (Powell et al., 1996b). **Foraging and Diet:** Active forager (Powell et al., 1996b); stomach contents include spiders (most important), other arachnids, a centipede, beetles (adults and larvae), roaches, flies (adults and larvae), hemipterans, ants, other hymenopterans, lepidopteran larvae, grasshoppers, arthropodan egg cases, snails, miscellaneous plant material, presumably adventitiously ingested (Nelson et al., 2001). **Predation:** *Hypsirhynchus parvifrons* (Henderson et al., 1987a); *Uromacer oxyrhynchus* (Henderson et al., 1987c); scorpion (*Rhopalurus princeps*, length 72 mm) consumed a female *A. lineolata* (SVL = 48 mm; Armas and Abud, 1992; Armas, 2001). **Size:** Male SVL to 59 mm, female SVL to 54 mm (Schwartz, 1965b). **Tail Autotomy:** 14 of 19 individuals with regenerated tails, not significantly related to size or sex (RP).

Conservation Status: Unknown.

Additional Pertinent Literature: Hartley et al. (1998).

Ameiva major Duméril and Bibron, 1839

Distribution: Thought to have occurred on Martinique, but Breuil (2002b) indicated instead that it was endemic to Les Îles de la Petite Terre near Guadeloupe.

Habitat and Natural History: Unknown.

Size: Maximum known SVL to 197 mm (Schwartz and Henderson, 1991).

Conservation Status: Listed as "extinct" in the IUCN Red List (World Conservation Monitoring Centre, 1996b); a hurricane might have been largely responsible for the extinction of the species (Breuil, 2002b).

Ameiva maynardii Garman, 1888

Distribution: Three currently recognized subspecies occur in the Bahama Islands (Great and Little Inagua islands).

Habitat: Xerophilic (Schwartz and Henderson, 1991); prefers sandy and loamy areas (absent in rocky areas), wherein it digs shallow burrows (rarely exceeding depth of 30–45 cm), on and behind beaches, especially common in the upper beach zone with sand, rock rubble, and sparse vegetation, about the ruins of old houses or partially destroyed walls (Noble and Klingel, 1932).

Behavior: "Exercises" by keeping body, hindlimbs, and tail motionless while continuing walking movements with forelimbs (Noble and Klingel, 1932). **Foraging and Diet:** Occasionally climbs trees to secure food (insects) as high as 1.8 m (Noble and Klingel, 1932). **Predation:** *Leiocephalus inaguae* and *Tropidophis canus* (Noble and Klingel, 1932). **Size:** Male SVL to 72 mm, female SVL to 70 mm (Schwartz and Henderson, 1991).

Conservation Status: Not assessed.

Ameiva pleii Duméril and Bibron, 1839

Distribution: Anguilla Bank (Anguilla, St.-Martin/St. Maarten, St.-Barthélemy, and satellites (but not Little Scrub Island off Anguilla).

Habitat: Mesophilic (Schwartz and Henderson, 1991); ground dwelling, on Anguilla, in coastal and inland scrub on sandy and rock soils, on St.-Barthélemy, in beach scrub, mangrove edges, xeric woods among limestone blocks near the summit of a hill, among rocks and cacti (*Opuntia*) on Île Forchue, on St.-Martin/St. Maarten, in xeric scrub and woods, along beaches in *Ipomoea*, in and about a garbage dump in a road across saltpan, about *Opuntia* clumps in grassland (Schwartz and Henderson, 1991); abundant along beaches in association with *Coccoloba* and *Ipomoea*, also in gardens, along rock walls, on dry hillsides with *Opuntia*, and in dry forests with *Pisonia, Eugenia, Capparis cynophallophora, Guajacum officinale, Croton flavens, Caesalpinia bonduc*, mangrove borders, human debris (Breuil, 2002b); particularly common on Anguilla around human habitations where it feeds on food scraps and garbage (Hodge et al., 2003); on St.-Martin/St. Maarten, on beaches, especially in association with *Ipomoea* sp., in dry scrub around *Opuntia*, in heavily disturbed situations such as streets, alleys, vacant lots, and garbage dumps (Powell et al., 2005); especially heavy individuals exploit animal food and the invertebrates attracted to it on the grounds of the St. Maarten Zoo (RP).

Activity: Active 0900–1500 h, with a peak from 1100 to 1400 h, large males emerge earlier and retreat later than smaller individuals in order to patrol home ranges for receptive females (Censky, 1995b); on Dog Island, emerges ~0900 h, forages until ~1300 h, with a few individuals reemerging from 1500 to 1700 h (Censky, 2003); generally active from early to midmorning until early to midafternoon, with some individuals emerging from burrows to forage again from late afternoon into early evening, retreats into self-excavated burrows at night and during inclement weather (Hodge et al., 2003; Powell et al., 2005). **Behavior:** Burrows apparently occupied by single individuals, aggressive encounters usually short lunges (40% of 120 observations) or chases, occasionally escalating (between equally matched adult males) into prolonged chases during which individuals may bipedal (40%) or engage in "arched dances" (16%), fights with vigorous biting are rare (2%), female-female agonistic encounters rare, larger females always displace smaller individuals, agonistic male-female encounters usually involve a "guarding" male approaching a female that jumps at or even "nips" the male, who always moves (Censky, 1995b). **Competition:** Hermit crabs (*Paguristes* sp.) outcompete lizards for food on Dog Island but not on Anguilla, where crab density much lower (Censky, 1995a). **Foraging and Diet:** Active forager (Schwartz and Henderson, 1991); invertebrates and fruits (Censky, 1995a); males "guarding" females spend less time foraging than those not "guarding" females (15.2 ± 10% versus 51.2 ± 6%), "guarded" females spend more time foraging than unguarded females (67.5 ± 9% versus 49.6 ± 18%; Censky, 1995b); food includes hymenopterans, neuropterans, hemipterans, lepidopteran adults and larvae, crickets, cockroaches, beetles (most frequently eaten), fly larvae, isopods, spiders, a solpugid, snails, crabs, *Anolis* eggs, seeds, and fruit, with sizes of prey not differing by sex (Censky, 1996); aggregated around and in a goat carcass, presumably eating maggots and possibly scavenging (A. M. White et al., 2002b); ascends tree trunks to ~1 m in search of land snails (Powell and Censky, 2002); actively explores cavities, cracks, and fissures in and under surface debris, relies heavily on chemical cues but may also hunt by sight, may dig to 5 cm into soil and litter to extract a worm or beetle larva, various insects and other small arthropods, *Sphaerodactylus*, extracted a hermit crab and snails from shells, fruits (including *Hippomane mancinella*) and may facilitate dissemination of some species (*Salanum racemosum*, *Malpighia* sp.), exploits human debris where lizards eat any organic matter as well as insects attracted to trash, eats chicken scraps, licked a discarded yogurt cup (Breuil, 2002b); primarily insects and other small invertebrates, may consume fruits, flowers, and animal prey, including anoles and their eggs (Hodge et al., 2003; Powell et al., 2005); competes for food scraps on a beach (K. Questel in Sánchez Muñoz, 2008); exploits animal food and invertebrates on the grounds of the St. Maarten Zoo (RP). **Growth:** Growth rates slowed with increasing SVL on both Anguilla and Dog Island, did not differ between sexes on Dog Island, but rates of growth for males higher than those for females on Anguilla (Censky, 1996). **Home Range:** Male home ranges (1,550.8 ± 566.1 m^2) larger than those of females (864.2 ± 503.7 m2), with home range sizes positively correlated with SVL, but sizes of home ranges did not differ by sex when corrected for SVL, home ranges of large males overlapped an average of 6 other male home ranges, number of overlapping female home ranges not correlated with male SVL, but strongly correlated with home range size, smaller males have large home ranges because they are frequently chased by larger males, average percent overlap by male home ranges on both male and female home ranges 67.19%, average percent overlap by female home ranges on both male and female home ranges 45.67% (Censky, 1995b). **Parasites:** Cestode *Oochoristica maccoyi* and nematodes *Parapharyngodon cubensis*, *Physaloptera squamatae*, and acuariid larvae (Goldberg and Bursey, 2000); prevalence of chigger mites (9.1–83.8%) in 5 populations on Anguilla varied significantly, but did not differ by sex at any site, extent of areas covered by mites was not significantly correlated with indices of condition nor did indices of condition of infected and mite-free lizards vary significantly, suggesting that mite infections are relatively asymptomatic (Brennan et al., 2009). **Population Size and Density:** 91 lizards/ha on Dog Island, 74 lizards/ha on Anguilla (Censky, 1996); on Tintamarre, to 1 lizard/10 m^2 (Breuil, 2002b). **Predation:** American Kestrel (*Falco sparvarius*), on Anguilla 12.5% (June–July) and 23.1% (March) by frequency of occurrence, 13.4% and 24% by biomass (McLaughlin and Roughgarden, 1989); also *Alsophis rijgersmaei*, Yellow-crowned Night Heron (*Nyctanassa violacea*), Pearly-eyed Thrashers (*Margarops fucatus*), rats, cats, and dogs (Censky, 1996); also mongooses (*Herpestes javanicus*) on St.-Martin/St. Maarten (Hodge et al., 2003; Powell et al., 2005). **Reproduction:** Seasonal, concentrated during May–October, with timing associated with day length and temperature, not all females reproductively active, females lay 1–7 eggs on Anguilla (mean 3.1) and 1–3 on Dog Island (mean 2.1), with larger females laying larger clutches (clutch sizes did not differ between Anguilla and Dog Island when corrected for female SVL), females may lay 3 clutches in a season on Anguilla, with later clutches equal in size to first clutches, but not Dog Island (also related to body size), juveniles with yolk-sac scars in June, July, and September (Censky, 1995a); male-male aggression common during the breeding season when males "guard" females throughout the receptive period of 1–4 days, with the largest males almost inevitably "winning," only those guarding females mated, 4 largest males involved in 84% of matings and presumably responsible for 88% of fertilized eggs, most matings (11 of 16) just before the female retreated into a burrow, 4 of 16 upon first emergence from a burrow, only one at midday, male makes the first approach with head low, moving along female's tail from tip to base, persisting in this or moving to the front of the female, once positioned anteriorly, "quivering" cloacal region and base of tail for several minutes, moves to base of tail and attempts to mount, if rejected, sequence repeated, if not, male mounts and copulates, one attempt at forced copulation, males sometimes cover burrows of females at night, only 30% of males "guard" females, spending an average of 7.9% of the season "guarding," but the largest males spend 14.9% of the reproductive season "guarding" females, enlarged follicles 8.4–10.9 mm, smallest ovulated egg 12.5 mm, second clutches within 16–19 days of first clutches and third clutches within 27–42 days of second clutches, eggs presumably laid in chambers off the main burrow, after mating, females disappear into burrows and are not active on the surface again for several days (Censky, 1995b); when "guarding" males removed, females rejected small males, large males may be better able to protect females from harassment and behavior of larger males is less harassing than that of smaller males, affording a female more foraging time (Censky, 1997). **Sex Ratio (M:F):** Does not vary substantially from 1:1 in nature (Censky, 1995b); 152:61 in a sample (A. M. White et al., 2002b) attributable to males spending more time active on the surface (E. J. Censky in A. M. White et al., 2002b). **Size:** Male SVL to 181 mm, female SVL to 139 mm (Censky and Paulson, 1992); smallest mature male 62 mm SVL, smallest mature female 64 mm SVL (Censky, 1995a), but not all females with 74 mm SVL mature (Censky, 1995b); sexual size dimorphism on Anguilla 1.37:1, 1.17:1 on Dog Island, with selection favoring large males that successfully "guard" receptive females, which occurs only on Anguilla (Censky, 1996). **Thermal Biology:** Invests considerable time and effort in thermoregulation, mean cloacal temperatures (36.6 ± 0.2 °C) significantly higher than corresponding substrate temperatures (31.5 ± 0.2 °C), both strongly correlated with time of day, temperatures of males and females do not differ, but females are on warmer substrates, body temperatures drop significantly during handling of animals, in the laboratory, *A. pleii* heated and cooled at rates comparable to those of melanistic *A. corax* (A. M. White et al., 2002b).

Conservation Status: Not formally assessed; populations on Anguilla have adapted well to human-mediated alterations of habitats and remain abundant, with no conservation measures necessary (Hodge et al., 2003); *A. pleii* remains abundant on St.-Barthélemy (Breuil, 2002b); populations on St.-Martin/St. Maarten are localized and absent from many areas of presumably suitable habitat, probably due to predation by the Indian Mongoose (*Herpestes javanicus*; Breuil, 2002b; Powell et al., 2005; Powell, 2006b); Powell and Henderson (2005a) suggested monitoring on St.-Martin/St. Maarten.

Additional Pertinent Literature: Censky (1998).

Ameiva pluvianotata Garman, 1887

Distribution: Montserrat.

Habitat: Mesophilic to xerophilic, ground dwelling, in mixed forest and pastures, in meadows, and on beaches (Schwartz and Henderson, 1991).

Activity: Diurnal (Schwartz and Henderson, 1991). **Foraging and Diet:** Active forager (Schwartz and Henderson, 1991). **Size:** Male SVL to 169 mm, female SVL to 158 mm (Schwartz and Henderson, 1991).

Conservation Status: Not assessed; Johnson (1988) recommended surveys to assess conservation status; has declined markedly in abundance since the Soufrière Hills volcano began erupting in 1995 (Malhotra and Thorpe, 1999).

Ameiva polops Cope, 1862

Distribution: St. Croix (almost certainly extirpated) and satellites, including Green Cay, Protestant Cay, Buck Island (introduced), and Ruth Island (introduced).

Habitat: Largely xerophilic (Schwartz and Henderson, 1991); forages among dead leaves on the ground (Heatwole and Torres, 1967); sandy areas with some trees, low cover, and patches of direct sunlight (Philibosian and Ruibal, 1971); on Green Cay, lizard density is positively correlated with tree density (Meier et al., 1993); on Protestant Cay, on beaches (dominated by *Coccoloba uvifera*) and in dry forest with abundant litter (McNair, 2003; McNair and Coles, 2003); smaller individuals use more exposed sites, adults are in canopied areas (McNair and Coles, 2003); on Green Cay, most abundant in forested areas, relatively rare on beaches, with the number of lizards positively associated with the number of shrub stems (McNair and Lombard, 2004); on artificial Ruth Island, more abundant in woodland than in scrub, absent from unvegetated coral rubble, highest counts next to temporary fish camps with boards, food cans, and other debris (McNair and Mackay, 2005).

Activity: Diurnal (Heatwole and Torres, 1967); overnights in burrows underground and under rocks, earliest emergence in March 0702 h (27 °C), mean emergence time 0850 ± 31 min (28 °C) and mean retreat time 1430 ± 54 min, latest active individual at 1757, in July and August, earliest emergence 0720 h, mean emergence time 0910 ± 43 min, mean retreat time 1730 h (Meier et al., 1993); active throughout the year (McNair and Coles, 2003). **Behavior:** 76% (forest) and 59% (beach) of individuals foraging, 22% and 38% thermoregulating, 0.5% and 0.2% engaged in agonistic interactions, 1% and 2% in nonforaging movements, and 0.5% and 0.8% in "maintenance" (Meier et al., 1993). **Foraging and Diet:** Active forager, extends tongue frequently, came to waxed paper that had contained a ham sandwich and repeatedly tried to feed, although no food or insects were visible, food includes ants (Heatwole and Torres, 1967); beach-litter amphipods that are abundant in beach debris (Beatty in Grant, 1937); hermit crab (*Coenobita clypeatus*; Philibosian and Ruibal, 1971); climbs to 0.3 m, consumes Limber Caper (*Capparis flexuosa*), annelids, amphipods (*Orchestia*, most frequently taken), orthopterans, earwigs, termites, beetles, lepidopterans, flies, hymenopterans (ants?), and eggs of *Sphaerodactylus beattyi* (Meier et al., 1993). **Home Range:** Maximum longitudinal axis in July 61 m, mean home range size on beach 190 ± 86 m^2, home ranges overlap extensively (Meier et al., 1993). **Population Size and Density:** "Maybe 50 individuals" on Protestant Cay, but the "population was much larger in the 1920's," "over a hundred" on Green Cay (G. A. Seaman in Baskin and Williams, 1966); estimated 200 lizards on Protestant Cay prior to 1968 (Philibosian and Ruibal, 1971); reduced to 50–100 animals after development of a hotel complex, "perhaps" 200 lizards on Green Cay, which became a National Wildlife Refuge in 1978 (Dodd, 1978); by 1989 (pre-Hurricane Hugo), ~530 lizards on Green Cay and ~60 on Protestant Cay, with 18% of the population on Green Cay composed of juveniles (Vilella, 1991); in 1987, estimated 421 ± 166 and 462 ± 200 lizards on Green Cay (Meier et al., 1993); on Protestant Cay, in 1987, present in all available habitats (P. Zwank in McNair, 2003); in 1994–

1996, population 53–88 lizards/ha (W. C. Knowles in McNair, 2003); 30 animals in 2002, 128 lizards/ha in inhabited areas, 25/ha on the entire island (McNair, 2003); 36 animals in November 2002 (McNair and Coles, 2003); estimated 183 lizards on Green Cay in 2002 (McNair and Lombard, 2004); estimated 60 lizards in 2003, up from 10 introduced on this artificial island in 1990 (McNair and Mackay, 2005). **Population Structure:** 8% juveniles on 20 June on Protestant Cay, 22% on 6 August, 18% and 33% during 2 surveys on 7 August (Meier et al., 1993); 20 adults/12 juveniles in spring, 19 adults/21 juveniles in winter (McNair and Coles, 2003). **Predation:** Presumably the Indian Mongoose (*Herpestes javanicus*; Baskin and Williams, 1966; Heatwole and Torres, 1967). **Reproduction:** Mating in August, juveniles 30–36 mm SVL in June, 30–40 mm SVL in July, 30–45 mm SVL in August, 35–50 mm SVL in October, umbilical scars lost at ~40 mm SVL, juvenile with SVL 33 mm marked 1 August measured 40 mm SVL on 3 October (Meier et al., 1993). **Size:** Male SVL to 64 mm (Grant, 1937), female SVL to 69 mm (Schwartz and Henderson, 1991); mean size of adults apparently decreased after introduction of the Indian Mongoose (*Herpestes javanicus*; G. A. Seaman in Baskin and Williams, 1966).

Conservation Status: Listed as "critically endangered" on the IUCN Red List (Nellis, 1996) and as "endangered" on the U.S. List of Endangered and Threatened Wildlife and Plants (Dodd, 1978; U.S. Fish & Wildlife Service, 2008); increasingly rare on St. Croix proper after introduction of the Indian Mongoose (*Herpestes javanicus*), apparently disappeared prior to 1964 (Seaman and Randall, 1962; G. A. Seaman in Baskin and Williams, 1966), although a small population remained as recently as 1967 (G. A. Seaman in Heatwole and Torres, 1967); introduced population on Buck Island thrived initially (Philibosian and Ruibal, 1971), but declined and disappeared by 1974, after the mongoose trapping program ended in 1970 (Philibosian and Yntema, 1976; Dodd, 1978); a species recovery plan exists (Furniss, 1984); "endangered" (Moreno, 1991d).

Additional Pertinent Literature: Dodd (1980).

Ameiva taeniura Cope, 1862

Distribution: Thirteen currently recognized subspecies occur on Hispaniola and satellites (Île de la Gonâve, Île-à-Vache, Îles Petite and Grande Cayemite, Gosse Caye, Isla Saona; elevational range, sea level to >1,700 m.

Habitat: Mesophilic (Schwartz and Henderson, 1991); along coastal areas in dense *Avicennia* thickets and in cultivated areas (Mertens, 1939); in shaded lowland and upland situations, the former sometimes as dense stands of *Acacia* in otherwise very xeric regions, along roads and in cultivated and natural openings in broadleaf forest, in a wooded but over-cut ravine, very mesic hammock woods, along the coast in shaded thornscrub, *Coccoloba*, and *Cocos*, in xeric scrub adjacent to a fence row of shade trees, in brush-covered rocks in a river flood plain, within drier portions of mangrove forests, and in mixed pine-hardwoods (Schwartz, 1967c); some populations survive in Cayes (Haiti) in yards and open parts of orchards and beside a freshwater trickle flowing into the bay (Olson, 1995); in xeric habitat on the Barahona Peninsula, *A. taeniura* used sites with more tree cover, more ground cover, and farther from clearings than sympatric *A. chrysolaema*, with differences in tree cover significant (Sproston et al., 1999).

Activity: Unimodal in June at xeric site on the Barahona Peninsula, with few lizards active before 1200 h and after 1600 h (Sproston et al., 1999). **Behavior:** Much more skittish than *A. chrysolaema* (Mertens, 1939). **Foraging and Diet:** Food includes roaches (most frequently taken), fly larvae (with roaches, in greatest numbers), one leafhopper, one lepidopteran larva, one orthopteran, and some plant material, present in greatest volume (Powell et al., 1989a); on the Barahona Peninsula, stomach contents were roaches, beetles (adults and larvae), flies, hemipterans, ants, termites (adults and eggs), lepidopteran larvae, mantids, orthopterans (most important), spiders, isopods, snails, plant matter, shed skin, and grit, with lizards taking larger prey than larger sympatric *A. chrysolaema* (Sproston et al., 1999). **Parasites:** Coccidian *Isospora jedlickai* (Huntington et al., 1997; note that the type-host was incorrectly listed as *A. chrysolaema*). **Predation:** *Uromacer frenatus*, *U. oxyrhynchus* (Henderson et al., 1987c). **Reproduction:** 5 eggs under a large flat rock, 2 measured 18.9 × 13.6 mm and 18.8 × 11.5 mm (Schwartz, 1967c); clutch size 3 (Hedges, in prep.). **Size:** Male SVL to 102 mm, female SVL to 103 mm (Schwartz, 1967c). **Thermal Biology:** Mean cloacal temperature 37.2 ± 0.4 °C at mean substrate temperature 34.6 ± 0.5 °C (Sproston et al., 1999).

Conservation Status: Not assessed.

Ameiva wetmorei Stejneger, 1913

Distribution: Southwestern Puerto Rico and adjacent satellites, including Isla Magueyes, Caja de Muertos, Isla Morrillito (= Platillo).

Habitat: Xerophilic (Schwartz and Henderson, 1991); inhabits area with mean annual rainfall <76 cm, more or less open deciduous forest and scrub, in *Opuntia* clumps and in noncoastal *Coccoloba*, retreats under rocks and logs (Heatwole and Torres, 1967; Rivero, 1998); where sympatric with *A. exsul*, "dominant" on more exposed ridges with scrubby vegetation, in very dry areas where *A. exsul* does not occur, *A. wetmorei* is abundant (Rivero, 1998).

Activity: Most active during the heat of the day (Rivero, 1998). **Predation:** *Borikenophis portoricensis* (M. Leal in Henderson and Sajdak, 1996). **Reproduction:** Seasonal, more closely associated with day length than rainfall or temperature, testes at maximal sizes February–August, females with eggs March–September, although individuals with enlarged ovarian follicles in all months except October, November, January, fat bodies reach maximal size at the onset of vitellogenesis, clutch size invariably 1 egg, number of enlarged ovarian follicles 1–3, probably produces 2 clutches of 1 egg per year, average investment per clutch 0.18 ± 0.037 of body mass (Rodríguez-Ramirez and Lewis, 1991). **Size:** Male SVL to 52 mm, female SVL to 45 mm (K. P. Schmidt, 1928); minimum size at maturity 81.7% of mean adult size in males, 85.7% in females (Rodríguez-Ramirez and Lewis, 1991). **Tail Autotomy:** Tails extremely fragile, 25 of 27 specimens with broken or regenerated tails (K. P. Schmidt, 1928). **Thermal Biology:** "Torpid" at 23 °C (Grant, 1931a).

Conservation Status: "Stable" (Moreno, 1991d).

Cnemidophorus vanzoi
Baskin and Williams, 1966

Distribution: Maria Islands, off the southeastern coast of St. Lucia; introduced to Praslin Island (also off coastal St. Lucia; Dickenson and Fa, 2000); may once have occurred on St. Lucia (D. W. Corke in Johnson, 1988).

Habitat: On Maria Major, in grassy areas at margins of woods, in woods, will climb thick stems of vines to lower branches of trees and lie along them (Baskin and Williams, 1966); grasses and Prickly Pear (*Opuntia* sp.) interspersed with small trees, will bask on low branches or rocks (Long, 1974); on the lower edge of tree growth among bushes, cacti, and rocks (Schwartz and Henderson, 1991); also in low scrub, will forage on beaches, on Maria Minor, primarily where grass grows, especially at the foot of low cliffs and on the beach (Corke, 1987); not seen in the beach area or at the base of the cliffs (Buley et al., 1997); on Praslin Island, the translocated population exhibits seasonal changes in habitat preference, although preferences are broad during the wet season, highest density in shrub habitat characterized by *Lantana camera*, mixed woodland characterized by Milkwood (*Tabernaemontana citrifolia*), during the dry season, highest density in shrub habitat, followed by Manchineel (*Hippomane mancinella*) woodland, thermoregulation largely dictates habitat use, as exposed habitats are avoided, especially during the dry season (Dickenson et al., 2001).

Abundance: On Praslin Island, mean population size 151.5 ± 25.9, abundance significantly different between wet and dry seasons (Dickenson and Fa, 2000). **Activity:** Sun loving, will disappear underground when a cloud moves across the sun (Long, 1974); peak activity 0900–1100 h (Corke, 1987). **Behavior:** Excavates burrows by scooping first with a few strokes of one forelimb, then with the other, using the pointed snout "like a wedge," shows little fear of humans, intensely curious, investigating any slight movement, will bipedal when frightened (Long, 1974). **Foraging and Diet:** Primarily insectivorous, but will eat fish, meat, and even ripe fruit (Long, 1974); scratches soil surfaces and leaf litter foraging for insects, readily takes termites (but not ants), male ate a scorpion, eats amphipods and fed on the carcass of a dead bird on the beach, fruits of *Ficus citrifoliella* and juveniles seen to pick up and carry bright red seeds of *Abrus praecatorius*, but not eat them, could be baited during the dry season with meat or soft fruit (Corke, 1987). **Longevity:** In captivity, 5 years, 3 months (R. Gibson in Slavens and Slavens, 2003). **Population Size and Density:** Estimated 425 lizards on Maria Major (Nichols and Maggio, 1977 in Corke, 1987); estimated 60–80 lizards in a circle of 20-m radius on Maria Major, estimated population size on the entire island ~740 individuals, population appears to be stable, but the population on Maria Minor appeared low (Corke, 1987). **Reproduction:** Lays eggs underground, juveniles rarely encountered (Long, 1974). **Sex Ratio (M:F):** On Praslin Island, mean ratio 1.2:1 (Dickenson and Fa, 2000). **Size:** Male SVL to 121 mm, female SVL to 95 mm (Corke, 1987); individuals on Maria Major are larger than those on Maria Minor (Corke, 1987). **Tail Autotomy:** 7 of 13 adult males and 5 of 25 females and juveniles showed evidence of loss and regrowth of parts of tails (Corke, 1987). **Thermal Biology:** In September, on Maria Major, one lizard with an internal transmitter, body temperature 27.5 °C early in the morning (0700–0730 h) when the lizard was in an underground burrow, to 41.0 °C at 1100 h on a sunny day when 15–20 m from the burrow (Buley et al., 1997). **Translocation:** The population introduced to Praslin Island apparently successfully colonized the island and no significant changes in animals' conditions or size occurred during the first 3 years after the translocation of 42 lizards from Maria Major (Dickenson and Fa, 2000).

Conservation Status: Considered "vulnerable" on the IUCN Red List (Gibson, 1996a); listed as "endangered" by U.S. Fish & Wildlife Service (2008).

Kentropyx borckiana Peters, 1869

Distribution: Barbados in the West Indies; elsewhere on Trinidad and in northern South America.

Habitat: On Barbados, beside a creek leading to a carenage (Schwartz and Henderson, 1991).

Reproduction: Unisexual (Cole et al., 1995). **Size:** Maximum SVL 100 mm (Schwartz and Henderson, 1991).

Conservation Status: Not assessed, presumably rare on Barbados.

19 Family Xantusiidae

Cricosaura typica Gundlach and Peters, 1863

Distribution: Eastern Cuba.

Habitat: Xerophilic, under rocks, stones, logs, debris in xeric forest and woods (Schwartz and Henderson, 1991); under rocks and *Cocos* debris, in root tangles, and in leaf litter by day (M. Černý, 1992); under rocks of various sizes (5.0 cm to 1.0 m at greatest length), sometimes with a covering of litter and humus (juveniles and subadults) or on bare ground (adults), soil under the rocks formed a series of small galleries (Estrada and Armas, 1998); under rocks (20–30 cm long) and fallen trunks in evergreen and semideciduous forests, leaf litter 5–10 cm deep (Fong Grillo et al., 1999).

Figure 22. *Cricosaura typica* (Xantusiidae) is the only species in the family that occurs in the West Indies, where it is endemic to eastern Cuba. Photograph by S. Blair Hedges.

Abundance: "Quite common" (Schwartz and Henderson, 1991). **Activity:** Nocturnal, easily collected by day (Schwartz and Henderson, 1991). **Behavior:** When a rock under which lizards are hiding is turned, they quickly retreat into cracks and crevices in loose soil, often succeeding in escaping (Z. Vogel, 1965b); like geckos, licks eyes with tongue (captive observation; Petzold, 1969a). **Defensive Mechanism:** Tail autotomy (Martínez Reyes and Moreno García, 2003). **Diet:** Fecal analysis revealed a diet of small invertebrates, including snails (*Opisthosiphon echinatum*), arachnids (spiders and mites), millipedes, and insects, among which were leafhoppers, termites (*Nasutitermes rippertii*), hymenopterans (mimarid wasps and ants, *Odontomachus haematoda insularis*, which were most abundant, *Strumigenys* sp., *Rogeria* sp., *Monomorium* sp.), and beetles, including curculionids and carabids (Armas et al., 1987); subsequent fecal analysis indicated a diet of 94% insects (of which 89% were hymenopterans, especially ants), 4% arachnids, and 3% miscellaneous material, most prey 3–4 mm in size, with 3–12 prey items/lizard (Estrada and Armas, 1998). **Injuries:** 32.6% of 42 lizards were missing toes, front or hind feet, or entire appendages (Fong Grillo et al., 1999); probably attributable to conflicts with conspecifics (Moreno, 1987). **Parasites:** Nematode *Skrjabinodon cricosaurae* (Baruš and Coy Otero, 1974; Coy Otero and Baruš, 1979b; Coy Otero and Lorenzo Hernández, 1982); prevalence of this nematode very high, 80% and 86% in two populations, with 1–28 parasites/host (Goldberg et al., 1998c). **Reproduction:** Smallest female with gonadal development 24.2 mm SVL, smallest male 29.9 mm SVL (Fong Grillo et al., 1999); clutch size one (Hedges, in prep.). **Size:** Male SVL to 37 mm, female SVL to 39 mm (Schwartz and Henderson, 1991). **Tail Autotomy:** Regenerated tails in 61.5% of adults and 50.0% of juveniles, possibly a result of fighting among conspecifics (Fong Grillo et al., 1999). **Thermal Relations:** Mean cloacal temperature of males (mean air, mean substrate) 30.6 °C (26.4 °C, 24.6 °C), and of females 31.4 °C (26.6 °C, 25.2 °C; Estrada and Armas, 1998).

Conservation Status: Considered "endangered" (Díaz et al., 1998a; Rodríguez Schettino and Chamizo Lara, 1998a).

Section 2 Squamata (amphisbaenians)

20 Family Amphisbaenidae

Amphisbaena bakeri Stejneger, 1904

Distribution: Western Puerto Rico; elevational range, sea level to ~600 m.

Habitat: Mesophilic (Schwartz and Henderson, 1991); in a small, grassy pasture in a doline bottom in karst topography under a rock on moist clayey soil, in a coffee-growing area in a rotten log along the edge of a clearing with short weedy vegetation (Thomas, 1966d).

Parasites: Nematode *Aplectana pusilla* (Dyer et al., 1999). **Size:** Adult SVL to 260 mm, tail to 21 mm (Thomas, 1966d).

Conservation Status: "Stable" (Moreno, 1991d).

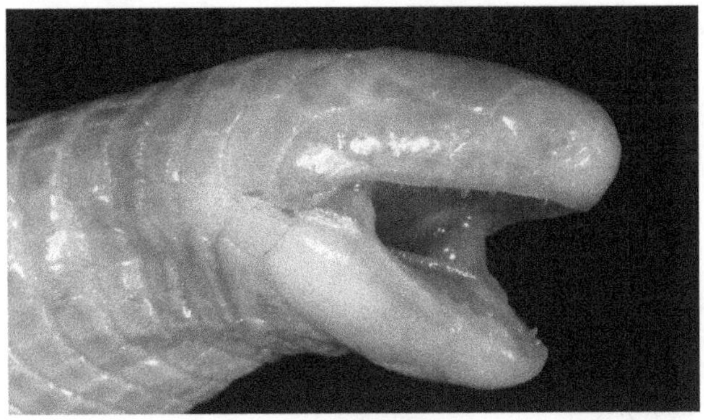

Figure 23. *Amphisbaena bakeri* (Amphisbaenidae) is a blind burrower with dentition that suggests an ability to attack and consume small vertebrates in addition to arthropods. Photograph by Alejandro J. Sánchez Muñoz.

Amphisbaena barbouri Gans and Alexander, 1962

Distribution: West central Cuba, coastal or near-coastal localities.

Habitat: Mesophilic (Schwartz and Henderson, 1991); under stones on lowland forest floor (Barbour and Ramsden, 1919); surrounded by ants under cardboard on the top of a cliff (Gans and Alexander, 1962).

Size: Adult SVL to 177 (Thomas and Hedges, 1998b).

Conservation Status: Not assessed.

Amphisbaena caeca Cuvier, 1829

Distribution: Puerto Rico, widespread on the main island, plus Isla Piñeros and Isla Vieques, including Cayo de Afuera; elevational range, sea level to ~670 m.

Habitat: Mesophilic (Schwartz and Henderson, 1991); on Puerto Rico, fossorial, in shaded situations, burrows in upper soil and humus layers beneath rocks, accumulations of dead vegetation, abandoned termite nests, and rotten logs, but apparently does not burrow in clay (Thomas and Gaa Kessler, 1996); in shaded habitats including the interior of upland coffee plantations, about 12 m from the ocean (shed skins in the same area), in coastal *Terminalia* woods, mesic woods next to a pasture, an exposed ravine slope, shady ravine woods (Thomas, 1966d); under stones and in ant nests (Stejneger, 1904); burrows in the ground (most uncovered by cultivation), one "about three inches beneath an ant's nest under a log" (K. P. Schmidt, 1928); in rotting wood, under rocks, under the stump of a fallen palm, in a *Cocos* plantation about 100 m from shore, under a pile of dead grass in a pasture, under cement slabs on a guava-wooded hillside (Schwartz and Henderson, 1991); in logs not in firm contact with the ground, on Cayo de Afuera, in woody, shaded (but not moist) habitat (Thomas, 1966d).

Foraging and Diet: In captivity, locates carrion by olfaction, 20 stomachs contained an oligochaete worm, 2 slugs, 12 gastropod eggs, 4 mites, 1 harvestman, 1 spider, 3 millipedes, 1 springtail, 6 termites (*Nasutitermes*), 1 silverfish, 5 adult beetles, 17 larval beetles, 3 adult ants, 12 ant pupae, 3 flies, 2 *Anolis* sp. (Thomas and Gaa Kessler, 1996). **Reproduction:** 3 eggs (1 beneath a termite nest, 2 under a log) with the largest 42 × 11 mm (K. P. Schmidt, 1928); clutch size 1 (Hedges, in prep.). **Size:** Adult SVL to 291 mm (male), tail to 21 mm (Thomas, 1966d).

Conservation Status: "Stable" (Moreno, 1991d).

Amphisbaena carlgansi Thomas and Hedges, 1998

Distribution: Cuba, only from Meseta de Cabo Cruz, Granma Province.

Habitat: Xerophilic (Schwartz and Henderson, 1991); scrub-covered, terraced limestone with vegetation in solution pits (Thomas and Hedges, 1998b).

Size: Adult SVL to 168 mm, tail to 14 mm (Thomas and Hedges, 1998b).

Conservation Status: Unknown.

Amphisbaena caudalis Cochran, 1928

Distribution: Hispaniola, Île Grande Cayemite and Presqu'île de Baradères (Haiti).

Habitat: Under limestone rocks in cultivated areas and semi-open horticultural woods (Thomas and Hedges, 2006).

Size: Adult SVL to 247 mm, tail 24 mm (Thomas and Hedges, 2006).

Conservation Status: Not assessed.

Amphisbaena cayemite Thomas and Hedges, 2006

Distribution: Hispaniola, only from the type locality near Anse à Macon, Département de la Grande Anse (Haiti).

Habitat: Under limestone rocks in cultivated areas and semi-open horticultural woods (Thomas and Hedges, 2006).

Size: SVL to 220 mm, tail 17 mm (Thomas and Hedges, 2006).

Conservation Status: Unknown.

Amphisbaena cubana Gundlach and Peters, 1878

Distribution: Disjunct range in eastern and westcentral Cuba, extreme western Cuba, and Isla de la Juventud.

Habitat: Mesophilic (Schwartz and Henderson, 1991); under rocks (Alayo Dalmau, 1951); under debris in open areas (Collette, 1961); under stones near a brook (Gans and Alexander, 1962); excavated by plows in cane fields (Barbour, 1914).

Behavior: Crawls backward as quickly as forward (Z. Vogel, 1965b). **Diet:** Ant and termite adults, larvae, and pupae, although dentition and the ability to inflict a painful bite suggest that other prey may be taken (Z. Vogel, 1965b). **Parasites:** Nematode *Parapharyngodon cubensis* (Coy Otero and Baruš, 1979b; Coy Otero and Lorenzo Hernández, 1982). **Predation:** *Arrhyton taeniatum* (Schwartz and Garrido, 1981). **Size:** Adult SVL to 206 (Thomas and Hedges, 1998b).

Conservation Status: Not assessed.

Amphisbaena fenestrata Cope, 1862

Distribution: British and U.S. Virgin Islands (but not known from Anegada or the St. Croix Bank).

Habitat: Mesophilic to semixerophilic (Schwartz and Henderson, 1991); on St. John, under dry sugarcane debris left in fields after the harvest, on St. Thomas, under stones in small holes on moist ground (Reinhardt and Lütken, 1862); on Virgin Gorda, shed skins under rocks in shady parts of low xeric woods, 2 individuals under rocks on a heavily wooded hillside after a morning rainfall (Thomas, 1966d); on Great Thatch Island, under a log above the beach (Lazell, 2005a); under rocks on a heavily wooded hillside, under bricks and stones associated with Danish plantation house ruins, under rocks in xeric woods, under rocks on the sides of a rather mesic ravine and on gradual slopes, under agriculturally terraced rock on an open, dry hillside (Schwartz and Henderson, 1991).

Behavior: Bites (O. Chalwell in Lazell, 2005a). **Mortality:** On Guana, roadkilled (J. Lazell, pers. comm.). **Predation:** On Guana, preyed upon by *Borikenophis portoricensis* (Barun and Perry, 2003). **Size:** Adult SVL to 242 mm, tail to 17 mm (Schwartz and Henderson, 1991).

Conservation Status: "Stable" (Moreno, 1991d).

Amphisbaena gonavensis Gans and Alexander, 1962

Distribution: Hispaniola, Île-de-la-Gonâve and Île-de-la-Petite Gonâve (Haiti).

Habitat: Xerophilic (Schwartz and Henderson, 1991).

Size: SVL to 210 mm, tail 14 mm (Thomas and Hedges, 2006).

Conservation Status: Unknown.

Additional Pertinent Literature: Powell (1992b).

Amphisbaena hyporissor Thomas, 1965

Distribution: Hispaniola, where two currently recognized subspecies (*A. h. hyporissor* and *A. h. leberi*) occur on the Barahona Peninsula (Dominican Republic).

Habitat: Xerophilic (Schwartz and Henderson, 1991); in xeric woods with abundant limestone, friable soil, and plant litter (Thomas, 1965c); under rocks of all sizes, shed skins sometimes under rocks with *Typhlops syntherus* and *Celestus agasepsoides* (RP).

Activity: Nocturnal, but not strictly so, when photoperiod reversed, did not reverse activity, under constant light, nocturnal rhythm persisted at least 2 days, then random, with a tendency toward 30-h activity periods during the last 4 days of the experiment, suggests endogenous circadian element can be altered by changes in light/dark cycles, relatively exposed areas used during nocturnal foraging would have to be avoided by day due to high ground temperatures, often thin soil, and the threat of desiccation (Thomas and Thomas, 1978). **Diet:** Stomach contents ($N = 21$) contained beetles (adults and larvae, the latter the most important prey item), lepidopteran larvae, termites, hymenopteran larvae, roaches, vertebrate tissue, wax, plant matter, and grit, the latter presumably adventitiously ingested (Cusumano and Powell, 1991); stomach contents ($N = 44$, 26 of which were empty) included adult beetles ($N = 4$, total volume 0.030 mm^3, in 3 stomachs), larval beetles (17, 1.510, 11), orthopteran (1, 0.030, 1), termites (7, 0.113, 3), hemipteran (1, 0.003, 1), roaches (2, 0.013, 2), lepidopteran larvae (15, 0.235, 14), dipteran larvae (2, 0.010, 2), hymenopteran larvae (4, 0.035, 2), ants (7, 0.010, 1), spider (1, 0.003, 1), isopod (1, 0.005, 1), unidentified vertebrate flesh (1, 0.100, 1), plus shed skin, plant matter, waxy material (possibly from subterranean insect nests), and grit, the latter 3 of which presumably adventitiously ingested (L. R. White et al., 1992a); fed voraciously on termites in captivity, also attacked and ate hatchling *Hypsirhynchus parvifrons* and consumed hamburger, the latter worked into the mouth by action of the stout, rasplike tongue, bites and twists viciously, suggesting ingestion of carrion or large prey items (Thomas, 1965e). **Size:** Adult SVL to 310 mm (Cusumano and Powell, 1991; L. R. White et al., 1992a).

Conservation Status: Not assessed.

Additional Pertinent Literature: Powell (1992b).

Amphisbaena innocens Weinland, 1862

Distribution: Hispaniola, Tiburon Peninsula east to the western Sierra de Baoruco, north into the Plaine-de-Cul-de-Sac and Montagnes du Trou d'Eau, plus Île Grande Cayemite.

Habitat: Mesophilic to xerophilic, under rocks along a road cut and in a semi-clearing, in pine forest (Schwartz and Henderson, 1991).

Predation: Land crabs (Schwartz and Henderson, 1991; Thomas and Gaa Kessler, 1996). **Reproduction:** 4 eggs under a rock on 28 July, shells granular, medium brown, very irregular in shape (pear-shaped, constricted cylinder, etc., 35.5 × 10.7 mm, 28.5 × 11.5 mm, 38.9 × 10.9 mm), contained well-developed embryos (Schwartz and Henderson, 1991); clutch size 2 (Hedges, in prep.). **Size:** Adult SVL to 262 mm, tail to 18 mm (Schwartz and Henderson, 1991).

Conservation Status: Unknown.

Amphisbaena leali Thomas and Hedges, 2006

Distribution: Hispaniola, only from the type locality in the Département de la Grande Anse, Haiti.

Habitat: Under limestone rocks in a mesic, hilly karst area with moderate tree cover (Thomas and Hedges, 2006).

Size: Adult SVL to 254 mm, tail to 24 mm (Thomas and Hedges, 2006).

Conservation Status: Unknown.

Amphisbaena manni Barbour, 1914

Distribution: Hispaniola, widespread on the main island, but absent from the Tiburon and Barahona peninsulas, also on Île-de-la-Tortue; elevational range, sea level to >1,200 m.

Habitat: Mesophilic, under logs, stones, rocks, a dry decayed log, in a fibrous palm log, a large wet palm log, under a rock at the base of a termitarium, in a pile of wet *Cocos* husks, in a floodplain, mesic coastal woods, lowland forest, open pasture, coffee groves, cacao groves, mesic rocky piedmont, a forested ravine, a beach backed by mangroves, and rainforest (Schwartz and Henderson, 1991); under rocks on a rainforest hillside (RP).

Activity: Less nocturnal than *A. hyporissor*, but the activity pattern more nocturnal than diurnal, inversion of 12L:12D had little effect on activity (Thomas and Thomas, 1978). **Parasites:** Coccidians *Eimeria witcheri* and *E. amphisbaeniarum* (Huntington et al., 1996). **Reproduction:** Eggs cylindrical, 8.5–10.5 × 26.5–28.0 mm, embryo (SVL ~100 mm) in twice-doubled position, attached to an elongate yolk mass (Gans and Alexander, 1962). **Size:** Adult SVL to 212 mm, tail to 22 mm (Schwartz and Henderson, 1991).

Conservation Status: Not assessed.

Amphisbaena schmidti Gans, 1964

Distribution: Puerto Rico, northwestern limestone region; elevational range, sea level to ~400 m.

Habitat: Xerophilic to semimesophilic (Schwartz and Henderson, 1991); under a fallen palm frond on humus, and possibly from excavated termite nests (Gans, 1964); in *Acacia* scrub, on moderately to well-exposed scrubby ravine sides, an exposed hillside pasture, high, open ravine woods, moderately exposed edges of coffee plantations, *Psidium* scrub at the rim of a doline pasture, a cleared doline slope, a moderately exposed roadside ditch, an open lake margin (Thomas, 1966d).

Size: Adult SVL to 240 mm, tail to 26 mm (Schwartz and Henderson, 1991).

Conservation Status: "Stable" (Moreno, 1991d).

Amphisbaena xera Thomas, 1966

Distribution: Southwestern Puerto Rico; elevational range, sea level to ~600 m.

Habitat: Xerophilic (Schwartz and Henderson, 1991); in xeric woods on limestone hills on sandy (with moderate plant content) to loamy soil covered with leaf litter, transitional area between xeric lowlands and mesic uplands near a small stream at the base of a hillside, in a patch of sandy, mesic streamside woods with Spiny Palms (*Acrocomia*) on adjacent hillsides, in a mesic wooded ravine surrounded by typically xeric to semixeric regions, also at night in xeric foothills under a rock not embedded in soil in an exposed area with very little vegetation (Thomas, 1966d).

Size: Adult SVL 245 mm, tail to 20 mm (Schwartz and Henderson, 1991).

Conservation Status: "Stable" (Moreno, 1991d).

21 Family Cadeidae

Figure 24. *Cadea blanoides* (Cadeidae) is one of two species in this endemic Cuban family. This individual was under a rock coiled around a clutch of two eggs. Photograph by S. Blair Hedges.

Cadea blanoides Stejneger, 1916

Distribution: Western Cuba and Isla de la Juventud (record from eastern Cuba questionable; Schwartz and Henderson, 1991).

Habitat: Mesophilic (Schwartz and Henderson, 1991); in manure piles, under planks and rocks (Gundlach, 1880); at times plowed up during cultivation (Barbour and Ramsden, 1919); beneath a rock at the top of a rock pile, the interstices of which were filled with moist soil, in a sinkhole cave (Zug and Schwartz, 1958); in an orchard (Garrido, 1980b); on Isla de la Juventud, beneath rocks in moist soil in a rock-strewn area (Zug and Schwartz, 1958).

Diet: Insectivorous (Gundlach, 1880). **Parasites:** Nematode *Aplectana rysavy* (Coy Otero and Baruš, 1979b; Coy Otero and Lorenzo Hernández, 1982). **Reproduction:** Clutch size 2, eggs laid under rocks (S. B. Hedges, in litt., 09.V.2008; figure 24). **Size:** Adult SVL 260 mm, tail 14 mm (Zug and Schwartz, 1958).

Conservation Status: Unknown.

Cadea palirostrata Dickerson, 1916

Distribution: Cuba, Isla de la Juventud.

Habitat: Mesophilic (Schwartz and Henderson, 1991).

Parasites: Nematode *Aplectana rysavy* (Baruš and Coy Otero, 1969a; Coy Otero, 1970; Coy Otero and Baruš, 1979b; Coy Otero and Lorenzo Hernández, 1982). **Size:** Adult SVL 280 mm, tail 12 mm (Schwartz and Henderson, 1991).

Conservation Status: Considered "vulnerable" (Rodríguez Schettino and Chamizo Lara, 1998a).

Section 3 Squamata (snakes)

22 Family Boidae

Figure 25. *Corallus cookii* (Boidae) is one of two species in the genus endemic to West Indian islands. Photograph by Robert Powell.

Boa nebulosa Lazell, 1964

Distribution: Dominica.

Habitat: Coastal xeric and littoral woodlands and montane rainforest (Malhotra and Thorpe, 1999); edges of banana fields (Lazell, 1964b); disturbed forest (Schwartz and Henderson, 1991); by day under a sheet of tin in scrub habitat within 100 m of the Caribbean Sea (RWH); in a compost pile in the Botanic Garden in Roseau (A. James, in litt., 14.VI.2007); dead on the road in an area of xeric scrub and sheer cliffs (RWH, RP); aggregations of 2–7 in rock piles, root systems of trees, and shallow undercuts in stream banks in the vicinity of hot springs and sulfur gas vents along mountain streams (Vandeventer, 1992).

Activity: More active by night than by day (Malhotra and Thorpe, 1999); basking on a termitarium at 1300 h (Tolson and Henderson, 1993). **Behavior:** Loud hissing and aggressive striking, aggregations of 3–12 in hollow logs, tree stumps (Lazell, 1964b); very aggressive in captivity (Binder and Lamp, 2007). **Foraging and Diet:** Diet includes introduced rodents (*Rattus*) and agoutis (*Dasyprocta noblei*), boa with agouti was 1.95 m SVL and weighed 5.9 kg (Schwartz and Henderson, 1991); will occasionally raid hen houses (Lazell, 1964b); *Iguana delicatissima* (Breuil, 2002b; Knapp et al., 2009); probing *I. delicatissima* burrows while actively foraging on Dominica (A. Mitchell, pers. comm.). **Miscellany:** Lesser Antillean Bullfinches (*Loxigilla noctis*) and Bananaquits (*Coereba flaveola*) congregated and appeared to be highly agitated (jumping from branch to branch and loudly vocalizing) around a person holding a live boa (RP). **Mortality:** Often dead on roads in the morning, especially on some northern coastal roads (Malhotra and Thorpe, 1999). **Parasites:** Ticks, tapeworms, roundworms, protozoans, and pentastomes (Vandeventer, 1992). **Predators:** *Leptodactylus fallax* (Lescure, 1979); humans (Day, 1990; Bonny, 2007). **Reproduction:** 16 nearly full-term young on 25 April, 1,496-mm female with 12 enlarged eggs on 12 March (Schwartz and Henderson, 1991); to 25 young in a single litter (James, 2004). **Size:** SVL to about 3.0 m (Schwartz and Henderson, 1991); females 190–260 cm total length, males 120–180 cm total length (Binder and Lamp, 2007).

Conservation Status: Listed in CITES Appendix II (UNEP-WCMC, 2008); local consumption of oil from *B. nebulosa* fat may sell for as much as EC $100 (~US $38) per 0.5-liter bottle (Day, 1990); boas are collected as needed for "medicinal" oil (Vandeventer, 1992); boas are routinely killed despite an awareness that they eat banana-damaging rats (Bonny, 2007); residents kill boas because they prey on chickens (Binder and Lamp, 2007).

Remarks: Price and Russo (1991) suggested elevation of *Boa constrictor nebulosa* to species status. The new combination was used twice without sufficient rationale in publications devoted largely to husbandry (Bosch, 1994; Walls, 1998). More recently, Bonny (2007) treated this taxon as a full species and provided substantial evidence. We concur with his decision.

Lazell (1964b) presented data regarding the distinctive (from *Boa c. constrictor*) head shape (prominent snout and "strikingly" convex canthus versus a straight or nearly straight canthus in *B. c. constrictor*), dorsal scale rows at midbody (59–69 versus 75–98 in *B. c. constrictor* and 65–75 in *B. orophias*), number of ventrals (258–273 versus 214–250 in *B. c. constrictor* and 270–288 in *B. orophias*), and dorsal saddles (32–35 versus usually <22 in *B. c. constrictor* and 27–31 in *B. orophias*). The dorsal ground color is very dark, clouded, and gray brown; loreal and subocular stripes are absent or obscure (these stripes tend to be obvious and sharply demarcated in *B. c. constrictor*).

Boa nebulosa (or its ancestor) probably has been isolated from mainland *B. constrictor* for at least two million years (late Pliocene–early Quaternary) and likely much longer (perhaps paralleling the endemic West Indian species of *Bothrops* with a lineage split dating back to the Miocene or early Pliocene; Wüster et al., 2002). The potential for gene flow between this species and *B. orophias* or *B. constrictor* is virtually nil.

Boa orophias Linnaeus, 1758

Distribution: St. Lucia; a reference to St. Kitts in Stöckl and Stöckl (2008) is in error.

Habitat: *Musa* groves (Lazell, 1964b); less likely than *Bothrops caribbaeus* to be in cultivated areas (Long, 1974); primary rainforest (Arendt and Anthony, 1987); uses caves and hollows, where snakes may be aggregated, also arboreal (even adults), may spend as much as 90% of its time in trees and bushes to heights ~12 m (R. Larket in Binder and Lamp, 2007).

Abundance: In 2002, boas routinely seen without intensive efforts (R. Larket in Binder and Lamp, 2007). **Activity:** May stay in one spot for long periods and may use the same den for years (Long, 1974); largely nocturnal, adults largely ground dwelling, but young (smaller) snakes may ascend as high as 12 m (Lazell, 1964b). **Behavior:** Hisses when aroused and strikes "savagely" (Lazell, 1964b); tames readily, but hisses and strikes when first captured (Long, 1974); relatively calm in nature, but very aggressive and unpredictable in captivity (Binder and Lamp, 2007); more aggressive in captivity than *Boa constrictor* (Stöckl and Stöckl, 2008). **Foraging and Diet:** Small mammals (including mongooses) and birds, which are reputed to give "peculiar warning calls" when they spot a snake (Long, 1974); diet includes bats (*Brachyphylla cavernarum*), prey mass/snake mass 0.17 (Arendt and Anthony, 1987). **Home Range:** Appears to remain in one area (Binder and Lamp, 2007). **Reproduction:** Live-bearer (Long, 1974); a breeding "ball" of one large female and 5 smaller males on St. Lucia in February (Malhotra and Thorpe, 1999). **Size:** SVL to about 3.0 m (Schwartz and Henderson, 1991); females 240–300 cm total length, males 170–220 cm total length (Binder and Lamp, 2007).

Conservation Status: Listed in CITES Appendix II (UNEP-WCMC, 2008); we do not believe that the pet trade substantively impacts West Indian populations of *Boa*; this species is considered beneficial by farmers who realize that they help control rodent populations (Lazell, 1964b); however, persecution continues by residents who do not distinguish between dangerous *Bothrops caribbaeus* and boas (Bonny, 2007); protected by law in St. Lucia (Binder and Lamp, 2007); habitat destruction is rampant and residents believe that boas hinder the profitable tourist industry and kill them (V. Kratzke in Stöckl and Stöckl, 2008).

Remarks: Stull (1935) recognized *Boa orophias* as a full species, but that combination fell into disfavor after Lazell (1964b), despite acknowledging its distinctive features, recognized it only as a subspecies of widely distributed *B. constrictor*. Price and Russo (1991) subsequently suggested re-elevation of *Boa constrictor orophias* to species status. More recently, the binomen was used twice without sufficient rationale in publications devoted largely to husbandry (Bosch, 1994; Walls, 1998). Still more recently, Bonny (2007) treated it as a full species and provided substantial evidence. We concur with his treatment of this taxon.

Lazell (1964b) presented data regarding the distinctive (from *Boa c. constrictor*) head shape (prominent snout and convex canthus versus a straight or nearly straight canthus in *B. c. constrictor*), dorsal scale rows at midbody (65–75 versus 75–98 in *B. c. constrictor* and 59–69 in *B. nebulosa*), number of ventrals (270–288 versus 214–250 in *B. c. constrictor* and 258–273 in *B. nebulosa*), and dorsal saddles (27–31 versus usually <22 in *B. c. constrictor* and 32–35 in *B. nebulosa*). The dorsal ground color is rich brown; the loreal stripe is largely obscure and the subocular stripe distinct and complete (these stripes both tend to be obvious and sharply demarcated in *B. c. constrictor*).

Boa orophias (or its ancestor) probably has been isolated from mainland *Boa constrictor* for at least two million years (late Pliocene-early Quaternary) and likely much longer (perhaps paralleling the endemic West Indian species of *Bothrops* with a lineage split dating back to the Miocene or early Pliocene; Wüster et al., 2002). The potential for gene flow between this species and *B. nebulosa* or *B. constrictor* is virtually nil.

Corallus cookii Gray, 1842

Distribution: St. Vincent.

Habitat: In a variety of habitats at elevations between sea level and 500 m, including secondary growth, mixed agriculture (mango, coconut, nutmeg, papaya, breadfruit, and cacao), rainforest (including strangler figs, *Ficus* sp., and tree ferns, *Cyathea* sp.), stands of pine (*Pinus caribbaea*) and bamboo (*Bambusa* sp.; Henderson, 2002); also in *Gliricidia sepium* (S. B. Powell et al., 2007); in residential areas, along roads with frequent vehicular traffic, in artificially illuminated areas, in gardens with orchard trees, foraging on chain-link fences, and perching on and in buildings (Henderson, 2002; S. B. Powell et al., 2007).

Abundance: Near Layou (St. Patrick Parish), in a heavily disturbed area of mixed agriculture encounter rate 4–6/person-hour (Henderson, 2002). **Activity:** Becomes active at dusk, activity probably terminated no later than dawn, and most likely well before (RWH). **Defense:** Strategies include retreat and hiding, balling posture, emitting an offensive odor, biting, and body rotation (RWH). **Dispersal:** A treeboa was swimming 1.5–2.0 km off Chateaubelair (St. David Parish, St. Vincent), when rescued it was "exhausted," but, upon recovery, attempted to bite (Henderson and Sajdak, 2003). **Foraging and Diet:** Based on 8 prey items, snakes 396–910 mm SVL take anoles (*Anolis griseus* and *A. trinitatis*), and those 655–1,374 mm SVL consume introduced rodents (*Mus* and *Rattus*), an unidentified bird was eaten by a snake 655 mm SVL, prey size ranged from 1 cm^3 (*A. trinitatis*) to 31 cm^3 (*A. griseus*), a *M. musculus* had a volume of 11 cm^3, this species undergoes an ontogenetic shift in diet from predominantly anoles to mostly rodents (Henderson, 2002); a snake approached a sleeping *A. trinitatis* from behind, in 10 min approached from ~15 cm to 2.0 cm, at which movement was imperceptible for 15 min, "virtually touching" the lizard before grabbing it by the nape and throwing a coil around it, swallowing completed 14 min later (Henderson et al., 2007). **Longevity:** In captivity, 14 years, 3 months (Bowler, 1977). **Reproduction:** Based on the appearance of snakes <500 mm SVL, birth occurs between September and March (Henderson, 2002); 5 neonates in close proximity to one another in the same tree on 6 June (one with 365 mm SVL and mass of 11.5 g), probably born within the previous several days and had not yet dispersed from the vicinity of the birth site (RWH). **Size:** Maximum SVL 1,374 mm in a female (Henderson, 1997).

Conservation Status: Listed on CITES Appendix II (UNEP-WCMC, 2008); although habitat destruction and alteration may have a deleterious impact, this is a fairly versatile species and will probably continue to do well in areas where clear-cutting of appropriate habitat has not occurred.

Corallus grenadensis Barbour, 1914

Distribution: Grenada Bank, Grenada and the Grenadines.

Habitat: Arboreal, in a wide range of habitats displaying expanses of contiguous crown vegetation from sea level to at least 525 m, including thorny scrub with large stands of *Acacia* with some *Agave* and cactus (*Cephalocereus*, *Opuntia*), mangrove (*Rhizophora mangle*) and mangrove associates (in situations sometimes completely surrounded by water), mixed agriculture (mango, cacao, coconut, citrus, breadfruit, banana, nutmeg, vines of passion fruit), Sea Grape (*Coccoloba*, on Carriacou), and rainforest, usually in edge situations (Henderson, 2002); may have benefited from creation of edge situations during early human settlements, uses perches and retreats in and on houses (Henderson, 2008); also in heavily disturbed residential areas where it will occasionally enter houses and outhouses (Daudin and de Silva, 2007); perch height, perch diameter, and distance from the distal end of a branch is positively correlated with snake SVL and mass (e.g., at Mt. Hartman Bay, snakes <600 mm SVL at mean perch height 2.4 ± 0.1 m, those 600–1,100 mm SVL 2.9 ± 0.4 m, and those >1,100 mm SVL 3.8 ± 1.0 m; Henderson, 2002; Henderson et al., 1998); different size classes exhibit differential use of a given habitat type (e.g., small snakes often associated with thorny vegetation, such as *Acacia*, whereas larger snakes are most frequently in fruit trees, such as mango, or mangroves, *Rhizophora*; Henderson, 2002; Henderson et al., 1998); on the Grenadines, in dry basal fronds of the bromeliad (*Tillandsia utriculata*) or cavities in tree trunks by day (Daudin and de Silva, 2007).

Abundance: Population density estimates ~4–69/ha, depending on site (Henderson, 2002). **Activity:** By day, sleeps/rests coiled on the distal ends of branches (often with an umbrella of vegetation providing cover and protection from direct sun) with only the prehensile tail wrapped around the branch, in mangroves selects branches over water (Henderson, 2002); foraging/traveling begins at dusk as light levels and air temperatures decrease, remains high 1900–2200 h, and then begins to decline, no foraging/traveling activity after 0530 h as dawn approaches (Henderson and Winstel, 1997). **Defense:** Defensive strategies include crypsis, apostatic selection, retreat and hiding, balling posture, emitting an offensive odor, biting, and body rotation (Henderson, 2002). **Dispersal:** A treeboa was swimming 3.5 km off Bequia, whether it would successfully reach landfall and whether St. Vincent or an island of the Grenada Bank is open to speculation (Henderson and Sajdak, 2003). **Foraging and Diet:** Lizards (*Anolis aeneus*, *A. richardii*, and *Iguana iguana*; Pendlebury, 1974; Henderson, 2002); birds (Bananaquit, *Coereba flaveola*), and introduced murid rodents (*Mus*, *Rattus*; Henderson, 1993, 2002, 2008); probably bats (RWH); ontogenetic shift in diet, anoles (*Anolis aeneus* and *A. richardii*) are the predominant prey of snakes <600 mm SVL, snakes 600–1,100 mm SVL in transition from anoles to rodents, and those >1,100 mm SVL prey almost exclusively on rodents, prey size 0.5–55 cm^3 correlated with SVL, midbody circumference, and head width, mean mass ratio (prey mass/snake mass) 0.11 ± 0.02 (0.01–0.30; N = 22; Henderson, 1993, 2002); sleeping anoles located by active foraging with frequent tongue flicks on branch and leaf surfaces, and often on the branches and leaves below that which supports the snake (Yorks et al., 2004); boas large enough to prey on rodents tend to be sit-and-wait foragers, often choosing ambush sites close to the ground (Henderson, 2002). **Growth:** Based on limited data, rates of growth 0.08–0.62 mm/day (Henderson, 2002). **Hurricane Response:** Two months after a Category 4 storm on Grenada (Hurricane Ivan, 2004), predominantly young-of-the-year boas foraging in crowns of trees that were on the ground and others foraging in leafless, standing trees (Henderson and Berg, 2005a). **Parasites:** Ticks (Henderson, 2002). **Predators:** On Grenada, opossums (*Didelphis marsupialis*) and, potentially, the mongoose (*Herpestes javanicus*; Henderson, 2002); potential predators on the Grenadines include opossums and feral cats (Daudin and de Silva, 2007). **Reproduction:** Two preserved females had 29 and 40 undeveloped eggs, another preserved female had 17 nearly full-term young, a preserved litter comprised 21 neonates, gravid or pregnant females in May–June, births likely in June–November (Henderson, 2002; Schwartz and Henderson, 1991). **Size:** Largest individual (female) with SVL 1,625 mm (Henderson, 1997). **Spatial Distribution:** Preliminary evidence suggests that spatial patterns (i.e., individual distances) of large adults may reduce potential competition for rodent prey that occurs at much lower densities than lizard prey, snakes <600 mm SVL at mean distance 18.1 ± 5.8 m, those 600–1,100 mm SVL 32.8 ± 6.6 m, those >1,100 mm SVL 158.5 ± 2.5 m (Henderson and Sajdak, 2002). **Thermal Biology:** Mean body temperature of foraging treeboas 25.6 ± 1.7 °C (20.8–28.6 °C) and mean air temperature 24.9 ± 1.40 °C (20.0–27.1 °C); boas at higher elevations foraged with significantly lower body temperatures than those at lower elevations (Henderson and Henderson, 1995).

Conservation Status: Listed on CITES Appendix II (UNEP-WCMC, 2008); although habitat destruction and alteration may have a deleterious impact, this is a fairly versatile species and will probably continue to do well in areas where clear-cutting of appropriate habitat has not occurred; however, decreasing numbers of snakes have been documented at some sites (Henderson and Berg, 2006).

Epicrates angulifer Cocteau and Bibron, 1840

Distribution: Cuba, including Isla de la Juventud and several satellite islands; elevational range, sea level to 325 m.

Habitat: Xerophytic scrub to broadleaf forest, often in trees (to 9.2 m coiled and resting by day), frequently associated with caves, rocky cliffs (and their talus), on roads at night (Sheplan and Schwartz, 1974); at 0200 h in a low tree in tangled brush and cactus ~23 m from the high tide mark, in an area of tall grass and scattered cactus, in a residential area (Lando and Williams, 1969); in orchards and small patches of forest in the proximity of human activity in rural areas (Estrada, 1994); on rocks at a river cut, in a bunker on the U.S. Naval Base at Guantanamo Bay, at the base of a *Corypha umbraculifera* palm (Schwartz and Henderson, 1991); on the U.S. Naval Base at Guantanamo Bay, in the sewer of a deserted fort killing *Rattus*, adjacent to salt flats, in bunkers and residential areas, crossing roads leading to storage areas at night (Tolson and Henderson, 1993); in a hollow Ironwood Tree (*Guaiacum sanctum*; Alberts, 2003); in hollow trunks of palms (6.2%), on the ground (9.4%), in trees (12.5%), often near houses, in an abandoned house (3.1%), and in caves (68.8%), equally divided between warm caves and those that are not, with caves providing refuges but also food (Morell Savall et al., 1998); over an 8-day period, 41 boas at Cueva de Majas (Hardy, 1957c); in a sample of 32 boas, 22 in caves (Berovides Álvarez and Carbonell Paneque, 1998); inactive during the day, resting coiled in trees >8 m off the ground (Díaz, 2006).

Abundance: During 22–24 December 1954 and 8–12 September 1955, 41 boas collected (including 35 from a bat-populated cave; Hardy, 1957c); 9 boas at the mouth of a cave (Ševčik, 1989); encounter rate 2.06 ± 1.74/h, highest in caves (Berovides Álvarez and Carbonell Paneque, 1998). **Activity:** Crepuscular and nocturnal (Z. Vogel, 1965c); primarily nocturnal, but may occasionally be active by day (Lando and Williams, 1969). **Behavior:** In captivity, reacted aggressively toward dipsadid snakes (*Caraiba andreae*, *Cubophis cantherigerus*), but not toward *Tropidophis melanurus* (Morell Savall et al., 1998). **Defense:** Besides biting, may expel a stream ("fétido") that may be a combination of musk and fecal matter (Garrido and Schwartz, 1969). **Foraging and Diet:** Primarily a sit-and-wait forager, but might actively enter hen houses for domestic fowl (Barbour and Ramsden, 1919; Sheplan and Schwartz, 1974; Estrada, 1994); hunts Cuban Hutias (*Capromys pilorides*) and bats, pillages bird nests, and readily takes domestic fowl (Z. Vogel, 1965c); diet includes snakes (*Tropidophis melanurus*; Viña Dávila and Armas, 1989), *Rattus rattus*, *Capromys* (Tolson and Henderson, 1993; Buide, 1966), and bats (Barbour and Ramsden, 1919), including *Brachyphylla*, *Mormoops*, and *Phyllonycteris* (Hardy, 1957c; Sheplan and Schwartz, 1974); young goat of 2.72 kg (Buide, 1966); "natural enemy" of the White-crowned Pigeon (*Columba leucocephala*; Godinez et al., 1987); nests and hatchlings of *Trachemys decussata* (Sampedro Marín and Montañez Huguez, 1989); eggs of Cave Swallows (*Hirundo fulva*; Mancina and Llanes Sosa, 1997); in captivity, took a variety of birds (*Passer domesticus*, *Gallus gallus*, *Columbina passerina*, *Zenaida macroura*, *Z. asiatica*, *Crotophaga ani*), mice (*Mus musculus*), rarely reptiles (*Anolis sagrei*, *A. allisoni*, *Ameiva auberi*), and only once a frog (*Osteopilus septentrionalis*), showing a strong preference for birds and mammals (Morell Savall et al., 1998). **Longevity:** In captivity, 22 years, 7 months (Bowler, 1977); >24 years (RWH). **Parasites:** Sporozoan red blood cell parasite *Haemogregarina* sp. (Zajicek and Mauri Méndez, 1969); cestode *Ophiotaenia nattereri* (Freze and Rysavy, 1976; Coy Otero and Lorenzo Hernández, 1982); ticks *Amblyomma quadricavum* (V. Černý, 1966, 1969; Guglielmone et al., 2003); *A. albopictum* (V. Černý, 1969; Coy Otero, 1999; Guglielmone et al., 2003); *Anocentor nitens* (Pérez Vigueras, 1934a; V. Černý, 1969). **Predators:** Humans, who "persecute" these snakes (Gundlach, 1880); also exploit snakes for meat, fat (for curing hemorrhoids, respiratory problems), and hides (for belts and shoes; Rivalta et al., 2003a; Díaz and Abreu Guerra, 2005; Fong Grillo et al., 2005a); snakes also killed because people fear them and believe them to be major predators of domestic chickens (Morell Savall et al., 1998). **Reproduction:** Viviparous (Schwartz and Henderson, 1991); males display ritualized combat behavior (Tolson, 1983; Tolson and Henderson, 1993); several males may engage a female at one time in order to mate (Gundlach, 1880); during courtship the male drapes his tail over the female or encloses her with a single coil (Tolson and Henderson, 1993); copulation in nature in April (Morell Savall et al., 1998); litter of 4 born 4 November, neonates on 17 November had total lengths of 600–610 mm and were able to exude a mild musk (Sheplan and Schwartz, 1974); in captivity, gestation 152–252 days, litter size 2–28, neonate body mass 880.0–203.6 g, relative clutch mass 0.145–0.415 (Tolson, 1983; Tolson and Henderson, 1993). **Size:** SVL approaches 4.0 m (although Gundlach, 1880, reported individuals of 4.6–6.4 m). **Social Behavior:** May aggregate in caves (Hardy, 1957c).

Conservation Status: Listed on CITES Appendix II (UNEP-WCMC, 2008); on the IUCN Red List as "near threatened" with extinction (Day and Tolson, 1996); also as at "lower risk" of extinction (Rodríguez Schettino and Chamizo Lara, 1998a).

Additional Pertinent Literature: Henderson and Arias (2001).

Epicrates chrysogaster Cope, 1871

Distribution: Bahama Islands, where three currently recognized subspecies occur on Acklins and Crooked islands (*E. c. schwartzi*), Great Inagua Island and Sheep Cay (*E. c. relicquus*), and in the Turks and Caicos Islands (*E. c. chrysogaster*).

Habitat: In barren, xeric situations as well as wooded areas and heavily disturbed *Cocos* groves, from inside a log in a *Cocos* grove on Great Inagua (Sheplan and Schwartz, 1974); in *Cocos* litter, garden plots, and around human habitations (Buden, 1975); copulating under a flat limestone slab and crawling on a stone wall in the ruins of a plantation building by night on North Caicos (Tolson and Henderson, 1993).

Activity: Nocturnal (Tolson and Henderson, 1993). **Diet:** Rodents (*Mus*, *Rattus*) and chicks and eggs of domestic fowl (*Gallus*; Schwartz and Henderson, 1991). **Longevity:** In captivity, 14 years, 2 months and still alive at last report in 1992 (Snider and Bowler, 1992). **Reproduction:** Pregnant female on 5 June on North Caicos gave birth to 28 young on 14–15 September (Buden, 1975); in captivity, North Caicos pair mated on 4 July 1976, the 294.6-g female gave birth to 9 young on 27 February 1977, gestation 238 days, neonate mass 7.3–8.4 g, relative clutch mass 0.241 (Tolson, 1980; Tolson and Henderson, 1993). **Size:** SVL to 1,310 mm (Sheplan and Schwartz, 1974).

Conservation Status: Listed on CITES Appendix II (UNEP-WCMC, 2008); protected in the Bahamas by the Wild Animal Protection Act of 1968 (Tolson and Henderson, 1993); Barbour (1941) suggested that the species (*E. c. relicquus*) had been extirpated from Great Inagua.

Epicrates exsul Netting and Goin, 1944

Distribution: Bahama Islands, Little Bahama Bank.

Habitat: On Great Abaco, in burned-over pinewoods and cutover scrub and pine covering porous limestone substrate (Sheplan and Schwartz, 1974); along the margin of a clearing in pinewoods (Netting and Goin, 1944); beneath debris at the high tide line (Tolson and Henderson, 1993).

Activity: Presumably nocturnal (RWH). **Longevity:** In captivity, 21 years, 8 months and still alive at last report in 1992 (Snider and Bowler, 1992). **Behavior:** Male-male combat in captivity (Tolson and Henderson, 1993). **Reproduction:** In captivity, 234-g female gave birth to 9 young on 20 August, neonates 4.5–7.3 g with SVL 231–280 mm, relative clutch mass 0.267, the female consumed 2 infertile eggs and a deformed stillborn neonate after parturition (Tolson and Henderson, 1993). **Size:** SVL to 810 mm (Sheplan and Schwartz, 1974). **Thermal Biology:** Gravid females presumably thermoregulate beneath debris at the high tide mark on Great Abaco (Schwartz and Henderson, 1991); on Great Abaco, emerges from cavities in coral to warm in the sun at the margin of a clearing (Netting and Goin, 1944).

Conservation Status: Listed on CITES Appendix II (UNEP-WCMC, 2008); although protected in the Bahamas under the Wild Animals Protection Act of 1968, often killed by local residents (Tolson and Henderson, 1993); an anecdotal account indicated extirpation on Stranger's Cay after the introduction of feral pigs (Netting and Goin, 1944).

Epicrates fordii Günther, 1861

Distribution: Hispaniola, where three currently recognized subspecies occur on much of Hispaniola, including Île de la Gonâve and Isla Saona (*E. f. fordii*), in northwestern Haiti (*E. f. agametus*), and on Île à Cabrit (*E. f. manototus*).

Habitat: In both xeric and mesic situations (Schwartz and Henderson, 1991); at 1.5 m in a fire-scarred cabbage palm, dead on the road in a *Musa* grove, dead on the road in an area irrigated for rice culture, and beneath rocks on grassy hillsides (Sheplan and Schwartz, 1974); in stems of *Agave* (Schwartz and Graham, 1980); on Isla Saona, at 1.2–1.5 m at mangrove edge (Henderson and Sajdak, 1983); on Île à Cabrit at the base of a grass tussock in *Acacia*-cactus scrub and another under a large dead cactus trunk on an exposed hillside (Schwartz, 1979d); beneath limestone rocks partially embedded in soil and from the interior of an *Aloe* clump bordering a garden in a populated area near Manneville, Haiti (Tolson and Henderson, 1993).

Activity: Nocturnal (Schwartz and Henderson, 1991). **Foraging and Diet:** May use both active and ambush foraging strategies, diet includes *Anolis* (4 of 6 prey items), *Mus* (2 of 6 prey items; Henderson et al., 1987b); *Rattus* (Sheplan and Schwartz, 1974); seen regularly near Hispaniolan Parrot (*Amazona ventralis*) nests, but no direct evidence of predation (Snyder et al., 1987). **Longevity:** In captivity, 24 years, 5 months and still alive at last report in 1992 (Snider and Bowler, 1992). **Reproduction:** Female from Frères had 7 young on 25 June, female from near Pétionville gravid on 16 June (Sheplan and Schwartz, 1974); 5 females from near Manneville gave birth 24–28 May, litter sizes 3–4, neonate mass 3.6–5.3 g, relative clutch mass 0.416–0.493, female from near Pétionville gave birth to 4 young on 24 August, neonate mass 3.8–4.3 g, relative clutch mass 0.263 (Tolson and Teubner, 1987; Tolson and Henderson, 1993); reproductive male and female >15 years old (Tolson, 1991a; Tolson and Henderson, 1993). **Size:** SVL to 860 mm (Sheplan and Schwartz, 1974).

Conservation Status: Listed on CITES Appendix II (UNEP-WCMC, 2008).

Additional Pertinent Literature: Henderson and Powell (2004a).

Epicrates gracilis Fischer, 1888

Distribution: Hispaniola, where two currently recognized subspecies occur north of the Plaine de Cul de Sac-Valle de Neiba (*E. g. gracilis*) and on the Tiburon Peninsula east to Port-au-Prince and Jacmel (*E. g. hapalus*).

Habitat: Lowland deciduous woods, frequently adjacent to bodies of water, 0.5–4.6 m above the ground in trees, bushes, and vine tangles, on a cutbank adjacent to a pasture, on a large *Ficus* trunk adjacent to a road in a grassy pasture with scattered bushes (Sheplan and Schwartz, 1974); 5 m high in vine tangles hanging from trees and telephone wires, 3 m high in *Acacia* scrub (RP).

Abundance: Locally abundant (Sheplan and Schwartz, 1974). **Activity:** Nocturnal and arboreal (Schwartz and Henderson, 1991); by day in a pile of rocks 6.1 m from a house in a cultivated area, in a limestone rock solution hole, in the knothole of a fence, crawling on a barbed-wire fence in early afternoon adjacent to an orange grove (Sheplan and Schwartz, 1974). **Foraging and Diet:** An active forager that preys predominantly on *Anolis* (6 of 6 prey items), mean prey volume 2.4 cm^3 (Henderson et al., 1987b); seen regularly near Hispaniolan Parrot (*Amazona ventralis*) nests, but no direct evidence of predation (Snyder et al., 1987). **Reproduction:** 730 mm SVL female from Limbé (Haiti) gave birth to 5 young on 22 September (Murphy et al., 1978); 2 females from Limbé each gave birth to 3 young and 2 infertile egg masses on 21 and 24 October, neonate mass 2.0–2.3 g, relative clutch mass 0.192–0.224 (Tolson and Henderson, 1993). **Size:** SVL to 905 mm (Sheplan and Schwartz, 1974).

Conservation Status: Listed on CITES Appendix II (UNEP-WCMC, 2008).

Additional Pertinent Literature: Henderson and Powell (2002).

Epicrates inornatus Reinhardt, 1843

Distribution: Puerto Rico; elevational range, sea level to 1,050 m.

Habitat: Rocky areas (Stejneger, 1904); xeric scrub (Tolson and Henderson, 1993); caves and plantations, not uncommon in urban and suburban areas (Schwartz and Henderson, 1991); in outbuildings and houses (Wiley, 2003; D. P. Reagan and C. P. Zucca, 1982 in Joglar, 2005b; Joglar, 2005b); "dormant" on rafters in buildings in the Caribbean National Forest, resting in pipes near El Verde (Reagan, 1984); in various forested situations, including wet montane forest, lowland wet forest, remnant coastal rainforest, mangrove forest, wet limestone karst, also pastureland with patches of exotic trees, and offshore cays (Wiley, 2003); at the Mata de Plátano Reserve bat cave, males spend more time under litter, in trees, and on the ground, while females spend more time inside the cave, on the ground, and on rock walls (Puente-Rolón and Bird-Picó, 2004); compared to random trees in the Luquillo Experimental Forest, trees in which boas were located had conspicuous vine cover, larger trunks, more crown contact with neighboring trees, were taller and surrounded by a higher density of understory vegetation, and had a lower percentage of canopy cover (Wunderle et al., 2004); in 1887 at Bayamón, "very common" along streams on tree branches and "easily taken by means of a stick bearing a noose of strong twine" (Garman, 1887).

Activity: Largely nocturnal, basks by day in light gaps and forest edge situations, as well as open areas in mahogany and kadam plantations, and on branches along forest streams (Reagan, 1984). **Behavior:** Considerably more aggressive than *E. m. monensis* (Rivero et al., 1982); defensive displays include hissing, throat inflation, biting, and coiling into a ball with head hidden in coils (Rivero, 1998). **Foraging and Diet:** Likely employs active and ambush foraging modes (Wunderle et al., 2004); 3–14 (mean 7) adult snakes (1.5–2.0 m total length) in foraging positions at a cave entrance feeding on bats (including *Brachyphylla cavernarum, Monphyllus redmani*), one observation of capture, constriction, and ingestion took 16 min (Rodríguez and Reagan, 1984); in captivity, *Ameiva exsul* (M. Leal in Thomas and Gaa Kessler, 1996); in captivity, *Bufo marinus*, which proved fatal to the snake (J. Wunderle in Thomas and Gaa Kessler, 1996); diet includes land crabs, lizards (*Anolis cuvieri, A. evermanni, A. gundlachi*), domestic fowl (*Gallus gallus*), Ground Doves (*Columbina passerina*), bats (*Monophyllus redmani*), and rats (*Rattus rattus*), with the latter accounting for 61.8% of prey items in a sample of 34, anoles taken by small (388 mm SVL) and large (1,780 mm SVL) boas, rats taken by boas 1,200–2,050 mm SVL (Wiley, 2003); ate an egg and a hatchling at a Cattle Egret (*Bubulcus ibis*) nest (Wiley, 2003); preys on the Puerto Rican Parrot (*Amazona vittata*; Wiley, 1981), although predation on parrots probable but not confirmed (Snyder et al., 1987); cannibalism (Acevedo-Torres et al., 2005); boas at the entrance of a bat cave from 1730–0600 h, but most activity occurred 1900–2400 h, boas hung 75% or more of their bodies from a vine, branch, or limestone outcropping in attempts to capture bats leaving the cave or active inside, at least 5 species of bats captured and consumed, as many as 4 bats captured by a single snake on a given night, average bat handling time 12.53 ± 6.5 minutes (Puente-Rolón and Bird-Picó, 2004); piracy, boa stealing bat from the boa that captured it (Rodríguez-Durán, 1996; Puente-Rolón and Bird-Picó, 2004); bat carrion (Rodríguez-Durán, 1996); freshly captured snake regurgitated a *Diploglossus pleii* (A. J. Sánchez Muñoz, in litt., 17.III.2008). **Home Range and Movements:** At Mata de Plátano Reserve, home ranges (minimum convex polygon) 138.9–18,380 m^2, mean female home range 7,890 ± 6,614 m^2, mean male home range 5,000 ± 4,102 m^2, mean distance/move 47.9 ± 18 m for females and 51.5 ± 24 m for males, mean distance traveled/day 83 ± 51 m for females and 99 ± 88 m for males, mean time spent at one location 37.4 ± 38.3 days (range 5–53) in males and 47 ± 34.7 days (2–96) in females (Puente-Rolón and Bird-Picó, 2004); in the Luquillo Experimental Forest, boa SVL and 95% adaptive kernel (AK) home range positively correlated for 18 boas, median AK home range size did not differ significantly between sexes (9.1 ha in males, 6.7 ha in females; Wunderle et al., 2004). **Hurricane Impact:** In the Luquillo Experimental Forest, post-Hurricane Georges (21 September 1998; sustained winds of 184 km/h), mean monthly movement increased (prehurricane 10.7 ± 1.1 m/d/mo; hurricane 14.9 ± 1.3 m/d/mo; posthurricane 16.8 ± 2.2 m/d/mo, home range sizes in males changed little between pre- and post-hurricane periods, but increased for 2 females, although this increase might be related to postparturition behavior (Wunderle et al., 2004). **Longevity:** In captivity, 23 years, 11 months (A. Wisnieski in Slavens and Slavens, 2003). **Mobbing:** Seven bird species (warblers, tanagers, Puerto Rican Bullfinches, Bananaquits) mobbed female boas in the Luquillo Experimental Forest, sometimes approaching as close as 25 cm (Mercado et al., 2002). **Mortality:** Roadkills (Reagan, 1984); more than a dozen boas dead on roads at elevations below 400 m, but a lack of observations above 400 m may be an artifact, as little time was spent driving high-elevation roads (Snyder et al., 1987). **Parasites:** Mites (Reagan, 1984); ticks (*Amblyomma* sp. or *Boophilus* sp.) and red mites (*Ornithodorus portoricensis*; F. J. Bird, 1994 in Joglar, 2005b). **Population Size**

and Density: At Cueva de los Culebrones, Mata de Plátano Reserve, 3–21 boas observed on a given night, with 1–10 seen at a cave entrance (Rodríguez-Durán, 1996); mean monthly density 5.6/ha at several sites between February 2001 and February 2002 (Rio-López and Aide, 2007). **Predation:** Potential predators include mongooses (*Herpestes javanicus*), cats (*Felis catus*), Red-tailed Hawks (*Buteo jamaicensis*), Pearly-eyed Thrashers (*Margarops fuscatus*), Red-legged Thrushes (*Turdus plumbeus*), Puerto Rican Lizard Cuckoos (*Saurothera vieilloti*), although some are restricted to taking small snakes (Reagan, 1984); other possible predators include pigs, toads, chickens, *Ameiva*, and other birds of prey (Rivero and Seguí-Crespo, 1992; Rivero, 1998); humans kill boas as they depredate poultry or for the supposed medicinal and aphrodisiac qualities of the oil rendered from their fat (Snyder et al., 1987; Wiley, 2003). **Reproduction:** In captivity, severe male-male aggression during the breeding season, females also may act aggressively toward conspecifics (Tolson and Henderson, 1993); females in reproductive condition late April–mid-August, nonreproductive females February–April and November (Wiley, 2003); pair captured on 17 May mated the same day while being transported in a box and the female contained 32 embryos on 20 July (Grant, 1932f); gestation 152–193 days, smallest gravid female 1,080 mm, 685 g (Wiley, 2003); litter size 12–32 (Tolson, 1992); neonate body mass 12.0–21.7 g, relative clutch mass 0.189–0.216 (Tolson, 1992; Tolson and Henderson, 1993); 1.5-m female gave birth to 15 live and 1 dead neonate, total lengths 385–430 mm (R. Joglar and N. Ríos in Joglar, 2005b). **Size:** To 2,050 mm SVL (Wiley, 2003).

Conservation Status: Listed on CITES Appendix I (UNEP-WCMC, 2008); on the IUCN Red List as "near threatened" with extinction (Tolson, 1996b); as "endangered" on the U.S. Fish & Wildlife Service list of threatened and endangered animals and plants (U.S. Fish & Wildlife Service, 2008); protected as "threatened" or "endangered" (Joglar, 2005b); although original forest on Puerto Rico is largely gone, as the rural human population moves to cities, land once used for agriculture has reverted to secondary forest, now the species is apparently common in suitable habitat, occurs in disturbed situations, and is not coveted in the pet trade, oil from the fat of this species is used by some rural Puerto Ricans as an aphrodisiac and as a remedy for rheumatism and arthritis, unknown numbers may be killed to extract the fat (Snyder et al., 1987; Rivero, 1998; Wiley, 2003; Joglar, 2005b); this species, however, may be "holding its own despite routine killing by humans when it is found" (Snyder et al., 1987); "threatened (currently endangered)" (Moreno, 1991d).

Epicrates monensis Zenneck, 1898

Distribution: Two currently recognized subspecies on Isla Mona (*E. m. monensis*) and the Puerto Rico Bank, including the U.S. and British Virgin Islands (*E. m. granti*); the two taxa likely are distinct at the species level.

Habitat: In dry subtropical forest throughout most of the range (Tolson, 1988, 1991b); in subtropical moist forest on Puerto Rico (Sierra de Luquillo; Tolson, 1996a); on Isla Mona (*E. m. monensis*), in open canopy depression forest and closed canopy coastal forest, 34 tree and shrub species, 3 vine species, and one bromeliad species used, boas <400 mm SVL at mean perch height 1.47 ± 0.15 m and diameter 12.3 ± 4.3 mm, most frequently in *Eugenia* sp., *Ficus citrifolia*, and *Tillandsia utriculata*, 401–500 mm SVL (young-of-the-year) at mean perch height and diameter 1.31 ± 0.16 m and 13.4 ± 3.8 mm, exploited 15 species of trees and shrubs (notably *F. citrifolia* and *Antirhea acutata*), mean perch height and diameter for subadults (501–600 mm SVL) 1.66 ± 0.24 m and 10.2 ± 2.9 mm, used 18 species of trees and shrubs (most in *F. citrifolia* and *A. acutata*), subadults (601–700 mm SVL) 1.75 ± 0.23 m and 24.4 ± 8.9 mm, used 14 species of trees and shrubs (mostly *Caesalpinea monensis* and *F. citrifolia*), adults (>700 mm SVL) 2.12 ± 0.40 m and 51.4 ± 12.1 mm, used 19 species of trees and shrubs (most frequently *Clusia rosea* and *F. citrifolia*), nearest neighbor tree or shrub distance in 77 of 82 captures zero (Tolson et al., 2007); *E. m. granti* often associated with *Coccoloba uvifera*, roadside *Acacia* and *Leucaena* (Tolson, 1996a); on Cayo Diablo, 51 of 79 active boas in *Coccoloba*, 10 in *Caesalpinea*, 9 in *Cassythia*, 7 in *Suriana*, and 2 in *Opuntia*, on St. Thomas, beneath a limestone slab in a housing development (Tolson, 1988, 1991b); in a stone wall and dead on the road (Nellis et al., 1983; Tolson, 1991b); at 2.4 m in a viny tangle (Sheplan and Schwartz, 1974); on Tortola, in the rafters of a house (RWH); on Great Camanoe (British Virgin Islands), hanging from a rock wall (B. S. Barker, pers. comm.); islands with high *Rattus* densities (~20 rats/ha; Buck Island, Cas Cay, Salt Cay) have lower *Ameiva* and *Anolis* densities and apparently no *E. monensis* despite suitable habitat (Tolson, 1988, 1991b).

Activity: Nocturnal (Tolson, 1988). **Behavior:** Even when handled, shows no signs of aggression (Rivero et al., 1982). **Foraging and Diet:** Individuals forage every night under most circumstances (Tolson, 1988); employs active and ambush foraging modes, subadult *E. m. monensis* prey on *Anolis monensis* and adults take *Rattus rattus* (Tolson et al., 2007); *A. monensis* taken by *E. m. monensis* (K. P. Schmidt, 1926); in *E. m. granti*, *Anolis cristatellus* is predominant prey, rarely *Ameiva* or endotherms (Tolson, 1988; Chandler and Tolson, 1990); on Cayo Diablo, greatest boa concentrations in areas with anole densities >60/100 m^2 (Tolson, 1988); 78.9% of sleeping anoles and 88.6% of foraging boas use *Coccoloba uvifera* (Chandler and Tolson, 1990). **Habitat Degradation:** Feral pigs (*Sus scrofa*) and goats (*Capra hircus*) threaten endemic plant communities and degrade boa habitat (Tolson, 1991b). **Hurricane Impact:** On Mona, Hurricane Georges (22 September 1998) altered capture heights of adults, prehurricane mean 3.18 ± 1.17 m, posthurricane 1.04 + 0.34 m, all large *Clusia rosea* in a depression site toppled (Tolson et al., 2007). **Longevity:** In captivity (for *E. monensis granti*), 9 years, 11 months and still alive at last report in 1995 (A. Odum in Slavens and Slavens, 2003). **Population Size:** On Cayo Diablo, >100 boas/ha at some sites (Tolson, 1988); on St. Thomas, 300–400 individuals (Tolson, 1991b). **Predation:** Rats (*Rattus*), domestic and feral cats (*Felis*), mongooses (*Herpestes*), potential natural predators include Yellow-crowned Night Herons (*Nyctanassa violacea*) and Puerto Rican Screech Owls (*Otus nudipes*), evidence circumstantial, but >17% of boas have wounds indicative of natural predation, *Borikenophis portoricensis* also may prey on boas (Tolson, 1991b). **Refugia:** Of 149 inactive snakes removed from refugia, 43% in *Cocos* or *Sabal* axils, 36% in termite nests, 21% under rocks or debris, 51% of boas taken from termite nests were females, >50% gravid (Tolson, 1988, 1991b). **Reintroductions:** Reintroductions successful on Cayo Diablo and Cayo Ratones on the Puerto Rico Bank (Tolson, 1996a). **Reproduction:** Copulation on 10 March, gravid females use termite nests or sun-baked debris to thermoregulate and are able to elevate body temperatures to >33° C (Tolson, 1988); gestation 132 days, litter size 2–10, neonate total length 206–280 mm, body mass 3.6–6.4 g, relative clutch mass 0.244–0.372 (Tolson and Henderson, 1993); female with SVL 918 mm gravid with 4 embryos (total length 246–256 mm) on 27 May (Joglar, 2005b). **Size:** SVL to 1,230 mm (Tolson et al., 2007). **Thermal Biology:** Gravid females use termite nests or sun-baked debris to thermoregulate, and may elevate their body temperatures to 33 °C (Tolson, 1991b).

Conservation Status: Listed on CITES Appendix I (UNEP-WCMC, 2008); *E. m. granti* as "endangered" and *E. m. monensis* as "threatened" on the U.S. Fish & Wildlife Service list of threatened and endangered animals and plants (U.S. Fish & Wildlife Service, 2008); "endangered" (Moreno, 1991d); protected as "threatened" or "endangered" (Joglar, 2005b).

Epicrates striatus Fischer, 1856

Distribution: Eight currently recognized subspecies occur in the Bahama Islands, Alligator Cay, Cat Island (*E. s. ailurus*), Bimini Islands (*E. s. fosteri*), Andros and Berry islands (*E. s. fowleri*), Ragged Islands (*E. s. mccraniei*), New Providence, Rose, Eleuthera, and Long islands, and the Exuma Cays (*E. s. strigilatus*), and on Hispaniola, main island, Île de la Gonâve, Isla Saona (*E. s. striatus*), Tiburon Peninsula, Île à Vache (*E. s. exagistus*), and Île de la Tortue (*E. s. warreni*); elevational range, sea level to 1,220 m.

Habitat: On Hispaniola, usually in mesic situations, but does occur in dry forests, in *Musa*, tangled vines at 2.1–2.7 m, in trees at 1.5–6.1 m, in a small roadside shrub, and in various other shrubs, trees, and bushes, dead on the road in xeric scrub (Sheplan and Schwartz, 1974); at night active on a tangled vine mass 2 m above the ground in a streamside forest in an otherwise xeric area, live and dead on the coastal road along the eastern Peninsula de Barahona (RP); in the Bahamas, on New Providence Island, in a bush in treeless scrub and in a tree in coppice (i.e., a stand of tropical hardwoods), on Little Ragged Island (Ragged Islands), in a cranny of a rock wall surrounded by a buttonwood (*Conocarpus*) swamp at dusk, on South Bimini Island, in coppice and crossing the road at night in mangroves, on Great Harbour Cay (Berry Islands), on roads during late morning, on Andros Island, under a large boat hatch lying on sand and leaves near a *Coccoloba-Casuarina* stand, at 2.5 m in an oleander tree, on an unpaved road in pinewoods at night (Sheplan and Schwartz, 1974); in scrub and in pine woodlands with open, partial, and closed understories (Knapp and Owens, 2004); on a trail at the edge of a buttonwood swamp in the Exuma Islands (Franz et al., 1993); by day on Hispaniola, in a rotten palm log adjacent to cacao, in a dry log, in a large hollow adjacent to a cornfield surrounded by high-canopied coffee plantation, in the straw roofing of a fallen shack, and in a rotten stump (Sheplan and Schwartz, 1974); also in a bird nest and clumps of vegetation growing on horizontal branches at 20 m (Schwartz and Henderson, 1991); by day in the South Andros area (Sand Cay, Linder Cay, and Mangrove Cay), refugia include being underground, in limestone holes, under leaf litter or in logs, and in water (Knapp and Owens, 2004).

Activity: Nocturnal (Schwartz and Henderson, 1991); although usually in arboreal situations, not infrequently encountered at ground level, including crossing roads, paved and otherwise (Sheplan and Schwartz, 1974); emerged in late afternoon from a burrow in loose sand 6.0 m from Caribbean Sea (Olson, 1990). **Defense:** Produces exceedingly foul-smelling musk and excretia when stressed (Powell, 2003b). **Foraging and Diet:** Probably employs active and ambush strategies, from close to or at ground level to at least 7.0 m, on Hispaniola, snakes <600 mm SVL prey on *Anolis* (75% of prey items for boas <800 mm SVL; mean boa SVL 572 ± 391 mm; mean anole size 4.6 ± 0.7 cm^3), those >800 mm SVL prey on birds (*Dulus*, *Quiscalus*, *Coereba*; mean boa SVL 822 ± 178 mm; mean bird size 50.0 ± 11.0 cm^3) and rodents (*Mus*, *Rattus*; mean boa SVL 1,166 ± 730 mm; mean rodent size 100.4 ± 13.5 cm^3; Henderson et al., 1987b); opportunistically preys on birds (*Amazona*, *Aratinga*, *Columba*, *Zenaida*, *Chrysolophus*, *Phasianus*, *Ortalis*) in a free-flight aviary (Ottenwalder, 1980; Henderson et al., 1987b); eats the large rodent *Plagiodontia aedium* and likely the insectivore *Solenodon paradoxus* (Henderson et al., 1987b); on South Bimini, an 860-mm SVL boa contained a Catbird (*Dumetella*; Sheplan and Schwartz, 1974); in the South Andros area, preys on hatchling *Cyclura cychlura*, with one boa consuming 3 from the same clutch on the same night (Knapp and Owens, 2004; Knapp et al., 2004b, 2005); in the Bahamas (Exuma Islands), a boa in an artificial well ingested adult *Osteopilus septentrionalis* 3 hours after sunset (Franz et al., 1993); also feeds on domestic fowl and house cats (Schwartz and Henderson, 1991); seen regularly near Hispaniolan Parrot (*Amazona ventralis*) nests, but no direct evidence of predation (Snyder et al., 1987); on Hispaniola, forages for bats at cave openings (J. A. Ottenwalder, pers. comm.); in captivity, female ate her own offspring (Hanlon, 1964). **Longevity:** In captivity, 22 years, 1 month (Bowler, 1977). **Mortality:** Eight dead boas hung from tree branches at the site of some kind of religious ritual near the Haitian border, frequently killed on roads (RP). **Movements:** In the South Andros area, *E. s. fowleri* moved 0–159 m/day, minimum convex polygon home ranges 0.03–2.0 ha (Knapp and Owens, 2004). **Parasites:** Ticks *Amblyomma albopictum* (Schulze, 1941; Morel, 1967; Guglielmone et al., 2003); *Amblyomma dissimile* on *E. s. fowleri* on Andros Island (Durden and Knapp, 2005). **Reproduction:** Female *E. s. striatus* collected between 13 June and 27 August gravid, number of embryos or fetuses ranged from 8 (in the smallest female, 1,313 mm SVL) to 31 (in the largest female, 1,672 mm), between those extremes, number of embryos 5, 9, 12, 14, 17, 18, and 19 (Sheplan and Schwartz, 1974); 7 females (SVL 1,420–1,635 mm) from Camp Perrin, Haiti (*E. s. exagistus*) on 24–26 July gravid with 7–18 fetuses, no correlation between female SVL and number of embryos, female (1,509 mm SVL) on 16 May on Andros with 35 slightly embryonated eggs (Sheplan and Schwartz, 1974; Schwartz and Henderson, 1991); female (1,676 mm total length) on New Providence on 19 June produced a litter of 51 on 16–18 September (Hanlon, 1964); in captivity, male-male combat (Tolson and Teubner, 1987); strong relationship between 4 stages of male reproductive behavior (precourtship, courtship, copulatory, postcopulatory) and serum testosterone levels (Teubner and Tolson, 1984); courtship may be protracted, sometimes in excess of 3 weeks (Tolson, 1980); gestation in captivity 192–224 days (Tolson and Henderson, 1993); litter sizes 6–51, neonate body mass 12.9–19.1 g, and relative clutch mass 0.160–0.319 (Tolson, 1992; Tolson and Henderson, 1993). **Size:** SVL to 2,330 mm (Henderson et al., 1987b).

Conservation Status: Listed on CITES Appendix II (UNEP-WCMC, 2008); *Epicrates s. fowleri* considered "endangered" by the Bahamas National Trust (Knapp and Owens, 2004); factors threatening boas include habitat loss, predation by feral cats, indiscriminate killing, and smuggling for the illegal pet trade (Knapp and Owens, 2004); likewise, populations of *E. s. fosteri* on South Bimini were heavily exploited for the pet trade in the United States, despite protection in the Bahama Islands (Tolson and Henderson, 1993); on Hispaniola, habitat loss is the primary threat, particularly at some sites in Haiti and the Dominican Republic in the mid-1980s (RWH).

Epicrates subflavus Stejneger, 1901

Distribution: Jamaica, widely distributed but disjunct, also Goat Island.

Habitat: In rock crevices, caves, and trees by day (Schwartz and Henderson, 1991); wet limestone forest and an overgrown *Musa* grove (Prior and Gibson, 1997); enters houses (Gosse, 1851).

Activity: Nocturnal, but at 2.0 m in a tree at 1500 h presumably foraging for birds (Prior and Gibson, 1997); will lie in the sun on rocks after heavy rains (Grant, 1940a). **Foraging and Diet:** Foraging behavior appears to involve active and ambush strategies, will forage on the ground, in trees (1.5–2.0 m), and among tank bromeliads (*Aechmea paniculigera*), presumably for birds (Prior and Gibson, 1997); forages from vegetation growing above a cave opening for emerging bats (Prior and Gibson, 1997; Dávalos and Eriksson, 2004); during 105 min, a snake made >200 unsuccessful strikes at bats emerging from a cave, also will enter homes to forage (Prior and Gibson, 1997); diet includes *Anolis valencienni* (Singhal et al., 2007); hatchling iguanas (*Cyclura collei*; Wilson and Van Veen, 2006); *Rattus rattus* and bats (Prior and Gibson, 1997); domestic fowl (Gosse, 1851); Black-billed Amazon Parrots (*Amazona agilis*; Koenig, 2001; Koenig et al., 2007); consumed contents of a parrot nest (Cruz and Gruber, 1980); juvenile female (940 mm SVL, 345 g) captured and ate (headfirst) a Jamaican Fruit Bat (*Artibeus jamaicensis*; Koenig and Schwartz,

2003). **Longevity:** In captivity, 24 years, 4 months and still alive at last report in 1997 (Curator of Reptiles, Dallas Zoo in Slavens and Slavens, 2003). **Parasites:** Ticks common on snakes from Goat Island (Grant, 1940a), and elsewhere on Jamaica (Gosse, 1851). **Predators:** Feral cats (Grant, 1940a). **Reproduction:** Breeding aggregations (balling behavior) occur (Gosse, 1851). **Size:** ~2.0 m (Schwartz and Henderson, 1991), but reported to attain 2.8 m (Grant, 1940a).

Conservation Status: Listed on CITES Appendix I (UNEP-WCMC, 2008); considered "vulnerable" on the IUCN Red List (Gibson, 1996h); listed as "endangered" on the U.S. Fish & Wildlife Service list of threatened and endangered animals and plants (U.S. Fish & Wildlife Service, 2008); apparently once common in many Jamaican parishes (Gosse, 1851), this boa has experienced a long history of decline since the colonization of Jamaica by Europeans; killed on sight and suffering habitat loss, but not thought to be in danger of immediate extinction (Oliver, 1982); habitat destruction via clearing of forests for agriculture and predation by introduced mammals have severely impacted populations (Tolson and Henderson, 1993).

23 Family Colubridae

Figure 26. The arboreal tendencies of *Mastigodryas bruesi* (Colubridae) may render it less vulnerable to predation by the introduced Indian Mongoose (*Herpestes javanicus*) than more terrestrial snakes and lizards. Photograph by Robert Powell.

Chironius vincenti Boulenger, 1891

Distribution: St. Vincent; elevational range 275–600 m.

Habitat: Wet forest, on shaded trails and, in early morning, in sunny clearings (Henderson and Haas, 1993).

Activity: Diurnal (Henderson and Haas, 1993). **Foraging and Diet:** Active forager, diet includes adult *Eleutherodactylus johnstonei* and *Eleutherodactylus* and *Pristimantis* eggs (Henderson et al., 1988b); adult *Pristimantis shrevei* (RWH). **Size:** SVL to 1,260 mm (Henderson and Haas, 1993).

Conservation Status: Considered "critically endangered" on the IUCN Red List (Henderson, 1996).

Additional Pertinent Literature: Henderson and Powell (1996c).

Mastigodryas bruesi Barbour, 1914

Distribution: St. Vincent, the Grenadines, and Grenada; introduced on Barbados.

Habitat: Generally in mesic situations, although habitat on several of the Grenadines is quite xeric, in wooded areas, either on the ground or in low vegetation (Schwartz and Henderson, 1991); 1 m high in the fork of a large tree on St. Vincent (RP); in developed, suburban habitats (e.g., St. George's, Grenada; Henderson and Berg, 2006); on Barbados, often associated with *Musa* and other fruit trees (Underwood et al., 1999); sleeps at night in bushes (Malhotra and Thorpe, 1999).

Activity: Diurnal, sleeps in trees to heights of at least 5 m (Schwartz and Henderson, 1991). **Behavior:** Elevates forepart of body and feigns strikes (RWH). **Foraging and Diet:** Presumably an active forager, diet includes *Eleutherodactylus* (25% by frequency of occurrence in a sample of 12 prey items), *Leptodactylus validus* (8.3%), *Anolis* (50%), and *Ameiva* (16.7%; Henderson and Crother, 1989). **Reproduction:** Oviparous (RWH). **Size:** SVL to 830 mm (Schwartz and Henderson, 1991).

Conservation Status: Unknown.

Additional Pertinent Literature: Greene et al. (2003).

Pantherophis guttatus Linnaeus, 1766

Distribution: Native to the eastern United States from New Jersey to Florida and west to Louisiana; in the West Indies, the species is known from Grand Cayman, St. Thomas, Tortola, Anguilla, St. Maarten, St.-Barthélemy, Antigua, and Martinique; apparently, reproductive populations have been established only on Grand Cayman (Seidel and Franz, 1994; Fred Burton, in litt., 26.V.2006) and St. Thomas (Perry et al., 2003a), but multiple snakes have been found in Martinique (Breuil, 2009).

Habitat: In the United States, in a variety of habitats (including edificarian; Conant and Collins, 1998); on Grand Cayman associated with shrubbery and trees imported from Miami, Florida (Fred Burton, in litt., 26.V.2006).

Diet: On Grand Cayman, includes *Anolis conspersus* (Seidel and Franz, 1994). **Reproduction:** Oviparous (RWH). **Size:** Total length to 1,829 mm (Conant and Collins, 1998).

Conservation Status: A species of "least concern" on the IUCN Red List (Hammerson, 2007c); as an introduced species, the hope is that additional reproductive populations do not become established in the West Indies; Seidel and Franz (1994) noted that the establishment of this species on West Indian islands is not unlikely as its native range includes southern Florida and the Florida Keys; should this happen, the impact on native species cannot be predicted.

Remarks: This species was formerly placed in the genus *Elaphe*; Burbrink and Lawson (2007) suggested that it be placed in the genus *Pituophis*.

Tantilla melanocephala Linnaeus, 1758

Distribution: Widely distributed on the Neotropical mainland and on Trinidad and Tobago; in the West Indies recorded from Mustique and Union Island in the St. Vincent Grenadines (Henderson and Powell, 2006).

Habitat: On Union Island, under rocks in moist, loose soil along a rock ledge in shade (J. Boone, in litt., 14.VIII.2008).

Size: Maximum size of West Indian snakes unknown, SVL to 120 mm on Trinidad and Tobago (Murphy, 1997), maximum total length to 486 mm in Venezuela (Roze, 1966).

Conservation Status: As an introduced species, the main concern is whether or not it negatively impacts native species; it will not compete with snake species native to the Grenadines, but it may compete with invertebrate-eating lizards.

Additional Pertinent Literature: L. D. Wilson (1992).

24 Family Dipsadidae

Figure 27. *Borikenophis portoricensis* (Dipsadidae) can elevate its head and spread its throat in a cobra-like defensive threat display. Photograph by Mel José Rivera.

Alsophis antiguae Parker, 1933

Distribution: Antigua; whether subfossil records from Barbuda (Pregill et al., 1994) apply to *A. antiguae* or *A. sajdaki* is uncertain.

Habitat: Unknown.

Activity: Probably diurnal (Schwartz and Henderson, 1991). **Diet:** Includes *Anolis* sp. and *Mus musculus* (Henderson and Sajdak, 1996). **Size:** SVL to ~700 mm SVL (RWH).

Conservation Status: Listed as "critically endangered" on the IUCN Red List (Day, 1996d); possibly extinct (Henderson et al., 1996a).

Additional Pertinent Literature: Henderson et al. (1996a).

Alsophis antillensis Schlegel, 1837

Distribution: Guadeloupe and Marie-Galante.

Habitat: Rocky wooded hillsides and areas of mixed agriculture (Henderson et al., 1992).

Activity: Diurnal (Schwartz and Henderson, 1991). **Diet:** Includes frogs (*Eleutherodactylus* spp.; 2 in a sample of 3 prey items) and lizards (*Anolis marmoratus*; 1 of 3; Henderson and Sajdak, 1996). **Size:** SVL to 910–930 mm (Schwartz and Henderson, 1991).

Conservation Status: Not assessed; extremely rare, possibly due to mongoose predation (Powell and Henderson, 2005a), but it has recently (2003) been observed on both Basse-Terre and Grande-Terre (Guadeloupe; B. Ibéné in Breuil, 2009).

Alsophis manselli Parker, 1933

Distribution: Montserrat.

Habitat: Rainforest, forest edge, dry scrub, and banana and coconut groves, crossing roads (Schwartz and Henderson, 1991).

Activity: Diurnal (RWH). **Diet:** Prey items included 6 *Eleutherodactylus johnstonei* and 2 *Anolis lividus*, including one frog and one lizard in the same stomach (Henderson and Sajdak, 1996). **Predator:** *Leptodactylus fallax* (Buley, 2003).

Reproduction: Copulation in a *Musa* grove on 6 April at 1715 h, 5 shelled eggs (10.3 × 19.3 mm), 510 mm hatchling on 6 August (Schwartz and Henderson, 1991). **Size:** SVL to at least 940 mm (RWH).

Conservation Status: Not assessed; the possible impact of recent volcanic eruptions is unknown.

Remarks: This species was formerly considered a subspecies of *Alsophis antillensis*.

Alsophis rijgersmaei Cope, 1869

Distribution: Anguilla Bank, Anguilla (including Scrub Island), St.-Martin/St. Maarten, St.-Barthélemy.

Habitat: Xerophilic to mesophilic (Schwartz and Henderson, 1991); rock-strewn fields and pastures, rock walls, and rock piles (Schwartz and Henderson, 1991; Powell et al., 2005); thorn forest and residential gardens (Powell et al., 2005); under embedded rocks by day and night (Schwartz and Henderson, 1991).

Activity: Diurnal, most activity in early morning and late afternoon, presumably to avoid being abroad during the hottest hours of the day (Powell et al., 2005). **Diet:** Three prey items all *Anolis gingivinus* (Henderson and Sajdak, 1996); also eats introduced frog species (*Eleutherodactylus johnstonei* and *Osteopilus septentrionalis*; Powell et al., 2005). **Size:** SVL to ~790 mm (Schwartz and Henderson, 1991).

Conservation Status: On the IUCN Red List as "endangered" (Day, 1996e); snakes remain fairly common on mongoose-free islands (Anguilla, Scrub, and St.-Barthélemy), but have been essentially extirpated from mongoose-infested St.-Martin/St. Maarten.

Alsophis rufiventris Duméril and Bibron, 1854

Distribution: Saba and St. Christopher Bank (St. Eustatius, St. Christopher, Nevis).

Habitat: Mesophilic (Schwartz and Henderson, 1991); in rainforest, forest scrub, forest edge, gardens, other heavily disturbed situations including edificarian (Rojer, 1997a, 1997b; Schwartz and Henderson, 1991); often associated with rocky hillsides (Savit et al., 2005); rock walls, forest paths, in leaf litter, under rocks, discarded plywood, or palm debris (Schwartz and Henderson, 1991).

Activity: Diurnal, on The Quill (St. Eustatius), activity appears to be bimodal, starts after 0700 h, rises markedly after 0900 h and remains high until 1200 h, after a midday depression, activity increases after 1500 h, peaks at 1600 h, and declines again after 1700 h (Savit et al., 2005). **Behavior:** On The Quill (St. Eustatius), 28% of time spent foraging, 21% moving steadily with no tongue flicks, 21% sprawled in mixed sun and shade, and less time sprawled or coiled in sun or shade, 8% spent with head elevated at ~45° angle to substrate with no tongue flicking (Heinz et al., 2004); possesses a mild venom that may elicit reactions in some humans, nonaggressive and rarely bites even when handled (Maley, 2004). **Foraging and Diet:** Ground-dwelling active forager that preys on frogs and lizards, based on 16 prey items from across the range, 5 (31.2%) *Eleutherodactylus johnstonei*, 10 (62.5%) *Anolis* (*A. sabanus*, *A. schwartzi*, *A.* sp.), and 1 (6.3%) *Ameiva erythrocephala*, one snake from Saba had 4 *E. johnstonei* froglets in its stomach (Schwartz and Henderson, 1991; Henderson and Sajdak, 1996); on The Quill highest concentration of snakes associated with high concentrations of *Anolis schwartzi*, which were closely associated with high concentrations of rocks (Savit et al., 2005); actively foraging in leaf litter and consuming reptilian eggs, presumably of *A. schwartzi* (Heinz et al., 2005). **Longevity:** In captivity, 2 years, 6 months and still alive at last report in 1990 (Aquarium/Reptile Curator, Milwaukee County Zoo in Slavens and Slavens, 2003). **Size:** SVL to 920 mm (Schwartz and Henderson, 1991). **Thermal Biology:** On Mt. Scenery (Saba), body temperatures on 3 February at 1410–1600 h 24.8–31.4 °C at air temperatures of 21.2–23.0 °C (Schwartz and Henderson, 1991); often in mottled sunlight (Heinz et al., 2004).

Conservation Status: Listed as "endangered" on the IUCN Red List (Day, 1996f); this species has been extirpated from St. Christopher and Nevis (Henderson, 2004; Sajdak and Henderson, 1991); both islands harbor mongooses; the species remains abundant on mongoose-free Saba and St. Eustatius (Maley, 2004; Powell and Henderson, 2005a).

Alsophis sajdaki Henderson, 1990

Distribution: Great Bird Island (off Antigua); (re)introduced to Rabbit Island and Green Island (Daltry, 2007); whether subfossil records from Barbuda (Pregill et al., 1994) represent this species or *A. antiguae* is uncertain.

Habitat: On Great Bird Island, favors habitat with a high percentage of canopy cover, leaf litter, and dense undergrowth (Daltry et al., 2001).

Activity: Diurnal (Daltry et al., 2001). **Foraging and Diet:** Lies in thin layers of leaf litter with only the head protruding to ambush passing lizards, based on fecal analysis from 73 individuals on Great Bird Island, 41.1% of fecal pellets contained scales of *Anolis* (mostly *A. wattsi*, but some *A.*

leachii), 30.1% *Ameiva griswoldi*, and 4.1% (all from juvenile snakes) *Sphaerodactylus elegantulus* (Daltry et al., 2001). **Movements:** Based on 40 days of radio tracking, activity range (minimum polygon) 4,425 m^2 (Daltry et al., 2001). **Size:** SVL to 711 mm (Henderson et al., 1996a).

Conservation Status: Listed (as *A. antiguae*) as "critically endangered" on the IUCN Red List (Day, 1996d).

Additional Pertinent Literature: Henderson et al. (1996a).

Remarks: This species was formerly considered a subspecies of *Alsophis antiguae*.

Alsophis sanctonum Barbour, 1915

Distribution: Iles des Saintes, Guadeloupe, where two currently recognized subspecies occur on Terre-de-Haut and Îlet-à-Cabrit (*A. s. sanctonum*) and Terre-de-Bas (*S. s. danforthi*).

Habitat: Mangrove edge, roadsides, gardens, wooded areas, preferentially semideciduous forest with *Pimenta racemosa*, *Pisonia subcordata*, *Hymenea courbaril*, and *Bursera simaruba*, generally in areas less frequented by humans (Breuil, 2002b).

Activity: Diurnal, particularly active in late morning and afternoon when *Anolis* is most active, active until 1915 h (Breuil, 2002b). **Foraging and Diet:** Moves slowly, exploring cavities with its tongue, feeds primarily on *Anolis terraealtae*, but has been observed eating eggs of *Iguana* and turtledoves (Breuil, 2002b); on Terre-de-Bas, a 45.0 g snake swallowed a 7.2 g *A. terraealtae* for a mass ratio (prey mass/predator mass) of 0.16 (Henderson and Sajdak, 1996). **Reproduction:** On Terre-de-Haut, newborn snakes in December and February (Breuil, 2002b). **Size:** SVL to ~900 mm (RWH).

Conservation Status: Not assessed.

Remarks: This species was formerly considered a subspecies of *Alsophis antillensis*.

Alsophis sibonius Cope, 1879

Distribution: Dominica.

Habitat: Xerophilic to mesophilic, in rainforest and edges, xeric scrub, coastal scrub, mountain pastures, mangrove edge, deciduous forest, fruit groves, often associated with rock-strewn areas (frequently on hillsides), rock walls and rock piles, *Cocos* trash, and roads (Schwartz and Henderson, 1991); also occasionally arboreal in shrubs and small trees (James, 2004; L. A. White et al., 2008; Muelleman et al., 2009); to heights >10 m (A. Mitchell, pers. comm.); tolerant of human-mediated alterations of habitat at Cabrits National Park, most snakes (73.9% in the morning and 81.3% in the afternoon) in full shade, 136 of 165 snakes associated with leaf litter (Muelleman et al., 2009).

Activity: Diurnal, most active in the morning and late afternoon (Malhotra and Thorpe, 1999); earliest activity at Cabrits National Park at 0606 h, latest at 1806 h, peaks at 0700–0800 h and 1600–1900 h, but activity extends throughout the day in cool, rainy weather (Muelleman et al., 2009); an individual in a foraging posture at a night light on hotel grounds (Powell and Henderson, 2008d); activity included foraging (32.3% of time), "traveling" (30.6%), sprawled motionless (28.2%), moving slowly (6.5%), and coiled (2.4%; L. A. White et al., 2008). **Behavior:** Rarely bites, but may release a foul-smelling cloacal secretion (Malhotra and Thorpe, 1999; James, 2004); often lies motionless in leaf litter, usually sprawled, rarely loosely coiled, responds to threats by either remaining motionless or fleeing, with rate of escape more rapid after a close approach, some individuals elevate the anterior body and assume an "S-shaped strike posture" (Muelleman et al., 2009); after responding to a close approach by an observer, a snake wriggled its tail tip while appearing to monitor the potential threat (L. A. White et al., 2008). **Foraging and Diet:** Sample of 6 prey items included 4 *Eleutherodactylus martinicensis*, 1 *Anolis oculatus*, and 1 *Ameiva fuscata* (Henderson and Sajdak, 1996); mainly *A. oculatus*, small rodents also may be eaten (Malhotra and Thorpe, 1999); appears to use both active and sit-and-wait foraging strategies (Daniells et al., 2008; L. A. White et al., 2008); active foraging consists of slow movement while tongue-flicking and nosing through the substrate (Muelleman et al., 2009); probes in leaf litter and small holes, consumes eggs, presumably those of *A. oculatus*, tongue-flicks and assumes a position with head elevated and adpressed against a tree trunk on which *A. oculatus* was present and from which a snake made an unsuccessful strike at an anole (L. A. White et al., 2008). **Longevity:** In captivity, 2 years, 6 months and still alive at last report in 1992 (Snider and Bowler, 1992). **Predation:** Frequently killed on sight by humans, although completely harmless (Malhotra and Thorpe, 1999); failed predation attempt by a subadult Broad-winged Hawk (*Buteo platypterus*) resulted in a snake defensively constricting the bird, which probably would have died except for human intervention (A. Mitchell and Prince, 2008). **Reproduction:** Clutch size 5 (Malhotra and Thorpe, 1999). **Size:** SVL to at least 905 mm (RWH).

Conservation Status: Unknown; can tolerate altered habitats when not persecuted (Powell and Henderson, 2008c, 2008d; Muelleman et al., 2009).

Remarks: This species was formerly considered a subspecies of *Alsophis antillensis*.

Arrhyton ainictum Schwartz and Garrido, 1981

Distribution: Eastern Cuba, presumably in the Sierra de Najasa.

Habitat: Mesophilic (Schwartz and Henderson, 1991).

Reproduction: Oviparous (RWH). **Size:** SVL to 363 mm (Schwartz and Garrido, 1981).

Conservation Status: Considered "endangered" (Rodríguez Schettino and Chamizo Lara, 1998a).

Arrhyton dolichura Werner, 1909

Distribution: Cuba, vicinity of La Habana (Schwartz and Garrido, 1981).

Habitat: Mesophilic (Schwartz and Henderson, 1991); under rocks by day (Garrido and Jaume, 1984).

Activity: Nocturnal (Garrido and Jaume, 1984). **Size:** SVL to 249 mm (Schwartz and Garrido, 1981).

Conservation Status: Considered "endangered" (Rodríguez Schettino and Chamizo Lara, 1998a).

Arrhyton procerum Hedges and Garrido, 1992

Distribution: Cuba, coastal Matanzas Province.

Habitat: In semixeric Sea Grape (*Coccoloba*) and palm thatch woods near a beach with many coral-limestone rocks in the area (Hedges and Garrido, 1992a).

Size: SVL to 207 mm based on one specimen (Hedges and Garrido, 1992a).

Conservation Status: Considered "endangered" (Rodríguez Schettino and Chamizo Lara, 1998a).

Arrhyton redimitum Cope, 1863

Distribution: Southeastern Cuba.

Habitat: Xerophilic to mesophilic (Schwartz and Henderson, 1991); in a rotting mat of palm fibers at the base of a fan palm in a palm thicket on a xeric, scrubby hillside (Schwartz, 1965a); under a stone ~15 cm below the surface, when exposed, the snake burrowed into loose soil and rocks, also under a large piece of coral (Lando and Williams, 1969); in a wooded area under a large sheet of plywood almost completely concealed under leaf mold and debris (Schwartz and Garrido, 1981).

Behavior: Efficient burrower (Lando and Williams, 1969). **Size:** SVL to 250 mm (Schwartz and Garrido, 1981).

Conservation Status: Unknown.

Remarks: Most recently published information on this species is under the name *Arrhyton landoi*, now considered a synonym of *A. redimitum*.

Arrhyton supernum Hedges and Garrido, 1992

Distribution: Eastern Cuba.

Habitat: Under rocks by day along a trail in a cacao grove and in an open pasture above a forested ravine (Hedges and Garrido, 1992a).

Size: SVL to 255 mm (Hedges and Garrido, 1992a).

Conservation Status: Considered "endangered" (Rodríguez Schettino and Chamizo Lara, 1998a).

Arrhyton taeniatum Günther, 1858

Distribution: Cuba, including Isla de la Juventud.

Habitat: Mesophilic (Schwartz and Henderson, 1991); ground-dwelling to fossorial in open areas under rocks in dry pastures, Guinea Grass (*Panicum maximum*), a clod of dirt (Schwartz and Garrido, 1981); turned up by plows or hidden under flat stones (Barbour and Ramsden, 1919); in an old termite nest on the beach (B. I. Crother, in litt., 26.II.2008).

Activity: Nocturnal (Barbour and Ramsden, 1919). **Foraging and Diet:** Active forager, diet includes *Amphisbaena cubana* (Schwartz and Garrido, 1981); *Typhlops* sp. (Schwartz and Henderson, 1991). **Size:** SVL to 448 mm (Schwartz and Garrido, 1981).

Conservation Status: Unknown.

Arrhyton tanyplectum Schwartz and Garrido, 1981

Distribution: Western Cuba.

Habitat: Mesophilic (Schwartz and Henderson, 1991); in a rural garden during heavy rain, under rocks at a cliff base (Schwartz and Garrido, 1981).

Activity: Nocturnal (Schwartz and Garrido, 1981). **Foraging and Diet:** Active forager (Schwartz and Garrido, 1981). **Size:** SVL to 306 mm (Schwartz and Garrido, 1981).

Conservation Status: Considered "endangered" (Rodríguez Schettino and Chamizo Lara, 1998a).

Arrhyton vittatum Gundlach in Peters, 1861

Distribution: Cuba (islandwide, but disjunct), including Isla de la Juventud.

Habitat: Mesophilic (Schwartz and Henderson, 1991); under rocks (Alayo Dalmau, 1951); crossing a road in semixeric forest by day, under a rock, ground dwelling (Schwartz and Garrido, 1981).

Activity: Nocturnal (Schwartz and Garrido, 1981). **Parasites:** Larval nematode Stomachidae gen. sp. (Baruš and Coy Otero, 1978; Coy Otero and Baruš, 1979b; Coy Otero and Lorenzo Hernández, 1982). **Size:** SVL to 207 mm (Schwartz and Garrido, 1981).

Conservation Status: Unknown.

Borikenophis portoricensis Reinhardt and Lütken, 1862

Distribution: Puerto Rico Bank, where six currently recognized subspecies occur on much of Puerto Rico and Cayo Santiago (*B. p. portoricensis*), in the British Virgin Islands (*B. p. anegadae*), on Isla Vieques (*B. p. aphantus*), on Buck Island (*B. p. nicholsi*), on Caja de Muertos, Platillo, and in southern Puerto Rico (*B. p. prymnus*), and on Isla Culebra, St. Thomas and satellites, Lovango Cay, Peter Island, Salt Island (*B. p. richardi*).

Habitat: Xerophilic to mesophilic, in various situations, including xeric scrub, gardens, rainforest, *Musa* groves, *Hippomane* woods (St. Thomas), *Cocos* trash (Virgin Gorda), open pasture, rock piles, under an active termitarium buried under *Cocos* fronds adjacent to a mangrove swamp (Schwartz and Henderson, 1991; Barun et al., 2007); on Tortola, along the beaches (Grant, 1932h); on Anegada, along a path through sandy, rocky habitat (Carey, 1972); occasionally edificarian (RWH); on Puerto Rico, in an arroyo behind the Coamo Springs Hotel, also under a rock at the base of a cliff behind the "bath house" (K. P. Schmidt, 1928); snakes readily enter standing water on Guana (Powell et al., 2006); primarily ground dwelling, but will ascend to 18–20 m in rainforest canopy (J. C. Gillingham and D. P. Reagan in Thomas and Gaa Kessler, 1996); on Guana Island (British Virgin Islands), a snake (SVL 404 mm) sleeping draped over branches on a bush ~75 cm above the ground (S. Valentine-Cooper, pers. comm.).

Activity: Diurnal, crossing a road at 0745 h (Schwartz and Henderson, 1991); on beaches in early morning on Tortola, seldom seen after 1000 h on Water Island (Grant, 1932h); forages only during the day on Puerto Rico (Clark and Gillingham, 1990); on Guana (British Virgin Islands), activity bimodal with a depression at the hottest time of day (Barun et al., 2007); also on Guana, exploits the night-light niche on hotel grounds after dark, preying on *Anolis*, itself using the artificial light to ambush prey (Perry and Lazell, 2000; Perry et al., 2008); on Anegada, on a trail 0700–1200 h (Carey, 1972). **Behavior:** Raises foreparts of the body and flattens dorsoventrally (Lazell, 2005a; Rodríguez-Robles, 2005); rarely attempts to bite humans, but will occasionally (K. P. Schmidt, 1928); venom has caused severe (but not fatal) reactions in humans (Heatwole and Banuchi, 1966; Perez-Rivera and Vélez, 1978; Rodríguez-Robles, 2005; K. Lindsay, pers. comm.); when handled on Guana Island, may rotate body rapidly and wave tail (RWH, RP). **Foraging and Diet:** Active forager (RWH, RP); pokes head into holes and even turns small rocks (A. J. Sánchez Muñoz, in litt., 17.III.2008); prey movement usually provokes an attack (Rodríguez-Robles, 1992); small prey grasped and swallowed almost immediately, for larger prey, venom from Duvernoy's gland injected by chewing on the prey with rear maxillary fangs repeatedly embedded until the prey is moribund (snake may use its body to restrain prey while the venom takes effect), *Ameiva* may be particularly susceptible to the venom, saliva with a high protein content, primary action of venom hemolytic, effects include subcutaneous hemorrhage in mammals and pulmonary and peritoneal hemorrhage in lizards (Thomas and Prieto Hernández, 1985); based on 44 feeding sequences on *Anolis cristatellus*, mean subduction time 834.1 sec and average handling time 883.4 sec ($N = 33$), lizards gaping and showing signs of respiratory distress within a few minutes after being seized suggests that the venom weakens but does not immediately immobilize or kill, on 11 occasions anoles struggled inside a snake to 414 sec after being completely engulfed, 60 of 61 anoles swallowed head first, constriction occasionally used to subdue prey (Rodríguez-Robles, 1992); faster digestion rates recorded for skin, forelimbs, liver, and lungs of envenomated anoles (*A. cristatellus*) versus non-envenomated anoles, digestion of envenomated anoles is faster than that of non-envenomated lizards (Rodríguez-Robles and Thomas, 1992); subduction time for *Anolis cristatellus* greater (mean 623.1 ± 151.0 sec) than for *Eleutherodactylus coqui* (242.2 ± 117.9 sec), but swallowing time faster for *A. cristatellus* (174.3 ± 18.8 sec) than for *E. coqui* (538.2 ± 159.2 sec), *Anolis* initially attacked on the trunk, frogs on head or limbs, venom usually not used on frogs (1 of 14 envenomated), but is used on anoles (13 of 17 envenomated), anoles always swallowed headfirst, but frogs either head, tail, or side-first, frogs swallowed and regurgitated in the field were still alive several days after being regurgitated (Rodríguez-Robles and Leal, 1993b); *Magliophis exiguus* and *Typhlops* sp. (Schwartz and Henderson, 1991); a sample of 45 prey items from Puerto Rico and satellite islands included 6 frogs (3 *Eleutherodactylus* sp., 3 *E. antillensis*), 38 lizards (14 *Anolis* sp., 10 *A. cristatellus*, 3 *A. pulchellus*, 2 *A. stratulus*, 2 *Ameiva* sp., 2 *A. exsul*, 3 *Sphaerodactylus macrolepis*), and 1 snake

(*Typhlops platycephalus*), a sample of 6 from the U.S. Virgin Islands included 2 *Sphaerodactylus macrolepis*, 2 *Anolis*, 1 *Iguana iguana*, and 1 *Ameiva exsul*, a sample of 7 prey items from the British Virgin Islands included 1 frog (*Eleutherodactylus* sp.) and 6 lizards (1 *Sphaerodactylus* sp., 1 *S. macrolepis*, 3 *Anolis* sp., and 1 *A. stratulus*; Henderson and Sajdak, 1996); also takes *Eleutherodactylus coqui*, *Sphaerodactylus townsendi*, *Anolis evermanni*, *Magliophis exiguus*, and *Rattus rattus* on Puerto Rico (Rodriguez-Robles and Leal, 1993a); *Ameiva wetmorei* (M. Leal in Henderson and Sajdak, 1996); *Sphaerodactylus nicholsi* (J. Rodríguez-Robles, 1995; Thomas and Gaa Kessler, 1996); mouse (M. Leal in Thomas and Gaa Kessler, 1996); snake from Water Island contained 3 "young" *I. iguana* (Grant, 1932h); baby *Cyclura pinguis* on Guana Island (Levering and Perry, 2003); on Guana, a 621-mm SVL, 88-g snake consumed a hatchling *Cyclura pinguis* of 40.4 g, a mass representing 45.9% of the snake's pre-meal mass (RWH, RP); cannibalism on Guana Island (Perry and Powell, 2008); on Congo Cay, "masticating" small, dried fish (*Hernagula* sp.) lost from a Brown Pelican (*Pelicanus occidentalis*) nest, on Little Saba Cay entered the nests of ground-nesting Zenaida Doves (*Zenaida aurita*) and positioned its head over 2 eggs while the adult stood nearby in a combative posture (the snake fled when disturbed; Norton, 1993); on Puerto Rico, *Anolis krugi* (A. J. Sánchez Muñoz in Horstink, 2004); also on Puerto Rico, snake entered the nest of a Bananaquit (*Coereba flaveola*) at a height of 2.5 m in a tree and captured a nestling (Pérez-Rivera and Laboy-Rivera, 1996); on Guana, *Amphisbaena fenestrata* (Barun and Perry, 2003); additional prey species include *Hemidactylus mabouia*, *Sphaerodactylus nicholsi*, *S. roosevelti*, *Diploglossus pleii*, and *Anolis gundlachi* (Rodríguez-Robles, 2005); scavenges discarded chicken bones (Perry et al., 2007b); juvenile *Bufo marinus* (A. J. Sánchez Muñoz, in litt., 17.III.2008); caudal luring in juveniles (Leal and Thomas, 1994); struck at a foraging *Ameiva exsul* from vegetation bordering an open sandy area near a beach on Guana Island, British Virgin Islands (RP). **Parasites:** Tick *Amblyomma arianae* (Keirans and Garris, 1986); subcutaneous helminths (A. J. Sánchez Muñoz, in litt., 17.III.2008). **Population Density:** Mark-recapture indicates that this species is phenomenally abundant on Guana, British Virgin Islands (Lazell, 1991, 2005; RWH, RP); 50/ha (and 3.05 kg of biomass/ha) in *Leucaena* plots on Guana (Rodda et al., 2001); on Puerto Rico, mean monthly densities 5.6–11.1/ha at several sites between February 2001 and February 2002 (Rios-López and Aide, 2007). **Predators:** Red-tailed Hawk (*Buteo jamaicensis*; Santana and Temple, 1988); mongoose (*Herpestes javanicus*; Perez-Rivera and Vélez, 1978); house cats, also *Anolis cristatellus* attempted to eat a juvenile (Thomas and Gaa Kessler, 1996); on Guana, a very high percentage of snakes have scars on the body and damaged tails (nearly 70%), probably attributable to unsuccessful predation attempts by Soldier Crabs (*Coenobita clypeatus*), larger snakes had a higher incidence of scarring and tail damage (Barun et al., 2007). **Reproduction:** Gravid March–April, clutch size 4–10 in females 522–612 mm SVL (Schwartz and Henderson, 1991); female produced 12 eggs in captivity on 8 May, hatched 28 June (M. Leal in Rodríguez Robles, 2005). **Size:** SVL to 923 mm (Schwartz and Henderson, 1991).

Conservation Status: "Stable" for *B. p. portoricensis*, *B. p. anegadae*, *B. p. prymnus*, "rare everywhere, protect habitat" for *B. p. nicholsi*, "decreasing in some areas, protect habitat" for *B. p. richardi*, and "endangered," possibly extinct for *B. p. aphantus* (Moreno, 1991d).

Remarks: This species was until recently placed in the genera *Alsophis* and *Ocyophis*.

Borikenophis sanctaecrucis Cope, 1863

Distribution: St. Croix (U.S. Virgin Islands).

Habitat: Xerophilic (Schwartz and Henderson, 1991).

Abundance: Prior to 1900, "the older generation remembers them as having been plentiful" (Grant, 1937b). **Activity:** Diurnal (Schwartz and Henderson, 1991). **Size:** SVL to 1,025 mm (Schwartz and Henderson, 1991).

Conservation Status: Designated "extinct" on the IUCN Red List (World Conservation Monitoring Centre, 1996f); "may be extinct" (Seaman and Randall, 1962); "endangered," possibly extinct (Moreno, 1991d).

Additional Pertinent Literature: Henderson and Powell (1996b).

Remarks: This species was formerly placed in the genus *Alsophis*.

Borikenophis variegatus Schmidt, 1926

Distribution: Isla Mona and possibly the species on Isla Desecheo.

Habitat: In *Cocos* trash (Schwartz and Henderson, 1991); beneath rubbish on the low terrace of cultivated land (K. P. Schmidt, 1928); 1 m high on the trunk of a *Casuarina* (Powell, 2003b).

Activity: Presumably diurnal (Schwartz and Henderson, 1991). **Diet:** *Hemidactylus mabouia* and *Sphaerodactylus monensis* (M. Leal in Henderson and Sajdak, 1996); *Ameiva alboguttata* (K. P. Schmidt, 1928; Heatwole and Torres, 1967); *Anolis desechensis* on Isla Desecheo (Meier and Noble, 1991). **Size:** SVL to 827 mm (Rivero, 1998).

Conservation Status: Not assessed.

Remarks: This species was formerly considered a subspecies of *Borikenophis portoricensis*, which until recently had been included in the genera *Alsophis* and *Ocyophis*.

Caraiba andreae Reinhardt and Lütken, 1862

Distribution: Cuba, where six currently recognized subspecies occur in central Cuba (*C. a. andreae*), Cayo Cantiles (*C. a. melopyrrha*), cayos Santa María and Guajaba (*C. a. morenoi*), Isla de la Juventud (*C. a. nebulata*), western Cuba (*C. a. orientalis*), and the Península de Guanahacabibes (*C. a. peninsulae*).

Habitat: Mesophilic and ground dwelling (Schwartz and Henderson, 1991); savannas, gardens, patios, cultivated areas (Domínguez Díaz and Moreno García, 2003; Kunz, 2005); wooded areas, under rocks (Barbour, 1916); at the side of a road in the sun (Garrido and Schwartz, 1968b); in dry scrub, coastal scrub, a mound at the edge of a cornfield and crawling in palm trash, in a pasture under a log with *Cubophis cantherigerus*, in palm forest, and on damp mud and logs (Schwartz and Henderson, 1991); on floors under houses where they come up through the walls (Barbour and Ramsden, 1919); in leaf litter, fissures between rocks, and in hollow trunks (Estrada, 1993b); under building materials (Edwards, 1997).

Activity: Diurnal (Schwartz and Henderson, 1991); basking in the sun at the side of a road (Garrido and Schwartz, 1968b). **Defensive Behavior:** Raises the anterior part of the body and flattens the neck dorsoventrally, voids cloacal contents (Domínguez Díaz and Moreno García, 2003); hides under leaf litter, vegetation, rocks, fissures between rocks, or in hollow logs (Estrada, 1993b). **Foraging and Diet:** Feeds mostly on *Eleutherodactylus* spp. (Z. Vogel, 1965c); active forager, diet includes *Eleutherodactylus* (41.2% by frequency in a sample of 17 prey items), *Anolis* (52.9%), and *Leiocephalus* (5.9%; Schwartz and Henderson, 1991); *Bufo peltocephalus* taken by a ~300 mm SVL snake, *E. atkinsi* by a 266 mm SVL snake, and *E. dimidiatus* (Fong Grillo, 2004a); takes *Anolis sagrei* and small rodents, swallows prey headfirst without killing it first (Domínguez Díaz and Moreno García, 2003). **Parasites:** Nematodes *Hastospiculum cubaense* (Baruš and Sonin, 1971; Baruš and Coy Otero, 1978; Coy Otero and Baruš, 1979b; Coy Otero and Lorenzo Hernández, 1982); *Kalicephalus costatus costatus*, *K. rectiphilus rectiphilus*, *Oswaldocruzia lenteixeirai*, and larval Stomachidae gen. sp. (Baruš and Coy Otero, 1978; Coy Otero and Baruš, 1979b; Coy Otero and Lorenzo Hernández, 1982); trematode *Infidum simile* (Groschaft and Coy Otero, 1978; Coy Otero and Lorenzo Hernández, 1982). **Reproduction:** Mating in May (Estrada, 1994); clutch of 3 eggs (6.0 × 19.3 mm) in April from a female 294 mm SVL (Schwartz and Henderson, 1991); 310 mm SVL female laid a clutch of 3 off-white eggs (30.1–31.2 × 8.1–9.2 mm) on 10 June, another clutch of 4 under a 250 × 200 mm rock (Fong Grillo, 2004b). **Size:** SVL to ~700 mm (Domínguez Díaz and Moreno García, 2003).

Conservation Status: Not assessed.

Remarks: This species was until recently placed in the genus *Antillophis*.

Clelia clelia Daudin, 1803

Distribution: West Indian distribution is restricted to Grenada (*C. c. groomei*); elsewhere widely distributed on the Neotropical mainland (Guatemala to northern Argentina).

Habitat: Wet lowlands (Greer, 1965).

Activity: Likely active by day or night (J. Lazell in Greer, 1965). **Foraging and Diet:** Active forager on the ground, West Indian specimens yielded *Rattus* (Schwartz and Henderson, 1991). **Reproduction:** Oviparous (RWH). **Size:** SVL to ~2 m (Schwartz and Henderson, 1991).

Conservation Status: Listed on CITES Appendix II (UNEP-WCMC, 2008); if the species survives on Grenada (and conversations with Grenadians in 2005–2007 suggest it does), it is extremely rare (Henderson, 2004; Powell and Henderson, 2005a).

Clelia errabunda Underwood, 1993

Distribution: St. Lucia.

Habitat: Unknown.

Diet: Other snakes (Long, 1974); includes *Bothrops caribbaeus* (Underwood, 1995). **Size:** SVL to 1,380 mm (Underwood, 1993).

Conservation Status: Extinct (Underwood, 1993); ironically, extinction is likely to have occurred as a consequence of persecution of snakes as a result of fear of being bitten by *B. caribbaeus* (Malhotra and Thorpe, 1999).

Cubophis cantherigerus Bibron, 1840

Distribution: Five currently recognized subspecies occur in western Cuba, Isla de la Juventud, Archipiélago de Sabana-Camagüey, Archipiélago de los Canarreos, Cayos de San Felipe (*C. c. cantherigerus*), central Cuba (*C. c. schwartzi*), eastern Cuba (*C. c. adspersus*, *C. c. pepei*), and Little Swan Island (*C. c. brooksi*).

Habitat: In fields, wooded groves, beach areas (Lando and Williams, 1969); on a mountainside, along a lake edge, under and among rocks, in trash and scrub in a clearing, under a log in a pasture (with *Caraiba andreae*), in a cave mouth under a log, in a flooded ditch in rainforest, in grass and among cacti and driftwood (Schwartz and Henderson, 1991); in residential gardens and patios and in city parks (Domínguez Díaz and Moreno García, 2003; Martínez Reyes and Hernández Marrero, 2003; Kunz, 2005); in a hollow in dogtooth limestone (Garrido and Schwartz, 1969); on dry, rocky surfaces around a hotel in Santiago (Petzold et al., 1970); on a sandy, rock-strewn beach (Neill, 1958); under a pile of grass clippings (B. I. Crother, in litt., 26.II.2008).

Activity: Diurnal, crossing a road at high noon, foraging in late morning or early afternoon in grass and cactus (Lando and Williams, 1969); chasing a *Leiocephalus* at 1400 h, at night seeks refuge under rocks, also under a concrete slab in a nursery, active at night on roads, crawling in a garden, and in the water at a pond edge (Schwartz and Henderson, 1991). **Behavior:** Bites when captured, but tames quickly in captivity (Z. Vogel, 1965c); defensive display includes raising forepart of body off the ground and flattening it dorsoventrally ("cobra-like"), expels foul-smelling cloacal fluids, bites, venom may cause a reaction in some humans (Domínguez Díaz and Moreno García, 2003; Jaume and Garrido, 1980; Neill, 1954). **Foraging and Diet:** Very quick, captures *Leiocephalus* spp. and subdues them by constriction or by pressing them against the substrate (Z. Vogel, 1965c); active forager that can subdue prey with a weak venom, but also capable of killing prey by constriction, 38 prey items from Cuban snakes included 5 frogs (*Eleutherodactylus* and *Osteopilus*; 13.2% of total prey items by frequency), 1 turtle (*Eretmochelys imbricata* [Crother, 1986]; 2.6%), 22 lizards (18 *Anolis*, 1 *Cyclura*, 1 *Leiocephalus*, 2 *Diploglossus*; 57.9%), 2 snakes (*Caraiba andreae*; 5.3%), 2 unidentified birds (5.3%), 5 rodents (*Mus musculus* and unidentified; 13.2%), and 1 bat (2.6%; Henderson and Sajdak, 1996); *Eleutherodactylus planirostris* (Alonso Bosch and Rodríguez Gómez, 2003b); other prey includes bats (*Phyllonycteris poeyi*; Hardy, 1957b), domestic fowl (*Gallus gallus*) and a snake (*Tropidophis* sp.; Novo Rodríguez and Arazoza Rodríguez, 1986); at 1400 h, pursuing *Leiocephalus cubensis* down the center of a road, passed between 2 biologists without concern (Schwartz and Henderson, 1991); mean prey size 6.5 cm^3 (Schwartz and Henderson, 1991). **Longevity:** In captivity, 4 years, 8 months and still alive at last report in 1992 (Snider and Bowler, 1992). **Parasites:** In Cuba, nematodes *Hastospiculum cubaense* (Baruš and Sonin, 1971; Baruš and Coy Otero, 1978; Coy Otero and Baruš, 1979b; Coy Otero and Lorenzo Hernández, 1982); *Abbreviata baracoa*, *Terranova caballeroi*, *Kalicephalus costatus alsophisi* (Baruš and Coy Otero, 1966; Baruš and Coy Otero, 1978; Coy Otero and Baruš, 1979b; Coy Otero and Lorenzo Hernández, 1982; Coy Otero, 1999); *K. rectiphilus rectiphilus*, *Oswaldocruzia lenteixeirai*, *Parapharyngodon cubensis*, and larval Stomachidae gen. sp. (Baruš and Coy Otero, 1978; Coy Otero and Baruš, 1979b; Coy Otero and Lorenzo Hernández, 1982); trematodes *Infidum simile* (Groschaft and Coy Otero, 1978; Coy Otero and Lorenzo Hernández, 1982) and *Ochetosoma adenodermis* (Pérez Vigueras, 1942, 1956; Coy Otero and Lorenzo Hernández, 1982); ticks *Amblyomma quadricavum* (V. Černý, 1966, 1969; Guglielmone et al., 2003); *A. albopictum* (Coy Otero, 1999). **Predation:** Barn Owl (*Tyto alba furcata*; Arredondo Antúnez and Chirino Flores, 2002). **Reproduction:** In Cuba, mating in May, oviposition 15–20 days after copulation, incubation in captivity 69–77 days, clutch size 10–24, positively correlated with female SVL, egg color yellowish white, egg size 13 × 25.7–19.75 × 31.8 mm, egg weight 3.3–6.0 g, hatchlings 210–270 mm total length, weight 2.6–4.4 g (Domínguez Díaz and Moreno García, 2003; Petzold, 1978); a female from Cayo Boca de Sagua (Archipiélago de Sabana-Camagüey) produced 13 adherent eggs (Garrido, 1973d); 58 oviductal eggs in a 1,333 mm SVL female (Domínguez Díaz and Moreno García, 2006a). **Size:** On Cuba, SVL to 1,333 mm (Domínguez Díaz and Moreno García, 2006a).

Conservation Status: "Of concern" (Buide et al., 1974); Cuban populations listed, but not assessed (Rodríguez Schettino and Chamizo Lara, 1998a); status unknown in the Swan Islands.

Remarks: This species was until recently placed in the genera *Alsophis* and *Ocyophis*.

Cubophis caymanus Garman, 1887

Distribution: Cayman Islands, Grand Cayman.

Habitat: Ground dwelling, but may ascend into bushes and trees (to 9.1 m; Grant, 1940c); "climbs well" (English, 1912); in scrubby logwood, evergreen woodlands, old fields, on sand near the ocean, coastal *Terminalia* and *Coccoloba*, brackish water ponds, mangrove swamps, around houses (Schwartz and Henderson, 1991; Seidel and Franz, 1994).

Activity: Diurnal (Schwartz and Henderson, 1991). **Defensive Behavior:** Flattens neck dorsoventrally and bites (Grant, 1940c). **Foraging and Diet:** "lives principally on frogs" (English, 1912); active forager that had consumed 5 *Eleutherodactylus*, and 16 *Anolis* in a sample of 21 prey items (Henderson and Sajdak, 1996); introduced *Iguana iguana* (T. Ebanks, in litt., 23.II.2009). **Longevity:** In captivity, 3 years, 8 months (Snider and Bowler, 1992). **Mortality:** Many killed by cars as they attempt to cross roads (Seidel and Franz, 1994). **Parasites:** Abdominal parasites and ticks (Grant, 1940c). **Reproduction:** Eggs laid May–June, hatch August–September (Grant, 1940c); female 617 mm SVL contained 9 shelled eggs (average 12 × 25 mm) in May (Seidel and Franz, 1994). **Shedding:** Many shedding under flat stones (Grant, 1940c). **Size:** SVL to 1,190 mm and mass to 1.25 kg (Seidel and Franz, 1994).

Conservation Status: Not assessed.

Remarks: This species was formerly considered a subspecies of *Cubophis cantherigerus*, which until recently had been included in the genera *Alsophis* and *Ocyophis*.

Cubophis fuscicauda Garman, 1888

Distribution: Cayman Islands, Cayman Brac.

Habitat: Unknown.

Diet: Five prey items included 1 *Sphaerodactylus*, 3 *Anolis*, and 1 *Mus musculus* (Henderson and Sajdak, 1996); also took a "honey creeper" and *Anolis sagrei* (Grant, 1940c). **Reproduction:** Two females, 693 mm and 641 mm SVL, had 8 and 9 eggs, respectively (Seidel and Franz, 1994). **Size:** SVL to 700 mm (Grant, 1940c).

Conservation Status: Unknown.

Remarks: This species was formerly considered a subspecies of *Cubophis cantherigerus*, which until recently had been included in the genera *Alsophis* and *Ocyophis*.

Cubophis ruttyi Grant, 1941

Distribution: Cayman Islands, Little Cayman Island.

Habitat: On the road in Black Mangrove (*Avicennia*; Seidel and Franz, 1994); presumably ground dwelling (RWH).

Activity: Presumably diurnal (Schwartz and Henderson, 1991). **Foraging and Diet:** Presumably an active forager (Schwartz and Henderson, 1991). **Reproduction:** Presumably oviparous (RWH). **Size:** SVL to at least 840 mm (Grant, 1940c).

Conservation Status: Unknown.

Remarks: This species was formerly considered a subspecies of *Cubophis cantherigerus*, which until recently had been placed in the genera *Alsophis* and *Ocyophis*.

Cubophis vudii Cope, 1863

Distribution: The Bahama Islands, where five currently recognized subspecies occur on the Great Bahama Bank (except Bimini Islands), Cat Island, Little San Salvador Island (*C. v. vudii*), Little Bahama Bank (*C. v. aterrimus*), Bimini Islands (*C. v. picticeps*), Crooked and Acklins islands (*C. v. raineyi*), and Great Inagua Island, including Sheep Cay (*C. v. utowanae*).

Habitat: Xerophilic to mesophilic, in palm scrub, hammocks (South Bimini Island, Andros Island), wooded areas, coppice (Cat Island), on beaches and burnt and cutover pine (Great Abaco Island), pinewoods (Andros Island), buttonwood and saw grass (Cat Island), lawns (Eleuthera), in and adjacent to coastal mangroves, and in an abandoned shack adjacent to pinewoods, under and among rocks, under *Cocos* trash and logs, under a board on sand (East Bimini Island; Schwartz and Henderson, 1991).

Abundance: On Little Ragged Island, abundant in sandy areas close to the shore in 1970 (L. Porras, in litt., 17.VII.2007). **Activity:** Diurnal, on a road at 0900 h (Great Abaco), prowling at dusk in rocks and weeds (Schwartz and Henderson, 1991); on Waderick Wells Cay, moving from a sandy beach to an adjoining limestone ridge at 1655 h (Franz and Dodd, 1994). **Dispersal:** Emerging from surf onto Alligator Cay (Exuma Cays), an island from which *Cubophis* was not known, the nearest island was 100 m away and separated by a channel of flowing water (Knapp, 2000c). **Foraging and Diet:** Ground-dwelling, active forager, but at 1.2 m in a bush in hillside scrub at midday on Cat Island (Schwartz and Henderson,1991); 87 prey items from across the range included one invertebrate (1.1% of the total sample by frequency; unidentified Phymatidae), 19 frogs (21.8%; *Eleutherodactylus planirostris, Osteopilus septentrionalis*; stomachs of several snakes contained multiple—5, 9, 18, 26—*O. septentrionalis* froglets), 61 lizards (70.1%; *Sphaerodactylus* sp., *S. notatus, Anolis* sp., *A. brunneus, A. sagrei, A. scriptus, Leiocephalus punctatus, Ameiva auberi, A. maynardii*), 4 snakes (4.6%; *C. vudii*), 1 unidentified bird (1.1%), 1 mammal (1.1%; *Mus musculus*; Henderson and Sajdak, 1996); on Eleuthera, a snake grabbed a Yellow Throat (*Geothlypis trichas*) after it was shot and fell into shallow water at a mangrove edge (Schwartz and Henderson, 1991); on Waderick Wells Cay, a 196 g snake ate a 34.5 g *Ameiva auberi*, which was 17.6% of the snake's mass (Franz and Dodd, 1994); preys on hatchling *Cyclura cychlura* on Andros Island (Knapp et al., 2004, 2005); gape limits appear to be responsible for failed predation attempts on hatchling *C. cychlura* (Knapp and Owens, 2008b). **Growth:** Over 484 days, an adult female grew 15 mm (0.031 mm/day; Franz and Dodd, 1994). **Miscellany:** A Bananaquit (*Coereba fla-*

veola) feigned injury to "lure" a snake from an apparent nest site (A. Mitchell, pers. comm.). **Parasites:** Trematode *Ochetosoma kansense* (Franz, 1976). **Reproduction:** Clutch size 4–15, gravid females taken April–July (480–675 mm SVL), eggs 8.6–15 × 19–29.3 mm (Schwartz and Henderson, 1991). **Size:** SVL to 840 mm (Franz and Dodd, 1994).

Conservation Status: Not assessed.

Remarks: This species was until recently placed in the genera *Alsophis* and *Ocyophis*.

Haitiophis anomalus Peters, 1893

Distribution: Hispaniola, widely distributed, but disjunct, also on Isla Beata and Île de la Tortue.

Habitat: Xerophilic, in open, thorn forest (Isla Beata), essentially ground dwelling, but in a mango tree on Île de la Tortue (Schwartz and Henderson, 1991).

Activity: Diurnal (Schwartz and Henderson, 1991). **Diet:** Includes *Anolis* and *Mus* (Henderson and Sajdak, 1996; Schwartz and Henderson, 1991). **Mortality:** Killed by native Dominicans (Thomas et al., 2007); 2 roadkills in the Valle de Neiba (RP). **Reproduction:** Female (1,770 mm SVL) with 22 eggs on 8 May (Cochran, 1941). **Size:** SVL to ~2,000 mm (total length to 2,743 mm; Thomas et al., 2007).

Conservation Status: Apparently a genuinely rare snake, despite many years of intensive fieldwork by many people on Hispaniola, very few specimens have been collected; however, the species is afforded no official protection; although not formally assessed, this species was included in a list of threatened Hispaniolan species with the annotation "data deficient" (Powell et al., 2000).

Additional Pertinent Literature: Powell and Henderson (1998a).

Remarks: This species was until recently placed in the genera *Alsophis* and *Ocyophis*.

Hypsirhynchus ater Gosse, 1851

Distribution: Jamaica, widely distributed, but disjunct.

Habitat: At forest edge, in leaf litter, and on rock walls, largely ground dwelling but will climb into trees (Gosse, 1851).

Activity: Diurnal (Gosse, 1851). **Behavior:** Defensive display includes dorsoventral flattening of the neck (Gosse, 1851). **Diet:** Includes *Anolis*, *Mabuya* (Gosse, 1851); also *Celestus* (Schwartz and Henderson, 1991). **Reproduction:** Oviparous (RWH). **Size:** SVL to 850 mm (Schwartz and Henderson, 1991).

Conservation Status: Although designated "critically endangered" on the IUCN Red List of Threatened Species (Gibson, 1996c), this species is possibly extinct (Henderson and Powell, 1996a; S. B. Hedges, in litt., 18.IV.2008); thought to persist in small numbers in the Cockpit Country (W. L. R. Oliver in Johnson, 1988).

Additional Pertinent Literature: Henderson and Powell (1996a).

Remarks: This species was until recently placed in the genus *Alsophis*.

Hypsirhynchus callilaemus Gosse, 1851

Distribution: Jamaica, islandwide; elevational range, sea level to >900 m.

Habitat: Ground dwelling, in dry limestone plains, pastures, and forests, under logs, *Cocos* trash, abandoned termitaria, and inside rotted wood (Grant, 1938, 1940a; Schwartz and Henderson, 1991); in loose soil under a rock in a dry wash (B. I. Crother, in litt., 26.II.2008).

Activity: Diurnal (Schwartz and Henderson, 1991). **Foraging and Diet:** Active forager that preys largely on *Anolis* (Schwartz and Henderson, 1991). **Size:** SVL to 465 mm (Schwartz and Henderson, 1991).

Conservation Status: Unknown; not quite extirpated, but rare due to mongoose predation (Barbour, 1914).

Remarks: This species was until recently placed in the genera *Arrhyton* and *Schwartzophis*.

Hypsirhynchus ferox Günther, 1858

Distribution: Hispaniola, where three currently recognized subspecies occur on most of the main island (except the Tiburon Peninsula; *H. f. ferox*), Isla Saona (*H. f. exedrus*), and Île de la Gonâve (*H. f. paracrousis*); elevational range, below sea level to >1,700 m.

Habitat: Mesophilic and xerophilic, in arid, cactus-studded lowlands to mesic lowlands and uplands and mesic coastal forest, by day under an *Agave* stump, a metal laundry tray, rocks, *Cocos* trash, a small log, a piece of flat concrete adjacent to the ruins of an abandoned house, actively crawling on the ground, ascending the trunk of a large tree in an Haitian oasis, in *Acacia* scrub, crossing a sandy trail at midday on Isla Saona, and basking in late morning on a fallen palm frond (Schwartz and Henderson, 1991); on a street in Barahona at dusk (Powell and Henderson, 2008c).

Activity: Diurnal (RWH). **Diet:** Includes *Anolis* (26.1% by frequency of occurrence in a sample of 46 prey items),

Leiocephalus (17.4%), *Ameiva* (52.2%), and *Celestus* (4.3%), ontogenetic shift in diet from primarily *Anolis* to larger ground-dwelling species (Henderson, 1984b). **Reproduction:** Clutches of 3–15 eggs (10.1–16.7 × 13.5–30.0 mm) in females 530–740 mm SVL on 19 May–22 September, *H. f. paracrousis* laid eggs on 16 May, hatched 7 August, same subspecies produced 2 eggs on 3 August, creamy color, very pliable, granular texture, 30.6 × 17.2 mm and 25.9 mm × 17.0 mm (Schwartz and Henderson, 1991). **Size:** SVL to 773 mm (Schwartz, 1971b).

Conservation Status: Unknown.

Additional Pertinent Literature: Sosa et al. (1995).

Hypsirhynchus funereus Cope, 1863

Distribution: Jamaica, widely distributed in moist lowlands; elevational range, sea level to > 600 m.

Habitat: Mesophilic, in ravine beds, open pastures, margins of *Musa* groves under logs and rocks (Schwartz and Henderson, 1991); under heaps of vegetation and rotting *Cocos* husks (Grant, 1938, 1940a); under rock piles in pastures (B. I. Crother, in litt., 26.II.2008).

Activity: Diurnal (Schwartz and Henderson, 1991). **Behavior:** "not so nervous or active as [*H.*] *callilaemus*" (Grant, 1938). **Foraging and Diet:** Active forager with a diet that includes *Eleutherodactylus* (eggs and adults; 41.7% by frequency in a sample of 24 prey items), *Sphaerodactylus* (8.3%), *Anolis* (50.0%), also *Typhlops* eggs (Schwartz and Henderson, 1991). **Parasites:** Infested with ticks under the chin and anterior to the vent (Grant, 1940a). **Reproduction:** Oviparous, eggs on 31 August, egg hatched 1 September, hatchling in a black ant nest under a log on 2 July (Schwartz and Henderson, 1991). **Size:** SVL to 479 mm (Schwartz and Henderson, 1991).

Conservation Status: Not assessed.

Remarks: This species was until recently placed in the genera *Arrhyton* and *Schwartzophis*.

Hypsirhynchus melanichnus Cope, 1863

Distribution: Hispaniola, widely distributed, but disjunct.

Habitat: Presumably ground dwelling (Schwartz and Henderson, 1991).

Activity: Presumably diurnal (Schwartz and Henderson, 1991). **Reproduction:** Presumably oviparous (RWH). **Size:** SVL to 630 mm (Schwartz and Henderson, 1991).

Conservation Status: This species has been afforded no official conservation status, but it is probably extinct (Powell and Henderson, 1998b); listed as "critically endangered" by Powell et al. (2000).

Additional Pertinent Literature: Powell and Henderson (1998b).

Remarks: This species was until recently placed in the genera *Alsophis* and *Ocyophis*.

Hypsirhynchus parvifrons Cope, 1863

Distribution: Hispaniola, where nine currently recognized subspecies occur on the Tiburon Peninsula, Île Grande Cayemite, and Grosse Caye (*H. p. parvifrons*), Île de la Gonâve and Île Petite Gonâve (*H. p. alleni*), Isla Beata and the Barahona Peninsula (*H. p. lincolni*), the Península de Samaná (*H. p. niger*), in the southeastern Dominican Republic and on Isla Catalina (*H. p. paraniger*), throughout much of the main island (*H. p. protenus*), on Île-à-Vache (*H. p. rosamondae*), on Isla Saona (*H. p. stygius*), and on Île de la Tortue (*H. p. tortuganus*); a presumed vagrant found on Little Inagua Island (Bahamas).

Habitat: Primarily mesophilic, in arid scrub to wet forest, ground dwelling, but taken under the bark of a *Cocos* palm at 1.5 m (Schwartz and Henderson, 1991); on Île de la Gonâve (420 m) in disturbed areas, limestone rocks adjacent to agricultural plots (Franz and Gicca, 1982); in grassy pastures, shaded Royal Palm–coffee groves, pine barrens, a xeric hammock, hillside pastures, in the dining room of the Hotel Taino (just west of Grand Goâve, Haiti), active in forest leaf litter, *Cocos* trash, rock-strewn fields, beach litter, around shallow temporary pools in thornscrub, in a breadfruit (*Artocarpus altilis*) grove and *Acacia* scrub, under rocks, root masses, logs, vegetal debris, palm chips, bamboo trash in a cacao grove, a fallen thatch roof, cardboard (on sand at a beach), small *Roystonea* fronds in dry coffee and cacao, logger's chips at a sawmill, and *Musa* trash, in a solution cavity in limestone rock (Schwartz and Henderson, 1991).

Activity: Diurnal (Schwartz and Henderson, 1991). **Behavior:** Defensive display includes raised forepart of body and dorsoventrally flattened neck (Thomas and Schwartz, 1965; Sajdak and Henderson, 1982); aggregations of 3–5 individuals coiled together on Île de la Gonâve (Franz and Gicca, 1982). **Foraging and Diet:** Active forager that probes leaf litter with its head (Franz and Gicca, 1982); diet includes *Osteopilus dominicensis* (5.5% by frequency in a sample of 199 prey items), *Eleutherodactylus abbotti* (5.5%), *Sphaerodactylus difficilis* (4.4%), *Anolis caudalis* (5.5%), *A. chlorocyanus* (3.3%), *A. coelestinus* (14.6%), *A. cybotes* (34.8%),

Ameiva chrysolaema (4.4%), *Mus musculus* (2.2%), and others, all *Anolis* combined accounted for 70.9% in frequency and 15.9% in volume, mean prey size 2.5 ± 0.3 cm^3 (Franz and Gicca, 1982; Henderson et al., 1987a); also takes *Celestus haetianus* and *Ameiva taeniura*, attempting to swallow a *Leiocephalus semilineatus* much too large for it, feeding on *Eleutherodactylus* at 1300 h in cacao (Schwartz and Henderson, 1991); *Sphaerodactylus difficilis* and *Celestus costatus* (J. W. Smith et al., 1994); swallowing an *Ameiva chrysolaema* at Las Charcas in coastal scrub (RP). **Parasites:** Nematode *Terranova caballeroi* (Powell et al., 1990f); trematodes in intestinal mesenteries and on body walls (Franz and Gicca, 1982). **Predation:** Approximately 13% of 89 snakes from Île de la Gonâve had stubbed tails, indicating encounters with potential predators (Franz and Gicca, 1982); predators include Ridgway's Hawk (*Buteo ridgwayi*; Wiley and Wiley, 1981). **Reproduction:** Oviparous, females from Isla Saona (sea level) laid 4 eggs 22 July (hatched 4 September) and 3 eggs 4 September (hatched 10 November), eggs 30–34 × 10–11 mm, mean SVL of 5 hatchlings 152 mm (Sajdak and Henderson, 1982); females from the Sierra de Baoruco (elevation ~1,020 m) in March–April had 3–8 shelled eggs (15.8–28.0 × 8.3–12.0 mm), clutch size (based on number of yolked ovarian follicles and oviductal eggs 3–10 (mean 5.2 ± 1.4, $N = 22$), clutch size and SVL positively correlated (Powell et al., 1991); on Île de la Gonâve, females with 3–9 maturing follicles or eggs (mean 5.1), number positively correlated with female SVL (Franz and Gicca, 1982); elsewhere, hatchlings on 6, 8, 10, 16, 19 August, sample of 26 clutches, 2–15 eggs (4.6–12.1 × 9.4–25.3 mm), clutch size correlated with female SVL (305–530 mm), March–December (Schwartz and Henderson, 1991). **Size:** SVL to 557 mm (Franz and Gicca, 1982). **Thermal Biology:** Body and adjacent air temperatures of 2 snakes on Isla Saona 29.9 °C (27.8 °C) and 31.2 °C (31.2 °C; Sajdak and Henderson, 1982).

Conservation Status: Not assessed.

Remarks: This species was until recently placed in the genus *Antillophis*.

Hypsirhynchus polylepis Buden, 1966

Distribution: Eastern Jamaica

Habitat: Mesophilic, in piles of *Cocos* trash, ground dwelling (Schwartz and Henderson, 1991); in and under rock piles (B. I. Crother, in litt., 26.II.2008).

Foraging and Diet: Active forager, diet includes *Eleutherodactylus* and *Anolis* (Schwartz and Henderson, 1991). **Reproduction:** Oviparous (Schwartz and Henderson, 1991). **Size:** SVL to 455 mm (Schwartz and Henderson, 1991).

Conservation Status: Unknown.

Remarks: This species was until recently placed in the genera *Arrhyton* and *Schwartzophis*.

Hypsirhynchus scalaris Cope, 1863

Distribution: Hispaniola, Tiburon Peninsula (Haiti); elevational range, sea level to at least 305 m.

Habitat: Probably mesophilic (Schwartz, 1971b); presumably ground dwelling (RWH).

Diet: Includes *Anolis* spp. and *Leiocephalus* sp. (RWH). **Reproduction:** Oviparous (RWH). **Size:** SVL to 592 mm (Schwartz, 1971b).

Conservation Status: Unknown.

Additional Pertinent Literature: Sosa et al. (1995).

Remarks: This species was formerly considered a subspecies of *Hypsirhynchus ferox*.

Ialtris agyrtes Schwartz and Rossman, 1976

Distribution: Hispaniola, Barahona Peninsula and Sierra de Neiba (Dominican Republic).

Habitat: Unknown; probably ground dwelling in xeric forest (Schwartz and Henderson, 1991).

Activity: Diurnal (Schwartz and Henderson, 1991). **Diet:** Includes *Typhlops pusillus* (Schwartz and Rossman, 1976). **Size:** SVL to 576 mm (Schwartz and Henderson, 1991).

Conservation Status: Although not formally assessed, this species was included as "endangered" in a list of threatened Hispaniolan species (Powell et al., 2000).

Additional Pertinent Literature: Powell and Henderson (1994a, 1994b).

Ialtris dorsalis Günther, 1858

Distribution: Hispaniola, widely distributed but disjunct on the main island, also on Île-à-Vache, Île de la Gonâve, Île de la Tortue.

Habitat: Ground dwelling, under a rock in mesic forest on a karst substrate, under a rock and in moist palm trash in a coffee grove (Schwartz and Henderson, 1991).

Activity: Diurnal (RWH). **Foraging and Diet:** Probably an active forager, diet includes *Osteopilus dominicensis* (1 in a sample of nine prey items), *Anolis* (2), *Celestus* (2), *Ameiva* (2), and *Rattus* (2), mean prey size 10.7 ± 4.7 cm^3 (Henderson, 1984a; Henderson et al., 1988a). **Mortality:** Killed by rural residents (RP). **Predator:** Ridgway's Hawk (*Buteo ridgwayi*; Wiley and Wiley, 1981). **Reproduction:** Female (SVL 660 mm SVL) with a clutch of 12 eggs (~28.7 × 15.7 mm) on 27 June (Schwartz and Henderson, 1991). **Size:** SVL to 990 mm (Schwartz and Henderson, 1991).

Conservation Status: Not assessed, but the tenuous nature of its habitat and its seeming rarity warrants special concern (Powell et al., 2000).

Additional Pertinent Literature: Powell and Henderson (1994a), Zippel et al. (1994).

Ialtris haetianus Cochran, 1935

Distribution: Hispaniola, where three currently recognized subspecies occur in the Massif de la Hotte (*I. h. haetianus*), the Sierra de Baoruco and southern versant of Massif de la Selle (*I. h. perfector*), and the Montagne Noire (*I. h. vaticinata*).

Habitat: Mesophilic (Schwartz and Henderson, 1991); beneath rocks and vegetable debris and a mat of cut vines in wet, pasturelike clearings surrounded by broadleaf mesic forest or coffee groves, under a pile of drying grass (Schwartz and Thomas, 1965; Henderson and Schwartz, 1986); a montane ravine with dense growth of bamboo (Schwartz and Thomas, 1965); rocky moist hillsides with little cover, around native houses in cutover cloud forest, in *Musa* trash, under a rotten log in a cloud forest clearing (Schwartz and Henderson, 1991); under surface debris in a recent clearcut (RP).

Foraging and Diet: As the diet includes anuran eggs, probably an active forager (RWH); diet composed almost exclusively of *Eleutherodactylus* (45 of 46 prey items, including 13 egg clutches and *E. abbotti*, *E. armstrongi*, *E. eunaster*, *E. glandulifer*, *E. glaphycompus*, *E. heminota*, *E. oxyrhyncus*, *E. sciagraphus*, and *Anolis armouri*, mean prey size 1.2 ± 0.3 cm^3 (Henderson and Schwartz, 1986). **Reproduction:** Females with SVL 235–310 mm with clutches of 1–7 eggs (5.9–7.0 × 13.0–27.6 mm) on 25–28 June, eggs under a roadside rock on 27 August, eggs desiccate very rapidly (Schwartz and Henderson, 1991). **Size:** SVL to 305 mm (Schwartz, 1986).

Conservation Status: Unknown.

Additional Pertinent Literature: Schwartz (1986).

Remarks: This species was until recently placed in the genus *Darlingtonia*.

Ialtris parishi Cochran, 1932

Distribution: Hispaniola, Île de la Tortue.

Habitat: Presumably ground dwelling (Schwartz and Henderson, 1991).

Reproduction: Presumably oviparous (RWH). **Size:** SVL to 745 mm (Schwartz and Rossman, 1976).

Conservation Status: Unknown, presumably a very rare snake.

Additional Pertinent Literature: Powell and Henderson (1994a, 1994c).

Liophis cursor Lacépède, 1789

Distribution: Martinique, including the 0.06 km^2 satellite Rocher du Diamant.

Habitat: Mesophilic, presumably ground dwelling (Schwartz and Henderson, 1991); on Rocher du Diamant, in thick vegetation on the north side of the island (Lazell, 1967).

Activity: Presumably diurnal (Schwartz and Henderson, 1991). **Foraging and Diet:** Presumably an active forager, stomach contents included a beetle (1 of 6 prey items in sample), *Eleutherodactylus* (2), and *Anolis roquet* (3; Schwartz and Henderson, 1991). **Reproduction:** Female with 5 shelled eggs on 8 February (Schwartz and Henderson, 1991). **Size:** SVL to 671 mm (Dixon, 1981).

Conservation Status: "Critically endangered" on the IUCN Red List (Breuil, 1996); not collected since 1962 (Honegger, 1981); the species is almost certainly extirpated from Martinique, if still extant on Rocher du Diamant, at a very low population density (Henderson, 2004; Powell and Henderson, 2005a).

Liophis juliae Cope, 1879

Distribution: Central Lesser Antilles, where three currently recognized subspecies occur on Dominica (*L. j. juliae*), Guadeloupe (*L. j. copeae*), and Marie-Galante (*L. j. mariae*).

Habitat: Mesophilic, ground dwelling, in rainforest, cutover hardwoods, in citrus, *Musa*, and *Cocos* groves, xeric forest (in *Cocos* trash), wooded area adjacent to a beach, berm vegetation, gardens, on rocks and logs, edificarian (Schwartz and Henderson, 1991); in Cabrits National Park (Dominica), on leaf litter or associated with rocks, 8 of 9 individuals in full shade, one in a sun-shade mosaic (Muelleman et al., 2009).

Activity: Diurnal (Schwartz and Henderson, 1991); possibly unimodal (all 9 snakes observed during a 3-week period active in the morning), activity increases on rainy days (Muelleman et al., 2009). **Behavior:** Gentle and inoffensive, but may release a foul-smelling cloacal secretion when handled (Malhotra and Thorpe, 1999); one of 4 individuals on a rainy morning in Cabrits National Park (Dominica) responded to disturbance by flattening its neck dorsoventrally (Muelleman et al., 2009). **Foraging and Diet:** Active forager, stomach contents included 1 unidentified insect (in a sample of 9 prey items), 7 *Eleutherodactylus martinicensis* (eggs and adults), and 1 *Anolis oculatus* (Henderson and Bourgeois, 1993). **Predator:** Taken from the crop of a Broad-winged Hawk (*Buteo platypterus*; Schwartz and Henderson, 1991). **Reproduction:** Clutches of 2 and 4 eggs from females 410 mm and 441 mm SVL (Schwartz and Henderson, 1991); communal nesting site with more than 100 eggs in a single chamber encountered 17 February in Dominica, possibly used by 20–30 females, just prior to hatching 3 eggs 26.7–27.0 mm long, mass 3.1–3.5 g, some eggs hatched 25 February to 25 March, 2 hatchlings with total lengths 185 mm (James and Henderson, 2004). **Size:** SVL to 458 mm (Dixon, 1981).

Conservation Status: Not assessed; the species is moderately common on Dominica, but very rare or extirpated on Guadeloupe and Marie-Galante (Henderson, 2004; Powell and Henderson, 2005a).

Liophis ornatus Garman, 1887

Distribution: St. Lucia and 0.09 km^2 Maria Major.

Habitat: Xerophilic to mesophilic, ground dwelling (Schwartz and Henderson, 1991); on Maria Major, in wooded habitat and scrub (Corke, 1987).

Abundance: One snake encountered during 117 hours of searching in 1997 (Buley et al., 1997). **Activity:** Diurnal (Schwartz and Henderson, 1991); snakes at 0600 h (Corke, 1987) and 1505 h (Buley et al., 1997), after or during heavy rain. **Foraging and Diet:** Likely feeds on lizards and frogs (Long, 1974); presumably an active forager (Schwartz and Henderson, 1991). **Reproduction:** Presumably oviparous (RWH). **Size:** SVL to 1,235 mm (Dixon, 1981) or 610 mm SVL (Buley et al., 1997).

Conservation Status: "Endangered" on the IUCN Red List (Gibson, 1996f); almost certainly extirpated from St. Lucia, it may still occur on Maria Major (Johnson, 1988; Malhotra and Thorpe, 1999; Henderson, 2004; Powell and Henderson, 2005a).

Liophis perfuscus Cope, 1863

Distribution: Barbados.

Habitat: Mesophilic and ground dwelling (Schwartz and Henderson, 1991).

Activity: Presumably diurnal (Schwartz and Henderson, 1991). **Foraging and Diet:** Active forager (Schwartz and Henderson, 1991). **Reproduction:** Oviparous (RWH). **Size:** SVL to 797 mm (Dixon, 1981).

Conservation Status: "Endangered" on the IUCN Red List (Carrington et al., 1996); the species is possibly extinct (Henderson, 2004; Powell and Henderson, 2005a); it has not been found on Barbados since 1963 (Underwood et al., 1999).

Magliophis exiguus Cope, 1863

Distribution: Puerto Rico Bank, islands east of Puerto Rico, Culebra, St. Thomas, Hassel Island, Tortola, Guana Island, Peter Island, Virgin Gorda (*M. e. exiguus*), and southern Puerto Rico (*M. e. subspadix*).

Habitat: In wooded and cultivated situations, cactus-*Acacia* woods, under rocks, logs, and dead vegetation(Schwartz and Henderson, 1991); on Tortola, "frequently met with by persons gardening or cultivating," two "torpid" individuals in winter in the cracks of an old wall (Grant, 1932h); among small stones and rocks between shaded buttresses at the base of a tree, by day under trash, in piles of *Cocos* fronds and husks, in cracks of an old wall in "winter" (Schwartz, 1967a); occasionally edificarian on Guana Island (RP).

Activity: Diurnal (Schwartz and Henderson, 1991; Thomas and Gaa Kessler, 1996; Joglar, 2005b); crossing a road on Virgin Gorda in dry woods during the heat of late morning, but also active at 2130 h on palm trash in light rain on Tortola (Schwartz, 1967a); foraging by both day and night in forest floor leaf litter (Clark and Gillingham, 1990). **Defensive Behavior:** On Guana Island, may wave tail, perhaps to distract a predator from the head (RP). **Foraging and Diet:** Active forager, diet includes *Eleutherodactylus* eggs, *Sphaerodactylus*, and *Anolis* (Schwartz and Henderson, 1991); on Puerto Rico, includes *Anolis cristatellus* and *A. pulchellus* (Thomas and Gaa Kessler, 1996); *A. gundlachi* (G. Rogowitz in Thomas and Gaa Kessler, 1996); *A. krugi* in captivity (A. J. Sánchez Muñoz in Horstink, 2004); *Anolis* eggshells regurgitated on Guana (Lazell, 2005a); after capture, snakes delivered "embedding" bites to either inject more venom or to assess the responsiveness of the lizard, anoles taken by snakes exhibited (in sequence) swallowing, gaping, and spasmodic breathing movements of the thorax, and these signs were repeated at intervals until lizards were immobile (Thomas and Leal, 1993). **Population Density:** At several sites on Puerto Rico, means 5.6–16.7/ha during February 2001–February 2002 (Rios-López and Aide, 2007). **Reproduction:** On Guana, 3 eggs laid on 16 July, eggs 21–22 × 8–8.5 mm, hatchlings 86–91 mm SVL (Lazell, 2005a); apparent copulation in March on Puerto Rico (Blair, 2007). **Size:** SVL to 325 mm (Thomas and Gaa Kessler, 1996).

Conservation Status: "Stable" (Moreno, 1991d).

Remarks: This species was until recently placed in the genus *Arrhyton*; *M. e. subspadix* may be distinct at the species level.

Magliophis stahli Stejneger, 1904

Distribution: Puerto Rico, all but the southern part of the island; elevational range, sea level to at least 550 m.

Habitat: Terrestrial and secretive, under logs, rocks, tree stumps, and in termite nests (Puente-Rolón, 2001); ground dwelling (Schwartz and Henderson, 1991); rainforest, coastal woods, dry woods, rocky pastures, around bases of rock walls, in gardens and yards, by day under trash piles, piles of *Cocos* fronds and husks, logs, and rocks (Schwartz, 1967a); under a log in a pasture, in a coffee plantation, on a road, under tree stumps (K. P. Schmidt, 1928).

Activity: Probably diurnal (Schwartz and Henderson, 1991); although observed feeding on an anole egg at night (Puente-Rolón, 2001). **Foraging and Diet:** Diet includes *Eleutherodactylus* (eggs, hatchlings, adults; 50% by frequency in a sample of 18 prey items), *Sphaerodactylus* (11.1%), and *Anolis* (38.9%; RWH); *Anolis pulchellus* (K. P. Schmidt, 1928); also *Amphisbaena* and adults and eggs of *Typhlops* (Rivero, 1998); 21 × 11 mm *Anolis* (probably *A. cuvieri*) egg (Puente-Rolón, 2001); after capture by a snake, *Anolis* lizards exhibit swallowing, gaping, and spasmodic breathing movements of the thorax until they are immobile (and apparently dead), autopsies of lizards revealed bloody froth in the lungs, trachea, and back of throat, snakes bite down forcefully, apparently to inject venom (Thomas and Leal, 1993). **Reproduction:** Eggs under a log in a pasture, in loose soil under a stump, under an abandoned termite nest in a coffee plantation (adherent clutches of 6 and 7), groups of 13 and 40 eggs suggest communal nesting, female with 6 well-developed eggs, egg size 21–25 × 12–15 mm, eggs white and finely striate (K. P. Schmidt, 1920, 1928). **Size:** SVL to 438 mm (Schwartz, 1967a).

Conservation Status: Not assessed.

Remarks: This species was formerly considered a subspecies of *Magliophis exiguus*, which had been placed in the genus *Arrhyton*.

Tretanorhinus variabilis Duméril and Bibron, 1854

Distribution: Five currently recognized subspecies occur on Cuba (*T. v. variabilis*, *T. v. binghami*, *T. v. wagleri*), Isla de la Juventud (*T. v. insulaepinorum*), and the Cayman Islands, Grand Cayman (*T. v. lewisi*).

Habitat: Mesophilic, largely aquatic (Schwartz and Henderson, 1991); on Cuba, in small streams, saltwater-filled ditches, around dock pilings in the ocean, in salt marshes (Neil, 1958, 1965); under a car tire on a river bank (Collette, 1961); in a puddle in a dirt road (Garrido and Schwartz, 1968b); in shallow rivers, lagoons, small lakes, in a stream cut through limestone in pinewoods, crossing a dirt road in woodland, by day under submerged rocks or in an accumulations of drift vegetation (Schwartz and Henderson, 1991); in a brackish tidal estuary (Grant, 1946); young individual captured "half a mile or more from the nearest small stream" (Barbour and Ramsden, 1919); on Grand Cayman, along rocky or muddy edges of freshwater or slightly brackish ponds in "cow wells," by day buried in loose flocculent organic material that forms on the bottom (Seidel and Franz, 1994); mostly hidden in mud, rather than in water, secured by bailing cow wells and searching bottom mud and rock crevices (Grant, 1940c); swimming along the margin of a cattail-clogged pool in a pasture (Schwartz and Henderson, 1991).

Activity: Nocturnal (Alayo Dalmau, 1951; Z. Vogel, 1965c); emerges shortly after dark (Neil, 1965). **Behavior:** When handled, rolls into balls with the head hidden inside the coils (Barbour and Ramsden, 1919; Seidel and Franz, 1994); responds to touch by repeatedly striking an object with its body (Mertens, 1946); will hide in vegetation along a bank or under rocks in a stream if it senses danger (Sampedro Marín and Rodríguez Schettino, 2003). **Foraging and Diet:** At night, majority in more quiet areas of a stream near the bank, "usually completely submerged, with their bodies looped around rocks in the bottom of the stream" (Schwartz and Ogren, 1956); move slowly in water or lie in wait for prey (Barbour and Ramsden, 1919); pursuing small fishes in saltwater of a deep ditch by "snapping randomly right and left amidst the darting schools," crawls "very slowly on the bottom, beneath about 2 feet of water" (Neil, 1965); takes small fish and frogs (Z. Vogel, 1965c); forages underwater in cow wells (Seidel and Franz, 1994); diet primarily fishes, usually cyprinodonts (Barbour and

Ramsden, 1919); in Cuba, regurgitated a *Rana catesbeiana* that had been swallowed rear first (Neil, 1965); snakes easily captured in minnow traps due to the presence of the poeciliid fish *Limia caymanensis* (Seidel and Franz, 1994); also includes aquatic invertebrates, larval frogs, and small lizards (Sampedro Marín and Rodríguez Schettino, 2003); one snake struck at and caught a small fish and finished swallowing it while being held (Schwartz and Ogren, 1956). **Longevity:** In captivity, 5 years, 2 months (S. Minton in Slavens and Slavens, 2003). **Parasites:** Larval nematode *Contracaecum* sp. (Baruš and Coy Otero, 1966, 1978; Coy Otero and Baruš, 1979b; Coy Otero and Lorenzo Hernández, 1982); cestodes *Ophiotaenia barbouri* (Pérez Vigueras, 1934c, 1935a; Coy Otero and Lorenzo Hernández, 1982); *O. viperis* (Freze and Rysavy, 1976; Coy Otero and Lorenzo Hernández, 1982); and *Sparganum* sp. (Pérez Vigueras, 1935a; Coy Otero and Lorenzo Hernández, 1982; Coy Otero, 1999). **Reproduction:** Cuban female laid 8 adherent eggs (33–36 × 16–18 mm) on 31 August, hatched 5 October, hatchlings 143–145 mm total length (Petzold, 1967); female on Grand Cayman on 17 July laid 8 eggs on 5 September, females in January and February nonreproductive, postreproductive in mid-August (Seidel and Franz, 1994). **Size:** SVL to ~800 mm (Schwartz and Henderson, 1991).

Conservation Status: Of concern (Buide et al., 1974); listed but not formally assessed (Rodríguez Schettino and Chamizo Lara, 1998a).

Uromacer catesbyi Schlegel, 1837

Distribution: Hispaniola, including satellite islands, where seven currently recognized subspecies occur on the Tiburon Peninsula (*U. c. catesbyi*), Île Grande Cayemite, presumably Île Petite Cayemite (*U. c. cereolineatus*), Île de la Gonâve (*U. c. frondicolor*), much of Haiti (*U. c. hariolatus*), Isla Saona (*U. c. inchausteguii*), Île-à-Vache (*U. c. insulaevaccarum*), and much of the Dominican Republic (*U. c. pampineus*).

Habitat: Mesophilic, but in a variety of habitats, including *Acacia* scrub, rainforest, and oases in xeric areas, highly arboreal, but will make long forays on the ground (Schwartz and Henderson, 1991); often in heavily disturbed situations in close proximity to human activity (RWH); uses heavier branches than *U. oxyrhynchus* on Isla Saona, mean diameter 10.2 ± 0.95 mm (3–16 mm) and height 2.1 ± 0.77 m (0.3–3.5 m) by day and 2.8 ± 0.13 m (1.0–5.3 m) by night (Henderson et al., 1981).

Activity: Diurnal (Schwartz and Henderson, 1991). **Behavior:** Likely male-male combat in nature (Schwartz, 1970e); gapes and feigns strikes when handled (RP). **Effects of Bite:** One enlarged rear maxillary tooth penetrated skin between thumb and forefinger causing instant burning pain that subsided within about 2 minutes (no subsequent swelling or discoloration; Schwartz and Henderson, 1991). **Foraging and Diet:** Exhibits active and sit-and-wait foraging, diet includes *Osteopilus pulchrilineatus* (1.6% by frequency of occurrence in a sample of 64 prey items), *Osteopilus dominicensis* (46.9%), *Anolis* spp. (45.3%), *Leiocephalus* spp. (6.3%), mean prey size 4.0 ± 1.0 cm^3 (Horn, 1969; Henderson et al., 1987c); Bananaquit (*Coereba flaveola*; M. A. Landestoy, in litt., 12.XI.2008). **Predation:** Snake (0.5 m) partially ingested by a Hispaniolan Lizard Cuckoo (*Coccyzus longirostris*) had coiled its tail tightly around a small sapling, effectively tethering the bird (T. H. White, 2006). **Reproduction:** Gravid females 12 April (520 mm SVL, 5 eggs, 9.0 × 21.2 mm; Schwartz and Henderson, 1991); female (SVL 746 mm) on 2 June laid 5 white oblong eggs (3.2–4.5 g, mean 3.7 g; 38.5–45.5 × 14.0–15.0 mm; mean 42.1 × 14.4 mm) on 7 June, hatched on 2 and 4 September, hatchlings 175–187 mm SVL and 301–315 mm total length (Bowersox et al., 1994a). **Size:** SVL to 830 mm (Schwartz, 1970e). **Thermal Biology:** Body temperatures 23.2–31.6 °C on Isla Saona, preferred temperature range probably 29–30 °C (Henderson, 1982).

Conservation Status: Not assessed.

Additional Pertinent Literature: Schwartz and Henderson (1984a, 1984b).

Uromacer frenatus Günther, 1865

Distribution: Hispaniola, including several satellite islands, where four currently recognized subspecies occur on the "south island" of Hispaniola, including Grosse Caye, with one record on the "North Island" at the base of the Sierra Martín García (*U. f. frenatus*, *U. f. chlorauges*), Île de la Gonâve (*U. f. dorsalis*), and Isla Beata (*U. f. wetmorei*).

Habitat: Mesophilic to xerophilic, *Acacia* forest, low bushes, and tall grass (Schwartz and Henderson, 1991); arboreal, but frequently on or near the ground (RWH, RP).

Activity: Diurnal (RWH, RP). **Foraging and Diet:** Sit-and-wait forager, strictly saurophagous with *Anolis* (70.4% by frequency in a sample of 108 prey items), *Leiocephalus* (10.2%), *Ameiva* (18.5%) especially important, mean prey size 2.8 ± 0.3 cm^3 (Henderson and Horn, 1983; Henderson et al., 1987c); geographic variation in diet, with *Ameiva chrysolaema* taken more frequently on Île de la Gonâve than elsewhere in the range (Henderson et al., 1987c); unfed neonates exhibit a strong preference for lizard prey extracts over anuran prey extract and distilled water (Henderson et al., 1983). **Reproduction:** Île de la Gonâve females deposited egg clutches late March and early April, clutch size 4–6 ($N = 2$ clutches), eggs hatched late June and early July, mean SVL of hatchlings 27.6 ± 0.5 cm, mean mass 2.5 ± 0.07 g (Schwartz and Henderson, 1991). **Size:** SVL to 930 mm (Schwartz and Henderson, 1991).

Conservation Status: Not assessed.

Additional Pertinent Literature: Schwartz and Henderson (1984a), Henderson and Schwartz (1984).

Uromacer oxyrhynchus Duméril and Bibron, 1854

Distribution: Hispaniola, essentially islandwide, including Île de la Tortue, Isla Saona, and Isla Catalina.

Habitat: Mesophilic to xerophilic habitats (Schwartz and Henderson, 1991); especially common in *Acacia* scrub, uses slender branches in low bushes and trees, on Isla Saona, mean perch diameter 4.9 ± 0.6 mm (2–10 mm), mean daytime perch height 1.7 ± 0.9 m (0.2–3.4 m), mean nighttime perch height 2.9 ± 0.2 m (0.7–10 m; Henderson et al., 1981); sleeps coiled on distal ends of branches (RWH).

Activity: Diurnal (Schwartz and Henderson, 1991); 94–95% of daylight hours relatively motionless, moving <3.0 m/h (RWH). **Foraging and Diet:** Sit-and-wait forager that preys almost exclusively on lizards, stenophagous for *Anolis*, diet includes *Anolis* (81.8% by frequency of occurrence in a sample of 44 prey items), *Leiocephalus* (6.8%), and *Ameiva* (11.4%), mean prey size 2.2 ± 0.3 cm^3 (Henderson et al., 1987c); grabbed a *Leiocephalus schreibersii* on the ground (RP); keratophagy in captivity (Groves and Altimari, 1977). **Longevity:** In captivity, 18 years, 11 months (Snider and Bowler, 1992). **Mortality:** Roadkills (RP). **Predator:** Ridgway's Hawk (*Buteo ridgwayi*; Wiley and Wiley, 1981). **Reproduction:** Gravid female 701 mm SVL with 5 eggs (Schwartz and Henderson, 1991). **Size:** SVL to ~1,500 mm (Schwartz and Henderson, 1991). **Thermal Biology:** Mean body temperature at 0730–0845 h 24.2 ± 0.2 °C (23.9–24.4 °C), mean body temperature 0845–1445 h 29.6 ± 0.4 °C (27.5–29.9 °C; Henderson, 1982).

Conservation Status: Not assessed.

Additional Pertinent Literature: Schwartz and Henderson (1984a, 1984c).

25 Family Leptotyphlopidae

Figure 28. Very few natural history data address species in the family Leptotyphlopidae, probably attributable primarily to the small size and secretive habits of these snakes. *Leptotyphlops breuili* occurs on St. Lucia. Photograph by S. Blair Hedges.

Leptotyphlops asbolepis Thomas, McDiarmid, and Thompson, 1985

Distribution: Hispaniola, Sierra Martín García (Dominican Republic); elevational range ~350–400 m.

Habitat: Mesophilic (Schwartz and Henderson, 1991); under limestone boulders in mesic forest (Thomas et al., 1985).

Size: SVL to 156 mm (Thomas et al., 1985).

Conservation Status: Unknown.

Leptotyphlops bilineatus Schlegel, 1844

Distribution: Martinique (including Ilet Chancel).

Habitat: In dry and wet forests, gardens, and fields (Breuil, 2009).

Reproduction: Oviparous, 2–6 elongated eggs laid in August–September (Currat, 1980). **Size:** Total length to 108 mm (SVL to 102 mm), total length of one known juvenile 60 mm (Hedges, 2008c).

Conservation Status: Unknown.

Leptotyphlops breuili Hedges, 2008

Distribution: St. Lucia (including Maria Major).

Habitat: On Maria Major, under rocks or in leaf litter in woodland (Buley et al., 1997); in forested areas on slopes (Hedges, 2008c).

Diet: Ants and termites (Long, 1974). **Reproduction:** Oviparous (Long, 1974). **Size:** Total length to 113 mm (SVL to 105 mm), total lengths of 4 known juveniles 59–85 mm (Hedges, 2008c).

Conservation Status: Not assessed; "development of forested areas on Saint Lucia, for housing, roads, tourism, and other activities, is likely to continue reducing the available habitat of this species" (Hedges, 2008c).

Leptotyphlops calypso Thomas, McDiarmid, and Thompson, 1985

Distribution: Hispaniola, Samaná Peninsula (Dominican Republic).

Habitat: Mesophilic (Schwartz and Henderson, 1991); in open pasture and mixed mesic cultivation (bananas, yams, coffee, coconuts, papaya) interspersed with scrubby to semiwooded areas, by day under large limestone rocks in open pasture, crawling on the floor of an outdoor kitchen (Thomas et al., 1985).

Size: SVL to 190 mm (Thomas et al., 1985).

Conservation Status: Unknown.

Leptotyphlops carlae Hedges, 2008

Distribution: Barbados.

Habitat: In soil under vegetable litter (Emsley, 1966); adjacent to a small patch of secondary forest, may depend on forested habitat (Hedges, 2008c).

Reproduction: Holotype contained a single, greatly elongated (2.2 × 13.7 mm) oviductal egg and one paratype contained a smaller, less-developed egg, clutch size presumably 1 (Hedges, 2008c). **Size:** Total length to 104 mm, SVL to 98 mm (Hedges, 2008c).

Conservation Status: Not assessed; potentially suitable habitat is probably no more than a few square kilometers and the species may face competition with invasive *Ramphotyphlops braminus* (Hedges, 2008c).

Leptotyphlops columbi Klauber, 1939

Distribution: Bahama Islands, San Salvador Island and satellites.

Habitat: In logs, known locally as "wood worms" (Schwartz and Henderson, 1991).

Size: SVL to 180 mm (Schwartz and Henderson, 1991).

Conservation Status: Unknown.

Leptotyphlops goudoti Duméril and Bibron, 1844

Distribution: In the West Indies, on the Swan Islands; elsewhere on the mainland from Mexico into northern South America and on some continental islands.

Habitat and Natural History: Unknown in the West Indies.

Size: SVL to ~160 mm (Schwartz and Henderson, 1991).

Conservation Status: Unknown.

Leptotyphlops leptepileptus Thomas, McDiarmid, and Thompson, 1985

Distribution: Hispaniola, southern Haiti.

Habitat: Mesophilic (Schwartz and Henderson, 1991); in piles of stream-worn cobbles in a mango-shaded rest area along a road (Thomas et al., 1985).

Size: SVL to 156 mm (Thomas et al., 1985).

Conservation Status: Unknown.

Leptotyphlops pyrites Thomas, 1965

Distribution: Hispaniola, Valle de Neiba and lowlands along the southern Dominican/Haitian border.

Habitat: Xerophilic to mesophilic, in very dry *Acacia* to moderately mesic situations such as coffee groves (Schwartz and Henderson, 1991).

Size: SVL to 165 mm (Schwartz and Henderson, 1991).

Conservation Status: Unknown.

26 Family Natricidae

Figure 29. Semiaquatic *Nerodia clarkii* (Natricidae) also occurs in the southeastern United States. Populations along the northern coast of Cuba are found in rivers, ponds, lagoons, and coastal mangrove stands. Photograph by S. Blair Hedges.

Nerodia clarkii Baird and Girard, 1853

Distribution: Northern coast of Cuba and the southeastern United States (including the Florida Keys); of two currently recognized subspecies, *N. c. compressicauda* occurs in Cuba.

Habitat: Semiaquatic, in rivers, ponds, and lagoons (Gundlach, 1880); in coastal mangroves (*Rhizophora mangle*; Jaume, 1965); basks on limbs and prop roots in the early morning (Neill, 1965); on a narrow mudflat, wet with salt water and between tide marks (Barbour and Ramsden, 1919); by day under rocks and logs, in 0.3 m of water during rain (Schwartz and Henderson, 1991).

Activity: Nocturnal (Schwartz and Henderson, 1991); may be active by day (Kwet, 1995). **Behavior:** In mangroves, submerges itself in salt or brackish water when startled (Jaume, 1965). **Diet:** Includes fishes (Gundlach, 1880; Schwartz and Henderson, 1991). **Reproduction:** Viviparous (RWH); female giving birth by day at water's edge in August (Kwet, 1995). **Size:** Total length to 933 mm (Conant and Collins, 1998).

Conservation Status: On the IUCN Red List as a species of "least concern" (Hammerson, 2007b).

27 Family Tropidophiidae

Figure 30. This *Tropidophis wrighti* (Tropidophiidae) was found foraging at night on ferns in Holguin Province, Cuba. Photograph by S. Blair Hedges.

Tropidophis bucculentus Cope, 1868

Distribution: Navassa Island.

Habitat: Unknown.

Diet: Includes *Celestus badius* (Thomas, 1966c). **Size:** Male SVL to 360 mm, female SVL to 596 mm (Hedges, 2002).

Conservation Status: Not assessed, possibly extinct (Powell, 1999a).

Additional Pertinent Literature: Powell (2002d).

Tropidophis canus Cope, 1868

Distribution: Bahama Islands, where three currently recognized subspecies occur on Great Inagua (*T. c. canus*), Andros Island (*T. c. androsi*), and Eleuthera Island, Eleuthera Cays, Long Island, Cat Island, Exuma Cays, Ragged Islands (*T. c. barbouri*).

Habitat: Mesophilic (Schwartz and Henderson, 1991); under flat stones and decayed wood and boards, coiled in a hollow and decayed coconut palm 2.0 m above the ground (Noble and Klingel, 1932).

Activity: Nocturnal, often emerges after rainfall (Noble and Klingel, 1932). **Diet:** Includes *Anolis scriptus*, *Ameiva maynardii*, and possibly beetles and spiders (Noble and Klingel, 1932). **Longevity:** In captivity, 16 years, 1 month and still alive at last report in 1992 (Snider and Bowler, 1992). **Size:** Male SVL to 363 mm, female SVL to 338 mm SVL (Hedges, 2002).

Conservation Status: Unknown.

Tropidophis caymanensis Battersby, 1938

Distribution: Cayman Islands, Grand Cayman.

Habitat: Xerophilic to mesophilic, wooded areas, in a logwood stump associated with a termitarium 23 m from the sea (Schwartz and Henderson, 1991); under rocks or in crevices and cracks in wood and decaying stumps, found in logs when cutting firewood (Grant, 1940c).

Abundance: Abundant after late summer (July, August) rains (Grant, 1940c). **Activity:** Nocturnal, spends daylight hours under cover (Grant, 1940c). **Defensive Behavior:** Coils into a tight knot (Grant, 1940c). **Foraging:** Ground-dwelling active forager (Schwartz and Henderson, 1991).

Reproduction: Presumably viviparous (RWH). **Size:** Male SVL to 470 mm, female SVL to 438 mm (Hedges, 2002).

Conservation Status: Unknown.

Tropidophis celiae Hedges, Estrada, and Diaz, 1999

Distribution: Northwestern Cuba.

Habitat: Sea Grape (*Coccoloba uvifera*) and dry scrub forest (Hedges et al., 1999).

Activity: Crawling on limestone rocks at 0300 h within 6.0 m of the ocean (Hedges et al., 1999). **Diet:** Includes *Eleutherodactylus blairhedgesi* (Hedges et al., 1999). **Size:** Female SVL to 344 mm (Hedges et al., 1999).

Conservation Status: Unknown.

Tropidophis curtus Garman, 1887

Distribution: Bahama Islands, where two currently recognized subspecies occur on New Providence Island, Bimini Islands, Cay Sal Bank, Eleuthera, Eleuthera Cays, Long Island, Cat Island, Exuma Cays, Ragged Islands, Andros Island.

Habitat: Ground dwelling, under flat stones near shore (North and South Bimini), in the vicinity of and in wells on the Exumas (Hecht et al., 1955); in hammock woods on Andros, in a rotting log and between sheeting standing around a house, under rocks, rock piles, palm fronds, at the base of an old termitarium (Schwartz and Henderson, 1991).

Defensive Behavior: Includes coiling of body into a compact ball with the head in the middle and autohemorrhaging from eyes and mouth (Hecht et al., 1955); undergoes a physiological color change associated with the light-dark cycle that may afford some protective advantage (Rehák, 1987). **Diet:** *Osteopilus septentrionalis* (6.7% by frequency of occurrence in a sample of 15 prey items), *Sphaerodactylus* (3.3%), and *Anolis* (80.0%; Schwartz and Henderson, 1991); juveniles hold the yellow-tipped tail erect, suggestive of caudal luring (Hecht et al., 1955). **Size:** Male SVL to 357 mm, female SVL to 354 mm (Hedges, 2002).

Conservation Status: Not assessed.

Tropidophis feicki Schwartz, 1957

Distribution: Western Cuba.

Habitat: Mesophilic, ground dwelling to arboreal, in wooded areas, on a rock wall, a rock cliff face, the roof of a cave 1.5 m above the ground, under a roadside rock, 1.5 m off ground in an 18 cm diameter tree, in trees and vines associated with cliff faces (1–1.5 m off the ground; Schwartz and Henderson, 1991); 4.6 m within a cave entrance, crawling into a hole in a limestone cliff (Schwartz, 1957); in mogotes (W. Gehrmann in Burger, 2004).

Activity: Nocturnal (Schwartz and Henderson, 1991). **Diet:** Eats geckos (*Hemidactylus*) in captivity (Burger, 2004). **Miscellany:** Undergoes daily color changes, from a light phase at night to a dark phase during the day, temperatures <17 °C result in changes to the transitional color phase at any time of day, photoperiod shown to influence color change (Rehák, 1987). **Reproduction:** Viviparous (Schwartz and Henderson, 1991). **Size:** Male SVL to 411 mm, female SVL to 448 mm (Hedges, 2002).

Conservation Status: Unknown.

Tropidophis fuscus Hedges and Garrido, 1992

Distribution: Eastern Cuba.

Habitat: Under a large log and a large rock by day in pinewoods (*Pinus cubensis*) on red, lateritic soil (Hedges and Garrido, 1992b); by day coiled under stones (one male partially underground) in pine (*Pinus cubensis*) forest and broadleaf woods growing in red lateritic soil, under a fallen tree trunk along the edge of a coffee plantation in semideciduous woodland (Fong Grillo, 2005).

Activity: Nocturnal (Schwartz and Henderson, 1991). **Color Change:** In the dark, snakes are very light, almost gray, with very dark dorsal blotches, by day, very dark dorsal coloration (Fong Grillo, 2005). **Parasites:** Nematodes (Hedges and Garrido, 1992b). **Reproduction:** Viviparous, female (304 mm SVL) with 4 or 5 well-developed young on 27 July (Hedges and Garrido, 1992b); female (258 mm SVL) gave birth to 3 young on 12 October, neonates 110–113 mm SVL, each weighed 0.75 g (Fong Grillo, 2005). **Size:** Male SVL to 287 mm, female SVL to 304 mm (Hedges, 2002).

Conservation Status: Considered "vulnerable" (Rodríguez Schettino and Chamizo Lara, 1998a).

Tropidophis galacelidus Schwartz and Garrido, 1975

Distribution: Southcentral Cuba.

Habitat: Mesophilic, ground dwelling (Schwartz and Henderson, 1991).

Activity: Nocturnal (Schwartz and Henderson, 1991). **Size:** Female SVL to 405 mm (Hedges, 2002).

Conservation Status: Unknown.

Tropidophis greenwayi Barbour and Shreve, 1936

Distribution: Caicos Islands, where two currently recognized subspecies (*T. g. greenwayi*, *T. g. lanthanus*) occur.

Habitat: Ground dwelling in rocky coppice, dense scrub, and vacant lots near towns, under limestone rocks, one under a small board, none under rocks embedded in soil, all on substrates of rocks or soil (Iverson, 1986).

Activity: Nocturnal (Schwartz and Henderson, 1991). **Behavior:** Secretive, defensive response includes coiling in a tight ball with the head hidden, smearing a foul-smelling cloacal discharge on coils, and autohemorrhaging (Iverson, 1986). **Diet:** *Sphaerodactylus caicosensis* (20% by occurrence in a sample of 10 prey items) and *Anolis scriptus* (80%; Iverson, 1986). **Growth Rate:** 7.4–9.2 mm/month in juveniles (Iverson, 1986). **Longevity:** In captivity, 10 years, 3 months (Snider and Bowler, 1992). **Reproduction:** Sexual maturity at 220–235 mm SVL and 9.5–10.5 g in females, ~195 mm and 8.0 g in males, ovarian cycle seasonal, mating probably in January-early February (when testes size is maximal), captive mating in late January (in copula for 3 h 25 min and 5 h), during coitus, male's head about one-third the distance down female's body, at one point the male coiled his tail around the female's tail, birth coincides with the onset of the rainy season in early June, brood size (oviductal eggs or embryos) 2–3 and positively correlated with female SVL (Iverson, 1986). **Seasonal Activity:** Apparent increase in activity in December–March may be associated with mating rather than seasonal rainfall patterns (Iverson, 1986). **Size:** Male SVL to 313 mm SVL (Hedges, 2002), female SVL to 332 mm (Iverson, 1986).

Conservation Status: Unknown.

Tropidophis haitianus Cope, 1879

Distribution: Hispaniola, where the combined ranges of three currently recognized subspecies are essentially islandwide, including Île de la Gonâve and Île de la Tortue.

Habitat: Mesophilic, wooded situations, including *Acacia*, creekbeds, mesic forests, cactus-*Acacia* scrub, grassy pastures, on a paved road at night, under boards, logs, bark, rocks, debris, and *Cocos* trash, in a termitarium, rock jumbles, and in water along the edge of a cement cattle tank (Schwartz and Henderson, 1991); often especially common along watercourses (RP).

Activity: Nocturnal (Schwartz and Henderson, 1991). **Defensive Behavior:** Coils into a tight ball with its head hidden, autohemorrhages from mouth and eyes in juveniles and adult males (D. D. Smith et al., 1991, 1993). **Foraging and Diet:** Active forager that is largely ground dwelling, but will ascend above ground level on occasion, diet includes *Eleutherodactylus* (25% by frequency of occurrence in a sample of 12 prey items), *Anolis* (58.3%), *Ameiva* (8.3%), and rodents (8.3%), also *Sphaerodactylus* and a juvenile *Anolis barahonae* (Schwartz and Henderson, 1991); with one *Osteopilus pulchrilineatus* in its mouth while constricting another (K. P. Schmidt, 1921). **Locomotion:** Sidewinding in juveniles (163 mm and 241 mm SVL) and an adult (520 mm SVL; D. D. Smith et al., 1991). **Longevity:** In captivity, 15 years, 4 months (Snider and Bowler, 1992). **Predator:** Ridgway's Hawk (*Buteo ridgwayi*; Wiley and Wiley, 1981). **Reproduction:** Gravid female 19 July, litter size 8 (Schwartz and Henderson, 1991); gravid female 7 June (D. D. Smith et al., 1991); litter size 8 (Hedges, in prep.). **Size:** Male SVL to 534 mm, female SVL to 552 mm (Hedges, 2002).

Conservation Status: Not assessed, can be locally abundant, but information regarding the impact of severe habitat degradation throughout much of the species' range is lacking.

Tropidophis hardyi Schwartz and Garrido, 1975

Distribution: Southcentral Cuba.

Habitat: Mesophilic (Schwartz and Henderson, 1991).

Activity: Presumably nocturnal (Schwartz and Henderson, 1991). **Reproduction:** Presumably viviparous (RWH). **Size:** Male SVL to 303 mm, female SVL to 334 mm (Hedges, 2002).

Conservation Status: Unknown.

Tropidophis hendersoni Hedges and Garrido, 2002

Distribution: Northern coast of eastern Cuba.

Habitat: Unknown.

Reproduction: Female in August had enlarged follicles (Hedges and Garrido, 2002). **Size:** Female SVL to 280 mm (based on one individual; Hedges and Garrido, 2002).

Conservation Status: Unknown.

Tropidophis jamaicensis Stull, 1928

Distribution: Southern Jamaica

Habitat: Mesophilic (Schwartz and Henderson, 1991); resting on limestone in the early morning and under debris (B. S. Wilson, in litt., 28.IV.2006; Wilson and Vogel, 2000); coiled under a flat rock (Barbour, 1910).

Activity: Presumably nocturnal (Schwartz and Henderson, 1991). **Defensive Behavior:** Includes autohemorrhaging (B. S. Wilson, in litt., 28.IV.2006). **Diet:** Includes *Anolis lineatopus* (Wilson and Vogel, 2000; B. S. Wilson, in litt., 28.IV.2006). **Reproduction:** Viviparous (Schwartz and Henderson, 1991). **Size:** Male SVL to 338 mm, female SVL to 306 mm (Hedges, 2002).

Conservation Status: Unknown.

Tropidophis maculatus Bibron, 1820

Distribution: Western Cuba, Isla de la Juventud.

Habitat: Mesophilic, ground dwelling, in scrub savanna (Schwartz and Henderson, 1991); beneath debris at the edge of weedy tidal flats near La Habana (Neill, 1958); under rocks and other debris (Collette, 1961).

Activity: Nocturnal (Schwartz and Henderson, 1991). **Diet:** Includes *Anolis alutaceus* and *A. angusticeps* (Collette, 1961). **Size:** Male SVL to 375 mm, female SVL to 372 mm (Domínguez Díaz and Moreno García, 2006b).

Conservation Status: Unknown.

Tropidophis melanurus Schlegel, 1837

Distribution: Cuba, where three currently recognized subspecies occur on the main island (*T. m. melanurus*, *T. m. dysodes*) and Isla de la Juventud (*T. m. ericksoni*).

Habitat: Mesophilic, in rainforest, woodlands, oases in xeric habitats, heavily disturbed situations (gardens), in *Musa*, on a patch of cliff face, in bases of palm (*Corypha umbraculifera*) petioles and axils, under a tire in a garden, crawling on a lawn, in a flooded rainforest ditch, at a drinking trough (Schwartz and Henderson, 1991); under rocks and other objects, in grasses or thick foliage (Gundlach, 1880); in a fallen bromeliad, in a tree at 9.1 m in a bromeliad, under a brookside rock, 2 on the ground in early morning, at night climbing among vines on a tree trunk at 1.0 m (Stull, 1928); by day among rocks (Garrido and Schwartz 1968b); under decaying vegetation, in the rolled up "boots" of fallen palm fronds, especially where bananas or mangoes occur in Royal Palm (*Roystonea*) groves with well-shaded, damp situations on the ground (Barbour and Ramsden, 1919); under palm logs in moist palm forest near the coast (Schwartz and Thomas, 1960); on highway on the Península de Zapata (Garrido, 1980b); an adult male in the crook of a tree 1.5 m high (Edwards, 1997); takes refuge in holes in tree trunks and crevices (Domínguez Díaz and Moreno García, 2003); often associated with humans (Kunz, 2005); on the Zapata Peninsula, takes refuge under palm fronds, logs, and stones by day, but forages on the ground at night, even close to houses (Díaz and Abreu Guerra, 2005).

Activity: Nocturnal, at night on rock outcroppings (Schwartz and Henderson, 1991); crossing paved roads at 2000 h and 2300 h (Lando and Williams, 1969); when lights are on, snakes are inactive and in the dark color phase, when lights are off, snakes are active and in the light color phase (Hedges et al., 1989). **Defensive Behavior:** Includes discharging an "extremely offensive" pungent "whitish slime" from the vent (Barbour and Ramsden, 1919); autohemorrhaging (Mertens, 1946); adults and a neonate coiled into balls with heads hidden (Tolson, 1982). **Foraging and Diet:** Active forager, diet includes frogs (13 in a sample of 26 prey items), lizards (9), birds (2), and rodents (2; Greene, 1983); constricting *Osteopilus septentrionalis* in 10 cm of water, lizards (*Anolis* spp.; Schwartz and Henderson, 1991); eating *Osteopilus* while in a small pond (Estrada, 1994). **Longevity:** In captivity, 2 years, 5 months (Biegler, 1966). **Parasites:** Nematodes *Hastospiculum cubaense* (Baruš and Sonin, 1971; Baruš and Coy Otero, 1978; Coy Otero and Baruš, 1979b; Coy Otero and Lorenzo Hernández, 1982); *Parapharyngodon cubensis* (Coy Otero and Baruš, 1979b; Coy Otero and Lorenzo Hernández, 1982); and larval Stomachidae gen. sp. (Baruš and Coy Otero, 1978; Coy Otero and Baruš, 1979b; Coy Otero and Lorenzo Hernández, 1982); trematode *Infidum simile* (Groschaft and Coy Otero, 1978; Coy Otero and Lorenzo Hernández, 1982). **Predator:** *Epicrates angulifer* (Viña Dávila and Armas, 1989). **Reproduction:** In captivity, minimum gestation 155 days, births on 2 and 26 February, newborn 176–196 mm total length and 2.0–3.3 g (Tolson, 1982); 950 mm SVL female with 36 ova with diameters ~7 mm (Grant, 1957). **Size:** Male SVL to 770 mm, female SVL to 957 mm (Hedges, 2002).

Conservation Status: Not assessed.

Tropidophis morenoi Hedges, Garrido, and Diaz, 2001

Distribution: North central Cuba.

Habitat and Natural History: Unknown.

Size: Female SVL to 359 mm (Domínguez Díaz et al., 2006b).

Conservation Status: Unknown.

Tropidophis nigriventris Bailey, 1937

Distribution: East central Cuba.

Habitat: Mesophilic, under rocks in woodland (Schwartz and Henderson, 1991).

Size: Male SVL to 184 mm, female SVL to 227 mm (Hedges, 2002).

Conservation Status: Considered "vulnerable" (Rodríguez Schettino and Chamizo Lara, 1998a).

Tropidophis pardalis Gundlach, 1840

Distribution: Cuba, widely distributed, but disjunct on the main island (apparently less abundant in the east), Archipiélago de Sabana-Camagüey, Isla de la Juventud.

Habitat: Mesophilic, parks, river banks, rainforest, under rocks in a cave mouth, in a rock pile in recently cut woods, on rocks in woodland, on the ground among dead palm fronds, under rocks (Schwartz and Henderson, 1991); wooded areas, shady wet forest (Petzold, 1969b); under rocks in a disturbed area near a building (Garrido and Schwartz, 1968b); under a roadside rock (Edwards, 1997).

Activity: Nocturnal (Schwartz and Henderson, 1991). **Defensive Behavior:** Autohemorrhaging (Domínguez Díaz and Moreno García, 2003); also coiling into a ball (Petzold, 1969b). **Foraging and Diet:** Ground dwelling, likely an active forager, stalking *Anolis* in grass at night (Schwartz and Henderson, 1991). **Parasites:** Cestode *Ophiotaenia habanensis* (Freze and Rysavy, 1976; Coy Otero and Lorenzo Hernández, 1982); nematodes *Oswaldocruzia lenteixeirai* and larval Stomachidae gen. sp. (Baruš and Coy Otero, 1978; Coy Otero and Baruš, 1979b; Coy Otero and Lorenzo Hernández, 1982). **Predator:** Great Lizard Cuckoo (*Saurothera merlini*; Garrido, 1976a). **Size:** Male SVL to 284 mm (Domínguez Díaz and Moreno García, 2006c), female SVL to 287 mm (Hedges, 2002).

Conservation Status: Not assessed.

Tropidophis parkeri Grant, 1940

Distribution: Cayman Islands, Little Cayman.

Habitat: Xerophilic to mesophilic, usually ground dwelling (Schwartz and Henderson, 1991); in "beach situations," often beneath fragments of limestone, within eroded chambers of rocks, among the leaves of dead *Agave*, 0.9 m above the ground in the leaf base of an epiphyte, 2.4 m above the ground in an outhouse roof, beneath rocks and Sea Grape (*Coccoloba uvifera*) litter, resting in a tin can (Thomas, 1963); in rotten logs (Grant, 1940c); under fallen *Cocos* fronds and husks at an abandoned home site on the beach (Seidel and Franz, 1994).

Activity: Nocturnal, crawling about at 2200 h (Grant, 1940c). **Defensive Behavior:** Will autohemorrhage and also coils into a tight ball with the head completely hidden (Thomas, 1963). **Diet:** Includes *Anolis sagrei* (Grant, 1940c); *Osteopilus septentrionalis* (Thomas, 1963). **Longevity:** In captivity, 25 years, 11 months and still alive at last report in 1993 (Curator of Reptiles, Dallas Zoo in Slavens and Slavens, 2003). **Size:** Male SVL to 422 mm, female SVL to 512 mm (Hedges, 2002).

Conservation Status: Unknown.

Tropidophis pilsbryi Bailey, 1937

Distribution: Eastern Cuba.

Habitat: Mesophilic, ground dwelling (Schwartz and Henderson, 1991); under a stone in secondary vegetation surrounded by pine forests and *Hibiscus* plantation, under a fallen trunk (5.0 m × 60 cm) in an open grassy area adjacent to secondary vegetation (A. Fong Grillo, in litt., 05.V.2006).

Activity: Nocturnal (Schwartz and Henderson, 1991). **Reproduction:** Viviparous (RWH). **Size:** Male SVL to 295 mm (Hedges, 2002), female SVL to 371 mm (Domínguez Díaz and Moreno García, 2005).

Conservation Status: Listed but not formally assessed (Rodríguez Schettino and Chamizo Lara, 1998a).

Tropidophis schwartzi Thomas, 1963

Distribution: Cayman Islands, Cayman Brac.

Habitat: Xerophilic to mesophilic (Schwartz and Henderson, 1991).

Activity: Presumably nocturnal (Schwartz and Henderson, 1991). **Reproduction:** Viviparous (RWH). **Size:** Male SVL to 385 mm, female SVL to 321 mm (Hedges, 2002).

Conservation Status: Unknown.

Tropidophis semicinctus Gundlach and Peters, 1865

Distribution: Western and central Cuba.

Habitat: Mesophilic (Schwartz and Henderson, 1991); arboreal (Burger, 2004); under a box (Stull, 1928); 2 under one stone with an *Arrhyton* (Darlington, 1927); beneath a pile of bricks by day (Schwartz and Ogren, 1956); neonates under rocks between a road and regenerated scrub (Edwards, 1997).

Activity: Nocturnal (Barbour and Ramsden, 1919). **Behavior:** Rolls into a ball with the head tucked inside and remains motionless (Barbour and Ramsden, 1919); "a very offensive anal secretion is produced" followed by autohemorrhaging from mouth and eyes (Darlington, 1927); neonates exhibit "balling" behavior (Edwards, 1997). **Parasites:** Nematode *Parapharyngodon cubensis* (Baruš and

Coy Otero, 1978; Coy Otero and Baruš, 1979b; Coy Otero and Lorenzo Hernández, 1982). **Reproduction:** Neonate 160 mm, 1.5 g (Petzold, 1969b). **Size:** Male SVL to 383 mm, female SVL to 408 mm (Hedges, 2002).

Conservation Status: Unknown.

Tropidophis spiritus Hedges and Garrido, 1999

Distribution: Central Cuba.

Habitat: Unknown.

Defensive Behavior: Includes autohemorrhaging (Domínguez Díaz and Moreno García, 2003). **Size:** Male SVL to 320 mm, female SVL to 372 mm (Hedges, 2002).

Conservation Status: Unknown.

Tropidophis stejnegeri Grant, 1940

Distribution: Northern Jamaica.

Habitat: Mesophilic (Schwartz and Henderson, 1991); under rotting *Cocos* husks, in bromeliads on the ground or low in trees (Schwartz, 1975).

Behavior: When handled, coils itself into a ball with the head inside (Grant, 1940a); oral autohemorrhaging when handled (Sánchez Muñoz, 2008). **Color Change:** Changes from a yellow dorsal ground color with a pattern at night to patternless black by day (Hedges et al., 1989). **Size:** Male SVL to 395 mm, female SVL to 529 mm (Hedges, 2002).

Conservation Status: Unknown.

Tropidophis stullae Grant, 1940

Distribution: Jamaica, Portland Point.

Habitat: Mesophilic (Schwartz and Henderson, 1991); under vegetable trash and rocks (Grant, 1940a).

Behavior: When handled, rolls into a ball with the head hidden in the coils (Grant, 1940a). **Size:** Male SVL to 260 mm, female SVL to 248 mm (Hedges, 2002).

Conservation Status: Unknown.

Tropidophis wrighti Stull, 1938

Distribution: Eastern and central Cuba.

Habitat: Mesophilic, in a shack in a pasture, 0.6 m above the ground, on an outhouse wall at night, 2.4 m above ground in *Coffea* (Schwartz and Henderson, 1991); under rocks (Alayo Dalmau, 1951); under a rock along a bank (Estrada, 1994); arboreal (Burger, 2004); under a rock in an ecotone between pine forest and xeric subthorny thicket (Fernández Velzáquez and Alonso Bosch, 2005); under stones, under pine bark (at 15 cm and 60 cm above the ground), under the bark of a dead pine trunk in sclerophyll rainforest and scrub vegetation (Charrascal; A. Fong Grillo, in litt., 05.V.2006); foraging at night on ferns (S. B. Hedges, in litt., 09.V.2008; figure 30).

Activity: Presumably nocturnal (Schwartz and Henderson, 1991); under cover at 0800–1800 h (A. Fong Grillo, in litt., 05.V.2006). **Parasites:** Nematodes *Parapharyngodon cubensis* (Coy Otero and Lorenzo Hernández, 1982). **Reproduction:** Viviparous, in October, 336 mm SVL female with a 154 mm (total length) neonate (Fernández Velzáquez and Alonso Bosch, 2005); on 27 August, 385 mm female gave birth to 3 young (2 live, 1 dead + 2 undeveloped ova) on 8 September, gestation at least 138 days, neonates 124–136 mm SVL (Fong Grillo, 2008). **Size:** Male SVL to 488 mm (Domínguez Díaz et al., 2005), female SVL to 336 mm (Fernández Velzáquez and Alonso Bosch, 2005); to 400 mm SVL (Domínguez Díaz and Moreno García, 2003).

Conservation Status: Considered at "lower risk" of extinction (Rodríguez Schettino and Chamizo Lara, 1998a).

Tropidophis xanthogaster Domínguez Díaz, Moreno, and Hedges, 2006

Distribution: Cuba, Guanahacabibes Peninsula.

Habitat: Inside a cave, under a rock close to a road in an area of dogtooth limestone (Domínguez Díaz et al., 2006a).

Size: Male SVL to 239 mm, single female 227 mm SVL (Domínguez Díaz et al., 2006a).

Conservation Status: Unknown.

28 Family Typhlopidae

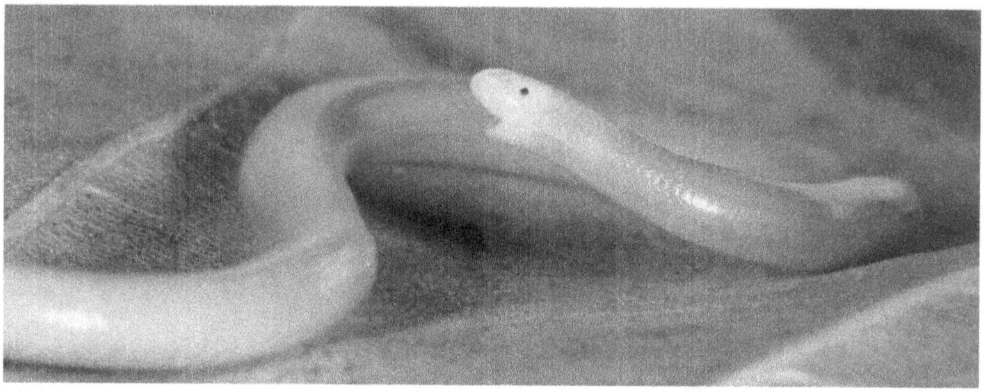

Figure 31. *Typhlops granti* (Typhlopidae) is gape-limited to feeding on very small prey such as termites (adults, larvae, and pupae) and other tiny arthropods. Photograph by Alejandro J. Sánchez Muñoz.

Ramphotyphlops braminus Daudin, 1803

Distribution: Native to the Indian subcontinent, but introduced widely (Lever, 2003), presumably with ornamental plants; in the West Indies, probably from Florida, reported from Grand Cayman, Anguilla, St.-Martin/St. Maarten, St.-Barthélemy, and Barbados.

Habitat: Burrower, probably restricted to "artificial" habitats, such as moist gardens around hotels and homes (Breuil, 2002b; Hodge et al., 2003; Powell et al., 2005); on Grand Cayman, under a plank half-buried in leaf litter under a Tamarind Tree next to a house, in a garden, and in the Botanic Park under a potted plant (Echternacht and Burton, 2003); on Anguilla, unearthed while digging holes for tree planting on hotel grounds (Censky and Hodge, 1997).

Diet: Includes termites, ants, their eggs and larvae (Powell et al., 2005). **Reproduction:** All-female parthenogenetic species (Breuil, 2002b; Hodge et al., 2003; Powell et al., 2005). **Size:** Total length to ~175 mm (Powell et al., 2005).

Conservation Status: The primary concern regarding this introduced species is any potential impact on native species; to date effects appear to be negligible.

Typhlops agoralionis Thomas and Hedges, 2007

Distribution: Hispaniola, Massif de la Hotte (Haiti); elevation ~435 m.

Habitat: Mesophilic (R. Thomas, in litt., 22.IV.2008).

Size: Total length to 193 mm (Thomas and Hedges, 2007).

Conservation Status: Unknown.

Typhlops anchaurus Thomas and Hedges, 2007

Distribution: Extreme eastern tip of Cuba.

Habitat: Semixeric situations (R. Thomas, in litt., 22.IV.2008).

Size: Total length of only known specimen 240 mm (Thomas and Hedges, 2007).

Conservation Status: Unknown.

Typhlops annae Breuil, 1999

Distribution: St.-Barthélemy.

Habitat: In vegetation at the edge of a road (Breuil, 1999).

Size: Total length of only known specimen (presumably a young animal) 110 mm (Breuil, 1999).

Conservation Status: Unknown.

Typhlops anousius Thomas and Hedges, 2007

Distribution: Eastern Cuba.

Habitat: In "an extremely xeric ravine along the coast" (Thomas and Hedges, 2007).

Size: Total length (based on 2 specimens) to 197 mm (Thomas and Hedges, 2007).

Conservation Status: Unknown.

Typhlops arator Thomas and Hedges, 2007

Distribution: Western Cuba.

Habitat: Under a large rock in a semi-open situation near the shore (R. Thomas, in litt., 22.IV.2008).

Size: Total length (based on a single specimen) to 460 mm (Thomas and Hedges, 2007).

Conservation Status: Unknown.

Typhlops biminiensis Richmond, 1955

Distribution: Bahama Islands (Great Bahama Bank, Cay Sal Bank) and Cuba, where widely distributed, but disjunct.

Habitat: Xerophilic (Schwartz and Henderson, 1991); on Cuba, coastal situations (Thomas, 1968b); crawling on a paved road through an area of mixed cactus and small shade trees and loamy soil at 0445 h, coiled in sand beneath sections of a fallen wall (Lando and Williams, 1969); under cardboard in the middle of an ant nest (Schwartz and Henderson, 1991); in an active termite nest (Tolson and Campbell, 1989); in the Bahamas, on Elbow Cay, in and around *Tournefourtia* shrubs growing in pockets of sandy soil in limestone substrate, with woody stems often occupied by the termite *Neotermes jouteli* (perhaps principal food on Elbow Cay), also under rocks close to the base of a termite-inhabited stem (Thomas, 1968b); on Great Inagua Island, beneath stacked railway ties in an area between buttonwood and scrub, also beneath a large rock embedded in dark soil at the edge of woody scrub (Thomas, 1968b); on Andros Island, in hammock woods, on South Bimini, type in a termite nest at the base of a thatch palm stump, also crawling on the surface after rain (Richmond, 1955); also on South Bimini, at the shaded base of a Gumbo-Limbo Tree (*Bursera*) in a log pile in sandy humus (ant colony nearby) at 1600 h, in dry leaves and sandy humus at the shaded border of coppice at 1000 h (Schwartz and Henderson, 1991); in a abandoned mesic nursery under potted plants, in termite-inhabited stumps and in termitaria (Schwartz and Henderson, 1991).

Activity: Active at night (2200 h) and early morning (0445 h; Lando and Williams, 1969); active during early dark hours in captivity (Thomas and Thomas, 1978). **Diet:** Includes termites (*Neotermes jouteli*) on Elbow Cay (Thomas, 1968b). **Size:** Total length to 363 mm (Schwartz and Henderson, 1991).

Conservation Status: Unknown.

Typhlops capitulatus Richmond, 1964

Distribution: Hispaniola, southern Haiti.

Habitat: Xerophilic to mesophilic, in partial clearings to wooded areas, in a shaded ravine, on a rocky hillside, under limestone rocks, a large rotted log, and moist fallen thatch in a cemetery (Schwartz and Henderson, 1991).

Size: Total length 267 mm (Schwartz and Henderson, 1991).

Conservation Status: Unknown.

Additional Pertinent Literature: Thomas and Powell (1995a).

Typhlops catapontus Thomas, 1966

Distribution: British Virgin Islands, Anegada and Necker Island.

Habitat: Under rocks in sandy substrate with thick humus, sheet metal, and other cover objects (Anegada), and under a rock on sand in *Coccoloba* woods (Necker; J. Lazell, in litt., 08.V.2006); on Anegada, beneath large, flat limestone slabs in heavy leaf litter in shade, inside a natural limestone cavity occupied by an adult *Cyclura pinguis* (Carey, 1972); climbing into arboreal termite nests (Lazell, 2006).

Diet: Termites (Lazell, 2006). **Size:** Total length to 265 mm (Thomas, 1965f).

Conservation Status: Unknown.

Typhlops caymanensis Sackett, 1940

Distribution: Cayman Islands, Grand Cayman.

Habitat: Xerophilic (Schwartz and Henderson, 1991); sandy situations, dense *Coccoloba* and *Terminalia* woods, in a dead fallen termite-inhabited branch of a leguminous tree, in a termite-inhabited stump in sandy beach-associated scrub, and in a termite nest (Thomas, 1968b); beach areas with *Cocos* (Seidel and Franz, 1994).

Size: Total length to 260 mm (Schwartz and Henderson, 1991).

Conservation Status: Unknown.

Typhlops contorhinus Thomas and Hedges, 2007

Distribution: Extreme eastern Cuba, near the Río Jauco.

Habitat: Only known specimen collected along a river in xeric habitat (Thomas and Hedges, 2007).

Size: Total length of only known specimen 316 mm (Thomas and Hedges, 2007).

Conservation Status: Unknown.

Typhlops dominicanus Stejneger, 1904

Distribution: Dominica.

Habitat: Mesophilic, in a pile of wet, decaying *Theobroma* husks in a well-shaded grove, under rocks in a lime grove and an adjacent pasture on dry soil, under a log in a heavily disturbed area (Schwartz and Henderson, 1991); in dry coastal areas (Johnson, 1988); under rotting logs (Malhotra and Thorpe, 1999).

Activity: Active on surface in early morning after heavy rain the previous night (Malhotra and Thorpe, 1999). **Predator:** *Leptodactylus fallax* (Brooks, 1982). **Size:** Total length to 385 mm (Schwartz and Henderson, 1991).

Conservation Status: Unknown.

Typhlops epactius Thomas, 1968

Distribution: Cayman Islands, Cayman Brac.

Habitat: Mesophilic (Schwartz and Henderson, 1991); scrub (dark soil) immediately behind a beach, a beach dune area with *Coccoloba* and *Cocos*, under a palm stump and rocks (Thomas, 1968b).

Size: 247 mm total length (Thomas, 1968b).

Conservation Status: Unknown.

Typhlops eperopeus Thomas and Hedges, 2007

Distribution: Hispaniola, Valle de Neiba to relatively high elevations in the eastern Sierra de Baoruco (Dominican Republic); elevational range, 7 m below sea level to 697 m.

Habitat: Xeric lowlands to upland mesic situations, adjacent to a mesic spring surrounded by xeric scrub (R. Thomas, in litt., 22.IV.2008).

Size: Total length to 281 mm (Thomas and Hedges, 2007).

Conservation Status: Unknown.

Typhlops geotomus Thomas, 1966

Distribution: St. Christopher, Nevis, Antigua, Barbuda.

Habitat: Xerophilic to mesophilic, on St. Christopher, under a large log in wet soil and leaf mold in palm rainforest (Schwartz and Henderson, 1991); fossorial, in soil or under rocks, logs, or decaying vegetation (Malhotra and Thorpe, 1999); on Nevis, in maritime *Coccoloba*, on Antigua, in an artificially mesic area by human habitation, under a compost pile, and beneath piles of palm trash, in a mesic ravine in low hills on the southern part of the island, on Great Bird Island (off Antigua), under a rock on a beach about 6.0 m from water's edge, on Barbuda, under a rock in a stone wall in an exposed situation (Thomas, 1966b).

Diet: Insectivorous (Malhotra and Thorpe, 1999). **Size:** Total length to 213 mm (Thomas, 1966b).

Conservation Status: Not assessed.

Typhlops golyathi Domínguez and Moreno, 2009

Distribution: Western Cuba.

Habitat: Under a stone in semideciduous and evergreen forests and mogotes (Domínguez Díaz and Moreno García, 2009).

Natural History: Unknown. **Size**: Total length of male 371 mm (Domínguez Díaz and Moreno García, 2009).

Conservation Status: Unknown.

Typhlops gonavensis Richmond, 1964

Distribution: Hispaniola, Île de la Gonâve (Haiti).

Habitat: Xerophilic to mesophilic (Schwartz and Henderson, 1991).

Size: Total length to 220 mm (Schwartz and Henderson, 1991).

Conservation Status: Unknown.

Additional Pertinent Literature: Thomas and Powell (1995b).

Typhlops granti Ruthven and Gaige, 1935

Distribution: Puerto Rico, Isla Caja de Muertos.

Habitat: Xerophilic, under rocks on rocky and tree-covered hillsides, under a large rock embedded in the earth near a large cap of dogtooth limestone, in sandy soil with some humus, 2 under the same rock (Schwartz and Henderson, 1991).

Size: Total length to 210 mm (Schwartz and Henderson, 1991).

Conservation Status: "Rare everywhere, protect habitat" (Moreno, 1991d).

Typhlops guadeloupensis Richmond, 1966

Distribution: Guadeloupe.

Habitat: In leaf litter (Currat, 1980); under a rock in a hillside pasture, in *Cocos* trash (Schwartz and Henderson, 1991).

Diet: Includes ants and termites (Breuil, 2002b). **Size:** Total length to 280–299 mm (Thomas, 1976).

Conservation Status: Unknown, but Johnson (1988) recommended surveys to establish distribution and abundance.

Typhlops hectus Thomas, 1974

Distribution: Southwestern Hispaniola and Île Grand Cayemite (Haiti); elevational range, sea level to ~800 m.

Habitat: Xerophilic to mesophilic, at the edge of a canefield, in wooded areas, shaded woods and ravines, semixeric scrub woods, open fields, and yards, under rotted stumps and logs, rocks, and limestone (Schwartz and Henderson, 1991).

Population Structure: Some populations appear to be entirely female or nearly so (Thomas, 1974; Thomas and Hedges, 2007). **Size:** Total length to 237 mm (Schwartz and Henderson, 1991).

Conservation Status: Unknown.

Additional Pertinent Literature: Thomas and Powell (1992).

Typhlops hypomethes Hedges and Thomas, 1991

Distribution: Puerto Rico, including Palominos, Piñeros, and Vieques.

Habitat: In coastal localities and associated habitats (Hedges and Thomas, 1991a).

Predation: *Anolis cristatellus* (Colón Archilla, 2009). **Size:** Total length to 270 mm (Hedges and Thomas, 1991a).

Conservation Status: "Stable" (Moreno, 1991d).

Typhlops jamaicensis Shaw, 1802

Distribution: Jamaica, widely distributed, but disjunct; elevational range, sea level to ~600 m.

Habitat: Xerophilic to mesophilic, in pastures, clearings surrounded by woods, *Cocos* plantations, along a road cut through deciduous forest, in and under *Cocos* husks, rocks (as many as 3 snakes under a single, large rock), logs, rotten stumps, in sand by *Coccoloba* roots, in a log crevice, in termitaria (to 7 snakes in a single nest), in rotted wood with *Hypsirhynchus* (Schwartz and Henderson, 1991); crawling into an old stone wall (Barbour, 1910); in and under rock piles (B. I. Crother, in litt., 26.II.2008).

Reproduction: Clutch size 2–4 (Schwartz and Henderson, 1991); eggs in termite nests (Gosse, 1851). **Size:** Total length to 445 mm (Schwartz and Henderson, 1991).

Conservation Status: Unknown.

Typhlops lumbricalis Linnaeus, 1758

Distribution: Cuba, including Isla de la Juventud, and Bahama Islands, Little Bahama Bank, Great Bahama Bank.

Habitat: Mesophilic (Schwartz and Henderson, 1991); in the Bahama Islands, on South Bimini, in shady coppice with soil of dark humus overlying a clayey, limestone-derived basal soil (Thomas, 1968b); on Abaco, under a large rock in a sweet potato field (Barbour, 1904); on Eleuthera, in xeric scrub in a sink at the base of a tamarind tree (loose red soil), along a pasture edge with scattered high trees, also in an area of low woods, on New Providence, in a wooded area (soil red, loamy, friable), in a rotten log (3.0 × 0.3 m) on the leafy floor of a mesic coppice, 1.0 m above the ground in a rotting petiole base still attached to a fan palm, on Great Exuma, in an open, cleared area under rocks in red soil (Schwartz and Henderson, 1991); on Cuba, in mesic, heavily cultivated areas (Thomas, 1968b); in a sugarcane field, under a large rock in most soil at the edge of a path, in a trash pile, under dead palm fronds on wet ground, under a log at the edge of a cornfield, under rocks in a pasture, in woodland, at a cliff base, on a muddy road at night (Schwartz and Henderson, 1991).

Activity: Presumably nocturnal (Schwartz and Henderson, 1991). **Behavior:** When captured, uses the terminal spike in an effort to escape (Z. Vogel, 1965b). **Diet:** Termites and ants (Z. Vogel, 1965b; RWH). **Parasites:** Nematode *Typhlopsia kratochvili* (Baruš and Coy Otero, 1978; Coy Otero and Baruš, 1979b; Coy Otero and Lorenzo Hernández, 1982).

Reproduction: Clutch size 2–4 (Hedges, in prep.). **Size:** Total length to 257 mm (Schwartz and Henderson, 1991).

Conservation Status: Unknown.

Typhlops monastus Thomas, 1966

Distribution: Montserrat.

Habitat: Xerophilic to mesophilic (Schwartz and Henderson, 1991); in open or not heavily wooded habitats, open pastureland, cultivated clearings, sparse woods, scrub transitional between more arid lowlands and the mesic interior, in open fern growth (Thomas, 1966b); in a dry log in a burned-over *Musa* grove (Schwartz and Henderson, 1991).

Reproduction: Eggs (19–21 × 8–9 mm) on 1 August under a large rock in moderately mesic woods, hatched 3 September (Schwartz and Henderson, 1991). **Size:** Total length to 258 mm (Thomas, 1966b).

Conservation Status: Not assessed, impact of recent volcanic eruptions on Montserrat is unknown.

Typhlops monensis Schmidt, 1926

Distribution: Isla Mona.

Habitat: On a mesic beach with introduced vegetation (Hernández-Prieto, 1987); in shady low woods, an open stand of *Casuarina* behind a beach, under rocks, in a rotten log, in dark soil or sandy loam (Schwartz and Henderson, 1991).

Predator: Snowy Egret (*Egretta thula*) at 1544 h (Hernández-Prieto, 1987). **Size:** Total length to 207 mm (Schwartz and Henderson, 1991).

Conservation Status: Considered "endangered" on the IUCN Red List (Tolson, 1996d); "stable" (Moreno, 1991d).

Typhlops naugus Thomas, 1965

Distribution: British Virgin Islands, Virgin Gorda.

Habitat: Semixeric woodlands (R. Thomas, in litt., 22.IV.2008).

Size: Total length to at least 243 mm (Thomas, 1965f).

Conservation Status: Unknown.

Typhlops notorachius Thomas and Hedges, 2007

Distribution: Southern coast of eastern Cuba.

Habitat and Natural History: Unknown.

Size: Total length to 301 mm (Thomas and Hedges, 2007).

Conservation Status: Unknown.

Typhlops paradoxus Thomas, 1968

Distribution: Bahama Islands, Great Inagua Island.

Habitat: Holotype taken in semixeric Black Olive (*Bucida buceras*) woods (R. Thomas, in litt., 22.IV.2008).

Size: Total length to 245 mm (Thomas, 1968b).

Conservation Status: Unknown.

Typhlops perimychus Thomas and Hedges, 2007

Distribution: Eastern Cuba, vicinity of Guantanamo Bay; elevational range, near sea level.

Habitat: Most common in an artificially mesic situation (a plant nursery) in xeric habitat (Thomas and Hedges, 2007).

Size: Total length to 280 mm (Thomas and Hedges, 2007).

Conservation Status: Unknown.

Typhlops platycephalus Duméril and Bibron, 1844

Distribution: Puerto Rico, throughout coastal regions and inland to higher elevations at Cayey, also on Vieques.

Habitat: In a coastal *Cocos* grove, congregation of 19 snakes (juveniles and adults, 143–247 mm total lengths) beneath a 3.7 × 0.3 m rotting *Cocos* trunk in a 10 × 15 cm deep depression some 8 cm below the surface of sandy soil, congregation possibly in response to drought conditions in Puerto Rico at the time (Thomas, 1965f); eurytopic, in forests, but probably restricted to more open areas (Thomas and Gaa Kessler, 1996).

Foraging and Diet: Diet consists primarily of ants and termites, although other invertebrates are consumed in smaller numbers, adult ants most common prey (worker ants most common), ant species in scats significantly associated with their frequency in the field, *Wasmannia auropunctata* most commonly consumed, but also *Odontomachus ruginodis* pupae and *Tetramorium bicarinatum*, fire ants (*Solenopsis geminata* and *S. wagneri*) also eaten, snakes enter ant colonies to feed, the termite *Nasustitermes costalis* most frequently consumed, with workers much more often than soldiers (Torres et al., 2000); in pasture-woodland ecotones in northern Puerto Rico, 68% ants, 17% termites (mostly *Nasutitermes*), plus oribatid mites, beetle larvae, and a snail (R. Thomas in Thomas and Gaa Kessler, 1996); in forested regions, consumes "rotten-wood insects, such as termites and minute beetle larvae" (Thomas and Gaa Kessler, 1996). **Population Density:** Mean monthly density 5.6/ha during February 2001–February 2002 (Rios-López and Aide, 2007). **Predator:** *Borikenophis portoricensis* (Henderson and Sajdak, 1996). **Size:** SVL to at least 337 mm (Thomas and Gaa Kessler, 1996).

Conservation Status: "Stable" (Moreno, 1991d).

Typhlops proancylops Thomas and Hedges, 2007

Distribution: Hispaniola, northern slopes of the Massif de la Selle (Haiti) and Sierra de Baoruco (Dominican Republic); elevational range ~300–600 m.

Habitat: Upland sites under rocks in mesic situations, dry gallery forest, and xeric woods (*Acacia* and *Bucida*; Thomas and Hedges, 2007).

Size: Total length to 243 mm (Thomas and Hedges, 2007).

Conservation Status: Unknown.

Typhlops pusillus Barbour, 1914

Distribution: Hispaniola, essentially islandwide, and satellites.

Habitat: Xerophilic to mesophilic, in cultivated areas, limestone woods and cliff, *Cocos* groves, shaded xeric situations, *Acacia*-cactus scrub, mesic *Acacia* scrub, banana, coffee, and cacao groves, open fields and pastures, floodplain, beach scrub, well-forested mesic areas, canopied rainforest, an exposed hillside in an old guava grove, a rocky pasture surrounded by rice fields, coastal palm groves, in a man-made rock pile of alluvial pebbles and stones, a rock pile in a shady pothole (Île Grand Cayemite), in loose debris of an abandoned termitarium on a sandy beach in *Coccoloba*, in deep humus and rock fragments (semishaded, late morning and early afternoon), in *Cocos* trash, under rocks, rotting logs and stumps (left over from charcoal cutting), cardboard, limestone rocks, termitarium with *Amphisbaena*, fallen tree cactus, palm slats, dry palm trash and thatch palm cuttings, and a fallen bromeliad (Schwartz and Henderson, 1991).

Abundance: Locally abundant, 955 specimens collected in 36 hours in the Dominican Republic (Schwartz and Henderson, 1991). **Activity:** Nocturnal (Thomas and Thomas, 1978). **Predator:** *Ialtris agyrtes* (Schwartz and Henderson, 1991). **Size:** Total length to 226 mm (Schwartz and Henderson, 1991).

Conservation Status: Unknown.

Additional Pertinent Literature: Thomas and Powell (1994a).

Typhlops richardii Duméril and Bibron, 1844

Distribution: Puerto Rico Bank, Virgin Islands (exclusive of Anegada and Virgin Gorda, but including St. Croix), west through Culebra (and associated cays), also Cayo Diablo and Surprise Cay.

Habitat: Xerophilic to mesophilic, mesic hillsides, wooded areas (Schwartz and Henderson, 1991); ubiquitous on Guana (British Virgin Islands), in shaded moist areas of deep soil on wooded flats and in ravines under boards and rocks, encountered when digging pit-trap holes or cleaning out filled pit traps, as deep as 35 cm below the surface (Lazell, 1991, 2005); with *Eleutherodactylus*, *Leptodactylus*, and *Amphisbaena* under a pile of debris at a rum mill on Tortola (Grant, 1932h); in and under *Cocos* trunks, burrows in loose dirt around old stumps, under rocks and boards around an abandoned salt mine (Schwartz and Henderson, 1991).

Activity: Nocturnal, on Cayo Icacos at 2030 h at 1.5 m above the ground climbing the trunk of a 30 cm dbh tree (*Metopium toxiferum*), apparently moving toward a termite (*Nasutitermes nigriceps*) nest 33 cm above it, snake waving its head from side to side and tongue-flicking as it climbed (Tolson and Campbell, 1989). **Defensive Behavior:** Tail spine used to give sharp jab (Schwartz and Henderson, 1991). **Foraging and Diet:** Active forager predominantly on termites (*Nasutitermes*; Tolson and Campbell, 1989). **Population Density:** To ~580/ha on Guana (Lazell, 1991); 300/ha in predominantly *Coccoloba* habitat on Guana (and 0.71 kg of biomass/ha) via an "inappropriate" sampling technique (Rodda et al., 2001). **Reproduction:** Eggs (elongated, rounded ends, smooth white surface) in loose soil around a tree stump, embryo 98 mm total length, smallest hatchling 114 mm total length (Schwartz and Henderson, 1991). **Shed Skin:** Appearance of "quills," not collapsing like other snake sheds (Grant, 1932h). **Size:** Total length to 342 mm (Schwartz and Henderson, 1991).

Conservation Status: "Extremely rare" on St. Croix (Seaman and Randall, 1962); "stable" (Moreno, 1991d).

Typhlops rostellatus Stejneger, 1904

Distribution: Northern Puerto Rico.

Habitat: Under fallen logs in coffee groves (K. P. Schmidt, 1928); rainforest, rocky hillsides, and open pasture (Schwartz and Henderson, 1991); abundant in closed-canopy forest (Thomas and Gaa Kessler, 1996).

Diet: Stomachs of 3 specimens contained 52% (156 total items) termites (*Parvitermes*), 28% ants, and 19% small beetle larvae (Thomas and Gaa Kessler, 1996). **Size:** Total length to 222 mm (Schwartz and Henderson, 1991); SVL to at least 202 mm (Thomas and Gaa Kessler, 1996).

Conservation Status: "Stable" (Moreno, 1991d).

Typhlops satelles Thomas and Hedges, 2007

Distribution: Cuba, inlet to the Bahía de Cienfuegos.

Habitat: Only 2 specimens in beach habitat at sea level (Thomas and Hedges, 2007).

Size: Total length to 350 mm (Thomas and Hedges, 2007).

Conservation Status: Unknown.

Typhlops schwartzi Thomas, 1989

Distribution: Eastern Hispaniola (Dominican Republic).

Habitat: In mesic woods, lowland deciduous forest, and a grassy pasture, under rocks and inside a rotten log (Schwartz and Henderson, 1991).

Size: Total length 326 mm (Schwartz and Henderson, 1991).

Conservation Status: Unknown.

Additional Pertinent Literature: Thomas and Powell (1994b).

Typhlops sulcatus Cope, 1868

Distribution: Southwestern Hispaniola, including Isla Alto Velo, Île de la Gonâve, Île Grande Cayemite), presumably also on Navassa.

Habitat: Xerophilic, in shaded, moist *Acacia* woods and *Acacia* scrub, a shaded xeric situation adjacent to *Cocos*, in organic trash in a garbage dump, under rocks and pieces of cement house foundation in an open grassy area, under roadside rocks, in a termite-ridden log, under a log on sand, under a piece of concrete adjacent to a river, in dry palm logs (3 in one log 2.4 × 0.3 m) on dry mud, under bark of a 6.1 m log (Schwartz and Henderson, 1991).

Reproduction: Clutch size 1 (Hedges, in prep.). **Size:** Total length to 319 mm (Schwartz and Henderson, 1991).

Conservation Status: Unknown.

Additional Pertinent Literature: Thomas and Powell (1994c).

Typhlops sylleptor Thomas and Hedges, 2007

Distribution: Hispaniola, northern Tiburon Peninsula (Haiti); elevational range ~375–420 m.

Habitat: In karst in low, hilly, mesic habitat (Thomas and Hedges, 2007).

Size: Total length to 214 mm (Thomas and Hedges, 2007).

Conservation Status: Unknown.

Typhlops syntherus Thomas, 1965

Distribution: Hispaniola, lowlands of the Península de Barahona (Dominican Republic).

Habitat: Xerophilic to mesophilic (Schwartz and Henderson, 1991); under limestone rocks in xeric woods with deciduous trees of moderate to large size, scattered smaller thorn trees, sometimes with an undergrowth of *Opuntia* and *Agave* (Thomas, 1965b); also in a *Musa* grove and associated trash in very wet soil, in a deep, wooded, shaded ravine in *Acacia* forest (Schwartz and Henderson, 1991); under cover in shade or diffuse sunlight, sometimes 2 under the same cover object (L. R. White et al., 1992a); under rocks and broken concrete from an old foundation along a stream <40 m from the ocean (RP).

Behavior: When captured, uses the terminal spike in an effort to escape (Powell, 2003b); some locals believe this can deliver a dangerous "sting" (RP). **Diet:** Predominantly termites, but also beetle adults and larvae, trophic niche breadth 0.1366 and mean food item size < 0.003 cm^3 (L. R. White et al., 1992a). **Size:** Total length to 209 mm (Schwartz and Henderson, 1991).

Conservation Status: Unknown.

Additional Pertinent Literature: L. R. White et al. (1992b).

Typhlops tasymicris Thomas, 1974

Distribution: Grenada, widely distributed, but apparently disjunct.

Habitat and Natural History: Unknown, but one of 2 known specimens in an area of mixed agriculture (melons, citrus, cacao, mangoes, bananas, breadfruit) and native vegetation near sea level (RWH).

Size: Total length to 196 mm (Wallach, 2000).

Conservation Status: Unknown, but Johnson (1988) recommended surveys to establish its distribution and abundance.

Additional Pertinent Literature: Yorks et al. (2003).

Typhlops tetrathyreus Thomas, 1989

Distribution: Hispaniola, Plaine de Cul de Sac and adjacent mountain slopes (Haiti); elevational range, near sea level to ~550 m.

Habitat: Limestone woods, cultivated areas with scattered tree cover, in a rotten log in an oasis, in manmade rock piles composed of alluvial pebbles and stones, rocks fairly well embedded and intermingled with slightly moist soil (Schwartz and Henderson, 1991).

Size: Total length to 273 mm (Thomas, 1989).

Conservation Status: Unknown.

Additional Pertinent Literature: Thomas and Powell (1994d).

Typhlops titanops Thomas, 1989

Distribution: Hispaniola, Massif de la Selle (Haiti) and the Sierra de Bahoruco (Dominican Republic); elevational range ~240–730 m.

Habitat: Mesic to semixeric woodland (R. Thomas, in litt., 22.IV.2008); under rocks (Schwartz and Henderson, 1991).

Size: Total length to 216 mm (Schwartz and Henderson, 1991).

Conservation Status: Unknown.

Additional Pertinent Literature: Thomas and Powell (1995c).

29 Family Viperidae

Figure 32. *Bothrops caribbaeus* (Viperidae) is one of only two species of West Indian snakes with bites that might be life threatening to humans. This juvenile was crossing a trail between an agricultural area and intact forest during pouring rain at dusk. Photograph by Rod D. Wittenberg.

Bothrops caribbaeus Garman, 1887

Distribution: St. Lucia, largely coastal; elevational range, sea level to ~200 m; the West Indian distribution has been attributed to Amerindians releasing them as "weapons" against their enemies (e.g., Rufz, 1859).

Habitat: Mesophilic (Schwartz and Henderson, 1991); disturbed areas, apparently common in rock and coconut husk piles (Lazell, 1964b; Long, 1974; Long et al., 1980; Reichling, 1991); a good climber (Long, 1974); among rocks along rivers, often in trees (to 6.0 m above the ground; Lazell, 1964b).

Activity: Nocturnal (Long, 1974; Long et al., 1980) or crepuscular (R. Wittenberg, in litt., 28.III.2008; figure 29); may lie stretched out on a road after a heavy shower (Long, 1974). **Behavior:** Defensive display involves feeble tail vibration and throat inflation (Lazell, 1964b). **Danger to Humans:** If not bitten in a large blood vessel, little danger exists for a loss of life if a mixture of lime juice, rum, and salt is imbibed, followed by intoxication and sleep (Tyler, 1849); "the only treatment of the slightest avail is to make the patient intoxicated as fast as possible," local treatments "nearly all consist of rum in which cockroaches, scorpions, centipedes, and such like have been steeped for some time" (Duncan, 1889); all but one bite (in a backyard) in agricultural areas, with victims engaged in weeding, cutting wood, harvesting crops, or tending animals, most bites from 0500–1200 h, suggesting that resting snakes had been disturbed, most during the rainy season when snakes may abandon flooded resting sites, 52.4% of bites were on feet, 21.4% on legs, and 23.8% on hands, common early symptoms were pain, numbness, and local swelling, 4 of 5 victims who exhibited copious bleeding died, causes of death included gangrene, irreversible shock, and circulatory collapse (Long et al., 1980); an adult male developed multiple cerebral infarctions following a bite (Numeric et al., 2002). **Foraging and Diet:** Appears to be essentially an opportunistic feeder and probably changes its diet ontogenetically, but primarily a rodent eater (Gosner, 1987); reports of cannibalism (Duncan, 1889); although birds and mongooses (Lazell, 1964b; Long, 1974), domestic fowl (Long et al., 1980), and even slugs (Long, 1974) have been reported in the diet, only *Mus*, *Rattus*, and *Didelphis* in stomachs (Schwartz and Henderson, 1991; RWH); reports of these snakes taking fish and crayfish are not verified, does not need to coil to strike, does not retain its hold on prey, withdrawing after a "swift stroke," following the scent of the "unfortunate animal" until it catches up with it (Long, 1974). **Longevity:** In captivity, 14 years, 5 months (K. Wright in

Slavens and Slavens, 2003). **Miscellany:** Fangs, "with a little manipulation," could be used as hypodermic needles (Duncan, 1889). **Parasites:** Ticks *Amblyomma dissimile* (R. Pinchon in Morel, 1967; Guglielmone et al., 2003). **Predation:** *Clelia errabunda* (Duncan, 1889; Underwood, 1995). **Reproduction:** Viviparous, to 60 young (Long, 1974); "as many as 25 active young" (Long et al., 1980); female 1,505 mm SVL with 37 small, coiled embryos on 16 April, females (895 mm and 1,192 mm SVL) on 23 August with nearly full-term young (Schwartz and Henderson, 1991). **Size:** Maximum SVL ~1,300 mm (Schwartz and Henderson, 1991).

Conservation Status: Not assessed.

Additional Pertinent Literature: Powell and Wittenberg (1998).

Remarks: West Indian populations once were considered conspecific with those on the mainland, despite early statements to the contrary (e.g., Rufz, 1859), some information pertaining to populations on Martinique and St. Lucia may be included in references to mainland taxa; also, because *B. caribbaeus* and *B. lanceolatus* were considered conspecific until Lazell (1964b) clarified the taxonomy, some historical references to "*B. lanceolatus*" (under one of several generic names) also might refer to *B. caribbaeus.*

Bothrops lanceolatus Lacépède, 1789

Distribution: Martinique, largely in wetter coastal areas, distribution discontinuous; an early report of the species on Petit Martinique in the Grenadines is undoubtedly in error (possibly based upon a misidentified *Corallus grenadensis*), whereas those of snakes on Dominica and Guadeloupe in Werner (1927) "are doubtful and need to be verified" (Campbell and Lamar, 2004); however, the presence of 2 specimens in the British Museum from Guadeloupe might be explained by "an unsuccessful effort made, during a plague of rats in Guadeloupe, to introduce the fer-de-lance there" (Heam, 1890); the West Indian distribution has been attributed to Amerindians releasing them as "weapons" against their enemies (e.g., Rufz, 1859).

Habitat: Mesophilic, markedly arboreal (Schwartz and Henderson, 1991); in rainforest, sugarcane fields, and banana plantations, common along rivers, in rock piles, and in and around human-made structures, including banana sheds and piles of crates (Pinchon, 1967); rocky hillsides, vine tangles at forest edge, may ascend to 20 m (Lazell, 1964b); a large adult killed in a very populated part of Fort-de-France (Breuil, 2009).

Activity: Nocturnal (Rufz, 1859; Pinchon, 1967). **Danger to Humans:** Bites and fatalities recorded as early as 1667, fatalities now uncommon, in 1859, a popular snakebite remedy (efficacious for people of "the highest station") consisted of a fistful of grass, a fistful of stinkweed, and a fistful of green tobacco, put in a mortar, press the juice out through linen, put 2 fingers in a glass with rum (according to taste; if victim prefers vin ordinaire, use at least 3 fingers of the liquid; Pinchon, 1967); alternately, little danger of loss of life if a mixture of lime juice, rum, and salt is imbibed, leading to intoxication and sleep (Tyler, 1849); someone was bitten every day, mortality about 25 per year, death may be delayed by up to 20 days and may ultimately result from secondary infections (M. Guyon in Rufz, 1859); more than 40 treatments, most herbal remedies using particular plants prepared as infusions, "rational" or medical treatment of bites includes application of a ligature followed by suction and cauterization, if the latter not possible, treat with a caustic agent, treatment less efficacious as time elapses and venom spreads (Rufz, 1859); "the yellow viper of Martinique . . . has frequently killed negroes at work in the plantations. M. Guyon saw several soldiers perish from its bite. Death may occur almost immediately, or in the course of twenty-four hours" (Tanner, 1862); death in as little as 15 minutes if bitten in the eye, most bites on extremities, natives and early European colonists used plants (*Aristolochia anguicida*, *Euphorbia pilulifora*, "Bois de Coulevres"—a vine, and *Dracontium scandens*) to treat bites, *Mikania guaco* also used in the 1700s and root of lemon tree in the 1800s, human saliva and frogs also were applied to bites before antivenom became available (De Lalung, 1934); septicemia attributable to *Aeromonas hydrophila* following a snakebite (Lhomme et al., 1993); an adult male developed an occipital infarction after a bite (Merle et al., 2005); thrombotic stroke despite antivenom treatment (L. Thomas et al., 2006); fatal thrombosis after a snakebite (Malbranques et al., 2008). **Diet:** Appears to be essentially an opportunistic feeder and probably changes its diet ontogenetically (Gosner, 1987); although domestic fowl (Rufz, 1859; De Lalung, 1934; B. Raboteau, 1968 in Gosner, 1987), terrestrial crabs (De Lalung, 1934; B. Raboteau, 1968 in Gosner, 1987), toads, anoles, opossums (De Lalung, 1934), frogs, birds (Pinchon, 1967), bats (Pinchon, 1967; B. Raboteau, 1968 in Gosner, 1987), and mongooses (Lazell, 1964b) have been reported in the diet, rodents (*Rattus*) are likely the primary prey (De Lalung, 1934; Pinchon, 1967; Gosner, 1987); probably does not consume anoles, frogs, crabs, and insects, as had been reported (Rufz, 1859); stomach contents have yielded only *Mus* and *Rattus* (Schwartz and Henderson, 1991); does not drink milk from a cow's teats (Pinchon, 1967). **Longevity:** In captivity, 13 years (Snider and Bowler, 1992). **Population Size:** Between 1 January 1960 and 31 December 1966, bounty was paid on 51,228 individuals, 8,314 in 1963 alone

(Pinchon, 1967). **Predators:** Humans, a 50-centime bounty was implemented by the administration of Martinique in the 1820s and as many as 2,396 snakes were killed in a single area over an 8-year period (M. Guyon in Rufz, 1859); a committee on control of vipers recommended mongooses, secretary birds, hedgehogs, and opossums (Rufz, 1859); pigs (*Sus scrofa*) and mongooses, carcasses consumed by ants (De Lalung, 1934); although mongooses are considered supreme snake predators, in staged bouts, a *B. lanceolatus* killed 3 mongooses, with each dying in <5 min, and a 1.5 m (total length) snake killed a mongoose in <15 sec, one mongoose caught a snake by its head and killed it (Breuil, 2009). **Reproduction:** Viviparous, mate January–February, gestation 6–7 months, birth June–July, litter sizes average 40, high of 67 (De Lalung, 1934); litter size 20–75, young appear July–October (Pinchon, 1967); litter size to 82 (P. Clery in Breuil, 2009). **Size:** Historical record of a snake almost 9 feet long and >5 inches in diameter (De Lalung, 1934); total lengths of 1.8 m (Pinchon, 1967), 2.0 m, and 2.98 m (Lazell, 1964b) have been reported; maximum SVL 1,580 mm (Schwartz and Henderson, 1991).

Conservation Status: Not assessed, but clearing of forest and bounties have noticeably reduced numbers (Pinchon, 1967); pesticides used against banana pests are ingested by rats and birds and may accumulate in snakes, causing physiological problems and eventually killing them (Breuil, 2009); Dowling (1965) had difficulty obtaining animals while on Martinique.

Remarks: West Indian populations once were considered conspecific with those on the mainland, despite early statements to the contrary (e.g., Rufz, 1859), some information pertaining to populations on Martinique and St. Lucia may be included in references to mainland taxa; also, because *B. caribbaeus* and *B. lanceolatus* were considered conspecific until Lazell (1964b) clarified the taxonomy, some historical references to "*B. lanceolatus*" (under one of several generic names) also might refer to *B. caribbaeus*.

Section 4 Testudines

30 Family Emydidae

Figure 33. The genetic integrity of endemic populations of West Indian sliders, such as this *Trachemys stejnegeri stejnegeri* (Emydidae), may be threatened by hybridization with introduced *T. scripta* from North America. Photograph by Alejandro J. Sánchez Muñoz.

Trachemys decorata Barbour and Carr, 1940

Distribution: Hispaniola, Valle de Neiba/Plaine de Cul de Sac, distal Tiburon Peninsula and western Haiti.

Habitat: Brackish and freshwater lakes and ponds of moderate depth, with soft substrates and abundant aquatic vegetation (e.g., *Eichhornia* sp. and *Ceratophyllum* sp.; Incháustegui, 1975; Ottenwalder, 1994); at Etang Saumâtre (Haiti), in shallow water in or around *Conocarpus* (Thorbjarnarson, 1988).

Growth: In captivity, hatchlings grew to 54 mm plastron length after 1 year, 70 mm after 2 years, and 93 mm after 3 years (Incháustegui, 1975). **Hybridization:** Mating with introduced *T. scripta* may dilute the gene pool on Hispaniola (Powell et al., 2000; Powell and Incháustegui, 2009). **Longevity:** In captivity, 13 years, 11 months (Curator of Reptiles, Chicago Zoological Park in Slavens and Slavens, 2003). **Predators:** In Haiti, *Crocodylus acutus* (Thorbjarnarson, 1988); in the Dominican Republic, exploited for human consumption (Schubert, 1993; Ottenwalder, 1994; Powell et al., 2000). **Reproduction:** Courtship and mating occur year-round, male strokes female's eyes and interorbital region with his vibrating foreclaws as he swims in a face-to-face position, copulation underwater, clutch size 6–18 (mean 11.7), eggs 40 × 26 mm, 1–4 clutches/year (Incháustegui, 1975); in captivity in 1993 (breeding program at the national zoo in Santo Domingo), 26 clutches and 233 hatchlings from 24 nests (with 25 additional eggs from 2 clutches still in incubation), in 1994, 5 clutches and 56 eggs (Ottenwalder, 1994). **Size:** Male carapace length to 220 mm, female carapace length to 340 mm (Seidel, 1988a).

Conservation Status: On the IUCN Red List as "vulnerable" (Tortoise & Freshwater Turtle Specialist Group, 1996b); the species was included in lists of threatened Hispaniolan species by Aquino (1983) and Powell et al. (2000), the latter noted that interbreeding with introduced *T. scripta* could compromise the genetic integrity of *T. decorata*.

Additional Pertinent Literature: Bickham (1980).

Trachemys decussata Gray, 1831

Distribution: Two currently recognized subspecies occur in eastern and central Cuba, including Archipiélago de Sabana-Camagüey (*T. d. decussata*), and western Cuba, Isla de la Juventud, and Grand Cayman and Cayman Brac, Cayman Islands (*T. d. angusta*).

Habitat: In Cuba, associated with freshwater marshes, rivers, and lakes (Gundlach, 1880); widespread in lakes and rivers (Barbour and Ramsden, 1919); common on rocks in the Río Almendares (Collette, 1961); on Grand Cayman, in freshwater (but slightly saline with 2–3% sea water) roadside ditches (mostly unshaded in the afternoon, surface temperatures routinely reaching 31 °C), with Water Lilies (*Nymphoides*), Duckweed (*Lemna*), and Muskwort (*Chara*), and sedges (*Rhyncospora* or *Stellata*) growing on the banks (Seidel, 1990); on Grand Cayman, also in brackish mangrove swamps (56–81% sea water) and small ponds (about 27% sea water; Seidel, 1990); occasionally in open sea water off Grand Cayman (F. J. Burton in Seidel and Franz, 1994).

Abundance: Apparently abundant on Cayo Coco, Cuba (Estrada, 1993b) and in the Zapata swamps (Fidenci, 2001; Díaz and Abreu Guerra, 2005). **Activity:** As ponds dry up, turtles leave and aestivate beneath leaf mold in woods, scrub areas, or tall grass (Gundlach, 1880; Barbour and Carr, 1940); on Grand Cayman, 2 sleeping in <0.3 m of water at night (Schwartz and Henderson, 1991); on Cuba, adult terrestrial activity peaks during April–June (when nesting occurs), larger adults more likely to be on land than smaller adults, juveniles less likely to be active out of water, but if so, do not venture as far from water as adults of both sexes and are active on land earlier in the night than adults (Sampedro Marín, 2001). **Behavior:** In Cuba, basks "in the torrid rays of a vertical sun" (Barbour and Ramsden, 1919); in *T. d. decussata*, aggressive behavior among basking individuals changes with time of day and air temperature, gaping and biting peaks at 0900–1100 h, while physical displacement of another turtle peaks at 1200–1400 h, males more inclined to gape, whereas females more frequently displace other individuals (Sampedro Marín, 2002); on Grand Cayman, suns at water's edge on emergent mangrove (*Avicennia germinans*) roots (Seidel and Franz, 1994). **Diving:** May hold breath ~10–15 min (Nieto Dopico, 1997). **Foraging and Diet:** Individuals of all sizes congregate to eat fruits (*Anona palustris, Spondias mombin*; Gundlach, 1880); primarily herbivorous on aquatic plants, occasional fruits, and, possibly, crayfish (Barbour and Carr, 1940); on Grand Cayman, omnivorous, feeding on aquatic plants and invertebrates (Seidel, 1990; Seidel and Franz, 1994); in Cuba, juveniles insectivorous, adult males consume vegetable material and dead fishes, and females take mollusks (Sampedro Marín and Rodríguez Schettino, 2003). **Growth:** Hatchlings 31.0–33.5 mm plastron length, 2–3 years 67.1–69.5 mm, 3–4 years 79.5–102.6 mm, female 4–5 years 110.2 mm, male 4–5 years 110.3–123.7 mm, female 5–6 years 135.0–149.0 mm, male 5–6 years 128.0–131.0 mm female 6–7 years 170.0 mm, female 7–8 years 183.0 mm, male >6 years 153.0–183.0 mm (Seidel, 1990). **Longevity:** On Grand Cayman (based on marked animals), female about 21 years old, male 27–30 years old, maximum longevity on Grand Cayman likely >30 years (Seidel, 2003). **Hybridization:** Turtles on Great Exuma Island exhibit a blend of features shared by *T. terrapen*, *T. stejnegeri*, and *T. decussata*, and may represent a hybrid swarm as a possible result of multiple introductions (Franz et al., 1993). **Movements:** In Cuba, adults make long excursions (of several kilometers) over land (Sampedro Marín and Rodríguez Schettino, 2003). **Parasites:** In Cuba, nematodes *Capillaria* sp., *Falcaustra mascula*, and larval *Contracaecum* sp. (Baruš and Moravec, 1967b; Coy Otero and Baruš, 1979b; Coy Otero and Lorenzo Hernández, 1982); *Serpinema trispinosus* and *Spiroxys contorta* (Pérez Vigueras, 1936; Baruš and Moravec, 1967b; Coy Otero and Baruš, 1979b; Coy Otero and Lorenzo Hernández, 1982); acanthocephalan *Neoechinorhynchus emydis* (Pérez Vigueras, 1936; Baruš and Moravec, 1967b; Coy Otero and Lorenzo Hernández, 1982); trematode *Neopolystoma orbiculare* (Baruš and Moravec, 1967b; Coy Otero and Lorenzo Hernández, 1982). **Population Density:** On Grand Cayman, 1,307/ha, but likely an inflated value (Seidel, 1990). **Population Structure:** On Grand Cayman, 58% of 110 turtles were juveniles (<110 mm plastron length), 28% females (112–255 mm), and 15% males (110–204 mm; Seidel, 1990). **Predators:** Humans, meat particularly popular during Lent, as the Catholic church does not consider it to be meat (Gundlach, 1880); historically popular as source of food in Cuba ("Their flesh is delicious"; Barbour and Ramsden, 1919); "One last word I must repeat and that is how extremely pleasant a pastime eating tortoises really is" (Barbour, 1934); at the Ciénaga de Zapata (Cuba), dogs and mongooses (*Herpestes javanicus*) are nest predators, as is *Epicrates angulifer*, which also might take hatchlings (Sampedro Marín and Montañez Huguez, 1989); *Crocodylus rhombifer* takes adults, and pigs (*Sus scrofa*) excavate nests and eat eggs (Nieto Dopico, 1997). **Reproduction:** In Cuba, 32.3% of females in reproductive condition in April, 57.5% in May, sexual maturity at a plastral length of 13.0 cm in males, 19.0 cm in females, mean of oviductal eggs 2.33 × 3.94 cm ($N = 26$; Sampedro Marín et al., 1983); nesting begins in March, oviductal eggs in August, sperm retention in females for >1 year, nesting from dark to midnight, may dig 3 or more "trial nests" before depositing eggs, most nests 2–3 m from water in typical marsh substrate, but some in drier areas along the edges of the swamp, range in inclination from horizontal to vertical, covered by excavated soil, mean number of eggs 8.12 ± 0.46 in April ($N = 123$), 6.00 ± 0.37 in July ($N = 139$), mean size of eggs 3.66 ± 0.05 × 2.39 ± 0.04 cm in April, 3.55 ± 0.07 × 2.33 ± 0.04 cm in July, mean diameter of nest aperture and chamber 7.98 ± 0.05 cm and 4.41 ± 0.04 cm in April ($N = 16$), 7.86 ± 0.06 cm and 4.14 ± 0.05 cm in July ($N = 20$), mean depth of nest 9.48 ± 0.14 cm in April, 9.15 ± 0.06 cm in July, mean incubation period 87 days (85–90 days; Sampedro Marín and Montañez Huguez, 1989); some eggs deposited in nests of *Crocodylus rhombifer* (A. Ramos Targarona in Sampedro Marín and Montañez Huguez, 1989); most female *T. d. decussata* with eggs during April–June, fewer during July–September (Sampedro Marín, 2001); on Grand Cayman, females with shelled eggs April–May (start of the rainy season), hatching in wet season (Seidel, 1990); on Grand Cayman, eggs in July, during courtship, males flutter their forefeet in front of the female while resting in water snout-to-snout (Grant, 1940c); after this behavior repeated several times, the female settles to the bottom while the male mounts her from behind (Seidel and Franz, 1994); eggs deposited in trenches or cavities dug with female's hindlimbs for that purpose and covered with soil, egg 45 × 25 mm, white with parchmentlike shell (Gundlach, 1880); mean plastron length of hatchlings in captivity ($N = 11$) 32.0 mm (31.0–33.5 mm), wild-caught hatchling 32.7 mm on 11 March (Seidel, 1990); on Grand Cayman, clutch size usually 6–7, eggs 38–41 mm long (Seidel and Franz, 1994); in the laboratory, 2 clutches from Grand Cayman incubated at 28–35 °C produced 11 hatchlings with plastron lengths 31.0–33.5 mm (Seidel, 1990); in the laboratory, 6 eggs from Grand Cayman incubated at 30–35 °C hatched after 72 days, average hatchling size 40 mm (Seidel and Franz, 1994); in Cuba, females may make several preliminary nests (possibly to confuse nest predators) before digging that which receives the eggs (diameter 3.6 cm and weight 12.5 g), incubation takes about 90 days (Sampedro Marín and Rodríguez Schettino, 2003). **Salinity Tolerance:** Increase in sodium influx with declining size (hatchlings demonstrate considerable net uptake of sodium in sea water compared with larger individuals), large individuals (>200 g) tolerated immersion in 100% seawater for periods of at least 10–24 days (Dunson and Seidel, 1986). **Size:** Male carapace length to 270 mm, female carapace length to 390 mm (Seidel, 1988a). **Thermal Biology:** In *T. d. decussata*, most basking behavior on partially submerged tree trunks and branches at 1100 h and 1400 h when air temperatures 26.0–28.1 °C and water temperatures 23.5–23.7 °C, basking ends when air temperature reaches ~28.5 °C (Sampedro Marín, 2002).

Conservation Status: Cuban populations considered at "lower risk" of extinction (Rodríguez Schettino and Chamizo Lara, 1998a); according to Barbour and Ramsden (1919), "if measures for their preservation are not taken it is probable that the march of time will see them exterminated"; in addition to general habitat destruction, factors such as water pollution (industrial, agricultural, residential) and damming of rivers and creeks may negatively affect populations (Fidenci, 2001); these turtles are overharvested for food and religious use or for sale as souvenirs or pets (Díaz and Abreu Guerra, 2005).

Additional Pertinent Literature: Seidel (1988b).

Trachemys scripta Schoepff, 1792

Distribution: United States and Mexico, introduced widely throughout the world (Lever, 2003); in the West Indies, introduced on Grand Cayman, Hispaniola, Puerto Rico, U.S. Virgin Islands (St. Thomas, St. John, St. Croix), British Virgin Islands (Tortola, Virgin Gorda), St. Eustatius, St.-Martin/St. Maarten, Guadeloupe (*T. s. elegans*), Marie-Galante, and Martinique.

Habitat: On Tortola, in a manmade pond that also had *Pseudemys nelsoni* (Owen et al., 2005); on St.-Martin/St. Maarten, in most freshwater habitats, including ornamental pools in gardens and resorts, may move into brackish water along the coasts (Powell et al., 2005); on Guadeloupe, habitat depends on where released, in flooded forests, occurs with *Pelusios castaneus* (Breuil, 2002b); in nearly every freshwater body on St. Thomas and St. Croix (U.S. Virgin Islands), including agricultural ponds and ornamental pools at resorts and golf courses, but not in stormwater drainages with occasional connections to the sea (Platenberg, 2007); in ornamental pools in hotels and ponds on golf courses (RP).

Activity: Diurnal on Guadeloupe (Currat, 1980; Breuil, 2002b). **Diet:** Largely carnivorous on Guadeloupe (Currat, 1980). **Hybridization:** May mate with native species of *Trachemys* (e.g., Powell et al., 2000; Powell and Incháustegui, 2009). **Reproduction:** On Guadeloupe, possibly not reproductive (Breuil, 2002b). **Size:** Carapace length in males to 165 mm, in females to 280 mm (Carr, 1952); West Indian populations probably smaller. **Thermal Biology:** On Guadeloupe, observed basking (Breuil, 2002b).

Conservation Status: As all West Indian populations of this species are introduced, the primary conservation concern is their potential impact on native species of *Trachemys* (e.g., competition for resources, hybridizing with native species; Powell et al., 2000; Powell and Incháustegui, 2009); importation into Puerto Rico has been illegal since 1970, albeit not for ecological reasons, but for fear of *Salmonella* poisoning (Joglar, 2005b).

Trachemys stejnegeri Schmidt, 1928

Distribution: Three currently recognized subspecies occur in the Bahama Islands, Great Inagua Island (*T. s. malonei*), Hispaniola (*T. s. vicina*, largely in the Dominican

Republic), and Puerto Rico, Isla Culebra, and possibly Isla Vieques (*T. s. stejnegeri*); introduced on Guadeloupe, Les Saintes (Terre-de-Haut and Terre-de-Bas), Marie-Galante, and Dominica (Fritz, 1991; Seidel, 1996; Breuil, 2002b).

Habitat: In the Bahamas, in small, often temporary ponds (Schwartz and Henderson, 1991); on Hispaniola, in rivers, ponds, and marshes throughout coastal lowlands (Seidel, 1988a); in the Dominican Republic, small lagoons (surrounded by small trees) with 3 predominant types of vegetation: *Typha* sp., *Eleocharis* sp., and *Ceratophyllum* sp., small fishes and aquatic birds common in the lagoons (Incháustegui, 1975); on Puerto Rico, in lagoons, pools, irrigation ditches, natural and artificial ponds, and rivulets in lowlands (Rivero, 1998); on Vieques, a "pet" in a cistern, according to local residents, these turtles are "numerous in the lagoons and are sold on the street for fifteen cents apiece" (Grant, 1932a); on Marie-Galante, in diverse aquatic situations (i.e., water bodies of all sizes with aquatic vegetation; Breuil, 2002b); on Les Saintes, in situations with floating vegetation (Terre-de-Bas), and in ponds usually devoid of floating vegetation (Terre-de-Haut), on Marie-Galante and Terre-de-Bas, associated with the aquatic plant *Achrostichium aureum* (Breuil, 2002b).

Activity: On Great Inagua, female at 1800 h crossing a gravel road in scrub (Schwartz and Henderson, 1991); lays eggs at night (Danforth, 1925). **Behavior:** On Marie-Galante, when ponds dry, buries itself and aestivates (Breuil, 2002b). **Competition:** On Marie-Galante, *T. scripta* is a potential threat (Breuil, 2002b). **Diet:** On Puerto Rico, omnivorous, feeds on aquatic plants and invertebrates (Seidel, 1990; Rivero, 1998); fruits of the Pond Apple (*Anona glabra*) are an important part of the diet of turtles in the Reserva Natural de Humacao (León and Joglar, 2005); on Guadeloupean Archipelago, feeds on aquatic vegetation, fruits, and carrion, at night on land, consumes mangoes, on Terre-de-Haut, eats food given to ducks (Breuil, 2002b). **Disease:** About 30 of 354 turtles examined had inflamed lacerations and/or porous plastra, the pathogen is unknown (León and Joglar, 2005). **Exposure to Saltwater:** After 4 days in saltwater, 2 of 5 turtles died after reintroduction to freshwater and all had "impaired" eyes (Grant and de Sola, 1934). **Hybridization:** New Providence and Andros Island support interbreeding "swarms" of *Trachemys stejnegeri* and *T. terrapen* (Schwartz and Henderson, 1991); likewise, turtles on Great Exuma Island exhibit a blend of features shared by *T. terrapen*, *T. stejnegeri*, and *T. decussata*, and may represent another hybrid swarm as a possible result of multiple introductions (Franz et al., 1993). **Mortality:** On Marie-Galante, killed on roads when moving on land at night (Breuil, 2002b). **Longevity:** In captivity, 8 years, 1 month (Bowler, 1977). **Population Structure:** Of 44 turtles collected in Puerto Rico, 8 were juveniles (<90 mm plastron length), 21 females (92–218 mm), and 15 males (90–150 mm; Seidel, 1990). **Predation:** By humans, at Monte Cristi (Dominican Republic) "this turtle was a staple article of food" (K. P. Schmidt, 1921); sold in a market at Cap Haïtien, Haiti (Seidel and Incháustegui, 1984); turtles collected in a national park sold along the roadside (Powell, 2003a); hunted with the aid of lights at night in Puerto Rico, sold in markets as food (K. P. Schmidt, 1928); heavily exploited for food in Haiti and the Dominican Republic (Ottenwalder, 1994); formerly consumed by humans on Marie-Galante, but the practice apparently has stopped (Breuil, 2002b); on Puerto Rico, ants (*Solenopsis* sp.) prey on eggs and hatchlings, but the mongoose (*Herpestes javanicus*) is the major predator of eggs and adults, introduced caimans (*Caiman crocodilus*) are possible predators of adults (León and Joglar, 2005). **Reproduction:** In the Bahamas, females dig saucer-shaped symmetrical nests ~6.0–7.5 cm deep and several centimeters in diameter, female may assist hatchlings exit the nest by scratching soil from the top of the nest in August (Hodson and Pearson, 1943); in the Dominican Republic, incubation 57–79 days (Incháustegui, 1975); on Puerto Rico, lays eggs in April (Danforth, 1925); on Puerto Rico, the reproductive season is April–July, copulation is perpetrated by the male grasping the female's carapace with claws of 4 feet and curving his tail down and beneath that of the female, eggs are laid in shallow excavations within 90–180 m of water, incubation 2–3 months (Rivero, 1998); at the Reserva Natural de Humacao, most nests in disturbed areas without or with little vegetation, only 13 of 114 nesting attempts completed, when disturbed, females return later and try again, excavate using both legs alternately, mean cavity depth 10.1 ± 3.2 cm, 4–15 (mean 10.1 ± 3.2) eggs/clutch, eggs 37.6 ± 2.6–25.1 ± 1.4 mm, large size may be attributable to retention of eggs in oviducts for some time before deposition, neonatal growth 0.16 mm/month for males and 0.12 mm/month for females (León and Joglar, 2005); on Marie-Galante, reproduction April–May, 6–20 eggs in nests dug by the female at pond's edge, incubation ~2 months (Breuil, 2002b); hatching in *T. s. malonei* July–August (K. Bjorndal in Groombridge, 1982). **Sex Ratio (M:F):** 1:1.89 at the Reserva Natural de Humacao, Puerto Rico (León and Joglar, 2005). **Size:** Male carapace length to 259 mm, female carapace length to 397 mm (León and Joglar, 2005). **Thermal Biology:** In Puerto Rico, basking on masses of vegetation, and one climbed into a dead bush to bask (Grant and De Sola, 1934); on Marie-Galante, basks on submerged objects, vegetation rafts, and land (Breuil, 2002b).

Conservation Status: Historically, nesting females were collected at night with the aid of lights and sold in markets for food (Danforth, 1925); on the IUCN Red List as "near threatened" with extinction (Tortoise and Freshwater Turtle Specialist Group, 1996c); however, several populations in the Dominican Republic have been extirpated due to habitat destruction, pollution, and subsistence hunting (Incháustegui, 1975); the species was included in lists of threatened Hispaniolan species by Aquino (1983) and Powell et al. (2000); the genetic integrity of Puerto Rican (León and Joglar, 2005) and Hispaniolan (Powell et al., 2000; Powell and Incháustegui, 2009) populations may be threatened by hybridization with introduced *T. scripta*; the subspecies *T. s. malonei* is listed as "endangered" by the U.S. Fish & Wildlife Service (2008); "stable" in Puerto Rico (Moreno, 1991d).

Additional Pertinent Literature: Seidel (1988c).

Trachemys terrapen Lacépède, 1788

Distribution: Jamaica and the Bahama Islands, Cat Island, Eleuthera Island, Andros Island; New Providence and Andros Island support interbreeding "swarms" of *Trachemys stejnegeri* and *T. terrapen* (Schwartz and Henderson, 1991); likewise, turtles on Great Exuma Island exhibit a blend of features shared by *T. terrapen*, *T. stejnegeri*, and *T. decussata*, and may represent another hybrid swarm as a possible result of multiple introductions (Franz et al., 1993); Bahamian populations are likely the result of human mediated introductions (Seidel, 1988a); population on Eleuthera present at least since 1954 (Schwartz and Henderson, 1991).

Habitat: On Jamaica, in permanent freshwater bodies such as rivers, ponds, streams, and swamps, on Cat Island, in fresh rain puddles and small temporary "ponds which dry up in the dry season during which time [turtles] hibernate in mud at the bottom of the puddles and pot holes where they work into the mud which extends under the rock at the shelving edges of the pools" (Barbour and Shreve, 1935); on Cat Island and Eleuthera, in small, temporary limestone sinks (Seidel, 1988a); on Cat Island, in a freshwater pond with a heavy growth of *Chara*, on Eleuthera, in a small pond (intermittently dry) with fairly dense growth of emergent and submergent aquatics (Schwartz and Henderson, 1991).

Diet: Frugivorous, on Cat Island, prefers Custard Apples (*Annona glabra*; Barbour and Shreve, 1935). **Longevity:** In captivity, 1 year, 8 months (Snider and Bowler, 1992). **Predators:** Humans, on Cat Island, locals bait hooks with Custard Apples and catch turtles during the day (Barbour and Shreve, 1935). **Reproduction:** In Jamaica, when courting in water, the male faces the female and holds his forelegs straight out in front with the backs of his feet on each side of the female's head, with the backs of long claws, he uses vibratory motions or rapid or very rapid slapping on the sides of the female's head, slapping more on the side she tries to turn, the female apparently keeps her eyes shut and "appears very much bored at his solicitude" (Grant, 1940a); female with 4 shelled eggs (24 × 39 mm) and 4 large yolks on 14 April (Grant, 1940a); in captivity, turtles from Jamaica copulate 8 February–4 September, copulation lasts 8–14 min, produce 3 clutches/year with mean 5.4 ± 0.4 eggs/clutch (N = 18 clutches), mean of third clutch (3.7 ± 0.2 eggs) smaller than the first 2 (6.3 ± 0.3 eggs), egg dimensions 43.6 ± 0.5 × 26.3 ± 0.1 mm (N = 34), hatchlings with mean carapace length 40.0 ± 0.9 mm (N = 8; Fritz, 1990). **Size:** Male carapace length to 200 mm, female carapace length to 270 mm (Seidel, 1988a).

Conservation Status: On the IUCN Red List as "vulnerable" (Tortoise and Freshwater Specialist Group, 1996d); listed as "endangered" by the U.S. Fish & Wildlife Service (2008).

Additional Pertinent Literature: Seidel (1988d).

31 Family Pelomedusidae

Figure 34. *Pelusios castaneus* (Pelomedusidae) is native to Africa; populations on Guadeloupe, where invasive Water Hyacinths (*Eichhornia crassipes*) provide favored habitat, may have originated from the 18th-century "triangular trade," when slaves were shipped from Africa to the West Indies. Photograph by Michel Breuil.

Pelusios castaneus Schweigger, 1812

Distribution: Native to Africa; in the West Indies, introduced to Guadeloupe (Basse-Terre and Grande-Terre), possibly during the era of "triangular trade" during the 18th century when slaves were shipped to the West Indies from western Africa (Lescure, 1983).

Habitat: A variety of aquatic situations (temporary or permanent), including pools as small as 20–30 m^2, with (*Eichhornia crassipes*, *Pistia stratoides*) or without aquatic vegetation, water depth variable, but usually 1–2 m, water bodies often in pastures with mango trees, but also in flooded forests (Breuil, 2002b).

Abundance: One pond, which dried up, contained at least 200 individuals (Breuil, 2002b). **Activity:** Nocturnal (Currat, 1980; Breuil, 2002b); remains immobile by day (Currat, 1980). **Foraging and Diet:** Primarily carnivorous, eats fish and amphibians, takes salad, bread, and ham in captivity (Currat, 1980); omnivorous, includes mangoes, algae, snails, will leave water to forage on land at dusk (Breuil, 2002b). **Mortality:** Often killed on roads (Breuil, 2002b). **Movements:** Will change water bodies voluntarily, but locals also move captives from one body of water to another (Breuil, 2002b). **Predation:** Mongooses (*Herpestes javanicus*) and humans (Currat, 1980). **Reproduction:** Nest in July (Breuil, 2002b). **Size:** Maximum length 22 cm (Breuil, 2002b). **Thermal Biology:** Basks on natural (clumps of vegetation) or artificial basking sites surrounded by water (Breuil, 2002b).

Conservation Status: As this is an introduced species, its conservation status is of concern only as it may impact native plants and animals.

32 Family Testudinidae

Figure 35. *Chelonoidis carbonaria* (Testudinidae) is widely distributed in the Lesser Antilles. Many populations probably are of mixed origins, descending from ancestors that arrived naturally via overwater dispersal or with Amerindian, early colonial, or recent pet-trade mediated human assistance. Photograph by John S. Parmerlee, Jr.

Chelonoidis carbonaria Spix, 1824

Distribution: Panama and Colombia to Paraguay, Brazil, and Argentina; in the West Indies, known from St. Thomas, Water Island, St. John, St. Croix, Tortola, Virgin Gorda, Guana (re-introduction?), Necker (re-introduction?), Peter Island, Anguilla, St.-Martin/St. Maarten (including Tintamarre), St.-Barthélemy, Barbuda, Antigua (including Green Island), Saba, Montserrat, Guadeloupe (Basse-Terre), Les Saintes, Iles de Petite Terre, Dominica (probably a recent introduction, a viable population unlikely; Malhotra and Thorpe, 1999), Martinique, Grenadines (Bequia, Balliceaux, Mustique, Canouan, Mayreau, Palm, Union, Carriacou), and Grenada; the status of some populations (e.g., Antigua; Malhotra and Thorpe, 1999) is uncertain; many West Indian populations probably are composed of descendants of ancestors that arrived naturally via overwater dispersal or mediated by humans, either Amerindians, early European colonists, or recently as a consequence of an active trade in live animals (e.g., Hodge et al., 2003; Powell et al., 2005).

Habitat: On Anguilla, in dry scrub woodlands (Hodge et al., 2003); on St.-Barthélemy, in disturbed woodland (RWH); in shaded ravines, thickets, and dry forest, in Croton thicket on Tintamarre (Breuil, 2002b); in leaf litter along a road on Mustique (RWH); in dense forest and scrub habitat in the U.S. Virgin Islands (Platenberg, 2007); apparently restricted to residential gardens (i.e., no feral populations occur) on Martinique (Breuil, 2009).

Abundance: The small (0.58 km^2) Carriacou satellite Isle de Large "probably has one of the highest concentrations . . . found anywhere in the region" (Daudin and de Silva, 2007); apparently still common on Antigua and Barbuda (Daltry, 2007). **Activity:** Diurnal (Currat, 1980); on St.-Barthélemy, peak activity 0600–0900 h and 1600–1700 h (Breuil, 2002b); active midmorning on St.-Barthélemy and on Mustique (RWH). **Behavior:** When threatened, retracts completely into its shell (Currat, 1980); on St.-Barthélemy, hides under low branches in dappled light where the carapace pattern of light and dark provides good camouflage (Breuil, 2002b); a juvenile on Guana quickly became tame when fed *Hibiscus* blossoms, grapes, or beetle grubs (Lazell, 2005b). **Diet:** Omnivorous, but primarily herbivorous, eats herbaceous material, fruits, worms, slugs, larvae, and carrion, accepts salads in captivity (Currat, 1980); on St.-Barthélemy, ate *Iguana* excrement (Breuil, 2002b); on Martinique, will eat most anything, including carrion (Pinchon, 1967); on Dominica, eating a banana (Krintler, 1986); a juvenile on Guana (British Virgin Islands) ate a *Tabebuia* blossom (Lazell, 2005b); in the Grenadines, takes fruit, flowers, carrion, and feces, which makes it scorned by locals as a food source (Daudin and de Silva, 2007). **Exploitation:** In Puerto Rico, known as *morrocollo* or "latrine cleaner" as turtles were kept for that purpose (Stahl in Grant, 1937a). **Growth:** On Guana, a juvenile (48 g, carapace length 60 mm) increased in size by 5% (to 58 g and 63 mm) in 4 weeks, an adult male added clear growth lamina (9 to 10) and increased in size from 24.5 to 26.8 cm (9.4%) in carapace length in 1 year (Lazell, 2005b). **Movements:** An adult female (carapace length 27.5 cm) on Guana remained in a 40 m^2 area for 3 weeks, whereas an adult male (26.8 cm) ranged over 4,743 m^2 (Lazell, 2005b). **Predators:** Mongooses (*Herpestes javanicus*; Currat, 1980); humans, on Saba, as recently as the early 20th century, turtles were routinely killed because they ate crops in agricultural plots (Powell et al., 2005); dogs may take small individuals (Hodge et al., 2003). **Reproduction:** Oviparous, lays 2–12 spherical eggs (diameter 3–4 cm, 10–20 g) in an excavation 10–20 cm deep, hatch after 3 months (Currat, 1980); on St.-Barthélemy, mate July–August, lays eggs after rains have started, produce 2–7 eggs (usually 5), in August, nest (circular, 10 cm in diameter, 7 cm deep, 1 cm of dirt covering eggs) in clay soil at the base of a wall with 4 eggs (another had been eaten by *Rattus*), hatchlings 45–46 mm carapace length, 54–56 g, another nest at the same site on 12 August, 4 hatchlings under 4 cm of dirt, 58–63 mm carapace length, 39–45 g, on Grand Terre, eggs hatch early August (Breuil, 2002b). **Size:** Male carapace length to 60 cm, female carapace length to ~40 cm (Ernst and Leuteritz, 1999a).

Conservation Status: Listed on CITES Appendix II (UNEP-WCMC, 2008); despite uncertainty regarding the origins of this species in the West Indies, but because at least some individuals may be descendants of animals that reached the islands naturally, populations should be monitored until genetic studies can confirm origins (Powell et al., 2005); numbers in the Grenadines dwindled in the 1960s and 1970s, when tortoises were shipped to Trinidad for food and to Martinique for food and to clean the feces from chicken farms (de Silva and Wilson, 2006).

Additional Pertinent Literature: Ernst and Leuteritz (1999a).

Chelonoidis denticulata Linnaeus, 1766

Distribution: South America east of the Andes, from Colombia and Venezuela to Bolivia and Brazil, Trinidad; in the West Indies, introduced to Guadeloupe and possibly to St. John (U.S. Virgin Islands), the latter not yet verified (Platenberg and Boulton, 2006); report of an introduction on Dominica (Tortoise and Freshwater Turtle Specialist Group, 1996a) is apparently in error.

Habitat and Natural History: Unknown in the West Indies (Breuil, 2002b).

Size: Maximum carapace length to 820 mm (on the Neotropical mainland; Pritchard and Trebbau, 1984).

Conservation Status: As an introduced species, its status in the West Indies is of concern only in terms of impact on native plants and animals; listed on CITES Appendix II (UNEP-WCMC, 2008); on the IUCN Red List as "vulnerable" (Tortoise and Freshwater Turtle Specialist Group, 1996a).

Additional Pertinent Literature: Ernst and Leuteritz (1999b).

Section 5 Crocodilia

33 Family Alligatoridae

Figure 36. *Caiman crocodilus* (Alligatoridae) has become established in Cuba and Puerto Rico. This juvenile was captured in a backyard in Caguas, Puerto Rico. Photograph by Luis O. Nieves.

Caiman crocodilus Linnaeus, 1758

Distribution: Central America and much of South America; in the West Indies, 9 individuals introduced into the Cienega de Lanier on Isla de la Juventud, Cuba in 1959 (*C. c. fuscus*; Varona, 1976); also introduced into Puerto Rico (*C. c. crocodilus* or *C. c. fuscus*).

Habitat: Aquatic, in ponds, streams, marshes, rivers, and occasionally in brackish estuaries (Schwartz and Henderson, 1991); on Puerto Rico, particularly abundant in Laguna de Tortuguero (Thomas and Joglar, 1996).

Abundance: On Isla de la Juventud, estimated at 40,000 (Soberon et al., 1996); in Laguna de Tortuguero (Puerto Rico), 14 juveniles captured in 1 hour (Joglar, 2005b). **Diet:** In Puerto Rico, the introduced fish *Tilapia* (Watlington, 1996). **Parasites:** Pentastomids (Rivero, 1998). **Predators:** On Puerto Rico, *Ameiva*, *Rattus*, and *Herpestes* may eat eggs, egrets and herons may take hatchlings, humans, as caiman "nuggets" are on some restaurant menus (Rivero, 1998). **Size:** Adults to about 225–250 cm depending on subspecies (Brazaitis, 1973).

Conservation Status: Listed on CITES Appendix II (UNEP-WCMC, 2008); a species of "least concern" on the IUCN Red List (Crocodile Specialist Group, 1996a); as an introduced species in the West Indies, however, its potential impact on native species is of primary concern.

34 Family Crocodylidae

Figure 37. *Crocodylus acutus* (Crocodylidae) can move very quickly when startled or when attacking prey. Photograph by Alejandro J. Sánchez Muñoz.

Crocodylus acutus Cuvier, 1807

Distribution: Southeastern United States through Central America to northwestern South America; in the West Indies, Cuba, including Archipiélago de los Canarreos, Jardines de la Reina, Archipiélago de Sabana-Camagüey, Cayos de San Felipe, Isla de la Juventud, Hispaniola, including Île-à-Vache and Île de la Gonâve, Jamaica, and as an occasional vagrant to the Cayman Islands (Little Cayman and Cayman Brac).

Habitat: In Cuba, in fresh, brackish, and salt water, common in coastal situations (Neill, 1958; Sampedro Marín and Rodríguez Schettino, 2003); on Hispaniola, in a variety of situations including mangroves, coastal areas, saline and freshwater lakes and rivers (Thorbjarnarson, 1988; Incháustegui, 2003b); in the Dominican Republic at Lago Enriquillo, in highly saline situation, with denuded shorelines but abundant sandy beaches for nesting (Incháustegui, 2003b); at Etang Saumâtre (Haiti), riparian vegetation is a virtual monoculture of Buttonwood (*Conocarpus erectus*), with varied aquatic vegetation comprising grasses, sedges, floating vines, and cattails (Thorbjarnarson, 1988); in Jamaica, in coastal situations and brackish lagoons (Grant, 1940a).

Abundance: Sixty or more observed from one vantage point at Lago Enriquillo in 1968, and gatherings of more than 125 adults seen in the late 1970s (Incháustegui, 2003b); by the mid-1990s, the population at Lago Enriquillo estimated at ~200 (Schubert and Santana, 1996). **Biomass:** In Haiti at Etang Saumâtre, total biomass estimated as 4,741.8 kg, a crude density biomass of 66.6 kg/km of shore, and an ecological density biomass of 101.7 kg/km (Thorbjarnarson, 1988). **Foraging and Diet:** In Haiti, adults sit on the bottom of the lake (in water 1–2 m deep) and make rapid sideways sweeps of the jaws to catch fish, also take fish trapped in gill nets (sit on the bottom at night next to nets), diet at Etang Saumâtre (Haiti), for crocodiles <0.5 m (N = 14) exclusively invertebrates (crustaceans in 85.7% of stomachs, beetles 78.6%, dragon- and damselflies 64.3%, and others), for crocodiles 0.5–0.9 m (N = 43) crustaceans 90.7% of stomachs, beetles 72.1%, dragon- and damselflies 46.5%, spiders 30.2%, fishes 11.6%, reptiles 2.3%, birds, 2.3%, mammals 11.6%, for crocodiles 0.9–1.8 m (N = 7) crustaceans 71.4%, dragon- and damselflies 42.9%, spiders 42.9%, birds 28.6%, mammals 14.3%; for crocodiles >1.8 m (N = 2) fishes 100%, birds 50%, among crustaceans, Fiddler Crabs (*Uca bergersii*) most common prey, fishes included *Limia* sp. and *Gambusia hispaniole*, reptiles included

Trachemys decorata, birds included *Dulus dominicus* and *Gallus gallus*, mammals included *Mus musculus*, *Rattus* sp., *Capra hircus*, and *Canis familiaris* (Thorbjarnarson, 1988); in Lago Enriquillo (Dominican Republic), diet includes the introduced fish *Tilapia* (Incháustegui, 2003b). **Growth:** At Etang Saumâtre, average hatchling growth rate over first 65–86 days 0.135 cm/day, mean growth rate 0.058 cm/day in crocodiles 0.3–0.9 m (Thorbjarnarson, 1988). **Hatchling Distribution:** From Isla Cabritos in highly saline Lago Enriquillo (Dominican Republic), salt-intolerant hatchlings are transported across the lake on the backs of their mothers to sources of freshwater on the main island (Incháustegui, 2003b). **Mortality:** Egg loss at Etang Saumâtre due to embryonic mortality 5.8–20.8% (Thorbjarnarson, 1988); at Lago Enriquillo, nests poached by humans (Schubert and Santana, 1996). **Nesting:** In the early 1970s at Lago Enriquillo, the average number of nests/season was ~150 (Incháustegui, 2003b); in 1990, 30 nests, and only 10 in 1991 (Schubert and Santana, 1996); at Etang Saumâtre, nesting beaches determined by (1) suitability of terrestrial habitat, (2) human population density, and (3) presence of nearby aquatic habitats protected from wave action, nests located on coralliferous limestone outcrops with deep sand that extended to the lakeshore, and in old charcoal mounds, mean height of nests above lake level 1.2 m at the beginning of the incubation period, the bottom of an average clutch was ~80 cm above lake level, mean hole depth 37.9 cm, soil moisture content (mean 6.62%) critical in nest choice, nest temperature over a 30 h period 29.3 ± 0.41 °C, females remain in the vicinity of nests during incubation (Thorbjarnarson, 1988). **Parasites:** In Cuba, trematodes *Pelaezia loossi* and *Acanthostomum americanum* (Pérez Vigueras, 1956; Coy Otero and Lorenzo Hernández, 1982); nematode *Dujardinascaris helicina* (Pérez Vigueras, 1936; Coy Otero and Lorenzo Hernández, 1982). **Population Density:** At Etang Saumâtre, mean density 6.3 crocodiles/km of shoreline, ecological density (i.e., eliminating habitats unsuitable for crocodiles) 9.6/km, at Lago Enriquillo, 2.58–3.88 adults/km, at Etang Saumâtre 0.98 adults/km (Thorbjarnarson, 1988); in the Zapata Swamp (Cuba), 1.2–19.6/km^2 (Ramos Targarona et al., 1994). **Population Size:** At Lago Enriquillo, 336–504 individuals, at Etang Saumâtre, 450 individuals, including all size classes (Thorbjarnarson, 1988). **Predators:** At Lago Enriquillo, Yellow-crowned Night Herons (*Nyctanassa violacea*) take young, nest predators include dogs and possibly mongooses (J. A. Ottenwalder in Thorbjarnarson, 1988); at Etang Saumâtre, humans are the greatest threat to adults, including the killing of males in order to collect penises as a possible aphrodisiac (Thorbjarnarson, 1988); at Lago Enriquillo, hunted for meat, fat for local pharmaceuticals, and penises, but also killed indiscriminately (Schubert and Santana, 1996); historically, the exploitation of crocodiles was officially sanctioned in the Dominican Republic (G. K. Noble in Powell, 2003a). **Reproduction:** At Lago Enriquillo, mating in December–January, eggs laid February–March (to 50 nests on a single nesting beach), peak of hatchling season is May–June (Incháustegui, 2003b); at Etang Saumâtre, courtship and mating in late December and early January, when "harems" consisting of one large male and several smaller adults (presumably females and possibly smaller males) form, nesting 22 January–12 February and hatching in late April–early May (incubation ~90 days), average clutch size 22.5 ± 2.7 eggs, fertility 70.6–100% within clutches, mean clutch mass 2.18 kg, about 4.4% of female body mass (Thorbjarnarson, 1988); in Cuba, clutch of 60 eggs (F. Sauvalle in Gundlach, 1880); eggs 73–81 × 42–47 mm, increase to 100 × 50 mm (Gundlach, 1880). **Size:** At Etang Saumâtre, males to 3.5 m total length, females to 2.4 m total length (Thorbjarnarson, 1988); in Cuba, to 3.8 m in nature and 4.05 m in captivity (Sampedro Marín and Rodríguez Schettino, 2003).

Conservation Status: On CITES Appendix I (Appendix II for Cuba; UNEP-WCMC, 2008); on the IUCN Red List as "vulnerable" (Crocodile Specialist Group, 1996b); this species was included in lists of threatened Hispaniolan species by Aquino (1983) and Powell et al. (2000); Cuban populations "vulnerable" (Rodríguez Schettino and Chamizo Lara, 1998a).

Additional Pertinent Literature: Ernst et al. (1999).

Crocodylus rhombifer Cuvier, 1807

Distribution: Cuba, including the Archipiélago de los Canarreos and Isla de la Juventud; historically in the Bahama Islands (Franz et al., 1995) and Cayman Islands (Morgan et al., 1993).

Habitat: Freshwater marshes and swamps, may enter brackish situations, common about beaches and mangrove swamps (Barbour and Ramsden, 1919); appears to prefer freshwater habitats, but may enter the sea (Ramos Targarona, 1998).

Behavior: More aggressive than *C. acutus* (Z. Vogel, 1965c). **Diet and Feeding:** If a food item is small, it is swallowed, if too large, it is shaken and pieces are swallowed, may allow food to putrefy to facilitate dismemberment, may go long periods without eating or drinking (Gundlach, 1880); primarily fishes and turtles (*Trachemys*), occasionally small mammals (Ramos Targarona et al., 1994); includes gar (*Lepisosteus*) and the rodent *Capromys* (de Sola, 1930); able to "jump" vertically and laterally out of the water, possibly to capture prey (Varona, 1986). **Exploitation:** Farmed for meat and hides in Cuba (Ramos Targarona, 1998; Rivalta et al., 2003a). **Longevity:** In captivity, 38 years, 2 months and still alive at last report in 1996 (J. Behler in Slavens and Slavens, 2003). **Parasites:** Trematodes *Pelaezia loossi, Acanthostomum americanum, Pseudoneodiplostomum groschafti*, and acanthocephalan *Hexaglandula mutabilis* (Groschaft and Baruš, 1970; Coy Otero and Lorenzo Hernández, 1982); nematodes *Contracaecum caballeroi, Dujardinascaris helicina*, and larval *Contracaecum* sp. and *Spiroxys* sp. (Groschaft and Baruš, 1970; Coy Otero and Baruš, 1979b; Coy Otero and Lorenzo Hernández, 1982). **Population Density:** Based on mark-recapture, 11.9–104.7/km^2 (Ramos Targarona et al., 1994) **Population Size:** 3,000–6,000 in 300 km^2 in the southwestern portion of the Zapata Swamp (Ramos Targarona et al., 1994). **Reproduction:** Females deposit eggs in trenches or holes near water and cover them with soil, eggs 78 × 52 mm (Gundlach, 1880); reproduces at about 2.0 m total length and 9–10 years of age, copulation in the water but near shore so the female can maintain contact with the bottom, nest dug by the female to a depth of 30 cm with her hind legs, females aggressively attend nests which may have 12–60 eggs, but usually 30–40 (Varona, 1986); in a captive population, sexual maturity at 6 years of age and at a size >183 cm, copulation February–April, females construct nests and defend them aggressively, mean 22 eggs/female, increasing with female size, incubation ~80 days (Ramos Targarona, 1998); often only ~50% of eggs develop (Nieto Dopico, 1997). **Sex Ratio:** Among adults, varied from approximately 1:1 to 90% females at different sites in the Zapata Swamp (Ramos Targarona et al., 1994). **Size:** May reach 4.9 m total length, but few exceed 3.5 m (Varona, 1966); "16½ feet" (Gundlach, 1880).

Conservation Status: On CITES Appendix I (UNEP-WCMC, 2008); on the IUCN Red List as "critically endangered" (Ramos Targarona et al., 2008); "endangered" (Crocodile Specialist Group, 1996c); as "endangered" (Rodríguez Schettino and Chamizo Lara, 1998a); listed as "endangered" by the U.S. Fish & Wildlife service (2008); this species has the smallest known natural distribution of any extant crocodilian (Ramos Targarona et al., 1994; Chamizo Lara et al., 2003); hybridization with *C. acutus* threatens the genetic purity of wild populations (Ramos Targarona et al., 1994, 2008; Díaz and Abreu Guerra, 2005); other current threats include illegal hunting for food, illegal sales of mounted specimens and live juveniles as pets (Díaz and Abreu Guerra, 2005); a reintroduction program on Isla de la Juventud (where the species was virtually extirpated by 1950) began in 1990 (Soberon et al., 1996).

Additional Pertinent Literature: Ross (1998).

Literature Cited

References with one or two authors are listed alphabetically, but those with three or more authors are listed chronologically after those with two to facilitate the location of specific references cited as "et al." in the text.

Please note that for Hispanic surnames (= *apellidos*) cited inconsistently in the literature, with the second surname sometimes omitted or abbreviated, we have consistently used the full second surname for all citations, both in the literature cited and in the text, if ever used in any cited publication.

Abreu, R. M., and J. de la Cruz. 1988. Algunos datos sobre la alimentación del Almiquí. Garciana 10: 2–3.

Acevedo-Torres, M. A., N. Rios-Lopez, and M. del Carmen Ruiz-Jaen. 2005. *Epicrates inornatus* (Puerto Rican Boa). Cannibalism. Herpetological Review 36: 195.

Acholonu, A. D. 1970. Studies on *Mesocoelium danforthi* Hoffmann, 1935 (Trematoda: Brachycoeliidae) and its distribution in the herpetofauna of the Greater Antilles. Journal of Parasitology 56 (supplement) 703: 391.

———. 1976. Helminth fauna of sauriens from Puerto Rico with observations on the life cycle of *Lueheia inscripta* (Westrumb, 1821) and description of *Allopharynx puertoricensis* sp. n. Proceedings of the Helminthological Society of Washington 43: 106–116.

Ackermann, T. 2006. Schreibers Glatkopfleguan *Leiocephalus schreibersii*. Münster, Germany: Natur und Tier.

Ackley, J. W., P. J. Muelleman, R. E. Carter, R. W. Henderson, and R. Powell. 2009. A rapid assessment of herpetofaunal diversity in variously altered habitats on Dominica. Applied Herpetology 6: 171–184.

Adams, C. 2004. Requiem for Godzilla. Iguana 11: 168–171.

Adamson, M. L. 1981. *Parapharyngodon osteopili* n. sp. (Pharyngodonidae: Oxyuroidea) and a revision of *Parapharyngodon* and *Thelandros*. Systematic Parasitology 3: 105–117.

Adolph, S. C., and J. Roughgarden. 1983. Foraging by passerine birds and *Anolis* lizards on St. Eustatius (Neth. Antilles): Implications for interclass competition, and predation. Oecologia 56: 313–317.

Ajayi, A. 2001. The territorial behavior of the Mountain Chicken. Unpublished report. Texas A&M University Study Abroad, College Station. http://dominica.tamu.edu/student%20projects/projects_by_year.htm.

Alayo Dalmau, P. 1951. Especies herpetológicas halladas en Santiago de Cuba. Boletin de Historia Natural Sociedad Felipe Poey 2: 106–110.

Alayón García, G. 1976. Araneidos depredados por anolinos. Miscelanea Zoológica 2: 1.

Alberts, A. C. 1993. The most endangered lizard in the world: The Jamaican Iguana, *Cyclura collei*. Vivarium 5: 12–14.

———. 1995. Use of statistical models based on radiographic measurements to predict oviposition dates and clutch size in rock iguanas (*Cyclura nubila*). Zoo Biology 14: 543–553.

———. 2003. Conserving the remarkable reptiles of Guantanamo Bay. In: R. W. Henderson and R. Powell (eds.), Islands and the Sea: Essays on Herpetological Exploration in the West Indies, pp. 67–73. Contributions to Herpetology, volume 20. Ithaca, N.Y.: Society for the Study of Amphibians and Reptiles.

———. 2004. Conservation strategies for West Indian Rock Iguanas (Genus *Cyclura*): Current efforts and future directions. Iguana 11: 212–223.

Alberts, A. C., and A. Perera. 1996. *Cyclura nubila* ssp. *nubila*. In: IUCN 2008. 2008 IUCN Red List of Threatened Species. http://www.iucnredlist.org.

Alberts, A. C., M. L. Oliva, M. B. Worley, S. R. Telford, Jr., P. J. Morris, and D. L. Janssen. 1998. The need for pre-release health screening in animal translocations: A case study of the Cuban Iguana (*Cyclura nubila*). Animal Conservation 1: 165–172.

Alberts, A. C., J. L. Lemm, T. D. Grant, and L. A. Jackintell. 2004. Testing the utility of headstarting conservation strategy for West Indian iguanas. In: A. C. Alberts, R. L. Carter, W. K. Hayes, and E. P. Martins (eds.), Iguanas: Biology and Conservation, pp. 210–219. Berkeley and Los Angeles: University of California Press.

Alexander, K. 2007. A photographic field guide to the reptiles and amphibians of Dominica, West Indies. Unpublished report. Texas A&M University Study Abroad, College Station. http://dominica.tamu.edu/student%20projects/projects_by_year.htm.

Alfau, C. A. de. 1980. Primera cita: *Foleyella brachyoptera* en *Bufo marinus*. Naturalista Postal (Santo Domingo, Dominican Republic) 80: 1.

Alfonso Alvarez, O., E. Morell Saval, R. Díaz Aguiar, R. Carbonell Paneque, F. Morera, and V. Berovides Alvarez. 1998. *Epicrates angulifer*. In: E. Pérez, E. Osa, Y. Matamoros, and

U. S. Seal (eds.), Report of Conservation Assessment and Management Plan Workshop for Selected Cuban Species II. Sección III. Reptiles. Apple Valley, Minn.: IUCN/SSC Conservation Breeding Specialist Group.

Allen, A. L., M. K. Albright, and R. Powell. 1994. Notes on the reproductive biology of *Anolis barahonae* (Lacertilia: Polychrotidae), a crown-dwelling giant anole from Hispaniola. Herpetological Natural History 2(2): 93–95.

Allen, E. R., and W. T. Neill. 1957. The gecko-like habits of *Anolis lucius*, a Cuban anole. Herpetologica 13: 246–247.

Alonso Bosch, R., and A. Rodríguez Gómez. 2001 (2000). Llamadas del repertorio vocal de *Eleutherodactylus eileenae* (Anura: Leptodactylidae) del Occidente de Cuba. Cuadernos de Herpetología 14: 123–130.

———. 2003a. Insospechados habitants de las penumbras. In: L. Rodríguez Schettino (ed.), Anfibios y Reptiles de Cuba, pp. 20–29. Vaasa, Finland: UPC Print.

———. 2003b. Entre la hojarasca bajo nuestros pies. In: L. Rodríguez Schettino (ed.), Anfibios y Reptiles de Cuba, pp. 30–37. Vaasa, Finland: UPC Print.

———. 2003c. Advertisement calls of Cuban toads of the genus *Bufo* (Anura, Bufonidae). Phyllomedusa 2: 75–82.

———. 2004. *Anolis lucius* (Coronel, Gorrita). Vocalization. Herpetological Review 35: 385–386.

Alonso Bosch, R., A. Rodríguez Gómez, and A. R. Estrada. 2001. Patrones de actividad acustica y trófica de machos cantors de *Eleutherodactylus eileenae* (Anura: Leptodactylidae). Revista Española de Herpetología 15: 45–52.

Alonso Bosch, R., A. Rodríguez Gómez, and A. Hernández. 2005. *Eleutherodactylus albipes* (NCN). Reproduction. Herpetological Review 36: 433.

Alonso Bosch, R., A. Rodríguez Gómez, and M. C. Mena. 2007. Guia Sonora de los Anfibios de Cuba. Barcelona, Spain: Alosa, Sons de la Natura.

Altig, R., and R. Lohoefener. 1982. *Rana grylio*. Catalogue of American Amphibians and Reptiles 286: 1–2.

Álvarez, A., N. Pérez-Buitrago, M. Andrade-Núñez, and M. García-Bermúdez. 2007. Problem assessment and control of the introduced Cuban Iguana (*Cyclura nubila nubila*) on Isla Magueyes. Iguana Specialist Group Newsletter 10(1): 11–12 (reprinted in Iguana 14: 241).

Amorín Ponce, J. A., L. Bacallao Mesa, O. Martínez, T. Piñeiro Cordero, and G. Fornaris. 2003. La Ciénaga de Zapata. Historia y Naturaleza. La Habana: Editorial Academia.

Andrews, R. 1976. Growth rate in island and mainland anoline lizards. Copeia 1976: 477–482.

———. 1979. Evolution of life histories: A comparison of *Anolis* lizards from matched island and mainland habitats. Breviora 454: 1–51.

Andrews, R., and A. S. Rand. 1974. Reproductive effort in anoline lizards. Ecology 55: 1317–1327.

Angulo, A. 2008a. *Peltophryne lemur*. In: IUCN 2008. 2008 IUCN Red List of Threatened Species. http://www.iucnredlist.org.

———. 2008b. *Eleutherodactylus antillensis*. In: IUCN 2008. 2008 IUCN Red List of Threatened Species. http://www.iucnredlist.org.

———. 2008c. *Eleutherodactylus hedricki*. In: IUCN 2008. 2008 IUCN Red List of Threatened Species. http://www.iucnredlist.org.

———. 2008d. *Eleutherodactylus jasperi*. In: IUCN 2008. 2008 IUCN Red List of Threatened Species. http://www.iucnredlist.org.

———. 2008e. *Eleutherodactylus monensis*. In: IUCN 2008. 2008 IUCN Red List of Threatened Species. http://www.iucnredlist.org.

———. 2008f. *Eleutherodactylus portoricensis*. In: IUCN 2008. 2008 IUCN Red List of Threatened Species. http://www.iucnredlist.org.

———. 2008g. *Eleutherodactylus richmondi*. In: IUCN 2008. 2008 IUCN Red List of Threatened Species. http://www.iucnredlist.org.

———. 2008h. *Eleutherodactylus rogersi*. In: IUCN 2008. 2008 IUCN Red List of Threatened Species. http://www.iucnredlist.org.

———. 2008i. *Eleutherodactylus wightmanae*. In: IUCN 2008. 2008 IUCN Red List of Threatened Species. http://www.iucnredlist.org.

Anonymous. 2006a. Durrell Wildlife finds Montserrat Galliwasp lizards thought to be extinct. The Montserrat Reporter Online (27 January 2006; reprinted in Iguana 13: 80).

———. 2006b. Fichas de especies y poster: *Anolis equestris*. Reptilia 62: 47–50.

Appler, H. 1963. Über Pflege und Zucht von *Anolis cuvieri* Merrem, dem Riesenanolis aus Portoriko. Die Aquarien- und Terrarienzeitschrift (DATZ) 16: 279–282.

Aquino, A. T. 1983. Anfibios y reptiles de la Hispaniola en peligro de extinción. Conferencias, Revista Cientifica de Divulgación, Museo Nacional de Historia Natural 1: 26–30.

Arendt, W. J. 1987. An observation of *Iguana iguana* feeding on eggs of the Cattle Egret (*Bubulcus ibis*) at Fox's Bay, Montserrat, West Indies: A case of predation or scavenging? Caribbean Journal of Science 22: 221–222.

Arendt, W. J., and D. Anthony. 1987. Bat predation by the St. Lucia Boa (*Boa constrictor orophias*). Caribbean Journal of Science 22: 219–220.

Arias, Y., S. Incháustegui, and E. Rupp. 2004. *Cyclura ricordii* on the Barahona Peninsula: A preliminary report. Iguana 11: 8–14.

Armas, L. F. de. 1987a. Notas sobre la alimentación de *Leiocephalus carinatus cayensis* (Sauria: Iguanidae). Poeyana 350: 1–7.

———. 2001. Frogs and lizards as prey of some Greater Antillean arachnids. Revista Ibérica de Aracnológia 3: 87–88.

Armas, L. F. de, and A. J. Abud. 1992. Depredación de vertebrados por escorpiones (Scorpiones: Buthidae) de República Dominicana. In: Comunicaciones Breves de Zoología, pp. 5–6. La Habana: Editorial Academia.

Armas, L. F. de, and G. Alayón García. 1986. Depredadores y parasitoides de *Argiope trifasciata* (Araneae: Araneidae) en el sur de La Habana. Ciencias Biológicas 16: 114–117.

———. 1987. Observaciones sobre la ecología trófica de una población de *Argiope trifasciata* (Araneae: Araneidae) en el sur de La Habana. Poeyana 344: 1–18.

Armas, L. F. de, A. Fong Grillo, and F. Rodríguez. 1999. Depredación del alacrán *Centruroides anchorellus* (Scorpiones: Buthidae) por la lagartija *Anolis porcatus* (Iguania: Polychridae). Cocuyo 9: 14.

Armas, L. F. de, A. Rans, and A. Torres Leyva. 1987. Primeras observaciones sobre la alimentación de *Cricosaura typica* (Sauria: Xantusiidae) en condiciones naturales. Miscelanea Zoológica 32: 1–2.

Arnold, D. L. 1980. Geographic variation in *Anolis brevirostris* (Sauria: Iguanidae) in Hispaniola. Breviora 461: 1–31.

Arredondo Antónez, C., and V. N. Chirino Flores. 2002. Consideraciones sobre la alimentación de *Tyto alba furcata* (Aves: Strigiformes) con implicaciones ecológicas en Cuba. Journal of Caribbean Ornithology 15: 16–24.

Audebaud, G., and E. Courmes. 1961 (1962). Ixodidés de la Guadeloupe. Présence d'*Amblyomma dissimile* (Koch, 1844). Rapport de l'Institut Pasteur de la Guadeloupe 1961: 90–91.

Auffenberg, W. 1982. Feeding strategy of the Caicos Ground Iguana, *Cyclura carinata*. In: G. M. Burghardt and A. S. Rand (eds.), Iguanas of the World, pp. 84–116. Park Ridge, N.J.: Noyes.

Austen, N. L. 1867. The crested anolis. Land and Water 4: 9.

Autumn, K., and J. B. Losos. 1997. Notes on jumping ability and thermal biology of the enigmatic anole *Chamaelinorops barbouri*. Journal of Herpetology 31: 442–444.

Avery, R. A. 1988. Observations on habitat utilization by the lizard *Anolis conspersus* on the island of Grand Cayman, West Indies. Amphibia-Reptilia 9: 417–420.

Ayala, S. C. 1977. Plasmodia of reptiles. In: J. P. Kreier (ed.), Parasitic Protozoa. Volume 3, pp. 267–309. New York: Academic Press.

———. 1978. Checklist, host index, and annotated bibliography of *Plasmodium* in reptiles. Journal of Protozoology 25: 87–100.

Ayala, S. C., and P. E. Hertz. 1981. Malaria infection in *Anolis* lizards on Martinique, Lesser Antilles. Revista do Instituto de Medicina Tropical de São Paulo 23: 12–17.

Bach, S., and P. Schönecker. 2008. Herpetologische Eindrücke bei der Leguansuche auf den Französischen Antillen: Martinique. Terraria 3: 71–80.

Baedermann, L. 1989. Vergleichende Beobachtungen bei Haltung und Zucht kubanischer Höhlenanolis. Sauria 11: 13–17.

Banbury, B. L., and Y. M. Ramos. 2005. The rock iguanas of Parque Nacional Isla Cabritos. Iguana 12: 256–261.

Banbury, B. L., Y. M. Ramos, R. Powell, and J. S. Parmerlee, Jr. 2000. The *Cyclura* of Parque Nacional Isla Cabritos. Iguana Times (Journal of the International Iguana Society) 8(2): 3–7.

Barber, D. 2008. Building conservation partnerships with zoos. Endangered Species Bulletin 33: 18–19.

Barbour, T. 1904. Batrachia and Reptilia from the Bahamas. Bulletin of the Museum of Comparative Zoölogy 46: 55–61.

———. 1910. Notes on the herpetology of Jamaica. Bulletin of the Museum of Comparative Zoölogy 52: 273–301 + 2 plates.

———. 1914. A contribution to the zoögeography of the West Indies, with especial reference to amphibians and reptiles. Memoirs of the Museum of Comparative Zoölogy 44: 209–359.

———. 1916. The reptiles and amphibians of the Isle of Pines. Annals of the Carnegie Museum 10: 297–308.

———. 1920. A note on *Xiphocercus*. Proceedings of the New England Zoölogical Club 7: 61–63.

———. 1921. Sphaerodactylus. Memoirs of the Museum of Comparative Zoölogy 47: 215–278 + 26 plates.

———. 1930a. Some faunistic changes in the Lesser Antilles. Proceedings of the New England Zoölogical Club 11: 73–85.

———. 1930b. A list of Antillean reptiles and amphibians. Zoologica 11: 61–116.

———. 1930c. The anoles. I. The forms known to occur on the Neotropical islands. Bulletin of the Museum of Comparative Zoölogy 70: 105–144.

———. 1934. Observations on Antillean tortoises and terrapins. Copeia 1934: 111–113.

———. 1941. A new boa from the Bahamas. Proceedings of the New England Zoölogical Club 18: 61–65.

———. 1945. A Naturalist in Cuba. Boston: Little, Brown.

Barbour, T., and A. F. Carr, 1940. Antillean terrapins. Memoirs of the Museum of Comparative Zoölogy 54: 318–415.

Barbour, T., and G. K. Noble. 1915. A revision of the lizards of the genus *Ameiva*. Bulletin of the Museum of Comparative Zoölogy 59: 417–479.

Barbour, T., and C. T. Ramsden. 1916. Catalogo de los reptiles y anfibios de la isla de Cuba. Memorias de la Sociedad Cubana de Historia Natural Felipe Poey 2: 124–143.

———. 1919. The herpetology of Cuba. Memoirs of the Museum of Comparative Zoölogy 47: 71–213.

Barbour, T., and B. Shreve. 1935. Concerning some Bahamian reptiles, with notes on the fauna. Proceedings of the Boston Society of Natural History 40: 347–366.

———. 1937. Novitates Cubanae. Bulletin of the Museum of Comparative Zoölogy 80: 377–387 + 4 plates.

Barnes, V., Jr. 1946. The birds of Mona Island, Puerto Rico. Auk 63: 318–327.

Barré, N., O. Lorvelec, and M. Breuil. 1997. Les oiseaux et les reptiles de Îles de la Petite Terre. Bilan d'un suivi écologique d'une année. Association pour l'Etude et la Protection des Vertébrés et Végétaux des Petites Antilles (AEVA) Rapport 16: 1–57.

Barun, A., and G. Perry. 2003. *Amphisbaena fenestrata* (Virgin Islands Amphisbaena). Predation. Herpetological Review 34: 244.

Barun, A., G. Perry, R. W. Henderson, and R. Powell. 2007. *Alsophis portoricensis anegadae* (Squamata: Colubridae): Morphometric characteristics, activity patterns, and habitat use. Copeia 2007: 93–100.

Baruš, V. 1972. Nemadodes parasitizing hosts of the genus *Eleutherodactylus* (Amphibia) from Cuba. Vestník Ceskoslovenské Spolecnosti Zoologické 36: 161–168.

———. 1973. Nematodes parasitizing hosts of the genus *Bufo* (Amphibia) in Cuba. Folia Parasitologica 20: 28–39.

Baruš, V., and A. Coy Otero. 1966. Nota sobre la helmintofauna de ofidios en Cuba. Descripción de tres especies nuevas de nemátodos. Poeyana 23: 1–16.

———. 1968. *Freitasia teixeirai* gen. n. et sp. n. and other nematodes parasitizing *Anolis equestris* (Squamata: Iguanidae). Folia Parasitologica 15: 41–54.

———. 1969a. Systematic survey of nematodes parasitizing lizards (Sauria) in Cuba. Helminthology 10: 329–346.

———. 1969b. Nemátodos del género *Parapharyngodon* Chatterji, 1933 (Oxyuridae) en Cuba. Torreia 7: 1–10.

———. 1974. Nematodes of the genera *Spauligodon*, *Skrjabinodon* and *Pharyngodon* (Oxyuridae) parasitizing Cuban lizards. Vestník Ceskoslovenské Spolecnosti Zoologické 38: 1–12.

———. 1978. Nematodes parasitizing Cuban snakes (Ophidia). Vestník Ceskoslovenské Spolecnosti Zoologické 42: 85–100.

Baruš, V., and F. Moravec. 1967a. Systematic studies of parasitic worms found in the host *Lepisosteus tritoechus* (Gynglimodi: Lepisosteidae) and *Hyla insulsa* (Ecaudata: Hylidae) from Cuba. Vestník Ceskoslovenské Spolecnosti Zoologické 31: 1–14.

———. 1967b. A survey of helminths from the Cuban turtle *Pseudemys decussata* Gray (Emydidae). Vestník Ceskoslovenské Spolecnosti Zoologické 31: 313–324.

Baruš, V., and M. D. Sonin. 1971. A new species of filarial nematode *Hastospiculum cubaense* sp. n. (Diplotriaenidae: Dicheilonematinae) from Cuban snakes. Folia Parasitologica 18: 187–190.

Baruš, V., A. Coy Otero, and O. H. Garrido. 1969. Helmintofauna de *Cyclura macleayi* Gray (Sauria: Iguanidae) en Cuba. Torreia 8: 1–20.

Baskin, J. N., and E. E. Williams. 1966. The Lesser Antillean *Ameiva* (Sauria, Teiidae). Studies on the Fauna of Curaçao and Other Caribbean Islands 23: 144–178.

Bauer, A. M., and A. P. Russell. 1992. The evolutionary significance of regional integumentary loss in island geckos: A complement to caudal autotomy. Ethology Ecology & Evolution 4: 343–358.

———. 1993a. *Aristelliger*. Catalogue of American Amphibians and Reptiles 565: 1–4.

———. 1993b. *Aristelliger barbouri*. Catalogue of American Amphibians and Reptiles 566: 1.

———. 1993c. *Aristelliger hechti*. Catalogue of American Amphibians and Reptiles 569: 1–2.

———. 1993d. *Aristelliger lar*. Catalogue of American Amphibians and Reptiles 570: 1–3.

———. 1993e. *Aristelliger praesignis*. Catalogue of American Amphibians and Reptiles 570: 1–4.

———. 2003. *Phyllodactylus wirshingi*. Catalogue of American Amphibians and Reptiles 770: 1–4.

Bauer, A. M., A. P. Russell, and R. E. Shadwick. 1989. Mechanical properties and morphological correlates of fragile skin in gekkonid lizards. Journal of Experimental Biology 145: 79–102.

———. 1992. Skin mechanics and morphology in *Sphaerodactylus roosevelti* (Reptilia: Gekkonidae). Herpetologica 48: 124–133.

Baum, T. 2008. Jamaika, seine Leguane und der Riesenanolis *Anolis garmani*. Reptilia 13: 73–80.

Bayless, B. 2005. Diet study of *Eleutherodactylus martinicensis* and *Anolis oculatus*. Unpublished report. Texas A&M University Study Abroad, College Station. http://dominica.tamu.edu/student%20projects/projects_by_year.htm.

Beard, K. H., and E. M. O'Neill. 2005. Infection of an invasive frog *Eleutherodactylus coqui* by the chytrid fungus *Batrachochytrium dendrobatidis* in Hawaii. Biological Conservation 126: 591–595.

Beard, K. H., A. K. Eschtruth, K. A. Vogt, D. J. Vogt, and F. N. Scatena. 2003a. The effects of the frog *Eleutherodactylus coqui* on invertebrates and ecosystem processes at two scales in the Luquillo Experimental Forest, Puerto Rico. Journal of Tropical Ecology 19: 607–617.

Beard, K. H., S. McCullough, and A. K. Eschtruth. 2003b. Quantitative assessment of habitat preferences for the Puerto Rican terrestrial frog, *Eleutherodactylus coqui*. Journal of Herpetology 37: 10–17.

Bech, R. 1979. Anolis im Terrarium. Aquarien Terrarien 26: 385–392.

———. 1985. Zur Haltung und Vermehrung von *Anolis allisoni* im Terrarium. Elaphe 4: 72–73.

———. 1986. Zur Haltung und Vermehrung von Ritteranolis im Terrarium. Aquarien Terrarien 33: 20–22.

Bello, R. E. 2000. *Anolis* sp. and *Gonatodes albogularis* (Yellow-headed gecko). Predation. Herpetological Review 31: 239–240.

Bels, V. L. 1984. Communication and agnostic [sic] behavior of *Anolis chlorocyanus* in captivity. In: Reptile Symposium on Captive Propagation and Husbandry, pp. 78–102. Thurmont, Md.: Zoological Consortium.

———. 1986. Analysis of the display-action-pattern of *Anolis chlorocyanus* (Sauria: Iguanidae). Copeia 1986: 963–970.

Bels, V. L., and I. Baltus. 1988. The influence of food items on the feeding cycle in *Anolis equestris* (Reptilia: Iguanidae). Copeia 1988: 479–481.

Bendon, J. 2003. The St. Lucian Green Iguana—A special case? Iguana 10: 71–78.

———. 2004. Magnificent iguanas of Isla Mona. Iguana 11: 98–106.

Benedix, J. H., Jr., and P. M. Narins. 1999. Competitive calling behavior by male treefrogs, *Eleutherodactylus coqui* (Anura: Leptodactylidae). Copeia 1999: 1118–1122.

Ben Slimane, B., M.-C. Durette-Desset, and A. G. Chabaud. 1995. Phénomènes de speciation chez les *Oswaldocruzia* (Nematoda: Trichostrongylina, Molineoidea) parasites de lézards de Guadeloupe. Parasite 2: 373–383.

Beovides-Casas, K., and C. A. Mancina. 2006. Natural history and morphometry of the Cuban Iguana (*Cyclura nubila* Gray, 1831) in Cayo Sijú, Cuba. Animal Biodiversity and Conservation 29: 1–8.

Bequaert, J. C. 1932. On the ornate nymphs of the tick genus *Amblyomma* (Acarina, Ixodidae). Zeitschrift für Parasitenkunde 4: 776–783.

Berovides Álvarez, V. 1980. Notas sobre la ecología de la iguana (*Cyclura nubila*) en Cayo del Rosario. Ciencias Biológicas 5: 112–115.

———. 1998. La iguana (*Cyclura nubila nubila*): Universo de interrogantes por descifrar. Flora y Fauna 2: 38–39.

Berovides Álvarez, V., and R. Carbonell Paneque. 1998. Morfometría y abundancia del majá de Santa María *Epicrates angulifer* (Ophidia, Boidae). In: E. Pérez, E. Osa, Y. Matamoros, and U. Seal (eds.), Taller para la Conservación, Análisis y Manejo Planificado de un Selección de Especies Cubanas II. Apple Valley, Minn.: IUCN/SSC Conservation Breeding Specialist Group.

Berovides Álvarez, V., and A. Sampedro Marín. 1980. Competición en especies de lagartos iguánidos de Cuba. Ciencias Biológicas 5: 115–122.

Berovides Álvarez, V., M. Morales Díaz, and J. Castillo Pérez. 2003. Morphometric and abundance data for the Cuban Iguana (*Cyclura nubila nubila*) on the Cays north of Carahata, Villa Clara. Iguana Specialist Group Newsletter 6(2): 8–9.

Beuttell, K., and J. B. Losos. 1999. Ecological morphology of Caribbean anoles. Herpetological Monographs 13: 1–28.

Bhajan, W. R. 1966. A comparative study of montane and lowland larval populations of *Leptodactylus albilabris* in Puerto Rico. Master of Science thesis, Michigan State University, Lansing.

Bickham, J. W. 1980. *Chrysemys decorata*. Catalogue of American Amphibians and Reptiles 235: 1–2.

Bickham, J. W., J. B. Iverson, J. F. Parham, H.-D. Philippen, A. G. J. Rhodin, H. B. Shaffer, P. Q. Spinks, and P. P. Van Dijk. 2007. An annotated list of modern turtle terminal taxa with comments on areas of taxonomic instability and recent change. Chelonian Research Monographs 4: 173–199.

Biegler, R. 1966. A survey of recent longevity records for reptiles and amphibians in zoos. International Zoo Yearbook 6: 487–493.

Binder, S., and A. Lamp. 2007. *Boa constrictor.* Die Abgottschlange. Münster, Germany: Natur und Tier.

Binns, J. 2003. Stout Iguanas: Historical perspectives and status report. Iguana 10: 39–48.

———. 2008a. The seven blues of May. Iguana 15: 66–76.

———. 2008b. Green Iguanas (*Iguana iguana*). Iguana 15(4): inside back cover.

Blair, C. 2007. *Arrhyton exiguum* (Puerto Rican Garden Snake). Reproduction. Herpetological Review 38: 466.

Blair, D., and West Indian Iguana Specialist Group. 2000. *Cyclura cychlura* ssp. *inornata*. In: IUCN 2008. 2008 IUCN Red List of Threatened Species. http://www.iucnredlist.org.

Blok, J. 1971. Verslag over gedrag en voortplanting van *Anolis bimaculatus sabanus* in het terrarium. Lacerta 30: 3–6.

Boal, C. 2008a. Observations of an Antillean Crested Hummingbird (*Orthorhyncus cristatus*) attacking Saddled Anoles (*Anolis stratulus*). Caribbean Journal of Ornithology 21: 48–49.

———. 2008b. Predation of a dwarf gecko (*Sphaerodactylus macrolepis*) by a Bridled Quail-Dove (*Geotrygon mystacea*). Caribbean Journal of Ornithology 21: 50–51.

Bofill, M., and A. R. Lewis. 1999. Detection of socially relevant chemical cues by the teiid lizard *Ameiva exsul*. Journal of Herpetology 33: 713–715.

Bogert, C. M. 1982. Investigations in the field and laboratory of behavioural themoregulation in an iguanid lizard. In: T. T. P. Gunawardana, L. Prematilleka, and R. Silva (eds.), P. E. P. Deriniyagala Commemorative Volume, pp. 90–102. Columbo, Sri Lanka: Lake House Investments.

Böker, H. 1939. Über einige Ergebnisse meiner Forschungen am Deutsch-Dominikanischen Tropenforschungsinstitut. Sobre algunos resultados de mis investigaciones in el Instituto Científico Domínico-Alemán. Veröffentlichungen des Deutsch-Dominikanischen Tropenforschungsinstituts (Publicaciones del Instituto Científico Domínico-Alemán) 1: 14–77 + plates II–XI.

Bolaños, F., F. Castro, C. Cortez, I. de la Riva, T. Grant, S. B. Hedges, W. R. Heyer, R. Ibáñez, E. la Marca, E. Lavilla, D. L. Silvano, S. Lötters, G. Parra Olea, S. Reichle, R. Reynolds, L. Rodríguez, G. Santos Barrera, N. J. Scott, C. Ubeda, A. Veloso, M. Wilkinson, and B. Young. 2008. Neotropics. In: S. N. Stuart, M. Hoffmann, J. S. Chanson, N. A. Cox, R. Berridge, P. Ramani, and B. E. Young (eds.), Threatened Amphibians of the World, pp. 92–99. Barcelona, Spain: Lynx Ediciones, with IUCN—The World Conservation Union, Conservation International, and NatureServe.

Bonny, K. 2007. Die Gattung *Boa*. Taxonomie und Fortpflanzung. Rheinstetten, Germany: KUS-Verlag.

Boone, J. 2005. Freilandbeobachtungen an Roosevelts Kugelfingergecko, *Sphaerodactylus roosevelti* Grant, 1931. Sauria 27(3): 3–7.

Bosch, H. 1994. *Boa constrictor*. Münster, Germany: Heselhaus & Schmidt Verlag.

Boulenger, G. A. 1885. Catalogue of the Lizards in the British Museum (Natural History). Codicote, Hertfordshire: Wheldon and Wesley.

Bowersox, S. R., S. Calderón, G. Cisper, R. S. Garcia, C. Huntington, A. Lathrop, L. Lenart, J. S. Parmerlee, Jr., R. Powell, A. Queral, D. D. Smith, S. P. Sowell, and K. C. Zippel. 1994a. Miscellaneous natural history notes on amphibians and reptiles from the Dominican Republic. Bulletin of the Chicago Herpetological Society 29: 54–55.

Bowersox, S. R., S. Calderón, R. Powell, J. S. Parmerlee, Jr., D. D. Smith, and A. Lathrop. 1994b. Nahrung eines Riesenanolis, *Anolis barahonae*, von Hispaniola, mit einer Zusammenfassung des Nahrungsspektrums westindischer Riesenanolis-Arten. Salamandra 30: 155–160.

Bowie, A., and R. Franz. 1974. *Cyrtosomum mega* sp. n. (Nematoda: Oxyuroidea) from the lizard, *Cyclura carinata*. Journal of Parasitology 60: 628–631.

Bowler, J. K. 1977. Longevity of reptiles and amphibians in North American collections. Society for the Study of Amphibians and Reptiles, Herpetological Circular 6: iv + 32 pp.

Boylan, T. 1984. Breeding the Rhinocerous Iguana, *Cyclura c. cornuta*, at Sydney Zoo. International Zoo Yearbook 23: 144–148.

———. 1985. Captive management of a population of Rhinoceros Iguanas *Cyclura cornuta cornuta* at Taronga Zoo, Sydney. In: G. Grigg, R. Shine, and H. Ehman (eds.), Biology of Australian Frogs and Reptiles, pp. 491–494. Sydney, New South Wales: Royal Zoological Society.

Bradley, K. A., and G. P. Gerber. 2005. Conservation of the Anegada Iguana (*Cyclura pinguis*). Iguana 12: 78–85.

Brandley, M. C., and K. de Queiroz. 2004. Phylogeny, ecomorphological evolution, and historical biogeography of the *Anolis cristatellus* series. Herpetological Monographs 18: 90–126.

Brazaitis, P. 1973. The identification of living crocodilians. Zoologica 58: 59–105.

Brennan, A. M., E. J. Censky, and R. Powell. 2009. Effects of chigger mite (Acari: Trombiculidae) infections on *Ameiva* (Squamata: Teiidae) from the Anguilla Bank. Contemporary Herpetology 2009(1): 1–3.

Breuil, M. 1996. *Liophis cursor*. In: IUCN 2008. 2008 IUCN Red List of Threatened Species. http://www.iucnredlist.org.

———. 1998. Lizard letters. Iguana Times (Journal of the International Iguana Society) 7: 45.

———. 1999. Nouvelle espèce de genre *Typhlops* (Serpentes, Typhlopidae) de l'île de Saint-Barthélemy, comparison avec les autres espèces des Petites Antilles. Bulletin Mensuel de la Société Linnéane de Lyon 68: 30–40.

———. 2000a. *Iguana delicatissima* and *Iguana iguana* in FWI. Summer 1999 field report. West Indian Iguana Specialist Group Newsletter 3(1): 4–5.

———. 2000b. Taxon reports. Lesser Antilles *Iguana delicatissima* and *Iguana iguana*. West Indian Iguana Specialist Group Newsletter 3(2): 4–5.

———. 2000c. Taxon reports. Lesser Antilles *Iguana delicatissima* and *Iguana iguana*. West Indian Iguana Specialist Group Newsletter 3(2): 11–13.

———. 2000d. Taxon reports (Lesser Antilles): *Iguana delicatissima* and *Iguana iguana*. Hybridization in the Guadeloupean Archipelago. West Indian Iguana Specialist Group Newsletter 3(2): 13–15.

———. 2000e. Iguana newsbriefs: Lesser Antilles iguanas. Iguana Times (Journal of the International Iguana Society) 8(3): 20.

———. 2001. Taxon reports. Lesser Antilles *Iguana delicatissima* and *Iguana iguana*. Iguana Specialist Group Newsletter 4(2): 9–11.

———. 2002a. Iguana newsbriefs: *Iguana delicatissima*. Iguana Times (Journal of the International Iguana Society) 9(1–2): 43.

———. 2002b. Histoire naturelle des amphibiens et reptiles terrestres de l'Archipel Guadeloupéen. Guadeloupe, Saint-Martin, Saint-Barthélemy. Patrimoines Naturels, Paris 54: 1–339.

———. 2003. In the footsteps of French naturalists, a "battle" of iguanas, and "improvements" in biodiversity. In: R. W. Henderson and R. Powell (eds.), Islands and the Sea: Essays on Herpetological Exploration in the West Indies, pp. 255–270. Contributions to Herpetology, volume 20. Ithaca, N.Y.: Society for the Study of Amphibians and Reptiles.

———. 2004. Les Amphibiens et Reptiles des Antilles. Abymes, Guadeloupe: PLB Editions.

———. 2009. The terrestrial herpetofauna of Martinique: Past, present, future. Applied Herpetology 6: 123–149.

Breuil, M., and M. Day. 1996. *Iguana delicatissima*. In: IUCN 2008. 2008 IUCN Red List of Threatened Species. http://www.iucnredlist.org.

Breuil, M., and B. Ibéné. 2008. Les hylidés envahissants dans les Antilles françaises et le peuplement batrachologique natural. Bulletin de la Société Herpétologique de France 125: 1–27.

Breuil, M., and C. Sastre. 1994. Végétation et reptiles de l'Archipel des Saintes (Antilles françaises). Bulletin des Amis du Muséum National d'Histoire Naturelle 178: 17–19.

Breuil, M., F. Guiougou, and B. Ibéné. 2007. Lesser Antillean Iguana (*Iguana delicatissima*). Iguana Specialist Group Newsletter 10(2): 15–17.

Breustedt, A. 1991. *Anolis equestris* Merrem. Sauria Supplement 13: 207–212.

Brooks, G. R., Jr. 1968. Body temperatures of three lizards from Dominica, West Indies. Herpetologica 24: 209–214.

———. 1981. Ecological significance of sexual dimorphism in body size in *Anolis lividus*. Program (abstract), p. 34. Memphis, Tenn.: Joint Annual Meeting of the Society for the Study of Amphibians and Reptiles and the Herpetologists' League.

———. 1982. An analysis of prey consumed by the anuran, *Leptodactylus fallax*, from Dominica, West Indies. Biotropica 14: 301–309.

Bryan, J. J., G. P. Gerber, M. E. Welch, and C. L. Stephen. 2007. Re-evaluating the taxonomic status of the Booby Cay Iguana, *Cyclura carinata bartschi*. Copeia 2007: 734–739.

Buckley, L. B., and J. Roughgarden. 2005a. Effect of species interactions on landscape abundance patterns. Journal of Animal Ecology 74: 1182–1194.

———. 2005b. Lizard habitat partitioning on islands: The interaction of local and landscape scales. Journal of Biogeography 32: 2113–2121.

Buckner, S., and D. Blair. 2000. Andros Island Iguana, *Cyclura cychlura cychlura*. In: A. Alberts (ed.), West Indian Iguanas: Status Survey and Conservation Action Plan, pp. 31–32. Gland, Switzerland and Cambridge, UK: International Union for the Conservation of Nature and Natural Resources.

Buden, D. W. 1974. Prey remains of Barn Owls in the southern Bahama Islands. Wilson Bulletin 86: 336–343.

———. 1975. Notes on *Epicrates chrysogaster* (Serpentes: Boidae) of the southern Bahamas, with the description of a new subspecies. Herpetologica 31: 167–177.

Buden, D. W., and A. Schwartz. 1969. Reptiles and birds of the Cay Sal Bank, Bahama Islands. Quarterly Journal of the Florida Academy of Sciences 31: 290–320.

Bui, H. T., R. Powell, D. D. Smith, J. S. Parmerlee, Jr., and A. Lathrop. 1992a. A new coccidian parasite (Apicomplexa: Eimeriorina) from *Anolis distichus* (Sauria: Polychridae) in the Dominican Republic. Journal of Parasitology 78: 784–785.

Bui, H. T., D. D. Smith, R. Powell, J. S. Parmerlee, Jr., and A. Lathrop. 1992b. Redescription of *Eimeria helenlevineae* (Apicomplexa: Eimeriidae) from *Hemidactylus brooki haitianus* (Sauria: Gekkonidae) from Hispaniola. Caribbean Journal of Science 28: 109–110.

Buide, M. S. 1951. Observations on habits of the Cuban Iguana. Herpetologica 7: 124.

———. 1966. Reptiles de la Península Hicacos. Poeyana, serie A 21: 1–21.

———. 1967. Lista de los anfibios y reptiles de Cuba. Torreia 1: 1–60.

———. 1985. Reptiles de Cuba. La Habana: Editorial Gente Nueva.

Buide, M. S., J. Fernández, F. García, O. H. Garrido, H. de los Santos, G. Silva, and L. Varona. 1974. Las Especies Amenazadas de Vertebrados Cubanos. La Habana: Academica de Ciencieas de Cuba.

Buley, K. R. 2003. *Leptodactylus fallax* (Mountain Chicken). Ophiophagy. Herpetological Review 34: 358–359.

Buley, K. R., K. A. Prior, and R. C. Gibson. 1997. In search of the St. Lucia Racer *Liophis ornatus*, with further observations on the reptile fauna of the Maria Islands, St. Lucia, West Indies. Dodo 33: 104–117.

Bullock, D. J., and P. G. H. Evans. 1990. The distribution, density and biomass of terrestrial reptiles in Dominica, West Indies. Journal of Zoology (London) 222: 421–443.

Bullock, D. J., H. M. Jury, and P. G. H. Evans. 1993. Foraging ecology in the lizard *Anolis oculatus* (Iguanidae) from Dominica, West Indies. Journal of Zoology (London) 230: 19–30.

Bundt, D. 2007. Terrarienhaltung von Anolis—ein Erfahrungsbericht am Beispiel von *Anolis bahorucoensis bahorucoensis*. Terraria 2: 18–31.

Bundy, D. A. P., P. Vogel, and E. A. Harris. 1987. Helminth parasites of Jamaican anoles (Reptilia: Iguanidae): A comparison of the helminth fauna of 6 *Anolis* species. Journal of Helminthology 61: 77–83.

Burbrink, F. T., and R. Lawson. 2007. How and when did Old World rat snakes disperse into the New World? Molecular Phylogenetics and Evolution 43: 173–189.

Burger, R. M. 2004. Boas del genero *Tropidophis*. Reptilia 48: 47–51.

Burns, J. K., C. A. Cunningham, R. A. Dupuis, M. N. Trask, J. S. Tulloch, R. Powell, J. S. Parmerlee, Jr., K. L. Kopecky, and M. L. Jolley. 1992. Lizards of the Cayos Siete Hermanos, Dominican Republic, Hispaniola. Bulletin of the Chicago Herpetological Society 27: 225–232.

Burrowes, P. A. 2000. Parental care and sexual selection in the Puerto Rican cave-dwelling frog, *Eleutherodactylus cooki*. Herpetologica 56: 375–386.

———. 2005. El demonio de Puerto Rico. In: R. L. Joglar (ed.), Biodiversidad de Puerto Rico. Vertebrados Terrestres y Ecosistemas, pp. 88–90. Serie de Historia Natural. San Juan: Editorial Instituto de Cultura Puertorriqueña.

Burrowes, P. A., and R. L. Joglar. 1991. A survey of the population status and ecological evaluation of three Puerto Rican frogs. In: J. A. Moreno (ed.), Status y Distribución de los Reptiles y Anfibios de la Región de Puerto Rico, pp. 42–46. Publicación Científica Miscelánea No. 1. San Juan: Departamento de Recursos Naturales de Puerto Rico.

———. 2005. El estado de los anfibios de Puerto Rico. In: R. L. Joglar (ed.), Biodiversidad de Puerto Rico. Vertebrados Terrestres y Ecosistemas, pp. 60–65. Serie de Historia Natural. San Juan: Editorial Instituto de Cultura Puertorriqueña.

Burrowes, P. A., R. L. Joglar, and D. E. Green. 2004. Potential causes for amphibian declines in Puerto Rico. Herpetologica 60: 141–154.

Burrowes, P. A., A. V. Longo, R. L. Joglar, and A. A. Cunningham. 2008. Geographic distribution of *Batrachochytrium dendrobatidis* in Puerto Rico. Herpetological Review 39: 321–324.

Bursey, C. R., and S. R. Goldberg. 1994. *Physalopteroides bahamensis* n. sp. (Nematoda: Spiruroidea) from the Cuban Treefrog *Osteopilus septentrionalis* (Hylidae) from San Salvador Island, Bahamas. Transactions of the American Microscopical Society 113: 169–176.

———. 1995. *Spauligodon caymanensis* sp. n. (Nematoda: Pharyngodonidae) from *Anolis conspersus* (Sauria: Polychridae) from Grand Cayman Island, British West Indies. Journal of the Helminthological Society of Washington 62: 183–187.

———. 1996. *Oochoristica maccoyi* n. sp. (Cestoda: Linstowiidae) from *Anolis gingivinus* (Sauria: Polychrotidae) collected in Anguilla, Lesser Antilles. Caribbean Journal of Science 32: 390–394.

———. 1998. Reclassification of *Spauligodon anolis* (Chitwood, 1934) Inglis 1968 as *Spauligodon anolis* (Chitwood, 1934) n. comb. (Nematoda: Pharyngodonidae) from *Anolis* lizards of the Caribbean. Journal of Parasitology 84: 819–822.

Burton, F. J. 2000. Grand Cayman Iguana, *Cyclura nubila lewisi*. In: A. Alberts (ed.), West Indian Iguanas: Status Survey and Conservation Action Plan, pp. 45–47. Gland, Switzerland and Cambridge, UK: International Union for the Conservation of Nature and Natural Resources.

———. 2004a. *Cyclura lewisi*. In: IUCN 2008. 2008 IUCN Red List of Threatened Species. http://www.iucnredlist.org.

———. 2004b. *Cyclura nubila lewisi* (Grand Cayman Blue Rock Iguana). Reproduction. Herpetological Review 35: 388–389.

———. 2004c. Revision to species of *Cyclura nubila lewisi*, the Grand Cayman Blue Iguana. Caribbean Journal of Science 40: 198–203.

———. 2005a. Restoring a new wild population of Blue Iguanas (*Cyclura lewisi*) in the Salina Reserve, Grand Cayman. Iguana 12: 166–174.

———. 2005b. Blue Iguana Recovery Program update. Iguana Specialist Group Newsletter 8(2): 2–3.

———. 2006. Blue Iguana Recovery Program. Iguana 13: 116–118.

———. 2007. Blue Iguana Recovery Program update. Iguana Specialist Group Newsletter 10(1): 2–3 (reprinted in Iguana 14: 231–232).

Burton, F. J., and Q. M. C. Bloxam (facilitators). 2007a. Turks and Caicos Iguana, *Cyclura carinata carinata*, Conservation and Management Plan, 2005–2009. IUCN/SSC Iguana Specialist Group. http://www.iucn-isg.org/recoveryplan/cover.php.

———. 2007b. Anegada Iguana, *Cyclura pinguis*, Species Recovery Plan, 2006–2010. IUCN/SSC Iguana Specialist Group. http://www.iucn-isg.org/recoveryplan/cover.php.

Burton, F. J., and A. C. Echternacht. 2003. *Bufo marinus*. Geographic Distribution. Herpetological Review 34: 257.

Butler, M. A., and J. B. Losos. 2002. Multivariate sexual dimorphism, sexual selection, and adaptation in Greater Antillean *Anolis* lizards. Ecological Monographs 72: 541–559.

Calderón, S. R., S. R. Bowersox, J. S. Parmerlee, Jr., and R. Powell. 1995. *Anolis barahonae*. Catalogue of American Amphibians and Reptiles 609: 1–3.

Calsbeek, R. 2008. An ecological twist on the morphology-performance-fitness axis. Evolutionary Ecology Research 10: 197–212.

———. 2009. Experimental evidence that competition and habitat use shape the individual fitness surface. Journal of Evolutionary Biology 22: 97–108.

Calsbeek, R., and C. Bonneaud. 2008. Postcopulatory fertilization bias as a form of cryptic sexual selection. Evolution 62: 1137–1148.

Calsbeek, R., and D. J. Irschick. 2007. The quick and the dead: Correlational selection on morphology, performance, and habitat use in island lizards. Evolution 61: 2493–2503.

Calsbeek, R., and E. Marnocha. 2006. Context dependent territory defense: The importance of habitat structure in *Anolis sagrei*. Ethology 112: 537–543.

Calsbeek, R., and T. B. Smith. 2007. Probing the adaptive landscape using experimental islands: Density-dependent natural selection on lizard body size. Evolution 61: 1052–1061.

———. 2008. Experimentally replicated disruptive selection on performance traits in a Caribbean lizard. Evolution 62: 478–484.

Calsbeek, R., J. H. Knouft, and T. B. Smith. 2006. Variation in scale numbers is consistent with ecologically based natural selection acting within and between lizard species. Evolutionary Ecology 20: 377–394.

Calsbeek, R., C. Bonneaud, S. Prabhu, N. Manoukis, and T. B. Smith. 2007a. Multiple paternity and sperm storage lead to increased genetic diversity in *Anolis* lizards. Evolutionary Ecology Research 9: 495–503.

Calsbeek, R., T. B. Smith, and C. Bardeleben. 2007b. Intraspecific variation in *Anolis sagrei* mirrors the adaptive radiation of Greater Antillean anoles. Biological Journal of the Linnean Society 90: 189–199.

Calsbeek, R., C. Bonneaud, and T. B. Smith. 2008. Differential fitness effects of immunocompetence and neighbourhood density in alternative female lizard morphs. Journal of Animal Ecology 77: 103–109.

Calsbeek, R., W. Buermann, and T. B. Smith. 2009. Parallel shifts in ecology and natural selection in an island lizard. BMC Evolutionary Biology 9(3): 12 pp.

Campbell, H. W., and C. K. Dodd, Jr. 1982. Culebra Island Giant Anole recovery plan. Atlanta: U.S. Fish and Wildlife

Service. http://ecos.fws.gov/docs/recovery_plans/1983/830128.pdf.

Campbell, J. A., and W. W. Lamar. 2004. The Venomous Reptiles of the Western Hemisphere. Volume I. Ithaca, New York and London: Comstock (Cornell University Press).

Carey, W. M. 1966. Observations on the ground iguana *Cyclura macleayi caymanensis* on Cayman Brac, West Indies. Herpetologica 22: 265–268.

———. 1972. The herpetology of Anegada, British Virgin Islands. Caribbean Journal of Science 12: 79–89.

———. 1975. The rock iguana, *Cyclura pinguis*, on Anegada, British Virgin Islands, with notes on *Cyclura ricordi* and *Cyclura cornuta* of Hispaniola. Bulletin of the Florida State Museum, Biological Sciences 19: 189–233.

———. 1976. Iguanas of the Exumas. Wildlife 18: 59–61.

Carpenter, C. C. 1982. The aggressive displays of iguanine lizards. In: G. M. Burghardt and A. S. Rand (eds.), Iguanas of the World, pp. 215–231. Park Ridge, N.J.: Noyes.

Carpenter, C. C., and J. C. Gillingham. 1984. Giant Centipede (*Scolopendra alternans*) attacks Marine Toad (*Bufo marinus*). Caribbean Journal of Science 20: 71–72.

———. 1987. Water hole fidelity in the Marine Toad, *Bufo marinus*. Journal of Herpetology 21: 158–161.

Carr, A. F. 1952. Handbook of Turtles. Ithaca, N.Y.: Comstock (Cornell University Press).

Carr, D. K., R. Powell, J. S. Parmerlee, Jr., and D. D. Smith. 1989. *Anolis whitemani whitemani*. Food habits. Herpetological Review 20: 49, 53.

Carrington, C. M. S., G. Evans, and J. Horrocks. 1996. *Liophis perfuscus*. In: IUCN 2008. 2008 IUCN Red List of Threatened Species. http://www.iucnredlist.org.

Carter, R. L., and W. K. Hayes. 1996a. *Cyclura rileyi*. In: IUCN 2008. 2008 IUCN Red List of Threatened Species. http://www.iucnredlist.org.

———. 1996b. *Cyclura rileyi* ssp. *cristata*. In: IUCN 2008. 2008 IUCN Red List of Threatened Species. http://www.iucnredlist.org.

———. 2004. Conservation of an endangered Bahamian rock iguana, II: Morphological variation and conservation priorities. In: A. C. Alberts, R. L. Carter, W. K. Hayes, and E. P. Martins (eds.), Iguanas: Biology and Conservation, pp. 258–273. Berkeley and Los Angeles: University of California Press.

Carter, R. L., W. K. Hayes, and West Indian Iguana Specialist Group. 1996a. *Cyclura rileyi* ssp. *nuchalis*. In: IUCN 2008. 2008 IUCN Red List of Threatened Species. http://www.iucnredlist.org.

———. 1996b. *Cyclura rileyi* ssp. *rileyi*. In: IUCN 2008. 2008 IUCN Red List of Threatened Species. http://www.iucnredlist.org.

Cast, E. E., M. E. Gifford, K. R. Schneider, A. J. Hardwick, J. S. Parmerlee, Jr., and R. Powell. 2000. Natural history of an anoline lizard community in the Sierra de Baoruco, República Dominicana. Caribbean Journal of Science 36: 258–266.

Catesby, M. 1743. The Natural History of Carolina, Florida and the Bahama Islands. Volume 2. London: Published by the author.

CBSG (Conservation Breeding Specialist Group). 2005. Population and Habitat Viability Analysis for the Puerto Rican Crested Toad. Final Report. Apple Valley, Minn.: IUCN/SSC Conservation Breeding Specialist Group.

Censky, E. J. 1989. *Eleutherodactylus johnstonei* (Salientia: Leptodactylidae) from Anguilla, West Indies. Caribbean Journal of Science 35: 229–230.

———. 1995a. Reproduction in two Lesser Antillean populations of *Ameiva plei* (Teiidae). Journal of Herpetology 29: 553–560.

———. 1995b. Mating strategy and reproductive success in the teiid lizard, *Ameiva plei*. Behaviour 132: 529–557.

———. 1996. The evolution of sexual size dimorphism in the teiid lizard *Ameiva plei*: A test of alternative hypotheses. In: R. Powell and R. W. Henderson (eds.), Contributions to West Indian Herpetology: A Tribute to Albert Schwartz, pp. 277–289. Contributions to Herpetology, volume 12. Ithaca, N.Y.: Society for the Study of Amphibians and Reptiles.

———. 1997. Female mate choice in the non-territorial lizard *Ameiva plei* (Teiidae). Behavioral Ecology and Sociobiology 40: 221–225.

———. 1998. *Ameiva plei*. Catalogue of American Amphibians and Reptiles 671: 1–3.

———. 2003. Dog Island days. In: R. W. Henderson and R. Powell (eds.), Islands and the Sea: Essays on Herpetological Exploration in the West Indies, pp. 233–239. Contributions to Herpetology, volume 20. Ithaca, N.Y.: Society for the Study of Amphibians and Reptiles.

Censky, E. J., and K. Hodge. 1997. *Ramphotyphlops braminus* (Brahminy Blind Snake). Geographic Distribution. Herpetological Review 28: 210.

Censky, E. J., and D. R. Paulson. 1992. Revision of the *Ameiva* (Reptilia: Teiidae) of the Anguilla Bank, West Indies. Annals of the Carnegie Museum 61: 177–195.

Censky, E. J., and R. Powell. 2001. Little black dragons of Little Scrub Island. Fauna 2(3): 24–31.

Censky, E. J., K. Hodge, and J. Dudley. 1998. Over-water dispersal of lizards due to hurricanes. Nature 395: 556.

Černý, M. 1992. Natur und Reptilien im östlichen Kuba. Die Aquarien- und Terrarienzeitschrift (DATZ) 45: 596–600.

———. 1994. Kubanische Kugelfingergeckos. Die Aquarien- und Terrarienzeitschrift (DATZ) 47: 656–659.

Černý, V. 1966. Nuevas garrapatas (Ixodoidea) en aves y reptiles de Cuba. Poeyana, Serie A, 26: 1–10.

———. 1969. The tick fauna of Cuba. Folia Parasitologica (Praha) 16: 279–284.

Chamizo Lara, A., L. Rodríguez Schettino, L. V. Moreno

García, M. Dominguez Díaz, and L. M. Díaz Beltrán. 2003. Gigantes and enanos. In: L. Rodríguez Schettino (ed.), Anfibios y Reptiles de Cuba, pp. 74–89. Vaasa, Finland: UPC Print.

Chandler, C. R., and P. J. Tolson. 1990. Habitat use by a boid snake, *Epicrates monensis*, and its anoline prey, *Anolis cristatellus*. Journal of Herpetology 24: 151–157.

Chaparro, J. C., J. B. Pramuk, and A. G. Gluesenkamp. 2007. A new species of arboreal *Rhinella* (Anura: Bufonidae) from cloud forest of southeastern Peru. Herpetologica 63: 203–212.

Cheng, H. Y. 1983. Predation, tail regeneration and reproduction in two sympatric geckos (*Sphaerodactylus*) in Hispaniola. Abstracts of the 26th Annual Meeting of the Society for the Study of Amphibians and Reptiles, p. 52. Tucson, Ariz.

Chitwood, B. G. 1934. Two new nematodes. Smithsonian Miscellaneous Collections 91: 1–4.

Christian, K. 1987. Aspects of the life history of Cuban Iguanas on Isla Magueyes, Puerto Rico. Caribbean Journal of Science 22: 159–164.

Christian, K., and W. T. Lawrence. 1991. Microclimate conditions in nests of the Cuban Iguana (*Cyclura nubila*). Biotropica 23: 287–293.

Christian, K., I. E. Clavijo, N. Cordero-López, E. E. Elías-Maldonado, M. A. Franco, M. V. Lugo-Ramírez, and M. Marengo. 1986. Thermoregulation and energetics of a population of Cuban Iguanas (*Cyclura nubila*) on Isla Magueyes, Puerto Rico. Copeia 1986: 65–69.

Christian, K., W. T. Lawrence, and H. L. Snell. 1991. Effect of soil moisture on yolk and fat distribution in hatchling lizards from natural nests. Comparative Biochemistry and Physiology 99A: 13–19.

Cintrón, G. 1970. Niche separation of tree frogs in the Luquillo Forest. In: H. T. Odum (ed.), A Tropical Rain Forest, pp. E51–E53. Springfield, Va.: U.S. Atomic Energy Commission.

Cisek, J. M., C. A. Cunningham, R. A. Dupuis, R. P. Seibolt, and R. Powell. 1990. *Wetmorena haitiana*. Food habits. Herpetological Review 21: 62.

Cisper, G. L., C. Huntington, D. D. Smith, R. Powell, J. S. Parmerlee, Jr., and A. Lathrop. 1995. Four new Coccidia (Apicomplexa: Eimeriidae) from anoles (Lacertilia: Polychrotidae) in the Dominican Republic. Journal of Parasitology 81: 252–255.

Clark, D. L., and J. C. Gillingham. 1990. Sleep-site fidelity in two Puerto Rican lizards. Animal Behaviour 39: 1138–1148.

Clifford, C. M., and G. M. Kohls. 1962. Description of the female of *Dermacentor latus* Cooley and of *Amblyomma albopictum* Neumann (Acarina, Ixodidae). Journal of Parasitology 48: 486–489.

Clough, G. C., and G. Fulk. 1971. The vertebrate fauna and the vegetation of East Plana Cay, Bahama Islands. Atoll Research Bulletin 138: 1–17.

Cochran, D. M. 1928. A new genus and species of lizard, *Hispaniolus pratensis*, from the Haitian Republic. Proceedings of the Biological Society of Washington 41: 49–52.

———. 1941. Herpetology of Hispaniola. Bulletin of the United States National Museum 177: vii + 398 pp.

Cochran, P. A. 1989. *Anolis sagrei* (Brown Anole). Behavior. Herpetological Review 20: 70.

Coenen, C. 1995. Observations on the Bahamian rock iguana of the Exumas. Bahamas Journal of Science 2: 8–14.

Cofresí-Sala, F. 1964. Notes on the distribution and redescription of *Mesocoelium danforthi* Hoffman, 1935 (Trematoda: Brachycoeliidae). Caribbean Journal of Science 4: 435–439.

Cole, C. J., H. C. Dessauer, C. R. Townsend, and M. G. Arnold. 1995. *Kentropyx borckiana* (Squamata, Teiidae): A unisexual lizard of hybrid origin in the Guiana region, South America. American Museum Novitates 3145: 1–23.

Collen, B., M. Ram, N. Dewhurst, V. Clausnitzer, V. Kalkman, N. Cumberlidge, and J. E. M Baillie. 2008. Broadening the coverage of biodiversity assessments. In: J.-C. Vié, C. Hilton-Taylor, and S. N. Stuart (eds.), The 2008 Review of The IUCN Red List of Threatened Species. Gland, Switzerland: International Union for the Conservation of Nature and Natural Resources. http://cmsdata.iucn.org/downloads/broadening_the_coverage_of_biodiversity_assessments.pdf.

Collette, B. B. 1961. Correlations between ecology and morphology in anoline lizards from Havana, Cuba and southern Florida. Bulletin of the Museum of Comparative Zoology 125: 137–162.

———. 1962. Arboreality, lamellae, and body size in two species of Cuban geckos. Copeia 1962: 644–645.

Colón Archilla, A. D. 2009. Turning the tables: Lizard eats snake. Reptiles & Amphibians 16: 28–29.

Conant, R., and J. T. Collins. 1998. A Field Guide to Reptiles and Amphibians of Eastern and Central North America. 3rd ed., expanded. New York: Houghton Mifflin.

Conners, S., J. Wasilewski, J. Burgess, and J. Bendon. 2005. Bartsch's Rock Iguana (*Cyclura carinata bartschi*) 2005 update. Iguana Specialist Group Newsletter 8(2): 13.

———. 2006. Bartsch's Rock Iguana (*Cyclura carinata bartschi*). Iguana 13: 128.

Conyers, J., and D. Wingate. 1996. *Anolis roosevelti*. In: IUCN 2008. 2008 IUCN Red List of Threatened Species http://www.iucnredlist.org.

Cooper, J. E. 1958. Ecological notes on some Cuban lizards. Herpetologica 14: 53–54.

Cooper, W. E., Jr. 2001. Multiple roles of tail display by the curly-tailed lizard *Leiocephalus carinatus*: Pursuit deterrent and deflective roles of a social signal. Ethology 107: 1137–1149.

———. 2005. Ecomorphological variation in foraging behaviour by Puerto Rican *Anolis* lizards. Journal of Zoology, London 265: 133–139.

———. 2006a. Risk factors affecting escape behaviour by Puerto Rican *Anolis* lizards. Canadian Journal of Zoology 84: 495–504.

———. 2006b. Dynamic risk assessment: Prey rapidly adjust flight initiation distance to changes in predator approach speed. Ethology 112: 858–864.

———. 2007. Escape and its relationship to pursuit-deterrent signaling in the Cuban Curly-tailed Lizard *Leiocephalus carinatus*. Herpetologica 63: 144–150.

Cooper, W. E., Jr., and K. A. Bradley. 2009. Prey chemical discrimination by a diploglossine lizard, the giant Hispaniolan galliwasp (*Celestus warreni*). Amphibia-Reptilia 30: 135–140.

Corke, D. 1987. Reptile conservation on the Maria Islands (St. Lucia, West Indies). Biological Conservation 40: 263–279.

———. 1992. The status and conservation needs of the terrestrial herpetofauna of the Windward Islands (West Indies). Biological Conservation 62: 47–58.

Courchamp, F., E. Angulo, P. Rivalan, R. J. Hall, L. Signoret, L. Bull, and Y. Meinard. 2006. Rarity value and species extinction: The anthropogenic Allee effect. PLoS Biology 4(12). http://biology.plosjournals.org/perlserv/?request=get-document&doi=10.1371%2Fjournal.pbio.0040415.

Cousens, P. N. 1956. Notes on the Jamaican and Cayman Island lizards of the genus *Celestus*. Breviora 56: 1–6.

Coy Otero, A. 1970. Contribución al conocimiento de la helmintofauna de los saurios cubanos. Ciencias Biológicas 4: 1–50.

———. 1979. Tres nuevos hispederos definitivos de nemátodos en Cuba. Miscelanea Zoológica 8: 1.

———. 1989. Nuevo género de tremátodo (Plagiorchiidae) parásito de reptiles cubanos. Poeyana 376: 1–5.

———. 1999. Parasites. In: L. Rodríguez Schettino (ed.), The Iguanid Lizards of Cuba, pp. 77–85. Gainesville: University Press of Florida.

Coy Otero, A., and V. Baruš. 1973a. Notes on nematodes of the genus *Cyrtosomum* (Atractidae) parasitic in Cuban lizards (Sauria). Folia Parasitologica 20: 297–305.

———. 1973b. New hosts for *Parapharyngodon cubensis* (Oxyuridae) of the families Gekkonidae and Iguanidae. Folia Parasitologica 20: 379–380.

———. 1979a. On the species *Trichospirura teixeirai* (Nematoda: Rhabdochonidae) parasitizing Cuban reptiles. Vestník Ceskoslovenské Spolecnosti Zoologické 43: 90–97.

———. 1979b. Nematodes parasitizing Cuban reptiles. Acta Scientiarum Naturalium Brno 13: 1–43.

———. 1980. Nueva especie del género *Physalopteroides* Wu et Liu, 1940 (Nematoda: Physalopteridae) en Cuba. Poeyana 200: 1–6.

Coy Otero, A., and N. Lorenzo Hernández. 1982. Lista de los helmintos parásitos de los vertebrados cubanos. Poeyana 235: 1–57.

Coy Otero, A., and L. Rodríguez Schettino. 1982. Nuevo hospedero definitivo para *Piratuba digiticauda* Lent et Freitas, 1941 (Nematoda: Splendidofilariidae) en Cuba. Miscelanea Zoológica 17: 3–4.

Coy Otero, A., and L. Ventosa. 1984. Nemátodos parásitos de anfibios cubanos. Poeyana 269: 1–20.

Coy Otero, A., L. Ventosa, and A. Quintana. 1980. Nuevo record de nemátoda para Cuba. Miscelanea Zoológica 9: 1.

Coy Otero, A., J. Espinosa, and A. Rams. 1987. Contribución al conocimiento de la fauna de Cayo Fragoso, costa norte de Cuba, Parte 1. Parásitos. Garciana 4: 1–3.

Crocodile Specialist Group. 1996a. *Caiman crocodilus*. In: IUCN 2008. 2008 IUCN Red List of Threatened Species. http://www.iucnredlist.org.

———. 1996b. *Crocodylus acutus*. In: IUCN 2007. 2007 IUCN Red List of Threatened Species. http://www.iucnredlist.org.

———. 1996c. *Crocodylus rhombifer*. In: IUCN 2008. 2008 IUCN Red List of Threatened Species. http://www.iucnredlist.org.

Crombie, R. I. 1977. A new species of frog of the genus *Eleutherodactylus* (Amphibia: Leptodactylidae) from the Cockpit Country of Jamaica. Proceedings of the Biological Society of Washington 90: 194–204.

———. 1986. Another new forest-dwelling frog (Leptodactylidae: *Eleutherodactylus*) from the Cockpit Country of Jamaica. Transactions of the San Diego Society of Natural History 21: 145–153.

Crombie, R. I., D. W. Steadman, and J. C. Barber. 1983 (1984). A preliminary survey of the vertebrates of Cabarita Island, St. Mary Parish, Jamaica. Atoll Research Bulletin 280: 1–12.

Crother, B. I. 1986. *Alsophis cantherigerus* (Cuban Racer). Diet. Herpetological Review 17: 47.

———. 2003. Man of the yard. In: R. W. Henderson and R. Powell (eds.), Islands and the Sea: Essays on Herpetological Exploration in the West Indies, pp. 109–111. Contributions to Herpetology, volume 20. Ithaca, N.Y.: Society for the Study of Amphibians and Reptiles.

Cruz, A., and S. Gruber. 1980. The distribution, ecology, and breeding biology of Jamaican Amazon parrots. In: R. F. Pasquier (ed.), Conservation of New World Parrots, pp. 103–131. International Council for Bird Preservation Technical Publication No. 1. Cambridge, UK: International Council for Bird Preservation.

Cubillas Hernández, S. Y., and V. Berovides Álvarez. 1991. Características de los refugios de la iguana de Cuba *Cyclura nubila*. Revista Biología 5: 85–87.

Cullen, D. J., and R. Powell. 1994. A comparison of food habits of a montane and a lowland population of *Anolis distichus* (Lacertilia: Polychrotidae) from the Dominican

Republic. Bulletin of the Maryland Herpetological Society 30: 62–66.

Cunningham, C. A., J. S. Tulloch, and R. Powell. 1993. Food habits of four Hispaniolan species of *Sphaerodactylus* (Lacertilia: Gekkonidae). Herpetological Natural History 1(2): 91–93.

Cunningham, C. A., R. Powell, and S. B. Hedges. 1998. *Eleutherodactylus chlorophenax*. Catalogue of American Amphibians and Reptiles 648: 1–2.

Curio, E. 1970. Jagdverhalten eines *Anolis* und das Farbmuster seiner Beute. Die Naturwissenschaften 57: 361–362.

Currat, P. 1980. Les Reptiles. Pointe-à-Pitre, Guadeloupe: Centre Departemental de Documentation Pédagogique.

Cusumano, M. A., and R. Powell. 1991. A note on the diet of *Amphisbaena gonavensis* in the Dominican Republic. Amphibia-Reptilia 12: 350–352.

Cusumano, M. A., D. J. Pflanz, and R. Powell. 1991. *Anolis whitemani* (NCN). Tail autotomy. Herpetological Review 22: 58.

Daltry, J. C. 1999. Status, Distribution, and Natural History of the Antiguan Racer (*Alsophis antiguae*): Field Research and Conservation Technical Report, 1995 to 1999. Prepublication Draft Report. Cambridge, UK: Fauna & Flora International Caribbean Programme.

———. 2007. An introduction to the herpetofauna of Antigua, Barbuda and Redonda, with some conservation recommendations. Applied Herpetology 4: 97–130.

Daltry, J. C., and G. Gray. 1999. Effects of volcanic activity on the endangered Mountain Chicken Frog (*Leptodactylus fallax*). Froglog 32: 1–2.

Daltry, J. C., Q. Bloxam, G. Cooper, M. L. Day, J. Hartley, M. Henry, K. Lindsay, and B. E. Smith. 2001. Five year's of conserving the 'world's rarest snake,' the Antiguan Racer *Alsophis antiguae*. Oryx 35: 119–127.

Danforth, S. T. 1925. Porto Rican herpetological notes. Copeia 1925: 76–79.

———. 1928. Bird observations in Jamaica during the summer of 1926. Auk 45: 480–491.

———. 1939. Birds of Guadeloupe and adjacent islands. Journal of Agriculture, University of Porto Rico 23: 9–46.

Daniells, E. A., J. W. Ackley, R. E. Carter, P. J. Muelleman, S. M. Rudman, P. A. Turk, N. J. Vélez Espinet, L. A. White, and N. N. Wyszynski. 2008. An annotated checklist of the amphibians and reptiles of Dominica, West Indies. Iguana 15: 130–141.

Darlington, P. J. 1927. Autohemorrhage in *Tropidophis semicinctus*. Bulletin of the Antivenin Institute of America 1: 59.

Daszak, P., and S. J. Ball. 1991. Five new species of *Eimeria* (Apicomplexa: Eimeriidae) from lizards. Systematic Parasitology 20: 141–147.

Daudin, J., and M. de Silva. 2007. An annotated checklist of the amphibians and terrestrial reptiles of the Grenadines with notes on their local natural history and conservation. Applied Herpetology 4: 163–175.

Dávalos, L. M., and R. Eriksson. *Epicrates subflavus* (Jamaican Boa). 2004. Foraging behavior. Herpetological Review 35: 66.

Dávila, N. V., and L. F. de Armas. 1989. Depradación de *Tropidophis melanurus* (Serpentes: Tropidophiidae) por *Epicrates angulifer* (Serpentes: Boidae). Miscelanea Zoológica 41: 2–3.

Davis, S. 2000. Genetic research needs. In: A. Alberts (ed.), West Indian Iguanas: Status Survey and Conservation Action Plan, pp. 89–91. Gland, Switzerland and Cambridge, UK: International Union for the Conservation of Nature and Natural Resources.

Davis, S. L., R. B. Davis, A. James, and B. C. P. Talyn. 2000. Reproductive behavior and larval development of *Leptodactylus fallax* in Dominica, West Indies. Herpetological Review 31: 217–220.

Day, M. 1990. The ecology and status of the Lesser Antillean Iguana on the island of Dominica, Lesser Antilles: Initial report. Unpublished report to the Commonwealth of Dominica, Rosseau.

———. 1996a. *Cyclura cornuta*. In: IUCN 2008. 2008 IUCN Red List of Threatened Species. http://www.iucnredlist.org.

———. 1996b. *Cyclura nubila*. In: IUCN 2008. 2008 IUCN Red List of Threatened Species. http://www.iucnredlist.org.

———. 1996c. *Diploglossus montisserrati*. In: IUCN 2008. 2008 IUCN Red List of Threatened Species. http://www.iucnredlist.org.

———. 1996d. *Alsophis antiguae*. In: IUCN 2008. 2008 IUCN Red List of Threatened Species. http://www.iucnredlist.org.

———. 1996e. *Alsophis rijersmai* [sic]. In: IUCN 2008. 2008 IUCN Red List of Threatened Species. http://www.iucnredlist.org.

———. 1996f. *Alsophis rufiventris*. In: IUCN 2008. 2008 IUCN Red List of Threatened Species. http://www.iucnredlist.org.

Day, M., and R. S. Thorpe. 1996. Population differentiation of *Iguana delicatissima* and *I. iguana* in the Lesser Antilles (abstract in English, Spanish, and French). In: R. Powell and R.W. Henderson (eds.), Contributions to West Indian Herpetology: A Tribute to Albert Schwartz, pp. 436–437. Contributions to Herpetology, volume 12. Ithaca, N.Y.: Society for the Study of Amphibians and Reptiles.

Day, M., and P. J. Tolson. 1996. *Epicrates angulifer*. In: IUCN 2008. 2008 IUCN Red List of Threatened Species. http://www.iucnredlist.org.

Day, M., M. Breuil, and S. Reichling. 2000. Lesser Antillean Iguana: *Iguana delicatissima*. In: A. Alberts (ed.), West Indian Iguanas: Status Survey and Conservation Action Plan, pp. 62–67. Gland, Switzerland and Cambridge, UK:

International Union for Conservation of Nature and Natural Resources.

Daza, J. D., P. A. Burrowes, and P. Medina. 2008. *Eleutherodactylus coqui* (Coquí). Predation. Herpetological Review 39: 459–460.

de la Cruz, J. 1973. Nueva especie de ácaro del género *Geckobia*, Megnin, 1878 (Acarina; Pterygosomidae) parásito de la *Tarentola americana* (Gray) de Cuba. Poeyana 102: 1–6.

———. 1984. Nueva especie de garrapata del género *Ornithodoros* (Acarina, Ixodoidea, Argasidae), parásita nasal de iguana *Cyclura nubila* (Sauria, Iguanidae) de Cuba. Poeyana 277: 1–6.

de la Cruz, J., and R. M. Abreu. 1986. El abuje o abujo, *Eutrombicula alfreddugesi* (Oudemans, 1910) (Acarina: Trombiculidae), un ácaro de importancia biomedical en las provincias orientales de Cuba. Revista Cubana de Medicina Tropical 38: 119–125.

De Lalung, H. 1934. Le Serpent de la Martinique. Sa Légende, ses Mœurs, ses Ennemis. Comment les Caraïbes et les Nègres Soignaient ses Piqûres. Essai de Médecine Rétrospective. Paris: Laboratoires Corbière.

Dent, J. N. 1956. Observations on the life history and development of *Leptodactylus albilabris*. Copeia 1956: 207–210.

de Silva, M., and D. Wilson. 2006. A Natural History of Mustique. Mustique Island: The Mustique Company.

de Sola, C. R. 1930. The Cuban Crocodile: An account of the species *Crocodilus rhombifer* Cuvier, with notes on its life history. Copeia 1930: 81–83.

Dial, R., and J. Roughgarden. 1995. Experimental removal of insectivores from rain forest canopy: Direct and indirect effects. Ecology 76: 1821–1834.

Dial, R., J. Roughgarden, and S. C. Tobin. 1994. Notes on the absolute abundance of canopy anoles, *Anolis cuvieri*, *A. stratulus*, and *A. evermanni* (Lacertilia: Polychridae) in the Luquillo Forest, Puerto Rico. Caribbean Journal of Science 30: 278–279.

Díaz, D. C. 1984. Recovery Plan for the Mona Iguana, *Cyclura stejnegeri*. Unpublished report. Atlanta: U.S. Fish and Widlife Service.

Díaz, L. M. 1999. Planarias y chinches acuáticas (Tricladia, Hemiptera), depredadores de la rana platanera *Osteopilus septentrionalis* (Anura: Hylidae). Cocuyo 9: 24–25.

———. 2004. *Eleutherodactylus simulans*. In: IUCN 2008. 2008 IUCN Red List of Threatened Species. http://www.iucnredlist.org.

———. 2006. Anfibios y Reptiles. En Cuba: Camaguey, Sierra de Cubitas. In: L. M. Díaz, W. S. Alverson, A. Barreto, and T. Wachter (eds.), Rapid Biological Inventories Report 08, pp. 48–50 + 164–165 + Figures A–B. Chicago: The Field Museum.

———. 2008a. *Eleutherodactylus michaelschmidi*. In: IUCN 2008. 2008 IUCN Red List of Threatened Species. http://www.iucnredlist.org.

———. 2008b. Pflege und Zucht des Kubanischen Gelbkopfgeckos *Gonatodes albogularis fuscus*. Aquaristik Fachmagazin 199: 90–91.

Díaz, L. M., and E. Abreu Guerra. 2005. Anfibios y Reptiles. En Cuba: Península de Zapata. Rapid Biological Inventories Report 07, pp. 50–53 + 126 + Figures 4A–L. Chicago: The Field Museum.

Díaz, L. M., and A. Cádiz. 2006. Pflege und Vermehrung von *Bufo longinasus* Stejneger, 1905. Ein Beitrag zur Erhaltung dieser Art. Aquaristik Fachmagazin 188: 18–21.

———. 2008. Guía taxonómica de los anfibios de Cuba. Abc Taxa 4: vi + 294 pp. + audio compact disk.

Díaz, L. M., and A. R. Estrada. 2000. The male and female vocalizations of the Cuban frog *Eleutherodactylus guanahacabibes* (Anura: Leptodactylidae). Caribbean Journal of Science 36: 328–331.

Díaz, L. M., and A. Fong Grillo. 2001. A new mottled frog of the genus *Eleutherodactylus* (Anura: Leptodactylidae) from eastern Cuba. Solenodon 1: 76–84.

Díaz, L. M., and S. B. Hedges. 2008. A new gecko of the genus *Tarentola* (Squamata: Gekkonidae) from eastern Cuba. Zootaxa 1743: 43–52.

———. 2009. First record of the genus *Aristelliger* (Squamata: Sphaerodactylidae) in Cuba, with the description of a new species. Zootaxa 2028: 31–40.

Díaz, L. M., A. R. Estrada, and L. V. Moreno. 1996. A new species of *Anolis* (Sauria: Iguanidae) from the Sierra de Trinidad, Sancti Spiritus, Cuba. Caribbean Journal of Science 32: 54–58.

Díaz, L. M., L. V. Moreno, and V. Berovides. 1998a. *Cricosaura typica*. In: E. Pérez, E. Osa, Y. Matamoros, and U. S. Seal (eds.), Report of Conservation Assessment and Management Plan Workshop for Selected Cuban Species II. Sección III. Reptiles. Apple Valley, Minn.: IUCN/SSC Conservation Breeding Specialist Group.

Díaz, L. M., N. Navarro, and O. H. Garrido. 1998b. Nueva especies de *Chamaeleolis* (Sauria: Iguanidae) de la Meseta de Cabo Cruz, Granma, Cuba. A new species of *Chamaeleolis* (Sauria: Iguanidae) from Meseta de Cabo Cruz, Granma, Cuba. Avicennia 8/9: 27–34.

Díaz, L. M., L. V. Moreno, and G. Begue. 2000. Descripción de las larvas de *Bufo taladai* (Anura: Bufonidae). Caribbean Journal of Science 36: 156–160.

Díaz, L. M., A. R. Estrada, and S. B. Hedges. 2001. A new riparial frog of the genus *Eleutherodactylus* (Anura: Leptodactylidae) from eastern Cuba. Caribbean Journal of Science 37: 63–71.

Díaz, L. M., A. Cádiz, and S. B. Hedges. 2003. A new grass frog from pine forests of western Cuba, and description of acoustic and pattern variation in *Eleutherodactylus varleyi* (Amphibia: Leptodactylidae). Caribbean Journal of Science 39: 176–188.

Díaz, L. M., A. Cádiz, and N. Navarro. 2005a. A new ground-dwelling frog of the genus *Eleutherodactylus* (Anura: Leptodactylidae) from eastern Cuba, and a reconsideration of the *E. dimidiatus* group. Caribbean Journal of Science 41: 307–318.

Díaz, L. M., A. Fong Grillo, N. Viña Dávila, and G. Knell. 2005b. Anfibios y Reptiles. En Cuba: Parque Nacional La Bayamesa. In: D. Maceira F., A. Fong Grillo, W. S. Alverson, and T. Wachter (eds.), Rapid Biological Inventories Report 13, pp. 72–76 + 228–231 + Figs. 6A–E. Chicago: The Field Museum.

Díaz, L. M., A. Cádiz, A. Chong, and A. Silva. 2007a. First report of chytridiomycosis in a dying toad (Anura: Bufonidae) from Cuba: A new conservation challenge for the island. EcoHealth 4: 172–175.

Díaz, L. M., A. Cádiz, and N. Navarro. 2007b. A new rock dwelling frog of the genus *Eleutherodactylus* (Amphibia: Leptodactylidae) from eastern Cuba, with comments on other species with similar habits. Zootaxa 1435: 51–68.

Diaz-Lameiro, A. M., R. Powell, and C. S. Berg. 2007. Colour and pattern polymorphism in *Pristimantis shrevei* and *Eleutherodactylus johnstonei* (Leptodactylidae) on St. Vincent, West Indies. Herpetological Bulletin 101: 18–25.

Díaz-Lameiro, A. M., H. Kaiser, and R. Powell. 2008. *Pristimantis shrevei*. Catalogue of American Amphibians and Reptiles 844: 1–4.

Dickenson, H. C., and J. E. Fa. 2000. Abundance, demographics and body condition of a translocated population of St. Lucia Whiptail Lizards (*Cnemidophorus vanzoi*). Journal of Zoology 251: 187–197.

Dickenson, H. C., J. E. Fa, and S. M. Lenton. 2001. Microhabitat use by a translocated population of St. Lucia Whiptail Lizards (*Cnemidophorus vanzoi*). Animal Conservation 4: 143–156.

Diesel, R., G. Bäurle, and P. Vogel. 1995. Cave breeding and froglet transport: A novel pattern of anuran brood care in the Jamaican frog, *Eleutherodactylus cundalli*. Copeia 1995: 354–360.

Dixon, J. R. 1962. The Leaf-toed Geckoes, genus *Phyllodactylus*, of northeastern South America. Southwestern Naturalist 7: 211–226.

———. 1973. A systematic review of the teiid lizards, genus *Bachia*, with remarks on *Heterodactylus* and *Anotosaurus*. Miscellaneous Publications of the University of Kansas Museum of Natural History 57: 1–47.

———. 1981. The Neotropical colubrid snake genus *Liophis*: The Eastern Caribbean complex. Copeia 1981: 296–304.

Dmi'el, R., G. Perry, and J. Lazell. 1997. Evaporative water loss in nine insular populations of the lizard *Anolis cristatellus* group in the British Virgin Islands. Biotropica 29: 111–116.

Dobson, A. P., and S. V. Pacala. 1992. The parasites of *Anolis* lizards in the northern Lesser Antilles. II. The structure of the parasite communities. Oecologica 92: 118–125.

Dobson, A. P., S. V. Pacala, J. D. Roughgarden, E. R. Carper, and E. A. Harris. 1992. The parasites of *Anolis* lizards in the northern Lesser Antilles. I. Patterns of distribution and abundance. Oecologia 91: 110–117.

Dodd, C. K., Jr. 1978. Island lizard in danger. National Parks & Conservation Magazine 52(8): 10–11.

———. 1980. *Ameiva polops*. Catalogue of American Amphibians and Reptiles 240: 1–2.

———. 1985. *Sphaerodactylus micropithecus*. Catalogue of American Amphibians and Reptiles 363: 1–2.

———. 1994. Culebra Island Giant Anole *Anolis roosevelti*. In: M. Emanoil (ed.), Encyclopedia of Endangered Species, pp. 813–814. Detroit: Gale Research.

———. 2003. Turquoise water, silver palms, and fluorescent green *Leiocephalus*. In: R. W. Henderson and R. Powell (eds.), Islands and the Sea: Essays on Herpetological Exploration in the West Indies, pp. 29–38. Contributions to Herpetology, volume 20. Ithaca, N.Y.: Society for the Study of Amphibians and Reptiles.

Dodd, C. K., Jr., and H. W. Campbell. 1982. *Anolis roosevelti*. Catalogue of American Amphibians and Reptiles 300: 1–2.

Dodd, C. K., Jr., and R. Franz. 1996. Species richness and biogeography of the herpetofauna in the Exuma Cays Land and Sea Park, Bahamas. In: R. Powell and R. W. Henderson (eds.), Contributions to West Indian Herpetology: A Tribute to Albert Schwartz, pp. 359–369. Contributions to Herpetology, volume 12. Ithaca, N.Y.: Society for the Study of Reptiles and Amphibians.

Dodd, C. K., Jr., and P. R. Ortiz R. 1983. An endemic gecko in the Caribbean. Oryx 17: 119–121.

———. 1984. Variation of dorsal pattern and scale counts in the Monito Gecko, *Sphaerodactylus micropithecus*. Copeia 1984: 768–770.

Domínguez Díaz, M., and L. V. Moreno García. 2003. Serpientes del suelo. In: L. Rodríguez Schettino (ed.), Anfibios y Reptiles de Cuba, pp. 98–109. Vaasa, Finland: UPC Print.

———. 2005. *Tropidophis pilsbryi* (NCN). Size record. Herpetological Review 36: 330.

———. 2006a. *Alsophis cantherigerus* (Jubo, Jubo de Sabana, Jubo Sabanero). Size record. Herpetological Review 37: 349.

———. 2006b. *Tropidophis maculatus* (NCN). Size record. Herpetological Review 37: 355–356.

———. 2006c. *Tropidophis pardalis* (NCN). Size record. Herpetological Review 37: 356.

———. 2009. Taxonomy of the Cuban blind snakes (Scolecophidia, Typhlopidae), with the description of a new species. Zootaxa 2028: 59–66.

Domínguez Díaz, M., L. V. Moreno García, and M. Sánchez. 2005. *Tropidophis wrighti* (NCN). Size record. Herpetological Review 36: 197.

Domínguez Díaz, M., L. V. Moreno García, and S. B. Hedges. 2006a. A new snake of the genus *Tropidophis* (Tropidophiidae) from the Gunahacabibes Peninsula of western Cuba. Amphibia-Reptilia 27: 427–432.

Domínguez Díaz, M., L. V. Moreno García, and L. Rodríguez Schettino. 2006b. *Tropidophis morenoi* (NCN). Size record. Herpetological Review 37: 356.

Domínguez Díaz, M., A. Sanz, N. Almaguer, and J. Chávez. 2006c. Seasonal reproduction in males of the Cuban lizard *Anolis lucius* (Polychrotidae). Boletín de la Sociedad Herpetológica Mexicana 14: 1–9.

Dosse, G. 1939. Über einige Nematoden aus *Cyclura carinata* R. Harlan, 1824 (*Macracis monhystera* v. Linstow, 1902; *Travassozolaimus travassosi* Vigueras, 1938; *Mammillomacracis cyclurae* n. g., n. sp.). Zeitschrift für Parasitenkunde 10: 694–703.

Dowling, H. G. 1965. The puzzle of the *Bothrops*; or, a tangle of serpents. Animal Kingdom, February: 18–21.

Drewry, G. E. 1970a. The role of amphibians in the ecology of Puerto Rican rain forest. In: R. G. Clements, G. E. Drewry, and R. J. Lavigne (eds.), The Rain Forest Project Annual Report, pp. 16–54. PRNC-147. Río Piedras: Puerto Rico Nuclear Center.

———. 1970b. Factors affecting activity of rain forest frog populations as measured by electrical recording of sound pressure levels. In: H. T. Odum (ed.), A Tropical Rain Forest, pp. E55–E68. Springfield, Va.: U.S. Atomic Energy Commission.

Drewry, G. E., and K. L. Jones. 1976. A new ovoviviparous frog, *Eleutherodactylus jasperi* (Amphibia, Anura, Leptodactylidae) from Puerto Rico. Journal of Herpetology 10: 161–165.

Drewry, G. E., and A. S. Rand. 1983. Characteristics of an acoustic community: Puerto Rican frogs of the genus *Eleutherodactylus*. Copeia 1983: 941–953.

Duellman, W. W., and R. I. Crombie. 1970. *Hyla septentrionalis*. Catalogue of American Amphibians and Reptiles 92: 1–4.

Duellman, W. E., and A. Schwartz. 1958. Amphibians and reptiles of southern Florida. Bulletin of the Florida State Museum 3: 181–324.

Duellman, W. E., and J. J. Wiens. 1993. Hylid frogs of the genus *Scinax* Wagler, 1830, in Amazonian Ecuador and Peru. Occasional Papers of the Museum of Natural History, University of Kansas 153: 1–57.

Duer, C. K., J. M. Cisek, and R. Powell. 1992. Food habits of *Osteopilus dominicensis* (Anura: Hylidae). Caribbean Journal of Science 28: 226–228.

Duméril, A.-M.-C., and G. Bibron. 1836. Erpétologie Générale ou Histoire Naturelle Complète des Reptiles. Volume 3. Paris: Librarie Encyclopédique de Roret.

Duncan, W. 1889. "Trigonocephalus Lanceolatus." Notes on the West Indian "Fer-de-Lance." Proceedings of the Bristol Naturalists' Society, Series 2, 6: 44–55.

Dunn, E. R. 1925. New frogs from Cuba. Occasional Papers of the Boston Society of Natural History 5: 163–166.

———. 1926a. Additional frogs from Cuba. Occasional Papers of the Boston Society of Natural History 5:209–215.

———. 1926b. The frogs of Jamaica. Proceedings of the Boston Society of Natural History 38: 111–130.

———. 1926c. The habits of *Sminthillus limbatus*. Copeia 1926: 141–143.

———. 1926d. Notes on Cuban anoles. Copeia 1926: 153–154.

———. 1926e. Notes on two Cuban frogs. Copeia 1926: 154–156.

Dunson, W. A., and M. E. Seidel. 1986. Salinity tolerance of estuarine and insular turtles (*Pseudemys nelsoni* and *Trachemys decussata*). Journal of Herpetology 20: 237–245.

Durden, L. A., and C. R. Knapp. 2005. Ticks parasitizing reptiles in the Bahamas. Medical and Veterinary Entomology 19: 326–328.

Dyer, W. G., L. Bunkley-Williams, and E. H. Williams, Jr. 1999. Two new Caribbean records of parasitic nematodes collected from reptiles in Puerto Rico: *Aplectana pusilla* in *Amphisbaena bakeri* and *Alaeuris vogelsangi* in *Iguana iguana*. Caribbean Journal of Science 35: 158–159.

Eales, J., R. S. Thorpe, and A. Malhotra. 2008. Weak founder effect signal in a recent introduction of Caribbean *Anolis*. Molecular Ecology 17: 1416–1426.

Easteal, S. 1986. *Bufo marinus*. Catalogue of American Amphibians and Reptiles 395: 1–4.

Eaton, J. M., K. G. Howard, and R. Powell. 2001. *Anolis carolinensis*. Geographic distribution. Herpetological Review 32: 118.

Eaton, J. M., S. C. Larimer, K. G. Howard, R. Powell, and J. S. Parmerlee, Jr. 2002. Population densities and ecological release of a solitary species: *Anolis gingivinus* on Anguilla, West Indies. Caribbean Journal of Science 38: 27–36.

Echternacht, A. C., and F. J. Burton. 2003. *Ramphotyphlops braminus*. Geographic Distribution. Herpetological Review 34: 265–266.

Echternacht, A. C., and G. P. Gerber. 2000. *Anolis conspersus* (Grand Cayman Blue-throated Anole). Nectivory. Herpetological Review 31: 173.

Edwards, E. 1997. The pearl of the Caribbean. The Reptilian Magazine 5(4): 47–52.

Elstrott, J., and D. J. Irschick. 2004. Evolutionary correlations among morphology, habitat use and clinging performance in Caribbean *Anolis* lizards. Biological Journal of the Linnean Society 83: 389–398.

Emsley, M. G., 1966. The confirmation of the blind snake *Leptotyphlops* in Barbados. Copeia 1966: 125.

Engeman, R. M., H. T. Smith, and B. Constantine. 2005. Invasive Green Iguanas as airstrike hazards at San Juan International Airport, Puerto Rico. Journal of Aviation-Aerospace Education and Research 14: 45–50.

English, T. M. S. 1912. Some notes on the natural history of Grand Cayman. Handbook of Jamaica for 1912: 598–600.

Ernst, C. H., and T. E. J. Leuteritz. 1999a. *Geochelone carbonaria*. Catalogue of American Amphibians and Reptiles 690: 1–7.

———. 1999b. *Geochelone denticulata*. Catalogue of American Amphibians and Reptiles 691: 1–6.

Ernst, C. H., F. D. Ross, and C. A. Ross. 1999. *Crocodylus acutus*. Catalogue of American Amphibians and Reptiles 700: 1–17.

Espeut, W. B. 1882. On the acclimatization of the Indian mungoos in Jamaica. Proceedings of the Zoological Society of London 1882: 712–714.

Estrada, A. R. 1987a. Sintopía de *Anolis argenteolus* y *A. lucius* en la Sierra de Najasa, Camagüey, Cuba. Miscelanea Zoológica 31: 3–4.

———. 1987b. *Anolis argillaceus* (Sauria: Iguanidae): Un nuevo caso de puestas comunales en anolis cubanos. Poeyana 353: 1–9.

———. 1987c. Los nidos terrestres de dos especies de anfibios cubanos del genero *Eleutherodactylus* (Anura: Leptodactylidae). Poeyana 352: 1–9.

———. 1988. Nido communal de *Anolis argenteolus* (Sauria: Iguanidae). Garciana 9: 3–4.

———. 1990. Las puestas de *Eleutherodactylus varians* (Gundlach et Peters). Revista Biologia 4: 163–167.

———. 1992. Comentarios sobre *Eleutherodactylus intermedius* Barbour et Shreve, 1937 de la Región Oriental de Cuba. Comunicaciones Breves de Zoología Academicia de Ciencias de Cuba 1992: 12–14.

———. 1993a. Herpetofauna del Archipiélago de los Canarreos, Cuba. Poeyana 431: 1–19.

———. 1993b. Anfibios y reptiles de Cayo Coco, Archipiélago de Sabana-Camagüey, Cuba. Poeyana 432: 1–21.

———. 1994. Herpetofauna de la Cuenca Banao-Higuanojo, Sancti Spíritus, Cuba. Revista de la Academia Colombiana de Ciencias 19: 353–360.

———. 1998. Anfibios y reptiles encontrados durante 1988 y 1989 en Cayo Paredón Grande, Archipiélago de Sabana-Camagüey, Cuba. Caribbean Journal of Science 34: 106–112.

———. 2002. Data on a clutch of *Anolis evermani* [sic] (Sauria: Iguanidae) from Puerto Rico. Solenodon 2: 53–54.

Estrada, A. R., and R. Alonso Bosch. 1997. Nueva especie del grupo *limbatus* (Leptodactylidae: *Eleutherodactylus*) de la region Oriental de Cuba. Caribbean Journal of Science 33: 41–44.

Estrada, A. R., and L. F. de Armas. 1998. Apuntes ecológicos sobre *Cricosaura typica* (Sauria: Xantusiidae) de Cuba. Caribbean Journal of Science 34: 157–160.

Estrada, A. R., and O. H. Garrido. 1990. Nueva subespecie de *Anolis jubar* (Lacertilia: Iguanidae) para Cayo Coco y la loma de Cunagua, Ciego de Avila, Cuba. Revista Biología 4: 71–79.

———. 1991. Dos nuevas especies de *Anolis* (Lacertilia: Iguanidae) de la región oriental de Cuba. Caribbean Journal of Science 27: 146–161.

Estrada, A. R., and S. B. Hedges. 1991. Nueva especie de *Eleutherodactylus* (Anura: Leptodactylidae) de la región Oriental de Cuba. Caribbean Journal of Science 27:139–145.

———. 1995. A new species of *Anolis* (Sauria: Iguanidae) from eastern Cuba. Caribbean Journal of Science 31: 65–72.

———. 1996a. A new frog of the genus *Eleutherodactylus* from eastern Cuba (Anura: Leptodactylidae). Herpetologica 52: 435–439.

———. 1996b. At the lower size limit in tetrapods: A new diminutive frog from Cuba (Leptodactylidae: *Eleutherodactylus*). Copeia 1996: 852–859.

———. 1997a. Nueva especie de *Eleutherodactylus* (Anura: Leptodactylidae) del Macizo Sagua-Baracoa, Cuba. Caribbean Journal of Science 33: 222–226.

———. 1997b. A new species of frog from the Meseta de Cabo Cruz, eastern Cuba (Leptodactylidae, *Eleutherodactylus*). Caribbean Journal of Science 33: 227–232.

———. 1997c. A new species of frog from the Sierra Maestra, Cuba (Leptodactylidae, *Eleutherodactylus*). Journal of Herpetology 31: 364–368.

———. 1998. Sistemática de las ranas ribereñas de Cuba (Leptodactylidae: *Eleutherodactylus*) con la descripción de una especie nueva. Caribbean Journal of Science 34: 218–230.

Estrada, A. R., and J. Novo Rodríguez. 1985a. Nueva especie de *Eleutherodactylus* del grupo *ricordii* (Anura: Leptodactylidae) del occidente de Cuba. Poeyana 303: 1–10.

———. 1985b. Nuevos datos sobre las puestas comunales de *Anolis bartschi* (Sauria: Iguanidae) en la Sierra de los Organos, Pinar del Río, Cuba. Ciencias Biológicas 15: 135–136.

———. 1986a. Subnicho estructural de *Anolis bartschi* en la Sierra de los Órganos, Pinar del Río, Cuba. Poeyana 316: 1–10.

———. 1986b. Subnicho estructural de *Anolis sagrai* [sic] en Cayo Inés de Soto, Cuba: Análisis intra- y extrapoblacional. Poeyana 320: 1–13.

———. 1987. Subnicho climático de *Anolis bartschi* (Sauria: Iguanidae). Poeyana 341: 1–19.

Estrada, A. R., and J. Novo Rodríguez, and L. V. Moreno. 1986. Las ranas del grupo *symingtoni*, género *Eleutherodactylus* (Anura: Leptodactylidae) de Cuba. Poeyana 329: 1–14.

———. 1987. Datos sobre una puesta de *Ameiva auberi* Cocteau (Sauria: Teiidae). Miscelanea Zoológica 30: 2–3.

Estrada, A. R., L. M. Díaz, and A. Rodríguez. 1997. Nueva especie de *Eleutherodactylus* (Anura: Leptodactylidae) del litoral norte de la Habana, Cuba. Revista Española de Herpetología 11: 19–24.

Etheridge, R. E. 1965. Fossil lizards from the Dominican Republic. Quarterly Journal of the Florida Academy of Sciences 30: 197–220.

Evans, L. T. 1938a. Cuban field studies on territoriality of the lizard, *Anolis sagrei*. Journal of Comparative Psychology 25: 97–125.

———. 1938b. Courtship behavior and sexual selection of *Anolis*. Journal of Comparative Psychology 26: 475–497.

———. 1953. Tail display in an iguanid lizard, *Leiocephalus carinatus*. Copeia 1953: 50–54.

Evans, M. A. 1989. Ecology and removal of introduced Rhesus Monkeys: Desecheo Island National Wildlife Refuge, Puerto Rico. Puerto Rico Journal of Health Sciences 8: 139–156.

Evans, M. A., H. J. Herbert, and K. Rohnke. 1991. Observations on the status of the herpetofauna of Desecheo Island National Wildlife Refuge, Puerto Rico. In: J. A. Moreno (ed.), Status y Distribución de los Reptiles y Anfibios de la Región de Puerto Rico, pp. 34–36. Publicación Científica Miscelánea No. 1. San Juan: Departamento de Recursos Naturales de Puerto Rico.

Eyerdam, W. J. 1954. With Dr. Eric Ekman in southern Haiti, in the summer of 1927. Field and Laboratory 22: 85–106.

Fa, J. E., S. B. Hedges, B. Ibéné, M. Breuil, R. Powell, and C. Magin. 2004. *Leptodactylus fallax*. In: IUCN 2008. 2008 IUCN Red List of Threatened Species. http://www.iucnredlist.org.

Faaborg, J. R., and W. J. Arendt. 1985. Wildlife Assessment in the Caribbean. Río Piedras, Puerto Rico: Institute of Tropical Forestry.

Faivovich, J., C. E. B. Haddad, P. C. A. Garcia, D. R. Frost, J. A. Campbell, and W. C. Wheeler. 2005. Systematic review of the frog family Hylidae with special reference to Hylinae: Phylogenetic analysis and taxonomic revision. Bulletin of the American Museum of Natural History 294: 1–240.

FAO (United Nations Food and Agriculture Organization). 2005. Global Forest Resources Assessment 2005: Progress Towards Sustainable Forest Management. Rome: United Nations Food and Agriculture Organization.

Fernández, E. 2007. Hispaniola. A Photographic Journey Through Island Biodiversity. Biodiversidad a Través de un Recorrido Fotográfico. Cambridge, Mass.: Belknap Press (Harvard University Press).

Fernández Méndez, I. 1998a. *Anolis pigmaequestris*. In: E. Pérez, E. Osa, Y. Matamoros, and U. S. Seal (eds.), Report of Conservation Assessment and Management Plan Workshop for Selected Cuban Species II. Sección III. Reptiles. Apple Valley, Minn.: IUCN/SSC Conservation Breeding Specialist Group.

———. 1998b. *Anolis equestris* ssp. In: E. Pérez, E. Osa, Y. Matamoros, and U. S. Seal (eds.), Report of Conservation Assessment and Management Plan Workshop for Selected Cuban Species II. Sección III. Reptiles. Apple Valley, Minn.: IUCN/SSC Conservation Breeding Specialist Group.

Fernández Méndez, I., and E. Manso Valdés. 1998. Anolis gigantes. Flora y Fauna 2: 43–45.

Fernández Méndez, I., O. Alonso, L. Rodríguez Schettino, and A. R. Chamizo Lara. 1998. *Anolis equestris potior*. In: E. Pérez, E. Osa, Y. Matamoros, and U. S. Seal (eds.), Report of Conservation Assessment and Management Plan Workshop for Selected Cuban Species II. Sección III. Reptiles. Apple Valley, Minn.: IUCN/SSC Conservation Breeding Specialist Group.

Fernández Milera, J. 1984. Agresividad de *Leiocephalus cubensis* Gray 1840 (Reptilia: Sauria: Iguanidae). Miscelanea Zoológica 22: 2.

Fernández Velzáquez, A., and R. Alonso Bosch. 2005. *Tropidophis wrighti* (NCN). Reproduction. Herpetological Review 36: 330.

Ferrara, S. 2003. *Anolis oculatus* habitat selection. Unpublished report. Texas A&M University Study Abroad, College Station. http://dominica.tamu.edu/student%20projects/projects_by_year.htm.

Fidenci, P. 2001. An overview of the Cuban turtle, *Trachemys decussata*, in Cuba. Herpetological Bulletin 80: 20–22.

Fiske, R. S., and H. Sigurdsson. 1982. Soufriere Volcano, St. Vincent: Observations of its 1979 eruption from the ground, aircraft, and satellites. Science 216: 1105–1126.

Fitch, H. S. 2003. Dewlaps and the lack thereof: Or, lizard studies in the Neotropics. In: R. W. Henderson and R. Powell (eds.), Islands and the Sea: Essays on Herpetological Exploration in the West Indies, pp. 153–157. Contributions to Herpetology, volume 20. Ithaca, N.Y.: Society for the Study of Amphibians and Reptiles.

Fitch, H. S., and R. W. Henderson. 1987. Ecological and ethological parameters in *Anolis bahorucoensis*, a species having rudimentary development of the dewlap. Amphibia-Reptilia 8: 69–80.

Fitch, H. S., R. W. Henderson, and H. Guarisco. 1989. Aspects of the ecology of an introduced anole: *Anolis cristatellus* in the Dominican Republic. Amphibia-Reptilia 10: 307–320.

Fläschendräger, A. 1986. Zur Haltung und Nachzucht kleinerer *Anolis*-Arten. Elaphe 8: 21–25.

———. 1988. *Anolis luciae* Garman, 1887 im Terrarium. Elaphe 10: 71–72.

———. 1990. *Anolis bahorucoensis*—ein bemerkenswerter Saumfinger aus der Dominikanischen Republik. Herpetofauna 12: 6–10.

———. 1998. Beobachtungen an drei Grasanolis-Arten im Biotop sowie Bemerkungen zu Haltung und Fortpflanzung im Terrarium. Herpetofauna 20: 19–28.

———. 1999a. Erste Nachzuchtergebnisse bei *Anolis reconditus* Underwood & Williams, 1959. Herpetofauna 21: 5–10.

———. 1999b. Erstnachweis von *Hemidactylus mabouia* (Moreau de Jonnes, 1818) für Jamaica. Herpetofauna 21: 18.

———. 2000. Zur Biologie, Haltung und Nachzucht von *Anolis* (Synonym: *Chamaelinorops*) *barbouri* (Schmidt, 1919). Herpetofauna 22: 5–10.

———. 2001a. Was sind *Anolis* und wo leben sie? Reptilia 6: 16–21.

———. 2001b. *Anolis*, ¿qué son y donde viven? Reptilia (Spain) 28: 16–22.

———. 2001c. What are anoles and where do they live? Reptilia (Spain) 15: 16–22.

———. 2002. Beobachtungen an *Anolis*-Arten in der Cordillera de Guaniguanico, Provinz Pinar del Río, West-Kuba. Herpetofauna 24: 5–18.

———. 2007. *Anolis loysianus*, ein stachelbeschuppter Rindenanolis aus Kuba. Terraria 2: 32–35.

Fläschendräger, A., and L. C. M. Wijffels. 1996. *Anolis*. Münster, Germany: Natur und Tier.

———. 2009. *Anolis*. 2. vollständig überarbeitete und erweitere Auflage. Münster, Germany: Natur und Tier.

Fleishman, L. J. 2000. Signal function, signal efficiency and the evolution of anoline lizard dewlap color. In: Y. Espmark, T. Admundsen, and G. Rosenqvist (eds.), Animal Signals: Signalling and Signal Design in Animal Communication, pp. 209–236. Trondheim, Norway: Tapir Academic Press.

Fleishman, L. J., E. R. Loew, and M. Leal. 1993. Ultraviolet vision in lizards. Nature 365: 397.

Fleishman, L. J., M. Bowman, D. Saunders, W. E. Miller, M. J. Rury, and E. R. Loew. 1997. The visual ecology of Puerto Rican anoline lizards: Habitat light and spectral sensitivity. Journal of Comparative Physiology 181: 446–460.

Flores, G., J. H. Lenzycki, and J. Palumbo, Jr. 1994. An ecological study of the endemic Hispaniolan anoline lizard, *Chamaelinorops barbouri* (Lacertilia: Iguanidae). Breviora 499: 1–23.

Floyd, H. B., and T. A. Jenssen. 1983. Food habits of the Jamaican lizard *Anolis opalinus*: Resource partitioning and seasonal effects examined. Copeia 1983: 319–331.

———. 1984. Prey diversity comparisons between stomach and hindgut of the lizard, *Anolis opalinus*. Journal of Herpetology 18: 204–205.

Fobes, T. M., R. Powell, J. S. Parmerlee, Jr., A. Lathrop, and D. D. Smith. 1992. Natural history of *Anolis cybotes* (Sauria: Polychridae) in a disturbed habitat in Barahona, Dominican Republic. Caribbean Journal of Science 28: 200–207.

Fobes, T. M., J. S. Parmerlee, Jr., and R. Powell. 1993. *Anolis cybotes*. Catalogue of American Amphibians and Reptiles 564: 1–5.

Fogarty, J. H., and F. J. Vilella. 2002. Population dynamics of *Eleutherodactylus coqui* in cordillera forest reserves of Puerto Rico. Journal of Herpetology 36: 193–201.

Fogarty, S. P., V. H. Zero, and R. Powell. 2004. Revisiting St. Eustatius: Estimating the population size of Lesser Antillean Iguanas, *Iguana delicatissima*. Iguana 11: 139–146.

Fong Grillo, A. 1999. Changes in amphibian composition in altered habitats of eastern Cuba. Froglog 36: 2.

———. 2002. *Anolis smallwoodi smallwoodi* (Chipojo). Maximum size. Herpetological Review 33: 204.

———. 2003. The pleasure of the adventure: Searching for reptiles in eastern Cuba. In: R. W. Henderson and R. Powell (eds.), Islands and the Sea: Essays on Herpetological Exploration in the West Indies, pp. 101–106. Contributions to Herpetology, volume 20. Ithaca, N.Y.: Society for the Study of Amphibians and Reptiles.

———. 2004a. *Antillophis andreae* (Black and White Racer, Jubito). Diet. Herpetological Review 35: 57.

———. 2004b. *Antillophis andreae* (Black and White Racer, Jubito). Reproduction. Herpetological Review 35: 57.

———. 2005. Variation in and natural history notes on *Tropidophis fuscus* (Serpentes: Tropidophiidae) from Cuba. Herpetological Review 36: 118–119.

———. 2006. Anfibios y reptiles. En Cuba: Pico Mogote. Rapid Biological Inventories Report 09, pp. 58–59 + 178–179 + Figures 5A–J. Chicago: The Field Museum.

———. 2008. *Tropidophis wrighti* (NCN). Neonates. Herpetological Review 39: 474.

Fong Grillo, A., and J. Blanco Ojeda. 2002. *Anolis smallwoodi* (Chipojo). Food habits. Herpetological Review 33: 204.

Fong Grillo, A., and E. B. del Castillo. 2002. *Leiocephalus macropus* (Bayoya). Saurophagy. Herpetological Review 33: 205–206.

Fong Grillo, A., and L. M. Díaz. 2004. Two new species of *Sphaerodactylus* (Sauria: Gekkonidae) from the southeastern coast of Cuba. Solenodon 4: 73–84.

Fong Grillo, A., and G. Garcés G. 2001. *Chamaeleolis chamaeleonides*. Notes on care and breeding. Reptilia 27: 49–51.

———. 2002a. Composición y variación estacional de la dieta de *Leiocephalus carinatus* (Sauria: Iguanidae) en Santiago de Cuba, Cuba. Boletín de la Sociedad Herpetológica Mexicana 10: 29–34.

———. 2002b. *Bufo taladai* (NCN). Diet. Herpetological Review 33: 302.

Fong Grillo, A., and O. H. Garrido. 2000. Nueva especie de *Anolis* (Sauria: Iguanidae) de la región norte de Cuba oriental. Revista de Biología Tropical 48: 665–670.

Fong Grillo, A., and J.-M. Hero. 2006. Population dynamics of the stream-dwelling frog *Eleutherodactylus cuneatus* on La Gran Piedra, eastern Cuba. Froglog 77: 2–4.

Fong Grillo, A., R. Vina M., and A. Arias B. 1999. Aspectos de la historia natural de *Cricosaura typica* (Sauria: Xantusiidae) de Cuba. Caribbean Journal of Science 35: 148–150.

Fong Grillo, A., L. M. Díaz, and N. Viña Dávila. 2005a. Anfibios y reptiles. En Cuba: Humboldt. In: A. Fong Grillo, D. Maceira F., W. S. Alverson, and J. M. Shopland (eds.), Rapid Biological Inventories Report 14, pp. 92–98 + 346–349. Chicago: The Field Museum.

Fong Grillo, A., R. Viña Dávila, and N. Viña Bayés. 2005b. Anfibios y reptiles de la Altiplanicie del Toldo. En Cuba: Humboldt. In: A. Fong Grillo, D. Maceira F., W. S. Alverson, and J. M. Shopland (eds.), Rapid Biological Inventories Report 14, pp. 98–101 + 350–351. Chicago: The Field Museum.

Font, E., and M. Kramer. 1989. A multivariate clustering approach to display repertoire analysis: Headbobbing in *Anolis equestris*. Amphibia-Reptilia 10: 331–344.

Fontenot, B. E., M. E. Gifford, and R. Powell. 2003. Seasonal variation in dietary preferences of a Hispaniolan anole, *Anolis longitibialis*. Herpetological Bulletin 86: 2–4.

Formanowicz, D. R., Jr., M. M. Stewart, F. H. Pough, and P.

F. Brussard. 1981. Predation by giant crab spiders on the Puerto Rican frog *Eleutherodatylus coqui*. Herpetologica 37: 125–129.

Fowler, H. W. 1918. Some amphibians and reptiles from Porto Rico and the Virgin Islands. Papers from the Department of Marine Biology of the Carnegie Institution of Washington 12: 1–15 + 1 plate.

Franz, R. 1976. A digenetic trematode from a West Indian racer. Florida Scientist 39: 1–2.

———. 2003. Wet mountains and mountain frogs. In: R. W. Henderson and R. Powell (eds.), Islands and the Sea: Essays on Herpetological Exploration in the West Indies, pp. 159–167. Contributions to Herpetology, volume 20. Ithaca, N.Y.: Society for the Study of Amphibians and Reptiles.

Franz, R., and D. Cordier. 1986. The Herpetofaunas of the Proposed National Parks in Southern Haiti. Gainesville: USAID/Haiti, University of Florida.

Franz, R., and C. K. Dodd, Jr. 1994. *Alsophis vudii vudii* (Brown Runner). Diet and growth. Herpetological Review 25: 28.

Franz, R., and D. F. Gicca. 1982. Observations on the Haitian snake *Antillophis parvifrons alleni*. Journal of Herpetology 16: 419–421.

Franz, R., C. K. Dodd, Jr., and D. W. Buden. 1993. Distributional records of amphibians and reptiles from the Exuma Islands, including the first reports of a freshwater turtle and an introduced gecko. Caribbean Journal of Science 29: 165–173.

Franz, R., G. S. Morgan, N. Albury, and S. D. Buckner. 1995. Fossil skeleton of a Cuban Crocodile (*Crocodylus rhombifer*) from a blue hole on Abaco, Bahamas. Caribbean Journal of Science 31: 149–152.

Freze, V. I., and B. Rysavy. 1976. Cestodes of the suborder Proteocephalata Spassky, 1957 (Cestoda: Pseudophyllidea) from Cuba, and description of a new species *Ophiotaenia habanensis* sp. n. Folia Parasitologica 23: 97–104.

Fritz, U. 1990. Haltung und Nachzucht der Jamaika-Schmuckschildkröte *Trachemys terrapen* (Lacepede, 1788) und Bemerkungen zur Fortpflanzungsstrategie von neotropischen Schmuckschildkröten der Gattung *Trachemys*. Salamandra 26: 1–18.

———. 1991. Contribution to the knowledge of the Hispaniola Slider, *Trachemys decorata* (Barbour and Carr, 1940). Sauria 1991(1): 11–14.

Frost, D. R., and G. A. Hammerson. 2007. *Anolis carolinensis*. In: IUCN 2008. 2008 IUCN Red List of Threatened Species. http://www.iucnredlist.org.

Frost, D. R., T. Grant, J. Faivovich, R. H. Bain, A. Haas, C. F. B. Haddad, R. O. de Sá, A. Channing, M. Wilkinson, S. C. Donnellan, C. J. Raxworthy, J. A. Campbell, B. L. Blotto, P. Moler, R. C. Drewes, R. A. Nussbaum, J. D. Lynch, D. M. Green, and W. C. Wheeler. 2006. The amphibian tree of life. Bulletin of the American Museum of Natural History 297: 1–370.

Furniss, S. B. 1984. Recovery plan for the St. Croix Ground Lizard, *Ameiva polops*. Atlanta: United States Fish and Wildlife Service.

Gallant, A. L., R. W. Klaver, G. S. Casper, and M. J. Lannoo. 2007. Global rates of habitat loss and implications for amphibian conservation. Copeia 2007: 967–979.

Gamble, T., A. M. Bauer, E. Greenbaum, and T. R. Jackman. 2007. Evidence for Gondwanan vicariance in an ancient clade of gecko lizards. Journal of Biogeography 35: 88–104.

———. 2008. Out of the blue: A novel, trans-Atlantic clade of geckos (Gekkota, Squamata). Zoologica Scripta 37: 355–366.

Gans, C. 1964. *Amphisbaena schmidti*, a third species of the genus from Puerto Rico (Amphisbaenia: Reptilia). Breviora 198: 1–12.

Gans, C., and A. A. Alexander. 1962. Studies on amphisbaenids (Amphisbaenia, Reptilia). 2. On the amphisbaenids of the Antilles. Bulletin of the Museum of Comparative Zoology 128: 67–158.

Garber, S. D. 1978. Opportunistic feeding behavior of *Anolis cristatellus* (Iguanidae: Reptilia) in Puerto Rico. Transactions of the Kansas Academy of Science 81: 79–80.

Garcea, R., and G. Gorman. 1968. A difference in male territorial display behavior in two sibling species of *Anolis*. Copeia 1968: 419–420.

García, G. 2008. Projekt "Berghuhn"—der Antillen-Ochsenfrosch (*Leptodactylus fallax*). Draco 9: 70–79.

García, G., A. Cunningham, D. Horton, T. Garner, A. Hyatt, S. Hengstberger, J. Lopez, A. Ogrodowczyk, C. Fenton, and J. E. Fa. 2007. Mountain Chickens *Leptodactylus fallax* and sympatric amphibians appear to be disease free on Montserrat. Oryx 41: 398–401.

García, M. A. 2005a. Mona Island Iguana, *Cyclura cornuta stejnegeri* update. Iguana Specialist Group Newsletter 8(2): 11.

———. 2005b. Introduced Cuban Iguana, *Cyclura nubila nubila* on Isla Magueyes. Iguana Specialist Group Newsletter 8(2): 11.

———. 2006a. Mona Island Iguana, *Cyclura cornuta stejnegeri*. Iguana 13: 125–126.

———. 2006b. *Cyclura nubila* on Isla Magueyes, Puerto Rico. Iguana 13: 126.

García, M. A., N. F. Perez-Buitrago, and T. Wiewandt. 2000. *Cyclura cornuta* ssp. *stejnegeri*. In: IUCN 2008. 2008 IUCN Red List of Threatened Species. http://www.iucnredlist.org.

García, M. A., C. E. Diez, and A. O. Alvarez. 2001. The impact of feral cats on Mona Island wildlife and recommendations for their control. Caribbean Journal of Science 37: 107–108.

García, M. A., N. F. Pérez-Buitrago, A. O. Álvarez, and P. J. Tolson. 2007. Survival, dispersal and reproduction of headstarted Mona Island Iguanas, *Cyclura cornuta stejnegeri*. Applied Herpetology 4: 357–363.

Garcia, R., A. Queral, R. Powell, J. S. Parmerlee, Jr., D. D. Smith, and A. Lathrop. 1994. Evidence of hybridization among green anoles (Lacertilia: Polychrotidae) from Hispaniola. Caribbean Journal of Science 30: 279–281.

García Díaz, J. 1967. Rediscovery of *Bufo lemur* Cope and additional records of reptiles from Puerto Rico. Stahlia 10: 1–6.

García Moll, A. 1978. Abundance studies on the *Anolis* lizards and insect populations of altitudinally different tropical forest habitats. Occasional Paper of the Center for Energy and Environment Research CEER-11. San Juan: University of Puerto Rico and U.S. Department of Energy.

Garman, S. 1887. On West Indian reptiles in the Museum of Comparative Zoology, at Cambridge, Mass. Proceedings of the American Philosophical Society 24: 278–286.

———. 1888. Reptiles and batrachians from the Caymans and the Bahamas. Bulletin of the Essex Institute 20: 1–13.

Garrick, L. D., R. L. Sutton, and J. W. Lang. 1985. Observations on the largest Jamaican tree frog, *Calyptahyla crucialis*. Caribbean Journal of Science 21: 159–162.

Garrido, O. H. 1967. Sobre el *Anolis rubribarbus* (Sauria: Iguanidae) en Cuba. Academia de Ciencias de Cuba, Trabajo Divulga 55: 1–7.

———. 1972. *Anolis bremeri* Barbour (Lacertilia: Iguanidae) en el occidente de Cuba e Isla de Pinos. Caribbean Journal of Science 7: 59–77.

———. 1973a. Nueva especie de *Leiocephalus* (Lacertilia, Iguanidae) para Cuba. Poeyana 116: 1–19.

———. 1973b. Anfibios, reptiles y aves de Cayo Real (Cayos San Felipe), Cuba. Poeyana 119: 1–50.

———. 1973c. Distribución y variación de *Anolis homolechis* Cope (Lacertilia: Iguanidae) en Cuba. Poeyana 120: 1–68.

———. 1973d. Anfibios, reptiles y aves del Archipiélago de Sabana-Camagüey, Cuba. Torreia, nueva serie 27: 1–72.

———. 1975a. Nuevos reptiles del Archipiélago Cubano. Poeyana 141: 1–58.

———. 1975b. Distribución y variación de *Anolis argillaceus* Cope (Lacertilia: Iguanidae) en Cuba. Poeyana 142: 1–28.

———. 1975c. Distribución y variación del complejo *Anolis cyanopleurus* (Lacertilia: Iguanidae) en Cuba. Poeyana 143: 1–60.

———. 1975d. Variación de *Anolis angusticeps* Hallowell (Lacertilia: Iguanidae) en el occidente de Cuba y en Isla de Pinos. Poeyana 144: 1–18.

———. 1976a. Aves y reptiles de Cayo Coco, Cuba. Miscelanea Zoológica 3: 3–4.

———. 1976b. Nota sobre *Deiroptyx vermiculatus* Duméril et Bibron (Lacertilia: Iguanidae). Miscelanea Zoológica 4: 1–2.

———. 1979. Nuevas subspecies de *Leiocephalus macropus* Cope (Lacertilia: Iguanidae) para Cuba. Poeyana 188: 1–16.

———. 1980a. Revisión del complejo *Anolis alutaceus* (Lacertilia: Iguanidae) y descripción de una nueva especie de Cuba. Poeyana 201: 1–41.

———. 1980b. Los vertebrados terrestres de la Península de Zapata. Poeyana 203: 1–49.

———. 1981. Nueva subespecie de *Anolis equestris* (Sauria: Iguanidae) para Cuba, con comentarios sobre la distribución y afinidades de otras poblaciones del complejo. Poeyana 232: 1–15.

———. 1982a. Descripción de una nueva especie cubana de *Chamaeleolis* (Lacertilia: Iguanidae), con notas sobre su comportamiento. Poeyana 236: 1–25.

———. 1982b. Nueva especie de *Anolis* (Lacertilia, Iguanidae) para Cuba. Doñana, Acta Vertebrata 9: 131–137.

———. 1983. Nueva especie de *Anolis* (Lacertilia: Iguanidae) de la Sierra del Turquino, Cuba. Caribbean Journal of Science 19: 71–76.

———. 1985. Nueva subespecie de *Anolis isolepis* (Lacertilia: Iguanidae) para Cuba. Doñana, Acta Vertebrata 12: 41–49.

———. 1988. Nueva especies para la ciencia de *Anolis* (Lacertilia: Iguanidae) de Cuba perteneciente al complejo *argillaceus*. Doñana, Acta Vertebrata 15: 45–57.

———. 1990. Nueva especies de *Anolis* de la sección beta (Lacertilia: Iguanidae) para Cuba. Revista Biología 4: 157–162.

———. 1991. Conducta de enfrentamiento, cortejo y copula en tres especies de anolinos cubanos (Sauria: Iguanidae). Caribbean Journal of Science 27: 169–172.

———. 2003. Discovery and rediscovery. In: R. W. Henderson and R. Powell (eds.), Islands and the Sea: Essays on Herpetological Exploration in the West Indies, pp. 59–66. Contributions to Herpetology, volume 20. Ithaca, N.Y.: Society for the Study of Amphibians and Reptiles.

Garrido, O. H., and A. R. Estrada. 1989. Nueva especie del complejo *Anolis alutaceus* (Lacertilia: Iguanidae) para Cuba. Revista Biología 3: 57–66.

Garrido, O. H., and S. B. Hedges. 1992. Three new grass anoles from Cuba (Squamata: Iguanidae). Caribbean Journal of Science 28: 21–29.

———. 2000. Elevation of *Anolis vanidicus rejectus* Garrido and Schwartz (Sauria: Iguanidae) to species status and designation of *Anolis mimus* Schwartz and Thomas as a synonym. Caribbean Journal of Science 36: 160–161.

———. 2001. A new anole from the northern slope of the Sierra Maestra in eastern Cuba (Squamata: Iguanidae). Journal of Herpetology 35: 378–383.

Garrido, O. H., and M. L. Jaume. 1984. Catálogo descriptivo de los anfibios y reptiles de Cuba. Doñana, Acta Vertebrata 11: 5–128.

Garrido, O. H., and L. V. Moreno. 1998. Nueva especie de *Anolis* (Lacertilia: Iguanidae) del Pico Turquino, Sierra

Maestra, Cuba. A new species of *Anolis* (Lacertilia: Iguanidae) from Pico Turquino, Sierra Maestra, Cuba. Avicennia 8/9: 35–40.

Garrido, O. H., and L. Pareta. 1994. Puestas comunales de *Anolis allisoni* (Lacertilia: Iguanidae) en Cuba. Revista Biología 8: 145–147.

Garrido, O. H., and A. Schwartz. 1968a. Cuban lizards of the genus *Chamaeleolis*. Quarterly Journal of the Florida Academy of Sciences 30: 197–220.

———. 1968b. Anfibios, reptiles y aves de la Península de Guanahacabibes, Cuba. Poeyana, serie A, 53: 1–68.

———. 1969. Anfibios, reptiles y aves de Cayo Cantiles. Poeyana 67: 1–44.

———. 1972. The Cuban *Anolis spectrum* complex (Sauria, Iguanidae). Proceedings of the Biological Society of Washington 85: 509–522.

———. 1982. A new species of *Sphaerodactylus* (Reptilia: Sauria: Gekkonidae) from eastern Cuba. Proceedings of the Biological Society of Washington 95: 392–397.

Garrido, O. H., A. R. Estrada, and A. Llanes. 1986. Anfibios, reptiles y aves de Cayo Guajaba, Archipiélago de Sabana-Camagüey, Cuba. Poeyana 328: 1–34.

Garrido, O. H., O. Pérez-Beato, and L. V. Moreno. 1991. Nueva especie de *Chamaeleolis* (Lacertilia: Iguanidae) para Cuba. Caribbean Journal of Science 27: 162–168.

Garrido, O. H., L. V. Moreno, and A. R. Estrada. 2001. Subespecies nuevas de reptiles de complejo *Anolis equestris* (Lacertilia: Iguanidae) para los cayos Las Brujas, Coco y Sabinal, Archipiélago de Sabana-Camagüey, Cuba. Solenodon 1: 55–65.

Gassett, J. W., T. H. Folk, K. J. Alexy, K. V. Miller, B. R. Chapman, F. L. Boyd, and D. I. Hall. 2000. Food habits of Cattle Egrets on St. Croix, U.S. Virgin Islands. Wilson Bulletin 112: 268–271.

Genet, K. S. 2002. Structural habitat and ecological overlap of the Puerto Rican lizards *Anolis cristatellus* and *A. cooki*, with comments on the long-term survival and conservation of *A. cooki*. Caribbean Journal of Science 38: 272–278.

Gerber, G. P. 1996a. *Cyclura carinata* ssp. *carinata*. In: IUCN 2008. 2008 IUCN Red List of Threatened Species. http://www.iucnredlist.org.

———. 1996b. *Cyclura nubila* ssp. *caymanensis*. In: IUCN 2008. 2008 IUCN Red List of Threatened Species. http://www.iucnredlist.org.

———. 1999. A review of intraguild predation and cannibalism in *Anolis*. In: J. B. Losos and M. Leal (eds.), *Anolis* Newsletter V, pp. 28–39. St. Louis, Mo.: Washington University.

———. 2000a. Lesser Caymans Iguana, *Cyclura nubila caymanensis*. In: A. Alberts (ed.), West Indian Iguanas: Status Survey and Conservation Action Plan, pp. 41–44. Gland, Switzerland and Cambridge, UK: International Unions for the Conservation of Nature and Natural Resources.

———. 2000b. Conservation of the Anegada Iguana, *Cyclura pinguis*. Field Research Report. San Diego: Unpublished report to the British Virgin Islands National Parks Trust, Flora and Fauna International, and the Zoological Society of San Diego.

———. 2004a. An update on the ecology and conservation of *Cyclura pinguis* on Anegada. Iguana 11: 23–26.

———. 2004b. *Cyclura carinata*. In: IUCN 2008. 2008 IUCN Red List of Threatened Species. http://www.iucnredlist.org.

Gerber, G. P., and A. C. Alberts. 2005. Turks and Caicos Iguana, 2005 update. Iguana Specialist Group Newsletter 8(2): 6–7.

———. 2006. Turks and Caicos Iguana. Iguana 13: 120–121.

Gerber, G. P., and J. Iverson. 2000. Turks and Caicos Iguana, *Cyclura carinata carinata*. In: A. Alberts (ed.), West Indian Iguanas: Status Survey and Conservation Action Plan, pp. 15–18. Gland, Switzerland and Cambridge, UK: International Union for the Conservation of Nature and Natural Resources.

Gerber, G. P., and West Indian Iguana Specialist Group. 2000. *Cyclura carinata* ssp. *bartschi*. In: IUCN 2008. 2008 IUCN Red List of Threatened Species. http://www.iucnredlist.org.

Gerber, G. P., T. D. Grant, A. C. Alberts, and M. A. Hostetter. 2002. *Cyclura nubila nubila* (Cuban Iguana). Carrion feeding. Herpetological Review 33: 133–134.

Gerber, G. P., L. Pagni, and K. Bradley. 2007. Anegada Iguana 2006 conservation and research update. Iguana Specialist Group Newsletter 10(1): 8–10 (reprinted in Iguana 14: 237–239).

Germano, J. M., J. M. Sander, R. W. Henderson, and R. Powell. 2003. Herpetofaunal communities in Grenada: A comparison of altered sites, with an annotated checklist of Grenadian amphibians and reptiles. Caribbean Journal of Science 39: 68–76.

Giannasi, N. C., R. S. Thorpe, and A. Malhotra. 1997. Introductions of *Anolis* species to the islands of St. Lucia, West Indies: Testing for hybrids using multivariate morphometrics. Journal of Herpetology 31: 586–589.

Gibson, R. 1996a. *Cnemidophorus vanzoi*. In: IUCN 2008. 2008 IUCN Red List of Threatened Species. http://www.iucnredlist.org.

———. 1996b. *Cyclura collei*. In: IUCN 2008. 2008 IUCN Red List of Threatened Species. http://www.iucnredlist.org.

———. 1996c. *Alsophis ater*. In: IUCN 2008. 2008 IUCN Red List of Threatened Species. http://www.iucnredlist.org.

———. 1996d. *Celestus duquesneyi*. In: IUCN 2008. 2008 IUCN Red List of Threatened Species. http://www.iucnredlist.org.

———. 1996e. *Celestus fowleri*. In: IUCN 2008. 2008 IUCN Red List of Threatened Species. http://www.iucnredlist.org.

———. 1996f. *Liophis ornatus*. In: IUCN 2008. 2008 IUCN

Red List of Threatened Species. http://www.iucnredlist.org.

———. 1996g. *Celestus microblepharis*. In: IUCN 2008. 2008 IUCN Red List of Threatened Species. http://www.iucnredlist.org.

———. 1996h. *Epicrates subflavus*. In: IUCN 2008. 2008 IUCN Red List of Threatened Species. http://www.iucnredlist.org.

———. 2001. Lesser Antilles Iguana (*Iguana delicatissima*). Dustbins, D3, diet, and determination: How to make baby *delicatissima*. Iguana Specialist Group Newsletter 4(1): 6–7.

Gibson, R. C., and K. R. Buley. 2004. Maternal care and obligatory oophagy in *Leptodactylus fallax*: A new reproductive mode in frogs. Copeia 2004: 128–135.

Gicca, D. 1980. The status and distribution of *Cyclura r. rileyi* (Reptilia: Iguanidae) a Bahamian rock iguana. Caribbean Journal of Science 16: 9–12.

Gifford, M. E., and R. Powell. 2002a. *Anolis longitibialis*. Catalogue of American Amphibians and Reptiles 749: 1–4.

———. 2002b. *Anolis sheplani*. Catalogue of American Amphibians and Reptiles 750: 1–2.

———. 2007. Sexual dimorphism and reproductive characteristics in five species of *Leiocephalus* lizards from the Dominican Republic. Journal of Herpetology 41: 521–527.

Gifford, M. E., R. Powell, and W. E. Steiner, Jr. 2000. Relationship of diet and prey availability in *Aristelliger cochranae*, a gecko from Navassa Island, West Indies. Caribbean Journal of Science 36: 323–326.

Gifford, M. E., Y. M. Ramos, R. Powell, and J. S. Parmerlee, Jr. 2002. Natural history of a saxicolous anole, *Anolis longitibialis* from Hispaniola. Herpetological Natural History 9: 15–20.

Gifford, M. E., R. Powell, A. Larson, and R. L. Gutberlet, Jr. 2004. Population structure and history of a phenotypically variable teiid lizard (*Ameiva chrysolaema*) from Hispaniola: The influence of a geographically complex island. Molecular Phylogenetics and Evolution 32: 735–478.

Gifford, M. E., A. Herrel, and D. L. Mahler. 2008. The evolution of locomotor morphology, performance, and anti-predator behaviour among populations of *Leiocephalus* lizards from the Dominican Republic. Biological Journal of the Linnean Society 93: 445–456.

Girdler, J. 2004. Microclimate regulation of *Anolis oculatus* on the island of Dominica. Unpublished report. Texas A&M University Study Abroad, College Station. http://dominica.tamu.edu/student%20projects/projects_by_year.htm.

Glor, R. E. 2003. Rediscovering the diversity of Dominican anoles. In: R. W. Henderson and R. Powell (eds.), Islands and the Sea: Essays on Herpetological Exploration in the West Indies, pp. 141–152. Contributions to Herpetology, volume 20. Ithaca, N.Y.: Society for the Study of Amphibians and Reptiles.

Glor, R. E., R. Powell, and J. S. Parmerlee, Jr. 1998. *Cyclura ricordii*. Catalogue of American Amphibians and Reptiles 657: 1–3.

———. 2000. *Cyclura cornuta*. Catalogue of American Amphibians and Reptiles 709: 1–6.

Glor, R. E., A. S. Flecker, M. F. Benard, and A. G. Power. 2001. Lizard diversity and agricultural disturbance in a Caribbean forest landscape. Biodiversity and Conservation 10: 711–723.

Glor, R. E., J. J. Kolbe, R. Powell, A. Larson, and J. B. Losos. 2003. Phylogenetic analysis of ecological and morphological diversification in Hispaniolan trunk-ground anoles (*Anolis cybotes* group). Evolution 57: 2383–2397.

Glor, R. E., M. E. Gifford, A. Larson, J. B. Losos, L. Rodríguez Schettino, A. R. Chamizo Lara, and T. R. Jackman. 2004. Partial island submergence and speciation in an adaptive radiation: A multilocus analysis of the Cuban green anoles. Proceedings of the Royal Society B 271: 2257–2265.

Godinez, E., M. Gómez, J. A. Puentes, and S. Vargas. 1987. Características reproductivas de *Columba leucocephala* en la Península de Guanacahabibes, Cuba. Poeyana 340: 1–8.

Goetz, M. 2008. The Cayman Sister Isles Iguana Project: Identifying the conservation needs of *Cyclura nubila caymanensis*. Iguana 15: 12–19.

———. 2009. Die Wirtelschwanzleguane der Kaimaninseln: Schutz and Erforschung von *Cyclura lewisi* und *Cyclura nubila caymanensis*. Terraria 4: 42–47.

Goin, C. J. 1947. Studies of the life history of *Eleutherodactylus ricordii planirostris* (Cope) in Florida. University of Florida Studies, Biological Science Series 4: xii + 66 pp.

———. 1953. Rediscovery of the frog *Litoria luteola* Gosse in Jamaica. Occasional Papers of the Institute of Jamaica 7: 1–4.

———. 1955. Description of a new subspecies of the frog *Eleutherodactylus ricordi* from the Bahama Islands. American Museum Novitates 1708: 1–7.

Goin, C. J., and B. W. Cooper. 1950. Notes on a collection of amphibians from Jamaica. Occasional Papers of the Institute of Jamaica 4: 1–9.

Goldberg, S. R., and C. R. Bursey. 1996a. Gastrointestinal helminths of the anole *Anolis oculatus* (Polychridae) from Dominica, Lesser Antilles. Journal of the Helminthological Society of Washington 63: 125–128.

———. 1996b. *Anolis maynardi* (NCN). Endoparasites. Herpetological Review 27: 19.

———. 2000. New helminth records for one teiid and four gekkonid lizard species from the Lesser Antilles. Caribbean Journal of Science 36: 342–344.

Goldberg, S. R., C. R. Bursey, and R. Tawil. 1994a. Gastrointestinal nematodes of the Cuban Treefrog, *Osteopilus septentrionalis* (Hylidae) from San Salvador Island, Bahamas. Journal of the Helminthological Society of Washington 61: 230–233.

———. 1994b. Helminth parasites of the Bark Anole, *Anolis*

distichus and the Brown Anole, *Anolis sagrei* (Polychridae) from Florida and the Bahamas. Caribbean Journal of Science 30: 275–277.

Goldberg, S. R., C. R. Bursey, and H. Cheam. 1995. Helminth parasites of three sympatric lizards from Grand Cayman Island, *Anolis conspersus*, *Anolis sagrei* (Polychrotidae) and *Leiocephalus carinatus* (Tropiduridae). Caribbean Journal of Science 31: 339–340.

Goldberg, S. R., C. R. Bursey, and L. N. Ajimine. 1996a. *Anolis scriptus* (Silver Key Anole). Endoparasites. Herpetological Review 27: 142.

———. 1996b. Gastrointestinal helminthes of *Anolis bimaculatus* (Polychrotidae) from the Lesser Antilles. Caribbean Journal of Science 32: 244–247.

Goldberg, S. R., C. R. Bursey, and H. Cheam. 1996c. Gastrointestinal helminths of six anole species: *Anolis armouri*, *A. barahonae*, *A. bahorucoensis*, *A. brevirostris*, *A. chlorocyanus* and *A. coelestinus* (Polychrotidae) from Hispaniola. Caribbean Journal of Science 32: 112–115.

Goldberg, S. R., H. J. Holshuh, and R. L. Joglar. 1996d. *Eleutherodactylus cooki* ("Guajon" or "Demonio de Puerto Rico"). Ectoparasites. Herpetological Review 27: 18.

Goldberg, S. R., C. R. Bursey, and H. Cheam. 1997a. Helminths of 12 species of *Anolis* lizards (Polychrotidae) from the Lesser Antilles, West Indies. Journal of the Helminthological Society of Washington 64: 248–257.

———. 1997b. Helminths of *Anolis acutus* (Sauria: Polychrotidae) from St. Croix, U.S. Virgin Islands. Journal of Parasitology 83: 530–531.

———. 1998a. Helminths of the lizard *Anolis cristatellus* (Polychrotidae) from the British Virgin Islands, West Indies. Journal of the Helminthological Society of Washington 65: 259–262.

———. 1998b. Helminths of six species of *Anolis* lizards (Polychrotidae) from Hispaniola, West Indies. Journal of Parasitology 84: 1291–1295.

Goldberg, S. R., C. R. Bursey, and A. Fong Grillo. 1998c. *Cricosaura typica* (Cuban Night Lizard). Endoparasites. Herpetological Review 29: 237–238.

Goldberg, S. R., C. R. Bursey, and H. Kaiser. 1998d. Gastrointestinal helminths of five species of *Eleutherodactylus* (Anura: Leptodactylidae) from the West Indies. Caribbean Journal of Science 34: 146–149.

Goldwasser, L., and J. Roughgarden. 1993. Construction and analysis of a large Caribbean food web. Ecology 74: 1216–1233.

González Bermúdez, F., and L. Rodríguez Schettino. 1982. Dato etoecológicos sobre *Anolis vermiculatus* (Sauria: Iguanidae). Poeyana 245: 1–18.

González Grau, A., N. Manójina, O. Valdés Lafont, and A. Hernández Marrero. 1989. Reporte de una cueva de calor en la Reserva de la Biosfera Sierra del Rosario. Miscelanea Zoológica 43: 3–4.

González Rossell, A., V. Berovides Álvarez, and M. A. Castañeira Colomé. 2001. Aspectos de morfometría, abundancia y alimentación de la Iguana Cubana (*Cyclura nubila nubila*) en el Archipiélago de los Canarreos, Cuba. Revista Biología 15: 98–104.

Goodman, R. M. 2003. Studying the world's most endangered rock iguana, *Cyclura nubila lewisi*. Iguana 10: 15–19.

———. 2007. Activity patterns and foraging behavior of the endangered Grand Cayman Blue Iguana, *Cyclura lewisi*. Caribbean Journal of Science 43: 73–86.

Goodman, R. M., and F. J. Burton. 2005. *Cyclura lewisi* (Grand Cayman Blue Iguana). Hatchlings. Herpetological Review 36: 176.

Goodman, R. M., F. J. Burton, and A. C. Echternacht. 2005a. Habitat use of the endangered iguana *Cyclura lewisi* in a human-modified landscape on Grand Cayman. Animal Conservation 8: 397–405.

Goodman, R. M., A. C. Echternacht, and F. J. Burton. 2005b. Spatial ecology of the endangered iguana, *Cyclura lewisi*, in a disturbed setting on Grand Cayman. Journal of Herpetology 39: 402–408.

Goodyear, N. C., and J. Lazell. 1994. Status of a relocated population of endangered *Iguana pinguis* on Guana Island, British Virgin Islands. Restoration Ecology 2: 43–50.

Gorman, G. C. 1968. The relationships of *Anolis* of the *roquet* species group (Sauria: Iguanidae)—III. Comparative study of display behavior. Breviora 284: 1–31.

———. 1969. Intermediate territorial display of a hybrid *Anolis* lizard (Sauria: Iguanidae). Zeitschrift für Tierpsychologie 26: 390–393.

———. 1976. Observations on the distribution of *Anolis extremus* (Sauria: Iguanidae) on St. Lucia, West Indies—a "colonizing" species. Herpetologica 32: 184–188.

———. 1977. Comments on ontogenetic color change in *Anolis cuvieri* (Reptilia, Lacertilia, Iguanidae). Journal of Herpetology 11: 221.

———. 1980. *Anolis occultus*, a small cryptic canopy lizard: Are there pair bonds? Caribbean Journal of Science 15: 29–31.

———. 2003. The mule lizards of Trinidad: Jungle dreams and backyard explorations. In: R. W. Henderson and R. Powell (eds.), Islands and the Sea: Essays on Herpetological Exploration in the West Indies, pp. 279–293. Contributions to Herpetology, volume 20. Ithaca, N.Y.: Society for the Study of Amphibians and Reptiles.

Gorman, G. C., and R. Harwood. 1977. Notes on population density, vagility, and activity patterns of the Puerto Rican grass lizard, *Anolis pulchellus* (Reptilia, Lacertilia, Iguanidae). Journal of Herpetology 11: 363–368.

Gorman, G. C., and S. Hillman. 1977. Physiological basis for climatic niche partitioning in two species of Puerto Rican *Anolis* (Reptilia, Lacertilia, Iguanidae). Journal of Herpetology 11: 337–340.

Gorman, G. C., and B. Stamm. 1975. The *Anolis* lizards of Mona, Redonda, and La Blanquilla: Chromosomes, relationships, and natural history notes. Journal of Herpetology 9: 197–205.

Gorman, G. C., R. B. Huey, and E. E. Williams. 1969. Cytotaxonomic studies on some unusual iguanid lizards assigned to the genera *Chamaeleolis*, *Polychrus*, and *Phenacosaurus*, with behavioral notes. Breviora 316: 1–17.

Gosner, K. L. 1987. Observations on Lesser Antillean pit vipers. Journal of Herpetology 21: 78–80.

Gosner, R. A., and L. L. Woolbright. 1995. Homing behavior in the Puerto Rican frog, *Eleutherodactylus coqui*. Journal of Herpetology 29: 481–484.

Gosse, P. H. 1850. Descriptions of a new genus and six new species of saurian reptiles. Annals of the Magazine of Natural History 2: 344–348.

———. 1851. A Naturalist's Sojourn in Jamaica. London: Longman, Brown, Green, & Longmans.

Goto, M. M., and M. A. Osborne. 1989. Nocturnal microhabitats of two Puerto Rican grass anoles, *Anolis pulchellus* and *Anolis krugi*. Journal of Herpetology 23: 79–81.

Government of Barbados. 2002. A national biodiversity strategy & action plan for Barbados. The Ministry of Physical Development and Environment, Barbados. http://www.biodiv.org/doc/world/bb/bb-nbsap-01-en.pdf.

Graham, E. D., Jr. 1981. A new species of lizard (*Sphaerodactylus*) from northwestern Haiti. Journal of Herpetology 15: 363–366.

Graham, K. S., and M. Morton. 2004. Taxon update: St. Lucia Iguana (*Iguana iguana*) update. Iguana Specialist Group Newsletter 7(2): 8.

———. 2005. Taxon update: St. Lucia Iguana (*Iguana iguana*) update. Iguana 12: 29–30.

Grant, C. 1931a. Notes on the ameivas of Puerto Rico. Copeia 1931: 61–62.

———. 1931b. The sphaerodactyls of Porto Rico, Culebra, and Mona islands. Journal of the Department of Agriculture of Porto Rico 15: 199–213.

———. 1932a. The herpetology of Vieques Island. Journal of the Department of Agriculture of Porto Rico 16: 37–39.

———. 1932b. The hemidactyls of the Porto Rico region. Journal of the Department of Agriculture of Porto Rico 16: 51–57.

———. 1932c. A new frog from Puerto Rico. Journal of the Department of Agriculture of Porto Rico 16: 145–148.

———. 1932d. Herpetological notes from the Puerto Rico area. Journal of the Department of Agriculture of Porto Rico 16: 161–165.

———. 1932e. A new frog from the Virgin Islands. Journal of the Department of Agriculture of Porto Rico 16: 325–326.

———. 1932f. Notes on the boas of Puerto Rico and Mona. Journal of the Department of Agriculture of Porto Rico 16: 327–329.

———. 1932g. The herpetology of St John and adjacent keys, U.S. Virgin Islands. Journal of the Department of Agriculture of Porto Rico 16: 331–334.

———. 1932h. Herpetology of Tortola; notes on Anegada and Virgin Gorda, British Virgin Islands. Journal of the Department of Agriculture of Porto Rico 16: 339–346.

———. 1937a. Two early Puerto Rican herpetologists. Journal of Agriculture, University of Porto Rico 21: 491–501.

———. 1937b. Herpetological notes with new species from the American and British Virgin Islands, 1936. Journal of Agriculture, University of Puerto Rico 21: 503–522.

———. 1938. The Jamaican *Dromicus funereus* (Cope) reestablished. Copeia 1938: 83–86.

———. 1939. Additional data on Jamaican snakes of the genus *Dromicus*. Copeia 1939: 105–106.

———. 1940a. The herpetology of Jamaica. II. The reptiles. Bulletin of the Institute of Jamaica, Science Series 1: 61–148.

———. 1940b. Notes on the reptiles and amphibians of Jamaica, with diagnoses of new species and subspecies. In: Jamaica Today, pp. 151–157. London: Hazell, Watson & Viney.

———. 1940c [1941]. The herpetology of the Cayman Islands. Bulletin of the Institute of Jamaica, Science Series 2: iv + 65 pp.

———. 1944a. New sphaerodactyls from Cuba and the Isle of Pines. Herpetologica 2: 118–125.

———. 1944b. *Cyclura stejnegeri*—A portrait. Herpetologica 2: 139–140.

———. 1946. Notes on *Tretanorhinus* of Cuba and the Isle of Pines. Journal of Agriculture, University of Puerto Rico 30: 102–117.

———. 1957. The black tailed *Tropidophis* (Reptilia: Serpentes). Herpetologica 13: 154.

———. 1959. Another new *Sphaerodactylus* from Guantanamo, Cuba. Herpetologica 15: 53.

Grant, C., and H. A. Beatty. 1944. Herpetological notes on St. Croix, Virgin Islands. Herpetologica 2: 110–113.

Grant, C., and C. R. de Sola. 1934. Antillean tortoises and terrapins: Distribution, status, and habits of *Testudo* and *Pseudemys*. Copeia 1934: 73–79.

Grant, T., and J. Lemm (authors not credited). 2007. Grand Cayman Iguanas hatch. Iguana Specialist Group Newsletter 10(2): 1–2.

Grant, T., D. R. Frost, J. P. Caldwell, R. Gagliardo, C. F. B. Haddad, P. J. R. Kok, D. B. Means, B. P. Noonan, W. E. Schargel, and W. C. Wheeler. 2006. Phylogenetic systematics of dart-poison frogs and their relatives (Amphibia: Athesphatanura: Dendrobatidae). Bulletin of the American Museum of Natural History 299: 1–262.

Gravelle, K., and C. A. Simon. 1980. Field observations on the use of the tongue—Jacobson's organ system in two iguanid lizards, *Sceloporus jarrovi* and *Anolis trinitatis*. Copeia 1980: 356–359.

Graves, G. R. 2006. Greater Antillean Grackle (*Quiscalus*

niger) preys on *Anolis grahami*. Journal of Caribbean Ornithology 19: 56–58.

Greene, B. T., D. T. Yorks, J. S. Parmerlee, Jr., R. Powell, and R. W. Henderson. 2002. Discovery of *Anolis sagrei* in Grenada with comments on its potential impact on native anoles. Caribbean Journal of Science 38: 270–272.

Greene, B. T., R. Powell, and R. W. Henderson. 2003. *Mastigodryas bruesi*. Catalogue of American Amphibians and Reptiles 777: 1–3.

Greene, H. W. 1983. Dietary correlates of the origin and radiation of snakes. American Zoologist 23: 431–441.

———. 1988. Antipredator mechanisms in reptiles. In: C. Gans and R. B. Huey (eds.), Biology of the Reptilia. Volume 16. Ecology B, pp. 1–152. New York: Alan R. Liss.

Greer, A. E. 1965. A new subspecies of *Clelia* (Serpentes: Colubridae) from the island of Grenada. Breviora 223: 1–6.

———. 1967. Notes on the mode of reproduction in anguid lizards. Herpetologica 23: 94–99.

Greve, J. H., and R. Powell. 1989. *Skrjabinoptera leiocephalorum* sp. n. (Nematoda: Physalopteroidea) in *Leiocephalus* spp. (Reptilia: Iguanidae) from Hispaniola. Journal of Parasitology 75: 677–679.

Groombridge, B. 1982. The IUCN Amphibia-Reptilia Red Data Book. Part I. Gland, Switzerland: International Union for the Conservation of Nature and Natural Resources.

Groome, J. R. 1970. A Natural History of the Island of Grenada. Arima, Trinidad: Caribbean Printers.

Groschaft, J., and V. Baruš. 1970. Studies on the helminth fauna of crocodiles in Cuba. Vestník Ceskoslovenské Spolecnosti Zoologické 34: 289–303.

Groschaft, J., and A. Coy Otero. 1978. Occurrence of *Infidum similis* Travassos, 1916 (Trematoda: Dicrocoelidae) in Cuban snakes. Vestník Ceskoslovenské Spolecnosti Zoologické 42: 105–107.

Groves, J. D., and W. Altimari. 1977. Keratophagy in the Slender Vine Snake, *Uromacer oxyrhynchus*. Herpetological Review 8: 124.

Guglielmone, A. A., A. Estrada-Peña, J. E. Keirans, and R. G. Robbins. 2003. Ticks (Acari: Ixodida) of the Neotropical Zoogeographic Region. International Consortium on Ticks and Tick-borne Diseases. Houten, The Netherlands: Atalanta.

Guillette, L. J., P. H. Jaffe, L. R. Franz, and A. Ramírez. 1988. The reproductive biology of the high elevation Haitian lizard *Wetmorena haitiana*. Program and Abstracts, Joint Meeting of the Herpetologists' League and the American Society of Ichthyologists and Herpetologists: 102.

Gundlach, J. 1880. Contribución á la Erpetologia Cubana. La Habana: G. Monteil.

Haast, W. E. 1969. Hatching Rhinoceros Iguanas *Cyclura cornuta* at the Miami Serpentarium. International Zoo Yearbook 9: 49.

Haberham, Z. L. 2003. Bij de voorplaat: *Anolis baleatus* (Cope 1861). Lacerta 61: 1–2.

Hammerson, G. A. 2007a. *Sphaerodactylus notatus*. In: IUCN 2008. 2008 IUCN Red List of Threatened Species. http://www.iucnredlist.org.

———. 2007b. *Nerodia clarkii*. In: IUCN 2008. 2008 IUCN Red List of Threatened Species. http://www.iucnredlist.org.

———. 2007c. *Pantherophis guttatus*. In: IUCN 2008. 2008 IUCN Red List of Threatened Species. http://www.iucnredlist.org.

Hammerson, G. A., and S. B. Hedges. 2004a. *Gastrophryne carolinensis*. In: IUCN 2008. 2008 IUCN Red List of Threatened Species. http://www.iucnredlist.org.

———. 2004b. *Lithobates sphenocephalus*. In: IUCN 2008. 2008 IUCN Red List of Threatened Species. http://www.iucnredlist.org.

———. 2004c. *Hyla cinerea*. In: IUCN 2008. 2008 IUCN Red List of Threatened Species. http://www.iucnredlist.org.

———. 2004d. *Hyla squirella*. In: IUCN 2008. 2008 IUCN Red List of Threatened Species. http://www.iucnredlist.org.

Hammerson, G. A., S. B. Hedges, and R. L. Joglar. 2008. *Lithobates grylio*. In: IUCN 2008. 2008 IUCN Red List of Threatened Species. http://www.iucnredlist.org.

Haneke, B. 1995. A study of Mona Rock Iguana (*Cyclura cornuta stejnegeri*) nesting sites on Mona Island, Puerto Rico. Iguana Times (Journal of the International Iguana Society) 4: 61–71.

Hanlon, R. W. 1964. Reproductive activity of the Bahaman boa (*Epicrates striatus*). Herpetologica 20: 143–144.

Harcourt, C., and J. Ottenwalder. 1996. Hispaniola. In: C. S. Harcourt and J. A. Sayer (eds.), The Conservation Atlas of Tropical Forests: The Americas, pp. 102–111. New York: Simon & Schuster.

Hardy, J. D., Jr. 1956. Notes on the Cuban Iguana. Herpetologica 12: 323–324.

———. 1957a. Observations on the life history of the Cuban lizard, *Anolis lucius*. Herpetologica 13: 241–245.

———. 1957b. A note on the feeding habits of the Cuban Racer, *Alsophis angulifer* (Bibron). Copeia 1957: 49–50.

———. 1957c. Bat predation by the Cuban Boa, *Epicrates angulifer* Bibron. Copeia 1957: 151–152.

———. 1958a. A new lizard of the genus *Leiocephalus* from Cuba (Squamata: Iguanidae). Journal of the Washington Academy of Sciences 48: 294–300.

———. 1958b. Tail prehension and related behavior in a New World lizard. Herpetologica 14: 205–206.

———. 1967. Geographic variation in the West Indian lizard, *Anolis angusticeps*, with the description of a new form, *Anolis angusticeps paternus*, subsp. nov., from the Isle of Pines, Cuba (Reptila: Iguanidae). Caribbean Journal of Science 6: 23–31.

Hardy, J. D., Jr., W. R. Heyer, S. B. Hedges, and R. Powell. 2004. *Leptodactylus validus*. In: IUCN 2008. 2008 IUCN Red List of Threatened Species. http://www.iucnredlist.org.

Hardy, L. M., C. J. Cole, and C. R. Townsend. 1989. Parthenogenetic reproduction in the Neotropical unisexual lizard, *Gymnophthalmus underwoodi* (Reptilia: Teiidae). Journal of Morphology 201: 215–234.

Harris, B. R., D. T. Yorks, C. A. Bohnert, J. S. Parmerlee, Jr., and R. Powell. 2004. Population densities and structural habitats in lowland populations of *Anolis* lizards on Grenada. Caribbean Journal of Science 40: 31–40.

Hartley, L. M., R. Powell, and J. S. Parmerlee, Jr. 1998. *Ameiva lineolata*. Catalogue of American Amphibians and Reptiles 654: 1–4.

Hartley, L. M., R. E. Glor, A. L. Sproston, R. Powell, and J. S. Parmerlee, Jr. 2000. Germination rates of seeds consumed by two species of Rock Iguanas (*Cyclura* spp.) in the Dominican Republic. Caribbean Journal of Science 36: 149–151.

Hass, C. A., S. B. Hedges, and L. R. Maxson. 1993. Molecular insights into the relationships and biogeography of West Indian anoline lizards. Biochemical Systematics and Ecology 21: 97–114.

Hassler, W. G. 1930. Digging for lizard nests. Collector's luck in the forests and deserts of the Dominican Republic. Natural History 30: 409–420.

Hayes, W. K. 2000a. San Salvador Iguana, *Cyclura rileyi rileyi*. In: A. Alberts (ed.), West Indian Iguanas: Status Survey and Conservation Action Plan, pp. 56–59. Gland, Switzerland and Cambridge, UK: International Union for the Conservation of Nature and Natural Resources.

———. 2000b. White Cay Iguana, *Cyclura rileyi cristata*. In: A. Alberts (ed.), West Indian Iguanas: Status Survey and Conservation Action Plan, pp. 59–60. Gland, Switzerland and Cambridge, UK: International Union for the Conservation of Nature and Natural Resources.

Hayes, W. K., and R. Montanucci. 2000. Acklins Iguana, *Cyclura rileyi nuchalis*. In: A. Alberts (ed.), West Indian Iguanas: Status Survey and Conservation Action Plan, pp. 60–62. Gland, Switzerland and Cambridge, UK: International Union for the Conservation of Nature and Natural Resources.

Hayes, W. K., R. L. Carter, S. Cyril, Jr., and B. Thornton. 2004. Conservation of an endangered Bahamian rock iguana, I: Population assessments, habitat restoration, and behavioral ecology. In: A. C. Alberts, R. L. Carter, W. K. Hayes, and E. P. Martins (eds.), Iguanas: Biology and Conservation, pp. 232–257. Berkeley and Los Angeles: University of California Press.

Heam, L. 1890. Two years in the French West Indies. London: Harper & Brothers.

Heatwole, H. 1961. Institute of Caribbean Studies field excursion to Isla Mona. Herpetology. Caribbean Journal of Science 1: 32–33.

———. 1968a. Relationship of escape behavior and camouflage in anoline lizards. Copeia 1968: 109–113.

———. 1968b. Herpetogeography of Puerto Rico. V. Description of a new species of *Sphaerodactylus* from Desecheo Island. Breviora 292: 1–6.

———. 1976. Herpetogeography of Puerto Rico. VII. Geographic variation in the *Anolis cristatellus* complex in Puerto Rico and the Virgin Islands. Occasional Papers of the Museum of Natural History, The University of Kansas 46: 1–18.

Heatwole, H., and I. B. Banuchi. 1966. Envenomation by the colubrid snake, *Alsophis portoricensis*. Herpetologica 22: 132–134.

Heatwole, H., and F. Torres. 1963. Escape of *Anolis cristatellus* by submerging in water. Herpetologica 19: 223.

———. 1967. Distribution and geographic variation of the ameivas of Puerto Rico and the Virgin Islands. Studies on the Fauna of Curaçao and Other Caribbean Islands 24: 64–111.

Heatwole, H., and J. E. N. Veron. 1977. Vital limit and evaporative water loss in lizards (Reptilia, Lacertilia): A critique and new data. Journal of Herpetology 11: 341–348.

Heatwole, H., E. Ortiz, A. M. Díaz-Collazo, and A. R. Jimínez-Vélez. 1962. Aquatic tendencies in the Puerto Rican pasture lizard, *Anolis pulchellus*. Herpetologica 17: 272–274.

Heatwole, H., D. S. Sade, and R. Hildreth. 1963. Herpetogeography of Puerto Rico. I. Herpetofauna of Cayo Santiago and Cayo Batata. Caribbean Journal of Science 3: 1–5.

Heatwole, H., S. Blasini de Austin, and R. Herrero. 1968. Heat tolerances of tadpoles of two species of tropical anurans. Comparative Biochemistry and Physiology 27: 807–815.

Heatwole, H., T.-H. Lin, E. Villalón, A. Muñiz, and A. Matta. 1969. Some aspects of the thermal ecology of Puerto Rican anoline lizards. Journal of Herpetology 3: 65–77.

Hecht, M. K. 1952. Natural selection in the lizard genus *Aristelliger*. Evolution 6: 112–124.

Hecht, M. K., V. Walters, and G. Ramm. 1955. Observations on the natural history of the Bahaman Pygmy Boa, *Tropidophis pardalis*, with notes on autohemorrhage. Copeia 1955: 249–251.

Heckel, D. G., and J. Roughgarden. 1979. A technique for estimating the size of lizard populations. Ecology 60: 966–975.

Hedges, S. B. 1987. Vocalization and habitat preference of the Jamaican treefrog, *Hyla marianae* (Anura, Hylidae). Caribbean Journal of Science 23: 380–384.

———. 1988. A new diminutive frog from Hispaniola (Leptodactylidae: *Eleutherodactylus*). Copeia 1988: 636–641.

———. 1992. A reconsideration of two montane species of *Eleutherodactylus* in Hispaniola (Anura: Leptodactylidae). Caribbean Journal of Science 28: 11–16.

———. 1993. Global amphibian declines: A perspective from the Caribbean. Biodiversity and Conservation 2: 290–303.

———. 1996. The origin of West Indian amphibians and reptiles. In: R. Powell and R. W. Henderson (eds.), Contributions to West Indian herpetology: A Tribute to Albert

Schwartz, pp. 95–127. Contributions to Herpetology, volume 12. Ithaca, N.Y.: Society for the Study of Amphibians and Reptiles.

———. 1999. Distribution patterns of amphibians in the West Indies. In: W. E. Duellman (ed.), Patterns of Distribution of Amphibians: A Global Perspective, pp. 211–254. Baltimore: Johns Hopkins University Press.

———. 2002. Morphological variation and the definition of species in the snake genus *Tropidophis* (Serpentes: Tropidophiidae). Bulletin of the Natural History Museum, Zoology Series 68: 83–90.

———. 2004a. *Eleutherodactylus griphus*. In: IUCN 2008. 2008 IUCN Red List of Threatened Species. http://www.iucnredlist.org.

———. 2004b. *Eleutherodactylus junori*. In: IUCN 2008. 2008 IUCN Red List of Threatened Species. http://www.iucnredlist.org.

———. 2004c. *Eleutherodactylus orcutti*. In: IUCN 2008. 2008 IUCN Red List of Threatened Species. http://www.iucnredlist.org.

———. 2004d. *Eleutherodactylus pentasyringos*. In: IUCN 2008. 2008 IUCN Red List of Threatened Species. http://www.iucnredlist.org.

———. 2006. An overview of the evolution and conservation of West Indian amphibians and reptiles. Applied Herpetology 3: 281–292.

———. 2008a. Caribherp: Database of West Indian Amphibians and Reptiles. http://evo.bio.psu.edu/caribherp.

———. 2008b. Threatened amphibians of the West Indies. In: S. N. Stuart, M. Hoffmann, J. S. Chanson, N. A. Cox, R. Berridge, P. Ramani, and B. E. Young (eds.), Threatened Amphibians of the World, pp. 104–105. Barcelona, Spain: Lynx Ediciones, with IUCN—The World Conservation Union, Conservation International, and NatureServe.

———. 2008c. At the lower size limit in snakes: Two new species of threadsnakes (Squamata: Leptotyphlopidae: *Leptotyphlops*) from the Lesser Antilles. Zootaxa 1841: 1–30.

———. In prep. Field Guide to West Indian Amphibians and Reptiles.

Hedges, S. B., and L. M. Díaz. 2004a. *Peltophryne cataulaciceps*. In: IUCN 2008. 2008 IUCN Red List of Threatened Species. http://www.iucnredlist.org.

———. 2004b. *Peltophryne empusa*. In: IUCN 2008. 2008 IUCN Red List of Threatened Species. http://www.iucnredlist.org.

———. 2004c. *Peltophryne fustiger*. In: IUCN 2008. 2008 IUCN Red List of Threatened Species. http://www.iucnredlist.org.

———. 2004d. *Peltophryne gundlachi*. In: IUCN 2008. 2008 IUCN Red List of Threatened Species. http://www.iucnredlist.org.

———. 2004e. *Peltophryne longinasus*. In: IUCN 2008. 2008 IUCN Red List of Threatened Species. http://www.iucnredlist.org.

———. 2004f. *Peltophryne peltocephala*. In: IUCN 2008. 2008 IUCN Red List of Threatened Species. http://www.iucnredlist.org.

———. 2004g. *Peltophryne taladai*. In: IUCN 2008. 2008 IUCN Red List of Threatened Species. http://www.iucnredlist.org.

———. 2004h. *Eleutherodactylus acmonis*. In: IUCN 2008. 2008 IUCN Red List of Threatened Species. http://www.iucnredlist.org.

———. 2004i. *Eleutherodactylus adelus*. In: IUCN 2008. 2008 IUCN Red List of Threatened Species. http://www.iucnredlist.org.

———. 2004j. *Eleutherodactylus albipes*. In: IUCN 2008. 2008 IUCN Red List of Threatened Species. http://www.iucnredlist.org.

———. 2004k. *Eleutherodactylus atkinsi*. In: IUCN 2008. 2008 IUCN Red List of Threatened Species. http://www.iucnredlist.org.

———. 2004l. *Eleutherodactylus auriculatus*. In: IUCN 2008. 2008 IUCN Red List of Threatened Species. http://www.iucnredlist.org.

———. 2004m. *Eleutherodactylus bartonsmithi*. In: IUCN 2008. 2008 IUCN Red List of Threatened Species. http://www.iucnredlist.org.

———. 2004n. *Eleutherodactylus blairhedgesi*. In: IUCN 2008. 2008 IUCN Red List of Threatened Species. http://www.iucnredlist.org.

———. 2004o. *Eleutherodactylus bresslerae*. In: IUCN 2008. 2008 IUCN Red List of Threatened Species. http://www.iucnredlist.org.

———. 2004p. *Eleutherodactylus casparii*. In: IUCN 2008. 2008 IUCN Red List of Threatened Species. http://www.iucnredlist.org.

———. 2004q. *Eleutherodactylus cubanus*. In: IUCN 2008. 2008 IUCN Red List of Threatened Species. http://www.iucnredlist.org.

———. 2004r. *Eleutherodactylus cuneatus*. In: IUCN 2008. 2008 IUCN Red List of Threatened Species. http://www.iucnredlist.org.

———. 2004s. *Eleutherodactylus dimidiatus*. In: IUCN 2008. 2008 IUCN Red List of Threatened Species. http://www.iucnredlist.org.

———. 2004t. *Eleutherodactylus eileenae*. In: IUCN 2008. 2008 IUCN Red List of Threatened Species. http://www.iucnredlist.org.

———. 2004u. *Eleutherodactylus emiliae*. In: IUCN 2008. 2008 IUCN Red List of Threatened Species. http://www.iucnredlist.org.

———. 2004v. *Eleutherodactylus etheridgei*. In: IUCN 2008. 2008 IUCN Red List of Threatened Species. http://www.iucnredlist.org.

———. 2004w. *Eleutherodactylus glamyrus*. In: IUCN 2008. 2008 IUCN Red List of Threatened Species. http://www.iucnredlist.org.

———. 2004x. *Eleutherodactylus goini*. In: IUCN 2008. 2008 IUCN Red List of Threatened Species. http://www.iucnredlist.org.
———. 2004y. *Eleutherodactylus greyi*. In: IUCN 2008. 2008 IUCN Red List of Threatened Species. http://www.iucnredlist.org.
———. 2004z. *Eleutherodactylus guanahacabibes*. In: IUCN 2008. 2008 IUCN Red List of Threatened Species. http://www.iucnredlist.org.
———. 2004aa. *Eleutherodactylus guantanamera*. In: IUCN 2008. 2008 IUCN Red List of Threatened Species. http://www.iucnredlist.org.
———. 2004bb. *Eleutherodactylus gundlachi*. In: IUCN 2008. 2008 IUCN Red List of Threatened Species. http://www.iucnredlist.org.
———. 2004cc. *Eleutherodactylus iberia*. In: IUCN 2008. 2008 IUCN Red List of Threatened Species. http://www.iucnredlist.org.
———. 2004dd. *Eleutherodactylus intermedius*. In: IUCN 2008. 2008 IUCN Red List of Threatened Species. http://www.iucnredlist.org.
———. 2004ee. *Eleutherodactylus ionthus*. In: IUCN 2008. 2008 IUCN Red List of Threatened Species. http://www.iucnredlist.org.
———. 2004ff. *Eleutherodactylus jaumei*. In: IUCN 2008. 2008 IUCN Red List of Threatened Species. http://www.iucnredlist.org.
———. 2004gg. *Eleutherodactylus klinikowskii*. In: IUCN 2008. 2008 IUCN Red List of Threatened Species. http://www.iucnredlist.org.
———. 2004hh. *Eleutherodactylus leberi*. In: IUCN 2008. 2008 IUCN Red List of Threatened Species. http://www.iucnredlist.org.
———. 2004ii. *Eleutherodactylus limbatus*. In: IUCN 2008. 2008 IUCN Red List of Threatened Species. http://www.iucnredlist.org.
———. 2004jj. *Eleutherodactylus mariposa*. In: IUCN 2008. 2008 IUCN Red List of Threatened Species. http://www.iucnredlist.org.
———. 2004kk. *Eleutherodactylus melacara*. In: IUCN 2008. 2008 IUCN Red List of Threatened Species. http://www.iucnredlist.org.
———. 2004ll. *Eleutherodactylus orientalis*. In: IUCN 2008. 2008 IUCN Red List of Threatened Species. http://www.iucnredlist.org.
———. 2004mm. *Eleutherodactylus pezopetrus*. In: IUCN 2008. 2008 IUCN Red List of Threatened Species. http://www.iucnredlist.org.
———. 2004nn. *Eleutherodactylus pinarensis*. In: IUCN 2008. 2008 IUCN Red List of Threatened Species. http://www.iucnredlist.org.
———. 2004oo. *Eleutherodactylus principalis*. In: IUCN 2008. 2008 IUCN Red List of Threatened Species. http://www.iucnredlist.org.
———. 2004pp. *Eleutherodactylus ricordii*. In: IUCN 2008. 2008 IUCN Red List of Threatened Species. http://www.iucnredlist.org.
———. 2004qq. *Eleutherodactylus riparius*. In: IUCN 2008. 2008 IUCN Red List of Threatened Species. http://www.iucnredlist.org.
———. 2004rr. *Eleutherodactylus rivularis*. In: IUCN 2008. 2008 IUCN Red List of Threatened Species. http://www.iucnredlist.org.
———. 2004ss. *Eleutherodactylus ronaldi*. In: IUCN 2008. 2008 IUCN Red List of Threatened Species. http://www.iucnredlist.org.
———. 2004tt. *Eleutherodactylus symingtoni*. In: IUCN 2008. 2008 IUCN Red List of Threatened Species. http://www.iucnredlist.org.
———. 2004uu. *Eleutherodactylus tetajulia*. In: IUCN 2008. 2008 IUCN Red List of Threatened Species. http://www.iucnredlist.org.
———. 2004vv. *Eleutherodactylus thomasi*. In: IUCN 2008. 2008 IUCN Red List of Threatened Species. http://www.iucnredlist.org.
———. 2004ww. *Eleutherodactylus toa*. In: IUCN 2008. 2008 IUCN Red List of Threatened Species. http://www.iucnredlist.org.
———. 2004xx. *Eleutherodactylus tonyi*. In: IUCN 2008. 2008 IUCN Red List of Threatened Species. http://www.iucnredlist.org.
———. 2004yy. *Eleutherodactylus turquinensis*. In: IUCN 2008. 2008 IUCN Red List of Threatened Species. http://www.iucnredlist.org.
———. 2004zz. *Eleutherodactylus varians*. In: IUCN 2008. 2008 IUCN Red List of Threatened Species. http://www.iucnredlist.org.
———. 2004aaa. *Eleutherodactylus varleyi*. In: IUCN 2008. 2008 IUCN Red List of Threatened Species. http://www.iucnredlist.org.
———. 2004bbb. *Eleutherodactylus zeus*. In: IUCN 2008. 2008 IUCN Red List of Threatened Species. http://www.iucnredlist.org.
———. 2004ccc. *Eleutherodactylus zugi*. In: IUCN 2008. 2008 IUCN Red List of Threatened Species. http://www.iucnredlist.org.
———. 2009. Amphibian conservation in the West Indies. In: H. H. Heatwole and J. W. Wilkenson (eds.), Status of Conservation and Decline of Amphibians: Western Hemisphere, in press. Chipping Norton, Australia: Surrey Beatty & Sons.
Hedges, S. B., and O. H. Garrido. 1992a. Cuban snakes of the genus *Arrhyton*: Two new species and a reconsideration of *A. redimitum* Cope. Herpetologica 48: 168–177.
———. 1992b. A new species of *Tropidophis* from Cuba (Serpentes: Tropidophiidae). Copeia 1992: 820–825.
———. 1993. A new species of gecko (*Sphaerodactylus*) from Central Cuba. Journal of Herpetology 27: 300–306.

———. 2002. A new snake of the genus *Tropidophis* (Tropidophiidae) from eastern Cuba. Journal of Herpetology 36: 157–161.

Hedges, S. B., and R. L. Joglar. 2004a. *Eleutherodactylus eneidae.* In: IUCN 2007. 2007 IUCN Red List of Threatened Species. http://www.iucnredlist.org.

———. 2004b. *Eleutherodactylus gryllus.* In: IUCN 2007. 2007 IUCN Red List of Threatened Species. http://www.iucnredlist.org.

———. 2004c. *Eleutherodactylus jasperi.* In: IUCN 2007. 2007 IUCN Red List of Threatened Species. http://www.iucnredlist.org.

———. 2004d. *Eleutherodactylus wightmanae.* In: IUCN 2007. 2007 IUCN Red List of Threatened Species. http://www.iucnredlist.org.

Hedges, S. B., and R. Powell. 1998a. *Eleutherodactylus amadeus.* Catalogue of American Amphibians and Reptiles 647: 1–2.

———. 1998b. *Eleutherodactylus parapelates.* Catalogue of American Amphibians and Reptiles 649: 1–2.

———. 1998c. *Eleutherodactylus thorectes.* Catalogue of American Amphibians and Reptiles 667: 1–2.

———. 2004a. *Eleutherodactylus amplinympha.* In: IUCN 2008. 2008 IUCN Red List of Threatened Species. http://www.iucnredlist.org.

———. 2004b. *Pristimantis euphronides.* In: IUCN 2008. 2008 IUCN Red List of Threatened Species. http://www.iucnredlist.org.

———. 2004c. *Pristimantis shrevei.* In: IUCN 2008. 2008 IUCN Red List of Threatened Species. http://www.iucnredlist.org.

Hedges, S. B., and N. Rios-López. 2008a. *Eleutherodactylus eneidae.* In: IUCN 2008. 2008 IUCN Red List of Threatened Species. http://www.iucnredlist.org.

———. 2008b. *Eleutherodactylus gryllus.* In: IUCN 2008. 2008 IUCN Red List of Threatened Species. http://www.iucnredlist.org.

———. 2008c. *Eleutherodactylus locustus.* In: IUCN 2008. 2008 IUCN Red List of Threatened Species. http://www.iucnredlist.org.

Hedges, S. B., and R. Thomas. 1987. A new burrowing frog from Hispaniola with comments on the *inoptatus* group of the genus *Eleutherodactylus* (Anura: Leptodactylidae). Herpetologica 43: 269–279.

———. 1989a. Supplement to West Indian Amphibians and Reptiles: A Check-list. Milwaukee Public Museum Contributions in Biology and Geology 77: 1–11.

———. 1989b. A new species of *Anolis* from the Sierra de Neiba, Hispaniola. Herpetologica 45: 330–336.

———. 1991a. Cryptic species of snakes (Typhlopidae: *Typhlops*) from the Puerto Rico Bank detected by protein electrophoresis. Herpetologica 47: 448–459.

———. 1991b. The importance of systematic research in the conservation of amphibian and reptile populations. In: J. A. Moreno (ed.), Status y Distribución de los Reptiles y Anfibios de la Región de Puerto Rico, pp. 56–61. Publicación Científica Miscelánea No. 1. San Juan: Departamento de Recursos Naturales de Puerto Rico.

———. 1992a. A new marsh-dwelling species of *Eleutherodactylus* from Haiti (Anura: Leptodactylidae. Caribbean Journal of Science 26: 191–195.

———. 1992b. Two new species of *Eleutherodactylus* from remnant cloud forest in Haiti (Anura: Leptodactylidae). Herpetologica 48: 351–358.

———. 2001. At the lower size limit in amniote vertebrates: A new diminutive lizard from the West Indies. Caribbean Journal of Science 37: 168–173.

———. 2004a. *Eleutherodactylus caribe.* In: IUCN 2008. 2008 IUCN Red List of Threatened Species. http://www.iucnredlist.org.

———. 2004b. *Eleutherodactylus corona.* In: IUCN 2008. 2008 IUCN Red List of Threatened Species. http://www.iucnredlist.org.

———. 2004c. *Eleutherodactylus dolomedes.* In: IUCN 2008. 2008 IUCN Red List of Threatened Species. http://www.iucnredlist.org.

———. 2004d. *Eleutherodactylus glandulifer.* In: IUCN 2008. 2008 IUCN Red List of Threatened Species. http://www.iucnredlist.org.

———. 2004e. *Eleutherodactylus glaphycompus.* In: IUCN 2008. 2008 IUCN Red List of Threatened Species. http://www.iucnredlist.org.

———. 2004f. *Eleutherodactylus grahami.* In: IUCN 2008. 2008 IUCN Red List of Threatened Species. http://www.iucnredlist.org.

———. 2004g. *Eleutherodactylus inoptatus.* In: IUCN 2008. 2008 IUCN Red List of Threatened Species. http://www.iucnredlist.org.

———. 2004h. *Eleutherodactylus lamprotes.* In: IUCN 2008. 2008 IUCN Red List of Threatened Species. http://www.iucnredlist.org.

———. 2004i. *Eleutherodactylus lentus.* In: IUCN 2008. 2008 IUCN Red List of Threatened Species. http://www.iucnredlist.org.

———. 2004j. *Eleutherodactylus lucioi.* In: IUCN 2008. 2008 IUCN Red List of Threatened Species. http://www.iucnredlist.org.

———. 2004k. *Eleutherodactylus oxyrhyncus.* In: IUCN 2008. 2008 IUCN Red List of Threatened Species. http://www.iucnredlist.org.

———. 2004l. *Eleutherodactylus paulsoni.* In: IUCN 2008. 2008 IUCN Red List of Threatened Species. http://www.iucnredlist.org.

———. 2004m. *Eleutherodactylus poolei.* In: IUCN 2008. 2008 IUCN Red List of Threatened Species. http://www.iucnredlist.org.

———. 2004n. *Eleutherodactylus rhodesi.* In: IUCN 2008. 2008 IUCN Red List of Threatened Species. http://www.iucnredlist.org.

———. 2004o. *Eleutherodactylus schwartzi*. In: IUCN 2008. 2008 IUCN Red List of Threatened Species. http://www.iucnredlist.org.

———. 2004p. *Eleutherodactylus sciagraphus*. In: IUCN 2008. 2008 IUCN Red List of Threatened Species. http://www.iucnredlist.org.

———. 2004q. *Eleutherodactylus semipalmatus*. In: IUCN 2008. 2008 IUCN Red List of Threatened Species. http://www.iucnredlist.org.

———. 2004r. *Eleutherodactylus ventrilineatus*. In: IUCN 2008. 2008 IUCN Red List of Threatened Species. http://www.iucnredlist.org.

———. 2004s. *Eleutherodactylus warreni*. In: IUCN 2008. 2008 IUCN Red List of Threatened Species. http://www.iucnredlist.org.

Hedges, S. B., R. Thomas, and R. Franz. 1987. A new species of *Eleutherodactylus* (Anura, Leptodactylidae) from the Massif de la Hotte, Haiti. Copeia 1987: 943–949.

Hedges, S. B., C. A. Hass, and T. K. Maugel. 1989. Physiological color change in snakes. Journal of Herpetology 23: 450–455.

Hedges, S. B., A. R. Estrada, and R. Thomas. 1992. Three new species of *Eleutherodactylus* from eastern Cuba, with notes on vocalizations of other species (Anura: Leptodactylidae). Herpetological Monographs 6: 68–83.

Hedges, S. B., L. González, and A. R. Estrada. 1995. Rediscovery of the Cuban frogs *Eleutherodactylus cubanus* and *E. turquinensis* (Anura: Leptodactylidae). Caribbean Journal of Science 31: 327–332.

Hedges, S. B., A. R. Estrada, and L. M. Diaz. 1999. New snake (*Tropidophis*) from western Cuba. Copeia 1999: 376–381.

Hedges, S. B., L. M. Díaz, B. Ibéné, R. L. Joglar, and R. Powell. 2004a. *Osteopilus septentrionalis*. In: IUCN 2008. 2008 IUCN Red List of Threatened Species. http://www.iucnredlist.org.

Hedges, S. B., L. M. Díaz, and R. Powell. 2004b. *Eleutherodactylus planirostris*. In: IUCN 2008. 2008 IUCN Red List of Threatened Species. http://www.iucnredlist.org.

Hedges, S. B., B. Ibéné, M. Breuil, and R. Powell. 2004c. *Allobates chalcopis*. In: IUCN 2008. 2008 IUCN Red List of Threatened Species. http://www.iucnredlist.org.

———. 2004d. *Eleutherodactylus barlagnei*. In: IUCN 2008. 2008 IUCN Red List of Threatened Species. http://www.iucnredlist.org.

———. 2004e. *Eleutherodactylus martinicensis*. In: IUCN 2008. 2008 IUCN Red List of Threatened Species. http://www.iucnredlist.org.

———. 2004f. *Eleutherodactylus pinchoni*. In: IUCN 2008. 2008 IUCN Red List of Threatened Species. http://www.iucnredlist.org.

Hedges, S. B., B. Ibéné, S. E. Koenig, E. La Marca, R. Ibáñez, and J. D. Hardy, Jr. 2004g. *Eleutherodactylus johnstonei*. In: IUCN 2007. 2007 IUCN Red List of Threatened Species. http://www.iucnredlist.org.

Hedges, S. B., S. J. Incháustegui, and M. Hernández. 2004h. *Eleutherodactylus pictissimus*. In: IUCN 2008. 2008 IUCN Red List of Threatened Species. http://www.iucnredlist.org.

———. 2004i. *Eleutherodactylus probolaeus*. In: IUCN 2008. 2008 IUCN Red List of Threatened Species. http://www.iucnredlist.org.

———. 2004j. *Osteopilus dominicensis*. In: IUCN 2008. 2008 IUCN Red List of Threatened Species. http://www.iucnredlist.org.

Hedges, S. B., S. J. Incháustegui, M. Hernández, and R. Powell. 2004k. *Peltophryne fluviatica*. In: IUCN 2008. 2008 IUCN Red List of Threatened Species. http://www.iucnredlist.org.

———. 2004l. *Peltophryne fracta*. In: IUCN 2008. 2008 IUCN Red List of Threatened Species. http://www.iucnredlist.org.

———. 2004m. *Peltophryne guentheri*. In: IUCN 2008. 2008 IUCN Red List of Threatened Species. http://www.iucnredlist.org.

———. 2004n. *Eleutherodactylus armstrongi*. In: IUCN 2008. 2008 IUCN Red List of Threatened Species. http://www.iucnredlist.org.

———. 2004o. *Eleutherodactylus audanti*. In: IUCN 2008. 2008 IUCN Red List of Threatened Species. http://www.iucnredlist.org.

———. 2004p. *Eleutherodactylus auriculatoides*. In: IUCN 2008. 2008 IUCN Red List of Threatened Species. http://www.iucnredlist.org.

———. 2004q. *Eleutherodactylus flavescens*. In: IUCN 2008. 2008 IUCN Red List of Threatened Species. http://www.iucnredlist.org.

———. 2004r. *Eleutherodactylus fowleri*. In: IUCN 2008. 2008 IUCN Red List of Threatened Species. http://www.iucnredlist.org.

———. 2004s. *Eleutherodactylus minutus*. In: IUCN 2008. 2008 IUCN Red List of Threatened Species. http://www.iucnredlist.org.

———. 2004t. *Eleutherodactylus montanus*. In: IUCN 2008. 2008 IUCN Red List of Threatened Species. http://www.iucnredlist.org.

———. 2004u. *Eleutherodactylus schmidti*. In: IUCN 2008. 2008 IUCN Red List of Threatened Species. http://www.iucnredlist.org.

———. 2004v. *Hypsiboas heilprini*. In: IUCN 2008. 2008 IUCN Red List of Threatened Species. http://www.iucnredlist.org.

———. 2004w. *Osteopilus pulchrilineatus*. In: IUCN 2008. 2008 IUCN Red List of Threatened Species. http://www.iucnredlist.org.

———. 2004x. *Osteopilus vastus*. In: IUCN 2008. 2008 IUCN Red List of Threatened Species. http://www.iucnredlist.org.

Hedges, S. B., S. J. Incháustegui, and R. Powell. 2004y. *Eleutherodactylus furcyensis*. In: IUCN 2008. 2008 IUCN Red List of Threatened Species. http://www.iucnredlist.org.

———. 2004z. *Eleutherodactylus haitianus*. In: IUCN 2008.

2008 IUCN Red List of Threatened Species. http://www.iucnredlist.org.

———. 2004aa. *Eleutherodactylus jugans*. In: IUCN 2008. 2008 IUCN Red List of Threatened Species. http://www.iucnredlist.org.

———. 2004bb. *Eleutherodactylus leoncei*. In: IUCN 2008. 2008 IUCN Red List of Threatened Species. http://www.iucnredlist.org.

———. 2004cc. *Eleutherodactylus nortoni*. In: IUCN 2008. 2008 IUCN Red List of Threatened Species. http://www.iucnredlist.org.

———. 2004dd. *Eleutherodactylus patriciae*. In: IUCN 2008. 2008 IUCN Red List of Threatened Species. http://www.iucnredlist.org.

———. 2004ee. *Eleutherodactylus pituinus*. In: IUCN 2008. 2008 IUCN Red List of Threatened Species. http://www.iucnredlist.org.

———. 2004ff. *Eleutherodactylus rufifemoralis*. In: IUCN 2008. 2008 IUCN Red List of Threatened Species. http://www.iucnredlist.org.

———. 2004gg. *Eleutherodactylus wetmorei*. In: IUCN 2008. 2008 IUCN Red List of Threatened Species. http://www.iucnredlist.org.

Hedges, S. B., S. J. Incháustegui, and R. Thomas. 2004hh. *Eleutherodactylus parabates*. In: IUCN 2008. 2008 IUCN Red List of Threatened Species. http://www.iucnredlist.org.

———. 2004ii. *Eleutherodactylus weinlandi*. In: IUCN 2008. 2008 IUCN Red List of Threatened Species. http://www.iucnredlist.org.

Hedges, S. B., S. J. Incháustegui, R. Thomas, and R. Powell. 2004jj. *Eleutherodactylus alcoae*. In: IUCN 2008. 2008 IUCN Red List of Threatened Species. http://www.iucnredlist.org.

———. 2004kk. *Eleutherodactylus heminota*. In: IUCN 2008. 2008 IUCN Red List of Threatened Species. http://www.iucnredlist.org.

———. 2004ll. *Eleutherodactylus hypostenor*. In: IUCN 2008. 2008 IUCN Red List of Threatened Species. http://www.iucnredlist.org.

———. 2004mm. *Eleutherodactylus ruthae*. In: IUCN 2008. 2008 IUCN Red List of Threatened Species. http://www.iucnredlist.org.

Hedges, S. B., R. L. Joglar, and R. Thomas. 2004nn. *Bufo lemur*. In: IUCN 2007. 2007 IUCN Red List of Threatened Species. http://www.iucnredlist.org.

———. 2004oo. *Eleutherodactylus brittoni*. In: IUCN 2007. 2007 IUCN Red List of Threatened Species. http://www.iucnredlist.org.

———. 2004pp. *Eleutherodactylus cooki*. In: IUCN 2007. 2007 IUCN Red List of Threatened Species. http://www.iucnredlist.org.

———. 2004qq. *Eleutherodactylus hedricki*. In: IUCN 2007. 2007 IUCN Red List of Threatened Species. http://www.iucnredlist.org.

———. 2004rr. *Eleutherodactylus karlschmidti*. In: IUCN 2007. 2007 IUCN Red List of Threatened Species. http://www.iucnredlist.org.

———. 2004ss. *Eleutherodactylus locustus*. In: IUCN 2007. 2007 IUCN Red List of Threatened Species. http://www.iucnredlist.org.

———. 2004tt. *Eleutherodactylus monensis*. In: IUCN 2007. 2007 IUCN Red List of Threatened Species. http://www.iucnredlist.org.

———. 2004uu. *Eleutherodactylus portoricensis*. In: IUCN 2007. 2007 IUCN Red List of Threatened Species. http://www.iucnredlist.org.

———. 2004vv. *Eleutherodactylus richmondi*. In: IUCN 2007. 2007 IUCN Red List of Threatened Species. http://www.iucnredlist.org.

———. 2004ww. *Eleutherodactylus unicolor*. In: IUCN 2007. 2007 IUCN Red List of Threatened Species. http://www.iucnredlist.org.

———. 2004xx. *Leptodactylus albilabris*. In: IUCN 2007. 2007 IUCN Red List of Threatened Species. http://www.iucnredlist.org.

Hedges, S. B., R. L. Joglar, R. Thomas, and R. Powell. 2004yy. *Eleutherodactylus antillensis*. In: IUCN 2007. 2007 IUCN Red List of Threatened Species. http://www.iucnredlist.org.

———. 2004zz. *Eleutherodactylus cochranae*. In: IUCN 2007. 2007 IUCN Red List of Threatened Species. http://www.iucnredlist.org.

Hedges, S. B., S. E. Koenig, and B. S. Wilson. 2004aaa. *Eleutherodactylus alticola*. In: IUCN 2008. 2008 IUCN Red List of Threatened Species. http://www.iucnredlist.org.

———. 2004bbb. *Eleutherodactylus andrewsi*. In: IUCN 2008. 2008 IUCN Red List of Threatened Species. http://www.iucnredlist.org.

———. 2004ccc. *Eleutherodactylus cavernicola*. In: IUCN 2008. 2008 IUCN Red List of Threatened Species. http://www.iucnredlist.org.

———. 2004ddd. *Eleutherodactylus cundalli*. In: IUCN 2008. 2008 IUCN Red List of Threatened Species. http://www.iucnredlist.org.

———. 2004eee. *Eleutherodactylus fuscus*. In: IUCN 2008. 2008 IUCN Red List of Threatened Species. http://www.iucnredlist.org.

———. 2004fff. *Eleutherodactylus glaucoreius*. In: IUCN 2008. 2008 IUCN Red List of Threatened Species. http://www.iucnredlist.org.

———. 2004ggg. *Eleutherodactylus gossei*. In: IUCN 2008. 2008 IUCN Red List of Threatened Species. http://www.iucnredlist.org.

———. 2004hhh. *Eleutherodactylus grabhami*. In: IUCN 2008. 2008 IUCN Red List of Threatened Species. http://www.iucnredlist.org.

———. 2004iii. *Eleutherodactylus jamaicensis*. In: IUCN 2008. 2008 IUCN Red List of Threatened Species. http://www.iucnredlist.org.

———. 2004jjj. *Eleutherodactylus luteolus*. In: IUCN 2008. 2008 IUCN Red List of Threatened Species. http://www.iucnredlist.org.
———. 2004kkk. *Eleutherodactylus nubicola*. In: IUCN 2008. 2008 IUCN Red List of Threatened Species. http://www.iucnredlist.org.
———. 2004lll. *Eleutherodactylus pantoni*. In: IUCN 2008. 2008 IUCN Red List of Threatened Species. http://www.iucnredlist.org.
———. 2004mmm. *Eleutherodactylus sisyphodemus*. In: IUCN 2008. 2008 IUCN Red List of Threatened Species. http://www.iucnredlist.org.
———. 2004nnn. *Osteopilus brunneus*. In: IUCN 2008. 2008 IUCN Red List of Threatened Species. http://www.iucnredlist.org.
———. 2004ooo. *Osteopilus crucialis*. In: IUCN 2008. 2008 IUCN Red List of Threatened Species. http://www.iucnredlist.org.
———. 2004ppp. *Osteopilus marianae*. In: IUCN 2008. 2008 IUCN Red List of Threatened Species. http://www.iucnredlist.org.
———. 2004qqq. *Osteopilus wilderi*. In: IUCN 2008. 2008 IUCN Red List of Threatened Species. http://www.iucnredlist.org.
Hedges, S. B., R. Thomas, S. J. Incháustegui, and R. Powell. 2004rrr. *Eleutherodactylus abbotti*. In: IUCN 2008. 2008 IUCN Red List of Threatened Species. http://www.iucnredlist.org.
Hedges, S. B., R. Thomas, and R. Powell. 2004sss. *Eleutherodactylus amadeus*. In: IUCN 2008. 2008 IUCN Red List of Threatened Species. http://www.iucnredlist.org.
———. 2004ttt. *Eleutherodactylus apostates*. In: IUCN 2008. 2008 IUCN Red List of Threatened Species. http://www.iucnredlist.org.
———. 2004uuu. *Eleutherodactylus bakeri*. In: IUCN 2008. 2008 IUCN Red List of Threatened Species. http://www.iucnredlist.org.
———. 2004vvv. *Eleutherodactylus brevirostris*. In: IUCN 2008. 2008 IUCN Red List of Threatened Species. http://www.iucnredlist.org.
———. 2004www. *Eleutherodactylus chlorophenax*. In: IUCN 2008. 2008 IUCN Red List of Threatened Species. http://www.iucnredlist.org.
———. 2004xxx. *Eleutherodactylus counouspeus*. In: IUCN 2008. 2008 IUCN Red List of Threatened Species. http://www.iucnredlist.org.
———. 2004yyy. *Eleutherodactylus darlingtoni*. In: IUCN 2008. 2008 IUCN Red List of Threatened Species. http://www.iucnredlist.org.
———. 2004zzz. *Eleutherodactylus eunaster*. In: IUCN 2008. 2008 IUCN Red List of Threatened Species. http://www.iucnredlist.org.
———. 2004aaaa. *Eleutherodactylus glanduliferoides*. In: IUCN 2008. 2008 IUCN Red List of Threatened Species. http://www.iucnredlist.org.
———. 2004bbbb. *Eleutherodactylus parapelates*. In: IUCN 2008. 2008 IUCN Red List of Threatened Species. http://www.iucnredlist.org.
———. 2004cccc. *Eleutherodactylus thorectes*. In: IUCN 2008. 2008 IUCN Red List of Threatened Species. http://www.iucnredlist.org.
Hedges, S. B., R. L. Joglar, R. Thomas, and R. Powell. 2006. *Eleutherodactylus coqui*. In: IUCN 2007. 2007 IUCN Red List of Threatened Species. http://www.iucnredlist.org.
Hedges, S. B., W. E. Duellman, and M. P. Heinicke. 2008a. New World direct-developing frogs (Anura: Terrarana): Molecular phylogeny, classification, biogeography, and conservation. Zootaxa 1737: 1–182.
Hedges, S. B., B. Ibéné, S. E. Koenig, E. La Marca, R. Ibáñez, and J. D. Hardy, Jr. 2008b. *Eleutherodactylus johnstonei*. In: IUCN 2008. 2008 IUCN Red List of Threatened Species. http://www.iucnredlist.org.
Hedges, S. B., R. L. Joglar, and R. Thomas. 2008c. *Eleutherodactylus karlschmidti*. In: IUCN 2008. 2008 IUCN Red List of Threatened Species. http://www.iucnredlist.org.
Hedges, S. B., A. Couloux, and N. Vidal. 2009. Molecular phylogeny, classification, and biogeography of West Indian racers of the Tribe Alsophiini (Squamata, Dipsadidae, Xenodontinae). Zootaxa 2067: 1–28.
Heinicke, M. P., W. E. Duellman, and S. B. Hedges. 2007. Major Caribbean and Central American frog faunas originated by ancient oceanic dispersal. Proceedings of the National Academy of Sciences 104: 10092–10097.
Heinz, H. M., A. J. Maley, A. Z. Savit, R. W. Henderson, and R. Powell. 2004. Behaviour and time allotment in the West Indian snake *Alsophis rufiventris* (Colubridae). Herpetological Bulletin 89: 22–25.
Heinz, H. M., A. Z. Savit, A. J. Maley, R. W. Henderson, and R. Powell. 2005. *Alsophis rufiventris* (Red-bellied Racer). Foraging and diet. Herpetological Review 36: 186–187.
Helmer, E. H., T. A. Kennaway, D. H. Pedreros, M. L. Clark, H. Marcano-Vega, L. L. Tieszen, T. R. Ruzycki, S. R. Schill, and C. M. S. Carrington. 2008. Land cover and forest formation distributions for St. Kitts, Nevis, St. Eustatius, Grenada and Barbados from decision tree classification of cloud-cleared satellite imagery. Caribbean Journal of Science 44: 175–198.
Henderson, R. W. 1982. Thermoregulation in an Hispaniolan tree snake, *Uromacer catesbyi*. Journal of Herpetology 16: 89–91.
———. 1984a. The diets of Hispaniolan colubrid snakes I. Introduction and prey genera. Oecologia 62: 234–239.
———. 1984b. The diet of the Hispaniolan snake *Hypsirhynchus ferox* (Colubridae). Amphibia-Reptilia 5: 367–371.
———. 1988a. *Diploglossus anelpistus*. Catalogue of American Amphibians and Reptiles 424: 1.

———. 1988b. *Diploglossus carraui*. Catalogue of American Amphibians and Reptiles 425: 1.
———. 1988c. *Diploglossus warreni*. Catalogue of American Amphibians and Reptiles 426: 1–2.
———. 1990. A new subspecies of *Alsophis antiguae* (Parker) from Great Bird Island (Antigua), Lesser Antilles. Caribbean Journal of Science 25: 119–122.
———. 1993. Foraging and diet in West Indian *Corallus enydris* (Serpentes: Boidae). Journal of Herpetology 27: 24–28.
———. 1996. *Chironius vincenti*. In: IUCN 2008. 2008 IUCN Red List of Threatened Species. http://www.iucnredlist.org.
———. 1997. A taxonomic review of the *Corallus hortulanus* complex of Neotropical tree boas. Caribbean Journal of Science 33: 198–221.
———. 2002. Neotropical Treeboas: Natural History of the *Corallus hortulanus* Complex. Malabar, Fla.: Krieger.
———. 2004. Lesser Antillean snake faunas: Distribution, ecology, and conservation concerns. Oryx 38: 311–320.
———. 2008. The shared history of treeboas (*Corallus grenadensis*) and humans on Grenada: A hypothetical excursion. Iguana 15: 198–203.
Henderson, R. W., and A. Arias B. 2001. *Epicrates angulifer*. Catalogue of American Amphibians and Reptiles 734: 1–4.
Henderson, R. W., and C. S. Berg. 2005a. A post-Hurricane Ivan assessment of frog and reptile populations on Grenada, West Indies. Herpetological Bulletin 91: 4–9.
———. 2005b. *Thecadactylus rapicauda* (Turnip-tailed Gecko). Diet. Herpetological Review 36: 454.
———. 2006. The herpetofauna of Grenada and the Grenada Grenadines: Conservation concerns. Applied Herpetology 3: 197–213.
Henderson, R. W., and R. W. Bourgeois. 1993. Notes on the diets of West Indian *Liophis* (Serpentes: Colubridae). Caribbean Journal of Science 29: 253–254.
Henderson, R. W., and B. I. Crother. 1989. Biogeographic patterns of predation in West Indian colubrid snakes. In: C. A. Woods (ed.), Biogeography of the West Indies: Past, Present, and Future, pp. 479–517. Gainesville, Fla.: Sandhill Crane Press.
Henderson, R. W., and G. T. Haas. 1993. Status of the West Indian snake *Chironius vincenti*. Oryx 27: 181–184.
Henderson, R. W., and K. F. Henderson. 1995. Altitudinal variation in body temperature in foraging tree boas (*Corallus enydris*) on Grenada. Caribbean Journal Science 31: 73–76.
Henderson, R. W., and R. M. Henderson. 2004. *Osteopilus vastus*. Catalogue of American Amphibians and Reptiles 786: 1–3.
Henderson, R. W., and H. S. Horn. 1983. The diet of the snake *Uromacer frenatus dorsalis* on Île de la Gonâve, Haiti. Journal of Herpetology 17: 409–412.
Henderson, R. W. and R. Powell. 1996a. *Alsophis ater*. Catalogue of American Amphibians and Reptiles 633: 1–2.
———. 1996b. *Alsophis sanctaecrucis*. Catalogue of American Amphibians and Reptiles 634: 1–2.
———. 1996c. *Chironius vincenti*. Catalogue of American Amphibians and Reptiles 635: 1–2.
———. 1999. West Indian herpetoecology. In: B. I. Crother (ed.), Caribbean Amphibians and Reptiles, pp. 223–268. San Diego: Academic Press.
———. 2001. Responses by the West Indian herpetofauna to human-influenced resources. Caribbean Journal of Science 37: 41–54.
———. 2002. *Epicrates gracilis*. Catalogue of American Amphibians and Reptiles 754: 1–4.
——— (eds.). 2003. Islands and the Sea: Essays on herpetological Exploration in the West Indies. Contributions to Herpetology, volume 20. Ithaca, N.Y.: Society for the Study of Amphibians and Reptiles.
———. 2004a. *Epicrates fordii*. Catalogue of American Amphibians and Reptiles 800: 1–3.
———. 2004b. Thomas Barbour and the *Utowana* voyages (1929–1934) in the West Indies. Bonner Zoologische Beiträge 52: 297–309.
———. 2005. Geographic distribution. *Anolis sagrei*. Herpetological Review 36: 467.
———. 2006. Geographic distribution. *Tantilla melanocephala* (NCN). Herpetological Review 37: 501.
Henderson, R. W., and R. A. Sajdak. 1983. Notes on reptiles from Isla Saona, República Dominicana. Florida Scientist 46: 59–61.
———. 1996. Diets of West Indian racers (Colubridae: *Alsophis*): Composition and biogeographic implications. In: R. Powell and R. W. Henderson (eds.), Contributions to West Indian Herpetology: A Tribute to Albert Schwartz, pp. 317–326. Contributions to Herpetology, volume 12. Ithaca, N.Y.: Society for the Study of Amphibians and Reptiles.
———. 2002. A preliminary look at the spatial distribution of treeboas at a site on Grenada, Herpetological Bulletin 81: 5–7.
———. 2003. *Corallus* spp. (Treeboas). Evidence of over-water dispersal. Herpetological Review 34: 63–64.
Henderson, R. W., and A. Schwartz. 1984. *Uromacer frenatus*. Catalogue of American Amphibians and Reptiles 357: 1–2.
———. 1986. The diet of the Hispaniolan colubrid snake, *Darlingtonia haetiana*. Copeia 1986: 529–531.
Henderson, R. W., and R. A. Winstel. 1997. Daily activity in tree boas (*Corallus grenadensis*) on Grenada. Herpetological Natural History 5: 175–180.
Henderson, R. W., M. H. Binder, and R. A. Sajdak. 1981. Ecological relationships of the tree snakes *Uromacer catesbyi* and *U. oxyrhynchus* (Colubridae) on Isla Saona, República Dominicana. Amphibia-Reptilia 2: 153–163.

Henderson, R. W., M. H. Binder, and G. M. Burghardt. 1983. Responses of neonate Hispaniolan vine snakes (*Uromacer frenatus*) to prey extracts. Herpetologica 39: 75–77.

Henderson, R. W., B. I. Crother, T. A. Noeske-Hallin, A. Schwartz, and C. R. Dethloff. 1987a. The diet of the Hispaniolan snake *Antillophis parvifrons*. Journal of Herpetology 21: 328–332.

Henderson, R. W., T. A. Noeske-Hallin, J. A. Ottenwalder, and A. Schwartz. 1987b. On the diet of the boa *Epicrates striatus* on Hispaniola, with notes on *E. fordi* and *E. gracilis*. Amphibia-Reptilia 8: 251–258.

Henderson, R. W., A. Schwartz, and T. A. Noeske-Hallin. 1987c. Food habits of three colubrid tree snakes (genus *Uromacer*) on Hispaniola. Herpetologica 43: 241–248.

Henderson, R. W., T. A. Noeske-Hallin, B. I. Crother, and A. Schwartz. 1988a. The diets of Hispaniolan colubrid snakes II. Prey species, prey size, and phylogeny. Herpetologica 44: 55–70.

Henderson, R. W., R. A. Sajdak, and R. M. Henderson. 1988b. The rediscovery of the West Indian colubrid snake *Chironius vincenti*. Amphibia-Reptilia 9: 415–422.

Henderson, R. W., J. Daudin, G. T. Haas, and T. J. McCarthy. 1992. Significant distribution records for some reptiles in the Lesser Antilles. Caribbean Journal of Science 28: 101–103.

Henderson, R. W., A. Delatte, and T. J. McCarthy. 1993. *Gekko gecko* (Sauria: Gekkonidae) established on Martinique, French West Indies. Caribbean Journal of Science 29: 128–129.

Henderson, R. W., R. Powell, J. C. Daltry, and M. L. Day. 1996a. *Alsophis antiguae*. Catalogue of American Amphibians and Reptiles 632: 1–3.

Henderson, R. W., A. R. Winstel, and J. Friesch. 1996b. *Corallus enydris* (Serpentes: Boidae) in the post-Columbian West Indies: New habitats, new prey species, and new predators. In: R. Powell and R. W. Henderson (eds.), Contributions to West Indian herpetology: A Tribute to Albert Schwartz, pp. 417–423. Contributions to Herpetology, volume 12. Ithaca, N.Y.: Society for the Study of Amphibians and Reptiles.

Henderson, R. W., R. A. Sajdak, and W. A. Winstel. 1998. Habitat utilization by the arboreal boa *Corallus grenadensis* in two ecologically disparate habitats on Grenada. Amphibia-Reptilia 19: 203–214.

Henderson, R. W., M. L. Treglia, and S. D. Powell. 2007. *Corallus cookii* (St. Vincent Treeboa). Foraging. Herpetological Review 38: 466.

Hensley, R. L. and R. Powell. 2006. *Sphaerodactylus sabanus*. Catalogue of American Amphibians and Reptiles 809: 1–3.

Hensley, R. L., S. M. Wissmann, R. Powell, and J. S. Parmerlee, Jr. 2004. Habitat preferences and abundance of dwarf geckos (*Sphaerodactylus*) on St. Eustatius, Netherlands Antilles. Caribbean Journal of Science 40: 427–429.

Hensley, R. L., A. Z. Savit, and R. Powell. 2006. *Anolis schwartzi*. Catalogue of American Amphibians and Reptiles 816: 1–5.

Hernández-Prieto, E. 1987. *Typhlops monensis* (Mona Island Blind Snake). Predation. Herpetological Review 18: 76.

Herrel, A., and V. Holanova. 2008. Cranial morphology and bite force in *Chamaeleolis* lizards: Adaptations to molluscivory? Zoology—Analysis of Complex Systems 111: 467–475.

Herrel, A., B. Vanhooydonck, R. Joachim, and D. J. Irschick. 2004. Frugivory in polychrotid lizards: Effects of body size. Oecologia 140: 160–168.

Herrel, A., R. Joachim, B. VanHooydonck, and D. J. Irschick. 2006. Ecological consequences of ontogenetic changes in head shape and bite performance in the Jamaican lizard *Anolis lineatopus*. Biological Journal of the Linnean Society 89: 443–454.

Herrel, A. B., B. Vanhooydonck, J. Porck, and D. J. Irschick. 2008. Anatomical basis of differences in locomotor behavior in *Anolis* lizards: A comparison between two ecomorphs. Bulletin of the Museum of Comparative Zoology 159: 213–238.

Herrmann, H.-J. 2006. Verfluchte Kobolde Kuba-Laubfrösche *Osteopilus septentrionalis*. Aquaristik Fachmagazin and Aquarium Heute 38: 22–23.

Hertz, P. E. 1976. *Anolis alumina*, new species of grass anole from the Barahona Peninsula of Hispaniola. Breviora 437: 1–19.

———. 1979a. Comparative thermal biology of sympatric grass anoles (*Anolis semilineatus* and *A. olssoni*) in lowland Hispaniola (Reptilia, Lacertilia, Iguanidae). Journal of Herpetology 13: 329–333.

———. 1979b. Sensitivity to high temperatures in three West Indian grass anoles (Sauria, Iguanidae), with a review of heat sensitivity in the genus *Anolis*. Comparative Biochemistry and Physiology 63A: 217–222.

———. 1980a. Comparative physiological ecology of the sibling species *Anolis cybotes* and *A. marcanoi*. Journal of Herpetology 14: 92–95.

———. 1980b. Responses to dehydration in *Anolis* lizards sampled along altitudinal transects. Copeia 1980: 440–446.

———. 1981. Adaptation to altitude in two West Indian anoles (Reptilia: Iguanidae): Field thermal biology and physiological ecology. Journal of Zoology (London) 195: 25–37.

———. 1983. Eurythermy and niche breadth in West Indian *Anolis* lizards: A reappraisal. In: A. G. J. Rhodin and K. Miyata (eds.), Advances in Herpetology and Evolutionary Biology: Essays in Honor of Ernest E. Williams, pp. 472–483. Cambridge, Mass.: Museum of Comparative Zoology, Harvard University.

———. 1992a. Evaluating thermal resource partitioning in sympatric lizards *Anolis cooki* and *A. cristatellus*: A field test using null hypotheses. Oecologia 90: 127–136.

———. 1992b. Temperature regulation in Puerto Rican *Ano-*

lis lizards: A field test using null hypotheses. Ecology 73: 1405–1417.

Hertz, P. E., and R. B. Huey. 1981. Compensation for altitudinal changes in the thermal environment by some *Anolis* lizards on Hispaniola. Ecology 62: 515–521.

Hertz, P. E., A. Arce-Hernández, J. Ramírez-Vásquez, W. Tirado-Rivera, and L. Vásquez-Vives. 1979. Geographical variation of heat sensitivity and water loss rates in the tropical lizard, *Anolis gundlachi*. Comparative Biochemistry and Physiology 63A: 947–953.

Hertz, P. E., R. B. Huey, and R. D. Stevenson. 1993. Evaluating temperature regulation by field-active ectotherms: The fallacy of the inappropriate question. American Naturalist 142: 796–818.

Hertz, P. E., L. J. Fleishman, and C. Armsby. 1994. The influence of light intensity and temperature on microhabitat selection in two *Anolis* lizards. Functional Ecology 8: 720–729.

Hess, N. E., and J. B. Losos. 1991. Interspecific aggression between *Anolis cristatellus* and *A. gundlachi*: Comparison of sympatric and allopatric populations. Journal of Herpetology 25: 256–259.

Heselhaus, R. 1990. Von einigen kubanischen *Anolis*. Die Aquarien- und Terrarienzeitschrift (DATZ) 43: 663–665.

Heselhaus, R., and M. Schmidt. 1990. Karibische Anolis. Münster, Germany: Herpetologischer Fachverlag (Terrarien Bibliotek).

Heyer, W. R. 1979. Systematics of the *pentadactylus* species group of the frog genus *Leptodactylus* (Amphibia: Leptodactylidae). Smithsonian Contributions to Zoology 301: iv + 43 pp.

Hicks, R. A. 1973. New studies on a montane lizard of Jamaica, *Anolis reconditus*. Breviora 404: 1–23.

Hicks, R. A., and R. L. Trivers. 1983. The social behavior of *Anolis valencienni*. In: A. G. J. Rhodin and K. Miyata (eds.), Advances in Herpetology and Evolutionary Biology: Essays in Honor of Ernest E. Williams, pp. 570–595. Cambridge, Mass.: Museum of Comparative Zoology, Harvard University.

Hicks, S. 2008. In memory of the seven blues. Iguana 15: 77.

Higham, T. E., M. S. Davenport, and B. C. Jayne. 2001. Maneuvering in an arboreal habitat: The effects of turning angle on the locomotion of three sympatric ecomorphs of *Anolis* lizards. Journal of Experimental Biology 204: 4141–4155.

Hilgenhof, R., and R. J. Hilgenhof. 2004. Zur Kenntnis, Haltung und Nachzucht des Schwertschwanz-Anolis (*Anolis valencienni*) (Dumeril & Bibron, 1837). Reptilia 9: 72–76.

Hilgenhof, R. J. 1998a. Eine herpetologische Reise nach Nordost-Jamaika unter besonderer Berücksichtigung der Gattung *Anolis* Daudin, 1802—Teil I. Sauria 20(2): 15–24.

———. 1998b. Eine herpetologische Reise nach Nordost-Jamaika unter besonderer Berücksichtigung der Gattung *Anolis* Daudin, 1802—Teil II. Sauria 20(3): 3–10.

Hill, W. C. O. 1954. Report of the Society's presector for the year 1953. Proceedings of the Zoological Society, London 124: 304–311.

Hillman, S. S., and G. C. Gorman. 1977. Water loss, desiccation tolerance, and survival under desiccating conditions in 11 species of Caribbean *Anolis*. Evolutionary and ecological implications. Oecologia 29: 105–116.

Hillman, S., G. C. Gorman, and R. Thomas. 1979. Water loss in *Anolis* lizards: Evidence for acclimation and intraspecific differences along a habitat gradient. Comparative Biochemistry and Physiology 62A: 491–494.

Hines, K. N., and J. B. Iverson. 2006a. Taxon reports: Allen Cays Iguana (*Cyclura cychlura inornata*). Iguana Specialist Group Newsletter 9(1): 10–11.

———. 2006b. Allen Cays Iguana (*Cyclura cychlura inornata*). Iguana 13: 279–281.

Hines, K. N., J. B. Iverson, and J. M. Valiulis. 2002a. *Cyclura cychlura inornata* (Allen Cays Rock Iguana). Bird predation. Herpetological Review 33: 306.

———. 2002b. *Leiocephalus carinatus* (Exuma Islands Curlytail Lizard). Interspecific aggression. Herpetological Review 33: 308.

Hite, J. L., C. A. Rodríguez Gomez, S. C. Larimer, A. M. Díaz-Lameiro, and R. Powell. 2008a. Anoles of St. Vincent (Squamata: Polychrotidae): Population densities and structural habitat use. Caribbean Journal of Science 44: 102–115.

Hite, J. L., D. S. Steinberg, and R. Powell. 2008b. *Sphaerodactylus kirbyi*. Catalogue of American Amphibians and Reptiles 852: 1–2.

Hoagland, D. B., G. R. Horst, and C. W. Kilpatrick. 1989. Biogeography and population biology of the mongoose in the West Indies. In: C. A. Woods (ed.), Biogeography of the West Indies: Past, Present, and Future, pp. 611–633. Gainesville, Fla.: Sandhill Crane Press.

Hodge, K. V. D., E. J. Censky, and R. Powell. 2003. The Reptiles and Amphibians of Anguilla, British West Indies. The Valley: Anguilla National Trust.

Hodson, L. A., and J. F. W. Pearson. 1943. Notes on the discovery and biology of two Bahaman fresh-water turtles of the genus *Pseudemys*. Quarterly Journal of the Florida Academy of Science 6: 17–23.

Hoffman, W. A. 1935. *Mesocoelium danforthi*, n. sp. (Dicrocoeliidae) from a lizard, *Celestus pleei*, in Puerto Rico. Proceedings of the Helminthological Society of Washington 2: 64.

Holanova, V. 2003. *Sphaerodactylus cinereus*. Reptilia (Spain) 41: 55–59.

Holanova, V., and J. Hribal. 2004. Dos anolis cubanos: *Anolis bartschi* y *Anolis lucius*. Reptilia (Spain) 47: 62–68.

———. 2005. *Anolis baracoae*. Reptilia (Spain) 53: 56–60.

Honegger, R. E. 1970. Eine Kröte erobert die Welt. Die heutige Verbreitung der Aga-Kröte (*Bufo marinus*). Natur und Museum 100: 447–453.

———. 1981. List of amphibians and reptiles either known or thought to be extinct since 1600. Biological Conservation 19: 141–158.

Horn, H. S. 1969. Polymorphism and evolution of the Hispaniolan snake genus *Uromacer* (Colubridae). Breviora 324: 1–23.

Horstink, Y. 2004. Een kleine grasanolis van Puerto Rico *Anolis krugi* Peters, 1876. Lacerta 62: 60–65.

Hottmar, V. 1991. Zur Erstnachzucht von *Chamaeleolis barbatus* Garrido, 1982, im Terrarium mit einer Übersicht über die Gattung *Chamaeleolis* auf Kuba. Herpetofauna 13: 29–33.

Howard, A. K., J. D. Forester, J. M. Ruder, and R. Powell. 1997. Diets of two syntopic frogs: *Eleutherodactylus abbotti* and *E. armstrongi* (Leptodactylidae) from the Sierra de Baoruco, Hispaniola. Herpetological Natural History 5: 77–82.

Howard, A. K., J. D. Forester, J. M. Ruder, J. S. Parmerlee, Jr., and R. Powell. 1999a. Natural history of a terrestrial Hispaniolan anole: *Anolis barbouri*. Journal of Herpetology 33: 702–706.

Howard, A. K., R. Powell, and J. S. Parmerlee, Jr. 1999b. *Anolis barbouri*. Catalogue of American Amphibians and Reptiles 692: 1–4.

Howard, K. G., J. S. Parmerlee, Jr., and R. Powell. 2001. Natural history of the edificarian geckos *Hemidactylus mabouia*, *Thecadactylus rapicauda*, and *Sphaerodactylus sputator* on Anguilla. Caribbean Journal of Science 37: 285–288.

Hower, L. M., and S. B. Hedges. 2001. Molecular phylogeny and biogeography of West Indian teiid lizards of the genus *Ameiva*. Caribbean Journal of Science 39: 298–306.

Hudson, R. D., and A. C. Alberts. 2004. The role of zoos in the conservation of West Indian iguanas. In: A. C. Alberts, R. L. Carter, W. K. Hayes, and E. P. Martins (eds.), Iguanas: Biology and Conservation, pp. 274–289. Berkeley and Los Angeles: University of California Press.

Huey, R. B. 1977. Egg retention in some high-altitude *Anolis* lizards. Copeia 1977: 373–375.

———. 1983. Natural variation in body temperature and physiological performance in a lizard (*Anolis cristatellus*). In: A. G. J. Rhodin and K. Miyata (eds.), Advances in Herpetology and Evolutionary Biology: Essays in Honor of Ernest E. Williams, pp. 484–490. Cambridge, Mass.: Museum of Comparative Zoology, Harvard University.

Huey, R. B., and P. T. Webster. 1975. Thermal biology of a solitary lizard: *Anolis marmoratus* of Guadeloupe, Lesser Antilles. Ecology 56: 445–452.

———. 1976. Thermal biology of *Anolis* lizards in a complex fauna: The *cristatellus* group on Puerto Rico. Ecology 57: 985–994.

Huntington, C., and G. L. Cisper. 1994. A new host and locality record for *Eimeria leiocephali* (Apicomplexa: Eimeriorina) with comments on the type-host. Caribbean Journal of Science 30: 152.

Huntington, C., G. Cisper, D. D. Smith, R. Powell, J. S. Parmerlee, Jr., and A. Lathrop. 1996. Two new *Eimeria* (Apicomplexa: Eimeriidae) from *Amphisbaena manni* (Amphisbaenia: Amphisbaenidae) in the Dominican Republic. Caribbean Journal of Science 32: 50–53.

Huntington, C., R. Powell, J. S. Parmerlee, Jr., G. L. Cisper, D. D. Smith, and A. Lathrop. 1997. Two new Coccidia (Apicomplexa: Eimeriorina) from *Ameiva* spp. (Lacertilia: Teiidae) in the Dominican Republic. Revista Brasileira Biologia 57: 11–14.

Huyghe, K., A. Herrel, B. Vanhooydonck, J. J. Meyers, and D. J. Irschick. 2007. Microhabitat use, diet, and performance data on the Hispaniolan twig anole, *Anolis sheplani*: Pushing the boundaries of morphospace. Zoology 110: 2–8.

Incháustegui M., S. J. 1975. Las tortugas dominicanas de agua dulce *Chrysemys decussata vicina* y *Chrysemys decorata* (Testudines, Emydidae). Anuario del Academia de Ciencias del República Dominicana 1: 139–278.

———. 2003a. Current status of *Cyclura ricordii*. Iguana 10: 4.

———. 2003b. Isla Cabritos: Paradise lost? In: R. W. Henderson and R. Powell (eds.), Islands and the Sea: Essays on Herpetological Exploration in the West Indies, pp. 191–200. Contributions to Herpetology, volume 20. Ithaca, N.Y.: Society for the Study of Amphibians and Reptiles.

Incháustegui M., S. J., A. Schwartz, and R. W. Henderson. 1985. Hispaniolan giant *Diploglossus* (Sauria: Anguidae): Description of a new species and notes on the ecology of *D. warreni*. Amphibia-Reptilia 6: 195–201.

Irschick, D. J. 2000a. Effects of behavior and ontogeny on the locomotor performance of a West Indian lizard *Anolis lineatopus*. Functional Ecology 14: 438–444.

———. 2000b. Comparative and behavioral analyses of preferred speed: *Anolis* lizards as a model system. Physiological and Biochemical Zoology 73: 428–437.

Irschick, D. J., and J. B. Losos. 1996. Morphology, ecology, and behavior of the twig anole *Anolis angusticeps*. In: R. Powell and R. W. Henderson (eds.), Contributions to West Indian Herpetology: A Tribute to Albert Schwartz, pp. 291–301. Contributions to Herpetology, volume 12. Ithaca, N.Y.: Society for the Study of Amphibians and Reptiles.

———. 1998. A comparative analysis of the ecological significance of maximal locomotor performance in Caribbean *Anolis* lizards. Evolution 52: 219–226.

———. 1999. Do lizards avoid habitats in which performance is submaximal? The relationship between sprinting capabilities and structural habitat use in Caribbean anoles. American Naturalist 154: 293–305.

Irschick, D. J., L. J. Vitt, P. A. Zani, and J. B. Losos. 1997. A comparison of evolutionary radiations in mainland and West Indian *Anolis* lizards. Ecology 78: 2191–2203.

Irschick, D. J., T. E. Macrini, S. Koruba, and J. Forman. 2000. Ontogenetic differences in morphology, habitat use, behavior and sprinting capacity in two West Indian *Anolis* lizard species. Journal of Herpetology 34: 444–451.

IUCN (International Union for Conservation of Nature and Natural Resources). 2001. 2001 Categories and Criteria (version 3.1). http://www.redlist.org/info/categories_criteria2001>.

———. 2008. 2008 IUCN Red List of Threatened Species. http://www.iucnredlist.org.

Iverson, J. B. 1978. The impact of feral cats and dogs on populations of the West Indian rock iguana, *Cyclura carinata*. Biological Conservation 14: 63–73.

———. 1979. Behavior and ecology of the rock iguana *Cyclura carinata*. Bulletin of the Florida State Museum, Biological Sciences 24: 175–358.

———. 1980. Colic modifications in iguanine lizards. Journal of Morphology 163: 79–93.

———. 1982. Adaptations to herbivory in iguanine lizards. In: G. M. Burghardt and A. S. Rand (eds.), Iguanas of the World, pp. 60–76. Park Ridge, N.J.: Noyes.

———. 1986. Notes on the natural history of the Caicos Islands Dwarf Boa, *Tropidophis greenwayi*. Caribbean Journal of Science 22: 191–198.

———. 2000. Allen's Cay Iguana, *Cyclura cychlura inornata*. In: A. Alberts (ed.), West Indian Iguanas: Status Survey and Conservation Action Plan, pp. 34–36. Gland, Switzerland and Cambridge, UK: International Union for Conservation of Nature and Natural Resources.

———. 2007a. Juvenile survival in the Allen Cays Rock Iguana (*Cyclura cychlura inornata*). Copeia 2007: 740–744.

———. 2007b. 2006 research update for Allen Cays Iguanas. Iguana Specialist Group Newsletter 10(1): 12 (reprinted in Iguana 14: 241–243).

———. 2007c. Bahamian iguanas (*Cyclura cychlura inornata* and *C. rileyi nuchalis*). Iguana Specialist Group Newsletter 10(2): 8–10.

———. 2008. Allen Cays Iguana (*Cyclura cychlura inornata*). Iguana Specialist Group Newsletter 11(2): 7–8.

Iverson, J. B. and M. R. Mamula. 1989. Natural growth in the Bahaman iguana *Cyclura cychlura*. Copeia 1989: 502–505.

Iverson, J. B., and G. R. Smith. 2006. *Leiocephalus psammodromus* (Turks and Caicos Curly-tail Lizard). Necrophagy. Herpetological Review 37: 345–346.

Iverson, J. B., K. N. Hines, and J. M. Valiulis. 2004a. The nesting ecology of the Allen Cays Rock Iguana, *Cyclura cychlura inornata* in the Bahamas. Herpetological Monographs 18: 1–36.

Iverson, J. B., G. R. Smith, and L. Pieper. 2004b. Factors affecting long-term growth of the Allen Cays Rock Iguana in the Bahamas. In: A. C. Alberts, R. L. Carter, W. K. Hayes, and E. P. Martins (eds.), Iguanas: Biology and Conservation, pp. 176–192. Berkeley and Los Angeles: University of California Press.

Iverson, J. B., D. P. Young, C. A. Young, and C. Schofield. 2005. *Cyclura cychlura inornata* (Allen Cays Iguana). Mortality. Herpetological Review 36: 175.

Iverson, J. B., S. J. Converse, G. R. Smith, and J. M. Valiulis. 2006a. Long-term trends in the demography of the Allen Cays Rock Iguana (*Cyclura cychlura inornata*): Human disturbance and density-dependent effects. Biological Conservation 132: 300–310.

Iverson, J. B., S. A. Pasachnik, C. R. Knapp, and S. D. Buckner. 2006b. *Cyclura cychlura*. Catalogue of American Amphibians and Reptiles 810: 1–8.

Jakowska, S., 1972. Lesions produced by ticks, *Amblyomma dissimile*, in *Bufo marinus* toads from the Dominican Republic [abstract]. American Zoologist 12: 731.

James, A. 2004. Flora and Fauna of Cabrits National Park, Dominica. Roseau: Forestry, Wildlife & Parks Division, Dominica.

James, A., and R. W. Henderson. 2004. Communal nesting site in the snake *Liophis juliae* in Dominica, West Indies. Caribbean Journal of Science 40: 263–265.

James, S. B., B. L. Raphael, V. Greco, and J. B. Iverson. 2006. Health assessment of Allen Cays Rock Iguana (*Cyclura cychlura inornata*). Journal of Herpetological Medicine and Surgery 16: 93–100.

Jaume, M. L. 1965. La mas rara culebra de Cuba. Museo "Felipe Poey" de la Academia de Ciencias de la República de Cuba, Trabajos de Divulgacion 33: 1–7.

Jaume, M. L., and O. H. Garrido. 1980. Notas sobre mordidas de jubo *Alsophis* (Serpentes: Colubridae). Miscelánea Zoologica 11: 2–3.

Jenkins, M., and C. Harcourt. 1996a. Cuba. In: C. S. Harcourt and J. A. Sayer (eds.), The Conservation Atlas of Tropical Forests: The Americas, pp. 95–101. New York: Simon & Schuster.

———. 1996b. Lesser Antilles. In: C. S. Harcourt and J. A. Sayer (eds.), The Conservation Atlas of Tropical Forests: The Americas, pp. 120–137. New York: Simon & Schuster.

Jenssen, T. A. 1973. Shift in the structural habitat of *Anolis opalinus* due to congeneric competition. Ecology 54: 863–869.

———. 1977. Morphological, behavioral and electrophoretic evidence of hybridization between the lizards *Anolis grahami* and *Anolis lineatopus neckeri* on Jamaica. Copeia 1977: 270–276.

———. 1979a. Display modifiers of *Anolis opalinus* (Lacertilia: Iguanidae). Herpetologica 35: 21–30.

———. 1979b. Display behavior of male *Anolis opalinus* (Sauria, Iguanidae): A case of weak display stereotypy. Animal Behaviour 27: 173–184.

———. 1981. Unusual display behavior by *Anolis grahami* from western Jamaica. Copeia 1981: 728–733.

———. 1983. Display behavior of two Haitian lizards, *Anolis cybotes* and *Anolis distichus*. In: A. G. J. Rhodin and K. Miyata (eds.), Advances in Herpetology and Evolutionary Biology: Essays in Honor of Ernest E. Williams, pp. 552–569. Cambridge, Mass.: Museum of Comparative Zoology, Harvard University.

———. 1990. *Anolis cooki*. Catalogue of American Amphibians and Reptiles 488: 1–2.

———. 1996. A test of assortative mating between sibling lizard species, *Anolis websteri* and *A. caudalis*, in Haiti. In: R. Powell and R. W. Henderson (eds.), Contributions to West Indian Herpetology: A Tribute to Albert Schwartz, pp. 303–315. Contributions to Herpetology, volume 12. Ithaca, N.Y.: Society for the Study of Amphibians and Reptiles.

———. 2002. Spatial awareness by the lizard *Anolis cristatellus*: Why should a non-ranging species demonstrate homing behavior? Herpetologica 58: 364–371.

Jenssen, T. A., and R. M. Andrews. 1984. Seasonal growth rates in the Jamaican lizard, *Anolis opalinus*. Journal of Herpetology 18: 338–341.

Jenssen, T. A., and P. C. Feely. 1991. Social behavior of the male anoline lizard *Chamaelinorops barbouri*, with a comparison to *Anolis*. Journal of Herpetology 25: 454–462.

Jenssen, T. A., and N. L. Gladson. 1984. A comparative display analysis of the *Anolis brevirostris* complex in Haiti. Journal of Herpetology 18: 217–230.

Jenssen, T. A., and D. L. Marcellini. 1986. *Leiocephalus schreibersii* (Curly-tailed Lizard). Tail autotomy. Herpetological Review 17: 89.

Jenssen, T. A., and S. C. Nuñez. 1994. Male and female reproductive cycles of the Jamaican lizard, *Anolis opalinus*. Copeia 1994: 767–780.

Jenssen, T. A., and B. Swenson. 1974. An ecological correlate of critical flicker-fusion frequencies from some *Anolis* lizards. Vision Research 14: 965–970.

Jenssen, T. A., D. L. Marcellini, C. A. Pague, and L. A. Jenssen. 1984. Competitive interference between the Puerto Rican lizards, *Anolis cooki* and *A. cristatellus*. Copeia 1984: 853–862.

Jenssen, T. A., D. L. Marcellini, and E. P. Smith. 1988. Seasonal micro-distribution of sympatric *Anolis* lizards in Haiti. Journal of Herpetology 22: 266–274.

Jenssen, T. A., D. L. Marcellini, K. A. Buhlmann, and P. H. Goforth. 1989. Differential infanticide by adult curly-tailed lizards, *Leiocephalus schreibersi*. Animal Behaviour 36: 1054–1061.

Jezkova, T., M. Leal, J. A. Rodríguez-Robles. 2009. Living together but remaining apart: Comparative phylogeography of *Anolis poncensis* and *A. cooki*, two lizards endemic to the aridlands of Puerto Rico. Biological Journal of the Linnean Society 96: 617–634.

Joglar, R. L. 1998. Los Coquíes de Puerto Rico: Su Historia Natural y Conservación. San Juan: Editorial de la Universidad de Puerto Rico.

———. 2005a. Anfibios. In: R. L. Joglar (ed.), Biodiversidad de Puerto Rico. Vertebrados Terrestres y Ecosistemas, pp. 39–96. Serie de Historia Natural. San Juan: Editorial Instituto de Cultura Puertorriqueña.

———. 2005b. Reptiles. In: R. L. Joglar (ed.), Biodiversidad de Puerto Rico. Vertebrados Terrestres y Ecosistemas, pp. 99–168. Serie de Historia Natural. San Juan: Editorial Instituto de Cultura Puertorriqueña.

Joglar, R. L., and P. A. Burrowes. 1991. El efecto del Huracan Hugo sobre una comunidad de anfibios en El Yunque, Puerto Rico y algunas recomendaciones para la proteccion de las especies del genero *Eleutherodactylus*. In: J. A. Moreno (ed.), Status y Distribución de los Reptiles y Anfibios de la Región de Puerto Rico, pp. 47–53. Publicación Científica Miscelánea No. 1. San Juan: Departamento de Recursos Naturales de Puerto Rico.

———. 1996. Declining amphibian populations in Puerto Rico. In: R. Powell and R. W. Henderson (eds.), Contributions to West Indian Herpetology: A Tribute to Albert Schwartz, pp. 371–380. Contributions to Herpetology, volume 12. Ithaca, N.Y.: Society for the Study of Amphibians and Reptiles.

———. 2005. Greater Antilles region progress report. Froglog 72: 1–2.

Joglar, R. L., and N. Rios-López. 1998. *Eleutherodactylus coqui* (Puerto Rican Coqui). Geographic distribution. Herpetological Review 29: 107.

Joglar, R. L., P. A. Burrowes, and N. Rios. 1996. Biology of the Puerto Rican cave-dwelling frog, *Eleutherodactylus cooki*, and some recommendations for conservation. In: R. Powell and R. W. Henderson (eds.), Contributions to West Indian Herpetology: A Tribute to Albert Schwartz, pp. 251–258. Contributions to Herpetology, volume 12. Ithaca, N.Y.: Society for the Study of Amphibians and Reptiles.

Joglar, R. L., P. A. Burrowes, D. Dávila, A. Rodríguez, A. López, A. Longo, and P. Medina. 2005a. *Eleutherodactylus wightmanae* (Coquí Melodioso). Reproduction. Herpetological Review 36: 433–434.

Joglar, R. L., V. Rentas, and F. C. Li. 2005b. Ranita de Labio Blanco (*Leptodactylus albilabris*). In: R. L. Joglar (ed.), Biodiversidad de Puerto Rico. Vertebrados Terrestres y Ecosistemas, pp. 74–88. Serie de Historia Natural. San Juan: Editorial Instituto de Cultura Puertorriqueña.

Joglar, R. L., A. O. Álvarez, T. M. Aide, D. Barber, P. A. Burrowes, M. A. García, A. León-Cardona, A. V. Longo, N. Pérez-Buitrago, A. Puente, N. Rios-López, and P. J. Tolson. 2007. Conserving the Puerto Rican herpetofauna. Applied Herpetology 4: 327–345.

Johnson, B. 1999. Recovery of the Puerto Rican Crested Toad. Endangered Species Bulletin 24(3): 8–9.

Johnson, M. A., R. Kirby, S. Wang, and J. B. Losos. 2006. What drives variation in habitat use by *Anolis* lizards: Habitat availability or selectivity? Canadian Journal of Zoology 84: 877–886.

Johnson, M. A., M. Leal, L. Rodríguez Schettino, A. Chamizo Lara, L. J. Revell, and J. B. Losos. 2008. A phylogenetic perspective on foraging mode evolution and habitat use in West Indian *Anolis* lizards. Animal Behaviour 75: 555–563.

Johnson, T. H. 1988. Biodiversity and conservation in the Caribbean: Profiles of selected islands. International Council for Bird Preservation Monograph 1: xvii + 144 pp.

Johnston, D. W. 1974. Food of the Barn Owl on Grand Cayman, B.W.I. Quarterly Journal of the Florida Academy of Science 35: 171–172.

———. 1975. Ecological analysis of the Cayman Island avifauna. Bulletin of the Florida State Museum 19: 235–300.

Jones, K. L. 1982. Prey patterns and trophic niche overlap in four species of Caribbean frogs, pp. 49–55. In: N. J. Scott (ed.), Herpetological Communities. Washington, D.C.: U.S. Department of the Interior, Fish and Wildlife Service, Wildlife Research Report 13.

Kaden, U. 1985. Die Haltung und Vermehrung von kubanischen Wasseranolis im Terrarium. Elaphe Sonderheft 7: 88–90.

Kaiser, H. 1992. The trade-mediated introduction of *Eleutherodactylus martinicensis* (Anura: Leptodactylidae) on St. Barthélemy, French Antilles, and its implications for Lesser Antillean biogeography. Journal of Herpetology 26: 264–273.

———. 1994. *Leptodactylus fallax*. Catalogue of American Amphibians and Reptiles 583: 1–3.

Kaiser, H., and R. Altig. 1994. The atypical tadpole of the dendrobatid frog, *Colostethus chalcopis*, from Martinique, French Antilles. Journal of Herpetology 28: 374–378.

Kaiser, H., and M. L. Day. 1995. Abiotic disturbances in the Lesser Antilles. Froglog 15: 1–2.

Kaiser, H., and J. D. Hardy, Jr. 1994a. *Eleutherodactylus johnstonei*. Catalogue of American Amphibians and Reptiles 581: 1–5.

———. 1994b. *Eleutherodactylus martinicensis*. Catalogue of American Amphibians and Reptiles 582: 1–4.

Kaiser, H., and R. W. Henderson. 1994. The conservation status of Lesser Antillean frogs. Herpetological Natural History 2(2): 41–56.

Kaiser, H., and R. Wagenseil. 1995. Colonization and distribution of *Eleutherodactylus johnstonei* Barbour (Anura: Leptodactylidae) on Dominica, West Indies. Caribbean Journal of Science 31: 341–344.

Kaiser, H., L. A. Coloma, and H. M. Gray. 1994a. A new species of *Colostethus* (Anura: Dendrobatidae) from Martinique, French Antilles. Herpetologica 50: 23–32.

Kaiser, H., D. M. Green, and M. Schmid. 1994b. Systematics and biogeography of Eastern Caribbean frogs (Leptodactylidae: *Eleutherodactylus*), with the description of a new species from Dominica. Canadian Journal of Zoology 72: 2217–2237.

Kaiser, H., J. T. Murdoch, R. Boistel, and M. Breuil. 2003. *Eleutherodactylus pinchoni*. Catalogue of American Amphibians and Reptiles 765: 1–3.

Kaiser, H., R. Boistel, and M. Breuil. 2004. *Eleutherodactylus barlagnei*. Catalogue of American Amphibians and Reptiles 784: 1–3.

Karch, P. 2002. Bemerkungen zur gattung *Chamaeleolis*. Iguana (Germany) 15(2): 11–12.

Keirans, J. E. 1985. *Amblyomma antillorum* Kohls, 1969 (Acari: Ixodidae): Description of the immature stages from the rock iguana, *Iguana pinguis* (Sauria: Iguanidae) in the British Virgin Islands. Proceedings of the Entomological Society of Washington 87: 821–825.

Keirans, J. E., and G. I. Garris. 1986. *Amblyomma arianae*, n. sp. (Acari: Ixodidae), a parasite of *Alsophis portoricensis* (Reptilia: Colubridae) in Puerto Rico. Journal of Medical Entomology 23: 622–625.

Kerr, A. M., R. Powell, and J. S. Parmerlee, Jr. 2005. *Ameiva erythrocephala* (Teiidae) on Sint Eustatius, Netherlands Antilles: Baseline data on a small population in a severely altered habitat. Caribbean Journal of Science 41: 162–169.

Kerr, A. M., V. H. Zero, and R. Powell. 2006. *Ameiva erythrocephala*. Catalogue of American Amphibians and Reptiles 817: 1–4.

Kiester, R. K., G. C. Gorman, and D. C. Arroyo. 1975. Habitat selection behavior of three species of *Anolis* lizards. Ecology 56: 220–225.

King, J. D., L. A. Rollins-Smith, P. F. Nielsen, A. John, and J. M. Conlon. 2005. Characterization of a peptide from skin secretions of male specimens of the frog, *Leptodactylus fallax* that stimulates aggression in male frogs. Peptides 26: 597–601.

King, W. 1962. Systematics of Lesser Antillean lizards of the genus *Sphaerodactylus*. Bulletin of the Florida State Museum 7: 1–52.

Klingel, G. C. 1929. Lizard hunting in the Black Republic. Natural History 29: 451–464.

Klug, M. 1995. Kugelfingergeckos der Gattung *Sphaerodactylus*. Die Aquarien- und Terrarienzeitschrift (DATZ) 48: 162.

Kluge, A. G. 1987. Cladistic relationships in the Gekkonoidea (Squamata, Sauria). Miscellaneous Publications of the Museum of Zoology, University of Michigan 173: iv + 54 pp.

———. 1995. Cladistic relationships of sphaerodactyl lizards. American Museum Novitates 3139: 1–23.

Knapp, C. R. 1995. A flora and fauna survey of Guana Cay, with an emphasis on its rock iguana. Bahamas Journal of Science 2: 2–7.

———. 2000a. Home range and intraspecific interactions of a translocated iguana population (*Cyclura cychlura inornata* Barbour and Noble). Caribbean Journal of Science 36: 250–257.

———. 2000b. Exuma Island Iguana, *Cyclura cychlura figginsi*. In: A. Alberts (ed.), West Indian Iguanas: Status Survey and Conservation Action Plan, pp. 32–34. Gland, Switzerland and Cambridge, UK: International Union for Conservation of Nature and Natural Resources.

———. 2000c. *Alsophis vudii* (Bahamian Brown Racer). Overwater dispersal. Herpetological Review 31: 244.

———. 2001a. Status of a translocated *Cyclura* iguana colony in the Bahamas. Journal of Herpetology 35: 239–248.

———. 2001b. *Leiocephalus carinatus virescens* (Exuma Islands Curlytail Lizard). Interspecific aggression. Herpetological Review 32: 104–105.

———. 2002. Exuma Island Iguana (*Cyclura cychlura figginsi*). Iguana Specialist Group Newsletter 5(1): 6–9.

———. 2004. 2003 research update for Exuma (*Cyclura cychlura figginsi*) and Andros (*C. c. cychlura*) islands, Bahamas. Iguana 11: 32–33.

———. 2005. Working to save the Andros Iguana. Iguana 12: 8–13.

———. 2007a. Ecology and conservation of the Lesser Antillean Iguana (*Iguana delicatissima*). Iguana 14: 222–225.

———. 2007b. Lesser Antillean Iguana (*Iguana delicatissima*). Iguana Specialist Group Newsletter 10(2): 13–14.

Knapp, C. R., and S. D. Buckner. 2004a. *Cyclura cychlura* ssp. *cychlura*. In: IUCN 2008. 2008 IUCN Red List of Threatened Species. http://www.iucnredlist.org.

———. 2004b. *Cyclura cychlura* ssp. *figginsi*. In: IUCN 2008. 2008 IUCN Red List of Threatened Species. http://www.iucnredlist.org.

Knapp, C. R., and K. N. Hines. 2006a. Taxon reports: Exuma Island Iguana (*Cyclura cychlura figginsi*). Iguana Specialist Group Newsletter 9(1): 12–13.

———. 2006b. Exuma Island Iguana (*Cyclura cychlura figginsi*). Iguana 13: 281–282.

Knapp, C. R., and C. L. Malone. 2003. Patterns of reproductive success and genetic variability in a translocated iguana population. Herpetologica 59: 195–202.

Knapp, C. R., and A. K. Owens. 2004. Diurnal refugia and novel ecological attributes of the Bahamian boa, *Epicrates striatus fowleri* (Boidae). Caribbean Journal of Science 40: 265–270.

———. 2005. Home range and habitat associations of a Bahamian iguana: Implications for conservation. Animal Conservation 8: 269–278.

———. 2008a. Nesting behavior and the use of termitaria by the Andros Iguana (*Cyclura cychlura cychlura*). Journal of Herpetology 42: 46–53.

———. 2008b. *Cyclura cychlura cychlura* (Andros Iguana). Attempted predation. Herpetological Review 39: 221–222.

Knapp, C. R., and L. Prince. 2008. *Eretmochelys imbricata* (Hawksbill Sea Turtle). Predation. Herpetological Review 39: 213–214.

Knapp, C. R., and S. Valeri. 2008. *Iguana delicatissima* (Lesser Antillean Iguana). Mortality. Herpetological Review 39: 227–228.

Knapp, C. R., S. D. Buckner, A. Feldman, and L. Roth. 1999. Status update and empirical field observations of the Andros Rock Iguana, *Cyclura cychlura cychlura*. Bahamas Journal of Science 7: 2–5.

Knapp, C. R., J. B. Iverson, and S. Buckner. 2004a. *Cyclura cychlura*. In: IUCN 2008. 2008 IUCN Red List of Threatened Species. http://www.iucnredlist.org.

Knapp, C. R., A. K. Owens, and C. Sheehy. 2004b. Taxon update: 2004 research update for *Cyclura cychlura cychlura* and *C. c. figginsi*. Iguana Specialist Group Newsletter 7(2): 2–3.

———. 2005. Taxon update: 2004 research update for *Cyclura cychlura cychlura* and *C. c. figginsi*. Iguana 12: 24–25.

Knapp, C. R., J. B. Iverson, and A. K. Owens. 2006. Geographic variation in nesting behavior and reproductive biology of an insular iguana (*Cyclura cychlura*). Canadian Journal of Zoology 84: 1566–1575.

Knapp, C. R., M. Greenaway, A. James, and L. Prince. 2009. *Boa constrictor* (Boa Constrictor). Diet. Herpetological Review 40: 229.

Koenig, S. E. 2001. The breeding biology of the Black-billed Parrot *Amazona agilis* and Yellow-billed Parrot *A. collaria* in Cockpit Country, Jamaica. Bird Conservation International 11: 205–225.

Koenig, S. E., and M. Schwartz. 2003. *Epicrates subflavus* (Jamaican Yellow Boa). Diet. Herpetological Review 34: 363–364.

Koenig, S. E., J. M. Wunderle, Jr., and E. C. Enkerlin-Hoeflich. 2007. Vines and canopy contact: A route for snake predation on parrot nests. Bird Conservation International 17: 79–91.

Kohls, G. M. 1969. A new species of *Amblyomma* from iguanas in the Caribbean (Acarina: Ixodidae). Journal of Medical Entomology 6: 439–442.

Kohls, G. M., D. E. Sonenshine, and C. M. Clifford. 1965. The systematics of the subfamily Ornithodorinae (Acarina: Argasidae). II. Identification of the larvae of the Western Hemisphere and descriptions of three new species. Annals of the Entomological Society of America 58: 331–364.

Kolbe, J. J., A. Larson, and J. B. Losos. 2007a. Differential admixture shapes morphological variation among invasive populations of the lizard *Anolis sagrei*. Molecular Ecology 16: 1579–1591.

Kolbe, J. J., R. E. Glor, L. Rodríguez Schettino, A. Chamizo Lara, A. Larson, and J. B. Losos. 2007b. Multiple sources, admixture, and genetic variation in introduced *Anolis* lizard populations. Conservation Biology 21: 1612–1625.

Kolbe, J. J., P. L. Colbert, and B. E. Smith. 2008. Niche relationships and interspecific interactions in Antiguan lizard communities. Copeia 2008: 261–272.

Königsted, B., and D. Königstedt. 1982. Haltung und Zucht des Kuba-Laubfroschs. Aquarien Terrarien 29: 280–282.

Kraus, F., and D. Cravalho. 2001. The risk to Hawai'i from snakes. Pacific Science 55: 409–417.

Kraus, F., E. W. Campbell, A. Allison, and T. Pratt. 1999. *Eleutherodactylus* frog introductions to Hawaii. Herpetological Review 30: 21–25.

Krintler, K. 1977. Pflege und Zucht des Jamaika-Anolis (*Anolis garmani*). Die Aquarien- und Terrarienzeitschrift (DATZ) 30: 424–426.

———. 1985. *Anolis garmani*. Sauria Supplement 7: 31–32.

———. 1986. Dominica—Herpetologische Kleinod in der Karibik. Herpetofauna 8: 26–30.

Krintler, K., and K. Dommaschk. 1980. Herpetologische Reise zur Karibikinsel Dominica. Die Aquarien- und Terrarienzeitschrift (DATZ) 30: 284–286.

Krüger, B. 2008. Herpetologisches zwischen Kolumbus' Landungspunkt und der Karibik: Kubas mittlerer Osten. Reptilia 13: 44–49.

Krysko, K. L., and A. P. Borgia. 2005. *Hemidactylus mabouia* (Cosmopolitan House Gecko). Geographic Distribution. Herpetological Review 36: 486.

Krysko, K. L., and T. M. Thomas. 2007. *Hemidactylus mabouia* (Amerafrican House Gecko). Geographic Distribution Herpetological Review 38: 351.

Kunz, K. 2005. Una herpetofauna fotogenica. Observaciones sobre las especies sinantropicas de Cuba. Reptilia (Spain) 56: 51–57.

Kwet, A. 1995. Amphibien und Reptilien auf Kuba. Die Aquarien- und Terrarienzeitschrift (DATZ) 48: 662–665.

Lachner, R. 1987. Inseln der Karibik, Landschaft und Tiere. Hannover, Germany: Landbuch-Verlag.

Lacy, K. E., and E. P. Martins. 2003. The effect of anthropogenic habitat usage on the social behaviour of a vulnerable species, *Cyclura nubila*. Animal Conservation 6: 3–9.

Lailvaux, S., and D. J. Irschick. 2007. The evolution of performance-based male fighting ability in Caribbean *Anolis* lizards. American Naturalist 170: 573–586.

Lambert, M. 1986. Herpetology in Jamaica with general reference to conservation. British Herpetological Bulletin 18: 3–8.

Lambrechts, D. 2005. *Anolis argenteolus* (Iguanidae: Polychrotinae): Een Cubaanse 'rots' anolis in het terrarium. Lacerta 63: 152–165.

Lando, R. V., and E. E. Williams. 1969. Notes on the herpetology of the U.S. Naval Base at Guantanamo Bay, Cuba. Studies on the Fauna of Curaçao and Other Caribbean Islands 31: 159–201.

Lane, C. S., S. P. Horn, and K. H. Orvis. 2008. The earliest evidence of Ostionoid maize agriculture from the interior of Hispaniola. Caribbean Journal of Science 44: 43–52.

Langerhans, R. B., J. H. Knouft, and J. B. Losos. 2006. Shared and unique features of diversification in Greater Antillean *Anolis* ecomorphs. Evolution 60: 362–369.

Lannoo, M. J., D. S. Townsend, and R. J. Wassersug. 1987. Larval life in the leaves: Arboreal tadpole types, with special attention to the morphology, ecology, and behavior of the oophagous *Osteopilus brunneus* (Hylidae) larva. Fieldiana Zoology, new series 38: 1–31.

Larimer, S. C., R. Powell, and J. S. Parmerlee, Jr. 2006. Effects of structural habitat on the escape behavior of the lizard, *Anolis gingivinus*. Amphibia-Reptilia 27: 569–574.

Laska, A. L. 1970. The structural niche of *Anolis scriptus* on Inagua. Breviora 349: 1–6.

Lavigne, R. J., and G. E. Drewry. 1970. Feeding behavior of the frogs and lizards in the tropical wet forest: Preliminary report, pp. 64–73. In: R. G. Clements, G. E. Drewry, and R. J. Lavigne (eds.), The Rain Forest Project Annual Report. PRNC-147. Rio Piedras: Puerto Rico Nuclear Center.

Lawler, H. E., and C. Norris. 1979. Breeding the Haitian Giant Galliwasp, *Diploglossus warreni* (Sauria: Anguinidae) at the Knoxville Zoological Park. Third Annual Symposium on Captive Propagation and Husbandry: 73–79.

Lazell, J. 1961. A new species of *Sphaerodactylus* from northern Haiti. Breviora 139: 1–5.

———. 1962. The anoles (Sauria: Iguanidae) of the eastern Caribbean. Part V. Geographic differences in *Anolis oculatus* on Dominica. Bulletin of the Museum of Comparative Zoology 127: 466–475, plates 1–3.

———. 1964a. The anoles (Sauria: Iguanidae) of the Guadeloupéen Archipelago. Bulletin of the Museum of Comparative Zoology 131: 359–401.

———. 1964b. The Lesser Antillean representatives of *Bothrops* and *Constrictor*. Bulletin of the Museum of Comparative Zoology 132: 245–273.

———. 1964c. The reptiles of Sombrero, West Indies. Copeia 1964: 716–718.

———. 1966. Contributions to the herpetology of Jamaica. Studies on *Anolis reconditus* Underwood and Williams. Bulletin of the Institute of Jamaica Science Series 18: 1–15.

———. 1967. Wiederentdeckung von zwei angeblich ausgestorbenen Schlangenarten der westindischen Inseln. Salamandra 3: 91–97.

———. 1972. The anoles (Sauria, Iguanidae) of the Lesser Antilles. Bulletin of the Museum of Comparative Zoology 143: 1–115.

———. 1973. The lizard genus *Iguana* in the Lesser Antilles. Bulletin of the Museum of Comparative Zoology 145: 1–28.

———. 1983. Biogeography of the herpetofauna of the British Virgin Islands, with description of a new anole (Sauria: Iguanidae). In: A. G. J. Rhodin and K. Miyata (eds.), Advances in Herpetology and Evolutionary Biology: Essays in Honor of Ernest E. Williams, pp. 99–117. Cambridge, Mass.: Museum of Comparative Zoology, Harvard University.

———. 1989. Wildlife of the Florida Keys. Washington, D.C.: Island Press.

———. 1991. The herpetofauna of Guana Island: Diversity, abundance, rarity, and conservation. In: J. A. Moreno (ed.), Status y Distribución de los Reptiles y Anfibios de la Región de Puerto Rico, pp. 28–33. Publicación Científica Miscelánea No. 1. San Juan: Departamento de Recursos Naturales de Puerto Rico.

———. 1994. A new *Sphaerodactylus* (Sauria: Gekkonidae) from Bequia, Grenada Bank, Lesser Antilles. Breviora 496: 1–20.

———. 1995. Natural Necker. The Conservation Agency Occasional Paper 2: 1–28.

———. 1997. The Stout Iguana of the British Virgin Islands. Iguana Times (Journal of the International Iguana Society) 6: 75–80.

———. 2000. Necker Island update. West Indian Iguana Specialist Group Newsletter 3(1): 2.

———. 2002. Restoring vertebrate animals in the British Virgin Islands. Ecological Restoration 20: 179–185.

———. 2005a. Island: Fact and Theory in Nature. Berkeley and Los Angeles: University of California Press.

———. 2005b. Tortoise tales. In: J. Lazell (ed.), Island: Fact and Theory in Nature, p. 321. Berkeley and Los Angeles: University of California Press.

———. 2006. Natural Necker: Thirty hectares of amazing biological diversity. The Conservation Agency, Occasional Papers 3: 1–56.

Lazell, J., and N. C. Mitchell. 1998. *Anolis cristatellus wileyae* (Virgin Islands Crested Anole). Herbivory. Herpetological Review 29: 237.

Lazell, J., and G. Perry. 1997. *Anolis cristatellus wileyae* (Virgin Islands Crested Anole). Frugivory. Herpetological Review 28: 150.

Le, M., C. J. Raxworthy, W. P. McCord, and L. Mertz. 2006. A molecular phylogeny of tortoises (Testudines: Testudinidae) based on mitochondrial and nuclear genes. Molecular Phylogenetics and Evolution 40: 517–531.

Leal, M. 1999. Honest signalling during prey-predator interactions in the lizard *Anolis cristatellus*. Animal Behaviour 58: 521–526.

Leal, M., and L. J. Fleishman. 2002. Evidence for habitat partitioning based on adaptation to environmental light in a pair of sympatric lizard species. Proceedings of the Royal Society B 269: 351–359.

———. 2004. Differences in visual signal design and detectability between allopatric populations of *Anolis* lizards. American Naturalist 163: 26–39.

Leal, M., and J. B. Losos. 2000. Behavior and ecology of the Cuban "Chipojo Bobos" *Chamaeleolis barbatus* and *C. porcus*. Journal of Herpetology 34: 318–322.

Leal, M., and J. A. Rodríguez-Robles. 1995. Antipredator responses of *Anolis cristatellus* (Sauria: Polychrotidae). Copeia 1995: 155–161.

———. 1997a. Signalling displays during predator-prey interactions in a Puerto Rican anole, *Anolis cristatellus*. Animal Behaviour 54: 1147–1154.

———. 1997b. Antipredator responses of the Puerto Rican Giant Anole, *Anolis cuvieri* (Sauria: Polychrotidae). Biotropica 29: 372–375.

Leal, M., and R. Thomas. 1992. *Eleutherodactylus coqui* (Puerto Rico Coquí). Prey. Herpetological Review 23: 79–80.

———. 1994. Notes on the feeding behavior and caudal luring by juvenile *Alsophis portoricensis* (Serpentes: Colubridae). Journal of Herpetology 28: 126–128.

Leal, M., J. A. Rodríguez-Robles, and J. B. Losos. 1998. An experimental study of interspecific interactions between two Puerto Rican *Anolis* lizards. Oecologia 117: 273–278.

Leal, M., A. K. Knox, and J. B. Losos. 2002. Lack of convergence in aquatic *Anolis* lizards. Evolution 56: 785–791.

Lebron, R., A. T. Batra, J. Bontempo, C. Buckley, M. Cron, L. Fenstermacher, C. Mahoney, L. Schmitt, and A. M. Bauer. 1995. *Leptodactylus albilabris* (Caribbean White-lipped Frog). Larval diet. Herpetological Review 26: 31.

Leclair, R., Jr. 1978. Water loss and microhabitats in three sympatric species of lizards (Reptilia, Lacertilia) from Martinique, West Indies. Journal of Herpetology 12: 177–182.

Leclair, R., Jr., and G. Provencher. 1988. Demographic traits of a small sphaerodactyline lizard from Martinique, West Indies. Program and Abstracts, Combined Herpetology Meetings, pp. 127–128.

Leidy, J. 1890 (1891). Notices of entozoa. Proceedings of the Academy of Natural Sciences of Philadelphia 1890: 410–418.

Lenart, L. A., R. Powell, J. S. Parmerlee, Jr., D. D. Smith, and A. Lathrop. 1994. The diet and a gastric parasite of *Anolis armouri*, a cybotoid anole from montane pine forests in southern Hispaniola. Herpetological Natural History 2(2): 97–100.

Lenart, L. A., J. S. Parmerlee, Jr., and R. Powell. 1995. *Anolis armouri*. Catalogue of American Amphibians and Reptiles 607: 1–3.

Lenart, L. A., R. Powell, J. S. Parmerlee, Jr., A. Lathrop, and D. D. Smith. 1997. Anoline diversity in three differentially altered habitats in the Sierra de Baoruco, República Dominicana, Hispaniola. Biotropica 29: 117–123.

León, A., and R. L. Joglar. 2005. La Jicotea Puertorriqueña (*Trachemys stejnegeri stejnegeri*). In: R. L. Joglar (ed.), Biodiversidad de Puerto Rico. Vertebrados Terrestres y Ecosistemas, pp. 181–190. Serie de Historia Natural. San Juan: Editorial Instituto de Cultura Puertorriqueña.

Lescure, J. 1979. Étude taxinomique et éco-éthologique d'un Amphibien des petites Antilles: *Leptodactylus fallax* Muller, 1926 (Leptodactylidae). Bulletin du Muséum National d'Histoire Naturelle, Paris 4, serie 1, section A, no. 3: 757–774.

———. 1983. Introductions passives et actives de reptiles et d'amphibiens dans les Antilles et les Guyanes. C. R. Societie Biogéographie 59: 59–70.

Leuck, B. E., K. W. Hughes, and H. Y. Cheng. 1990. Social displays of experimentally paired dwarf geckos (*Sphaerodactylus clenchi*). Journal of Herpetology 24: 416–418.

Leussink, J. A. 1958. Nematodes of the genus *Ozalaimus* in West Indian iguanas. Studies on the Fauna of Curaçao and Other Caribbean Islands 8: 127–145.

Lever, C. 2003. Naturalized Amphibians and Reptiles of the World. Oxford, UK: Oxford University Press.

Levering, K., and G. Perry. 2003. *Cyclura pinguis* (Stout Iguana, Anegada Rock Iguana). Juvenile predation. Herpetological Review 34: 367–368.

Lewis, A. R. 1983. Preliminary estimates of density and individual growth rates for a population of the Puerto Rican Ground Lizard. In: R. A. Pérez-Rivera, L. O. Nieves-Rivera, and E. Ortiz-Corp's (eds.), Memorias del Cuarto Simposio sobre la Fauna de Puerto Rico y el Caribe, pp. 58–73. Humacao: Universidad de Puerto Rico, Colegio Universitario de Humacao.

———. 1986. Body size and growth in two populations of the Puerto Rican Ground Lizard (Teiidae). Journal of Herpetology 20: 190–195.

———. 1989. Diet selection and depression of prey abundance by an intensively foraging lizard. Journal of Herpetology 23: 164–170.

Lewis, A. R., and J. F. Saliva. 1987. Effects of sex and size on home range, dominance, and activity budgets in *Ameiva exsul* (Lacertilia: Teiidae). Herpetologica 43: 374–383.

Lewis, A. R., G. Tirado, A. Aponte, J. Blanco, and J. Sepulveda. 1996. Social control of male reproductive success in *Ameiva exsul* (abstract). In: R. Powell and R. W. Henderson (eds.), Contributions to West Indian Herpetology: A Tribute to Albert Schwartz, pp. 445–449. Contributions to Herpetology, volume 12. Ithaca, N.Y.: Society for the Study of Amphibians and Reptiles.

Lewis, A. R., G. Tirado, J. A. Sepulveda. 2000. Body size and paternity in a teiid lizard (*Ameiva exsul*). Journal of Herpetology 34: 110–120.

Lewis, C. B. 1940. The Cayman Islands and marine turtle. In: C. Grant (ed.), The Herpetology of the Cayman Islands. Bulletin of the Institute of Jamaica, Science Series 2: 1–65.

———. 1944. Notes on *Cyclura*. Herpetologica 2: 93–98.

Lewis, D. S., P. Vogel, and B. S. Wilson. 2008. Survival, dispersal, and home range of headstarted Jamaican Iguanas, *Cyclura collei*. Caribbean Journal of Science 44: 423–425.

Lewis, T. R. 2001. Reptile observations on the Maria Islands: Two satellite islands of St. Lucia (West Indies). Herpetological Bulletin 77: 25–28.

Lhomme, V., R. Barberteguy, J. Jouannelle, and L. Elisabeth. 1993. Septicémie à *Aeromonas hydrophila* après morsure de serpent (*Bothrops lanceolatus*). A propos d'un cas martiniquais. Médicine et Maladies Infectieuses 23: 808–815.

Licht, P. 1974. Response of *Anolis* lizards to food supplementation in nature. Copeia 1974: 215–221.

Licht, P., and G. C. Gorman. 1970. Reproductive and fat cycles in Caribbean *Anolis* lizards. University of California Publications in Zoology 95: 1–52.

———. 1975. Altitudinal effects on the seasonal testis cycles of tropical *Anolis* lizards. Copeia 1975: 496–504.

Lindsay, K., and B. Cooper. 2008. *Osteopilus septentrionalis* becomes established on Antigua, West Indies. Applied Herpetology 5: 96–98.

Linstow, O. F. B. v. 1902. *Atractis cruciata* und *Oxyuris monhystera*, zwei neue Nematoden aus *Metopoceros cornutus*. Zentralblatt Bakteriologie I. Abteilung Originale 31: 28–32.

Lister, B. C. 1976. The nature of niche expansion in West Indian *Anolis* lizards. I. Ecological consequences of reduced competition. Evolution 30: 659–676.

———. 1981. Seasonal niche relationships of rain forest anoles. Ecology 62: 1548–1560.

Long, E. G. 1974. The Serpent's Tale. Reptiles and Amphibians of St. Lucia. Iounaloa Series No. 2. The Morne, St. Lucia: University of West Indies Extra Mural Department.

Long, E. G., D. Brill, and M. Fevrier. 1980. Snakebite in a rural community in St. Lucia. Castries, Saint Lucia: Unpublished report to the Ministry of Health, Housing, and Labor.

Lopez, P. T. 1996. Mate selection in the Puerto Rican frog, *Eleutherodactylus coqui*. In: R. Powell and R. W. Henderson (eds.), Contributions to West Indian Herpetology: A Tribute to Albert Schwartz, pp. 241–250. Contributions to Herpetology, volume 12. Ithaca, N.Y.: Society for the Study of Amphibians and Reptiles.

López-Ortiz, R., and A. R. Lewis. 2002. Seasonal abundance of hatchlings and gravid females of *Sphaerodactylus nicholsi* in Cabo Rojo, Puerto Rico. Journal of Herpetology 36: 276–280.

———. 2004. Habitat selection by *Sphaerodactylus nicholsi* (Squamata: Gekkonidae) in Cabo Rojo, Puerto Rico. Herpetologica 60: 438–444.

Lorvelec, O., A. Levesque, G. Leblond, M.-È. Jaffard, N. Barré, P. Feldmann, M. Pascal, and C. Pavis. 2000. Suive écologique des reptiles, oiseaux et mammifères aux Îles de la Petite Terre (commune de la Désirade, Guadeloupe). Années 1998 et 1999. Association pour l'Etude et la Protection des Vertébrés et Végétaux des Petites Antilles (AEVA) Rapport 24: 1–104 + 8 maps + 8 plates.

Lorvelec, O., A. Levesque, N. Barré, P. Feldmann, G. LeBlond, M.-E. Jaffard, M. Pascal, and C. Pavis. 2002. Évolution de la densité de population de l'iguane de Petites Antilles (*Iguana delicatissima*) dans ls Réserve Naturelle des Îlets de la Petite Terre (Guadeloupe) entre 1995 et 2002. Evolution of the population density of the Lesser Antillean Iguana (*Iguana delicatissima*) of the Îlets de la Petite Terre Natural Reserve (Guadeloupe, French West Indies): 1995–2002. Evolución de la densidad poblacional de la iguana de las Antillas Menores (*Iguana delicatissima*) en la Reserva Natural de las Islas de Petite Terre (Guadeloupe) entre 1995 y 2002. [Abstracts]. Journal of Caribbean Ornithology 15: 78–79.

Lorvelec, O., M. Pascal, D. Pavis, and P. Feldmann. 2007. Amphibians and reptiles of the French West Indies: Inventory, threats and conservation. Applied Herpetology 4: 131–161.

Losos, J. B. 1985a. Male aggressive behavior in a pair of sympatric sibling species. Breviora 484: 1–30.

———. 1985b. An experimental demonstration of the species-recognition role of *Anolis* dewlap color. Copeia 1985: 905–910.

———. 1990a. Ecomorphology, performance capability, and

scaling of West Indian *Anolis* lizards: An evolutionary analysis. Ecological Monographs 60: 369–388.
———. 1990b. Concordant evolution of locomotor behaviour, display rate, and morphology in *Anolis* lizards. Animal Behaviour 39: 879–890.
———. 2009. Lizards in an Evolutionary Tree: Ecology and Adaptive Radiation of *Anolis*. Berkeley and Los Angeles: University of California Press.
Losos, J. B., and K. de Queiroz. 1997a. Evolutionary consequences of ecological release in Caribbean *Anolis* lizards. Biological Journal of the Linnaean Society 61: 459–483.
———. 1997b. Darwin's lizards. Natural History 106: 34–39.
Losos, J. B., and D. J. Irschick. 1996. The effect of perch diameter on the escape behavior of *Anolis* lizards: Laboratory-based predictions and field tests. Animal Behaviour 51: 593–602.
Losos, J. B., and B. Sinervo. 1989. The effects of morphology and perch diameter on sprint performance of *Anolis* lizards. Journal of Experimental Biology 145: 23–30.
Losos, J. B., and D. A. Spiller. 1999. Differential colonization success and asymmetrical interactions between two lizard species. Ecology 80: 252–258.
———. 2005. *Anolis smaragdinus* (Bahamian Green Anole). Dispersal. Herpetological Review 36: 315–316.
Losos, J. B., M. R. Gannon, W. J. Pfeiffer, and R. B. Waide. 1990. Notes on the ecology and behavior of *Anolis cuvieri* (Lacertilia: Iguanidae) in Puerto Rico. Caribbean Journal of Science 26: 65–66.
Losos, J. B., J. C. Marks, and T. W. Schoener. 1993. Habitat use and ecological interactions of an introduced and a native species of *Anolis* lizard on Grand Cayman, with a review of the outcomes of anole introductions. Oecologia 95: 525–532.
Losos, J. B., D. J. Irschick, and T. W. Schoener. 1994. Adaptation and constraint in the evolution of specialization of Bahamian *Anolis* lizards. Evolution 48: 1786–1798.
Losos, J. B., T. W. Schoener, and D. A. Spiller. 2003. Effect of immersion in seawater on egg survival in the lizard *Anolis sagrei*. Oecologia 137: 360–362.
———. 2004. Predator-induced behaviour shifts and natural selection in field-experimental lizard populations. Nature 432: 505–508.
Losos, J. B., T. W. Schoener, R. B. Langerhans, and D. A. Spiller. 2006. Rapid temporal reversal in predator-driven natural selection. Science 314: 1111.
Loughridge, R. 2006. Ecomorphs of *Anolis* on Dominica: The relationship of morphology to performance. Unpublished report. Texas A&M University Study Abroad, College Station. http://dominica.tamu.edu/student%20projects/projects_by_year.htm.
Lu, W. 2009. *Anolis cristatellus* (Crested Anole). Avian predation. Herpetological Review 40: 219–220.
Lynch, J. D. 1991. Three replacement names for pre-occupied names in the genus *Eleutherodactylus* (Amphibia: Leptodactylidae). Copeia 1991: 1138–1139.
Lynn, W. G. 1937. Two new frogs from Jamaica. Herpetologica 1: 88–91.
———. 1940. The Herpetology of Jamaica. I. The Amphibians. Bulletin of the Institute of Jamaica, Science Series 1: 1–60.
———. 1954. Description of a new frog of the genus *Eleutherodactylus* from Jamaica, B.W.I. Journal of the Washington Academy of Science 44: 400–402.
———. 1957. Notes on a collection of reptiles and amphibians from Antigua, B. W. I. Herpetologica 13: 53–56.
———. 1958. Some amphibians from Haiti and a new subspecies of *Eleutherodactylus schmidti*. Herpetologica 14: 153–157.
Lynn, W. G., and J. N. Dent. 1942. A new frog of the genus *Eleutherodactylus* from Jamaica. Herpetologica 2: 71–74.
———. 1943. Notes on Jamaican amphibians. Copeia 1943: 234–242.
Lynxwiler, J. R., and J. S. Parmerlee, Jr. 1993. *Aristelliger cochranae*. Catalogue of American Amphibians and Reptiles 567: 1–3.
Lynxwiler, J. R., R. Powell, J. S. Parmerlee, Jr., A. Lathrop, and D. D. Smith. 1991. Notes on the natural history of *Hemidactylus brookii haitianus* (Sauria: Gekkonidae) from the Dominican Republic. Dactylus 1: 2–9.
MacDonald, E. A., N. M. Czekala, G. P. Gerber, and A. C. Alberts. 2007. Diurnal and seasonal patterns in corticosterone in the Turks and Caicos Iguana (*Cyclura carinata carinata*). Caribbean Journal of Science 43: 266–272.
Macedonia, J. M., and J. A. Stamps. 1994. Species recognition in *Anolis grahami* (Sauria, Iguanidae): Evidence from responses to video playbacks of conspecific and heterospecific displays. Ethology 98: 246–264.
Macedonia, J. M., C. S. Evans, and J. B. Losos. 1994. Male *Anolis* lizards discriminate video-recorded conspecific and heterospecific displays. Animal Behaviour 47: 1220–1223.
Macedonia, J. M., S. James, L. W. Wittle, and D. L. Clark. 2000. Skin pigments and coloration in the Jamaican radiation of *Anolis* lizards. Journal of Herpetology 34: 99–109.
Macko, J., and B. Rysavy. 1968. *Diorchia longihamulus* sp. n. (Hymenolepididae), a new cestode from *Gailinula chloropus cerceris* (Railiformes). Folia Parasitologica 15: 267–270.
MacLean, W. P. 1982. Reptiles and Amphibians of the Virgin Islands. London: MacMillan Caribbean.
———. 1985. Water loss rates of *Sphaerodactylus parthenopion* (Reptilia: Gekkonidae), the smallest amniote vertebrate. Comparative Biochemistry and Physiology 82A: 759–761.
MacLean, W. P., and R. D. Holt. 1979. Distributional patterns in St. Croix *Sphaerodactylus* lizards: The taxon cycle in action. Biotropica 11: 189–195.

Macrini, T., and D. J. Irschick. 1998. An intraspecific analysis of trade-offs in sprinting performance in a West Indian lizard (*Anolis lineatopus*). Biological Journal of the Linnean Society 63: 579–591.

Mahon, R., and K. Aiken. 1977. The establishment of the North American Bullfrog, *Rana catesbeiana* (Amphibia, Anura, Ranidae) in Jamaica. Journal of Herpetology 11: 197–199.

Maillard, J.-F., and M. Breuil. 2007. The conservation of *Iguana delicatissima* in Martinique and a future action plan for the French West Indies. Iguana Specialist Group Newsletter 10(1): 15 (reprinted in Iguana 14: 245–246).

Malbranques, S., M. D. Piercecchi-Marti, L. Thomas, C. Barbey, D. Courcier, B. Bucher, A. Ridarch, D. Smadja, and D. A. Warrell. 2008. Fatal diffuse thrombotic microangiopathy after a bite by the "Fer-de-Lance" pit viper (*Bothrops lanceolatus*) of Martinique. American Journal of Tropical Medicine and Hygiene 78: 856–861.

Maldonado Capriles, J., and S. Medina Gaud. 1977. The ticks in Puerto Rico (Arachnida: Acarina). Journal of Agriculture of the University of Puerto Rico 61: 402–404.

Maley, A. J. 2004. Species profile: *Alsophis rufiventris*. Iguana 11: 147.

Maley, A. J. A. Z. Savit, H. M. Heinz, R. Powell, and R. W. Henderson. 2006. *Alsophis rufiventris*. Catalogue of American Amphibians and Reptiles 818: 1–4.

Malhotra, A., and R. S. Thorpe. 1991a. Microgeographic variation in *Anolis oculatus* on the island of Dominica, West Indies. Journal of Evolutionary Biology 4: 321–335.

———. 1991b. Experimental detection of rapid evolutionary response in natural lizard populations. Nature 353: 347–348.

———. 1992. *Anolis oculatus*. Catalogue of American Amphibians and Reptiles 540: 1–4.

———. 1993. An experimental field study of a eurytopic anole, *Anolis oculatus*. Journal of the Zoological Society of London 229: 163–170.

———. 1995. *Ameiva fuscata*. Catalogue of American Amphibians and Reptiles 606: 1–3.

———. 1997a. Size and shape variation in a Lesser Antillean anole, *Anolis oculatus* (Sauria: Iguanidae) in relation to habitat. Biological Journal of the Linnean Society 60: 53–72.

———. 1997b. Microgeographic variation in scalation of *Anolis oculatus* (Dominica, West Indies): A multivariate analysis. Herpetologica 53: 49–62.

———. 1999. Reptiles & Amphibians of the Eastern Caribbean. London: Macmillan Education.

Malhotra, A., R. S. Thorpe, E. Hypolite, and A. James. 2007. A report on the status of the herpetofauna of the Commonwealth of Dominica, West Indies. Applied Herpetology 4: 177–194.

Mallery, C. S., Jr., M. A. Marcum, R. Powell, J. S. Parmerlee, Jr., and R. W. Henderson. 2007. Herpetofaunal communities of the leeward slopes and coasts of St. Vincent: A comparison of sites variously altered by human activity. Applied Herpetology 4: 313–325.

Malone, C. L., and S. K. Davis. 2004. Genetic contributions to Caribbean iguana conservation. In: A. C. Alberts, R. L. Carter, W. K. Hayes, and E. P. Martins (eds.), Iguanas: Biology and Conservation, pp. 45–57. Berkeley and Los Angeles: University of California Press.

Malone, C. L., T. Wheeler, J. F. Taylor, and S. K. Davis. 2000. Phylogeography of the Caribbean rock iguana (*Cyclura*): Implications for conservation and insights on the biogeographic history of the West Indies. Molecular Phylogenetics and Evolution 17: 269–279.

Mancina, C. A., and A. Llanes Sosa. 1997. Indicios de depredación de huevos de *Hirundo fulva* (Passeriformes: Hirundinidae) por *Epicrates angulifer* (Serpentes: Boidae). Journal of Caribbean Ornithology 10: 95–96.

Manójina, N., A. González, and R. M. Abreu. 1989. Datos sobre alimentación de la Jutía Conga (*Capromys pilorides*) en Guanahacabibes. Poeyana 369: 1–13.

Marcellini, D. L., and T. A. Jenssen. 1983. A character for differentiating the sympatric lizards *Anolis cooki* and *Anolis cristatellus*. Herpetological Review 14: 113–114.

———. 1989. Thermal ecology of the tropical iguanid lizard, *Leiocephalus schreibersi*. American Midland Naturalist 122: 44–50.

———. 1991. Avoidance learning by the tropical iguanid lizard, *Leiocephalus schreibersii*: Implications for antipredator behavior. Journal of Herpetology 25: 238–241.

Marcellini, D. L., and L. Rodríguez Schettino. 1987. Notes on the natural history of the unusual Cuban lizard, *Anolis lucius*. Herpetological Review 18: 52–53.

Marcellini, D. L., T. A. Jenssen, and C. A. Pague. 1985. Distribution of the lizard *Anolis cooki*, with comments on its possible future extinction. Herpetological Review 16: 99–102.

Marcum, M. A., M. A. Powell, A. J. Muensch, H. F. Arnold, and R. Powell. 2008. Social behaviour of the dwarf gecko *Sphaerodactylus vincenti vincenti* on St. Vincent, Lesser Antilles. Salamandra 44: 15–22.

Maria, R., J. Ramer, T. Reichard, P. J. Tolson, and M. M. Christopher. 2007. Biochemical reference intervals and intestinal microflora of free-ranging Ricord's Iguanas (*Cyclura ricordii*). Journal of Zoo and Wildlife Medicine 38: 414–419.

Martin, C. 2001. The abundance of *Eleutherodactylus martinices* [sic] from a water source. Unpublished report. Texas A&M University Study Abroad, College Station. http://dominica.tamu.edu/student%20projects/projects_by_year.htm.

Martínez Reyes, M. 1994. Aspectos reproductivos de *Leiocephalus cubensis cubensis* (Iguania: Tropiduridae) en una localidad de Ciudad de La Habana, Cuba. Ciencias Biológicas 27: 83–89.

———. 1995. Saurios de la Reserva de la Biosfera "Sierra del Rosario" Pinar del Río, Cuba. Evaluación ecológica de tres comunidades. Investigaciones Geográficas Boletín 30: 50–77.

Martínez Reyes, M., and I. Fernández García. 1994. Hábitat y alimentación de *Leiocephalus cubensis cubensis* (Iguania: Tropiduridae) en una localidad de Ciudad de La Habana, Cuba. Ciencias Biológicas 26: 21–30.

Martínez Reyes, M., and A. Hernández Marrero. 2003. Moradores de jardines, casas y cultivos. In: L. Rodríguez Schettino (ed.), Anfibios y Reptiles de Cuba, pp. 138–143. Vaasa, Finland: UPC Print.

Martínez Reyes, M., and L. V. Moreno García. 2003. Lagartos habitantes de los suelos. In: L. Rodriguez Schettino (ed.), Anfibios y Reptiles de Cuba, pp. 90–97. Vaasa, Finland: UPC Print.

Martínez Reyes, M., and L. Rodríguez Schettino. 1987. Canibalismo en *Leiocephalus carinatus* (Gray) (Sauria: Iguanidae). Miscelanea Zoológica 29: 1–2.

Martínez Reyes, M., A. Estrada, and J. Novo R. 1990. Aspectos ecológicos y reproductivos de *Leiocephalus s. stictigaster* (Sauria: Iguanidae) en la Península de Guanahacabibes, Cuba. Poeyana 403: 1–20.

Martínez Rivera, C. C., A. González N., M. Bertrand, and J. Acosta. 2003. *Hemidactylus mabouia* (Sauria: Gekkonidae), host of *Geckobia hemidactyli* (Actinedida: Pterygosomatidae), throughout the Caribbean and South America. Caribbean Journal of Science 39: 321–326.

Martins, E. P., and J. Lamont. 1998. Estimating ancestral states of a communicative display: A comparative study of *Cyclura* rock iguanas. Animal Behaviour 55: 1685–1706.

Mattingly, W. B., and B. D. Jayne. 2004. Resource use in arboreal habitats: Structure affects locomotion of four ecomorphs of *Anolis* lizards. Ecology 85: 1111–1124.

———. 2005. The choice of arboreal escape paths and its consequences for the locomotor behaviour of four species of *Anolis* lizards. Animal Behaviour 70: 1239–1250.

Mayer, G. C. 1989. Deterministic Aspects of Community Structure in West Indian Amphibians and Reptiles. Ph.D. thesis, Harvard University, Cambridge, Mass.

Mayer, G. C., and J. Lazell. 2000. A new species of *Mabuya* (Sauria: Scincidae) from the British Virgin Islands. Proceedings of the Biological Society of Washington 113: 871–886.

McCoy, C. J. 1970. *Hemidactylus turcicus*. Catalogue of American Amphibians and Reptiles 87: 1–2.

McFarlane, D. A., and K. L. Garrett. 1989. The prey of Common Barn Owls (*Tyto alba*) in dry limestone scrub forest of southern Jamaica. Caribbean Journal of Science 25: 21–23.

McGinnity, D., and R. Powell. 2004a. *Celestus anelpistus*. In: IUCN 2008. 2008 IUCN Red List of Threatened Species. http://www.iucnredlist.org.

———. 2004b. *Celestus warreni*. In: IUCN 2008. 2008 IUCN Red List of Threatened Species. http://www.iucnredlist.org.

McKeown, S. 1998. Notes on a newly established frog, *Eleutherodactylus coqui*, in the Hawaiian Islands. Bulletin of the Chicago Herpetological Society 33: 30–31.

McLaughlin, J. F., and J. Roughgarden. 1989. Avian predation on *Anolis* lizards in the northeastern Caribbean: An inter-island contrast. Ecology 70: 617–628.

McManus, J. J., and D. W. Nellis. 1973. Temperature and metabolism of a tropical lizard, *Anolis acutus*. Comparative Biochemistry and Physiology 45A: 403–410.

McNair, D. B. 2003. Population estimate, habitat associations, and conservation of the St. Croix Ground Lizard *Ameiva polops* at Protestant Cay, United States Virgin Islands. Caribbean Journal of Science 39: 94–99.

McNair, D. B., and W. Coles. 2003. Response of the St. Croix Ground Lizard *Ameiva polops* population to severe local disturbance of critical habitat at Protestant Cay: Before-and-after comparison. Caribbean Journal of Science 39: 392–398.

McNair, D. B., and C. D. Lombard. 2004. Population estimates, habitat associations, and management of *Ameiva polops* (Cope) at Green Cay, United States Virgin Islands. Caribbean Journal of Science 40: 353–361.

McNair, D. B., and A. Mackay. 2005. Population estimates and management of *Ameiva polops* (Cope) at Ruth Island, United States Virgin Islands. Caribbean Journal of Science 41: 352–357.

Medina Díaz, P., and R. L. Joglar. 2008. *Eleutherodactylus richmondi* (Coquí Caoba). Reproduction. Herpetological Review 39: 460.

Medina Díaz, P., H. M. Heinz, J. S. Parmerlee, Jr., and R. Powell. 2005. Population densities and structural habitats of *Anolis* lizards on St. Eustatius, Netherlands Antilles. Caribbean Journal of Science 41: 296–306.

Meier, A. J., and R. E. Noble. 1990. Notes on the status and habits of the Desecheo Gecko, *Sphaerodactylus levinsi*. Journal of Herpetology 24: 426–428.

———. 1991. Notes on the natural history of *Anolis desechensis*. Florida Field Naturalist 19: 17–18.

Meier, A. J., R. E. Noble, and S. L. Rathbun. 1993. Population status and notes on the biology of the St. Croix Ground Lizard on Green Cay (St. Croix, U.S. Virgin Islands). Caribbean Journal of Science 29: 147–152.

Mercado, J. E., E. Terranova, and J. M. Wunderle, Jr. 2002. Avian mobbing of the Puerto Rican Boa (*Epicrates inornatus*). Caribbean Journal of Science 38: 125–126.

Merle, H., A. Donnio, L. Ayeboua, Y. Plumelle, D. Smadja, and L. Thomas. 2005. Occipital infarction revealed by quadranopsia following snakebite by *Bothrops lanceolatus*. American Journal of Tropical Medicine and Hygiene 73: 583–585.

Mertens, R. 1939. Herpetologische Ergebnisse einer Reise

nach der Insel Hispaniola, Westindien. Abhandlungen der Senckenbergischen Naturforschenden Gesellschaft 449: 1–84.

———. 1946. Die Warn- und Droh-Reaktionen der Reptilien. Abhandlungen der Senckenbergischen Naturforschende Gesellschaft 471: 1–108.

Meshaka, W. E., Jr. 1995. Geographic Distribution. *Hemidactylus garnotii* (Indo-Pacific Gecko). Herpetological Review 26: 108.

———. 1996a. Diet and colonization of buildings by the Cuban Treefrog, *Osteopilus septentrionalis* (Anura: Hylidae). Caribbean Journal of Science 32: 59–63.

———. 1996b. *Osteopilus septentrionalis* (Cuban Treefrog). Maximum Size. Herpetological Review 27: 74.

———. 1996c. Occurrence of the nematode *Skrjabinoptera scelopori* in the Cuban Treefrog, *Osteopilus septentrionalis*: Mainland and island comparisons. In: R. Powell and R. W. Henderson (eds.), Contributions to West Indian Herpetology: A Tribute to Albert Schwartz, p. 271–276. Contributions to Herpetology, volume 12. Ithaca, N.Y.: Society for the Study of Amphibians and Reptiles.

Mettrick, D. F., and L. C. Dunkley 1968. Observations on the occurrence, growth, and morphological variation of the trematode, *Mesocoelium danforthi* Hoffman, 1935, in Jamaica. Caribbean Journal of Science 8: 71–94.

Micco, S. M., G. J. Lahey, R. A. Sosa, R. Powell, E. J. Censky, and J. S. Parmerlee, Jr. 1997. Natural history of *Leiocephalus barahonensis* (Tropiduridae) on the Península de Barahona, Hispaniola: An examination of two populations. Herpetological Natural History 5: 147–156.

Michael, S. F. 1995. Captive breeding of two species of *Eleutherodactylus* (Anura: Leptodactylidae) from Puerto Rico, with notes on behavior in captivity. Herpetological Review 26: 27–29.

———. 1996. Courtship calls of three species of *Eleutherodactylus* from Puerto Rico (Anura: Leptodactylidae). Herpetologica 52: 116–120.

Miller, T. J. 1985. Husbandry and breeding of the Puerto Rican Toad (*Peltophryne lemur*) with comments on its natural history. Zoo Biology 4: 281–286.

Milton, T. H., and T. A. Jenssen. 1979. Description and significance of vocalization by *Anolis grahami* (Sauria: Iguanidae). Copeia 1979: 481–489.

Ministry of Physical Development and Environment, Barbados. 2001. Barbados State of the Environment Report 2000. United Nations Environment Programme for the Ministry of Physical Development and Environment, Barbados. http://www.pnuma.org/dewalac/PDFs/Barbados.pdf.

Miralles, A. 2005. The identity of *Lacertus mabouya* Lacepède, 1788, with description of a neotype. Herpetologica 61: 46–53.

Miranda-Castro, L., A. R. Puente-Rolón, and S. Vega-Castillo. 2000. First list of the vertebrates of Los Tres Picachos State Forest, Puerto Rico, with data on relative abundance and altitudinal distribution. Caribbean Journal of Science 36: 117–126.

Mitchell, A., and L. Prince. 2008. A predation attempt gone awry: Defensive constriction of a predator by a Dominican Racer. Iguana 15: 154–156.

Mitchell, N. C. 1996. *Cyclura pinguis*. In: IUCN 2008. 2008 IUCN Red List of Threatened Species. http://www.iucnredlist.org.

———. 1999. Effect of introduced ungulates on density, dietary preferences, home range, and physical condition of the iguana *Cyclura pinguis* on Anegada. Herpetologica 55: 7–17.

———. 2000a. Anegada Island Iguana, *Cyclura pinguis*. In: A. Alberts (ed.), West Indian Iguanas: Status Survey and Conservation Action Plan, pp. 47–49. Gland, Switzerland and Cambridge, UK: International Union for Conservation of Nature and Natural Resources.

———. 2000b. Anegada Iguana. In: R. P. Reading and B. Miller (eds.), Endangered Animals: A Reference Guide to Conflicting Issues, pp. 22–27. Westport, Conn.: Greenwood.

———. 2003. Another chance for the Turks and Caicos Rock Iguana. Iguana 10: 10–14.

Mitchell, N. C., R. Haeffner, V. Veer, M. Fulford-Gardner, W. Clerveaux, C. R. Veitch, and G. Mitchell. 2002. Cat eradication and the restoration of endangered iguanas (*Cyclura carinata*) on Long Cay, Caicos Bank, Turks and Caicos Islands, British West Indies. In: C. R. Veitch and M. N. Clout (eds.), Turning the Tide: The Eradication of Invasive Species, pp. 206–212. Gland, Switzerland: IUCN/SSC Invasive Species Specialist Group, International Union for Conservation of Nature and Natural Resources.

Moermond, T. C. 1979a. Habitat constraints on the behavior, morphology, and community structure of *Anolis* lizards. Ecology 60: 152–164.

———. 1979b. The influence of habitat structure on *Anolis* foraging behavior. Behaviour 70: 147–167.

———. 1981. Prey-attack behavior of *Anolis* lizards. Zeitschrift für Tierpsychologie 56: 128–136.

Montañez Huguez, L., A. Sampedro Marín, and M. Alvarez Rodríguez. 1996. Utilización diferenciada de los recursos espacio y alimento por juveniles y adultos de *Rana catesbeiana* en la Ciénaga de Zapata. Revista Biología 10: 21–26.

Moravec, F., and H. Kaiser. 1994. Description of *Cosmocerca longispicula* sp. nov. (Nematoda: Cosmocercidae), a parasite of a dendrobatid frog from Martinique, French Antilles. Parisitological Research 80: 29–32.

———. 1995. Helminth parasites from West Indian frogs, with descriptions of two new species. Caribbean Journal of Science 31: 252–268.

Morel, P. C. 1967. Les tiques des animaux sauvages des Antilles (Acariens, Ixodoidea). Acarologia, Paris 9: 341–352.

Morell Savall, E., R. Díaz Aguiar, and O. Alfonso Álvarez. 1998. El Majá de Santa María (*Epicrates angulifer*): La boa de la mayor de las Antillas. Flora y Fauna 2: 40–42.

Moreno, J. A. 1991a. Introduccion. Introduction. In: J. A. Moreno (ed.), Status y Distribución de los Reptiles y Anfibios de la Región de Puerto Rico, pp. 3–6. Publicación Científica Miscelánea No. 1. San Juan: Departamento de Recursos Naturales de Puerto Rico.

———. 1991b. Accounts of those species considered to be of concern. In: J. A. Moreno (ed.), Status y Distribución de los Reptiles y Anfibios de la Región de Puerto Rico, pp. 9–10. Publicación Científica Miscelánea No. 1. San Juan: Departamento de Recursos Naturales de Puerto Rico.

———. 1991c. Status survey of the Golden Coqui *Eleutherodactylus jasperi*. In: J. A. Moreno (ed.), Status y Distribución de los Reptiles y Anfibios de la Región de Puerto Rico, pp. 37–41. Publicación Científica Miscelánea No. 1. San Juan: Departamento de Recursos Naturales de Puerto Rico.

———. 1991d. Appendix II. Status recommendations/recomendaciones de designation. In: J. A. Moreno (ed.), Status y Distribución de los Reptiles y Anfibios de la Región de Puerto Rico, pp. 65–67. Publicación Científica Miscelánea No. 1. San Juan: Departamento de Recursos Naturales de Puerto Rico.

———. 1995. Interim Report—Mona Island Iguana Survey. Mayagüez, Puerto Rico: U.S. Fish and Wildlife Service, Caribbean Field Office.

Moreno, L. V. 1987. Primeras observaciones sobre *Cricosaura typica* (Squamata: Xantusiidae) en cautiverio. Ciencias Biológicas 17: 104–108.

Moreno, L. V., and V. Rivalta. 2007. Especie nueva de sapo del género *Bufo* (Anura: Bufonidae) del Península de Zapata, Cuba. Solenodon 6: 60–69.

Morgan, G. S., and C. A. Woods. 1986. Extinction and zoogeography of West Indian land mammals. Biological Journal of the Linnaean Society 28: 167–203.

Morgan, G. S., R. Franz, and R. I. Crombie. 1993. The Cuban Crocodile, *Crocodylus rhombifer*, from late Quaternary fossil deposits on Grand Cayman. Caribbean Journal of Science 29: 153–164.

Morrison, W., L. Valentine, M. Kerehgyarto, and A. Burklund. 2002. Anole density and biomass in Dominica. Unpublished report. Texas A&M University Study Abroad, College Station. http://dominica.tamu.edu/student%20projects/projects_by_year.htm.

Morton, M. 2007. St. Lucia Iguana 2006 project update. Iguana Specialist Group Newsletter 10(1): 16–17 (reprinted in Iguana 14: 246–247).

Moster, J. A., R. Powell, J. S. Parmerlee, Jr., D. D. Smith, and A. Lathrop. 1992. Natural history notes on a small population of *Anolis brevirostris* (Sauria: Polychridae) from altered habitat in the Dominican Republic. Bulletin of the Maryland Herpetological Society 28: 150–161.

Moster, J. A., J. S. Parmerlee, Jr., and R. Powell. 1995. *Anolis brevirostris*. Catalogue of American Amphibians and Reptiles 610: 1–4.

Mudde, P. 1999. Cubaans grasdwergje *Anolis alutaceus*. Onder het Palmblad 2: 30.

Muelleman, P. J., L. A. White, R. W. Henderson, and R. Powell. 2009. Activity patterns of *Alsophis sibonius* and *Liophis juliae* (Dipsadidae) in Cabrits National Park, Dominica, West Indies. South American Journal of Herpetology 4: 55–60.

Murphy, J. B. 1969. Notes on iguanids and varanids in a mixed exhibit at Dallas Zoo. International Zoo Yearbook 9: 39–41.

Murphy, J. B., D. G. Barker, and B. W. Tryon. 1978. Miscellaneous notes on the reproductive biology of reptiles. 2. Eleven species of the family Boidae, genera *Candoia*, *Corallus*, *Epicrates*, and *Python*. Journal of Herpetology 12: 385–390.

Murphy, J. C. 1997. Amphibians and Reptiles of Trinidad and Tobago. Malabar, Fla.: Krieger.

Myers, C. W. 1971. Central American lizards related to *Anolis pentaprion*: Two new species from the Cordillera de Talamanca. American Museum Novitates 2471: 1–40.

Myers, N., R. A. Mittermeier, C. G. Mittermeier, G. A. B. da Fonseca, and J. Kent. 2000. Biodiversity hotspots for conservation priorities. Nature 403: 853–858.

Narins, P. M., and R. R. Capranica. 1976. Sexual differences in the auditory system of the tree frog *Eleutherodactylus coqui*. Science 192: 378–380.

Narins, P. M., and S. L. Smith. 1986. Clinal variation in anuran advertisement calls: Basis for acoustic isolation? Behavioral Ecology and Sociobiology 19: 135–141.

Nava, S. S. 2004. Microhabitat Selection, Resource Partitioning, and Evaporative Water Loss by Dwarf Geckos (*Sphaerodactylus*) on Puerto Rico. Master's thesis, University of Texas-El Paso.

Nava, S. S., C. R. Lindsay, R. W. Henderson, and R. Powell. 2001. Microhabitat, activity, and density of a dwarf gecko (*Sphaerodactylus parvus*) on Anguilla, West Indies. Amphibia-Reptilia 22: 455–464.

Nava, S. S., C. R. Lindsay, R. Powell, and R. W. Henderson. 2002. *Sphaerodactylus parvus*. Catalogue of American Amphibians and Reptiles 752: 1–2.

Navarro Pacheco, N., and O. H. Garrido. 2004. Especie nueva de *Anolis* (Sauria: Lacertilia: Iguanidae) de la region suroriental de Cuba. Solenodon 4: 85–90.

Navarro Pacheco, N., A. Fernández V., and O. H. Garrido. 2001. Reconsideración taxonómica de *Anolis centralis litoralis* y descripción de una especie nueva del grupo *argillaceus* (Sauria: Iguanidae) para Cuba. Solenodon 1: 66–75.

Neill, W. T. 1954. Evidence of venom in snakes of the genera *Alsophis* and *Rhadinea*. Copeia 1954: 59.

———. 1958. The occurrence of amphibians and reptiles in saltwater areas, and a bibliography. Bulletin of Marine Sciences of the Gulf and Caribbean 8: 1–97.

———. 1965. Notes on aquatic snakes, *Natrix* and *Tretanorhinus*, in Cuba. Herpetologica 21: 62–67.

Neill, W. T., and E. R. Allen. 1957. *Deiroptyx*—Cuba's reptilian oddity. Nature Magazine 50: 39–41, 52.

Neils, A., and C. Bugbee. 2007. *Rana catesbeiana* (American Bullfrog). Diet. Herpetological Review 38: 443.

Nellis, D. W. 1996. *Ameiva polops*. In: IUCN 2008. 2008 IUCN Red List of Threatened Species. http://www.iucnredlist.org.

Nellis, D. W., R. L. Norton, and W. P. Mac Lean. 1983. On the biogeography of the Virgin Islands Boa, *Epicrates monensis granti*. Journal of Herpetology 17: 413–417.

Nelson, C. E. 1972. *Gastrophyrne carolinensis*. Catalogue of American Amphibians and Reptiles 120: 1–4.

Nelson, S. E., and R. Powell, R. 2002. *Leiocephalus semilineatus*. Catalogue of American Amphibians and Reptiles 751: 1–2.

Nelson, S. E., B. L. Banbury, R. A. Sosa, and R. Powell. 2001. Natural history of *Leiocephalus semilineatus* in association with sympatric *Leiocephalus schreibersii* and *Ameiva lineolata*. Contemporary Herpetology 2001(1): 1–6 + 4 figures + 2 tables http://www.contemporaryherpetology.org/ch/2001/1/index.htm.

Netting, M. G., and C. J. Goin. 1944. Another new boa of the genus *Epicrates* from the Bahamas. Annals of the Carnegie Museum 30: 71–76.

Neumann, L. G. 1899. Révision de la famille des Ixodidés. III. Ixodae. Memoire de la Société Zoologique de France 12: 107–294.

———. 1901. Révision de la famille des Ixodidés. IV. Memoire de la Société Zoologique de France 14: 249–372.

Newsom, L. A., and E. S. Wing. 2004. On Land and Sea: Native American Uses of Biological Resources in the West Indies. Tuscaloosa: University of Alabama Press.

Newstead, R. 1909. Ticks and other blood-sucking Arthropoda of Jamaica. Annals of Tropical Medicine and Parasitology 3: 421–468.

Nicholson, K. L., S. M. Torrence, D. M. Ghioca, J. Bhattacharjee, A. E. Andrei, J. Owen, N. J. A. Radke, and G. Perry. 2005. The influence of temperature and humidity on activity patterns of the lizards *Anolis stratulus* and *Ameiva exsul* in the British Virgin Islands. Caribbean Journal of Science 41: 870–873.

Nieto Dopico, A. 1997. La Fauna Silvestre de la Ciénaga de Zapata. La Habana: Editorial Científico-Técnica.

Nieves, L. O. 2007. Predation on a Tarantula (*Cyrtopholis bartholomaei*) by a Puerto Rican Ground Lizard (*Ameiva exsul*). Iguana 14: 226–227.

Noble, G. K. 1923a. In pursuit of the giant tree frog. Natural History 23: 105–116.

———. 1923b. Field studies of Dominican tree frogs and their haunts. Natural History 23: 117–121.

———. 1923c. Trailing the Rhinoceros Iguana. Natural History 23: 541–558.

———. 1926. The hatching process in *Alytes*, *Eleutherodactylus* and other amphibians. American Museum Novitates 229: 1–7.

———. 1927. The value of life history data in the study of the evolution of the Amphibia. Annals of the New York Academy of Science 30: 31–128.

Noble, G. K., and H. T. Bradley. 1933. The mating behavior of lizards; its bearing on the theory of sexual selection. Annals of the New York Academy of Sciences 35: 25–100.

Noble, G. K., and W. G. Hassler. 1933. Two new species of frogs, five new species and a new race of lizards from the Dominican Republic. American Museum Novitates 652:1–17.

Noble, G. K., and G. C. Klingel. 1932. The reptiles of Great Inagua Island, British West Indies. American Museum Novitates 549: 1–25.

Noble, G. K., and H. K. Teale. 1930. The courtship of some iguanid and teiid lizards. Copeia 1930: 54–56.

Norton, R. L. 1993. *Alsophis portoricensis richardii* (Ground Snake). Feeding. Herpetological Review 24: 34.

———. 1997. First report of *Peltophryne lemur* from St. John, U.S. Virgin Islands. Bulletin of the Chicago Herpetological Society 32: 106–107.

———. 1998. Comments on habitat and threats to populations of the Puerto Rican Crested Toad (*Peltophryne lemur*) in the Virgin Islands. British Herpetological Society Bulletin 62: 20–22.

Novo Rodríguez, J. 1985a. Nido communal de *Anolis angusticeps* (Sauria: Iguanidae) en Cayo Francés, Cuba. Miscelanea Zoológica 26: 3–4.

———. 1985b. Datos reproductivos sobre *Tarentola americana* (Sauria: Gekkonidae) en Cuba. Miscelanea Zoológica 28: 2–3.

Novo Rodríguez, J., and F. de Arazoza Rodríguez. 1986. Cría en cautiverio de *Alsophis cantherigerus* (Ophidia: Colubridae). Reporte Investigación del Instituto de Zoología 30: 1–15.

Novo Rodríguez, J., and A. R. Estrada. 1986. Ciclo reproductivo de *Anolis bartschi* (Sauria: Iguanidae). Poeyana 318: 1–5.

Novo Rodríguez, J., A. R. Estrada, and G. Alayón. 1985. *Eleutherodactylus* (Anura: Leptodactylidae) depredado por un aranéido. Miscelanea Zoológica 28: 1–2.

Novo Rodríguez, J., A. R. Estrada, and L. V. Moreno. 1987. Apuntes sobre la reproducción de *Eleutherodactylus atkinsi* (Anura: Leptodactylidae). Miscelanea Zoológica 29: 2–3.

Numeric, P., V. Moravie, M. Didier, D. Chatot-Henry, S. Cirille, B. Bucher, and L. Thomas. 2002. Multiple cerebral infarctions following a snakebite by *Bothrops caribbaeus*. American Journal of Tropical Medicine and Hygiene 67: 287–288.

Ober, F. A. 1879. Camps in the Caribees. Boston: Lee and Shepard.

Ober, L. D. 1968. Reproduction and growth rates in the His-

paniolan anguid *Diploglossus stenurus* Cope. Herpetologica 24: 326–327.
———. 1970. Reproduction in the anguid lizard, *Diploglossus curtissi aporus* Schwartz. Herpetologica 26: 275.
Ogrodowczyk, A., P. Murrain, L. Martin, and R. P. Young. 2006. Recent observations of the Montserrat Galliwasp, *Diploglossus montisserrati*. Herpetological Bulletin 98: 9–11.
Olexa, A. 1972. *Anolis loysiana* von Kuba. Aquaterra 9(3): 28–31.
Oliver, J. A. 1948. The anoline lizards of Bimini, Bahamas. American Museum Novitates 1383: 1–36.
Oliver, W. L. R. 1982. The Coney and the Yellow Snake: The distribution and status of the Jamaican Hutia *Geocapromys brownii* and the Jamaican Boa *Epicrates subflavus*. Dodo 19: 6–33.
Olson, R. E. 1990. Herpetological observations on Tiburon Peninsula, Haiti. Bulletin of the Maryland Herpetological Society 26: 135–152.
———. 1995. Microhabitat in ecologically devastated areas. Bulletin of the Maryland Herpetological Society 31: 93–95.
Ord, T. J. 2008. Dawn and dusk "chorus" in visually communicating Jamaican anole lizards. American Naturalist 172: 585–592.
Ord, T. J., and J. A. Stamps. 2008. Alert signals enhance animal communication in "noisy" environments. Proceedings of the National Academy of Sciences 105: 18830–18835.
Ord, T. J., R. A. Peters, B. Clucas, and J. A. Stamps. 2007. Lizards speed up visual displays in noisy motion habitats. Proceedings of the Royal Society B 274: 1057–1062.
Orrell, K., and T. A. Jenssen. 1998. Display behavior of *Anolis bahoruconensis*: An anole with a diminutive dewlap. Caribbean Journal of Science 34: 113–125.
Ortiz, P. R. 1990. Status and distribution of *Anolis cooki* (Reptilia, Sauria, Iguanidae). Acta Cientifica 4: 157–159.
———. 1991. Status and distribution of *Anolis cooki* (Reptilia, Sauria, Iguanidae). In: J. A. Moreno (ed.), Status y Distribución de los Reptiles y Anfibios de la Región de Puerto Rico, pp. 20–22. Publicación Científica Miscelánea No. 1. San Juan: Departamento de Recursos Naturales de Puerto Rico.
Ortiz, P. R., and T. A. Jenssen. 1982. Interspecific aggression between lizard competitors, *Anolis cooki* and *Anolis cristatellus*. Zeitschrift für Tierpsychologie 60: 227–238.
Otero, A. R. 1950. Sobre la alimentación de *Anolis porcatus*. Boletín de Historia Natural 1: 186–187.
Ottenwalder, J. A. 1980. *Epicrates striatus* como predador de aves. Naturalista Postal (Santo Domingo, Dominican Republic) 21: 1–2.
———. 1994. Conservation program for the South Hispaniolan Slider Turtle (*Trachemys decorata*) in the Dominican Republic, 1994. Summary progress report prepared for the 1994 AZA/Chelonian Advisory Group Meeting. New Orleans: International Herpetological Symposium.
———. 1996a. *Cyclura cornuta* ssp. *cornuta*. In: IUCN 2008. 2008 IUCN Red List of Threatened Species. http://www.iucnredlist.org.
———. 1996b. *Cyclura ricordi*. In: IUCN 2008. 2008 IUCN Red List of Threatened Species. http://www.iucnredlist.org.
———. 2000a. Rhinoceros Iguana, *Cyclura cornuta cornuta*. In: A. Alberts (ed.), West Indian Iguanas: Status Survey and Conservation Action Plan, pp. 22–27. Gland, Switzerland and Cambridge, UK: International Union for Conservation of Nature and Natural Resources.
———. 2000b. Ricord's Iguana, *Cyclura ricordi*. In: A. Alberts (ed.), West Indian Iguanas: Status Survey and Conservation Action Plan, pp. 51–54. Gland, Switzerland and Cambridge, UK: International Union for Conservation of Nature and Natural Resources.
Ottenwalder, J. A., and R. Powell. 2002. Brief notes on the status of Hispaniolan Rhinoceros Iguanas, *Cyclura cornuta*. Iguana Times (Journal of the International Iguana Society) 9: 78.
Ovaska, K. 1991a. Reproductive phenology, population structure, and habitat use of the frog *Eleutherodactylus johnstonei* in Barbados, West Indies. Journal of Herpetology 25: 424–430.
———. 1991b. Diet of the frog *Eleutherodactylus johnstonei* (Leptodactylidae) in Barbados, West Indies. Journal of Herpetology 25: 486–488.
———. 1992. Short- and long-term movements of the frog *Eleutherodactylus johnstonei* in Barbados, West Indies. Copeia 1992: 569–573.
———. 2005. The frog (*Eleutherodactylus antillensis*). In: J. Lazell, Island: Fact and Theory in Nature, pp. 175–180. Berkeley and Los Angeles: University of California Press.
Ovaska, K., and J. Caldbeck. 1997a. Courtship behavior and vocalizations of the frogs *Eleutherodactylus antillensis* and *E. cochranae* on the British Virgin Islands. Journal of Herpetology 31: 149–155.
———. 1997b. Vocal behaviour of the frog *Eleutherodactylus antillensis* on the British Virgin Islands. Animal Behaviour 54: 181–188.
Ovaska, K., and A. R. Estrada. 2003. *Eleutherodactylus antillensis* (Coqui Churi). Reproduction. Herpetological Review 34: 229.
Ovaska, K., and W. Hunte. 1992. Male mating behavior of the frog *Eleutherodactylus johnstonei* (Leptodactylidae) in Barbados, West Indies. Herpetologica 48: 40–49.
Ovaska, K., J. Caldbeck, and J. Lazell. 1998. *Eleutherodactylus schwartzi* (NCN). Reproduction. Herpetological Review 29: 97.
———. 2000. New records and distributional and ecological notes of leptodactylid frogs, *Leptodactylus* and *Eleutherodactylus*, from the British Virgin Islands. Breviora 508: 1–25.

Owen, J., and G. Perry. 2005. *Anolis cristatellus wileyae* (Virgin Islands Crested Anole). Saurophagy. Herpetological Review 36: 444.

Owen, J., G. Perry, J. Lazell, and C. Petrovic. 2005. Geographic Distribution. *Pseudemys nelsoni*. Herpetological Review 36: 466.

Owen, J., G. Perry, J. Lazell, C. Petrovic, and J. Egelhoff. 2006. *Osteopilus septentrionalis* (Cuban Tree Frog). Colonization of the British Virgin Islands. Herpetological Review 37: 74–75.

Owens, A. K., and C. R. Knapp. 2004. *Leiocephalus carinatus coryi* (NCN). Nest location. Herpetological Review 35: 392–393.

———. 2007. *Cyclura cychlura cychlura* (Andros Iguana). Scoliosus; kyphosis. Herpetological Review 38: 454–455.

Pacala, S. W., and J. Roughgarden. 1982. Resource partitioning and interspecific competition in two two-species insular *Anolis* lizard communities. Science 217: 444–446.

———. 1984. Control of arthropod abundance by *Anolis* lizards on St. Eustatius (Neth. Antilles). Oecologia 64: 160–162.

———. 1985. Populations experiments with the *Anolis* lizards of St. Maarten and St. Eustatius. Ecology 66: 129–141.

Paine, F. L. 1984. The husbandry, management & reproduction of the Puerto Rican Crested Toad (*Bufo lemur*). In: R. Hahn (ed.), Proceedings of the 8th International Symposium on Captive Propagation and Husbandry, pp. 59–73. Thurmont, Md.: Zoological Consortium.

Panton, E. S. 1928. Eight months with the green lizard or the history of *Anolis garmani*. Kingston, Jamaica: Times Printery.

Parmerlee, J. S., Jr., R. Powell, D. D. Smith, and A. Lathrop. 1992. Unusual behavior in the Cuban Green Anole, *Anolis porcatus* (Sauria: Polychridae). Bulletin of the Chicago Herpetological Society 27: 118.

Parsons, K. J. 1995. *Peltophryne guentheri* (NCN). Diet. Herpetological Review 26: 202.

Parsons, K. J., R. Powell, and J. H. Greve. 1995. *Peltophryne guentheri* (NCN). Parasitism. Herpetological Review 26: 31–32.

Pasachnik, S. A., J. J. Shew, J. H. Townsend, and R. Powell. 2002. *Iguana delicatissima*. Activity. Herpetological Review 33: 51–52.

Pasachnik, S. A., M. Breuil, and R. Powell. 2006. *Iguana delicatissima*. Catalogue of American Amphibians and Reptiles 811: 1–14.

Paulson, D. R. 1966. New records of birds from the Bahama Islands. Journal of the Academy of Natural Sciences of Philadelphia 4: 242–251.

Pendlebury, G. B. 1972. Nesting sites, eggs and young of *Hemidactylus mabouia* from Carriacou, West Indies. Herpetological Review 4: 203.

———. 1974. Stomach and intestine contents of *Corallus enydris*; a comparison of island and mainland specimens. Journal of Herpetology 8: 241–244.

Pereira, H. M., S. R. Loarie, and J. Roughgarden. 2002. Monogamy, polygyny and interspecific interactions in the lizards *Anolis pogus* and *Anolis gingivinus*. Caribbean Journal of Science 38: 132–136.

Perera, A. 1984. Aspectos de la ecomorfología de *Cyclura nubila* (Sauria: Iguanidae). Ciencias Biológicas 11: 129–141.

———. 1985a. Datos sobre la abundancia y actividad de *Cyclura nubila* (Sauria: Iguanidae) en los alrededores de Cayo Largo del Sur, Cuba. Poeyana 288: 1–17.

———. 1985b. Datos sobre la dieta de *Cyclura nubila* (Sauria: Iguanidae) en los alrededores de Cayo Largo del Sur, Cuba. Poeyana 291: 1–12.

———. 2000. Cuban Iguana, *Cyclura nubila nubila*. In: A. Alberts (ed.), West Indian Iguanas: Status Survey and Conservation Action Plan, pp. 36–39. Gland, Switzerland and Cambridge, UK: International Union for Conservation of Nature and Natural Resources.

Perez, M. 1951. The food of *Rana catesbeiana* Show [sic] in Puerto Rico. Herpetologica 7: 102–104.

Pérez-Beato, O. 1997. A new subspecies of *Anolis porcatus* (Sauria: Polychrotidae) from western Cuba. Revista Biología Tropical 44(3)/45(1): 295–299.

Pérez-Buitrago, N. F., and A. Sabat. 2000. Population status of the rock ground iguana (*Cyclura cornuta stejnegeri*) in Mona Island, Puerto Rico. Acta Científica 14: 67–76.

———. 2007. Natal dispersal, home range and habitat use of hatchlings of the Mona Island Iguana (*Cyclura cornuta stejnegeri*). Applied Herpetology 4: 365–376.

Pérez-Buitrago, N. F., A. O. Álvarez, and M. A. García. 2006. Cannibalism in an introduced population of *Cyclura nubila nubila* on Isla Magueyes, Puerto Rico. Iguana 13: 206–208.

Pérez-Buitrago, N. F., K. Rosas, S. Funk, M. A. García, A. Álvarez, and O. McMillan. 2007a. Mona Island Iguana reports on ecology, conservation, and blindness. Iguana Specialist Group Newsletter 10(1): 10–11 (reprinted in Iguana 14: 239–240).

Pérez-Buitrago, N. F., A. Sabat, S. M. Funk, M. A. García, A. O. Álvarez, and W. O. McMillan. 2007b. Spatial ecology of the Mona Island Iguana *Cyclura cornuta stejnegeri* in an undisturbed environment. Applied Herpetology 4: 347–355.

Pérez-Buitrago, N. F., M. A. García, A. Sabat, J. Delgado, A. Álvarez, O. McMillan, and S. M. Funk. 2008. Do headstart programs work? Survival and body condition in headstarted Mona Island Iguanas *Cyclura cornuta stejnegeri*. Endangered Species Research 6: 55–65.

Pérez-Rivera, R. A. 1985. Nota sobre el habitat, los hábitos alimentarios y los depredadores del lagarto *Anolis cuvieri* (Lacertilia: Iguanidae) de Puerto Rico. Caribbean Journal of Science 21: 101–103.

Pérez-Rivera, R. A., and M. A. Laboy-Rivera. 1996. Aspectos de la conducta de un individuo de *Alsophis portoricensis* y adultos de Reinita (*Coereba flaveola*) durante un episodio de depradación. Caribbean Journal of Science 32: 432–433.

Pérez-Rivera, R. A., and A. Molina-Opio. 2008. *Ameiva exsul* (Puerto Rican Ground Lizard). Diet. Herpetological Review 39: 219–220.

Pérez-Rivera, R. A., and E. Nadal D. 1996. Use of bird nests by *Eleutherodactylus* frogs in the Carite Forest, Puerto Rico. Caribbean Journal of Science 32: 201–205.

Pérez-Rivera, R. A., and M. J. Vélez, Jr. 1978. Notas sobre algunas culebras de Puerto Rico. Science-Ciencia 6: 68–73.

Pérez Vigueras, I. 1934a. On the ticks of Cuba, with description of a new species, *Amblyomma torrei*, from *Cyclura macleayi* Gray. Psyche 41: 13–18.

———. 1934b. Descripción de la hembra de *Amblyomma albopictum* Neumann, 1899. Revista Universidad de La Habana 1: 128–129.

———. 1934c. *Ophiotaenia barbouri* n. sp. (Cestoda), parásito de *Tretanorhinus variabilis* (Reptilia). Memorias de la Sociedad Cubana de Historia Natural "Felipe Poey" 8: 231–234.

———. 1935a. Notas sobre la fauna parasitológica de Cuba. Parte I. Vermes. C-Cestoda. Memorias de la Sociedad Cubana de Historia Natural "Felipe Poey" 9: 59–66.

———. 1935b. Sobre la validez de la especie *Atractis cruciata* Linstow, 1902. Revista Parasitología Clínica y Laboratorio La Habana 1: 188–190.

———. 1936. Notas sobre la fauna parasitológica de Cuba. Parte I. Vermes. II. Acanthocephala-Nemathelminthes. Memorias de la Sociedad Cubana de Historia Natural "Felipe Poey" 10: 53–86.

———. 1938. Nota sobre algunos nemátodos parásitos nuevos de Cuba. In: Livro jubilar do Professor Lauro Travassos: Editado para Commemorar a 25 Anniversario de suas Actividades Scientificas (1913–1938), pp. 501–508. Rio de Janeiro: Typographia do Instituto Oswaldo Cruz.

———. 1940. Nota sobre algunas especies nuevas de tremátodos y sobre otras poco conocidas. Revista Universidad de La Habana 28–29: 1–28.

———. 1942. Notas helmintológicas. Revista Universidad de La Habana 40–42: 193–213.

———. 1956. Contribución al conocimiento de la fauna helmintológica cubana. Memorias de la Sociedad Cubana de Historia Natural "Felipe Poey" 23: 1–36.

Perkins, S. L. 2001. Phylogeography of Caribbean lizard malaria: Tracing the history of vector-borne parasites. Journal of Evolutionary Biology 14: 34–45.

Perry, G. 1999. The evolution of search modes: Ecological versus phylogenetic perspectives. American Naturalist 153: 98–109.

———. 2005. The lizard genus *Anolis*. In: J. Lazell, Island: Fact and Theory in Nature, pp. 186–190. Berkeley and Los Angeles: University of California Press.

———. 2006. *Eleutherodactylus johnstonei*. Habitat. Herpetological Review 37: 443–444.

Perry, G., and R. N. Fisher. 2006. Night lights and reptiles: Observed and potential effects. In: C. Rich and T. Longcore (eds.), Ecological Consequences of Artificial Night Lighting, pp. 169–191. Washington, D.C.: Island Press.

Perry, G., and G. P. Gerber. 2006. Conservation of amphibians and reptiles in the British Virgin Islands: Status and patterns. Applied Herpetology 3: 237–256.

Perry, G., and J. Lazell. 1997. *Anolis stratulus* (Saddled Anole). Nectivory. Herpetological Review 28: 150–151.

———. 2000. *Liophis portoricensis anegadae* (NCN). Night-light hunting. Herpetological Review 31: 247.

———. 2006. *Anolis pulchellus* (Grass Anole). Nectivory. Herpetological Review 37: 218–219.

Perry, G., and N. Mitchell. 2003. Guana and Necker Island population assessments 2002. Iguana 10: 49.

Perry, G., and R. Powell. 2008. *Alsophis portoricensis anegadae* (Racer). Cannibalism. Herpetological Review 39: 465.

Perry, G., R. Dmi'el, and J. Lazell. 1999. Evaporative water loss in insular populations of the *Anolis cristatellus* group (Reptilia: Sauria) in the British Virgin Islands. II. The effect of drought. Biotropica 31: 337–342.

———. 2000. Evaporative water loss in insular populations of the *Anolis cristatellus* group (Reptilia: Sauria) in the British Virgin Islands. III. Response to the end of drought and a common garden experiment. Biotropica 32: 722–728.

Perry, G., J. Pierce, D. Griffin, G. Van Buurt, and J. Lazell. 2003a. Geographic distribution. *Elaphe guttata guttata*. Herpetological Review 34: 264–265.

Perry, G., K. Levering, and N. Mitchell. 2003b. *Cyclura pinguis* (Stout Iguana, Anegada Rock Iguana). Juvenile behavior. Herpetological Review 34: 367.

Perry, G., K. LeVering, I. Girard, and T. Garland, Jr. 2004. Locomotor performance and social dominance in male *Anolis cristatellus*. Animal Behaviour 67: 37–47.

Perry, G., R. Powell, and H. Watson. 2006. Keeping invasive species off Guana Island, British Virgin Islands. Iguana 13: 272–277.

Perry, G., J. Lazell, K. LeVering, and N. Mitchell. 2007a. Body size and timing of reproduction in the highly endangered Stout Iguana, *Cyclura pinguis*, in the British Virgin Islands. Caribbean Journal of Science 43: 155–159.

Perry, G., K. LeVering, A. Barun, and J. Lazell. 2007b. *Alsophis portoricensis anegadae* (Puerto Rican Racer). Scavenging. Herpetological Review 38: 203.

Perry, G., B. W. Buchanan, R. N. Fisher, M. Salmon, and S. E. Wise. 2008. Effects of artificial night lighting on amphibians and reptiles in urban environments. In: J. C. Mitchell, R. E. Jung, and B. Bartholomew (eds.), Urban Herpetology, pp. 239–256. Herpetological Conservation, volume 3.

Salt Lake City, Utah: Society for the Study of Amphibians and Reptiles.

Persons, M. H., L. J. Fleishman, M. A. Frye, and M. E. Stimphil. 1999. Sensory response patterns and the evolution of visual signal design in anoline lizards. Journal of Comparative Physiology 184: 585–607.

Peters, G. 1970. Zur Taxonomie und Zoogeographie der kubanischen anolinen Eidechsen (Reptilia, Iguanidae). Mitteilungen des Zoologisches Museum in Berlin 46: 197–234.

———. 1974. Notizen über die Batrachofauna der Insel Kuba. Mitteilungen des Zoologisches Museum in Berlin 50: 299–322.

Peters, G., and R. Schubert. 1968. Theoretische und praktische Aspekte einer biologischen Expedition durch Kuba. Biologie in der Schule 8/9: 337–347 + figures 1–16.

Peters, G., and A. Schwartz. 1972. Ein neuer, Bromelien bewohnender Kugelfingergecko (Gekkonidae: *Sphaerodactylus*) aus Oriente/Cuba. Mitteilungen des Zoologisches Museum in Berlin 48: 393–399.

Peters, W. 1876. Über eine von Hrn. Viceconsul L. Krug und Dr. J. Gundlach auf der Insel Puertorico gemachte Sammlung von Säugethieren und Amphibien, so wie über die Entwicklung eines Batrachiers, *Hylodes martinicensis* Dum. Bibr., ohne Metamorphose. Monatsberichte der Königlichen Preussische Akademie der Wissenschaften zu Berlin 1876: 703–714.

Petzold, H.-G. 1962. Successful breeding of *Leiocephalus carinatus* Gray. International Zoo Yearbook 4: 97–98.

———. 1967. Some remarks on the breeding biology and the keeping of *Tretanorhinus variabilis*, a water snake of Cuba. Herpetologica 23: 242–246.

———. 1969a. *Cricosaura typica* Gundlach & Peters, eine Herpetologische Kostbarkeit aus Kuba. Die Aquarien- und Terrarienzeitschrift (DATZ) 22: 82–85.

———. 1969b. Zur Haltung und Fortpflanzungsbiologie einiger kubanischer Schlangen im Tierpark Berlin. Salamandra 5: 124–140.

———. 1978. Zur Haltung und Fortpflanzungsbiologie der Kuba-Schlanknatter, *Alsophis cantherigerus* (Bibron 1840), im Terrarium. Zoologischer Garten 39: 304–322.

———. 1981. *Anolis bimaculatus sabanus* Garman, 1887, Pantheranolis. Aquarium-Terrarum 28: 287.

Petzold, H.-G., H. A. Pederzani, and H. Szidat. 1970. Einige Beobachtungen zur Biologie des kubanischen Rollschwanzleguans, *Leiocephalus carinatus*. Zoologischer Garten 39: 304–322.

Pflanz, D. J., M. A. Cusumano, and R. Powell. 1991. Notes on reproduction in a xeric-adapted anole, *Anolis whitemani* (Sauria: Polychridae) from Hispaniola. Journal of Herpetology 25: 491–493.

Philbosian, R. 1975. Territorial behavior and population regulation in the lizards *Anolis acutus* and *A. cristatellus*. Copeia 1975: 428–444.

Philibosian, R., and R. Ruibal. 1971. Conservation of the lizard *Ameiva polops* in the Virgin Islands. Herpetologica 27: 450–454.

Philibosian, R., and J. A. Yntema. 1976. Records and status of some reptiles and amphibians in the Virgin Islands. I. 1968–1975. Herpetologica 32: 81–85.

———. 1977. Annotated Checklist of the Birds, Mammals, Reptiles, and Amphibians of the Virgin Islands and Puerto Rico. Christiansted, St. Croix: Information Services.

Pinchon, R. 1967. Quelques aspects de la nature aux Antilles. Fort-de-France, Martinique: Imprimeurs à Caen.

Platenberg, R. J. 2007. Impacts of introduced species on an island ecosystem: Non-native reptiles and amphibians in the U.S. Virgin Islands. In: G. W. Witmer, W. C. Pitt, and K. A. Fagerstone (eds.), Managing Vertebrate Invasive Species: Proceedings of an International Symposium, pp. 168–174. Fort Collins, Colo.: USDA APHIS Wildlife Services, National Wildlife Research Center.

Platenberg, R. J., and R. H. Boulon, Jr. 2006. Conservation status of reptiles and amphibians in the U.S. Virgin Islands. Applied Herpetology 3: 215–235.

Poche, A. J., Jr., R. Powell, and R. W. Henderson. 2005. Sleep-site selection and fidelity in Grenadian anoles (Reptilia: Squamata: Polychrotidae). Schlafplatzwahl und -treue bei Echsen der Gattung *Anolis* aus Grenada (Reptilia: Squamata: Polychrotidae). Herpetozoa 18: 3–10.

Pough, F. H., T. L. Taigen, M. M. Stewart, and P. F. Brussard. 1983. Behavioral modification of evaporative water loss by a Puerto Rican frog. Ecology 64: 244–252.

Pounds, J. A., and M. L. Crump. 1994. Amphibian declines and climate disturbance: The case of the Golden Toad and the Harlequin Frog. Conservation Biology 8: 72–85.

Powell, G. L., and A. P. Russell. 1992. Locomotor correlates of ecomorph designation in *Anolis*—an examination of 3 sympatric species from Jamaica. Canadian Journal of Zoology 70: 725–739.

Powell, M. A., R. Powell, and R. W. Henderson. 2006. *Anolis sabanus*. Catalogue of American Amphibians and Reptiles 815: 1–5.

Powell, R. 1990a. *Anolis porcatus*. Behavior. Herpetological Review 21: 93.

———. 1990b. *Hemidactylus palaichthus*. Catalogue of American Amphibians and Reptiles 468: 1.

———. 1992a. *Peltophryne guentheri*. Catalogue of American Amphibians and Reptiles 535: 1–4.

———. 1992b. *Amphisbaena gonavensis*. Catalogue of American Amphibians and Reptiles 539: 1–3.

———. 1992c. *Anolis porcatus*. Catalogue of American Amphibians and Reptiles 541: 1–5.

———. 1999a. Herpetology of Navassa Island. Caribbean Journal of Science 35: 1–13.

———. 1999b. *Anolis longiceps*. Catalogue of American Amphibians and Reptiles 693: 1–2.

———. 1999c. *Celestus badius*. Catalogue of American Amphibians and Reptiles 694: 1–2.
———. 1999d. *Leiocephalus barahonensis*. Catalogue of American Amphibians and Reptiles 695: 1–4.
———. 1999e. *Leiocephalus eremitus*. Catalogue of American Amphibians and Reptiles 696: 1–2.
———. 1999f. *Sphaerodactylus becki*. Catalogue of American Amphibians and Reptiles 697: 1–2.
———. 2000. *Cyclura onchiopsis*. Catalogue of American Amphibians and Reptiles 710: 1–3.
———. 2002a. Species profile: *Leiocephalus personatus*. Iguana Times (Journal of the International Iguana Society) 9: 80.
———. 2002b. Geographic distribution: *Anolis carolinensis*. Herpetological Review 33: 321.
———. 2002c. *Anolis aliniger*. Catalogue of American Amphibians and Reptiles 748: 1–3.
———. 2002d. *Tropidophis bucculentus*. Catalogue of American Amphibians and Reptiles 760: 1–3.
———. 2003a. Exploitation of reptiles in the West Indies: A long history. Iguana 10: 67–70.
———. 2003b. As good as it gets. In: R. W. Henderson and R. Powell (eds.), Islands and the Sea: Essays on Herpetological Exploration in the West Indies, pp. 169–183. Contributions to Herpetology, volume 20. Ithaca, N.Y.: Society for the Study of Amphibians and Reptiles.
———. 2004a. Lesser Antillean iguanas, *Iguana delicatissima* and *I. iguana*. Iguana 11: 37.
———. 2004b. Species profile: *Leiocephalus semilineatus*. Iguana 11: 97.
———. 2004c. Species profile: The ground lizards of Isla Mona: *Ameiva alboguttata*. Iguana 11: 107.
———. 2004d. Species profile: Saw-scaled Curlytail: *Leiocephalus carinatus*. Iguana 11: 153.
———. 2004e. Conservation of iguanas (*Iguana delicatissima* and *I. iguana*) in the Lesser Antilles. Iguana 11: 238–246.
———. 2006a. Lizard warfare. Iguana 13: 22–23.
———. 2006b. Conservation of the herpetofauna on the Dutch Windward Islands: St. Eustatius, Saba, and St. Maarten. Applied Herpetology 3: 293–306.
———. 2007a. Collecting animals from nature. Iguana 14: 47–51.
———. 2007b. Geographic Distribution. *Osteopilus septentrionalis* (Cuban Treefrog). Herpetological Review 38: 215.
Powell, R., and D. K. Carr. 1990. *Anolis whitemani*. Catalogue of American Amphibians and Reptiles 462: 1–2.
Powell, R., and E. J. Censky. 2002. *Ameiva alboguttata*. Arboreal activity. Herpetological Review 33: 50.
Powell, R., and R. E. Glor. 2000. *Cyclura stejnegeri*. Catalogue of American Amphibians and Reptiles 711: 1–4.
Powell, R., and R. W. Henderson. 1992. *Anolis gingivinus*. Nocturnal activity. Herpetological Review 23: 117.
———. 1994a. *Ialtris*. Catalogue of American Amphibians and Reptiles 590: 1–2.
———. 1994b. *Ialtris agyrtes*. Catalogue of American Amphibians and Reptiles 591: 1–2.
———. 1994c. *Ialtris parishi*. Catalogue of American Amphibians and Reptiles 593: 1–2.
———. 1998a. *Alsophis anomalus*. Catalogue of American Amphibians and Reptiles 659: 1–2.
———. 1998b. *Alsophis melanichnus*. Catalogue of American Amphibians and Reptiles 660: 1–2.
———. 1999. Addenda to the checklist of West Indian amphibians and reptiles. Herpetological Review 30: 137–139.
———. 2003a. A second set of addenda to the checklist of West Indian amphibians and reptiles. Herpetological Review 34: 341–345.
———. 2003b. The taxonomic and conservation status of giant Hispaniolan *Celestus* (Anguidae). Caribbean Journal of Science 39: 237–240.
———. 2005a. Conservation status of Lesser Antillean Reptiles. Iguana 12: 62–77.
———. 2005b. A new species of *Gonatodes* (Squamata: Gekkonidae) from the West Indies. Caribbean Journal of Science 41: 709–715.
———. 2006. Geographic distribution. *Sphaerodactylus kirbyi*. Herpetological Review 37: 495–496.
———. 2007. The St. Vincent (Lesser Antilles) herpetofauna: Conservation concerns. Applied Herpetology 4: 295–312.
———. 2008a. Avian predators of West Indian reptiles. Iguana 15: 8–11.
———. 2008b. *Sphaerodactylus cochranae*. Catalogue of American Amphibians and Reptiles 851: 1–2.
———. 2008c. Urban herpetology in the West Indies. In: J. C. Mitchell, R. E. Jung Brown, and B. Bartholomew (eds.), Urban Herpetology, pp. 389-404. Herpetological Conservation, volume 3. Salt Lake City, Utah: Society for the Study of Amphibians and Reptiles.
———. 2008d. Exploitation of the night-light niche by a Dominican Racer. Iguana 15: 156–157.
Powell, R., and S. J. Incháustegui. 2009. Conservation of the herpetofauna of the Dominican Republic. Applied Herpetology 6: 103–122.
Powell, R., and K. Lindsay. 1999. Geographic distribution. *Gymnophthalmus underwoodi*. Herpetological Review 30: 110.
Powell, R., and S. A. Maxey. 1990. *Hemidactylus brookii*. Catalogue of American Amphibians and Reptiles 493: 1–3.
Powell, R., and J. A. Neland. 1998. *Anolis eugenegrahami*. Catalogue of American Amphibians and Reptiles 672: 1–2.
Powell, R. and J. S. Parmerlee, Jr. 1991. *Anolis chlorocyanus*. Behavior. Herpetological Review 22: 130.
———. 1992. In the spotlight: *Aristelliger lar* Cope, 1862. Dactylus 1(3): 37–38.
———. 1993a. *Anolis chlorocyanus*. Aquatic activity. Herpetological Review 24: 59.
———. 1993b. In the spotlight: *Hemidactylus haitianus* Meer-

warth 1901, an endemic West Indian house gecko. Dactylus 2: 54–55.

Powell, R., and G. K. Pregill. 1991. *Peltophryne fluviatica*. Catalogue of American Amphibians and Reptiles 507: 1–2.

Powell, R., and R. D. Wittenberg. 1998. *Bothrops caribbaeus*. Catalogue of American Amphibians and Reptiles 676: 1–4.

Powell, R., S. S. Duer, J. S. Parmerlee, Jr., and D. D. Smith. 1989a. *Ameiva taeniura*. Food habits. Herpetological Review 20: 49.

Powell, R., J. S. Parmerlee, Jr., P. Ward, and D. D. Smith. 1989b. Geographic distribution: *Anolis whitemani whitemani*. Herpetological Review 20: 12.

Powell, R., S. G. Thornhill, and D. D. Smith. 1989c. Geographic Distribution: *Sphaerodactylus a. altavelensis* (NCN). Herpetological Review 20: 13.

Powell, R., P. J. Hall, J. H. Greve, and D. D. Smith. 1990a. Occurrence of *Trichospirura teixeirai* (Spiurida: Rhabdochonidae) in *Hemidactylus brookii haitianus* (Sauria: Gekkonidae) from Hispaniola. Journal of the Helminthological Society of Washington 57: 74–75.

Powell, R., P. J. Hall, and J. H. Greve. 1990b. Occurrence of *Skrjabinoptera leiocephalorum* (Spirurida: Physalopteridae) in *Leiocephalus* spp. (Sauria: Iguanidae) from Hispaniola. Journal of the Helminthological Society of Washington 57: 75–77.

Powell, R., J. S. Parmerlee, Jr., M. A. Rice, and D. D. Smith. 1990c. Ecological observations of *Hemidactylus brookii haitianus* Meerwarth (Sauria: Gekkonidae) from Hispaniola. Caribbean Journal of Science 26: 67–70.

Powell, R., D. J. Pflanz, J. H. Greve, and D. D. Smith. 1990d. *Leiocephalus semilineatus*. Parasitism. Herpetological Review 21: 60–61.

Powell, R., D. D. Smith, J. S. Parmerlee, Jr., C. V. Taylor, Jr., and M. L. Jolley. 1990e. Range expansion by an introduced anole: *Anolis porcatus* in the Dominican Republic. Amphibia-Reptilia 11: 421–425.

Powell, R., S. A. Maxey, and J. H. Greve. 1990f. The occurrence of *Terranova caballeroi* (Nematoda: Ascarididae) in *Antillophis parvifrons protenus* (Serpentes: Colubridae) from Hispaniola. Caribbean Journal of Science 26:72.

Powell, R., S. A. Maxey, J. S Parmerlee, Jr., and D. D. Smith. 1991. Notes on the reproductive biology of a montane population of *Antillophis parvifrons protenus* (Serpentes: Colubridae) from the Dominican Republic. Journal of Herpetology 25: 121–122.

Powell, R., R. J. Passaro, and R. W. Henderson. 1992. Noteworthy herpetological records from Saint [sic] Maarten, Netherlands Antilles. Caribbean Journal of Science 28: 234–235.

Powell, R., P. J. Hall, D. D. Smith, and J. Riley. 1993. The occurrence of *Raillietiella* sp. (Pentastomida: Cephalobaenida) in *Hemidactylus haitianus* (Lacertilia: Gekkonidae) from Hispaniola. Dactylus 2: 51–53.

Powell, R., R. W. Henderson, K. Adler, and H. A. Dundee. 1996a. An annotated checklist of West Indian amphibians and reptiles. In: R. Powell and R. W. Henderson (eds.), Contributions to West Indian Herpetology: A Tribute to Albert Schwartz, pp. 51–93. Contributions in Herpetology, volume 12. Ithaca, N.Y.: Society for the Study of Amphibians and Reptiles.

Powell, R., J. S. Parmerlee, Jr., and D. D. Smith. 1996b. Evidence of spatial niche partitioning by a Hispaniolan lizard community in a xeric habitat. In: R. Powell and R. W. Henderson (eds.), Contributions to West Indian Herpetology: A Tribute to Albert Schwartz, pp. 317–326. Contributions to Herpetology, volume 12. Ithaca, N.Y.: Society for the Study of Amphibians and Reptiles.

Powell, R., J. H. Greve, and A. K. Howard. 1998a. Hispaniolan *Eleutherodactylus* (Anura: Leptodactylidae) as hosts of immature *Terranova* (Nematoda: Ascarididae), with notes on additional nematodes. Caribbean Journal of Science 34: 155–157.

Powell, R., R. I. Crombie, and H. E. A. Boos. 1998b. *Hemidactylus mabouia*. Catalogue of American Amphibians and Reptiles 674: 1–11.

Powell, R., J. A. Ottenwalder, and S. J. Incháustegui. 1999. The Hispaniolan herpetofauna: Diversity, endemism, and historical perspectives, with comments on Navassa Island. In: B. I. Crother (ed.), Caribbean Amphibians and Reptiles, pp. 93–168. San Diego: Academic Press.

Powell, R., J. A. Ottenwalder, S. J. Incháustegui, R. W. Henderson, and R. E. Glor. 2000. Amphibians and reptiles of the Dominican Republic: Species of special concern. Oryx 34: 118–128.

Powell, R., D. M. Nieves, M. E. Gifford, and B. E. Fontenot. 2002. The "Rhino Factory" at Manati Park. Iguana Times (Journal of the International Iguana Society) 9: 75–77, 79, 81, back cover.

Powell, R., R. W. Henderson, and J. S. Parmerlee, Jr. 2005. Reptiles and Amphibians of the Dutch Caribbean: St. Eustatius, Saba, and St. Maarten. Gallows Bay, St. Eustatius, Netherlands Antilles: St. Eustatius National Parks Foundation.

Powell, R., G. Perry, R. W. Henderson, and A. Barun. 2006. *Alsophis portoricensis anegadae* (NCN). Aquatic activity. Herpetological Review 37: 228–229.

Powell, S. B., M. L. Treglia, R. W. Henderson, and R. Powell. 2007. Treeboas in the West Indies: Responses of *Corallus cookii* and *C. grenadensis* to disturbed habitats, pp. 375–386. In: R. W. Henderson and R. Powell (eds.), Biology of the Boas and Pythons. Eagle Mountain, Utah: Eagle Mountain Publishing.

Pregill, G. K. 1992. Systematics of the West Indian lizard genus *Leiocephalus* (Squamata: Iguania: Tropiduridae). University of Kansas Museum of Natural History Miscellaneous Publication 84: 1–69.

Pregill, G. K., D. W. Steadman, and D. R. Watters. 1994.

Late Quaternary vertebrate faunas of the Lesser Antilles: Historical components of Caribbean biogeography. Bulletin of the Carnegie Museum of Natural History 30: iv + 51 pp.

Price, R. M., and P. Russo. 1991. Revisionary comments on the genus *Boa* with the description of a new subspecies of *Boa constrictor* from Peru. The Snake 23: 29–35.

Prior, K. A., and R. C. Gibson. 1997. Observations on the foraging behavior of the Jamaican Boa, *Epicrates subflavus*. Herpetological Review 28: 72–73.

Pritchard, P. C. H., and P. Trebbau. 1984. The Turtles of Venezuela. Contributions to Herpetology, volume 2. Ithaca, N.Y.: Society for the Study of Amphibians and Reptiles.

Proctor, G. R. 1950. Results of the Catherwood-Chaplin West Indies Expedition, 1948. Part I. Plants of Cayo Largo (Cuba), San Andrés and Providencia. Proceedings of the Academy of Natural Sciences of Philadelphia 102: 27–42, Plate 1, Maps 1–2.

Puente-Rolón, A. R. 2001. *Arrhyton exiguum* (Puerto Rican Garden Snake). Diet. Herpetological Review 32: 261.

Puente-Rolón, A. R., and F. J. Bird-Picó. 2004. Foraging behavior, home range, movements and activity patterns of *Epicrates inornatus* (Boidae) at Mata de Plátano Reserve in Arecibo, Puerto Rico. Caribbean Journal of Science 40: 343–352.

Queral, A., R. Garcia, R. Powell, J. S. Parmerlee, Jr., D. D. Smith, and A. Lathrop. 1995. Agonistic responses by a grass anole, *Anolis olssoni* from the Dominican Republic, to male conspecifics. Amphibia-Reptilia 16: 313–321.

Quesnel, V. C., and J. S. Kenny. 1959. The anoles of the eastern Caribbean (Sauria, Iguanidae). Pt. II. Two sibling species in Trinidad. Bulletin of the Museum of Comparative Zoology 121: 189–191.

Quick, B. 2007. A field guide to the herpetofauna on Dominica, W.I. Unpublished report. Texas A&M University Study Abroad, College Station. http://dominica.tamu.edu/student%20projects/projects_by_year.htm.

Rabb, G. B., and E. B. Hayden, Jr. 1957. The Van Voast-American Museum of Natural History Bahama Islands Expedition. Record of the expedition and general features of the islands. American Museum Novitates 1836: 1–53.

Raetzsch, C. F. 2004. Size and morphology of *Anolis oculatus* along an elevation gradient at selected areas of Dominica. Unpublished report. Texas A&M University Study Abroad, College Station. http://dominica.tamu.edu/student%20projects/projects_by_year.htm.

Ramer, J. 2004. A survey of Ricord's Iguanas (*Cyclura ricordii*) and Rhinoceros Iguanas (*Cyclura cornuta cornuta*) in Isla Cabritos National Park, Dominican Republic 2003: A preliminary report. Iguana 11: 88–95.

Ramos, Y. M., and R. Powell. 2001a. *Anolis chlorocyanus*. Catalogue of American Amphibians and Reptiles 728: 1–6.

———. 2001b. *Anolis coelestinus*. Catalogue of American Amphibians and Reptiles 729: 1–5.

Ramos Targarona, R. 1998. Manejo en cautiverio en el zoocriadero de la Ciénaga de Zapata. Flora y Fauna 2: 10–15.

Ramos Targarona, R., V. de Buffrenil, and J. P. Ross. 1994. Current status of the Cuban Crocodile, *Crocodylus rhombifer*, in the wild. In: Crocodiles. Proceedings of the 12th Working Meeting of the IUCN/SSC Crocodilian Specialist Group. Vol. 1, pp. 113–139. Gland, Switzerland: International Union for the Conservation of Nature and Natural Resources.

Ramos Targarona, R., R. Rodríguez Soberón, L. Cotayo, M. Alonso Tabet, and J. Thorbjarnarson. 2008. *Crocodylus rhombifer*. In: IUCN 2008. 2008 IUCN Red List of Threatened Species. http://www.iucnredlist.org.

Rand, A. S. 1961. Notes on Hispaniolan herpetology. 4. *Anolis koopmani*, new species, from the southwestern peninsula of Haiti. Breviora 137: 1–4.

———. 1962. Notes on Hispaniolan herpetology. 5. The natural history of three sympatric species of *Anolis*. Breviora 154: 1–15.

———. 1964a. Ecological distribution in anoline lizards of Puerto Rico. Ecology 45: 745–752.

———. 1964b. Inverse relationship between temperature and shyness in the lizard *Anolis lineatopus*. Ecology 45: 863–864.

———. 1967a. Ecology and social organization in the iguanid lizard *Anolis lineatopus*. Proceedings of the United States National Museum 122: 1–79.

———. 1967b. The ecological distribution of the anoline lizards around Kingston, Jamaica. Breviora 272: 1–18.

———. 1967c. Communal egg laying in anoline lizards. Herpetologica 23: 227–230.

Rand, A. S., and R. Andrews. 1975. Adult color dimorphism and juvenile pattern in *Anolis cuvieri*. Journal of Herpetology 9: 257–260.

Rand, A. S., and E. E. Williams. 1969. The anoles of La Palma: Aspects of their ecological relationships. Breviora 327: 1–18.

Read, C. P., and Y. U. Amrein. 1953. North American nematodes of the genus *Pharyngodon* Diesing (Oxyuridea). Journal of Parasitology 39: 365–370.

Reagan, D. P. 1984. Ecology of the Puerto Rican Boa (*Epicrates inornatus*) in the Luquillo Mountains of Puerto Rico. Caribbean Journal of Science 20: 119–127.

———. 1986. Foraging behavior of *Anolis stratulus* in a Puerto Rican rain forest. Biotropica 18: 157–160.

———. 1991. The response of *Anolis* lizards to hurricane-induced habitat changes in a Puerto Rican rain forest. Biotropica 23: 468–474.

———. 1992. Congeneric species distribution and abundance in a three-dimensional habitat: The rain forest anoles of Puerto Rico. Copeia 1992: 392–403.

———. 1995. Lizard ecology in the canopy of an island rain forest. In: M. D. Lowman and N. Nadkarni (eds.), Forest Canopies, pp. 149–164. San Diego: Academic Press.

———. 1996a. Anoline lizards. In: D. P. Reagan and R. B. Waide (eds.), The Food Web of a Tropical Rain Forest, pp. 321–345. Chicago: University of Chicago Press.

———. 1996b. The role of amphibians and reptiles in a West Indian rain forest food web. In: R. Powell and R. W. Henderson (eds.), Contributions to West Indian Herpetology: A Tribute to Albert Schwartz, pp. 217–227. Contributions to Herpetology, volume 12. Ithaca, N.Y.: Society for the study of Amphibians and Reptiles.

Reagan, D. P., and R. B. Waide (eds.). 1996. The Food Web of a Tropical Rain Forest. Chicago: University of Chicago Press.

Reagan, D. P., G. R. Camilo, and R. B. Waide. 1996. The community food web: Major properties and patterns of organization. In: D. P. Reagan and R. B. Waide (eds.), The Food Web of a Tropical Rain Forest, pp. 461–510. Chicago: University of Chicago Press.

Regal, P. J. 1978. Behavioral differences between reptiles and mammals: An analysis of activity and mental capabilities. In: N. Greenberg and P. D. MacLean (eds.), Behavior and Neurology of Lizards: An Interdisciplinary Colloquium, pp. 183–202. Rockville, Md.: National Institute of Mental Health.

Regalado, R. 1997. Social behavior of the Ashy Gecko (*Sphaerodactylus elegans* Macleay): Repertoire and sex recognition. Herpetological Natural History 5: 41–52.

———. 1998. Approach distance and escape behavior of three species of Cuban *Anolis* (Squamata: Polychrotidae). Caribbean Journal of Science 34: 211–217.

———. 2003. Social behavior and sex recognition in the Puerto Rican dwarf gecko *Sphaerodactylus nicholsi*. Caribbean Journal of Science 39: 77–93.

———. 2006. Reproduction and growth of seven species of dwarf gecko, *Sphaerodactylus* (Gekkonidae), in captivity. Herpetological Review 37: 13–20.

Regalado, R., and O. H. Garrido. 1993. Diferencias en el comportamiento social de dos especies gemelas de anolinos cubanos. Caribbean Journal of Science 29: 18–23.

Rehák, I. 1987. Color change in the snake *Tropidophis feicki* (Reptilia: Squamata: Tropidophiidae). Vestník Ceskoslovenské Spolecnosti Zoologické 51: 300–303.

Reichling, S. B. 1991. The serpents of St. Lucia, *Bothrops caribbaeus* and *Boa constrictor orophias*. Vivarium 3(3): 7–9, 22

———. 2000. The status of the Lesser Antillean Iguana on Sint Eustatius. Iguana Times (Journal of the International Iguana Society) 8(1): 3–6.

Reinhardt, J., and C. F. Lütken. 1862. Bidrag til vestindiske Öriges og navnligen til de danskvestindiske Öers Herpetologie. Videns Meddelelser fra den naturhistoriske Forening Kjöbenhavn 10–18: 153–291.

Reisinger, M. 2003. Zur Haltung und Zucht von *Anolis baracoae* Schwartz 1964. Iguana-Rundschreiben 16: 23–27.

Reubold, E. 1983. Antillen-Pfeiffrösche der Gattung *Eleutherodactylus*. Die Aquarien- und Terrarienzeitschrift (DATZ) 36: 113–116.

Reyes Campos, N. B. 1971. Observaciones sobre la conducta del coquí, *Eleutherodactylus coqui*. Caribbean Journal of Science 11: 209–210.

Reynolds, R. G. 2009. *Leiocephalus psammodromus* (Turks and Caicos Curly-tailed Lizard). Conspecific necrophagy. Herpetological Review 40: 88–89.

Richmond, N. D. 1955. The blind snakes (*Typhlops*) of Bimini, Bahama Islands, British West Indies, with description of a new species. American Museum Novitates 1734: 1–7.

Riley, J., R. Powell, and D. D. Smith. 1991. Further observations of blunt-hooked pentastomids belonging to the genus *Raillietiella* Sambon, 1910 infecting *Hemidactylus brookii* (Sauria: Gekkonidae) in Africa and the Caribbean: Comparison with closely related *Raillietiella* spp. from an African skink (*Mabuya perrotetii*). Systematic Parasitology 20: 47–57.

Rios-López, N. 1999. Geographic distribution. *Scinax rubra*. Herpetological Review 30: 232.

———. 2004. *Anolis stratulus* (Saddled Anole). Extrafloral herbivory. Herpetological Review 35: 386.

———. 2008a. Effects of increased salinity on tadpoles of two anurans from a Caribbean coastal wetland in relation to their natural abundance. Amphibia-Reptilia 29: 7–18.

———. 2008b. *Leptodactylus albilabris*. In: IUCN 2008. 2008 IUCN Red List of Threatened Species. http://www.iucnredlist.org.

———. 2008c. *Eleutherodactylus brittoni*. In: IUCN 2008. 2008 IUCN Red List of Threatened Species. http://www.iucnredlist.org.

———. 2008d. *Eleutherodactylus cochranae*. In: IUCN 2008. 2008 IUCN Red List of Threatened Species. http://www.iucnredlist.org.

———. 2008e. *Eleutherodactylus cooki*. In: IUCN 2008. 2008 IUCN Red List of Threatened Species. http://www.iucnredlist.org.

———. 2008f. *Eleutherodactylus coqui*. In: IUCN 2008. 2008 IUCN Red List of Threatened Species. http://www.iucnredlist.org.

———. 2008g. *Eleutherodactylus juanariveroi*. In: IUCN 2008. 2008 IUCN Red List of Threatened Species. http://www.iucnredlist.org.

———. 2008h. *Eleutherodactylus unicolor*. In: IUCN 2008. 2008 IUCN Red List of Threatened Species. http://www.iucnredlist.org.

Rios-López, N., and T. M. Aide. 2007. Herpetofaunal dynamics during secondary succession. Herpetologica 63: 35–50.

Rios-López, N., and R. L. Joglar. 1999. Geographic distribution. *Rana grylio*. Herpetological Review 30: 231–232.

Rios-López, N., and A. R. Puente-Colón. 2007. *Anolis cuvieri* (Puerto Rican Giant Anole). Reproductive ecology. Herpetological Review 38: 73–75.

Rios-López, N., and R. Thomas. 2007. A new species of palustrine *Eleutherodactylus* (Anura: Leptodactylidae) from Puerto Rico. Zootaxa 1512: 51–64.

Rivalta G., V. 2008. *Peltophryne florentinoi*. In: IUCN 2008. 2008 IUCN Red List of Threatened Species. http://www.iucnredlist.org.

Rivalta G., V., A. Chamizo L., L. V. Moreno G., A. Sampedro M., and A. Torres B. 2003a. Mitos, creencias populares y usos. In: L. Rodríguez Schettino (ed.), Anfibios y Reptiles de Cuba, pp. 144–155. Vaasa, Finland: UPC Print.

Rivalta G., V., A. Sampedro M., and L. V. Moreno G. 2003b. Misteriosos pobladores de la naturaleza. In: L. Rodríguez Schettino (ed.), Anfibios y Reptiles de Cuba, pp. 56–63. Vaasa, Finland: UPC Print.

Rivalta G., V., L. Rodríguez Schettino, A. Torres Barboza, M. Martínez Reyes, and J. Larramendi Joa. 2004. *Bufo peltocephalus* (Cuban Toad). Diurnal retreat. Herpetological Review 35: 157.

Rivera-Vélez, N., and A. R. Lewis. 1994. Threshold temperatures and the thermal cycle of a heliothermic lizard. Journal of Herpetology 28: 1–6.

Rivero, J. A. 1959. Two new species of *Eleutherodactylus* from Puerto Rico. Breviora 185: 1–6.

———. 1963. *Eleutherodactylus hedricki*, a new species of frog from Puerto Rico (Salientia, Leptodactylidae). Breviora 185: 1–7.

———. 1998. Los Anfibios y Reptiles de Puerto Rico, segunda edición revisada. The Amphibians and Reptiles of Puerto Rico, second edition revised. San Juan: University of Puerto Rico.

Rivero, J. A., and R. Joglar. 1979. *Eleutherodactylus cochranae*. Herpetological Review 10: 101.

Rivero, J. A., and D. Seguí Crespo. 1992. Anfibios y Reptiles en Nuestro Folklore. Quebradillas, Puerto Rico: Imprenta San Rafael.

Rivero, J. A., J. Maldonado, and H. Mayorga. 1963. On the habits and food of *Eleutherodactylus karlschmidti* Grant. Caribbean Journal of Science 3: 25–27.

Rivero, J. A., H. Mayorga, E. Estremera, and I. Izquierdo. 1980. Sobre el *Bufo lemur*. Caribbean Journal of Science 15: 33–40.

Rivero, J. A., R. L. Joglar, and I. Vazquez. 1982. Cinco nuevos ejemplares del culebrón de Mona, *Epicrates m. monensis* Ophidia, Boidae). Caribbean Journal of Science 17: 7–14.

Robinson, L. E. 1926. The genus *Amblyomma*. In: G. H. F. Nuttall, C. Warburton, and L. E. Robinson (eds.), Ticks. A Monograph of the Ixodoidea, Part 4, pp. 155–188. Cambridge: Cambridge University Press.

Rodda, G. H., and J. A. Savidge. 2007. Biology and impacts of Pacific island invasive species. 2. *Boiga irregularis*, the Brown Tree Snake (Reptilia: Colubridae). Pacific Science 61: 307–324.

Rodda, G. H., G. Perry, R. Rondeau, and J. Lazell. 2001. The densest terrestrial vertebrate. Journal of Tropical Ecology 17: 331–338.

Rodríguez, A., and R. Alonso. 2003a. Ranas arborícolas. In: L. Rodríguez Schettino (ed.), Anfibios y Reptiles de Cuba, pp. 10–19. Vaasa, Finland: UPC Print.

———. 2003b. Ranas diminutas. In: L. Rodríguez Schettino (ed.), Anfibios y Reptiles de Cuba, pp. 50–55. Vaasa, Finland: UPC Print.

Rodríguez, G. A., and D. P. Reagan. 1984. Bat predation by the Puerto Rican Boa, *Epicrates inornatus*. Copeia 1984: 219–220.

Rodríguez-Durán, A. 1996. Foraging ecology of the Puerto Rican Boa (*Epicrates inornatus*): Bat predation, carrion feeding, and piracy. Journal of Herpetology 30: 533–536.

Rodríguez Gómez, A., and R. Alonso Bosch. 2000. Threatened amphibians of Cuba. Froglog 37: 5–6.

Rodríguez-Ramirez, J., and A. R. Lewis. 1991. Reproduction in the Puerto Rican teiids *Ameiva exsul* and *A. wetmorei*. Herpetologica 47: 395–403.

Rodríguez-Robles, J. A. 1992. Notes on the feeding behavior of the Puerto Rican Racer, *Alsophis portoricensis* (Serpentes: Colubridae). Journal of Herpetology 26: 100–102.

———. 2005. La culebra común (*Alsophis portoricensis*). In: R. L. Joglar (ed.), Biodiversidad de Puerto Rico. Vertebrados Terrestres y Ecosistemas, pp. 169–180. Serie de Historia Natural. San Juan: Editorial Instituto de Cultura Puertorriqueña.

Rodríguez-Robles, J. A., and H. W. Greene. 1996. Ecological patterns in Greater Antillean macrostomatan snake assemblages, with comments on body-size evolution in *Epicrates* (Boidae). In: R. Powell and R. W. Henderson (eds.), Contributions to West Indian Herpetology: A Tribute to Albert Schwartz, pp. 339–357. Contributions to Herpetology, volume 12. Ithaca, N.Y.: Society for the Study of Amphibians and Reptiles.

Rodríguez-Robles, J. A., and M. Leal. 1993a. *Alsophis portoricensis* (Puerto Rican Racer). Diet. Herpetological Review 24: 150–151.

———. 1993b. Effects of prey type on the feeding behavior of *Alsophis portoricensis* (Serpentes: Colubridae). Journal of Herpetology 27: 163–168.

Rodríguez-Robles, J. A., and R. Thomas. 1992. Venom function in the Puerto Rican Racer, *Alsophis portoricensis* (Serpentes: Colubridae). Copeia 1992: 62–68.

Rodríguez-Robles, J. A., M. Leal, and J. B. Losos. 2005. Habitat selection by the Puerto Rican Yellow-chinned Anole, *Anolis gundlachi*. Canadian Journal of Zoology 83: 983–988.

Rodríguez-Robles, J. A., T. Jezkova, and M. A. García. 2007. Evolutionary relationships and biogeography of *Anolis desechensis* and *Anolis monensis*, two lizards endemic to small islands in the eastern Caribbean Sea. Journal of Biogeography 34: 1546–1558.

Rodríguez-Robles, J. A., T. Jezkova, and M. Leal. 2008. Genetic structuring in the threatened "Lagartijo del Bosque Seco" (*Anolis cooki*) from Puerto Rico. Molecular Phylogenetics and Evolution 46: 503–514.

Rodríguez Schettino, L. 1999a. Introduction. In: L. Rodríguez Schettino (ed.), The Iguanid Lizards of Cuba, pp. 1–16. Gainesville: University Press of Florida.

———. 1999b. Systematic accounts of the species. In: L. Rodríguez Schettino (ed.), The Iguanid Lizards of Cuba, pp. 104–380. Gainesville: University Press of Florida.

———. 2001. *Anolis loysianus* (NCN). Diet. Herpetological Review 32: 252–253.

———. 2003a. Anfibios y Reptiles de Cuba. Vaasa, Finland: UPC Print.

———. 2003b. My love affair with Cuban reptiles. In: R. W. Henderson and R. Powell (eds.), Islands and the Sea: Essays on Herpetological Exploration in the West Indies, pp. 93–99. Contributions to Herpetology, volume 20. Ithaca, N.Y.: Society for the Study of Amphibians and Reptiles.

Rodríguez Schettino, L., and A. R. Chamizo Lara. 1998a. Reptiles cubanos con algún grado de amenaza de extinción. Poeyana 463: 1–8.

———. 1998b. *Anolis vermiculatus*. In: E. Pérez, E. Osa, Y. Matamoros, and U. S. Seal (eds.), Report of Conservation Assessment and Management Plan Workshop for Selected Cuban Species II. Sección III. Reptiles. Apple Valley, Minn.: IUCN/SSC Conservation Breeding Specialist Group.

———. 2001. *Anolis mestrei* (NCN). Thermal habitat. Herpetological Review 32: 253.

Rodríguez Schettino, L., and K. de Queiroz. 2002a. *Anolis homolechis homolechis* (NCN). Thermal biology and microhabitat. Herpetological Review 33: 304.

———. 2002b. *Anolis sagrei sagrei* (Cuban Brown Lizard; Chino; Torito). Habitat use and thermal biology. Herpetological Review 33: 305.

Rodríguez Schettino, L., and M. Lizana. 1997. Historia natural del Lagarto Caimán Cubano, *Anolis vermiculatus* (Iguania: Polychridae). Boletín de la Asociación Herpetológica Española 8: 23–26.

Rodríguez Schettino, L., and M. Martínez Reyes. 1992. Hábitos alimentarios de *Anolis bartschi* (Sauria: Iguanidae) en San Vicente, Pinar del Río, Cuba. Ciencias Biológicas 23: 30–40.

———. 1994. Características tróficas de una población de *Anolis lucius* (Iguania: Polychridae) en la costa septentrional de Cuba. Avicennia 1: 67–77.

———. 1996. Algunos aspectas de la ecología trófica de *Anolis argenteolus* (Sauria: Polychridae) en una población de la costa suroriental de Cuba. Biotropica 28: 252–257.

Rodríguez Schettino, L., and J. Novo Rodríguez. 1985. Nuevos datos etoecológicos sobre *Anolis vermiculatus* (Sauria: Iguanidae). Poeyana 296: 1–11.

Rodríguez Schettino, L., and M. de J. Valderrama Puente. 1986. Algunos aspectos del nicho estructural y climático de *Anolis lucius* (Sauria: Iguanidae). Poeyana 319: 1–12.

Rodríguez Schettino, L., D. L. Marcellini, and J. Novo Rodríguez. 1987. Algunos aspectos ecológicos sobre *Anolis vermiculatus* (Sauria: Iguanidae). Poeyana 343: 1–9.

Rodríguez Schettino, L., M. Martínez Reyes, and L. V. Moreno García. 1999. Ecology and behavior. In: L. Rodríguez Schettino (ed.), The Iguanid Lizards of Cuba, pp. 36–58. Gainesville: University Press of Florida.

Rodríguez Schettino, L., A. Torres Barboza, and A. Hernández Marrero. 2003. Trepadores excelentes. In: L. Rodríguez Schettino (ed.), Anfibios y Reptiles de Cuba, pp. 110–127. Vaasa, Finland: UPC Print.

Rodríguez Soberón, R. 1998a. *Crocodylus rhombifer*. In: E. Pérez, E. Osa, Y. Matamoros, and U. S. Seal (eds.), Report of Conservation Assessment and Management Plan Workshop for Selected Cuban Species II. Sección III. Reptiles. Apple Valley, Minn.: IUCN/SSC Conservation Breeding Specialist Group.

———. 1998b. *Crocodylus acutus*. In: E. Pérez, E. Osa, Y. Matamoros, and U. S. Seal (eds.), Report of Conservation Assessment and Management Plan Workshop for Selected Cuban Species II. Sección III. Reptiles. Apple Valley, Minn.: IUCN/SSC Conservation Breeding Specialist Group.

Rogner, M. 2000. Haltung, Pflege und Vermehrung von *Anolis cybotes*. Draco 1: 50–53.

———. 2002. Kuba: Auf der Suche nach *Anolis smallwoodi*. Reptilia 7: 46–50.

———. 2007. *Anolis*-Haltung, Pflege und Vermehrung. Terraria 2: 4–13.

Rogowitz, G. L. 1996. Evaluation of thermal acclimation and altitudinal variation of metabolism in a Neotropical lizard, *Anolis gundlachi*. Copeia 1996: 535–542.

Rojer, A. 1997a. Biological inventory of Saba. Unpublished Report 96-10. Curaçao, Netherlands Antilles: Carmabi Foundation.

———. 1997b. Biological inventory of St. Eustatius. Unpublished Report 96-10. Curaçao, Netherlands Antilles: Carmabi Foundation.

Röll, B. 2006. Kugelfingergeckos *Sphaerodactylus elegans* und *Sphaerodactylus torrei*. Münster, Germany: Natur und Tier.

———. 2009. Neotropische Taggeckos. Münster, Germany: Natur und Tier.

Rolle, F. J. 1963. Life history of the Red-legged Thrush (*Mimocichla plumbea ardodiacea*) in Puerto Rico. Studies on the Fauna of Curaçao and Other Caribbean Islands 14: 1–40.

Roming, U. 1993. Zur Biologie und Haltung des Kuba-Laubfrosches—ein Zuchtbericht. Elaphe (neue Folge) 1(2): 16–18.

Rose, B. 1982. Food intake and reproduction in *Anolis acutus*. Copeia 1982: 322–330.

Ross, F. D. 1998. *Crocodylus rhombifer*. Catalogue of American Amphibians and Reptiles 680: 1–18.

Rothblum, L. M., J. W. Watkins, and T. A. Jenssen. 1979. A learning paradigm and the behavioral demonstration of audition for the lizard, *Anolis grahami*. Copeia 1979: 490–494.

Roughgarden, J. 1974. Niche width: Biogeographic patterns among *Anolis* lizard populations. American Naturalist 108: 429–442.

———. 1986. A comparison of food-limited and space-limited animal competition communities. In: J. Diamond and T. J. Case (eds.), Community Ecology, pp. 492–516. New York: Harper & Row.

———. 1995. *Anolis* Lizards of the Caribbean: Ecology, Evolution, and Plate Tectonics. Oxford, UK: Oxford University Press.

Roughgarden, J., and E. Fuentes. 1977. The environmental determinants of size in solitary populations of West Indian *Anolis* lizards. Oikos 29: 44–51.

Roughgarden, J., W. Porter, and D. Heckel. 1981. Resource partitioning of space and its relationship to body temperature in *Anolis* lizard populations. Oecologia 50: 256–264.

Roughgarden, J., D. Heckel, and E. R. Fuentes. 1983a. Coevolutionary theory and the biogeography and community structure of *Anolis*. In: R. B. Huey, E. R. Pianka, and T. W. Schoener (eds.), Lizard Ecology: Studies of a Model Organism, pp. 371–410. Cambridge, Mass.: Harvard University Press.

Roughgarden, J., J. Rummel, and S. Pacala. 1983b. Experimental evidence of strong present-day competition between *Anolis* populations of the Anguilla Bank—a preliminary report. In: A. G. J. Rhodin and K. Miyata (eds.), Advances in Herpetology and Evolutionary Biology: Essays in Honor of Ernest E. Williams, pp. 499–506. Cambridge, Mass.: Museum of Comparative Zoology, Harvard University.

Roughgarden, J., S. Pacala, and J. Rummel. 1984. Strong present-day competition between the *Anolis* lizard populations of St. Maarten (Neth. Antilles). In: B. Shorrocks (ed.), Evolutionary Ecology, pp. 203–220. Oxford: Blackwell Science.

Roze, J. A. 1966. La Taxonomía y Zoogeografía de los Ofidios en Venezuela. Caracas: Ediciones de la Biblioteca de la Universidad Central de Venezuela.

Ruder, J. M., R. Powell, and J. S. Parmerlee, Jr. 1995. *Eleutherodactylus rufifemoralis*. Catalogue of American Amphibians and Reptiles 601: 1–2.

Rudman, S. M., R. Powell, and J. S. Parmerlee, Jr. 2009. *Ameiva fuscata* (Dominican Ground Lizard). Arboreal activity and diet. Herpetological Review 40: 219.

Rufz, E. 1859. Enquête suer le Serpent de la Martinique. Paris: Chez Germer Bailliere.

Ruibal, R. 1946. A new *Sphaerodactylus* from the Dominican Republic. American Museum Novitates 1308: 1–4.

———. 1957. A preliminary investigation of the ecology and taxonomy of Cuban lizards. Yearbook of the American Philosophical Society 1957: 256–258.

———. 1961. Thermal relations of five species of tropical lizards. Evolution 15: 98–111.

———. 1964. An annotated checklist and key to the anoline lizards of Cuba. Bulletin of the Museum of Comparative Zoology 130: 475–520.

———. 1967. Evolution and behavior in West Indian anoles. In: W. W. Milstead (ed.), Lizard Ecology, A Symposium, pp. 116–140. Columbia: University of Missouri Press.

Ruibal, R., and R. Philibosian. 1970. Eurythermy and niche expansion in lizards. Copeia 1970: 645–653.

———. 1974a. Aggression in the lizard *Anolis acutus*. Copeia 1974: 349–357.

———. 1974b. The population ecology of the lizard *Anolis acutus*. Ecology 55: 525–537.

Ruibal, R., and E. E. Williams. 1961a. Two sympatric Cuban anoles of the *carolinensis* group. Bulletin of the Museum of Comparative Zoology 125: 183–208.

———. 1961b. The taxonomy of the *Anolis homolechis* complex of Cuba. Bulletin of the Museum of Comparative Zoology 125: 211–246.

Ruibal, R., R. Philbosian, and J. L. Adkins. 1972. Reproductive cycle and growth in the lizard *Anolis acutus*. Copeia 1972: 509–518.

Ruiz García, F. N. 1975. Observaciones ecológicas sobre *Anolis homolechis* (Lacertilia: Iguanidae) en Cuba. Miscelanea Zoológica 1: 4.

———. 1977. Observaciones sobre *Anolis sagrei* Duméril et Bibron (Lacertilia: Iguanidae) en cautiverio. Miscelanea Zoológica 6: 2.

———. 1987. Anfibios de Cuba. Habana, Cuba: Editorial Gente Nueva.

Ruiz Urquiola, A., O. H. Garrido, T. González Blanco, G. Espinosa López, and M. E. Ibarra Martín. 1998a. *Chamaeleolis barbatus*. In: E. Pérez, E. Osa, Y. Matamoros, and U. S. Seal (eds.), Report of Conservation Assessment and Management Plan Workshop for Selected Cuban Species II. Sección III. Reptiles. Apple Valley, Minn.: IUCN/SSC Conservation Breeding Specialist Group.

———. 1998b. *Chamaeleolis guamuhaya*. In: E. Pérez, E. Osa, Y. Matamoros, and U. S. Seal (eds.), Report of Conservation Assessment and Management Plan Workshop for Selected Cuban Species II. Sección III. Reptiles. Apple Valley, Minn.: IUCN/SSC Conservation Breeding Specialist Group.

Rummel, J. D., and J. Roughgarden. 1985. Effects of reduced perch-height separation on competition between two *Anolis* lizards. Ecology 66: 430–444.

Rupp, E., S. Incháustegui, and Y. Arias. 2005. Conservation of *Cyclura ricordii* in the southwestern Dominican Republic and a brief history of the Grupo Jaragua. Iguana 12: 222–230.

———. 2007. Preliminary report on the distribution and status of *Cyclura ricordii* along the southern shore of Lago Enriquillo. Iguana 14: 2–7.

———. 2008. Conserving *Cyclura ricordii* 2007. Iguana 15: 2–7.

Russell, A. P., and A. M. Bauer. 1991a. *Anolis garmani*. Catalogue of American Amphibians and Reptiles 513: 1–3.

———. 1991b. *Anolis grahami*. Catalogue of American Amphibians and Reptiles 514: 1–4.

———. 1991c. *Anolis opalinus*. Catalogue of American Amphibians and Reptiles 515: 1–2.

———. 2002. *Thecadactylus rapicauda*. Catalogue of American Amphibians and Reptiles 753: 1–16.

Sajdak, R. A., and R. W. Henderson. 1982. Notes on the eggs and young of *Antillophis parvifrons stygius* (Reptilia, Serpentes, Colubridae). Florida Scientist 45: 200–204.

———. 1991. Status of West Indian racers in the Lesser Antilles. Oryx 25: 33–38.

Salvato, R. 1977. In: E. E. Williams (ed.), The Third *Anolis* Newsletter, p. 96. Cambridge, Mass.: Museum of Comparative Zoology, Harvard University.

Sampedro Marín, A. 1998. *Trachemys decussata*. In: E. Pérez, E. Osa, Y. Matamoros, and U. S. Seal (eds.), Report of Conservation Assessment and Management Plan Workshop for Selected Cuban Species II. Sección III. Reptiles. Apple Valley, Minn.: IUCN/SSC Conservation Breeding Specialist Group.

———. 2001. Actividad terrestre de *Trachemys decussata decussata* (Chelonia: Emydidae) en la Ciénaga de Zapata, Cuba. Revista Biología 15: 111–116.

———. 2002. Actividad termorreguladora de *Trachemys decussata decussata* (Chelonia: Emydidae) en una localidad de la Ciénaga de Zapata, Cuba. Revista Biología 16: 19–26.

Sampedro Marín, A., and V. Berovides Álvarez. 1982. Biología reproductiva del sapo común (*Bufo peltocephalus*). Reportes Investigaciónes de Instituto de Zoologia 8: 1–15.

Sampedro Marín, A., and L. Montañez Huguez. 1989. Estratégia Reproductiva de la Jicotea Cubana (*Pseudemys decussata*) en la Ciénaga de Zapata. La Habana: Editorial Academia.

Sampedro Marín, A., and L. Rodríguez Schettino. 2003. Reptiles acuáticos. In: L. Rodríguez Schettino (ed.), Anfibios y Reptiles de Cuba, pp. 128–137. Vaasa, Finland: UPC Print.

Sampedro Marín, A., V. Berovides Álvarez, and O. Torres Fundora. 1979. Hábitat, alimentación, y actividad de dos especies de *Leiocephalus* (Sauria: Iguanidae) en dos localidades de la región suroriental de Cuba. Ciencias Biológicas 3: 129–139.

Sampedro Marín, A., V. Berovides Álvarez, and L. Rodríguez Schettino. 1982a. Algunos aspectos ecológicos sobre dos especies cubanas del género *Anolis* (Sauria: Iguanidae). Ciencias Biológicas 7: 87–103.

Sampedro Marín, A., V. Berovides Álvarez, and O. Torres Fundora. 1982b. Hábitos alimentarios y actividad de *Bufo peltocephalus* Tschudi (Amphibia: Bufonidae) en el Jardín Botánico de Cienfuegos. Poeyana 233: 1–14.

Sampedro Marín, A., V. Berovides Álvarez, A. Perera Puga, and P. Lorente Díaz. 1983. Datos morphométricos y ecológicos de *Pseudemys decussata* (Reptilia: Chelonia) en la Cienega de Zapata, Cuba. Reporte de Investigación del Instituto de Zoologia. Academia de Ciencias de Cuba 8: 1–15.

Sampedro Marín, A., L. Montañez Huguez, and O. Suárez Boado. 1985. Alimentación de *Rana catesbeiana* en dos zonas de captura de Cuba. Ciencias Biológicas 13: 59–66.

Sampedro Marín, A., L. Montañez Huguez, and M. Alvarez Rodríguez. 1992. Dinámica del crecimiento en larvas de *Rana catesbeiana* del Jardín Botánico Nacional de Cuba. Revista del Jardín Botánico Nacional 13: 115–121.

Sampedro Marin, A., V. Rivalta R., and L. M. Díaz B. 2003. Ranas acuáticas y ribereñas. In: L. Rodríguez Schettino (ed.), Anfibios y Reptiles de Cuba, pp. 38–43. Vaasa, Finland: UPC Print.

Sánchez Muñoz, A. J. 2008. Father Sanchez's Web Site of West Indian Natural History. http://www.kingsnake.com/westindian/.

Sander, J. M., H. Kaiser, and R. Powell. 2003. *Eleutherodactylus euphronides*. Catalogue of American Amphibians and Reptiles 764: 1–3.

Sanger, T. J., J. B. Losos, and J. J. Gibson-Brown. 2008. A developmental staging series for the lizard genus *Anolis*: A new system for the integration of evolution, development, and ecology. Journal of Morphology 269: 129–137.

Sanlley Castro, C., and J. Duval. 1979. Reproducción en cautividad de iguanas del género *Cyclura*. ZOODOM 3(2): 13–18.

Santana, C. E., and S. A. Temple. 1988. Breeding biology and diet of Red-tailed Hawks in Puerto Rico. Biotropica 20: 151–160.

Santos-Barrera, G., G. A. Hammerson, S. B. Hedges, R. L. Joglar, S. J. Incháustegui, L. Kuangyang, C. Wenhao, G. Huiqing, S. Haitao, A. Diesmos, D. Iskandar, P. P. van Dijk, M. Matsui, B. Schmidt, C. Miaud, and Iñi. 2006. *Lithobates catesbeianus*. In: IUCN 2008. 2008 IUCN Red List of Threatened Species. http://www.iucnredlist.org.

Sanz Ochotorena, A. 2000. Artículo de revisión. Ciclos reproductivos en reptiles (género *Anolis*). Revista Biología 14: 97–106.

Sanz Ochotorena, A., and M. C. Uribe Aranzábal. 1999. Ciclo gonadal y de los cuerpos grasos de *Anolis sagrei* (Sauria: Iguanidae) en Ciudad de La Habana. Revista Biología 13: 26–32.

Sanz Ochotorena, A., M. C. Uribe Aranzábal, R. M. Coro Antich, and F. Torres Barbosa. 2000. Ciclo reproductivo de *Anolis porcatus* (Sauria: Polychrotidae) en Ciudad de La Habana. Revista Biología 14: 114–120.

Sanz Ochotorena, A., M. C. Uribe Aranzábal, and L. J. Guillette, Jr. 2005. Seasonal gametogenic cycles in a Cuban tropical lizard, *Anolis porcatus*. Journal of Herpetology 39: 443–454 [this article was published with the senior author's name appearing as "Ochotorena" instead of "Sanz Ochotorena"].

Savit, A. Z., A. J. Maley, H. M. Heinz, R. W. Henderson, and R. Powell. 2005. Distribution and activity periods of *Also-*

phis rufiventris (Colubridae) on The Quill, St. Eustatius, Netherlands Antilles. Amphibia-Reptilia 26: 418–421.

Schall, J. J. 1992. Parasite-mediated competition in *Anolis* lizards. Oecologia 92: 58–64.

Schall, J. J., and A. R. Pearson. 2000. Body condition of a Puerto Rican anole, *Anolis gundlachi*: Effect of a malaria parasite and weather variation. Journal of Herpetology 34: 489–491.

Schall, J. J., and C. M. Staats. 2002. Virulence of lizard malaria: Three species of *Plasmodium* infecting *Anolis sabanus*, the endemic anole of Saba, Netherlands Antilles. Copeia 2002: 39–43.

Schall, J. J., and S. P. Vogt. 1993. Distribution of malaria in *Anolis* lizards of the Luquillo Forest, Puerto Rico: Implications for host community ecology. Biotropica 25: 229–235.

Schardt, M. 1998a. *Iguana delicatissima* im Freiland. Reptilia 3(2): 28–32.

———. 1998b. Freilanduntersuchungen zur Lebensweise und Ökologie von *Iguana iguana* (Linnaeus, 1758) auf der Karibikinsel Terre de Haute, mit einer Bemerkung zum sympatrischen Vorkommen mit *Iguana delicatissima* Laurenti, 1768. Sauria 20(3): 15–21.

Schell, P. T., J. S. Parmerlee, Jr., and R. Powell. 1993a. *Ameiva chrysolaema*. Catalogue of American Amphibians and Reptiles 563: 1–6.

Schell, P. T., R. Powell, J. S. Parmerlee, Jr., A. Lathrop, and D. D. Smith. 1993b. Notes on the natural history of *Ameiva chrysolaema* (Sauria: Teiidae) from Barahona, Dominican Republic. Copeia 1993: 859–862.

Schirra, K. J. 1993. *Anolis christophei* Williams. Sauria Supplement 15: 261–264.

Schlaepfer, M. A., C. Hoover, and C. K. Dodd, Jr. 2005. Challenges in evaluating the impact of the trade in amphibians and reptiles on wild populations. BioScience 55: 256–264.

Schmidt, J. 1979. Warum eigentlich immer *Anolis carolinensis*? Herpetofauna 1(3): 26–29.

———. 1980a. Warum eigentlich immer *Anolis carolinensis*? Teil 2. Herpetofauna 2(5): 22–25.

———. 1980b. Warum eigentlich immer *Anolis carolinensis*? Teil 3. Herpetofauna 2(8): 24–26.

———. 1981. Warum eigentlich immer *Anolis carolinensis*? Teil 4. Herpetofauna 3(12): 6–8.

———. 1994. Nachzucht von *Cyclura nubila nubila* (Kuba-Wirtelschwanzleguan). Elaphe (neue Folge) 2(2): 21–23.

———. 2009. Der Exuma-Wirtelschwanzleguan, *Cyclura cychlura figginsi*: Biologie, Haltung, und Erstnachzucht. Terraria 4: 26–35.

Schmidt, K. P. 1919. Descriptions of new amphibians and reptiles from Santo Domingo and Navassa. Bulletin of the American Museum of Natural History 41: 519–525.

———. 1920. Contribution to the herpetology of Porto Rico. Annals of the New York Academy of Sciences 28: 167–200.

———. 1921. Notes on the herpetology of Santo Domingo. Bulletin of the American Museum of Natural History 44: 7–20.

———. 1926. The herpetology of Mona Island, West Indies. Publications of the Field Museum of Natural History, Zoology 12: 217–224.

———. 1928. Amphibians and reptiles of Porto Rico, with a list of those reported from the Virgin Islands. New York Academy of Sciences 10, Part 1: 1–160 + plates I–IV.

Schmidt, M. 1993. Martinique: Die Heimat der schönen *Anolis roquet*. Die Aquarien- und Terrarienzeitschrift (DATZ) 46: 458–461.

Schneider, K. R., R. Powell, and J. S. Parmerlee, Jr. 2000. Escape behavior of *Anolis* lizards from the Sierra de Baoruco, Dominican Republic. Caribbean Journal of Science 36: 321–323.

Schoener, A., and T. W. Schoener. 1984. Experiments on dispersal: Short-term flotation of insular anoles, with a review of similar abilities in other terrestrial animals. Oecologia 63: 289–294.

Schoener, T. W. 1967. The ecological significance of sexual dimorphism in size of the lizard *Anolis conspersus*. Science 155: 474–478.

———. 1968. The *Anolis* lizards of Bimini: Resource partitioning in a complex fauna. Ecology 49: 704–726.

———. 1970. Nonsynchronous spatial overlap of lizards in patchy habitats. Ecology 51: 408–418.

———. 1975. Presence and absence of habitat shift in some widespread lizard species. Ecological Monographs 45: 233–258.

———. 1976. Habitat shift in widespread *Anolis* lizard species. National Geographic Society Research Reports, 1968 Projects: 369–378.

———. 1977. Competition and the niche. In: C. Gans and D. W. Tinkle (eds.), Biology of the Reptilia. Volume 7. Ecology and Behaviour A, pp. 35–136. London: Academic Press.

———. 1979. Feeding, spacing, and growth in the four species of Bimini *Anolis* lizards. National Geographic Society Research Reports, 1970 Projects: 479–485.

———. 1988. Testing for non-randomness in sizes and habitats of West Indian lizards: Choice of species pool affects conclusions from null models. Evolutionary Ecology 2: 1–26.

Schoener, T. W., and G. C. Gorman. 1968. Some niche differences in three Lesser Antillean lizards of the genus *Anolis*. Ecology 49: 819–830.

Schoener, T. W., and A. Schoener. 1971a. Structural habitats of West Indian *Anolis* lizards. I. Lowland Jamaica. Breviora 368: 1–53.

———. 1971b. Structural habitats of West Indian *Anolis* lizards. II. Puerto Rican uplands. Breviora 375: 1–39.

———. 1976. The ecological context of female pattern poly-

morphism in the lizard *Anolis sagrei*. Evolution 30: 650–658.
———. 1978a. Inverse relation of survival of lizards with island size and avifaunal richness. Nature 274: 685–687.
———. 1978b. Estimating and interpreting body-size growth in some *Anolis* lizards. Copeia 1978: 390–405.
———. 1980a. Densities, sex ratios, and population structure in four species of Bahamian *Anolis* lizards. Journal of Animal Ecology 49: 19–53.
———. 1980b. Ecological and demographic correlates of injury rates in some Bahamian *Anolis* lizards. Copeia 1980: 839–850.
———. 1982a. The ecological correlates of survival in some Bahamian *Anolis* lizards. Oikos 39: 1–16.
———. 1982b. Intraspecific variation in home-range size in some *Anolis* lizards. Ecology 63: 809–823.
———. 1983a. On the voluntary departure of lizards from very small islands. In: A. G. J. Rhodin and K. Miyata (eds.), Advances in Herpetology and Evolutionary Biology: Essays in Honor of Ernest E. Williams, pp. 491–498. Cambridge, Mass.: Museum of Comparative Zoology, Harvard University.
———. 1983b. Distribution of vertebrates on some very small islands. I. Occurrence sequences of individual species. Journal of Animal Ecology 52: 209–235.
Schoener, T. W., J. B. Slade, and C. H. Stinson. 1982. Diet and sexual dimorphism in the very catholic lizard genus, *Leiocephalus* of the Bahamas. Oecologia 53: 160–169.
Schoener, T. W., D. A. Spiller, and J. B. Losos. 2001. Natural restoration of a species-area relation for a lizard after a hurricane. Science 294: 1525–1528.
———. 2002. Predation of a common *Anolis* lizard: Can the food-web effects of a devastating predator be reversed? Ecological Monographs 72: 383–407.
Schoener, T. W., J. B. Losos, and D. A. Spiller. 2005. Island biogeography of populations: An introduced species transforms survival patterns. Science 310: 1807–1809.
Schreiber, M. C., R. Powell, J. S. Parmerlee, Jr., A. Lathrop, and D. D. Smith. 1993. Natural history of a small population of *Leiocephalus schreibersii* (Sauria: Tropiduridae) from altered habitat in the Dominican Republic. Florida Scientist 56: 18–27.
Schreiber, M. C., R. Powell, and J. S. Parmerlee, Jr. 1995. *Leiocephalus schreibersii*. Catalogue of American Amphibians and Reptiles 613: 1–4.
Schröter, G. 1986. Anfänge der Pflege und Zucht von *Anolis porcatus*. Aquarien Terrarien 33: 411–415.
Schubert, A. 1993. Conservation of biological diversity in the Dominican Republic. Oryx 27: 115–121.
Schubert, A., and G. Santana. 1996. Conservation of the American Crocodile (*Crocodylus acutus*) in the Dominican Republic. In: R. Powell and R. W. Henderson (eds.), Contributions to West Indian Herpetology: A Tribute to Albert Schwartz, pp. 425–433. Contributions to Herpetology, volume 12. Ithaca, N.Y.: Society for the Study of Amphibians and Reptiles.
Schulze, P. 1941. Ein neues *Amblyomma* und ein neues *Aponomma* mit Augenrudimenten aus Haiti. Zoologische Anzeigen 133: 225–229.
Schwartz, A. 1957. A new species of boa (genus *Tropidophis*) from western Cuba. American Museum Novitates 1839: 1–8.
———. 1958a. Another new large *Eleutherodactylus* (Amphibia: Leptodactylidae) from western Cuba. Proceedings of the Biological Society of Washington 71: 37–42.
———. 1958b. Four new frogs of the genus *Eleutherodactylus* (Leptodactylidae) from Cuba. American Museum Novitates 1873: 1–20.
———. 1958c. Another new *Eleutherodactylus* (Amphibia: Leptodactylidae) from western Cuba. Journal of the Washington Academy of Sciences 48: 127–131.
———. 1958d. A new frog of the *auriculatus* group of the genus *Eleutherodactylus* from western Cuba. Herpetologica 14: 69–77.
———. 1959a. A new species of toad, *Bufo cataulaciceps*, from the Isla de Pinos and western Cuba. Proceedings of the Biological Society of Washington 72: 109–120.
———. 1959b. A new species of frog of the *Eleutherodactylus ricordi* group from central Cuba. American Museum Novitates 1926: 1–16.
———. 1959c. The Cuban lizards of the species *Leiocephalus carinatus* (Gray). Reading Public Museum and Art Gallery Scientific Publications 10: 1–47.
———. 1959d. The status of *Eleutherodactylus pinarensis* and a new species of the genus from western Cuba. Herpetologica 15: 61–69.
———. 1959e. Variation in lizards of the *Leiocephalus cubensis* complex in Cuba and the Isla de Pinos. Bulletin of the Florida State Museum 4: 97–143.
———. 1960a. Nine new Cuban frogs of the genus *Eleutherodactylus*. Reading Public Museum and Art Gallery Scientific Publications 11: 1–50.
———. 1960b. The large toads of Cuba. Proceedings of the Biological Society of Washington 73: 45–56.
———. 1960c. Variation in the Cuban lizard *Leiocephalus raviceps* Cope. Proceedings of the Biological Society of Washington 73: 67–82.
———. 1964a. A review of *Sphaerodactylus vincenti* on the southern Windward Islands. Caribbean Journal of Science 4: 391–409.
———. 1964b. Two new species of *Eleutherodactylus* from the eastern Cordillera Central of the "Republica Dominicana." Caribbean Journal of Science 4: 473–484.
———. 1964c. *Anolis equestris* in Oriente Province, Cuba. Bulletin of the Museum of Comparative Zoology 131: 405–428.

———. 1964d. Three new species of frogs (Leptodactylidae, *Eleutherodactylus*) from Hispaniola. Breviora 208: 1–15.
———. 1965a. A review of the colubrid snake genus *Arrhyton* with the description of a new subspecies from southern Oriente Province, Cuba. Proceedings of the Biological Society of Washington 78: 99–114.
———. 1965b. The *Ameiva* (Reptilia: Teiidae) of Hispaniola I. *Ameiva lineolata* Dumeril and Bibron. Caribbean Journal of Science 5: 45–57.
———. 1965c. The *Leiocephalus* (Lacertilia, Iguanidae) of Hispaniola I. *Leiocephalus melanochlorus* Cope. Journal of the Ohio Herpetological Society 5: 39–48.
———. 1965d. A new Cuban *Eleutherodactylus* of the *auriculatus* group. Herpetologica 21: 27–31.
———. 1965e. Variation and natural history of *Eleutherodactylus ruthae* on Hispaniola. Bulletin of the Museum of Comparative Zoology 132: 481–508.
———. 1966. The relationships of four small Hispaniolan *Eleutherodactylus* (Leptodactylidae). Bulletin of the Museum of Comparative Zoology 133: 369–399.
———. 1967a. A review of the genus *Dromicus* in Puerto Rico and the Virgin Islands. Stahlia 9: 1–14 + 4 figures.
———. 1967b. Frogs of the genus *Eleutherodactylus* in the Lesser Antilles. Studies on the Fauna of Curaçao and Other Caribbean Islands 24: 1–62.
———. 1967c. The *Ameiva* (Lacertilia, Teiidae) of Hispaniola. III. *Ameiva taeniura* Cope. Bulletin of the Museum of Comparative Zoology 135: 345–375.
———. 1967d. The *Leiocephalus* (Lacertilia, Iguanidae) of Hispaniola. II. The *Leiocephalus personatus* complex. Tulane Studies in Zoology 14: 1–53.
———. 1967e. The *Leiocephalus* (Lacertilia, Iguanidae) of the southern Bahama Islands. Annals of the Carnegie Museum 39: 153–184.
———. 1968a. Geographic variation in the Hispaniolan frog *Eleutherodactylus wetmorei* Cochran. Breviora 290: 1–13.
———. 1968b. Geographic variation in the New World gekkonid lizard *Tarentola americana* Gray. Proceedings of the Biological Society of Washington 81: 123–142.
———. 1968c. The geckos (*Sphaerodactylus*) of the southern Bahama Islands. Annals of the Carnegie Museum 39: 227–271.
———. 1968d. Geographic variation in *Anolis distichus* Cope (Lacertilia, Iguanidae) in the Bahama Islands and Hispaniola. Bulletin of the Museum of Comparative Zoology 137: 255–309.
———. 1968e. The *Leiocephalus* (Lacertilia, Iguanidae) of Hispaniola. III. *Leiocephalus schreibersi*, *L. semilineatus*, and *L. pratensis*. Journal of Herpetology 24: 39–63.
———. 1968f. The Cuban lizards of the *Anolis homolechis* complex. Tulane Studies in Zoology 14: 140–184.
———. 1968g. Two new subspecies of *Ameiva* (Lacertilia, Teiidae) from Hispaniola. Herpetologica 24: 21–28.
———. 1969a. The Antillean *Eleutherodactylus* of the *auriculatus* group. Studies on the Fauna of Curaçao and Other Caribbean Islands 114: 99–115.
———. 1969b. A review of the Hispaniolan lizard *Anolis coelestinus* Cope. Caribbean Journal of Science 9: 33–38.
———. 1970a. A new species of large *Diploglossus* (Sauria, Anguidae) from Hispaniola. Proceedings of the Biological Society of Washington 82: 777–788.
———. 1970b. *Sphaerodactylus notatus*. Catalogue of American Amphibians and Reptiles 90: 1–2.
———. 1970c. The subspecies of *Eleutherodactylus schmidti* Noble (Anura: Leptodactylidae). Caribbean Journal of Science 10: 109–118.
———. 1970d. A systematic review of *Ameiva auberi* Cocteau (Reptilia, Teiidae) in Cuba and the Bahamas. I. The Cuban subspecies. III. Discussion. Annals of the Carnegie Museum 41: 45–117, 152–168.
———. 1970e. A systematic review of *Uromacer catesbyi* Schlegel (Serpentes, Colubridae). Tulane Studies in Zoology and Botany 16: 131–149.
———. 1971a. A new species of bromeliad-inhabiting galliwasp (Sauria: Anguidae) from Jamaica. Breviora 371: 1–10.
———. 1971b. A systematic account of the Hispaniolan snake genus *Hypsirhynchus*. Studies on the Fauna of Curaçao and Other Caribbean Islands 35: 63–94.
———. 1971c. *Anolis distichus*. Catalogue of American Amphibians and Reptiles 108: 1–4.
———. 1972. The native toads (Anura, Bufonidae) of Hispaniola. Journal of Herpetology 6: 217–231.
———. 1973a. A new subspecies of *Ameiva chrysolaema* (Sauria, Teiidae) from Haiti. Herpetologica 29: 101–105.
———. 1973b. Six new species of *Eleutherodactylus* (Anura, Leptodactylidae) from Hispaniola. Journal of Herpetology 7: 249–273.
———. 1973c. A new species of montane *Anolis* (Sauria, Iguanidae) from Hispaniola. Annals of the Carnegie Museum of Natural History 44: 183–195.
———. 1974a. *Eleutherodactylus planirostris*. Catalogue of American Amphibians and Reptiles 154: 1–4.
———. 1974b. A new species of primitive *Anolis* (Sauria, Iguanidae) from the Sierra de Baoruco, Hispaniola. Breviora 423: 1–19.
———. 1974c. An analysis of variation in the Hispaniolan giant anole, *Anolis ricordi* Duméril and Bibron. Bulletin of the Museum of Comparative Zoology 146: 89–146.
———. 1975. Variation in the Antillean boid snake *Tropidophis haitianus* Cope. Journal of Herpetology 9: 303–311.
———. 1976a. A new species of *Sphaerodactylus* (Sauria, Gekkonidae) from the República Dominicana. Florida Scientist 39: 65–70.
———. 1976b. Variation and relationships of some Hispaniolan frogs (Leptodactylidae, *Eleutherodactylus*) of the *ri-*

cordi group. Bulletin of the Florida State Museum, Biological Sciences 21: 1–46.

———. 1976c. Two new species of Hispaniolan *Eleutherodactylus* (Leptodactylidae). Herpetologica 32: 163–171.

———. 1977a. A new subspecies of *Anolis baleatus* (Sauria: Iguanidae) from Isla Saona, República Dominicana. Florida Scientist 40: 401–405.

———. 1977b. The geckoes (Sauria, Gekkonidae) of the genus *Sphaerodactylus* of the Dominican Península de Barahona, Hispaniola. Proceedings of the Biological Society of Washington 90: 243–254.

———. 1977c. *Eleutherodactylus abbotti*. Catalogue of American Amphibians and Reptiles 191: 1–2.

———. 1977d. *Eleutherodactylus alcoae*. Catalogue of American Amphibians and Reptiles 192: 1–2.

———. 1977e. *Eleutherodactylus apostates*. Catalogue of American Amphibians and Reptiles 201: 1.

———. 1977f. A new subspecies of *Eleutherodactylus wetmorei* Cochran (Anura: Leptodactylidae) from northern Haiti. Herpetologica 33: 66–72.

———. 1978a. *Eleutherodactylus armstrongi*. Catalogue of American Amphibians and Reptiles 208: 1–2.

———. 1978b. The Hispaniolan *Anolis* (Reptilia, Lacertilia, Iguanidae) of the *hendersoni* complex. Journal of Herpetology 12: 355–370.

———. 1978c. A new species of aquatic *Anolis* (Sauria, Iguanidae) from Hispaniola. Annals of the Carnegie Museum 47: 261–279.

———. 1979a. A new species of *Eleutherodactylus* (Amphibia, Anura, Leptodactylidae) from northwestern Haiti, Hispaniola. Journal of Herpetology 13: 199–202.

———. 1979b. The status of Greater Antillean *Phyllodactylus* (Reptilia, Gekkonidae). Journal of Herpetology 13: 419–426.

———. 1979c. *Eleutherodactylus audanti*. Catalogue of American Amphibians and Reptiles 224: 1–2.

———. 1979d. The herpetofauna of Île à Cabrit, Haiti, with the description of two new subspecies. Herpetologica 35: 248–255.

———. 1979e. A new species of cybotoid anole (Sauria, Iguanidae) from Hispaniola. Breviora 451: 1–27.

———. 1979f. A new species of *Leiocephalus* (Reptilia: Iguanidae) from Hispaniola. Proceedings of the Biological Society of Washington 92: 272–279.

———. 1980a. *Eleutherodactylus auriculatoides*. Catalogue of American Amphibians and Reptiles 246: 1–2.

———. 1980b. *Eleutherodactylus bakeri*. Catalogue of American Amphibians and Reptiles 247: 1.

———. 1980c. Two new species of *Eleutherodactylus* (Amphibia, Anura, Leptodactylidae) from Hispaniola. Annals of the Carnegie Museum 49: 103–112.

———. 1980d. *Eleutherodactylus brevirostris*. Catalogue of American Amphibians and Reptiles 259: 1.

———. 1980e. *Eleutherodactylus counouspeus*. Catalogue of American Amphibians and Reptiles 260: 1–2.

———. 1980f. Variation in Hispaniolan *Anolis whitemani* Williams. Journal of Herpetology 14: 399–406.

———. 1981. Variation in Hispaniolan *Anolis olssoni* Schmidt (Reptilia: Sauria: Iguanidae). Milwaukee Public Museum Contributions in Biology and Geology 47: 1–21.

———. 1982a. *Eleutherodactylus darlingtoni*. Catalogue of American Amphibians and Reptiles 284: 1.

———. 1982b. *Eleutherodactylus eunaster*. Catalogue of American Amphibians and Reptiles 285: 1.

———. 1982c. *Eleutherodactylus flavescens*. Catalogue of American Amphibians and Reptiles 296: 1–2.

———. 1982d. *Eleutherodactylus fowleri*. Catalogue of American Amphibians and Reptiles 297: 1.

———. 1983a. The *difficilis* complex of *Sphaerodactylus* (Sauria, Gekkonidae) of Hispaniola. Part 1. *Sphaerodactylus difficilis, S. clenchi*, and *S. lazelli*. Bulletin of the Carnegie Museum of Natural History 22: 5–30.

———. 1983b. *Eleutherodactylus furcyensis*. Catalogue of American Amphibians and Reptiles 309: 1.

———. 1983c. *Eleutherodactylus glandulifer*. Catalogue of American Amphibians and Reptiles 310: 1.

———. 1983d. *Eleutherodactylus glanduliferoides*. Catalogue of American Amphibians and Reptiles 322: 1.

———. 1983e. *Eleutherodactylus glaphycompus*. Catalogue of American Amphibians and Reptiles 323: 1.

———. 1983f. *Eleutherodactylus haitianus*. Catalogue of American Amphibians and Reptiles 335: 1–2.

———. 1983g. *Eleutherodactylus heminota*. Catalogue of American Amphibians and Reptiles 336: 1.

———. 1984. *Eleutherodactylus hypostenor*. Catalogue of American Amphibians and Reptiles 346: 1.

———. 1986. *Darlingtonia, D. haetiana*. Catalogue of American Amphibians and Reptiles 390: 1–2.

———. 1989. A review of the cybotoid anoles (Reptilia: Sauria: Iguanidae) from Hispaniola. Milwaukee Public Museum Contributions in Biology and Geology 78: 1–32.

Schwartz, A., and M. Carey. 1977. Systematics and evolution in the West Indian iguanid genus *Cyclura*. Studies on the Fauna of Curaçao and Other Caribbean Islands 53: 15–97.

Schwartz, A., and R. I. Crombie. 1975. A new species of the genus *Aristelliger* (Sauria: Gekkonidae) from the Caicos Islands. Proceedings of the Biological Society of Washington 88: 305–313.

Schwartz, A., and D. C. Fowler. 1973. The Anura of Jamaica: A progress report. Studies on the Fauna of Curaçao and Caribbean Islands 43: 50–142.

Schwartz, A., and R. Franz. 1976. A new species of *Sphaerodactylus* (Sauria: Gekkonidae) from Hispaniola. Proceedings of the Biological Society of Washington 88: 367–372.

Schwartz, A., and O. H. Garrido. 1967. A review of the Cuban lizard *Leiocephalus macropus* Cope. Reading Public Museum and Art Gallery Scientific Publications 14: 1–41.

———. 1968a. An undescribed subspecies of *Leiocephalus raviceps* Cope (Sauria: Iguanidae) from western Cuba. Proceedings of the Biological Society of Washington 81: 23–30.

———. 1968b. Four new subspecies of *Leiocephalus stictigaster* from Cuba. National Museum of Canada Natural History Papers 37: 1–23.

———. 1972. The lizards of the *Anolis equestris* complex in Cuba. Studies on the Fauna of Curaçao and Other Caribbean Islands 39: 1–86.

———. 1981. A review of the Cuban members of the genus *Arrhyton* (Reptilia, Serpentes, Colubridae). Annals of the Carnegie Museum of Natural History 50: 207–230.

———. 1985. The Cuban lizards of the genus *Sphaerodactylus* (Sauria, Gekkonidae). Milwaukee Public Museum Contributions in Biology and Geology 62: 1–67.

Schwartz, A., and E. D. Graham, Jr. 1980. The *shrevei* group of Hispaniolan *Sphaerodactylus* (Reptilia, Gekkonidae). Tulane Studies in Zoology and Botany 22: 1–15.

Schwartz, A., and R. W. Henderson. 1982. *Anolis cybotes* (Reptilia, Iguanidae): The eastern Hispaniolan populations. Milwaukee Public Museum Contributions in Biology and Geology 49: 1–8.

———. 1984a. *Uromacer*. Catalogue of American Amphibians and Reptiles 355: 1.

———. 1984b. *Uromacer catesbyi*. Catalogue of American Amphibians and Reptiles 356: 1–2.

———. 1984c. *Uromacer oxyrhynchus*. Catalogue of American Amphibians and Reptiles 358: 1–2.

———. 1988. West Indian amphibians and reptiles: A checklist. Milwaukee Public Museum Contributions in Biology and Geology 74: 1–264.

———. 1991. Amphibians and Reptiles of the West Indies: Descriptions, Distributions, and Natural History. Gainesville: University of Florida Press.

Schwartz, A., and S. J. Incháustegui. 1976. A new species of *Diploglossus* (Reptilia, Lacertilia, Anguidae) from Hispaniola. Journal of Herpetology 10: 241–246.

———. 1980. The endemic Hispaniolan lizard genus *Chamaelinorops*. Journal of Herpetology 14: 51–56.

Schwartz, A., and J. F. Jacobs. 1989. A new subspecies of *Celestus costatus* (Sauria: Anguidae) from the República Dominicana. Journal of Herpetology 23: 193–195.

Schwartz, A., and R. F. Klinikowski. 1966. The *Ameiva* (Lacertilia, Teiidae) of Hispaniola. II. Geographic variation in *Ameiva chrysolaema* Cope. Bulletin of the Museum of Comparative Zoology 133: 427–487.

Schwartz, A., and L. H. Ogren. 1956. A collection of reptiles and amphibians from Cuba, with the description of two new forms. Herpetologica 12: 91–110.

Schwartz, A., and D. A. Rossman. 1976. A review of the Hispaniolan colubrid snake genus *Ialtris*. Studies on the Fauna of Curaçao and Other Caribbean Islands 50: 76–102.

Schwartz, A., and R. Thomas. 1960. Four new snakes (*Tropidophis, Dromicus, Alsophis*) from the Isla de Pinos and Cuba. Herpetologica 16: 73–90.

———. 1965. The genus *Darlingtonia* (Serpentes) in Hispaniola, including a new species from the Dominican Republic. Breviora 229: 1–10.

———. 1969. A review of *Anolis angusticeps* in the West Indies. Quarterly Journal of the Florida Academy of Sciences 31: 51–69.

———. 1975. A check-list of West Indian amphibians and reptiles. Carnegie Museum of Natural History Special Publication 1: 1–216.

———. 1977. Two new species of *Sphaerodactylus* (Reptilia, Lacertilia, Gekkonidae) from Hispaniola. Journal of Herpetology 11: 61–66.

Schwartz, A., R. Thomas, and L. D. Ober. 1978. First supplement to a check-list of West Indian amphibians and reptiles. Carnegie Museum of Natural History, Special Publication 5: 1–35.

Schwartz, A., E. D. Graham, Jr., and J. J. Duval. 1979. A new species of *Diploglossus* (Sauria: Anguidae) from Hispaniola. Proceedings of the Biological Society of Washington 92: 1–9.

Schwenk, K., and G. C. Mayer. 1991. Tongue display in anoles and its evolutionary basis. In: J. B. Losos and G. C. Mayer (eds.), *Anolis* Newsletter IV, pp. 131–140. Washington, D.C.: National Museum of Natural History.

SEA/DVS (Secretaría de Estado de Agricultura/Departamento de Vida Silvestre). 1992. Reconocimiento y Evaluación de los Recursos Naturales en el Bahoruco Oriental. Santo Domingo, Dominican Republic: Secretaría de Estado de Agricultura, Departamento de Vida Silvestre.

Seaman, G. A. 1961. Mammals, Reptiles & Amphibians of the Virgin Islands. Christiansted, St. Croix: Brodhurst's Printery.

Seaman, G. A., and J. E. Randall. 1962. The mongoose as a predator in the Virgin Islands. Journal of Mammalogy 43: 544–546.

Seidel, M. E. 1988a. Revision of the West Indian emydid turtles (Testudines). American Museum Novitates 2918: 1–41.

———. 1988b. *Trachemys decussata*. Catalogue of American Amphibians and Reptiles 440: 1–3.

———. 1988c. *Trachemys stejnegeri*. Catalogue of American Amphibians and Reptiles 441: 1–3.

———. 1988d. *Trachemys terrapen*. Catalogue of American Amphibians and Reptiles 442: 1–2.

———. 1990. Growth and population characteristics of the slider turtle, *Trachemys decussata*, on Grand Cayman Island. Journal of Herpetology 24: 191–196.

———. 1996. Current status of biogeography of the West Indian turtles in the genus *Trachemys* (Emydidae). In: R.

Powell and R. W. Henderson (eds.), Contributions to West Indian Herpetology: A Tribute to Albert Schwartz, pp. 169–174. Contributions to Herpetology, volume 12. Ithaca, N.Y.: Society for the Study of Amphibians and Reptiles.

———. 2003. *Trachemys decussata angusta*. (North Antillean Slider). Longevity. Herpetological Review 34: 363–364.

Seidel, M. E., and R. Franz. 1994. Amphibians and reptiles (exclusive of marine turtles) of the Cayman Islands. In: M. A. Brunt and J. E. Davies (eds.), The Cayman Islands: Natural History and Biogeography, pp. 407–433. Dordrecht, The Netherlands: Kluwer Academic Publishers.

Seidel, M. E., and S. J. Incháustegui M. 1984. Status of the trachemyd turtles (Testudines: Emydidae) on Hispaniola. Journal of Herpetology 18: 468–479.

Self, J. T., and J. García Díaz. 1961. *Raillietiella* (*Heymonsia*) *hemidactyli* Heft, 1934, from *Hemidactylus mabouia* in Puerto Rico, with a correction of the identity of *R. hebtihamata* Self and Kuntz, 1961. Journal of Parasitology 47: 912.

Ševčik, J. 1989. Die Reptilien des Escambray-Gebirges auf Kuba. Die Aquarien- und Terrarienzeitschrift (DATZ) 42: 731–734.

———. 1990. Ritteranolis in der Natur und im Terrarium. Die Aquarien- und Terrarienzeitschrift (DATZ) 43: 156–158.

Sexto, L. 1989. Los lagartos sí comen huevo. Bohemia 81: 36–38.

Shafir, S., and J. Roughgarden. 1994. Instrumental discrimination conditioning of *Anolis cristatellus* in the field with food as a reward. Caribbean Journal of Science 30: 228–233.

———. 1998. Testing predictions of foraging theory for a sit-and-wait forager, *Anolis gingivinus*. Behavioral Ecology 9: 74–84.

Shaw, C. E. 1969. Breeding the Rhinoceros Iguana *Cyclura cornuta cornuta* at San Diego Zoo. International Zoo Yearbook 9: 45–48.

Shaw, G. 1802. General Zoology or Systematic Natural History. Volume III. Amphibia. London: Thomas Davison.

Sheplan, B. R., and A. Schwartz. 1974. Hispaniolan boas of the genus *Epicrates* (Serpentes, Boidae) and their Antillean relationships. Annals of the Carnegie Museum 45: 57–143.

Shew, J. J., E. J. Censky, and R. Powell. 2002a. *Ameiva corvina*. Catalogue of American Amphibians and Reptiles 747: 1–2.

Shew, J. J., S. C. Larimer, R. Powell, and J. S. Parmerlee, Jr. 2002b. Sleeping patterns and sleep-site fidelity of *Anolis gingivinus* on Anguilla. Caribbean Journal of Science 38: 136–138.

Shreve, B. 1936. A new *Anolis* and new Amphibia from Haiti. Proceedings of the New England Zoölogical Club 15: 93–99.

———. 1968. The *notatus* group of *Sphaerodactylus* (Sauria, Gekkonidae) in Hispaniola. Breviora 280: 1–28.

Shreve, B., and E. E. Williams. 1963. The frogs. In: E. E. Williams, B. Shreve, and P. S. Humphrey. 1963. The herpetology of the Port-au-Prince region and Gonave Island, Haiti, pp. 302–341. Parts I and II. Bulletin of the Museum of Comparative Zoology 129: 291–342.

Sifers, S. M., M. L. Yeska, Y. M. Ramos, R. Powell, and J. S. Parmerlee, Jr. 2001. *Anolis* lizards restricted to altered edge habitats in a Hispaniolan cloud forest. Caribbean Journal of Science 37: 55–62.

Silva Lee, A. 1984. Chipojos, Bayoyas y Camaleones. La Habana: Editorial Científico Técnica.

Silva Rodríguez, A., and A. R. Estrada. 1984. Ciclos reproductivos de dos lagartos género *Anolis* (*A. homolechis* y *A. allogus*) en la Estación Ecológica Sierra del Rosario. Ciências Biológicas 12: 81–89.

———. 1994. Subnicho trófico de dos especies de *Anolis* (Iguania: Polycridae [sic]) en la Sierra del Rosario, Pinar del Río, Cuba. Ciencias Biológicas 26: 13–20.

Silva Rodríguez, A., V. Berovides Alvarez, and A. Estrada Acosta. 1982. Sitios de puesta comunales de *Anolis bartschi* (Sauria: Iguanidae). Miscelanea Zoológica 15: 1.

Simmons, P. M., B. T. Greene, K. E. Williamson, R. Powell, and J. S. Parmerlee, Jr. 2005. Ecological interactions within a lizard community on Grenada. Herpetologica 61: 124–134.

Singhal, S., M. A. Johnson, and J. T. Ladner. 2007. The behavioral ecology of sleep: Natural sleeping site choice in three *Anolis* lizard species. Behaviour 144: 1033–1052.

Skála, M. 1989. *Anolis bartschi*, chov a odchov. Akvárium Terárium 32(5): 26–27.

Slavens, F., and K. Slavens. 2003. Reptiles and amphibians in captivity–longevity–home page. http://www.pondturtle.com/longev.html.

Smith, B. E., and R. E. Baum. 2000. Surveys of the Lizards *Anolis wattsi* and *Ameiva griswoldi* on Antiguan Offshore Islands I: Summer 1999. Antiguan Racer Conservation Project Report No. 5, Pre-publication Draft Report. Spearfish, S.Dak.: Black Hills State University.

Smith, B. E., and P. L. Colbert. 2002. Surveys of the Lizard *Anolis wattsi* on Antiguan Offshore Islands III: Summer 2001. Antiguan Racer Conservation Project Report No. 7. Spearfish, S.Dak.: Black Hills State University.

Smith, B. E., R. E. Baum, J. L. Massie, and O. Davis. 2001. Surveys of the Lizards *Anolis wattsi* and *Ameiva griswoldi* on Antiguan Offshore Islands II: Summer 2000. Antiguan Racer Conservation Project Report No. 6. Spearfish, S.Dak.: Black Hills State University.

Smith, B. E., O. Davis, and N. S. Bartscher. 2002. Surveys of the lizard *Ameiva griswoldi* on Antiguan offshore islands III: Summer 2001. Antiguan Racer Conservation Project Report No. 8. Spearfish, S.Dak.: Black Hills State University.

Smith, D. D., J. S. Parmerlee, Jr., R. Powell, L. R. White, and A.

Lathrop. 1991. *Tropidophis haitianus* (NCN). Locomotion. Herpetological Review 22: 132–133.

Smith, D. D., D. J. Pflanz, and R. Powell. 1993. Observations of autohemorrhaging in *Tropidophis haitianus*, *Rhinocheilus lecontei*, *Heterodon platyrhinos*, and *Nerodia erythrogaster*. Herpetological Review 24: 130–131.

Smith, D. D., P. T. Schell, R. Powell, and J. S. Parmerlee, Jr. 1994. Pharyngeal myiasis by sarcophagid larvae (Diptera) in *Ameiva chrysolaema* (Sauria: Teiidae) from the Dominican Republic. Caribbean Journal of Science 30: 148–149.

Smith, G. R. 1992. Sexual dimorphism in the curly-tailed lizard, *Leiocephalus psammodromus*. Caribbean Journal of Science 28: 99–101.

———. 1994. Observations on activity in two lizards (*Anolis scriptus* and *Leiocephalus psammodromus*). Bulletin of the Maryland Herpetological Society 30: 120–125.

———. 1995. Observations on perch use in two lizards (*Anolis scriptus* and *Leiocephalus psammodromus*). Herpetological Journal 5: 285–286.

Smith, G. R., and J. B. Iverson. 1993. Reproduction in the curly-tailed lizard, *Leiocephalus psammodromus* from the Caicos Islands. Canadian Journal of Zoology 71: 2147–2151.

———. 2006. Changes in sex ratio over time in the endangered iguana *Cyclura cychlura inornata*. Canadian Journal of Zoology 84: 1522–1527.

Smith, G. R., and A. M. Nickel. 2002. *Leiocephalus macropus*, *Leiocephalus raviceps*, and *Leiocephalus stictigaster*. Clutch Size. Herpetological Review 33: 308.

Smith, G. R., L. Pieper, and J. B. Iverson. 2008. *Leiocephalus carinatus* (Northern Curly-tail Lizard). Sap feeding. Herpetological Review 39: 228.

Smith, H. M., and E. T. Willis, Jr. 1955. Interspecific variation in compression of tail in a Cuban lizard. Herpetologica 11: 86–88.

Smith, H. M., G. Sinelnik, J. D. Fawcett, and R. E. Jones. 1972. A unique reproductive cycle in *Anolis* and its relatives. Bulletin of the Philadelphia Herpetological Society 20: 28–30.

———. 1973. A survey of the chronology of ovulation in anoline lizard genera. Transactions of the Kansas Academy of Science 75: 107–120.

Smith, J. W., R. Powell, J. S. Parmerlee, Jr., D. D. Smith, and A. Lathrop. 1994. Natural history notes on a population of *Anolis olssoni* (Sauria: Polychrotidae) from the Dominican Republic. Bulletin of the Maryland Herpetological Society 30: 67–75.

Smith, J. W., J. S. Parmerlee, Jr., and R. Powell. 1995. *Anolis olssoni*. Catalogue of American Amphibians and Reptiles 611: 1–5.

Smith, M. L., S. B. Hedges, W. Buck, A. Hemphill, S. Incháustegui, M. Ivie, D. Martina, M. Maunder, and J. F. Ortega. 2005. Caribbean islands. In: R. A. Mittermeier, P. Robles-Gil, M. Hoffmann, C. G. Mittermeier, J. Pilgrim, T. Brooks, and G. A. B. Fonseca (eds.), Hotspots Revisited: Earth's Biologically Richest and Most Endangered Terrestrial Ecoregions, pp. 112–118. Mexico City: CEMEX.

Snider, A. T., and J. K. Bowler. 1992. Longevity of reptiles and amphibians in North American collections. 2nd ed. Society for the Study of Amphibians and Reptiles, Herpetological Circular 21: iv + 40 pp.

Snyder, G. K. 1975. Respiratory metabolism and evaporative water loss in a small tropical lizard. Journal of Comparative Physiology 104: 13–18.

———. 1979. Water loss and oxygen consumption in tropical *Sphaerodactylus*. Oecologia 38: 107–110.

Snyder, N. F. R., J. W. Wiley, and C. B. Kepler. 1987. The Parrots of Luquillo: Natural History and Conservation of the Puerto Rican Parrot. Los Angeles: Western Foundation of Vertebrate Zoology.

Soberon, R., R. Ramos, W. McMahan, and P. Ross. 1996. Cuba: Reproduction of Cuban Crocodile on the Isla of Pines. Crocodile Specialist Group Newsletter 15(3): 10–11.

Socarrás, A. A., J. de la Cruz, G. Garcés G., and A. Ruiz. 1988. Saurofagia en *Anolis* (Sauria: Iguanidae). Miscelanea Zoológica 38: 4.

Solís, F., R. Ibáñez, G. Hammerson, S. B. Hedges, A. Diesmos, M. Matsui, J.-M. Hero, S. Richards, L. A. Coloma, S. Ron, E. La Marca, J. D. Hardy, Jr., and R. Powell. 2004a. *Bufo marinus*. In: IUCN 2007. 2007 IUCN Red List of Threatened Species. http://www.iucnredlist.org.

Solís, F., R. Ibáñez, C. Jaramillo, Q. Fuenmayor, C. Azevedo-Ramos, E. La Marca, L. A. Coloma, S. Ron, J. D. Hardy, Jr., S. B. Hedges, B. Ibéné, M. Breuil, and R. Powell. 2004b. *Scinax ruber*. In: IUCN 2007. 2007 IUCN Red List of Threatened Species. http://www.iucnredlist.org.

Solís, F., R. Ibáñez, G. Hammerson, S. B. Hedges, A. Diesmos, M. Matsui, J.-M. Hero, S. Richards, L. A. Coloma, S. Ron, E. La Marca, J. D. Hardy, Jr., R. Powell, F. Bolaños, and G. Chaves. 2008a. *Rhinella marina*. In: IUCN 2008. 2008 IUCN Red List of Threatened Species. http://www.iucnredlist.org.

———. 2008b. *Scinax ruber*. In: IUCN 2008. 2008 IUCN Red List of Threatened Species. http://www.iucnredlist.org.

Somma, C. A., and G. R. Brooks. 1976. Reproduction in *Anolis oculatus*, *Ameiva fuscata* and *Mabuya mabouya* from Dominica. Copeia 1976: 249–256.

Sosa, R. A., R. W. Henderson, and R. Powell. 1995. *Hypsirhynchus*, *H. ferox*. Catalogue of American Amphibians and Reptiles 617: 1–4.

Sowell, S. P., J. S. Parmerlee, Jr., and R. Powell. 1995. *Anolis bahorucoensis*. Catalogue of American Amphibians and Reptiles 608: 1–3.

Spiller, D. A., and T. W. Schoener. 1998. Lizards reduce spider species richness by excluding rare species. Ecology 79: 503–516.

Spiller, D. A., J. B. Losos, and T. W. Schoener. 1998. Impact of a catastrophic hurricane on island populations. Science 281: 695–697.

Sprackland, R. G., and G. N. Swinney. 1998. A new species of giant gecko of the genus *Tarentola* (Reptilia: Squamata: Gekkonidae) from Jamaica. Journal of Zoology, London 245: 73–78.

Sproston, A. L., J. H. Greve, and R. Powell. 1997. *Ameiva chrysolaema ficta*. Parasitism. Herpetological Review 28: 201.

Sproston, A. L., R. Powell, and J. S. Parmerlee, Jr. 1998. *Ameiva leberi*. Catalogue of American Amphibians and Reptiles 670: 1–2.

Sproston, A. L., R. E. Glor, L. M. Hartley, E. J. Censky, R. Powell, and J. S. Parmerlee, Jr. 1999. Niche differences among three sympatric species of *Ameiva* (Reptilia: Teiidae) on Hispaniola. Journal of Herpetology 33: 131–136.

Staats, C. M., and J. J. Schall. 1996a. Distribution and abundance of malarial parasites of the endemic *Anolis* lizard of Saba Island, Netherlands Antilles. Journal of Parasitology 82: 409–413.

———. 1996b. Malarial parasites (*Plasmodium*) of *Anolis* lizards: Biogeography in the Lesser Antilles. Biotropica 28: 388–393.

Staats, C. M., T. Bartolotta, and J. J. Schall. 1997. Body size and perching behavior of *Anolis sabanus*, the solitary anole of Saba, Netherlands Antilles. Herpetological Natural History 5: 157–163.

Stamps, J. A. 1973. Displays and social organization in female *Anolis aeneus*. Copeia 1973: 264–272.

———. 1975. Courtship patterns, estrus periods and reproductive condition in a lizard, *Anolis aeneus*. Physiology and Behavior 14: 531–535.

———. 1976a. Rainfall, activity, and social behaviour in the lizard *Anolis aeneus*. Animal Behaviour 24: 603–608.

———. 1976b. Egg retention, rainfall and egg laying in a tropical lizard *Anolis aeneus*. Copeia 1976: 759–764.

———. 1977a. The relationship between resource competition, risk, and aggression in a tropical territorial lizard. Ecology 58: 349–358.

———. 1977b. Rainfall, moisture and dry season growth rates in *Anolis aeneus*. Copeia 1977: 415–419.

———. 1977c. The function of the survey posture in *Anolis* lizards. Copeia 1977: 756–758.

———. 1978. A field study of the ontogeny of social behavior in the lizard *Anolis aeneus*. Behaviour 66: 1–31.

———. 1983a. The relationship between ontogenetic habitat shift, competition and predation avoidance in a juvenile lizard (*Anolis aeneus*). Behavioral Ecology and Sociobiology 12: 19–33.

———. 1983b. Territoriality and the defence of the predator-refuges in juvenile lizards. Animal Behaviour 31: 857–870.

———. 1983c. Sexual selection, sexual dimorphism, and territoriality. In: R. B. Huey, E. R. Pianka, and T. W. Schoener (eds.), Lizard Ecology: Studies of a Model Organism, pp. 169–204. Cambridge, Mass.: Harvard University Press.

———. 1984a. The growth costs of territorial overlap: Experiments with juvenile lizards (*Anolis aeneus*). Behavioral Ecology and Sociobiology 15: 115–119.

———. 1984b. Rank-dependent compromises between growth and predator protection in lizard dominance hierarchies. Animal Behaviour 32: 1101–1107.

———. 1987a. Conspecifics as cues to territory quality: A preference of juvenile lizards (*Anolis aeneus*) for previously used territories. American Naturalist 129: 629–642.

———. 1987b. The effect of familiarity with a neighborhood on territorial acquisition. Behavioral Ecology and Sociobiology 21: 273–277.

———. 1988a. Conspecific attraction and aggregation in territorial species. American Naturalist 131: 329–347.

———. 1988b. The effect of body size on habitat and territory choice in juvenile lizards. Herpetologica 44: 369–376.

———. 1990a. The effect of contender pressure on territory size and overlap in seasonally territorial species. American Naturalist 135: 614–632.

———. 1990b. Starter homes for young lizards. Natural History, October 1990: 40–44.

———. 1991. The effect of conspecifics on habitat selection in territorial species. Behavioral Ecology and Sociobiology 28: 29–36.

———. 1992. Simultaneous versus sequential settlement in territorial species. American Naturalist 139: 1070–1088.

———. 1999. Relationships between female density and sexual size dimorphism in samples of *Anolis sagrei*. Copeia 1999: 760–765.

Stamps, J. A., and G. W. Barlow. 1973. Variation and stereotypy in the displays of *Anolis aeneus* (Sauria: Iguanidae). Behaviour 47: 67–94.

Stamps, J. A., and D. P. Crews. 1976. Seasonal changes in reproduction and social behavior in the lizard *Anolis aeneus*. Copeia 1976: 467–476.

Stamps, J. A., and P. K. Eason. 1989. Relationship between spacing behavior and growth rates: A field study of lizard feeding territories. Behavioral Ecology and Sociobiology 25: 99–107.

Stamps, J. A., and V. V. Krishnan. 1994a. Territory acquisition in lizards: I. First encounters. Animal Behaviour 47: 1375–1385.

———. 1994b. Territory acquisition in lizards: II. Establishing social and spatial relationships. Animal Behaviour 47: 1387–1400.

———. 1995. Territory acquisition in lizards: III. Competing for space. Animal Behaviour 49: 679–693.

———. 1997. The function of fights in territory establishment. American Naturalist 150: 393–405.

———. 1998. Territory acquisition in lizards. IV. Obtaining

high status and exclusive home ranges. Animal Behaviour 55: 461–472.

Stamps, J. A., and S. K. Tanaka. 1981a. The influence of food and water on growth rates in a tropical lizard (*Anolis aeneus*). Ecology 61: 33–40.

———. 1981b. The relationship between food and social behavior in juvenile lizards (*Anolis aeneus*). Copeia 1981: 422–434.

Stamps, J. A., and K. Tollestrup. 1984. Prospective resource defense in a territorial species. American Naturalist 123: 99–114.

Stamps, J. A., S. K. Tanaka, and V. V. Krishnan. 1981. The relationship between selectivity and food abundance in a juvenile lizard. Ecology 62: 1079–1092.

Steinberg, D. S., S. D. Powell, R. Powell, J. S. Parmerlee, Jr., and R. W. Henderson. 2007. Population densities, water-loss rates, and diets of *Sphaerodactylus vincenti* on St. Vincent, West Indies. Journal of Herpetology 41: 330–336.

Steinberg, D. S., J. L. Hite, R. Powell, and R. W. Henderson. 2008. *Sphaerodactylus vincenti*. Catalogue of American Amphibians and Reptiles 853: 1–6.

Stejneger, L. 1903. A new species of large iguana from the Bahama Islands. Proceedings of the Biological Society of Washington 16: 129–132.

———. 1904. The herpetology of Porto Rico. Report of the United States National Museum for the Year 1902: 551–724.

———. 1917. Cuban amphibians and reptiles collected for the United States National Museum from 1899 to 1902. Proceedings of the United States National Museum 53: 259–291.

Stephen, C. 2005. Genetic studies update. Iguana Specialist Group Newsletter 8(2): 12–13.

———. 2006. Genetic studies update. Iguana 13: 127–128.

———. 2007. Update on iguanine phylogeography and phylogenetics. Iguana Specialist Group Newsletter 10(1): 20 (reprinted in Iguana 14: 251).

Stettler, P. H. 1978. Handbuch der Terrarienkunde. Stuttgart, Germany: Franckh-Verlag.

Stevens, M., and G. Waldmann. 2001. Animal Biodiversity of the Lesser Antillean Island of Montserrat (British West Indies): An Annotated Checklist of Terrestrial and Freshwater Animals. Archiv zoologischer Publikationen, Band 6. Nümbrecht, Germany: Martina Galunder-Verlag.

Stevenson, P. H. 1991. *Anolis lividus* (NCN). Feeding. Herpetological Review 22: 57–58.

Stewart, M. M. 1977. The role of introduced species in a Jamaican frog community. Actas del IV Simposium Internacional de Ecología Tropical, Tomo 1: 111–146.

———. 1985. Arboreal habitat use and parachuting by a subtropical forest frog. Journal of Herpetology 19: 391–401.

———. 1995. Climate driven population fluctuations in rain forest frogs. Journal of Herpetology 29: 437–446.

Stewart, M. M., and P. J. Bishop. 1994. Effects of increased sound level of advertisement calls on calling male frogs, *Eleutherodactylus coqui*. Journal of Herpetology 28: 46–53.

Stewart, M. M., and G. E. Martin. 1980. Coconut husk-piles—a unique habitat for Jamaican terrestrial frogs. Biotropica 12: 107–116.

Stewart, M. M., and F. H. Pough. 1983. Population density of tropical forest frogs: Relation to retreat sites. Science 221: 570–572.

Stewart, M. M., and A. S. Rand. 1991. Vocalizations and the defense of retreat sites by male and female frogs, *Eleutherodactylus coqui*. Copeia 1991: 1013–1024.

———. 1992. Diel variation in the use of aggressive calls by the frog *Eleutherodactylus coqui*. Herpetologica 48: 49–56.

Stewart, M. M., and L. L. Woolbright. 1996. Amphibians. In: D. P. Reagan and R. B. Waide (eds.), The Food Web of a Tropical Rain Forest, pp. 273–320. Chicago: University of Chicago Press.

Stöckl, H., and E. Stöckl. 2008. Die Abgottschlange *Boa constrictor*. Münster, Germany: Natur und Tier.

Stuart, S. N. 2006. *Eleutherodactylus maestrensis*. In: IUCN 2008. 2008 IUCN Red List of Threatened Species. http://www.iucnredlist.org.

Stuart, S. N., M. Hoffmann, J. S. Chanson, N. A. Cox, R. Berridge, P. Ramani, and B. E. Young (eds.). 2008. Threatened Amphibians of the World. Barcelona, Spain: Lynx Ediciones, with IUCN-The World Conservation Union, Conservation International, and NatureServe.

Stull, O. G. 1928. A revision of the genus *Tropidophis*. Occasional Papers of the Museum of Zoology, University of Michigan 195: 1–49 + 3 plates.

———. 1935. A checklist of the family Boidae. Proceedings of the Boston Society of Natural History 40: 387–408.

Sutcliffe, R. 1952. Results of the Catherwood-Chaplin West Indies expedition, 1948. Part VI. Amphibia and Reptilia. Notulae National Academy of Natural Sciences, Philadelphia 243: 1–8.

Szidat, H. 1969. Zucht und Haltung kubanischer *Anolis* im Terrarium 1. Aquarien Terrarien 16: 194–197.

Taigen, T. L., F. H. Pough, and M. M. Stewart. 1984. Water balance of terrestrial anuran (*Eleutherodactylus coqui*) eggs: Importance of parental care. Ecology 65: 248–255.

Tanner, T. H. 1862. Memoranda on Poisons. 2nd edition, enlarged and improved. London: Henry Renshaw.

Tarbox, G. 2005. Habitat selection by Dominica anoles. Unpublished report. Texas A&M University Study Abroad, College Station. http://dominica.tamu.edu/student%20-projects/projects_by_year.htm.

Taylor, E. H. 1963. The lizards of Thailand. University of Kansas Science Bulletin 44: 687–1077.

Teale, H. K. 1930. Incubation period of *Hemidactylus brooki* Gray and *Sphaerodactylus difficilis* Barbour. Copeia 1930: 52.

Telford, S. R. 1975. Saurian malaria in the Caribbean; *Plas-*

modium azurophilum, sp. nov., a malarial parasite with schizogony and gametogony in both red and while blood cells. International Journal of Parasitology 5: 383–394.

Ten Eyck, G. R., and L. E. Kruger. 2002. *Eleutherodactylus coqui* (Puerto Rican Coqui). Immobilization. Herpetological Review 33: 46.

Tenlén, R. 1997a. Mona-saaren leguaani, *Cyclura cornuta stejnegeri*, Barbour and Noble 1916. Herpetomania 6(3): 5–13.

———. 1997b. Vähä-Antillien Leguaani *Iguana delicatissima* Laurenti 1768: Lajin tila St. Eustatius. Herpetomania 6(4): 5–24.

Tenlén, R., and L. C. M. Wijffels. 1999. Veldwaarnemingen aan de anolis van het eiland Mona, *Anolis monensis*. Lacerta 57: 143–148.

Teubner, V. A., and P. J. Tolson. 1984. Behavioral and endocrine aspects of the reproductive cycle in the Haitian boa *Epicrates striatus* (abstract). American Zoologist 24: 114A.

Thomas, K. R., and R. Thomas. 1978. Locomotor activity responses to photoperiod in four West Indian fossorial squamates of the genera *Amphisbaena* and *Typhlops* (Reptilia, Lacertilia). Journal of Herpetology 12: 35–41.

Thomas, L., N. Chausson, J. Uzan, S. Kaidomar, R. Vignes, Y. Plumelle, B. Bucher, and D. Smadja. 2006. Thrombotic stroke following snake bites by the "Fer-de-Lance" *Bothrops lanceolatus* in Martinique despite antivenom treatment: A report of three recent cases. Toxicon 48: 23–28.

Thomas, R. 1963. Cayman Islands *Tropidophis* (Reptilia, Serpentes). Breviora 195: 1–8.

———. 1964. The races of *Sphaerodactylus fantasticus* Dumeril and Bibron in the Lesser Antilles. Caribbean Journal of Science 4: 373–390.

———. 1965a. A new anole (Sauria, Iguanidae) from Puerto Rico. Part II. Field observations on *Anolis occultus* Williams and Rivero. Breviora 231: 10–16.

———. 1965b. A new species of *Typhlops* from the Barahona Peninsula of Hispaniola. Copeia 1965: 436–439.

———. 1965c. Two new species of *Amphisbaena* (Amphisbaenia, Reptilia) from the Barahona Peninsula of Hispaniola. Breviora 215: 1–14.

———. 1965d. A congregation of the blind snake, *Typhlops richardi*. Herpetologica 21: 309.

———. 1965e. The feeding habits of captive amphisbaenids. Herpetologica 21: 238.

———. 1965f. A reassessment of the Virgin Islands *Typhlops* with the description of two new subspecies. Revista de Biología Tropical 13: 187–201.

———. 1965g. The races of *Sphaerodactylus fantasticus* Dumeril and Bibron in the Lesser Antilles. Caribbean Journal of Science 4: 373–390.

———. 1966a. New species of Antillean *Eleutherodactylus*. Quarterly Journal of the Florida Academy of Science 28: 375–391.

———. 1966b. Leeward Islands *Typhlops* (Reptilia, Serpentes). Proceedings of the Biological Society of Washington 79: 255–266.

———. 1966c. A reassessment of the herpetofauna of Navassa Island. Journal of the Ohio Herpetological Society 5: 73–89.

———. 1966d. Additional notes on the amphisbaenids of Greater Puerto Rico. Breviora 249: 1–23.

———. 1966e. A new Hispaniolan gecko. Breviora 253: 1–5.

———. 1968a. Notes on Antillean geckos (*Sphaerodactylus*). Herpetologica 24: 46–60.

———. 1968b. The *Typhlops biminiensis* group of Antillean blind snakes. Copeia 1968: 713–722.

———. 1971. A new species of *Diploglossus* (Sauria: Anguidae) from Hispaniola. Occasional Papers of the Museum of Zoology, Louisiana State University 40: 1–9.

———. 1974. A new species of *Typhlops* (Serpentes: Typhlopidae) from Hispaniola. Proceedings of the Biological Society of Washington 87: 11–18.

———. 1975. The *argus* group of West Indian *Sphaerodactylus* (Sauria: Gekkonidae). Herpetologica 31: 177–195.

———. 1976. Systematics of Antillean Blind Snakes of the Genus *Typhlops* (Serpentes: Typhlopidae). Ph.D. dissertation, Louisiana State University, Baton Rouge.

———. 1982. A new dwarf *Sphaerodactylus* from Haiti (Lacertilia: Gekkonidae). Proceedings of the Biological Society of Washington 95: 81–88.

———. 1989. The relationships of Antillean *Typhlops* (Serpentes: Typhlopidae) and the descriptions of three new species. In: C. A. Woods (ed.), Biogeography of the West Indies: Past, Present, and Future, pp. 409–432. Gainesville, Fla.: Sandhill Crane Press.

———. 1999. The Puerto Rico area. In: B. I. Crother (ed.), Caribbean Amphibians and Reptiles, pp. 169–179. San Diego: Academic Press.

Thomas, R., and A. Gaa Kessler. 1996. Nonanoline reptiles. In: D. P. Reagan, and R. B. Waide (eds.), The Food Web of a Tropical Rain Forest, pp. 347–362. Chicago: University of Chicago Press.

Thomas, R., and S. B. Hedges. 1988. Two new geckos (*Sphaerodactylus*) from the Sierra Martin Garcia of Hispaniola. Herpetologica 44: 96–104.

———. 1989. A new *Celestus* (Sauria: Anguidae) from the Chaine de la Selle of Haiti. Copeia 1989: 886–891.

———. 1991. Rediscovery and description of the Hispaniolan lizard *Anolis darlingtoni* (Sauria: Iguanidae). Caribbean Journal of Science 27: 90–93.

———. 1992. An unusual new *Sphaerodactylus* from Hispaniola (Squamata: Gekkonidae). Journal of Herpetology 26: 289–292.

———. 1993. A new banded *Sphaerodactylus* from eastern Hispaniola (Squamata: Gekkonidae). Herpetologica 49: 350–354.

———. 1998a. New anguid lizard (*Diploglossus*) from Cuba. Copeia 1998: 97–103.

———. 1998b. A new amphisbaenian from Cuba. Journal of Herpetology 32: 92–96.

———. 1998c. A new gecko from the Sierra de Neiba of Hispaniola (Squamata: Gekkonidae: *Sphaerodactylus*). Herpetologica 54: 333–336.

———. 2006. Two new species of *Amphisbaena* (Reptilia: Squamata: Amphisbaenidae) from the Tiburon Peninsula of Haiti. Caribbean Journal of Science 42: 208–219.

———. 2007. Eleven new species of snakes of the genus *Typhlops* (Serpentes: Typhlopidae) from Hispaniola and Cuba. Zootaxa 1400: 1–26.

Thomas, R., and R. Joglar. 1996. The herpetology of Puerto Rico, past present and future. In: J. C. Figueroa Colon (ed.), The Scientific Survey of Puerto Rico and Virgin Islands. An Eighty-year Reassessment of the Island's Natural History, pp. 181–200. Annals of the New York Academy of Science, Volume 766.

Thomas, R., and M. Leal. 1993. Feeding envenomation by *Arrhyton exiguum* (Serpentes: Colubridae). Journal of Herpetology 27: 107–109.

Thomas, R., and R. Powell. 1992. *Typhlops hectus*. Catalogue of American Amphibians and Reptiles 550: 1–2.

———. 1994a. *Typhlops pusillus*. Catalogue of American Amphibians and Reptiles 595: 1–2.

———. 1994b. *Typhlops schwartzi*. Catalogue of American Amphibians and Reptiles 596: 1–2.

———. 1994c. *Typhlops sulcatus*. Catalogue of American Amphibians and Reptiles 597: 1–2.

———. 1994d. *Typhlops tetrathyreus*. Catalogue of American Amphibians and Reptiles 598: 1–2.

———. 1995a. *Typhlops capitulatus*. Catalogue of American Amphibians and Reptiles 618: 1–2.

———. 1995b. *Typhlops gonavensis*. Catalogue of American Amphibians and Reptiles 619: 1–2.

———. 1995c. *Typhlops titanops*. Catalogue of American Amphibians and Reptiles 620: 1–2.

Thomas, R., and J. A. Prieto Hernández. 1985. The use of venom by the Puerto Rican snake, *Alsophis portoricensis*. Decimo Simposio Recursos Naturales 1983: 13–22a.

Thomas, R., and A. Schwartz. 1965. Hispaniolan snakes of the genus *Dromicus* (Colubridae). Revista de Biología Tropical 13: 59–83.

———. 1966. *Sphaerodactylus* (Gekkonidae) in the Greater Puerto Rico Region. Bulletin of the Florida State Museum 10: 193–260.

———. 1967. The *monticola* group of the lizard genus *Anolis* in Hispaniola. Breviora 261: 1–27.

———. 1977. Three new species of *Sphaerodactylus* (Sauria: Gekkonidae) from Hispaniola. Annals of the Carnegie Museum 46: 33–43.

———. 1983a. The *difficilis* complex of *Sphaerodactylus* (Sauria, Gekkonidae) of Hispaniola. Part 2. *Sphaerodactylus savagei*, *S. cochranae*, *S. darlingtoni*, *S. armstrongi*, *S. streptophorus*, and conclusions. Bulletin of Carnegie Museum of Natural History 22: 31–60.

———. 1983b. Variation in Hispaniolan *Sphaerodactylus* (Sauria: Gekkonidae). In: A. G. J. Rhodin and K. Miyata (eds.), Advances in Herpetology and Evolutionary Biology, pp. 86–98. Cambridge, Mass.: Museum of Comparative Zoology, Harvard University.

Thomas, R., and K. R. Thomas. 1977. Distributional records of amphibians and reptiles from Puerto Rico. Herpetological Review 8: 40.

Thomas, R., R. W. McDiarmid, and F. G. Thompson. 1985. Three new species of thread snakes (Serpentes: Leptotyphlopidae) from Hispaniola. Proceedings of the Biological Society of Washington 98: 204–220.

Thomas, R., S. B. Hedges, and O. H. Garrido. 1992. Two new species of *Sphaerodactylus* from eastern Cuba (Squamata: Gekkonidae). Herpetologica 48: 358–367.

———. 1998. A new gecko (*Sphaerodactylus*) from the Sierra Maestra of Cuba. Journal of Herpetology 32: 66–69.

Thomas, R., R. W. Henderson, R. Powell, and P. Genero Rodríguez. 2007. *Alsophis anomalus* (Hispaniolan Brown Racer). Maximum Size. Herpetological Review 38: 338–339.

Thompson, R. L. 1996. Larval habitat, ecology, and parental investment of *Osteopilus brunneus* (Hylidae). In: R. Powell and R. W. Henderson (eds.), Contributions to West Indian Herpetology: A Tribute to Albert Schwartz, pp. 259–269. Contributions to Herpetology, volume 12. Ithaca, N.Y.: Society for the Study of Amphibians and Reptiles.

Thorbjarnarson, J. B. 1988. The status and ecology of the American Crocodile in Haiti. Bulletin of the Florida State Museum, Biological Sciences 33: 1–86.

Thorpe, R. S. 2002. Analysis of color spectra in comparative evolutionary studies: Molecular phylogeny and habitat adaptation in the St. Vincent anole (*Anolis trinitatis*). Systematic Biology 51: 554–569.

Thorpe, R. S., and A. G. Stenson. 2003. Phylogeny, paraphyly and ecological adaptation of the colour and pattern in the *Anolis roquet* complex on Martinique. Molecular Ecology 12: 117–132.

Thorpe, R. S., A. Malhotra, A. Stenson, and J. T. Reardon. 2004. Adaptation and speciation in Lesser Antillean anoles. In: U. Dieckmann, M. Doebeli, J. A. J. Metz, and D. Tautz (eds.), Adaptive Speciation, pp. 324–342. Cambridge: Cambridge University Press.

Thorpe, R. S., A. G. Jones, A. Malhotra, and Y. Surget-Groba. 2008a. Adaptive radiation in Lesser Antillean lizards: Molecular phylogenetics and species recognition in the Lesser Antillean dwarf gecko complex, *Sphaerodactylus fantasticus*. Molecular Ecology 17: 1489–1504.

Thorpe, R. S., Y. Surget-Groba, and H. Johansson. 2008b. The relative importance of ecology and geographic isolation for speciation in anoles. Philosophical Transactions of the Royal Society B: Biological Sciences 363: 3071–3081.

Tilman, D., R. M. May, C. L. Lehman, and M. A. Nowak. 2002. Habitat destruction and the extinction debt. Nature 371: 65–66.

Timmermann, A., B. Dalsgaard, J. M. Olesen, L. H. Andersen, and A. M. Martín González. 2008. *Anolis aeneus* (Grenadian Bush Anole). *Anolis richardii* (Grenadian Tree Anole). Nectivory/pollination. Herpetological Review 39: 84–85.

Tirado, G., and A. R. Lewis. 1997. Use of RAPD-PCR for parentage analysis in the teiid lizard *Ameiva exsul*. Journal of Herpetology 31: 436–440.

Tolson, P. J. 1980. Comparative reproductive behavior of four species of snakes in the boid genus *Epicrates*. In: Proceedings of the 4th Reptile Symposium on Captive Propagation and Husbandry, pp. 87–97. Thurmont, Md.: Zoological Consortium.

———. 1982. Captive propagation and husbandry of the Cuban Dwarf Boa, *Tropidophis melanurus* at the Museum of Zoology, the University of Michigan. In: Proceedings of the 5th Reptile Symposium on Captive Propagation and Husbandry, pp. 43–46. Thurmont, Md.: Zoological Consortium.

———. 1983. Captive propagation and husbandry of the Cuban Boa, *Epicrates angulifer*. In: Proceedings of the 6th Reptile Symposium on Captive Propagation and Husbandry, pp. 285–292. Thurmont, Md.: Zoological Consortium.

———. 1988. Critical habitat, predator pressures, and the management of *Epicrates monensis* (Serpentes: Boidae) on the Puerto Rico Bank: A multivariate analysis. In: R. C. Szaro, K. E. Severson, and D. R. Patton (eds.), Management of Amphibians, Reptiles, and Small Mammals in North America, pp. 228–238. General Technical Report RM-166. Fort Collins, Colo.: United States Department of Agriculture (USDA) Forest Service.

———. 1991a. *Epicrates* (West Indian Boa). Reproductive Longevity. Herpetological Review 22: 100.

———. 1991b. The conservation status of *Epicrates monensis* (Serpentes: Boidae) on the Puerto Rico Bank. In: J. A. Moreno (ed.), Status y Distribución de los Reptiles y Anfibios de la Región de Puerto Rico, pp. 11–19. Publicación Científica Miscelánea No. 1. San Juan: Departamento de Recursos Naturales de Puerto Rico.

———. 1992. The reproductive biology of the Neotropical boid genus *Epicrates* (Serpentes: Boidae). In: W. C. Hamlett (ed.), Reproductive Biology of South American Vertebrates, pp. 165–178. New York: Springer-Verlag.

———. 1996a. Conservation of *Epicrates monensis* on the satellite islands of Puerto Rico. In: R. Powell and R. W. Henderson (eds.), Contributions to West Indian herpetology: A Tribute to Albert Schwartz, pp. 407–416. Contributions to Herpetology, volume 12. Ithaca, N.Y.: Society for the Study of Amphibians and Reptiles.

———. 1996b. *Epicrates inornatus*. In: IUCN 2008. 2008 IUCN Red List of Threatened Species. http://www.iucnredlist.org.

———. 1996c. *Sphaerodactylus micropithecus*. In: IUCN 2008. 2008 IUCN Red List of Threatened Species. http://www.iucnredlist.org.

———. 1996d. *Typhlops monensis*. In: IUCN 2008. 2008 IUCN Red List of Threatened Species. http://www.iucnredlist.org.

Tolson, P. J., and E. W. Campbell III. 1989. *Typhlops richardi* (Richard's Blind Snake). Arboreal activity. Herpetological Review 20: 75.

Tolson, P. J., and R. W. Henderson. 1993. The Natural History of West Indian Boas. Taunton, England: R & A Publishing.

———. 2006. An overview of snake conservation in the West Indies. Applied Herpetology 3: 345–356.

Tolson, P. J., and V. A. Teubner. 1987. The role of social manipulation and environmental cycling in propagation of the boid genus *Epicrates*: Lessons from the field and laboratory. American Association of Zoological Parks and Aquariums Regional Conference Proceedings 1987: 606–613.

Tolson, P. J., M. A. García, and C. L. Ellsworth. 2007. Habitat use by the Mona Boa (*Epicrates monensis monensis*) on Isla Mona, West Indies. In: R. W. Henderson and R. Powell (eds.), Biology of the Boas and Pythons, pp. 118–126. Eagle Mountain, Utah: Eagle Mountain Publishing.

Toro, E., A. Herrel, and D. J. Irschick. 2004. The evolution of jumping performance in Caribbean *Anolis* lizards: Solutions to biomechanical trade-offs. American Naturalist 163: 844–856.

———. 2006. Movement control strategies during jumping in a lizard (*Anolis valencienni*). Journal of Biomechanics 39: 2014–2019.

Torres, J. A., R. Thomas, M. Leal, and T. Gush. 2000. Ant and termite predation by the tropical blindsnake *Typhlops platycephalus*. Insectes Sociaux 47: 1–6.

Tortoise & Freshwater Turtle Specialist Group. 1996a. *Geochelone denticulata*. In: IUCN 2008. 2008 IUCN Red List of Threatened Species. http://www.iucnredlist.org.

———. 1996b. *Trachemys decorata*. In: IUCN 2008. 2008 IUCN Red List of Threatened Species. http://www.iucnredlist.org.

———. 1996c. *Trachemys stejnegeri*. In: IUCN 2008. 2008 IUCN Red List of Threatened Species. http://www.iucnredlist.org.

———. 1996d. *Trachemys terrapen*. In: IUCN 2008. 2008 IUCN Red List of Threatened Species. http://www.iucnredlist.org.

Townsend, D. S. 1986. The costs of male parental care and its evolution in a Neotropical frog. Behavioral Ecology and Sociobiology 19: 187–195.

———. 1989. The consequences of microhabitat choice for male reproductive success in a tropical frog (*Eleutherodactylus coqui*). Herpetologica 45: 451–458.

———. 1996. Patterns of parental care in frogs of the genus *Eleutherodactylus*. In: R. Powell and R. W. Henderson (eds.), Contributions to West Indian Herpetology: A Tribute to Albert Schwartz, pp. 229–239. Contributions to Herpetology, volume 12. Ithaca, N.Y.: Society for the Study of Amphibians and Reptiles.

Townsend, D. S., and M. M. Stewart. 1986a. Courtship and mating behavior of a Puerto Rican frog, *Eleutherodactylus coqui*. Herpetologica 42: 165–170.

———. 1986b. The effect of temperature on direct development in a terrestrial-breeding, Neotropical frog. Copeia 1986: 520–523.

———. 1994. Reproductive ecology of the Puerto Rican frog *Eleutherodactylus coqui*. Journal of Herpetology 28: 34–40.

Townsend, D. S., M. M. Stewart, F. H. Pough, and B. F. Brussard. 1981. Internal fertilization in an oviparous frog. Science 212: 469–471.

Townsend, D. S., M. M. Stewart, and F. H. Pough. 1984. Male parental care and its adaptive significance in a Neotropical frog. Animal Behaviour 32: 421–431.

Townsend, J. H., J. M. Eaton, R. Powell, J. S. Parmerlee, Jr., and R. W. Henderson. 2000. Cuban Treefrogs (*Osteopilus septentrionalis*) in Anguilla, Lesser Antilles. Caribbean Journal of Science 36: 326–328.

Townson, S. 1980. West Indian iguanas of the genus *Cyclura*: The threat of habitat destruction in the Cayman islands. British Journal of Herpetology 6: 101–104.

Trefaut Rodrigues, M., U. Caramaschi, and A. Mijares. 2004. *Scinax x-signatus*. In: IUCN 2008. 2008 IUCN Red List of Threatened Species. http://www.iucnredlist.org.

Treglia, M. L. 2006. An annotated checklist of the amphibians and reptiles of St. Vincent, West Indies. Iguana 13: 251–262.

Treglia, M. L., A. J. Muensch, R. Powell, and J. S. Parmerlee, Jr. 2008. Invasive *Anolis sagrei* on St. Vincent and its potential impact on perch heights of *Anolis trinitatis*. Caribbean Journal of Science 44: 251–256.

Trivers, R. L. 1972. Parental investment and sexual selection. In: B. Campbell (ed.), Sexual Selection and the Descent of Man 1871–1971, pp. 136–179. Chicago: Aldine.

———. 1976. Sexual selection and resource-accruing abilities in *Anolis garmani*. Evolution 30: 253–269.

Trueb, L., and M. J. Tyler. 1974. Systematics and evolution of the Greater Antillean hylid frogs. Occasional Papers of the Museum of Natural History University of Kansas 24: 1–60.

Turner, F. B., and C. S. Gist. 1970. Observations of lizards and tree frogs in an irradiated Puerto Rican forest. In: H. T. Odum (ed.), A Tropical Rain Forest, pp. E25–E49. Springfield, Va.: U.S. Atomic Energy Commission.

Tyler, Lieut. 1849. Notes on the serpents of St. Lucia. Proceedings of the Zoological Society of London 1849: 100–104.

Underwood, G. 1959a. A new Jamaican galliwasp (Sauria, Anguidae). Breviora 102: 1–13.

———. 1959b. The anoles of the eastern Caribbean (Sauria, Iguanidae). Pt. III. Revisionary notes. Bulletin of the Museum of Comparative Zoology 121: 191–226.

———. 1962. Reptiles of the eastern Caribbean. Caribbean Affairs (new series) 1: [2] + 192 pp.

———. 1964. An anguid lizard from the Leeward Islands. Breviora 200: 1–10.

———. 1993. A new snake from St. Lucia, West Indies. Bulletin of the Natural History Museum (Zoology) 59: 1–9.

———. 1995. A Tale of Old Serpents. Castries: St. Lucia National Trust.

Underwood, G., and E. E. Williams. 1959. The anoline lizards of Jamaica. Bulletin of the Institute of Jamaica Science Series 9: 1–48.

Underwood, G., J. A. Horrocks, and J. C. Daltry. 1999. A new snake from Barbados. Journal of the Barbados Museum and Historical Society 45: 67–75.

UNEP-WCMC (United Nations Environment Programme World Conservation Monitoring Centre). 2008. UNEP-WCMC Species Database: CITES-Listed Species http://www.cites.org/eng/resources/species.html.

United Nations. 2007. World Urbanization Prospects: The 2007 Revision. Highlights. http://www.un.org/esa/population/publications/wup2007/2007WUP_Highlights_web.pdf.

USFWS (United States Fish & Wildlife Service). 2008. Species Information Threatened and Endangered Animals and Plants. http://www.fws.gov/endangered/wildlife.html.

Utiger, U., B. Schätti, and N. Helfenberger. 2005. The Oriental colubrine genus *Coelognathus* Fitzinger, 1843 and classification of Old and New World racers and ratsnakes (Reptilia, Squamata, Colubridae, Colubrinae). Russian Journal of Herpetology 12: 39–60.

Valderrama Puente, M. de J., and L. Rodríguez Schettino. 1988. Algunas characterísticas reproductivas de *Anolis lucius* (Sauria: Iguanidae). Poeyana 358: 1–15.

Valdés de la Osa, A. 1978. Utilazación del nido de *Speotyto cunicularia* (Aves: Strigiformes) como refugio por *Bufo peltocephalus* (Amphibia: Bufonidae). Miscelanea Zoológica 7: 4.

———. 1989. Notas sobre *Sminthillus limbatus* (Cope). Caribbean Journal of Science 25: 96–98.

Valdés de la Osa, A., and F. N. Ruíz García. 1979. Manifestación del patrón conductual de aumento de volumen en bufónidos cubanos. Miscelanea Zoológica 8: 3–4.

———. 1980a. Consideraciones sistemáticas sobre *Bufo longinasus* (Anura: Bufonidae) de Cuba y descripción de una nueva subespecie. Poeyana 206: 1–33.

———. 1980b. Caso de canabalismo en *Eleutherodactylus cuneatus* (Cope) (Salientia: Leptodactylidae). Miscelanea Zoológica 6: 4.

Vale, A. 2002. A characterization, correlation, and comparison of Tree Gecko and Tink Frog vocalizations of Dominica. Unpublished report. Texas A&M University Study Abroad, College Station. http://dominica.tamu.edu/student%20projects/projects_by_year.htm.

Valentine, L. 2002. Morphological variation in *Anolis oculatus* between Dominican habitats. Unpublished report. Texas A&M University Study Abroad, College Station. http://dominica.tamu.edu/student%20projects/projects_by_year.htm.

Vales, M., A. Alvarez, L. Montes, and A. Ávila (eds.). 1998. Estudio Nacional sobre la Diversidad Biológica en la República de Cuba. Madrid: CESYTA, S.L.

Valido, A. 2006. *Anolis allisoni* (Allison's Anole/Cameleón Azul). Nectar feeding. Herpetological Review 37: 461.

Valiulis, J. M., K. N. Hines, and J. B. Iverson. 2003. Adventures of the iguana patrol. In: R. W. Henderson and R. Powell (eds.), Islands and the Sea: Essays on Herpetological Exploration in the West Indies, pp. 39–48. Contributions to Herpetology, volume 20. Ithaca, N.Y.: Society for the Study of Amphibians and Reptiles.

Van Berkum, F., F. H. Pough, M. M. Stewart, and P. F. Brussard. 1982. Altitudinal and interspecific differences in the rehydration abilities of Puerto Rican frogs (*Eleutherodactylus*). Physiological Zoology 55: 130–136.

van Buel, H., and R. Powell. 2006. Geographic distribution: *Gymnophthalmus underwoodi*. Herpetological Review 37: 494.

Van Veen, R., and B. Wilson. 2004. Taxon update: Jamaican Iguana (*Cyclura collei*) recovery project, 2004 update. Iguana Specialist Group Newsletter 7(2): 6–7.

———. 2005. Taxon update: Jamaican Iguana (*Cyclura collei*) recovery project, 2004 update. Iguana 12: 28–29.

Vandeventer, T. L. 1992. In search of the Tete'chein: Observations on the natural history of *Boa constrictor nebulosus*. Proceedings of the 16th International Herpetological Symposium on Captive Husbandry & Propagation, pp. 1–9. St. Louis: International Herpetological Symposium (see also http://www.boa-subspecies.com/subspecies/nebulosus.htm).

VanHooydonck, B., A. Herrel, R. Van Damme, and D. J. Irschick. 2005. Does dewlap size predict male bite performance in Jamaican *Anolis* lizards? Functional Ecology 19: 38–42.

———. 2006a. The quick and the fast: The evolution of acceleration capacity in *Anolis* lizards. Evolution 60: 2137–2147.

VanHooydonck, B., A. Herrel, and D. J. Irschick. 2006b. Out on a limb: The differential effect of substrate diameter on acceleration capacity in *Anolis* lizards. Journal of Experi- [illegible] 4515–4523.

———. 2007. Determinants of intersexual differences in escape behaviour in *Anolis* lizards: A comparative approach. Integrative and Comparative Biology 47: 200–210.

Vanhooydonck, B., A. B. Herrel, J. J. Meyers, and D. J. Irschick. 2009. What determines dewlap diversity in *Anolis* lizards? An among-island comparison. Journal of Evolutionary Biology 22: 293–305.

Vanzolini, P. E., and E. E. Williams. 1962. Jamaican and Hispaniolan *Gonatodes* and allied forms (Sauria, Gekkonidae). Bulletin of the Museum of Comparative Zoology 127: 479–498.

Vargas-Salinas, F. 2005. *Bufo marinus* (Cane Toad). Amplexus displacement. Herpetological Review 36: 431–432.

———. 2006. Sexual size dimorphism in the Cuban Treefrog *Osteopilus septentrionalis*. Amphibia-Reptilia 27: 419–426.

———. 2007. Breeding behavior of the Cane Toad *Bufo marinus* (Bufonidae): A successfully invasive species. Herpetological Review 38: 12–17.

Varona, L. S. 1966. Notas sobre los cocodrilos de Cuba y descripción de una especie del Pleistoceno. Poeyana, serie A 16: 1–34.

———. 1976. *Caiman crocodilus* (Reptilia: Alligatoridae) en Cuba. Miscelanea Zoológica 5: 1.

———. 1986. Algunos datos sobre la etología de *Crocodylus rhombifer* (Reptilia: Crocodylidae). Poeyana 313: 1–7.

Varona, L. S., and O. H. Garrido. 1970. Vertebrados de los Cayos de San Felipe, Cuba, incluyendo una nueva especie de jutía. Poeyana, serie A 75: 1–25.

Vergner, I. 1998. Gecko-Zwerge aus der Gattung *Sphaerodactylus*. Die Aquarien- und Terrarienzeitschrift (DATZ) 51: 516–519.

Vergner, I., and P. Polák. 1990. Beobachtungen am Kuba-Rotkehlanolis *Anolis porcatus* Gray, 1840 und Bemerkungen zu *Anolis allisoni* Barbour, 1928. Herpetofauna 12: 6–10.

———. 1996. Kubanische Höhlenanolis (*Anolis lucius*-Gruppe) in der Natur und im Terrarium. Herpetofauna 18: 5–12.

Verstegen, L. 1988. Mijn ervaringen met de anolis *Anolis roquet summus*. Lacerta 46: 178–180.

Vidal, N., A.-S. Delmas, P. David, C. Cruaud, A. Couloux, and S. B. Hedges. 2007. The phylogeny and classification of caenophidian snakes inferred from seven nuclear protein-coding genes. Comptes Rendus Biologies 330: 182–187.

Vidal, N., A. Azvolinsky, C. Cruaud, and S. B. Hedges. 2008. Origin of tropical American burrowing reptiles by transatlantic rafting. Biology Letters 4: 115–118.

Viella, F. J. 1991. Comments on the process of introducing the St. Croix Ground Lizard *Ameiva polops* into new and former habitats. In: J. A. Moreno (ed.), Status y Distribución de los Reptiles y Anfibios de la Región de Puerto Rico, pp. 23–25. Publicación Científica Miscelánea No. 1. San Juan: Departamento de Recursos Naturales de Puerto Rico.

———. 1998. Biology of the mongoose (*Herpestes javanicus*) in a rain forest of Puerto Rico. Biotropica 30: 120–125.

Villa, J., and D. S. Townsend. 1983. Viable frog eggs eaten by phorid fly larvae. Journal of Herpetology 17: 278–281.

Villanueva-Rivera, L. J., and R. L. Joglar. 2001. *Eleutherodactylus cochranae* (Coqui Pitito). Reproduction. Herpetological Review 32: 182.

Villanueva-Rivera, L. J., R. L. Joglar, and F. C. Li-Objio. 2000. *Eleutherodactylus coqui* (Coqui). Predation. Herpetological Review 31: 100.

Villard, P. 1999. Le Pic de la Guadeloupe. Paris: La Société d'Études Ornithologiques de France.

Viña Dávila, N., and L. F. de Armas. 1989. Depradación de *Tropidophis melanurus* (Serpentes: Tropidophidae) por *Epicrates angulifer* (Serpentes: Boidae). Miscelanea Zoológica 41: 2–3.

Viña Dávila, N., A. M. Botta, G. Garcés, N. Viña Bayés, and A. Ruiz. 1984. Análisis de la variación altitudinal de los vertebrados presentes en la cuenca del Río Palma Mocha en la vertiente sur de la Sierra Maestra, pp. 104–115. In: M. L. García (ed.), Anuario Biología 1984. Santiago de Cuba: Dirección de Información Científico-Técnica, Universidad de Oriente.

Vinnikov, K. Y., A. Robock, R. J. Stouffer, J. E. Wals, C. L. Parkinson, D. J. Cavalieri, J. F. B. Mitchell, D. Garrett, and V. F. Zakharov. 1999. Global warming and northern hemisphere sea ice extent. Science 286: 1934–1937.

Vit, Z. 1981. Der kubanische Gelbkopfgecko—*Gonatodes albogularis fuscus* (Hallowell, 1855). Aquarien Terrarien 28: 68–69.

Vitt, L. J., and P. A. Zani. 1997. Ecology of the nocturnal lizard *Thecadactylus rapicauda* (Sauria: Gekkonidae) in the Amazon Region. Herpetologica 53: 165–179.

Vogel, P. 1983. On an ant-lizard interaction. Oecologia 58: 269–271.

———. 1984. Seasonal hatchling recruitment and juvenile growth of the lizard *Anolis lineatopus*. Copeia 1984: 747–757.

———. 1992. Conservation strategy for the Jamaican Iguana and its habitat. Unpublished report. Jamaican Iguana Research and Conservation Group, Kingston.

———. 1994. Evidence of reproduction in a remnant population of the endangered Jamaican Iguana, *Cyclura collei* (Lacertilia: Iguanidae). Caribbean Journal of Science 30: 234–241.

———. 1996. Harmful prey aversion in hatchling *Anolis lineatopus*: Learning the good things first (abstract). In: R. Powell and R. W. Henderson (eds.), Contributions to West Indian Herpetology: A Tribute to Albert Schwartz, pp. 441–442. Contributions to Herpetology, volume 12. Ithaca, N.Y.: Society for the Study of Amphibians and Reptiles.

———. 2000a. Jamaican Iguana, *Cyclura collei*. In: A. Alberts (ed.), West Indian Iguanas: Status Survey and Conservation Action Plan, pp. 19–22. Gland, Switzerland and Cambridge, UK: International Union for the Conservation of Nature and Natural Resources.

———. 2000b. Wirbeltiere (Vertebrata) und Bromelien. Die Bromelie 2: 32–37.

Vogel, P., and D. A. B. Bundy. 1987. Helminth parasites of Jamaican anoles (Reptilia: Iguanidae): Variation in prevalence and intensity with host age and sex in a population of *Anolis lineatopus*. Parasitology 94: 399–404.

Vogel, P., and F. von Brockhusen-Holzer. 1984. Ants as prey of juvenile *Anolis lineatopus* (Rept., Iguanidae) in prey choice experiments. Zeitschrift für Tierpsychologie 65: 66–76.

Vogel, P., R. Nelson, and R. Kerr. 1996. Conservation strategy for the Jamaican Iguana. In: R. Powell and R. W. Henderson (eds.), Contributions to West Indian Herpetology: A Tribute to Albert Schwartz, pp. 395–406. Contributions to Herpetology, volume 12. Ithaca, N.Y.: Society for the Study of Amphibians and Reptiles.

Vogel, Z. 1965a. Herpetologische Beobachtungen auf Kuba. Aquarien Terrarien 12: 234–239 (reprinted in Dutch as Herpetologische waarnemingen op Cuba I. Lacerta 24: 54–55 and Herpetologische waarnemingen op Cuba II. Lacerta 24: 59–60).

———. 1965b. Herpetologische Beobachtungen auf Kuba (II). Aquarien Terrarien 12: 304–305 (reprinted in Dutch as Herpetologische waarnemingen op Cuba III. Lacerta 24: 69–70).

———. 1965c. Herpetologische Beobachtungen auf Kuba (III). Aquarien Terrarien 12: 340–343 (reprinted in Dutch as Herpetologische waarnemingen op Cuba IV. Lacerta 24: 75–77).

———. 1965d. Herpetologische Beobachtungen auf Kuba (IV). Froschfang auf Kuba. Aquarien Terrarien 12: 420–422 (reprinted in Dutch as Herpetologische waarnemingen op Cuba V [slot]. Lacerta 25: 5–7).

Waide, R. B., and D. P. Reagan. 1996. The rain forest setting. In: D. P. Reagan and R. B. Waide (eds.), The Food Web of a Tropical Rain Forest, pp. 1–16. Chicago: University of Chicago Press.

Wake, M. H. 1978. The reproductive biology of *Eleutherodactylus jasperi* (Amphibia, Anura, Leptodactylidae), with comments on the evolution of live-bearing systems. Journal of Herpetology 12: 121–133.

Walker, L. R., D. J. Lodge, N. V. L. Brokaw, and R. B. Waide. 1991. An introduction to hurricanes in the Caribbean. Biotropica 23: 313–316.

Wallach, V. 2000. *Typhlops tasymicris*. Maximum length and second known specimen. Herpetological Review 31: 180.

Walls, J. G. 1998. The Living Boas. Neptune City, N.J.: T.F.H. Publications.

Walton, A. C. 1940. Notes on amphibian parasites. Proceedings of the Helminthological Society of Washington 7: 87–91.

Wasilewski, J. 1998. Booby Cay update. Iguana Times (Journal of the International Iguana Society) 7: 29–30.

Watlington, F. 1996. Caiman in Puerto Rico. Crocodile Specialist Group Newsletter 17(3): 11–12.

Weaver, J. D., H. Heatwole, J. R. Gorham, and F. J. Rolle. 1961. Institute of Caribbean Studies field excursion to Isla Mona. Caribbean Journal of Science 1: 30–35.

Webster, T. P. 1969. Ecological observations on *Anolis occultus* Williams and Rivero (Sauria, Iguanidae). Breviora 312: 1–5.

———. 1977. Hybridization of Hispaniolan lizards in the *Anolis distichus* species group. In: E. E. Williams (ed.), The Third *Anolis* Newsletter, pp. 166–172. Cambridge, Mass.: Museum of Comparative Zoology, Harvard University.

Webster, T. P., and J. M. Burns. 1973. Dewlap color variation and electrophoretically detected sibling species in a Haitian lizard, *Anolis brevirostris*. Evolution 27: 368–377.

Weiss, A. J., and S. B. Hedges. 2007. Molecular phylogeny and biogeography of the Antillean geckos *Phyllodactylus wirshingi*, *Tarentola americana*, and *Hemidactylus haitianus* (Reptilia, Squamata). Molecular Phylogenetics and Evolution 45: 409–416.

Weldon, P. J., B. J. Demeter, and R. Rosscoe. 1993. A survey of shed skin-eating (dermatophagy) in amphibians and reptiles. Journal of Herpetology 27: 219–228.

Welzel, A. 1981. Die Nachzucht von *Anolis oculatus montanus* im Terrarium. Herpetofauna 13: 22–24.

Werner, F. 1927. Neue oder wenig bekannte Schlangen aus dem Wiener naturhistorischen Staatsmuseum. III. Teil. Sitzungsberichten der Akademie der Wissenschaften in Wien. Mathematisch-naturwissenschaftliche Klasse, Abteilung I, 135: 243–257.

Werner, Y. L. 1969. Eye size in geckos of various ecological types (Reptilia: Gekkonidae and Sphaerodactylidae). Israel Journal of Zoology 18: 291–316.

Werning, H. 2009. Eine der seltensten Echsen der Welt: Der Anegada-Wirtelschwanzleguan. Terraria 4: 34–41.

Westerman, J. H. 1953. Nature preservation in the Caribbean. A review of literature on the destruction and preservation of flora and fauna in the Caribbean area. Publication of the Foundation for Scientific Research in Surinam and the Netherlands Antilles 9: 1–107.

Wetmore, A. 1916. Birds of Porto Rico. United States Department of Agriculture Bulletin 326: 1–140.

Wetmore, A., and B. H. Swales. 1931. The birds of Haiti and the Dominican Republic. Bulletin of the United States National Museum 155: iv + 483 pp.

White, A. M., E. J. Censky, and R. Powell. 2002a. *Ameiva corax*. Catalogue of American Amphibians and Reptiles 746: 1–2.

White, A. M., R. Powell, and E. J. Censky. 2002b. On the thermal biology of *Ameiva* (Teiidae) from the Anguilla Bank, West Indies: Does melanism matter? Amphibia-Reptilia 23: 517–523.

White, L. A., P. J. Muelleman, R. W. Henderson, and R. Powell. 2008 (2009). Behaviour, time management, and foraging modes of a West Indian racer, *Alsophis sibonius*. Herpetological Bulletin 106: 20–25.

White, L. R., and R. Powell. 1996. *Celestus agasepsoides*. Catalogue of American Amphibians and Reptiles 627: 1–2.

White, L. R., R. Powell, J. S. Parmerlee, Jr., A. Lathrop, and D. D. Smith. 1992a. Food habits of three syntopic reptiles from the Barahona Peninsula, Hispaniola. Journal of Herpetology 26: 518–520.

White, L. R., J. S. Parmerlee, Jr., and R. Powell. 1992b. *Typhlops syntherus*. Catalogue of American Amphibians and Reptiles 551: 1–2.

White, T. H., Jr. 2006. Hispaniolan Lizard-Cuckoo (*Coccyzus longirostris*) tethered by Common Green Snake (*Uromacer catesbyi*). Journal of Caribbean Ornithology 19: 52–53.

Whittick, R. J. 1939. On some ticks belonging to the genera *Aponomma* and *Amblyomma*. Parasitology 31: 434–441.

Wiewandt, T. A. 1977. Ecology, Behavior, and Management of the Mona Island Ground Iguana, *Cyclura stejnegeri*. Ph.D. thesis, Cornell University, Ithaca, N.Y.

———. 1979. La Gran Iguana de Mona: On a small island in the Caribbean lives a very large lizard. Natural History 88: 57–65.

Wiewandt, T. A., and M. García. 2000. Mona Island Iguana, *Cyclura cornuta stejnegeri*. In: A. Alberts (ed.), West Indian Iguanas: Status Survey and Conservation Action Plan, pp. 27–31. Gland, Switzerland and Cambridge, UK: International Union for Conservation of Nature and Natural Resources.

Wijffels, L. C. M. 1960. *Anolis*, Gedrag en reacties. Lacerta 18(11): 82–86.

———. 1962. Grasanolissen. Lacerta 20: 27–29.

———. 1964. *Anolis bimaculatus gingivinus*. Lacerta 22: 74.

———. 1971. Enige notities over *Anolis bimaculatus sabanus* Garman, 1887 en zijn natuurlijk voorkomen. Lacerta 30: 7–9.

———. 1972. Die kehlwammenlosen Anolis Kubas, *Anolis bartschi* (Cochran, 1928) und *Anolis vermiculatus* Duméril & Bibron, 1837. Aqua Terra 9(3): 33–36.

———. 1975. Ein seltener Saumfinger von Puerto Rico, *Anolis cuvieri*. Das Aquarium 67: 32–33.

———. 1976. De Antillenleguaan. Lacerta 34: 135–136.

———. 1984. Over Sint Eustatius en de anolissen daar. Bulletin de Nederlandse Studiegroep Anolissen 7(4): 5–7.

———. 1997a. Een nieuwe ontmoeting met de Antillenleguaan (*Iguana delicatissima*). Lacerta 55: 124–128.

———. 1997b. *Anolis bimaculatus* als huisgenoot. Lacerta 55: 201–205.

———. 1997c. *Anolis aeneus*: Een geslaagd emigrant een bijdrage over emigratie bij anolissen. Lacerta 56: 39–43.

———. 2001. Zur Variation bei „Le Roquet," dem *Anolis* von Martinique. Reptilia 6: 35–43.

Wilcox, K., J. Y. Carter, and L. V. Wilcox, Jr. 1973. Range extension of *Cyclura figginsi* Barbour in the Bahamas. Caribbean Journal of Science 13: 211–213.

Wiley, J. W. 1981. The Puerto Rican Amazon (*Amazona vittata*): Its decline and the program for its conservation. In:

R. F. Pasquier (ed.), Conservation of New World Parrots, pp. 133–160. International Council for Bird Preservation Technical Publication No. 1. Washington, D.C.: Smithsonian Institution Press.
———. 2003. Habitat association, size, stomach contents, and reproductive condition of Puerto Rican Boas (*Epicrates inornatus*). Caribbean Journal of Science 39: 189–194.
Wiley, J. W., and B. N. Wiley. 1981. Breeding season ecology and behavior of Ridgway's Hawk (*Buteo ridgwayi*). Condor 83: 132–151.
Williams, E. E. 1961. Notes on Hispaniolan herpetology. 3. The evolution and relationships of the *Anolis semilineatus* group. Breviora 136: 1–8.
———. 1962a. Notes on Hispaniolan herpetology 6. The giant anoles. Breviora 155: 1–15.
———. 1962b. Notes on the herpetology of Hispaniola. 7. New material of two poorly known anoles: *Anolis monticola* Shreve and *Anolis christophei* Williams. Breviora 164: 1–11.
———. 1962c. The anoles (Sauria: Iguanidae) of the eastern Caribbean. Part IV. The anoles of the northern Leewards, Anguilla to Montserrat: New data and a new species. Bulletin of the Museum of Comparative Zoology 127: 453–465.
———. 1963a. Notes on Hispaniolan herpetology. 8. The forms related to *Anolis hendersoni* Cochran. Breviora 186: 1–13.
———. 1963b. *Anolis whitemani*, new species from Hispaniola (Sauria, Iguanidae). Breviora 197: 1–8.
———. 1964. Remarks on the relationships of the reptiles and amphibians of the Cayman Islands. Occasional Papers on Mollusks, Museum of Comparative Zoology 2: 383–384.
———. 1965a. The species of Hispaniolan green anoles (Sauria, Iguanidae). Breviora 227: 1–16.
———. 1965b. Hispaniolan giant anoles (Sauria, Iguanidae): New data and a new subspecies. Breviora 232: 1–7.
———. 1969. The ecology of colonization as seen in the zoogeography of anoline lizards on small islands. Quarterly Review of Biology 44: 345–389.
———. 1972. The origin of faunas. Evolution of lizard congeners in a complex island fauna: A trial analysis. Evolutionary Biology 6: 47–89.
———. 1975. *Anolis marcanoi* new species: Sibling to *Anolis cybotes*: Description and field evidence. Breviora 439: 1–9.
———. 1983. Ecomorphs, faunas, island size, and diverse end points in island radiations of *Anolis*. In: R. B. Huey, E. R. Pianka, and T. W. Schoener (eds.), Lizard Ecology: Studies of a Model Organism, pp. 326–370. Cambridge, Mass.: Harvard University Press.
———. 1999. Over 300 years of collecting in the Caribbean. In: B. I. Crother (ed.), Caribbean Amphibians and Reptiles, pp. 1–30. San Diego: Academic Press.
Williams, E. E., and S. M. Case. 1986. Interactions among members of the *Anolis distichus* complex in and near the Sierra de Baoruco, Dominican Republic. Journal of Herpetology 20: 535–546.
Williams, E. E., and M. Hecht. 1955. "Sunglasses" in two anoline lizards from Cuba. Science 122: 691–692.
Williams, E. E., and A. S. Rand. 1961. Notes on Hispaniolan herpetology. 2. A review of the *Anolis semilineatus* group with the description of *Anolis cochranae*, new species. Breviora 135: 1–11.
———. 1969. *Anolis insolitus*, a new dwarf anole of zoogeographic importance from the mountains of the Dominican Republic. Breviora 326: 1–21.
———. 1977. Species recognition, dewlap function and faunal size. American Zoologist 17: 261–270.
Williams, E. E., and J. A. Rivero. 1965. A new anole (Sauria, Iguanidae) from Puerto Rico. Part I. Description. Breviora 231: 1–9.
Williams, E. E., and T. P. Webster. 1974. *Anolis rupinae* new species a syntopic sibling of *A. monticola* Shreve. Breviora 429: 1–22.
Williams, E. E., B. Shreve, and P. S. Humphrey. 1963. The herpetology of the Port-au-Prince region and Gonave Island, Haiti. Parts I and II. Bulletin of the Museum of Comparative Zoology 129: 291–342.
Williamson, K. E., and R. Powell. 2004. *Gymnophthalmus underwoodi*. Catalogue of American Amphibians and Reptiles 793: 1–5.
Williamson, K. E., A. J. Poche, Jr., B. T. Greene, B. R. Harris, J. M. Germano, P. M. Simmons, D. T. Yorks, R. Powell, J. S. Parmerlee, Jr., and R. W. Henderson. 2002. Herpetofauna of Hog Island, Grenada. Herpetological Bulletin 82: 26–29.
Wilson, B. S. 2003. Of mongooses and lizards. In: R. W. Henderson and R. Powell (eds.), Islands and the Sea: Essays on Herpetological Exploration in the West Indies, pp. 113–119. Contributions to Herpetology, volume 20. Ithaca, N.Y.: Society for the Study of Amphibians and Reptiles.
Wilson, B. S., and R. Van Veen. 2005. Jamaican Iguana Recovery Project, 2005 update. Iguana Specialist Group Newsletter 8(2): 4–6.
———. 2006. Jamaican Iguana Recovery Project. Iguana 13: 119–120.
Wilson, B. S., and P. Vogel. 2000. A survey of the herpetofauna of the Hellshire Hills, Jamaica, including the rediscovery of the Blue-tailed Galliwasp (*Celestus duquesneyi* Grant). Caribbean Journal of Science 36: 244–249.
Wilson, B. S., A. C. Alberts, K. Graham, R. Hudson, R. Kerr, D. Lewis, N. Lung, R. Nelson, N. Thompson, J. L. Kunna, and P. Vogel. 2004a. Survival and reproduction of repatriated Jamaican Iguanas: Headstarting as a viable conservation strategy. In: A. C. Alberts, R. L. Carter, W. K. Hayes, and E. P. Martins (eds.), Iguanas: Biology and Conservation, pp. 220–231. Berkeley and Los Angeles: University of California Press.
Wilson, B. S., O. F. Robinson, and P. Vogel. 2004b. Status of the Jamaican Iguana (*Cyclura collei*): Assessing 15 years of conservation effort. Iguana 11: 224–231.

Wilson, B. S., J. A. Horrocks, and A. Hailey. 2006. Conservation of insular herpetofaunas in the West Indies. Applied Herpetology 3: 181–195.

Wilson, E. O. 1957. Behavior of the Cuban lizard *Chamaeleolis chamaeleontides* (Duméril and Bibron) in captivity. Copeia 1957: 145.

———. 1994. Naturalist. New York: Island Press.

Wilson, L. D. 1992. *Tantilla melanocephala*. Catalogue of American Amphibians and Reptiles 547: 1–3.

Wilson, S. M. 1997. Introduction to the study of the indigenous people of the Caribbean. In: S. M. Wilson (ed.), The Indigenous People of the Caribbean, pp. 1–8. Gainesville: University Press of Florida.

Wing, E. S. 1989. Human exploitation of animal resources in the Caribbean. In: C. A. Woods (ed.), Biogeography of the West Indies: Past, Present, and Future, pp. 137–152. Gainesville, Fla.: Sandhill Crane Press.

———. 2001. Native American use of animals in the Caribbean. In: C. A. Woods and F. E. Sergile (eds.), Biogeography of the West Indies, pp. 481–518. 2nd ed. Boca Raton, Fla.: CRC Press.

Wissmann, S. M., R. L. Hensley, R. Powell, and J. S. Parmerlee, Jr. 2005. Social behaviour in the dwarf geckos *Sphaerodactylus sabanus* and *S. sputator* from St. Eustatius, Netherlands Antilles. Salamandra 41: 45–50.

Wolcott, G. N. 1923. The food of Porto Rican lizards. Journal of the Department of Agriculture of Porto Rico 7: 5–38.

———. 1937. What the giant Surinam Toad, *Bufo marinus* L., is eating now in Puerto Rico. Journal of Agriculture of the University of Puerto Rico 21: 79–84.

———. 1953. The food of the mongoose (*Herpestes javanicus auropunctatus*) in St. Croix and Puerto Rico. Journal of the Department of Agriculture of Porto Rico 37: 241–247.

Wolman, W. 2002. Tink Frog density. Unpublished report. Texas A&M University Study Abroad, College Station. http://dominica.tamu.edu/student%20projects/projects_by_year.htm.

Woodley, J. D. 1980. Survival of the Jamaican Iguana, *Cyclura collei*. Journal of Herpetology 14: 45–48.

Woods, C. A. 1989. The biogeography of West Indian rodents. In: C. A. Woods (ed.), Biogeography of the West Indies: Past, Present, and Future, pp. 741–798. Gainesville, Fla.: Sandhill Crane Press.

Woolbright, L. L. 1985. Patterns of nocturnal movement and calling by the tropical frog *Eleutherodactylus coqui*. Herpetologica 41: 1–9.

———. 1989. Sexual dimorphism in *Eleutherodactylus coqui*: Selection pressures and growth rates. Herpetologica 45: 68–74.

———. 1991. The impact of Hurricane Hugo on forest frogs in Puerto Rico. Biotropica 23: 462–467.

———. 1996. Disturbance influences long-term population patterns in the Puerto Rican frog, *Eleutherodactylus coqui* (Anura: Leptodactylidae). Biotropica 28: 493–501.

———. 1997. Local extinctions of anuran amphibians in the Luquillo Experimental Forest of northeastern Puerto Rico. Journal of Herpetology 31: 572–576.

Woolbright, L. L., and M. M. Stewart. 1987. Foraging success of the tropical frog, *Eleutherodactylus coqui*: The cost of calling. Copeia 1987: 69–75.

Woolbright, L. L., A. H. Hara, C. M. Jacobsen, W. J. Mautz, and F. L. Benevides, Jr. 2006. Population densities of the Coqui, *Eleutherodactylus coqui* (Anura: Leptodactylidae) in newly invaded Hawaii and in native Puerto Rico. Journal of Herpetology 40: 122–126.

World Conservation Monitoring Centre. 1996a. *Ameiva cineracea*. In: IUCN 2008. 2008 IUCN Red List of Threatened Species. http://www.iucnredlist.org.

———. 1996b. *Ameiva major*. In: IUCN 2008. 2008 IUCN Red List of Threatened Species. http://www.iucnredlist.org.

———. 1996c. *Leiocephalus eremitus*. In: IUCN 2008. 2008 IUCN Red List of Threatened Species. http://www.iucnredlist.org.

———. 1996d. *Leiocephalus herminieri*. In: IUCN 2008. 2008 IUCN Red List of Threatened Species. http://www.iucnredlist.org.

———. 1996e. *Celestus occiduus*. In: IUCN 2008. 2008 IUCN Red List of Threatened Species. http://www.iucnredlist.org.

———. 1996f. *Alsophis sanctaecrucis*. In: IUCN 2008. 2008 IUCN Red List of Threatened Species. http://www.iucnredlist.org.

Wright, A. H. 1932. Life-histories of the frogs of the Okefinokee Swamp, Georgia. North American Salientia (Anura) no. 2. New York: MacMillan.

Wright, S. J. 1981. Extinction-mediated competition: The *Anolis* lizards and insectivorous birds of the West Indies. American Naturalist 117: 181–192.

Wunderle, J. M., Jr. 1981. Avian predation upon *Anolis* lizards on Grenada, West Indies. Herpetologica 37: 104–108.

Wunderle, J. M., Jr., J. E. Mercado, B. Parresol, and E. Terranova. 2004. Spatial ecology of Puerto Rican Boas (*Epicrates inornatus*) in a hurricane impacted forest. Biotropica 36: 555–571.

Wüster, W., R. S. Thorpe, M. da Graça Salomão, L. Thomas, G. Puorto, R. D. G. Theakston, and D. A. Warrell. 2002. Origin and phylogenetic position of the Lesser Antillean species of *Bothrops* (Serpentes: Viperidae): Biogeographical and medical implications. Bulletin of the Natural History Museum of London (Zoology) 68: 101–106.

Yanek, K., W. R. Heyer, and R. O. de Sa. 2006. Genetic resolution of the enigmatic Lesser Antillean distribution of the frog *Leptodactylus validus* (Anura, Leptodactylidae). South American Journal of Herpetology 1: 192–201.

Yeska, M. L., R. Powell, and J. S. Parmerlee, Jr. 2000. The lizards of Cayo Pisaje, Dominican Republic, Hispaniola. Herpetological Review 31: 18–20.

Yorks, D. T., R. W. Henderson, and R. Powell. 2003. *Typhlops tasymicris*. Catalogue of American Amphibians and Reptiles 780: 1–2.

Yorks, D. T., K. E. Williamson, R. W. Henderson, R. Powell, and J. S. Parmerlee, Jr. 2004. Foraging behavior in the arboreal boid *Corallus grenadensis*. Studies of the Neotropical Fauna and Environment 38: 167–172.

Young, B. E., S. N. Stuart, J. S. Chanson, N. A. Cox, and T. M. Boucher. 2004. Disappearing Jewels: The Status of New World Amphibians. Arlington, Va.: NatureServe.

Zaher, H., F. Gobbi Grazziotin, J. E. Cadle, R. W. Murphy, J. C. de Moura-Leite, and S. L. Bonatto. 2009. Molecular phylogeny of advanced snakes (Serpentes, Caenophidia) with an emphasis on South American xenodontines: A revised classification and descriptions of new taxa. Papéis Avulsos de Zoologia 49: 115–153.

Zajicek, D., and M. Mauri Méndez. 1969. Hemoparásitos de algunos animals de Cuba. Poeyana 66: 1–10.

Zani, P. W., S. I. Guttman, and R. Powell. 1993. The genetic relations of *Anolis cristatellus* (Sauria: Polychridae) from Hispaniola and Puerto Rico. Caribbean Journal of Science 29: 250–253.

Zeilfelder, I., and U. Bartelt. 2006. Der Bahama-Anolis *Anolis sagrei*. Münster, Germany: Natur und Tier.

Zernikow, W. 1987. *Anolis argenteolus*—Verhalten und Problematik der Nachzucht. Elaphe 9(3): 44–45.

Zippel, K. C. 2009. Zoos play a vital role in amphibian conservation. http://amphibiaweb.org/declines/zoo/index.html.

Zippel, K. C., J. S. Parmerlee, Jr., and R. Powell. 1994. *Ialtris dorsalis*. Catalogue of American Amphibians and Reptiles 592: 1–3.

Zippel, K. C., R. Powell, J. S. Parmerlee, Jr., S. Monks, A. Lathrop, and D. D. Smith. 1996. The distribution of larval *Eutrombicula alfreddugesi* (Acari: Trombiculidae) infesting *Anolis* lizards (Lacertilia: Polychrotidae) from different habitats on Hispaniola. Caribbean Journal of Science 32: 43–49.

Zoological Society of London. 2008. Chytridiomycosis and the Caribbean. http://www.zsl.org/conservation/animals/amphibians/mountain-chicken,830,AR.html.

Zug, G. R., and A. Schwartz. 1958. Variation in the species of *Cadea* (Amphisbaenidae), and a record of *C. blanoides* from Isla de Pinos. Herpetologica 14: 176–179.

Index to Scientific Names of Amphibians and Reptiles

A **boldface** page number refers to the primary account for that species. An *italicized* number refers to an illustration. If an illustration is on the same page as the primary account, the boldfaced page number is ***italicized***.

Allobates, ix, 6
  chalcopis, ix, 14, ***25***
Alsophis, xv, 7, 8, 9, 14, 20, 133, 136, 361, 362, 363, 364, 365, 366
  antiguae, xv, 10, 14, 21,231, 289, **356**, 357, 358
  antillensis, xv, 18, 21, 60, 243, **356**, 357, 358
  manselli, xv, 18, 93, 236, **356–57**
  rijgersmaei, xv, 18, 216, 311, 315, 330, **357**
  rufiventris, xv, 10, 18, 21, 60, 180, 267, 275, 315, 324, **357**
  sajdaki, xv, 14, 231, 289, 328, 356, **357–58**
  sanctonum, xv, 18, 65, 283, **358**
  sibonius, xv, 18, 65, 248, **358**
Ameiva, xv, 7, 9, 12, 14, 19, 20, 102, 198, 201, 202, 311, 321, 327, 350, 354, 360, 366, 368, 371, 372, 378, 403
  alboguttata, xv, 14, 244, **319**, 361
  ameiva, xv, 21, 159, 262, ***319***, **320**
  atrata, xv, 14, 141, **320**
  auberi, xv, 19, 271, **320–21**, 347, 364
  chrysolaema, xv, 101, 138, 145, 148, 188, 237, **321–22**, 332, 367, 371
  cineracea, xv, 21, **322**
  corax, xv, 14, 15, **322–23**, 323, 331
  corvina, xv, 14, **323**
  desechensis, xv, **323–24**
  dorsalis, xv, **324**
  erythrocephala, xv, 180, 275, 315, **324**, 357
  exsul, xv, 8, 18, 194, 281, 307, 309, **324–27**, 333, 349, 360, 361
  fuscata, xv, 133, 322, **327**, 358
  griswoldi, xv, 289, 290, **328**, 358
  leberi, 322
  lineolata, xv, **328**
  major, xv, 14, 21, **328–29**
  maynardii, xv, 142, **329**, 364, 376
  pleii, xv, 8, 133, 216, 254, 311, 315, 323, 325, **329–31**
  pluvianotata, xv, **331**
  polops, xv, 299, **331–32**
  taeniura, xv, 321, 322, **332**, 367
  wetmorei, xv, **332–33**, 361
Amphisbaena, xv, 7, 16, 20, 370, 387
  bakeri, xv, ***337***
  barbouri, xv, **337**
  caeca, xv, **337**
  carlgansi, xv, **338**
  caudalis, xv, **338**
  cayemite, xv, **338**
  cubana, xv, **338**, 359
  fenestrata, xv, **338**, 361
  gonavensis, xv, **338**
  hyporissor, xv, **338–39**, 339
  innocens, xv, **339**
  leali, xv, **339**
  manni, xv, **339**
  schmidti, xv, **339–40**
  xera, xv, **340**
Anolis, xii, 6, 9, 12, 14, 15, 19, 20, 86, 93, 106, 108, 155, 156, 197, 288, 290, 311, 321, 325, 329, 337, 348, 350, 351, 354, 356, 357, 358, 360, 361, 362, 363, 364, 365, 366, 367, 368, 369, 370, 371, 372, 377, 378, 379, 380
  acutus, xii, **157–58**
  aeneus, xii, 6, 8, **158–62**, 262, 263, 284, 320, 346
  agueroi, xii, **162**
  ahli, xii, **162–63**
  alayoni, xii, **163**
  alfaroi, xii, **163**
  aliniger, xii, **163–64**
  allisoni, xii, **164–65**, 256, 257, 269, 347
  allogus, xii, *157*, **165–66**, 262
  altavelensis, xii, **166**
  altitudinalis, xii, **166**
  alumina, xii, **166**
  alutaceus, xii, **166–67**, 379
  anfiloquioi, xii, **167–68**
  angusticeps, xii, **168–70**, 219, 252, 379
  argenteolus, xii, **170–71**, 226, 239
  argillaceus, xii, **171–72**
  armouri, xii, **172**, 368
  bahorucoensis, xii, **172–74**, 188, 199, 203, 223
  baleatus, xii, **174**
  baracoae, xii, **174–75**
  barahonae, xii, 174, **175**, 378
  barbatus, xii, **175–77**, 257
  barbouri, xii, **177–78**, 188, 200, 204
  bartschi, xii, **178–79**
  bimaculatus, xii, 2, **179–81**, 231, 275, 315
  birama, xii, **181**
  bremeri, xii, **181**, 272, 273

breslini, xii, **181**
brevirostris, xii, **181–82**, 183, 185, 204, 250, 290, 291
brunneus, xii, 142, 146, **183**, 364
carolinensis, xii, **183**, 217, 279
caudalis, xii, 182, **183**, 290, 366
centralis, xii, **183–84**, 235
chamaeleonides, xii, 179, **184–85**
chlorocyanus, xii, 182, **185–86**, 187, 188, 193, 204, 256, 366
christophei, xii, **186**
chrysops, xii, **186**
clivicola, xii, **186–87**
coelestinus, xii, 173, 177, 185, 186, **187–88**, 199, 200, 204, 322, 366
confusus, xii, **188**
conspersus, xii, 11, **188–89**, 217, 270, 355
cooki, xii, **189–90**, 191, 192, 195
cristatellus, xii, 6, 11, 18, 19, 20, 44, 110, 185, 189, **190–96**, 199, 201, 203, 204, 207, 210, 222, 230,244, 246, 247, 249, 281, 283, 309, 325, 350, 360, 361, 369, 385
cupeyalensis, xii, 163, **196**
cuvieri, xii, 194, **196–97**, 230, 259, 282, 349, 370
cyanopleurus, xii, 37, **197–98**
cybotes, xii, 6, 172, 173, 177, 188, 193, **198–201**, 204, 241, 242, 256, 290, 366
darlingtoni, xii, **201**
delafuentei, xii, **201**
desechensis, xii, **201**, 361
desiradei, xii, **201–2**
distichus, xii, 8, 148, 177, 182, 188, 191, 193, 194, 200, **202–5**
dolichocephalus, xii, **205**, 223
equestris, xii, 18, **205–7**, 241, 256
ernestwilliamsi, xiii, **207**
etheridgei, xiii, **207**
eugenegrahami, xiii, **207–8**
evermanni, xiii, 15, 18, 43, 190, 192, 195, 197, **208–10**, 222, 230, 282, 283, 349, 361
extremus, xiii, **210–11**, 238
fairchildi, xiii, **211**
ferreus, xiii, **211**
fowleri, xiii, **211**
fugitivus, xiii, **211–12**
garmani, xiii, **212–13**, 232, 234
garridoi, xiii, **213–14**
gingivinus, xiii, 110, 155, **214–16**, 253, 254, 311, 315, 357
grahami, xiii, 212, **216–18**, 232, 233, 234, 251, 268, 286
griseus, xiii, **218**, 284, 345
guafe, xiii, **218–19**
guamuhaya, xiii, **219**
guazuma, xiii, **219**
gundlachi, xiii, 6, 15, 18, 43, 44, 191, 192, 193, 195, 197, 209, 210, **219–23**, 230, 282, 283, 306, 349, 361, 369
haetianus, xiii, **223**
hendersoni, xiii, **223**
homolechis, xiii, **223–26**, 235, 241, 256, 268, 270
imias, xiii, 171, **226**
incredulus, xiii, **226**
inexpectatus, xiii, **226–27**
insolitus, xiii, **227**
isolepis, xiii, **227**
juangundlachi, xiii, **227–28**
jubar, xiii, 223, **228–29**
kahouannensis, xiii, **229**
koopmani, xiii, **229**
krugi, xiii, 193, 195, 197, 210, 221, **229–30**, 258, 259, 277, 283, 361, 369
leachii, xiii, **230–31**, 289, 290, 358
lineatopus, xiii, 6, 8, 212, 217, 218, **231–34**, 250, 251, 261, 286, 379
litoralis, xiii, 184, **234–35**
lividus, xiii, **235–36**, 356
longiceps, xiii, 14, 15, **236**
longitibialis, xiii, 138, **236–37**, 321
loysianus, xiii, **237**
luciae, xiii, 210, 218, **237–38**
lucius, xiii, **238–40**
luteogularis, xiii, 179, 225, **240–41**
luteosignifer, xiii, 139, **241**
macilentus, xiii, **241**
marcanoi, xiii, 199, **241–42**
marmoratus, xiii, 18, **242–43**, 305, 356
marron, xiii, **243**
maynardii, xiii, **243**, 298
mestrei, xiii, **243–44**
mimus, 262
monensis, xiii, 190, **244–45**, 319, 350
monticola, xiii, **245**
noblei, xiii, **245**
nubilis, xiii, **245–46**
occultus, xiii, **246**, 285
oculatus, xiii, 11, 193, **246–49**, 327, 358, 369
olssoni, xiii, **249–50**, 291
opalinus, xiii, 218, 232, **250–51**, 286
ophiolepis, xiii, **251–52**
oporinus, xiii, **252**
paternus, xiii, **252**
pigmaequestris, xiii, **253**
placidus, xiii, **253**
pogus, xiii, 214, 215, 216, **253–54**, 311, 315
poncensis, xiii, **254–55**
porcatus, xiii, 18, 88, 164, 185, 206, 225, **255–57**, 269
porcus, xiii, **257–58**
pulchellus, xiii, 197, 229, 230, 254, **258–60**, 277, 307, 360, 369, 370
pumilus, xiii, **260**
quadriocellifer, xiii, **260–61**
reconditus, xiii, 231, **261**
rejectus, xiii, 165, **261–62**

richardii, xiii, 158, 159, 160, 161, 162, **262–63**, 320, 346
ricordii, xiii, 174, **263–64**
rimarum, xiii, **264**
roosevelti, xiii, 21, **264**
roquet, xiii, 238, **264–66**, 368
rubribarbus, xiii, **266**
ruibali, xiii, **266**
rupinae, xiii, **266**
sabanus, xiii, **266–67**, 357
sagrei, xiii, 2, 6, 8, 11, 19, 88, 139, 140, 141, 181, 188, 189, 204, 217, 224, 225, 231, 235, 255, **267–74**, 279, 284, 303, 347, 362, 364, 380
schwartzi, xiii, 180, 181, **274–75**, 315, 357
scriptus, xiii, 141, 146, **275–76**, 364, 376, 378
semilineatus, xiii, **276–77**
sheplani, xiii, **277**
shrevei, xiii, **277**
singularis, xiii, **277–78**
smallwoodi, xiii, 18, **278**
smaragdinus, xiii, 268, 270, 271, 273, **278–79**
spectrum, xiii, **279–80**
strahmi, xiii, **280**
stratulus, xiii, 8, 14, 192, 195, 197, 209, 210, 221, 222, 230, 259, **280–83**, 309, 360, 361
terraealtae, xiii, **283**, 358
terueli, xiii, **283**
toldo, xiii, **283**
trinitatis, xiii, 161, 218, **283–84**, 345
valencienni, xiii, 217, 233, 234, 246, 251, **284–86**, 352
vanidicus, xiii, **286**
vermiculatus, xiii, **287–88**
vescus, xiii, **288–89**
wattsi, xiii, 231, **289–90**, 328, 357
websteri, xiii, 182, 183, **290**
whitemani, xiii, 182, 199, 250, **290–91**
Antillophis, 362, 367
Aristelliger, xiv, 6, 20
barbouri, xiv, **294**
cochranae, xiv, 14, **294–95**
expectatus, xiv, **295**
hechti, xiv, **295**
lar, xiv, **295–96**
praesignis, xiv, **296**
reyesi, xiv, **296**
Arrhyton, xvi, 7, 8, 19, 20, 365, 366, 367, 370, 380
ainictum, xvi, **359**
dolichura, xvi, **359**
landoi, 359
procerum, xvi, **359**
redimitum, xvi, **359**
supernum, xvi, **359**
taeniatum, xvi, 20, 338, **359**
tanyplectum, xvi, **359**
vittatum, xvi, **359–60**

Bachia, xi, 6, 8
heteropus, xi, **112**
Basiliscus, 287
Boa, xv, 7, 17
constrictor, 343, 344
nebulosa, *ii*, xv, 12, 20, 93, 133, **343–44**
orophias, xv, **344**
Boiga
irregularis, 11
Borikenophis, xvi, 7, 8, 20
portoricensis, xvi, 8, 9, 18, 20, 36, 44, 92, 126, 135, 192, 193, 194, 196, 197, 210, 222, 230, 259, 282, 307, 309, 317, 325, 326, 333, 338, 351, *356*, **360–61**, 362, 387
sanctaecrucis, xvi, 21, **361**
variegatus, xvi, 201, 244, 308, 319, **361–62**
Bothrops, xvii, 7, 344
caribbaeus, xvii, 20, 344, 362, ***390,*** **391**, 392
lanceolatus, xvii, 12, 391, **391–92**
Bufo, ix, 6, 8, 11
cataulaciceps, ix, **26**
empusus, ix, **26–27**, 87
florentinoi, ix, **27**
fluviaticus, ix, 21, **27**
fractus, ix, **27**
fustiger, ix, **27–28**
guentheri, ix, **28**
gundlachi, ix, **28–29**
lemur, ix, *26*, **29**
longinasus, ix, 10, **29–30**
marinus, ix, 8, 10, 18, 19, 20, **30–31**, 91, 349, 361
peltocephalus, ix, 8, 20, 28, **31–32**, 362
taladai, ix, 32, **32**

Cadea, xv, 7
blanoides, xv, ***341***
palirostrata, xv, **341**
Caiman, xvii, 7
crocodilus, xvii, 397, ***403***
Caraiba, xvi, 7
andreae, xvi, 18, 20, 57, 176, 273, 288, 346, **362**, 363
Celestus, xi, 6, 8, 9, 19, 20, 365, 366, 368
agasepsoides, xi, **101**, 322, 339
anelpistus, xi, 21, **101–2**, 102
badius, xi, 14, *101*, **102**, 376
barbouri, xi, **102**
carraui, 106
costatus, xi, **102**, 106, 137, 299, 367
crusculus, xi,**102–3**
curtissi, xi, **103**
darlingtoni, xi, **103**
duquesneyi, xi, **103**
fowleri, xi, **103**
haetianus, xi, **103–4**, 367
hewardii, xi, **104**

macrotus, xi, **104**
maculatus, xi, **104**
marcanoi, xi, **104**
microblepharis, xi, **104–5**
molesworthi, xi, **105**
occiduus, xi, 21, **105**
sepsoides, xi, **105**
stenurus, xi, **105**
warreni, xi, 102, **105–6**
Chamaeleolis, 162, 176, 184, 219, 258
Chamaelinorops, 178
Chaunus, 31
Chelonoidis, xvii, 7, 19
carbonaria, xvii, 12, ***400***, **401**
denticulata, xvii, **401**
Chironius, xv, 7, 19
vincenti, xv, 8, 18, 98, **354**
Clelia, xvi, 7
clelia, xvi, 21, **362**
errabunda, xvi, 20, 21, **362**, 391
Cnemidophorus, xv, 7
vanzoi, xv, 14, **333–34**
Colostethus, 25
Corallus, xv, 7, 16, 17, 20
cookii, xv, 8, 218, 284, *343*, **345**
grenadensis, xv, 8, 12, 15, 136, 161, 262, 263, **345–46**, 391
Cricosaura, xv, 7
typica, xv, ***335***
Crocodylus, xvii, 7
acutus, xvii, 3, 8, 9, 10, 12, 13, 21, 394, ***404***, **405**, 406
intermedius, 2
rhombifer, xvii, 8, 12, 13, 395, 396, **405–6**
Cubophis, xvi, 7, 8, 20, 140, 256, 364
cantherigerus, xvi, 18, 20, 72, 88, 106, 179, 256, 261, 273, 288, 346, **362–63**, 364
caymanus, xvi, 18, 72, 120, 136, 189, **363–64**
fuscicauda, xvi, 124, 241, 273, 298, **364**
ruttyi, xvi, **364**
vudii, xvi, 18, 20, 72, 88, 118, 183, 273, 276, 310, **364–65**
Cyclura, xi, 1, 6, 8, 9, 12, 16, 17, 19, 20, 363
carinata, xi, 8, 10, **114–15**
collei, xi, **115–16**, 352
cornuta, xi, 21, **116–17**, 120, 122
cychlura, xi, 8, **117–19**, 140, 352, 364
lewisi, xi, 8, *114*, **120–21**, 122, 123
nubila, xi, 8, 120, **121–25**
onchiopsis, xi, 6, **125**
pinguis, xi, 8, 10, 12, 16, **125–27**, 361, 383
ricordii, xi, 117, **127**
rileyi, xii, 8, **127–29**, 301
stejnegeri, xii, 8, 120, 122, **129–31**

Darlingtonia, 368
Diploglossus, xi, 6, 9, 20, 363
delasagra, xi, **106**
garridoi, xi, **106**
montisserrati, xi, 14, **106–7**
nigropunctatus, xi, **107**
pleii, xi, **107**, 349, 361

Elaphe, 355
guttata, 11
Eleutherodactylus, ix, 6, 8, 9, 14, 18, 20, 43, 86, 93, 98, 108, 156, 193, 200, 221, 242, 288, 306, 354, 356, 360, 361, 362, 363, 364, 366, 367, 368, 369, 370, 387, 378
abbotti, ix, 17, 20, **33**, 177, 366, 368
acmonis, ix, **33–34**
adelus, ix, **34**
albipes, ix, **34**
alcoae, ix, 20, **34**
alticola, ix, **34–35**
amadeus, ix, **35**
amplinympha, ix, **35**
andrewsi, ix, **35**
antillensis, ix, 8, 20, **35–36**, 360
aporostegus, ix, **36**
apostates, ix, **36–37**
armstrongi, ix, 20, **37**, 177, 368
atkinsi, ix, 20, **37**, 48, 362
audanti, ix, **37–38**, 65, 67
auriculatoides, ix, **38**
auriculatus, ix, **38**, 299
bakeri, ix, *33*, **38–39**
barlagnei, ix, **39**
bartonsmithi, ix, **39**
blairhedgesi, ix, 20, **39–40**, 377
bothroboans, ix, **40**
bresslerae, ix, **40**
brevirostris, ix, **40**
brittoni, ix, **40**
caribe, ix, **41**
casparii, ix, **41**
cavernicola, ix, **41**
chlorophenax, ix, **41**
cochranae, ix, 18, **41–42**
cooki, ix, 8, **42–43**
coqui, ix, 8, 14, 16, 18, 19, 20, 36, **43–45**, 210, 221, 222, 293, 360, 361
corona, ix, **46**
counouspeus, ix, **46**
cubanus, ix, **46**
cundalli, ix, **46–47**, 67
cuneatus, ix, **47**, 75
darlingtoni, ix, **47**
dimidiatus, ix, 20, 37, **48**, 362
diplasius, ix, **48**
dolomedes, ix, **48**
eileenae, ix, **48–49**

emiliae, ix, **49**
eneidae, ix, 21, **49**
erythroproctus, ix, **49**
etheridgei, ix, **50**
eunaster, ix, 20, **50**, 368
flavescens, ix, **50**
fowleri, ix, **50**
furcyensis, ix, **51**
fuscus, ix, **51**
glamyrus, ix, **51**
glandulifer, ix, 20, **51–52**, 368
glanduliferoides, ix, **52**
glaphycompus, ix, 20, **52**, 368
glaucoreius, ix, **52**
goini, ix, **52–53**
gossei, ix, **53**
grabhami, ix, **53**
grahami, x, **53**
greyi, x, **53–54**
griphus, x, **54**
gryllus, x, **54**
guanahacabibes, x, **54–55**
guantanamera, x, **55**
gundlachi, x, **55**
haitianus, x, **55**
hedricki, x, 14, **55–56**
heminota, x, 20, **56**, 368
hypostenor, x, 20, **56**
iberia, x, 1, 18, 20, **56–57**
inoptatus, x, 20, **57**
intermedius, x, **57**
ionthus, x, **57–58**
jamaicensis, x, **58**
jasperi, x, 18, 21, **58**
jaumei, x, **58**
johnstonei, x, 8, 10, 16, 17, 18, 19, 20, 21, 39, **59–60**, 71, 89, 98, 156, 354, 356, 357
juanariveroi, x, **60**
jugans, x, **60**
junori, x, **60–61**
karlschmidti, x, 21, **61**
klinikowskii, x, **61**
lamprotes, x, **61–62**
leberi, x, **62**
lentus, x, **62**
leoncei, x, **62**
limbatus, x, **62–63**
limbensis, x, **63**
locustus, x, **63**
lucioi, x, **63**
luteolus, x, **63–64**
maestrensis, x, **64**
mariposa, x, **64**
martinicensis, x, 20, 39, **64–65**, 89, 358, 369
melacara, x, **65**
melatrigonum, x, 38, **65**
michaelschmidi, x, **65–66**
minutus, x, **66**
monensis, x, **66**
montanus, x, **66**
nortoni, x, **66**
notidodes, x, 38, **67**
nubicola, x, **67**
olibrus, x, **67**
orcutti, x, 21, **67**
orientalis, x, **68**
oxyrhyncus, x, 20, *33*, **68**, 368
pantoni, x, 18, 67, **68**
parabates, x, **68–69**
paralius, x, **69**
parapelates, x, **69**
patriciae, x, **69**
paulsoni, x, **69**
pentasyringos, x, **70**
pezopetrus, x, **70**
pictissimus, x, **70**
pinarensis, x, **70–71**
pinchoni, x, **71**
pituinus, x, **71**
planirostris, x, 18, 20, 38, **71–72**, 75, 88, 363, 364
poolei, x, **72**
portoricensis, x, 15, 18, 45, 56, **72–73**
principalis, x, **73**
probolaeus, x, **73**
rhodesi, x, **73**
richmondi, x, 14, 18, **73–74**
ricordii, x, **74**
riparius, x, 47, **74–75**
rivularis, x, **75**
rogersi, x, **75**
ronaldi, x, **75–76**
rucillensis, x, **76**
rufifemoralis, x, **76**
ruthae, x, 20, 36, **76**, 80
schmidti, x, 63, **76**
schwartzi, x, **76–77**
sciagraphus, x, 20, **77**, 368
semipalmatus, x, 21, **77**
sierramaestrae, 47
simulans, x, **77**
sisyphodemus, x, **77–78**
sommeri, x, **78**
staurometopon, x, **78**
symingtoni, x, **78**
tetajulia, x, **78**
thomasi, x, **79**
thorectes, x, **79**
toa, x, **79**

tonyi, x, **79–80**
turquinensis, x, **80**
tychathrous, x, **80**
unicolor, x, **80**
varians, x, 67, 78, **80–81**
varleyi, x, **81**
ventrilineatus, x, **81**
warreni, x, **81**
weinlandi, x, 69, **82**
wetmorei, x, 78, **82**
wightmanae, x, 18, 44, 56, **82–83**
zeus, x, **83**
zugi, x, **83**
Epicrates, xv, 7, 16, 17, 20
angulifer, xv, 8, 12, 16, 20, 179, **346–47**, 379, 395
chrysogaster, xv, **347**
exsul, xv, 10, **347–48**
fordii, xv, 200, **348**
gracilis, xv, 200, 205, **348**
inornatus, xv, 8, 15, 20, 44, 107, 197, 210, 222, 326, **349–50**
monensis, xv, 8, 10, 19, 191, 194, 244, 349, **350–51**
striatus, xv, 12, 18, 88, 118, 188, 200, 280, **351–52**
subflavus, xv, 10, 116, 286, **352–53**
Eretmochelys
imbricata, 20, 327, 363

Gastrophryne, xi, 6
carolinensis, xi, 18, 88, ***95***
Gekko, xi, 6
gecko, xi, **108**
Gonatodes, xiv, 6
albogularis, xiv, **297**
daudini, xiv, 14, **297**
Gymnophthalmus, xi, 6
pleii, xi, **112–13**
underwoodi, xi, *112*, **113**

Haitiophis, xvi, 7
anomalus, xvi, 149, **365**
Hemidactylus, xi, 6, 19, 20, 108, 153, 155, 156, 377
angulatus, xi, **108–9**
garnotii, xi, **109**
haitianus, 109
mabouia, xi, 19, *108*, **109–10**, 155, 156, 193, 311, 315, 361
palaichthus, xi, **110–11**
turcicus, xi, **111**
Hyla, xi, 6, 85
cinerea, xi, **84**
crucifer, 2
squirella, xi, **84**
Hypsiboas, xi, 6
heilprini, xi, **84–85**
Hypsirhynchus, xvi, 7, 8, 20, 385
ater, xvi, 21, **365**
callilaemus, xvi, 234, **365**, 366
ferox, xvi, 149, 322, **365–66**, 367
funereus, xvi, 18, **366**
melanichnus, xvi, 21, **366**
parvifrons, xvi, 18, 20, 33, 34, 37, 56, 57, 86, 102, 104, 145, 172, 177, 183, 186, 188, 200, 205, 223, 277, 298, 301, 302, 304, 322, 328, 339, **366–67**
polylepis, xvi, 18, **367**
scalaris, xvi, **367**

Ialtris, xvi, 7, 8, 20
agyrtes, xvi, 20, **367**, 387
dorsalis, xvi, **367–68**
haetianus, xvi, 18, 33, 34, 50, 52, 56, 172, **368**
parishi, xvi, **368**
Iguana, xii, 1, 6, 12, 19, 20, 358, 401
delicatissima, xii, 9, 11, 17, **131–34**, 135, 247, 327, 343
iguana, xii, 3, 9, 11, 15, 16, 132, 133, **134–36**, 326, 346, 361, 364

Kentropyx, xv, 7
borckiana, xv, **334**

Leiocephalus, xii, 6, 9, 12, 19, 20, 198, 202, 321, 362, 363, 366, 367, 371, 372
altavelensis, xii, **137**
barahonensis, xii, **137–38**, 237, 321
carinatus, xii, 87, **138–40**, 141, 144, 241, 269, 270, 273, 274
cubensis, xii, **141–42**, 272, 273, 363
cuneus, xii, 21, **142**
endomychus, xii, **142**
eremitus, xii, 21, **142**
greenwayi, xii, **142**, 183
herminieri, xii, 21, **142**
inaguae, xii, **142–43**, 329
loxogrammus, xii, **143**
lunatus, xii, *137*, **143**, 148
macropus, xii, **143–44**, 147, 228
melanochlorus, xii, **144**
onaneyi, xii, 21, **145**
personatus, xii, **145–46**, 148
pratensis, xii, **146**
psammodromus, xii, **146**, 276
punctatus, xii, **146**, 183, 364
raviceps, xii, 144, **146–47**
rhutidira, xii, **147**
schreibersii, xii, **147–50**, 151, 372
semilineatus, xii, 148, **150–51**, 367
stictigaster, xii, 144, **151–52**
vinculum, xii, **152**
Leptodactylus, xi, 6, 387
albilabris, xi, 10, 18, 20, 87, **91–92**, 97
fallax, xi, 9, 10, 12, 13, 16, 18, 20, **92–94**, 343, 356, 384
validus, xi, 20, *91*, **94**, 354

Leptotyphlops, xvi, 7, 8, 19, 20
asbolepis, xvi, **373**
bilineatus, xvi, **373**
breuili, xvi, ***373***
calypso, xvi, **373–74**
carlae, xvi, 1, 11, **374**
columbi, xvi, **374**
goudoti, xvi, **374**
leptepileptus, xvi, **374**
pyrites, xvi, **374**
Liophis, xvi, 7, 9, 14, 20, 21
cursor, xvi, 14, 18, 265, **368**
juliae, xvi, 18, 248, **368–69**
ornatus, xvi, 14, **369**
perfuscus, xvi, 21, **369**
Lithobates, 97

Mabuya, xiv, 6, 9, 20, 293, 365
lineolata, xiv, **292**
mabouya, xiv, ***292*, 293**
macleani, xiv, 14, **293**
sloanei, xiv, 18, 44, **293**
Magliophis, xvi, 7, 8, 20
exiguus, xvi, 18, 20, 44, 194, 222, 230, 259, 360, 361, **369–70**, 370
stahli, xvi, 18, 20, 197, 259, **370**
Mastigodryas, xv, 7, 19
bruesi, xv, 18, 94, ***354***

Nerodia, xvi, 7
clarkii, xvi, ***375***

Ocyophis, 361, 362, 363, 364, 365, 366
Ophisaurus
ventralis, 2
Osteopilus, xi, 6, 8, 11, 363, 379
brunneus, xi, 8, **85**
crucialis, xi, **85**
dominicensis, xi, 18, 20, **85–86**, 97, 366, 367, 371
marianae, xi, **86**
pulchrilineatus, xi, 20, **86**, 371, 378
septentrionalis, xi, 3, 8, 10, 11, 17, 18, 19, 20, 36, *84*, **86–89**, 91, 95, 141, 155, 206, 256, 273, 278, 347, 352, 357, 364, 377, 379, 380
vastus, xi, **89**
wilderi, xi, **89**

Pantherophis, xv, 7, 189
guttatus, xv, 11, **355**
Peltophryne, 26, 28, 29, 30, 32
Pelusios, xvii, 7, 19
castaneus, xvii, 396, ***399***
Phrynops, 2
geoffroanus, 2
Phyllodactylus, xii, 6, 153
hispaniolae, xii, **153**
pulcher, xii, **153**
sommeri, xii, **153–54**
wirshingi, xii, 153, **154**
Pituophis, 355
Pristimantis, xi, 6, 10, 354
euphronides, xi, 8, 10, 15, **98**
shrevei, xi, 8, 16, **98**, 354
Pseudemys
nelsoni, 396

Ramphotyphlops, xvi, 7
braminus, xvi, 3, 11, 374, **382**
Rana, xi, 6
catesbeiana, xi, 3, 18, 19, 86, 92, ***96***, **97**, 371
grylio, xi, **97**
sphenocephala, xi, **97**
Rhinella, 31

Schwartzophis, 365, 366, 367
Scinax, xi, 6
ruber, xi, **90**
cf. *x-signatus*, xi, **90**
Sphaerodactylus, xiv, 6, 8, 14, 19, 20, 39, 193, 199, 200, 242, 283, 313, 329, 361, 364, 366, 369, 370, 377, 378
altavelensis, xiv, **297–98**
argivus, xiv, **298**
argus, xiv, **298**
ariasae, xiv, 1, 20, **298**
armasi, xiv, **299**
armstrongi, xiv, 102, **299**
asterulus, xiv, **299**
beattyi, xiv, **299**, 331
becki, xiv, 14, **299**
bromeliarum, xiv, 38, **299**
caicosensis, xiv, **300**, 378
callocricus, xiv, **300**
celicara, xiv, **300**
cinereus, xiv, **300**
clenchi, xiv, **300**
cochranae, xiv, **300–301**
copei, xiv, **301**
corticola, xiv, 128, **301**
cricoderus, xiv, **301**
cryphius, xiv, **301**
dacnicolor, xiv, **301–2**
darlingtoni, xiv, **302**
difficilis, xiv, **302**, 366, 367
dimorphicus, xiv, **302–3**
docimus, xiv, **303**
elasmorhynchus, xiv, **303**
elegans, xiv, **303–4**
elegantulus, xiv, **304**, 358

epiurus, xiv, **304**
fantasticus, xiv, **304–5**, 312
gaigeae, xiv, **305**
gilvitorques, xiv, 21, **305**
goniorhynchus, xiv, 18, **305**
inaguae, xiv, **305**
intermedius, xiv, **305–6**
kirbyi, xiv, **306**
klauberi, xiv, 18, 44, 221, **306**
ladae, xiv, **306**
lazelli, xiv, **306**
leucaster, xiv, **306**
levinsi, xiv, **306–7**
macrolepis, xiv, 1, 19, 259, *294*, **307–8**, 325, 360, 361
mariguanae, xiv, **308**
microlepis, xiv, **308**
micropithecus, xiv, 10, **308**
monensis, xiv, **308**, 361
nicholsi, xiv, **308–9**, 313, 361
nigropunctatus, xiv, **309**, 316
notatus, xiv, **309–10**, 364
nycteropus, xiv, **310**
ocoae, xiv, **310**
oliveri, xiv, **310**
omoglaux, xiv, **310**
oxyrhinus, xiv, **310**
parkeri, xiv, **311**
parthenopion, xiv, **311**
parvus, xiv, 215, 254, **311**, 315
perissodactylius, xiv, **311**
phyzacinus, xiv, **311–12**
pimienta, xiv, **312**
plummeri, xiv, **312**
ramsdeni, xiv, **312**
randi, xiv, **312**
rhabdotus, xiv, **312**
richardi, xiv, **313**
richardsoni, xiv, **313**
roosevelti, xiv, 309, **313**, 361
ruibali, xiv, **313**
sabanus, xiv, **313–14**
samanensis, xiv, **314**
savagei, xiv, **314**
scaber, xiv, **314**
schuberti, xiv, **314**
schwartzi, xiv, **314**
semasiops, xiv, **314–15**
shrevei, xiv, **315**
siboney, xiv, **315**
sommeri, xiv, **315**
sputator, xiv, 155, 311, **315–16**
storeyae, xiv, **316**
streptophorus, xiv, **316**
thompsoni, xiv, **316**
torrei, xiv, 309, **316**
townsendi, xiv, **316–17**, 361
underwoodi, xv, **317**
vincenti, xv, 113, **317–18**
williamsi, xv, **318**
zygaena, xv, **318**

Tantilla, xv, 7
melanocephala, xv, **355**
Tarentola, xii, 6
albertschwartzi, xii, 21, **154**
americana, xii, 87, 141, **154**, 155, 238
crombiei, xii, **154–55**
Thamnophis, 296
Thecadactylus, xii, 6
rapicauda, xii, 18, *153*, **155–56**, 311, 315
Trachemys, xvii, 7, 14, 19, 396, 405
decorata, xvii, **394**, 405
decussata, xvii, 13, 347, **394–96**, 397, 398
scripta, xvii, 10, 394, **396**, 397, 398
stejnegeri, xvii, 12, *394*, 395, **396–98**, 398
terrapen, xvii, 395, 397, **398**
Tretanorhinus, xvi, 7
variabilis, xvi, 18, 97, 288, **370–71**
Tropidophis, xvi, 7, 14, 20, 39, 175, 179, 256, 363
bucculentus, xvi, 10, 21, **376**
canus, xvi, 276, 329, **376**
caymanensis, xvi, **376–77**
celiae, xvi, 18, **377**
curtus, xvi, 18, **377**
feicki, xvi, **377**
fuscus, xvi, **377**
galacelidus, xvi, **377**
greenwayi, xvi, 276, 300, **378**
haetianus, xvi, 18, 175, 205, 314, **378**
hardyi, xvi, **378**
hendersoni, xvi, **378**
jamaicensis, xvi, 234, **378–79**
maculatus, xvi, 167, 170, **379**
melanurus, xvi, 18, 20, 176, 179, 184, 257, 288, 346, 347, **379**
morenoi, xvi, **379**
nigriventris, xvi, **379–80**
pardalis, xvi, **380**
parkeri, xvi, 18, 273, **380**
pilsbryi, xvi, **380**
schwartzi, xvi, **380**
semicinctus, xvi, **380–81**
spiritus, xvi, **381**
stejnegeri, xvi, **381**
stullae, xvi, **381**
wrighti, xvi, *376*, **381**
xanthogaster, xvi, **381**
Typhlops, xvi, 7, 8, 14, 16, 19, 20, 359, 360, 366, 370
agoralionis, xvi, **382**

anchaurus, xvi, **382**
annae, xvi, 14, **382**
anousius, xvi, **383**
arator, xvi, **383**
biminiensis, xvii, **383**
capitulatus, xvii, **383**
catapontus, xvii, **383**
caymanensis, xvii, **383**
contorhinus, xvii, **384**
dominicanus, xvii, 18, 93, **384**
epactius, xvii, **384**
eperopeus, xvii, **384**
geotomus, xvii, **384**
golyathi, xvii, **384**
gonavensis, xvii, **384**
granti, xvii, *382*, **384–85**
guadeloupensis, xvii, **385**
hectus, xvii, **385**
hypomethes, xvii, 193, **385**
jamaicensis, xvii, **385**
lumbricalis, xvii, **385–86**
monastus, xvii, **386**
monensis, xvii, **386**
naugus, xvii, **386**
notorachius, xvii, **386**
paradoxus, xvii, **386**
perimychus, xvii, **386**
platycephalus, xvii, 20, 361, **386–87**
proancylops, xvii, **387**
pusillus, xvii, 20, 367, **387**
richardii, xvii, **387**
rostellatus, xvii, **387–88**
satelles, xvii, **388**
schwartzi, xvii, **388**
sulcatus, xvii, **388**
sylleptor, xvii, **388**
syntherus, xvii, 339, **388**
tasymicris, xvii, **388–89**
tetrathyreus, xvii, **389**
titanops, xvii, **389**

Uromacer, xvi, 7, 8, 20
catesbyi, xvi, 18, 86, 143, 144, 186, 188, 200, 205, 277, 280, **371**
frenatus, xvi, 144, 145, 152, 186, 188, 200, 205, 250, 277, 304, 322, 332, **371–72**
oxyrhynchus, xvi, 12, 20, 149, 183, 186, 200, 205, 250, 277, 328, 332, 371, **372**

Robert W. Henderson is senior curator of herpetology at the Milwaukee Public Museum. In addition to having published many articles in scientific journals, he has written or edited twelve books, including *Amphibians and Reptiles of the West Indies: Descriptions, Distributions, and Natural History* (with Albert Schwartz), *Neotropical Treeboas: Natural History of the* Corallus hortulanus *Complex*, and *Biology of the Boas and Pythons* (coedited with Robert Powell).

Robert Powell is professor of biology at Avila University in Kansas City, Missouri. Among hundreds of publications (many with students), he has written or edited seven books, including *A Key to the Amphibians and Reptiles of the Continental United States and Canada* (with Joseph T. Collins and Errol D. Hooper Jr.), *The Reptiles and Amphibians of Anguilla, British West Indies* (with Karim V. D. Hodge and Ellen J. Censky), and *Reptiles and Amphibians of the Dutch Caribbean: St. Eustatius, Saba, and St. Maarten* (with Robert W. Henderson and John S. Parmerlee Jr.).

Combined the authors have nearly 60 years of experience studying reptiles and amphibians in the West Indies, with Powell focusing mostly on lizards and Henderson on the snakes (especially treeboas and racers) that eat them.

www.ingramcontent.com/pod-product-compliance
Lightning Source LLC
Chambersburg PA
CBHW081217180926
51613CB00004B/262